T0230807

# Event-Based Control and Signal Processing

# Embedded Systems

**Series Editor**

Richard Zurawski
*SA Corporation, San Francisco, California, USA*

Event-Based Control and Signal Processing, *edited by Marek Miskowicz*

Real-Time Embedded Systems: Open-Source Operating Systems Perspective, *Ivan Cibrario Bertolotti and Gabriele Manduchi*

Time-Triggered Communication, *edited by Roman Obermaisser*

Communication Architectures for Systems-on-Chip, *edited by José L. Ayala*

# Event-Based Control and Signal Processing

Edited by Marek Miśkowicz

CRC Press
Taylor & Francis Group
Boca Raton London New York

CRC Press is an imprint of the
Taylor & Francis Group, an **informa** business

CRC Press
Taylor & Francis Group
6000 Broken Sound Parkway NW, Suite 300
Boca Raton, FL 33487-2742

First issued in paperback 2017



ISBN-13: 978-1-4822-5655-0 (hbk)
ISBN-13: 978-1-138-89318-4 (pbk)



**Visit the Taylor & Francis Web site at**
**http://www.taylorandfrancis.com**

**and the CRC Press Web site at**
**http://www.crcpress.com**

# Contents

# Contributors

**Adolfo Anta**
General Electric
Munich, Germany

**Panos J. Antsaklis**
University of Notre Dame
Notre Dame, IN, USA

**Manuel Berenguel**
University of Almería
ceiA3-CIESOL, Almería, Spain

**Brigitte Bidégaray-Fesquet**
University Grenoble Alpes, CNRS, LJK
Grenoble, France

**Dieter Brückmann**
University of Wuppertal
Wuppertal, Germany

**Azime Can-Cimino**
University of Pittsburgh
Pittsburgh, PA, USA

**Christos G. Cassandras**
Boston University
Boston, MA, USA

**Luis F. Chaparro**
University of Pittsburgh
Pittsburgh, PA, USA

**Yu Chen**
Columbia University
New York, NY, USA

**Jorge Cortés**
University of California
San Diego, CA, USA

**M.C.F. (Tijs) Donkers**
Eindhoven University of Technology
Eindhoven, Netherlands

**Sebastián Dormido**
UNED
Madrid, Spain

**Laurent Fesquet**
University Grenoble Alpes, CNRS, TIMA
Grenoble, France

**Anqi Fu**
Delft University of Technology
Delft, Netherlands

**Eloy Garcia**
Infoscitex Corp.
Dayton, OH, USA

**Peter Gawthrop**
University of Melbourne
Melbourne, Victoria, Australia

**Henrik Gollee**
University of Glasgow
Glasgow, UK

**Luiz Carlos Paiva Gouveia**
University of Glasgow
Glasgow, UK

**José Luis Guzmán**
University of Almería
ceiA3-CIESOL, Almería, Spain

**Alister Hamilton**
University of Edinburgh
Edinburgh, UK

**Uwe D. Hanebeck**
Karlsruhe Institute of Technology
Karlsruhe, Germany

**W.P. Maurice H. Heemels**
Eindhoven University of Technology
Eindhoven, Netherlands

**Burkhard Hensel**
Dresden University of Technology
Dresden, Germany

**Alfred O. Hero**
University of Michigan
Ann Arbor, MI, USA

**Klaus Kabitzsch**
Dresden University of Technology
Dresden, Germany

**Thomas Jacob Koickal**
Beach Theory
Sasthamangalam, Trivandrum, India

**Karsten Konrad**
University of Wuppertal
Wuppertal, Germany

**Dariusz Kościelnik**
AGH University of Science and Technology
Kraków, Poland

**Maria Kurchuk**
Pragma Securities
New York, NY, USA

**Walter Lang**
University of Siegen
Siegen, Germany

**Mircea Lazar**
Eindhoven University of Technology
Eindhoven, Netherlands

**Michael D. Lemmon**
University of Notre Dame
Notre Dame, IN, USA

**Lichun Li**
Georgia Institute of Technology
Atlanta, GA, USA

**Ian Loram**
Manchester Metropolitan University
Manchester, UK

**Jan Lunze**
Ruhr-University Bochum
Bochum, Germany

**Pablo Martinez-Nuevo**
Massachusetts Institute of Technology
Cambridge, MA, USA

**Manuel Mazo Jr.**
Delft University of Technology
Delft, Netherlands

**Michael J. McCourt**
University of Florida
Shalimar, FL, USA

**Marek Miśkowicz**
AGH University of Science and Technology
Kraków, Poland

**George V. Moustakides**
University of Patras
Rio, Greece

**Dragan Nešić**
University of Melbourne
Parkville, Australia

**Benjamin Noack**
Karlsruhe Institute of Technology
Karlsruhe, Germany

**Steven M. Nowick**
Columbia University
New York, NY, USA

**Cameron Nowzari**
University of Pennsylvania
Philadelphia, PA, USA

**Roman Obermaisser**
University of Siegen
Siegen, Germany

**Sharvil Patil**
Columbia University
New York, NY, USA

**Mirosław Pawlak**
University of Manitoba
Winnipeg, MB, Canada

**Andrzej Pawlowski**
UNED
Madrid, Spain

**Joern Ploennigs**
IBM Research–Ireland
Dublin, Ireland

**Romain Postoyan**
University of Lorraine
CRAN UMR 7039, CNRS
Nancy, France

**Dominik Rzepka**
AGH University of Science and Technology
Kraków, Poland

**Bob Schell**
Analog Devices Inc.
Somerset, NJ, USA

**Joris Sijs**
TNO Technical Sciences
The Hague, Netherlands

**Christian Stöcker**
Ruhr-University Bochum
Bochum, Germany

**Paulo Tabuada**
University of California
Los Angeles, CA, USA

**Andrew R. Teel**
University of California
Santa Barbara, CA, USA

**Nguyen T. Thao**
City College of New York
New York, NY, USA

**Yannis Tsividis**
Columbia University
New York, NY, USA

**Volodymyr Vasyutynskyy**
SAP SE
Research & Innovation
Dresden, Germany

**Christos Vezyrtzis**
IBM
Yorktown Heights, NY, USA

**Xiaodong Wang**
Columbia University
New York, NY, USA

**Thomas Werthwein**
University of Wuppertal
Wuppertal, Germany

**Yasin Yilmaz**
University of Michigan
Ann Arbor, MI, USA

# Acknowledgments

Preparation of the book is an extensive task requiring complex coordination and assistance from authors and the publisher. There are many people I have to acknowledge and thank for their contribution in this work at various stages of its evolution.

First of all, I wish to thank Karl Johan Åström, the pioneer of modern event-based control, for his insightful Foreword to the book. Richard Zurawski, Editor of the CRC Press Embedded Systems Series, shared with me his experience on the process of publishing edited books for what I am most obliged. Furthermore, I would like to thank Jan Lunze and Yannis Tsividis for their extended involvement in the book preparation, constructive advice, and especially for the contribution of survey chapters beginning Part I and Part II of the book.

I am very grateful to authors who took the lead in the preparation of chapters: Panos J. Antsaklis, Dieter Brückmann, Christos G. Cassandras, Luis F. Chaparro, Yu Chen, Jorge Cortés, Sebastián Dormido, Laurent Fesquet, Peter Gawthrop, Alister Hamilton, Maurice Heemels, Klaus Kabitzsch, Michael D. Lemmon, Manuel Mazo, Roman Obermaisser, Joris Sijs, Nguyen T. Thao, and Yasin Yilmaz.

I am indebted to Nora Konopka, Publisher with CRC Press/Taylor & Francis, for giving me the opportunity to prepare that book. Hayley Ruggieri, Michele Smith, Jennifer Stair, and John Gandour of the CRC Press/Taylor & Francis offered great assistance throughout the whole process. I also thank Viswanath Prasanna at Datapage for his patience and constructive cooperation.

Finally, I express my deep gratitude to all the authors who contributed to the book. I believe the book will stimulate research in the areas covered, as well as fruitful interaction between event-based control and signal processing communities.

Marek Miśkowicz

# *Foreword*

Control and signal processing are performed very differently in man and machine. In biological systems it is done by interconnecting neurons that communicate by pulses. The systems are analog bio-chemical reactions, execution is parallel. Emulation of neurons in silicon was pioneered by Carver Mead. Some of his designs have shown to give very efficient processing of images particularly when sensing information generates pulses. Engineering applications of control and signal processing were originally implemented by mechanical or electrical analog computing, which was also analog and asynchronous. The signals were typically continuous.

The advent of the digital computer was a major paradigm shift, the analog signals were sampled and converted to digital form, control and signal processing were executed by numerical algorithms, the results were converted to analog form and applied to the physical devices by actuators. A very powerful theory was developed for analysis and design by introducing proper abstractions. The computer based systems for control and signal processing were traditionally based on a central periodic clock, which controlled sampling, computing and actuation. For physical processes modeled as linear time-invariant systems the approach leads to closed loop systems that are linear but periodic. The periodic nature can be avoided by considering the behaviour at times that are synchronised with the sampling, resulting in the *stroboscopic model*. The system can then be described by difference equations with constant coefficients. Traditional sampled data theory based on the stroboscopic model is powerful and simple. It has been used extensively in practical applications of control and signal processing. Periodic sampling also matches the time-triggered model of real time software which is simple and easy to implement safely. The combination of computers and sampled data theory is a very powerful tool for design and implementation of engineered systems. Today it is the standard way of implementing control and signal processing.

In spite of the tremendous success of sampled-data theory, it has some disadvantages. There are severe difficulties when dealing with systems having multiple sampling rates and systems with distributed computing. With multi-rate sampling, the complexity of the system depends critically on the ratio of the sampling rates. For distributed systems the theory requires that the clocks are synchronised. Sampling jitter, lost samples and delays create significant difficulties in networked, distributed systems on computer controlled systems.

Event-based sampling is an alternative to periodic sampling. Signals are then sampled only when significant events occurs, for example, when a measured signal exceeds a limit or when an encoder signal changes. Event-based control has many conceptual advantages. Control is not executed unless it is required (control by exception). Event-based control is also useful in situations when control actions are expensive; changes of production rates can be very costly when controlling manufacturing complexes. Acquiring information in computer systems and networks can be costly because it consumes computer resources.

Even if periodic sampling and sampled-data theory will remain a standard method to analyse and design systems for control and signal processing, there are many problems where event-based control is natural. Event-based control may also bridge the gap between biological and engineered systems. Recently there has been an increased interest in event-based control and papers have appeared in many different fields. The theory currently available is far from what is needed for analysis and design, an equivalent of the theory sampled-data systems is missing. The results are scattered among different publications. Being interested in event-based control, I am therefore very pleased that this book, which covers views of different authors, is published. I hope that it will stimulate further research.

Karl Johan Åstrœm
*Lund University, Sweden*

# Introduction

Control and signal processing belong to the most prominent areas in science and engineering that have a great impact on the development of modern technology. To benefit from continuing progress of computing in the past century, most control and signal processing systems have been designed in the discrete-time domain using digital electronics. Because of the existence of mature theory, well-established design methods, and easier implementation, the classical digital control, signal processing, and electronic instrumentation are based on *periodic sampling*. Indeed, the analysis and design of control systems sampled equidistantly in time is relatively simple since the closed-loop system can be then described by difference equations with constant coefficients for linear time-invariant processes. As time is a critical issue in real-time systems, the adoption of time-driven model for sampling, estimation, control, and optimization harmonizes well with the development of time-triggered real-time software architectures. On the other hand, the recovery of signals sampled uniformly in time requires just low-pass filtering which promotes practical implementations of classical signal processing systems.

Although synchronous architectures of circuits and systems have monopolized electrical engineering since the 1960s, they are definitely suboptimal in terms of resource utilization. Sampling the signal and transmitting the samples over the communication channel, as well as occupying the central processing unit to perform computations when the signal does not change significantly, are evident waste of resources. Nowadays, the efficient utilization of communication, computing, and energy capabilities become a critical issue in various application domains because many systems have become increasingly networked, wireless, and spatially distributed. The demand for the economic use of computational and communication resources becomes indispensable as the digital revolution drives the development and deployment of pervasive sensing systems producing huge amount of data whose acquisition, storage, and transmission is a tremendous challenge. On the other hand, the postulate of providing ubiquitous wireless connectivity between smart devices with scarce energy resources poses a requirement to manage efficiently the constrained energy budget.

The response to these challenges claims the reformulation of answers to fundamental questions referred to discrete-time systems: "when to sample," "when to update control action," or "when to transmit information."

One of the generic answers to these questions is covered by adoption of the *event-based paradigm* that represents a model of calls for resources only if it is really necessary in system operation. In event-based systems, any activity is triggered as a response to *events* defined usually as a *significant change* of the state of relevance. Most "man-made" products, such as computer networks, complex software architectures, and production processes are natural, discrete event systems as all activity in these systems is driven by asynchronous occurrences of events.

The event-based paradigm was an origin of fundamental concepts of classical computer networking. In particular, the random access protocols for shared communication channels (ALOHA and CSMA) and the most successful commercial technology in local area networks (Ethernet) have been inherently based on the event-triggered model. The same refers to CSMA/CA scheme adopted in IEEE 802.11 standard to wireless local area network technology. Many fieldbus technologies for industrial communication are event-based controller area network. A representative example of the commercial platform designed consistently on the basis of event-based architecture is LonWorks (Local Operating Networks, ISO/IEC 14908), where the event-driven model is involved in task scheduling, strategy of data reporting, application programming, explicit flow control, and protocol stack design. The other well-known, event-driven communication architecture is controller area network, which is based on a message-oriented protocol of CSMA type. Perhaps, the most spectacular example of a widespread modern event-based system is Internet, the operation of which, as a global network, is dictated by events, such as error detection, message transmission and reception, and segmentation and reassembling. The rigorous theory for study of discrete event dynamic systems was developed in the 1980s [1] (see Chapter 2).

The notion of "event" is also an efficient abstraction for computer programming and data processing. Most graphical user interface (GUI) applications designed to be event-driven and reactive to certain kinds of events (e.g., mouse button clicks) while ignoring others (e.g., mouse movement). Many current trends in computing are event-based, for example, event stream processing, complex event processing, smart metering, or the Internet of Things. Finally, the event-based paradigm is one of the underlying concepts adapted to various aspects of human activity (e.g., event-driven business process management or event-driven marketing) as "events define our world."

The past decade has seen increasing scientific interest in event-based paradigm applied for continuous-time systems in a wide spectrum of engineering disciplines, including control, communication, signal processing, and electronic instrumentation. The significant research effort was initiated by publishing a few important research works independently in the context of control [2–4] and signal processing [5–9] at the turn of the century. These works released great research creativity and gave rise to a systematic development of new event-triggered approaches for continuous-time systems.

The concepts to involve the event-based paradigm in control, communication, and signal processing are in general not new. The event-based systems with continuous-time signals exist in nature. In particular, the hand control system is event driven because human hand movements are synchronized to input signals rather than to an internal clock (see also Chapter 14).

Event-based control is the dominating principle in biological neurons that process information using energy-efficient, highly parallel, event-driven architectures as opposed to clock-driven serial processing adopted in computers. The observation that the brain operates on event-based analog principles of neural computation that are fundamentally different from digital techniques in traditional computing was the motivation for the development of *neuromorphic engineering*. Event-based neuromorphic technologies inspired by the architecture of the brain (see [10]) was pioneered by the group of Mead at Caltech that proposed to integrate biologically inspired electronic sensors with analog circuits, as well as to introduce an address-event-based asynchronous, continuous-time communications protocol (see also Chapter 21).

Early event-based systems (termed also as *aperiodic* or *asynchronous*) have been designed in feedback control, data transmission, and signal processing at least since the 1950s. The first idea to adopt a *significant change of the signal* as a criterion for triggering the sampling of continuous-time signals, termed later as *Lebesgue sampling* [3], is attributed to Ellis [11] and appeared in the context of control although it does not refer explicitly to the concept of "event." In signal processing, the independent idea equivalent to Lebesgue sampling provided the ground for the invention of asynchronous delta modulation in the 1960s [12]. In a parallel line of research, the framework for the class of adaptive (time-triggered) sampling designed under event-based criteria was evolved during the 1960s and 1970s (see Chapter 3). In particular, the *constant absolute-difference criterion* introduced in [13] is the adaptive (time-triggered) sampling equivalent of the (event-triggered) Lebesgue sampling. Since the recent decade, the adaptive sampling has been applied in the *self-triggered control* strategy to cope with event unexpectedness and to avoid event detection by the use of prediction (see Chapter 10). Nowadays, the Lebesgue sampling, known also in industrial communication as *send-on-delta* scheme [14], is used in many industrial applications, including building automation, street lighting, and smart metering, and has been extended among others to spatiotemporal sampling adopted in sensor networks. The notion of the "event" in the technical context comes neither from control nor from signal processing but originated in computing in the 1970s [15].

In signal processing, the event-based scheme similar to the Lebesgue sampling was rediscovered by Mark and Todd at the beginning of 1980s and introduced under the name *level-crossing sampling* [16]. The inspiration to adopt the level-crossings as a sampling criterion was derived not from the control theory but from studies on the *level-crossing problem* in stochastic processes (see Chapter 3). The level-crossing rate is a function of the signal bandwidth, which for bandlimited Gaussian processes were proven by Rice in his monumental study published in the 1940s [17]. Hence, the level-crossing rate provides an estimate of local bandwidth for time-varying signals (see Chapter 23). In parallel, at least since the 1990s, event-based sampling has been used to solve problems in statistical signal processing (see Chapter 20). The adoption of the level-triggered sampling concept to hypothesis testing for distributed systems via Sequential Probability Ratio Test (SPRT) first appeared in [18] although without any reference to the event-based sampling. The potential of level-crossing scheme for signal processing applications has been significantly extended because of an observation made by Tsividis that the level-crossing sampling in continuous time results in no aliasing [6,7]. This observation was the ground for the formulation of a new concept of continuous-time digital signal processing (CT-DSP) based essentially on amplitude quantization, and dealing with binary signals in continuous time, which is complementary to conventional clocked DSP (see Chapters 15, 17, and 18).

In recent research studies, the concept of "event" evolves to address a wide range of design objectives and application requirements. The answer to the question "when to sample" depends on the technical context. In particular, the developments of new event-based sampling criteria for state estimation focused on reducing state uncertainty (e.g., variance-based triggering, or Kullback–Leibler divergence) (see Chapter 13) give rise to the extension of the traditional concept of event originated in computing: an event is now regarded not only to a significant change of the state, but may define a *significant change of the state uncertainty*.

The primary reason for the increasing interest in event-based design of circuits and systems is its superiority in the resource-constrained applications. The main benefit provided by adoption of event-based strategies in networked sensor and control systems consists in decreasing the rate of messages transferred through the shared communication channel (see Chapter 3), which for wireless transmission allows to minimize power consumption.

On the other hand, the reduction of communication traffic is the most effective method to minimize the network-induced imperfections on the control performance (i.e., packet delay, delay jitter, and packet losses). In this way, the adoption of the event-triggered strategy meets one of the fundamental criteria of the networked control system design. This criterion is aimed to achieve performance close to that of the conventional point-to-point digital control system while still retaining well-known advantages provided by networking (e.g., high flexibility, efficient reconfigurability, reduction of installation and maintenance costs).

The other strong motivation behind the development of event-based methodology is that the networked control systems with nonzero transmission delays and packet losses are event-based. Therefore, the control loops established over networks with varying transmission delays and packet losses can no longer be modeled by conventional discrete-time control theory, even if data are processed and transmitted with a fixed sampling rate. Hence, the strategies for event-based control are expected to be similar to strategies for networked control.

Finally, the appropriability of event-based control is not restricted only to economizing the use of digital networks but relates also to a class of control problems in which the independent variable is different from time. For example, it is more efficiently to sample the crankshaft angle position in the value domain instead of in time (see Chapter 1).

In parallel to the development of the event-triggered communication and control strategies, the event-based paradigm has been adopted during the past decade in signal processing and electronic circuitry design as an alternative to the conventional approach based on periodic sampling. The motivation behind event-based signal processing is that the modern nanoscale VLSI technology promotes fast circuit operation but requires low supply voltage (even below 1 V). The latter makes the fine quantization of the amplitude increasingly difficult as the voltage increment corresponding to the least significant bit may become close to noise level. In the event-based signal processing mode, the periodic sampling is substituted by discretization in the amplitude (e.g., level crossings), whereas the quantization process (if any) is moved from the amplitude to the time domain. This allows for the removal of the global clock from system architecture and ensures benefits from growing time resolution provided by the downscaling of the feature size of VLSI technology. The technique of encoding signals in time instead of in amplitude is expected to be further improved with advances in chip fabrication technology. Finally, the event-based signal processing systems are characterized in general by activity-dependent power consumption, and energy savings on the circuit level.

In response to challenges of nanoscale VLSI design, the development of event-based signal processing techniques has resulted in the introduction of a new class of analog-to-digital converters and digital filters that use event-triggered sampling (see Chapter 22), as well as a novel concept of CT-DSP. According to Tsividis' classification, the CT-DSP may be regarded as a "missing category" among well-established signal processing modes: continuous-time continuous-amplitude (analog processing), discrete-time discrete-amplitude (digital processing), and discrete-time continuous-amplitude processing techniques (see Chapters 15, 17, and 18).

Although major concepts and ideas in both technical areas are similar, so far the event-based control and event-based signal processing have evolved independently, being explored by separate research communities in parallel directions.

But control and signal processing are closely related. Many controllers can be viewed as special kind of signal processors that convert an input and a feedback into a control signal. The history of engineering provides notable examples of successful interactions between communities working on control and signal processing domains. In particular, the term "feedback," which is an ancient idea and a central concept of control, was used for the first time in the context of signal processing to express boosting the gain of an electronic amplifier [19]. On the other hand, the Kalman filter, which is now a critical component of many signal processing systems, has been developed and enhanced by engineers involved in the design of control systems [17]. One of main objectives of this book is to provide interactions between both research areas because concepts and ideas migrate between the fields.

Marek Miśkowicz

## Bibliography


[1] C. G. Cassandras, S. Lafortune, *Introduction to Discrete Event Systems*. Springer, New York, 2008.

[2] K. E. Årzén, A simple event-based PID controller. In *Proceedings of IFAC World Congress*, 1999, vol. 18, pp. 423–428.

[3] K. J. Åstrom, B. Bernhardsson, Comparison of periodic and event based sampling for firstorder stochastic systems. In *Proceedings of IFAC World Congress*, 1999, pp. 301–306.

[4] W. P. M. H. Heemels, R. J. A. Gorter, A. van Zijl, P. P. J. van den Bosch, S. Weiland, W. H. A. Hendrix, et al., Asynchronous measurement and control: A case study on motor synchronization. *Control Eng. Pract.*, vol. 7, pp. 1467–1482, 1999.

[5] N. Sayiner, H. V. Sorensen, T. R. Viswanathan, A level-crossing sampling scheme for A/D conversion. *IEEE Trans. Circuits Syst. II: Analog Digital Signal Process*, vol. 43, no. 4, pp. 335–339, 1996.

[6] Y. Tsividis, Continuous-time digital signal processing. *Electron. Lett.*, vol. 39, no. 21, pp. 1551–1552, 2003.

[7] Y. Tsividis, Digital signal processing in continuous time: A possibility for avoiding aliasing and reducing quantization error. In *Proceedings of IEEE International Conference on Acoustics, Speech, and Signal Processing ICASSP 2004*, 2004, pp. 589–592.

[8] E. Allier, G. Sicard, L. Fesquet, M. Renaudin, A new class of asynchronous A/D converters based on time quantization. In *Proceedings of IEEE International Symposium on Asynchronous Circuits and Systems ASYNC 2003*, 2003, pp. 196–205.

[9] F. Aeschlimann, E. Allier, L. Fesquet, M. Renaudin, Asynchronous FIR filters: Towards a new digital processing chain. In *Proceedings of International Symposium on Asynchronous Circuits and Systems ASYNC 2004*, 2004, pp. 198–206.

[10] S. C. Liu, *Event-Based Neuromorphic Systems*. Wiley, 2015.

[11] P. H. Ellis, Extension of phase plane analysis to quantized systems. *IRE Trans. Autom. Control*, vol. 4, pp. 43–59, 1959.

[12] H. Inose, T. Aoki, K. Watanabe, Asynchronous delta-modulation system. *Electron. Lett.*, vol. 2, no. 3, pp. 95–96, 1966.

[13] J. Mitchell, W. McDaniel, Adaptive sampling technique. *IEEE Trans. Autom. Control*, vol. 14, no. 2, pp. 200–201, 1969.

[14] M. Miśkowicz, Send-on-delta concept: An *event-based* data reporting strategy *Sensors*, vol. 6, pp. 49–63, 2006.

[15] R. E. Nance, The time and state relationships in simulation modeling. *Commun. ACM*, vol. 24, no. 4, pp. 173–179, 1981.

[16] J. W. Mark, T. D. Todd, A nonuniform sampling approach to data compression. *IEEE Trans. Commun.*, vol. 29, no. 1, pp. 24–32, 1981.

[17] S. O. Rice, Mathematical analysis of random noise. *Bell System Tech. J.*, vol. 24, no. 1, pp. 46–156, 1945.

[18] A. M. Hussain, Multisensor distributed sequential detection. *IEEE Trans. Aerosp. Electron. Syst.*, vol. 30, no. 3, pp. 698–708, 1994.

[19] W. S. Levine, Signal processing for control. In *Handbook of Signal Processing Systems*, S. S. Bhattacharyya, E. F. Deprettere, R. Leupers, J. Takala, Eds. Springer, New York, 2010.

# About the Book

The objective of this book is to provide an up-to-date picture of the developments in event-based control and signal processing, especially in the perspective of networked sensor and control systems.

Despite significant research efforts, the event-based control and signal processing systems are far from being mature and systematically developed technical domains. Instead, they are the subject of scientific contributions at the cutting edge of modern science and engineering. The book does not pretend to be an exhaustive presentation of the state-of-the-art in both areas. It is rather aimed to be a collection of studies that reflect major research directions. Due to a broad range of topics, the mathematical notation is not uniform throughout the book, but is rather adapted by the authors to capture the content of particular chapters in the clearest way possible.

The Introduction to the book outlines the broad range of technical fields that adopt the events as an underlying concept of system design and provides a historical background of scientific and engineering developments of event-based systems. The present section includes short presentations of the book contributions.

The book contains 23 chapters divided in two parts: Event-Based Control and Event-Based Signal Processing. Over 60 authors representing leading research teams from all over the world have contributed to this work. Each part of the book begins with a survey chapter that covers the topic discussed: Chapter 1 for "Event-Based Control" and Chapter 15 for "Event-Based Signal Processing."

Most chapters included in Part I, "Event-Based Control," describe the research problems in the perspective of networked control systems and address research topics on event-driven control and optimization of hybrid systems, decentralized control, event-based proportional-integral-derivative (PID) regulators, periodic event-triggered control (PETC), event-based state estimation, self-triggered and team-triggered control, event-triggered and time-triggered real-time computing and communication architectures for embedded systems, intermittent control, model-based event-triggered control, and event-based generalized predictive control.

Chapter 1 is a tutorial survey on the main methods of analysis and design of event-based control and includes links to the other chapters in the book. It has been emphasized in particular that the event-based control may be viewed as a combination of feedforward and feedback control in the sense that the feedback loop is closed only at event times while remaining open between consecutive events. Although this is the case with all the discrete-time control systems, the average time interval between event-triggered control updates is in general much longer than the sampling period in the classical approach to sampled-data control. Therefore, in the event-triggered control, modeling the plan behavior only at the sampling instants, as in classical sampled-data control theory, is insufficient and requires integration of the continuous-time model for time intervals between events.

Chapter 2 focuses on the issue of how to apply the event-driven paradigm to control and optimization of hybrid systems, including both time-driven and event-driven components. The general-purpose control and optimization problem is carried out by evaluation (or estimation for stochastic environment) of gradient information related to a given performance metric using infinitesimal perturbation analysis (IPA). The adoption of IPA calculus to control and optimize hybrid systems results in a set of simple and computationally efficient, event-driven iterative equations and is advantageous because the size of the event space of the system model is much smaller than of the state space.

Chapter 3 provides a survey on event-triggered sampling schemes applied both in control and signal processing systems, including threshold-based sampling, Lyapunov sampling, reference-crossing sampling, cost-aware sampling, quality-of-service-based sampling, and adaptive sampling. Furthermore, the chapter contains derivation of the analytical formulae referred to the mean sampling rate for threshold-based sampling schemes (send-on-delta, send-on-area, and send-on-energy). It is shown that the *total variation* of the sampled signal is an essential measure to approximate the number of samples produced by the send-on-delta (Lebesgue sampling) scheme. Furthermore, Chapter 3 addresses the problem of evaluating the reduction of event-based sampling rate compared with periodic sampling. As proved, the lower bound for send-on-delta reduction of communication (guaranteed reduction) is determined by the ratio of the peak-to-mean slope of the sampled signal and does not depend on the parameter of the sampling algorithm (threshold size).

Chapter 4 gives a comparative analysis of event-triggered and time-triggered real-time computing and communication architectures for embedded control applications in terms of composability, real-time requirements, flexibility, and heterogeneous timing models. The attention of the reader is focused on time-triggered architectures as a complementary framework to event-triggered models in implementation of real-time systems. Finally, the chapter concludes with a case study of a control application realized alternatively as the event-triggered and as the time-triggered system and their comparative analysis based on experimental results.

Chapter 5 extends the event-based state-feedback approach introduced in technical literature to the distributed event-based control of physically interconnected systems. To avoid the realization of the distributed state feedback with continuous communications between subsystems, it is proposed to introduce interactions between the components of the networked controller only at the event times although the feedback of the local state information is still continuous. Two event-based communication schemes initiated by a local controller have been adopted: the transmission of local state information to and the request for information from the neighboring subsystems. The chapter concludes with a demonstration of the distributed event-based state-feedback approach for a thermofluid process by simulation and experimental results.

Chapter 6 focuses on the analysis of PETC systems that combine the benefits of time-triggered and event-triggered control strategies. In PETC, the sensor is sampled equidistantly and event occurrence is verified at every sampling time to decide on the transmission of a new measurement to a controller. The PETC can be referred to the concept of globally asynchronous locally asynchronous (GALS) model of computation introduced in the 1980s and adopted as the reference architecture for digital electronics, including system-on-chip design. Furthermore, the PETC suits practical implementations in embedded software architectures and is used, for example, in event-based PID controllers (see Chapter 12), LonWorks commercial control networking platform (ISO/IEC 14908), and EnOcean energy harvesting wireless technology.

The chapter focuses on approaches to PETC that include a formal analysis framework, which applies for continuous-time plants and incorporates intersample behavior. The analysis first addresses the case of plants modeled by linear continuous-time systems and is further extended to preliminary results for plants with nonlinear dynamics. The chapter concludes with a discussion of open research problems in the area of PETC.

In Chapter 7, a decentralized wireless networked control architecture is considered where the sensors providing measurements to the controller have only a partial view of the full state of the system. The objective of this study is to minimize the energy consumption of the system not only by reducing the communication but also by decreasing the time the sensors have to listen for messages from the other nodes. Two techniques of decentralized conditions are examined. The first technique requires synchronous transmission of measurements from the sensors, while the second one proposes the use of asynchronous measurements. Both techniques are finally compared through some examples followed by a discussion of their benefits and shortcomings. By employing asynchronous updates, the sensor nodes do not need to stay listening for updates from other sensors thus potentially bringing some additional energy savings. The chapter concludes with ideas for the future and ongoing extensions of the proposed approach.

Chapter 8 focuses on the practical implementation of event-based strategy using the generalized predictive control (GPC) scheme, which is the most popular model predictive control (MPC) algorithm in industry. To circumvent complexity of MPC implementation, a straightforward algorithm that includes event-based capabilities in a GPC controller is provided both for sensor and actuator deadbands. The chapter concludes with a simulation study, where the event-based control schemes are evaluated for the greenhouse temperature control problem that highlights the advantages of the proposed event-based GPC controllers.

Chapter 9 presents the model-based event-triggered (MB-ET) control framework for networked systems. The MB-ET control makes explicit use of existing knowledge of the plant dynamics built into its mathematical model to enhance the performance of the system and to use network communication resources more efficiently. The implementation of an event-based scheme in the model-based approach to networked control systems represents a very intuitive way of saving network resources and has many advantages compared with the classical periodic implementation (e.g., robustness to packet dropouts).

In Chapter 10, the extensions to event-triggered control called *self-triggered* and *team triggered control* are presented. Unlike the event-triggered execution model that consists in taking an action when a relevant event occurs, in the self-triggered control strategy, an appropriate action is scheduled for a time when an appropriate event is predicted to happen. Unlike the event-triggered strategy, the self-triggered control does not require continuous access to some state or output of the system in order to monitor the possibility of an event being triggered, which can be especially costly to implement over networked systems. However, the self-triggered algorithms can provide conservative decisions on when actions should take place to properly guarantee the desired stability results. The team-triggered control approach summarized in Chapter 10 combines ideas from both event- triggered and self-triggered controls to help overcome these limitations.

Chapter 11 addresses the problem of conditions under which the event-triggered system is *efficiently attentive* in the sense that the intersampling interval gets longer when the system approaches the equilibrium. The problem of efficient attention is formulated from the perspective of networked control systems when the communication channel transmits data in discrete packets of a fixed size that are received after a finite delay. The actual bandwidth needed to transmit the sampled information is defined as a *stabilizing instantaneous bit rate* that equals the number of bits used to encode each sampled measurement and the maximum delay with which the sampled-data system is known to be asymptotically stable. The chapter provides a sufficient condition for event-triggered systems to be efficiently attentive. Furthermore, the study introduces a dynamic quantization policy and identifies the maximum acceptable transmission delay for which the system is assured to be locally input-to-state stable and efficiently attentive.

The objective of Chapter 12 is to give an overview of the state-of-the-art on event-based PID control. The introduction of the event-based PID controller by Årzén in 1999 was one of the stimuli for increasing research interest in event-based control. The PID regulators are the backbone of most industrial control systems involved in more than 90% of industrial control loops. The chapter synthesizes research on event-based PID control and provides practical guidelines on design and tuning of event-based PID controllers that allow for an adjustable trade-off between high control performance and low message rate.

Chapter 13 refers to event-based state estimation discussed from the perspective of distributed networked systems that benefit from the reduction in resource utilization as the sensors transmit the measurements to the estimator according to the event-triggered framework (see also Chapter 20). The send-on-delta sampling, predictive sampling (send-on-delta with prediction), and matched sampling are analyzed in this chapter in terms of their applicability to state estimation. The matched sampling uses the threshold-based scheme in relation to the Kullback–Leibler divergence that shows how much information has been lost since the recent measurement update. Unlike the send-on-delta scheme that is aimed to detect the difference between two measurement values, the Kullback–Leibler divergence is used for comparison of two probability distributions representing state estimates. The approaches discussed in the chapter adopt not only events to state estimation but utilize also *implicit information* defined as *a lack of event* to improve system performance between events. Hence, Chapter 13 belongs to a small number of works on event-based control that propose to utilize the implicit information to improve system performance (see also Section 3.2.9 in Chapter 3).

Chapter 14 gives an extensive overview of the current state-of-the-art of the event-based intermittent control discussed in the context of physiological systems. Intermittent control has a long history in the physiological literature and appears in various forms in engineering literature. In particular, intermittent control has been shown to provide an event-driven basis for human control systems associated with balance and motion control. The main motivation to use intermittent control in the context of event-based systems is the observation that the zero-order hold is inappropriate for event-based systems as the intersample interval varies with time and is determined by the event times.

The intermittent approach avoids these issues by replacing the zero-order hold by the *system-matched hold* (SMH), a type of generalized hold based on system states, which explicitly generates an open-loop intersample control trajectory using the underlying continuous-time closed-loop control system. Furthermore, the SMH-based intermittent controller is associated with a separation principle similar to that of the underlying continuous-time controller, which, however, is not valid for zero-order hold. An important consequence of this separation principle is that neither the design of the SMH nor the stability of the closed-loop system is dependent on sample interval. Hence, the SMH is particularly appropriate when sample times are unpredictable or nonuniform, possibly arising from an event-driven design. Finally, at the end of the chapter, challenging problems for relevant future research in intermittent control have been indicated.

In Part II, "Event-Based Signal Processing," the technical content focuses on topics such as continuous-time digital signal processing, its spectral properties and hardware implementations, statistical event-based signal processing in distributed detection and estimation, asynchronous spike event coding technique, perfect signal recovery, including nonstationary signals based on event-triggered samples, event-based digital filters, and local bandwidth estimation by level-crossing sampling for signal reconstruction at local Nyquist rate.

Chapter 15 opens part of the book related to event-based signal processing. This chapter discusses research on event-based, continuous-time signal acquisition and digital signal processing presented as an alternative to well-known classical analog (continuous both in time and in amplitude) and classical digital (discrete in time and in amplitude) signal processing modes. The continuous-time digital signal processing is characterized by no aliasing, immediate response to input changes, lower in-band quantization error, and activity-dependent power dissipation. Although the level-crossing sampling has been known for decades, earlier works seem not to have recognized aliasing as its property in continuous-time, perhaps because in those references the time was assumed discretized. Although the discussion in the chapter is concentrated on the level-crossing sampling realized in continuous time, most of the principles presented are valid for other types of event-based sampling, and for the technique that uses discretized time. The chapter also discusses design challenges of CT-DSP systems and reports experimental results from recent test chips, operating at kilohertz (kHz) to gigahertz (GHz) signal frequencies.

In the CT-DSP mode discussed in Chapter 15, the method of signal reconstruction from the time instants of amplitude-related events by the zero-order hold is focused on feasibility of practical implementation rather than perfectibility of signal recovery. The theoretical issue of perfect signal reconstruction from the discrete events (e.g., level crossings) requires the use of advanced methods of linear algebra. The recovery of signals from event-based samples may use methods developed for more general class of nonuniform sampling. Although a number of surveys have been published in the area of nonuniform sampling and its applications, this literature is, however, addressed more toward applied mathematicians rather than engineers because of its focus on advanced knowledge on functional and complex analysis. Chapter 16 tries to bridge the gap in the scientific literature and is aimed to be a tutorial on advanced mathematical issues referred to processing of nonuniformly sampled signals for engineers and keeps a connection with the constraints of real implementations. The problems of perfect signal recovery of nonuniformly sampled signals are addressed also in Chapters 19 and 23.

Chapter 17 refers to spectral analysis of continuous-time ADC and DSP and, thus, is closely related to Chapter 15. Chapter 17 discusses the spectral features of the quantization error introduced by continuous-time amplitude quantization. In particular, it is shown that the amplitude quantization in continuous time can be analyzed as a memoryless nonlinear transformation of continuous signals, which yield spectral results that are harmonic based, rather than noise oriented. Furthermore, the chapter addresses the effects of fine quantization of the time axis by uniform sampling of piecewise constant signals needed, for example, for interface compatibility with standard data processing.

Chapter 18 contains concepts for hardware-efficient design of CT-DSP. The most critical point in the implementation of the CT-DSP is the realization of time-continuous delay elements that require considerable amount of hardware. In the chapter, several new concepts developed recently for the implementation of delay elements, which enable significant hardware reduction, have been discussed. In particular, the study includes the proposition of a new CT base cell for the CT delay element built of a cascade of cells that allow the chip area be reduced by up to 90% at the cost of negligible minor performance losses. It is also shown that the hardware requirements can be reduced by applying asynchronous sigma-delta modulation (ASDM) for quantization. Finally, filter architectures especially well-suited for event-driven signal processing are proposed. The presented design methods help to evolve event-driven systems based on continuous-time DSPs toward commercial application.

Chapter 19 discusses the issues related to asynchronous data acquisition and processing of nonstationary signals. The nonstationary signals occur in many applications, such as biomedical and sensor networking. As follows from the Wold-Cramer spectrum, the distribution of the power of a nonstationary signal is a function of time and frequency; therefore, the sampling of such signals needs to be signal dependent to avoid collection of unnecessary samples when the signal is quiescent. In the chapter, sampling and reconstruction of nonstationary signals using the level-crossing sampling and ASDM are discussed (see Chapters 16 and 23 that also address the signal recovery problems). It is argued that level-crossing is a suitable sampling scheme for nonstationary signals, as it has the advantage of not requiring bandlimiting constraint. On the other hand, the prolate spheroidal wave or the Slepian function is more appropriate than the classical sinc functions for reconstruction of nonstationary signals as the former have better time and frequency localization than the latter. Unlike the level-crossing sampler, the ASDM encodes the signal only into sample times without amplitude quantization but requires the condition that the input signal must be bandlimited. To improve signal encoding and processing, the method of decomposition of a nonstationary signal using the ASDM by latticing the time–frequency plane, as well as the modified ASDM architecture, are discussed.

Chapter 20 addresses statistical signal processing problems by the use of event-based strategy (see also Chapter 13). First, the decentralized detection problem based on statistical hypothesis testing is considered, in which a number of distributed sensors under energy and bandwidth constraints report sequentially a summary of their discrete-time observations to a fusion center, which makes a decision satisfying certain performance constraints. The level-triggered sampling has been already used to transmit effectively the sufficient local statistics in decentralized systems for many applications (e.g., spectrum sensing and channel estimation in cognitive radio networks, target detection in wireless sensor networks, and power quality monitoring in power grids). In the context of the decentralized detection problem, the chapter discusses the event-based information transmission schemes at the sensor nodes and the sequential decision function at the fusion center. The second part of the chapter, by the use of the level-triggered sampling, addresses the decentralized estimation of linear regression parameters, which is one of the common tasks in wireless sensor networks (e.g., in system identification or estimation of wireless multiple access channel coefficients). It is shown that the decentralized detection and estimation schemes based on level-triggered sampling significantly outperform the schemes based on conventional uniform sampling in terms of average stopping time.

Chapter 21 focuses on a technique of signal processing named the asynchronous spike event coding (ASEC) scheme associated with the communication method known as the address event representation (AER) and realized via an asynchronous digital channel. The ASEC consists of analog-to-spike and spike-to-analog coding schemes based on asynchronous delta modulation. The ASEC is suitable to perform a series of arithmetic operations for analog computing (i.e., negation, amplification, modulus, summation and subtraction, multiplication and division, and averaging). Although digital circuits are more robust and flexible than their analog counterparts, analog designs usually present a more efficient solution than digital circuits because they are usually smaller, consume less power, and provide a higher processing speed. The AER communication protocol originates from ideas to replicate concepts of signal processing in biological neural systems into VLSI systems. The applications of ASEC are programmable analog arrays (complementary to FPGAs) for low-power computing or remote sensing, mixed-mode architectures, and neuromorphic systems.

Chapter 22 focuses on synthesis and design of discrete-time filters using the level-crossing sampling. It presents several filtering strategies, such as finite impulse-response (FIR) and infinite impulse-response (IIR) filters. Furthermore, a novel strategy of filter synthesis in the frequency domain based on the level-crossing sampling referred to the transfer function is introduced. This novel strategy allows for a significant reduction of the number of coefficients in filter design. Finally, the architecture for hardware implementation of the FIR filter based on an asynchronous, data-driven logic is provided. The designed circuits are event-driven and able to process data only when new samples are triggered, which results both in a reduction of system activity and average data rate on the filter output.

In Chapter 23, the problem of perfect recovery of bandlimited signals based on local bandwidth is considered. In this sense, the content in Chapter 23 is related to the content in Chapters 16 and 19. In general, the signal processing aware of local bandwidth refers to a class of signals, whose local spectral content strongly varies with time. These signals, represented, for example, by FM signals, radar, EEG, or neuronal activity, do not change significantly their values during some long time intervals, followed by rapid aberrations over short time intervals. By exploiting local properties, these signals can be sampled faster when the local bandwidth becomes higher and slower in regions of lower local bandwidth, which provides the potential for more efficient resource utilization. The method which is used for this purpose as discussed in Chapter 23 is the level-crossing sampling because the mean level-crossing rate depends on the power spectral density of the signal. The estimates of the local bandwidth are obtained by evaluating the number of level crossings in time windows of a finite size. The chapter develops a methodology for estimation of local bandwidth by level-crossing sampling for signals modeled by Gaussian stochastic processes. Finally, the recovery of the signal from the level crossings based on the methods suited for irregular samples combined with time warping is presented.

Marek Miśkowicz

# Part I

# Event-Based Control

# 1

## Event-Based Control: Introduction and Survey

**Jan Lunze**
*Ruhr-University Bochum*
*Bochum, Germany*

**CONTENTS**



**ABSTRACT** This chapter gives an introduction to the aims and to the analysis and design methods for event-based control, surveys the current research topics, and gives references to the following chapters of this book.

### 1.1 Introduction to Event-Based Control

#### 1.1.1 Basic Event-Based Control Configuration

Event-based control is a new control method that closes the feedback loop only if an event indicates that the control error exceeds a tolerable bound and triggers a data transmission from the sensors to the controllers and the actuators. It is currently being developed as an important means for reducing the load of the digital communication networks that are used to implement feedback control. This chapter explains the main ideas of event-based control, surveys analysis and design methods, and gives an overview of the problems and solutions that are presented in the following chapters of this book.

Feedback is the main principle of control. It enables a controller

- To stabilize an unstable plant
- To attenuate disturbances
- To change the dynamical properties of a plant
- To tolerate uncertainties of the model that has been used to design the controller

This principle has been elaborated in detail for continuous-time systems and for implementing controllers in a discrete-time setting as sampled-data controllers with sufficiently high sampling rate.

If the reasons for information feedback mentioned above do not exist in an application, feedforward control can be used, which does not necessitate any communication network and, moreover, leads to a quicker system response. In many situations, feedforward control is combined with feedback control in a two-degrees-of-freedom configuration to take advantage of the characteristics of both control methods. This combination shows that an information feedback is necessary only to the extend of instability of the plant, to the magnitude of the effects of disturbances, to the necessity to change the dynamics of the plant, and to the size of uncertainties in the model used for control design. Event-based control combines feedforward and feedback control in the sense that only at the event times is the feedback loop closed, and between consecutive event times, the control is carried out in an open-loop fashion. Hence, event-based control lives in the range between both control methods. It chooses the sampling instances online in dependence on the current plant behavior and in doing so it answers the question when information feedback is necessary to reach a desired level of performance.

**Basic event-based control configuration.** Figure 1.1 shows the basic configuration of event-based control. A digital communication network is used to implement the data flow from the event generator toward the controller and from the controller toward the control input generator. The dashed lines indicate that this data flow occurs only at the event time instants $t_k$ $(k = 0, 1, \ldots)$. In the typical situation considered in event-based control, the communication network is not used exclusively for the control task at hand, but also for other tasks; hence, it may introduce a time delay or even loss of information into the control loop in dependence on the overall data traffic. To emphasize this situation, digital networks are drawn as clouds in the figures of this chapter.

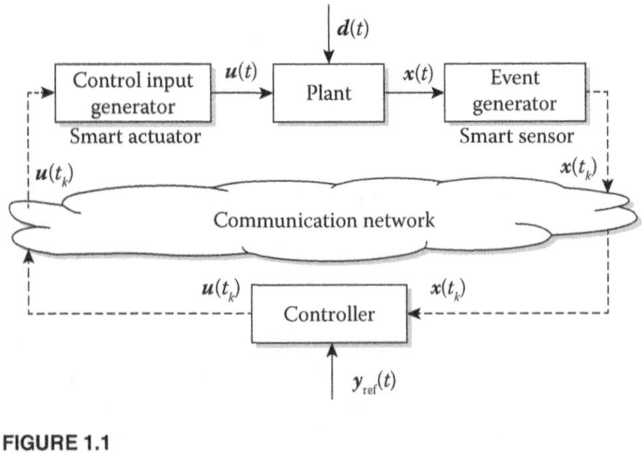


**FIGURE 1.1**
Event-based control loop.

There are three components to be designed for event-based control:

- **The event generator** evaluates continuously the current behavior of the plant, which is expressed in the measured state $x(t)$ or output $y(t)$, generates events, and initiates at the event times $t_k$ the data transfer of the current state $x(t_k)$ or output $y(t_k)$ toward the controller.

- **The controller** acts at the time instants $t_k$ and determines the current input $u(t_k)$ of the plant in dependence on the new information $x(t_k)$ or $y(t_k)$ and the reference trajectory $y_{\text{ref}}(t)$.

- **The control input generator** uses the values $u(t_k)$ of the control input received from the controller to generate the continuous-time input $u(t)$ for the time interval $t \in [t_k, t_{k+1})$ before it receives the next input value at time $t_{k+1}$.

In general, the event generator and the control input generator are, like the controller, dynamical components. They are implemented on smart sensors or smart actuators, which have enough computing capabilities to solve the equations that describe the event-generation rules or determine the progress of the input signals, respectively. In literature, zero-order holds have been widely used as control input generators, although this simple element limits the effectiveness of event-based control considerably. For the investigation of the control loop shown in Figure 1.1, the controller is often integrated into the control input generator, and the control loop is simplified accordingly.

**Motivation for event-based control.** The main motivation of the huge current research interest in event-based control results from the aim to adapt the usage of all elements in a control loop to the current needs. In particular, the usage of the communication network should be considerably reduced in time intervals, in which the plant is in the surroundings of a steady state and the network can be released for other tasks. Likewise, the sensors, the actuators, and the controller hardware should be used only if necessary to save energy in battery-powered equipment. All these reasons for event-based control are summarized in the popular term of a *resource-aware design* of the control loop.

In addition, there are several reasons why event-based control is an excellent option:

- **Asynchronous communication and real-time computation:** The main digital components of a control loop work in asynchronous modes and necessitate synchronization steps whenever two of them have to communicate information.

Hence, if these components are not specifically scheduled for control purposes, the standard assumptions of sampled-data control with fixed sampling instants and delay-free transmissions are not satisfied, and a more detailed analysis needs to be done by methods of event-based control. Due to such technological circumstances, field bus systems, asynchronous transfer mode networks, or wireless communication networks may not be available if needed by the control loop, and even the sampling theorem can be violated.

- **Working principles of technological systems:** Several technologies require adaptation of the control systems to the timing regime of the plant, for example, internal combustion engines, DC (direct current)/DC converters, or satellites. For example, the control of internal combustion engines does not use the time as the independent variable, but instead uses the crankshaft angle, the progress of which depends on the engine speed and not on the clock time. Sampling should not be invoked by a clock, but by the movement of the engine.
- **Organization of real-time programming:** Shared utilization of computing units in concurrent programming can be made much easier if the real-time constraints are not given by strict timing rules, but by "soft" bounds. Then, the average performance of the computing system is improved at the price of shifting the sampling instants of the controllers.
- **State-dependent measurement principles:** There are several measurement principles that lead to nonperiodic sensor information. For example, the positions of rotating axes are measured by equidistant markers, which appear in the sensor quicker or slower in accordance with the rotation speed. Quantization of measurement signals may lead likewise to nonperiodic sensor information.

These and further situations show that the motivation for event-based control is not restricted to economizing the use of digital networks and, hence, will not be overcome by future technological progress in communication.

### 1.1.2 Event-Based versus Sampled-Data Control

For the digital implementation of control loops, sampled-data control with the sample-and-hold configuration shown in Figure 1.2 has been elaborated during the last decades and numerous modeling, design, and implementation methods and tools are available. As Figure 1.2 shows, this standard configuration is similar to event-based control with the event generator replaced by the sampler and the control input generator by a hold element. The main difference is in the fact that in sampled-data control, both the sampler and the hold are triggered by a clock, which determines the time instants

$$t_k = kT_s, \quad k = 0, 1, \ldots, \tag{1.1}$$

for a reasonably chosen sampling period $T_s$. The periodic sampling occurs after the elapse of a time, not in dependence upon the behavior of the plant. Therefore, in the top part of Figure 1.3, the event instants are

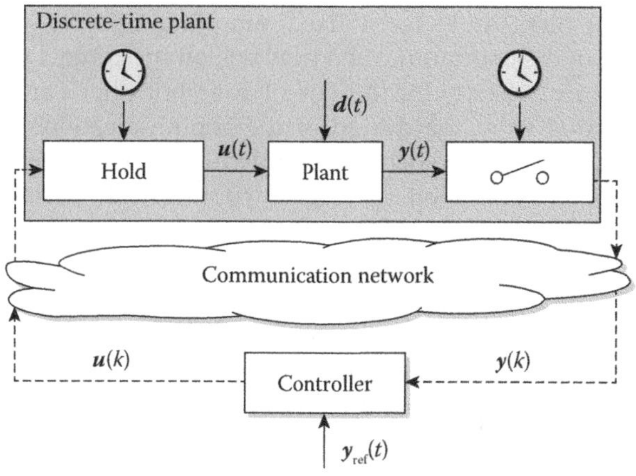


**FIGURE 1.2**
Sampled-data control loop.

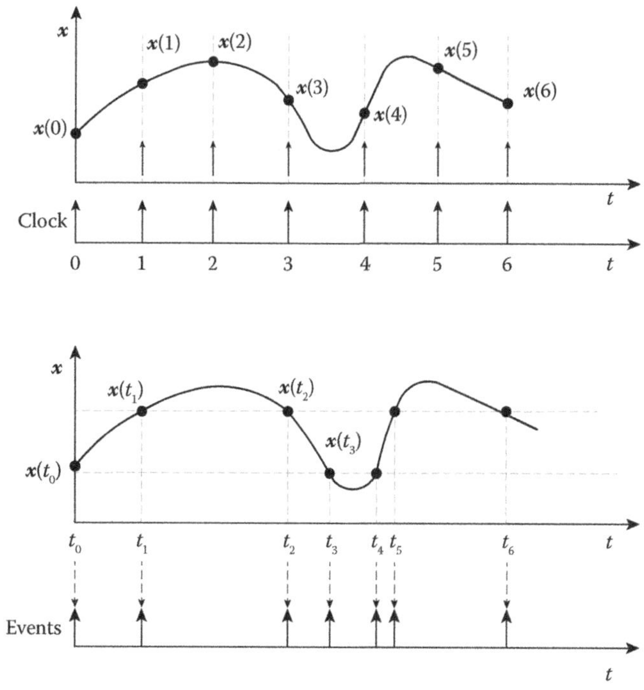


**FIGURE 1.3**
Periodic versus event-triggered sampling.

equidistant on the time axis, and the lines from the event markers toward the measurement values $x(k)$ are drawn as arrows from the clock axis toward the curve showing the time evolution of the state $x(t)$.

The usefulness of sampled-data control results from the fact that the continuous-time plant together with the sampler and the hold can be represented in a compact form by a discrete-time state-space model or for linear systems by a transfer function in the $z$-domain. This model is not an approximation but a precise representation of the input–output behavior of the plant at the time instances (1.1), which are enumerated by $k$ $(k = 0, 1, \ldots)$. It includes the dynamics of the plant and—for a fixed sampling time $T_s$—the effect of the communication network on the control performance. Hence, the design of the controller can be separated from the design of the communication network, and a comprehensive sampled-data control theory has been elaborated for digital control with periodic sampling.

Under these circumstances, it is necessary to clearly determine why a new theory of event-based control is necessary and useful. The starting point for this argumentation is shown in the bottom part of Figure 1.3. In event-based control, the system itself decides when to sample next. Therefore, the dashed arrows between the measurement points and the events are oriented from the curve of $x(t)$ toward the event markers. In the figure, two event thresholds are drawn as horizontal dashed lines. Whenever the signal crosses one of these thresholds, sampling is invoked. Obviously, the sampling frequency is adapted to the behavior of the plant with quicker sampling in time intervals with quick dynamics and slower sampling in time intervals with slow state changes. To emphasize the difference between periodic and event-based sampling, one can characterize the periodic sampling as *feedforward sampling* and the event-based approach as *feedback sampling*, where the sampling itself influences the selection of the next sampling instant.

There is a severe consequence of the step from periodic toward event-based sampling: Sampled-data control theory is no longer applicable, because its fundamental assumption of a periodic sampling is violated. As the sampling time is not predictable and the sampling frequency changes from one event to the next, there is no chance to represent the plant behavior at the sampling instants in a compact form as in sampled-data control theory, but the continuous-time model has to be integrated for the current sampling time $T_s(k) = t_{k+1} - t_k$. As shown in Section 1.2.3 in more detail, the closed-loop system represents a hybrid dynamical system, and this characteristic leads to rather complex analysis and design tasks.

Event-based control leads to new theoretical problems. The event generator, the controller, and the control input generator have to be designed so as to answer the following fundamental questions:

- When should information be transmitted from the sensors via the controller toward the actuators?
- How should the control input be generated between two consecutive event instants?

If these components are chosen appropriately, the main question of event-based control can be answered:

- Which improvements can be obtained by event-based control in comparison to sampled-data control with respect to an effective operation of all components in the control loop?

The answer to this question depends, of course, on the practical circumstances of the control loop under investigation, which determine whether the usage of the communication network, the activities of the sensors or the actuators, or the complexity of the control algorithm bring about the main restrictions on the design and the implementation of the controller.

Event-based control leads to more complex online computational problems than sampled-data control and should, therefore, be used only for the reasons mentioned in Section 1.1.1. Otherwise, sampled-data control, which is easier to design and implement, should be used. A more detailed comparison of event-triggered versus time triggered system can be found in Chapter 3.

### 1.1.3 Architecture of Event-Based Control Loops

This section discusses three different architectures in which event-based control has been investigated in the past.

**Single-loop event-based control.** The basic structure shown in Figure 1.1 applies to systems, where, with respect to the communication structure, all sensors and all actuators are lumped together in a single node each. Consequently, the current measurement data are collected at the sensor node and sent simultaneously to the controller. Analogously, the updated values of all input signals of the plant are sent together from the controller toward the actuator node. If the communication network introduces some time delay into the feedback loop, then all output data or input data are delayed simultaneously.

**Multivariable event-based control.** Figure 1.4 shows a multivariable control loop, where all the actuators $A_i$ $(i = 1, 2, \ldots, m)$ and all the sensors $S_i$ $(i = 1, 2, \ldots, r)$ are drawn as single units because they represent separate nodes in the communication structure of the overall

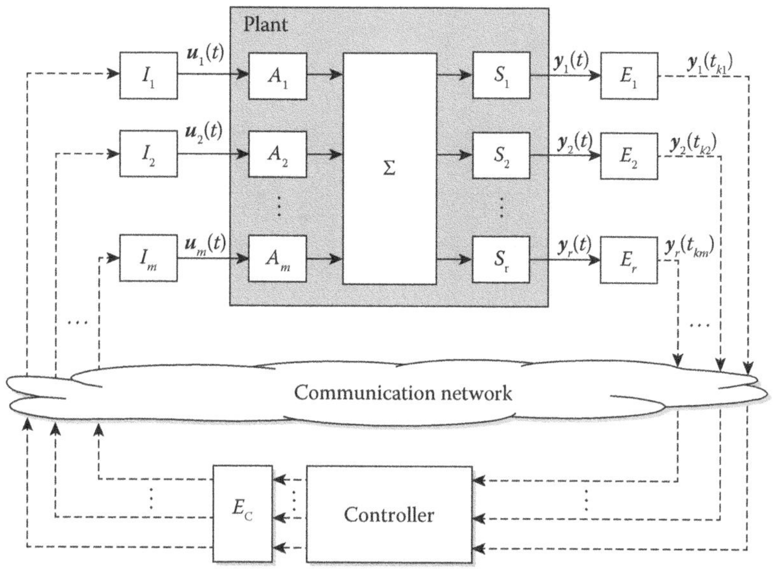


**FIGURE 1.4**
Multivariable event-based control.

system. Hence, separate event generators $E_i$ and control input generators $I_i$ have to be placed at these units.

From the methodological point of view, this structure brings about additional problems. First, the event generators $E_i$ have to make their decision to send or not to send data to the controller based on the measurement of only the single output that the associated sensor $S_i$ generates (decentralized event triggering rules). This output alone does not represent the state of the plant and, hence, makes it difficult to decide whether the new measurement is important for the controller and, thus, has to be communicated. This difficulty can be avoided if the sensor nodes are connected to form a sensor network, where each sensor receives the measurement information of all other sensors. This connection can be implemented by the digital network used for the communication with the controller or by a separate network [2]. Then, the event generators can find a consensus about the next event time.

Second, as the sensor nodes decide independently of each other when to send information, the communication in the overall system is no longer synchronized and the control algorithm has to cope with sensor data arriving asynchronously. In the formal representation, the event times are denoted by $t_{k_i}$ with separate counters $k_i$ $(i = 1, 2, \ldots, m)$ for the $m$ sensor nodes.

Third, it may be reasonable to place another event generator $E_C$ at the controller node, because the separate decisions about sending measurement data may not imply the necessity to send control input data from the controller toward the actuators. Even if the sensor data have changed considerably and the event generators at the sensor nodes have invoked new data exchanges, the changes of the input data may remain below the threshold for sending new information from the controller toward the actuators.

**Decentralized and distributed event-based control.** The third architecture refers to interconnected plants that consist of several subsystems $\Sigma_i$, $(i = 1, 2, \ldots, N)$, which are physically coupled via an interconnection block $\boldsymbol{K}$ (Figure 1.5). In decentralized event-based control, separate control loops around the subsystems are implemented, which are drawn in the figure with the event generator $E_i$ and the control input generator $I_i$ for the subsystem $\Sigma_i$. For simplicity of presentation, the control station $C_i$ of this control loop is thought to be an element of the control input generator $I_i$ and, hence, not explicitly shown. In this situation, the communication network closes the separate loops by sending the state $\boldsymbol{x}_i(t)$ of the subsystem (or the subsystem output) from the event generator $E_i$ toward the control input generator $I_i$ of the same subsystem, which uses this new information to update its input, for example, according to the control law

$$\boldsymbol{u}_i(t) = -\boldsymbol{K}_i \boldsymbol{x}_i(t_{k_i}), \quad t_{k_i} \le t < t_{k_i+1}.$$

The design problem for the decentralized controller includes the "normal" problem of choosing separate controllers $\boldsymbol{K}_i$ for interconnected subsystems that is known from the theory of continuous decentralized control [27] and the new problem of respecting the restrictions of the event-based implementation of these decentralized control stations [30,32,43,45]. The latter

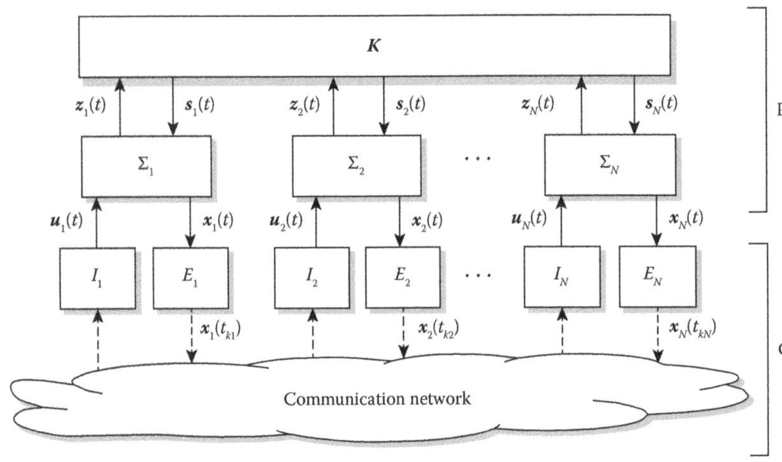


**FIGURE 1.5**
Decentralized event-based control.

problem includes the asynchronous activation of communication links by the event generators of the different loops, which is symbolized by the indexed counters $k_i$ for the event times.

In distributed event-based control, the overall system architecture is the same as in Figure 1.5, but now the communication network submits information from the event generator $E_i$ not only to the control input generator $I_i$ of the same subsystem but also to neighboring subsystems. Hence, the input to the block $I_i$, which is shown as a dashed arrow, includes state information $x_j(t_{k_j})$ originating in neighboring subsystems $\Sigma_j$ and arriving at the event times $t_{k_j}$, which are determined by the event generators $E_j$ of these subsystems. Therefore, the control input of the $i$th subsystem may depend upon several or even upon all other subsystem states, which is shown in the example control law

$$u_i(t) = -(K_{i1}, K_{i1}, \ldots, K_{iN}) \begin{pmatrix} x_1(t_{k_1}) \\ x_2(t_{k_2}) \\ \vdots \\ x_N(t_{k_N}) \end{pmatrix},$$
$$\max\{t_{k_1}, \ldots, t_{k_N}\} \leq t < \min\{t_{k_1+1}, \ldots, t_{k_N+1}\},$$

in which $u_i$ depends upon all subsystem states [11,47–49].

In both the decentralized and the distributed event-based control structure, the asynchronous data transfer poses the main problems in the modeling of the closed-loop system and the design of the controller. In comparison to multivariable event-based control, the structural constraints of the control law lead to further difficulties. In a more general setting, this structure can be dealt with as a set of computing units that are distributed over the control loop and are connected by a packet-based network. The question to be answered is as follows: How can the control problem be solved in such a structure with as few communication activities as possible?

New methods to cope with these problems are presented in Chapter 5 for distributed event-based control and in Chapter 7 for implementing a centralized controller with decentralized event generators.

## 1.2 Components of Event-Based Control Loops

This section describes in more detail the components of event-based control loops and points to the main assumptions made in literature when elaborating their working principles. This explanation is focused on the basic configuration shown in Figure 1.1.

### 1.2.1 Event Generator

The main task of the event generator $E$ is to evaluate the current measurement data in order to decide at which time instant an information feedback from the sensors via the controller toward the actuators is necessary. The rule for this decision is often formulated as a condition

$$h(t_k, x(t_k), x(t), \bar{e}) = 0, \quad k = 0, 1, \ldots, \tag{1.2}$$

with $\bar{e}$ being an event threshold, $t_k$ describing the last event time, $x(t_k)$ the state at the last event time, and $x(t)$ the current state of the plant. For any time $t > t_k$, the condition $h$ depends only on the state $x(t)$. If the state $x(t)$ makes the function $h$ vanish, the next event time $t = t_{k+1}$ is reached.

From the methodological point of view, it is important to see that the time point $t_{k+1}$ is only implicitly described by the condition (1.2). Even if an explicit solution of the plant model for the state $x(t)$ can be obtained, Equation 1.2 can usually not be transformed into an explicit expression for $t_{k+1}$. In the online implementation, the event generator takes the current measurement vector $x(t)$, evaluates the function $h$, and in case this function is (nearly) vanishing, invokes a data transfer.

There is an important difference between event-based signal processing described in the second part of this book and event-based control. The introductory chapter on event-based data acquisition and digital signal processing shows that level-crossing sampling adapts the sampling frequency to the properties of the signal to be transmitted. The aim is to set sampling points in such a way that the progress of the signal can be reconstructed by the receiver.

In contrast to this setting, event-based control has to focus on the behavior of the closed-loop system. The event generator should send the next signal value if this value is necessary to ensure a certain performance of the closed-loop system. Consequently, the event generation is not based on the evaluation of a single measurement, but on the evaluation of features describing the closed-loop performance with and without sending the new information.

**Event generation to ensure stability.** To explain the event generation for feedback purposes in more detail, a basic idea from [1] is briefly outlined here (Figure 1.6). Consider the linear plant

$$\Sigma: \begin{cases} \dot{x}(t) = Ax(t) + Bu(t), & x(0) = x_0 \\ y(t) = Cx(t) \end{cases}, \tag{1.3}$$

with the $n$-dimensional state vector $x$ and apply, in an event-based fashion, the state feedback

$$u(t) = -Kx(t), \tag{1.4}$$

which, in the continuous implementation, leads to an asymptotically stable closed-loop system

$$\bar{\Sigma}: \begin{cases} \dot{x}(t) = \underbrace{(A - BK)}_{\bar{A}} x(t), & x(0) = x_0 \\ y(t) = Cx(t) \end{cases}. \tag{1.5}$$

The stability can be proved by means of the Lyapunov function $V(x(t)) = x^{\mathrm{T}}(t)Px(t)$, where $P$ is the symmetric, positive definite solution of the Lyapunov equation

$$\bar{A}^{\mathrm{T}}P + P\bar{A} = -Q,$$

for some positive definite matrix $Q$.

Now, the controller should be implemented in an event-based fashion with a zero-order hold as the control input generator:

$$u(t) = -Kx(t_k), \quad t_k \le t < t_{k+1}. \tag{1.6}$$

The event generator should supervise the behavior of the system $\Sigma$ under the control (1.6) in order to ensure that the Lyapunov function $V(x(t))$ decays at least with the rate $\lambda$:

$$V(x(t)) \le S(t) = V(x(t_k))e^{-\lambda(t-t_k)}.$$

This requirement can be formulated with a tolerance of $\bar{e}$ as an event function (1.2):

$$h(t_k, x(t_k), x(t), \bar{e}) = x^{\mathrm{T}}(t)Px(t) - x^{\mathrm{T}}(t_k)Px(t_k)e^{-\lambda(t-t_k)} - \bar{e} = 0, \tag{1.7}$$

that can be tested online for the current measured state $x(t)$. If the equality is satisfied, the next event time $t_{k+1}$ is reached.

Figure 1.7 illustrates this method of the event generation. If a continuous controller were used, the temporal evolution of $V(x(t))$ followed the lowest curve, which

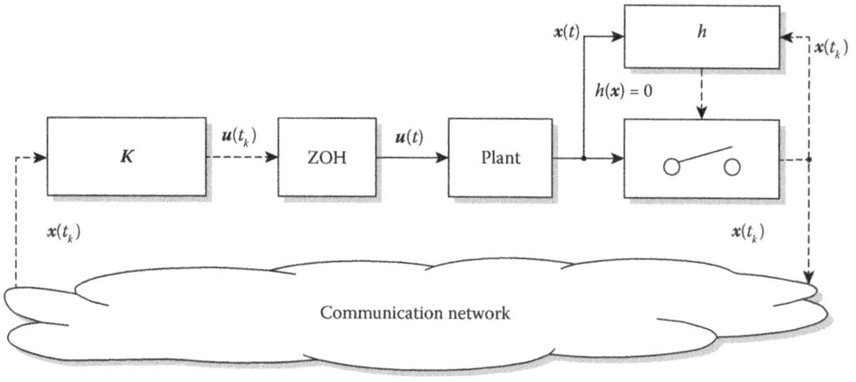


**FIGURE 1.6**
Event-based control loop showing the event generation mechanism.

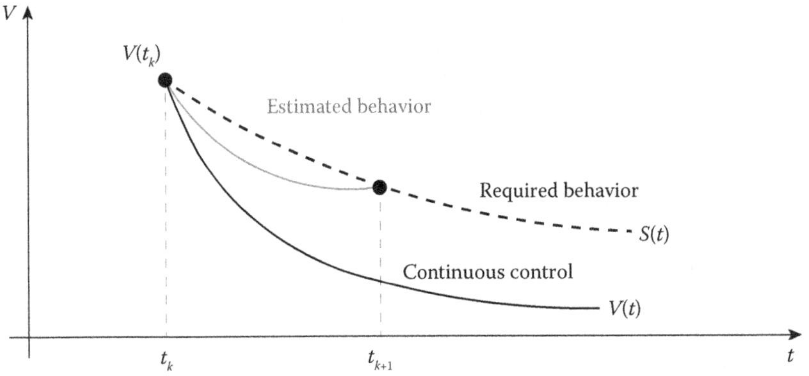


**FIGURE 1.7**
Determination of the next sampling instant.

has the largest decay rate of the three lines shown in the figure. This line is denoted as $V(t)$ because it illustrates how the Lyapunov function decreases from its value $V(\boldsymbol{x}(t_k))$ at the event time $t_k$ over the time interval $t > t_k$.

With event-based feedback (1.6), the same decay rate cannot be required, because less information is sent around the control loop. A reasonable requirement is to claim that the Lyapunov function should at least decay as the function $S(t)$, which is drawn as dashed curve in the figure. For the event-based controller, the Lyapunov function $V$ has for a time $t$, which is only a little bit greater than $t_k$ a similar decay rate as for a continuous controller, because the inputs $\boldsymbol{u}(t)$ generated by the continuous controller (1.4) and by the event-based controller (1.6) are nearly the same. Later on, the decay rate decreases for the event-based controller. A crossing of the middle curve showing the Lyapunov function for the event-based controller with the dashed line occurs, and this point defines the next event time $t_{k+1}$. The threshold $\bar{e}$ in Equation 1.7 acts as a bound for the tolerance that is accepted in the comparison of the two curves.

The important issue is the fact that the rule of the event generator for the determination of the next event instant depends not only on a single measurement signal but also on the Lyapunov function characterizing the behavior of the closed-loop system. Furthermore, this method illustrates that typically no fixed threshold is used in the event condition but a changing border like the function $S(t)$ in the figure. Hence, many event generators include dynamical models of the closed-loop system to determine the novelty of the measurement information and the necessity to transfer this information. This is one of the main reasons for the higher computational load of event-based control in comparison with sampled-data control.

**Minimum attention control.** Although there are multiple aims that are used to find appropriate methods for event generation, the main idea is to reduce the sampling frequency whenever possible. An early attempt to find such an event generation rule goes back to 1997 [9], which tries to limit the number of changes of the input value $\boldsymbol{u}(t_k)$. The notion of "minimum attention control" results from the aim to minimize the attention that the controller has to spend to the plant behavior during operation.

This idea has been developed further [15], and new results are presented in Chapter 11 on efficiently attentive systems, for which the intersampling interval gets longer when the system approaches its operating point.

### 1.2.2 Input Signal Generator

The input signal generator has the task to generate a continuous-time input $\boldsymbol{u}(t)$ for a given input value $\boldsymbol{u}(t_k)$ over the entire time interval $t \in [t_k, t_{k+1})$ until the next event time occurs. In sampled-data control, usually zero-order holds are used to perform this task. The rationale behind this choice is, in addition to the simplicity of its implementation, the fact that sampled-data control uses a sufficiently high sampling frequency so that the piecewise constant input signals generated by a zero-order hold approximate a continuous-time input signal with sufficient accuracy.

This argument does not hold for event-based control, because over a long time interval, a constant input signal is not adequate for control [4,51]. For linear systems, the model of the overall system (e.g., Equation 1.5) shows that the typical form of the input is a sum of exponential functions like in the equation

$$\boldsymbol{u}(t) = -\boldsymbol{K}e^{\bar{\boldsymbol{A}}(t-t_k)}\boldsymbol{x}(t_k), \quad (1.8)$$

for the continuous controller (1.4). This is the reason why in the recent literature, the focus has moved from the use of zero-order holds as input-generating to more involved input-generating schemes.

The necessity to sample in the feedback loop is minimized if the model of the closed-loop system (1.5) is used to generate the input to the plant:

$$I: \begin{cases} \dot{x}_s(t) = \bar{A}x_s(t), & x_s(t_0) = x(t_0) \\ u(t) = -Kx_s(t) \end{cases}. \tag{1.9}$$

In order to avoid confusion, the state of this component has been renamed to $x_s(t)$. After the state of the control input generator (1.9) has been set to the plant state $x(t_0)$ at the first event time $t_0$, the model (1.9) precisely generates the input $u(t)$, $(t \geq t_0)$ given in Equation 1.8 for $k = 0$ for the continuous state-feedback controller (1.4). Hence, the dynamics of the event-based and the continuous feedback loop are the same. Deviations occur, in practice, due to disturbances or model uncertainties, which bring the argumentation back to the list of reasons for feedback control mentioned in Section 1.1.1. Only if disturbances or model uncertainties are considered, is feedback necessary. In the structure considered here, the activation of the feedback means to reset the state $x_s(t_k)$ at the next event time $t_k$. It is surprising to see that, nevertheless, many papers on event-based control in the recent literature consider neither disturbances nor model uncertainties in their problem setting.

Chapter 14 gives a survey on *intermittent control*, which is a technique to generate the continuous control input between two consecutive sampling instants by means of a dynamical model. The control input generator is similar to Equation 1.9. Similarly, Chapter 9 is concerned with a dynamical control input generator for which the closed-loop system tolerates time delay and quantization effects.

### 1.2.3 Characteristics of the Closed-Loop System

This section summarizes properties of event-based control loops that occur due to the structure of the components or that have to be considered, tested, or proved in the analysis of event-based systems.

**Combination of open-loop and closed-loop control.** An important aspect of event-based control can be found in the fact that the input signal $u(t)$ to the plant is generated alternately in an open-loop and a closed-loop fashion (Figure 1.8). Only at the event times $t_k$ is the feedback loop closed. In the time between two consecutive event times, the control input is generated in dependence upon the data received until the last event time and, usually, by means of a dynamical plant model.

This situation does not differ from sampled-data control, but here the time interval between two consecutive events should be as long as possible and is, in general, not negligible. Therefore, the analysis of the event-based control loop has to take into consideration this time interval and cannot be restricted to the event time points.

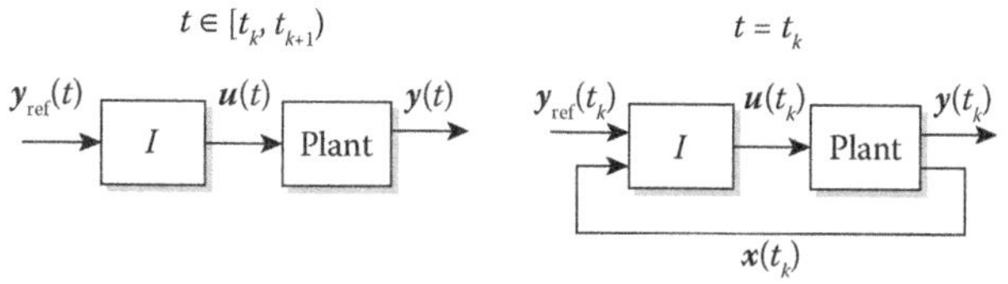


**FIGURE 1.8**
Open-loop versus closed-loop control in event-based control loops.

As a direct consequence of these structural differences, the plant state $x(t_k)$ has to be available for the control input generator $I$ in order to enable this component to generate a reasonable control input in the time interval $t \in [t_k, t_{k+1})$ between two consecutive event instants. Consequently, most of the event-based control methods developed in the literature assume that the whole state vector $x$ is measurable and transmitted at event times. If the measurement is restricted to some output $y$, an observer is typically used to reconstruct the state.

**Hybrid systems character.** The information feedback in event-based control loops usually results in state jumps in the controller or in the control input generator, or in switching among several controllers that have been designed for different sampling frequencies. Hence, the event-based control loop represents a hybrid dynamical system [28].

To illustrate this fact, consider the system

$$\Sigma_d: \begin{cases} \dot{x}(t) = Ax(t) + Bu(t) + Ed(t) & x(0) = x_0 \\ y(t) = Cx(t) \end{cases}, \tag{1.10}$$

with disturbance $d(t)$ together with the controller (1.9). At the event time $t_k$, the state of the control input generator jumps, which gives the overall system the characteristics of an *impulsive system*. Impulsive systems are specific forms of hybrid systems, where state jumps occur but the dynamics are the same before and after the jumps [17]. Such systems can be represented by a state equation together with a jump condition:

$$\text{Impulsive system}: \begin{cases} \dot{x}(t) = f(x(t), d(t)) & \text{for } x(t) \notin \mathcal{D} \\ x(t^+) = \Phi(x(t)) & \text{for } x(t) \in \mathcal{D} \end{cases}.$$

For the event-based systems considered here, the state equation results from the models (1.10) and (1.9) of the plant and the controller, the jump set $\mathcal{D}$ represents the event-generation mechanism, and the reset function $\Phi$ shows how the state of the control input generator is reset at event times $t_k$. As long as the state $x(t)$ does not satisfy the event condition (e.g., Equation 1.2), the state runs according to the differential equation in the first line of the model. If the event condition is satisfied, the current state $x(t)$ belongs to the jump set $\mathcal{D}$, and the state is reset as indicated in the second line. The symbol for the time just after the reset is $t^+$.

This reformulation of the event-based control loop as an impulsive system makes it possible to apply methods that have been specifically developed for this system class to the analysis of event-based control (cf. [12, 14,41] and Chapter 6). This way of analysis and control design is important for multivariable event-based control and distributed event-based control, because in these architectures the different sensors or control loops, respectively, generate events asynchronously; hence, numerous state jumps occur in different parts of the closed-loop system at different times.

Figure 1.9 shows experimental results to illustrate this phenomenon. For two coupled reactors, decentralized event-based controllers react independently of one another for their associated subsystem. Whereas in the right part of the figure, events occur only in the first time interval of $[0, 50]$ s, the left part shows events for the whole time horizon. Such asynchronous state jumps bring about severe problems for the analysis of the system. The "noisy" curves represent the measured liquid level or temperature, whereas the curves with the jumps show the behavior of the internal model (1.9) of the control input generator. Jumps at the event times show the resetting of the state of the control input generator.

**Minimum interevent time.** An important analysis aim is to show that there is a guaranteed minimum time span between the events in all possible situations (disturbances, model uncertainties, etc.). This aim is important for two reasons. First, the motivation for event-based control is to reduce the information exchange within a control loop, and one has to prove that with a specific event-generation scheme, this task is really fulfilled. Second, for hybrid systems, the phenomenon of a Zeno behavior is known, which means that the system generates an infinite number of events within a finite time interval. Of course, in practice, this situation seems to be artificial, and the effects of technological systems like short time delays that are neglected in the plant model prevent the Zeno behavior from occurring in reality. However, Zeno behavior has also to be avoided for the model of the event-based control loop, because this phenomenon indicates that the event-generation mechanism is not selected appropriately.

The usual way to prove the suitability of the event-based control and, simultaneously, avoid Zeno behavior is to derive a lower bound for the interevent time $T_s(k) = t_{k+1} - t_k$. Equation 1.7 shows that the event rules have to be carefully chosen. If in this equation $\bar{e}$ is set to zero, the event condition $h(\boldsymbol{x}(t)) = 0$ is satisfied trivially, because the equality $\boldsymbol{x}(t) = \boldsymbol{x}(t_k)$ leads to $h(t_k, \boldsymbol{x}(t_k), \boldsymbol{x}(t_k), 0) = 0$ and, hence, an instantaneous invocation of the next data transfer. With a positive bound $\bar{e}$, a minimum interevent time can be proved to exist.

**Stability of the control loop.** Stability as a prerequisite for the implementation of any controller has to be tested for the event-based control loop with specific

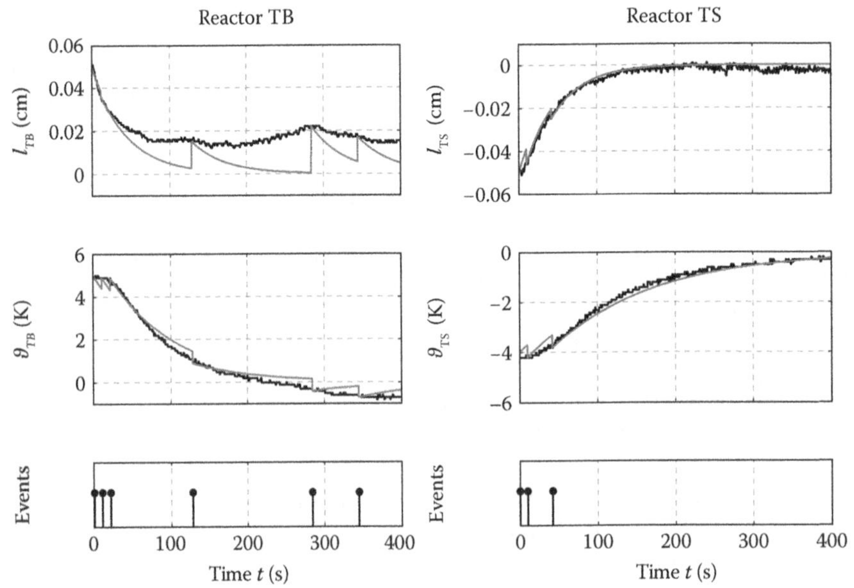


**FIGURE 1.9**
Asynchronous events in distributed event-based control loops.

methods. In many schemes, asymptotic stability can no longer be reached, because the event-triggering mechanism includes some tolerance between the current state and the last communicated state as, for example, in the inequality

$$|\boldsymbol{x}(t) - \boldsymbol{x}(t_k)| > \bar{e},$$

which represents a tolerance of the size $\bar{e}$. Then, the overall system does not reach its equilibrium asymptotically, although the continuous control loop would do so.

In this situation, "practical stability" is required. The controller should hold the state $\boldsymbol{x}(t)$ of the plant inside a set $\Omega_d$ for all initial states $\boldsymbol{x}_0 \in \Omega_1$ and all bounded disturbances $\boldsymbol{d}(t)$:

$$\boldsymbol{x}(t) \in \Omega_d \ \forall t \geq T(\boldsymbol{x}_0),\ \boldsymbol{x}_0 \in \Omega_1,\ \boldsymbol{d}(t) \in [\boldsymbol{d}_{\min}, \boldsymbol{d}_{\max}].$$

Then, the set $\Omega_d$ is called *robust positively invariant*, and the system state is said to be "ultimately bounded."

A further property of event-based control loops, which results likewise from some tolerance that is built into the event-triggering rule, is the insensitivity to small disturbances. This property means that for sufficiently small disturbances, no information feedback is invoked.

**Tolerance of event-based control with respect to network imperfections.** The communication networks may introduce time delays, packet loss, and other imperfections into the control loop, and an analysis has to show which kind of such phenomena can be tolerated by a specific event-based control method [26]. This situation shows that one has to understand which information is available at which node in the network at which time. For example, if the communication of the state information $\boldsymbol{x}(t_k)$ is delayed, the control input generator may use the former information $\boldsymbol{x}(t_{k-1})$ for generating the current input $\boldsymbol{u}(t)$, whereas the event generator assumes that the input is in accordance with the new information $\boldsymbol{x}(t_k)$ that it has sent toward the control input generator. To send acknowledgement information may avoid this situation, but this leads to more data traffic and more complicated working principles of the control loop.

## 1.3 Approaches to Event-Based Control

Several different methods for event-based control have been proposed in recent years, which can be distinguished with respect to the answers given to these questions. Some of them have been published under different names like *event-driven sampling* [19], *event-based sampling* [4,5], *event-triggered control*, *Lebesgue sampling* [6], *deadband control* [36], *level-crossing control* [23], *minimum attention control* [9], *self-triggered control* [1], or *send-on-delta control* [33]. The following paragraphs classify these approaches.

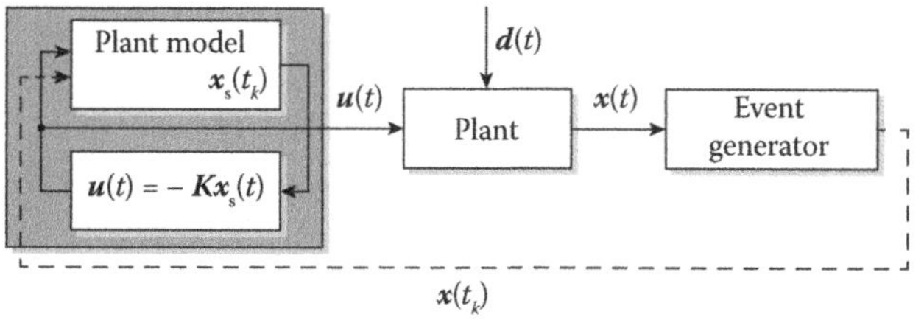


**FIGURE 1.10**
Event-based state feedback.

### 1.3.1 Emulation-Based Design Methods

A well-developed approach is to take a continuous feedback and to implement this controller in an event-based fashion. As the aim is to get a bounded deviation of the behavior of the event-based control loop from the behavior of the continuous loop, this strategy of the event-based control is called emulation-based.

As an example of this method, consider the *event-based state feedback* shown in Figure 1.10 [29]. A linear plant (1.3) is combined with a state feedback (1.4) so as to get a tolerable deviation of the event-based control loop from the continuous control loop (1.5). The events are generated in dependence on the deviation of the state $\boldsymbol{x}_s(t)$ of the continuous closed-loop model from the measured state $\boldsymbol{x}(t)$:

$$h(t_k, \boldsymbol{x}(t_k), \boldsymbol{x}(t), \bar{e}) = |\boldsymbol{x}(t) - \boldsymbol{x}(t_k)| - \bar{e} = 0.$$

At the event time $t_k$ the difference $\boldsymbol{x}(t) - \boldsymbol{x}(t_k)$ vanishes, and it increases afterward due to the effect of the disturbance, which only affects the plant, not the model. The threshold $\bar{e}$ describes the tolerable deviation of both states. It can be shown that the deviation of the behavior of the event-based control loop from the behavior of the continuous control loop for the same disturbance $\boldsymbol{d}(t)$ is bounded, and that this bound decreases if the threshold $\bar{e}$ is chosen to be smaller. Hence, the event-based loop mimics the continuous feedback loop with adjustable precision, where, however, the price for improvement of the approximation is an increased communication effort.

Many event-based control methods have been developed as emulation-based controllers [1,12,24,29,31,41,44].

### 1.3.2 Switching among Different Sampling Frequencies

The second strategy of event-based control considers sampled-data control loops with different sampling frequencies. An interesting question to ask is under what

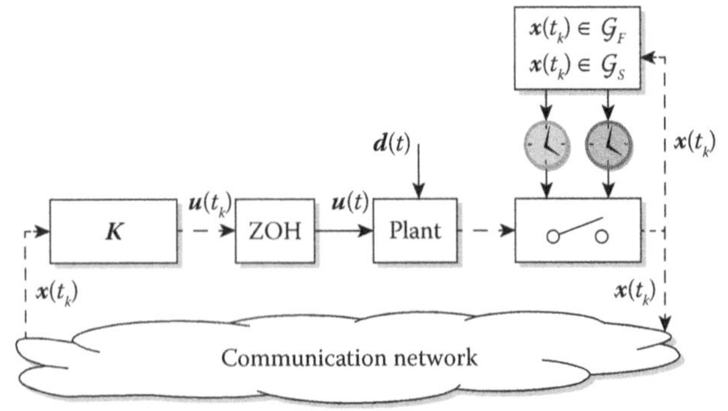


**FIGURE 1.11**
Switching between two sampling frequencies.

condition it is possible to get better loop performance with less sampling events.

To illustrate the approach, consider a sampled-data control loop with two sampling times $T_{\text{slow}}$ and $T_{\text{fast}}$ (Figure 1.11). The objective function

$$J = \int_0^\infty x^{\mathrm{T}}(t)Qx(t) + u^{\mathrm{T}}(t)Ru(t)\,\mathrm{d}t,$$

has to be brought into a discrete-time form for the sampling time used:

$$J = \sum_{k=1}^{\infty} \begin{pmatrix} x(k) \\ u(k) \end{pmatrix}^{\mathrm{T}} N \begin{pmatrix} x(k) \\ u(k) \end{pmatrix} = x_0^{\mathrm{T}} P x_0,$$

where $N$ is an appropriate matrix that depends upon the sampling time. In [2] the performance of the system with the two sampling times $T_{\text{slow}}$ and $T_{\text{fast}}$ is compared with a system with constant sampling time $T_{\text{nominal}}$ with $T_{\text{slow}} > T_{\text{nominal}} > T_{\text{fast}}$. For a linear plant (1.3), the optimal controller is a state feedback with feedback matrix $K_{\text{fast}}$, $K_{\text{nominal}}$, or $K_{\text{slow}}$, which results from the solutions $P_{\text{fast}}$, $P_{\text{nominal}}$, or $P_{\text{slow}}$ of the corresponding algebraic Ricatti equation.

To illustrate that it is possible to improve the performance with less sampling over the long time horizon, assume that the undisturbed plant is in the state $x_0$ and should be brought into the equilibrium state $\bar{x} = \mathbf{0}$. Then, it has been shown in [2] that the controller should start with the fast sampling and switch to the slow sampling if the current state $x(t)$ satisfies the event rule

$$x^{\mathrm{T}}(t)(P_{\text{slow}} - P_{\text{fast}})x(t) - x_0^{\mathrm{T}}(P_{\text{nominal}} - P_{\text{fast}})x_0 = 0.$$

The set of all states $x(t)$ satisfying this relation is denoted by $\mathcal{G}_{\text{slow}}$. Hence, if $x(t_k) \in \mathcal{G}_{\text{slow}}$ holds, the controller switches from the fast toward the slow sampling. It can be shown that over the infinite time horizon, this strategy leads to a smaller value of the objective function $J$ and to less sampling instances.

If a disturbance $d(t)$ occurs, another set $\mathcal{G}_{\text{fast}}$ has to be defined such that the controller switches from the slow to the fast sampling if the current state satisfies the switching rule $x(t) \in \mathcal{G}_{\text{fast}}$. Then, the controller adapts to the effect of the disturbance and switches back to the slow sampling if this effect has been sufficiently attenuated.

A similar approach is described in Chapter 8, where a set of controllers is used, which are designed offline for different sampling times. A deadband is used for the control error to select the current sampling time and, hence, the appropriate controller from the predefined set.

### 1.3.3 Self-Triggered Control

In self-triggered control, the plant state is not continuously supervised by the event generator, but the next event time $t_{k+1}$ is determined by the event generator at the event time $t_k$ in advance [1,50]. Compared to event-based control, this approach has the advantage that the sensors can "sleep" until the predicted next sampling instant.

To find the prediction of $t_{k+1}$, the event condition (1.2) has to be reformulated so as to find the time

$$t_{k+1} = \min\{t > t_k : h(t_k, x(t_k), x(t), \bar{e}) = 0\}.$$

For the undisturbed plant (1.3) subject to the control input (1.6):

$$\dot{x}(t) = Ax(t) - BKx(t_k),$$

the state trajectory $x(t)$, $(t \geq t_k)$ can be predicted and the time $t_{k+1}$ determined. The problem lies in the situation where the plant is subject to some disturbance. As usual, the disturbance is not known in advance, and one has to consider a restricted class of disturbances when determining the next event time, for example, a set of bounded disturbances with a known magnitude bound. This way of solution leads often to rather conservative results, which means that the predicted interevent time $T_s(k) = t_{k+1} - t_k$ is much slower than it is in event-based control with online evaluation of the event condition.

Chapter 10 elaborates a new method for self-triggered control of multi-agent systems, where the overall system should ensure a prescribed decay rate for the common Lyapunov function.

### 1.3.4 Simultaneous Design of the Sampler and the Controller

One can expect to get the best possible event-based controller if the event generation strategy is designed together with the control law. This way of solution has been elaborated, for example, in [34,35] by formulating the design task as a stochastic optimization problem, where the common objective function penalize the state and input to the plant (like in linear-quadratic regulator theory) as well as the transmission of the plant state in the feedback loop. This problem is rather complex, because it is known from stochastic linear systems that the resulting strategy is, in general, nonlinear and includes signaling. Signaling means that the event generator sends the state information $x(t_k)$ toward the controller, and sends further information, which is included in the fact to send or not to send the state information. On the other hand, the controller can modulate its input to the plant to send some information while using the plant as a communication channel to the event generator.

An interesting result of these investigation shows under what conditions the design problem reduces to the emulation-based approach, where first the controller is designed, and second an event generation strategy is developed for the event-based implementation of the controller. Such conditions require that the policies of the event generator and of the controller be restricted to be deterministic. Then, the overall problem can be decomposed into the two problems that can be solved sequentially.

### 1.3.5 Event-Based State Estimation

The ideas explained so far for event-based control can also be used for state estimation, where the measured input and output signals are not transmitted to the estimator at equidistant time intervals but in dependence upon their changes. This problem differs from event-based signal processing that is dealt with in the second part of this book, because now the input and the output of a dynamical system are processed together with a dynamical model of the system under consideration. The aim is to maintain a small deviation of the estimated state $\hat{x}(t)$ from the true state $x(t)$.

Chapter 13 presents a state estimation method with event-based measurement updates, which illustrate this problem.

### 1.3.6 Event-Based Control of Interconnected Systems

Additional problems occur if the event-based control strategies are not considered for a single-loop system, but are considered instead for an interconnected system, where the events occur asynchronously in the subsystems. From a theoretical viewpoint, interconnected systems pose the most interesting problems to be solved. Therefore, multiloop systems, event-based control of multi-agent systems, or distributed event-based control of systems has been the subject of intensive research for many years, as shown in [11,30,47–49] and Chapter 5.

## 1.4 Implementation Issues

So far, the overall system was investigated as a continuous-time system in which information is exchanged at any discrete time points that are fixed by an event generator. If the event generator and the controller or the control input generator should be implemented, one has to consider these components in a discrete-time version. The following outlines the problems to be solved.

**Sampled-data event-based control.** For the implementation, one has to go back to the sampled-data framework of control, because the components to be implemented work in a discrete-time fashion. The event-based character can be retained for the communication step, whereas the rules for the event generator and the control input generator have to be transformed into their sampled-data versions. Usually, these components can be realized with a sufficiently high sampling rate so that a quasi-continuous behavior is obtained.

*Periodic event-triggered control* has emerged as a topic of interest, as it results from these implementation issues. At each sampling instant, the event generator is invoked to decide whether information has to be transmitted. If not, the event generator waits for the next sampling instant; otherwise, it sends information and, by doing so, starts the control algorithm for updating the control input. The idea for periodic event-based control is not new [18,19], but Chapter 6 shows that there are still interesting problems to be solved.

**Asynchronous components in an event-based feedback loop.** In network systems, clock synchronization poses a severe problem. In contrast, the theory of event-based control assumes that the timing regime is the same for all components considered. In particular, it is often taken for granted that an information packet, which is sent at time $t_k$, arrives at the same time at the controller and that the controller can determine the control input based on precise information about the elapsed sampling time $T_s(k) = t_{k+1} - t_k$. What happens if there is a time delay in the communication of the data from the event generator toward the control input generator or if the clocks used in both components show different times is illustrated in Figure 1.12. At time $t_k$, the event generator measures the current state $x(t_k)$ and sends the state toward the control input generator. Simultaneously, it resets the state $x_e(t_k)$ of its internal

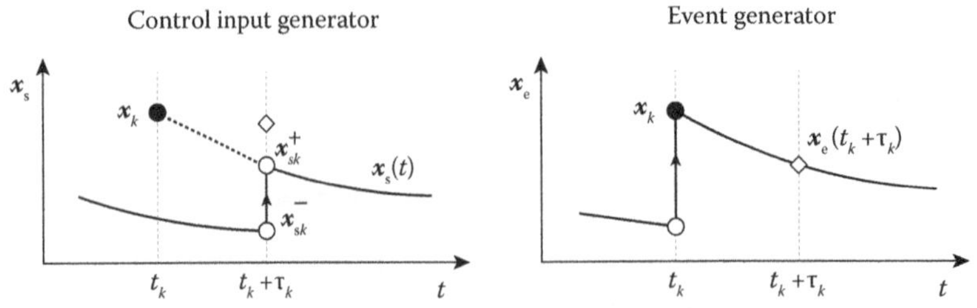


**FIGURE 1.12**
State reset at time $t_k$ or $t_k + \tau_k$.

model of the plant. The control input generator receives the measured information after a time delay of $\tau_k$ at time $t_k + \tau_k$, and the question arises, to which value it should now reset the state $x_s(t_k + \tau_k)$ of its internal model. The aim is to reinstall the synchrony of both internal models, and this requires that the state $x_s$ does not get the new value $x(t_k)$ but the value $x_{sk}^+$ which lies on the trajectory $x_s(t)$ of the model that the event generator uses. This strategy requires finding methods for determining the right value for $x_s$ such that the equality $x_s(t_k + \tau_k) = x_e(t_k + \tau_k)$ holds.

**Scheduling.** Event-based control reduces the load of the communication network only if the scheduling of the network is taken into account. If a fixed allocation protocol is used, event-based data transmission can only decide whether a data packet should be sent or not, but the empty slot in the packet flow cannot be used by other users of the network. Consequently, one of the main advantages of event-based control compared to periodic sampling becomes effective only if online scheduling methods are applied.

The best performance can be expected if a full-level event-based processing is reached, where sensing, communication, computation, and actuation are all included in the online scheduling [46]. Then, no time delays occur when data packets are waiting to be processed and no additional delays have to be considered when analyzing the closed-loop system. On the other hand, the more different control loops are closed over the same network, the worse is the average performance, even if event-based control methods are applied [10,22].

A detailed comparison of event-based and time-triggered control has been described in [7,8] for the ALOHA protocol. It shows that the performance of event-based control is better than that of sampled-data control as long as a congestion in the communication network does not introduce long time delays.

In [38] it has been shown that event-based triggering is better than periodic control if the loss probability is smaller than 0.25 for independent and identically distributed random packet loss and level-triggered sampling.

The best situation is found in self-triggered control, where the next event time $t_{k+1}$ is known well in advance and can be used to schedule the processes mentioned above.

**Experimental evaluation of event-based control.** Theoretical results are only valid as long as the main assumptions of the underlying theory are satisfied. It is, therefore, important that several applications of event-based control strategies have shown that the promises of this control method can, at least partly, be kept. As an example, Figure 1.13 shows the result of an experiment described in [25]. The middle part of the figure shows the tank level and the liquid temperature of a thermo-fluid process, where the "noisy" curves represent measurements, and the other curves the state of the control input generator (1.9). A typical sampling period for this process would be $T_s = 10$ s, which would lead to about 200 sampling instants in the time interval shown. In the bottom part of the figure, one can see that with event-based sampling only 10 events occur, which leads to a reduction of the sampling effort by a factor of 20. The experiment has been done with industry-like components and shows that event-based control can really reduce the control and communication effort. Similar results are reported in [21,40].

The application of event-based control techniques also leads to important extensions of the methodology, which are necessary for specific areas. In particular, in the majority of papers, event-based control is investigated for linear static feedback, but for set-point following or disturbance attenuation, the usual extensions with integral or other dynamical parts are necessary. Investigations along this line are reported, for example, in [25,39,40] and in Chapter 12.

Finally, it should be mentioned that experiments with periodic and event-based sampling have been the starting point of the current wave of research in event-triggered and self-triggered control. References [3] and [5] have been two of the first papers that demonstrated with the use of interesting examples, that an adaptation of the sampling rate to the state of the plant or the effect

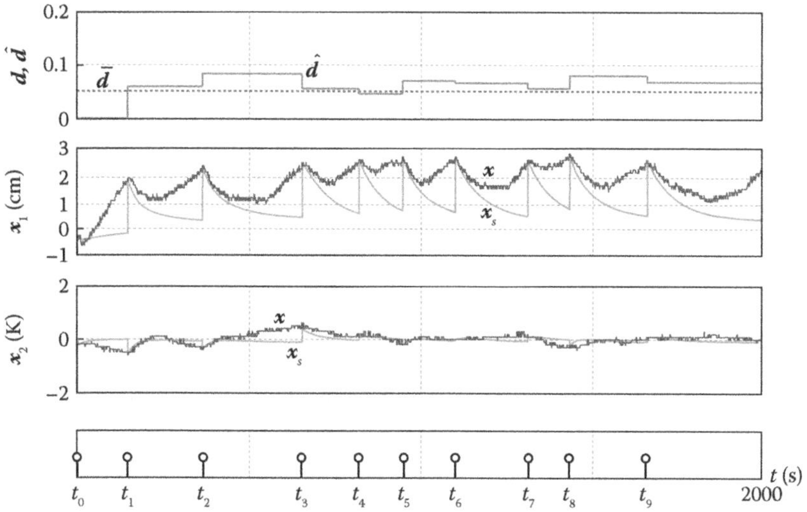


**FIGURE 1.13**
Experimental evaluation of event-based state-feedback control.

of the disturbance cannot only save sampling efforts but even improve the performance of the closed-loop system.

## 1.5 Extension of the Basic Principles and Open Problems

This section summarizes several extensions and open problems and shows that the field of event-based control is rich in interesting problems just waiting to be solved.

The main ideas of event-based control have been surveyed in this chapter for linear systems with static state feedback. With respect to these restrictions, the chapter does not reproduce the state-of-the-art in the expected way, because, from the beginnings of research into this topic, literature was not restricted to linear systems but has tackled many of the problems mentioned above directly for nonlinear systems. For such systems, stability is usually investigated in the sense of input-to-state stability, and the performance is described by gains between the disturbance input and the state or the output.

To assume that the state vector $x(t)$ is measurable is, in contrast, a common assumption, which has to be released for applications. Observers are the usual means for reconstructing the state, and event-based methods can also be applied as outlined in Section 1.3.5 and described in more detail in Chapter 13.

Another important extension to be investigated concerns the computational complexity of event-based control. Literature has posed the problem to find a trade-off between communication and computation in the sense that event-based control with fewer sampling instants usually requires more computation to make full use of the information included in the available data. It is, however, an open problem how the complexity of the event generation and the control input generation can be restricted (e.g., with respect to the restricted computing capabilities in the car industry).

Besides these theoretical issues, the application of the theory of event-based control is still marginal, and experimental evaluations showing where event-based control outperforms the classical periodic sampled-data implementation of continuous control are still to come.

**Open problems.** The following outlines three lines of research that can give new direction in the field of event-based control.

- **Complete event-based control loop.** Event-based control has been considered in the literature mainly in the one-degree-of-freedom structure shown in Figure 1.1 with only a feedback part in the loop. In applications, often control techniques with more degrees of freedom are used in order to cope with different control tasks for the same plant. In particular, feedback control to attenuate disturbances is combined with feedforward control for a quick reference following and rejection of measurable disturbances. In multi-agent systems, further communications are used to coordinate several subsystems. In such extended structures, several components work and communicate and may do so in an event-based fashion. The question of what sampling strategies are adequate for such systems is quite open.

- **Event-based control and quantization.** Event-based control has been developed as a means to cope with bandwidth-restricted networks. An alternative way is quantization that reduces the information to be sent over a channel and, hence, the necessary bandwidth for its communication. The question is how both methods can be combined. Can a higher sampling rate compensate for the information loss due to quantization? As Chapter 15 shows, level-crossing sampling with a zero-order hold can be equivalently represented as quantization without sampling. Can one use this relation for event-based control?

- **Protocol design for event-based control.** Assume that the network protocol can be designed for control purposes. What would be the ideal protocol for event-based control loops? This question may seem to be irrelevant because network protocols will most probably not be designed for control purposes, because the market is thought to be too small. However, the answer to this question can lead to new event-based control mechanisms or configurations, if one compares what would be optimal for control with what control engineers get as tools or hardware components for the implementation of their control loops.

## Acknowledgment

This work has been supported by the German Research Foundation (DFG) within the priority program, "Control Theory of Digitally Networked Dynamic Systems."

## Bibliography

[1] A. Anta, P. Tabuada, To sample or not to sample: Self-triggered control of nonlinear systems, *IEEE Trans. Autom. Control*, vol. 55, pp. 2030–2042, 2010.

[2] J. Araujo, *Design, Implementation and Validation of Resource-Aware and Resilient Wireless Networked Control Systems*, PhD Thesis, KTH Stockholm, 2014.

[3] K.-E. Arzen, A simple event-based PID controller, *IFAC Congress*, Beijing, 1999, pp. 423–428.

[4] K. Aström, Event based control, In A. Astolfi and L. Marconi (Eds.), *Analysis and Design of Nonlinear Control Systems*, Springer-Verlag, Berlin, 2008, pp. 127–147.

[5] K. Aström, B. Bernhardsson, Comparison of periodic and event-based sampling for first order stochastic systems, *IFAC Congress*, Beijing, 1999, pp. 301–306.

[6] K. Aström, B. Bernhardsson, Comparison of Riemann and Lebesgue sampling for first order stochastic systems, *IEEE Conf. on Decision and Control*, Las Vegas, NV, 2002, pp. 2011–2016.

[7] R. Blind, F. Allgöwer, Analysis of networked event-based control with a shared communication medium: Part I—pure ALOHA; Part II—slotted ALOHA, *18th IFAC World Congress*, Milan, 2011, pp. 10092–10097 and pp. 8830–8835.

[8] R. Blind, F. Allgöwer, On the optimal sending rate for networked control systems with a shared communication medium, *Joint 50th IEEE Conf. on Decision and Control and European Control Conf.*, Orlando, FL, 2011, pp. 4704–4709.

[9] R. W. Brockett, Minimum attention control, *Conf. on Decision and Control*, San Diego, CA, 1997, pp. 2628–2632.

[10] A. Cervin, T. Henningsson, Scheduling of event-triggered controllers on a shared network, *IEEE Conf. on Decision and Control*, Cancun, Mexico, 2008, pp. 3601–3606.

[11] C. De Persis, R. Sailer, F. Wirth, On a small-gain approach to distributed event-triggered control. *IFAC Congress*, Milan, 2011.

[12] M. C. F. Donkers, *Networked and Event-Triggered Control Systems*, Technische Universiteit Eindhoven, 2011.

[13] M. C. F. Donkers, M. P. M. H. Heemels, N. van de Wouw, L. L. Hetel, Stability analysis of networked control systems using a switched linear systems approach, *IEEE Trans. Autom. Control*, vol. 56, pp. 2101–2115, 2011.

[14] M. C. F. Donkers, M. P. M. H. Heemels, Output-based event-triggered control with guaranteed $\mathcal{L}_\infty$-gain and improved and decentralised event-triggering, *IEEE Trans. Autom. Control*, vol. 57, no. 6, pp. 1362–1376, 2012.

[15] M. C. F. Donkers, P. Tabuada, M. P. M. H. Heemels, Minimum attention control for linear systems: a linear programming approach, *Discrete Event Dyn. Sys.*, vol. 24, pp. 199–218, 2011.

[16] L. Grüne, S. Hirche, O. Junge, P. Koltai, D. Lehmann, J. Lunze, A. Molin, R. Sailer, M. Sigurani, C. Stöcker, F. Wirth, Event-based control, in J. Lunze (Ed.): *Control Theory of Digitally Networked Dynamic Systems*, Springer-Verlag, Heidelberg, Germany, 2014.

[17] W. Haddad, V. Chellaboina, S. Nersesov, *Impulsive and Hybrid Dynamical Systems: Stability, Dissipativity, and Control*, Princeton University Press, Princeton, NJ, 2006.

[18] W. P. M. H. Heemels, M. C. F. Donkers, Model-based periodic event-triggered control for linear systems, *Automatica*, vol. 49, pp. 698–711, 2013.

[19] W. P. M. H. Heemels, J. Sandee, P. P. J. Van den Bosch, Analysis of event-driven controllers for linear systems, *Int. J. Control*, vol. 81, pp. 571–590, 2007.

[20] W. P. M. H. Heemels, M. C. F. Donkers, A. R. Teel, Periodic event-triggered control for linear systems, *IEEE Trans. Autom. Control*, vol. 58, pp. 847–861, 2013.

[21] W. P. M. H. Heemels, R. Gorter, A. van Zijl, P. v.d. Bosch, S. Weiland, W. Hendrix, M. Vonder, Asynchronous measurement and control: A case study on motor synchronisation, *Control Eng. Prac.*, vol. 7, pp. 1467–1482, 1999.

[22] T. Henningsson, E. Johannesson, A. Cervin, Sporadic event-based control of first-order linear stochastic systems, *Automatica*, vol. 44, pp. 2890–2895, 2008.

[23] E. Kofman, J. H. Braslavsky, Level crossing sampling in feedback stabilization under data-rate constraints, *IEEE Conf. on Decision and Control*, San Diego, CA, 2006, pp. 4423–4428.

[24] D. Lehmann, *Event-Based State-Feedback Control*, Logos-Verlag, Berlin, 2011.

[25] D. Lehmann, J. Lunze, Extension and experimental evaluation of an event-based state-feedback approach, *Control Eng. Prac.*, vol. 19, pp. 101–112, 2011.

[26] D. Lehmann, J. Lunze, Event-based control with communication delays and packet losses, *Int. J. Control*, vol. 85, no. 5, pp. 563–577, 2012.

[27] J. Lunze, *Feedback Control of Large Scale Systems*, Prentice Hall, Upper Saddle River, NJ, 1992.

[28] J. Lunze, F. Lamnabhi-Lagarrigue (Eds.), *Handbook of Hybrid Systems Control—Theory, Tools, Applications*, Cambridge University Press, Cambridge, UK, 2009.

[29] J. Lunze, D. Lehmann, A state-feedback approach to event-based control, *Automatica*, vol. 46, pp. 211–215, 2010.

[30] M. Mazo, P. Tabuada, Decentralized event-triggered control over wireless sensor/actuator networks, *IEEE Trans. Autom. Control*, vol. 56, pp. 2456–2461, 2011.

[31] M. Mazo, A. Anta, P. Tabuada, An ISS self-triggered implementation of linear controllers, *Automatica*, vol. 46, pp. 1310–1314, 2010.

[32] M. Mazo, M. Cao, Asynchronous decentralized event-triggered control, *Automatica*, vol. 50, pp. 3197–3203, 2014.

[33] M. Miskowicz, Send-on-delta concept: An event-based data reporting strategy, *Sensors*, vol. 6, pp. 49–63, 2002.

[34] A. Molin, S. Hirche, On the optimality of certainty equivalence for event-triggered control systems, *IEEE Trans. Automat. Control*, vol. 58, pp. 470–474, 2013.

[35] A. Molin, S. Hirche, A bi-level approach for the design of event-triggered control systems over a shared network, *Discrete Event Dyn. Sys.*, vol. 24, pp. 153–171, 2014.

[36] P. G. Otanez, J. G. Moyne, D. M. Tilbury, Using deadbands to reduce communication in networked control systems, *American Control Conf.*, Anchorage, AK, 2002, pp. 3015–3020.

[37] M. Rabi, K. H. Johansson, M. Johansson, Optimal stopping for event-triggered sensing and actuation, *IEEE Conf. on Decision and Control*, Cancun, Mexico, 2008, pp. 3607–3612.

[38] M. Rabi, K. H. Johansson, Scheduling packets for event-triggered control, *European Control Conf.*, Budapest, 2009.

[39] A. Ruiz, J. E. Jimenez, J. Sanchez, S. Dormido, Design of event-based PI-P controllers using interactive tools, *Control Eng. Prac.*, vol. 32, pp. 183–202, 2014.

[40] M. Sigurani, C. Stöcker, L. Grüne, J. Lunze, Experimental evaluation of two complementary decentralized event-based control methods, *Control Eng. Prac.*, vol. 35, pp. 22–34, 2015.

[41] C. Stöcker, *Event-Based State-Feedback Control of Physically Interconnected Systems*, Logos-Verlag, Berlin, 2014.

[42] C. Stöcker, J. Lunze, Event-based control of input-output linearizable systems, *IFAC World Congress*, 2011, pp. 10062–10067.

[43] C. Stöcker, D. Vey, J. Lunze, Decentralized event-based control: Stability analysis and experimental evaluation, *Nonlinear Anal. Hybrid Sys.*, vol. 10, pp. 141–155, 2013.

[44] P. Tabuada, Event-triggered real-time scheduling of stabilizing control tasks, *IEEE Trans. Autom. Control*, vol. 52, pp. 1680–1685, 2007.

[45] P. Tallapragada, N. Chopra, Decentralized event-triggering for control of nonlinear systems, *IEEE Trans. Autom. Control*, vol. 59, pp. 3312–3324, 2014.

[46] M. Velasco, P. Mari, E. Bini, Control-driven tasks: Modeling and analysis, *Real-Time Systems Symposium*, 2008, pp. 280–290.

[47] X. Wang, M. D. Lemmon, Event-triggered broadcasting across distributed networked control systems, In *Proc. American Control Conference*, 2008, pp. 3139–3144.

[48] X. Wang, M. D. Lemmon, Event-triggering in distributed networked systems with data dropouts and delays, in R. Majumdar, P. Tabuada (Eds.), *Hybrid Systems: Computation and Control*, Springer, 2009, pp. 366–380.

[49] X. Wang, M. D. Lemmon, Event-triggering in distributed networked control systems, *IEEE Trans. Autom. Control*, vol. 56, pp. 586–601, 2011.

[50] X. Wang, M. Lemmon, Self-triggered feedback control systems with finite-gain $L_2$ stability, *IEEE Trans.*, vol. AC-54, pp. 452–467, 2009.

[51] P. V. Zhivoglyadov, R. H. Middleton, Networked control design for linear systems, *Automatica*, vol. 39, pp. 743–750, 2003.

# 2

# *Event-Driven Control and Optimization in Hybrid Systems*

**Christos G. Cassandras**
*Boston University*
*Boston, MA, USA*

**CONTENTS**



**ABSTRACT** The event-driven paradigm offers an alternative complementary approach to the time-driven paradigm for modeling, sampling, estimation, control, and optimization. This is largely a consequence of systems being increasingly networked, wireless, and consisting of distributed communicating components. The key idea is that control actions need not be dictated by time steps taken by a "clock"; rather, an action should be triggered by an "event," which may be a well-defined condition on the system state, including the possibility of a simple time step, or a random state transition. In this chapter, the event-driven paradigm is applied to control and optimization problems encountered in the general setting of hybrid systems where controllers are parameterized and the parameters are adaptively tuned online based on observable data. We present a general approach for evaluating (or estimating in the case of a stochastic system) gradients of performance metrics with respect to various parameters based on the infinitesimal perturbation analysis (IPA) theory originally developed for discrete event systems (DESs) and now adapted to hybrid systems. This results in an "IPA calculus," which amounts to a set of simple, event-driven iterative equations. The event-driven nature of this approach implies its scalability in the size of an event set, as opposed to the system state space. We also show how the event-based IPA calculus may be used in multi-agent systems for determining optimal agent trajectories without any detailed knowledge of environmental randomness.

## 2.1 Introduction

The history of modeling and analysis of dynamic systems is founded on the time-driven paradigm provided by a theoretical framework based on differential (or difference) equations. In this paradigm, time is an independent variable, and as it evolves, so does the state of the system. Conceptually, we postulate the existence of an underlying "clock," and with every "clock tick" a state update is performed, including the case where no change in the state occurs. The methodologies developed for sampling, estimation, communication, control, and optimization of dynamic systems have also evolved based on the same *time-driven* principle. Advances in digital technologies that occurred in the 1970s and beyond have facilitated the implementation of this paradigm with digital clocks embedded in hardware and used to drive processes for data collection or for the actuation of devices employed for control purposes.

As systems have become increasingly networked, wireless, and distributed, the universal value of this point of view has understandably come to question. While it is always possible to postulate an underlying clock with time steps dictating state transitions, it may not be feasible to guarantee the synchronization of all components of a distributed system to such a clock, and it is not efficient to trigger actions with every time step when such actions may be unnecessary. The *event-driven* paradigm offers an alternative, complementary look at modeling, control, communication, and optimization. The key idea is that a clock should not be assumed to dictate actions simply because a time step is taken; rather, an action should be triggered by an "event" specified as a well-defined condition on the system state or as a consequence of environmental uncertainties that result in random state transitions. Observing that such an event could actually be defined to be the occurrence of a "clock tick," it follows that this framework may in fact incorporate time-driven methods as well. On the other hand, defining the proper "events" requires more sophisticated techniques compared to simply reacting to time steps.

The motivation for this alternative event-driven view is multifaceted. For starters, there are many natural DESs where the only changes in their state are dictated by event occurrences. The Internet is a prime example, where "events" are defined by packet transmissions and receptions at various nodes, causing changes in the contents of various queues. For such systems, a time-driven modeling approach may not only be inefficient, but also potentially erroneous, as it cannot deal with events designed to occur concurrently in time. The development of a rigorous theory for the study of DES in the 1980s (see, e.g., [1–5]) paved the way for event-based models of certain classes of dynamic systems and spurred new concepts and techniques for control and optimization. By the early 1990s, it became evident that many interesting dynamic systems are in fact "hybrid" in nature, i.e., at least some of their state transitions are caused by (possibly controllable) events [6–12]. This has been reinforced by technological advances through which sensing and actuating devices are embedded into systems allowing physical processes to interface with such devices which are inherently event driven. A good example is the modern automobile where an event induced by a device that senses slippery road conditions may trigger the operation of an antilock braking system, thus changing the operating dynamics of the actual vehicle. More recently, the term "cyber–physical system" [13] has emerged to describe the hybrid structure of systems where some components operate as physical processes modeled through time-driven dynamics, while other components (mostly digital devices empowered by software) operate in event-driven mode.

Moreover, many systems of interest are now networked and spatially distributed. In such settings, especially when energy-constrained wireless devices are involved, frequent communication among system components can be inefficient, unnecessary, and sometimes infeasible. Thus, rather than imposing a rigid time-driven communication mechanism, it is reasonable to seek instead to define specific events that dictate when a particular node in a network needs to exchange information with one or more other nodes. In other words, we seek to complement synchronous operating mechanisms with asynchronous ones, which can dramatically reduce communication overhead without sacrificing adherence to design specifications and desired performance objectives. When, in addition, the environment is stochastic, significant changes in the operation of a system are the result of random event occurrences, so that, once again, understanding the implications of such events and reacting to them is crucial. Besides their modeling potential, it is also important to note that event-driven approaches to fundamental processes such as sampling, estimation, and control possess important properties related to variance reduction and robustness of control policies to modeling uncertainties. These properties render them particularly attractive, compared to time-driven alternatives.

While the importance of event-driven behavior in dynamic systems was recognized as part of the development of DES and then hybrid systems, more recently there have been significant advances in applying event-driven methods (also referred to as "event-based" and "event-triggered") to classical feedback control systems; see [14–18] and references therein. For example, in [15] a controller for a linear system is designed to update control values only when a specific error measure (e.g., for tracking or stabilization purposes) exceeds a given threshold, while refraining from any updates otherwise. It is also shown how such controllers may be tuned and how bounds may be computed in conjunction with known techniques from linear system theory. Trade-offs between interevent times and controller performance are further studied in [19]. As another example, in [18] an event-driven approach termed "self-triggered control" determines instants when the state should be sampled and control actions taken for some classes of nonlinear control systems. Benefits of event-driven mechanisms for estimation purposes are considered in [20,21]. In [20], for instance, an event-based sampling mechanism is studied where a signal is sampled only when measurements exceed a certain threshold, and it is shown that this approach outperforms a classical periodic sampling process at least in the case of some simple systems.

In distributed systems, event-driven mechanisms have the advantage of significantly reducing communication among networked components without affecting desired performance objectives (see [22–27]). For instance, Trimpe and D'Andrea [25] consider the problem of estimating the state of a linear system based on information communicated from spatially distributed sensors. In this case, each sensor computes the measurement prediction variance, and the event-driven process of transmitting information is defined by events such that this variance exceeds a given threshold. A scenario where sensors may be subject to malicious attacks is considered in [28], where event-driven methods are shown to lead to computationally advantageous state reconstruction techniques. It should be noted that in all such problems, one can combine event-driven and time-driven methods, as in [24] where a control scheme combining periodic (time-driven) and event-driven control is used for linear systems to update and communicate sensor and actuation data only when necessary in the sense of maintaining a satisfactory closed-loop performance. It is shown that this goal is attainable with a substantial reduction in communication over the underlying network. Along the same lines, combining event-driven and time-driven sensor information it is shown in [29] that stability can be guaranteed where the former methods alone may fail to do so.

In multi-agent systems, on the other hand, the goal is for networked components to cooperatively maximize (or minimize) a given objective; it is shown in [23] that an event-driven scheme can still achieve the optimization objective while drastically reducing communication (hence, prolonging the lifetime of a wireless network), even when delays are present (as long as they are bounded). Event-driven approaches are also attractive in receding horizon control, where it is computationally inefficient to reevaluate a control value over small time increments as opposed to event occurrences defining appropriate planning horizons for the controller (e.g., see [30]).

In the remainder of this chapter, we limit ourselves to discussing how the event-driven paradigm is applied to control and optimization problems encountered in the general setting of hybrid systems. In particular, we consider a general-purpose control and optimization framework where controllers are parameterized and the parameters are adaptively tuned online based on observable data. One way to systematically carry out this process is through gradient information pertaining to given performance measures with respect to these parameters, so as to iteratively adjust their values. When the environment is stochastic, this entails generating gradient estimates with desirable properties such as unbiasedness. This gradient evaluation/estimation approach is based on the IPA theory [1,31] originally developed for DES and now adapted to hybrid system where it results in an "IPA calculus" [32], which amounts to a set of simple, event-driven iterative equations. In this approach, the gradient evaluation/estimation procedure is based on directly observable data, and it is entirely event driven. This makes it computationally efficient, since it reduces a potentially complex process to a finite number of actions. More importantly perhaps, this approach has two key benefits that address the need for scalable methods in large-scale systems and the difficulty of obtaining accurate models especially in stochastic settings. First, being event driven, it is scalable in the size of the event space and not the state space of the system model. As a rule, the former is much smaller than the latter. Second, it can be shown that the gradient information is often independent of model parameters, which may be unknown or hard to estimate. In stochastic environments, this implies that complex control and optimization problems can be solved with little or no knowledge of the noise or random processes affecting the underlying system dynamics.

This chapter is organized as follows. A general online control and optimization framework for hybrid systems is presented in Section 2.2, whose centerpiece is a methodology used for evaluating (or estimating in the stochastic case) a gradient of an objective function with respect to controllable parameters. This event-driven methodology, based on IPA, is described in Section 2.3. In Section 2.4, three key properties of IPA are presented and illustrated through examples. In Section 2.4, an application to multi-agent systems is given. In particular, we consider cooperating agents that carry out a persistent monitoring mission in simple one-dimensional environments and formulate this mission as an optimal control problem. Its solution results in agents operating as hybrid systems with parameterized trajectories. Thus, using the event-based IPA calculus, we describe how optimal trajectories can be obtained online without any detailed knowledge of environmental randomness.

## 2.2 A Control and Optimization Framework for Hybrid Systems

A hybrid system consists of both time-driven and event-driven components [33]. The modeling, control, and optimization of these systems is quite challenging. In particular, the performance of a stochastic hybrid system (SHS) is generally hard to estimate because of the absence of closed-form expressions capturing the dependence of interesting performance metrics on various design or control parameters. Most approaches rely on approximations and/or using computationally taxing methods, often involving dynamic programming techniques. The inherent computational complexity of these approaches, however, makes them unsuitable for online control and optimization. Yet, in some cases, the structure of a dynamic optimization problem solution can be shown to be of parametric form, thus reducing it to a parametric optimization problem. As an example, in a linear quadratic Gaussian setting, optimal feedback policies simply depend on gain parameters to be selected subject to certain constraints. Even when this is not provably the case, one can still define parametric families of solutions which can be optimized and yield near-optimal or at least vastly improved solutions relative to ad hoc policies often adopted. For instance, it is common in solutions based on dynamic programming [34] to approximate cost-to-go functions through parameterized function families and then iterate over the parameters involved seeking near-optimal solutions for otherwise intractable problems.

With this motivation in mind, we consider a general-purpose framework as shown in Figure 2.1. The starting point is to assume that we can observe state trajectories of a given hybrid system and measure a performance (or cost) metric denoted by $L(\theta)$, where $\theta$ is a parameter vector. This vector characterizes a controller (as shown in Figure 2.1) but may also include design or model parameters. The premise here is that the system is too complex for a closed-form expression of $L(\theta)$ to be available, but that it is possible to measure it over a given time window. In the case of a stochastic environment, the observable state trajectory is a sample path of a SHS, so that $L(\theta)$ is a sample function, and performance is measured through $E[L(\theta)]$ with the expectation defined in the context of a suitable probability space. In addition to $L(\theta)$, we assume that all or part of the system state is observed, with possible noisy measurements. Thus, randomness may enter through the system process or the measurement process or both.

The next step in Figure 2.1 is the evaluation of the gradient $\nabla L(\theta)$. In the stochastic case, $\nabla L(\theta)$ is a random variable that serves as an estimate (obtained over a given time window) of $\nabla E[L(\theta)]$. Note that we require $\nabla L(\theta)$ to be evaluated based on available data observed from a *single* state trajectory (or sample path) of the hybrid system. This is in contrast to standard derivative approximation or estimation methods for $\frac{dL(\theta)}{d\theta}$ based on finite differences of the form $\frac{L(\theta+\Delta\theta)-L(\theta)}{\Delta\theta}$. Such methods require two state trajectories under $\theta$ and $\theta+\Delta\theta$, respectively, and are vulnerable to numerical problems when $\Delta\theta$ is selected to be small so as to increase the accuracy of the derivative approximation.

The final step then is to make use of $\nabla L(\theta)$ in a gradient-based adaptation mechanism of the general form $\theta_{n+1} = \theta_n + \eta_n \nabla L(\theta)$, where $n = 1, 2, \ldots$ counts the iterations over which this process evolves, and $\{\eta_n\}$ is a step size sequence which is appropriately selected to ensure convergence of the controllable parameter sequence $\{\theta_n\}$ under proper stationarity assumptions. After each iteration, the controller is adjusted, which obviously affects the behavior of the system, and the process repeats. Clearly, in a stochastic setting there is no guarantee of stationarity conditions, and this framework is simply one where the controller is perpetually seeking to improve system performance.

The cornerstone of this *online* framework is the evaluation of $\nabla L(\theta)$ based *only* on data obtained from the observed state trajectory. The theory of IPA [32,35] provides the foundations for this to be possible. Moreover, in the stochastic case where $\nabla L(\theta)$ becomes an estimate of $\nabla E[L(\theta)]$, it is important that this estimate possess desirable properties such as unbiasedness, without which the ultimate goal of achieving optimality cannot be provably attained. As we see in the next section, it is possible to evaluate $\nabla L(\theta)$ for virtually arbitrary SHS through a simple systematic event-driven procedure we refer to as the "IPA calculus." In addition, this gradient is characterized by several attractive properties under mild technical conditions.

In order to formally apply IPA and subsequent control and optimization methods to hybrid systems, we need to establish a general modeling framework. We use a standard definition of a hybrid automaton [33].

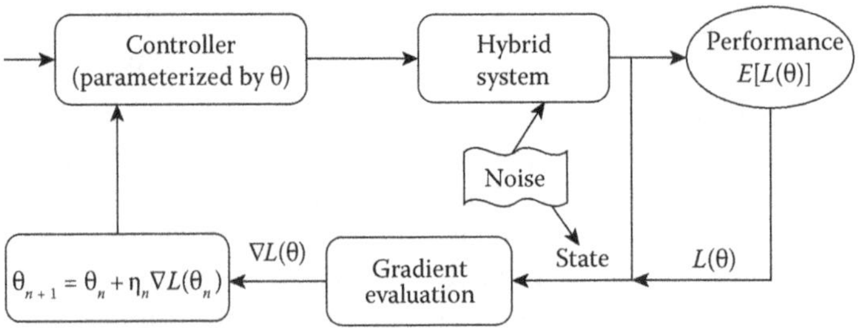


**FIGURE 2.1**
Online control and optimization framework for hybrid systems.

Thus, let $q \in Q$ (a countable set) denote the discrete state (or mode) and $x \in X \subseteq \mathbb{R}^n$ denote the continuous state of the hybrid system. Let $\upsilon \in \Upsilon$ (a countable set) denote a discrete control input and $u \in U \subseteq \mathbb{R}^m$ a continuous control input. Similarly, let $\delta \in \Delta$ (a countable set) denote a discrete disturbance input and $d \in D \subseteq \mathbb{R}^p$ a continuous disturbance input. The state evolution is determined by means of

- A vector field $f : Q \times X \times U \times D \to X$
- An invariant (or domain) set $Inv : Q \times \Upsilon \times \Delta \to 2^X$
- A guard set $Guard : Q \times Q \times \Upsilon \times \Delta \to 2^X$
- A reset function $r : Q \times Q \times X \times \Upsilon \times \Delta \to X$

A trajectory or sample path of such a system consists of a sequence of intervals of continuous evolution followed by a discrete transition. The system remains at a discrete state $q$ as long as the continuous (time-driven) state $x$ does not leave the set $Inv(q, \upsilon, \delta)$. If, before reaching $Inv(q, \upsilon, \delta)$, $x$ reaches a set $Guard(q, q', \upsilon, \delta)$ for some $q' \in Q$, a discrete transition is allowed to take place. If this transition does take place, the state instantaneously resets to $(q', x')$, where $x'$ is determined by the reset map $r(q, q', x, \upsilon, \delta)$. Changes in the discrete controls $\upsilon$ and disturbances $\delta$ are discrete events that either *enable* a transition from $q$ to $q'$ when $x \in Guard(q, q', \upsilon, \delta)$ or *force* a transition out of $q$ by making sure $x \notin Inv(q, \upsilon, \delta)$. We also use $\mathcal{E}$ to denote the set of all events that cause discrete state transitions and will classify events in a manner that suits the purposes of perturbation analysis. In what follows, we provide an overview of the "IPA calculus" and refer the reader to [32] and [36] for more details.

## 2.3 IPA: Event-Driven IPA Calculus

In this section, we describe the general framework for IPA as presented in [37] and generalized in [32] and [36]. Let $\theta \in \Theta \subset \mathbb{R}^l$ be a global variable, henceforth called the *control parameter*, where $\Theta$ is a given compact, convex set. This may include system design parameters, parameters of an input process, or parameters that characterize a policy used in controlling this system. The disturbance input $d \in D$ encompasses various random processes that affect the evolution of the state $(q, x)$ so that, in general, we can deal with a SHS. We will assume that all such processes are defined over a common probability space, $(\Omega, \mathcal{F}, P)$. Let us fix a particular value of the parameter $\theta \in \Theta$ and study a resulting sample path of the SHS. Over such a sample path, let $\tau_k(\theta)$, $k = 1, 2, \ldots$, denote the occurrence times of the discrete events in increasing order, and define $\tau_0(\theta) = 0$ for convenience. We will use the notation $\tau_k$ instead of $\tau_k(\theta)$ when no confusion arises. The continuous state is also generally a function of $\theta$, as well as of $t$, and is thus denoted by $x(\theta, t)$. Over an interval $[\tau_k(\theta), \tau_{k+1}(\theta))$, the system is at some mode during which the time-driven state satisfies:

$$\dot{x} = f_k(x, \theta, t), \tag{2.1}$$

where $\dot{x}$ denotes $\frac{\partial x}{\partial t}$. Note that we suppress the dependence of $f_k$ on the inputs $u \in U$ and $d \in D$ and stress instead its dependence on the parameter $\theta$ which may generally affect either $u$ or $d$ or both. The purpose of perturbation analysis is to study how changes in $\theta$ influence the state $x(\theta, t)$ and the event times $\tau_k(\theta)$ and, ultimately, how they influence interesting performance metrics that are generally expressed in terms of these variables. The following assumption guarantees that (2.1) has a unique solution w.p.1 for a given initial boundary condition $x(\theta, \tau_k)$ at time $\tau_k(\theta)$.

**ASSUMPTION 2.1** W.p.1, there exists a finite set of points $t_j \in [\tau_k(\theta), \tau_{k+1}(\theta))$, $j = 1, 2, \ldots$, which are independent of $\theta$, such that the function $f_k$ is continuously differentiable on $\mathbb{R}^n \times \Theta \times ([\tau_k(\theta), \tau_{k+1}(\theta)) \setminus \{t_1, t_2 \ldots\})$. Moreover, there exists a random number $K > 0$ such that $E[K] < \infty$ and the norm of the first derivative of $f_k$ on $\mathbb{R}^n \times \Theta \times ([\tau_k(\theta), \tau_{k+1}(\theta)) \setminus \{t_1, t_2 \ldots\})$ is bounded from above by $K$.

An event occurring at time $\tau_{k+1}(\theta)$ triggers a change in the mode of the system, which may also result in new dynamics represented by $f_{k+1}$, although this may not always be the case; for example, two modes may be distinct because the state $x(\theta, t)$ enters a new region where the system's performance is measured differently without altering its time-driven dynamics (i.e., $f_{k+1} = f_k$). The event times $\{\tau_k(\theta)\}$ play an important role in defining the interactions between the time-driven and event-driven dynamics of the system.

We now classify events that define the set $\mathcal{E}$ as follows:

- *Exogenous*: An event is exogenous if it causes a discrete state transition at time $\tau_k$ independent of the controllable vector $\theta$ and satisfies $\frac{d\tau_k}{d\theta} = 0$. Exogenous events typically correspond to uncontrolled random changes in input processes.

- *Endogenous*: An event occurring at time $\tau_k$ is endogenous if there exists a continuously differentiable function $g_k : \mathbb{R}^n \times \Theta \to \mathbb{R}$ such that

$$\tau_k = \min\{t > \tau_{k-1} \ : \ g_k(x(\theta, t), \theta) = 0\}. \tag{2.2}$$

The function $g_k$ is normally associated with an invariant or a guard condition in a hybrid automaton model.

- *Induced*: An event at time $\tau_k$ is induced if it is triggered by the occurrence of another event at time $\tau_m \leq \tau_k$. The triggering event may be exogenous, endogenous, or itself an induced event. The events that trigger induced events are identified by a subset of the event set, $\mathcal{E}_I \subseteq \mathcal{E}$.

Although this event classification is sufficiently general, recent work has shown that in some cases, it is convenient to introduce further event distinctions [38]. Moreover, it has been shown in [36] that an explicit event classification is in fact unnecessary if one is willing to appropriately extend the definition of the hybrid automaton described earlier. However, for the rest of this chapter, we only make use of the above classification.

Next, consider a performance function of the control parameter $\theta$:

$$J(\theta; x(\theta,0), T) = E[L(\theta; x(\theta,0), T)],$$

where $L(\theta; x(\theta,0), T)$ is a sample function of interest evaluated in the interval $[0,T]$ with initial conditions $x(\theta,0)$. For simplicity, we write $J(\theta)$ and $L(\theta)$. Suppose that there are $N$ events, with occurrence times generally dependent on $\theta$, during the time interval $[0,T]$ and define $\tau_0 = 0$ and $\tau_{N+1} = T$. Let $L_k : \mathbb{R}^n \times \Theta \times \mathbb{R}^+ \to \mathbb{R}$ be a function satisfying Assumption 2.1 and define $L(\theta)$ by

$$L(\theta) = \sum_{k=0}^{N} \int_{\tau_k}^{\tau_{k+1}} L_k(x,\theta,t)dt, \tag{2.3}$$

where we reiterate that $x = x(\theta,t)$ is a function of $\theta$ and $t$. We also point out that the restriction of the definition of $J(\theta)$ to a finite horizon $T$ which is independent of $\theta$ is made merely for the sake of simplicity.

Returning to Figure 2.1 and considering (for the sake of generality) the stochastic setting, the ultimate goal of the iterative process shown is to maximize $E_\omega[L(\theta,\omega)]$, where we use $\omega$ to emphasize dependence on a sample path $\omega$ of a SHS (clearly, this is reduced to $L(\theta)$ in the deterministic case). Achieving such optimality is possible under standard ergodicity conditions imposed on the underlying stochastic processes, as well as the assumption that a single global optimum exists; otherwise, the gradient-based approach is simply continuously attempting to improve the observed performance $L(\theta,\omega)$. Thus, we are interested in estimating the gradient

$$\frac{dJ(\theta)}{d\theta} = \frac{dE_\omega[L(\theta,\omega)]}{d\theta},$$

by evaluating $\frac{dL(\theta,\omega)}{d\theta}$ based on *directly observed data*. We obtain $\theta^*$ (under the conditions mentioned above) by optimizing $J(\theta)$ through an iterative scheme of the form

$$\theta_{n+1} = \theta_n - \eta_n H_n(\theta_n; x(\theta,0), T, \omega_n), \quad n = 0,1,\ldots, \tag{2.4}$$

where $\{\eta_n\}$ is a step size sequence and $H_n(\theta_n; x(\theta,0), T, \omega_n)$ is the estimate of $\frac{dJ(\theta)}{d\theta}$ at $\theta = \theta_n$. In using IPA, $H_n(\theta_n; x(\theta,0), T, \omega_n)$ is the sample derivative $\frac{dL(\theta,\omega)}{d\theta}$, which is an unbiased estimate of $\frac{dJ(\theta)}{d\theta}$ if the condition (dropping the symbol $\omega$ for simplicity)

$$E\left[\frac{dL(\theta)}{d\theta}\right] = \frac{d}{d\theta}E[L(\theta)] = \frac{dJ(\theta)}{d\theta}, \tag{2.5}$$

is satisfied, which turns out to be the case under mild technical conditions to be discussed later. The conditions under which algorithms of the form (2.4) converge are well-known (e.g., see [39]). Moreover, in addition to being unbiased, it can be shown that such gradient estimates are independent of the probability laws of the stochastic processes involved and require minimal information from the observed sample path.

The process through which IPA evaluates $\frac{dL(\theta)}{d\theta}$ is based on analyzing how changes in $\theta$ influence the state $x(\theta,t)$ and the event times $\tau_k(\theta)$. In turn, this provides information on how $L(\theta)$ is affected, because it is generally expressed in terms of these variables. Given $\theta = [\theta_1,\ldots,\theta_l]^{\mathrm{T}}$, we use the Jacobian matrix notation:

$$x'(\theta,t) \equiv \frac{\partial x(\theta,t)}{\partial\theta}, \quad \tau_k' \equiv \frac{\partial \tau_k(\theta)}{\partial\theta}, \quad k = 1,\ldots,K,$$

for all state and event time derivatives. For simplicity of notation, we omit $\theta$ from the arguments of the functions above unless it is essential to stress this dependence. It is shown in [32] that $x'(t)$ satisfies

$$\frac{d}{dt}x'(t) = \frac{\partial f_k(t)}{\partial x}x'(t) + \frac{\partial f_k(t)}{\partial\theta}, \tag{2.6}$$

for $t \in [\tau_k(\theta), \tau_{k+1}(\theta))$ with boundary condition

$$x'(\tau_k^+) = x'(\tau_k^-) + [f_{k-1}(\tau_k^-) - f_k(\tau_k^+)]\tau_k', \tag{2.7}$$

for $k = 0,\ldots,K$. We note that whereas $x(t)$ is often continuous in $t$, $x'(t)$ may be discontinuous in $t$ at the event times $\tau_k$; hence, the left and right limits above are generally different. If $x(t)$ is not continuous in $t$ at $t = \tau_k(\theta)$, the value of $x(\tau_k^+)$ is determined by the reset function $r(q,q',x,\upsilon,\delta)$ discussed earlier and

$$x'(\tau_k^+) = \frac{dr(q,q',x,\upsilon,\delta)}{d\theta}. \tag{2.8}$$

Furthermore, once the initial condition $x'(\tau_k^+)$ is given, the linearized state trajectory $\{x'(t)\}$ can be computed in the interval $t \in [\tau_k(\theta), \tau_{k+1}(\theta))$ by solving (2.6) to obtain

$$x'(t) = e^{\int_{\tau_k}^{t} \frac{\partial f_k(u)}{\partial x} du} \left[ \int_{\tau_k}^{t} \frac{\partial f_k(v)}{\partial \theta} e^{-\int_{\tau_k}^{t} \frac{\partial f_k(u)}{\partial x} du} dv + \xi_k \right], \tag{2.9}$$

with the constant $\xi_k$ determined from $x'(\tau_k^+)$ in either (2.7) or (2.8).

In order to complete the evaluation of $x'(\tau_k^+)$ in (2.7), we need to also determine $\tau_k'$. Based on the event classification above, $\tau_k' = 0$ if the event at $\tau_k(\theta)$ is exogenous and

$$\tau_k' = -\left[\frac{\partial g_k}{\partial x} f_k(\tau_k^-)\right]^{-1} \left(\frac{\partial g_k}{\partial \theta} + \frac{\partial g_k}{\partial x} x'(\tau_k^-)\right), \tag{2.10}$$

if the event at $\tau_k(\theta)$ is endogenous, that is, $g_k(x(\theta, \tau_k), \theta) = 0$, defined as long as $\frac{\partial g_k}{\partial x} f_k(\tau_k^-) \neq 0$. (Details may be found in [32].) Finally, if an induced event occurs at $t = \tau_k$ and is triggered by an event at $\tau_m \leq \tau_k$, the value of $\tau_k'$ depends on the derivative $\tau_m'$. The event induced at $\tau_m$ will occur at some time $\tau_m + w(\tau_m)$, where $w(\tau_m)$ is a (generally random) variable which is dependent on the continuous and discrete states $x(\tau_m)$ and $q(\tau_m)$, respectively. This implies the need for additional state variables, denoted by $y_m(\theta, t)$, $m = 1, 2, \ldots$, associated with events occurring at times $\tau_m$, $m = 1, 2 \ldots$. The role of each such a state variable is to provide a "timer" activated when a triggering event occurs. Triggering events are identified as belonging to a set $\mathcal{E}_I \subseteq \mathcal{E}$ and letting $e_k$ denote the event occurring at $\tau_k$. Then, define $F_k = \{m : e_m \in \mathcal{E}_I, m \leq k\}$ to be the set of all indices with corresponding triggering events up to $\tau_k$. Omitting the dependence on $\theta$ for simplicity, the dynamics of $y_m(t)$ are then given by

$$\dot{y}_m(t) = \begin{cases} -C(t) & \tau_m \leq t < \tau_m + w(\tau_m),\ m \in F_m \\ 0 & \text{otherwise} \end{cases}, \tag{2.11}$$

$$y_m(\tau_m^+) = \begin{cases} y_0 & y_m(\tau_m^-) = 0,\ m \in F_m \\ 0 & \text{otherwise} \end{cases},$$

where $y_0$ is an initial value for the timer $y_m(t)$, which decreases at a "clock rate" $C(t) > 0$ until $y_m(\tau_m + w(\tau_m)) = 0$ and the associated induced event takes place. Clearly, these state variables are only used for induced events, so that $y_m(t) = 0$ unless $m \in F_m$. The value of $y_0$ may depend on $\theta$ or on the continuous and discrete states $x(\tau_m)$ and $q(\tau_m)$, while the clock rate $C(t)$ may depend on $x(t)$ and $q(t)$ in general, and possibly $\theta$. However, in most simple cases where we are interested in modeling an induced event to occur at time $\tau_m + w(\tau_m)$, we have $y_0 = w(\tau_m)$ and $C(t) = 1$—that is, the timer simply counts down for a total of $w(\tau_m)$ time units until the induced event takes place. Henceforth, we will consider $y_m(t)$, $m = 1, 2, \ldots$, as part of the continuous state of the SHS, and we set

$$y_m'(t) \equiv \frac{\partial y_m(t)}{\partial \theta}, \quad m = 1, \ldots, N. \tag{2.12}$$

For the common case where $y_0$ is independent of $\theta$ and $C(t)$ is a constant $c > 0$ in (2.11), Lemma 2.1 facilitates the computation of $\tau_k'$ for an induced event occurring at $\tau_k$. Its proof is given in [32].

**Lemma 2.1**

If in (2.11), $y_0$ is independent of $\theta$ and $C(t) = c > 0$ (constant), then $\tau_k' = \tau_m'$.

With the inclusion of the state variables $y_m(t)$, $m = 1, \ldots, N$, the derivatives $x'(t)$, $\tau_k'$, and $y_m'(t)$ can be evaluated through (2.6)–(2.11) and this set of equations is what we refer to as the "IPA calculus." In general, this evaluation is recursive over the event (mode switching) index $k = 0, 1, \ldots$. In other words, *the IPA estimation process is entirely event driven.* For a large class of problems, the SHS of interest does not involve induced events, and the state does not experience discontinuities when a mode-switching event occurs. In this case, the IPA calculus reduces to the application of three equations:

1. Equation 2.9:

$$x'(t) = e^{\int_{\tau_k}^{t} \frac{\partial f_k(u)}{\partial x} du} \left[ \int_{\tau_k}^{t} \frac{\partial f_k(v)}{\partial \theta} e^{-\int_{\tau_k}^{t} \frac{\partial f_k(u)}{\partial x} du} dv + \xi_k \right],$$

which describes how the state derivative $x'(t)$ evolves over $[\tau_k(\theta), \tau_{k+1}(\theta))$.

2. Equation 2.7:

$$x'(\tau_k^+) = x'(\tau_k^-) + [f_{k-1}(\tau_k^-) - f_k(\tau_k^+)]\tau_k',$$

which specifies the initial condition $\xi_k$ in (2.9).

3. Either $\tau_k' = 0$ or Equation 2.10:

$$\tau_k' = -\left[\frac{\partial g_k}{\partial x} f_k(\tau_k^-)\right]^{-1} \left(\frac{\partial g_k}{\partial \theta} + \frac{\partial g_k}{\partial x} x'(\tau_k^-)\right),$$

depending on the event type at $\tau_k(\theta)$, which specifies the event time derivative present in (2.7).

From a computational standpoint, the IPA derivative evaluation process takes place iteratively at each event defining a mode transition at some time instant $\tau_k(\theta)$. At this point in time, we have at our disposal the value of $x'(\tau_{k-1}^+)$ from the previous iteration, which specifies $\xi_{k-1}$

in (2.9) applied for all $t \in [\tau_{k-1}(\theta), \tau_k(\theta))$. Therefore, setting $t = \tau_k(\theta)$ in (2.9) we also have at our disposal the value of $x'(\tau_k^-)$. Next, depending on whether the event is exogenous or endogenous, the value of $\tau_k'$ can be obtained: it is either $\tau_k' = 0$ or given by (2.10) since $x'(\tau_k^-)$ is known. Finally, we obtain $x'(\tau_k^+)$ using (2.7). At this point, one can wait until the next event occurs at $\tau_{k+1}(\theta)$ and repeat the process which can, therefore, be seen to be entirely *event driven*.

The last step in the IPA process involves using the IPA calculus in order to evaluate the IPA derivative $dL/d\theta$. This is accomplished by taking derivatives in (2.3) with respect to $\theta$:

$$\frac{dL(\theta)}{d\theta} = \sum_{k=0}^{N} \frac{d}{d\theta} \int_{\tau_k}^{\tau_{k+1}} L_k(x,\theta,t)dt. \tag{2.13}$$

Applying the Leibnitz rule, we obtain, for every $k = 0, \ldots, N$,

$$\begin{aligned} &\frac{d}{d\theta} \int_{\tau_k}^{\tau_{k+1}} L_k(x,\theta,t)dt \\ &= \int_{\tau_k}^{\tau_{k+1}} \left[ \frac{\partial L_k}{\partial x}(x,\theta,t)x'(t) + \frac{\partial L_k}{\partial \theta}(x,\theta,t) \right] dt \\ &\quad + L_k(x(\tau_{k+1}),\theta,\tau_{k+1})\tau_{k+1}' - L_k(x(\tau_k),\theta,\tau_k)\tau_k', \end{aligned} \tag{2.14}$$

where $x'(t)$ and $\tau_k'$ are determined through (2.6)–(2.10). What makes IPA appealing is the simple form the right-hand-side in Equation 2.14 often assumes. As we will see, under certain commonly encountered conditions, this expression is further simplified by eliminating the integral term.

## 2.4 IPA Properties

In this section, we identify three key properties of IPA. The first one is important in ensuring that when IPA involves estimates of gradients, these estimates are unbiased under mild conditions. The second is a robustness property of IPA derivatives in the sense that they do not depend on specific probabilistic characterizations of any stochastic processes involved in the hybrid automaton model of a SHS. This property holds under certain sufficient conditions which are easy to check. Finally, under conditions pertaining to the switching function $g_k(x,\theta)$, which we have used to define endogenous events, the event-driven IPA derivative evaluation or estimation process includes some events that have the property of allowing us to decompose an observed state trajectory into cycles, thus greatly simplifying the overall computational effort.

### 2.4.1 Unbiasedness

We begin by returning to the issue of unbiasedness of the sample derivatives $\frac{dL(\theta)}{d\theta}$ derived using the IPA calculus described in the last section. In particular, the IPA derivative $\frac{dL(\theta)}{d\theta}$ is an unbiased estimate of the performance (or cost) derivative $\frac{dJ(\theta)}{d\theta}$ if the condition (2.5) holds. In a pure DES, the IPA derivative satisfies this condition for a relatively limited class of systems (see [1,31]). This has motivated the development of more sophisticated perturbation analysis methods that can still guarantee unbiasedness at the expense of additional information to be collected from the observed sample path or additional assumptions regarding the statistical properties of some of the random processes involved. However, in a SHS, the technical conditions required to guarantee the validity of (2.5) are almost always applicable.

The following result has been established in [40] regarding the unbiasedness of IPA:

**Theorem 2.1**

Suppose that the following conditions are in force: (1) For every $\theta \in \Theta$, the derivative $\frac{dL(\theta)}{d\theta}$ exists w.p.1. (2) W.p.1, the function $L(\theta)$ is Lipschitz continuous on $\Theta$, and the Lipschitz constant has a finite first moment. Then, for a fixed $\theta \in \Theta$, the derivative $\frac{dJ(\theta)}{d\theta}$ exists, and the IPA derivative $\frac{dL(\theta)}{d\theta}$ is unbiased.

The crucial assumption for Theorem 2.1 is the continuity of the sample function $L(\theta)$, which in many SHSs is guaranteed in a straightforward manner. Differentiability w.p.1 at a given $\theta \in \Theta$ often follows from mild technical assumptions on the probability law underlying the system, such as the exclusion of co-occurrence of multiple events (see [41]). Lipschitz continuity of $L(\theta)$ generally follows from upper boundedness of $|\frac{dL(\theta)}{d\theta}|$ by an absolutely integrable random variable, generally a weak assumption. In light of these observations, the proofs of unbiasedness of IPA have become standardized, and the assumptions in Theorem 2.1 can be verified fairly easily from the context of a particular problem.

### 2.4.2 Robustness to Stochastic Model Uncertainties

Next, we turn our attention to properties of the estimators obtained through the IPA calculus which render them, under certain conditions, particularly simple and efficient to implement with minimal information required about the underlying SHS dynamics.

The first question we address is related to $\frac{dL(\theta)}{d\theta}$ in (2.13), which, as seen in (2.14), generally depends on information accumulated over all $t \in [\tau_k, \tau_{k+1})$. It is, however, often the case that it depends *only* on information related to the event times $\tau_k$, $\tau_{k+1}$, resulting in an IPA estimator that is simple to implement. Using the notation $L'_k(x,t,\theta) \equiv \frac{dL_k(x,t,\theta)}{d\theta}$, we can rewrite $\frac{dL(\theta)}{d\theta}$ in (2.13) as

$$\frac{dL(\theta)}{d\theta} = \sum_k \Big[ \tau'_{k+1} \cdot L_k(\tau^+_{k+1}) - \tau'_k \cdot L_k(\tau^+_k) + \int_{\tau_k}^{\tau_{k+1}} L'_k(x,t,\theta)dt \Big]. \qquad (2.15)$$

The following theorem provides two sufficient conditions under which $\frac{dL(\theta)}{d\theta}$ involves only the event time derivatives $\tau'_k$, $\tau'_{k+1}$ and the "local" performance $L_k(\tau^+_{k+1})$, $L_k(\tau^+_k)$, which is obviously easy to observe. The proof of this result is given in [42].

**Theorem 2.2**

If condition (C1) or (C2) below holds, then $\frac{dL(\theta)}{d\theta}$ depends only on information available at event times $\{\tau_k\}$, $k = 0, 1, \ldots$.

(C1) $L_k(x,t,\theta)$ is independent of $t$ over $[\tau_k, \tau_{k+1})$ for all $k = 0, 1, \ldots$.

(C2) $L_k(x,t,\theta)$ is only a function of $x$, and the following condition holds for all $t \in [\tau_k, \tau_{k+1})$, $k = 0, 1, \ldots$:

$$\frac{d}{dt}\frac{\partial L_k}{\partial x} = \frac{d}{dt}\frac{\partial f_k}{\partial x} = \frac{d}{dt}\frac{\partial f_k}{\partial \theta} = 0. \qquad (2.16)$$

The implication of Theorem 2.2 is that (2.15), under either (C1) or (C2), reduces to

$$\frac{dL(\theta)}{d\theta} = \sum_k [\tau'_{k+1} \cdot L_k(\tau^+_{k+1}) - \tau'_k \cdot L_k(\tau^+_k)],$$

and involves *only* directly observable performance sample values at event times along with event time derivatives which are either zero (for exogenous events) or given by (2.10). The conditions in Theorem 2.2 are surprisingly easy to satisfy as the following example illustrates.

**EXAMPLE 2.1** Consider a SHS whose time-driven dynamics at all modes are linear and of the form

$$\dot{x} = a_k x(t) + b_k u_k(\theta,t) + w_k(t), \quad t \in [\tau_{k-1}(\theta), \tau_k(\theta)),$$

where $u_k(\theta,t)$ is a control used in the system mode over $[\tau_k(\theta), \tau_{k+1}(\theta))$, which depends on a parameter $\theta$ and $w_k(t)$ is some random process for which no further information is provided. Writing $f_k = a_k x(t) + b_k u_k(\theta,t) + w_k(t)$, we can immediately see that $\frac{\partial f_k}{\partial x} = a_k$ and $\frac{\partial f_k}{\partial \theta} = \frac{\partial u_k(\theta,t)}{\partial \theta}$; hence, the second of the three parts of (C2) is satisfied—that is, $\frac{d}{dt}\frac{\partial f_k}{\partial x} = 0$. Further, suppose that the dependence of $u_k(\theta,t)$ on $t$ is such that $\frac{\partial u_k(\theta,t)}{\partial \theta}$ is also independent of $t$; this is true, for instance, if $u_k(\theta,t) = u_k(\theta)$, that is, the control is fixed at that mode, or if $u_k(\theta,t) = \gamma(\theta)t$, in which case $\frac{d}{dt}\frac{\partial f_k}{\partial \theta} = 0$, and the last part of (C2) is also satisfied. Finally, consider a performance metric of the form

$$J(\theta) = E\left[\sum_{k=0}^{N} \int_{\tau_k}^{\tau_{k+1}} L_k(x,\theta,t)dt\right] = E\left[\sum_{k=0}^{N} \int_{\tau_k}^{\tau_{k+1}} x(t)dt\right],$$

where we have $\frac{\partial L_k}{\partial x} = 1$, thus satisfying also the first part of (C2). It is worthwhile pointing out that the IPA calculus here provides unbiased estimates of $\frac{dJ(\theta)}{d\theta}$ without any information regarding the noise process $w_k(t)$. Although this seems surprising at first, the fact is that the effect of the noise is captured through the values of the observable event times $\tau_k(\theta)$ and the observed performance values $L_k(\tau^+_k)$ at these event times only: modeling information about $w_k(t)$ is traded against observations made online at event times only. In other words, while the noise information is crucial if one is interested in the actual performance $\int_{\tau_k}^{\tau_{k+1}} L_k(x,\theta,t)dt$ over an interval $[\tau_{k-1}(\theta), \tau_k(\theta))$, such information is not always required to estimate the *sensitivity* of the performance $\int_{\tau_k}^{\tau_{k+1}} L_k(x,\theta,t)dt$ with respect to $\theta$.

We refer to the property reflected by Theorem 2.2 as "robustness" of IPA derivative estimators with respect to any noise process affecting the time-driven dynamics of the system. Clearly, that would not be the case if, for instance, the performance metric involved $x^2(t)$ instead of $x(t)$; then, $\frac{\partial L_k}{\partial x} = 2x(t)$ and the integral term in (2.15) would have to be included in the evaluation of $\frac{dL(\theta)}{d\theta}$. Although this increases the computational burden of the IPA evaluation procedure and requires the collection of sample data for $w_k(t)$, note that it still requires no prior modeling information regarding this random process.

Thus, one need not have a detailed model (captured by $f_{k-1}$) to describe the state behavior through $\dot{x} = f_{k-1}(x,\theta,t)$, $t \in [\tau_{k-1}, \tau_k)$ in order to estimate the effect of $\theta$ on this behavior. This explains why simple abstractions of a complex stochastic system are often adequate to perform sensitivity analysis and optimization, as long as the event times corresponding to discrete state transitions are accurately observed and the local system behavior at these event times, for example, $x'(\tau^+_k)$ in (2.7), can also be measured or calculated.

### 2.4.3 State Trajectory Decomposition

The final IPA property we discuss is related to the discontinuity in $x'(t)$ at event times, described in (2.7). This happens when endogenous events occur, since for exogenous events we have $\tau'_k = 0$. The next theorem identifies a simple condition under which $x'(\tau_k^+)$ is independent of the dynamics $f$ before the event at $\tau_k$. This implies that we can evaluate the sensitivity of the state with respect to $\theta$ without any knowledge of the state trajectory in the interval $[\tau_{k-1}, \tau_k)$ prior to this event. Moreover, under an additional condition, we obtain $x'(\tau_k^+) = 0$, implying that the effect of $\theta$ is "forgotten," and one can reset the perturbation process. This allows us to decompose an observed state trajectory (or sample path) into "reset cycles," greatly simplifying the IPA process. The proof of the next result is also given in [42].

**Theorem 2.3**

Suppose an endogenous event occurs at $\tau_k(\theta)$ with a switching function $g(x, \theta)$. If $f_k(\tau_k^+) = 0$, then $x'(\tau_k^+)$ is independent of $f_{k-1}$. If, in addition, $\frac{\partial g}{\partial \theta} = 0$, then $x'(\tau_k^+) = 0$.

The condition $f_k(\tau_k^+) = 0$ typically indicates a saturation effect or the state reaching a boundary that cannot be crossed, for example, when the state is constrained to be nonnegative. This often arises in stochastic flow systems used to model how parts are processed in manufacturing systems or how packets are transmitted and received through a communication network [43,44]. In such cases, the conditions of both Theorems 2.1 and 2.2 are frequently satisfied since (1) common performance metrics such as workload or overflow rates satisfy (2.16) and (2) flow systems involve nonnegative continuous states and are constrained by capacities that give rise to dynamics of the form $\dot{x} = 0$. This class of SHS is also referred to as stochastic flow models, and the simplicity of the IPA derivatives in this case has been thoroughly analyzed, for example, see [35,45]. We present an illustrative example below.

**EXAMPLE 2.2** Consider the fluid single-queue system shown in Figure 2.2, where the arrival-rate process $\{\alpha(t)\}$ and the service-rate process $\{\beta(t)\}$ are random processes (possibly correlated) defined on a common probability space. The queue has a finite buffer, $\{x(t)\}$ denotes the buffer workload (amount of fluid in the buffer), and $\{\gamma(t)\}$ denotes the overflow of excess fluid when the buffer is full. Let the controllable parameter $\theta$ be the buffer size, and consider the sample performance function to be the loss volume during a given horizon interval $[0, T]$, namely,

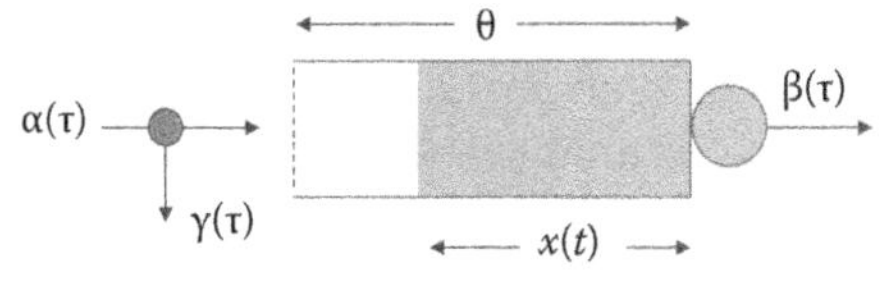


**FIGURE 2.2**
A simple fluid queue system for Example 2.2.

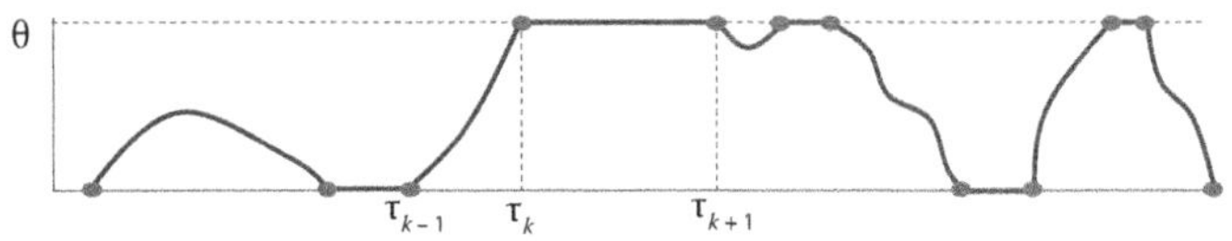


**FIGURE 2.3**
A typical sample path of the system in Figure 2.2.

$$L(\theta) = \int_0^T \gamma(\theta, t)dt. \tag{2.17}$$

We assume that $\alpha(t)$ and $\beta(t)$ are independent of $\theta$, and note that the buffer workload and overflow processes certainly depend upon $\theta$; hence, they are denoted by $\{x(\theta, t)\}$ and $\{\gamma(\theta, t)\}$, respectively. The only other assumptions we make on the arrival process and service process are that, w.p.1, $\alpha(t)$ and $\beta(t)$ are piecewise continuously differentiable in $t$ (but need not be continuous), and the terms $\int_0^T \alpha(t)dt$ and $\int_0^T \beta(t)dt$ have finite first moments. In addition, to satisfy the first condition of Theorem 2.1, we assume that w.p.1 no two events can occur at the same time (unless one induces the other), thus ensuring the existence of $\frac{dL(\theta)}{d\theta}$.

The time-driven dynamics in this SHS are given by

$$\dot{x}(\theta, t) = \begin{cases} 0 & \text{if } x(\theta, t) = 0, \alpha(t) \leq \beta(t) \\ 0 & \text{if } x(\theta, t) = \theta, \alpha(t) \geq \beta(t). \\ \alpha(t) - \beta(t) & \text{otherwise} \end{cases} \tag{2.18}$$

A typical sample path of the process $\{x(\theta, t)\}$ is shown in Figure 2.3. Observe that there are two endogenous events in this system: the first is when $x(\theta, t)$ increases and reaches the value $x(\theta, t) = \theta$ (as happens at time $\tau_k$ in Figure 2.3) and the second is when $x(\theta, t)$ decreases and reaches the value $x(\theta, t) = 0$. Thus, we see that the sample path is partitioned into intervals over which $x(\theta, t) = 0$, termed empty periods (EPs) since the fluid queue in Figure 2.2 is empty, and intervals over which $x(\theta, t) > 0$, termed nonempty periods (NEPs).

We can immediately see that Theorem 2.3 applies here for endogenous events with $g(x, \theta) = x$, which occur when an EP starts at some event time $\tau_k$. Since $\frac{\partial g}{\partial \theta} = 0$ and $f_k(\tau_k^+) = 0$ from (2.18), it follows that $x'(\tau_k^+) = 0$

and remains at this value throughout every EP. Therefore, the effect of the parameter $\theta$ in this case need only be analyzed over NEPs.

Next, observe that $\gamma(\theta, t) > 0$ only when $x(\theta, t) = \theta$. We refer to any such interval as a full period (FP) since the fluid queue in Figure 2.2 is full, and note that we can write $L(\theta)$ in (2.17) as

$$L(\theta) = \sum_{k \in \Psi_T} \int_{\tau_k}^{\tau_{k+1}} [\alpha(t) - \beta(t)]dt,$$

where $\Psi_T = \{k : x(\theta, t) = \theta \text{ for all } t \in [\tau_k(\theta), \tau_{k+1}(\theta))\}$ is the set of all FPs in the observed sample path over $[0, T]$. It follows that

$$\frac{dL(\theta)}{d\theta} = \sum_{k \in \Psi_T} [\alpha(\tau_{k+1}^-) - \beta(\tau_{k+1}^-)]\tau'_{k+1} - \sum_{k \in \Psi_T} [\alpha(\tau_k^+) - \beta(\tau_k^+)]\tau'_k, \quad (2.19)$$

and this is a case where condition (C2) of Theorem 2.2 holds: $\frac{d}{dt}\frac{\partial L_k}{\partial x} = \frac{d}{dt}[\alpha(t) - \beta(t)] = 0$ and $\frac{d}{dt}\frac{\partial f_k}{\partial x} = \frac{d}{dt}\frac{\partial f_k}{\partial \theta} = 0$ since $f_k = \alpha(t) - \beta(t)$ from (2.18). Thus, the evaluation of $\frac{dL(\theta)}{d\theta}$ reduces to the evaluation of $\tau'_{k+1}$ and $\tau'_k$ at the end and start, respectively, of every FP. Observing that $\tau'_{k+1} = 0$ since the end of a FP is an exogenous event depending only on a change in sign of $[\alpha(t) - \beta(t)]$ from nonnegative to strictly negative, it only remains to use the IPA calculus to evaluate $\tau'_k$ for every endogenous event such that $g(x(\theta, \tau_k), \theta) = x - \theta$. Applying (2.10) gives:

$$\tau'_k = \frac{1 - x'(\tau_k^-)}{\alpha(\tau_k^-) - \beta(\tau_k^-)}.$$

The value of $x'(\tau_k^-)$ is obtained using (2.9) over the interval $[\tau_{k-1}(\theta), \tau_k(\theta))$:

$$x'(\tau_k^-) = e^{\int_{\tau_{k-1}}^{\tau_k^-} \frac{\partial f_k(u)}{\partial x} du} \times \left[ \int_{\tau_{k-1}}^{\tau_k^-} \frac{\partial f_k(v)}{\partial \theta} e^{-\int_{\tau_{k-1}}^{\tau_k^-} \frac{\partial f_k(u)}{\partial x} du} dv + x'(\tau_{k-1}^+) \right],$$

where $\frac{\partial f_k(u)}{\partial x} = \frac{\partial f_k(u)}{\partial x} = 0$ and $\frac{\partial f_k(v)}{\partial \theta} = 0$. Moreover, using (2.7) at $t = \tau_{k-1}$, we have $x'(\tau_{k-1}^+) = x'(\tau_{k-1}^-) + [f_{k-1}(\tau_{k-1}^-) - f_k(\tau_{k-1}^+)]\tau'_{k-1} = 0$, since the start of a NEP is an exogenous event so that $\tau'_{k-1} = 0$ and $x'(\tau_{k-1}^-) = 0$ as explained earlier. Thus, $x'(\tau_k^-) = 0$, yielding

$$\tau'_k = \frac{1}{\alpha(\tau_k^-) - \beta(\tau_k^-)}.$$

Recalling our assumption that w.p.1 no two events can occur at the same time, $\alpha(t)$ and $\beta(t)$ can experience no discontinuities (exogenous events) at $t = \tau_k$ when the endogenous event $x(\theta, t) = \theta$ takes place, that is, $\alpha(\tau_k^-) - \beta(\tau_k^-) = \alpha(\tau_k^+) - \beta(\tau_k^+) = \alpha(\tau_k) - \beta(\tau_k)$. Then, returning to (2.19) we get

$$\frac{dL(\theta)}{d\theta} = -\sum_{k \in \Psi_T} \frac{\alpha(t) - \beta(t)}{\alpha(t) - \beta(t)} = -|\Psi_T|,$$

where $|\Psi_T|$ is simply the number of observed NEPs that include a "lossy" interval over which $x(\theta, t) = \theta$. Observe that this expression for $\frac{dL(\theta)}{d\theta}$ does not depend in any functional way on the details of the arrival or service rate processes. Furthermore, it is simple to compute, and in fact amounts to a simple counting process.

## 2.5 Event-Driven Optimization in Multi-Agent Systems

Multi-agent systems are commonly modeled as hybrid systems with time-driven dynamics describing the motion of the agents or the evolution of physical processes in a given environment, while event-driven behavior characterizes events that may occur randomly (e.g., an agent failure) or in accordance to control policies (e.g., an agent stopping to sense the environment or to change directions). As such, a multi-agent system can be studied in the context of Figure 2.1 with parameterized controllers aiming to meet certain specifications or to optimize a given performance metric. In some cases, the solution of a multi-agent *dynamic* optimization problem is reduced to a policy that is naturally parametric. Therefore, the adaptive scheme in Figure 2.1 provides a solution that is (at least locally) optimal. In this section, we present a problem known as "persistent monitoring," which commonly arises in multi-agent systems and where the event-driven approach we have described can be used.

Persistent monitoring tasks arise when agents must monitor a dynamically changing environment that cannot be fully covered by a stationary team of available agents. Thus, all areas of a given mission space must be visited infinitely often. The main challenge in designing control strategies in this case is in balancing the presence of agents in the changing environment so that it is covered over time optimally (in some well-defined sense) while still satisfying sensing and motion constraints. Examples of persistent monitoring missions include surveillance, patrol missions with unmanned vehicles, and environmental applications where routine sampling of an area is involved. Control and motion planning for this problem have been studied in the literature, for example, see [46–49]. We limit ourselves here to reviewing the optimal control formulation in [49] for a simple one-dimensional mission space taken to be an interval $[0, L] \subset \mathbb{R}$. Assuming $N$ mobile agents, let their positions at time $t$ be $s_n(t) \in [0, L]$, $n = 1, \ldots, N$, following the dynamics

$$\dot{s}_n(t) = g_n(s_n) + b_n u_n(t), \tag{2.20}$$

where $u_n(t)$ is the controllable agent speed constrained by $|u_n(t)| \leq 1$, $n = 1, \ldots, N$, that is, we assume that the agent can control its direction and speed. Without loss of generality, after some rescaling with the size of the mission space $L$, we further assume that the speed is constrained by $|u_n(t)| \leq 1, n = 1, \ldots, N$. For the sake of generality, we include the additional constraint,

$$a \leq s(t) \leq b, \; a \geq 0, \; b \leq L, \tag{2.21}$$

over all $t$ to allow for mission spaces where the agents may not reach the endpoints of $[0, L]$, possibly due to the presence of obstacles. We associate with every point $x \in [0, L]$ a function $p_n(x, s_n)$ that measures the probability that an event at location $x$ is detected by agent $n$. We also assume that $p_n(x, s_n) = 1$ if $x = s_n$, and that $p_n(x, s_n)$ is monotonically nonincreasing in the distance $|x - s_n|$ between $x$ and $s_n$, thus capturing the reduced effectiveness of a sensor over its range which we consider to be finite and denoted by $r_n$. Therefore, we set $p_n(x, s_n) = 0$ when $|x - s_n| > r_n$.

Next, consider a partition of $[0, L]$ into $M$ intervals whose center points are $\alpha_i = \frac{(2i-1)L}{2M}$, $i = 1, \ldots, M$. We associate a time-varying measure of uncertainty with each point $\alpha_i$, which we denote by $R_i(t)$. Without loss of generality, we assume $0 \leq \alpha_1 \leq \cdots \leq \alpha_M \leq L$ and, to simplify notation, we set $p_n(x, s_n(t)) = p_{n,i}(s_n(t))$ for all $x \in [\alpha_i - \frac{L}{2M}, \alpha_i + \frac{L}{2M}]$. Therefore, the joint probability of detecting an event at location $x \in [\alpha_i - \frac{L}{2M}, \alpha_i + \frac{L}{2M}]$ by all the $N$ agents simultaneously (assuming detection independence) is

$$P_i(\mathbf{s}(t)) = 1 - \prod_{n=1}^{N} [1 - p_{n,i}(s_n(t))], \tag{2.22}$$

where we set $\mathbf{s}(t) = [s_1(t), \ldots, s_N(t)]^{\mathrm{T}}$. We define uncertainty functions $R_i(t)$ associated with the intervals $[\alpha_i - \frac{L}{2M}, \alpha_i + \frac{L}{2M}]$, $i = 1, \ldots, M$, so that they have the following properties: (1) $R_i(t)$ increases with a prespecified rate $A_i$ if $P_i(\mathbf{s}(t)) = 0$, (2) $R_i(t)$ decreases with a fixed rate $B$ if $P_i(\mathbf{s}(t)) = 1$, and (3) $R_i(t) \geq 0$ for all $t$. It is then natural to model uncertainty so that its decrease is proportional to the probability of detection. In particular, we model the dynamics of $R_i(t), i = 1, \ldots, M$, as follows:

$$\dot{R}_i(t) = \begin{cases} 0 & \text{if } R_i(t) = 0, \; A_i \leq BP_i(\mathbf{s}(t)) \\ A_i - BP_i(\mathbf{s}(t)) & \text{otherwise} \end{cases}, \tag{2.23}$$

where we assume that initial conditions $R_i(0)$, $i = 1, \ldots, M$, are given and that $B > A_i > 0$ (thus, the uncertainty strictly decreases when there is perfect sensing $P_i(\mathbf{s}(t)) = 1$.) Note that $A_i$ represents the rate at which uncertainty increases at $\alpha_i$ which may be random. We will start with the assumption that the value of $A_i$ is known and will see how the robustness property of the IPA calculus (Theorem 2.2) allows us to easily generalize the analysis to random processes $\{A_i(t)\}$ describing uncertainty levels at different points in the mission space.

The goal of the optimal persistent monitoring problem is to control the movement of the $N$ agents through $u_n(t)$ in (2.20) so that the cumulative uncertainty over all sensing points $\{\alpha_i\}$, $i = 1, \ldots, M$, is minimized over a fixed time horizon $T$. Thus, setting $\mathbf{u}(t) = [u_1(t), \ldots, u_N(t)]$, we aim to solve the following optimal control problem:

$$\min_{\mathbf{u}(t)} J = \frac{1}{T} \int_0^T \sum_{i=1}^{M} R_i(t) dt, \tag{2.24}$$

subject to the agent dynamics (2.20), uncertainty dynamics (2.23), control constraint $|u_n(t)| \leq 1$, $t \in [0, T]$, and state constraints (2.21), $t \in [0, T]$.

Using a standard calculus of variations analysis, it is shown in [49] that the optimal trajectory of each agent $n$ is to move at full speed, that is, $u_n(t)$, until it reaches some switching point, dwell on the switching point for some time (possibly zero), and then switch directions. Consequently, each agent's optimal trajectory is fully described by a vector of switching points $\theta_n = [\theta_{n,1}, \ldots, \theta_{n,\Gamma_n}]^{\mathrm{T}}$ and $w_n = [w_{n,1} \ldots, w_{n,\Gamma_n}]^{\mathrm{T}}$, where $\theta_{n,\xi}$ is the $\xi$th control switching point and $w_{n,\xi}$ is the waiting time for this agent at the $\xi$th switching point. Note that $\Gamma_n$ is generally not known a priori and depends on the time horizon $T$. It follows that the behavior of the agents operating under optimal control is fully described by hybrid dynamics, and the problem is reduced to a *parametric* optimization one, where $\theta_n$ and $w_n$ need to be optimized for all $n = 1, \ldots, N$. This enables the use of the IPA calculus and, in particular, the use of the three equations, (2.9), (2.7), and (2.10), which ultimately leads to an evaluation of the gradient $\nabla J(\theta, w)$ with $J(\theta, w)$ in (2.24) now viewed as a function of the parameter vectors $\theta, w$.

In order to apply IPA to this hybrid system, we begin by identifying the events that cause discrete state transitions from one operating mode of an agent to another. Looking at the uncertainty dynamics (2.23), we define an event at time $\tau_k$ such that $\dot{R}_i(t)$ switches from $\dot{R}_i(t) = 0$ to $\dot{R}_i(t) = A_i - BP_i(\mathbf{s}(t))$ or an event such that $\dot{R}_i(t)$ switches from $\dot{R}_i(t) = A_i - BP_i(\mathbf{s}(t))$ to $\dot{R}_i(t) = 0$. In addition, since an optimal agent trajectory experiences switches of its control $u_n(t)$ from $\pm 1$ to 0 (the agent comes to rest before changing direction) or from 0 to $\pm 1$, we define events associated with each such action that affects the dynamics in (2.20). Denoting by $\tau_k(\theta, w)$ the occurrence time of any of these events, it is easy to obtain from (2.24):

$$\nabla J(\theta, w) = \frac{1}{T} \sum_{i=1}^{M} \sum_{k=0}^{K} \int_{\tau_k(\theta,w)}^{\tau_{k+1}(\theta,w)} \nabla R_i(t) dt,$$

which depends entirely on $\nabla R_i(t)$. Let us define the function

$$G_{n,i}(t) = B \prod_{d \neq n} (1 - p_i(s_d(t))) \left( \frac{\partial p_i(s_n)}{\partial s_n} \right) (t - \tau_k), \tag{2.25}$$

for all $t \in [\tau_k(\theta, w), \tau_{k+1}(\theta, w))$ and observe that it depends only on the sensing model $p_i(s_n(t))$ and the uncertainty model parameter $B$. Applying the IPA calculus (details are provided in [49]), we can then obtain

$$\frac{\partial R_i}{\partial \theta_{n,\xi}}(t) = \frac{\partial R_i(\tau_k^+)}{\partial \theta_{n,\xi}} - \begin{cases} 0 & \text{if } R_i(t) = 0, A_i < BP_i(\mathbf{s}(t)) \\ G_{n,i}(t) \frac{\partial s_n(\tau_k^+)}{\partial \theta_{n,\xi}} & \text{otherwise} \end{cases}, \tag{2.26}$$

and

$$\frac{\partial R_i}{\partial w_{n,\xi}}(t) = \frac{\partial R_i(\tau_k^+)}{\partial w_{n,\xi}} - \begin{cases} 0 & \text{if } R_i(t) = 0, A_i < BP_i(\mathbf{s}(t)) \\ G_{n,i}(t) \frac{\partial s_n(\tau_k^+)}{\partial w_{n,\xi}} & \text{otherwise} \end{cases}, \tag{2.27}$$

for all $n = 1, \ldots, N$ and $\xi = 1, \ldots, \Gamma_n$. It remains to derive event-driven iterative expressions for $\frac{\partial R_i(\tau_k^+)}{\partial \theta_{n,\xi}}$, $\frac{\partial R_i(\tau_k^+)}{\partial w_{n,\xi}}$ and $\frac{\partial s_n(\tau_k^+)}{\partial w_{n,\xi}}$, $\frac{\partial s_n(\tau_k^+)}{\partial \theta_{n,\xi}}$ above. These are given as follows (see [49] for details):

1. If an event at time $\tau_k$ is such that $\dot{R}_i(t)$ switches from $\dot{R}_i(t) = 0$ to $\dot{R}_i(t) = A_i - BP_i(\mathbf{s}(t))$, then $\nabla s_n(\tau_k^+) = \nabla s_n(\tau_k^-)$ and $\nabla R_i(\tau_k^+) = \nabla R_i(\tau_k^-)$ for all $n = 1, \ldots, N$.
2. If an event at time $\tau_k$ is such that $\dot{R}_i(t)$ switches from $\dot{R}_i(t) = A_i - BP_i(\mathbf{s}(t))$ to $\dot{R}_i(t) = 0$ (i.e., $R_i(\tau_k)$ becomes zero), then $\nabla s_n(\tau_k^+) = \nabla s_n(\tau_k^-)$ and $\nabla R_i(\tau_k^+) = 0$.
3. If an event at time $\tau_k$ is such that $u_n(t)$ switches from $\pm 1$ to 0, or from 0 to $\pm 1$, we need the components of $\nabla s_n(\tau_k^+)$ in (2.26) and (2.27) which are obtained as follows. First, for $\frac{\partial s_n(\tau_k^+)}{\partial \theta_{n,\xi}}$, if an event at time $\tau_k$ is such that $u_n(t)$ switches from $\pm 1$ to 0, then $\frac{\partial s_n}{\partial \theta_{n,\xi}}(\tau_k^+) = 1$ and

$$\frac{\partial s_n(\tau_k^+)}{\partial \theta_{n,j}} = \begin{cases} 0, & \text{if } j \neq \xi \\ 1, & \text{if } j = \xi \end{cases}, \quad j < \xi.$$

If on the other hand, $u_n(t)$ switches from 0 to $\pm 1$, then $\frac{\partial \tau_k}{\partial \theta_{n,\xi}} = -sgn(u(\tau_k^+))$ and

$$\frac{\partial s_n(\tau_k^+)}{\partial \theta_{n,j}} = \begin{cases} \frac{\partial s_n}{\partial \theta_{n,j}}(\tau_k^-) + 2, & \text{if } u_n(\tau_k^+) = 1, j \text{ even,} \\ & \text{or } u_n(\tau_k^+) = -1, j \text{ odd} \\ \frac{\partial s_n}{\partial \theta_{n,j}}(\tau_k^-) - 2, & \text{if } u_n(\tau_k^+) = 1, j \text{ odd,} \\ & \text{or } u_n(\tau_k^+) = -1, j \text{ even} \end{cases}, \quad j < \xi,$$

Finally, for $\frac{\partial s_n(\tau_k^+)}{\partial w_{n,\xi}}$, we have

$$\frac{\partial s_n(\tau_k^+)}{\partial w_{n,j}} = \begin{cases} 0, & \text{if } u_n(\tau_k^-) = \pm 1, u_n(\tau_k^+) = 0 \\ \mp 1, & \text{if } u_n(\tau_k^-) = 0, u_n(\tau_k^+) = \pm 1 \end{cases}.$$

In summary, this provides an event-driven procedure for evaluating $\nabla J(\theta, w)$ and proceeding with a gradient-based algorithm as shown in Figure 2.1 to determine optimal agent trajectories online or at least improve on current ones.

Furthermore, let us return to the case of stochastic environmental uncertainties manifested through random processes $\{A_i(t)\}$ in (2.23). Observe that the evaluation of $\nabla R_i(t)$, hence $\nabla J(\theta, w)$, is *independent* of $A_i$, $i = 1, \ldots, M$; in particular, note that $A_i$ does not appear in the function $G_{n,i}(t)$ in (2.25) or in any of the expressions for $\frac{\partial s_n(\tau_k^+)}{\partial \theta_{n,j}}$, $\frac{\partial s_n(\tau_k^+)}{\partial w_{n,j}}$. In fact, the dependence of $\nabla R_i(t)$ on $A_i$ manifests itself through the event times $\tau_k$, $k = 1, \ldots, K$, that do affect this evaluation, but they, unlike $A_i$ which may be unknown, are directly observable during the gradient evaluation process. This, once again is an example of the IPA robustness property discussed in Section 4.2. Extensive numerical examples of how agent trajectories are adjusted online for the persistent monitoring problem may be found in [49].

Extending this analysis from one-dimensional to two-dimensional mission spaces no longer yields optimal trajectories which are parametric in nature, as shown in [50]. However, one can represent an agent trajectory in terms of general function families characterized by a set of parameters that may be optimized based on an objective function such as (2.24) extended to two-dimensional environments. In particular, we may view each agent's trajectory as represented by parametric equations

$$s_n^x(t) = f(\Upsilon_n, \rho_n(t)), \quad s_n^y(t) = g(\Upsilon_n, \rho_n(t)), \qquad (2.28)$$

for all agents $n = 1, \ldots, N$. Here, $\Upsilon_n = [\Upsilon_n^1, \Upsilon_n^2, \ldots, \Upsilon_n^\Gamma]^T$ is the vector of parameters through which we control the shapes and locations of the $n$th agent trajectory, and $\Gamma$ is this vector's dimension. The agent position over time is controlled by a function $\rho_n(t)$ dependent on the agent dynamics. We can then formulate a problem such as

$$\min_{\Upsilon_n,\, n=1,\ldots,N} \quad J = \int_0^T \sum_{i=1}^M R_i(\Upsilon_1, \ldots, \Upsilon_N, t)dt,$$

which involves optimization over the controllable parameter vectors $\Upsilon_n, n = 1, \ldots, N$, characterizing each agent trajectory and placing once again the problem in the general framework of Figure 2.1.

## 2.6 Conclusions

Glancing into the future of systems and control theory, the main challenges one sees involve larger and ever more distributed wirelessly networked structures in application areas spanning cooperative multi-agent systems, energy allocation and management, and transportation, among many others. Barring any unexpected dramatic developments in battery technology, limited energy resources in wireless settings will have to largely dictate how control strategies are designed and implemented so as to carefully optimize this limitation. Taking this point of view, the event-driven paradigm offers an alternative to the time-driven paradigm for modeling, sampling, estimation, control, and optimization, not to supplant it but rather complement it. In hybrid systems, this approach is consistent with the event-driven nature of IPA which offers a general-purpose process for evaluating or estimating (in the case of stochastic systems) gradients of performance metrics. Such information can then be used on line so as to maintain a desirable system performance and, under appropriate conditions, lead to the solution of optimization problems in applications ranging from multi-agent systems to resource allocation in manufacturing, computer networks, and transportation systems.

## Acknowledgment

The author's work was supported in part by the National Science Foundation (NSF) under grants CNS-12339021 and IIP-1430145, by the Air Force Office of Scientific Research (AFOSR) under grant FA9550-12-1-0113, by the Office of Naval Research (ONR) under grant N00014-09-1-1051, and by the Army Research Office (ARO) under grant W911NF-11-1-0227.

## Bibliography

[1] C. G. Cassandras, S. Lafortune, *Introduction to Discrete Event Systems*, 2nd ed., Springer, New York 2008.

[2] Y. C. Ho, C. G. Cassandras, A new approach to the analysis of discrete event dynamic systems, *Automatica*, vol. 19, pp. 149–167, 1983.

[3] P. J. Ramadge, W. M. Wonham, The control of discrete event systems, *Proceedings of the IEEE*, vol. 77, no. 1, pp. 81–98, 1989.

[4] F. Baccelli, G. Cohen, G. J. Olsder, J. P. Quadrat, *Synchronization and Linearity*. Wiley, New York 1992.

[5] J. O. Moody, P. J. Antsaklis, *Supervisory Control of Discrete Event Systems Using Petri Nets*. Kluwer Academic, Dordrecht, the Netherlands 1998.

[6] M. S. Branicky, V. S. Borkar, S. K. Mitter, A unified framework for hybrid control: Model and optimal control theory, *IEEE Trans. Autom. Control*, vol. 43, no. 1, pp. 31–45, 1998.

[7] P. Antsaklis, W. Kohn, M. Lemmon, A. Nerode, and S. Sastry (Eds.), *Hybrid Systems*. Springer-Verlag, New York 1998.

[8] C. G. Cassandras, D. L. Pepyne, Y. Wardi, Optimal control of a class of hybrid systems, *IEEE Trans. Autom. Control*, vol. 46, no. 3, pp. 398–415, 2001.

[9] P. Zhang, C. G. Cassandras, An improved forward algorithm for optimal control of a class of hybrid systems, *IEEE Trans. Autom. Control*, vol. 47, no. 10, pp. 1735–1739, 2002.

[10] M. Lemmon, K. X. He, I. Markovsky, Supervisory hybrid systems, *IEEE Control Systems Magazine*, vol. 19, no. 4, pp. 42–55, 1999.

[11] H. J. Sussmann, A maximum principle for hybrid optimal control problems, in *Proceedings of 38th IEEE Conf. on Decision and Control*, December, 1999, pp. 425–430.

[12] A. Alur, T. A. Henzinger, E. D. Sontag (Eds.), *Hybrid Systems*. Springer-Verlag, New York 1996.

[13] Special issue on goals and challenges in cyber physical systems research, *IEEE Trans. Autom. Control*, vol. 59, no. 12, pp. 3117–3379, 2014.

[14] K. E. Arzen, A simple event based PID controller, in *Proceedings of 14th IFAC World Congress*, 2002, pp. 423–428.

[15] W. P. Heemels, J. H. Sandee, P. P. Bosch, Analysis of event-driven controllers for linear systems, *Int. J. Control*, vol. 81, no. 4, p. 571, 2008.

[16] J. Lunze, D. Lehmann, A state-feedback approach to event-based control, *Automatica*, vol. 46, no. 1, pp. 211–215, 2010.

[17] P. Tabuada, Event-triggered real-time scheduling of stabilizing control tasks, *IEEE Trans. Autom. Control*, vol. 52, no. 9, pp. 1680–1685, 2007.

[18] A. Anta, P. Tabuada, To sample or not to sample: Self-triggered control for nonlinear systems, *IEEE Trans. Autom. Control*, vol. 55, no. 9, pp. 2030–2042, 2010.

[19] V. S. Dolk, D. P. Borgers, W. P. M. H. Heemels, Dynamic event-triggered control: Tradeoffs between transmission intervals and performance, in *Proceedings 35th IEEE Conf. on Decision and Control*, 2014, pp. 2765–2770.

[20] K. J. Astrom, B. M. Bernhardsson, Comparison of Riemann and Lebesgue sampling for first order stochastic systems, in *Proceedings of 41st IEEE Conf. on Decision and Control*, 2002, pp. 2011–2016.

[21] T. Shima, S. Rasmussen, P. Chandler, UAV Team decision and control using efficient collaborative estimation, *ASME J. Dyn. Syst. Meas. Cont.* vol. 129, no. 5, pp. 609–619, 2007.

[22] X. Wang, M. Lemmon, Event-triggering in distributed networked control systems, *IEEE Trans. Autom. Control*, vol. 56, no. 3, pp. 586–601, 2011.

[23] M. Zhong, C. G. Cassandras, Asynchronous distributed optimization with event-driven communication, *IEEE Trans. Autom. Control*, vol. 55, no. 12, pp. 2735–2750, 2010.

[24] W. P. M. H. Heemels, M. C. F. Donkers, A. R. Teel, Periodic event-triggered control for linear systems, *IEEE Trans. Autom. Control*, vol. 58, no. 4, pp. 847–861, April 2013.

[25] S. Trimpe, R. D'Andrea, Event-based state estimation with variance-based triggering, *IEEE Trans. Autom. Control*, vol. 49, no. 12, pp. 3266–3281, 2014.

[26] E. Garcia, P. J. Antsaklis, Event-triggered output feedback stabilization of networked systems with external disturbance, in *Proceedings of 53rd IEEE Conf. on Decision and Control*, 2014, pp. 3572–3577.

[27] T. Liu, M. Cao, D. J. Hill, Distributed event-triggered control for output synchronization of dynamical networks with non-identical nodes, in *Proceedings of 53rd IEEE Conf. on Decision and Control*, 2014, pp. 3554–3559.

[28] Y. Shoukry, P. Tabuada, Event-triggered projected Luenberger observer for linear systems under sparse sensor attacks, in *Proceedings of 53rd IEEE Conf. on Decision and Control*, 2014, pp. 3548–3553.

[29] S. Trimpe, Stability analysis of distributed event-based state estimation, in *Proceedings of 53rd IEEE Conf. on Decision and Control*, 2014, pp. 2013–2018.

[30] Y. Khazaeni, C. G. Cassandras, A new event-driven cooperative receding horizon controller for multi-agent systems in uncertain environments, in *Proceedings of 53rd IEEE Conf. on Decision and Control*, 2014, pp. 2770–2775.

[31] Y. C. Ho, X. Cao, *Perturbation Analysis of Discrete Event Dynamic Systems*. Kluwer Academic, Dordrecht, the Netherlands 1991.

[32] C. G. Cassandras, Y. Wardi, C. G. Panayiotou, C. Yao, Perturbation analysis and optimization of stochastic hybrid systems, *Eur. J. Control*, vol. 16, no. 6, pp. 642–664, 2010.

[33] C. G. Cassandras, J. Lygeros (Eds.), *Stochastic Hybrid Systems*. CRC Press, Boca Raton, FL 2007.

[34] W. B. Powell, *Approximate Dynamic Programming* (2nd ed.). John Wiley and Sons, New York.

[35] C. G. Cassandras, Y. Wardi, B. Melamed, G. Sun, C. G. Panayiotou, Perturbation analysis for on-line control and optimization of stochastic fluid models, *IEEE Trans. Autom. Control*, vol. 47, no. 8, pp. 1234–1248, 2002.

[36] A. Kebarighotbi, C. G. Cassandras, A general framework for modeling and online optimization

of stochastic hybrid systems, in *Proceedings of 4th IFAC Conference on Analysis and Design of Hybrid Systems*, 2012.

[37] Y. Wardi, R. Adams, B. Melamed, A unified approach to infinitesimal perturbation analysis in stochastic flow models: The single-stage case, *IEEE Trans. Autom. Control*, vol. 55, no. 1, pp. 89–103, 2010.

[38] Y. Wardi, A. Giua, C. Seatzu, IPA for continuous stochastic marked graphs, *Automatica*, vol. 49, no. 5, pp. 1204–1215, 2013.

[39] H. J. Kushner, G. G. Yin, *Stochastic Approximation Algorithms and Applications*. Springer-Verlag, New York 1997.

[40] R. Rubinstein, *Monte Carlo Optimization, Simulation and Sensitivity of Queueing Networks*. John Wiley and Sons, New York 1986.

[41] C. Yao, C. G. Cassandras, Perturbation analysis and optimization of multiclass multiobjective stochastic flow models, *J. Discrete Event Dyn. Syst.*, vol. 21, no. 2, pp. 219–256, 2011.

[42] C. Yao, C. G. Cassandras, Perturbation analysis of stochastic hybrid systems and applications to resource contention games, *Frontiers of Electrical and Electronic Engineering in China*, vol. 6, no. 3, pp. 453–467, 2011.

[43] H. Yu, C. G. Cassandras, Perturbation analysis for production control and optimization of manufacturing systems, *Automatica*, vol. 40, pp. 945–956, 2004.

[44] C. G. Cassandras, Stochastic flow systems: Modeling and sensitivity analysis, in *Stochastic Hybrid Systems* (C. G. Cassandras, J. Lygeros, Eds.). Taylor and Francis, 2006, pp. 139–167.

[45] G. Sun, C. G. Cassandras, C. G. Panayiotou, Perturbation analysis of multiclass stochastic fluid models, *J. Discrete Event Dyn. Syst.: Theory Appl.*, vol. 14, no. 3, pp. 267–307, 2004.

[46] I. Rekleitis, V. Lee-Shue, A. New, H. Choset, Limited communication, multi-robot team based coverage, in *IEEE Int. Conf. on Robotics and Automation*, vol. 4, 2004, pp. 3462–3468.

[47] N. Nigam, I. Kroo, Persistent surveillance using multiple unmanned air vehicles, in *IEEE Aerospace Conf.*, 2008, pp. 1–14.

[48] S. L. Smith, M. Schwager, D. Rus, Persistent monitoring of changing environments using a robot with limited range sensing, in *Proc. of IEEE Conf. on Robotics and Automation*, pp. 5448–5455, 2011.

[49] C. G. Cassandras, X. Lin, X. C. Ding, An optimal control approach to the multi-agent persistent monitoring problem, *IEEE Trans. on Automatic Control*, vol. 58, no. 4, pp. 947–961, 2013.

[50] X. Lin, C. G. Cassandras, Trajectory optimization for multi-agent persistent monitoring in two-dimensional spaces, in *Proc. of 53rd IEEE Conf. Decision and Control*, 2014, pp. 3719–3724.

# 3

# *Reducing Communication by Event-Triggered Sampling*

**Marek Miśkowicz**
*AGH University of Science and Technology*
*Kraków, Poland*

**CONTENTS**



**ABSTRACT** Event-based sampling is a core of event-triggered control and signal processing of continuous-time signals. By flexible definition of the triggering conditions, the sampling released by events can address a wide range of design objectives and application requirements. The present study is a survey of event-triggered sampling schemes both in control and signal processing aimed to emphasize the motivation and wide technical context of event-based sampling criteria design.

## 3.1 Introduction

The convergence of computing, communication, and control enables integration of computational resources with the physical world under a concept of cyber-physical systems and is a major catalyst for future technology development. The core of cyber-physical systems is networking that enables coordination between smart sensing devices and computational entities. Due to high flexibility, easier mobility, and network expansion, the cyber-physical systems benefit from ubiquitous wireless connectivity. The postulate of providing wireless communication of a limited bandwidth between smart devices creates new challenges for networked sensor and control systems design, as the wireless transmission is energy expensive, while interconnected devices are often battery powered. The development of networked systems under constraint communication and energy resources has pushed a research interest toward concepts of establishing interactions between system components that are as rare as possible. This mechanism fostered to system design an adoption of the *event-based paradigm* represented by a model of calls for resources only if it is really necessary.

The event-paradigm in the context of control, communication, and signal processing is not new, as "probably one of the earliest issues to confront with respect to the control system design is when to sample the system so as to reduce the data needing to be transported over the network" [1]. Early event-based systems (termed also *aperiodic* or *asynchronous*) have been known in feedback control, data transmission, and signal processing at least since the 1950s, see [2,3]. Due to easier implementation and the existence of mature theory of digital systems based on periodic sampling, the event-based design strategy failed to compete with synchronous architectures that have monopolized electrical engineering since the early 1960s. In subsequent decades, the development of event-driven systems had attracted rather limited attention of the research community, although questions related to periodic versus aperiodic control, uniform versus nonuniform sampling in signal processing, or synchronous versus asynchronous circuit implementations have appeared since the origins of computer technology. The renaissance of interest in exploring the event-based paradigm was evoked more than a decade ago due to the publishing of a few important research works independently in the disciplines of control [4–6] and signal processing [7–11]. These works released great research creativity and gave rise to a systematic development of new event-triggered approaches (see [2,12], and other chapters in this book).

### 3.1.1 Event-Triggered Communication in Control

In the context of modern networked sensor and control systems, the event-based paradigm is aimed first at avoiding transmission of redundant data, thereby reducing the rate of messages transferred through the shared communication channel. The promise of saving communication bandwidth and energy resources is the primary motivation that stimulates research interest on event-triggered control strategies in networked systems. The other strong motivation behind development of event-based methodology is that the control loops established over networks with varying transmission delays and packet losses can no longer be modeled by conventional discrete-time control theory, even if data are processed and transmitted with a fixed sampling rate [5,12,42,64]. Thus, the progress in the event-triggered control strategies is enforced by the convergence of control and networking. The formulation of the event-based control methodology provides the fundamentals to model the operation of the networked control systems with nonzero transmission delays and packet losses. Perhaps, a more important benefit is that the development of approaches to the event-triggered control allows for relaxation of the control strategy based on equidistant sampling and for benefits to be gained from sporadic interactions between sensors and actuators at a lower rate.

Finally, the effect of reducing communication by the event-triggered data reporting model is the most effective method to minimize the network-induced imperfections on the control performance (i.e., packet delay, delay jitter, and packet losses) [13]. In this way, the adoption of the event-triggered strategy meets one of the fundamental criteria of the networked control system design aimed to achieve performance close to that of the conventional point-to-point digital control system. At the same time, control over networks provides well-known advantages over classical control (e.g., high flexibility and resource utilization, efficient reconfigurability, and reduction of installation and maintenance costs) and has been identified as one of the key directions for future developments of control systems [14].

### 3.1.2 Event-Based Signal Processing

During the last decade, in parallel to development of the event-triggered communication and control strategies, the event-based paradigm has been adopted in signal processing and electronic circuitry design as an alternative to the conventional approach based on periodic sampling. The strong motivation behind this alternative is that the modern nanoscale very large scale integration (VLSI) technology promotes fast circuit operation but requires low supply voltage (even below 1 V), which makes the fine quantization of the amplitude increasingly difficult. In the event-based signal processing mode, the periodic sampling is substituted by the discretization in the amplitude domain (e.g., by level-crossings), and the quantization process is moved from the amplitude to the time domain. This allows benefit to be gained from growing time resolution provided by the downscaling of the feature size of VLSI technology. The other reason for increasing interest in event-based signal processing is the promise to achieve activity-dependent power consumption and, thereby, save energy on the circuit level [15].

In response to challenges of nanoscale VLSI design, the development of the event-based signal processing techniques has resulted in introducing a new class of analog-to-digital converters and digital filters that use event-triggered sampling [10,11,16], as well as a novel concept of continuous-time digital signal processing (CT-DSP) [8,9,17].

The CT-DSP, based essentially on amplitude quantization and processing of binary signals in continuous time, is a complementary approach to conventional clocked DSP. Furthermore, the CT-DSP may be regarded as a "missing category" among well-established signal processing modes: continuous-time continuous-amplitude (analog processing), discrete-time discrete-amplitude (digital processing), and discrete-time continuous-amplitude processing techniques. Due to a lack of time discretization, the CT-DSP is characterized by no aliasing. The other significant advantages of this signal processing mode are the lower electromagnetic interference and minimized power consumption when the input does not contain novel information [17]. The application profile of the event-based signal processing techniques is related to battery-powered portable devices with low energy budget designed for wireless sensor networking and Internet of Things.

Despite different technical contexts and methodologies, both event-triggered control and signal processing essentially adopt the same sampling techniques, which are signal-dependent instead of based on the progression of time.

## 3.2 Event-Based Sampling Classes

Event-based sampling is a core of event-triggered control and signal processing of continuous-time signals. Furthermore, event-driven sampling specifies criteria for interactions between distributed system components, which is a critical issue in wireless communication systems. By flexible definition of the triggering conditions, the event-based sampling can address a wide range of design objectives and application requirements.

The fundamental postulate for formulation of event-based sampling criteria for monitoring and control systems is to achieve the satisfactory system performance at a low rate of events. Wireless transmission costs much more energy than information processing. Therefore, it is critical to maximize the usefulness of every bit transmitted or received [18], even if it requires much computational effort. In other words, it is reasonable to develop advanced event-triggered algorithms that provide discrete-time representation of the signal at low average sampling rates, even if these algorithms are complex.

On the other hand, in signal processing, another objective of event-based sampling is to eliminate the clock from system architecture, to benefit from increasing time resolution provided by nanoscale VLSI technology, or avoid aliasing. The present study is a survey of event-triggered sampling schemes in control and signal processing aimed to emphasize the motivation and wide technical context of event-based sampling criteria design.

### 3.2.1 Origins of Concept of Event

The notion of the *event* originates from computing domain and was formulated in the 1970s. An overview of early definitions of the event in the context of discrete-event digital simulation is summarized in [19]. Most concepts refer to the event as a change of the state of an object occurring at an instant. An event is said to be *determined*, if the condition on event occurrence can be expressed strictly as a function of time. Otherwise, the event is regarded as *contingent*. According to the classical definition widely adopted in real-time computer and communication systems in the 1990s, "an event is a significant change of a state of a system" [20,21].

Control theory was perhaps the first discipline that adopted the concept of event to define a criterion for triggering the sampling of continuous-time signals, although without an explicit reference to the notion of the event. In 1959, Ellis noticed, that "the most suitable sampling is by transmission of only significant data, as the new value obtained when the signal is changed by a given increment". Consequently, "it is not necessary to sample periodically, but only when quantized data change from one possible value to the next" [22]. As seen, Ellis defined the triggering condition for sampling as the occurrence of the significant change of the signal, which is consistent with definitions of the event adopted later in the computing domain. In the seminal work, Åstrom proposed for the criterion specified by Ellis the name *Lebesgue sampling* due to a clear connotation to the Lebesgue method of integration [5]. Using the same convention of terminology, the equidistant sampling has been called *Riemann sampling* [5]. Depending on the context, the Lebesgue sampling is also referred to as the *send-on-delta* concept [23], *deadbands* [13], *level-crossing sampling with hysteresis* [24], or *constant amplitude difference sampling criterion* [25]. The variety of existing terminology shows that it is really a generic concept adapted to a broad spectrum of technology and applications.

### 3.2.2 Threshold-Based Sampling

A class of event-based sampling schemes that may be referred literally to the concept of the event as a significant change of the particular signal parameter can be termed the *threshold-based sampling* that in the context of communication is known as the *send-on-delta* scheme [23]. According to the threshold-based (Lebesgue sampling, send-on-delta) criterion, an input signal is sampled when it deviates by a certain confidence interval (additive increment or decrement) (Figure 3.1). The size of the threshold (delta) defines a trade-off between system performance and the rate of samples. The maximum time interval between consecutive samples is theoretically unbounded.

The send-on-delta communication strategy is used in many industrial applications, for example, in LonWorks (ISO/IEC 14908), the commercial platform for networked control systems applied in building automation, street lighting, and smart metering [26], as well as in EnOcean energy harvesting wireless technology. Furthermore, the send-on-delta paradigm has been extended to spatiotemporal sampling, for example, as the basis of the Collaborative Event Driven Energy Efficient Protocol for sensor networks [27].

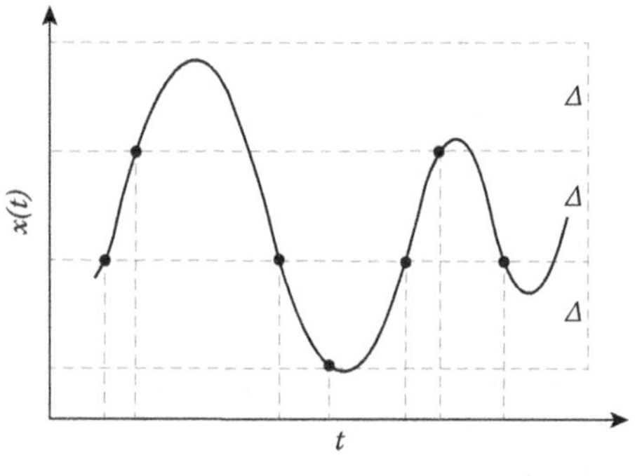


**FIGURE 3.1**
Send-on-delta sampling criterion.

In the basic version, the deviation is referred to the value of the most recent sample (zero-order hold) (Figure 3.1). In an advanced version, the threshold is related to the predicted value of the sampled signal (first-order or higher-order hold) [28–31]. The latter allows samples to be taken more rarely but the sampling device requires more information to be known about the signal (i.e., the time-derivatives) at the instant of the last sample.

Several extensions of send-on-delta criterion have been proposed: the integral sampling (Figure 3.2) [32–34] (called also area-triggered sampling, or *send-on-area* scheme), or the sampling triggered by error energy changes (*send-on-energy*) [35,36]. Due to the integration of any intersampling error, both the send-on-area and send-on-energy algorithms provide better conditions for effective sample triggering and help to avoid a lack of samples during long time intervals.

In the generalized version of the threshold-based sampling, a plant is sampled when the difference between the real process and its model is greater than a prespecified threshold [37].

### 3.2.3 Threshold-Based Criteria for State Estimation Uncertainty

The threshold-based sampling criteria (send-on-delta, send-on-area) are also used in the distributed state estimation when the state of a dynamic system is not measured directly by discrete values but is estimated with the probability density functions. The objective of the estimator is to keep the estimation uncertainty bounded. Although the use of the threshold-based sampling reduces the transmission of messages between the sensor and the estimator, compared to periodic updates, these transmissions occur not necessarily "when it is required." A demand to keep bounded the difference between the actual measurement and its prediction does not directly address the estimator's performance. The latter is based primarily on the estimation uncertainty. Recently, several novel event-based sampling criteria for state estimators have been proposed. These sampling schemes link the decision "when to transmit" to its contribution to the estimator performance [38–41].

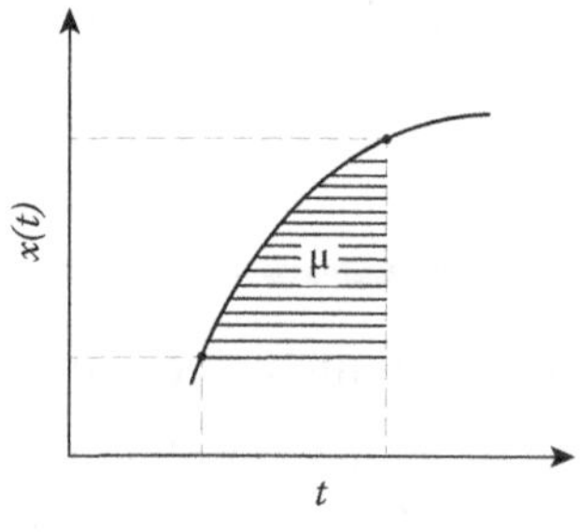


**FIGURE 3.2**
Send-on-area sampling criterion.

According to threshold-based sampling focused not on the sampled signal but on estimation uncertainty, a new measurement is not communicated as long as the state estimation confidence level is "good enough." The sampling is triggered only if the estimation uncertainty exceeds a tolerable upper bound. In the *matched sampling*, the sensor uses the Kullback–Leibler divergence $D_{KL}$ for triggering events [38,41]. The $D_{KL}$ is a measure of the difference between two probability distributions: $p_1(x)$ as a "true" reference distribution of state $x$ and $p_2(x)$ as the distribution related to the predicted state. More specifically, in the context of the state estimation, the $D_{KL}$ divergence shows how much information is lost when the predicted state defined by $p_2(x)$ is adopted instead of "true" reference distribution $p_1(x)$. The Kullback–Leibler divergence $D_{KL}(t)$ is a function of time due to temporal evolution of $p_1(x)$ and $p_2(x)$. As soon as $D_{KL}(t)$ crosses the prespecified threshold, a new sample is sent to the estimator, and the state of the system is updated in order to reduce the state uncertainty.

In [39,40], with the *variance-based sampling* criterion, new measurements transmitted from the sensor to the estimator are referred to prediction variance that grows in time when no measurement is available. A new measurement is broadcast if the prediction variance exceeds a tolerable bound, which indicates that the uncertainty of predicting the state is too large. As soon as the predefined threshold is exceeded, a new measurement is transmitted and the estimator variance drops. The comprehensive discussion of the event-based state estimation is included in [39,41].

It is worth noting that the development of new sampling criteria based on the state estimator uncertainty has given rise to the extension of the traditional concept of event originated in computing. Depending on the context, an event is now regarded not only to a significant change of the state of an object but may also define a *significant change of the state uncertainty*.

### 3.2.4 Lyapunov Sampling

In the context of control systems, the class of threshold-based sampling represents event-triggered sampling algorithms oriented to update the control law when needed to achieve a desired performance. The scheme related to the threshold-based schemes is *Lyapunov sampling*, introduced in [44] and explored further in [45–47], which consists of enforcing control update when required from the stability point of view.

The aim of the use of Lyapunov sampling is to drive the system to a neighborhood of the equilibrium. To guarantee the control system stability, Lyapunov sampling refers to variations of the Lyapunov function. The plant is sampled when the system trajectory sufficiently changes. Thus, a discretization is defined not in the amplitude but in the energy space domain, because the particular values of the Lyapunov function constitute a set of contour curves of constant energy in the state space. Each curve forms a closed region around the equilibrium point, and sampling is triggered when the system trajectory crosses contour curves from outside to inside, that is, when it migrates toward a decrease of system energy (Figure 3.3).

The next sample that triggers the control update is captured when the value of the Lyapunov function decreases with scaling coefficient $\eta$ compared to the actual sample, where $0 < \eta < 1$.

The tunable design parameter of Lyapunov sampling, $\eta$, represents the system energy decay ratio and can be referred to the threshold (delta) in the send-on-delta sampling. The difference is that the threshold in send-on-delta scheme ($\Delta$) is an additive parameter. Instead, the energy decay $\eta$ in Lyapunov sampling is defined as a multiplicative scaling coefficient, and the samples are triggered only if the energy of the system decreases ($\eta < 1$). The parameter $\eta$ determines the trade-off between the average rate of sampling frequency and control performance. For small values of $\eta$, the next sample is taken when the Lyapunov function decreases more significantly. Conversely, closer contour curves triggering events and more frequent sampling occur with the values of $\eta$ close to one.

If the parameter $\eta$ is constrained such that $0 < \eta < 1$, the system energy decreases with every new sample, and the trajectory successively traverses contour curves until the equilibrium point is reached. This is equivalent to the stability in the discrete Lyapunov sense. However, the requirement $0 < \eta < 1$ does not guarantee that the continuous-time trajectory is also stable, because a new sample might be never captured (i.e., Lyapunov function may not decrease sufficiently). The latter may happen if the system energy increases before the trajectory achieves the next contour curve [44]. Intuitively, the sampling intervals cannot be too long to preserve stability of the continuous-time system. As proved in [44], the stability of the continuous-time system trajectory is guaranteed if $\eta$ is higher than a certain value defined by the shape of the Lyapunov function: $0 < \eta^* < \eta < 1$. Thus, $\eta^*$ is regarded as the minimum value of $\eta$ that guarantees system stability. Further discussion on Lyapunov sampling is provided in Section 3.6. Recent experimental investigation of Lyapunov sampling for adaptive tracking controllers in robot manipulators is reported in [47].

### 3.2.5 Reference-Crossing Sampling

In signal processing, event-based sampling is considered mainly in the perspective of signal reconstruction.

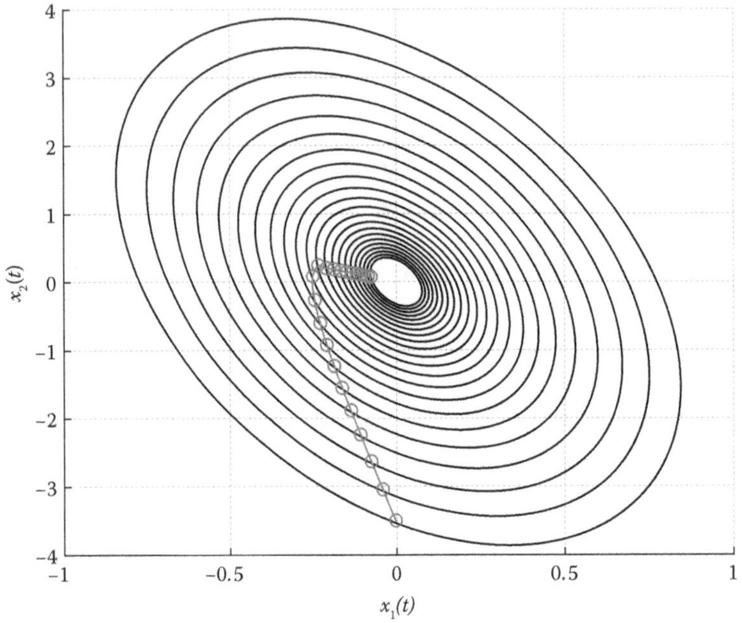


**FIGURE 3.3**
Lyapunov sampling as system trajectory-crossings scheme.

In order to recover the original signal perfectly from samples irregularly distributed in time, the mean sampling rate must exceed the Nyquist rate, which is twice higher than the highest frequency component in the signal spectrum ($B$) [48–50]. The survey of mathematical background related to signal recovery based on nonuniform sampling, which the event-triggered sampling belongs to, is included in [51].

A class of event-based criteria relevant to many applications in the event-based signal processing is the *reference-crossing sampling*. According to this criterion, the signal is sampled when it crosses a prespecified reference function. This sampling criterion is thus not referred to an explicit (additive or multiplicative) signal change but is dedicated to match a certain reference. The example is the sine-wave crossing sampling (Figure 3.4) [52,53].

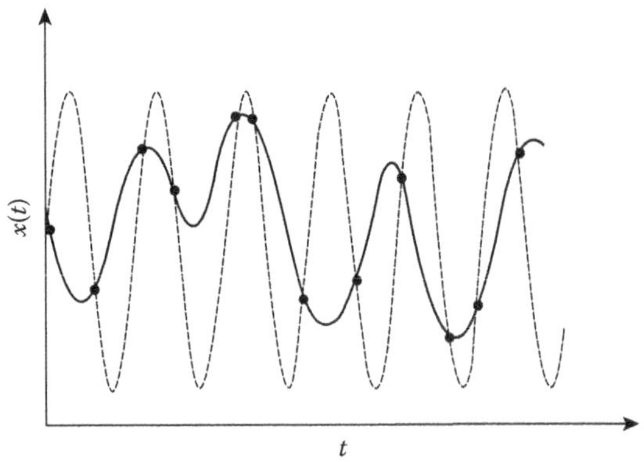


**FIGURE 3.4**
Sine-wave crossing sampling.

#### 3.2.5.1 Level-Crossing Sampling

In the simplest version of the reference-crossing sampling, the reference function can be defined as a single reference level disposed in the amplitude domain. Studies on the level-crossings problem for stochastic processes is one of the classical research topics in mathematics, with many applications to mechanics and signal processing [54]. For example, in the pioneered work published in 1945, Rice provided the formula for the mean rate of level-crossings for Gaussian processes [55].

The rate of crossings of the single level is usually lower than the Nyquist rate ($2B$). For example, the mean rate of zero-crossing for Gaussian process bandlimited to the bandwidth $B$ (Hz) equals $1.15B$, and the rate of crossings of level $L \neq 0$ is even lower [56]. Therefore, for the level-crossing sampling, multiple reference levels are used in practice (Figure 3.5) [7–11,56–63]. The work by Mark and Todd was the first study on level-crossing sampling with multiple reference levels addressed to the context of signal processing [57]. The level-crossing sampling is the most popular sampling scheme in event-based signal processing adopted to design of asynchronous analog-to-digital converters [7,10,60,61], and digital filters [11,16], as well as to

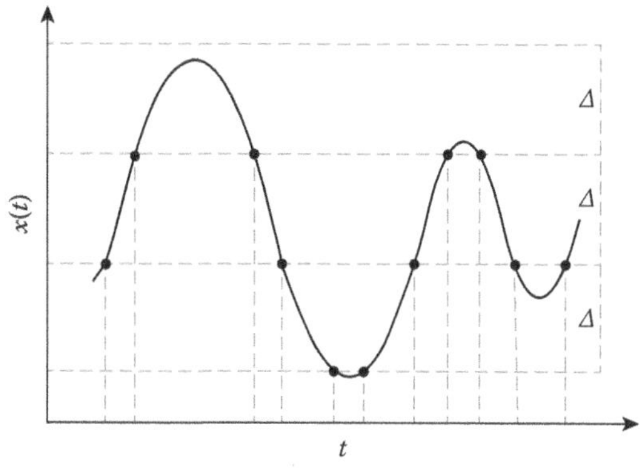


**FIGURE 3.5**
Level-crossing sampling with multiple reference levels.

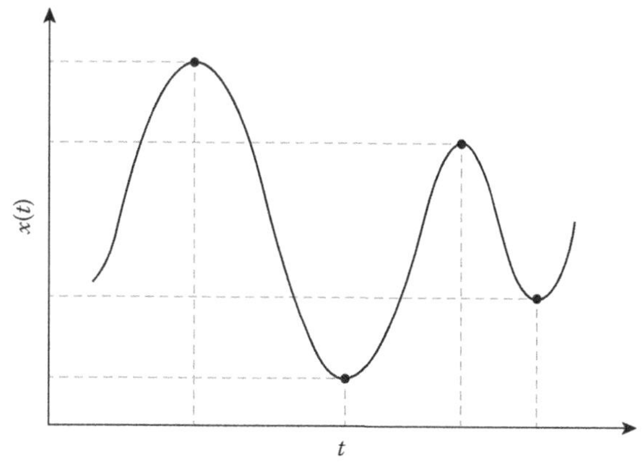


**FIGURE 3.6**
Extremum sampling.

CT-DSP [8,9,15,17,43,65]. In most scenarios of level-crossing sampling, the reference levels are uniformly distributed in the amplitude domain. However, the approaches to adaptive [60] or optimal distribution of levels [61] have also been proposed.

If the hysteresis is adopted to triggering level-crossings (i.e., the repeated crossings of the same level do not trigger new sampling operations), then such level-crossing criterion matches the threshold-based sampling (send-on-delta scheme). However, sometimes both terms are used interchangeably.

The mean rate of level-crossing sampling is a function of power spectral density of the sampled signal and depends on the signal bandwidth. This feature is intuitively comprehensible as the rate of level-crossings is higher when the signal varies quickly, and lower when it changes slowly. For bandlimited signals modeled by a stationary Gaussian process, the mean rate of level-crossing sampling is directly proportional to the signal bandwidth. In other words, the level-crossing scheme provides faster sampling when the local bandwidth becomes higher, and respectively slower in regions of lower local bandwidth. In this way, the level-crossings sampling provides dynamic estimation of time-varying local signal bandwidth. The approach to local bandwidth estimation and signal reconstruction by the use of level-crossing sampling is presented in [56] in this book.

#### 3.2.5.2 Extremum Sampling

By extending the concept of level-crossings to the transforms of input signal, new sampling criteria might be formulated. For example, the zero-crossings of the first derivative defines the *extremum sampling* (*mini–max* or *peak sampling*), which relies on capturing signal values at its local extrema (Figure 3.6) [66–68]. Extremum sampling allows estimation of dynamically time-varying signal bandwidths similarly as level-crossings. However, unlike level-crossing, extremum sampling enables recovery of the original signal only at half of the Nyquist rate [68], because it is two-channel sampling—that is, the samples provide information both on signal value and on zeros of its first time-derivative. For recovery of the signal based on irregular derivative sampling, see [69].

#### 3.2.5.3 Sampling Based on Timing

The reference-crossing sampling provides the framework for *time encoding* [70–72], also called *sampling based on timing* [73], a class of discrete-time signal representation complementary to classical sampling triggered by a timer. The concept of sampling by timing consists in recording the value of the signal at a preset time instant, instead of recording the time at which the function takes on a preset value. The objective of sampling based on timing is to represent the continuous-time signals fully in the time domain by a sequence of time intervals that include information on signal values. In this sense, the level-crossings do not allow signals to be encoded only in the time domain, because the information which level has been crossed is also required [72].

Sampling by timing is an invertible signal transformation, provided that the density of time intervals in general exceeds the Nyquist rate. The perfect recovery of the input signal from time-encoded output based on frame theory has been formulated in [70] and addressed in several later research studies.

As mentioned before, the motivation behind time encoding of signals is a desire to benefit from increasingly better time resolution of nanoscale VLSI circuits technology, while the coding of signals in the amplitude domain becomes more difficult due to the necessity to reduce supply voltage. The sampling by timing appears in nature and in biologically inspired circuit architectures (e.g., in integrate-and-fire neurons) that are usually based on integration. In various neuron models, the state transition (sampling) is triggered when an integral of input signal reaches a prespecified reference level [3,74–76]. In this sense, the time encoding realized in the integrate neurons is based on level-crossings of the input signal integral. The time encoding referred to circuit-based integration has been adopted to time-based analog-to-digital conversion, see, for example, [77–80].

### 3.2.6 Cost-Aware Sampling

The measures related to the input signal may not be the only factor that determines the conditions for triggering the sampling operations. By extension of the concept of the event, the sampling criteria may address the combination of the control performance and sampling cost. The control performance is defined by the threshold size or the density of reference levels. The sampling cost may address various application requirements, for example, the allocated fraction of communication bandwidth, energy consumption, or output buffer length. The fraction of bandwidth required for transmission of the samples is represented by the mean sampling rate, which is proportional to energy consumption for wireless communication.

The first work on cost-aware sampling was published by Dormido et al. in the 1970s in the context of adaptive sampling [81]. The cost-aware sampling criteria may balance the control performance and resource utilization, for example, by "discouraging" the triggering of a new sample when the previous sample was taken not long ago, even if the object has changed its state significantly [81,82].

### 3.2.7 QoS-Based Sampling

The other class of sampling schemes proposed to be used in the sensor and networked control systems is focused on the quality of service (QoS) of the network [83]. QoS performance metrics typically include time delays, throughput, and packet loss. Given that the bandwidth usage and the control performance are linked together in networked control systems, a sampling rate that is too fast due to its contribution to the total network traffic can significantly increase the transmission delays and packet losses, and thereby degrade the control performance. Unlike in classical point-to-point control systems, the performance of control loops in the networked control systems paradoxically could be increased with slower sampling, because the traffic reduction improves QoS metrics [13].

The idea to adapt the sampling interval to varying network conditions (e.g., congestion, length of delays, and wireless channel quality) is presented in [83]. In the QoS-based sampling scheme, a sampling interval is extended if the network is overloaded, and vice versa. The actual sampling rate can be dependent on the current state of QoS metrics, or a weighted mixture of the network conditions and closed-loop control performance. The QoS-based sampling scheme should be classified to the adaptive sampling category, as the samples are triggered by a timer, but the adaptation mechanism is event-triggered and related to the state of QoS performance metrics.

### 3.2.8 Adaptive Sampling

One of the principles of the event-based paradigm is that the events in event-triggered architectures occur unexpectedly—that is, they are *contingent* according to the classification provided in [19,21]. The attribute of unexpectedness of event occurrences is a great challenge for event-based systems design, because it precludes a priori allocation of suitable resources to handle them.

The well-known idea to cope with event unexpectedness relies on a prediction of a *contingent* event and converting it to a *determined* event, since the latter is predictable and can be expressed strictly as a function of time. The instant of occurrence of a determined event is an approximate of an arrival time of a corresponding contingent event. In the context of the sampling strategy, this idea comes down to a conversion of event-based to adaptive sampling under the same criterion. More specifically, the current sampling interval in adaptive sampling is assigned according to the belief that the sampled signal will meet the assumed criterion (e.g., a change by delta) at the moment of the next sampling instant. The adaptive sampling system does not need to use event detectors, because the new sample is not captured when an event is detected, but it is predicted based on the expansion of the signal at the actual sampling instant. For example, according to the adaptive sampling based on the constant absolute-difference criterion (send-on-delta), the next sample is taken after the time $T_{i+1} = \Delta/\dot{x}(t_i)$, where $\dot{x}(t_i)$ is the value of the first time-derivative at the instant $t_i$ of the actual sample, and $\Delta$ is the sampling threshold (Figure 3.1) [84]. For the adaptive sampling based on send-on-area criterion (integral-difference law), the next sampling time is $T_{i+1} = \mu/\sqrt{|\dot{x}(t_i)|}$, where $\mu$ is the assumed sampling threshold (Figure 3.2) [85]. The explicit difference between the event-triggered and adaptive sampling on the example of the send-on-delta criterion is illustrated explicitly in Figure 3.7. If the signal is convex (second time-derivative is negative), then the adaptive sampling ($t'_{i-1}$) is triggered earlier than the corresponding event-based sampling $t'_i$ (Figure 3.7), and vice versa.

To summarize, the essential difference between the adaptive sampling and the event-based sampling is that the former is based on the *time-triggered* strategy

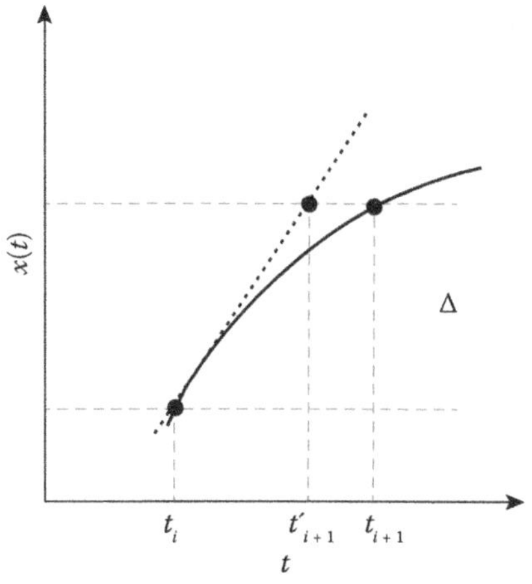


**FIGURE 3.7**
Comparison of event-triggered sampling and adaptive sampling under send-on-delta criterion.

where the sampling instants are controlled by the timer, and the latter belongs to the *event-triggered* systems where the sampling operations are determined only by signal amplitude variations, rather than by the progression of time.

The criteria for adaptive sampling [84–89] correspond to these for event-based sampling [5,22,23,32–36], and have been formulated in the 1960s and 1970s. In particular, Mitchell and McDaniel introduced the constant absolute-difference criterion that is the (adaptive sampling) equivalent of the pure (event-based) level-crossing/send-on-delta algorithm [84]. Dorf et al. proposed the constant *integral-difference* sampling law [85], which corresponds to the integral criterion (send-on-area) [32–34]. A unified approach to a design of adaptive sampling laws based on the same integral-difference criterion of different thresholds has been derived and proposed in [87]. In subsequent work, the proposition of generalized criteria for adaptive sampling has been introduced [88]. These criteria combine a sampling resolution with a sample cost modeled by the proposed cost functions, and [88] is the first work on cost-aware sampling design.

Adaptive sampling has been adopted in the *self-triggered control* strategy proposed in [89] and extensively investigated in technical literature, see [90]. In self-triggered control, the time of the next update is precomputed at the instant of the actual control update based on predictions using previously received data and knowledge of the plant dynamics. Thus, the controller computes simultaneously both the actual control input and the sampling period for the next update. The difference is that the prediction of the sampling period realized in early adaptive control systems was based on continuous estimation of the state of the plant, while it is computed in discrete time by software in modern self-triggered control systems.

Among other applications, adaptive sampling is a useful technique to determine the sleeping time of wireless sensor nodes between sampling instants in order to decrease energy consumption [91].

### 3.2.9 Implicit Information on Intersample Behavior

All the event-based sampling algorithms produce samples irregularly in time. However, compared to nonuniform sampling, the use of event-triggered schemes provides some implicit information on behavior of the signal between sampling instants [3], because "as long as no event is triggered, the actual measurement does not fulfill the event-sampling criterion [94], or rephrased equivalently "it is not true that there is zero relevant information between threshold crossings; on the contrary, there is a valuable piece of information, namely that the output has not crossed a threshold" [92]. Furthermore, in [93], the implicit information as a lack of event occurrence is termed the "negative information."

In particular, in the send-on-delta scheme, the implicit information means that the sampled signal does not vary between sampling instants more than by the threshold (delta) [5,22,23]. In the extremum sampling, it is known implicitly that the first time-derivative of the signal does not change its sign between sampling instants [66–68]. This extra information on intersample signal behavior might be additionally exploited to improve system performance. However, in the common strategy adopted in the existing research studies on event-based systems, the updates are restricted to event occurrences, for example, when a state of the plant varies by the threshold. The implicit information on intersample signal behavior to improve system performance in the context of control has been used only in a few works [92–94], see also [41] in this book. In these works, the implicit information is used to improve the state estimation by continuing updates between threshold crossings based on the known fact that the output lies within the threshold.

In the context of event-based signal processing, to the best of the author's knowledge, the implicit information on intersample behavior has not been used. The adoption of extra knowledge of the signal between event-triggered sampling instants to improve system performance seems to be one of the major unused potentials of the event-based paradigm.

### 3.2.10 Event Detection

In signal processing, and in many approaches to event-based control, the events that trigger sampling are detected in continuous time. The continuous verification

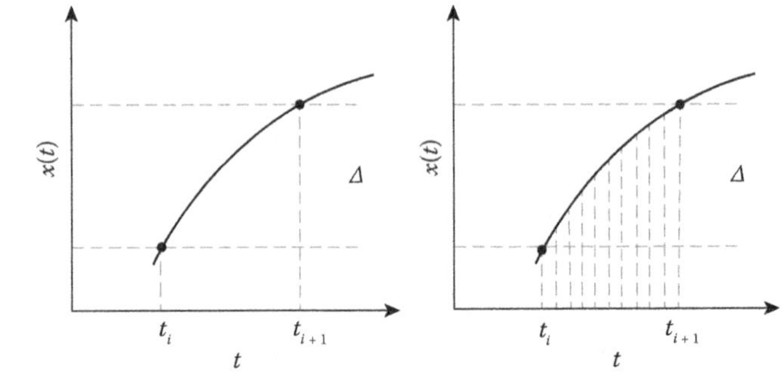


**FIGURE 3.8**
Continuous-time versus discrete-time event detection.

of an event-triggering condition is the basis of *continuous event-triggered control* (CETC) [95] and requires the use of analog or mixed signal (analog and digital) instrumentation (Figure 3.8a).

The alternative technique of event detection combines the event and time-triggered strategies, where the events are detected in discrete time on the basis of equidistant samples (Figure 3.8b). The implementation of even-triggered control on the top of periodically sampled data is referred to as the *periodic event-triggered control* (PETC) strategy [95]. In PETC, the sensor is sampled equidistantly, and event occurrence is verified at every sampling time in order to decide whether or not to transmit a new measurement to a controller. Thus, in PETC, the periodicity of control system components is relaxed in order to establish the event-driven communication between them.

The idea to build the event-based strategy on top of time-triggered architectures has been proposed and adopted in several approaches in control and real-time applications (e.g., in [4,96]). The PETC strategy can be referred to as the globally asynchronous locally asynchronous (GALS) model of computation introduced in the 1980s [97] and adopted as the reference architecture for digital electronics, including system-on-chip design [98]. The PETC is convenient for practical applications. In particular, the implementation of the event-triggered communication and processing activities on top of periodically sampled data has been applied in the send-on-delta scheme in LonWorks networked control systems platform (ISO/IEC 14908) [26] and in EnOcean energy harvesting wireless technology.

## 3.3 Send-on-Delta Data Reporting Strategy

Most event-based sampling schemes applied to control, communication, and signal processing involve triggering the sampling operations by temporal variations of the signal value.

### 3.3.1 Definition of Send-on-Delta Scheme

In the *send-on-delta sampling scheme* (*Lebesgue sampling*), the most generic sampling scheme, a continuous-time signal $x(t)$ is sampled when its value deviates from the value included in the most recent sample by an interval of confidence $\pm\Delta$ [5,22,23]. The parameter $\Delta > 0$ is called the *threshold*, or *delta* (Figures 3.1 and 3.9b):

$$|x(t_{i+1}) - x(t_i)| = \Delta, \tag{3.1}$$

where $x(t_i)$ is the $i$th signal sample. The size of the threshold (delta) defines a trade-off between system performance and the rate of samples: the smaller the threshold $\Delta$, the higher is the sampling rate.

By comparison, in the *periodic sampling*, the signal is sampled equidistantly in time. The sampling period $T$ is usually adjusted to the fastest expected change of a signal, as follows (Figure 3.9a):

$$T = \frac{\Delta}{|\dot{x}(t)|_{\max}}, \tag{3.2}$$

where $|\dot{x}(t)|_{\max}$ is the *maximum slope* of the sampled signal defined as the maximum of the absolute value of the first signal derivative in relation to time, and $\Delta$ is the maximum difference between values in the consecutive samples acceptable by the application. In [28], the send-on-delta scheme with prediction as an enhanced version of the pure send-on-delta principle has been introduced. This allows a further reduction of the number of samples. In the prediction-based send-on-delta scheme, $x(t)$ is sampled, when its value deviates from the value predicted on the basis of the most recent sample by a threshold $\pm\Delta$ (Figure 3.9c and d),

$$|x(t_{i+1}) - \hat{x}(t_{i+1})| = \Delta, \tag{3.3}$$

where $\hat{x}(t_{i+1})$ denotes the value predicted for the instant $t_{i+1}$ on the basis of the time-derivative(s) by the use of truncated Taylor series expanded at $t_i$ as follows:

$$\hat{x}(t_{i+1}) = x(t_i) + \dot{x}(t_i)(t_{i+1} - t_i) + \frac{\ddot{x}(t_i)}{2}(t_{i+1} - t_i)^2 + \cdots, \tag{3.4}$$

where $\dot{x}(t_i)$ and $\ddot{x}(t_i)$ are, respectively, the first and second time-derivatives of the signal at $t_i$. The send-on-delta scheme with zero-order prediction corresponds to the pure send-on-delta algorithm (Figure 3.9b):

$$\hat{x}_0(t_{i+1}) = x(t_i). \tag{3.5}$$

### 3.3.2 Illustrative Comparison of Number of Samples

It is intuitively comprehensible that the send-on-delta scheme reduces the mean sampling rate compared to the

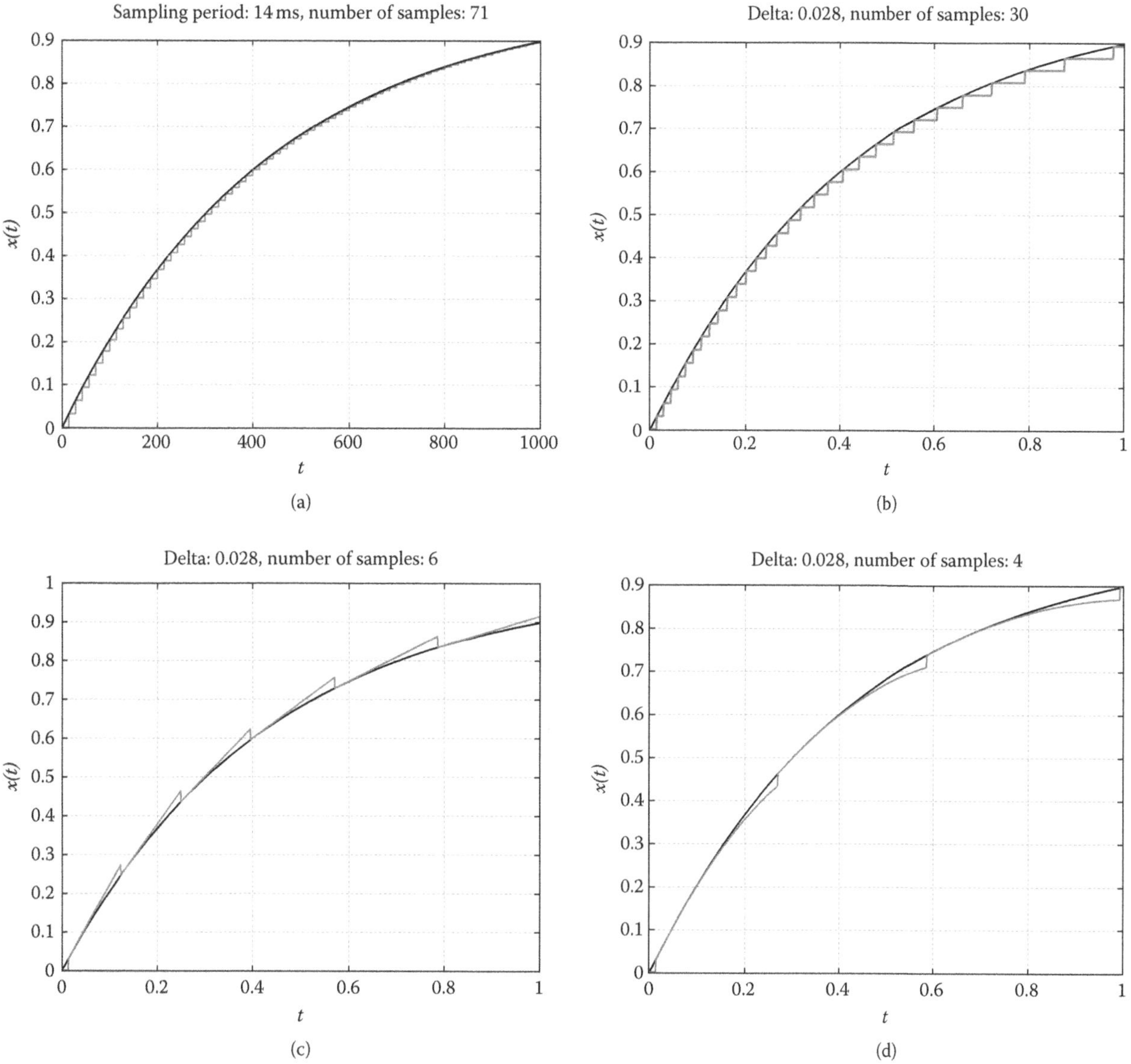


**FIGURE 3.9**
Step response of the first-order control system sampled: (a) periodically (71 samples), (b) according to send-on-delta scheme (28 samples), (c) with send-on-delta scheme and linear prediction (6 samples), and (d) with send-on-delta scheme and quadratic prediction (4 samples).

periodic sampling with the same maximum sampling error. The problem of comparative analysis of event-based sampling algorithms has been studied in several papers [99–102].

Figure 3.9a through d illustrates a comparison of the number of samples between periodic sampling and send-on-delta schemes. The step-response of the first-order control system as the sampled signal is used. In order to make the comparison illustrative, the maximum sampling error (delta) is set to 0.028, and the time constant of the system is 0.42 s. In uniform sampling, the sampling period of 14 ms is selected in order to keep the difference between values of the consecutive samples not higher than 0.028. The number of samples is evaluated during the settling time of the step response (i.e., as long as the step response reaches 90% of its final value).

As expected, the number of samples in the send-on-delta scheme (28 samples, Figure 3.9a) is lower compared to the classical periodic sampling (71 samples, Figure 3.9b). The use of linear and quadratic prediction contributes to further reduction of the number of samples (6 samples with linear prediction and just 4 samples with quadratic prediction). The reduction of the number of samples is higher than one order of magnitude in the prediction-based send-on-delta algorithms compared to equidistant sampling. However, this reduction is obtained at the price of a necessity to detect changes of the signal and also to sample its time-derivatives.

## 3.4 Send-on-Delta Bandwidth Evaluation

Since the send-on-delta sampling is triggered by asynchronous events, the time is a *dependent variable* in system modeling. The average requirements of networked system communication bandwidth for a single control loop are defined by the mean number of send-on-delta samples per unit of time.

### 3.4.1 Total Variation of Sampled Signal

As will be shown, the signal measure called the *total variation* of the sampled signal is essential to estimate the number of samples produced by the send-on-delta scheme.

In mathematics, the *total variation* of a function $x(t)$ on the time interval $[a,b]$ denoted by $V_a^b(x)$ is formally defined as the supremum on the following sum [103]:

$$V_a^b(x) = \sup \sum_{j=1}^{k} \left| x(t_j) - x(t_{j-1}) \right|, \tag{3.6}$$

for every partition of the interval $[a,b]$ into subintervals $(t_{j-1}, t_j)$; $j = 1, 2, \ldots, k$; and $t_0 = a$, $t_k = b$.

If $x(t)$ is continuous, the total variation $V_a^b(x)$ might be interpreted as the vertical component (ordinate) of the arc-length of the $x(t)$ graph, or alternatively, as the sum of all consecutive peak-to-valley differences.

For continuous signals, the total variation $V_a^b(x)$ of $x(t)$ on the time interval $[a,b]$ is

$$V_a^b = \int_a^b |\dot{x}(t)| dt. \tag{3.7}$$

The graphical interpretation of the total variation is illustrated in Figure 3.10. In the presented scenario, the total variation $V_{t_i}^{t_{i+7}}(x)$ between instants $t_i$ and $t_{i+7}$ is a sum $V_{t_i}^{t_{i+7}}(x) = V_{t_i}^{t_a}(x) + V_{t_a}^{t_b}(x) + V_{t_b}^{t_c}(x) + V_{t_c}^{t_d}(x) + V_{t_d}^{t_{i+7}}(x)$. Each component of this sum is represented by a corresponding arrow in Figure 3.10.

### 3.4.2 Total Signal Variation within Send-on-Delta Sampling Interval

Let us consider the total variation of the sampled signal between sampling instants in the send-on-delta scheme (Figure 3.10). If the signal $x(t)$ does not contain any local extrema between consecutive send-on-delta instants $t_i$ and $t_{i+1}$, then the derivative $\dot{x}(t)$ does not change its sign within $(t_i, t_{i+1})$. Then, the total variation $V_{t_i}^{t_{i+1}}(x) = \Delta$, because

$$V_{t_i}^{t_{i+1}}(x) = \int_{t_i}^{t_{i+1}} |\dot{x}(t)| dt = \left| \int_{t_i}^{t_{i+1}} \dot{x}(t) dt \right| = \Delta. \tag{3.8}$$

If the sampled signal $x(t)$ contains local extrema between consecutive send-on-delta instants which in Figure 3.10 occurs for $(t_{i+1}, t_{i+2})$, then $\dot{x}(t)$ changes its sign, and

$$V_{t_{i+1}}^{t_{i+2}}(x) > \Delta, \tag{3.9}$$

because

$$\int_{t_{i+1}}^{t_{i+2}} |\dot{x}(t)| dt > \left| \int_{t_{i+1}}^{t_{i+2}} \dot{x}(t) dt \right|. \tag{3.10}$$

In particular, if one local extremum occurs within $(t_{i+1}, t_{i+2})$, then [104]

$$\Delta < V_{t_{i+1}}^{t_{i+2}}(x) < 3\Delta. \tag{3.11}$$

Note that referring to Figure 3.10, it is evident that $V_{t_{i+1}}^{t_{i+2}}(x) \cong 3\Delta$ and $V_{t_{i+3}}^{t_{i+4}}(x) \cong \Delta$, which illustrates lower

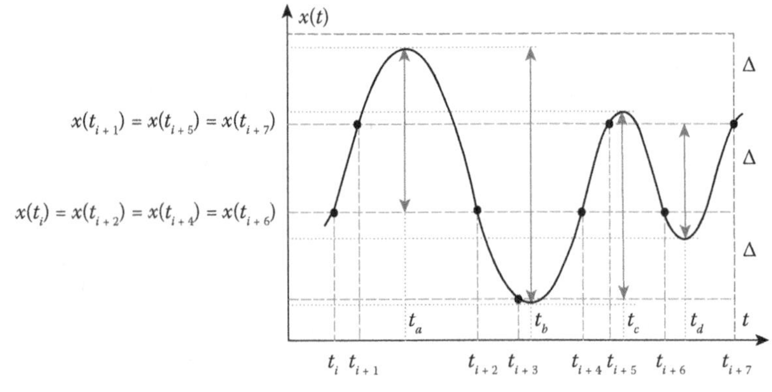


**FIGURE 3.10**
Total variation of a continuous-time signal.

and upper bounds for signal variation in the send-on-delta sampling interval with a single local extremum. If the send-on-delta sampling interval $(t_k, t_{k+1})$ contains a number of $m$ local extrema, then

$$\Delta < V_{t_k}^{t_{k+1}}(x) < (2m+1)\Delta. \tag{3.12}$$

To summarize, the total variation during a number of $n$ send-on-delta intervals $(t_i, t_{i+n})$ is lower bounded because $V_{t_i}^{t_{i+n}}(x) \geq n\Delta$. Therefore, the number of samples triggered during a certain time interval $[a,b]$ according to the send-on-delta scheme is not higher than $V_a^b(x)/\Delta$. In particular, if the signal $x(t)$ is a monotone function of time in $[a,b]$, then the number of samples $n$ is *exactly* equal to $V_a^b(x)/\Delta$:

$$n = V_a^b(x)/\Delta, \tag{3.13}$$

provided that the fractional part of this number is neglected.

Finally, in a general case, the expression $V_a^b(x)/\Delta$ might be regarded as an upper bound on the number $n$ of send-on-delta samples for the continuous signal $x(t)$ during $[a,b]$:

$$n \leq \frac{V_a^b(x)}{\Delta}. \tag{3.14}$$

Taking into account asymptotic conditions, it could be easily shown that the number $n$ of send-on-delta samples approaches $V_a^b(x)/\Delta$ if the threshold $\Delta$ becomes small [104]:

$$\lim_{\Delta \to 0} n = \frac{V_a^b(x)}{\Delta}. \tag{3.15}$$

**EXAMPLE 3.1** When the set point is changed by $x_0$ in the first-order control system, then the step response is given by $x(t) = x_0(1 - e^{-t/\tau})$, which is a monotone signal presented in Figure 3.9a through d. On the basis of (3.7), it is clear that $V_0^\infty(x) = x_0$, and the number of send-on-delta samples during the time $[0, \infty)$ when the control system migrates to the new set point according to (3.10) is

$$n = \frac{x_0}{\Delta}. \tag{3.16}$$

### 3.4.3 Mean Rate of Send-on-Delta Sampling

Taking into account (3.7) and (3.14), the mean send-on-delta sampling rate $\lambda$ during the time interval $[a,b]$ defined as $\lambda = n/(b-a)$ is constrained by $\lambda'$:

$$\lambda \leq \frac{\overline{|\dot{x}(t)|}}{\Delta} = \lambda', \tag{3.17}$$

where $\overline{|\dot{x}(t)|}$ is the *mean slope* of the signal—that is, the mean of the absolute value of the first time-derivative of $x(t)$ in the time interval $[a,b]$:

$$\overline{|\dot{x}(t)|} = \frac{1}{b-a}\int_a^b |\dot{x}(t)|dt, \tag{3.18}$$

or alternatively,

$$\overline{|\dot{x}(t)|} = \frac{V_a^b(x)}{b-a}. \tag{3.19}$$

For small $\Delta$, the mean send-on-delta sampling rate $\lambda$ approaches $\overline{|\dot{x}(t)|}/\Delta$ as follows from (3.15):

$$\lim_{\Delta \to 0} \lambda = \frac{\overline{|\dot{x}(t)|}}{\Delta} = \lambda'. \tag{3.20}$$

Since the mean slope of the signal can be calculated on the basis of (3.18) analytically or numerically either for deterministic or stochastic signals, the formula (3.17) provides a simple expression for the evaluation of the upper bound for the mean sampling rate produced by the send-on-delta scheme. In the context of a networked control system, this allows estimation of the bandwidth for transferring messages in a single control loop.

### 3.4.4 Mean Send-on-Delta Rate for Gaussian Processes

If $x(t)$ is a random process with a mean square value $\sigma_x$, bandlimited to $\omega_c = 2\pi f_c$, and with Gaussian amplitude probability density function $f(x)$ given by

$$f(x) = (1/\sqrt{2\pi}\sigma_x)\exp\left(\frac{-x^2}{2\sigma_x^2}\right), \tag{3.21}$$

the mean slope of the $x(t)$ is [105]

$$\overline{|\dot{x}(t)|} = \sqrt{\frac{8\pi}{3}} f_c \sigma_x. \tag{3.22}$$

On the basis of (3.17), the upper bound for the mean send-on-delta sampling rate for the bandlimited Gaussian random process $x(t)$ is

$$\lambda \leq \sqrt{\frac{8\pi}{3}} f_c \hat{\sigma}_x, \tag{3.23}$$

where $\hat{\sigma}_x = \sigma_x/\Delta$ is the mean square value normalized to the sampling threshold $\Delta$.

Let us introduce the measure of *spectral density of the send-on-delta sampling rate* as the ratio of the mean sampling rate $\lambda$ and the Gaussian input process cutoff frequency $f_c$:

$$\hat{\lambda} = \lambda / f_c. \tag{3.24}$$

The measure $\hat{\lambda}$ has intuitive interpretation as the number of samples generated in unit of time by a sensor using a send-on-delta scheme for the Gaussian process of unit bandwidth 1 Hz. On the basis of (3.23), the spectral density of the send-on-delta sampling rate of the bandlimited Gaussian process is upper bounded as follows [105]:

$$\hat{\lambda} \leq 2.89\hat{\sigma}_x. \tag{3.25}$$

**EXAMPLE 3.2** For $\Delta = 0.5\sigma_x$: $\hat{\lambda} \leq 5.78$ which means that for Gaussian bandlimited process with the cutoff frequency $f_c = 1$ Hz, the send-on-delta scheme produces no more than 5.78 samples per second in average.

### 3.4.5 Reducing Sampling Rate by Send-on-Delta

As illustrated in Figure 3.9a through d, the number of samples produced by the send-on-delta scheme is much lower compared to the periodic sampling. The analytical evaluation of sampling rate reduction has been studied in [23]. The reduction of the sampling rate might be defined as the ratio of the sampling frequency in the periodic sampling $\lambda_T = 1/T$ and the mean sampling rate $\lambda$ in the send-on-delta scheme for the same maximum sampling error ($\Delta$):

$$p = \lambda_T/\lambda. \tag{3.26}$$

According to (3.2), the frequency of periodic sampling is $\lambda_T = |\dot{x}(t)|_{\max}/\Delta$. Since $\lambda$ is upper bounded by $\overline{|\dot{x}(t)|}/\Delta$ according to (3.17), then the reduction of the sampling rate is limited by the following constraint:

$$p \leq \frac{|\dot{x}(t)|_{\max}}{\overline{|\dot{x}(t)|}}. \tag{3.27}$$

The expression (3.27) becomes an equality on the basis of (3.13) if the sampled signal $x(t)$ is a monotone function of time. Moreover, for small threshold $\Delta$, the reduction of the sampling rate reaches the minimum:

$$p_{min} = \lim_{\Delta \to 0} p = \frac{|\dot{x}(t)|_{\max}}{\overline{|\dot{x}(t)|}}. \tag{3.28}$$

By giving the lower limit, the formula shown in (3.28) provides a conservative evaluation of sampling rate decrease and might be interpreted as *guaranteed reduction* $p_{min}$. An interesting point is that $p_{min}$ is only a function of the sampled signal $x(t)$ and does not depend on the parameter of the sampling algorithm ($\Delta$).

Reference [23] contains the analytical formulae for the guaranteed reduction computed for time responses of the first- and the second-order control systems. For example, the guaranteed effectiveness for the pure harmonic signal is $\pi/2 = 1.57$. In [104], it is shown that the guaranteed reduction for bandlimited Gaussian random process $x(t)$ equals $p_{min} = \sqrt{4.5\pi} \cong 3.76$, and is independent of the process bandwidth and the mean square value.

**EXAMPLE 3.3** The reduction $p$ of the sampling rate for the step response of the first-order control system given by $x(t) = x_0(1 - e^{-t/\tau})$ during the time interval $(0, b)$ evaluated according to (3.27) equals [23]

$$p = b/[\tau(1 - e^{-b/\tau})]. \tag{3.29}$$

The relevant step response $x(t) = x_0(1 - e^{-t/\tau})$ is presented in Figure 3.9a through 3.9d. The reduction of the sampling rate on the basis of simulation results illustrated in Figure 3.10a and 3.10b is $71/28 \cong 2.54$. The reduction $p$ calculated according to (3.26) with $\tau = 0.42$ s and $b = 1$ s is $\cong 2.62$, which agrees with simulation results despite rounding errors ($\pm 1$ sample). Note that (3.29) represents not only guaranteed but even exact sampling reduction, because the relevant step response is monotone in time.

## 3.5 Extensions of Send-on-Delta Scheme

Both the level-crossing sampling and send-on-delta schemes are based on triggering sampling operations when the signal reaches certain values in the amplitude domain or exceeds an assumed linear sampling error. As mentioned, several extensions of send-on-delta criterion have been proposed to the other signal measures: the integral sampling [32–34] (area-triggered sampling, or send-on-area scheme), or the sampling triggered by error energy changes (send-on-energy) [35,36].

### 3.5.1 Send-on-Area Criterion

A signal $x(t)$ is sampled according to the uniform integral criterion (send-on-area) if the integral of the absolute error (*IAE*)—that is, the integral of the absolute difference between the current signal value $x(t)$ and the value included in the most recent sample $x(t_i)$, accumulated over an actual sampling interval—reaches a threshold $\delta > 0$ [32,33]:

$$\int_{t_i}^{t_{i+1}} |x(t) - x(t_i)| dt = \delta. \tag{3.30}$$

In the basic version, the zero-order hold is used to keep the value of the most recent sample between sampling instants, but the prediction may also be adopted to reduce further the number of samples as carried out in the predictive send-on-delta scheme, see [28–30].

There are a few reasons to use an event-based integral criterion in control systems. First, the integrated absolute sampling error is a useful measure to trigger a control update, because a summation of the absolute temporal sampling errors counteracts mutual compensation of positive and negative errors. Second, the performance of control systems is usually defined as the integral of the absolute value of the error (*IAE*), which directly corresponds to the integral sampling criterion definition.

The mean sampling rate $m$ in the integral sampling scheme with threshold $\delta$ is approximated as [32,33]

$$m \cong \frac{1}{\sqrt{2\delta}} \overline{\sqrt{|\dot{x}(t)|}}, \tag{3.31}$$

where $\overline{\sqrt{|\dot{x}(t)|}}$ is the mean of the square root of the signal derivative absolute value:

$$\overline{\sqrt{|\dot{x}(t)|}} = \frac{\int_a^b \sqrt{|\dot{x}(t)|}dt}{b-a}. \tag{3.32}$$

The approximation shown in (3.32) is more accurate if the sampling threshold $\delta$ is low.

### 3.5.2 Send-on-Energy Criterion

A further extension of send-on-delta and send-on-area schemes is the event-triggered criterion based on the energy of the sampling error that may be called the *send-on-energy* paradigm [35,36]. A signal $x(t)$ is sampled at the instant $t_{i+1}$ according to the error energy criterion if the energy of a difference between the signal value and the value included in the most recent sample $x(t_i)$ accumulated during the actual sampling interval reaches a threshold $\zeta > 0$

$$\int_{t_i}^{t_{i+1}} [x(t) - x(t_i)]^2 dt = \zeta. \tag{3.33}$$

The difference between send-on-area and send-on-energy is that the successive sampling error values are squared before integration in the error energy–based sampling criterion. Compared to the integral sampling, the error energy criterion gives more weight to extreme sampling error values. The mean sampling rate $s$ in the send-on-energy sampling scheme with the threshold $\zeta$ is [35,36]:

$$s \cong \frac{1}{\sqrt[3]{3\zeta}} \overline{\sqrt[3]{[\dot{x}(t)]^2}}, \tag{3.34}$$

where $\overline{\sqrt[3]{[\dot{x}(t)]^2}}$ denotes the mean of the cubic root of the signal derivative square in the time interval $[a,b]$:

$$\overline{\sqrt[3]{[\dot{x}(t)]^2}} = \frac{\int_a^b \sqrt[3]{[\dot{x}(t)]^2}dt}{b-a}. \tag{3.35}$$

The asymptotic reductions (i.e., reductions for small thresholds) of the sampling rate in send-on-area or send-on-energy compared to the periodic sampling derived analytically in [23,33,36] are given by an expression:

$$\chi_\infty(\alpha) = \frac{[\dot{x}(t)_{\max}]^\alpha}{\overline{[\dot{x}(t)]^\alpha}}, \tag{3.36}$$

where $\alpha = 1/2$ for send-on-area and $\alpha = 2/3$ for send-on-energy schemes. It may be shown that the asymptotic reduction of sampling rate for send-on-area is lower than for send-on-delta but higher from the send-on-energy:

$$p_{min} \leq \chi_\infty(\alpha = 1/2) \leq \chi_\infty(\alpha = 2/3). \tag{3.37}$$

As the price of higher sampling reduction, the quality of approximation of the sampled signal by the send-on-delta is lower than for the send-on-area and send-on-energy criteria. References [33] and [36] include the analytical formulae of asymptotic reductions for send-on-area or send-on-energy schemes for step responses of the first- and second-order control systems.

## 3.6 Lyapunov Sampling

As introduced in Section 3.2.4, the event-triggered criterion suggested to drive the control system to the neighborhood of the equilibrium is Lyapunov sampling based on the trajectory-crossings scheme. The Lyapunov sampling is stable provided that the energy decay ratio meets the following constraint: $0 < \eta^* < \eta < 1$. In this section, the exemplified scenario of Lyapunov sampling will be analyzed based on [44] and [46].

### 3.6.1 Stable Lyapunov Sampling

Following [44], consider a continuous-time control system:

$$\dot{x}(t) = f(x(t), u(t)), \tag{3.38}$$

with the control law

$$u(t) = k(x(t_i)) \quad \text{for } t \in (t_i, t_{i+1}),$$

updated by a feedback controller $k$ at instants $t_0, t_1, \ldots, t_i, \ldots$ and with the control signal kept constant during each sampling interval. Taking into account (3.38), the closed-loop system is

$$\dot{x}(t) = f(x(t), u(t_i)). \tag{3.39}$$

Let $V$ be a Lyapunov function of the control system and $t_i$ be the instant of taking the current sample. The next sample that triggers the control update is captured at the time instant $t_{i+1}$ when the following condition is met:

$$V(x(t_{i+1})) = \eta V(x(t_i)) \quad \text{where } 0 < \eta < 1. \tag{3.40}$$

The tunable design parameter of Lyapunov sampling $\eta$ represents the system energy decay ratio. Intuitively, the sampling intervals cannot to be too long to preserve the stability of the continuous-time system. As proved in [44], the stability of the continuous system trajectory is preserved if $\eta$ is lower bounded by $\eta^*$:

$$0 < \eta^* < \eta < 1, \tag{3.41}$$

where $\eta^*$ is calculated as

$$\eta^* = \max_{x_0} \frac{V^*(x_0)}{V(x_0)}, \tag{3.42}$$

and

$$V^*(x_0) = \min_t V(x(t, x_0)); \quad t \geq 0. \tag{3.43}$$

The Lyapunov sampling with $\eta$ such that $\eta^* < \eta < 1$ is called the *stable Lyapunov sampling* algorithm in the literature [44]. The setup of the control system with Lyapunov sampling will be illustrated by an example of a double integrator system analyzed originally in [44] and [46]:

$$\begin{aligned} \dot{x}(t) &= Ax(t) + Bu(t), \\ y(t) &= Cx(t) \end{aligned} \tag{3.44}$$

with

$$A = \begin{bmatrix} 0 & 1 \\ 0 & 0 \end{bmatrix} \quad \text{and} \quad B = \begin{bmatrix} 0 \\ 1 \end{bmatrix}, \tag{3.45}$$

for the initial conditions

$$x_0 = \begin{bmatrix} 0 \\ -3 \end{bmatrix}, \tag{3.46}$$

with the control law

$$u(t) = -Kx(t). \tag{3.47}$$

The Lyapunov function for the system is

$$V(x) = x^T(t)Px(t), \tag{3.48}$$

where $P$ is a positive definite matrix that makes the system stable. In the double integrator example,

$$P = \begin{bmatrix} 01.1455 & 0.1 \\ 0.1 & 0.0545 \end{bmatrix} \quad \text{and} \quad K = \begin{bmatrix} 10 & 11 \end{bmatrix}. \tag{3.49}$$

For the considered double integrator system, the value $\eta^*$ guaranteeing the stability of Lyapunov sampling equals 0.7818 [44]. On the basis of the results of the reproduced experiment reported originally in [44,46], double integrator system trajectories are shown for stable and unstable Lyapunov sampling in Figure 3.11a and b, respectively.

Figure 3.11a presents a system trajectory in plane $(x_1, x_2)$ with Lyapunov sampling for $\eta = 0.8$. The system is stable in the Lyapunov sense, because its energy decreases with each sample. The continuous control system is also stable since the system trajectory tends to the origin. As reported in [46], the performance of the double integrator system with Lyapunov sampling is similar to that with periodic sampling, although the number of samples is reduced by around 91%.

Figure 3.11b illustrates the system trajectory with Lyapunov sampling for $\eta = 0.65$. The lower value of $\eta$ promises to take less samples. However, the system is unstable. After a capture of seven samples, the triggering condition is no longer satisfied, and a new sample is never taken (see Figure 3.11b). The trajectory diverges because a nonupdated control signal is unable to drive the system to equilibrium.

### 3.6.2 Less-Conservative Lyapunov Sampling

The lower bound $\eta^*$ stated by (3.42) for the energy decay ratio $\eta$ in Lyapunov sampling is sufficient but not a necessary condition for system stability. Moreover, the condition (3.41) is quite conservative, and the use of $\eta$ values close to one enforces relatively frequent sampling that drives the discrete system into a quasi-periodic regime. For many initial conditions, the dynamics of the system is stable even for $\eta < \eta^*$ [46]. Furthermore, as argued in [46], a calculation of $\eta^*$ is computationally complex, especially for nonlinear systems, and the existence of $\eta^*$ is not always guaranteed.

To address these limitations, the less-conservative Lyapunov sampling with the gain factor $\eta$ varying dynamically during the evolution of the system is studied in [46]. The parameter $\eta$ is decreased ad hoc (even below $\eta^*$) when the system is stable, and is increased as soon as it becomes unstable. This allows for the number of samples to be reduced almost twice than with

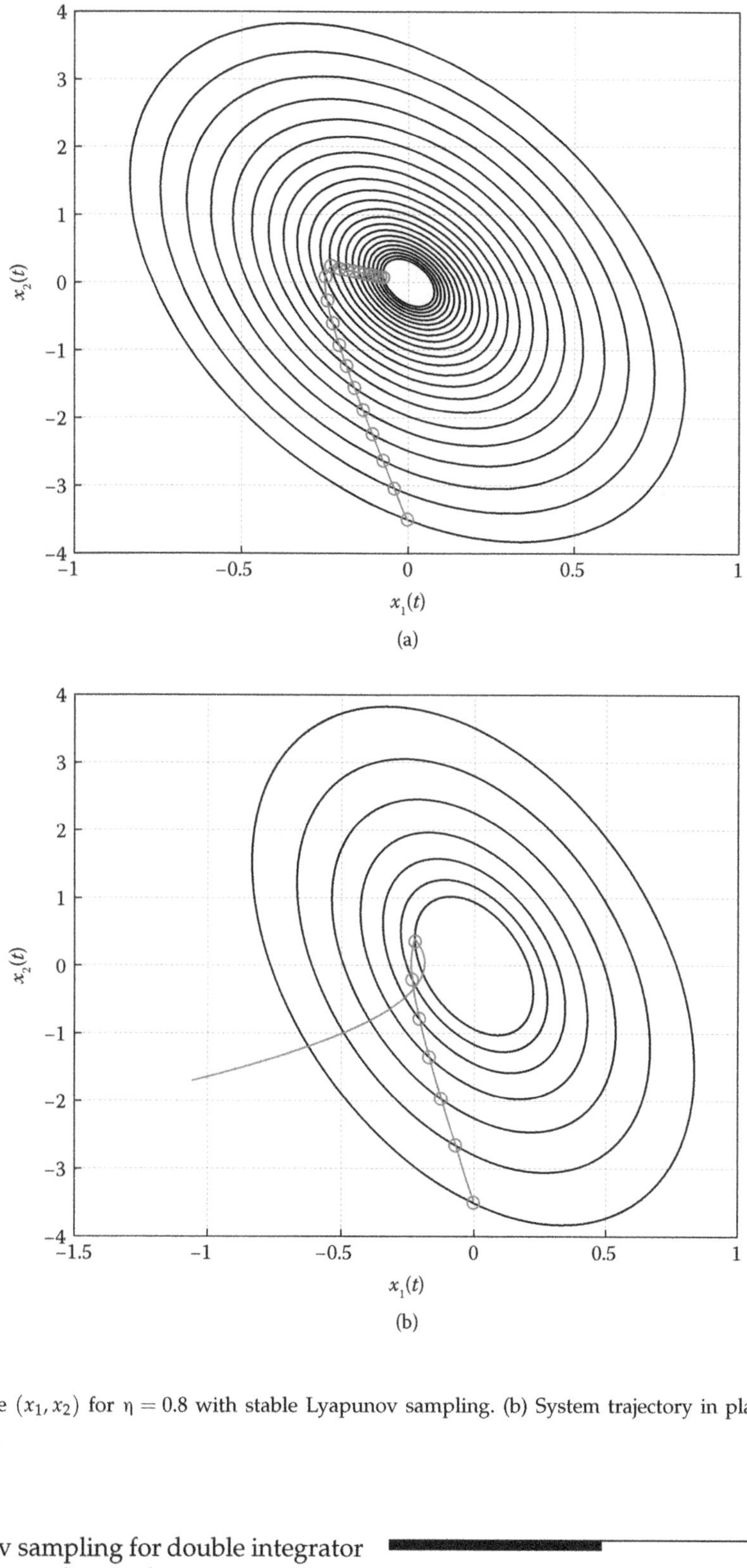


**FIGURE 3.11**
(a) System trajectory in plane $(x_1, x_2)$ for $\eta = 0.8$ with stable Lyapunov sampling. (b) System trajectory in plane $(x_1, x_2)$ for $\eta = 0.65$ with unstable Lyapunov sampling.

original stable Lyapunov sampling for double integrator systems while offering similar performance. Alternatively, to save system stability, the Lyapunov sampling has been proposed to be augmented with "safety sampling" that enforces a sudden control update as soon as the system energy increases [46].

## 3.7 Conclusions

Event-based sampling and communication are emerging research topics driven mainly by the fast development of event-triggered monitoring and control and its implementation in networked sensor and control systems, especially those based on a wireless link. The fundamental postulate for formulation of event-based sampling criteria for control is to achieve satisfactory system performance at a low rate of events. The overview of research studies shows that in many application scenarios, the event-based sampling schemes in control can provide around one order of magnitude of traffic reduction compared to conventional periodic sampling, while still keeping similar system performance. On the other hand, in signal processing, the objective of event-based sampling is to eliminate the clock from system architecture, benefit from increasing time resolution provided by nanoscale VLSI technology, or avoid aliasing. The present study is a survey of event-triggered sampling schemes, both in control and signal processing, aimed at emphasizing the motivation and wide technical context of event-based sampling criteria design.

## Acknowledgment

This work was supported by the Polish National Center of Science under grant DEC-2012/05/E/ST7/01143.

## Bibliography

[1] K. J. Åstrom, P. R. Kumar, Control: A perspective, *Automatica*, vol. 50, no. 1, pp. 3–43, 2014.

[2] W. P. M. H. Heemels, K. H. Johansson, P. Tabuada, An introduction to event-triggered and self-triggered control, *Proceedings of the IEEE Conf. on Decision and Control CDC 2012*, 2012, pp. 3270–3285.

[3] K. J. Åstrom, Event-based control, in *Analysis and Design of Nonlinear Control Systems* (A. Astolfi, L. Marconi, Eds.). Springer, New York 2008, pp. 127–147.

[4] K. E. Arzén, A simple event-based PID controller, *Proceedings of IFAC World Congress*, vol. 18, 1999, pp. 423–428.

[5] K. J. Åstrom, B. Bernhardsson, Comparison of periodic and event based sampling for first-order stochastic systems, *Proceedings of IFAC World Congress*, 1999, pp. 301–306.

[6] W. P. M. H. Heemels, R. J. A. Gorter, A. van Zijl, P. P. J. van den Bosch, S. Weiland, W. H. A. Hendrix, et al., Asynchronous measurement and control: A case study on motor synchronization, *Control Eng. Pract.*, vol. 7, pp. 1467–1482, 1999.

[7] N. Sayiner, H. V. Sorensen, T. R. Viswanathan, A level-crossing sampling scheme for A/D conversion, *IEEE Trans. Circuits Syst. II: Analog Digital Signal Process.*, vol. 43, no. 4, pp. 335–339, 1996.

[8] Y. Tsividis, Continuous-time digital signal processing, *Electron. Lett.*, vol. 39, no. 21, pp. 1551–1552, 2003.

[9] Y. Tsividis, Digital signal processing in continuous time: A possibility for avoiding aliasing and reducing quantization error, *Proceedings of IEEE International Conf. on Acoustics, Speech, and Signal Processing ICASSP 2004*, 2004, pp. 589–592.

[10] E. Allier, G. Sicard, L. Fesquet, M. Renaudin, A new class of asynchronous A/D converters based on time quantization, *Proceedings of IEEE International Symposium on Asynchronous Circuits and Systems ASYNC 2003*, 2003, pp. 196–205.

[11] F. Aeschlimann, E. Allier, L. Fesquet, M. Renaudin, Asynchronous FIR filters: Towards a new digital processing chain, *Proceedings of International Symposium on Asynchronous Circuits and Systems ASYNC 2004*, 2004, pp. 198–206.

[12] L. Grune, S. Hirche, O. Junge, P. Koltai, D. Lehmann, J. Lunze, et al., Event-based control, in *Control Theory of Digitally Networked Dynamic Systems* (J. Lunze, Ed.), Springer, New York 2014, pp. 169–261.

[13] P. G. Otanez, J. R. Moyne, D. M. Tilbury, Using deadbands to reduce communication in networked control systems, *Proceedings of American Control Conf., ACC 2002*, vol. 4, 2002, pp. 3015–3020.

[14] R. M. Murray, K. J. Astrom, S. P. Boyd, R. W. Brockett, G. Stein, Control in an information rich world, *IEEE Control Systems Magazine*, vol. 23, no. 2, pp. 20–33, 2003.

[15] Y. Tsividis, Event-driven data acquisition and digital signal processing: A tutorial, *IEEE Trans. Circuits Syst. II: Express Briefs*, vol. 57, no. 8, pp. 577–581, 2010.

[16] L. Fesquet, B. Bidégaray-Fesquet, Digital filtering with non-uniformly sampled data: From the algorithm to the implementation, in *Event-Based*

*Control and Signal Processing* (M. Miskowicz, Ed.), CRC Press, Boca Raton, FL 2015, pp. 515–528.

[17] Y. Tsividis, M. Kurchuk, P. Martinez-Nuevo, S. M. Nowick, S. Patil, B. Schell, C. Vezyrtzis, Event-based data acquisition and digital signal processing in continuous time, in *Event-Based Control and Signal Processing* (M. Miskowicz, Ed.), CRC Press, Boca Raton, FL 2015, pp. 353–378.

[18] J. Elson, D. Estrin, An address-free architecture for dynamic sensor networks. Technical Report 00-724. University of Southern California, Computer Science Department, Los Angeles, CA 2000.

[19] R. E. Nance, The time and state relationships in simulation modeling, *Commun. ACM*, vol. 24, no. 4, pp. 173–179, 1981.

[20] H. Kopetz, *Real-time systems. Design principles for distributed embedded applications*, Springer, New York 2011.

[21] R. Obermaisser, *Event-triggered and time-triggered control paradigms*, Springer, New York 2004.

[22] P. H. Ellis, Extension of phase plane analysis to quantized systems, *IRE Trans. Autom. Control*, vol. 4, pp. 43–59, 1959.

[23] M. Miskowicz, Send-on-delta concept: An event-based data reporting strategy, *Sensors*, vol. 6, pp. 49–63, 2006.

[24] E. Kofman, J. H. Braslavsky, Level crossing sampling in feedback stabilization under data-rate constraints, *Proceedings of IEEE Conf. on Decision and Control CDC2006* 2006, pp. 4423–4428.

[25] M. de la Sen, J. C. Soto, A. Ibeas, Stability and limit oscillations of a control event-based sampling criterion, *J. Appl. Math*, 2012, pp. 1–25.

[26] M. Miskowicz, R. Golański, LON technology in wireless sensor networking applications, *Sensors*, vol. 6, no. 1, pp. 30–48, 2006.

[27] M. Andelic, S. Berber, A. Swain, Collaborative Event Driven Energy Efficient Protocol (CEDEEP), *IEEE Wireless Commun. Lett.*, vol. 2, pp. 231–234, 2013.

[28] Y. S. Suh, Send-on-delta sensor data transmission with a linear predictor, *Sensors*, vol. 7, pp. 537–547, 2007.

[29] D. Bernardini, A. Bemporad, Energy-aware robust model predictive control based on wireless sensor feedback, *Proceedings of IEEE Conf. on Decision and Control CDC 2008*, 2008, pp. 3342–3347.

[30] K. Staszek, S. Koryciak, M. Miskowicz, Performance of send-on-delta sampling schemes with prediction, *Proceedings of IEEE International Symposium on Industrial Electronics ISIE 2011*, 2011, pp. 2037–2042.

[31] F. Xia, Z. Xu, L. Yao, W. Sun, M. Li, Prediction-based data transmission for energy conservation in wireless body sensors, *Proceedings of Wireless Internet Conf. WICON 2010*, pp. 1–9, 2010.

[32] M. Miskowicz, The event-triggered integral criterion for sensor sampling, *Proceedings of IEEE International Symposium on Industrial Electronics ISIE 2005*, vol. 3, 2005, pp. 1061–1066.

[33] M. Miskowicz, Asymptotic effectiveness of the event-based sampling according to the integral criterion, *Sensors*, vol. 7, pp. 16–37, 2007.

[34] V. H. Nguyen, Y. S. Suh, Networked estimation with an area-triggered transmission method, *Sensors*, vol. 8, pp. 897–909, 2008.

[35] M. Miskowicz, Sampling of signals in energy domain, *Proceedings of IEEE Conf. on Emerging Technologies and Factory Automation ETFA 2005*, vol. 1, 2005, pp. 263–266.

[36] M. Miskowicz, Efficiency of event-based sampling according to error energy criterion, *Sensors*, vol. 10, no. 3, pp. 2242–2261, 2010.

[37] J. Sánchez, A. Visioli, S. Dormido, Event-based PID control, in *PID Control in the Third Millennium. Advances in Industrial Control* (R. Villanova, A. Visioli, Eds.), Springer, New York 2012, pp. 495–526.

[38] J. W. Marck, J. Sijs, Relevant sampling applied to event-based state-estimation, *Proceedings of the International Conf. on Sensor Technologies and Applications SENSORCOMM 2010*, 2010, pp. 618–624.

[39] S. Trimpe, *Distributed and Event-Based State Estimation and Control*, PhD Thesis, ETH Zürich, Switzerland 2013.

[40] S. Trimpe, R. D'Andrea, Event-based state estimation with variance-based triggering, *IEEE Trans. Autom. Control*, vol. 59, no. 12, pp. 3266–3281, 2014.

[41] J. Sijs, B. Noack, M. Lazar, U. D. Hanebeck, Time-periodic state estimation with event-based measurement updates, in *Event-Based Control and Signal Processing* (M. Miskowicz, Ed.), CRC Press, Boca Raton, FL 2015, pp. 261–280.

[42] J. Lunze, Event-based control: Introduction and survey, in *Event-Based Control and Signal Processing* (M. Miskowicz, Ed.), CRC Press, Boca Raton, FL 2015, pp. 3–20.

[43] Y. Chen, M. Kurchuk, N. T. Thao, Y. Tsividis, Spectral analysis of continuous-time ADC and DSP, in *Event-Based Control and Signal Processing* (M. Miskowicz, Ed.), CRC Press, Boca Raton, FL 2015, pp. 409–420.

[44] M. Velasco, P. Martí, E. Bini, On Lyapunov sampling for event-driven controllers, *Proceedings of Joint IEEE Conf. on Decision and Control and Chinese Control Conf. CDC/CCC 2009*, 2009, pp. 6238–6243.

[45] J. Yépez, C. Lozoya, M. Velasco, P. Marti, J. M. Fuertes, Preliminary approach to Lyapunov sampling in CAN-based networked control systems, *Proceedings of Annual Conf. of IEEE Industrial Electronics IECON 2009*, 2009, pp. 3033–3038.

[46] S. Durand, N. Marchand, J. F. Guerrero Castellanos, Simple Lyapunov sampling for event-driven control, *Proceedings of the IFAC World Congress*, pp. 8724–8730, 2011.

[47] P. Tallapragada, N. Chopra, Lyapunov based sampling for adaptive tracking control in robot manipulators: An experimental comparison. Experimental robotics. in *Springer Tracts in Advanced Robotics*, vol. 88, Springer, New York 2013, pp. 683–698.

[48] J. Yen, On nonuniform sampling of bandwidth-limited signals, *IRE Trans. Circuit Theory*, vol. 3, no. 4, pp. 251–257, 1956.

[49] F. J. Beutler, Error-free recovery of signals from irregularly spaced samples, *Siam Rev.*, vol. 8, no. 3, pp. 328–335, 1966.

[50] H. G. Feichtinger, K. Gröchenig, Theory and practice of irregular sampling, in *Wavelets: Mathematics and Applications* (J. J. Benedetto, M. W. Frazier, Eds.), CRC Press, Boca Raton, FL 1994, pp. 305–363.

[51] N. T. Thao, Event-based data acquisition and reconstruction—Mathematical background, in *Event-Based Control and Signal Processing* (M. Miskowicz, Ed.), CRC Press, Boca Raton, FL 2015, pp. 379–408.

[52] I. Bar-David, An implicit sampling theorem for bounded band-limited functions, *Inf. Control*, vol. 24, pp. 36–44, 1974.

[53] J. Selva, Efficient sampling of band-limited signals from sine wave crossings, *IEEE Trans. Signal Process.*, vol. 60, no. 1, pp. 503–508, 2012.

[54] I. F. Blake, W. C. Lindsey, Level-crossing problems for random processes, *IEEE Trans. Inf. Theory*, vol. 3, pp. 295–315, 1973.

[55] S. O. Rice, Mathematical analysis of random noise, *Bell System Tech. J*, vol. 24, no. 1, pp. 46–156, 1945.

[56] D. Rzepka, M. Pawlak, D. Kościelnik, M. Miskowicz, Reconstruction of varying bandwidth signals from event-triggered samples, in *Event-Based Control and Signal Processing* (M. Miskowicz, Ed.), CRC Press, Boca Raton, FL 2015, pp. 529–546.

[57] J. W. Mark, T. D. Todd, A nonuniform sampling approach to data compression, *IEEE Trans. Commun.*, vol. 29, no. 1, pp. 24–32, 1981.

[58] C. Vezyrtzis, Y. Tsividis, Processing of signals using level-crossing sampling, *Proceedings of IEEE International Symposium on Circuits and Systems ISCAS 2009*, 2009, pp. 2293–2296.

[59] S. Senay, L. F. Chaparro, M. Sun, R. J. Sclabassi, Adaptive level-crossing sampling and reconstruction, *Proceedings of European Signal Processing Conf. EUSIPCO 2010*, pp. 196–1300, 2010.

[60] K. M. Guan, S. S Kozat, A. C. Singer, Adaptive reference levels in a level-crossing analog-to-digital converter, *EURASIP J. Adv. Signal Process.*, Article No. 183, 2008.

[61] S. S. Kozat, K. M. Guan, A. C. Singer, Tracking the best level set in a level-crossing analog-to-digital converter, *Digital Signal Process.*, vol. 23, no. 1, pp. 478–487, 2013.

[62] M. Greitans, R. Shavelis, Speech sampling by level-crossing and its reconstruction using spline-based filtering, *Proceedings of International Workshop on Systems, Signals and Image Processing 2007*, 2007, pp. 292–295.

[63] A. Can-Cimino, L. F. Chaparro, Asynchronous processing of nonstationary signals, in *Event-Based Control and Signal Processing* (M. Miskowicz, Ed.), CRC Press, Boca Raton, FL 2015, pp. 441–456.

[64] C. G. Cassandras, Event-driven control and optimization in hybrid systems, in *Event-Based Control and Signal Processing* (M. Miskowicz, Ed.), CRC Press, Boca Raton, FL 2015, pp. 21–36.

[65] D. Brückmann, K. Konrad, T. Werthwein, Concepts for hardware efficient implementation of continuous time digital signal processing, in *Event-Based Control and Signal Processing* (M. Miskowicz, Ed.), CRC Press, Boca Raton, FL 2015, pp. 421–440.

[66] M. Greitans, R. Shavelis, L. Fesquet, T. Beyrouthy, Combined peak and level-crossing sampling scheme, *Proceedings of International Conf. on Sampling Theory and Applications SampTA 2011*, pp. 1–4, 2011.

[67] I. Homjakovs, M. Hashimoto, T. Hirose, T. Onoye, Signal-dependent analog-to-digital conversion based on MINIMAX sampling, *IEICE Trans. Fundam. Electron. Commun. Comput. Sci.*, vol. E96-A, no. 2, pp. 459–468, 2013.

[68] D. Rzepka, M. Miskowicz, Recovery of varying-bandwidth signal from samples of its extrema, *Signal Processing: Algorithms, Architectures, Arrangements, and Applications SPA 2013*, 2013, pp. 143–148.

[69] D. Rzepka, M. Miskowicz, A. Gryboś, D. Kościelnik, Recovery of bandlimited signal based on nonuniform derivative sampling, *Proceedings of International Conf. on Sampling Theory and Applications SampTA* 2013, pp. 1–5, 2013.

[70] A. A. Lazar, L. T. Tóth, Perfect recovery and sensitivity analysis of time encoded bandlimited signals, *IEEE Trans. Circuits Syst. I: Regular Papers*, vol. 52, no. 10, pp. 2060–2073, 2005.

[71] A. A. Lazar, E. A. Pnevmatikakis, Video time encoding machines, *IEEE Trans. Neural Networks*, vol. 22, no. 3, pp. 461–473, 2011.

[72] A. A. Lazar, E. K. Simonyi, L. T. Toth, Time encoding of bandlimited signals, an overview, *Proceedings of the Conf. on Telecommunication Systems, Modeling and Analysis*, November 2005.

[73] G. David, M. Vetterli, Sampling based on timing: Time encoding machines on shift-invariant subspaces, *Appl. Comput. Harmon. Anal.*, vol. 36, no. 1, pp. 63–78, 2014.

[74] L. C. Gouveia, T. Koickal, A. Hamilton, Spike event coding scheme, in *Event-Based Control and Signal Processing* (M. Miskowicz, Ed.), CRC Press, Boca Raton, FL 2015, pp. 487–514.

[75] M. Rastogi, A. S. Alvarado, J. G. Harris, J. C. Principe, Integrate and fire circuit as an ADC replacement, *Proceedings of IEEE International Symposium on Circuits and Systems ISCAS 2011*, 2011, pp. 2421–2424.

[76] A. S. Alvarado, M. Rastogi, J. G. Harris, J. C. Principe, The integrate-and-fire sampler: A special type of asynchronous Σ-Δ modulator, *Proceedings of IEEE International Symposium on Circuits and Systems ISCAS 2011*, 2011, pp. 2031–2034.

[77] J. Daniels, W. Dehaene, M. S. J. Steyaert, A. Wiesbauer, A/D conversion using asynchronous Delta-Sigma modulation and time-to-digital conversion, *IEEE Trans. Circuits Syst. II: Express Briefs*, vol. 57, no. 9, pp. 2404–2412, 2010.

[78] D. Kościelnik, M. Miskowicz, Asynchronous Sigma-Delta analog-to-digital converter based on the charge pump integrator, *Analog Integr. Circuits Signal Process*, vol. 55, pp. 223–238, 2008.

[79] L. Hernandez, E. Prefasi, Analog-to-digital conversion using noise shaping and time encoding, *IEEE Trans. Circuits Syst. I: Regular Papers*, vol. 55, no. 7, pp. 2026–2037, 2008.

[80] J. Daniels, W. Dehaene, M. S. J. Steyaert, A. Wiesbauer, A/D conversion using asynchronous Delta-Sigma modulation and time-to-digital conversion, *IEEE Trans. Circuits Syst. II: Express Briefs*, vol. 57, no. 9, pp. 2404–2412, 2010.

[81] S. Dormido, M. de la Sen, M. Mellado, Criterios generales de determinacion de leyes de muestreo adaptive (General criteria for determination of adaptive sampling laws), *Revista de Informatica y Automatica*, vol. 38, pp. 13–29, 1978.

[82] M. Miskowicz, The event-triggered sampling optimization criterion for distributed networked monitoring and control systems, *Proceedings of IEEE International Conf. on Industrial Technology ICIT 2003*, vol. 2, 2003, pp. 1083–1088.

[83] J. Colandairaj, G. W. Irwin, W. G. Scanlon, Wireless networked control systems with QoS-based sampling, *IET Control Theory Appl.*, vol. 1, no. 1, pp. 430–438, 2007.

[84] J. Mitchell, W. McDaniel, Adaptive sampling technique, *IEEE Trans. Autom. Control*, vol. 14, no. 2, pp. 200–201, 1969.

[85] R. C. Dorf, M. C. Farren, C. A. Phillips, Adaptive sampling for sampled-data control systems, *IEEE Trans. Autom. Control*, vol. 7, no. 1, pp. 38–47, 1962.

[86] T. C. Hsia, Comparisons of adaptive sampling control laws, *IEEE Trans. Autom. Control*, vol. 17, no. 6, pp. 830–831, 1974.

[87] T. C. Hsia, Analytic design of adaptive sampling control laws, *IEEE Trans. Autom. Control*, vol. 19, no. 1, pp. 39–42, 1974.

[88] W. P. M. H. Heemels, R. Postoyan, M. C. F. (Tijs) Donkers, A. R. Teel, A. Anta, P. Tabuada, D. Nešić, Periodic Event-Triggered Control, in *Event-Based Control and Signal Processing* (M. Miskowicz, Ed.), CRC Press, Boca Raton, FL 2015, pp. 203–220.

[89] M. Velasco, P. Marti, J. M. Fuertes, The self-triggered task model for real-time control systems, *Proceedings of 24th IEEE Real-Time Systems Symposium RTSS 2003*, pp. 1–4, 2003.

[90] C. Nowzari, J. Cortés, Self-triggered and team-triggered control of networked cyber-physical systems, in *Event-Based Control and Signal Processing* (M. Miskowicz, Ed.), CRC Press, Boca Raton, FL 2015, pp. 203–220.

[91] J. Ploennigs, V. Vasyutynskyy, K. Kabitzsch, Comparative study of energy-efficient sampling approaches for wireless control networks, *IEEE Trans. Ind. Inf.*, vol. 6, no. 3, pp. 416–424, 2010.

[92] M. G. Cea, G. C. Goodwin, Event based sampling in non-linear filtering, *Control Eng. Pract.*, vol. 20, no. 10, pp. 963–971, 2012.

[93] J. Sijs, M. Lazar, On event based state estimation, in *Hybrid Systems: Computation and Control*, Springer, New York 2009, pp. 336–350.

[94] J. Sijs, B. Noack, U. D. Hanebeck, Event-based state estimation with negative information, *Proceedings of IEEE International Conf. on Information Fusion—FUSION 2013*, 2013, pp. 2192–2199.

[95] W. P. M. H. Heemels, M. C. F. Donkers, Model-based periodic event-triggered control for linear systems, *Automatica*, vol. 49, pp. 698–711, 2013.

[96] F. de Paoli, F. Tisato, On the complementary nature of event-driven and time-driven models, *Control Eng. Pract.*, vol. 4, no. 6, pp. 847–854, 1996.

[97] D. M. Chapiro, Globally-asynchronous locally-synchronous systems, PhD Thesis, Stanford University, CA 1984.

[98] M. Krstić, E. Grass, F. K. Gurkaynak, P. Vivet, Globally asynchronous, locally synchronous circuits: Overview and outlook, *IEEE Design Test*, vol. 24, no. 5, pp. 430–441, 2007.

[99] J. Ploennigs, V. Vasyutynskyy, K. Kabitzsch, Comparison of energy-efficient sampling methods for WSNs in building automation scenarios, *Proceedings of IEEE International Conf. on Emerging Technologies and Factory Automation ETFA 2009*, 2009, pp. 1–8.

[100] J. Sánchez, M. A. Guarnes, S. Dormido, On the application of different event-based sampling strategies to the control of a simple industrial process, *Sensors*, vol. 9, pp. 6795–6818, 2009.

[101] V. Vasyutynskyy, K. Kabitzsch, Towards comparison of deadband sampling types, *Proceedings of IEEE International Symposium on Industrial Electronics ISIE 2007*, 2007, pp. 2899–2904.

[102] M. Miskowicz, Event-based sampling strategies in networked control systems, *Proceedings of IEEE International Workshop on Factory Communication Systems WFCS 2014*, 2014, pp. 1–10.

[103] F. Riesz, B. Nagy, *Functional Analysis*, Dover, New York 1990.

[104] M. Miskowicz, Bandwidth requirements for event-driven observations of continuous-time variable, *Proceedings of IFAC Workshop on Discrete Event Systems*, 2004, pp. 475–480.

[105] M. Miskowicz, Efficiency of level-crossing sampling for bandlimited Gaussian random processes, *Proceedings of IEEE International Workshop on Factory Communication Systems WFCS 2006*, 2006, pp. 137–142.

# 4

# *Event-Triggered versus Time-Triggered Real-Time Systems: A Comparative Study*

**Roman Obermaisser**
*University of Siegen*
*Siegen, Germany*

**Walter Lang**
*University of Siegen*
*Siegen, Germany*

**CONTENTS**



**ABSTRACT** This chapter compares event-triggered and time-triggered systems, both of which can be used for the realization of an embedded control application. Inherent properties of time-triggered systems such as composability, real-time support, and diagnosis are well-suited for safety-relevant control applications. Therefore, we provide an overview of time-triggered architectures at chip level and at cluster level. The chapter finishes with an example of a control application, which is realized as an event-triggered system and as a time-triggered system in order to experimentally compare the respective behaviors.

## 4.1 Event-Triggered and Time-Triggered Control

Based on the source of temporal control signals, one can distinguish time-triggered and event-triggered systems. Time-triggered systems are designed for periodic activities, where all computational and communication activities are initiated at predetermined global points in time. The temporal behavior of a time-triggered communication network is controlled solely by the progression of time. Likewise, time-triggered operating systems perform scheduling decisions at predefined points in time according to schedule tables. Event-triggered systems, on the other hand, initiate activities whenever significant events occur. Any state change of relevance for a given application (e.g., external interrupt, software exception) can trigger a communication or computational activity.

In time-triggered systems, contention is resolved using the temporal coordination of resource access based on time division multiple access (TDMA). TDMA statically divides the capacity of a resource into a number of slots and assigns unique slots to every component.

In case of time-triggered networks, the communication activities of every component are controlled by a time-triggered communication schedule. The schedule specifies the temporal pattern of message transmissions (i.e., at what points in time components send and receive messages). A sequence of sending slots, which allows every component in an ensemble of $n$ components to send exactly once, is called a TDMA round. The sequence of the different TDMA rounds forms the cluster cycle and determines the periodicity of the time-triggered communication.

The a priori knowledge about the times of message exchanges enables the communication network to operate autonomously. Hence, the correct temporal behavior of the communication network is independent of the temporal behavior of the application software and can be verified in isolation.

Likewise, time-triggered operating systems and hypervisors use TDMA to manage access to processors. For example, XtratuM [1], LynxOS [2], and VxWorks [3] support fixed cyclic scheduling of partitions according to ARINC-653 [4].

Time-triggered systems typically employ implicit synchronization [5]. Components agree a priori on the global points in time when resources are accessed, thereby avoiding race conditions, interference, and inconsistency. In case of networks, a component's ability to handle received messages can be ensured at design time by fixing the points in time of communication activities. This implicit flow control [5] does not require acknowledgment mechanisms and is well-suited for multicast interactions, because a unidirectional data flow involves only a unidirectional control flow [6].

## 4.2 Requirements of Real-Time Architectures in Control Applications

This section discusses the fundamental requirements of an architecture for control applications. These requirements include composability, real-time support, flexibility, support for heterogeneous timing models, diagnosis, clock synchronization, and fault tolerance.

### 4.2.1 Composability

In many engineering disciplines, large systems are built from prefabricated components with known and validated properties. Components are connected via stable, understandable, and standardized interfaces. The system engineer has knowledge about the global properties of the components as they relate to the system functions and of the detailed specification of the component interfaces. Knowledge about the internal design and implementation of the components is neither needed nor available in many cases. Composability deals with all issues that relate to the component-based design of large systems.

Composability is a concept that relates to the ease of building systems out of subsystems [7]. We assume that subsystems have certain properties that have been validated at the subsystem level. A system, that is, a composition of subsystems, is considered composable with respect to a certain property, if this property, given that it has been established at the subsystem level, is not invalidated by the integration. Examples of such properties are timeliness or certification.

Composability is defined as the stability of component properties across integration. Composability is necessary for correctness-by-construction of component-based systems [8]. Temporal composability is an instantiation of the general notion of composability. A system is temporally composable, if temporal correctness is not refuted by the system integration [9]. For example, temporal correctness is essential to preserve the stability of a control application on the integration of additional components.

A necessary condition for temporal composability is that if $n$ components are already integrated, the integration of component $n+1$ will not disturb the correct operation of the $n$ already integrated components. This condition guarantees that the integration activity is linear and not circular. It has stringent implications for the management of the resources. Time-triggered systems satisfy this condition, because the allocation of resources occurs at design time (i.e., prior to system integration).

In an event-triggered system, however, which manages the resources dynamically, it must be ensured that even at the critical instant, that is, when all components request the resources at the same instant, the specified timeliness of all resource requests can be satisfied. Otherwise, failures will occur sporadically with a failure rate that is increasing with the number of integrated components. When resources are multiplexed between components, then each newly integrated component will affect the temporal properties of the system (e.g., network latencies, processor execution time) of already integrated components.

Time-triggered systems support invariant temporal properties during an incremental integration process. The resources are statically assigned to each component, and each newly integrated component simply exploits a time slot that has already been reserved at design time via a static schedule.

### 4.2.2 Real-Time Requirements

Many control applications are real-time systems, where the achievement of control stability and safety depends on the completion of activities (like reading sensor values, performing computations, conducting communication activities, implementing actuator control) in bounded time. In real-time systems, missed deadlines represent system failures with the potential for consequences as serious as in the case of providing incorrect results.

In the case of networks, important performance attributes are the bandwidth, the network delay, and the variability of the network delay (i.e., communication jitter). The bandwidth is a measure of the available communication resources expressed in bits/second. The bandwidth is an important parameter as it determines the types of functions that can be handled and the number of messages and components that can be handled by the communication network.

The network delay denotes the time difference between the production of a message at a sending component and the reception of the last bit of the message at a receiving component. Depending on whether the communication protocol is time triggered or event triggered, the network delay exhibits different characteristics. In a time-triggered system, the send instants of all components are periodically recurring instants, which are globally planned in the system and defined with respect to the global time base. The network delay is independent of the traffic from other components and depends solely on the relationship between the transmission request instant of a message and the preplanned send instants according to the time-triggered schedule. Furthermore, since the next send instant of every component is known a priori, a component can synchronize locally the production of periodic messages with the send instant and thus minimize the network delay of a message.

In an event-triggered system, the network delay of a message depends on the state of the communication system at the transmission request instant of a message. If the communication network is idle, the message transmission can start immediately, leading to a minimal transport delay. If the channel is busy, then the transport delay depends on the media access strategy implemented in the communication protocol. For example, in the Carrier Sense Multiple Access/Collision Detection (CSMA/CD) protocol of the Ethernet [10], components wait for a random delay before attempting transmission again. In the Carrier Sense Multiple Access/Collision Avoidance (CSMA/CA) protocol of the Controller Area Network (CAN) [11], the network delay of a message depends on its priority relative to the priorities of other pending messages. Hence, in an event-triggered network, the network delay of a message is a global property that depends on the traffic patterns of all components.

A bounded network delay with a minimum variability is important in many embedded applications. Achievement of control stability in real-time applications depends on the completion of actions such as reading sensor values, performing computations, and performing actuator control within bounded time. Hard real-time systems must guarantee bounded response times even in the case of peak load and fault scenarios. For example, in drive-by-wire applications, the dynamics for steered wheels in closed control loops enforce computer delays of less than 2 ms [12]. Taking the vehicle dynamics into account, a transient outage-time of the steering system must not exceed 50 ms [12].

While control algorithms can compensate known delays, delay jitter (i.e., the difference between the maximum and minimum value of delay) brings an additional uncertainty into a control loop, which has an adverse effect on the quality of control [5]. Delay jitter represents an uncertainty about the instant a physical entity was observed and can be expressed as an additional error in the value domain. In case of low jitter or a global timebase with good precision, state estimation techniques allow compensation for a known delay between the time of observation and the time of use of a real-time image. State estimation introduces models of physical quantities to compute the probable state of a variable at a future point in time.

### 4.2.3 Flexibility

Many systems are faced with the need to incorporate new application functionalities using new or changed components, for example, upgrade of a factory automation system with new Computerized Numerical Control (CNC) machines. Likewise, control applications need to cope with changes in processors and network technologies, which is also known as the challenge of technology obsolescence.

Event-triggered systems provide more flexibility in coping with new components and technological changes. For example, an additional component sending messages on a CAN bus does not require any changes in already existing nodes. In contrast, time-triggered systems require the computation of new schedules and TDMA schemes for the entire system.

However, the increased flexibility of event-triggered systems does not prevent undesirable implications such as missed deadlines after upgrades to the system. Although a change is functionally transparent to existing components, the resulting effects on extrafunctional properties and real-time performance can induce considerable efforts for revalidation and recertification.

### 4.2.4 Heterogeneous Timing Models

Due to the respective advantages of event-triggered and time-triggered systems, support for both control paradigms is desirable. The rationale of these integrated architectures is the effective covering of mixed-criticality systems, in which a safety-critical subsystem exploits time-triggered services, and a nonsafety-critical subsystem can exploit event-triggered services. Thereby, the system can support different, possibly contradicting requirements from different application subsystems.

For this reason, several communication protocols integrating event-triggered and time-triggered control have been developed [13–16]. These protocols differ at the level at which the integration takes place (i.e., either at the media access control [MAC] layer or above), and the basic operational principles for separating event-triggered and time-triggered traffic.

Communication protocols for the integration of event-triggered and time-triggered control can be classified, depending on whether the integration occurs on the MAC layer or through an overlay network.

In case of MAC layer integration, the start and end instants of the periodic time-triggered message transmissions, as well as the sending components are specified at design time. For this class of messages, contention is resolved statically. All time-triggered message transmissions follow an a priori defined schedule, which repeats itself with a fixed round length, which is the least common multiple of all time-triggered message periods. Within each round, the time intervals that are not consumed by time-triggered message exchanges are available for event-triggered communication. Consequently, time is divided into two types of slots: event-triggered and time-triggered. The time-triggered slots are used for periodic preplanned message transmissions, while event-triggered slots are located in between the time-triggered slots. In event-triggered slots, message exchanges depend on external control, and the start instants of message transmissions can vary. Furthermore, event-triggered slots can be assigned to multiple (or all) components of the system. For this reason, the MAC layer needs to support the dynamic resolving of contention when more than one component intends to transmit a message. During event-triggered slots, a subprotocol (e.g., CSMA/CA, CSMA/CD) takes over that is not required during time-triggered slots in which contention is prevented by design.

Event-triggered overlay networks based on a time-triggered communication protocol are another solution for the integration of the two control paradigms by layering event-triggered communication on top of time-triggered communication. The MAC protocol is TDMA—that is, time is divided into slots and each slot is statically assigned to a component that exclusively sends messages during this slot. A subset of the slots in each communication round is used for the construction of event-triggered overlay networks. Event-triggered overlay networks have been established for different event-triggered protocols [15,17].

### 4.2.5 Diagnosis

In order to achieve the required level of safety, it is required for many control applications that errors in the systems reliably be detected and isolated in bounded time.

In general, error detection mechanisms can be implemented at the architectural level, or within the application itself. The advantage of implementations at the architectural level is that they are built in a generic way and, thus, can be verified and certified. Furthermore, they can be implemented directly in hardware (e.g., within the communication controller), which can reduce the error detection latency and relieves the host CPU from computational overhead for error detection. Nevertheless, according to Salzer's end-to-end argument [18], every safety-critical system must contain additional end-to-end error detection mechanisms at the application level, in order to cover the entire controlled process.

Error detection mechanisms can be realized based on syntactic checks and semantic checks that are implemented at the protocol level in many time-triggered protocols, and error detection by active redundancy which is usually implemented at a higher level.

The syntactic checks are targeted at the syntax of the received messages. Protocols usually check for the satisfaction of specific constraints defined in the message format. Examples are start and end sequences of the entire message or cyclic redundancy checks (CRCs).

Semantic checks can be implemented efficiently in time-triggered protocols, due to the a-priori knowledge of communication patterns. Examples of semantic errors that can be detected by using this knowledge are omission failures, invalid sender identifiers, and violation of slot boundaries.

Value errors in a message's payload cannot be detected by the above-mentioned mechanisms, if the payload CRC and the message timing are valid. Such value errors can be systematically detected by active redundancy, which means that the sending component is replicated, and the values generated by the redundant senders are compared at the receivers.

### 4.2.6 Clock Synchronization

Due to clock drifts, the clock times in an ensemble of clocks will drift apart, if clocks are not periodically resynchronized. Clock synchronization is concerned with bringing the values of clocks in close relation with respect to each other. A measure for the quality of synchronization is the precision. An ensemble of clocks that are synchronized to each other with a specified precision offers a global time. The global time of the ensemble is represented by the local time of each clock, which serves as an approximation of the global time.

If the counter value is read by a component at a particular point in time, it is guaranteed that this value can only differ by one tick from the value that is read at any other correct component at the same point in time. This property is also known as the reasonableness of the global time, which states that the precision of the local clocks at the components is lower than the granularity of the global time base. Optionally, the global time base can be synchronized with an external time source (e.g., GPS or global time base of another integration level). In this case, the accuracy denotes the maximum deviation of any local clock to the external reference time.

The global time base allows the temporal coordination of distributed activities (e.g., synchronized messages), as well as the interrelation of timestamps assigned at different components.

### 4.2.7 Fault Tolerance

Fault detection, fault containment, and fault masking are three fault-tolerance techniques that can be realized to be consecutively building upon each other. These techniques differ in the incurred overhead and in their utility for realizing a reliable system. For example, fault detection allows the application to react to faults using an application-specific strategy (e.g., entering a safe state).

A fault containment region (FCR) is the boundary of the immediate impact of a fault. In conformance with the fault-error-failure chain introduced by Laprie [19], one can distinguish between faults that cause the failure of an FCR (e.g., design of the hardware or software of the FCR, operational fault of the FCR) and faults at the system level. The latter type of fault is a failure of an FCR, which could propagate to other FCRs through a sent message that deviates from the specification. If the transmission instant of the message violates the specifications, we speak of a message timing failure. A message value failure means that the data structure contained in a message is incorrect.

Such a failure of an FCR can be tolerated by distributed fault-tolerance mechanisms. The masking of failures of FCRs is denoted as error containment [20], because it avoids error propagation by the flow of erroneous messages. The error detection mechanisms must be part of different FCRs than the message sender [21]. Otherwise, the error detection mechanism may be impacted by the same fault that caused the message failure.

For example, a common approach for masking component failures is N-modular redundancy (NMR) [22]. The $N$ replicas receive the same requests and provide the same service. The output of all replicas is provided to a voting mechanism, which selects one of the results (e.g., based on majority) or transforms the results to a single one (average voter). The most frequently used $N$-modular configuration is triple modular redundancy (TMR).

We denote three replicas in a TMR configuration of a fault-tolerant unit (FTU). In addition, we consider the voter at the input of a component and the component itself as a self-contained unit, which receives the replicated inputs and performs voting by itself without relying on an external voter. We call this behavior incoming voting.

An important prerequisite for error containment are independent FCRs. Common-mode failures are failures of multiple FCRs, which are correlated and occur due to a common cause. Common-mode failures occur when the assumption of the independence of FCRs is compromised. They can result from replicated design faults or from common operational faults such as a massive transient disturbance. Common-mode failures of the replicas in an FRU must be avoided, because any correlation in the instants of the failures of FCRs significantly decreases the reliability improvements that can be achieved by NMR [23].

Replica determinism has to be supported by the architecture to ensure that the replicas of an FRU produce the same outputs in defined time intervals. A time-triggered communication system addresses key issues of replica determinism. In particular, a time-triggered communication system supports replica determinism by exploiting the global time base in conjunction with preplanned communication and computational schedules. Computational activities are triggered after the last message of a set of input messages has been received by all replicas of an FRU. This instant is a priori known due to the predefined time-triggered schedules. Thus, each replica wakes up at the same global tick, and operates on the same set of input messages. The alignment of communication and computational activities on the global time base ensures temporal predictability and avoids race conditions.

## 4.3 Communication Protocols

This section gives an overview of time-triggered protocols at the chip level and in distributed systems. In addition, we discuss how the requirements of control applications are addressed in the protocols (cf. overview in Table 4.1).

**TABLE 4.1**
Time-Triggered Protocols

| | Protocol | Composability | Real-Time (Bounded Latency and Jitter) | Timing Models | Diagnosis | Global Time | Fault Containment | Error Containment |
|---|---|---|---|---|---|---|---|---|
| On-chip | TTNoC | Yes | Yes | Periodic and sporadic | Local error detection | Yes | Yes | Yes |
| | AEthereal | Yes | Yes | Periodic and sporadic | Local error detection | No | Yes | No |
| Off-chip | TTP | Yes | Yes | Periodic | Membership | Yes | Yes | Yes |
| | TTE | Yes (time-triggered traffic) | Yes (time-triggered traffic) | Periodic, sporadic, aperiodic | Local error detection | Yes | Yes | Yes |
| | TTCAN | Yes (time-triggered traffic) | Yes (time-triggered traffic) | Periodic, sporadic, aperiodic | Local error detection | Yes | Yes | Yes |

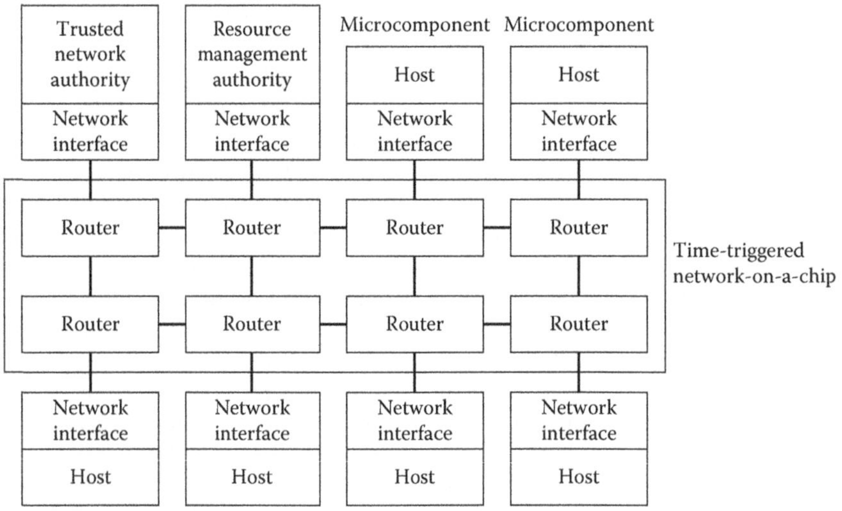


**FIGURE 4.1**
GENESYS MPSoC with example NoC topology.

### 4.3.1 Time-Triggered Network-on-a-Chip

The time-triggered network-on-a-chip (TTNoC) is part of the GENESYS Multi-Processor Systems-on-a-Chip (MPSoC) architecture [24]. This network-on-a-chip (NoC) interconnects multiple, possibly heterogeneous IP blocks called microcomponents. The MPSoC introduces a trusted subsystem, which ensures that a fault (e.g., a software fault) within the host of a microcomponent cannot lead to a violation of the microcomponent's temporal interface specification in a way that the communication between other microcomponents would be disrupted. For this reason, the trusted subsystem prevents a faulty microcomponent from sending messages during the sending slots of any other microcomponent. Figure 4.1 shows a GENESYS MPSoC with a mesh-based NoC topology.

Furthermore, the GENESYS MPSoC architecture supports integrated resource management. For this purpose, dedicated architectural elements called the Trusted Network Authority (TNA) and the Resource Management Authority (RMA) accept resource allocation requests from the microcomponents and reconfigure the MPSoC, for example, by dynamically updating the time-triggered communication schedule of the NoC and switching between power modes.

The TTNoC interconnects the microcomponents of an MPSoC. The purposes of the TTNoC encompass clock synchronization for the establishment of a global time base, as well as the predictable transport of periodic and sporadic messages.

**Clock Synchronization.** The TTNoC performs clock synchronization in order to provide a global time base for all microcomponents despite the existence of multiple clock domains. The TTNoC is based on a uniform time format for all configurations, which has been standardized by the Object Management Group (OMG) in the smart transducer interface standard [25].

The time format of the NoC is a binary time-format that is based on the physical second. Fractions of a second are represented as 32 negative powers of two (down to about 232 ps), and full seconds are presented in 32 positive powers of two (up to about 136 years). This time format is closely related to the time-format of the General Positioning System (GPS) time and takes the epoch from GPS. In case there is no external synchronization [26], the epoch starts with the power-up instant.

**Predictable Transport of Messages.** Using TDMA, the available bandwidth of the NoC is divided into periodic conflict-free sending slots. We distinguish between two utilizations of a periodic time-triggered sending slot by a microcomponent. A sending slot can be used for the periodic transmission of messages or the sporadic transmission of messages. In the latter case, a message is sent only if the sender must transmit a new event to the receiver.

The allocation of sending slots to microcomponents occurs using a communication primitive called pulsed data stream [27]. A pulsed data stream is a time-triggered periodic unidirectional data stream that transports data in pulses with a defined length from one sender to $n$ a priori identified receivers at a specified phase of every cycle of a periodic control system.

The pulsed behavior of the communication network enables the efficient transmission of large data in applications requiring a temporal alignment of sender and receiver. Temporal alignment of sender and receiver is required in applications where a short latency between sender and receiver is demanded. This is typical for many real-time systems.

For example, consider a control loop realized by three microcomponents performing sensor data acquisition (A), processing of the control algorithm (C), and actuator operating (E), as is schematically depicted in Figure 4.2. In this application, temporal alignment between sensor data transmission (B) and the start of the processing of the control algorithm (cf. instant 3 in Figure 4.2) as well as between the transmission of the control value (D) and the start of actuator output (cf. instant 5) is vital to reduce the end-to-end latency of the control loop, which is an important quality characteristic. By specifying two pulsed data streams corresponding to (B) and (D) in Figure 4.2, efficient, temporally aligned data transmission can be achieved.

Contrary to the on-chip communication system of the introduced MPSoC architecture, many existing NoCs provide only a guaranteed bandwidth to the individual senders without support for temporal alignment. The resulting consequences are as follows:

- The short latency cannot be guaranteed.
- A high bandwidth has to be granted to the sender throughout the entire period of the control cycle, although it is only required for a short interval.
- The communication system has to be periodically reconfigured in order to free and reallocate the nonused communication resources.

Similarly, in a fault-tolerant system that masks failures by TMR, a high bandwidth communication service is required for short intervals to exchange and vote on the state data of the replicated channels. A real-time communication network should consider these pulsed communication requirements and provide appropriate services.

Furthermore, the GENESYS MPSoC provides an extended functionality with respect to message ordering. The GENESYS MPSoC guarantees consistent delivery order across multiple channels. Consistent delivery order means that any two microcomponents will see the same sequence of reception events within the intersection of the two sets of messages that are received by the two microcomponents. Consistent delivery ordering is a prerequisite for replica deterministic behavior of microcomponents, which is required for the transparent masking of hardware errors by TMR [28].

### 4.3.2 Æthereal

COMPSOC [29] is an architectural approach for mixed-criticality integration on a multicore chip. The main interest of COMPSOC is to establish composability and predictability. Hardware resources, such as processor tiles, NoC, and memory tiles are virtualized to ensure temporal and spatial isolation. The result is a set of virtual platforms connected via virtual wires based on the

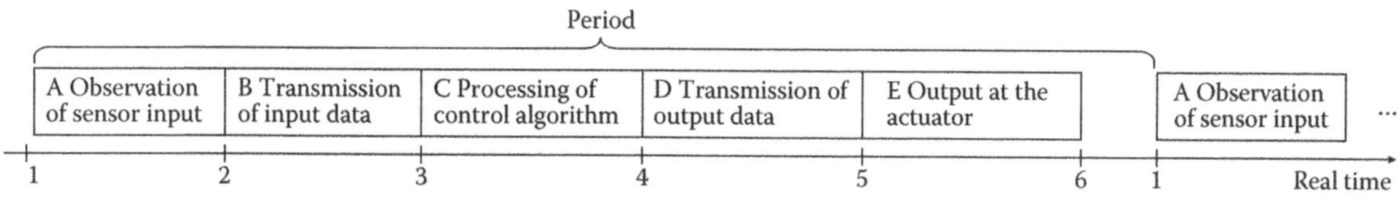


**FIGURE 4.2**
Cyclic time-triggered control activities.

Æthereal NoC [30]. A static TDMA scheme with preemption is used for processor and NoC scheduling, and shared resources have a bounded worst-case execution time (WCET) to ensure determinism. CoMPSOC is independent from the model of computation used by the application, but a reconfiguration of hardware resources at runtime is not supported.

Analogous to the time-triggered NoC, the TDMA scheme is employed to avoid conflicts on the shared interconnects and to provide encapsulation and timing guarantees for the individual communication channels. The objective of Æthereal is to establish resource guarantees with respect to bandwidth and latency.

The Æthereal architecture [31] combines guaranteed services (such as guaranteed throughput and bounded latency) with best-effort services. Guaranteed services aid in the compositional design of robust SoCs, while best-effort services increase the resource efficiency by exploiting the NoC capacity that is left unused by the guaranteed services.

The constituting elements of the architecture are network interfaces (NIs) and routers which are interconnected by links. The NIs translate the protocol of the attached IP core to the NoC-internal packet-based protocol. The routers transport messages between the NIs.

Æthereal offers a shared-memory abstraction to the attached IP modules [32] and employs a transaction-based master/slave protocol. The transaction-based model was chosen to ensure backward compatibility to existing on-chip network protocols like AXI [33] or OCP [34].

In the Æthereal NoC, the signals of an IP core with a standardized interface (e.g., AXI or OCP) are sequentialized in request and response messages, which are transported by the NoC in the form of packets. The communication services are provided via so-called connections that are composed out of multiple unidirectional peer-to-peer channels (e.g., a typical peer-to-peer OCP connection would consist of a request channel and a reply channel; a multicast or narrow-cast connection can be implemented by a collection of peer-to-peer connections, one for each master-slave pair).

Channels offer two types of service classes: guaranteed throughput (GT), and best effort (BE). GT channels give guarantees on minimum bandwidth and maximum latency by using a TDMA scheme. The TDMA scheme is based on a table with a given number of time slots (e.g., 128 slots). In each slot, a network interface can read and write at most one block of data. Given the duration of a slot, the size of a block that can be transferred within one slot and the number of slots in the table, a slots corresponds to a given bandwidth $B$. Therefore, reserving $N$ slots for a channel results in a total bandwidth of $N * B$. The granularity with which the bandwidth of a channel can be reserved equals $1/S$ of the maximum channel bandwidth, where $S$ is the number of slots in the TDMA table.

Within a single channel, temporal ordering of messages is guaranteed. This means that all messages are received in the same order as they were sent. Since the NI treats different channels as different entities, ordering guarantees are provided only within a single channel. Across different channels, message reordering is possible.

Æthereal provides a shared memory abstraction to the attached IP cores via a transaction-based master/slave protocol as it is used in OCP or AXI. These protocols define low-level signals like address signals, data signals, interrupt signals, reset signals, or clock signals. Examples where such protocols are typically employed are at the interfaces of processors, memory subsystems, or bus bridges.

### 4.3.3 Time-Triggered Protocol

The Time-Triggered Protocol (TTP) is a communication protocol for distributed fault-tolerant real-time systems. It is designed for applications with stringent requirements concerning safety and availability, such as avionic, automotive, industrial control, and railway systems. TTP was initially named TTP/C and later renamed to TTP. The initial name of the communication protocol originated from the classification of communication protocols of the Society of Automotive Engineers (SAE), which distinguishes four classes of in-vehicle networks based on the performance. TTP/C satisfies the highest performance requirements in this classification of in-vehicle networks and is suitable for network classes C and above.

TTP provides a consistent distributed computing base [35] in order to ease the construction of reliable distributed applications. Given the assumptions of the fault hypothesis, TTP guarantees that all correct nodes perceive messages consistently in the value and time domains. In addition, TTP provides consistent information about the operational state of all nodes in the cluster. For example, in the automotive domain these properties would reduce efforts for the realization of a safety-critical brake-by-wire application with four braking nodes. Given the consistent information about inputs and node failures, each of the nodes can adjust the braking force to compensate for the failure of other braking nodes. In contrast, the design of distributed algorithms becomes more complex [36] if nodes cannot be certain that every other node works on the same data. In such a case the agreement problem has to be solved at the application level.

The communication services of TTP support the predictable message transport with a small variability of the latency. The smallest unit of transmission and MAC on the TTP network is a TDMA slot. A TDMA slot is a time interval with a fixed duration that can be used by a node to broadcast a message to all other nodes. A sequence of TDMA slots is called a TDMA round. The cluster cycle defines a pattern of periodically recurring TDMA rounds.

The time of the TDMA slots and the allocation of TDMA slots to nodes is specified in a static data structure called the Message Descriptor List (MEDL). The MEDL is the central configuration data structure in the TTP. Each node possesses its own MEDL, which reflects the node's communication actions (e.g., sending of messages, clock synchronization) and parameters (e.g., delays to other nodes). At design time, TTP development tools are used to temporally align the MEDLs of the different nodes with respect to the global time base. For example, the period and phase of a message transmission are aligned with the respective message receptions taking into account propagation delays and jitter.

The fault-tolerant clock synchronization maintains a specified precision and accuracy of the global time base, which is initially established by the restart and startup services when transiting from asynchronous to synchronous operation. The fault-tolerant average (FTA) algorithm [26] is used for clock synchronization in TTP. The FTA algorithm computes the convergence function for the clock synchronization within a single TDMA round. It is designed to tolerate $k$ Byzantine faults in a system with $N$ nodes. Therefore, the FTA algorithm bounds the error that can be introduced by arbitrary faulty nodes. These nodes can provide inconsistent information to the other nodes.

The diagnostic services provide the application with feedback about the operational state of the nodes and the network using a consistent membership vector. A node $A$ considers another node $B$ as operational, if node $A$ has correctly received the message that was sent by node $B$ prior to the membership point. In case redundant communication channels are used, the reception on one of the channels is sufficient in order to consider a sender to be operational. The diagnostic services in conjunction with the a priori knowledge about the permitted behavior of nodes is the basis for the fault isolation services of TTP.

The TTP was designed to isolate and tolerate an arbitrary failure of a single node during synchronized operation [37]. After the error detection and the isolation of the node, a consecutive failure can be handled. Given fast error detection and isolation mechanisms, such a single fault hypothesis is considered to be suitable in many safety-critical systems [38]. The fault hypothesis assumes an arbitrary failure mode of a single node.

In order to tolerate timing failures, a TTP cluster uses local or central bus guardians. In addition, the bus guardian protects the cluster against slightly off-specification faults [39], which can lead to ambiguous results at the receiver nodes.

A local bus guardian is associated with a single TTP node and can be physically implemented as a separate device or within the TTP node (e.g., on the silicon die of the TTP communication controller or as a separate chip). The local bus guardian uses the a priori knowledge about the time-triggered communication schedule in order to ensure fail-silence of the respective node. If the node intends to send outside the preassigned transmission slot in the TDMA scheme, the local bus guardian cuts off the node from the network. In order to avoid common mode failures of the guardian and the node, the TTP suggests the provision of an independent external clock source for the local bus guardian.

The central bus guardian is always implemented as a separate device, which protects the TDMA slots of all attached TTP nodes. An advantage compared to the local bus guardians is the higher resilience against spatial proximity faults and the ability to handle slightly off-specification faults.

### 4.3.4 Time-Triggered Ethernet

A Time-Triggered Ethernet (TTEthernet) [40] network consists of a set of nodes and switches, which are interconnected using bidirectional communication links. TTEthernet combines different types of communication on the same network. A service layer is built on top of IEEE 802.3, thereby complementing layer two of the Open System Interconnection model [40].

TTEthernet supports synchronous communication using so-called time-triggered messages. Each participant of the system is configured offline with preassigned time slots based on a global time base. This network-access method based on TDMA offers a predictable transmission behavior without queuing in the switches and achieves low latency and low jitter.

While time-triggered periodic messages are always sent with a corresponding period and phase, time-triggered sporadic messages are assigned a periodic time slot but only transmitted if the host actually requests a transmission [41]. These messages can be used to transmit high-priority information about occurring events, but are transmitted only if the event really happens. If sporadic messages are not sent, the unused bandwidth of the communication slot can be used for other message types. Transmission request times for sporadic messages are unknown, but it is known that a minimum time interval exists between successive transmission requests. The knowledge about this minimum time interval can be used to compute the minimum period of the periodic communication slot.

The bandwidth that is either not assigned to time-triggered messages or assigned but not used is free for asynchronous message transmissions. TTEthernet defines two types of asynchronous messages: rate constrained and best effort. Rate-constrained messages are based on the AFDX protocol and intended for the transmission of data with less stringent real-time requirements [42]. Rate-constrained messages support bounded latencies but incur higher jitter compared to time-triggered messages. Best-effort messages are based on standard Ethernet and provide no real-time guarantees.

The different types of messages are associated with priorities in TTEthernet. Time-triggered messages have the highest priority, whereas best-effort messages are assigned the lowest priority. Using these priorities, TTEthernet supports three mechanisms to resolve collisions between the different types of messages [40,43]:

- **Shuffling.** If a low-priority message is being transmitted while a high-priority message arrives, the high-priority message will wait until the low-priority message is finished. That means that the jitter for the high-priority message is increased by the maximum transmission delay of a low-priority message. Shuffling is resource efficient but results in a degradation of the real-time quality.

- **Timely Block.** According to the time-triggered schedule, the switch knows in advance the transmission times of the time-triggered messages. Timely block means that the switch reserves guarding windows before every transmission time of a time-triggered message. This guarding window has a duration that is equal to the maximum transmission time of a lower-priority message. In the guarding window, the switch will not start the transmission of a lower-priority message to ensure that time-triggered messages are not delayed. The jitter for high-priority messages will be close to zero. Timely block ensures high real-time quality with a near constant delay. However, resource inefficiency occurs when the maximum size of low-priority messages is high or unknown [41].

- **Preemption.** If a high-priority message arrives while a low-priority message is being relayed by a switch, the switch stops the transmission of the low-priority message and relays the high-priority message. That means that the switch introduces an almost constant and a priori known latency for high-priority messages. However, the truncation of messages is resource inefficient and results in a low network utilization. Also, corrupt messages result from the truncation, which can be indistinguishable to the consequences of hardware faults. The consequence is a diagnostic deficiency.

The TTEthernet message format is fully compliant to the Ethernet message format. However, the destination address field in TTEthernet is interpreted differently depending on the traffic type. In best-effort traffic, the format for destination addresses as standardized in IEEE 802.3 is used. In time-triggered and rate-constrained traffic, the destination address is subdivided into a constant 32-bit field and a 16-bit field called the *virtual-link identifier*. TTEthernet communication is structured into virtual links, each of which offers a unidirectional connection from one node to one or more destination nodes. The constant field can be defined by the user but should be fixed for all time-triggered and rate-constrained traffic. This constant field is also denoted as the CT marker [42]. The least two significant bits of the first octet of the constant field must be equal to one, since rate-constrained and time-triggered messages are multicast messages.

### 4.3.5 Time-Triggered Controller Area Network

The Time-Triggered CAN (TTCAN) [44] protocol is an example of time-triggered services that are layered on top of an event-triggered protocol. TTCAN is based on the CAN data link layer as specified in ISO 11898 [11].

TTCAN is a master/slave protocol that uses a Time Master (TM) for initiating communication rounds. TTCAN employs multiple time masters for supporting fault tolerance. The time master periodically sends a reference message, which is recognized by clients via its identifier. The period between two successive reference messages is called *basic cycle*. The reference message contains information about the current time of the time master as well as the number of the cycle.

In order to improve the precision of the global time base, slave nodes measure the duration of the basic cycle and compare this measured duration with the values contained in the reference messages. The difference between these measured duration and the nominal duration (as indicated by the time master) is used for drift correction.

The basic cycle can consist of three types of windows, namely, exclusive time windows, free time windows, and arbitrating time windows.

1. **An exclusive time window** is dedicated to a single periodic message. A statically defined node sends the periodic message within an exclusive time window. An offline design tool ensures that no collisions occur.

2. **The arbitrating time windows** are dedicated to spontaneous messages. Multiple nodes can compete for transmitting a message within an arbitrating time window. The bitwise arbitration mechanism of CAN decides which message actually succeeds.

3. **Free time windows** are reserved for further extensions. When required, free time windows can be reconfigured as exclusive or arbitrating time windows.

Retransmissions are prohibited in all three types of windows.

In order to tolerate the failure of a single-channel bus, channel redundancy has been proposed for TTCAN [45]. This solution uses gateway nodes, which are connected to multiple channel busses, in order to synchronize the global time of the different busses and maintain a consistent schedule. However, no analysis of the dependencies between the redundant busses has been formed. For example, due to the absence of bus guardians, a babbling idiot failure [46] of a gateway node has the potential to disrupt communication across multiple redundant busses.

The CAN protocol, which forms the basis for TTCAN, has built-in error detection mechanisms such as CRC codes to detect faults on the communication channel or error counters.

In addition, the a priori knowledge of the time-triggered send and receive instants can be used to establish a membership service. Although such a service is not mandatory in TTCAN, [47] describes the design and implementation of a membership service on top of TTCAN.

## 4.4 Example of a Control Application Based on Time-Triggered and Event-Triggered Control

This section examines a rotary inverted pendulum controlled with several microcontrollers that communicate over a CAN bus. We compare the differences between time-triggered and event-triggered CAN communication with and without load. On every microcontroller, the operating system FreeRTOS with tasks and queues is used for the implementation of the control system.

The structure of the section is at first a short description of the rotary inverted pendulum model followed by a graph that shows the temporal behavior for a simple swing-up algorithm followed by an algorithm to hold the pendulum in an upright position. The following sections describe the printed circuit board (PCB) for control (cf. Control PCB in Figure 4.3) with the connections between the four microcontrollers, the sensors, the human–machine interface, and the motor controls with the help of a block diagram. A short description of the used TTCAN library and its configuration on all microcontrollers is described in the next section followed by the monitored signals on the CAN bus. The behavior of the arm and pendulum angles over time for small disruptions as well as the overall behavior of the control system with TTCAN communication are discussed in the next section.

Figure 4.3 shows the inverted rotary pendulum (IRP) model from a top-left point of view. The microcontroller-based control PCB is located at the bottom left border of the picture on the ground plate of the model foundation. There are two digital rotary encoders for the measurement of the pendulum bar angle (pendulum sensor) and the arm angle (arm sensor) with the rod for

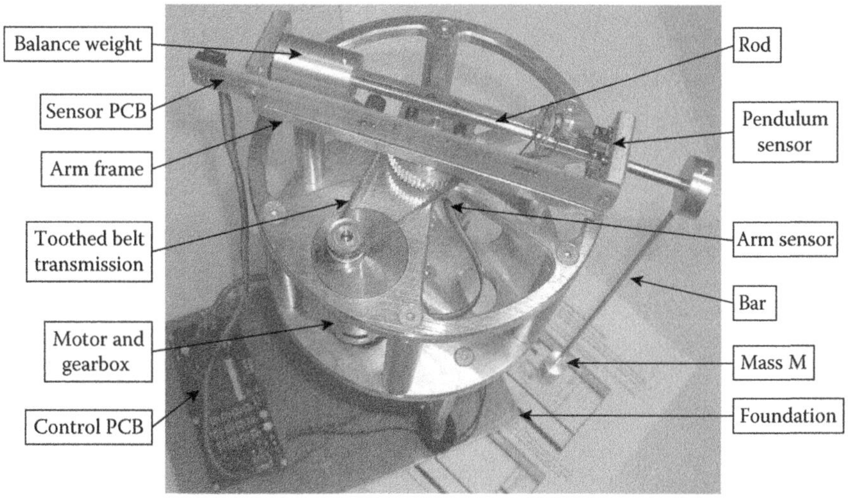


**FIGURE 4.3**
Inverted rotary pendulum.

mounting the pendulum bar with mass M at the end. The rotary encoder (pendulum sensor) for the pendulum bar angle sends its signals with the help of the sensor PCB over a flat ribbon cable to the control PCB. The measurement of the arm angle is also done with a rotary encoder (arm sensor) that is located below the toothed belt (called the toothed belt transmission) in the foundation of the IRP model. This encoder is directly connected with a flat ribbon cable to the control PCB, because its location is fixed in the foundation. The angle of the arm frame is driven by a DC motor over a first built-in gearbox (cf. motor and gearbox in Figure 4.3) in the motor case and an additional toothed belt transmission. The balance weight on the rod of the pendulum axis helps to balance the arm frame.

The graph in Figure 4.4 shows at the horizontal axis (time line) 13 s after power, on the start of the pendulum movement with a simple swing-up algorithm. At the time near 23 s, it holds the pendulum in the upright position with an appropriate algorithm derived from the state-space method of control theory. This is a typical behavior of the IRP model in combination with a well-designed control algorithm without disruptions. The stability of the pendulum angle is remarkable, which is manifested by a short time for swing-up with nearly constant 180 degrees as well as the small arm angle variability (less than $\pm 5$ degrees) for holding the pendulum in the upright position. The control algorithm is based on four states in the system:

- Pendulum angle
- Pendulum angle velocity
- Arm angle
- Arm angle velocity

These states are determined by a feedback system realized in the form of the distributed control algorithm based on the control PCB (see Figure 4.3) with the four microcontrollers and the TTCAN communication. The output signal of this feedback system is sent back to the IRP model for driving via a H-bridge driver and pulse width modulation the DC-motor movement.

Figure 4.5 shows the block diagram of the control PCB. On the left side, there is the first microcontroller where the angle sensors are connected and the pendulum and arm angles are calculated. These values are sent over the CAN bus driver (cf. angle box in the communication plan for TTCAN in Figure 4.6) to all other microcontrollers. Both values are the base for building the differential values over time of the arm and pendulum angles and in a final step the velocity of both angles. These four states (measured and calculated values) are the input signals of the feedback system in the form of the distributed control algorithm described in the last section. Microcontroller 2 transforms the binary values into a human-readable form (ASCII) and shows the measured values on a small liquid-crystal display (LCD). It is responsible for generating the first CAN message (reference message) as a time master. Its crystal-stabilized local time is sent to all other CAN bus participants, which will synchronize their own local time according to this time-master message. Microcontroller 3 reads the push-button matrix and sends its information over the CAN bus for controlling the sequence of the control

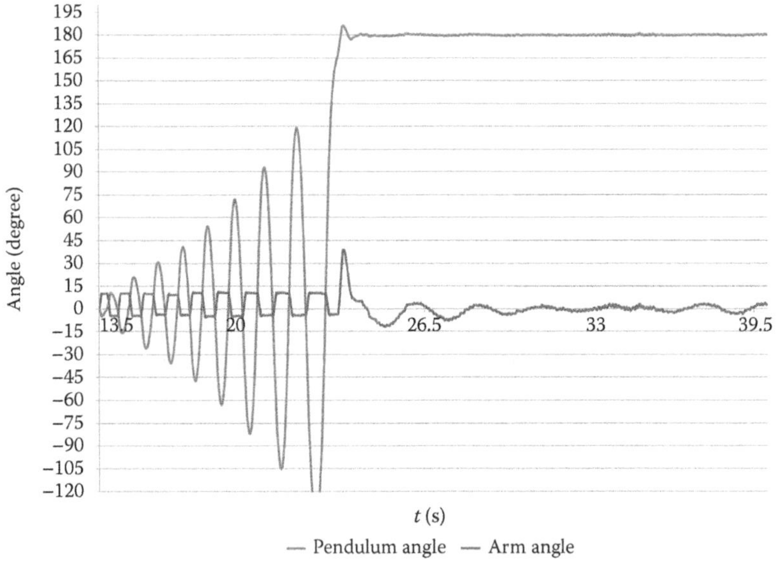


**FIGURE 4.4**
Pendulum swing-up and hold in upright position.

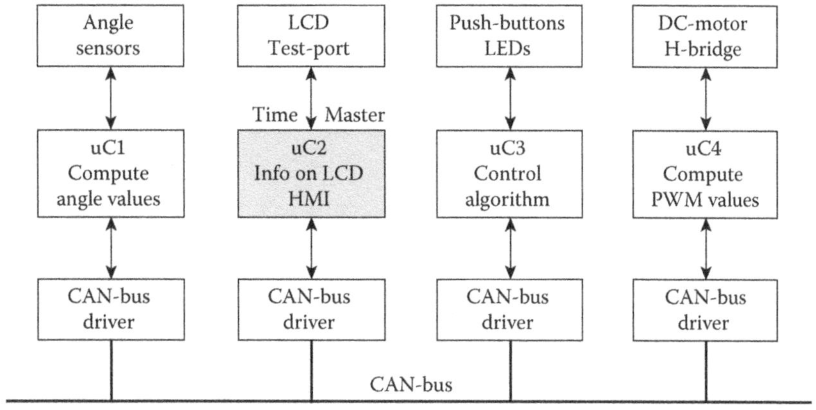


**FIGURE 4.5**
Block diagram of control PCB.

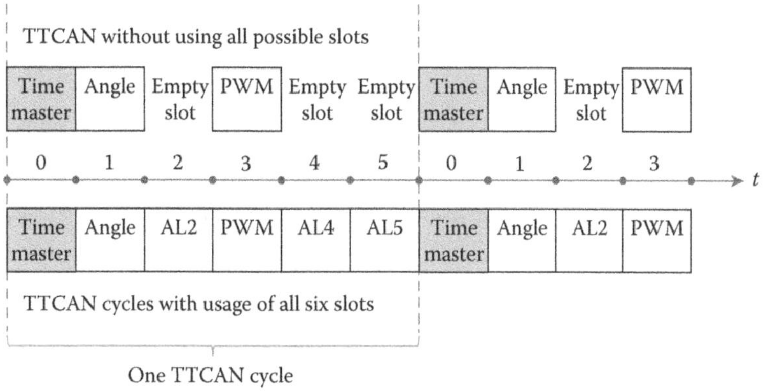


**FIGURE 4.6**
Communication plan for TTCAN.

algorithm (i.e., swing-up part, hold in upright position, and stop), and it realizes the control algorithm of the feedback system for the movement of the DC-motor in the IRP model. This information is also sent via the CAN bus [cf. Pulse-Width Modulation (PWM) box in the communication plan for TTCAN in Figure 4.6] for driving the DC-motor and showing the values on the small LCD. The last microcontroller (uC4) controls the revolution speed of the DC-motor with the help of a pulse width modulation algorithm. The software of all microcontrollers is based on FreeRTOS for controlling the time sequence of different tasks. FreeRTOS is also used to execute the sequence of activities according to the TTCAN communication plan (see Figure 4.6) such as sending CAN messages, receiving values via CAN messages with specific IDs, starting tasks, and synchronizing the local timers.

Figure 4.6 shows the TTCAN communication plan of the algorithm for controlling the IRP model. The upper part shows the first implementation with empty slots of the TTCAN communication cycle with a duration of six slots. The implementation of the feedback component with many matrix operations based on floating point values needs a lot of computation time on the used 8-bit microcontroller architecture (AT90CAN128 from Atmel). The implementation of the TTCAN management functions on every microcontroller (e.g., high-priority TTCAN scheduler task), the RTOS computations for the message queues, and the static arrays for the TTCAN plan require a significant amount of computation time. The TTCAN plan results in a very stable control system, as one can see in Figure 4.4.

The plan shows a TTCAN cycle with six slots indexed from zero to five. In slot number zero, the reference message with the local time of the time master microcontroller (uC2) is sent over the CAN bus with the hexadecimal value 0x000 as the CAN bus ID. All receiving microcontrollers synchronize their own FreeRTOS timers to the time in the reference message. This synchronized time is needed in order to see a constant time offset (see Figure 4.7) between the reference message and all other messages in the TTCAN

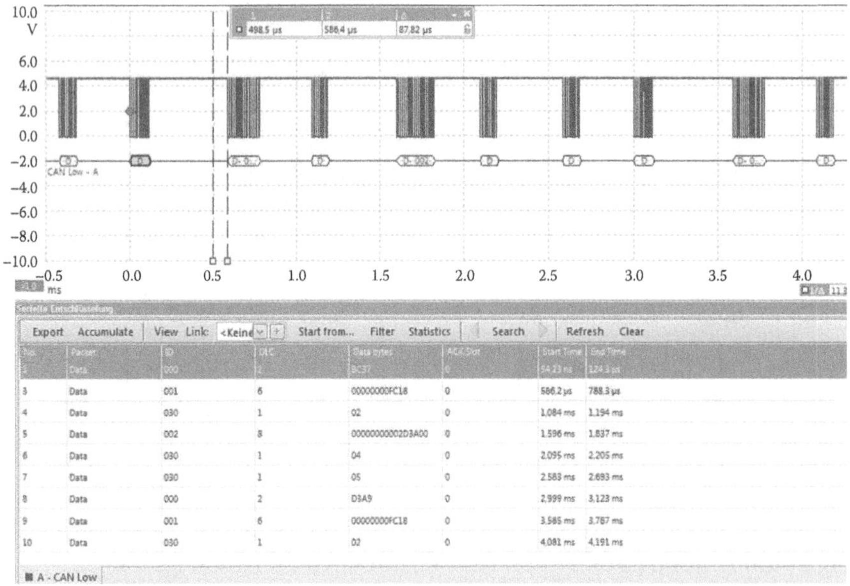


**FIGURE 4.7**
TTCAN signal and serial decoding of CAN messages.

communication. The configuration of FreeRTOS establishes on all microcontrollers a ticker period of half a millisecond. This ticker period determines the invocation of the FreeRTOS scheduler within the timer interrupt service routine. The task TTCAN-Scheduler with the highest priority is started every tick and increments the slot number (i.e., the index of the TTCAN schedule array). Then, the array with the TTCAN schedule will be checked for starting time-triggered activities, like sending the angles via the CAN bus. The lower part of Figure 4.6 shows a TTCAN schedule where all possible slots are used for sending CAN messages. In principle, it is the same plan as in the upper part, but some artificial loads (AL2, AL4, and AL5) are added in the empty slots in order to evaluate the temporal composability.

Figure 4.7 shows an example of a received signal at a microcontroller RXCAN pin of the CAN–bus interface. The marked D box at the time 0.0 ms under the CAN bus signal shows the time master message. All of the following messages have an offset of approximately 90 $\mu$s as indicated by the vertical rulers. The microcontrollers need computation time for synchronizing their local clocks, and therefore, all CAN messages after the time master message have a constant time delay in all slots except slot zero. The signal graph from the RXCAN pin shows that every 0.5 ms a CAN message is placed on the bus. The TTCAN cycle with six messages ends at 3.0 ms with the next time master message. The lower part of Figure 4.7 shows the decoded values of every TTCAN message. The start of the sequence begins with a time master message with CAN-ID 0x000 (see row 2 in the lower part of Figure 4.7). This message is followed by a sensor angle CAN message with ID 0x001, push-button events, and a PWM value for the motor with ID 0x002. All messages with ID 0x030 are artificial load messages for composability evaluations that transport in the data section of the CAN message only the sequence index of the TTCAN plan. In row 8, the next TTCAN cycle starts with a time master message that is easily identifiable in the ID column through the ID value 0x000.

Figure 4.8 shows the alteration by a disruption of both angles in the upright position of the pendulum. The disruption is similar to a finger snip force against the mass M at the upper end of the pendulum. You can see the effect near the time line values of 2.05 and 2.47 s as a local buckling at the axis of the pendulum angle. In contrast to the angle graph over time in Figure 4.4, the value domain of the arm angle is now in the range between +20 and −8 degrees.

The final part of this section is a short comparison to the behavior of the same model (IRP) with event-triggered CAN communication. The measured angles for the pendulum axis and the arm axis are sent in the same period (every 3 ms) as in the time-triggered CAN communication. The time point of all following CAN messages depends on this start-up message except for the artificial loads that follow later.

The graph in Figure 4.9 shows at the horizontal axis 13 s after power-on, the start of pendulum movement with a simple swing-up algorithm. To guarantee now a soft change from swing-up movement to the hold in upright position, the motor is driven with less

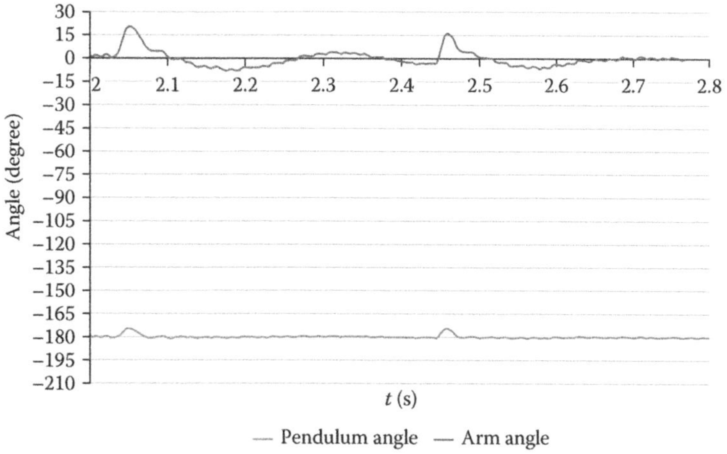


**FIGURE 4.8**
Angle behavior on a minor disruption.

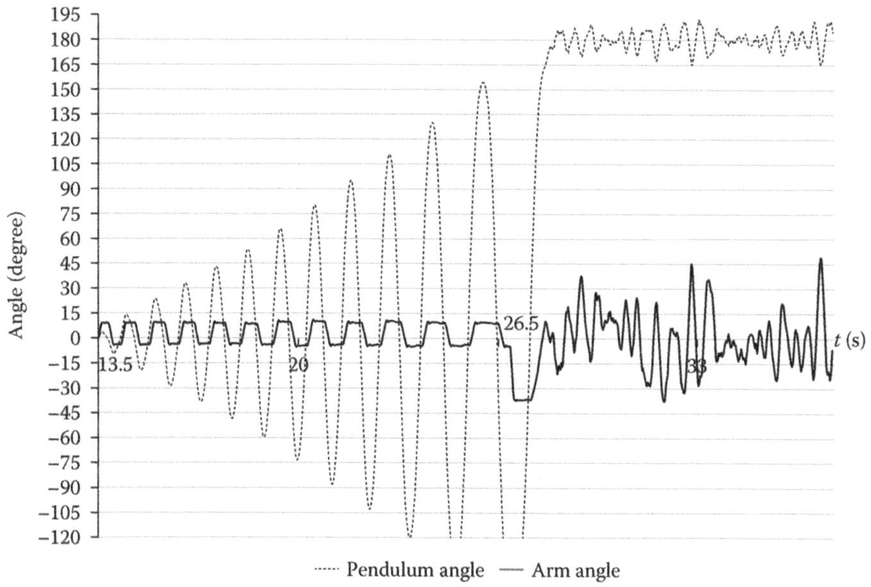


**FIGURE 4.9**
Swing-up and hold with event-triggered CAN communication.

power during the swing-up phase. In comparison with Figure 4.4 (10 s swing-up duration), it now needs a longer duration of approximately 12 s. The next interesting thing to see is the value domain of the arm angle. With the use of TTCAN communication, it is less than ±5 degrees, and now with the asynchronous CAN communication, one sees peaks up to 45 degrees as well as the pendulum angle within ±15 degrees. In general, the model behavior is near the limit of stability, and small disruptions produce angle values greater than the defined boundaries. The result is a system shut down.

Figure 4.10 shows a received signal at a microcontroller RXCAN pin of the CAN bus interface (the dashed signal CAN Low, A) with event-triggered CAN communication. The marked D box at the time 0.0 ms under the blue CAN bus signal shows the angle message (ID 0x11) from the sensor microcontroller. At the right part at the first vertical ruler is a CAN message placed with the ID 0x22, which includes the most recent output from the control algorithm. The level change at channel B (solid signal curve in Figure 4.10) at the second vertical ruler marks the time amount for receiving the CAN message with the event-triggered CAN library. The difference between first and second ruler is around 0.6 ms. Additional information is given through no level change after the second CAN message that contains also

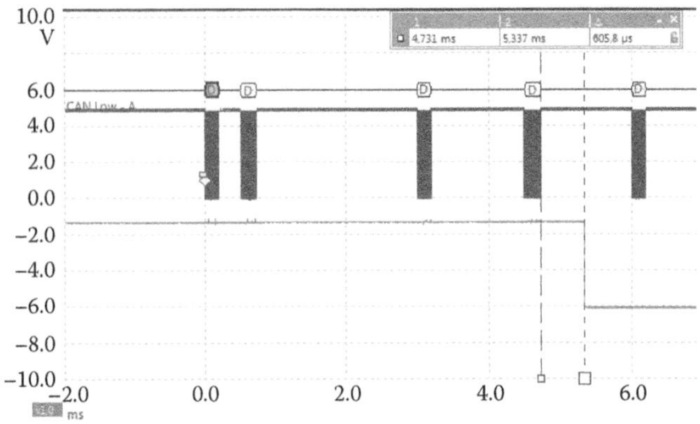


**FIGURE 4.10**
Event-triggered CAN signal without artificial loads.

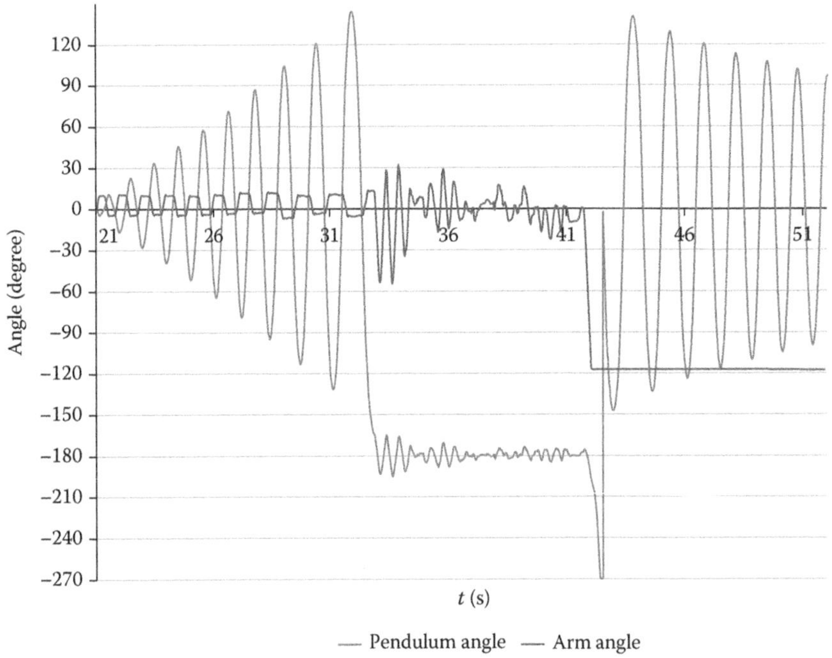


**FIGURE 4.11**
Reaching the arm boundary with a small disruption.

output for controlling the motor. This is a typical miss of information in event-triggered communication. The code for generating the signal of channel B is based on the comparison of the ID from the last received CAN message against the hexadecimal value 0x22 (defined ID for motor control messages). If this is the case, pin 6 of the 8-bit test port F is XORed with a logical 1 (binary 01000000). This leads to the level change that is shown in Figure 4.10 below the CAN signal.

The last section shows the behavior of the model and the signals on the CAN bus with artificial load in the case of event-triggered CAN communication. In Figure 4.11, we can see the swing-up of the pendulum followed by a barely stable behavior of the IRP model. Near 42 s on the time line, there is an event (same finger snip disruption as in Figure 4.8). The control algorithm reacts so hard that the angle boundaries of the arm angle are out of its limits. The boundaries of the arm angle are calibrated to $\pm 108$ degrees for mechanical reasons. When the arm angle exceeds these values, the IRP model is stopped for safety reasons in order not to harm the cables of the pendulum angle sensor. The rest of the graph shows the slow swing-down of the pendulum.

Figure 4.12 depicts the event-triggered communication on the CAN bus. Here, we can see the artificial load messages and the IRP model data messages in combination on the bus. Artificial loads can be identified by IDs greater than the value 0x100, for example, ID 0x401

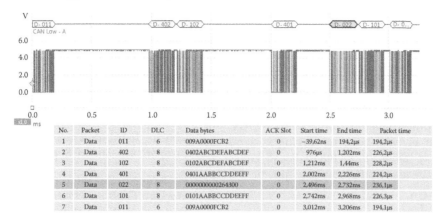

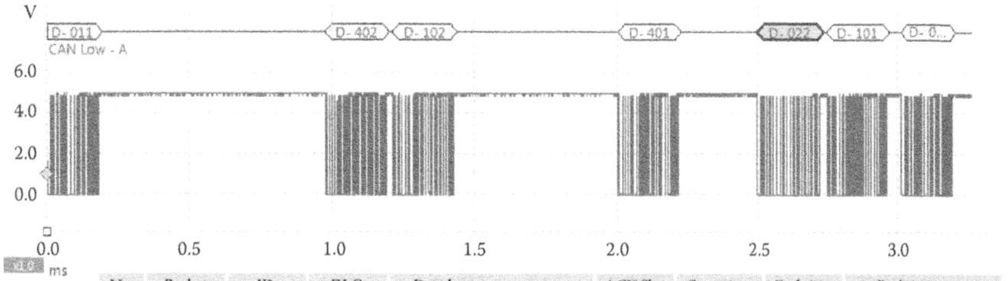


| No. | Packet | ID | DLC | Data bytes | ACK Slot | Start time | End time | Packet time |
|---|---|---|---|---|---|---|---|---|
| 1 | Data | 011 | 6 | 009A0000FCB2 | 0 | −39,62ns | 194,2μs | 194,2μs |
| 2 | Data | 402 | 8 | 0402ABCDEFABCDEF | 0 | 976μs | 1,202ms | 226,2μs |
| 3 | Data | 102 | 8 | 0102ABCDEFABCDEF | 0 | 1,212ms | 1,44ms | 228,2μs |
| 4 | Data | 401 | 8 | 0401AABBCCDDEEFF | 0 | 2,002ms | 2,226ms | 224,2μs |
| 5 | Data | 022 | 8 | 0000000000264300 | 0 | 2,496ms | 2,732ms | 236,1μs |
| 6 | Data | 101 | 8 | 0101AABBCCDDEEFF | 0 | 2,742ms | 2,968ms | 226,3μs |
| 7 | Data | 011 | 6 | 009A0000FCB2 | 0 | 3,012ms | 3,206ms | 194,1μs |

**FIGURE 4.12**
Event-triggered CAN signal with artificial loads and serial decoding.

before the marked box with the grey background in the table is below the measured CAN signal. The model data such as the measured angles (cf. Nos. 1 and 7 with ID 0x011) and the response of the control algorithm (cf. No. 5 with ID 0x022) are sent every 3 ms in case the CAN bus is idle at the moment.

In summary, the example system includes a complex control algorithm for controlling an unstable system (IRP) with multiple microcontrollers, and there are no problems with the control function when we use TTCAN communication between several microcontrollers with and without additional loads. The same model realized with event-triggered CAN communication is less stable and leads to an unstable system behavior with a comparable load, as used in the TTCAN communication.

## Bibliography


[1] A. Crespo, I. Ripoll, M. Masmano, Partitioned embedded architecture based on hypervisor: The xtratum approach, in *EDCC*, IEEE Computer Society, 2010, pp. 67–72.

[2] Lynuxworks, LynxOS User's Guide, Release 4.0, DOC-0453-02, 2005.

[3] P. Parkinson, L. Kinnan, Safety-Critical Software Development for Integrated Modular Avionics, Whitepaper, Wind River Systems, 2007.

[4] Aeronautical Radio, Inc., *ARINC Specification 653-2: Avionics Application Software Standard Interface.* Aeronautical Radio, Inc., Annapolis, MD, 2007.

[5] H. Kopetz, *Real-time systems—Design Principles for Distributed Embedded Applications* (2nd ed.). Springer, New York, 2011.

[6] H. Kopetz, Elementary versus composite interfaces in distributed real-time systems, in *Proc. of the International Symposium on Autonomous Decentralized Systems* (Tokyo, Japan), March 1999.

[7] *ARTEMIS Final Report on Reference Designs and Architectures—Constraints and Requirements.* ARTEMIS (Advanced Research and Technology for EMbedded Intelligence and Systems) Strategic Research Agenda, 2006. https://artemis-ia.eu/publication/download/639-sra-reference-designs-and-architectures-2.

[8] J. Sifakis, A framework for component-based construction, in *Proc. of Third IEEE International Conf. on Software Engineering and Formal Methods (SEFM05)*, September 2005, pp. 293–300.

[9] H. Kopetz, R. Obermaisser, Temporal composability, *Comput. Control Eng. J.*, vol. 13, pp. 156–162, August 2002.

[10] IEEE, IEEE standard 802.3—carrier sense multiple access with collision detect (CSMA/CD) access method and physical layer, tech. rep., IEEE, 2000.

[11] International Standardization Organisation (ISO), *Road vehicles—Interchange of Digital Information—Controller Area Network (CAN) for High-Speed Communication, ISO 11898*, 1993.

[12] G. Heiner, T. Thurner, Time-triggered architecture for safety-related distributed real-time systems in transportation systems, in *Proceedings of the Twenty-Eighth Annual International Symposium on Fault-Tolerant Computing*, June 1998, pp. 402–407.

[13] FlexRay Consortium. BMW AG, DaimlerChrysler AG, General Motors Corporation, Freescale GmbH, Philips GmbH, Robert Bosch GmbH, Volkswagen AG., *FlexRay Communications System Protocol Specification Version 2.1*, May 2005.

[14] P. Pedreiras, L. Almeida, Combining event-triggered and time-triggered traffic in FTT-CAN: Analysis of the asynchronous messaging system, in *Proceedings of Third IEEE International Workshop on Factory Communication Systems*, pp. 67–75, September 2000.

[15] R. Obermaisser, CAN Emulation in a time-triggered environment, in *Proceedings of the 2002 IEEE International Symposium on Industrial Electronics (ISIE)*, vol. 1, 2002, pp. 270–275.

[16] H. Kopetz, A. Ademaj, P. Grillinger, K. Steinhammer, The time-triggered ethernet (TTE) design, in *Proceedings of Eighth IEEE International Symposium on Object-Oriented Real-Time Distributed Computing (ISORC)*, pp. 22–33, May 2005.

[17] R. Benesch, TCP für die time-triggered architecture, Master's thesis, Technische Universität Wien, Institut für Technische Informatik, Vienna, Austria, June 2004.

[18] J. Saltzer, D. Reed, D. Clark, End-to-end arguments in system design, *ACM Trans. Comput. Syst. (TOCS)*, vol. 2, 1984, pp. 277–288.

[19] A. Avizienis, J. Laprie, B. Randell, Fundamental concepts of dependability, Research Report 01-145, LAAS-CNRS, Toulouse, France, April 2001.

[20] J. Lala, R. Harper, Architectural principles for safety-critical real-time applications, in *Proceedings of the IEEE*, vol. 82, pp. 25–40, January 1994.

[21] H. Kopetz, Fault containment and error detection in the time-triggered architecture, in *Proceedings of the Sixth International Symposium on Autonomous Decentralized Systems*, pp. 139–146, April 2003.

[22] P. Lee, T. Anderson, *Fault Tolerance Principles and Practice*, vol. 3 of *Dependable Computing and Fault-Tolerant Systems*. Springer-Verlag, New York, 1990.

[23] M. Hecht, D. Tang, H. Hecht, Quantitative reliability and availability assessment for critical systems including software, in *Proceedings of the 12th Annual Conf. on Computer Assurance* (Gaitherburg, Maryland), pp. 147–158, June 1997.

[24] R. Obermaisser, H. Kopetz, C. Paukovits, A cross-domain multi-processor system-on-a-chip for embedded real-time systems, *IEEE Trans. Ind. Inf.*, vol. 6, no. 4, pp. 548–567, 2010.

[25] OMG, Smart Transducers Interface, Specification ptc/2002-05-01, Object Management Group, May 2002. Available at http://www.omg.org/.

[26] H. Kopetz, W. Ochsenreiter, Clock synchronization in distributed real-time systems, *IEEE Trans. Comput.*, vol. 36, no. 8, pp. 933–940, 1987.

[27] H. Kopetz, Pulsed data streams, in *IFIP TC 10 Working Conf. on Distributed and Parallel Embedded Systems (DIPES 2006)*, (Braga, Portugal), Springer, New York, October 2006, pp. 105–124.

[28] S. Poledna, Replica determinism in distributed real-time systems: A brief survey, *Real-Time Systems*, vol. 6, 1994, pp. 289–316.

[29] K. Goossens, A. Azevedo, K. Chandrasekar, M. D. Gomony, S. Goossens, M. Koedam, Y. Li, D. Mirzoyan, A. Molnos, A. B. Nejad, A. Nelson, S. Sinha, Virtual execution platforms for mixed-time-criticality systems: The compsoc architecture and design flow, *SIGBED Rev.*, vol. 10, October 2013, pp. 23–34.

[30] K. Goossens, A. Hansson, The aethereal network on chip after ten years: Goals, evolution, lessons, and future, in *Proceedings of the 47th Design Automation Conf.*, DAC'10, (New York, NY), ACM, 2010, pp. 306–311.

[31] K. Goossens, J. Dielissen, A. Radulescu, The aethereal network on chip: Concepts, architectures, and implementations, *IEEE Des. Test Comput.*, vol. 22, no. 5, pp. 414–421, 2005.

[32] A. Radulescu, J. Dielissen, K. Goossens, E. Rijpkema, P. Wielage, An efficient on-chip network interface offering guaranteed services, shared-memory abstraction, and flexible network configuration, in *Proceedings of Design, Automation and Test in Europe Conference and Exhibition*, vol. 2, pp. 878–883, 2004.

[33] ARM, AXI protocol specification, 2004.

[34] O.-I. Association, Open core protocol specification 2.1, 2005.

[35] H. Kopetz, Time-triggered real-time computing, *Ann. Rev. Control*, vol. 27, no. 1, pp. 3–13, 2003.

[36] L. Lamport, R. Shostak, M. Pease, The byzantine generals problem, *ACM Trans. Program. Lang. Syst. (TOPLAS)*, vol. 4, no. 3, pp. 382–401, 1982.

[37] H. Kopetz, G. Bauer, S. Poledna, Tolerating arbitrary node failures in the time-triggered architecture, in *Proceedings of the SAE 2001 World Congress* (Detroit, MI), March 2001.

[38] R. Obermaisser, P. Peti, A fault hypothesis for integrated architectures, in *Proceedings of the Fourth International Workshop on Intelligent Solutions in Embedded Systems*, June 2006.

[39] A. Ademaj, Slightly-off-specification failures in the time-triggered architecture, in *Proceedings of the Seventh IEEE International High-Level Design Validation and Test Workshop* (Washington, DC), IEEE Computer Society, 2002, p. 7.

[40] R. Obermaisser, *Time-Triggered Communication*. Taylor and Francis, Boca Raton, FL 2011.

[41] K. Steinhammer, *Design of an FPGA-Based Time-Triggered Ethernet System*. PhD thesis, Technische Universität Wien, Austria, 2006.

[42] AS-6802—Time-Triggered Ethernet, November 2011.

[43] W. Steiner, G. Bauer, B. Hall, M. Paulitsch, S. Varadarajan, TTEthernet Dataflow Concept, in *Proceedings of the 8th IEEE International Symposium on Network Computing and Applications*, pp. 319–322, 2009.

[44] T. Führer, B. Müller, W. Dieterle, F. Hartwich, R. Hugel, Time-triggered CAN - TTCAN: Time-triggered communication on CAN, in *Proceedings of the Sixth International CAN Conf. (ICC6)*, (Torino, Italy), 2000.

[45] B. Müller, T. Führer, F. Hartwich, R. Hugel, H. Weiler, Fault tolerant TTCAN networks, in *Proceedings of the Eighth International CAN Conf. (iCC)*, 2002.

[46] C. Temple, Avoiding the babbling-idiot failure in a time-triggered communication system, in *Proceedings of the Symposium on Fault-Tolerant Computing*, 1998, pp. 218–227.

[47] C. Bergenhem, J. Karlsson, C. Archer, A. Sjöblom, Implementation results of a configurable membership protocol for active safety systems, in *Proceedings of the 12th Pacific Rim International Symposium on Dependable Computing (PRDC'06)*, pp. 387–388.

# 5

# *Distributed Event-Based State-Feedback Control*

**Jan Lunze**
*Ruhr-University Bochum*
*Bochum, Germany*

**Christian Stöcker**
*Ruhr-University Bochum*
*Bochum, Germany*

**CONTENTS**



**ABSTRACT** This chapter presents an approach to distributed control that combines continuous and event-based state feedback and aims at the suppression of the disturbance propagation through interconnected systems. First, a new method for the design of distributed state-feedback controllers is proposed, and second, the implementation of the distributed state-feedback law in an event-based manner is presented. At the event times, the event-based controllers transmit information to or request information from the neighboring subsystems, which is a new communication pattern. The distributed event-based state-feedback approach is tested on a thermofluid process through an experiment.

## 5.1 Introduction

In modern control systems, the feedback is increasingly closed through a digital communication network. Such systems, where the information exchange between sensors, controllers, and actuators occurs via a shared communication medium, are referred to as networked control systems [10,11,24]. The communication network allows a flexible distribution of information and, thus, opens the possibility of new control structures, which can hardly be realized in a conventional control system with fixed wiring. On the other hand, real networks have limited bandwidth, and in order to avoid network

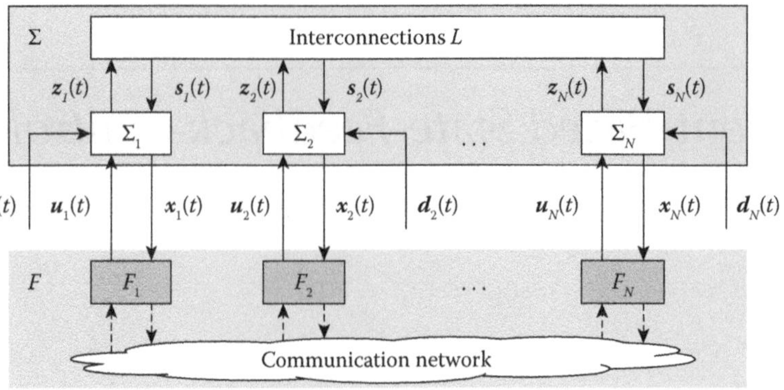


**FIGURE 5.1**
Event-based control system: The plant $\Sigma$ is composed of the interconnected subsystems $\Sigma_i$ $(i = 1, \ldots, N)$ that are controlled by the event-based control units $F_i$ $(i = 1, \ldots, N)$, which communicate over a network at the event times.

congestion, the communication should be kept to the minimum required to solve the respective control task, which has motivated the investigation of event-based control methods.

This section presents a new approach to the distributed event-based control of interconnected linear systems. The investigated control system is illustrated in Figure 5.1. The overall plant is composed of $N$ subsystems $\Sigma_i$ $(i = 1, \ldots, N)$, which are physically interconnected, and the subsystem $\Sigma_i$ is controlled by local control unit $F_i$. The overall event-based controller consists of the $N$ control units $F_i$ $(i = 1, \ldots, N)$ and the communication network. The main control goal is disturbance attenuation and the suppression of the propagation of the disturbance throughout the network.

Distributed event-based control has been previously investigated in [3,25,27] for nonlinear systems and in [2,7] for linear systems. Extensions of these approaches to imperfect communication between the control units considering delays and packet losses has been studied in [8] for linear systems and in [26] for nonlinear systems. In all these references, the stability of the overall system is ensured using small-gain arguments that claim the subsystems to be weakly connected in some sense.

In contrast to the existing literature, the proposed control approach combines continuous and event-based feedback. It is assumed that the control unit $F_i$ continuously measured the local state information $\boldsymbol{x}_i(t)$ but has access to the state information $\boldsymbol{x}_j(t)$ of some subsystem $\Sigma_j$ $(j \neq i)$ through the network at discrete time instants only. The novelty of this control approach is the use of two kinds of events that trigger either the transmission or the request of information, which leads to a new type of event-based control. Moreover, it is shown that the event-based control approach tolerates arbitrarily strong interconnections between neighboring subsystems and still guarantees the stability of the overall control system. In this way, the existing literature on distributed event-based control is extended.

## 5.2 Preliminaries

### 5.2.1 Notation

$\mathbb{R}$ and $\mathbb{R}_+$ denote the set of real numbers or the set of positive real numbers, respectively. $\mathbb{N}$ is the set of natural numbers, and $\mathbb{N}_0 = \mathbb{N} \cup \{0\}$. Throughout this chapter, scalars are denoted by italic letters $(s \in \mathbb{R})$, vectors by bold italic letters $(\boldsymbol{x} \in \mathbb{R}^n)$, and matrices by uppercase bold italic letters $(\boldsymbol{A} \in \mathbb{R}^{n \times n})$.

If $\boldsymbol{x}$ is a signal, the value of $\boldsymbol{x}$ at time $t \in \mathbb{R}_+$ is represented by $\boldsymbol{x}(t)$. Moreover,

$$\boldsymbol{x}(t^+) = \lim_{s \downarrow t} \boldsymbol{x}(s),$$

represents the limit of $\boldsymbol{x}(t)$ taken from above.

The transpose of a vector $\boldsymbol{x}$ or a matrix $\boldsymbol{A}$ is denoted by $\boldsymbol{x}^\top$ or $\boldsymbol{A}^\top$, respectively. $\boldsymbol{I}_n$ denotes the identity matrix of size $n$. $\boldsymbol{O}_{n \times m}$ is the zero matrix with $n$ rows and $m$ columns, and $\boldsymbol{0}_n$ represents the zero vector of dimension $n$. The dimensions of these matrices and vectors are omitted if they are clear from the context.

Consider the matrices $\boldsymbol{A}_1, \ldots, \boldsymbol{A}_N$. The notation $\boldsymbol{A} = \text{diag}(\boldsymbol{A}_1, \ldots, \boldsymbol{A}_N)$ is used to denote a block diagonal matrix:

$$\boldsymbol{A} = \text{diag}(\boldsymbol{A}_1, \ldots, \boldsymbol{A}_N) = \begin{pmatrix} \boldsymbol{A}_1 & & \\ & \ddots & \\ & & \boldsymbol{A}_N \end{pmatrix}.$$

The $i$th eigenvalue of a square matrix $\boldsymbol{A} \in \mathbb{R}^{n \times n}$ is denoted by $\lambda_i(\boldsymbol{A})$. The matrix $\boldsymbol{A}$ is called Hurwitz (or stable) if $\text{Re}(\lambda_i(A)) < 0$ holds for all $i = 1, \ldots, n$, where $\text{Re}(\cdot)$ denotes the real part of the indicated number.

Consider a time-dependent matrix $G(t)$ and vector $u(t)$. The asterisk $(*)$ is used to denote the convolution operator, for example,

$$G * u = \int_0^t G(t-\tau)u(\tau)d\tau.$$

The inverse of a square matrix $H \in \mathbb{R}^{n\times n}$ is symbolized by $H^{-1}$. For a nonsquare matrix $H^{n\times m}$ that has full rank, $H^+$ is the pseudoinverse that is defined by

$$H^+ = \begin{cases} \left(H^\top H\right)^{-1} H^\top, & \text{for } n > m \\ H^\top \left(H^\top H\right)^{-1}, & \text{for } m > n \end{cases}.$$

For $n > m$, $H^+$ denotes the left inverse $(H^+H = I_m)$, and for $m > n$, $H^+$ is the right inverse $(HH^+ = I_n)$ [1].

For two vectors $v, w \in \mathbb{R}^n$ the relation $v > w$ $(v \geq w)$ holds element-wise; that is, $v_i > w_i$ $(v_i \geq w_i)$ is true for all $i = 1, \ldots, n$, where $v_i$ and $w_i$ refer to the $i$th element of the vectors $v$ or $w$, respectively. Accordingly, for two matrices $V, W \in \mathbb{R}^{n\times m}$ where $V = (v_{ij})$ and $W = (w_{ij})$ are composed of the elements $v_{ij}$ and $w_{ij}$ for $i = 1, \ldots, n$ and $j = 1, \ldots m$, the relation $V > W$ $(V \geq W)$ refers to $v_{ij} > w_{ij}$ $(v_{ij} \geq w_{ij})$. For a scalar $s$, $|s|$ denotes the absolute value. For a vector $x \in \mathbb{R}^n$ or a matrix $A = (a_{ij}) \in \mathbb{R}^{n\times m}$, the $|\cdot|$-operator holds element-wise, that is,

$$|x| = \begin{pmatrix} |x_1| \\ \vdots \\ |x_n| \end{pmatrix}, \qquad |A| = \begin{pmatrix} |a_{11}| & \ldots & |a_{1m}| \\ \vdots & \ddots & \vdots \\ |a_{n1}| & \ldots & |a_{nm}| \end{pmatrix},$$

where $\|x\|$ and $\|A\|$ denote an arbitrary vector norm and the induced matrix norm according to

$$\|x\|_p := \left(\sum_{i=1}^{n} |x_i|^p\right)^{\frac{1}{p}}, \qquad \|A\|_p := \max_{x\neq 0} \frac{\|Ax\|_p}{\|x\|_p},$$

with the real number $p \geq 1$. Sets are denoted by calligraphic letters $(\mathcal{A} \subset \mathbb{R}^n)$. For the compact set $\mathcal{A} \subset \mathbb{R}^n$,

$$\|x\|_{\mathcal{A}} := \inf\left\{\|x - z\| \;\middle|\; z \in \mathcal{A}\right\},$$

denotes the point-to-set distance from $x \in \mathbb{R}^n$ to $\mathcal{A}$ [18].

### 5.2.2 Nonnegative Matrices and M-Matrices

**DEFINITION 5.1** A matrix $A = (a_{ij})$ is called nonnegative $(A \geq O)$ if all elements of $A$ are real and nonnegative $(a_{ij} \geq 0)$.

**DEFINITION 5.2** A matrix $A \in \mathbb{R}^{n\times n}$ is called reducible if there exists a permutation matrix $P$ that transforms $A$ to a block upper triangular matrix:

$$PAP^\top = \begin{pmatrix} \tilde{A}_{11} & O \\ \tilde{A}_{21} & \tilde{A}_{22} \end{pmatrix},$$

where $\tilde{A}_{11}$ and $\tilde{A}_{22}$ are square matrices. Otherwise, $A$ is called irreducible.

**Theorem 5.1: Theorem A1.1 in [15], Perron–Frobenius theorem**

Every irreducible nonnegative matrix $A \in \mathbb{R}_+^{n\times n}$ has a positive eigenvalue $\lambda_P(A)$ that is not exceeded by any other eigenvalue $\lambda_i(A)$

$$\lambda_P(A) \geq |\lambda_i(A)|.$$

$\lambda_P(A)$ is called the *Perron root* of $A$.

**DEFINITION 5.3** A matrix $P = (p_{ij}), P \in \mathbb{R}^{n\times n}$ is said to be an M-matrix if $p_{ij} \leq 0$ holds for all $i \neq j$ and all eigenvalues of $P$ have positive real part.

**Theorem 5.2: Theorem A1.3 in [15]**

A matrix $P = (p_{ij}), P \in \mathbb{R}^{n\times n}$ with $p_{ij} \leq 0$ for all $i \neq j$ is an M-matrix if and only if $P$ is nonsingular and $P^{-1}$ is nonnegative.

The next theorem makes a relationship between nonnegative matrices and M-matrices.

**Theorem 5.3: Theorem A1.5 in [15]**

If $A \in \mathbb{R}_+^{n\times n}$ is a nonnegative matrix, then the matrix $P = \mu I_n - A$ with $\mu \in \mathbb{R}_+$ is an M-matrix if and only if the relation

$$\mu > \lambda_P(A),$$

holds.

### 5.2.3 Models

**Plant.** The overall plant is represented by the linear state-space model:

$$\Sigma: \quad \dot{x}(t) = Ax(t) + Bu(t) + Ed(t), \quad x(0) = x_0, \tag{5.1}$$

where $x \in \mathbb{R}^n$, $u \in \mathbb{R}^m$, and $d \in \mathcal{D}$ denote the state, the control input, and the disturbance, respectively.

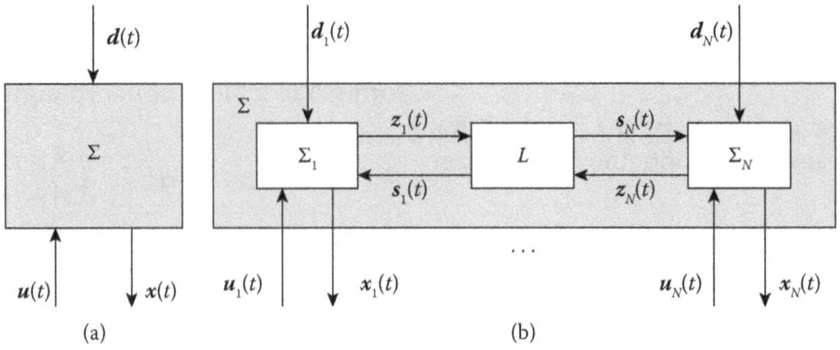


**FIGURE 5.2**
Structure of the overall system with (a) a central sensor unit and an actuator unit and (b) a sensor unit and an actuator unit for each subsystem.

Concerning the properties of the system (5.1), the following assumptions are made, unless otherwise stated:

**A** 5.1 The plant dynamics are accurately known.

**A** 5.2 The state $x(t)$ is measurable.

**A** 5.3 The disturbance $d(t)$ is bounded according to

$$d(t) \in \mathcal{D} := \left\{ d \in \mathbb{R}^p \,|\, |d| \leq \bar{d} \right\}, \quad \forall\, t \geq 0, \quad (5.2)$$

where $\bar{d} \in \mathbb{R}^p_+$ denotes the bound on the magnitude of the disturbance $d$.

Equation 5.1 is referred to as the *unstructured model* [15], because it barely gives an insight into the interconnections and the dynamics of the subsystems that form the overall plant. The model (5.1) is useful if the overall system has a central sensor unit that collects all measurements and a centralized actuator unit that drives all actuators of the plant (Figure 5.2a). The next section introduces a model that reflects the internal structure of the overall plant in more detail.

**Interconnected subsystems.** In this chapter, the overall plant is considered to be composed of $N$ physically interconnected subsystems. The subsystem $i \in \mathcal{N} = \{1, \ldots, N\}$ is represented by the linear state-space model:

$$\Sigma_i : \begin{cases} \dot{x}_i(t) = A_i x_i(t) + B_i u_i(t) + E_i d_i(t) + E_{si} s_i(t), \\ \qquad x_i(0) = x_{0i} \\ z_i(t) = C_{zi} x_i(t) \end{cases}, \quad (5.3)$$

where $x_i \in \mathbb{R}^{n_i}$, $u_i \in \mathbb{R}^{m_i}$, $d_i \in \mathcal{D}_i$, $s_i \in \mathbb{R}^{q_i}$, and $z_i \in \mathbb{R}^{r_i}$ denote the state, the control input, the disturbance, the coupling input, and the coupling output, respectively. The subsystem $\Sigma_i$ is interconnected with the remaining subsystems according to the relation

$$s_i(t) = \sum_{j=1}^{N} L_{ij} z_j(t), \quad (5.4)$$

where the matrix $L_{ij} \in \mathbb{R}^{q_i \times r_j}$ represents the couplings from some subsystem $\Sigma_j$ to subsystem $\Sigma_i$. The models (5.3) and (5.4) is called the *interconnection-oriented model* [15]. Concerning this model, the following assumptions are made:

**A** 5.4 The pair $(A_i, B_i)$ is controllable for each $i \in \mathcal{N}$.

**A** 5.5 $L_{ii} = O$ holds for all $i \in \mathcal{N}$, that is, the coupling input $s_i(t)$ does not directly depend on the coupling output $z_i(t)$.

The last assumption is weak and can always be fulfilled by modeling all internal dynamics in the matrix $A_i$ for all $i \in \mathcal{N}$. Note that the assumptions **A** 5.1 to **A** 5.3 imply that

- The dynamics of the subsystems are accurately known.
- The subsystem state $x_i(t)$ is measurable for each $i \in \mathcal{N}$.
- The local disturbance $d_i(t)$ is bounded according to

$$d_i(t) \in \mathcal{D}_i := \left\{ d_i \in \mathbb{R}^{p_i} \,|\, |d_i| \leq \bar{d}_i \right\}, \quad \forall\, t \geq 0, \quad (5.5)$$

  with $\bar{d}_i \in \mathbb{R}^{p_i}_+$ representing the bound on the local disturbance $d_i(t)$.

The unstructured model (5.1) can be determined from the interconnection-oriented models (5.3) and (5.4) by taking

$$A = \text{diag}\,(A_1, \ldots, A_N) + \text{diag}\,(E_{s1}, \ldots, E_{sN})\; L\, \text{diag}\,(C_{z1}, \ldots, C_{zN}), \quad (5.6)$$

$$B = \text{diag}\,(B_1, \ldots, B_N), \quad (5.7)$$

$$E = \text{diag}\,(E_1, \ldots, E_N), \quad (5.8)$$

with the overall interconnection matrix

$$L = \begin{pmatrix} O & L_{12} & \cdots & L_{1N} \\ L_{21} & O & \cdots & L_{2N} \\ \vdots & \vdots & \ddots & \vdots \\ L_{N1} & L_{N2} & \cdots & O \end{pmatrix}, \quad (5.9)$$

and by assembling the signal vectors according to

$$x(t) = \left(x_1^\top(t) \;\; \ldots \;\; x_N^\top(t)\right)^\top,$$
$$u(t) = \left(u_1^\top(t) \;\; \ldots \;\; u_N^\top(t)\right)^\top,$$
$$d(t) = \left(d_1^\top(t) \;\; \ldots \;\; d_N^\top(t)\right)^\top.$$

**Communication network.** Throughout this chapter, the communication network is assumed to be ideal in the following sense:

**A** 5.6 The transmission of information over the communication network happens instantaneously, without delays and packet losses. Multiple transmitters can simultaneously send information without collisions occurring.

The communication network is considered to transmit information much faster compared to the dynamical behavior of the control system, which from a control theoretic perspective justifies the assumption **A** 5.6.

### 5.2.4 Comparison Systems

The stability analysis in this section makes use of comparison systems that yield upper bounds on the signals of the respective subsystems.

Consider the isolated subsystem (5.3) (with $s_i(t) \equiv 0$):

$$\dot{x}_i(t) = A_i x_i(t) + B_i u_i(t) + E_i d_i(t), \quad x_i(0) = x_{0i}, \quad (5.10)$$

that has the state trajectory

$$x_i(t) = e^{A_i t} x_{0i} + G_{xui} * u_i + G_{xdi} * d_i, \quad \forall\, t \geq 0,$$

where

$$G_{xui}(t) = e^{A_i t} B_i, \quad (5.11)$$

$$G_{xdi}(t) = e^{A_i t} E_i, \quad (5.12)$$

denote the impulse-response matrices describing the impact of the control input $u_i(t)$ or the disturbance $d_i(t)$, respectively, on the state $x_i(t)$.

**DEFINITION 5.4** The system

$$r_{xi}(t) = \bar{F}_i(t)\,|x_{0i}| + \bar{G}_{xui} * |u_i| + \bar{G}_{xdi} * |d_i|, \quad (5.13)$$

with $r_{xi} \in \mathbb{R}^{n_i}$ is called a *comparison system* of subsystem (5.10) if it satisfies the inequality

$$r_{xi}(t) \geq |x_i(t)|, \quad \forall\, t \geq 0,$$

for arbitrary but bounded inputs $u_i(t)$ and $d_i(t)$.

A method for finding a comparison system is given in the following lemma.

**Lemma 5.1: Lemma 8.3 in [15]**

The system (5.13) is a *comparison system* of the system (5.10) if and only if the matrix $\bar{F}_i(t)$ and the impulse-response matrices $\bar{G}_{xui}(t)$ and $\bar{G}_{xdi}(t)$ satisfy the relations

$$\bar{F}_i(t) \geq \left| e^{A_i t} \right|, \quad \bar{G}_{xui}(t) \geq |G_{xui}(t)|,$$
$$\bar{G}_{xdi}(t) \geq |G_{xdi}(t)|, \quad \forall\, t \geq 0,$$

where the $G_{xui}(t)$ and $G_{xdi}(t)$ are defined in Equation 5.11.

### 5.2.5 Practical Stability

Event-based control systems are hybrid dynamical systems and, more specifically, belong to the class of impulsive systems [5,22]. The notion of stability that is used here for these kinds of systems is the stability with respect to compact sets [4,6].

**DEFINITION 5.5** Consider the system (5.1) and a compact set $\mathcal{A} \subset \mathbb{R}^n$.

- The set $\mathcal{A}$ is stable for the system (5.1) if for each $\varepsilon > 0$ there exists $\delta > 0$ such that $\|x(0)\|_{\mathcal{A}} \leq \delta$ implies $\|x(t)\|_{\mathcal{A}} \leq \varepsilon$ for all $t \geq 0$.
- The set $\mathcal{A}$ is globally attractive for the system (5.1) with (5.2) if

$$\lim_{t \to \infty} \|x(t)\|_{\mathcal{A}} = 0,$$

holds for all $x(0) \in \mathbb{R}^n$.

- The set $\mathcal{A}$ is globally asymptotically stable for the system (5.1) if it is stable and globally attractive.

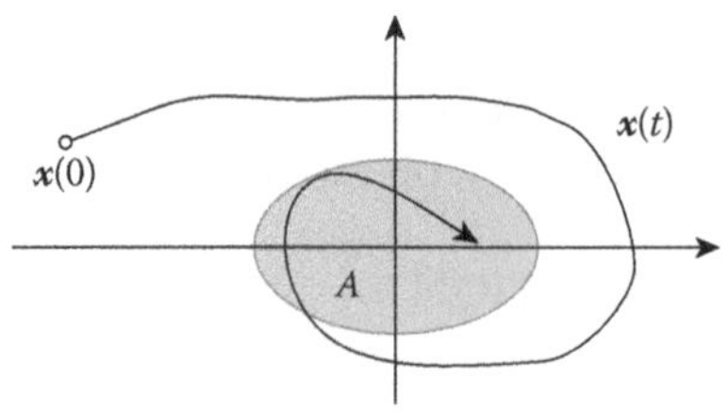


**FIGURE 5.3**
Practical stability with respect to the set $\mathcal{A}$.

This definition claims that for any bounded initial condition $x(0)$, the state trajectory $x(t)$ remains bounded for all $t \geq 0$ and eventually converges to the set $\mathcal{A}$ for $t \to \infty$.

**DEFINITION 5.6** If $\mathcal{A}$ is a globally asymptotically stable set for the system (5.1), then (5.1) is said to be *practically stable with respect to the set* $\mathcal{A}$.

A graphical interpretation of this stability notion is given in Figure 5.3, which shows the trajectory $x(t)$ of a system of the form (5.1) in the two-dimensional state space. The trajectory $x(t)$ is bounded to the set $\mathcal{A}$ for $t \to \infty$. The size of $\mathcal{A}$ depends on the disturbance magnitude $\bar{d}$ and, in case of event-based state feedback, also on some parameters of the event-based controller, as will be explained in more detail later.

## 5.3 State-Feedback Approach to Event-Based Control

A model-based approach to event-based state feedback has been presented first in [17]. This approach shows in a plain way the basic idea of how a dynamic model can be used for the control input generation and the determination of the event times in event-based control. For this purpose, the basic approach [17] is summarized in the following, before this idea is developed further in Section 5.4 to the distributed event-based state-feedback control of physically interconnected systems.

### 5.3.1 Basic Idea

The state-feedback approach to event-based control published in [17] is grounded on the consideration that feedback control, as opposed to feedforward control, is necessary in three situations:

- An unstable plant needs to be stabilized.
- The plant is inaccurately known so that the controller needs to react to model uncertainties.
- Unknown disturbances need to be attenuated.

In order to answer the question of, at which time instants a control loop needs to be closed, the focus in [17] is on disturbance attenuation, while the plant model is assumed to be known with negligible uncertainties. Hence, the main reason for a feedback communication is an intolerable effect of the disturbance $d(t)$ on the system performance. Further publications that have followed up on [17] have extended the event-based state-feedback approach to deal with model uncertainties [13], delays, and packet losses in the feedback communication [14] and nonlinear plant dynamics [19–21].

The structure of the event-based control loop investigated in [17] is illustrated in Figure 5.4. The plant is represented by the model (5.1), assuming that it provides a central sensor unit and a central actuator unit. Regarding the plant, the assumptions **A** 5.1–**A** 5.3 are made. The *event generator* $E$ determines the event times $t_k$ ($k = 0, 1, 2, \ldots$), at which the current state $x(t_k)$ is transmitted to the *control input generator* $C$ that uses the received information to update the generation of the control input $u(t)$. The solid arrows represent continuous information links, whereas the dashed line symbolizes a communication that occurs at the event times only. The feedback communication via the network is assumed to be ideal as specified in **A** 5.6.

The basic idea of the event-based state-feedback approach is to design the event generator $E$ and the control input generator $C$ such that the event-based control loop imitates the disturbance behavior of a continuous state-feedback loop with adjustable approximation

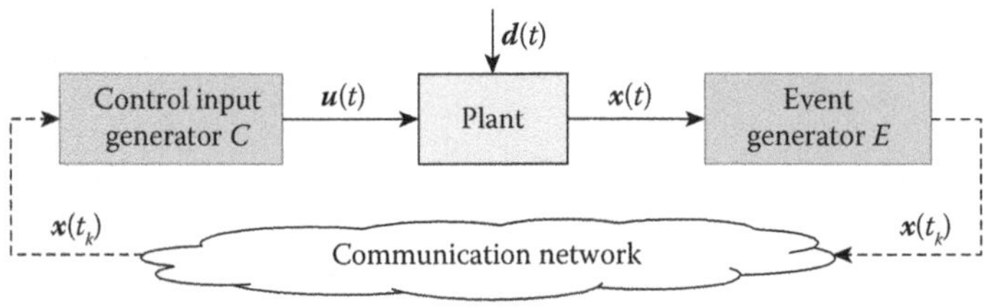


**FIGURE 5.4**
Event-based state-feedback loop.

error. In the following, this continuous control system is referred to as the *reference system* that is represented by the model

$$\Sigma_{\mathrm{r}}: \ \dot{x}_{\mathrm{r}}(t) = \underbrace{(A - BK)}_{=:\bar{A}} x_{\mathrm{r}}(t) + Ed(t), \quad x_{\mathrm{r}}(0) = x_0, \tag{5.14}$$

with the state $x_{\mathrm{r}} \in \mathbb{R}^n$. The state of the reference system is marked by the subscript r in order to distinguish it from the state of the event-based control system introduced later. The state-feedback gain $K$ in (5.14) is assumed to be designed such that the system $\Sigma_{\mathrm{r}}$ has a desired disturbance behavior; hence, the state $x_{\mathrm{r}}(t)$ of the reference system (5.14) is bounded.

**Proposition 5.1**

Given that the disturbance $d(t)$ is bounded as in (5.2), the reference system (5.14) is practically stable with respect to the set

$$\mathcal{A}_{\mathrm{r}} := \{x_{\mathrm{r}} \in \mathbb{R}^n \mid \|x_{\mathrm{r}}\| \leq b_{\mathrm{r}}\}, \tag{5.15}$$

with the bound

$$b_{\mathrm{r}} = \|\bar{d}\| \cdot \int_0^\infty \left\| \mathrm{e}^{\bar{A}\tau} E \right\| \mathrm{d}\tau. \tag{5.16}$$

**PROOF** Equation 5.14 yields

$$x_{\mathrm{r}}(t) = \mathrm{e}^{\bar{A}t} x_0 + \int_0^t \mathrm{e}^{\bar{A}(t-\tau)} Ed(\tau)\mathrm{d}\tau.$$

For an arbitrary large but bounded initial state $x_0$ and the maximum disturbance magnitude $\bar{d}$, the norm of the state $\|x_{\mathrm{r}}(t)\|$ is bounded by

$$\limsup_{t\to\infty} \|x_{\mathrm{r}}(t)\| \leq \|\bar{d}\| \cdot \int_0^\infty \left\| \mathrm{e}^{\bar{A}\tau} E \right\| \mathrm{d}\tau =: b_{\mathrm{r}},$$

for $t \to \infty$. ■

### 5.3.2 Components of the Event-Based Controller

This section describes the method for the design of the event-based controller proposed in [17]. The event-based controller consists of the following components:

- The event generator $E$ that determines what information is transmitted via the feedback link at which time instants $t_k$.
- The control input generator $C$ that generates the control input $u(t)$ in a feedforward manner in between consecutive event times $t_{k-1}$ and $t_k$, and it updates the determination of the input signal $u(t)$ using the information received at the event times $t_k$.

The communication network is also considered to be part of the event-based controller. However, the network is assumed to be ideal as stated in **A** 5.6; thus, the transmission of data is not particularly considered.

**Control input generator *C*.** The control input generator C is described by

$$C: \begin{cases} \Sigma_{\mathrm{s}}: \ \dot{x}_{\mathrm{s}}(t) = \bar{A}x_{\mathrm{s}}(t) + E\hat{d}_k, \quad x_{\mathrm{s}}(t_k^+) = x(t_k) \\ \qquad u(t) = -Kx_{\mathrm{s}}(t) \end{cases}. \tag{5.17}$$

It includes a model $\Sigma_{\mathrm{s}}$ of the reference system (5.14) with the model state $x_{\mathrm{s}} \in \mathbb{R}^n$, which is reset at the event times $t_k$ to the current plant state $x(t_k)$. Then, $\hat{d}_k$ denotes an estimate of the disturbance $d(t)$, which is determined at the event time $t_k$. A method for the disturbance estimate is presented in Section 5.3.2. The control input $u(t)$ is generated using the model state $x_{\mathrm{s}}(t)$ and is applied to the plant.

**Event generator *E*.** The event generator $E$ also includes a model $\Sigma_{\mathrm{s}}$ of the reference system (5.14) and it determines the event times $t_k$ as follows

$$E: \begin{cases} \Sigma_{\mathrm{s}}: \ \dot{x}_{\mathrm{s}}(t) = \bar{A}x_{\mathrm{s}}(t) + E\hat{d}_k, \quad x_{\mathrm{s}}(t_k^+) = x(t_k) \\ \qquad t_0 = 0 \\ \qquad t_{k+1} := \inf\{t > t_k \mid \|x(t) - x_{\mathrm{s}}(t)\| = \bar{e}\} \end{cases}. \tag{5.18}$$

Note that the state $x_{\mathrm{s}}(t)$ of the model $\Sigma_{\mathrm{s}}$ that is used by both the control input generator C and the event generator $E$ is a prediction of the plant state $x(t)$. Given that assumption **A** 5.1 holds true and that $x_{\mathrm{s}}(t_k^+) = x(t_k)$, a deviation between $x(t)$ and $x_{\mathrm{s}}(t)$ for $t > t_k$ is only caused by a difference between the actual disturbance $d(t)$ and its estimate $\hat{d}_k$. The event generator continuously monitors the difference state

$$x_\Delta(t) := x(t) - x_{\mathrm{s}}(t), \tag{5.19}$$

and triggers an event, whenever the norm of this difference attains some event threshold $\bar{e} \in \mathbb{R}_+$. The initial event at time $t_0 = 0$ is generated regardless of the triggering condition. At the event time $t_k$ $(k = 0, 1, 2, \ldots)$, the event generator $E$ transmits the current plant state $x(t_k)$ and a new disturbance estimate $\hat{d}_k$ over the communication network to the control input generator $C$. The state information $x(t_k)$ is used in both components at the event time $t_k$ in order to reset the model state $x_{\mathrm{s}}(t)$, as indicated in (5.17) and (5.18).

**Disturbance estimation.** The state-feedback approach to event-based control [17] works with any disturbance estimation method that yields bounded estimates $\hat{d}_k$, including the trivial estimation $\hat{d}_k \equiv \mathbf{0}$ for all $k \in \mathbb{N}_0$. This section presents an estimation method, which is based on the assumption that the disturbance $d(t)$ is constant in the time interval $t \in [t_{k-1}, t_k)$:

$$d(t) = d_c, \quad \text{for } t \in [t_{k-1}, t_k),$$

where $d_c \in \mathcal{D}$ denotes the disturbance magnitude. The idea of this estimation method is to determine the disturbance magnitude $d_c$ at the event time $t_k$ and to use this value as the disturbance estimation $\hat{d}_k$ for the time $t \geq t_k$ until the next event occurs.

The disturbance estimation method is given by the following recursion:

$$\hat{d}_0 = \mathbf{0}, \tag{5.20}$$

$$\hat{d}_k = \hat{d}_{k-1} + \left(A^{-1}\left(e^{A(t_k - t_{k-1})} - I_n\right)E\right)^{+} x_{\Delta}(t_k), \tag{5.21}$$

where the initial estimation $\hat{d}_0$ is chosen to be zero if no information about the disturbance is available. The pseudoinverse in (5.20) exists if $p \leq n$ holds, that is, the dimension $p$ of the disturbance vector $d(t)$ does not exceed the dimension $n$ of the state vector $x(t)$.

### 5.3.3 Main Properties of the Event-Based State Feedback

**Deviation between the behavior of the reference system and the event-based control loop.** The following result shows that the event-based state-feedback approach [17] has the ability to approximate the disturbance behavior of the continuous state-feedback system (5.14) with arbitrary precision.

**Theorem 5.4: Theorem 1 in [17]**

The approximation error $e(t) = x(t) - x_r(t)$ between the behavior of the event-based state-feedback loop (5.1), (5.17), and (5.18) and the reference system (5.14) is bounded from above by

$$\|e(t)\| \leq \bar{e} \cdot \int_0^{\infty} \left\| e^{\bar{A}t} BK \right\| dt =: e_{\max}, \tag{5.22}$$

for all $t \geq 0$.

Theorem 5.4 shows that the deviation between the disturbance rejection behavior of the event-based control system (5.1), (5.17), and (5.18) and the reference system (5.14) depends linearly on the event threshold $\bar{e}$. Hence, the behavior of the reference system is better approximated by the event-based state-feedback loop if the threshold $\bar{e}$ is reduced, and vice versa. Theorem 5.4 can also be regarded as a proof of stability for the event-based state-feedback loop, since the stability of the reference system (5.14) together with the bound (5.22) implies the boundedness of the state $x(t)$ of the event-based control loop.

In [12] it has been shown that the event-based state-feedback loop (5.1), (5.17), and (5.18) is practically stable with respect to the set

$$\mathcal{A} := \{x \in \mathbb{R}^n \mid \|x\| \leq b\},$$

with the bound

$$b = \bar{e} \cdot \int_0^{\infty} \left\| e^{\bar{A}\tau} BK \right\| d\tau + \|\bar{d}\| \cdot \int_0^{\infty} \left\| e^{\bar{A}\tau} E \right\| d\tau. \tag{5.23}$$

A comparison of (5.23) with the bound $b_r$ for the reference system given in (5.15) implies that $b = b_r + e_{\max}$ holds. Hence, the bound $b_r$ on the set $\mathcal{A}_r$ for the reference system is extended by the bound (5.22) if event-based state feedback is applied instead of continuous state feedback.

**Minimum interevent time.** The minimum time that elapses in between two consecutive events is referred to as the *minimum interevent time* $T_{\min}$. The event-based state-feedback approach [17] ensures that the minimum interevent time is bounded from below according to $T_{\min} \geq \bar{T}$.

**Theorem 5.5: Theorem 2 in [17]**

For any bounded disturbance, the minimum interevent time $T_{\min}$ of the event-based state-feedback loop (5.1), (5.17), and (5.18) is bounded from below by

$$\bar{T} = \arg\min_{t} \left\{ \int_0^t \left\| e^{A\tau} E \right\| d\tau = \frac{\bar{e}}{\bar{d}_{\Delta}} \right\}, \tag{5.24}$$

where $\bar{d}_{\Delta} \geq \left\| d(t) - \hat{d}_k \right\|$ denotes the maximum deviation between the disturbance $d(t)$ and the estimates $\hat{d}_k$ for all $t \geq 0$ and all $k \in \mathbb{N}_0$.

A direct inference of Theorem 5.5 is that no Zeno behavior can occur in the event-based state-feedback loop—that is, the number of events does not accumulate to infinity in finite time [16]. This result also shows that the minimum interevent time gets smaller if the event threshold $\bar{e}$ is reduced, and vice versa. This conclusion is in accordance with the expectation that a more precise approximation of the reference system's behavior by the event-based state feedback comes at the cost of a more frequent communication.

## 5.4 Disturbance Rejection with Event-Based Communication Using Information Requests

This section extends the event-based state-feedback approach [17] to the distributed event-based control of physically interconnected systems. The motivation for this extension is twofold:

1. The event-based state-feedback approach [17] requires the plant to have a centralized sensor unit that provides the overall state vector $\boldsymbol{x}(t)$, and a centralized actuator unit that updates all control inputs synchronously. Such centralized sensor and actuator units can generally not be presumed to exist, and particularly not in large-scale systems which are composed of interconnected systems (cf. Figure 5.2). From this consideration emerges the need for a decentralized controller structure.
2. For many large-scale systems, a stabilizing decentralized controller can only be found if the interconnections between the subsystems are sufficiently weak [9,23]. Moreover, there are applications where the control performance that can be achieved by means of decentralized control does not meet the requirements. In these cases, a distributed controller, where the control units do not use only local information but also information from neighboring subsystems, might be preferred in order to accomplish the control task.

This chapter presents a novel method for the implementation of a distributed state-feedback law in an event-based manner.

Section 5.4.1 introduces the notion of the *approximate model* and the *extended subsystem model*. These models are used in Section 5.4.2, which proposes a method for the design of a continuous distributed state feedback. The realization of the resulting distributed state feedback requires continuous communications between several subsystems, which would lead to a heavy load of the network and, thus, is undesired. A solution to this problem is presented in Section 5.4.3 in the form of a method for the implementation of the distributed state feedback in an event-based fashion, where the feedback of the local state information is continuous, and the communication between the components of the networked controller occurs at the event times only. The approximation of the behavior of the control system with distributed continuous state feedback by the event-based control is accomplished by a novel triggering mechanism that triggers either the transmission of information to or the request of information from the neighboring subsystems. Section 5.5 presents the results of an experiment that investigates the practical application of the distributed event-based state feedback to a thermofluid process.

### 5.4.1 Approximate Model and Extended Subsystem Model

The underlying idea of the subsequently proposed modeling approach is to obtain a model that not only describes the behavior of a single subsystem $\Sigma_i$ but also includes the interaction of $\Sigma_i$ with its neighboring subsystems and their controllers. The behavior of the neighboring subsystems can generally not be described precisely; hence, it is approximated by an *approximate model* $\Sigma_{\mathrm{a}i}$. The subsystem $\Sigma_i$ augmented with the approximate model $\Sigma_{\mathrm{a}i}$ then forms the *extended subsystem* $\Sigma_{\mathrm{e}i}$, which is used later for the design of a distributed state-feedback controller.

**Approximate model.** Regarding the interconnected subsystems (5.3) and (5.4), the following definitions are made:

**DEFINITION 5.7** Subsystem $\Sigma_j$ is called the *predecessor* of subsystem $\Sigma_i$ if $\|\boldsymbol{L}_{ij}\| > 0$ holds. In the following,

$$\mathcal{P}_i := \{j \in \mathcal{N} \setminus \{i\} \,|\, \|\boldsymbol{L}_{ij}\| > 0\},$$

denotes the set of those subsystems $\Sigma_j$ which directly affect $\Sigma_i$ through the coupling input $\boldsymbol{s}_i(t)$.

**DEFINITION 5.8** Subsystem $\Sigma_j$ is called the *successor* of subsystem $\Sigma_i$ if $\|\boldsymbol{L}_{ji}\| > 0$ holds. Hereafter,

$$\mathcal{S}_i := \{j \in \mathcal{N} \setminus \{i\} \,|\, \|\boldsymbol{L}_{ji}\| > 0\},$$

denotes the set of subsystems $\Sigma_j$, which the subsystem $\Sigma_i$ directly affects through the coupling output $\boldsymbol{z}_i(t)$.

To understand the notion of the approximate model, consider Figure 5.5. From Equation 5.4, it follows that the coupling input $\boldsymbol{s}_i(t)$ aggregates the influence of the remaining subsystems together with their controllers on $\Sigma_i$ (Figure 5.5a). Assume that, from the viewpoint of subsystem $\Sigma_i$, the relation between the coupling output

$z_i(t)$ and the coupling input $s_i(t)$ is approximately described by the *approximate model*:

$$\Sigma_{ai}: \begin{cases} \dot{x}_{ai}(t) = A_{ai}x_{ai}(t) + B_{ai}z_i(t) + E_{ai}d_{ai}(t) + F_{ai}f_i(t), \\ \qquad x_{ai}(0) = x_{a0i}, \\ s_i(t) = C_{ai}x_{ai}(t), \\ v_i(t) = H_{ai}x_{ai}(t) + D_{ai}z_i(t), \end{cases} \tag{5.25}$$

where $x_{ai} \in \mathbb{R}^{n_{ai}}, d_{ai} \in \mathbb{R}^{p_{ai}}, f_i \in \mathbb{R}^{\nu_i}$, and $v_i \in \mathbb{R}^{\mu_i}$ denote the state, the disturbance, the residual output, and the residual input, respectively (Figure 5.5b). The disturbance $d_{ai}(t)$ is assumed to be bounded by

$$|d_{ai}(t)| \leq \bar{d}_{ai}, \quad \forall\, t \geq 0. \tag{5.26}$$

In the following, the state $x_{ai}(t)$ of the approximate model $\Sigma_{ai}$ is considered to be directly related with the states $x_j(t)$ $(j \in \mathcal{P}_i)$ of the predecessors of $\Sigma_i$ in the sense that the approximate model $\Sigma_{ai}$ can be obtained by the linear transformation

$$x_{ai}(t) = \sum_{j \in \mathcal{P}_i} T_{ij}x_j(t), \quad \forall\, t \geq 0, \tag{5.27}$$

with $T_{ij} \in \mathbb{R}^{n_{ai} \times n_j}$. Moreover, the relation between the subsystem states $x_j(t)$ $(j \in \mathcal{P}_i)$ and the approximate model state $x_{ai}(t)$ is assumed to be bijective. Hence, for each $j \in \mathcal{P}_i$ there exists a matrix $C_{ji} \in \mathbb{R}^{n_j \times n_{ai}}$ that satisfies

$$x_j(t) = C_{ji}x_{ai}(t), \quad \forall\, t \geq 0. \tag{5.28}$$

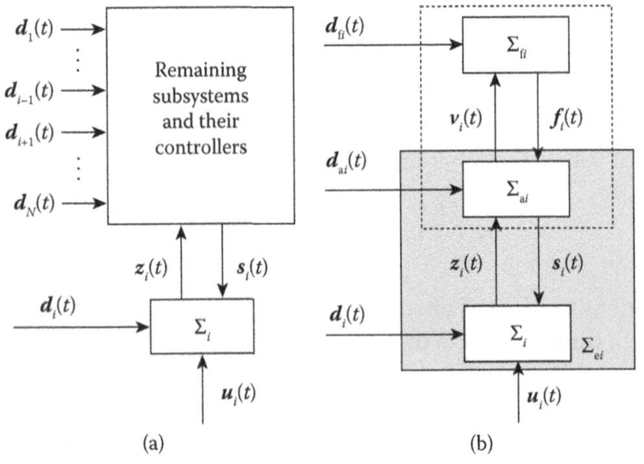


**FIGURE 5.5**
Interconnection of subsystem $\Sigma_i$ and the remaining control loops: (a) general structure and (b) decomposition of the remaining controlled subsystems into approximate model $\Sigma_{ai}$ and residual model $\Sigma_{fi}$.

Note that Equations 5.27 and 5.28 imply

$$C_{ji}T_{ip} = \begin{cases} I_{n_j}, & \text{if } j = p \\ O, & \text{if } j \neq p \end{cases}.$$

The model $\Sigma_{ai}$ approximately describes the behavior of the predecessor subsystems $\Sigma_j$ $(j \in \mathcal{P}_i)$ together with their controllers. This consideration leads to the following assumption:

**A** 7.1 The matrix $A_{ai}$ is Hurwitz for all $i \in \mathcal{N}$.

The mismatch between the behavior of the overall control system except for subsystem $\Sigma_i$ and the approximate model (5.25) is expressed by the *residual model*:

$$\Sigma_{fi}: f_i(t) = G_{fdi} * d_{fi} + G_{fvi} * v_i, \tag{5.29}$$

where $d_{fi} \in \mathbb{R}^{p_{fi}}$ denotes the disturbance that is assumed to be bounded by

$$|d_{fi}(t)| \leq \bar{d}_{fi}, \quad \forall\, t \geq 0. \tag{5.30}$$

Note that the approximate model $\Sigma_{ai}$ together with the residual model $\Sigma_{fi}$ represent the behavior of the remaining subsystems and their controllers (Figure 5.5b). Hereafter, the residual model $\Sigma_{fi}$ is not assumed to be known exactly but instead is described by some upper bounds $\bar{G}_{fdi}(t)$ and $\bar{G}_{fvi}(t)$ that satisfy the relations

$$\bar{G}_{fdi}(t) \geq |G_{fdi}(t)|, \quad \bar{G}_{fvi}(t) \geq |G_{fvi}(t)|, \tag{5.31}$$

for all $t \geq 0$.

**Extended subsystem model.** Consider the subsystem $\Sigma_i$ augmented with the approximate model $\Sigma_{ai}$ that yields the *extended subsystem*:

$$\Sigma_{ei}: \begin{cases} \dot{x}_{ei}(t) = A_{ei}x_{ei}(t) + B_{ei}u_i(t) + E_{ei}d_{ei}(t) + F_{ei}f_i(t) \\ x_{ei}(0) = x_{e0i} = \begin{pmatrix} x_{0i}^\top & x_{a0i}^\top \end{pmatrix}^\top \\ v_i(t) = H_{ei}x_{ei}(t), \end{cases} \tag{5.32}$$

with the state $x_{ei} = (x_i^\top \ x_{ai}^\top)^\top \in \mathbb{R}^{n_i + n_{ai}}$, the composite disturbance vector $d_{ei} = (d_i^\top \ d_{ai}^\top)^\top$, and the matrices

$$A_{ei} = \begin{pmatrix} A_i & E_{si}C_{ai} \\ B_{ai}C_{zi} & A_{ai} \end{pmatrix}, \quad B_{ei} = \begin{pmatrix} B_i \\ O \end{pmatrix},$$

$$E_{ei} = \begin{pmatrix} E_i & O \\ O & E_{ai} \end{pmatrix}, \quad F_{ei} = \begin{pmatrix} O \\ F_{ai} \end{pmatrix}, \tag{5.33}$$

$$H_{ei} = \begin{pmatrix} D_{ai}C_{zi} & H_{ai} \end{pmatrix}. \tag{5.34}$$

The boundedness of the disturbances $d_i(t)$ and $d_{ai}(t)$ implies the boundedness of $d_{ei}(t)$. With Equations 5.5 and 5.26, the bound

$$|d_{ei}(t)| \leq \begin{pmatrix} \bar{d}_i^\top & \bar{d}_{ai}^\top \end{pmatrix}^\top =: \bar{d}_{ei}, \quad \forall\, t \geq 0, \tag{5.35}$$

on the composite disturbance $d_{ei}(t)$ follows.

### 5.4.2 A Distributed State-Feedback Design Method

This section proposes a new approach to the design of a distributed continuous state-feedback law for physically interconnected systems, which is summarized in Algorithm 5.1. The main feature of the proposed design method is that the distributed state feedback can be designed for each subsystem $\Sigma_i$ without having exact knowledge about the overall system. The stability of the overall control system is tested by means of a condition that is presented in Theorem 5.7.

**Distributed state-feedback design.** The following proposes an approach to the design of a distributed state-feedback controller, which is carried out for each subsystem $\Sigma_i$ of the interconnected subsystems (5.3) and (5.4) separately. This design approach requires the extended subsystem model (5.32) to be given, in order to determine the state feedback

$$u_i(t) = -K_{ei}x_{ei}(t) = -\begin{pmatrix} K_{di} & K_{ai} \end{pmatrix}\begin{pmatrix} x_i(t) \\ x_{ai}(t) \end{pmatrix}. \quad (5.36)$$

The state-feedback gain $K_{ei}$ is assumed to be chosen for the extended subsystem (5.32) with $f_i(t) \equiv 0$ such that the closed-loop system

$$\bar{\Sigma}_{ei}: \begin{cases} \dot{x}_{ei}(t) = \bar{A}_{ei}x_{ei}(t) + E_{ei}d_{ei}(t), & x_{ei}(0) = \begin{pmatrix} x_{0i}^\top & x_{a0i}^\top \end{pmatrix}^\top \\ v_i(t) = H_{ei}x_{ei}(t) \end{cases}, \quad (5.37)$$

has a desired behavior that includes the minimum requirement that the matrix

$$\bar{A}_{ei} := A_{ei} - B_{ei}K_{ei} = \begin{pmatrix} A_i - B_iK_{di} & E_{si}C_{ai} - B_iK_{ai} \\ B_{ai}C_{zi} & A_{ai} \end{pmatrix}, \quad (5.38)$$

is stable. With the transformation (5.27), the control law (5.36) can be rewritten as

$$u_i(t) = -K_{di}x_i(t) - \sum_{j\in\mathcal{P}_i} K_{ij}x_j(t), \quad (5.39)$$

with

$$K_{ij} := K_{ai}T_{ij}.$$

This reveals that (5.36) is a distributed state feedback, where the control input $u_i(t)$ is a function of the local state $x_i(t)$ as well as of the states $x_j(t)$ $(j \in \mathcal{P}_i)$ of the predecessor subsystems of $\Sigma_i$.

Note that the proposed design approach neglects the impact of the residual model (5.29) since $f_i(t) \equiv 0$ is presumed. Hence, the design of the distributed state feedback only guarantees the stability of the isolated controlled extended subsystems. The next section derives a condition under which the design of the distributed state-feedback controller for the extended subsystem with $f_i(t) \equiv 0$ ensures the stability of the overall control system.

**Stability of the overall control system.** This section derives a condition to check the stability of the extended subsystem (5.32) with the distributed state feedback (5.39) taking the interconnection to the residual models (5.29) and (5.31) into account. The result is then extended to a method for the stability analysis of the overall control system.

The stability analysis method that is presented in this section makes use of the comparison system:

$$r_{xi}(t) = V_{x0i}(t)\,|x_{e0i}| + V_{xdi} * |d_{ei}| + V_{xfi} * |f_i| \geq |x_{ei}(t)|, \quad (5.40)$$

$$r_{vi}(t) = V_{v0i}(t)\,|x_{e0i}| + V_{vdi} * |d_{ei}| + V_{vfi} * |f_i| \geq |v_i(t)|, \quad (5.41)$$

for the extended subsystem (5.32) with the distributed controller (5.39), where

$$V_{x0i}(t) = \left|e^{\bar{A}_{ei}t}\right|, \qquad V_{xdi}(t) = \left|e^{\bar{A}_{ei}t}E_{ei}\right|,$$
$$V_{xfi}(t) = \left|e^{\bar{A}_{ei}t}F_{ei}\right|,$$
$$V_{v0i}(t) = \left|H_{ei}e^{\bar{A}_{ei}t}\right|, \quad V_{vdi}(t) = \left|H_{ei}e^{\bar{A}_{ei}t}E_{ei}\right|,$$
$$V_{vfi}(t) = \left|H_{ei}e^{\bar{A}_{ei}t}F_{ei}\right|. \quad (5.42)$$

The following lemma gives a condition that can be used to test the stability of the interconnection of the extended subsystem (5.32) with the distributed state feedback (5.39) and the residual models (5.29) and (5.31).

**Theorem 5.6**

Consider the interconnection of the extended subsystem (5.32) with the distributed state feedback (5.39) together with the residual model $\Sigma_{fi}$ as defined in (5.29) with the bounds (5.31). The interconnection of these systems is practically stable if the condition

$$\lambda_P\left(\int_0^\infty \bar{G}_{fvi}(t)\mathrm{d}t \int_0^\infty V_{vfi}(t)\mathrm{d}t\right) < 1, \quad (5.43)$$

is satisfied.

**PROOF** The proposed analysis method makes use of comparison systems, which are explained in Section 5.2.4. The following equations

$$r_{xi}(t) = V_{x0i}(t)\,|x_{e0i}| + V_{xdi} * |d_{ei}| + V_{xfi} * |f_i| \geq |x_{ei}(t)|, \quad (5.44)$$

$$r_{vi}(t) = V_{v0i}(t)\,|x_{e0i}| + V_{vdi} * |d_{ei}| + V_{vfi} * |f_i| \geq |v_i(t)|, \quad (5.45)$$

with

$$V_{x0i}(t) = \left|e^{\bar{A}_{ei}t}\right|, \qquad V_{xdi}(t) = \left|e^{\bar{A}_{ei}t}E_{ei}\right|,$$
$$V_{xfi}(t) = \left|e^{\bar{A}_{ei}t}F_{ei}\right|,$$
$$V_{v0i}(t) = \left|H_{ei}e^{\bar{A}_{ei}t}\right|, \quad V_{vdi}(t) = \left|H_{ei}e^{\bar{A}_{ei}t}E_{ei}\right|,$$
$$V_{vfi}(t) = \left|H_{ei}e^{\bar{A}_{ei}t}F_{ei}\right|,$$

represent a comparison system for the extended subsystem (5.32) with the distributed state feedback (5.39). In the following analysis, the assumption $x_{e0i} = \mathbf{0}$ is made, which is no loss of generality for the presented stability criterion but simplifies the derivation. Consider the interconnection of the comparison system (5.44) and the residual systems (5.29) and (5.31), which yields

$$r_{fi}(t) = \bar{G}_{fvi} * V_{vdi} * |d_{ei}| + \bar{G}_{fdi} * |d_{fi}| + \bar{G}_{fvi} * V_{vfi} * |f_i| \geq |f_i(t)|. \tag{5.46}$$

Equation 5.46 is an implicit bound on the signal $|f_i(t)|$. An explicit bound on $|f_i(t)|$ in terms of the signal $r_{vi}(t)$ can be obtained by means of the comparison principle [15], where the basic idea is to find an impulse-response matrix $G_i(t)$ that describes the input/output behavior of the gray highlighted block in Figure 5.6:

$$r_{fi}(t) = G_i * (\bar{G}_{fvi} * V_{vdi} * |d_{ei}| + \bar{G}_{fdi} * |d_{fi}|) \geq |f_i(t)|, \quad \forall\, t \geq 0. \tag{5.47}$$

The comparison principle says that the impulse-response matrix

$$G_i(t) = \delta(t)I + \bar{G}_{fvi} * V_{vfi} * G_i, \tag{5.48}$$

exists, which means the inequality

$$\int_0^\infty G_i(t)dt = I + \int_0^\infty \bar{G}_{fvi} * V_{vfi} * G_i dt = I + \int_0^\infty \bar{G}_{fvi}(t)dt \int_0^\infty V_{vfi}(t)dt \int_0^\infty G_i(t)dt < \infty,$$

holds true if the condition

$$\lambda_P\left(\int_0^\infty \bar{G}_{fvi}(t)dt \int_0^\infty V_{vfi}(t)dt\right) < 1, \tag{5.49}$$

is satisfied. On condition that relation (5.49) is fulfilled, the signal $r_{fi}(t)$ remains bounded for all $t \geq 0$ which implies that $f_i(t)$ is bounded, as well. Given that the disturbance inputs $d_{fi}(t)$ and $d_{ei}(t)$ are bounded by (5.30) or (5.35), respectively, and that the matrix $\bar{A}_{ei}$ in (5.42) is Hurwitz by definition, the output $r_{xi}(t)$ of the comparison system (5.40) is bounded for all $t \geq 0$. Since $r_{xi}(t)$ represents a bound on the subsystem state $x_i(t)$, the stability of overall comparison system implies the stability of (5.32) with the distributed state feedback (5.39) together with the residual model (5.29). ∎

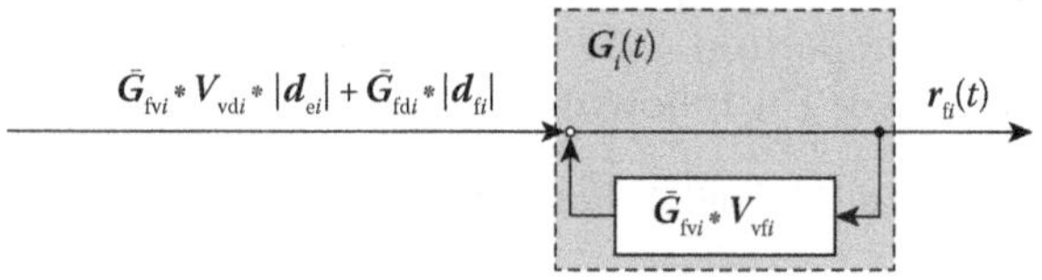


**FIGURE 5.6**
Block diagram representing Equation 5.46.

The relation (5.43) is interpreted as presented in Figure 5.6.

> The condition (5.43) can be considered as a small-gain theorem that claims that the extended subsystem (5.32) and the residual model (5.29) are weakly coupled. On the other hand, the condition (5.43) does not impose any restrictions on the interconnection of subsystem $\Sigma_i$ and its neighboring subsystems $\Sigma_j$ ($j \in \mathcal{N}_i$). These systems do not have to satisfy any small-gain condition but can be strongly interconnected.

Note that the residual model (5.29) together with the extended subsystem (5.32) with the distributed state feedback (5.39) describe the behavior of the overall control system from the perspective of subsystem $\Sigma_i$. Hence, the practical stability of the overall control system is implied by the practical stability of the interconnection of residual model (5.29) and the extended subsystem (5.32) with the controller (5.39) for all $i \in \mathcal{N}$. Theorem 5.7 makes the result on the practical stability of the overall control system more precise and determines the set $\mathcal{A}_{ri}$ to which the state $x_i(t)$ is bounded for $t \to \infty$. Hereafter, it is assumed that the condition (5.43) is satisfied, which implies that the matrix

$$\int_0^\infty G_i(t)dt := \left(I - \int_0^\infty \bar{G}_{fvi}(t)dt \int_0^\infty V_{vfi}(t)dt\right)^{-1},$$

exists and is nonnegative.

**Theorem 5.7**

Consider the interconnection of the controlled extended subsystems (5.32) and (5.39) with the residual models (5.29) and (5.31) and assume that the condition (5.43) is satisfied for all $i \in \mathcal{N}$. Then, the subsystem $\Sigma_i$ ($i \in \mathcal{N}$) is practically stable with respect to the set

$$\mathcal{A}_{ri} := \{x_i \in \mathbb{R}^{n_i} \mid |x_i| \leq b_{ri}\}, \tag{5.50}$$

with the ultimate bound

$$b_{ri} = (I_{n_i} \quad O)\left(\int_0^\infty V_{xdi}(t)dt \cdot \bar{d}_{ei} + \int_0^\infty V_{xfi}(t)dt \cdot \bar{f}_i\right), \tag{5.51}$$

where

$$\bar{f}_i = \int_0^\infty G_i(t)\mathrm{d}t\Big(\int_0^\infty \bar{G}_{\mathrm{fv}i}(t)\mathrm{d}t \int_0^\infty V_{\mathrm{vd}i}(t)\mathrm{d}t \cdot \bar{d}_{\mathrm{e}i} + \int_0^\infty \bar{G}_{\mathrm{fd}i}(t)\mathrm{d}t \cdot \bar{d}_{\mathrm{f}i}\Big),$$

is the bound on the maximum residual output $f_i(t)$ for all $t \geq 0$.

**PROOF** To begin with the proof, observe that the state $x_{\mathrm{e}i}(t)$ of each controlled extended subsystems (5.32) and (5.39) is bounded if the disturbance $d_{\mathrm{e}i}(t)$ as well as the signal $f_i(t)$ are bounded. While the former is bounded by definition (cf. Equation 5.35) the latter is known to be bounded if the condition (5.43) holds, which is fulfilled by assumption.

The next analysis derives the asymptotically stable set $\mathcal{A}_{\mathrm{r}i}$ for the subsystems $\Sigma_i$. Consider Equation 5.44 from which

$$\begin{aligned}|x_i(t)| &= (I_{n_i} \;\; O)\, |x_{\mathrm{e}i}(t)| \\ &\leq (I_{n_i} \;\; O)\, (V_{\mathrm{x}0i}(t)\, |x_{\mathrm{e}0i}| + V_{\mathrm{xd}i} * \bar{d}_{\mathrm{e}i} + V_{\mathrm{xf}i} * |f_i|),\end{aligned}$$

follows, where the disturbance $d_{\mathrm{e}i}$ has been replaced by the bound given in (5.35). Note that for $t \to \infty$, the term that depends on the initial state $x_{\mathrm{e}0i}$ vanishes. Hence, from the previous relation, the bound

$$\begin{aligned}&\limsup_{t\to\infty} |x_i(t)| \\ &\leq (I_{n_i} \;\; O)\Big(\int_0^\infty V_{\mathrm{xd}i}(t)\mathrm{d}t \cdot \bar{d}_{\mathrm{e}i} + \int_0^\infty V_{\mathrm{xf}i}(t)\mathrm{d}t \cdot \bar{f}_i\Big) \\ &=: b_{\mathrm{r}i},\end{aligned} \tag{5.52}$$

follows with

$$\limsup_{t\to\infty} |f_i(t)| \leq \bar{f}_i(t).$$

From Equation 5.52, it can be inferred that subsystem $\Sigma_i$ is practically stable with respect to the set

$$\mathcal{A}_{\mathrm{r}i} := \{x_i \in \mathbb{R}^{n_i} \mid |x_i| \leq b_{\mathrm{r}i}\}. \tag{5.53}$$

The aim of the following analysis is to specify the bound $\bar{f}_i$. Therefore, consider Equations 5.29 and 5.45, which yield

$$\begin{aligned}r_{\mathrm{f}i}(t) &= \bar{G}_{\mathrm{fv}i} * (V_{\mathrm{v}0i}\, |x_{\mathrm{e}0i}| + V_{\mathrm{vd}i} * |d_{\mathrm{e}i}| + V_{\mathrm{vf}i} * |f_i|) \\ &\quad + \bar{G}_{\mathrm{fd}i} * |d_{\mathrm{f}i}| \geq |f_i(t)|.\end{aligned}$$

Using the comparison principle, this bound can be restated in the explicit form

$$\begin{aligned}r_{\mathrm{f}i}(t) &= G_i * (\bar{G}_{\mathrm{fv}i} * (V_{\mathrm{v}0i}\, |x_{\mathrm{e}0i}| + V_{\mathrm{vd}i} * |d_{\mathrm{e}i}|) \\ &\quad + \bar{G}_{\mathrm{fd}i} * |d_{\mathrm{f}i}|) \geq |f_i(t)|,\end{aligned}$$

with the impulse-response matrix $G_i$ given in (5.48). From the last relation, the bound

$$\begin{aligned}&\limsup_{t\to\infty} r_{\mathrm{f}i}(t) \\ &= \int_0^\infty G_i(t)\mathrm{d}t\Big(\int_0^\infty \bar{G}_{\mathrm{fv}i}(t)\mathrm{d}t \int_0^\infty V_{\mathrm{vd}i}(t)\mathrm{d}t \cdot \bar{d}_{\mathrm{e}i} \\ &\quad + \int_0^\infty \bar{G}_{\mathrm{fd}i}(t)\mathrm{d}t \cdot \bar{d}_{\mathrm{f}i}\Big) =: \bar{f}_i,\end{aligned} \tag{5.54}$$

follows.

In summary, for $t \to \infty$, the state $x_i(t)$ asymptotically converges to the set $\mathcal{A}_{\mathrm{r}i}$ defined in (5.53) with the ultimate bound $b_{\mathrm{r}i}$ given in (5.52). The bound $b_{\mathrm{r}i}$ is a function of the disturbance bound $\bar{d}_{\mathrm{e}i}$ according to (5.35) and of the bound $\bar{f}_i$ given in (5.54). This completes the proof of Theorem 5.7. ■

Theorem 5.7 shows that the size of the set $\mathcal{A}_{\mathrm{r}i}$ depends on the disturbance bounds $\bar{d}_{\mathrm{e}i}$ and $\bar{d}_{\mathrm{f}i}$, which together represent a bound on the overall disturbance vector $d(t)$.

**Design algorithm.** The method for the design of the distributed continuous state-feedback controller is summarized in the following algorithm.

---

**Algorithm 5.1:** Design of a distributed continuous state-feedback controller

---

**Given:** Interconnected systems (5.3) and (5.4).
Do for each $i \in \mathcal{N}$:

1. Identify the set $\mathcal{P}_i$ and determine an approximate model (5.25) and the corresponding residual models (5.29) and (5.31). Compose the subsystem model (5.3) and the approximate model (5.25) to form the extended subsystem model (5.32).
2. Determine the state-feedback gain $K_{\mathrm{e}i}$ as in (5.36) such that the controlled extended subsystem has a desired behavior, implying that the matrix $\bar{A}_{\mathrm{e}i}$ given in (5.38) is Hurwitz.
3. Check whether the condition (5.43) is satisfied. If (5.43) does not hold, stop (the stability of the overall control system cannot be guaranteed).

**Result:** A state-feedback gain $K_{\mathrm{e}i}$ for each $i \in \mathcal{N}$ can be implemented in a distributed manner as shown in (5.39). The subsystem $\Sigma_i$ is practically stable with respect to the set $\mathcal{A}_{\mathrm{r}i}$ given in (5.50) with the ultimate bound (5.51).

---

Note that the design and the stability analysis of the distributed state feedback can be carried out for each $\Sigma_i$ $(i \in \mathcal{N})$ without the knowledge of the overall system model. For the design, the subsystem model $\Sigma_i$ and the approximate model $\Sigma_{ai}$ are required only. In order to analyze the stability of the control system, the coarse model information (5.31) of the residual model $\Sigma_{fi}$ is required.

The implementation of the distributed state feedback as proposed in (5.39) requires continuous information links from all subsystems $\Sigma_j$ $(j \in \mathcal{P}_i)$ to subsystem $\Sigma_i$. However, continuous communication between several subsystems is in many technical systems neither realizable nor desirable, due to high installation effort and maintenance costs. The following section proposes a method for the implementation of the distributed state-feedback law (5.39) in an event-based fashion, where a communication between subsystem $\Sigma_i$ and its neighboring subsystems $\Sigma_j$ $(j \in \mathcal{P}_i)$ is induced at the event time instants only.

### 5.4.3 Event-Based Implementation of a Distributed State-Feedback Controller

This section proposes a method for the implementation of a distributed control law of the form (5.39) in an event-based manner, where information is exchanged between subsystems at event time instants only. This implementation method obviates the need for a continuous communication among subsystems, and it guarantees that the behavior of the event-based control systems approximates the behavior of the control system with distributed continuous state feedback. Following the idea of [17], it will be shown that the deviation between the behavior of these two systems can be made arbitrarily small.

The main result of this section is a new approach to the distributed event-based state-feedback control that can be made to approximate the behavior of the control system with continuous distributed state feedback with arbitrary precision. This feature is guaranteed by means of a new kind of event-based control where two types of events are triggered: Events that are triggered due to the *transmit-condition* induce the transmission of local state information $x_i$ from controller $F_i$ to the controllers $F_j$ of the neighboring subsystems. In addition to this, a second event condition, referred to as *request-condition*, leads to the request of information from the controllers $F_j$ of the neighboring subsystems $\Sigma_j$ by the controller $F_i$.

**Basic idea.** The main idea of the proposed implementation method is as follows: The controller $F_i$ generates a prediction $\tilde{x}_j^i(t)$ of the state $x_j(t)$ for all $j \in \mathcal{P}_i$ which is used to determine the control input $u_i(t)$ according to

$$u_i(t) = -K_{\mathrm{d}i}x_i(t) - \sum_{j \in \mathcal{P}_i} K_{ij}\tilde{x}_j^i(t). \tag{5.55}$$

This state-feedback law is adapted from (5.39) by replacing the state $x_j(t)$ by its prediction $\tilde{x}_j^i(t)$. Note that the proposed approach to the distributed event-based control allows for different local control units $F_i$ and $F_l$ to determine different predictions $\tilde{x}_j^i(t)$ or $\tilde{x}_j^l(t)$, respectively, of the same state $x_j(t)$ of subsystem $\Sigma_j$. These signals are distinguished in the following by means of the superscript.

It will be shown that this implementation yields a stable overall control system if the difference

$$\left|x_{\Delta j}^i(t)\right| = \left|x_j(t) - \tilde{x}_j^i(t)\right|,$$

is bounded for all $t \geq 0$ and all $i, j \in \mathcal{N}$. This difference, however, can neither be monitored by the controller $F_i$ nor by the controller $F_j$, because the former knows only the prediction $\tilde{x}_j^i(t)$, and the latter has access to the subsystem state $x_j(t)$, only. This problem is solved in the proposed implementation method by a new event-triggering mechanism in the controllers $F_i$. The event generator $E_i$ in $F_i$ generates two different kinds of events: The first one leads to a transmission of the current subsystem state $x_i(t)$ to the controllers $F_j$ of the neighboring subsystems $\Sigma_j$, whereas the second one induces the request, denoted by $R$, of the current state information $x_j(t)$ from all neighboring subsystems $\Sigma_j$. In this way, the proposed implementation method introduces a new kind of event-based control, where information is not only sent but also requested by the event generators.

**Reference system.** Assume that for the interconnected subsystems (5.3) and (5.4), a distributed state-feedback controller is given, which has been designed according to Algorithm 5.1. The subsystem $\Sigma_i$ together with the distributed continuous state feedback (5.39) is described by the state-space model:

$$\Sigma_{\mathrm{r}i}: \begin{cases} \dot{x}_{\mathrm{r}i}(t) = \bar{A}_i x_{\mathrm{r}i}(t) - B_i \sum_{j \in \mathcal{P}_i} K_{ij} x_{\mathrm{r}j}(t) + E_i d_i(t) \\ \qquad + E_{\mathrm{s}i} s_{\mathrm{r}i}(t) \\ x_{\mathrm{r}i}(0) = x_{0i} \\ z_{\mathrm{r}i}(t) = C_{\mathrm{z}i} x_{\mathrm{r}i}(t) \end{cases} \tag{5.56}$$

where $\bar{A}_i := (A_i - B_i K_{\mathrm{d}i})$. Hereafter, $\Sigma_{ri}$ is referred to as the reference subsystem. The signals are indicated with $r$ to distinguish them from the corresponding signals in the event-based control system that is investigated later.

The interconnection of the reference subsystems $\Sigma_{ri}$ is given by

$$s_{ri}(t) = \sum_{j=1}^{N} L_{ij} z_{rj}(t), \tag{5.57}$$

according to Equation 5.4. Equations 5.56 and 5.57 yield the overall reference system

$$\Sigma_r: \ \dot{x}_r(t) = \underbrace{(A - B(K_d + \bar{K}))}_{:= \bar{A}} x_r(t) + Ed(t),$$
$$x_r(0) = x_0, \tag{5.58}$$

with the matrices $A$, $B$, $E$ given in (5.6) and the state-feedback gains

$$K_d = \operatorname{diag}(K_{d1}, \ldots, K_{dN}), \tag{5.59}$$

and

$$\bar{K} = \begin{pmatrix} O & K_{12} & \ldots & K_{1N} \\ K_{21} & O & \ldots & K_{2N} \\ \vdots & \vdots & \ddots & \vdots \\ K_{N1} & K_{N2} & \ldots & O \end{pmatrix}, \tag{5.60}$$

where $K_{ij} = O$ holds if $j \notin \mathcal{P}_i$. The overall control system (5.58) is stable since the distributed state-feedback gain (5.59) and (5.60) is the result of Algorithm 5.1, which includes the validation of the stability condition (5.43). Consequently, the subsystems (5.56) are practically stable with respect to the set $\mathcal{A}_{ri}$ given in (5.50) and (5.51).

The aim of the event-based controller that is to be designed subsequently is to approximate the behavior of the reference systems (5.56) and (5.57) with adjustable precision.

**Information transmissions and requests.** The event-based control approach that is investigated in this section works with a triggering mechanism that distinguishes between two kinds of events that lead to either the transmission of local state information or the request of information from neighboring subsystems. The event conditions, based on which these events are triggered, are called *transmit-condition* and *request-condition*. This event triggering scheme contrasts with the one of event-based control methods that are proposed in the previous chapters and, thus, necessitates an extended notion of the triggering time instants.

According to the subsequently presented event-based control approach, the local control unit $F_i$ sends and receives information. Both the sending and the reception of information can occur in two situations, which are explained next:

- The local control unit $F_i$ transmits the current state $x_i(t_{k_i})$ if at time $t_{k_i}$ its transmit-condition is fulfilled. On the other hand, $F_i$ sends the state $x_i(t_{r_j(i)})$ at time $t_{r_j(i)}$ when the controller $F_j$ requests this information from $F_i$.
- If at time $t_{r_i(j)}$ the request-condition of the local control unit $F_i$ is satisfied, it requests (and consequently receives) the information $x_j(t_{r_i(j)})$ from the controller $F_j$. The controller $F_i$ also receives information if the controller $F_j$ decides at time $t_{k_j}$ to transmit the state $x_j(t_{k_j})$ to the $F_i$.

In the following, $r_i(j)$ denotes the counter for the events at which the controller $F_i$ requests information from the controller $F_j$.

> From the viewpoint of the controller $F_i$, the transmission and the reception of information are either induced by its own triggering conditions (at the times $t_{k_i}$ or $t_{r_i(j)}$) or enforced by the neighboring controllers (at the times $t_{r_j(i)}$ or $t_{k_j}$).

**Networked controller.** The networked controller consists of the local components $F_i$ and the communication network as shown in Figure 5.1. Each control unit $F_i$ includes a control input generator $C_i$ that generates the control input $u_i(t)$ and an event generator $E_i$ that determines the time instants $t_{k_i}$ and $t_{r_i(j)}$ at which the current local state $x_i(t_{k_i})$ is communicated to the controllers $F_j$ of the successor subsystems $\Sigma_j$ $(j \in \mathcal{S}_i)$ or at which the current state $x_j(t_{r_i(j)})$ is requested from the controllers $F_j$ of the predecessor subsystems $\Sigma_j$ $(j \in \mathcal{P}_i)$, respectively. Figure 5.7 depicts the structure of the local control unit $F_i$. $R$ denotes a message that initiates the request of information, and $t_\star$ represents different event times that are specified later. The components $C_i$ and $E_i$ are explained in the following.

The control input generator $C_i$ determines the control input $u_i(t)$ for subsystem $\Sigma_i$ using the model

$$C_i: \begin{cases} \tilde{\Sigma}_{ai}: \begin{cases} \frac{d}{dt}\tilde{x}_{ai}(t) = A_{ai}\tilde{x}_{ai}(t) + B_{ai} z_i(t) \\ \tilde{x}_{ai}(t_\star^+) = \sum\limits_{p \in \mathcal{P}_i \setminus \{j\}} T_{ip} C_{pi} \tilde{x}_{ai}(t_\star) \\ \qquad + T_{ij} x_j(t_\star), \quad j \in \mathcal{P}_i \end{cases} \\ u_i(t) = -K_i x_i(t) - K_{ai}\tilde{x}_{ai}(t). \end{cases} \tag{5.61}$$

The structure of this generator is illustrated in Figure 5.8. The state $\tilde{x}_{ai}(t)$ is a prediction of the actual approximate model state $x_{ai}(t)$ that is generated by means of the model $\tilde{\Sigma}_{ai}$. Note that $\tilde{\Sigma}_{ai}$ differs from the approximate model $\Sigma_{ai}$ in that the disturbance $d_{ai}(t)$ and the

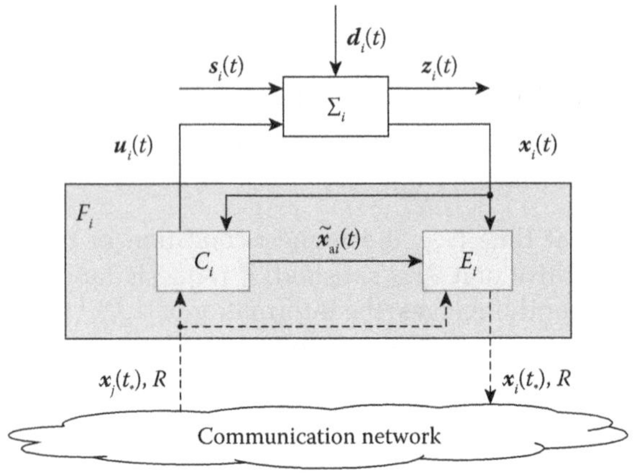


**FIGURE 5.7**
Component $F_i$ of the networked controller.

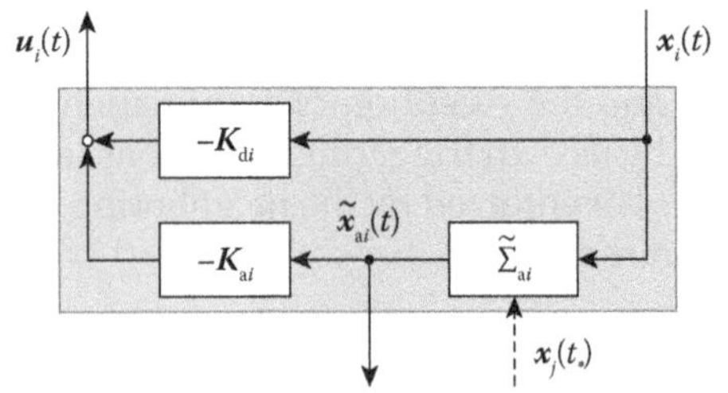


**FIGURE 5.8**
Control input generator $C_i$ of the control unit $F_i$.

residual output $f_i(t)$ are omitted, since these signals are unknown to the local controller $F_i$. The generation of the control input $u_i(t)$ is equivalent with Equation 5.55, since

$$K_{ai}\tilde{x}_{ai}(t) = \sum_{j\in\mathcal{P}_i} K_{ai}T_{ij}\tilde{x}_j^i(t) = \sum_{j\in\mathcal{P}_i} K_{ij}\tilde{x}_j^i(t).$$

The state $\tilde{x}_{ai}$ is reinitialized at the event times $t_\star$, where the $\star$-symbol is a placeholder either for $k_j$ or for $r_i(j)$. That is, $\tilde{x}_{ai}$ is reinitialized whenever $F_i$ receives current state information $x_j(t_\star)$ from $F_j$ ($j \in \mathcal{P}_i$) which occurs in two situations:

- At time $t = t_{k_j}$, the event generator $E_j$ decides to transmit the current state $x_j(t_{k_j})$ to all controllers $F_i$ of the successor subsystems $\Sigma_j$—that is, to all $F_i$ with $i \in \mathcal{S}_j$.

- At time $t = t_{r_i(j)}$, $E_i$ requests current state information from the controller $F_j$ of the predecessor subsystem $\Sigma_j$—that is, from $F_j$ with $j \in \mathcal{P}_i$.

The two event times $t_{k_j}$ and $t_{r_i(j)}$ are determined based on conditions that are defined in the next paragraph.

The task of the event generator $E_i$ is to bound the deviation

$$\left|x_{\Delta j}^i(t)\right| = \left|x_j(t) - C_{ji}\tilde{x}_{ai}(t)\right| = \left|x_j(t) - \tilde{x}_j^i(t)\right|, \quad (5.62)$$

between the actual subsystem state $x_j(t)$ and the prediction $C_{ji}\tilde{x}_{ai}(t) = \tilde{x}_j^i(t)$, applied for the control input generation (5.61). However, the difference (5.62) cannot be monitored by any of the event generators $E_i$ or $E_j$, which exclusively have access to the prediction $\tilde{x}_{ai}(t)$ or to the subsystem state $x_j(t)$, respectively. This consideration gives rise to the idea that the boundedness of (5.62) can be accomplished by a cooperation of the event generators $E_i$ and $E_j$, which is explained in the following.

Consider that both $E_i$ and $E_j$ include the model

$$\Sigma_{cj}: \ \dot{x}_{cj}(t) = \bar{A}_j x_{cj}(t), \quad x_{cj}(t_\star^+) = x_j(t_\star) \quad (\text{if } t_\star = t_{k_j} \text{ or } t_\star = t_{r_i(j)}), \quad (5.63)$$

with the matrix $\bar{A}_j = (A_j - B_jK_{dj})$. The state $x_{cj}(t)$ is used in both event generators $E_i$ and $E_j$ as a comparison signal as follows:

- Event generator $E_j$ triggers an event at time $t_{k_j}$ whenever the *transmit-condition*

$$\left|x_j(t) - x_{cj}(t)\right| \not< \alpha_j\bar{e}_j, \quad (5.64)$$

is satisfied, where $\bar{e}_j \in \mathbb{R}_+^{n_j}$ denotes the event threshold vector and $\alpha_j \in (0,1)$ is a weighting factor.

- The event generator $E_i$ triggers an event at time $t_{r_i(j)}$ which induces the request of the state $x_j(t_{r_i(j)})$ from $F_j$ whenever the *request-condition*

$$\left|C_{ji}\tilde{x}_{ai}(t) - x_{cj}(t)\right| \not< (1-\alpha_j)\bar{e}_j, \quad j \in \mathcal{P}_i, \quad (5.65)$$

is fulfilled, with the same event threshold vector $\bar{e}_j \in \mathbb{R}_+^{n_j}$ as in (5.64). That means, at time $t_{r_i(j)}$, the event generator $E_i$ sends a request $R$ to the controller $F_j$ in order to call this controller to transmit the current state $x_j(t_{r_i(j)})$.

The event generator $E_j$ of the controller $F_j$ responds to the triggering of the events caused by the transmit-condition (5.64) or by the request-condition (5.65) with the same action. It sends the current state $x_j(t_\star)$ (with $t_\star = t_{k_j}$ or $t_\star = t_{r_i(j)}$ if the transmit-condition (5.64) or the request-condition (5.65), respectively, has led to the event triggering) to all $F_p$ with $p \in \mathcal{S}_j$ (that is, to all controllers of the successor subsystems $\Sigma_p$ of subsystem $\Sigma_j$, which includes $F_i$). The received information is then used to reinitialize the state of the models (5.61) in all $C_p$

with $p \in \mathcal{S}_j$, as well as the state of the models (5.63) in $E_j$ and in all $E_p$ with $p \in \mathcal{S}_j$. Due to the event triggering and the state resets, the relations

$$|\boldsymbol{x}_i(t) - \boldsymbol{x}_{ci}(t)| \le \alpha_i \bar{\boldsymbol{e}}_i, \quad (5.66)$$

$$\left|\boldsymbol{C}_{ji}\tilde{\boldsymbol{x}}_{ai}(t) - \boldsymbol{x}_{cj}(t)\right| \le (1-\alpha_j)\bar{\boldsymbol{e}}_j, \quad (5.67)$$

hold for all $i \in \mathcal{N}$ and all $j \in \mathcal{P}_i$. By virtue of (5.66) the state difference (5.62) is bounded by

$$\begin{aligned}\left|\boldsymbol{x}^i_{\Delta j}(t)\right| &= \left|\boldsymbol{x}_j(t) - \boldsymbol{x}_{cj}(t) + \boldsymbol{x}_{cj}(t) - \boldsymbol{C}_{ji}\tilde{\boldsymbol{x}}_{ai}(t)\right| \\ &\le \left|\boldsymbol{x}_j(t) - \boldsymbol{x}_{cj}(t)\right| + \left|\boldsymbol{C}_{ji}\tilde{\boldsymbol{x}}_{ai}(t) - \boldsymbol{x}_{cj}(t)\right| \\ &\le \alpha_j\bar{\boldsymbol{e}}_j + (1-\alpha_j)\bar{\boldsymbol{e}}_j \\ &= \bar{\boldsymbol{e}}_j. \end{aligned} \quad (5.68)$$

In summary, the event generator $E_i$ is represented by

$$E_i : \begin{cases} \Sigma_{cj} \text{ as described in (5.63), for all } j \in \mathcal{P}_i \cup \{i\} \\ t_{k_i+1} := \inf\left\{t > \hat{t}_{\mathrm{Tx}i} \;\middle|\; |\boldsymbol{x}_i(t) - \boldsymbol{x}_{ci}(t)| \not\le \alpha_i\bar{\boldsymbol{e}}_i\right\} \\ t_{r_i(j)+1} := \inf\left\{t > \hat{t}_{\mathrm{Rx}i(j)} \;\middle|\; \left|\boldsymbol{C}_{ji}\tilde{\boldsymbol{x}}_{ai}(t) - \boldsymbol{x}_{cj}(t)\right| \right. \\ \qquad \left. \not\le (1-\alpha_j)\bar{\boldsymbol{e}}_j\right\}, \ \forall j \in \mathcal{P}_i \end{cases} \quad (5.69)$$

In (5.69) the time

$$\hat{t}_{\mathrm{Tx}i} := \max\left\{t_{k_i}, \max_{j \in \mathcal{S}_i}\left\{t_{r_j(i)}\right\}\right\},$$

denotes the last time at which $E_i$ has transmitted the state $\boldsymbol{x}_i$ to the controllers $F_j$ of the successor subsystems $\Sigma_j$, caused by either the violation of the transmit-condition or the request of information from one of the controllers $F_j$ with $j \in \mathcal{S}_i$. Accordingly, the time

$$\hat{t}_{\mathrm{Rx}i(j)} := \max\left\{t_{r_i(j)}, t_{k_j}\right\},$$

denotes the last time at which $E_i$ has received information about the state $\boldsymbol{x}_j$ either by request or due to the triggering of the transmit-condition in $E_j$.

Figure 5.9 illustrates the structure of the event generator $E_i$. Note that the logic that determines the time instants at which the current state $\boldsymbol{x}_i(t_\star)$ (with $t_\star = t_{k_i}$ or $t_\star = t_{r_j}$) is sent to the successor subsystems is decoupled from the logic that decides when to request state information from the predecessor subsystems. The latter is represented in the gray block for the example of requesting information from some subsystem $\Sigma_j$ ($j \in \mathcal{P}_i$). Note that this logic, highlighted in the gray block, is implemented in $E_i$ for each $j \in \mathcal{P}_i$.

The communication policy leads to a topology for the information exchange that can be characterized as follows:

> The event generator $E_i$ of the controller $F_i$ transmits the state information $\boldsymbol{x}_i$ only to the controllers $F_j$ of the successor subsystems $\Sigma_j$ ($\in \mathcal{S}_i$) of $\Sigma_i$, whereas $F_i$ receives the state information $\boldsymbol{x}_j$ only from the controllers $F_j$ of the predecessor subsystems $\Sigma_j$ ($j \in \mathcal{P}_i$) of $\Sigma_i$.

### 5.4.4 Analysis of the Distributed Event-Based Control

**Approximation of the reference system behavior.** This section shows that the deviation between the behavior of the reference systems (5.56) and (5.57) and the control systems (5.3) and (5.4) with the distributed event-based state feedback (5.61) and (5.69) is bounded, and the maximum deviation can be adjusted by means of an appropriate choice of the event threshold vectors $\bar{\boldsymbol{e}}_j$ for all $j \in \mathcal{N}$. This result is used to determine the asymptotically stable set $\mathcal{A}_i$ for each subsystem $\Sigma_i$.

The following theorem gives an upper bound for the deviation of the behavior of the event-based control systems (5.3), (5.4), (5.61), and (5.69) and the reference systems (5.56) and (5.57).

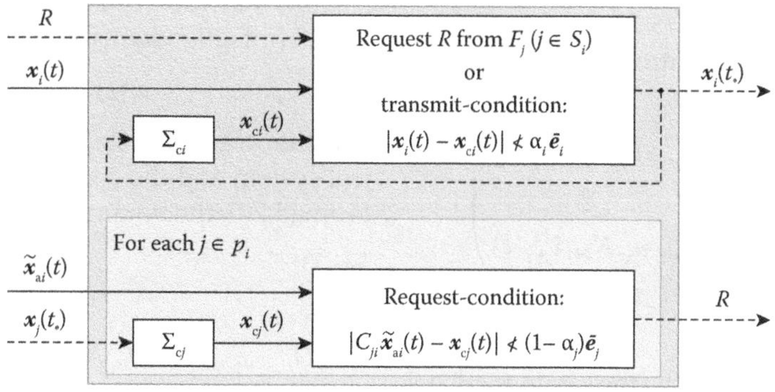


**FIGURE 5.9**
Event generator $E_i$ of the control unit $F_i$.

**Theorem 5.8**

Consider the event-based control systems (5.3), (5.4), (5.61), and (5.69) and the reference systems (5.56) and (5.57) with the plant states $x(t)$ and $x_r(t)$, respectively. The deviation $e(t) = x(t) - x_r(t)$ is bounded from above by

$$|e(t)| \le e_{\max} := \int_0^\infty \left|e^{\bar{A}t}B\right| dt \cdot |\bar{K}| \begin{pmatrix} \bar{e}_1 \\ \vdots \\ \bar{e}_N \end{pmatrix}, \tag{5.70}$$

for all $t \ge 0$ with the matrix $\bar{K}$ given in (5.60).

**PROOF** The subsystem $\Sigma_i$ represented by (5.3) with the control input that is generated by the control input generator (5.61) is described by the state-space model:

$$\dot{x}_i(t) = \bar{A}_i x_i(t) - B_i \sum_{j\in\mathcal{P}_i} K_{ij}\tilde{x}_j^i(t) + E_i d_i(t) + E_{si}s_i(t),$$
$$x_i(0) = x_{0i},$$
$$z_i(t) = C_{zi}x_i(t),$$

with the matrix $\bar{A}_i = (A_j - B_i K_{di})$. The state $x_i(t)$ depends on the difference states $x_{\Delta j}^i(t) = x_j(t) - \tilde{x}_j^i(t)$ for all $j \in \mathcal{P}_i$, which can be seen by reformulating the previous model as

$$\begin{aligned} \dot{x}_i(t) &= \bar{A}_i x_i(t) - B_i \sum_{j\in\mathcal{P}_i} K_{ij}x_j(t) \\ &\quad + B_i \sum_{j\in\mathcal{P}_i} K_{ij}(x_j(t) - \tilde{x}_j^i(t)) + E_i d_i(t) \\ &\quad + E_{si}s_i(t), \end{aligned} \tag{5.71}$$
$$x_i(0) = x_{0i}, \tag{5.72}$$
$$z_i(t) = C_{zi}x_i(t). \tag{5.73}$$

The interconnection of the models (5.71) according to the relation (5.4) yields the overall control system model:

$$\dot{x}(t) = \bar{A}x(t) + Ed(t) + B \begin{pmatrix} \sum_{j\in\mathcal{P}_1} K_{1j}\tilde{x}_{\Delta j}^1(t) \\ \vdots \\ \sum_{j\in\mathcal{P}_N} K_{Nj}\tilde{x}_{\Delta j}^N(t) \end{pmatrix},$$
$$x(0) = x_0, \tag{5.74}$$

where $\bar{A} = A - B(K_d + \bar{K})$ with the state-feedback gains $K_d$ and $\bar{K}$ as given in (5.59) and (5.60), respectively, and the matrices $A$, $B$, $E$ according to Equation 5.6.

Now consider the deviation $e(t) = x(t) - x_r(t)$. With Equations 5.58 and 5.74, the deviation $e(t)$ is described by the model

$$\dot{e}(t) = \bar{A}e(t) + B \begin{pmatrix} \sum_{j\in\mathcal{P}_1} K_{1j}\tilde{x}_{\Delta j}^1(t) \\ \vdots \\ \sum_{j\in\mathcal{P}_N} K_{Nj}\tilde{x}_{\Delta j}^N(t) \end{pmatrix}, \quad e(0) = 0,$$

which yields

$$e(t) = \int_0^t e^{\bar{A}(t-\tau)}B \begin{pmatrix} \sum_{j\in\mathcal{P}_1} K_{1j}\tilde{x}_{\Delta j}^1(\tau) \\ \vdots \\ \sum_{j\in\mathcal{P}_N} K_{Nj}\tilde{x}_{\Delta j}^N(\tau) \end{pmatrix} d\tau.$$

The last equation is used to derive a bound on the maximum deviation $|e(t)|$ by means of the following estimation:

$$\begin{aligned} |e(t)| &\le \int_0^t \left|e^{\bar{A}(t-\tau)}B\right| \begin{pmatrix} \sum_{j\in\mathcal{P}_1} |K_{1j}| \left|\tilde{x}_{\Delta j}^1(\tau)\right| \\ \vdots \\ \sum_{j\in\mathcal{P}_N} |K_{Nj}| \left|\tilde{x}_{\Delta j}^N(\tau)\right| \end{pmatrix} d\tau \\ &\le \int_0^\infty \left|e^{\bar{A}t}B\right| dt \begin{pmatrix} \sum_{j\in\mathcal{P}_1} |K_{1j}|\, \bar{e}_j \\ \vdots \\ \sum_{j\in\mathcal{P}_N} |K_{Nj}|\, \bar{e}_j \end{pmatrix} \\ &=: e_{\max}, \end{aligned} \tag{5.75}$$

where the relation (5.68) (for each $j \in \mathcal{N}$) has been applied. Consequently, Equation 5.75 can be reformulated as (5.70), which completes the proof. ∎

Theorem 5.8 shows that the systems (5.3) and (5.4) with the distributed event-based state feedback (5.61) and (5.69) can be made to approximate the behavior of the reference systems (5.56) and (5.57) with adjustable accuracy, by appropriately setting the event thresholds $\bar{e}_i$ for all $i \in \mathcal{N}$. Interestingly, the weighting factors $\alpha_i$ ($i \in \mathcal{N}$) that balance the event threshold vector $\bar{e}_i$ in the transmit- and request-conditions (5.64) and (5.65) do not affect the maximum deviation (5.70) in any way.

Note that the bound (5.70) holds element-wise and, thus, implicitly expresses also a bound on the deviation

$$e_i(t) = x_i(t) - x_{ri}(t),$$

between the subsystem states of the event-based control system and the corresponding reference subsystem. From (5.70), the bound

$$\begin{aligned} &|e_i(t)| \\ &\le e_{\max i} \\ &= \begin{pmatrix} O_{n_i\times n_1} & \dots & O_{n_i\times n_{i-1}} & I_{n_i} & O_{n_i\times n_{i+1}} & \dots & O_{n_i\times n_N} \end{pmatrix} \\ &\quad \times e_{\max}, \end{aligned} \tag{5.76}$$

follows.

The result (5.76) can be combined with the results (5.50) and (5.51) on the asymptotically stable set $\mathcal{A}_{ri}$ for the reference systems (5.56) in order to infer the asymptotically stable set $\mathcal{A}_i$ for the subsystems $\Sigma_i$ with distributed event-based state feedback.

**Corollary 5.1**

Consider the event-based control systems (5.3), (5.4), (5.61), and (5.69). Each subsystem $\Sigma_i$ is practically stable with respect to the set

$$\mathcal{A}_i := \left\{ \boldsymbol{x}_i \in \mathbb{R}^{n_i} \mid |\boldsymbol{x}_i| \leq \boldsymbol{b}_i \right\}, \qquad (5.77)$$

with the bound

$$\boldsymbol{b}_i = \boldsymbol{b}_{ri} + \boldsymbol{e}_{\max i}, \qquad (5.78)$$

where $\boldsymbol{b}_{ri}$, given in (5.51), is the bound for the reference subsystem (5.56) and $\boldsymbol{e}_{\max i}$, given in (5.76), denotes the maximum deviation between the behavior of subsystem $\Sigma_i$ subject to the distributed event-based controller (5.61) and (5.69) and the reference subsystem (5.56).

Corollary 5.1 can be interpreted as follows: Given that the vector $\boldsymbol{e}_{\max}$ can be freely manipulated by the choice of the event threshold vectors $\bar{\boldsymbol{e}}_i$ for all $i \in \mathcal{N}$, the difference between the sets $\mathcal{A}_{ri}$ and $\mathcal{A}_i$ can be arbitrarily adjusted, as well.

**Minimum interevent times.** This section investigates the minimum interevent times for events that are triggered by the transmit-condition (5.64) and the request-condition (5.65). The following analysis is based on the assumption that the state $\boldsymbol{x}_i(t)$ of subsystem $\Sigma_i$ is contained within the set $\mathcal{A}_i$ for all $i \in \mathcal{N}$.

The following theorem states that the minimum interevent time

$$T_{\min,\mathrm{TC}i} := \min_{k_i} \left( t_{k_i+1} - t_{k_i} \right), \quad \forall\, k_i \in \mathbb{N}_0, \qquad (5.79)$$

for two consecutive events that are triggered according to the transmit-condition (5.64) by the controller $F_i$ is bounded from below by some time $\bar{T}_{\mathrm{TC}i}$.

**Theorem 5.9**

The minimum interevent time $T_{\min,\mathrm{TC}i}$ defined in (5.79) is bounded from below by

$$T_{\min,\mathrm{TC}i} \geq \bar{T}_{\mathrm{TC}i}, \quad \forall\, t \geq 0,$$

where the time $\bar{T}_{\mathrm{TC}i}$ is given by

$$\bar{T}_{\mathrm{TC}i} := \arg\min_t \left\{ \int_0^t \left| \mathrm{e}^{\bar{\boldsymbol{A}}_i \tau} \right| \mathrm{d}\tau \left( \sum_{j \in \mathcal{P}_i} |\boldsymbol{B}_i \boldsymbol{K}_{ij}| \left( \boldsymbol{b}_j + \bar{\boldsymbol{e}}_j \right) + \sum_{j \in \mathcal{P}_i} |\boldsymbol{E}_{si} \boldsymbol{L}_{ij} \boldsymbol{C}_{zj}|\, \boldsymbol{b}_j + |\boldsymbol{E}_i|\, \bar{\boldsymbol{d}}_i \right) = \alpha_i \bar{\boldsymbol{e}}_i \right\}, \qquad (5.80)$$

with the bounds $\boldsymbol{b}_j$ given in (5.78).

**PROOF** Consider the transmit-condition (5.64) and observe that the difference

$$\boldsymbol{\delta}_{ci}(t) := \boldsymbol{x}_i(t) - \boldsymbol{x}_{ci}(t),$$

evolves for $t \geq \hat{t}_{\mathrm{Tx}i}$ according to the state-space model

$$\frac{\mathrm{d}}{\mathrm{d}t}\boldsymbol{\delta}_{ci}(t) = \bar{\boldsymbol{A}}_i \boldsymbol{\delta}_{ci}(t) - \sum_{j \in \mathcal{P}_i} \boldsymbol{B}_i \boldsymbol{K}_{ij} \tilde{\boldsymbol{x}}_j^i(t) + \boldsymbol{E}_i \boldsymbol{d}_i(t) + \boldsymbol{E}_{si} \boldsymbol{s}_i(t),$$
$$\boldsymbol{\delta}_{ci}(\hat{t}_{\mathrm{Tx}i}) = \boldsymbol{0}, \qquad (5.81)$$

With $\hat{t}_{\mathrm{TX}i} = t_{k_i}$, Equation 5.81 yields

$$\boldsymbol{\delta}_{ci}(t) = \int_{t_{k_i}}^t \mathrm{e}^{\bar{\boldsymbol{A}}_i (t-\tau)} \times \left( -\sum_{j \in \mathcal{P}_i} \boldsymbol{B}_i \boldsymbol{K}_{ij} \tilde{\boldsymbol{x}}_j^i(\tau) + \boldsymbol{E}_i \boldsymbol{d}_i(\tau) + \boldsymbol{E}_{si} \boldsymbol{s}_i(\tau) \right) \mathrm{d}\tau.$$

With Equation 5.4 and the relation $\boldsymbol{z}_j(t) = \boldsymbol{C}_{zj} \boldsymbol{x}_j(t)$, the previous equation can be restated as

$$\boldsymbol{\delta}_{ci}(t) = \int_{t_{k_i}}^t \mathrm{e}^{\bar{\boldsymbol{A}}_i (t-\tau)} \left( -\sum_{j \in \mathcal{P}_i} \boldsymbol{B}_i \boldsymbol{K}_{ij} \tilde{\boldsymbol{x}}_j^i(\tau) + \boldsymbol{E}_i \boldsymbol{d}_i(\tau) + \sum_{j \in \mathcal{P}_i} \boldsymbol{E}_{si} \boldsymbol{L}_{ij} \boldsymbol{C}_{zj} \boldsymbol{x}_j(\tau) \right) \mathrm{d}\tau.$$

In order to analyze the minimum time that elapses in between the two consecutive events $k_i$ and $k_i + 1$, assume that no information is requested from $F_i$ after $t_{k_i}$ until the next event at time $t_{k_i+1}$ occurs. Recall that an event is triggered according to the transmit-condition (5.64) whenever $|\boldsymbol{\delta}_{ci}(t)| = |\boldsymbol{x}_i(t) - \boldsymbol{x}_{ci}(t)| = \alpha_i \bar{\boldsymbol{e}}_i$ holds. Hence, the minimum interevent time $T_{\min,\mathrm{TC}i}$ is the minimum time $t$ for which

$$\left| \int_0^t \mathrm{e}^{\bar{\boldsymbol{A}}_i (t-\tau)} \left( -\sum_{j \in \mathcal{P}_i} \boldsymbol{B}_i \boldsymbol{K}_{ij} \tilde{\boldsymbol{x}}_j^i(\tau) + \boldsymbol{E}_i \boldsymbol{d}_i(\tau) + \sum_{j \in \mathcal{P}_i} \boldsymbol{E}_{si} \boldsymbol{L}_{ij} \boldsymbol{C}_{zj} \boldsymbol{x}_j(\tau) \right) \mathrm{d}\tau \right| = \alpha_i \bar{\boldsymbol{e}}, \qquad (5.82)$$

is satisfied. The following analysis derives a bound on that minimum interevent time. To this end, it is assumed that the state $x_i(t)$ is bounded to the set $\mathcal{A}_i$ given in (5.77) for all $i \in \mathcal{N}$, which implies that

$$|x_i(t)| \le b_i, \tag{5.83}$$

holds where the bound $b_i$ is given in (5.78). By virtue of the event-triggering mechanism and the state reset, the prediction $\tilde{x}^i_j(t)$ always remains in a bounded surrounding of the subsystems state $x_j(t)$. Hence, from Equation 5.68 the relation

$$\left|\tilde{x}^i_j(t)\right| \le |x_j(t)| + \bar{e}_j \le b_j + \bar{e}_j, \tag{5.84}$$

follows. Taking account of the bound (5.5) on the disturbance $d_i(t)$, the left-hand side of Equation 5.82 is bounded by

$$\begin{aligned}
&\left| \int_0^t e^{\bar{A}_i(t-\tau)} \Big( -\sum_{j\in\mathcal{P}_i} B_i K_{ij} \tilde{x}^i_j(\tau) + E_i d_i(\tau) \\
&\quad + \sum_{j\in\mathcal{P}_i} E_{si} L_{ij} C_{zj} x_j(\tau) \Big) d\tau \right| \\
&\quad \le \left| \int_0^t e^{\bar{A}_i\tau} \right| d\tau \Big( \sum_{j\in\mathcal{P}_i} |B_i K_{ij}| (b_j + \bar{e}_j) \\
&\quad + \sum_{j\in\mathcal{P}_i} |E_{si} L_{ij} C_{zj}|\, b_j + |E_i|\, \bar{d}_i \Big).
\end{aligned}$$

Consequently, for the minimum interevent time $T_{\min,\mathrm{TC}i}$ the relation

$$\bar{T}_{\mathrm{TC}i} \le T_{\min,\mathrm{TC}i},$$

holds, where

$$\begin{aligned}
&\bar{T}_{\mathrm{TC}i} \\
&:= \arg\min_t \left\{ \int_0^t \left| e^{\bar{A}_i\tau} \right| d\tau \Big( \sum_{j\in\mathcal{P}_i} |B_i K_{ij}| (b_j + \bar{e}_j) \right. \\
&\quad \left. + \sum_{j\in\mathcal{P}_i} |E_{si} L_{ij} C_{zj}|\, b_j + |E_i|\, \bar{d}_i \Big) = \alpha_i \bar{e}_i \right\},
\end{aligned}$$

which completes the proof. ∎

The next theorem presents a bound on the minimum interevent time

$$T^i_{\min,\mathrm{RC}j} := \min_{r_i(j)} \left( t_{r_i(j)+1} - t_{r_i(j)} \right),$$

for two consecutive information requests from controller $F_j$, triggered according to the request-condition (5.65) by the controller $F_i$.

**Theorem 5.10**

The minimum time $T^i_{\min,\mathrm{RC}j}$ that elapses in between two consecutive events at which the controller $F_i$ requests information from controller $F_j$ is bounded from below by

$$T^i_{\min,\mathrm{RC}j} \ge \bar{T}^i_{\mathrm{RC}j}, \quad \forall\, t \ge 0.$$

The time $\bar{T}^i_{\mathrm{RC}j}$ is given by

$$\begin{aligned}
&\bar{T}^i_{\mathrm{RC}j} \\
&:= \arg\min_t \left\{ \int_0^t \left| e^{\bar{A}_j\tau} \right| d\tau \Big( \sum_{p\in\mathcal{P}_i\setminus\{j\}} |C_{ji} A_{ai} T_{ip}| (b_p + \bar{e}_p) \right. \\
&\quad \left. + |C_{ji} B_{ai} C_{zi}|\, b_i \Big) = (1-\alpha_j)\bar{e}_j \right\},
\end{aligned} \tag{5.85}$$

with the bounds $b_i$ as in (5.78).

**PROOF** In order to investigate the difference,

$$\delta^i_{cj}(t) := C_{ji}\tilde{x}_{ai}(t) - x_{cj}(t), \tag{5.86}$$

that is monitored in the request-condition (5.65), consider the state-space model

$$\begin{aligned}
\frac{d}{dt} C_{ji}\tilde{x}_{ai}(t) &= C_{ji} A_{ai} \tilde{x}_{ai}(t) + C_{ji} B_{ai} C_{zi} x_i(t), \\
C_{ji}\tilde{x}_{ai}(\hat{t}_{\mathrm{R}xi(j)}) &= x_j(\hat{t}_{\mathrm{R}xi(j)}),
\end{aligned}$$

which follows from (5.61). With Equation 5.27, this model can be restated as

$$\begin{aligned}
\frac{d}{dt} C_{ji}\tilde{x}_{ai}(t) &= \bar{A}_j \tilde{x}^i_j(t) + \sum_{p\in\mathcal{P}_i\setminus\{j\}} C_{ji} A_{ai} T_{ip} \tilde{x}^i_p(t) \\
&\quad + C_{ji} B_{ai} C_{zi} x_i(t), \\
C_{ji}\tilde{x}_{ai}(\hat{t}_{\mathrm{R}xi(j)}) &= x_j(\hat{t}_{\mathrm{R}xi(j)}),
\end{aligned}$$

where the assumption is made that the matrices $T_{ij}$ and $C_{ji}$ are chosen such that

$$\bar{A}_j = C_{ji} A_{ai} T_{ij},$$

holds. This model together with Equation 5.63 yield the state-space model

$$\begin{aligned}
\frac{d}{dt}\delta^i_{cj}(t) &= \bar{A}_j \delta^i_{cj}(t) + \sum_{p\in\mathcal{P}_i\setminus\{j\}} C_{ji} A_{ai} T_{ip} \tilde{x}^i_p(t) \\
&\quad + C_{ji} B_{ai} C_{zi} x_i(t), \quad \delta^i_{cj}(\hat{t}_{\mathrm{R}xi(j)}) = 0,
\end{aligned} \tag{5.87}$$

that describes the behavior of the difference (5.86) for $t \ge \hat{t}_{\mathrm{R}xi(j)}$. With $\hat{t}_{\mathrm{R}xi(j)} = t_{r_i(j)}$, Equation 5.87 results in

$$\begin{aligned}
\delta^i_{cj}(t) &= \int_{t_{r_i(j)}}^t e^{\bar{A}_j(t-\tau)} \Big( \sum_{p\in\mathcal{P}_i\setminus\{j\}} C_{ji} A_{ai} T_{ip} \tilde{x}^i_p(\tau) \\
&\quad + C_{ji} B_{ai} C_{zi} x_i(\tau) \Big) d\tau.
\end{aligned}$$

In the following, it is assumed that the controller $F_j$ does not trigger an event according to its transmission-condition (5.64) in between the times $t_{r_i(j)}$ and $t_{r_i+1(j)}$ in order to identify a minimum time that elapses in between two consecutive events that trigger the request of information from $F_j$ by the controller $F_i$. Note that this information request is triggered whenever the equality $\left|\delta^i_{cj}(t)\right| = \left|C_{ji}\tilde{x}_{ai}(t) - x_{cj}(t)\right| = (1-\alpha_j)\bar{e}_j$ is satisfied in at least one element. The minimum interevent time $T^i_{\min,RCj}$ is given by the minimum time $t$ for which

$$\left| \int_0^t e^{\bar{A}_j(t-\tau)} \left( \sum_{p\in\mathcal{P}_i\setminus\{j\}} C_{ji}A_{ai}T_{ip}\tilde{x}^i_p(\tau) + C_{ji}B_{ai}C_{zi}x_i(\tau) \right) d\tau \right| = 1 - \alpha_j\bar{e}_j,$$

holds. Following the same arguments as in the previous proof, a bound

$$\bar{T}^i_{RCj} \leq T^i_{\min,RCj},$$

on that minimum interevent time is given by

$$\bar{T}^i_{RCj} := \arg\min_t \left\{ \int_0^t \left| e^{\bar{A}_j\tau} \right| d\tau \left( \sum_{p\in\mathcal{P}_i\setminus\{j\}} \left|C_{ji}A_{ai}T_{ip}\right| (b_p + \bar{e}_p) + \left|C_{ji}B_{ai}C_{zi}\right| b_i \right) = (1-\alpha_j)\bar{e}_j \right\},$$

which concludes the proof. ■

## 5.5 Example: Distributed Event-Based Control of the Thermofluid Process

This section demonstrates, using the example of a thermofluid process, how the distributed state feedback is designed according to Algorithm 5.1 and how the obtained control law is implemented in an event-based fashion. Moreover, the results of a simulation and an experiment illustrate the behavior of the distributed event-based state-feedback approach.

### 5.5.1 Demonstration Process

The event-based control approaches presented in this thesis are tested, and the analytical results are quantitatively evaluated through simulations and experiments on a demonstration process realized at the pilot plant at the Institute of Automation and Computer Control at Ruhr-University Bochum, Germany (Figure 5.10). The plant includes four cylindrical storage tanks, three batch reactors, and a buffer tank which are connected over a complex pipe system, and it is constructed with standard industrial components including more than 70 sensors and 80 actuators.

**Process description.** The experimental setup of the process is illustrated in Figure 5.11. The main components are the two reactors TB and TS used to realize continuous flow processes. Reactor TB is connected to the storage tank $T_1$ from where the inflow can be controlled by means of the valve angle $u_{T1}(t)$. Via the pump

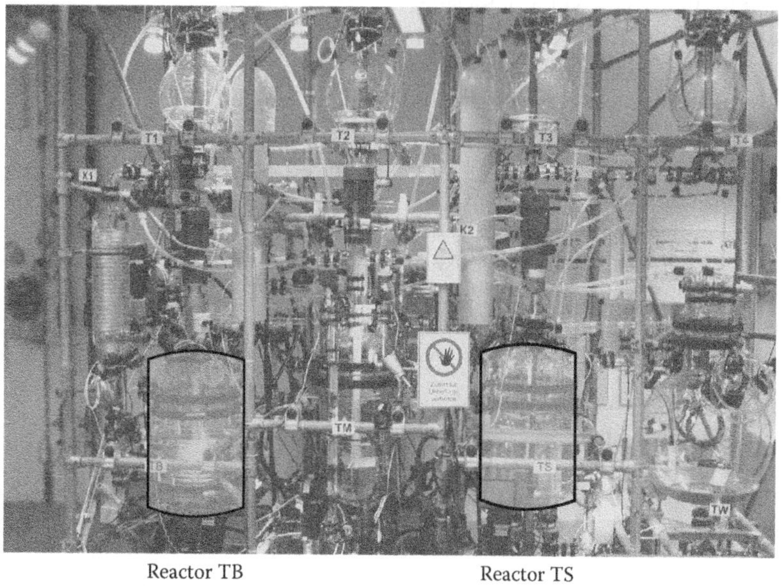


**FIGURE 5.10**
Pilot plant, where the reactors that are used for the considered process are highlighted.

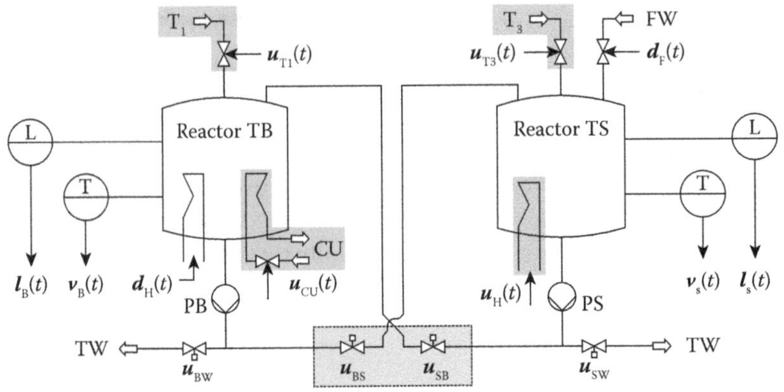


**FIGURE 5.11**
Experimental setup of the continuous flow process.

PB a part of the outflow is pumped out into the buffer tank TW (and is not used further in the process), while the remaining outflow is conducted to the reactor TS. The temperature $\vartheta_{\mathrm{B}}(t)$ of the water in reactor TB is influenced by the cooling unit (CU) using the input $u_{\mathrm{CU}}(t)$ or by the heating rods that are driven by the signal $d_{\mathrm{H}}(t)$. The inflow from the storage tank $T_3$ to the reactor TS can be adjusted by means of the opening angle $u_{\mathrm{T3}}(t)$. Reactor TS is additionally fed by the fresh water supply (FW) from where the inflow is controlled by means of the valve angle $d_{\mathrm{F}}(t)$. Equivalently to reactor TB, the outflow of reactor TS is split, and one part is conveyed via the pump PS to TW, and the other part is pumped to the reactor TB. The temperature $\vartheta_{\mathrm{TS}}(t)$ of the liquid in reactor TS can be increased by the heating rods that are controlled by the signal $u_{\mathrm{H}}(t)$. The signals $d_{\mathrm{H}}(t)$ and $d_{\mathrm{F}}(t)$ are disturbance inputs that are used to define test scenarios with reproducible disturbance characteristics, whereas the signals $u_{\mathrm{T1}}(t)$, $u_{\mathrm{CU}}(t)$, $u_{\mathrm{T3}}(t)$, and $u_{\mathrm{H}}(t)$ are control inputs. Both reactor TB and reactor TS are equipped with sensors that continuously measure the level $l_{\mathrm{B}}(t)$ or $l_{\mathrm{S}}(t)$ and the temperature $\vartheta_{\mathrm{B}}(t)$ or $\vartheta_{\mathrm{S}}(t)$, respectively.

The two reactors are coupled by the flow from reactor TB to reactor TS, and vice versa, where the coupling strength can be adjusted by means of the valve angles $u_{\mathrm{BS}}$ and $u_{\mathrm{SB}}$. The ratio of the volume that is used for the coupling of the systems and the outflow to TW is set by the valve angles $u_{\mathrm{BW}}$ and $u_{\mathrm{SW}}$.

**Model.** The linearized overall system is subdivided into four scalar subsystems with the states

$$x_1(t) = l_{\mathrm{B}}(t), \quad x_2(t) = \vartheta_{\mathrm{B}}(t),$$
$$x_3(t) = l_{\mathrm{S}}(t), \quad x_4(t) = \vartheta_{\mathrm{S}}(t).$$

The overall system is composed of the subsystems (5.3) with the following parameters:

Parameters of $\Sigma_1$ :

$$\begin{cases} A_1 = -10^{-3} \cdot 5.74, \quad B_1 = 10^{-3} \cdot 2.30, \quad E_1 = 0, \\ E_{\mathrm{s}1} = 10^{-3} \cdot 2.42, \quad C_{\mathrm{z}1} = 1. \end{cases} \tag{5.88}$$

Parameters of $\Sigma_2$ :

$$\begin{cases} A_2 = -10^{-3} \cdot 8.58, \; B_2 = -10^{-3} \cdot 38.9, \; E_2 = 0.169, \\ E_{\mathrm{s}2} = 10^{-3}\begin{pmatrix} -34.5 & 43.9 & 5.44 \end{pmatrix}, \quad C_{\mathrm{z}2} = 1. \end{cases} \tag{5.89}$$

Parameters of $\Sigma_3$ :

$$\begin{cases} A_3 = -10^{-3} \cdot 5.00, \quad B_3 = 10^{-3} \cdot 2.59, \\ E_3 = 10^{-3} \cdot 1.16, \quad E_{\mathrm{s}3} = 10^{-3} \cdot 2.85, \quad C_{\mathrm{z}3} = 1. \end{cases} \tag{5.90}$$

Parameters of $\Sigma_4$ :

$$\begin{cases} A_4 = -10^{-3} \cdot 5.58, \quad B_4 = 10^{-3} \cdot 35.0, \\ E_4 = -10^{-3} \cdot 20.7, \\ E_{\mathrm{s}4} = 10^{-3}\begin{pmatrix} -46.5 & 5.58 & 39.2 \end{pmatrix}, \quad C_{\mathrm{z}4} = 1. \end{cases} \tag{5.91}$$

The interconnection between the subsystems is represented by Equation 5.4 with

$$\begin{pmatrix} s_1(t) \\ s_2(t) \\ s_3(t) \\ s_4(t) \end{pmatrix} = \underbrace{\begin{pmatrix} 0 & 0 & 1 & 0 \\ \begin{pmatrix} 1 \\ 0 \\ 0 \end{pmatrix} & \begin{pmatrix} 0 \\ 0 \\ 0 \end{pmatrix} & \begin{pmatrix} 0 \\ 1 \\ 0 \end{pmatrix} & \begin{pmatrix} 0 \\ 0 \\ 1 \end{pmatrix} \\ 1 & 0 & 0 & 0 \\ \begin{pmatrix} 1 \\ 0 \\ 0 \end{pmatrix} & \begin{pmatrix} 0 \\ 1 \\ 0 \end{pmatrix} & \begin{pmatrix} 0 \\ 0 \\ 1 \end{pmatrix} & \begin{pmatrix} 0 \\ 0 \\ 0 \end{pmatrix} \end{pmatrix}}_{= L} \begin{pmatrix} z_1(t) \\ z_2(t) \\ z_3(t) \\ z_4(t) \end{pmatrix}. \tag{5.92}$$

### 5.5.2 Design of the Distributed Event-Based Controller

This section shows how a state-feedback law is determined first, which is afterward implemented in an event-based manner.

**Distributed state-feedback design.** According to step 1 in Algorithm 5.1, an approximate model $\Sigma_{ai}$ together with the bounds $\bar{G}_{fvi}(t)$ and $\bar{G}_{fdi}(t)$ on the impulse-response matrices for the residual models $\Sigma_{fi}$ must be determined for each subsystem $\Sigma_i$ $(i = 1, \ldots, 4)$. In this example, the approximate models are designed under the assumption that subsystem $\Sigma_i$ is controlled by a continuous state feedback, whereas, in fact, the control inputs are determined according to a distributed state feedback. The mismatch between the approximate model $\Sigma_{ai}$ and the model that describes the behavior of the neighboring subsystems of $\Sigma_i$ is aggregated in the residual model $\Sigma_{fi}$. For subsystem $\Sigma_1$, the approximate model with the corresponding residual model is presented by the state-space model:

$$\Sigma_{a1}: \begin{cases} \dot{x}_{a1}(t) = (A_3 - B_3 K_{d3})\, x_{a1}(t) \\ \qquad + E_3 d_{a1}(t) + E_{s3} L_{31} C_{z1} z_1(t) + 1 \cdot f_1(t) \\ s_1(t) = x_{a1}(t) \\ v_1(t) = z_1(t), \end{cases}$$

where $d_{a1}(t) = d_F(t)$ holds, and the residual model $\Sigma_{f1}$ is represented by

$$\Sigma_{f1}: \quad f_1(t) = -B_3 K_{a3} \cdot v_1(t),$$

where $K_{a3} = K_{31}$ holds. The remaining approximate models and residual models are determined accordingly.

In the second step, the state-feedback gains $K_{ei}$ are to be designed such that the controlled extended subsystems have a desired behavior. In this example, these matrices are chosen as follows:

$$K_{e1} = \begin{pmatrix} 3.00 & 1.05 \end{pmatrix},$$
$$K_{e2} = \begin{pmatrix} -0.70 & 0.89 & -1.13 & -0.14 \end{pmatrix},$$
$$K_{e3} = \begin{pmatrix} 1.00 & 1.10 \end{pmatrix}, \; K_{e4} = \begin{pmatrix} 0.60 & -1.33 & 0.16 & 1.12 \end{pmatrix}. \tag{5.93}$$

The structure of the distributed state-feedback law is better conveyed by restating these matrices in the form used in Equation 5.39 which yields

$$K_{d1} = 3.00, \qquad K_{13} = 1.05, \tag{5.94}$$
$$K_{21} = 0.89, \quad K_{d2} = -0.70, \quad K_{23} = -1.13, \quad K_{24} = -0.14, \tag{5.95}$$
$$K_{31} = 1.10, \qquad K_{d3} = 1.00, \tag{5.96}$$
$$K_{41} = -1.33, \quad K_{42} = 0.16, \quad K_{43} = 1.12, \quad K_{d4} = 0.60. \tag{5.97}$$

Note that the structure of the overall state-feedback gain is implicitly predetermined in the proposed design method by the interconnections of the subsystems. For the considered system, this design procedure leads to the fact that the controllers $F_2$ and $F_4$ of subsystems $\Sigma_2$ or $\Sigma_4$, respectively, use state information of all remaining subsystems.

In the final step, the stability of the overall control system is analyzed by means of the condition (5.43). The analysis results are summarized as follows:

$$\lambda_P\left(\int_0^\infty \bar{G}_{fv1}(t)\mathrm{d}t \int_0^\infty \left|H_{ei}e^{\bar{A}_{e1}t}F_{e1}\right| \mathrm{d}t\right) = 1.48 \cdot 10^{-4} < 1,$$
$$\lambda_P\left(\int_0^\infty \bar{G}_{fv2}(t)\mathrm{d}t \int_0^\infty \left|H_{e2}e^{\bar{A}_{e2}t}F_{e2}\right| \mathrm{d}t\right) = 0.37 < 1,$$
$$\lambda_P\left(\int_0^\infty \bar{G}_{fv3}(t)\mathrm{d}t \int_0^\infty \left|H_{e3}e^{\bar{A}_{e3}t}F_{e3}\right| \mathrm{d}t\right) = 2.51 \cdot 10^{-5} < 1,$$
$$\lambda_P\left(\int_0^\infty \bar{G}_{fv4}(t)\mathrm{d}t \int_0^\infty \left|H_{e4}e^{\bar{A}_{e4}t}F_{e4}\right| \mathrm{d}t\right) = 0.37 < 1.$$

Consequently, the distributed state-feedback controller (5.94) guarantees the stability of the overall control system. According to Theorem 5.7, each subsystem is practically stable with respect to the set $\mathcal{A}_{ri}$ defined in (5.50) with the bounds

$$b_{r1} = 0.00, \quad b_{r2} = 1.20, \quad b_{r3} = 6.11, \quad b_{r4} = 0.31. \tag{5.98}$$

The result $b_{r1} = 0$ shows that in the distributed continuous state-feedback system, the disturbance and couplings do not affect state $x_1(t)$; thus, this state eventually converges to the origin.

**Event-Based Implementation of the Distributed State Feedback.** For the implementation of the derived state feedback in an event-based manner, the event thresholds $\bar{e}_i$ and the weighting factor $\alpha_i$ for $i = 1, \ldots, 4$ are chosen as follows:

$$\begin{aligned} \bar{e}_1 = \bar{e}_3 = 0.035, &\qquad \alpha_1 = \alpha_3 = 0.85, \\ \bar{e}_2 = \bar{e}_4 = 3.2, &\qquad \alpha_2 = \alpha_4 = 0.94. \end{aligned} \tag{5.99}$$

According to Theorem 5.8, the maximum deviation between the distributed event-based state-feedback systems and the continuous reference system results in

$$e_{\max} = \begin{pmatrix} 0.67 & 0.51 & 1.31 & 0.79 \end{pmatrix}^\top. \tag{5.100}$$

By virtue of Corollary 5.1, this result together with the bounds (5.98) implies the practical stability of the subsystems $\Sigma_i$ with respect to the sets $\mathcal{A}_i$ given in (5.77) with

$$b_1 = 0.67, \quad b_2 = 1.71, \quad b_3 = 7.43, \quad b_4 = 1.10. \quad (5.101)$$

The following table summarizes the results on the bounds $\bar{T}_{\text{TC}i}$ and $\bar{T}^i_{\text{RC}j}$ on the minimum interevent times for events triggered due to the transmit-condition (5.64) or the request-condition (5.65). These times are obtained according to Theorems 5.9 and 5.10. The '—' symbolizes that $\Sigma_j$ is not a predecessor of $\Sigma_i$; hence, $F_i$ never requests information from $F_j$.

**TABLE 5.1**
Minimum Interevent Times (MIET)

| $F_i$ | MIET $\bar{T}_{\text{TC}i}$ in s for Transmissions | MIET $\bar{T}^i_{\text{RC}j}$ in s for Requests of Information from $F_j$ | | | |
|---|---|---|---|---|---|
| | | $j=1$ | $j=2$ | $j=3$ | $j=4$ |
| $F_1$ | 0.83 | – | – | 2.78 | – |
| $F_2$ | 4.21 | 0.29 | – | 2.64 | 0.58 |
| $F_3$ | 6.97 | 0.29 | – | – | – |
| $F_4$ | 4.62 | 0.29 | 5.97 | 0.65 | – |

The presented analysis results are evaluated by means of a simulation and an example in the next section.

### 5.5.3 Experimental Results

The following presents the results of an experiment, demonstrating the disturbance behavior of the distributed event-based state-feedback approach.

The experiment investigates the scenario where the overall system is perturbed in the time intervals $t \in [200, 600]$ s by the disturbance $d_{\text{H}}(t)$ and in $t \in [1000, 1400]$ s by the disturbance $d_{\text{F}}(t)$. The behavior of the overall plant with the distributed event-based controller subject to these disturbances is illustrated in Figure 5.12. It shows the disturbance behavior of each subsystem where from top to bottom the following signals are depicted: the disturbance $d_{\text{H}}(t)$ or $d_{\text{F}}(t)$, the subsystem state $x_i(t)$, the control input $u_i(t)$, and the time instants where information is received ($\text{Rx}_i$) and transmitted ($\text{Tx}_i$), indicated by stems. Regarding the reception of information, the amplitude of the stems refers to the subsystem that has transmitted the information.

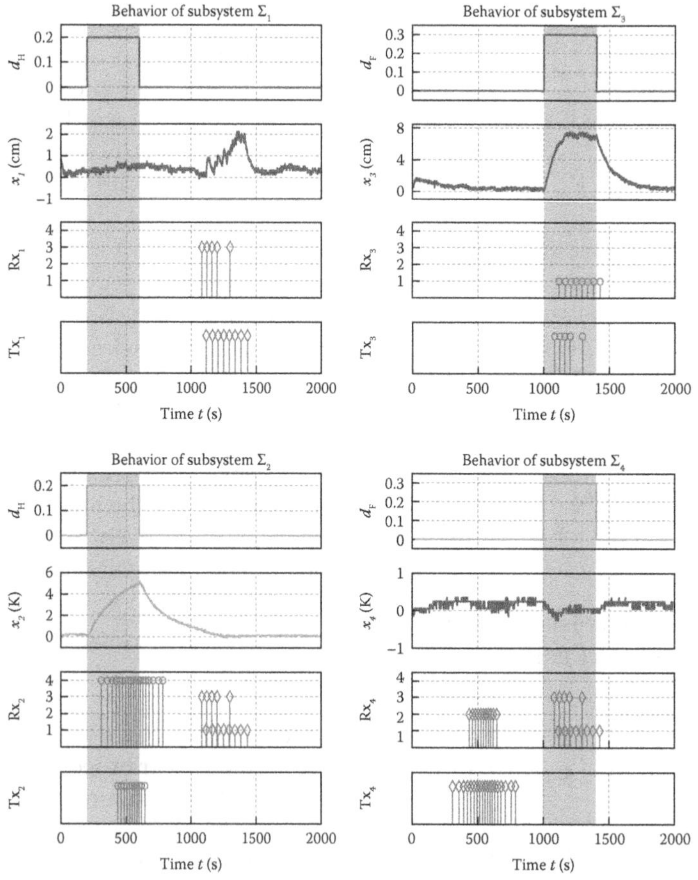


**FIGURE 5.12**
Experimental results for the disturbance behavior of the distributed event-based state feedback.

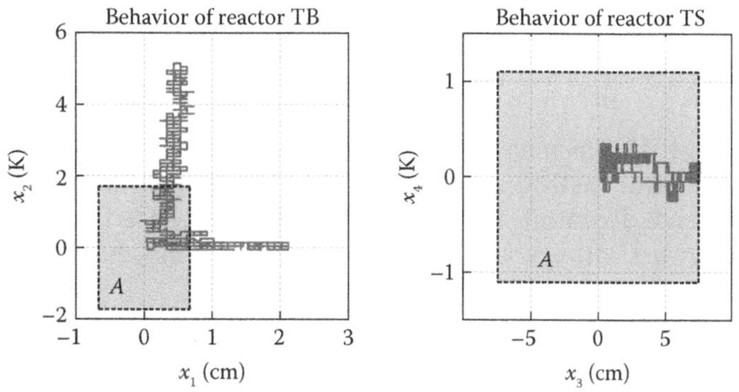


**FIGURE 5.13**
Trajectory of the overall system state in state space in the experiment.

For both $\mathrm{Rx}_i$ and $\mathrm{Tx}_i$ a stem with a circle denotes that the request-condition (5.65) or the transmit-condition (5.64) of the controller $F_i$ is satisfied and, thus, leads to the triggering of the respective event. For the receptions ($\mathrm{Rx}_i$), a stem with a diamond indicates that another control unit has transmitted information, because its local transmit-condition was satisfied or its state was requested by a third controller. For the transmissions ($\mathrm{Tx}_i$), a stem with a diamond shows that the controller $F_i$ was requested to transmit the local state by another controller.

The experiment shows how the communication is adapted to the behavior of the system. No event is triggered when the overall system is undisturbed, except for the times immediately after the disturbances vanish. In this experiment, the transmit-condition (5.64) is only satisfied in the controller $F_3$, whereas the request of information is only induced by the controllers $F_2$ and $F_3$. In this investigation, the minimum interevent times for events that are triggered according to the transmit-condition (5.64) or the request-condition (5.65) are

$$\begin{aligned} T_{\mathrm{min,TC3}} &\approx 90\,\mathrm{s} \ > \ 6.97\,\mathrm{s} = \bar{T}_{\mathrm{TC3}}, \\ T^2_{\mathrm{min,RC4}} &\approx 0.7\,\mathrm{s} \ > \ 0.58\,\mathrm{s} = \bar{T}^2_{\mathrm{RC4}}, \\ T^3_{\mathrm{min,RC1}} &\approx 49\,\mathrm{s} \ > \ 0.29\,\mathrm{s} = \bar{T}^3_{\mathrm{RC1}}. \end{aligned}$$

For these interevent times, the bounds that are listed in Table 5.1 hold true. In total, 43 events are triggered, which cause a transmission of information within 2000 s. Hence, from the communication aspect, the distributed event-based controller outperforms a conventional discrete-time controller (with appropriately chosen sampling time $h = 10\,\mathrm{s}$). Moreover, the experimental results also demonstrate that the proposed control approach is robust with respect to model uncertainties.

Figure 5.13 evaluates to what extend the analysis results, presented in the previous section, are valid for trajectories that are measured in the experiment. It shows that the bounds $b_1 = 0.67$ and $b_2 = 1.71$ do not hold, whereas the bounds $b_3 = 7.43$ and $b_4 = 1.10$ are not exceeded by the states $x_3(t)$ or $x_4(t)$, respectively. The reason why $x_2(t)$ grows much larger than the bound $b_2$ is again the saturation of the control input $u_2(t)$. The exceeding of the state $x_1(t)$ above the bound $b_1$ can be ascribed to unmodeled nonlinear dynamics of various components of the plant.

In summary, the experiment has shown that events are only triggered by the distributed event-based state-feedback controller whenever the plant is substantially affected by an exogenous disturbance and no feedback communication is required in the unperturbed case. This observation implies that the effect of the couplings between the subsystems has a minor impact on the triggering of events, since these dynamics are taken into account by using approximate models in the design of the local control units. In consequence, the proposed distributed event-based state-feedback method is suitable for the application to interconnected systems, where neighboring subsystems are strongly coupled.

## Bibliography

[1] D. S. Bernstein, *Matrix Mathematics: Theory, Facts and Formulas.* Princeton University Press, Princeton, NJ, 2009.

[2] C. De Persis, R. Sailer, F. Wirth, On inter-sampling times for event-triggered large-scale linear systems. In *Proceedings IEEE Conf. on Decision and Control*, 2013, pp. 5301–5306.

[3] C. De Persis, R. Sailer, F. Wirth, Parsimonious event-triggered distributed control: A Zeno free approach. *Automatica*, vol. 49, no. 7, pp. 2116–2124, 2013.

[4] M. C. F. Donkers, *Networked and Event-Triggered Control Systems*. PhD thesis, Eindhoven University of Technology, 2011.

[5] M. C. F. Donkers, W. P. M. H. Heemels, Output-based event-triggered control with guaranteed $\mathcal{L}_\infty$-gain and improved and decentralised event-triggering. *IEEE Trans. Autom. Control*, vol. 57, no. 6, pp. 1362–1376, 2012.

[6] R. Goebel, R. G. Sanfelice, A. R. Teel, Hybrid dynamical systems. *IEEE Control Syst. Magazine*, vol. 29, no. 2, pp. 28–93, 2009.

[7] M. Guinaldo, D. V. Dimarogonas, K. H. Johansson, J. Sánchez, S. Dormido, Distributed event-based control for interconnected linear systems. In *Proceedings IEEE Conf. on Decision and Control*, 2011, pp. 2553–2558.

[8] M. Guinaldo, D. Lehmann, J. Sánchez, S. Dormido, K. H. Johansson, Distributed event-triggered control with network delays and packet losses. In *Proceedings IEEE Conf. on Decision and Control*, 2012, pp. 1–6.

[9] W. P. M. H. Heemels, M. C. F. Donkers, Model-based periodic event-triggered control for linear systems. *Automatica*, vol. 49, no. 3, pp. 698–711, 2013.

[10] J. P. Hespanha, P. Naghshtabrizi, Y. Xu, A survey of recent results in networked control systems. *Proceedings of the IEEE*, vol. 95, no. 1, pp. 138–162, 2007.

[11] K.-D. Kim, P. R. Kumar, Cyber-physical systems: A perspective at the centennial. *Proceedings of the IEEE*, vol. 100, pp. 1287–1308, 2012.

[12] D. Lehmann, *Event-Based State-Feedback Control*. Logos-Verlag, Berlin, 2011.

[13] D. Lehmann, J. Lunze, Extension and experimental evaluation of an event-based state-feedback approach. *Contr. Eng. Practice*, vol. 19, no. 2, pp. 101–112, 2011.

[14] D. Lehmann, J. Lunze, Event-based control with communication delays and packet losses. *Int. J. Control*, vol. 85, no. 5, pp. 563–577, 2012.

[15] J. Lunze, *Feedback Control of Large-Scale Systems*. Prentice Hall, London, 1992.

[16] J. Lunze, F. Lamnabhi-Lagarrigue (Eds.), *Handbook of Hybrid Systems Control—Theory, Tools, Applications*. Cambridge University Press, Cambridge, 2009.

[17] J. Lunze, D. Lehmann, A state-feedback approach to event-based control. *Automatica*, vol. 46, no. 1, pp. 211–215, 2010.

[18] E. D. Sontag, Y. Wang, New characterizations of input-to-state stability. *IEEE Trans. Autom. Control*, vol. 41, no. 9, pp. 1283–1294, 1996.

[19] C. Stöcker, J. Lunze, Event-based control of input-output linearizable systems. In *Proceedings 18th IFAC World Congress*, 2011, pp. 10062–10067.

[20] C. Stöcker, J. Lunze, Event-based control of nonlinear systems: An input-output linearization approach. In *Proceedings Joint IEEE Conf. on Decision and Control and European Control Conf.*, 2011, pp. 2541–2546.

[21] C. Stöcker, J. Lunze, Event-based feedback control of disturbed input-affine systems. *J. Appl. Math. Mech.*, vol. 94, no. 4, pp. 290–302, 2014.

[22] C. Stöcker, J. Lunze, Input-to-state stability of event-based state-feedback control. In *Proceedings of the 13th European Control Conference*, 2013, pp. 49–54.

[23] C. Stöcker, D. Vey, J. Lunze, Decentralized event-based control: Stability analysis and experimental evaluation. *Nonlinear Anal. Hybrid Syst.*, vol. 10, pp. 141–155, 2013.

[24] Y. Tipsuwan, M.-Y. Chow, Control methodologies in networked control systems. *Contr. Eng. Practice*, vol. 11, no. 10, pp. 1099–1111, 2003.

[25] X. Wang, M. D. Lemmon, Event-triggered broadcasting across distributed networked control systems. In *Proceedings of the American Control Conference*, 2008, pp. 3139–3144.

[26] X. Wang, M. D. Lemmon, Event-triggering in distributed networked systems with data dropouts and delays, in *Hybrid Systems: Computation and Control* (R. Majumdar, P. Tabuada, Eds.), Springer: Berlin Heidelberg, 2009, pp. 366–380.

[27] X. Wang, M. D. Lemmon, Event-triggering in distributed networked control systems. *IEEE Trans. Autom. Control*, vol. 56, no. 3, pp. 586–601, 2011.

# 6

# *Periodic Event-Triggered Control*

**W. P. Maurice H. Heemels**
*Eindhoven University of Technology*
*Eindhoven, Netherlands*

**Romain Postoyan**
*University of Lorraine, CRAN UMR 7039, CNRS*
*Nancy, France*

**M. C. F. (Tijs) Donkers**
*Eindhoven University of Technology*
*Eindhoven, Netherlands*

**Andrew R. Teel**
*University of California*
*Santa Barbara, CA, USA*

**Adolfo Anta**
*General Electric*
*Munich, Germany*

**Paulo Tabuada**
*University of California*
*Los Angeles, CA, USA*

**Dragan Nešić**
*University of Melbourne*
*Parkville, Australia*

**CONTENTS**

**ABSTRACT** Recent developments in computer and communication technologies are leading to an increasingly networked and wireless world. This raises new challenging questions in the context of networked control systems, especially when the computation, communication, and energy resources of the system are limited. To efficiently use the available resources, it is desirable to limit the control actions to instances when the system really needs attention. Unfortunately, the classical time-triggered control paradigm is based on performing sensing and actuation actions periodically in time (irrespective of the state of the system) rather than when the system needs attention. Therefore, it is of interest to consider event-triggered control (ETC) as an alternative paradigm as it is more natural to trigger control actions based on the system state, output, or other available information. ETC can thus be seen as the introduction of feedback in the sensing, communication, and actuation processes. To facilitate an easy implementation of ETC, we propose to combine the principles and particularly the benefits of ETC and classical periodic time-triggered control. The idea is to periodically evaluate the triggering condition and to decide, at every sampling instant, whether the feedback loop needs to be closed. This leads to the periodic event-triggered control (PETC) systems. In this chapter, we discuss PETC strategies, their benefits, and two analysis and design frameworks for linear and nonlinear plants, respectively.

## 6.1 Introduction

In many digital control applications, the control task consists of sampling the outputs of the plant and computing and implementing new actuator signals. Typically, the control task is executed periodically, since this allows the closed-loop system to be analyzed and the controller to be designed using the well-developed theory on sampled-data systems. Although periodic sampling is preferred from an analysis and design point of view, it is sometimes less appropriate from a resource utilization point of view. Namely, executing the control task at times when no disturbances are acting on the system and the system is operating desirably is clearly a waste of resources. This is especially disadvantageous in applications where the measured outputs and/or the actuator signals have to be transmitted over a shared (and possibly wireless) network with limited bandwidth and energy-constrained wireless links. To mitigate the unnecessary waste of communication resources, it is of interest to consider an alternative control paradigm, namely, event-triggered control (ETC), which was proposed in the late 1990s, see [1–5] and [6] for a recent overview. Various ETC strategies have been proposed since then, see, for example, [7–18]. In ETC, the control task is executed after the occurrence of an event, generated by some well-designed event-triggering condition, rather than the elapse of a certain fixed period of time, as in conventional periodic sampled-data control. This can be seen as bringing feedback to the sensing, communication, and actuation processes, as opposed to "open-loop" sensing and actuation as in time-triggered periodic control. By using feedback principles, ETC is capable of significantly reducing the number of control task executions, while retaining a satisfactory closed-loop performance.

The main difference between the aforecited papers [1–5,7–18] and the ETC strategy that will be discussed in this chapter is that in the former, the event-triggering condition has to be monitored continuously, while in the latter, the event-triggering condition is evaluated only periodically, and at every sampling instant it is decided whether or not to transmit new measurements and control signals. The resulting control strategy aims at striking a balance between periodic time-triggered control on the one hand and event-triggered control on the other hand; therefore, we coined the term *periodic event-triggered control* (PETC) in [19,20] for this class of ETC. For the existing approaches that require monitoring of the event-triggering conditions continuously, we will use the term *continuous event-triggered control* (CETC). By mixing ideas from ETC and periodic sampled-data control, the benefits of reduced resource utilization are preserved in PETC as transmissions and controller computations are *not* performed periodically, even though the event-triggering conditions are evaluated only periodically. The latter aspect leads to several benefits, including a guaranteed minimum interevent time of (at least) the sampling interval of the event-triggering condition. Furthermore, as already mentioned, the event-triggering condition has to be verified only at periodic sampling instants, making PETC better suited for practical implementations as it can be implemented in more standard time-sliced embedded software architectures. In fact, in many cases CETC will typically be implemented using a discretized version based on a sufficiently high sampling period resulting in a PETC strategy (the results of [21] may be applied in this case to analyze stability of the resulting closed-loop system). This fact provides further motivation for a more direct analysis and design of PETC instead of obtaining them in a final implementation stage as a discretized approximation of a CETC strategy.

Initial work in the direction of PETC was taken in [2,7,8,22], which focused on restricted classes of systems, controllers, and/or (different) event-triggering conditions without providing a general analysis framework. Recently, the interest in what we call here PETC is growing, see, for example, [20,23–26] and [27, Sec. 4.5], although these approaches start from a discrete-time plant model instead of a continuous-time plant, as we do here. In this chapter, the focus is on approaches to PETC that include a formal analysis framework, which, moreover, apply for continuous-time plants and incorporate intersample behavior in the analysis. We first address the case of plants modeled by linear continuous-time systems. Afterward, we present preliminary results in the case where the plant dynamics is nonlinear. The presented results are a summary of our works in [19] and in [28], in which the interested reader will find all the proofs as well as further developments.

The chapter is organized as follows. We first introduce the PETC paradigm in Section 6.2. We then model PETC systems as impulsive systems in Section 6.3. Results for linear plants are presented in Section 6.4, and the case of nonlinear systems is addressed in Section 6.5. Section 6.6 concludes the chapter with a summary as well as a list of open problems.

### Nomenclature

Let $\mathbb{R} := (-\infty, \infty)$, $\mathbb{R}_+ := [0, \infty)$, $\mathbb{N} := \{1, 2, \ldots\}$, and $\mathbb{N}_0 := \{0, 1, 2, \ldots\}$. For a vector $x \in \mathbb{R}^n$, we denote by $\|x\| := \sqrt{x^\top x}$ its 2-norm. The distance of a vector $x$ to a set $\mathcal{A} \subset \mathbb{R}^n$ is denoted by $\|x\|_\mathcal{A} := \inf\{\|x - y\| \mid y \in \mathcal{A}\}$. For a real symmetric matrix $A \in \mathbb{R}^{n\times n}$, $\lambda_{\max}(A)$ denotes the maximum eigenvalue of $A$. For a matrix $A \in \mathbb{R}^{n\times m}$, we denote by $A^\top \in \mathbb{R}^{m\times n}$ the transpose of $A$, and by $\|A\| := \sqrt{\lambda_{\max}(A^\top A)}$ its induced 2-norm. For the sake of brevity, we sometimes write symmetric matrices of the form $\begin{bmatrix} A & B \\ B^\top & C \end{bmatrix}$ as $\begin{bmatrix} A & B \\ \star & C \end{bmatrix}$. We call a matrix $P \in \mathbb{R}^{n\times n}$ positive definite, and write $P \succ 0$, if $P$ is symmetric and $x^\top P x > 0$ for all $x \neq 0$. Similarly, we use $P \succeq 0$, $P \prec 0$, and $P \preceq 0$ to denote that $P$ is positive semidefinite, negative definite, and negative semidefinite, respectively. The notations $I$ and $0$ respectively stand for the identity matrix and the null matrix, whose dimensions depend on the context. For a locally integrable signal $w : \mathbb{R}_+ \to \mathbb{R}^n$, we denote by $\|w\|_{\mathcal{L}_2} := (\int_0^\infty \|w(t)\|^2 \mathrm{d}t)^{1/2}$ its $\mathcal{L}_2$-norm, provided the integral is finite. Furthermore, we define the set of all locally integrable signals with a finite $\mathcal{L}_2$-norm as $\mathcal{L}_2$. For a signal $w : \mathbb{R}_+ \to \mathbb{R}^n$, we denote the limit from below at time $t \in \mathbb{R}_+$ by $w^+(t) := \lim_{s\uparrow t} w(s)$. The solution $z$ of a time-invariant dynamical system at time $t \geqslant 0$ starting with the initial condition $z(0) = z_0$ will be denoted $z(t, z_0)$ or simply $z(t)$ when the initial state is clear from the context. The notation $\lfloor \cdot \rfloor$ stands for the floor function.

## 6.2 Periodic ETC Systems

In this section, we introduce the PETC paradigm. To do so, let us consider a plant whose dynamics is given by

$$\tfrac{\mathrm{d}}{\mathrm{d}t}x = f(x, u, w), \tag{6.1}$$

where $x \in \mathbb{R}^{n_x}$ denotes the state of the plant, $u \in \mathbb{R}^{n_u}$ is the input applied to the plant, and $w \in \mathbb{R}^{n_w}$ is an unknown disturbance.

In a conventional sampled-data state-feedback setting, the input $u$ is given by

$$u(t) = K(x(t_k)), \quad \text{for} \quad t \in (t_k, t_{k+1}], \tag{6.2}$$

where $t_k$, $k \in \mathbb{N}$, are the sampling instants, which are periodic in the sense that $t_k = kh$, $k \in \mathbb{N}$, for some properly chosen sampling interval $h > 0$. Hence, at each sampling instant, the state measurement is sent to the controller, which computes a new control input that is immediately applied to the plant.

The setup is different in PETC. In PETC, the sampled state measurement $x(t_k)$ is used to evaluate a criterion at each $t_k = kh$, $k \in \mathbb{N}$ for some $h > 0$, based on which it is decided (typically at the smart sensor) whether the feedback loop needs to be closed. In that way, a new control input is not necessarily periodically applied to the plant as in traditional sampled-data settings, even though the state is sampled at every $t_k$, $k \in \mathbb{N}$. This has the advantage of reducing the usage of the communication channel and of the controller computation resources, as well as the number of control input updates. The latter allows limiting the actuators, wear, and reducing the actuators, energy consumption, in some applications. As a consequence, the controller in PETC is given by

$$u(t) = K(\hat{x}(t)), \text{ for } t \in \mathbb{R}_+, \tag{6.3}$$

where $\hat{x}$ is a left-continuous signal* given for $t \in (t_k, t_{k+1}]$, $k \in \mathbb{N}$, by

$$\hat{x}(t) = \begin{cases} x(t_k), & \text{when } C(x(t_k), \hat{x}(t_k)) > 0 \\ \hat{x}(t_k), & \text{when } C(x(t_k), \hat{x}(t_k)) \leqslant 0 \end{cases}, \tag{6.4}$$

and some initial value for $\hat{x}(0)$. Considering the configuration in Figure 6.1, the value $\hat{x}(t)$ can be interpreted as the most recently transmitted measurement of the state $x$ to the controller at time $t$. Whether or not new state measurements are transmitted to the controller is based on the event-triggering criterion $\mathcal{C} : \mathbb{R}^{n_\xi} \to \mathbb{R}$ with $n_\xi := 2n_x$. In particular, if at time $t_k$ it holds that $\mathcal{C}(x(t_k), \hat{x}(t_k)) > 0$, the state $x(t_k)$ is transmitted over the network to the controller, and $\hat{x}$ and the control value $u$ are updated accordingly. In case $\mathcal{C}(x(t_k), \hat{x}(t_k)) \leqslant 0$, no new state information is sent to the controller, in which case the input $u$ is not updated and kept the same for (at least) another sampling interval, implying that no control computations are needed and no

*A signal $x : \mathbb{R}_+ \to \mathbb{R}^n$ is called left-continuous, if for all $t > 0$, $\lim_{s\uparrow t} x(s) = x(t)$.

new state measurements and control values have to be transmitted.

Contrary to CETC, we see that the triggering condition is evaluated only every $h$ units of time (and not continuously for all time $t \in \mathbb{R}_+$). Intuitively, we might want to design the criterion $\mathcal{C}$ as in CETC and to select the sampling period $h$ sufficiently small to obtain a PETC strategy which (approximately) preserves the properties ensured in CETC. Indeed, we know from [21] that if a disturbance-free CETC system is such that its origin (or, more generally, a compact set) is uniformly globally asymptotically stable, then the corresponding emulated PETC system preserves this property semiglobally and practically with fast sampling, under mild conditions as we will recall in Section 6.5.2. This way of addressing PETC may exhibit some limitations, as it may require very fast sampling of the state, which may not be possible to achieve because of the limited hardware capacities. Furthermore, we might want to work with "non-small" sampling periods in order to reduce the usage of the computation and communication resources. As such, there is a strong need for systematic methods to construct PETC strategies that appropriately take into account the features of the paradigm. The objective of this chapter is to address this challenge. We present in the next sections analysis and design results for systems (6.1), (6.3), and (6.4) such that desired stability or performance guarantees are satisfied, while the number of transmissions between the plant and the controller is kept small.

## 6.3 Impulsive System Formulation

The system described in the previous section combines continuous-time dynamics (6.1) with discrete-time phenomena (6.3) and (6.4). It is therefore natural to model PETC systems as impulsive systems (see [29]). An impulsive system is a system that combines the "flow" of the continuous dynamics with the discrete "jumps" occurring at each sampling instant.

We define $\xi := [x^\top \ \hat{x}^\top]^\top \in \mathbb{R}^{n_\xi}$, with $n_\xi = 2n_x$, and

$$g(\xi, w) := \begin{bmatrix} f(x, K(\hat{x}), w) \\ 0 \end{bmatrix}, \ J_1 := \begin{bmatrix} I & 0 \\ I & 0 \end{bmatrix}, \ J_2 := \begin{bmatrix} I & 0 \\ 0 & I \end{bmatrix}, \tag{6.5}$$

to arrive at an impulsive system given by

$$\frac{d}{dt}\begin{bmatrix} \xi \\ \tau \end{bmatrix} = \begin{bmatrix} g(\xi, w) \\ 1 \end{bmatrix}, \ \text{when } \tau \in [0, h], \tag{6.6}$$

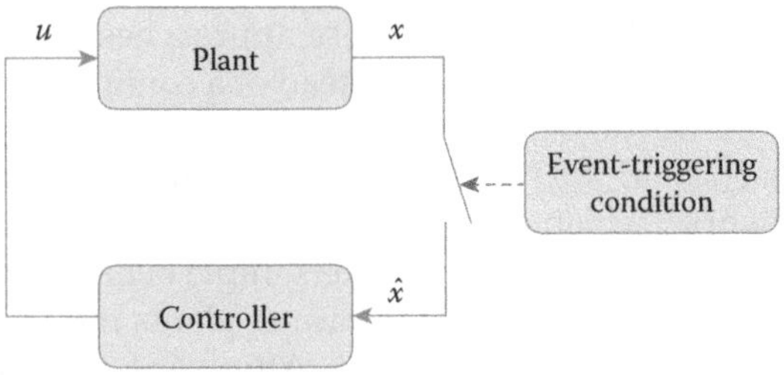


**FIGURE 6.1**
Periodic event-triggered control schematic.

$$\begin{bmatrix} \xi^+ \\ \tau^+ \end{bmatrix} = \begin{cases} \begin{bmatrix} J_1\xi \\ 0 \end{bmatrix}, & \text{when } \mathcal{C}(\xi) > 0, \ \tau = h \\ \begin{bmatrix} J_2\xi \\ 0 \end{bmatrix}, & \text{when } \mathcal{C}(\xi) \leqslant 0, \ \tau = h \end{cases} \tag{6.7}$$

where the state $\tau$ keeps track of the time elapsed since the last sampling instant. Between two successive sampling instants, $\xi$ and $\tau$ are given by the (standard) solutions to the ordinary differential equation (6.6), and these experience a jump dictated by (6.7) at every sampling instant. When the event-based condition is not satisfied, only $\tau$ is reset to 0, while $x$ and $\hat{x}$ are unchanged. In the other case, $\hat{x}$ and $\tau$ are, respectively, reset to $x$ and 0, which corresponds to a new control input being applied to the plant.

In what follows, we use the impulsive model to analyze the PETC system for both the case that the plant and controller are linear (Section 6.4), or the case that they are nonlinear (Section 6.5).

## 6.4 Analysis for Linear Systems

In this section, we analyze stability and performance of the PETC systems with linear dynamics. Hence, the plant model is given by (6.1) with $f(x, u, w) = A^p x + B^p u + B^w w$ and the feedback law by (6.2) with $K(x) = Kx$, where $A^p$, $B^p$, $B^w$, and $K$ are matrices of appropriate dimensions. This leads to the PETC (6.6)–(6.7) with

$$g(\xi, w) = \bar{A}\xi + \bar{B}w,$$
$$\text{with } \bar{A} := \begin{bmatrix} A^p & B^pK \\ 0 & 0 \end{bmatrix}, \ \bar{B} := \begin{bmatrix} B^w \\ 0 \end{bmatrix}. \tag{6.8}$$

Moreover, we focus on *quadratic* event-triggering conditions, i.e., $\mathcal{C}$ in (6.4) and (6.7), which is defined as

$$\mathcal{C}(\xi(t_k)) = \xi^\top(t_k) Q \xi(t_k), \tag{6.9}$$

for some symmetric matrix $Q \in \mathbb{R}^{n_\xi \times n_\xi}$. This choice is justified by the fact that various existing event-triggering conditions, including the ones in [11,12,16,30–33], that have been applied in the context of CETC, can be written as quadratic event-triggering conditions for PETC as in (6.9) (see [19] for more details).

We now make precise what we mean by stability and performance. Subsequently, we present two different approaches: (1) direct analysis of the impulsive system and (2) indirect analysis of the impulsive system using a discretization.

### 6.4.1 Problem Statement

Let us define the notion of global exponential stability and $\mathcal{L}_2$-performance, where the latter definition is adopted from [34].

**DEFINITION 6.1** The PETC system, given by (6.1), (6.2), and (6.3) is said to be *globally exponentially stable* (GES), if there exist $c > 0$ and $\rho > 0$ such that for any initial condition $\xi(0) = \boldsymbol{\xi}_0 \in \mathbb{R}^{n_\xi}$, all corresponding solutions to (6.6)–(6.7) with $\tau(0) \in [0, h]$ and $w = 0$ satisfy $\|\xi(t)\| \leqslant ce^{-\rho t}\|\boldsymbol{\xi}_0\|$ for all $t \in \mathbb{R}_+$ and some (lower bound on the) decay rate $\rho$.

Let us now define the $\mathcal{L}_2$-gain of a system, for which we introduce a performance variable $z \in \mathbb{R}^{n_z}$ given by

$$z = \bar{C}\xi + \bar{D}w, \tag{6.10}$$

where $\bar{C}$ and $\bar{D}$ are appropriately chosen matrices given by the considered problem. For instance, when $\bar{C} = [I_{n_x}\ 0_{n_x \times n_x}]$ and $\bar{D}$ is equal to $0_{n_x \times n_w}$, we simply have that $z = x$.

**DEFINITION 6.2** The PETC system, given by (6.1), (6.2), (6.3), and (6.10) is said to have an $\mathcal{L}_2$-gain from $w$ to $z$ smaller than or equal to $\gamma$, where $\gamma \in \mathbb{R}_+$, if there is a function $\delta : \mathbb{R}^{n_\xi} \to \mathbb{R}_+$ such that for any $w \in \mathcal{L}_2$, any initial state $\xi(0) = \boldsymbol{\xi}_0 \in \mathbb{R}^{n_\xi}$ and $\tau(0) \in [0, h]$, the corresponding solution to (6.6), (6.7), and (6.10) satisfies

$$\|z\|_{\mathcal{L}_2} \leqslant \delta(\boldsymbol{\xi}_0) + \gamma\|w\|_{\mathcal{L}_2}. \tag{6.11}$$

Equation 6.11 is a robustness property, and the gain $\gamma$ serves as a measure of the system's ability to attenuate the effect of the disturbance $w$ on $z$. Loosely speaking, small $\gamma$ indicates small impact of $w$ on $z$.

### 6.4.2 Stability and Performance of the Linear Impulsive System

We analyze the stability and the $\mathcal{L}_2$-gain of the impulsive system model (6.6)–(6.7) using techniques from Lyapunov stability analysis [34]. In short, the theory states that if an energy function (a so-called Lyapunov or storage function) can be found that satisfies certain properties, stability and a certain $\mathcal{L}_2$-gain can be guaranteed. In particular, we consider a Lyapunov function of the form

$$V(\xi, \tau) := \xi^\top P(\tau)\xi, \tag{6.12}$$

for $\xi \in \mathbb{R}^{n_\xi}$ and $\tau \in [0, h]$, where $P : [0, h] \to \mathbb{R}^{n_\xi \times n_\xi}$ with $P(\tau) \succ 0$, for $\tau \in [0, h]$. This function proves stability and a certain $\mathcal{L}_2$-gain from $w$ to $z$ if it satisfies

$$\tfrac{d}{dt}V \leqslant -2\rho V - \gamma^{-2}\|z\|^2 + \|w\|^2, \tag{6.13}$$

during the flow (6.6) and

$$V(J_1\xi, 0) \leqslant V(\xi, h), \text{ for all } \xi \text{ with } \xi^\top Q\xi > 0, \tag{6.14}$$

$$V(J_2\xi, 0) \leqslant V(\xi, h), \text{ for all } \xi \text{ with } \xi^\top Q\xi \leqslant 0, \tag{6.15}$$

during the jumps (6.7) of the impulsive system (6.6)–(6.7). Equation 6.13 indicates that along the solutions, the energy of the system decreases up to the perturbating term $\|w\|$ during the flow, and (6.14) indicates that the energy in the system does not increase during the jumps.

The main result presented below will provide a computable condition in the form of a linear matrix inequality (LMI) to verify if a function (6.12) exists that satisfies (6.13) and (6.14). Note that LMIs can be efficiently tested using optimization software, such as Yalmip [35]. We introduce the Hamiltonian matrix, given by

$$H := \begin{bmatrix} \bar{A} + \rho I + \bar{B}\bar{D}^\top L\bar{C} & \gamma^2\bar{B}(\gamma^2 I - \bar{D}^\top\bar{D})^{-1}\bar{B}^\top \\ -\bar{C}^\top L\bar{C} & -(\bar{A} + \rho I + \bar{B}\bar{D}^\top L\bar{C})^\top \end{bmatrix}, \tag{6.16}$$

with $L := (\gamma^2 I - \bar{D}\bar{D}^\top)^{-1}$. The matrix $L$ has to be positive definite, which can be guaranteed by taking $\gamma > \sqrt{\lambda_{\max}(\bar{D}^\top\bar{D})}$. In addition, we introduce the matrix exponential

$$F(\tau) := e^{-H\tau} = \begin{bmatrix} F_{11}(\tau) & F_{12}(\tau) \\ F_{21}(\tau) & F_{22}(\tau) \end{bmatrix}, \tag{6.17}$$

Besides this, we need the following technical assumption.

**ASSUMPTION 6.1** $F_{11}(\tau)$ is invertible for all $\tau \in [0, h]$.

Assumption 6.1 is always satisfied for a sufficiently small sampling period $h$. Namely, $F(\tau) = e^{-H\tau}$ is a continuous function, and we have that $F_{11}(0) = I$. Let us also introduce the notation $\bar{F}_{11} := F_{11}(h)$, $\bar{F}_{12} := F_{12}(h)$, $\bar{F}_{21} := F_{21}(h)$, and $\bar{F}_{22} := F_{22}(h)$, and a matrix $\bar{S}$ that satisfies $\bar{S}\bar{S}^\top := -\bar{F}_{11}^{-1}\bar{F}_{12}$. Such a matrix $\bar{S}$ exists under Assumption 6.1 because this assumption ensures that the matrix $-\bar{F}_{11}^{-1}\bar{F}_{12}$ is positive semidefinite.

**Theorem 6.1**

Consider the impulsive system (6.6)–(6.7) and let $\rho > 0$, $\gamma > \sqrt{\lambda_{\max}(\bar{D}^\top \bar{D})}$, and Assumption 6.1 hold. Suppose that there exist a matrix $P \succ 0$, and scalars $\mu_i \geqslant 0$, $i \in \{1, 2\}$, such that for $i \in \{1, 2\}$,

$$\begin{bmatrix} P + (-1)^i \mu_i Q & J_i^\top \bar{F}_{11}^{-\top} P \bar{S} & J_i^\top (\bar{F}_{11}^{-\top} P \bar{F}_{11}^{-1} + \bar{F}_{21} \bar{F}_{11}^{-1}) \\ \star & I - \bar{S}^\top P \bar{S} & 0 \\ \star & \star & \bar{F}_{11}^{-\top} P \bar{F}_{11}^{-1} + \bar{F}_{21} \bar{F}_{11}^{-1} \end{bmatrix} \succ 0. \tag{6.18}$$

Then, the PETC system (6.6)–(6.7) is GES with decay rate $\rho$ (when $w = 0$) and has an $\mathcal{L}_2$-gain from $w$ to $z$ smaller than or equal to $\gamma$.

The results of Theorem 6.1 guarantee both GES (for $w = 0$) and an upper bound on the $\mathcal{L}_2$-gain.

**REMARK 6.1** Recently, extensions to the above results were provided in [36]. Instead of adopting timer-dependent *quadratic* storage functions $V(\xi, \tau) = \xi^\top P(\tau) \xi$, in [36] more versatile storage functions were used of the *piecewise quadratic* form $V(\xi, \tau, \omega) = \xi^\top P_i(\tau) \xi$, $i \in \{1, \ldots, N\}$, where $i$ is determined by the region $\Omega_i$, $i \in \{1, \ldots, N\}$, in which the state $\xi$ is after $h - \tau$ time units (i.e., at the next jump time) that depends on the disturbance signal $\omega$. The regions $\Omega_1, \ldots, \Omega_N$ form a partition of the state-space $\mathbb{R}^{n_\xi}$. As such, the value of the storage function depends on future disturbance values, see [36] for more details. This approach leads to less conservative LMI conditions than the ones presented above.

### 6.4.3 A Piecewise Linear System Approach to Stability Analysis

In case disturbances are absent (i.e., $w = 0$), less conservative conditions for GES can be obtained than by using Theorem 6.1. These conditions can be obtained by discretizing the impulsive system (6.6)–(6.7) at the sampling instants $t_k = kh$, $k \in \mathbb{N}$, where we take* $\tau(0) = h$ and $w = 0$, resulting in a discrete-time piecewise-linear (PWL) model. By defining the state variable $\xi_k := \xi(t_k)$ (and assuming $\xi$ to be left-continuous), the discretization leads to the bimodal PWL model

$$\xi_{k+1} = \begin{cases} e^{\bar{A}h} J_1 \xi_k, & \text{when } \xi_k^\top Q \xi_k > 0 \\ e^{\bar{A}h} J_2 \xi_k, & \text{when } \xi_k^\top Q \xi_k \leqslant 0 \end{cases}. \tag{6.19}$$

*Note that $\tau(0)$ is allowed to take any value in $[0, h]$ in the stability definition (Definition 6.1), while in the discretization we take $\tau(0) = h$. Due to the linearity of the flow dynamics (6.6) and the fact that $\tau(0)$ lies in a bounded set, it is straightforward to see that GES for initial conditions with $\tau(0) = h$ implies GES for all initial conditions with $\tau(0) \in [0, h]$.

Using the PWL model (6.19) and a piecewise quadratic (PWQ) Lyapunov function of the form

$$V(\xi) = \begin{cases} \xi^\top P_1 \xi, & \text{when } \xi^\top Q \xi > 0 \\ \xi^\top P_2 \xi, & \text{when } \xi^\top Q \xi \leqslant 0 \end{cases}, \tag{6.20}$$

we can guarantee GES of the PETC system given by (6.1), (6.3), (6.4), and (6.8) under the conditions given next.

**Theorem 6.2**

The PETC system (6.6)–(6.7) is GES with decay rate $\rho$, if there exist matrices $P_1$, $P_2$, and scalars $\alpha_{ij} \geqslant 0$, $\beta_{ij} \geqslant 0$, and $\kappa_i \geqslant 0$, $i, j \in \{1, 2\}$, satisfying

$$\begin{aligned} e^{-2\rho h} P_i - (e^{\bar{A}h} J_i)^\top P_j e^{\bar{A}h} J_i + (-1)^i \alpha_{ij} Q \\ + (-1)^j \beta_{ij} (e^{\bar{A}h} J_i)^\top Q e^{\bar{A}h} J_i \succeq 0, \end{aligned} \tag{6.21}$$

for all $i, j \in \{1, 2\}$, and

$$P_i + (-1)^i \kappa_i Q \succ 0, \tag{6.22}$$

for all $i \in \{1, 2\}$.

When comparing the two different analysis approaches, two observations can be made. The first observation is that the direct analysis of the impulsive system allows us to analyze the $\mathcal{L}_2$-gain from $w$ to $z$, contrary to the indirect analysis using the PWL system. Second, the indirect analysis approach using the PWL system is relevant since, when comparing it to the direct analysis of the impulsive system, we can show that for stability analysis (when $w = 0$), the PWL system approach never yields more conservative results than the impulsive system approach, as is formally proven in [19].

**REMARK 6.2** This section is devoted to the analysis of the stability and the performance of linear PETC systems. The results can also be used to design the controllers as well as the triggering condition and the sampling period. The interested reader can consult Section IV in [19] for detailed explanations.

**REMARK 6.3** In [7], PETC closed-loop systems were analyzed with $\mathcal{C}(x(t_k), \hat{x}(t_k)) = \|x(t_k)\| - \delta$ with $\delta > 0$ some absolute threshold. Hence, the control value $u$ is updated to $Kx(t_k)$ only when $\|x(t_k)\| > \delta$, while in a region close to the origin, i.e., when $\|x(t_k)\| \leqslant \delta$, no updates of the control value take place at the sampling instants $t_k = kh$, $k \in \mathbb{N}$. For linear systems with bounded disturbances, techniques were presented in [7] to prove ultimate boundedness/practical stability, and calculate the ultimate bound $\Pi$ to which eventually all state trajectories converge (irrespective of the disturbance signal).

**REMARK 6.4** Extensions of the above analysis framework to output-based PETC with decentralized event triggering (instead of state-based PETC with centralized triggering conditions) can be found in [20]. Model-based (state-based and output-based) PETC controllers are considered in [19]. Model-based PETC controllers exploit model knowledge to obtain better predictions $\hat{x}$ of the true state $x$ in between sensor-to-controller communication than just holding the previous value as in (6.4). This can further enhance communication savings between the sensor and the controller. Similar techniques can also be applied to reduce the number of communications between the controller and the actuator.

**REMARK 6.5** For some networked control systems (NCSs), it is natural to periodically switch between time-triggered sampling and PETC. Examples include NCS with FlexRay (see [37]). FlexRay is a communication protocol developed by the automotive industry, which has the feature to switch between static and dynamic segments, during which the transmissions are, respectively, time triggered or event triggered. While the implementation and therefore the model differ in this case, the results of this section can be applied to analyze stability.

### 6.4.4 Numerical Example

We illustrate the presented theory using a numerical example. Let us consider the example from [12] with plant dynamics (6.1) given by

$$\frac{d}{dt}x = \begin{bmatrix} 0 & 1 \\ -2 & 3 \end{bmatrix} x + \begin{bmatrix} 0 \\ 1 \end{bmatrix} u + \begin{bmatrix} 1 \\ 0 \end{bmatrix} w, \tag{6.23}$$

and state-feedback controller (6.3), where we take $K(x) = [1 \;\; -4]\,x$ and $t_k = kh$, $k \in \mathbb{N}$, with sampling interval $h = 0.05$. We consider the event-triggering conditions given by

$$C(x, \hat{x}) = \|K\hat{x} - Kx\| - \sigma\|Kx\|, \tag{6.24}$$

for some value $\sigma > 0$. This can be equivalently written in the form of (6.9), by choosing

$$Q = \begin{bmatrix} (1-\sigma^2)K^\top K & -K^\top K \\ -K^\top K & K^\top K \end{bmatrix}. \tag{6.25}$$

For this PETC system, we will apply both approaches for stability analysis (for $w = 0$), and the impulsive system approach for performance analysis. We aim at constructing the largest value of $\sigma$ in (6.24) such that GES or a certain $\mathcal{L}_2$-gain can be guaranteed. The reason for striving for large values of $\sigma$ is that then large (minimum) interevent times are obtained, due to the form of (6.24).

For the event-triggering condition (6.24), the PWL system approach yields a maximum value for $\sigma$ of $\sigma_{\text{PWL}} = 0.2550$ (using Theorem 6.2), while still guaranteeing GES of the PETC system. The impulsive system approach gives a maximum value of $\sigma_{\text{IS}} = 0.2532$. Hence, as expected based on the discussion at the end of Section 6.4.3 indicating that the PWL system approach is less conservative than the impulsive system approach, see [19], we see that $\sigma_{\text{IS}} \leqslant \sigma_{\text{PWL}}$, although the values are rather close.

When analyzing the $\mathcal{L}_2$-gain from the disturbance $w$ to the output variable $z$ as in (6.10) where $z = [0\,1\,0\,0]\xi$, we obtain Figure 6.2a, in which the smallest upper bound on the $\mathcal{L}_2$-gain that can be guaranteed on the basis of Theorem 6.1 is given as a function of $\sigma$. This figure clearly demonstrates that better guarantees on the control performance (i.e., smaller $\gamma$) necessitate more updates (i.e., smaller $\sigma$), allowing us to make trade-offs between these two competing objectives, see also the discussion regarding Figure 6.2d. An important design observation is related to the fact that for $\sigma \to 0$, we recover the $\mathcal{L}_2$-gain for the periodic sampled-data system, given by (6.1) of the controller (6.2) with sampling interval $h = 0.05$ and $t_k = kh$, $k \in \mathbb{N}$. Hence, this indicates that an *emulation-based* design can be obtained by synthesizing first a state-feedback gain $K$ in a periodic time-triggered implementation of the feedback control given by $u(t_k) = Kx(t_k)$, $k \in \mathbb{N}$ (related to $\sigma = 0$), resulting in a small $\mathcal{L}_2$-gain of the closed-loop sampled-data control loop (using the techniques in, e.g., [38]). Next the PETC controller values of $\sigma > 0$ can be selected to reduce the number of communications and updates of control input, while still guaranteeing a small value of the guaranteed $\mathcal{L}_2$-gain according to Figure 6.2a and d.

Figure 6.2b shows the response of the performance output $z$ of the PETC system with $\sigma = 0.2$, initial condition $\xi_0 = [1\,0\,0\,0]^\top$ and a disturbance $w$ as also depicted in Figure 6.2b. For the same situation, Figure 6.2c shows the evolution of the interevent times. We see interevent times ranging from $h = 0.05$ up to 0.85 (17 times the sampling interval $h$), indicating a significant reduction in the number of transmissions. To more clearly illustrate this reduction, Figure 6.2d depicts the number of transmissions for this given initial condition and disturbance, as a function of $\sigma$. Using this figure and Figure 6.2a, it can be shown that the increase of the guaranteed $\mathcal{L}_2$-gain, through an increased $\sigma$, leads to fewer transmissions, which demonstrates the trade-off between the closed-loop performance and the number of transmissions that have to be made. Conclusively, using the PETC instead of the periodic sampled-data controller for this example

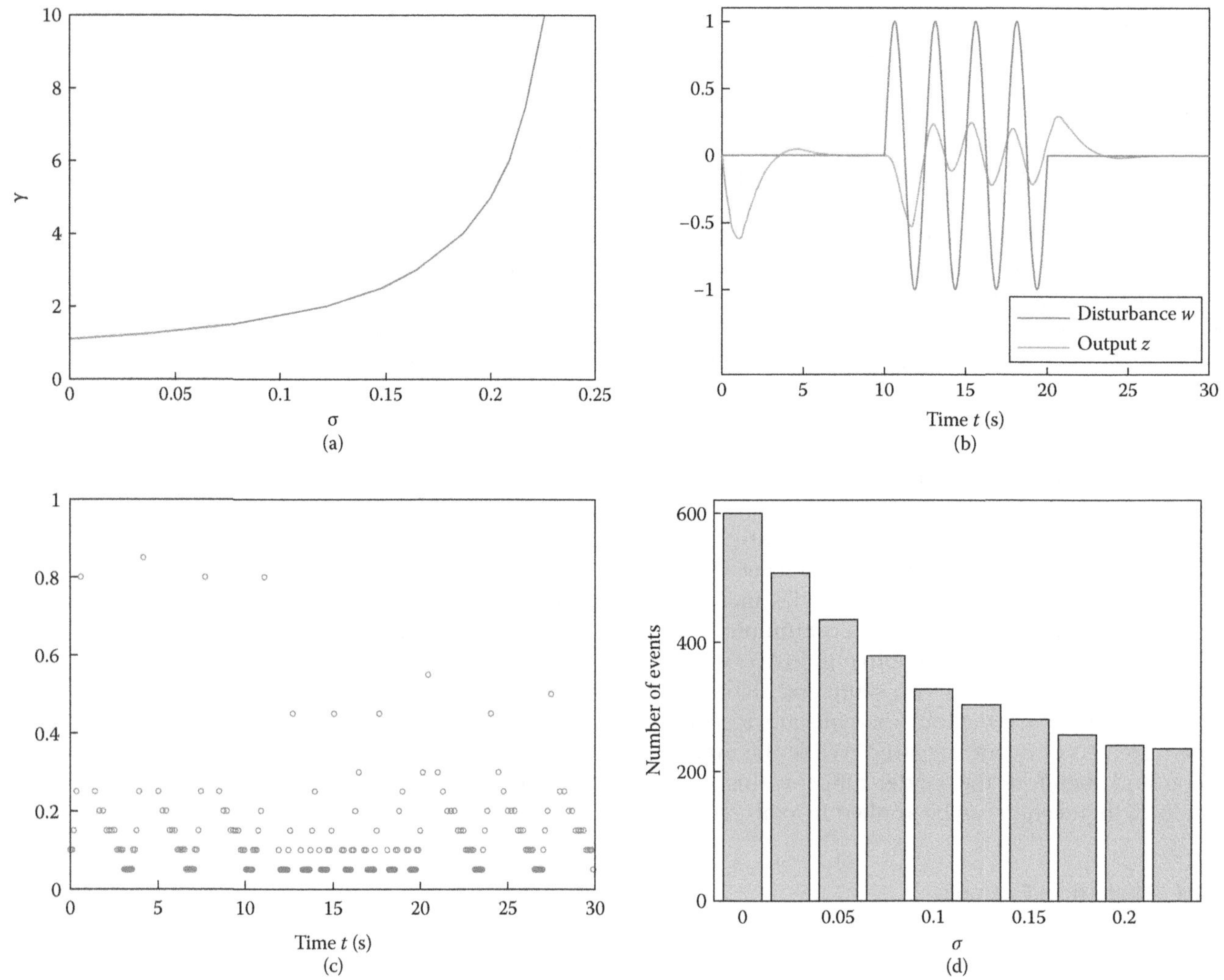


**FIGURE 6.2**

Figures corresponding to the numerical example of Section 6.4.4. (a) Upper-bound $\mathcal{L}_2$-gain as a function of $\sigma$. (b) The evolution of the disturbances $w$ and the output $z$ as a function of time for $\sigma = 0.2$. (c) The interevent times as a function of time for $\sigma = 0.2$. (d) Number of events as a function of $\sigma$.

yields a significant reduction in the number of transmissions/controller computations, while still preserving closed-loop stability and performance to some degree.

## 6.5 Design and Analysis for Nonlinear Systems

Let us now address the case where the plant dynamics is described by a nonlinear ordinary differential equation, and we ignore the possible presence of external disturbance $w$ for simplicity. As a consequence, (6.1) becomes

$$\tfrac{\mathrm{d}}{\mathrm{d}t}x = f(x, u, 0), \tag{6.26}$$

where $u$ is given by (6.3)–(6.4).

### 6.5.1 Problem Statement

The generalization of the results of Section 6.4 to nonlinear systems is a difficult task, and we will a priori not be able to derive similar easily computable criteria to verify the stability properties of the corresponding PETC systems. We therefore address the design of the sampling interval $h$ and of the triggering condition $\mathcal{C}$ from a different angle compared to Section 6.4.

We start by assuming that we already designed a *continuous* event-triggered controller and our objective is to design $h$ and $\mathcal{C}$ to preserve the properties of CETC. We thus assume that we know a mapping $K : \mathbb{R}^{n_x} \to \mathbb{R}^{n_u}$ and a criterion $\widetilde{\mathcal{C}} : \mathbb{R}^{2n_x} \to \mathbb{R}$, which are used to generate the control input. The corresponding transmission instants are denoted by $\tilde{t}_k$, $k \in \mathbb{N}_0$, and are defined by

$$\tilde{t}_{k+1} = \inf\left\{t > \tilde{t}_k \mid \tilde{C}(x(t), x(\tilde{t}_k)) \geqslant 0\right\}, \quad \tilde{t}_0 = 0. \tag{6.27}$$

The control input is thus given by

$$u(t) = K(\tilde{x}(t)), \quad \text{for} \quad t \in \mathbb{R}_+, \tag{6.28}$$

where* $\tilde{x}(t) = x(\tilde{t}_k)$ for $t \in (\tilde{t}_k, \tilde{t}_{k+1}]$. Note that $\tilde{C}$ is evaluated at any $t \in \mathbb{R}_+$ in (6.27) contrary to (6.4). The continuous event-triggered controller guarantees that, for all time† $t \geqslant 0$,

$$\tilde{C}(\xi(t)) \leqslant 0, \tag{6.29}$$

where $\xi := [x^\top \tilde{x}^\top]^\top$.

In the following, we first apply the results of [21] to show that, if (6.29) implies the global asymptotic stability of a given compact set, then this property is semiglobally and practically preserved in the context of PETC where the adjustable parameter is the sampling period $h$, under mild regularity conditions. We then present an alternative approach, which consists in redesigning the continuous event-triggering condition $\tilde{C}$ for PETC in order to recover the same properties as for CETC. In this case, we provide an explicit bound on the sampling period $h$.

### 6.5.2 Emulation

We model the overall CETC system as an impulsive system (like in Section 6.3)

$$\begin{array}{lll} \frac{d}{dt}\xi = g(\xi) & \text{when} & \tilde{C}(\xi) \leqslant 0 \\ \xi^+ = J_1\xi & \text{when} & \tilde{C}(\xi) \geqslant 0 \end{array}, \tag{6.30}$$

where $g(\xi) = [f(x, K(\tilde{x}), 0)^\top, 0^\top]^\top$ (with some abuse of notation with respect to (6.5)) and $J_1$ is defined in Section 6.3. We do not use strict inequalities in (6.30) to define the regions of the state space where the system flows and jumps, contrary to (6.6)–(6.7). This is justified by the fact that we want to work with a flow set, $\{\xi \mid \tilde{C}(\xi) \leqslant 0\}$, and a jump set, $\{\xi \mid \tilde{C}(\xi) \geqslant 0\}$, which are closed in order to apply the results of [21]. (We assume below that $\tilde{C}$ is continuous for this purpose.) When the state is in the intersection of the flow set and the jump set, the corresponding solution can either jump or flow, if flowing keeps the solution in the flow set. We make the following assumptions on the system (6.30).

**ASSUMPTION 6.2** The solutions to (6.30) do not undergo two consecutive jumps, i.e., $\tilde{t}_k < \tilde{t}_{k+1}$ for any $k \in \mathbb{N}_0$, and are defined for all positive time.

*We use the notation $\tilde{x}$ instead of $\hat{x}$ to avoid any confusion with the PETC setup.

†We assume that the solutions to the corresponding system are defined for all positive time, for any initial condition.

Assumption 6.2 can be relaxed by allowing two consecutive jumps, even Zeno phenomenon for the CETC system. In this case, a different concept of solutions is required as defined in [29], see for more detail [21].

**ASSUMPTION 6.3** The vector field $g$ and the scalar field $\tilde{C}$ are continuous.

We suppose that (6.29) ensures the global asymptotic stability of a given compact set $\mathcal{A} \subset \mathbb{R}^{2n_x}$, as formalized below.

**ASSUMPTION 6.4** The following holds for the CETC system.

(1) For each $\varepsilon > 0$, there exists $\delta > 0$ such that each solution $\xi$ starting at $\xi_0 \in \mathcal{A} + \delta\mathbb{B}$, where $\mathbb{B}$ is the unit ball of $\mathbb{R}^{2n_x}$, satisfies $\|\xi(t)\|_{\mathcal{A}} \leqslant \varepsilon$ for all $t \geqslant 0$.

(2) There exists $\mu > 0$ such that any solution starting in $\mathcal{A} + \mu\mathbb{B}$ satisfies $\|\xi(t)\|_{\mathcal{A}} \to 0$ as $t \to \infty$.

Set stability extends the classical notion of stability of an equilibrium point to a set. Essentially, a set is stable if a solution which starts close to it (in terms of the distance to this set) remains close to it [see item (1) of Assumption 6.4]; it is attractive if any solution converges toward this set [see item (1) of Assumption 6.4]. A set is asymptotically stable if it satisfies both properties. Set stability is fundamental in many control theoretic problems, see, for example, Chapter 3.1 in [29], or [39,40]. Many existing continuous event-triggered controllers satisfy Assumption 6.4 as shown in [18]. Examples include the techniques in [1,2,12–14] to mention a few.

We need to slightly modify the impulsive model (6.6)–(6.7) of the PETC system as follows, in order to apply the results of [21]

$$\frac{d}{dt}\begin{bmatrix}\xi \\ \tau\end{bmatrix} = \begin{bmatrix} g(\xi) \\ 1\end{bmatrix} \quad \text{when } \tau \in [0, h]$$

$$\begin{bmatrix}\xi^+ \\ \tau^+\end{bmatrix} \in \begin{cases} \left\{\begin{bmatrix} J_1\xi \\ 0\end{bmatrix}\right\}, \\ \text{when } \tilde{C}(\xi) > 0,\ \tau = h \\ \left\{\begin{bmatrix} J_2\xi \\ 0\end{bmatrix}\right\}, \\ \text{when } \tilde{C}(\xi) < 0,\ \tau = h \\ \left\{\begin{bmatrix} J_1\xi \\ 0\end{bmatrix}, \begin{bmatrix} J_2\xi \\ 0\end{bmatrix}\right\}, \\ \text{when } \tilde{C}(\xi) = 0,\ \tau = h. \end{cases} \tag{6.31}$$

The difference with (6.6)–(6.7) is that, when $\tau = h$ and $\tilde{C}(\xi) = 0$, we can either have a transmission (i.e., $\xi$ is reset to $[x^\top\ x^\top]^\top$) or not (i.e., $\xi$ remains unchanged). Hence, system (6.31) generates more solutions than (6.6)–(6.7). However, the results presented in Section 6.4 also apply when (6.31) (with linear plant) is used instead of (6.6)–(6.7). Furthermore, the jump map of (6.31) is outer semicontinuous (see Definition 5.9 in [29]), which is essential to apply [21]. Proposition 6.1 below follows from Theorem 5.2 in [21].

**Proposition 6.1**

Consider the PETC system (6.31) and suppose Assumptions 6.2 through 6.4 hold. For any compact set $\Delta \subset \mathbb{R}^{2n_x}$ and any $\varepsilon > 0$, there exists $h^*$ such that for any $h \in (0, h^*)$, any solution $[\xi^\top, \tau]^\top$ with $\xi(0) \in \Delta$, there exists $T \geqslant 0$ such that $\xi(t) \in \mathcal{A} + \varepsilon\mathbb{B}$ for all $t \geqslant T$.

Proposition 6.1 shows that the global asymptotic stability of $\mathcal{A}$ in Assumption 6.4 is semiglobally and practically preserved for the emulated PETC system, by adjusting the sampling period. It is possible to derive stronger stability guarantees for the PETC system such as the (semi)global asymptotic stability of $\mathcal{A}$, for instance, under additional assumptions on the CETC system.

### 6.5.3 Redesign

The results above are general, but they do not provide an explicit bound on the sampling period $h$, which is important in practice. Furthermore, in some cases, we would like to exactly (and not approximately) preserve the properties of CETC, which may not necessarily be those stated in Assumption 6.2, but may be some performance guarantees, for instance. We present an alternative approach to design PETC controllers for nonlinear systems for this purpose. Contrary to Section 6.5.2, we do not *emulate* the CETC controller, but we redesign the triggering criterion (but not the feedback law), and we provide an upper-bound on the sampling period $h$ to guarantee that $\widetilde{C}$ remains nonpositive for the PETC system.

We suppose that inequality (6.29) ensures the satisfaction of a desired stability or performance property. Consider the following example to be more concrete. In [12], a continuous event-triggering law of the form $\beta(\|x - \hat{x}\|) \geqslant \sigma\alpha(\|x\|)$ with* $\beta, \alpha \in \mathcal{K}_\infty$ and $\sigma \in (0, 1)$ is designed. We obtain $\widetilde{C}(x, \hat{x}) = \beta(\|x - \hat{x}\|) - \sigma\alpha(\|x\|)$ in this case. This triggering law is shown to ensure the global asymptotic stability of the origin of the nonlinear systems (6.26), (6.27), and (6.28) in [12], under some conditions on $f, K, \beta, \alpha$. In other words, $\widetilde{C}$ nonpositive along the system solutions implies that the origin of the closed-loop system is globally asymptotically stable. Similarly, the conditions of the form $\|x - \hat{x}\| \geqslant \varepsilon$ used in [1,2,13,14] and $\|x - \hat{x}\| \geqslant \delta\|x\| + \varepsilon$ in [16] to practically stabilize the origin of the corresponding CETC system give $\widetilde{C}(x, \hat{x}) = \|x - \hat{x}\| - \varepsilon$ and $\widetilde{C}(x, \hat{x}) = \|x - \hat{x}\| - \delta\|x\| - \varepsilon$, respectively. By reducing the properties of CETC to the satisfaction of (6.29), we cover a range of situations in a unified way.

We make the following assumption on the CETC system (which was not needed in Section 6.5.2).

**ASSUMPTION 6.5** Consider the CETC system (6.30), it holds that

$$T := \inf\{t > 0 \mid \widetilde{C}(\xi(t, [x_0^\top\ x_0^\top]^\top)) \geqslant 0,\\ [x_0^\top x_0^\top]^\top \in \Omega\} > 0, \tag{6.32}$$

where $\xi(t, [x_0^\top\ x_0^\top]^\top)$ is the solution to $\frac{d}{dt}\xi = g(\xi)$ at time $t$ initialized at $[x_0^\top\ x_0^\top]^\top$, and $\Omega \subseteq \mathbb{R}^{n_\xi}$ is bounded and forward invariant† for the CETC system (6.30).

Assumption 6.5 means that there exists a uniform minimum intertransmission time for the CETC system in the set $\Omega$. This condition is reasonable as most available event-triggering schemes of the literature ensure the existence of a uniform minimum amount of time between two transmissions over a given operating set $\Omega$, see [18]. The set $\Omega$ can be determined using the level set of some Lyapunov function when investigating stabilization problems, for example.

We have seen that under the PETC strategy, the input can be updated only whenever the triggering condition is evaluated—i.e., every $h$ units of time. Hence, it is reasonable to select the sampling interval to be less than the minimum intertransmission time of the CETC system (which does exist in view of Assumption 6.5). In that way, after a jump, we know that $\widetilde{C}$ will remain nonpositive at least until the next sampling instant. Therefore, we select $h$ such that

$$0 < h < T, \tag{6.33}$$

where $T$ is defined in (6.32). Estimates of $T$ are generally given in the analysis of the CETC system to prove the existence of a positive minimal interevent time.

We aim at guaranteeing that $\widetilde{C}$ remains nonpositive along the solutions to the CETC system. Hence, we would like to verify at $t_k$, $k \in \mathbb{N}_0$, whether the condition $\widetilde{C}(\xi(t)) > 0$ *may be satisfied* for $t \in [t_k, t_{k+1}]$

*A function $\beta : \mathbb{R}_+ \to \mathbb{R}_+$ is of class $\mathcal{K}_\infty$ if it is continuous, zero at zero, strictly increasing, and unbounded.

†The set $\Omega$ is forward invariant for the CETC system if $\xi_0 \in \Omega$ implies that the corresponding solution $\xi$, with $\xi(t_0) = \xi_0$ and $t_0 \in \mathbb{R}_+$, lies in $\Omega$ for all time larger than $t_0$.

(recall that $\xi = [x^\top \ \hat{x}^\top]^\top$). May $\widetilde{C}(\xi(t))$ be positive at some $t \in [t_k, t_{k+1}]$ (without updating the control action), a jump must occur when $t = t_k$ in order to guarantee $\widetilde{C}(\xi(t)) \leqslant 0$ for all $t \in [t_k, t_{k+1}]$. To determine at time $t_k$, $k \in \mathbb{N}$, whether the condition $\widetilde{C}(\xi(t)) \leqslant 0$ may be violated for some $t \in [t_k, t_{k+1}]$, the evolution of the triggering function $\widetilde{C}$ along the solutions to $\frac{\mathrm{d}}{\mathrm{d}t}\xi = g(\xi)$ needs to be analyzed. This point is addressed by resorting to similar techniques as in [41]. We make the following assumption for this purpose, which is stronger than Assumption 6.3.

**ASSUMPTION 6.6** The functions $g$ and $\widetilde{C}$ are $p$-times continuously differentiable where $p \in \mathbb{N}$ and the real numbers $c, \varsigma_j$ for $j \in \{0, 1, \dots, p-1\}$ satisfy

$$\mathcal{L}_g^p \widetilde{C}(\xi) \leqslant \sum_{j=0}^{p-1} \varsigma_j \mathcal{L}_g^j \widetilde{C}(\xi) + c, \tag{6.34}$$

for any $\xi \in \Omega$, where we have denoted the $j$th Lie derivative of $\widetilde{C}$ along the closed-loop dynamics $g$ as $\mathcal{L}_g^j \widetilde{C}$, with $\mathcal{L}_g^0 \widetilde{C} = \widetilde{C}$, $(\mathcal{L}_g \widetilde{C})(\xi) = \frac{\partial \widetilde{C}}{\partial \xi} g(\xi)$, and $\mathcal{L}_g^j \widetilde{C} = \mathcal{L}_g(\mathcal{L}_g^{j-1}\widetilde{C})$ for $j \geqslant 1$.

Inequality (6.34) always holds when $g$ and $\widetilde{C}$ are $p$-times continuously differentiable, as it suffices to take $c = \max_{\xi \in \Omega} \mathcal{L}_g^p \widetilde{C}(\xi)$ and $\varsigma_j = 0$ for $j \in \{0, 1, \dots, p-1\}$ to ensure (6.34) (recall that $\Omega$ is bounded in view of Assumption 6.5). However, this particular choice may lead to conservative results as explained below.

Assumption 6.6 allows to bound the evolution of $\widetilde{C}$ by a linear differential equation for which the analytical solution can be computed as stated in the lemma below, which directly follows from Lemma V.2 in [41].

**Lemma 6.1**

Under Assumption 6.6, for all solutions to $\frac{\mathrm{d}}{\mathrm{d}t}\xi = g(\xi)$ with initial condition $\boldsymbol{\xi}_0 \in \Omega$ such that $\xi(t, \boldsymbol{\xi}_0) \in \Omega$ for any $t \in [0, h]$, it holds that $\widetilde{C}(\xi(t, \boldsymbol{\xi}_0)) \leqslant y_1(t, \boldsymbol{y}_0)$ for any $t \in [0, h]$, where $y_1$ is the first component of the solution to the linear differential equation

$$\begin{cases} \frac{\mathrm{d}}{\mathrm{d}t} y_j &= y_{j+1}, \quad j \in \{1, 2, \dots, p-1\} \\ \frac{\mathrm{d}}{\mathrm{d}t} y_p &= \sum_{j=0}^{p-1} \varsigma_j y_{j+1} + y_{p+1} \\ \frac{\mathrm{d}}{\mathrm{d}t} y_{p+1} &= 0, \end{cases} \tag{6.35}$$

with $\boldsymbol{y}_0 = (\widetilde{C}(\boldsymbol{\xi}_0), \mathcal{L}_g \widetilde{C}(\boldsymbol{\xi}_0), \dots, \mathcal{L}_g^{p-1} \widetilde{C}(\boldsymbol{\xi}_0), c)$.

In that way, for a given state $\boldsymbol{\xi}_0 \in \Omega$ and $t \in [0, h]$, if $y_1(t, \boldsymbol{y}_0)$ is positive, then Lemma 6.1 implies that $\widetilde{C}(\xi(t, \boldsymbol{\xi}_0))$ *may* be positive. On the other hand, if $y_1(t, \boldsymbol{y}_0)$ is nonpositive, Lemma 6.1 ensures that $\widetilde{C}(t, \boldsymbol{\xi}_0))$ is nonpositive. We can therefore evaluate online $y_1(t, \boldsymbol{y}_0)$ for $t \in [0, h]$ and verify whether it takes a positive value, in which case a transmission occurs at $t_k$, otherwise that is not necessary. The analytic expression of $y_1(t, \boldsymbol{y}_0)$ is given by

$$y_1(t, \boldsymbol{\xi}_0) := C_p e^{A_p t} \begin{bmatrix} \widetilde{C}(\boldsymbol{\xi}_0) \\ \mathcal{L}_g \widetilde{C}(\boldsymbol{\xi}_0) \\ \vdots \\ \mathcal{L}_g^{p-1} \widetilde{C}(\boldsymbol{\xi}_0) \\ c \end{bmatrix}, \tag{6.36}$$

with

$$C_p := \begin{bmatrix} 1 & 0 & \dots & 0 \end{bmatrix}$$
$$A_p := \begin{bmatrix} 0 & 1 & 0 & \dots & 0 & 0 & 0 \\ 0 & 0 & 1 & \dots & 0 & 0 & 0 \\ \vdots & & & \ddots & & \vdots & \vdots \\ 0 & 0 & 0 & \dots & 1 & 0 & 0 \\ 0 & 0 & 0 & \dots & 0 & 1 & 0 \\ \varsigma_0 & \varsigma_1 & \varsigma_2 & \cdots & \varsigma_{p-2} & \varsigma_{p-1} & 1 \\ 0 & 0 & 0 & \dots & 0 & 0 & 0 \end{bmatrix}. \tag{6.37}$$

Hence, we define $C(\xi)$ for any $\xi \in \Omega$ as

$$C(\xi) := \max_{t \in [0,h]} y_1(t, \xi). \tag{6.38}$$

Every $h$ units of time, the current state $\xi$ is measured, and we verify whether $C(\xi)$ is positive, in which case the control input is updated. Conversely, if $C(\xi)$ is nonpositive, then the control input is not updated. It has to be noticed that we do not need to verify the triggering condition for the next $\lfloor \frac{T}{h} \rfloor$ sampling instants following a control input update according to Assumption 6.5, which allows us to further reduce computations.

**REMARK 6.6** The evaluation of $y_1(t, \xi)$ for any $t \in [0, h]$ in (6.38) involves an infinite number of conditions, which may be computationally infeasible. This shortcoming can be avoided by using convex overapproximation techniques, see [42]. The idea is to overapproximate $y_1(t, \xi)$ for $t \in [0, h]$. In that way, the control input is updated whenever the derived upper-bound is positive; otherwise, no update is needed. Note that these bounds can get as close as we want to $y_1(t, \xi)$, at the price of more computation at each sampling instant, see [42] for more detail.

The proposition below states that to choose $h$ such that (6.33) holds and $C$ as in (6.38) ensures that $\widetilde{C}$ will be nonpositive along the solutions to (6.6)–(6.7) as desired.

**Proposition 6.2**

Consider system (6.6)–(6.7) with $h$ which satisfies (6.33) and $C$ defined in (6.38) and suppose Assumptions 6.5 and 6.6 hold. Then for any solution $[\xi^\top(t)\ \tau^\top(t)]^\top$ for which $(\xi_0, \tau_0) \in \Omega \times \mathbb{R}_+, \widetilde{C}(\xi(t)) \leqslant 0$ for any $t \in \mathbb{R}_+$.

### 6.5.4 Numerical Example

We consider the rigid body previously studied in [43]. The model is given by

$$\begin{cases} \frac{d}{dt}x_1 = & u_1 \\ \frac{d}{dt}x_2 = & u_2 \quad , \\ \frac{d}{dt}x_3 = & x_1x_2 \end{cases} \tag{6.39}$$

and we consider the controller synthesized in [43] in order to stabilize the origin, which is given by

$$\begin{cases} u_1 = -x_1x_2 - 2x_2x_3 - x_1 - x_3 \\ u_2 = 2x_1x_2x_3 + 3x_3^2 - x_2. \end{cases} \tag{6.40}$$

The implementation of the controller on a digital platform leads to the following closed-loop system:

$$\begin{cases} \frac{d}{dt}x_1 = -\hat{x}_1\hat{x}_2 - 2\hat{x}_2\hat{x}_3 - \hat{x}_1 - \hat{x}_3 \\ \frac{d}{dt}x_2 = 2\hat{x}_1\hat{x}_2\hat{x}_3 + 3\hat{x}_3^2 - \hat{x}_2 \\ \frac{d}{dt}x_3 = x_1x_2. \end{cases} \tag{6.41}$$

In order to stabilize the origin of (6.41), we take the triggering condition as in [41,44]

$$\widetilde{C}(x,\hat{x}) \quad = \quad \|\hat{x} - x\|^2 - 0.79^2\sigma^2\|x\|^2 \tag{6.42}$$

**TABLE 6.1**
Average Intertransmission Times for 100 Points

| CETC | PETC | |
|---|---|---|
| | $h = 0.01$ | $h = 0.02$ |
| 0.3488 | 0.3440 | 0.3376 |

with $\sigma = 0.8$, which is obtained using the Lyapunov function

$$V(x) \quad = \quad \tfrac{1}{2}(x_1 + x_3)^2 + \tfrac{1}{2}(x_2 - x_3^2)^2 + x_3^2. \tag{6.43}$$

We design the PETC strategy by following the procedure in Section 6.5.3. Assumption 6.5 is satisfied with $T = 0.08$ (which has been determined numerically) for $\Omega = \{(x,\hat{x}) \mid V(x) \leqslant 5\}\backslash\{0\}$. Regarding Assumption 6.6, we note that the system vector fields and the triggering condition are smooth. In addition, (6.34) is verified with* $p = 3, \varsigma_0 = -748.4986, \varsigma_1 = -1.0008, \varsigma_2 = 4.3166$, and $c = 0$. We can thus apply the method presented in Section 6.5.3 as all the conditions of Proposition 6.2 are ensured. We have selected $h < T$. Table 6.1 provides the average intertransmission times for 100 points in $\Omega$ whose $x$-components are equally spaced along the sphere centered at 0 and of radius 1 and $\hat{x}(0) = x(0)$. PETC generates intertransmission times that are smaller than in CETC as expected. Moreover, we expect the average intertransmission time to increase when the sampling interval $h$ decreases as suggested by Table 6.1.

Assumption 6.4 is verified with $\mathcal{A} = \{0\}$ (in view of [18]) and Assumption 6.3 is also guaranteed. We can therefore also apply the emulation results of Section 6.5.2. To compare the strategies obtained by Sections 6.5.2 and 6.5.3, we plotted the evolution of $\widetilde{C}$ in both cases in Figure 6.3 with $h = 0.079$ and the

*SOSTools [45] was used to compute $\varsigma_i$ and $\chi_i$ for $i \in \{1,2,3\}$.

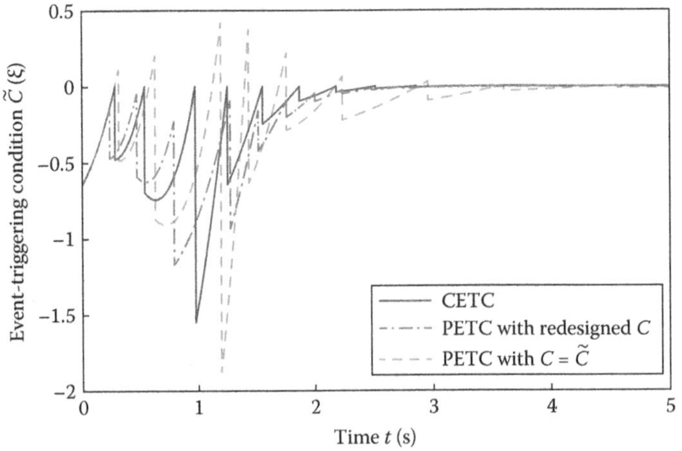


**FIGURE 6.3**
Evolution of $\widetilde{C}$.

same initial conditions. We see that $\widetilde{C}$ remains nonpositive all the time with the redesigned triggering condition, which implies that the periodic event-triggered controller ensures the same specification as the event-triggered controller, while $\widetilde{C}$ often reaches positive values with the emulated triggering law.

## 6.6 Conclusions, Open Problems, and Outlook

In this chapter, we discussed PETC as a class of ETC strategies that combines the benefits of periodic time-triggered control and event-triggered control. The PETC strategy is based on the idea of having an event-triggering condition that is verified only periodically, instead of continuously as in most existing ETC schemes. The periodic verification allows for a straightforward implementation in standard time-sliced embedded system architectures. Moreover, the strategy has an inherently guaranteed minimum interevent time of (at least) one sampling interval of the event-triggering condition, which is easy to tune directly.

We presented an analysis and design framework for linear systems and controllers, as well as preliminary results for nonlinear systems. Although we focused in the first case on static state-feedback controllers and centralized event-triggering conditions, extensions exist to dynamic output-feedback controllers and decentralized event generators, see [19]. Also model-based versions that can further enhance communication savings are available, see [20]. A distinctive difference between the linear and nonlinear results is that an emulation-based design for the former requires a well-designed time-triggered periodic controller (with a small $\mathcal{L}_2$ gain, e.g., synthesized using the tools in [38]), while the nonlinear part uses a well-designed continuous event-triggered controller as a starting point.

Several problems are still open in the area of PETC. First, obtaining tighter estimates for stability boundaries and performance guarantees (e.g., $\mathcal{L}_2$-gains), minimal interevent times, and average interevent times is needed. These are hard problems in general as we have shown in this chapter that PETC strategies result in closed-loop systems that are inherently of a hybrid nature, and it is hard to obtain nonconservative analysis and design tools in this context. One recent example providing improvements for the linear PETC framework regarding the determination of $\mathcal{L}_2$-gains is [36], see Remark 6.1. Moreover, in [46] a new lifting-based perspective is taken on the characterization of the $\mathcal{L}_2$-gain of the closed-loop PETC system, and, in fact, it is shown that the $\mathcal{L}_2$-gain of (6.6)–(6.7) is smaller than one (and the system is internally stable) if and only if the $\ell_2$-gain of a corresponding discrete-time piecewise linear system is smaller than one (and the system is internally stable). This new perspective on the PETC analysis yields an exact characterization of the $\mathcal{L}_2$-gain (and stability) that leads to significantly less conservative conditions.

Second, in the linear context, extensions to the case of output-feedback and decentralized triggering exist, see [19], and for the nonlinear context these extensions are mostly open. Also the consideration of PETC strategies for nonlinear systems with disturbances requires attention. Given these (and many other) open problems, it is fair to say that the system theory for ETC is far from being mature, certainly compared to the vast literature on time-triggered (periodic) sampled-data control. This calls for further theoretical research on ETC in general and PETC in particular.

Given the potential of ETC in saving valuable system's resources (computational time, communication bandwidth, battery power, etc.), while still preserving important closed-loop properties, as demonstrated through various numerical examples in the literature (including the two in this chapter), it is rather striking that the number of experimental and industrial applications is still rather small. To foster the further development of ETC in the future, it is therefore important to validate these strategies in practice. Getting feedback from industry will certainly raise new important theoretical questions. As such, many challenges are ahead of us both in theory and practice in this fledgling field of research.

## Acknowledgment

This work was supported by the Innovational Research Incentives Scheme under the VICI grant "Wireless control systems: A new frontier in automation" (No. 11382) awarded by NWO (The Netherlands Organization for Scientific Research) and STW (Dutch Science Foundation), the ANR under the grant COMPACS (ANR-13-BS03-0004-02), NSF grant ECCS-1232035, AFOSR grant FA9550-12-1-0127, NSF award 1239085, and the Australian Research Council under the Discovery Projects and Future Fellowship schemes.

## Bibliography

[1] K. J. Åström and B. M. Bernhardsson. Comparison of periodic and event based sampling for first order stochastic systems. In *Proceedings of the IFAC World Congress*, Beijing, China, pages 301–306, July 5–9, 1999.

[2] K.-E. Arzén. A simple event-based PID controller. In *Preprints IFAC World Conference*, Beijing, China, volume 18, pages 423–428, July 5–9, 1999.

[3] W. P. M. H. Heemels, R. J. A. Gorter, A. van Zijl, P. P. J. van den Bosch, S. Weiland, W. H. A. Hendrix, and M. R. Vonder. Asynchronous measurement and control: A case study on motor synchronisation. *Control Engineering Practice*, 7:1467–1482, 1999.

[4] E. Hendricks, M. Jensen, A. Chevalier, and T. Vesterholm. Problems in event based engine control. In *Proceedings of the American Control Conference*, Baltimore, volume 2, pages 1585–1587, 1994.

[5] W. H. Kwon, Y. H. Kim, S. J. Lee, and K.-N. Paek. Event-based modeling and control for the burn-through point in sintering processes. *IEEE Transactions on Control Systems Technology*, 7:31–41, 1999.

[6] W. P. M. H. Heemels, K. H. Johansson, and P. Tabuada. An introduction to event-triggered and self-triggered control. In *51th IEEE Conference on Decision and Control Conference (CDC), 2012*, Hawaï, pages 3270–3285, December 2012.

[7] W. P. M. H. Heemels, J. H. Sandee, and P. P. J. van den Bosch. Analysis of event-driven controllers for linear systems. *International Journal of Control*, 81:571–590, 2008.

[8] T. Henningsson, E. Johannesson, and A. Cervin. Sporadic event-based control of first-order linear stochastic systems. *Automatica*, 44:2890–2895, 2008.

[9] J. Lunze and D. Lehmann. A state-feedback approach to event-based control. *Automatica*, 46:211–215, 2010.

[10] P. J. Gawthrop and L. B. Wang. Event-driven intermittent control. *International Journal of Control*, 82:2235–2248, 2009.

[11] X. Wang and M. D. Lemmon. Event-triggering in distributed networked systems with data dropouts and delays. *IEEE Transactions on Automatic Control*, 56:586–601, 2011.

[12] P. Tabuada. Event-triggered real-time scheduling of stabilizing control tasks. *IEEE Transactions on Automatic Control*, 52:1680–1685, 2007.

[13] P. G. Otanez, J. R. Moyne, and D. M. Tilbury. Using deadbands to reduce communication in networked control systems. In *Proceedings of the American Control Conference*, Anchorage, pages 3015–3020, May 8–10, 2002.

[14] M. Miskowicz. Send-on-delta concept: An event-based data-reporting strategy. *Sensors*, 6:49–63, 2006.

[15] E. Kofman and J. H. Braslavsky. Level crossing sampling in feedback stabilization under data-rate constraints. In *Proceedings of the IEEE Conference Decision & Control*, pages 4423–4428, San Diego, CA, December 13–15, 2006.

[16] M. C. F. Donkers and W. P. M. H. Heemels. Output-based event-triggered control with guaranteed $\mathcal{L}_\infty$-gain and improved and decentralised event-triggering. *IEEE Transactions on Automatic Control*, 57(6):1362–1376, 2012.

[17] D. Lehmann and J. Lunze. Extension and experimental evaluation of an event-based state-feedback approach. *Control Engineering Practice*, 19:101–112, 2011.

[18] R. Postoyan, P. Tabuada, D. Nešić, and A. Anta. A framework for the event-triggered stabilization of nonlinear systems. *IEEE Transactions on Automatic Control*, 60(4):982–996, 2015.

[19] W. P. M. H. Heemels, M. C. F. Donkers, and A. R. Teel. Periodic event-triggered control for linear systems. *IEEE Transactions on Automatic Control*, 58(4):847–861, 2013.

[20] W. P. M. H. Heemels and M. C. F. Donkers. Model-based periodic event-triggered control for linear systems. *Automatica*, 49(3):698–711, 2013.

[21] R. G. Sanfelice and A. R. Teel. Lyapunov analysis of sampled-and-hold hybrid feedbacks. In *IEEE Conference on Decision and Control*, pages 4879–4884, San Diego, CA, 2006.

[22] J. K. Yook, D. M. Tilbury, and N. R. Soparkar. Trading computation for bandwidth: Reducing communication in distributed control systems using state estimators. *IEEE Transactions on Control Systems Technology*, 10(4):503–518, 2002.

[23] A. Eqtami, V. Dimarogonas, and K. J. Kyriakopoulos. Event-triggered control for discrete-time systems. In *Proceedings of the American Control Conference (ACC)*, pages 4719–4724, Baltimore, MD, 2010.

[24] R. Cogill. Event-based control using quadratic approximate value functions. In *Joint IEEE Conference on Decision and Control and Chinese Control Conference*, pages 5883–5888, Shanghai, China, December 15–18, 2009.

[25] L. Li and M. Lemmon. Weakly coupled event triggered output feedback system in wireless networked control systems. In *Allerton Conference on Communication, Control and Computing*, Urbana, IL, pages 572–579, 2011.

[26] A. Molin and S. Hirche. Structural characterization of optimal event-based controllers for linear stochastic systems. In *Proceedings of the IEEE Conference Decision and Control*, Atlanta, pages 3227–3233, December 15–17, 2010.

[27] D. Lehmann. *Event-Based State-Feedback Control*. Logos Verlag, Berlin, 2011.

[28] R. Postoyan, A. Anta, W. P. M. H. Heemels, P. Tabuada, and D. Nešić. Periodic event-triggered control for nonlinear systems. In *IEEE Conference on Decision and Control (CDC)*, Florence, Italy, pages 7397–7402, December 10–13, 2013.

[29] R. Goebel, R. G. Sanfelice, and A. R. Teel. *Hybrid Dynamical Systems*. Princeton University Press, Princeton, NJ, 2012.

[30] X. Wang and M. Lemmon. Self-triggered feedback control systems with finite-gain $\mathcal{L}_2$ stability. *IEEE Transactions on Automatic Control*, 45:452–467, 2009.

[31] M. Velasco, P. Marti, and E. Bini. On Lyapunov sampling for event-driven controllers. In *Proceedings of the IEEE Conference Decision & Control*, Shangaï, China, pages 6238–6243, December 15–18, 2009.

[32] X. Wang and M. D. Lemmon. Event design in event-triggered feedback control systems. In *Proceedings of the IEEE Conference Decision & Control*, Cancun, Mexico, pages 2105–2110, December 9–11, 2008.

[33] M. Mazo Jr., A. Anta, and P. Tabuada. An ISS self-triggered implementation of linear controllers. *Automatica*, 46:1310–1314, 2010.

[34] A. van der Schaft. *$\mathcal{L}_2$-Gain and Passivity Techniques in Nonlinear Control*. Springer-Verlag: Berlin Heidelberg, 2000.

[35] J. Löfberg. YALMIP: A toolbox for modeling and optimization in MATLAB. In *Proceedings of the CACSD Conference*, Taipei, Taiwan, pages 284–289, 2004.

[36] S. J. L. M. van Loon, W. P. M. H. Heemels, and A. R. Teel. Improved $\mathcal{L}_2$-gain analysis for a class of hybrid systems with applications to reset and event-triggered control. In *Proceedings of the IEEE Conference Decision and Control*, LA, pages 1221–1226, December 15–17, 2014.

[37] FlexRay Consortium. Flexray communications system-protocol specification. *Version*, 2(1):198–207, 2005.

[38] T. Chen and B. A. Francis. *Optimal Sampled-Data Control Systems*. Springer-Verlag, 1995.

[39] A. R. Teel and L. Praly. A smooth Lyapunov function from a class-estimate involving two positive semidefinite functions. *ESAIM: Control, Optimisation and Calculus of Variations*, 5(1):313–367, 2000.

[40] Y. Lin, E. D., Sontag, and Y. Wang. A smooth converse Lyapunov theorem for robust stability. *SIAM Journal on Control and Optimization*, 34(1):124–160, 1996.

[41] A. Anta and P. Tabuada. Exploiting isochrony in self-triggered control. *IEEE Transactions on Automatic Control*, 57(4):950–962, 2012.

[42] W. P. M. H. Heemels, N. van de Wouw, R. Gielen, M. C. F. Donkers, L. Hetel, S. Olaru, M. Lazar, J. Daafouz, and S. I. Niculescu. Comparison of overapproximation methods for stability analysis of networked control systems. In *Proceedings of the Conference Hybrid Systems: Computation and Control*, Stockholm, Sweden, pages 181–190, April 12–16, 2010.

[43] C. I. Byrnes and A. Isidori. New results and examples in nonlinear feedback stabilization. *Systems & Control Letters*, 12(5):437–442, 1989.

[44] A. Anta and P. Tabuada. To sample or not to sample: Self-triggered control for nonlinear systems. *IEEE Transactions on Automatic Control*, 55(9):2030–2042, 2010.

[45] S. Prajna, A. Papachristodoulou, P. Seiler, and P. A. Parrilo. Sostools: Sum of squares optimization toolbox for MATLAB. http://www.cds.caltech.edu/sostools, 2004.

[46] W. P. M. H. Heemels, G. Dullerud, and A. R. Teel. A lifting approach to L2-gain analysis of periodic event-triggered and switching sampled-data control systems. *Proceeding of the IEEE Conference on Decision and Control (CDC) 2015*, Osaka, Japan.

# 7

# *Decentralized Event-Triggered Controller Implementations*

**Manuel Mazo Jr.**
*Delft University of Technology*
*Delft, Netherlands*

**Anqi Fu**
*Delft University of Technology*
*Delft, Netherlands*

**CONTENTS**



**ABSTRACT** The triggering conditions found in the event-triggered control literature usually depend on a relation between the whole vector of states and the last measurements employed in the control. However, in networked implementations of control systems, it is often the case that sensors are not collocated with each other, therefore impeding the application of such event-triggered control techniques. In this chapter, motivated by the use of wireless communication networks in control systems' implementations, we present two alternative solutions to this problem. We proposed two types of decentralized conditions. The first technique requires synchronous transmission of measurements from all sensors, while the second proposes the use of asynchronous measurements. Furthermore, we introduce some naive communication protocols that are used, together with a realistic model of energy consumption on wireless sensor networks, to assess the energy usage of these two types of implementations.

## 7.1 Introduction

In the current chapter, we address the implementation of controllers in which the updates of the control action are event triggered. We consider architectures in which the sensors providing measurements to the controller are not collocated (neither among them nor with the controller). Most event-triggered implementations of controllers rely on triggering conditions that depend on the whole state vector, and thus, in the scenario, we consider monitoring the triggering conditions is either unfeasible or requires a prohibitive amount of communication exchange between the different sensors and the controller.

The study of the proposed scenarios is mainly motivated by networked implementations of control loops, in which a number of different sensors must provide measurements to recompute the control action. In such scenarios, in order to allow more control loops to share the same resources, access to the communication medium (regardless of the use of a wired or wireless physical layer) should be reduced. Furthermore, if the physical layer of the communication medium is a wireless channel, reducing its use has a direct impact on the energy consumption of each of the different nodes involved: sensors, actuators, and computation nodes (controllers). It is important to remark that in such wireless implementations, it is equally (if not more) important to reduce the amount of time employed in listening for messages in the channel as in transmitting messages.

Following this line of reasoning, after formalizing the system models and communication architecture under consideration, we introduce a decentralized strategy aimed at the reduction of message transmissions in the network. We discuss briefly strategies that could be employed to reduce the amount of time that the different nodes of the implementation need to spend listening for messages from other nodes. Next, we present an approach that addresses directly the reduction of listening time from nodes. Finally, we compare the two techniques through some examples, discuss the benefits and shortcomings of the two techniques, and conclude the chapter with some ideas for future and ongoing extensions of these techniques.

## 7.2 Preliminaries

We denote the positive real numbers by $\mathbb{R}^+$ and by $\mathbb{R}_0^+ = \mathbb{R}^+ \cup \{0\}$. We use $\mathbb{N}_0$ to denote the natural numbers including zero and $\mathbb{N}^+ = \mathbb{N}_0 \backslash \{0\}$. The usual Euclidean ($l_2$) vector norm is represented by $|\cdot|$. When applied to a matrix $|\cdot|$ denotes the $l_2$ induced matrix norm. A symmetric matrix $P \in \mathbb{R}^{n\times n}$ is said to be positive definite, denoted by $P > 0$, whenever $x^T P x > 0$ for all $x \neq 0$, $x \in \mathbb{R}^n$. By $\lambda_m(P), \lambda_M(P)$ we denote the minimum and maximum eigenvalues of $P$, respectively. A function $f : \mathbb{R}^n \to \mathbb{R}^m$ is said to be locally Lipschitz if for every compact set $S \subset \mathbb{R}^n$ there exists a constant $L \in \mathbb{R}_0^+$ such that $|f(x) - f(y)| \leq L|x - y|$, $\forall x, y \in S$. For a function $f : \mathbb{R}^n \to \mathbb{R}^n$, we denote by $f_i : \mathbb{R}^n \to \mathbb{R}$ the function whose image is the projection of $f$ on its $i$th coordinate, that is, $f_i(x) = \Pi_i(f(x))$. Consequently, given a Lipschitz continuous function $f$, we also denote by $L_{f_i}$ the Lipschitz constant of $f_i$. A function $\gamma : [0, a[ \to \mathbb{R}_0^+$, is of class $\mathcal{K}$ if it is continuous, strictly increasing, and $\gamma(0) = 0$; if furthermore $a = \infty$ and $\gamma(s) \to \infty$ as $s \to \infty$, then $\gamma$ is said to be of class $\mathcal{K}_\infty$. A continuous function $\beta : \mathbb{R}_0^+ \times \mathbb{R}_0^+ \to \mathbb{R}_0^+$ is of class $\mathcal{KL}$ if $\beta(\cdot, \tau)$ is of class $\mathcal{K}$ for each fixed $\tau \geq 0$ and for each fixed $s \geq 0$, $\beta(s, \tau)$ is decreasing with respect to $\tau$ and $\beta(s, \tau) \to 0$ for $\tau \to \infty$. Given an essentially bounded function $\delta : \mathbb{R}_0^+ \to \mathbb{R}^m$, we denote by $\|\delta\|_\infty$ the $\mathcal{L}_\infty$ norm, that is, $\|\delta\|_\infty = \text{ess sup}_{t\in\mathbb{R}_0^+}\{|\delta(t)|\}$. We also use the shorthand $\dot{V}(x, u)$ to denote the Lie derivative $\nabla V \cdot f(x, u)$, and $\circ$ to denote function composition, that is, $f \circ g(t) = f(g(t))$.

The notion of Input-to-State stability (ISS) [3] will be central to our discussion:

**DEFINITION 7.1: Input-to-state stability** A control system $\dot{\xi} = f(\xi, \upsilon)$ is said to be (uniformly globally) input-to-state stable (ISS) with respect to $\upsilon$ if there exist $\beta \in \mathcal{KL}, \gamma \in \mathcal{K}_\infty$ such that for any $t_0 \in \mathbb{R}_0^+$ the following holds:

$$\forall \xi(t_0) \in \mathbb{R}^n, \ \|\upsilon\|_\infty < \infty,$$
$$|\xi(t)| \leq \beta(|\xi(t_0)|, t - t_0) + \gamma(\|\upsilon\|_\infty), \ \forall t \geq t_0.$$

Rather than using this definition, we use an alternative characterization of ISS systems by virtue of the following fact: A system is ISS if and only if there exists an ISS Lyapunov function [3]:

**DEFINITION 7.2: ISS Lyapunov function** A continuously differentiable function $V : \mathbb{R}^n \to \mathbb{R}_0^+$ is said to be an ISS Lyapunov function for the closed-loop system $\dot{\xi} = f(\xi, \upsilon)$ if there exist class $\mathcal{K}_\infty$ functions $\underline{\alpha}$, $\overline{\alpha}$, $\alpha_v$, and $\alpha_e$ such that for all $x \in \mathbb{R}^n$ and $u \in \mathbb{R}^m$ the following is satisfied:

$$\underline{\alpha}(|x|) \leq V(x) \leq \overline{\alpha}(|x|),$$
$$\nabla V \cdot f(x, u) \leq -\alpha_v \circ V(x) + \alpha_e(|u|). \qquad (7.1)$$

Finally, we also employ the following Lemma in some of our arguments:

**Lemma 7.1: [16]**

Given two $\mathcal{K}_\infty$ functions $\alpha_1$ and $\alpha_2$, there exists some constant $L < \infty$ such that

$$\limsup_{s\to 0} \frac{\alpha_1(s)}{\alpha_2(s)} \leq L,$$

if and only if for all $S < \infty$ there exists a positive $\kappa < \infty$ such that

$$\forall s \in ]0, S],\ \alpha_1(s) \leq \kappa\alpha_2(s).$$

**PROOF** The necessity side of the equivalence is trivial; thus, we concentrate on the sufficiency part. By assumption, we know that the limit superior of the ratio of the functions tends to $L$ as $s \to 0$, and therefore, $\forall\epsilon > 0$, $\exists\delta > 0$ such that $\alpha_1(s)/\alpha_2(s) < L + \epsilon$ for all $s \in ]0, \delta[$. As $\alpha_1, \alpha_2 \in \mathcal{K}_\infty$, we know that in any compact set excluding the origin, the function $\alpha_1(s)/\alpha_2(s)$ is continuous and therefore attains a maximum, implying that there exists a positive $M \in \mathbb{R}^+$ such that $\alpha_1(s)/\alpha_2(s) < M$, $\forall s \in [\delta, S]$, $0 < \delta < S$. Putting these two results together, we have that $\forall s \in ]0, S]$, $S < \infty$, $\alpha_1(s) \leq \kappa\alpha_2(s)$, where $\kappa = \max\{L + \epsilon, M\}$. ■

Let us quickly revisit the event-triggering mechanism proposed in [21], which serves as a starting point for the techniques we propose in what follows. Consider a control system:

$$\dot{\xi}(t) = f(\xi(t), \upsilon(t)), \qquad \forall t \in \mathbb{R}_0^+, \tag{7.2}$$

where $\xi : \mathbb{R}_0^+ \to \mathbb{R}^n$ and $\upsilon : \mathbb{R}_0^+ \to \mathbb{R}^m$. Assume the following:

**ASSUMPTION 7.1: Lipschitz continuity** The functions $f$ and $k$, defining the dynamics and controller of the system, are locally Lipschitz.

This assumption guarantees the (not necessarily global) existence and uniqueness of solutions of the closed-loop system; and:

**ASSUMPTION 7.2: ISS w.r.t. measurement errors** There exists a controller $k : \mathbb{R}^n \to \mathbb{R}^m$ for system (7.2) such that the closed-loop system

$$\dot{\xi}(t) = f(\xi(t), k(\xi(t) + \varepsilon(t))), \qquad \forall t \in \mathbb{R}_0^+, \tag{7.3}$$

is ISS with respect to measurement errors $\varepsilon$. Furthermore, assume the knowledge of an ISS-Lyapunov function $V$ certifying that the system is ISS, that is, satisfying (7.1).

Let the control law $k(\xi)$ be applied in a sample-and-hold fashion:

$$\begin{aligned} \dot{\xi}(t) &= f(\xi(t), \upsilon(t)), \\ \upsilon(t) &= k(\xi(t_k)),\ t \in [t_k, t_{k+1}[, \end{aligned} \tag{7.4}$$

where $\{t_k\}_{k\in\mathbb{N}_0^+}$ is a divergent sequence of update times. The stability of the resulting hybrid system depends now on the selection of the sequence of update times $\{t_k\}_{k\in\mathbb{N}_0^+}$. In traditional time-triggered implementations, the sequence is simply defined by $t_k = kT$, where $T$ is a predefined sampling time. In contrast, an event-triggered controller implementation defines the sequence implicitly as the time instants at which some condition is violated, which in general results in aperiodic schemes of execution. The following is a typical scheme to design triggering conditions resulting in asymptotic stability of the event-triggered closed loop.

Define an auxiliary signal $\varepsilon : \mathbb{R}_0^+ \to \mathbb{R}^n$ as $\varepsilon(t) = \xi(t_k) - \xi(t)$ for $t \in [t_k, t_{k+1}[$, and regard it as a measurement error. By doing so, one can rewrite (7.4) for $t \in [t_k, t_{k+1}[$ as

$$\begin{aligned} \dot{\xi}(t) &= f(\xi(t), k(\xi(t) + \varepsilon(t))), \\ \dot{\varepsilon}(t) &= -f(\xi(t), k(\xi(t) + \varepsilon(t))),\ \varepsilon(t_k) = 0. \end{aligned}$$

Hence, as (7.3) is ISS with respect to measurement errors $\varepsilon$, by enforcing

$$\gamma(|\varepsilon(t)|) \leq \rho\alpha(|\xi(t)|),\ \forall t > 0,\ \rho \in ]0, 1[, \tag{7.5}$$

one can guarantee that

$$\frac{\partial V}{\partial x} f(x, k(x + e)) \leq -(1 - \rho)\alpha(|x|),\ \forall x, e \in \mathbb{R}^n,$$

and asymptotic stability of the closed loop follows.

Further assume that:

**ASSUMPTION 7.3** The operation of system (7.4) is confined to some compact set $S \subseteq \mathbb{R}^n$, and $\alpha^{-1}$ and $\gamma$ are Lipschitz continuous on $S$.

Then, the inequality (7.5) can be replaced by the simpler inequality $|\varepsilon(t)|^2 \leq \sigma|\xi(t)|^2$, for a suitably chosen $\sigma > 0$. We refer the interested reader to [21] for details on how to select $\sigma$. Hence, if the sequence of update times $\{t_k\}_{k\in\mathbb{N}_0^+}$ is such that

$$|\varepsilon(t)|^2 \leq \sigma|\xi(t)|^2, \quad t \in [t_k, t_{k+1}[, \tag{7.6}$$

the sample-and-hold implementation (7.4) is guaranteed to render the closed-loop system asymptotically stable. Note that enforcing this condition is achieved by simply closing the loop, that is, resetting to zero $\varepsilon(t_k) = \xi(t_k) - \xi(t_k) = 0$, at the instants of time:

$$t_k = \min\{t > t_{k-1} \,|\, |\varepsilon(t)|^2 \geq \sigma|\xi(t)|^2\}. \tag{7.7}$$

It was also shown in [21] (Theorem 3.1) that there is a non-zero minimum time that must always elapse after a triggering event until the next event is triggered, and that lower bounds can be explicitly computed. We denote a computable lower bound for such minimum time by $\tau_{\min}$. Furthermore, delays between the event generation and the application of an updated control signal can be accommodated by making the triggering conditions more conservative. We reproduce here Corollary IV.1 (with adjusted notation) from [21], which summarizes these results in the case of linear systems*:

**Corollary 7.1: [21]**

Let $\dot{\xi} = A\xi + B\upsilon$ be a linear control system, let $\upsilon = K\xi$ be a linear control law rendering the closed-loop system globally asymptotically stable, and assume that $\Delta\tau = 0$. For any initial condition in $\mathbb{R}^n$, the interevent times $\{t_{k+1} - t_k\}_{k\in\mathbb{N}}$ implicitly defined by the execution rule $|\varepsilon| = \sqrt{\sigma}|\xi|$ are lower bounded by the time $\tau_{\min}$ satisfying

$$\phi(\tau_{\min}, 0) = \sqrt{\sigma},$$

where $\phi(t, \phi_0)$ is the solution of

$$\dot{\phi} = |A + BK| + (|A + BK| + |BK|)\phi + |BK|\phi^2,$$

satisfying $\phi(0, \phi_0) = \phi_0$. Furthermore, for $\Delta\tau > 0$ and for any desired $\sqrt{\sigma} > 0$, the execution rule $|\varepsilon| = \sqrt{\sigma'}|\xi|$ with

$$\frac{\Delta\tau|[A + BK|BK]|(\sqrt{\sigma}+1)}{1 - \Delta\tau|[A + BK|BK]|(\sqrt{\sigma}+1)} \leq \sqrt{\sigma'} \leq \phi(-\Delta\tau, \sqrt{\sigma}),$$

enforces for any $k \in \mathbb{N}$ and for any $t \in [t_k + \Delta\tau, t_{k+1} + \Delta\tau[$ the following inequality:

$$|\varepsilon(t)| \leq \sqrt{\sigma}|\xi(t)|,$$

with interexecution times bounded by $\tau_{\min} = \Delta\tau + \tau$, where time $\tau$ satisfies

$$\phi\left(\tau, \frac{\Delta\tau|[A + BK|BK]|(\sqrt{\sigma}+1)}{1 - \Delta\tau|[A + BK|BK]|(\sqrt{\sigma}+1)}\right) = \sqrt{\sigma'}.$$

For completeness, we also provide the solution to the differential Riccati equation for $\phi$:

$$\phi(t, \phi_0) = -\frac{1}{2\alpha_2}(\alpha_1 - \Theta\tan(\frac{\Theta}{2}(t + C))),$$

$$C = \frac{2}{\Theta}\arctan\frac{2\alpha_2\phi_0 + \alpha_1}{\Theta}, \Theta = \sqrt{4\alpha_0\alpha_2 - \alpha_1^2},$$

where $\alpha_0 = |A + BK|$, $\alpha_1 = |A + BK| + |BK|$, and $\alpha_2 = |BK|$.

*Similar results can be obtained for nonlinear input affine systems under mild assumptions on the flow (e.g., being norm bounded from above). For details see, for example, [6].

## 7.3 System Architecture and Problem Statement

We consider systems consisting of three types of nodes interconnected through a shared communications channel: sensors, actuators, and controllers. Sensors and actuators are in direct contact with the plant to be controlled, the first ones to acquire measurements of the state, and the latter to set the values of the controllable inputs of the plant. We consider only centralized control computing situations—that is, each control loop has actions that are computed at a single controller node that collects data from all sensors and sends actions to all actuators. Additionally, a network may contain relay nodes employed to enlarge the area covered by the network. A graphical representation of one such generic network is given in Figure 7.1, in which sensor and actuator nodes are indicated by circles and diamonds, respectively, next to which their respective sensed or actuated signal is indicated, a central computing node is indicated with a box, and the remainder nodes are simple relay nodes to maintain connectivity. We have in mind, thus, tree-shaped networks in which sensors and actuators are leave nodes, the controller sits at the root, and all other intermediate nodes can only be relay nodes. The reason for this last restriction is to allow sensor (and possibly actuator) nodes to go to an idle state, in which they do not listen or transmit, more often than other nodes in the network without disrupting the network connectivity. This assumption is related to the energy constraints that we describe in the next two paragraphs.

The goal of the techniques that will be described in the remainder of the chapter is to reduce the amount of communication required between the different nodes of the network to stabilize a control plant. Attaining such a goal may be interesting to reduce congestion on the shared network, thus freeing it up for other communication tasks. Additionally, we consider situations in which the shared network is constructed over a wireless channel. In this case, the reduction of communication

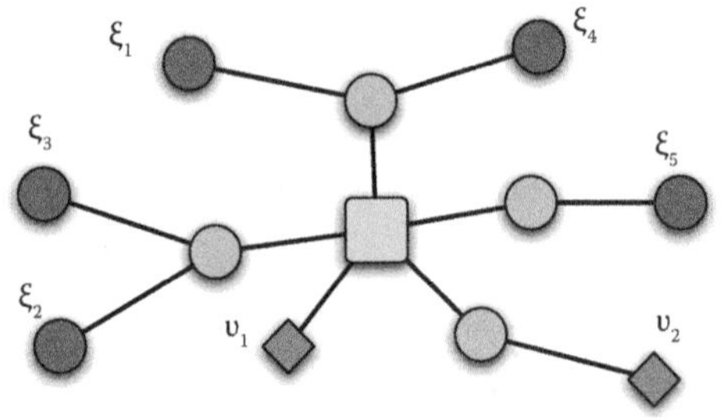


**FIGURE 7.1**
A generic control network, with a centralized computing (controller) node.

between nodes could be exploited to also reduce energy consumption on the network nodes, some of which may be battery powered, and thus extend the network's lifetime.

We consider that actuation nodes, controller nodes, and possible relay nodes are not as energy constrained as sensor nodes, and thus concentrate our efforts on energy savings at the sensor nodes. This is justified by the following considerations (taken as architectural assumptions): sensor nodes are often located in hard-to-reach areas, which limit their possible size and thus the size of the batteries supplying energy to them; actuators, while they can also have a restrictive placing, need larger energy reserves to act on the plant, which usually renders any communication consumption negligible; other nodes of the network necessary to relay packets can be equipped with sufficient energy reserves or even wired to a power source.

An important remark is due at this point: the techniques that we present in the following sections are limited to implementations of state-feedback controllers, and thus the following is one of the limiting assumptions of both techniques we present:

**ASSUMPTION 7.4: State measurements** Each (and all) of the states of the system are directly measurable by some sensor.

Note that this does not restrict each sensor node to sense only a single entry of the state vector.

Our objective is to propose controller implementations for systems of the form (7.2) satisfying Assumptions 7.4, 7.2, and 7.1. In particular, we are interested in finding stabilizing sample-and-hold implementations of a controller $\upsilon(t) = k(\xi(t))$ such that updates can be performed transmitting aperiodic measurements from sensors, and if possible, we would like to do so while reducing the amount of necessary transmissions. This problem can be formalized as follows:

**PROBLEM 7.1** Given system (7.2) and a controller $k : \mathbb{R}^n \to \mathbb{R}^m$ satisfying Assumptions 7.4, 7.2, and 7.1, find sequences of update times $\{t^i_{r_i}\}$, $r_i \in \mathbb{N}_0$ for each sensor $i = 1, \ldots, n$ such that a sample-and-hold controller implementation:

$$\upsilon_j(t) = k_j(\hat{\xi}(t)), \tag{7.8}$$

$$\hat{\xi}_i(t) = \xi_i(t^i_{r_i}),\ t \in [t^i_{r_i}, t^i_{r_i+1}[,\ \forall i = 1, \ldots, n, \tag{7.9}$$

renders the closed-loop system uniformly globally asymptotically stable (UGAS), that is, satisfying that there exists $\beta \in \mathcal{KL}$ such that for any $t_0 \geq 0$:

$$\forall \xi(t_0) \in \mathbb{R}^n,\ |\xi(t)| \leq \beta(|\xi(t_0)|, t - t_0),\ \forall t \geq t_0.$$

Note that a solution to this problem is already provided by the technique reviewed in Section 7.2, with the additional Assumption 7.3. However, such a solution requires that some node of the network has access to the full state vector in order to check the condition (7.7) which establishes when the controller needs to be updated with fresh measurements. In the architectures, we are considering the main challenge is therefore how to "decentralize" the decision of triggering new updates among the different sensors of the network, which only have a partial view of the full state of the system. Another similar problem, somehow complementary to the one we addressed here, is to coordinate the triggering of different subsystems weakly coupled with each other. For the interested reader we refer on that topic to the work of Wang and Lemmon in [24] and references therein.

Let us summarize this discussion by introducing the following additional assumption and a concrete reformulation of the problem we solve in the remainder.

**ASSUMPTION 7.5: Decentralized sensing** There is no single sensor in the network capable of measuring the whole state vector of the system (7.2).

**PROBLEM 7.2** Solve Problem 7.1 under the additional Assumption 7.5.

In the remainder of this chapter, we provide two alternative solutions to the problem we just described: in Section 7.4 a solution first introduced in [17] is presented in which $\{t^i_{r_i}\} = \{t^j_{r_j}\}$ for all $i, j \in [1, n]$, that is, measurements from all different sensors are acquired synchronized in time; Section 7.5 provides a solution, originally described in [16], in which this synchronicity requirement for the state vector employed in the controller computation is removed.

## 7.4 Decentralized Triggering of Synchronous Updates

For simplicity of presentation, we will consider in what follows a scenario in which each state variable is measured by a different sensor. Nonetheless, we will discuss at the end of the section how to apply these same ideas to more generic decentralized scenarios.

As discussed in the previous section, because of Assumption 7.5 no sensor can evaluate condition (7.6), since it requires the knowledge of the full state vector $\xi(t)$. The solution we propose is to employ a set of simple conditions that each sensor can check locally to decide when to trigger a controller update.

Then, whenever such a triggering event happens at one sensor the transmission of fresh measurements from all the sensors to the controller is initiated.

To show how to arrive to the local triggering conditions let us start by introducing a set of parameters: $\theta_1, \theta_2, \ldots, \theta_n \in \mathbb{R}$ such that $\sum_{i=1}^{n} \theta_i = 0$. Employing these parameters one can rewrite inequality (7.6) as

$$\sum_{i=1}^{n} \left(\varepsilon_i^2(t) - \sigma\xi_i^2(t)\right) \leq 0 = \sum_{i=1}^{n} \theta_i,$$

where $\varepsilon_i$ and $\xi_i$ denote the $i$th coordinates of $\varepsilon$ and $\xi$, respectively. Then, observing the following trivial implication

$$\bigwedge_{i=1}^{n} \left(\varepsilon_i^2(t) - \sigma\xi_i^2(t) \leq \theta_i\right) \Rightarrow |\varepsilon(t)|^2 \leq \sigma|\xi(t)|^2, \quad (7.10)$$

suggests the use of

$$\varepsilon_i^2(t) - \sigma\xi_i^2(t) \leq \theta_i, \quad (7.11)$$

as local event-triggering conditions.

Employing these local conditions (7.11) whenever one of them reaches equality the controller is recomputed. If the time elapsed between two events generated by the proposed scheme is smaller than the minimum time $\tau_{\min}$ between updates of the centralized event-triggered implementation (see the end of Section 7.2), the second event is discarded and the controller update is scheduled $\tau_{\min}$ units of time after the previous update. Waiting until $\tau_{\min}$ time has elapsed is justified because our implementation is merely trying to mimic a centralized triggering generation, but as will be discussed, the decentralization is achieved at a cost of being conservative.

This decentralization approach is in general conservative, meaning that times between updates will be shorter than in the centralised case. This conservativeness stems from the fact that (7.10) is not an equivalence but only an implication. The reader might wonder what the purpose of introducing the vector of parameters $\theta = [\theta_1\ \theta_2 \ldots \theta_n]^T$ is. In fact, the main purpose of these parameters is to aid in reducing the mentioned conservatism and thus reducing utilization of the communication network. To achieve this reduction, we allow the vector $\theta$ to change every time the control input is updated. To reflect the time-varying nature of the parameters, from here on we show explicitly this time dependence of $\theta$ by writing $\theta(k)$ to denote its value between the update instants $t_k$ and $t_{k+1}$. Note, that regardless of this time varying behavior, as long as $\theta$ satisfies $\sum_{i=1}^{n} \theta_i(k) = 0$, the stability of the closed loop is guaranteed independently of the specific value that $\theta$ takes and the rules used to update $\theta$. Some possible rules for the adaption of $\theta$ aiming at reducing the conservativeness of the decentralization are discussed in the next subsection.

The previous discussion is now summarized in the following proposition:

**Proposition 7.1: [17]**

Let Assumptions 7.1, 7.2, and 7.3 hold. For any choice of $\theta$ satisfying:

$$\sum_{i=1}^{n} \theta_i(k) = 0, \ \forall k \in \mathbb{N}_0^+,$$

the sequence of update times $\{t_k\}_{k\in\mathbb{N}_0^+}$ given by

$$t_{k+1} = t_k + \max\{\tau_{\min}, \min_{i=1,\ldots,n} \tau_i(\xi(t_k))\},$$

$$\tau_i(\xi(t_k)) = \min\{\tau \in \mathbb{R}_0^+ \mid \varepsilon_i^2(t_k+\tau) - \sigma\xi_i^2(t_k+\tau) = \theta_i(k)\},$$

renders the system (7.4) asymptotically stable.

### 7.4.1 Adaption Rules

With the exception of some very special types of systems [7],* finding a value of $\theta$ that for a given initial condition maximizes the time until the next event is generated is a nontrivial problem. Thus, rather than providing any optimal solution, we suggest a family of heuristics to adjust the vector $\theta$ whenever the control input is updated.

We need to introduce a new concept:

**DEFINITION 7.3: Decision gap** Let the *decision gap* at sensor $i$ at time $t \in [t_k, t_{k+1}[$ be defined as

$$G_i(t) = \varepsilon_i^2(t) - \sigma\xi_i^2(t) - \theta_i(k),$$

The family of heuristics is parametrized by an *equalization time* $t_e$ and an *approximation order* $q$, and we attempt to equalize the decision gap at time $t_e$.

For the *equalization time* $t_e : \mathbb{N}_0 \to \mathbb{R}^+$, we suggest the use of one of the following two choices:

- Constant and equal to the minimum time between controller updates $t_e(k) = \tau_{\min}$
- Previous time between updates $t_e(k) = t_k - t_{k-1}$

The *approximation order* is the order of the Taylor expansion that is used to estimate the decision gap at

*In that paper, the optimal values of $\theta$ are computed for diagonalized linear systems for the purpose of self-triggered control, see, for example, [6,15] for an introduction to that related topic.

the equalization time $t_e$:

$$\hat{G}_i(t_k + t_e) = \hat{\varepsilon}_i^2(t_k + t_e) - \sigma\hat{\xi}_i^2(t_k + t_e) - \theta_i(k),$$

where for $t \in [t_k, t_{k+1}[$

$$\hat{\xi}_i(t) = \xi_i(t_k) + \dot{\xi}_i(t_k)(t - t_k) + \frac{1}{2}\ddot{\xi}_i(t_k)(t - t_k)^2 + \dots + \frac{1}{q!}\xi_i^{(q)}(t_k)(t - t_k)^q,$$

$$\hat{\varepsilon}_i(t) = 0 - \dot{\xi}_i(t_k)(t - t_k) - \frac{1}{2}\ddot{\xi}_i(t_k)(t - t_k)^2 - \dots - \frac{1}{q!}\xi_i^{(q)}(t_k)(t - t_k)^q,$$

using the fact that $\dot{\varepsilon} = -\dot{\xi}$ and $\varepsilon(t_k) = 0$.

Once a choice for equalization time $t_e$ and an approximation order $q$ is selected, one can compute the vector $\theta(k) \in \mathbb{R}^n$ to satisfy the following conditions:

$$\hat{G}_i(t_k + t_e) = \hat{G}_j(t_k + t_e) \qquad \forall i, j \in \{1, 2, \dots, n\},$$
$$\sum_{i=1}^{n} \theta_i(k) = 0. \tag{7.12}$$

Note that solving for $\theta$, once the estimates $\hat{\xi}$ and $\hat{\varepsilon}$ have been computed, only requires to solve a system of $n$ linear equations:

$$\begin{bmatrix} 1 & -1 & 0 & 0 & \dots & 0 \\ 0 & 1 & -1 & 0 & \dots & 0 \\ 0 & 0 & \ddots & \ddots & 0 & 0 \\ 0 & 0 & 0 & \dots & 1 & -1 \\ 1 & 1 & 1 & \dots & 1 & 1 \end{bmatrix} \begin{bmatrix} \theta_1(k) \\ \theta_2(k) \\ \vdots \\ \theta_{n-1}(k) \\ \theta_n(k) \end{bmatrix} = \begin{bmatrix} \delta_{12}(t_k + t_e) \\ \delta_{23}(t_k + t_e) \\ \vdots \\ \delta_{(n-1)n}(t_k + t_e) \\ 0 \end{bmatrix}, \tag{7.13}$$

$$\delta_{ij}(t) = \left(\hat{\varepsilon}_i^2(t) - \sigma\hat{\xi}_i^2(t)\right) - \left(\hat{\varepsilon}_j^2(t) - \sigma\hat{\xi}_j^2(t)\right).$$

A solution $\theta$ to these equations could result in that for some sensor $i$, the following holds $-\sigma\xi_i^2(t_k) > \theta_i(k)$. Employing such a $\theta$ results in an immediate violation of the triggering condition at $t = t_k$, that is, $\tau_i(\xi(t_k))$ would be zero. Thus, in practice, when one encounters this situation, $\theta$ is reset to some default value such as the zero vector. Note that we propose that the computation of $\theta$ is made by the controller node for convenience, as it has access to $\xi(t_k)$. Nonetheless, if there would be any other node of the network that at each controller update also has access to the whole set of sensor measurements, such a node could also compute the updates of $\theta$.

Figure 7.2 shows a cartoon illustration of the operation of this heuristic algorithm. In the figure, $\theta$ is initialized

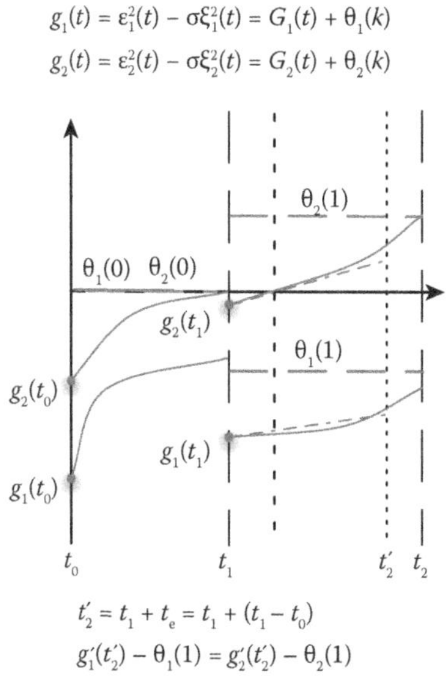


**FIGURE 7.2**
$\theta$ adaption.

to a zero value, and after $t_1 - t_0$ seconds, the second sensor has reached equality on its triggering condition. To compute the next values of $\theta$ the controller computes the estimate $\hat{\varepsilon}_i^2(t) - \hat{\sigma}\xi_i^2(t)$, $i = 1, 2$ at time $t_2' = 2t_1 - t_0$, that is, with $t_e$ equal to the last time between updates $t_1 - t_0$. The estimate is computed through a first-order Taylor approximation, that is, $q = 1$, as represented by the dash-dotted lines approximating the actual trajectory of the decision gap. Then, the new values of $\theta_1$ and $\theta_2$ are computed so that the estimates of the decision gap at $t_2'$ are equal. If one does not update the value of $\theta = 0$, it is indicated with a vertical dashed line the time at which the next update would be triggered, which can be seen to be much earlier compared to the time $t_2$ at which the next update is triggered when using the adapted new $\theta$ values.

The choice of $t_e$ and $q$ has a great impact on the amount of actuation required: equalizing at times $t_k + t_e$ as close as possible to the ideal next update time $t_{k+1}$ (the one that would result from a centralized event-triggered implementation) provides larger times between updates; but a large $t_e$ leads, in general, to poor estimates of the state of the plant at time $t_k + t_e$ and thus degrades the equalization of the gaps. Thus, employing small $t_e$ and simultaneously $t_k + t_e$ close to the ideal $t_{k+1}$ can be an impossible task, namely, when the time between controller updates is large. The effect of the order of approximation $q$ depends heavily on $t_e$, and in order to get an improvement of the estimates one might need to use prohibitively large $q$ values. An heuristic requiring relatively low computational effort (as long as $q$ is not too large) that provides good results in several case studies performed by the authors is given by Algorithm 7.1.

**Algorithm 7.1:** The $\theta$-adaptation heuristic algorithm.

**Input**: $q$, $t_{k-1}$, $t_k$, $\tau_{\min}$, $\xi(t_k)$
**Output**: $\theta(k)$
$t_e := t_k - t_{k-1}$;
Compute $\theta(k)$ according to Equation 7.12;
**if** $\exists i \in \{1,2,\ldots,n\}$ *such that* $-\sigma\xi_i^2(t_k) > \theta_i(k)$ **then**
  $t_e := \tau_{\min}$;
  Compute $\theta(k)$ according to Equation 7.12;
  **if** $\exists i \in \{1,2,\ldots,n\}$ *such that* $-\sigma\xi_i^2(t_k) > \theta_i(k)$ **then**
    $\theta(k) := 0$;
  **end**
**end**

We assumed at the beginning of the section that each node measured a single state of the system. In practice, one may encounter scenarios in which one sensor has access to several (but not all) states of the plant. Then, one can apply the same approach with a slight modification to the local triggering rules as follows:

$$|\bar{\varepsilon}_i(t)|^2 - \sigma|\xi_i(t)|^2 \le \theta_i,$$

where $\xi_i(t)$ is now the vector of states sensed at node $i$, $\bar{\varepsilon}_i(t)$ is its corresponding error vector, and $\theta_i$ is a scalar.

Let us illustrate the effect of the proposed decentralization of the event-triggering mechanism with an example. We only provide a brief description of the system and the results; for more details, we refer the reader to the original sources of this example in [12] and [17].

**EXAMPLE 7.1** Consider the quadruple-tank system as described in [12]. This is a multi-input multi-output nonlinear system consisting of four water tanks as shown in Figure 7.3. The state of the plant is composed of the water levels of the tanks: $\xi_1$, $\xi_2$, $\xi_3$, and $\xi_4$. Two inputs are available: $\upsilon_1$ and $\upsilon_2$, the input flows to the tanks. The fixed parameters $\gamma_1$ and $\gamma_2$ control how the flow is divided into the four tanks. The goal is to stabilize the levels $\xi_1$ and $\xi_2$ of the lower tanks at some specified values $x_1^*$ and $x_2^*$.

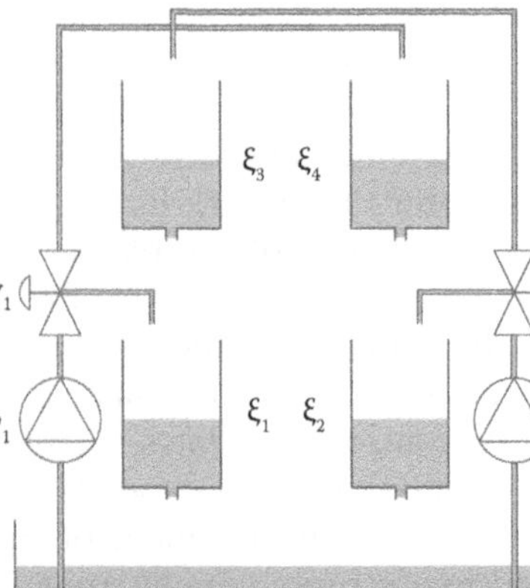


**FIGURE 7.3**
The quadruple-tank system.

The system dynamics are given by the equation:

$$\dot{\xi}(t) = f(\xi(t)) + g_c\upsilon,$$

with

$$f(x) = \begin{bmatrix} -\frac{a_1\sqrt{2gx_1}}{A1} + \frac{a_3\sqrt{2gx_3}}{A1} \\ -\frac{a_2\sqrt{2gx_2}}{A2} + \frac{a_4\sqrt{2gx_4}}{A2} \\ -\frac{a_3\sqrt{2gx_3}}{A3} \\ -\frac{a_4\sqrt{2gx_4}}{A4} \end{bmatrix},$$

$$g_c = \begin{bmatrix} \frac{\gamma_1}{A1} & 0 \\ 0 & \frac{\gamma_2}{A2} \\ 0 & \frac{1-\gamma_2}{A3} \\ \frac{1-\gamma_1}{A4} & 0 \end{bmatrix},$$

and $g$ denoting gravity's acceleration and $A_i$ and $a_i$ denoting the cross sections of the $i$th tank and outlet hole, respectively.

The controller design from [12] extends the dynamics with two additional auxiliary state variables $\xi_5$ and $\xi_6$. These are nonlinear integrators that the controller employs to achieve zero steady-state offset and evolve according to

$$\dot{\xi}_5(t) = k_{I1}a_1\sqrt{2g}\left(\sqrt{\xi_1(t)} - \sqrt{x_1^*}\right),$$

$$\dot{\xi}_6(t) = k_{I2}a_2\sqrt{2g}\left(\sqrt{\xi_2(t)} - \sqrt{x_2^*}\right),$$

where $k_{I1}$ and $k_{I2}$ are design parameters of the controller. With this additional dynamics, stabilizing the extended system implies that in steady state, $\xi_1$ and $\xi_2$ converge to the desired values $x_1^*$ and $x_2^*$. Assuming that the sensors measuring $\xi_1$ and $\xi_2$ also compute $\xi_5$ and $\xi_6$ locally and sufficiently fast, we can consider $\xi_5$ and $\xi_6$ as regular state variables.

Then, the controller from [12] results in the following state-feedback law:

$$\upsilon(t) = -K(\xi(t) - x^*) + u^*, \tag{7.14}$$

with

$$u^* = \begin{bmatrix} \gamma_1 & 1-\gamma_2 \\ 1-\gamma_1 & \gamma_2 \end{bmatrix}^{-1} \begin{bmatrix} a_1\sqrt{2gx_1^*} \\ a_2\sqrt{2gx_2^*} \end{bmatrix},$$

and $K = QP$, where $Q$ is a positive definite matrix and $P$ is given by

$$P = \begin{bmatrix} \gamma_1 k_1 & (1-\gamma_1)k_2 & 0 & (1-\gamma_1)k_4 & \gamma_1 k_1 & (1-\gamma_1)k_2 \\ (1-\gamma_2)k_1 & \gamma_2 k_2 & (1-\gamma_2)k_3 & 0 & (1-\gamma_2)k_1 & \gamma_2 k_2 \end{bmatrix},$$

with $k_1$, $k_2$, $k_3$, and $k_4$ being design parameters of the controller.

Consider the following function:

$$H_d(x) = \frac{1}{2}(x - x^*)^T P^T Q P(x - x^*) - u^{*T} P x \quad (7.15)$$
$$+ \sum_{i=1}^{4} \frac{2}{3} k_i a_i x_i^{3/2} \sqrt{2g} + k_1 a_1 x_5 \sqrt{2 g x_1^*}$$
$$+ k_2 a_2 x_6 \sqrt{2 g x_2^*},$$

which is positive definite and has a global minimum at $x^*$, as a candidate ISS Lyapunov function with respect to $\varepsilon$ for the system with the selected controller. The following bound shows that indeed $H_d$ is an ISS Lyapunov function:

$$\frac{d}{dt} H_d(\xi) \leq -\lambda_m(R)|\nabla H_d(\xi)|^2 + |\nabla H_d(\xi)||g_c' K||\varepsilon|.$$

Furthermore, from that equation, one can design the following as a triggering condition:

$$|\nabla H_d(\xi)||g_c' K||\varepsilon| \leq \rho \lambda_m(R)|\nabla H_d(\xi)|^2, \ \rho \in ]0, 1[.$$

And finally, assuming that the system is confined to a compact set containing a neighborhood of $x^*$, $|\nabla H_d(\xi)|$ can be bounded as $|\nabla H_d(\xi)| \geq \rho_m |\xi - x^*|$ to arrive to a triggering rule, ensuring asymptotic stability, of the desired form

$$|\varepsilon(t)|^2 \leq \sigma |\xi(t) - x^*|^2, \ \sigma = \left(\rho_m \rho \frac{\lambda_m(R)}{|g_c' K|}\right)^2 > 0.$$

In what follows we show some simulations of this plant and controller, for an initial states (13, 12, 5, 9) and $x_1^* = 15$ and $x_2^* = 13$, with the implementation described in this section. Employing the same parameters of the plant and the controller as in [12], and assuming that the system operates in the compact set $S = \{x \in \mathbb{R}^6 \mid 1 \leq x_i \leq 20,\ i = 1, \ldots, 4;\ 0 \leq x_i \leq 20,\ i = 5, 6\}$, one can take $\rho_m = 0.14$, which for a choice of $\rho = 0.25$ results in $\sigma = 0.0054^2$. A bound for the minimum time between controller updates is given by $\tau_{\min} = 0.1$ ms. In the implementation we employed Algorithm 7.1 with $q = 1$ and employed a combined single triggering condition for the pairs $\xi_1, \xi_5$ and $\xi_2, \xi_6$. Figure 7.4 shows a comparison with a centralized implementation (first row), our proposal (second row), and a decentralized implementation without adaption, with $\theta(k) = 0$ for all $k \in \mathbb{N}$ (last row). In the first, second, and third columns of the figure, the time between controller updates, the evolution of the ratio $\varepsilon/\xi$ versus $\sqrt{\sigma}$ and the state trajectories, are shown, respectively.

Figure 7.5 illustrates the evolution of the adaptation vector $\theta$ for the adaptive decentralized event-triggered implementation.

Although the three implementations produce almost undistinguishable state trajectories, the efficiency of them presents quite some differences. As expected, a centralized event-triggered implementation employs far less updates and produces larger times between updates than a decentralized event-triggered implementation without adaption. However, although Algorithm 7.1 does not completely recover the performance of the centralized event-triggered implementation, it produces results fairly close in terms of number of updates and time between them.

### 7.4.2 Listening Time Reduction and Delays

It is clear, especially by looking at Example 7.1, that the main benefit of employing event-triggered controllers, reducing transmissions of measurements, is retained by the proposed decentralization of the triggering conditions. However, while reducing the amount of information that needs to be transmitted from sensors to actuators, the proposed technique suggests that sensor nodes need to continuously listen for events triggered at other nodes, or at least to requests for measurements arriving from the controller (once this receives a triggering signal). This is a shortcoming of the current technique, as the power drawn by radios listening is usually as large as (if not larger than) that consumed while transmitting. In practice, wireless sensor nodes usually have their radio modules asleep most of the time and are periodically awakened according to a time division medium access (TDMA) protocol in order to address this problem. Current proposals of protocols for control over wireless networks, like WirelessHART, are also typically based on TDMA in order to provide noninterference and strict delay guarantees. Our technique can be adjusted to be implemented over a TDMA schedule forcing the radios of the sensor nodes to be asleep in certain slots. One can then regard the effect of this energy-saving mechanism as a bounded and known delay between the generation of an event and the corresponding effect in the control signal. Then, the same approach to reduce $\sigma$ from Corollary 7.1 can be applied to accommodate these medium-access induced (and any other possible) delays in our proposal.

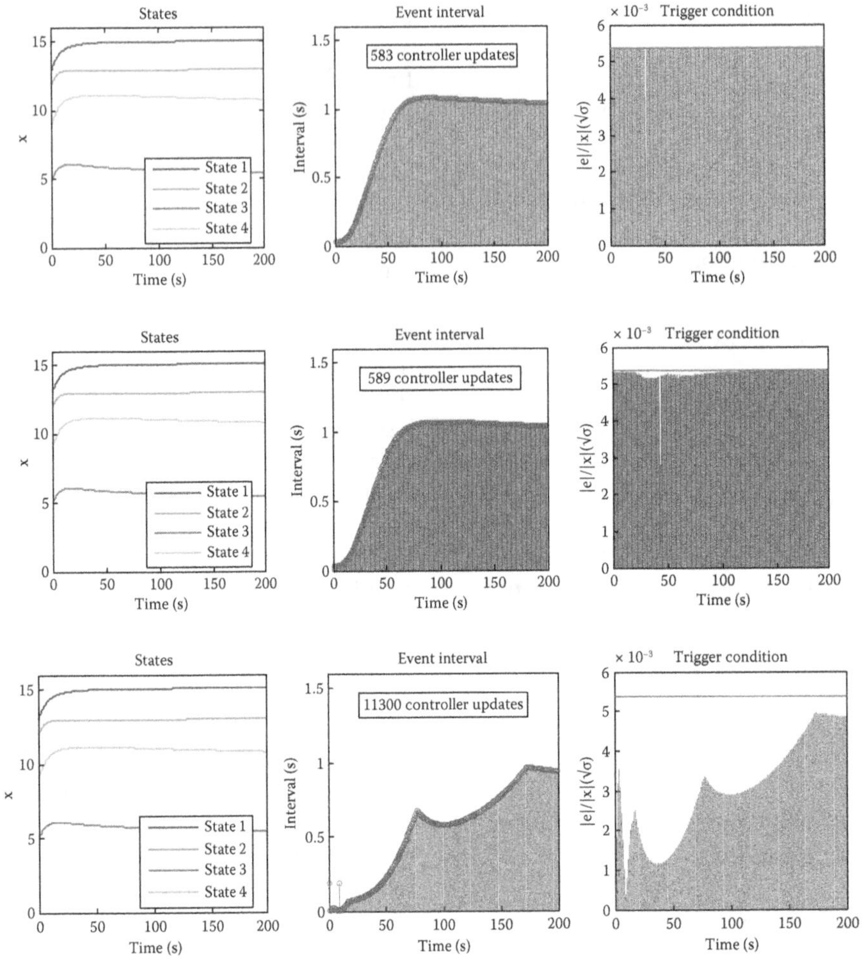


**FIGURE 7.4**
Evolution of the states, times between updates, and evolution of the triggering condition for the centralized event-triggering implementation (first row), decentralized event-triggering implementation with adaptation (second row), and decentralized event-triggering implementation without adaptation (third row).

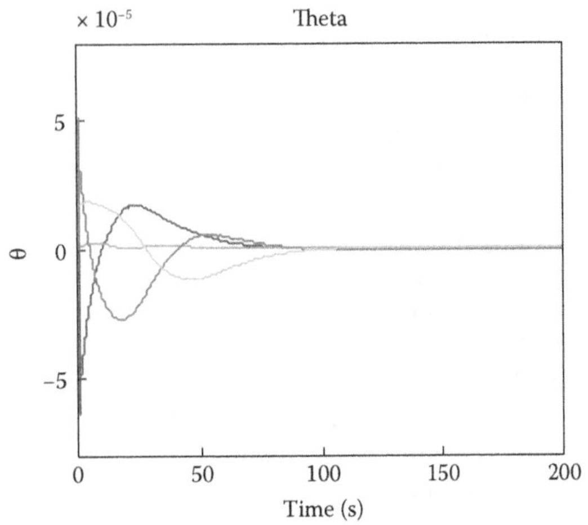


**FIGURE 7.5**
Adaptation parameter vector evolution for the adaptive decentralized event-triggered implementation.

## 7.5 Decentralized Triggering of Asynchronous Updates

We investigate now the possibility of implementing decentralized triggering in a manner in which the updates of the controller are performed with asynchronous measurements, that is, $\{t^i_{r_i}\}$ does not need to be identical to $\{t^j_{r_j}\}$ for $j \neq i$. This means that the controller is updated with measurements of the state vector containing entries not synchronized in time, thus the "asynchronous" name. By employing this asynchronous update, sensor nodes do not need to stay listening for updates from other sensors, as in the technique of Section 7.4, thus potentially bringing some additional energy savings.

We consider event-triggering rules defining implicitly, and independently, the sequences of update times $\{t^i_{r_i}\}$ for each sensor $i$. In particular, we begin by considering simple triggering conditions as

$$t^i_{r_i} := \min\{t > t^i_{r_i-1} \,|\, \varepsilon_i^2(t) = \eta_i\}, \tag{7.16}$$

where $\eta_i > 0$ are design parameters. As before, the effect of the sample-and-hold mechanism is represented as a measurement error at each sensor for all $i = 1, \ldots, n$ as

$$\varepsilon_i(t) = \xi_i(t^i_{r_i}) - \xi_i(t),\ t \in [t^i_{r_i}, t^i_{r_i+1}[,\ r_i \in \mathbb{N}_0.$$

Let us introduce the variable $\eta \in \mathbb{R}$ as

$$\eta = \sqrt{\sum_{i=1}^{n} \eta_i}, \tag{7.17}$$

which can be considered as a design parameter that once specified restricts the choices of $\eta_i$ to be used at each sensor. Then, the local parameters $\eta_i$ are defined through an appropriate scaling:

$$\eta_i := \omega_i^2 \eta^2,\ |\omega| = 1, \tag{7.18}$$

with $\omega_i$ as design constants introduced for analysis purposes only. Using this newly defined variable, the update rule (7.16) implies that $|\varepsilon(t)| \le \eta$ (with equality attained only when all sensors trigger simultaneously).

The following lemma will be central in the rest of this section:

**Lemma 7.2: [16] Intertransmission times bound**

If Assumptions 7.1 and 7.2 hold, for any $\eta > 0$, a lower bound for the minimum time between transmissions of the sensor $i$, for $i \in \{1, \ldots, n\}$, for all time $t \ge t_0$, is given by

$$\tau_i^* := L_{f_i}^{-1} \omega_i \frac{\eta}{\eta + \underline{\alpha}^{-1}(\max\{V(\xi(t_0)), \alpha_v^{-1} \circ \alpha_e(\eta)\})}, \tag{7.19}$$

where $L_{f_i}$ denotes the Lipschitz constant of the function $f_i(x, k(x+e))$ for $|x| \le \underline{\alpha}^{-1}(\max\{V(\xi(t_0)), \alpha_v^{-1} \circ \alpha_e(\eta)\})$ and $|e| \le \eta$.

**PROOF** Let us denote in what follows by: $S(y,z) = \{(x,e) \in \mathbb{R}^{n \times n} \,|\, V(x) \le y,\ |e| \le z\}$ and by $\overline{f}_i(y,z) = \max_{(x,e) \in S(y,z)} |f_i(x, k(x+e))|$. From Assumption 7.2, we have that $|e| \le \eta,\ V(x) \ge \alpha_v^{-1} \circ \alpha_e(\eta) \Rightarrow \dot{V}(x,e) \le 0$ and thus, $\tilde{S} := S(\max\{V(\xi(t_0)), \alpha_v^{-1} \circ \alpha_e(\eta)\}, \eta)$ is forward invariant. Recall that the minimum time between events at a sensor is given by the time it takes for $|\varepsilon_i|$ to evolve from the value $|\varepsilon_i(t^i_{k_i})| = 0$ to $|\varepsilon_i(t^{i-}_{k_i+1})| = \sqrt{\eta_i}$, and thus,* $\tau_i \ge \sqrt{\eta_i}(\max_{\tilde{S}} \frac{d}{dt}|\varepsilon_i|)^{-1}$. Therefore, all that needs to be proved is the existence of an upper bound on the rate of change of $|\varepsilon_i|$. One can trivially bound the evolution of $|\varepsilon_i|$ as $\frac{d}{dt}|\varepsilon_i| \le |\dot{\varepsilon}_i| = |f_i(\xi, k(\xi + \varepsilon))|$, and the maximum rate of change of $|\varepsilon_i|$ in $\tilde{S}$ by $\overline{f}_i(\max\{V(\xi(t_0)), \alpha_v^{-1} \circ \alpha_e(\eta)\}, \eta)$. Note that the existence of such a maximum is guaranteed by the continuity of the maps $f$ and $k$ and the compactness of the set $\tilde{S}$. Assumption 7.1 implies that $f_i(x, k(x+e))$ is also locally Lipschitz, and thus one can further bound $\overline{f}_i(\max\{V(\xi(t_0)), \alpha_v^{-1} \circ \alpha_e(\eta)\}, \eta) \le L_{f_i}\left(\underline{\alpha}^{-1}(\max\{V(\xi(t_0)), \alpha_v^{-1} \circ \alpha_e(\eta)\}) + \eta\right)$. Finally, recalling that $\eta_i = \omega_i^2 \eta^2$, a lower bound for the intertransmission times is given by (7.19) which proves the statement. ■

Employing a constant threshold value $\eta$ establishes a trade-off between the size of the intertransmission times and the size of the set to which the system converges. In order to achieve asymptotic stability, we allow the parameter $\eta$ to change over time and converge to zero. In particular, we consider an update policy for $\eta(t^c_{r_c})$ given by

$$\eta(t) = \eta(t^c_{r_c}),\ t \in [t^c_{r_c}, t^c_{r_c+1}[,$$
$$\eta(t^c_{r_c+1}) = \mu\eta(t^c_{r_c}), \tag{7.20}$$

for some $\mu \in ]0.5, 1[$ and with $\{t^c_{r_c}\}$ being a divergent sequence of times (to be defined later) with $t^c_0 = t_0$. The local update rules to be used are adjusted to incorporate these time-varying thresholds:

$$t^i_{r_i} := \min\{t > t^i_{r_i-1} \,|\, \varepsilon_i^2(t) = \eta_i(t)\},$$
$$\eta_i(t) := \omega_i^2 \eta(t)^2,\ |\omega| = 1. \tag{7.21}$$

Given the update policy (7.20), one can also design an event-triggered policy to decide the sequence of times $\{t^c_{r_c}\}$ such that the system is rendered asymptotically stable. The resulting overall implementation that we propose contains two independent triggering mechanisms to activate

- **Sensor-to-controller communication:** Sensors send measurements to the controller whenever the local condition (7.21) is violated. The update of the control commands is done with the measurements as they arrive in an asynchronous fashion.
- **Controller-to-sensor communication:** The controller orders the sensors to reduce the threshold used in their triggering condition, according to (7.20), when the system has "slowed down"

*The time derivative of $|\varepsilon_i|$ is defined for almost all $t$ (excluding the instants $\{t_{k_i}\}$), which is sufficient to bound the time between events.

enough to guarantee that the intersample times remain bounded from below. The controller checks this condition only in a periodic fashion, with some period $\tau^c$ (a design parameter), and therefore the sensors only need to listen at those time instants.

One of the features of our proposal, as we show later in this section, is that it enables implementations requiring only the exchange of one bit of information between a sensor and a controller, and vice versa, with the exception of the transmission of the initial state of the system at $t_0$.

The mechanism to trigger sensor-to-controller communication is already provided by (7.21). We focus now on designing an appropriate triggering mechanism for the communication from controller to sensors. The following lemma establishes some requirements to construct this new triggering mechanism.

**Lemma 7.3: [16]**

The closed-loop systems (7.2), (7.8), (7.20), (7.21) is UGAS if Assumptions 7.1 and 7.2 are satisfied and the following two conditions hold:

- $\{\eta(t^c_{r_c})\}$ is a monotonically decreasing sequence with $\lim_{r_c\to\infty}\eta(t^c_{r_c})\to 0$
- There exists $\kappa > 0$ such that for all $t^c_{r_c}$ $\frac{\underline{\alpha}^{-1}(\max\{V(\xi(t^c_{r_c})),\alpha_v^{-1}\circ\alpha_e(\eta(t^c_{r_c}))\})}{\eta(t^c_{r_c})} \leq \kappa < \infty$

**PROOF** In view of Lemma 7.2, the second condition of this lemma guarantees that there exists a minimum time between events at each sensor when both events fall in an open time interval $]t^c_{r_c}, t^c_{r_c+1}[$, that is, $t^i_{r_i+1} - t^i_{r_i} > \tau^*_i$ for all $i = 1,\ldots,n$ and $t^i_{r_i}, t^i_{r_i+1} \in ]t^c_{r_c}, t^c_{r_c+1}[$. It could happen, however, that some sensor update coincides with an update of the thresholds, that is, $t^i_{r_i+1} = t^c_{r_c+1}$, which could lead to two arbitrarily close events of sensor $i$. Similarly, events from two different sensors could be generated arbitrarily close to each other. Nonetheless, as the sequence $\{t^c_{r_c}\}$ is divergent (by assumption), and there is a finite number of sensors, none of these two effects can lead to Zeno executions.

The second condition of this lemma also implies that at $t^c_{r_c}$ either $V(\xi(t^c_{r_c})) \leq \alpha_v^{-1}\circ\alpha_e(\eta(t^c_{r_c}))$ or $V(\xi(t^c_{r_c})) \leq \underline{\alpha}(\kappa\eta(t^c_{r_c}))$. From Assumption 7.2, we have that for all $t \in [t^c_{r_c}, t^c_{r_c+1}[$ the following bound holds:

$$V(\xi(t)) \leq \max\{V(\xi(t^c_{r_c})),\alpha_v^{-1}\circ\alpha_e(\eta(t^c_{r_c}))\}$$
$$\leq \max\{\underline{\alpha}(\kappa\eta(t^c_{r_c})),\alpha_v^{-1}\circ\alpha_e(\eta(t^c_{r_c}))\}.$$

Thus, using definition (7.20) results in $V(\xi(t)) \leq \gamma_V(\eta(t)), \forall t \geq t_0$, where $\gamma_V \in \mathcal{K}_\infty$ is the function: $\gamma_V(s) = \max\{\alpha_v^{-1}\circ\alpha_e(s),\underline{\alpha}(\kappa s)\}$.

Next, we notice that the first condition of this lemma implies that $\exists\beta_\eta \in \mathcal{KL}$ such that $\eta(t) \leq \beta_\eta(\eta(t^c_0), t - t^c_0)$ for all $t \geq t^c_0$. Putting together these last two bounds, and assuming that the initial threshold is selected as $\eta(t^c_0) = \kappa_0 V(\xi(t^c_0))$, for some constant $\kappa_0 \in ]0,\infty[$, one can conclude that

$$V(\xi(t)) \leq \gamma_V(\beta_\eta(\kappa_0 V(t^c_0), t - t^c_0)),\ \forall t \geq t^c_0.$$

Finally, this last bound guarantees that

$$|\xi(t)| \leq \underline{\alpha}^{-1}(\gamma_V(\beta_\eta(\kappa_0\overline{\alpha}(|\xi(t^c_0)|), t - t^c_0))), \quad (7.22)$$
$$:= \beta(|\xi(t^c_0)|, t - t^c_0),\ \forall t \geq t^c_0. \quad (7.23)$$

with $\beta \in \mathcal{KL}$, which finalizes the proof. ■

**REMARK 7.1** Lemma 7.3 rules out the occurrence of Zeno behavior but does not establish a minimum time between transmissions of the same sensor. These bounds are provided later in Proposition 7.2. Similarly, the occurrence of arbitrarily close transmissions from different sensors is also not addressed by this lemma. A solution to this is discussed in Section 7.5.1.

Note that the first of the conditions established by this lemma is already guaranteed by employing the threshold update rule (7.20). The following assumption will help us devise a strategy satisfying the second condition. This is achieved at the cost of restricting the type of ISS controllers amenable to our proposed implementation.

**ASSUMPTION 7.6** For some $\epsilon > 1$, the ISS closed-loop system (7.3) satisfies the following property:

$$\limsup_{s\to 0}\underline{\alpha}^{-1}\circ\overline{\alpha}(\underline{\alpha}^{-1}\circ\epsilon\alpha_v^{-1}\circ\alpha_e(s) + 2s)s^{-1} < \infty. \quad (7.24)$$

**REMARK 7.2** This assumption, as well as Assumptions 7.1 and 7.2, are automatically satisfied by linear systems with a stabilizing linear state-feedback controller and the usual (ISS) quadratic Lyapunov function.

Let us introduce a simple illustrative example to provide some intuition about how this condition looks in practice. Note that we employ a scalar system as an example and thus this is not really representative of the benefits in a decentralized setting. More illustrative examples of the overall performance are provided later.

**EXAMPLE 7.2** Consider the system:

$$\dot{\xi}(t) = \mathtt{sat}(\upsilon(t)),$$

with a controller affected by measurement errors: $\upsilon(t) = -\xi(t) - \varepsilon(t)$, where the function $\mathtt{sat}(s)$ is the saturation function, saturating at values $|s| > 1$. In [20] it is shown that $V(x) = \frac{|x|^3}{3} + \frac{|x|^2}{2}$ is an ISS Lyapunov function for the system, with $\alpha_e(s) = 2s^2$ and $\alpha_v(s) = \alpha_x \circ \overline{\alpha}^{-1}(s)$, where $\alpha_x(s) = s^2/2$. Furthermore, we can also set $\overline{\alpha}(s) = \underline{\alpha}(s) = s^3/3 + s^2/2$. Then, noting that $\epsilon\overline{\alpha}(s) < \overline{\alpha}(\epsilon s)$, $\forall \epsilon > 1$,

$$\underline{\alpha}^{-1} \circ \overline{\alpha}(\underline{\alpha}^{-1} \circ \epsilon\alpha_v^{-1} \circ \alpha_e(s) + 2s) < \epsilon\alpha_x^{-1} \circ \alpha_e(s) + 2s = (2\epsilon + 2)s,$$

and we can conclude that any $\rho > 4$ will guarantee asymptotic stability to the origin. Furthermore, selecting, for example, $\rho = 4.1$, $\mu = 0.82$, and $\tau^c = 1\,\text{s}$, results in a minimum time between sensor transmissions of 0.04 s, after accounting for the effect of an aggregated (communication/actuation) delay of 0.002 s. Figure 7.6 shows a simulation in which it can be seen how the system is stabilized while respecting the lower bound for the intertransmission times.

Remember now that $\hat{\xi}$, defined in (7.9), is the vector formed by asynchronous measurements of the state entries that the controller is using to compute the input to the system. One can thus compute the following upper bound at the controller (which has access to all the latest measurements transmitted in the network):

$$|\overline{\xi}|(t) := |\hat{\xi}(t)| + \eta(t^c_{r_c}) \geq |\hat{\xi}(t) - \varepsilon(t)| = |\xi(t)|,\ \forall t \in [t^c_{r_c}, t^c_{r_c+1}[, \tag{7.25}$$

which also satisfies the bound $|\overline{\xi}|(t) \leq |\xi(t)| + 2\eta(t^c_{r_c})$.

Making use of this bound, the following theorem proposes a condition to trigger the update of sensor thresholds guaranteeing UGAS of the closed-loop system:

**Theorem 7.2: [16] UGAS**

Consider the closed-loop systems (7.2), (7.8), (7.21) with the threshold update rule (7.20) and satisfying Assumptions 7.1, 7.2, and 7.6. Let $\tau_c > 0$ be a design parameter. The sequence of threshold update times $\{t^c_{r_c}\}$ implicitly defined by

$$t^c_{r_c+1} := \min\{t = t^c_{r_c} + r\tau^c \mid r \in \mathbb{N}^+,\ |\overline{\xi}|(t) \leq \overline{\alpha}^{-1} \circ \underline{\alpha}(\rho\eta(t^c_{r_c}))\}, \tag{7.26}$$

with any $\rho < \infty$ satisfying

$$\underline{\alpha}^{-1} \circ \overline{\alpha}(\underline{\alpha}^{-1} \circ \epsilon\alpha_v^{-1} \circ \alpha_e(s) + 2s) \leq \rho s, \tag{7.27}$$

for all $s \in ]0, \eta(t_0)]$ and some $\epsilon > 1$ renders the closed-loop system UGAS.

**PROOF** We use Lemma 7.3 to show the desired result. The first itemized condition of the lemma is satisfied by the employment of the update rule (7.20) with a constant $\mu \in ]0, 1[$ if we can show that the sequence $\{t^c_{r_c}\}$

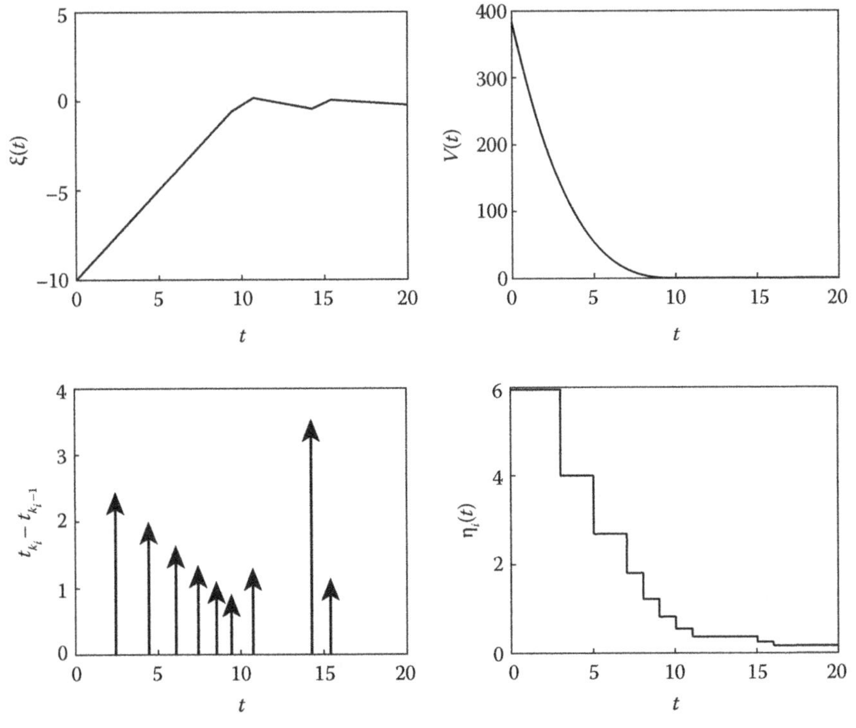


**FIGURE 7.6**
State trajectory, Lyapunov function evolution, events generated at the sensor, and evolution of the threshold.

is divergent. Thus, we must show that this sequence is divergent and that the second itemized condition in the lemma also holds.

First, we show that starting from some time $t^c_{r_c}$ there always exists some time $T \geq t^c_{r_c}$ such that for all $t \geq T$ $|\hat{\xi}|(t) \leq \overline{\alpha}^{-1} \circ \underline{\alpha}(\rho\eta(t^c_{r_c}))$. Showing this guarantees that $\{t^c_{r_c}\}$ is a divergent sequence. From Assumption 7.2, we know that for every $\bar{\epsilon} > 0$ there exists some $T \geq t^c_{r_c}$ such that $\underline{\alpha}(|\xi(t)|) \leq V(\xi(t)) \leq \alpha_v^{-1} \circ \alpha_e(\eta(t^c_{r_c})) + \bar{\epsilon}$ for every $t \geq T$. Let $\bar{\epsilon} = (\epsilon - 1)\alpha_v^{-1} \circ \alpha_e(\eta(t^c_{r_c}))$, for some $\epsilon > 1$, then

$$|\xi(t)| \leq \underline{\alpha}^{-1}(\epsilon\alpha_v^{-1} \circ \alpha_e(\eta(t^c_{r_c}))), \forall t > T,$$

and thus, we have that the proposed norm estimator (7.25) satisfies the bound:

$$|\hat{\xi}|(t) \leq \underline{\alpha}^{-1}(\epsilon\alpha_v^{-1} \circ \alpha_e(\eta(t^c_{r_c}))) + 2\eta(t^c_{r_c}), \forall t > T.$$

Therefore, if there exists a $\rho > 0$ satisfying (7.27) for some $\epsilon > 1$ and for all $s \in ]0, \eta(t_0)]$, a triggering event will eventually happen. Finally, from Assumption 7.6 and Lemma 7.1, one can conclude that such a $\rho < \infty$ exists.

The second condition of Lemma 7.3 is easier to prove. We start remarking that with $\rho$ so that (7.27) holds, as $\mu < 1, \epsilon > 1$ and $\underline{\alpha}^{-1} \circ \overline{\alpha}(s) \geq s$ for all $s$, we also have

$$\frac{\rho}{\mu}s \geq \underline{\alpha}^{-1} \circ \alpha_v^{-1} \circ \alpha_e(s), \quad (7.28)$$

for all $s \in ]0, \eta(t_0)]$. Thus, (7.28) and the triggering condition (7.26), as $V(\xi(t^c_{r_c})) \leq \overline{\alpha}(|\hat{\xi}|(t^c_{r_c}))$, guarantee that the following holds:

$$\underline{\alpha}^{-1}(\max\{V(\xi(t^c_{r_c})), \alpha_v^{-1} \circ \alpha_e(\eta(t^c_{r_c}))\}) \leq \frac{\rho}{\mu}\eta(t^c_{r_c}),$$

at all times $t^c_{r_c} > t^c_0$. Therefore,

$$\kappa := \max\left\{\frac{\rho}{\mu}, \frac{\underline{\alpha}^{-1} \circ V(\xi(t^c_0))}{\eta(t^c_0)}\right\} \quad (7.29)$$
$$\geq \frac{\underline{\alpha}^{-1}(\max\{V(\xi(t^c_{r_c})), \alpha_v^{-1} \circ \alpha_e(\eta(t^c_{r_c}))\})}{\eta(t^c_{r_c})},$$

for all $t^c_{r_c}$, which concludes the proof. ■

**REMARK 7.3** Assumption 7.6 is the most restrictive of the assumptions introduced this far. In general, it may be hard to find a controller that satisfies the property; however, in practice one can also disregard this condition. If one is only interested in stabilizing the system up to a certain precision, one can compute the Lyapunov level set to which converging guarantees such precision. By knowing that level set, one can compute the value of $\eta$ guaranteeing convergence to that level set [according to (7.1)]. If one denotes that value of the minimum threshold by $\eta_m$, then it is enough to guarantee (7.24) for $s \in [\eta_m, \eta(t^c_0)]$, $\eta_m > 0$. This can always be satisfied, given as for every $\alpha \in \mathcal{K}_\infty$ there always exists $\kappa < \infty$ such that $\alpha(s) \leq \kappa s$ for all $s \in [\eta_m, \eta(t^c_0)]$. An implementation for this type of "practical stability" would thus start with a larger threshold value than $\eta_m$ and keep decreasing it as in the proposed implementation, but once the minimum level $\eta_m$ is reached, it will stop updating the threshold. To attain the same level of stabilization precision, one could also just employ $\eta_m$ as the threshold from the beginning, but if the system has faster dynamics when it is farther from the origin (as happens with many nonlinear systems) that would result in much shorter intertransmission times.

The implementation we just described requires only the exchange of one bit after any event:

- To recover the value of a sensor after a threshold crossing, it is necessary only to know the previous value of the sensor and the sign of the error $\varepsilon_i$ when it crossed the threshold:

$$\hat{\xi}_i(t^i_{r_i}) = \hat{\xi}_i(t^i_{r_i-1}) + \texttt{sign}(\varepsilon_i(t^i_{r_i}))\sqrt{\eta_i(t^c_{r_c})}. \quad (7.30)$$

- Similarly, messages from the controller to the sensors, commanding a reduction of the thresholds, can be indicated with a single bit.

If one employs this one-bit implementation, bounds for the time between updates of a sensor valid globally (not only between threshold updates, but also across such updates) can be obtained as follows:

**Proposition 7.2: [16] Intertransmission time bounds**

The controller implementation from Theorem 7.2 with controller updates (7.30), $\tau^c \geq \max_{i\in[1,n]}\left\{\frac{\mu L_{f_i}^{-1}\omega_i}{\mu+\rho}\right\}$ and

$$\eta(t^c_0) \geq \frac{\mu}{\rho}\underline{\alpha}^{-1} \circ V(\xi(t^c_0)), \quad (7.31)$$

guarantees that a minimum time between events at each sensor is given, for all $t \geq t^c_0$, by

$$t^i_{r_i} - t^i_{r_i-1} \geq \tau^*_i \geq (2\mu - 1)\frac{L_{f_i}^{-1}\omega_i}{\mu + \rho} > 0, \quad (7.32)$$

where $L_{f_i}$ is the Lipschitz constant of the function $f_i(x, k(x + e))$ for $|x| \leq \underline{\alpha}^{-1}(\max\{V(\xi(t^c_0)), \alpha_v^{-1} \circ \alpha_e(\eta(t^c_0))\})$ and $|e| \leq \eta(t^c_0)$.

**PROOF** Theorem 7.2, by means of Lemma 7.3, guarantees that

$$t^i_{r_i} - t^i_{r_i-1} \geq \tau^b_i \geq \frac{L_{f_i}^{-1}\omega_i}{1+\kappa}, \forall t^i_{r_i}, t^i_{r_i-1} \in ]t^c_{r_c}, t^c_{r_c+1}[,$$

with (see proof of Theorem 7.2) $\kappa := \frac{\rho}{\mu}$ when (7.31) holds. However, it can happen that some sensors automatically violate their triggering condition when their local threshold is reduced, that is, some $t^i_{r_i} = t^c_{r_c}$. This can lead to two possible problematic situations: that $t^i_{r_i} - t^i_{r_i-1} < \tau^b_i$ and/or that $t^i_{r_i+1} - t^i_{r_i} < \tau^b_i$. In the first case, one can always bound $t^i_{r_i} - t^i_{r_i-1} \geq \mu\tau^i_b$ following the reasoning in the proof of Lemma 7.2 with the same bound for the system speed but to reach a threshold $|\varepsilon_i(t^{i-}_{r_i})| = \mu\sqrt{\eta_i(t^c_{r_c-1})}$, as (7.26) guarantees that no more than one threshold update can occur simultaneously. Note that by employing $\tau^c > \max_i\{\tau^b_i\}$, one also guarantees that threshold updates do not trigger sensor updates closer than $\tau^b_i$. In the second case, the source of the problem is the update of $\hat{\xi}$ following (7.30). When $|\varepsilon_i(t^{i-}_{r_i})| > \sqrt{\eta_i(t^c_{r_c})}$, the controller is updated with a value

$$\hat{\xi}_i(t^i_{r_i}) = \hat{\xi}_i(t^i_{r_i-1}) + \mathtt{sign}(\varepsilon_i(t^i_{r_i}))\sqrt{\eta_i(t^c_{r_c})}. \qquad (7.33)$$

Thus, updating the local error accordingly as $\varepsilon_i(t^i_{r_i}) := \hat{\xi}_i(t^i_{r_i}) - \xi_i(t^i_{r_i})$ results in an error satisfying $|\varepsilon_i(t^i_{r_i})| \leq (\frac{1}{\mu} - 1)\sqrt{\eta_i(t^c_{r_c})}$, and not necessarily equal to zero. Reasoning again as in the proof of Lemma 7.2, but now computing the time it takes $|\varepsilon_i|$ to go from a value of $(\frac{1}{\mu} - 1)\sqrt{\eta_i(t^c_{r_c})}$ to $\sqrt{\eta_i(t^c_{r_c})}$, one can show that $t^i_{r_i+1} - t^i_{r_i} \geq \left(2 - \frac{1}{\mu}\right)\tau^b_i$, whenever $t^i_{r_i} = t^c_{r_c}$. Finally, realizing that $\mu > 2 - \frac{1}{\mu} \geq 0$ for all $\mu \in ]0.5, 1[$ concludes the proof. ■

### 7.5.1 Time between Actuation Updates and Delays

The effect of transmission and computation speed limitations introduce delays in any control loop implementation. Additionally, in the presented asynchronous implementation, two controller updates could be required arbitrarily close to each other when they are generated by different sensors. To address this problem, one can consider two complementary solutions at the cost of introducing an additional delay: (1) To use a slotted time-division multiple access communication between sensors and controller, automatically, this means that controller updates can only occur at the frequency imposed by the size of the slots. (2) On top of this, one can impose a periodic update of the actuators, that is, allow changes on the controller inputs only at instants at $t = k\tau_{ua}$, for some period $\tau_{ua}$ satisfying $\tau_{ua} < \min_i\{\tau^*_i\}$.

Thus, it is important to briefly discuss how delays can be accommodated by the asynchronous implementation we just described. Similar to the synchronous case, the approach is to make the triggering conditions more conservative. Consider delays occurring between the event generation at the sensors and its effect being reflected in the control inputs applied to the system. Event-triggered techniques deal with delays of this type by controlling the magnitude of the virtual error $\varepsilon$ due to sampling. As has been illustrated across this chapter, as long as the magnitude of this error signal is successfully kept within certain margins, the controller implementation is stable. Looking carefully at the definitions, one can notice that this error signal is defined at the plant side. Therefore, in the presence of delays, while the sensors send new measurements trying to keep $|\varepsilon_i(t)| = |\xi_i(t^i_{r_i}) - \xi_i(t)| \leq \sqrt{\eta_i}$, the value of the error at the plant-side $\hat{\varepsilon}_i(t)$ might be different. This error $\hat{\varepsilon}_i(t)$ can be defined as

$$\hat{\varepsilon}_i(t) = \xi(t^i_{r_i-1}) - \xi(t) \qquad t \in [t^i_{r_i}, t^i_{r_i} + \Delta\tau^i_{r_i}[, \qquad (7.34)$$

$$\hat{\varepsilon}_i(t) = \varepsilon_i(t) \qquad t \in [t^i_{r_i} + \Delta\tau^i_{r_i}, t^i_{r_i+1}[, \qquad (7.35)$$

where $\Delta\tau^i_{r_i}$ denotes the delay between the time $t^i_{r_i}$, at which a measurement is transmitted, and the time $t^i_{r_i} + \Delta\tau^i_{r_i}$, at which the controller is updated with that new measurement. Thus, to retain asymptotic stability, $|\hat{\varepsilon}(t)|$ needs to be kept below the threshold $\eta$. Looking at the proof of Lemma 7.2, we know that the maximum speed $\bar{v}^e_i$ of the error signal is always bounded from above by $\bar{v}^e_i \leq L_{f_i}(\kappa + 1)\eta$, with $L_{f_i}$ as in Proposition 7.2. Assume now that the delays for every sensor are bounded from above by some known quantity $\Delta\tau$, that is, $\Delta\tau^i_{r_i} \leq \Delta\tau$. To appropriately accommodate for the delays, one needs to select a new threshold $\bar{\eta} := r_\eta\eta < \eta$ in such a way that

$$\bar{v}^e_i\Delta\tau + \sqrt{\bar{\eta}_i} \leq \sqrt{\eta_i},$$
$$\bar{v}^e_i\Delta\tau \leq \sqrt{\bar{\eta}_i}. \qquad (7.36)$$

The first condition makes sure that by keeping $|\varepsilon(t)| \leq \bar{\eta}$ for $t \in [t^i_{r_i-1}, t^i_{r_i}]$, the error at the plant side remains appropriately bounded: $|\hat{\varepsilon}(t)| \leq \eta$ in the interval $t \in [t^i_{r_i}, t^i_{r_i} + \Delta\tau]$. The second condition makes sure that after one transmission is triggered the next transmission is not triggered before the actuation is updated with the measurement from the first transmission.* From the two conditions (7.36), one can easily verify that $r_\eta$ must satisfy

$$\frac{\bar{v}^e_i\Delta\tau}{\sqrt{\eta_i}} \leq r_\eta \leq 1 - \frac{\bar{v}^e_i\Delta\tau}{\sqrt{\eta_i}}. \qquad (7.37)$$

*If one assumes that measurements are time-tagged and the controller discards old measurements, that may arrive later than more recent ones due to the variable delays, this second condition can be ignored resulting in less restrictive conditions on the admissible delays.

From this last relation, one can also deduce that the maximum allowable delay needs to satisfy

$$\Delta\tau \leq \frac{1}{2}\frac{\sqrt{\eta_i}}{\bar{v}_i^e}. \quad (7.38)$$

A sufficient condition for (7.38) to hold is given by

$$\Delta\tau \leq \frac{1}{2}\frac{L_{f_i}^{-1}\omega_i}{1+\kappa}, \quad (7.39)$$

and similarly for (7.37) to hold by

$$\Delta\tau\frac{1+\kappa}{L_{f_i}^{-1}\omega_i} \leq r_\eta \leq 1 - \Delta\tau\frac{1+\kappa}{L_{f_i}^{-1}\omega_i}. \quad (7.40)$$

Remember that $\kappa := \frac{\rho}{\mu}$ when (7.31) holds, and thus one can also use condition (7.39) to design the implementation so that a certain amount of delay can be accommodated for. Note also that the more conservative the estimates of $\kappa$ and $L_{f_i}$ that are employed, the more conservative the values of tolerable delays will be. Finally, as the size of the thresholds are reduced, the minimum interevent times will also reduce proportionally resulting in new bounds:

$$t_{r_i}^i - t_{r_i-1}^i \geq \tau_i^* \geq r_\eta\,(2\mu - 1)\,\frac{L_{f_i}^{-1}\omega_i}{\mu+\rho} > 0. \quad (7.41)$$

We provide now an example (borrowed from [16]) that illustrates a typical execution of the proposed controller implementation showing the asynchrony of measurements and the evolution of the thresholds.

**EXAMPLE 7.3** Consider a nonlinear system of the form

$$\dot{\xi}(t) = A\xi + B(g(\xi(t)) + \upsilon(t)),$$

where $g$ is a nonlinear locally Lipschitz function. Consider a controller affected by measurement errors:

$$\upsilon(t) = -g(\xi(t) + \varepsilon(t)) - K(\xi(t) + \varepsilon(t)),$$

with $K$ such that $A_c = A - BK$ is Hurwitz.

Let $V(x) = x^T Px$, where $PA_c + A_c^T P = -I$, be the candidate ISS-Lyapunov function for the system. In this case, one can employ the following functions: $\overline{\alpha}(s) = \lambda_M(P)s^2$, $\underline{\alpha}(s) = \lambda_m(P)s^2$, $\alpha_x(s) = a_x s^2$, and $\alpha_v(s) = \alpha_x \circ \overline{\alpha}^{-1}(s)$ with $\alpha_e(s) = a_e s^2$ to obtain the necessary ISS bounds. Observe that $\epsilon\overline{\alpha}(s) \leq \overline{\alpha}(\epsilon s)$, $\forall\epsilon > 1$, then

$$(\underline{\alpha}^{-1} \circ \epsilon\alpha_v^{-1} \circ \alpha_e(s) + 2s) < (\lambda_M(P)/\lambda_m(P))\epsilon\alpha_x^{-1} \circ \alpha_e(s) + 2s.$$

This shows that Assumption 7.6 is satisfied, and we get the condition

$$\rho > \frac{\lambda_M(P)}{\lambda_m(P)}\sqrt{\frac{a_e}{a_x}} + 2\sqrt{\frac{\lambda_M(P)}{\lambda_m(P)}},$$

where $a_x = \frac{1}{2}$ and $a_e = 2(|PBK| + L_g|PB|)^2$ with $L_g$ the Lipschitz constant of $g$ in the compact determined by $|x| \leq \underline{\alpha}^{-1}(\max\{V(\xi(t_0)), \alpha_v^{-1} \circ \alpha_e(\eta(t_0))\}) + \eta(t_0)$. In this case, $L_{f_i}$ can be taken as $L_{f_i} = \max\{|A_c|, |BK| + |B|L_g\}$.

In the simulation results that we show, the system is defined by

$$A = \begin{bmatrix} 1.5 & 0 & 7 & -5 \\ -0.5 & -4 & 0 & 0.5 \\ 1 & 4 & -6 & 6 \\ 0 & 4 & 1 & -2 \end{bmatrix}, \quad B = \begin{bmatrix} 0 & 0 \\ 5 & 0 \\ 1 & -3 \\ 1 & 0 \end{bmatrix},$$

$$g(x) = \begin{bmatrix} x_2^2 \\ \sin(x_3) \end{bmatrix}, \quad K = \begin{bmatrix} 0.1 & -0.2 & 0 & -0.2 \\ 1.5 & -0.2 & 0 & 0 \end{bmatrix}.$$

Figure 7.7 shows the results when $\mu = 0.9$, $\rho = 3106$, $\tau^c = 0.1\,\text{s}$, $\omega_i = [0.58\,0.33\,0.67\,0.30]^T$, $\xi(0) = [2\,1.6\,0.8\,1.2]^T$, $\varepsilon(0) = 0$, and $\eta(0) = 2.5 \times 10^{-3}$. Having initial conditions in $|\xi(t_0)| \leq 3$ imposes a $\rho > 3075$. Then, assuming the maximum delay introduced is $\Delta\tau = 6\,\mu\text{s}$ results in a minimum time between transmissions $\tau_i^* \geq 17\,\mu\text{s}$. These bounds are clearly conservative, as can be seen in the simulation results in which the minimum intertransmission time observed is $68\,\mu$s. Furthermore, the average intertransmission time during the 5 seconds of the simulation is one order of magnitude larger.

## 7.6 Energy Consumption Simulations

We have discussed in the introduction and in Section 7.4.2 the effect of listening time on the power consumption of the sensor nodes. In general, it is very hard to analytically compare the two alternative implementations we present in this chapter. This is mainly due to the fact that predicting the type of pattern of triggering is a tough problem, and the energy consumption is tightly connected to those triggering patterns. However, we argue that an asynchronous implementation generally results in less time spent by

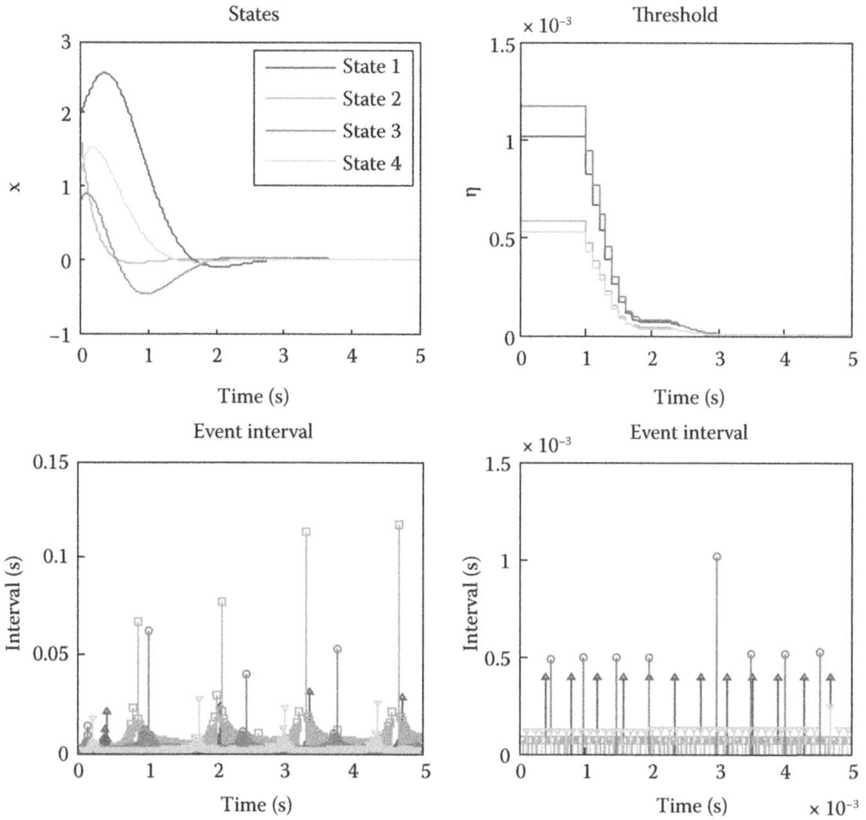


**FIGURE 7.7**
State trajectory, evolution of the thresholds, and events generated at the sensors.

the sensors listening for messages, and thus, one would hope that it also results in a lower energy consumption.

To check this intuitive hypothesis, we construct a simulator capable of tracking the energy consumed by the sensors in a simulation depending on three possible states of the sensor: transmitting, listening, or idle (neither transmitting nor listening). Using this simulator, we perform some experiments to compare a decentralized synchronous implementation with an asynchronous one providing a similar performance (in terms of decay rate of the Lyapunov function). In order for this simulation to be somehow realistic, we need to first come up with a TDMA scheme for each of the implementations. These are described in the following section.

### 7.6.1 TDMA Schemes

Two different simplified TDMA schemes are followed in each of the two proposed implementations. These schemes are far from being actually implementable, as they ignore, for example, overheads needed for synchronization and error correction. Furthermore, we do not consider the necessary modifications in the case of a multi-hop network. Thus, the type of network we are considering is essentially a star network with every sensor and actuator capable of direct communication with the controller, which is at the center of the network, see Figure 7.8.

Nonetheless, these simplified scenarios give us an abstraction sufficient to explore the type of energy consumption that each of the proposed schemes will exhibit in reality, if appropriate custom-made protocols are constructed. We briefly describe the two schemes as follows.

#### 7.6.1.1 Synchronous Event Triggering

Figure 7.9 shows the type of periodic hyperframes employed in the synchronous implementation. Two kinds of hyperframes are considered, depending on whether an event is triggered or not. The hyperframes are defined by the following subframes, slots, and respective parameters:

- **Notification (NOT) slot (length $\tau_{NOT}$).** In this slot, all sensors can notify the controller of an event. When there is an event, the sensor node that triggered the event switches to transmission (TX) mode to notify the controller node. The node is switched to idle mode immediately after to save

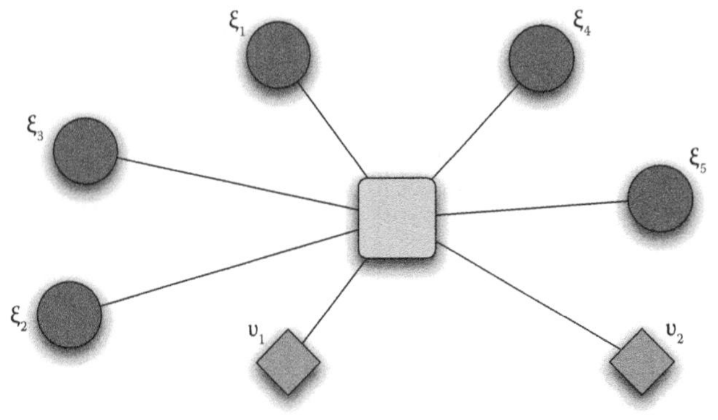


**FIGURE 7.8**
A star topology control network with a centralized computing (controller) node.

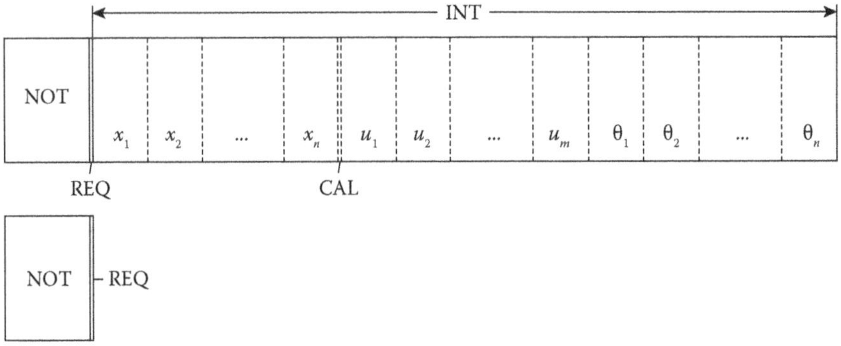


**FIGURE 7.9**
Synchronous event-triggering hyperframes.

energy. Meanwhile, other nodes remain idle if they do not trigger an event.

- **Request (REQ) slot (length $\tau_{REQ}$).** In this slot all sensor nodes switch to the listening mode (RX) to be able to receive a request of measurements from the controller node, if an event was notified in the NOT slot.

- **Intercommunication (INT) frame (length $\tau_{INT}$).** During this period, the sensors switch to TX mode to send their current measurements to the controller one at a time. Next, the actuators receive the new control commands $u_i$, while the sensors are idle. And finally all the sensor nodes switch to RX mode to receive the newest local parameters $\theta_i$.

- **$x_i$ slot (length $\tau_{x_i}$).** Slot employed to transmit measurements of sensor $i$ to the controller.

- **$u_i$ slot (length $\tau_{u_i}$).** Slot employed to transmit controller computed inputs to actuator $i$.

- **$\theta_i$ slot (length $\tau_{\theta_i}$).** Slot employed to transmit updated thresholds to sensor $i$.

- **Calculation (CAL) slot (length $\tau_{CAL}$).** Free slot to allow the computation of updated $u$ and $\theta$ values.

If a request is transmitted during the request (REQ) slot, the communication continues with an intercommunication (INT) frame; otherwise, the system skips that frame and a new hyperframe starts again with a notification period.

In terms of energy consumption, the sensor nodes are in listening mode both at the REQ slot and, if an even occurred, in their respective $\theta_i$ slot of the INT frame. Otherwise sensors are idle, except when they trigger an event, in which case they transmit for some amount of time in the NOT slot.

The length of the notification slot must satisfy the following condition: $\tau_{NOT} \geq \tau_{delay}$, where $\tau_{delay}$ is the transmission time associated with 1 bit, which is determined by the rate $k$ of the radio chip. We take as reference radio systems based on the IEEE 802.15.4 standard, which provides a maximum speed of 250 kbps. Using this rate value results in $\tau_{delay} = 4 \times 10^{-6}$ s. Similarly, 1 bit is enough to perform a request for measurements (REQ). $\tau_{INT}$ is determined by the number of bits employed to quantify the measurements and $\theta$, and the amount of time $\tau_{CAL}$. We assume the controller to be

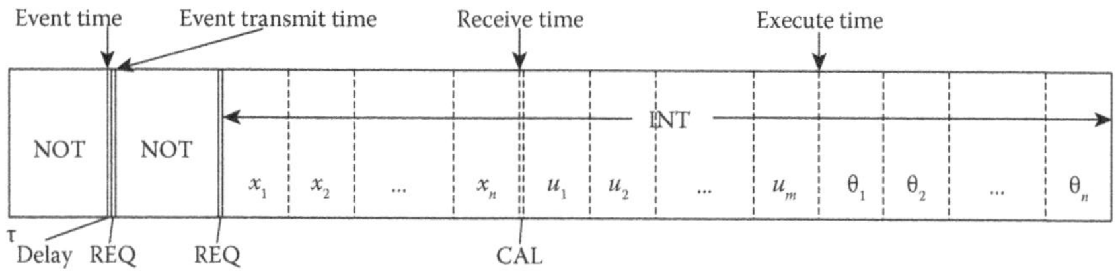


**FIGURE 7.10**
Synchronous event-triggering maximum delay.

sufficiently powerful so that $\tau_{CAL}$ can be neglected. We assume the use of analog-to-digital converters (ADC) employing 16 bits: $B_{AD} = 16$. Similarly, we assume $\theta$ and $u$ are also codified with 16 bits: $B_\theta = 16$, $B_u = 16$. In a system with $n$ sensors and $m$ inputs, this results in $\tau_{INT} = \frac{B_{AD} \times n}{k} + \frac{B_\theta \times n}{k} + \frac{B_u \times m}{k} = 1.28 \times 10^{-4} n + 6.4 \times 10^{-5} m$ s.

Assuming sufficiently good synchronization and given a pre-fixed schedule, the identity of the sensors could be inferred by the slot they employed to transmit, which allows us to reduce any additional addressing overheads. Thus, under all these considerations and the TDMA scheme described, the maximum delay that can appear between the generation of an event and the update of the controller is

$$\Delta\tau = 2\tau_{REQ} + \tau_{NOT} + \sum_{i=1}^{n} \tau_{x_i} + \sum_{i=1}^{m} \tau_{u_i} + \tau_{CAL} + \tau_{delay}.$$

Note that this holds for as long as $\tau_{NOT} + \tau_{REQ} + \tau_{INT} < \tau_{min}$, which ensures that no events can be generated during the transmissions of $u$ and $\theta$. This maximum delay is computed considering a worst-case scenario in which an event takes place $\tau_{delay}$ seconds in advance of the end of the notification period, meaning it cannot complete the notification in time. Thus, the event has to wait until the next NOT slot, and the updated control action cannot be computed until all measurements are collected. An illustration of this situation is shown in Figure 7.10.

A remark is needed at this point. The sensor that generates the event sends also the measurement taken at the request moment instead of at the actual time the event was triggered. Otherwise, this measurement would not be synchronized with the rest of the measurements. Note that this does not affect any of the previous delay analyses, as essentially this means that part of the delay is not harmful, as the measurement employed is even more recent than would have been otherwise.

#### 7.6.1.2 Asynchronous Event Triggering

Figure 7.11 shows the type of periodic frame employed in the asynchronous implementation. In this case, the type of frames employed are simpler and essentially contain a sequence of slots reserved for communication from each sensor to the controller, and slots for the communication from the controller to each of the actuators, followed by a broadcast slot to indicate if an update of the thresholds is needed. Note that additionally an initialization frame would be used in practice each time a change of the initial condition is made (e.g., because a change of a reference command at a higher control level) to transmit with a number of bits (e.g., 16) the whole state of the system. We do not discuss such a frame in detail as it is not relevant for the overall energy consumption comparison or to compute the delays.

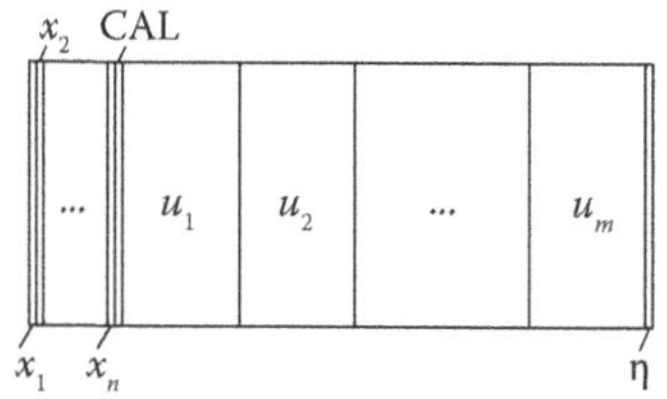


**FIGURE 7.11**
Asynchronous event-triggering frame.

The type of frames we consider are defined by the following slots and respective parameters:

- **$x_i$ slot (length $\tau_{x_i}$).** Slot employed to notify the controller of an event at sensor $i$ and the sign of the deviation with respect to the last stored value, as employed in Equation 7.30.
- **$u_i$ slot (length $\tau_{u_i}$).** Slot employed to transmit controller computed inputs to actuator $i$.
- **Calculation (CAL) slot (length $\tau_{CAL}$).** Free slot to allow the computation of updated $u$ and check if the thresholds need to be updated.
- **$\eta$ slot (length $\tau_\eta$).** Slot reserved to broadcast signals commanding the update of $\eta$.

All $\tau_{x_i}$, $\tau_{u_i}$, and $\tau_\eta$ are design parameters that should be greater than $\tau_{delay}$.

In this communication scheme, sensor nodes only need to listen during the slot reserved for the broadcast of $\eta$. The rest of the time the sensor nodes remain idle, except for the times when they need to transmit an event in their transmission slot. In fact, sensors do not even need to listen to the controller in each of the

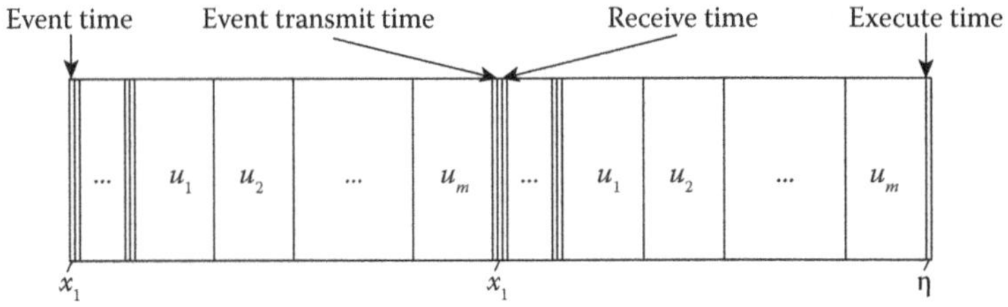


**FIGURE 7.12**
Asynchronous event-triggering maximum delay.

$\eta$ broadcast slots. The sensors only need to listen to the controller every $\tau^c$ units of time, which is selected as a multiple of the complete frame size, that is, $\tau^c = r\tau_f$ for some $r \in \mathbb{N}$, where

$$\tau_f := \sum_{i=1}^{n} \tau_{x_i} + \sum_{i=1}^{m} \tau_{u_i} + \tau_\eta + \tau_{\mathrm{CAL}}.$$

That means that given $r$, the sensors only need to listen in the $\eta$ slot of every $r$th frame.

The maximum delay that this scheme introduces is given by

$$\Delta\tau = 2\tau_f - \tau_\eta,$$

as illustrated in Figure 7.12 for a worst-case scenario in which an event is generated right at the time at which the slot reserved to the sensor triggering begins. This forces the sensor to wait until the next frame to send its notification, which is finally executed in the actuator right before the next $\eta$ update slot.

### 7.6.2 Controller Implementations

We employ the benchmark linear system originally coming from a batch reactor model [23]:

$$\dot{\xi} = A\xi + B\upsilon,$$

with the linear controller

$$u = Kx,$$

where

$$A = \begin{bmatrix} 1.38 & -0.20 & 6.71 & -5.67 \\ -0.58 & -4.29 & 0 & 0.67 \\ 1.06 & 4.27 & -6.65 & 5.89 \\ 0.04 & 4.27 & 1.34 & -2.10 \end{bmatrix},$$

$$B = \begin{bmatrix} 0 & 0 \\ 5.67 & 0 \\ 1.13 & -3.14 \\ 1.13 & 0 \end{bmatrix},$$

$$K = \begin{bmatrix} 0.1006 & -0.2469 & -0.0952 & -0.2447 \\ 1.4099 & -0.1966 & 0.0139 & 0.0823 \end{bmatrix},$$

For $Q = I$ the resulting $P$ defining the Lyapunov function $V(x) = x^T Px$ is given by

$$P = \begin{bmatrix} 0.5781 & -0.0267 & 0.3803 & -0.3948 \\ -0.0267 & 0.2809 & 0.0629 & 0.2011 \\ 0.3803 & 0.0629 & 0.4024 & -0.2279 \\ -0.3948 & 0.2011 & -0.2279 & 0.5780 \end{bmatrix}.$$

Note that even though this system is supposed to represent a batch reactor which usually exhibits rather slow dynamics, because of the state-feedback controller employed, the stabilization speed is unusually fast. Furthermore, we do not investigate whether the employed initial conditions are realistic or not. Thus, one should see this example as a purely academic one, and not think of it in terms of a physically realistic system.

For the two controller implementations described in the previous sections, we designed their relevant parameters as detailed in the following:

#### *7.6.2.1 Synchronous Event-Triggered Controller*

We start by finding the fastest synchronous event-triggered controller (SETC) implementation possible, that is, minimum possible $\sigma$, given the TDMA scheme described previously. We consider two different combinations of TDMA parameters resulting also in different maximum delays, as summarized in Table 7.1. Given this maximum delay, one can now compute the minimum value of $\sigma$ so that a smaller $\sigma'$ can be found to compensate for the delay $\Delta\tau$, that is, the constraints imposed in Corollary 7.1 can be satisfied. We find this lower bound on $\sigma$ through an iterative fixed point algorithm, described in Algorithm 7.2, where $\phi$ is the solution to the differential equation provided at the end of Section 7.2. An upper bound on $\sigma$ can also be computed following the results in [21]. In order to study the effect of enlarging the idle time of sensors in the SETC implementation, we computed also the range of $\sigma$ for the case when $\tau_{\mathrm{NOT}} = 0.0005$ s. Finally, in all the SETC implementations, a Taylor approximation order $q = 2$ is employed. The resulting range of possible $\sigma$ values is shown in Table 7.2 for the two different TDMA designs considered.

**TABLE 7.1**
TDMA Parameters (μs)

| Method | $\tau_{NOT}$ | $\tau_{REQ}$ | $\tau_{INT}$ | $\tau_{x_i}$ | $\tau_{u_i}$ | $\tau_{\theta_i}$ | $\tau_{\eta}$ | $\tau_f$ | $\Delta\tau$ |
|---|---|---|---|---|---|---|---|---|---|
| SETC(1) | 100 | 4 | 640 | 64 | 64 | 64 | — | 744 | 496 |
| SETC(2) | 500 | 4 | 640 | 64 | 64 | 64 | — | 1144 | 896 |
| AETC(1) | — | — | — | 4 | 64 | — | 4 | 148 | 292 |
| AETC(2) | — | — | — | 4 | 64 | — | 4 | 148 | 292 |

**TABLE 7.2**
Control Parameters

| Method | $\sqrt{\sigma}$ | $\rho$ | $\tau^c(s)$ | $\mu$ |
|---|---|---|---|---|
| SETC(1) | $]0.0138, 0.1831[$ | — | — | — |
| SETC(2) | $]0.0252, 0.1831[$ | — | — | — |
| AETC(1) | — | $]92.7915, 98.1302[$ | 0.111 | 0.9 |
| AETC(2) | — | $]92.7915, 98.1302[$ | 0.222 | 0.9 |

**Algorithm 7.2:** SETC parameter selection algorithm.

**Input**: $A$, $B$, $K$, $\Delta\tau$, precision
**Output**: $\sigma$
$i = 0$;
$\sigma(i) = 0$;
**while** $\sigma(i) - \sigma(i-1) >$ precision **do**
  $\sqrt{\sigma'} = \frac{\Delta\tau|[A+BK|BK]|(\sqrt{\sigma(i)}+1)}{1-\Delta\tau|[A+BK|BK]|(\sqrt{\sigma(i)}+1)}$;
  $t_{\sigma'} = \min\{t|\phi(t_{\sigma'}, 0) = \sqrt{\sigma'}\}$;
  $t_\sigma = t_{\sigma'} + \Delta\tau$;
  $i = i + 1$;
  $\sqrt{\sigma(i)} = \phi(t_\sigma, 0)$;
**end**

#### 7.6.2.2 Asynchronous Event-Triggered Controller

Similar to the SETC design, we first find the fastest asynchronous event-triggered controller (AETC) implementation once the maximum delay imposed by the choice of TDMA parameters selected, summarized in Table 7.1. In the AETC case, the design of the parameters is a bit more complex given that we need to design simultaneously $\rho$, $\mu$, $\tau^c$, and $\eta(0)$. We split this decision in several steps: first, we fixed the values of $\mu$ and $\tau^c$; and next we assume that $\eta(t_0) \geq \frac{\mu}{\rho}\underline{\alpha}^{-1} \circ V(\xi(0))$ and thus $\kappa = \frac{\rho}{\mu}$. These two decisions reduce then the design decision of $\rho$ to values satisfying the condition in Theorem 7.2 and the restriction imposed by the maximum delay as given by (7.39) by means of $\kappa$. Assume that $\omega$ parameters are found so that $L_{f_i}^{-1}\omega_i = L_{f_j}^{-1}\omega_j = L$ for all $i, j = 1, \ldots, n$. Then, (7.39) can be equivalently rewritten, by replacing $\kappa = \frac{\rho}{\mu}$ by

$$\rho \leq \mu\left(\frac{L}{2\Delta\tau} - 1\right). \tag{7.42}$$

To compute the lower bound, we employed Equation 7.39 with $\epsilon = 1$, and optimizing over different ISS bounds computed by tuning the parameters $a$ and $b$ satisfying $ab = |PBK|$:

$$\begin{aligned}\dot{V}(x) &= \dot{x}^{\mathrm{T}}Px + x^{\mathrm{T}}P\dot{x} = -x^{\mathrm{T}}Qx + e^{\mathrm{T}}BK^{\mathrm{T}}Px + x^{\mathrm{T}}PBKe \\ &\leq -\lambda_{\min}(Q)|x|^2 + 2ab|x||e| \\ &\leq -\frac{\lambda_{\min}(Q) - a^2}{\lambda_{\max}(P)}V(x) + b^2|e|^2.\end{aligned}$$

The resulting limits for the choices of $\rho$ are shown in Table 7.2. In general, for larger values of $\rho$ threshold, updates are triggered more frequently, thus resulting in a faster system, but the range of delays that can be tolerated is reduced, and the energy consumption is likely to be higher. We test both extremes in our simulations.

One can easily compute the Lipschitz constant for the closed-loop dynamics of each state variable from $A$ and $B$ and $K$ resulting in $L_{f_1} = 8.8948$, $L_{f_2} = 5.7603$, $L_{f_3} = 10.3322$, and $L_{f_4} = 4.8082$. The design parameters $\omega_i$ making $L_{f_i}^{-1}\omega_i = L_{f_j}^{-1}\omega_j$ for all $i, j = 1, \ldots, n$, result in $\omega = [0.5716\ 0.3702\ 0.6639\ 0.3090\ ]^T$.

### 7.6.3 Results

We perform simulations on four different implementation designs: (1) SETC(1) with $\sqrt{\sigma} = 0.014$; (2) SETC(2) with $\sqrt{\sigma} = 0.18$; (3) AETC(1) with $\rho = 98$; and (4) AETC(2) with $\rho = 93$.

Two different types of energy consumption profiles are employed in the simulations: CC2420 [1] and CC2530 [2], whose most relevant parameters are presented in Table 7.3.

**TABLE 7.3**
CC2420 and CC2530 Parameters

| Operation | CC2420 | CC2530 | Unit |
|---|---|---|---|
| Idle | 0.426 | 0.2 | mA |
| Listening to channel | 18.8 | 24.3 | mA |
| Receive | 18.8 | 24.3 | mA |
| Transmit | 17.4 (P=0 dBm) | 28.7 (P=1 dBm) | mA |

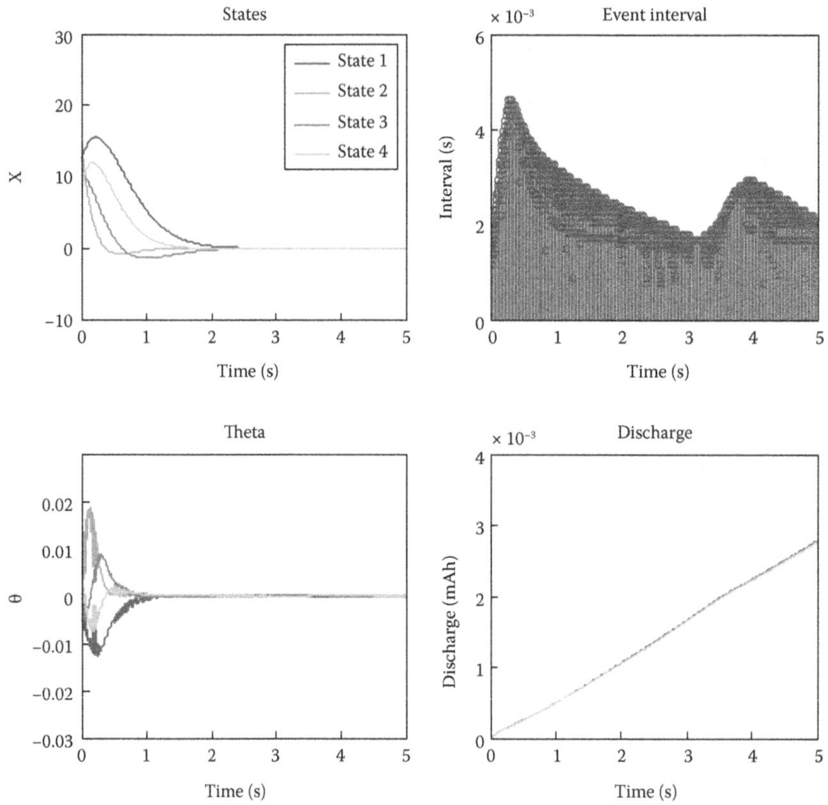


**FIGURE 7.13**
SETC(1), CC2420, $x(0) = x_a$.

Finally, we also perform simulations with the following two different initial conditions: $x(0) = x_a = [12\ 15\ 10\ 8]^T$, with $\eta(0) = \frac{\mu}{\rho}\sqrt{\frac{V(0)}{\lambda_{\min}(P)}} = 0.5285$; and $x(0) = x_b = [18\ 2\ 14\ 3]^T$, with $\eta(0) = 0.6605$.

#### 7.6.3.1 Parameters from TI CC2420

Employing the CC2420 chipset, we compare first what should be our fastest SETC and AETC implementations: SETC(1) and AETC(1). We see in Figures 7.13 and 7.14 that indeed both implementations stabilize the plant at a similar rate, see also Figure 7.15 showing a comparison of the Lyapunov functions evolution. However, we can clearly observe how the energy consumption of the asynchronous implementation is considerably lower.

We also include in Figure 7.16 a simulation result of the centralized event-trigger controller with the same delay and value of $\sigma$ as the decentralized SETC to verify that the decentralized implementation is operating very closely to the centralized one.

In order to show that these results are not coincidental, we tried several initial conditions in which similar patterns could be observed. For completeness, we provide in Figures 7.17 through 7.19 the results when $x(0) = x_b$.

In order to reduce the energy consumption in the synchronous implementation, we performed two modifications: enlarged the NOT slot to reduce listening times and reduce the speed of the implementation by increasing $\sigma$. These modifications resulted in implementation SETC(2). As can be seen in Figures 7.20 and 7.21, the energy consumption was greatly reduced as expected, at the cost of some slight performance degradation.

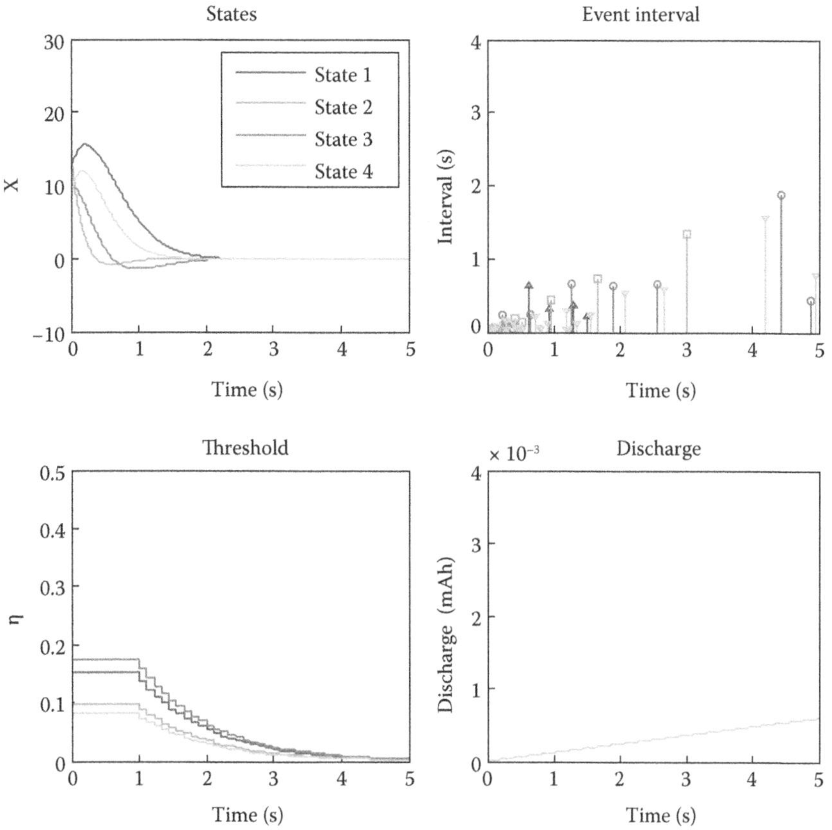


**FIGURE 7.14**
AETC(1), CC2420, $x(0) = x_a$.

Finally, we simulated the slower asynchronous implementation AETC(2) to see how much further energy consumption could be reduced in the asynchronous implementation without losing much performance. The result of this simulation can be found in Figure 7.22.

#### 7.6.3.2 Parameters from TI CC2530

To investigate further the effect of the cost of listening, we repeated the experiments with SETC(1) and AETC(1) employing the radio chip CC2530, which consumes less while idle and more both when transmitting and listening. As expected, the asynchronous implementation managed to gain more benefit from this change, in terms of energy consumption, than the synchronous implementation. The results of this comparison are shown in Figures 7.23 and 7.24.

## 7.7 Discussion

We presented two different schemes to decentralize the process of triggering events in event-triggered controllers so that no simultaneous centralized monitoring of all sensor measurements is needed. These strategies are thus well suited for systems where several sensors are not collocated with each other. Driven by the fact that wireless sensor nodes consume a large portion of their energy reserves when listening for transmissions from other nodes in the network, we investigated and proposed solutions that reduce the amount of time sensor nodes stay in a listening mode.

The first technique we proposed triggers synchronous transmissions of measurements and forces the radios of the sensors to go idle periodically in order to reduce listening times. This is done at the cost of introducing an artificial delay that can be accounted for by conservatively triggering, which usually produces more frequent updates. The second technique relies on allowing updates of the control with measurements from different sensors not synchronized in time. This allows sensors to operate without having to listen to other nodes very frequently, thus reducing automatically energy consumption on listening. However, by the nature of this later implementation, in general it can tolerate smaller delays than the first implementation technique.

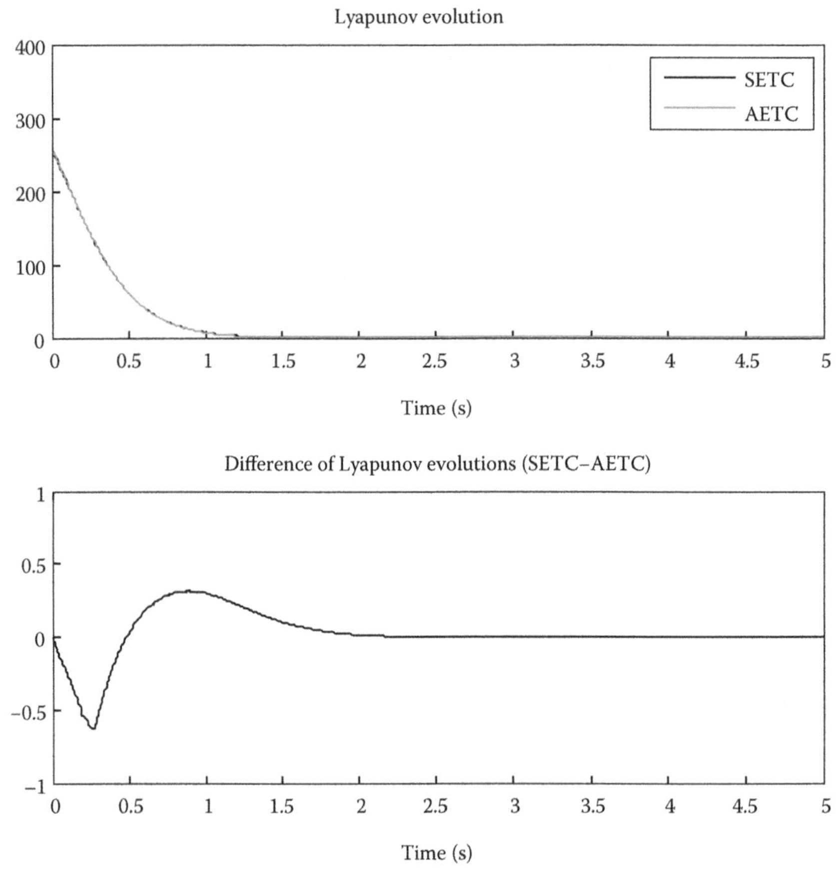


**FIGURE 7.15**
Lyapunov evolution for SETC(1) versus AETC(1). CC2420, $x(0) = x_a$.

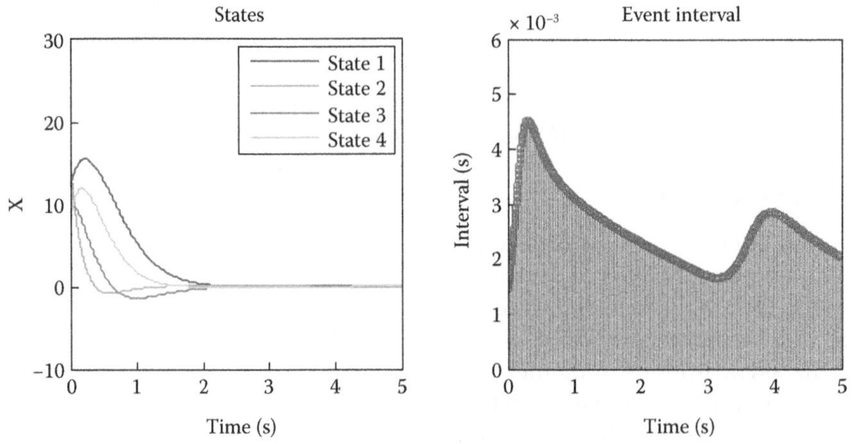


**FIGURE 7.16**
Centralized event-triggered controller, $x(0) = x_a$.

Alternative implementations could also be devised if one considers the use of wake-up radios [5,11,22], that is, radio chips that have a very low power consumption while in a stand-by mode in which, while they cannot receive messages, they can be remotely awakened. Such hardware would allow synchronous implementations with very small delays due to channel access scheduling, and consuming much less energy than in the form we presented. Similarly, wake-up radios could reduce even further the energy consumption of the

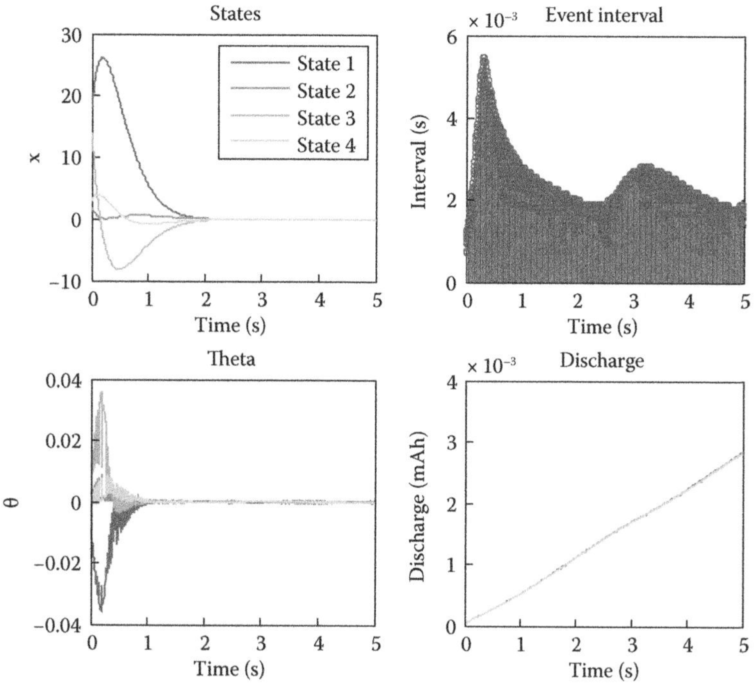


**FIGURE 7.17**
SETC(1), CC2420, $x(0) = x_b$.

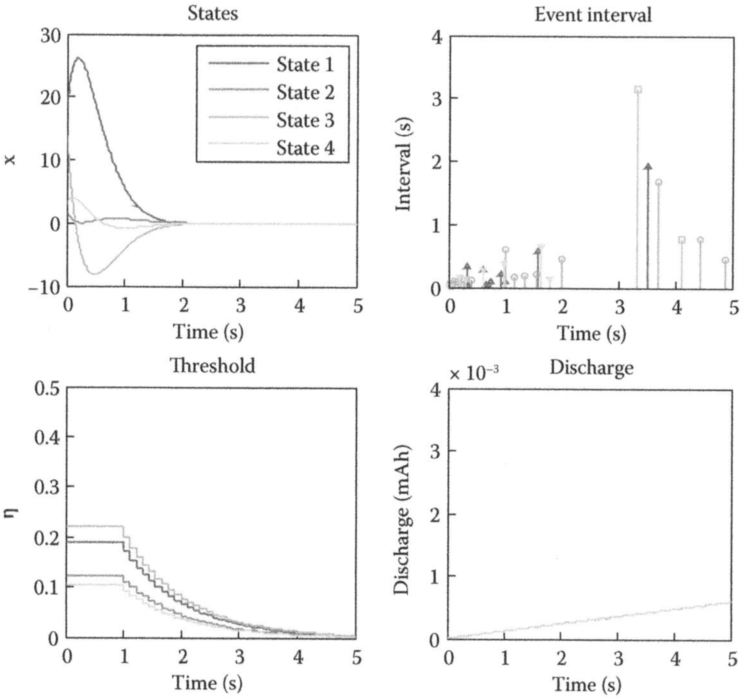


**FIGURE 7.18**
AETC(1), CC2420, $x(0) = x_b$.

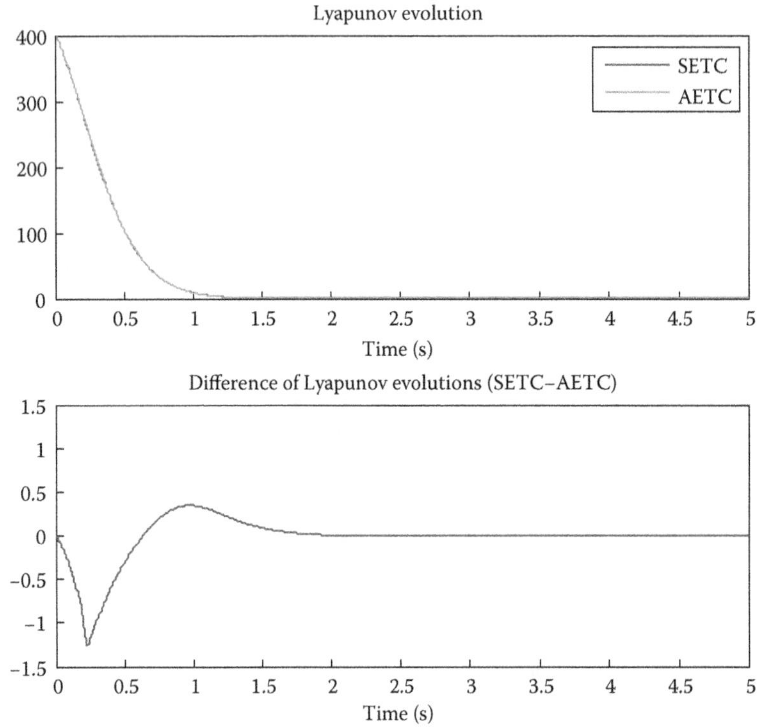


**FIGURE 7.19**
Lyapunov evolution for SETC(1) versus AETC(1). CC2420, $x(0) = x_b$.

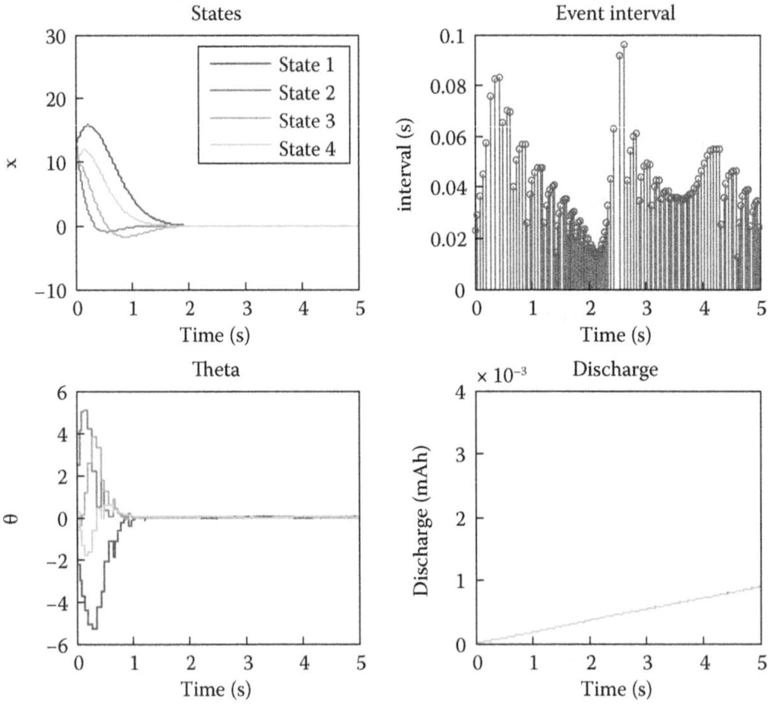


**FIGURE 7.20**
SETC(2), CC2420, $x(0) = x_a$.

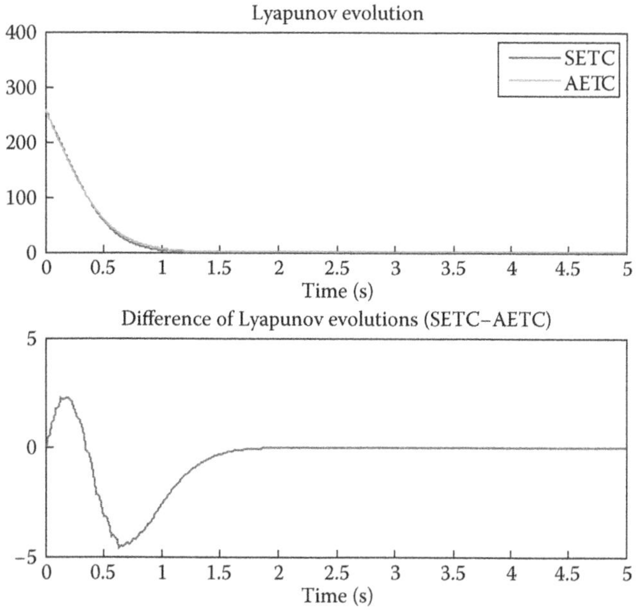


**FIGURE 7.21**
Lyapunov evolution for SETC(2) versus AETC(1). CC2420, $x(0) = x_a$.

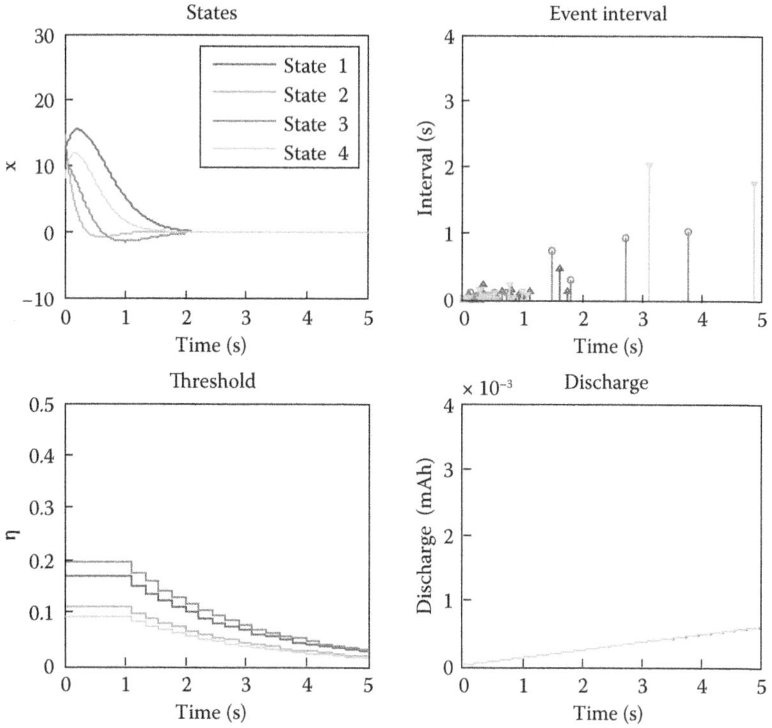


**FIGURE 7.22**
AETC(2), CC2420, $x(0) = x_a$.

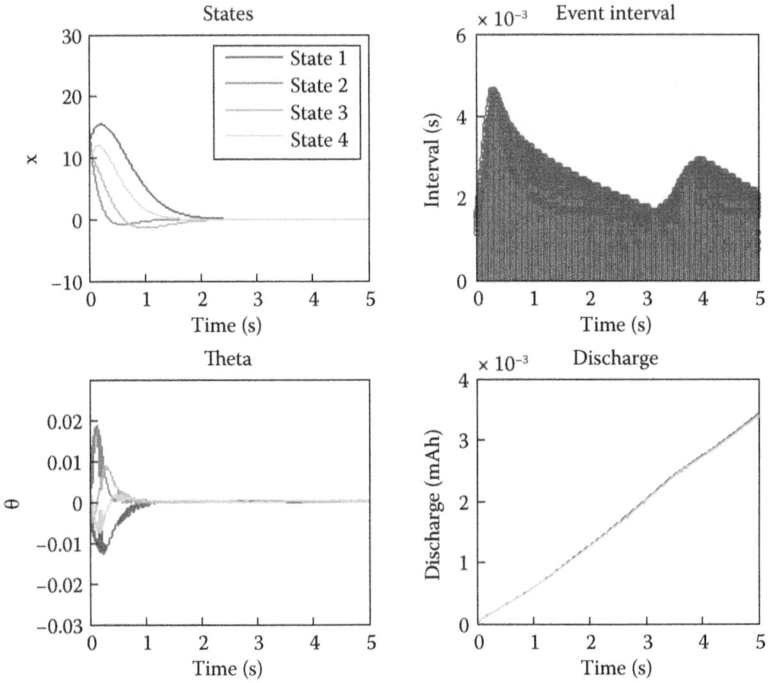


**FIGURE 7.23**
SETC(1), CC2530, $x(0) = x_a$.

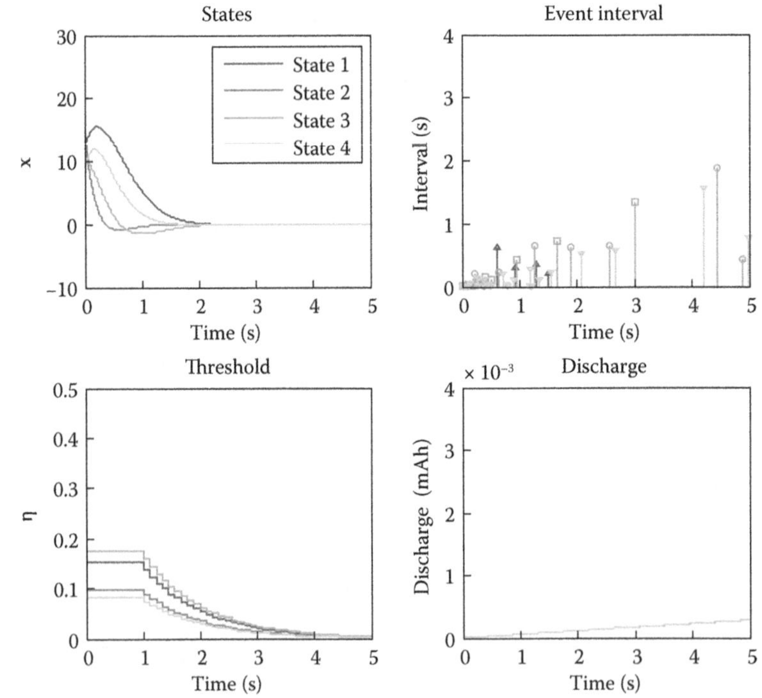


**FIGURE 7.24**
AETC(1), CC2530, $x(0) = x_a$.

asynchronous implementations. Whether with or without making use of wake-up radios, it would be interesting to see actual energy consumption comparisons based on real experimental data. To this end, appropriate new communication protocols that can exploit the benefits of the proposed implementations need to be designed.

Further work needs to also be done to address some of the limitations of the current proposals. The main limiting assumption of the proposed techniques when dealing with nonlinear plants is the requirement of having a controller rendering the system ISS on compacts with respect to measurement errors. It is important to remark that this property does not need to be globally satisfied but only in the compact of interest for the system operation (as our results are semiglobal in nature), which relaxes drastically the requirement and makes it also easier to satisfy [10]. Nonetheless, investigating controller designs satisfying such a property, as well as the more involved one from Assumption 7.6, is an interesting problem in its own right. One of the main drawbacks of the proposed implementations, even in the linear case, is that they can only be applied to state-feedback controllers. It would be interesting to combine these ideas with some of the work that is available addressing event-triggered output-feedback controllers, as, for example, [9]. As indicated earlier on, the proposed implementations are complementary to implementations proposed for more distributed systems with loosely interacting local control loops [24]. Once more, it would be interesting to see both techniques combined in a real application. As with most of the work on event-triggered control, the study of co-design of controller and implementation is largely an open issue, with some recent exceptions in the case of linear controllers, for example, [4]. Investigating such co-design and numerical methods, for example, employing LMIs [8], to further improve the triggering conditions or parameters of the implementations are another promising venue of future work. Additionally, further energy savings can be envisaged by combining these event-triggered strategies with self-triggered approaches [6]. A promising approach along these lines can be found in another chapter of this book, inspired in [18].

Finally, we believe it is worth investigating in more detail the possibilities the asynchronous approach brings by reducing the size of the exchanged messages to a single bit. In the present chapter, we have already remarked on the benefits in terms of energy savings that a smaller payload brings, an already well-known fact [25]. The careful reader may have also noticed the beneficial impact reduced payloads also have in reducing the medium access-induced delays. We believe that by using this type of one-bit strategy (heavily inspired in previous work on level sampling [13], dead-band control [19], and dynamic quantization [14]), it may be possible to produce more reliable communication protocols while keeping transmission delays within reasonable limits for control applications. This can be achieved by, for example, introducing in a smart fashion redundancy in the transmitted messages.

## Bibliography


[1] CC2420 Data Sheet. http://www.ti.com/lit/ds/symlink/cc2420.pdf.

[2] CC2530 Data Sheet. http://www.ti.com/lit/ds/symlink/cc2530.pdf.

[3] A. Agrachev, A. S. Morse, E. Sontag, H. Sussmann, and V. Utkin. Input to state stability: Basic concepts and results. In *Nonlinear and Optimal Control Theory, volume 1932 of Lecture Notes in Mathematics* (P. Nistri and G. Stefani, Eds.), pages 163–220. Springer, Berlin, 2008.

[4] S. Al-Areqi, D. Gorges, S. Reimann, and S. Liu. Event-based control and scheduling codesign of networked embedded control systems. In *American Control Conference (ACC), 2013*, pages 5299–5304, June 2013.

[5] J. Ansari, D. Pankin, and P. Mähönen. Radio-triggered wake-ups with addressing capabilities for extremely low power sensor network applications. *International Journal of Wireless Information Networks*, 16(3):118–130, 2009.

[6] A. Anta and P. Tabuada. To sample or not to sample: Self-triggered control for nonlinear systems. *IEEE Transactions on Automatic Control*, 55:2030–2042, 2010.

[7] J. Araújo, H. Fawzi, M. Mazo Jr., P. Tabuada, and K. H. Johansson. An improved self-triggered implementation for linear controllers. In *3rd IFAC Workshop on Distributed Estimation and Control in Networked Systems, volume 3(1) of Estimation and Control of Networked Systems* (F. Bullo, J. Cortes, J. Hespanha, and P. Tabuada, Eds.), NY, pages 37–42. International Federation of Automatic Control, 2012.

[8] M. Donkers, W. Heemels, N. van de Wouw, and L. Hetel. Stability analysis of networked control systems using a switched linear systems approach.

*IEEE Transactions on Automatic Control*, 56(9):2101–2115, 2011.

[9] M. C. F. Donkers and W. Heemels. Output-based event-triggered control with guaranteed l-gain and improved and decentralized event-triggering. *IEEE Transactions on Automatic Control*, 57(6):1362–1376, 2012.

[10] R. Freeman. Global internal stabilizability does not imply global external stabilizability for small sensor disturbances. *IEEE Transactions on Automatic Control*, 40(12):2119–2122, 1995.

[11] L. Gu and J. A. Stankovic. Radio-triggered wake-up capability for sensor networks. In *Proceedings of RTAS*, pages 27–36, May 25–28, 2004.

[12] J. Johnsen and F. Allgöwer. Interconnection and damping assignment passivity-based control of a four-tank system. In *Lagrangian and Hamiltonian Methods for Nonlinear Control 2006* (F. Bullo and K. Fujimoto, Eds.), Springer-Verlag, Heidelberg: Berlin, pages 111–122, 2007.

[13] E. Kofman and J. H. Braslavsky. Level crossing sampling in feedback stabilization under data-rate constraints. In *Decision and Control, 2006 45th IEEE Conference on*, pages 4423–4428, December 2006.

[14] D. Liberzon. Hybrid feedback stabilization of systems with quantized signals. *Automatica*, 39(9):1543–1554, 2003.

[15] M. Mazo Jr., A. Anta, and P. Tabuada. An ISS self-triggered implementation of linear controller. *Automatica*, 46:1310–1314, 2010.

[16] M. Mazo Jr. and M. Cao. Asynchronous decentralized event-triggered control. *Automatica*, 50(12): 3197–3203, 2014, doi: 10.1016/j.automatica.2014.10.029.

[17] M. Mazo Jr. and P. Tabuada. Decentralized event-triggered control over wireless sensor/actuator networks. *IEEE Transactions on Automatic Control, Special Issue on Wireless Sensor Actuator Networks*, 56(10):2456–2461, 2011.

[18] C. Nowzari and J. Cortés. Robust team-triggered coordination of networked cyberphysical systems. In *Control of Cyber-Physical Systems* (D. C. Tarraf, Ed.), Switzerland, pages 317–336. Springer, 2013.

[19] P. G. Otanez, J. R. Moyne, and D. M. Tilbury. Using deadbands to reduce communication in networked control systems. In *American Control Conference, 2002. Proceedings of the 2002*, volume 4, pages 3015–3020, 2002.

[20] E. Sontag. State-space and i/o stability for nonlinear systems. In *Feedback Control, Nonlinear Systems, and Complexity* (B. A. Francis and A. R. Tannenbaum, Eds.), Springer-Verlag Berlin Heidelberg: Berlin, pages 215–235, 1995.

[21] P. Tabuada. Event-triggered real-time scheduling of stabilizing control tasks. *IEEE Transactions on Automatic Control*, 52(9):1680–1685, 2007.

[22] B. Van der Doorn, W. Kavelaars, and K. Langendoen. A prototype low-cost wakeup radio for the 868 MHz band. *International Journal of Sensor Networks*, 5(1):22–32, 2009.

[23] G. C. Walsh and H. Ye. Scheduling of networked control systems. *Control Systems, IEEE*, 21(1):57–65, 2001.

[24] X. Wang and M. D. Lemmon. Event-triggering in distributed networked control systems. *IEEE Transactions on Automatic Control*, 56(3):586–601, 2011.

[25] W. Ye, J. Heidemann, and D. Estrin. An energy-efficient mac protocol for wireless sensor networks. In *INFOCOM 2002. Twenty-First Annual Joint Conference of the IEEE Computer and Communications Societies. Proceedings. IEEE*, volume 3, pages 1567–1576, 2002.

# 8

# *Event-Based Generalized Predictive Control*

**Andrzej Pawlowski**
*UNED*
*Madrid, Spain*

**José Luis Guzmán**
*University of Almería*
*ceiA3-CIESOL, Almería, Spain*

**Manuel Berenguel**
*University of Almería*
*ceiA3-CIESOL, Almería, Spain*

**Sebastián Dormido**
*UNED*
*Madrid, Spain*

**CONTENTS**

**ABSTRACT** This chapter describes a predictive event-based control algorithm focused on practical issues. The generalized predictive control scheme is used as a predictive controller, and sensor and actuator deadbands are included in the design procedure to provide event-based capabilities. The first configuration adapts a control structure to deal with asynchronous measurements; the second approach is used for asynchronous process updates. The presented ideas, for an event-based control scheme, conserve all of the well-known advantages of the adaptive sampling techniques applied to process sampling and updating. The objective of this combination is to reduce the control effort, whereas the control system precision is maintained at an acceptable level. The presented algorithm allows us to obtain a direct trade-off between performance and number of actuator events. The diurnal greenhouse temperature control problem is used as the benchmark to evaluate the different control algorithm configurations.

## 8.1 Introduction

Most industrial control loops are implemented as computer-based controlled systems, which are governed by a periodic sampling time. The main reason for using this solution is that the sampled-data theory in computer-based control is well established and is simple to implement [6]. However, there are many situations in the real world where a periodic sampling time does not make sense. That is, it is not always necessary to check the process and to compute the control law continuously, since changes in real systems do not follow any periodic pattern (although there are some exceptions for pure periodic systems). This is the case with biological systems, energy-based systems, or networked systems. Those processes are characterized for being in an equilibrium state, and system disturbances come in a sporadic way because of electrical stimulation, demand for energy, requests via the network, etc. [4]. For these processes, event-based control and event-based sampling are presented as ideal solutions, and for that reason they have become popular in the control community over the past few years [3,5,12,15,22–24,28,29,32,48,49,51,52,59,60]. With these approaches, the samples, and thus the computation of the control law, are calculated in an aperiodic way, where the control strategy is now governed by events associated with relevant changes in the process.

The main advantages of event-based control are quite remarkable from a practical point of view. One advantage is the reduction of resource utilization, for instance, actuator waste for mechanical or electromechanical systems. When the controller is event triggered, the control actions will be applied to the process in an asynchronous way and only when it is really necessary. However, with classical control systems, new control actions are produced every sampling instant, and even for very small errors, constant movements (position changes) of the actuator are required. On the contrary, the application of event-based control for distributed control systems (which are very common in industry) can reduce the information exchange between different control system agents [31,44]. This fact is even more important if the communication is performed through computer networks, where the network structure is shared with other tasks [42].

Another important benefit that characterizes the event-based control systems is the relaxed requirements on the sampling period. This problem is very important in all computer-implemented control systems [2]. The correct selection of the best sampling period for a digital control system is a compromise between many design aspects [1]. Lower sampling rates are usually characterized by their elevated cost; a slow sampling time directly reduces the hardware requirements and the overall cost of process automatization. A slower sample rate makes it possible for a slower computer to achieve a given control function or provides greater capability from a given computer. Factors that may require an increase in the sample rate are command signal tracking effectiveness, sensitivity to plant disturbance rejection, and sensitivity to plant parameter variations.

Therefore, it seems that the event-based control idea has more common sense than the traditional solution based on periodic sampling time—one must do something only if it is really necessary. So why are event-based approaches not being widely used in industry? The main reason is that it is necessary to propose new event-based control approaches or to adapt the current control algorithms to the event-based framework. Note that with an event-based strategy, the sampling time is not periodic; thus, it is necessary to study and to analyze how the control techniques should be modified to account for that [13]. On the contrary, its main drawback currently is that there is no well-supported theory as for computer-based controller systems, and there are still many open research problems regarding tuning and stability issues [4]. For that reason, the main contributions in this field recently have been focused on the adaptation, or modification, of well-known control strategies to be used within an event-based framework. See, for instance, [51], where a typical two-degrees-of-freedom control scheme based on a PI controller was modified to work based on events; [3], where a simple event-based PID control was proposed; or [29], where a new version of the state-feedback approach was presented for event-based control.

There are many active research topics related to the event-based control, where model predictive control (MPC) techniques are frequently used [8,18,25,36,46,54]. Most attention has been given to networked control systems where event-based MPC controllers have been used to reduce the amount of information exchanged through a network of control system nodes [7,26,30,47]. Moreover, the MPC properties are used to compensate for typical phenomena related to computer networks, such as variable delay, packet loss, and so on.

As is well known, in event-based control systems, the signal samples are triggered in an asynchronous way due to their own signal dynamics as well as asynchronous and delayed measurements. Usually, special requirements, or limitations on sampling-rate issues, appear in the distributed control system. The reason is that typical implementation of modern control systems involves communication networks connecting the different control system agents [11,14,25,44,57,58]. In some cases, a dedicated control network is required to achieve the necessary communication bandwidth and reliability. However, for economical reasons, a common network structure is desired for multiple services. In these cases, however, more complex algorithms for networked control systems (NCS) must be used to compensate for the typical issues in controlling networks. NCS, with compensation for random transmission delays using networked predictive control, was studied in [25]. Control systems based on this idea generally assume that future control sequences for all possible time delays are generated on the controller side and are afterward transmitted to the process side. The new assumption presented in that work is that the control signal is selected according to the plant output rather than the time-delay measurement. The obtained results show improvements in control performance when compared with the classical predictive controller. The authors in [47] study a predictive control architecture for plants with random disturbances, where a network is placed between the control system agents. The network imposes a constraint on the expected bit-rate and is affected by random independent dropouts. The controller explicitly incorporates bit-rate and network constraints by sending a quantized finite-horizon plant input sequence. In [26], the experimental study of an event-based input–output control scheme is presented. The sporadic control approach [24] is adopted and implemented in a networked environment. The main reported advantage is the reduction in the control rate, allowing a less saturated communication network. The main results were based on the real implementation, and experimental tests, in a networked system for a DC motor control, using a distributed control scheme with a TCP/IP protocol for the communication link. Other benefits of event-based control system implementation considering design and experimental MPC validation for a hybrid dynamical process with wireless sensors are presented in [7]. The control architecture is based on the reference governor approach, where the process is actuated by a local controller. A hybrid MPC algorithm is running on the remote base station, which sends optimal setpoints. The experimental results show that the proposed framework satisfies all requirements and presents good performance evaluated for a hybrid system in a networked control system.

All previously presented event-based and networked control systems are focused on the events and communications between the sensor and controller nodes. As a consequence of this assumption, the communication rates between the controller and the actuator nodes are determined by the controller-input events. However, the issue related to actuators has received less attention. Recently, some problems related to controller–actuator links were considered in certain research works. The control algorithm described in [19] takes the communication medium-access constraint into account and analyzes the actuator assignment problem. Medium-access constraints arise when the network involves a large number of sensors and actuators. In that work, the medium access of actuators is event driven, and only a subset of the actuators is assigned at each time step to receive control signals from the controller with a bounded time-varying transmission delay. A resource-aware distributed control system for spatially distributed processes is presented in [61]. The developed framework aims to enforce closed-loop stability with minimal information transfer over the network. The number of transmissions is limited by the use of models to compute the necessary control actions and system states using observers. The robustness of the control system against packet dropouts is described in [34], applying a packetized control paradigm, where each transmitted control packet contains tentative future plant input values. The main objective of that work is to derive sparse control signals for efficient encoding, whereas compressed sensing aims at decoding sparse signals.

Therefore, in the literature, there are several solutions dealing with event-based MPC. However, most of them are presented for specific applications such as NCS, to solve changing delay problems, or are based on complicated modifications of the classical MPC algorithm. On the contrary, most available solutions are quite difficult to implement as they require clock synchronization for each control system node. Thus, this chapter presents a new solution to the problem, where a straightforward algorithm to include event-based features in an MPC controller is provided both for sensor and actuator deadbands. The event-based control system is built using the generalized predictive control (GPC) algorithm since it is considered the most popular MPC algorithm in industry [10].

## 8.2 Event-Based Control System

The event-based control structure is shown in Figure 8.1, where $C$ represents an event-based controller and $P(s)$ the controlled process. In this configuration, two types of events can be generated from *u-based* and *y-based* conditions. In the developed application, the actuator possesses a zero-order hold (ZOH) so that the current control action is maintained until the arrival of a new one.

The *u-based* criterion is used to trigger the input-side event, $E_u$, consisting of the transmission of a new control action, $u(t_k)$, if it is different enough (bigger than $\beta_u$) with respect to the previous control action. On the contrary, the *y-based* condition will trigger the output side event, $E_y$, when the difference between the reference $w(t)$ and the process output is out of the limit $\beta_y$. The deadband sampling/updating (based on the level crossing technique) is used for *y-based* and *u-based* conditions.

Since the *u-based* condition is related to the actuator and the *y-based* condition to the sensor, two approaches to perform event-based control strategies can be established:

- Sensor deadband approach
- Actuator deadband approach

Both configurations can be combined to compose an event-based control system with the desired characteristics. Both approaches are presented in detail, maintaining the universal formulation that permits it to be used in a wide range of control techniques.

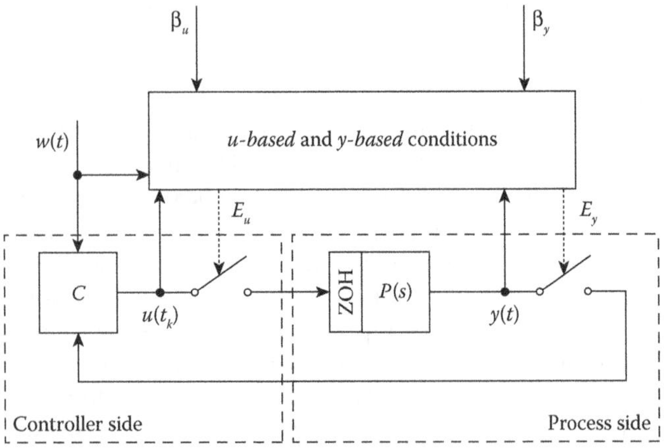


**FIGURE 8.1**
Event-based control approach.

The main ideas of these approaches are detailed one after the other.

## 8.3 Sensor Deadband Approach

When events are coming only from the sensor, an event-based controller consists of two parts: an event detector and a controller [4]. The event detector indicates to the controller when a new control signal has to be calculated, caused by a new event. The complete event-based control structure for this approach, including the process, the actuator, the controller, and the event generator, is shown in Figure 8.2. In this configuration, the event-based controller is composed of a set of controllers, in such a way that one of them will be selected according to the time instant when a new event is detected, such as is described below. This control scheme operates using the following ideas:

- The process is sampled using a constant sampling time $T_{base}$ at the event generator block, whereas the control action is computed and applied to the process using a variable sampling time $T_f$, which is determined by an event occurrence.
- $T_f$ is a multiple of $T_{base}$ ($T_f = fT_{base}$, $f \in [1, n_{max}]$) and verifies $T_f \leq T_{max}$, $T_{max} = n_{max}T_{base}$ being the maximum sampling time value. This maximum sampling time will be chosen to maintain minimum performance and stability margins.
- $T_{base}$ and $T_{max}$ are defined by considering process data and closed-loop specifications, following classical methods for the sampling time choice.

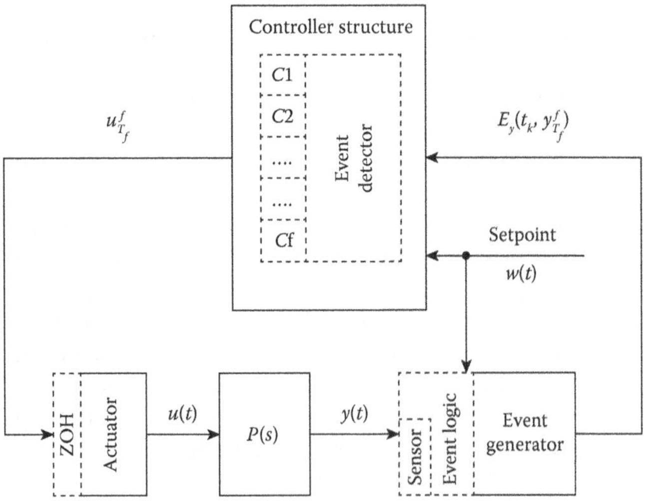


**FIGURE 8.2**
Proposed event-based control scheme.

- After applying a control action at time $t$, the process output is monitored by the event generator block at each base sampling time, $T_{base}$. This information is used by the event detector block, which verifies if the process output satisfies certain specific conditions (*y-based* condition and $T_f \leq T_{max}$). If some of those conditions are met, an event is generated with a sampling period $T_f$, and a new control action is computed. Otherwise, the control action is only computed by a timing event, at $t_k = t_s + T_{max}$, where $t_k$ is the current discrete time instant and $t_s$ is the discrete time instant corresponding to the last acquired sample in event-based sampling.

- It has to be pointed out that, according to the previous description, the control actions will be computed based on a variable sampling time, $T_f$. For that reason, a set of controllers is used, where each controller can be designed for a specific sampling time $T_f = fT_{base}, f \in [1, n_{max}]$. Thus, when an event is detected, the controller associated to sampling time of the current event occurrence will be selected. Conversely, the controller can be designed for continuous time using a tuning procedure corresponding to the control technique used, and the obtained continuous time controller is discretized for each sampling frequency from $T_f$.

The following sections describe in detail the fundamental issues of signal sampling for the proposed strategy.

### 8.3.1 Signal Sampling for Event-Based Control

As discussed above, in an event-based control system with sensor deadband, the control actions are executed in an asynchronous way. The sampling period is governed by system events, and these events are determined according to the event-based sampling techniques described in [33,43].

As can be seen in Figure 8.2, this event-based sampling is governed by the event generator block. This block continuously checks several conditions in the process variables to detect possible events and thus determines the current closed-loop sampling time.

The event generator block includes two different kinds of conditions in order to generate new events. When one of those conditions become true, a new event is generated and then the current signal values of the process variables are transmitted to the controller block, which is used to compute a new control action.

The first types of a conditions are those focused on checking the process variables. These conditions are based on the level-crossing sampling technique [33,43], that is to say, a new event is considered when the absolute value between two signals is higher than a specific limit $\beta_y$. For instance, in the case of setpoint tracking, the condition would be as follows:

$$|w(k) - y(k)| > \beta_y, \tag{8.1}$$

trying to detect that the process output, $y(k)$, is tracking the reference, $w(k)$, within a specific tolerance $\beta_y$.

The second condition is a time condition used for stability and performance reasons. This condition defines that the maximum period of time between two control signal computations, and thus between two consecutive events, is given by $T_{max}$. Hence, this condition is represented as

$$t_k - t_s \geq T_{max}. \tag{8.2}$$

These conditions are checked with the smallest sampling rate $T_{base}$, where the detection of an event will be given within a variable sampling time $T_f = fT_{base}$, $f \in [1, n_{max}]$. It has to be highlighted that this variable sampling period determines the current closed-loop sampling time to be used in the computation of the new control action.

Figure 8.3 shows several examples of how events are generated in the proposed control structure, according to the conditions described above. As can be seen, the continuous time signal $y(t)$ is checked in every $T_{base}$, and new events are generated if the current signal value is outside the established sensitivity band $w \pm \beta_y$. For instance, the events $E_{t_k}$, $E_{t_{k+1}}$,$E_{t_{k+2}}$, and $E_{t_{k+5}}$ are generated from an error condition such as that proposed in (8.1). On the contrary, the events $E_{t_{k+3}}$ and $E_{t_{k+4}}$ correspond to the time limit cycle condition, described by (8.2). In this case, the event is governed by the time condition since there is no other event within that time period.

Note that events are generated at time instant multiples of $T_{base}$, and the effective sampling time for the control loop is $T_f$, computed as the time between two consecutive events. This shows that during rapid and high transients (during setpoint changes and when load disturbances affect the process output), the control action is executed with a small sampling time, near $T_{base}$, allowing rapid closed-loop responses. Conversely, in steady-state conditions, the control action is updated with the maximal sampling time $T_{max}$, minimizing energy and actuator movements but allowing the controller to correct small deviations from the setpoint. Therefore, when the changes in the monitored signals are relatively small and slow, the number of computed control actions is significantly smaller than in a periodic sampling scheme. This is the principal advantage of the proposed event-based control scheme.

It has to be pointed out that it is assumed that the transmission time, $T_T$, between the control system

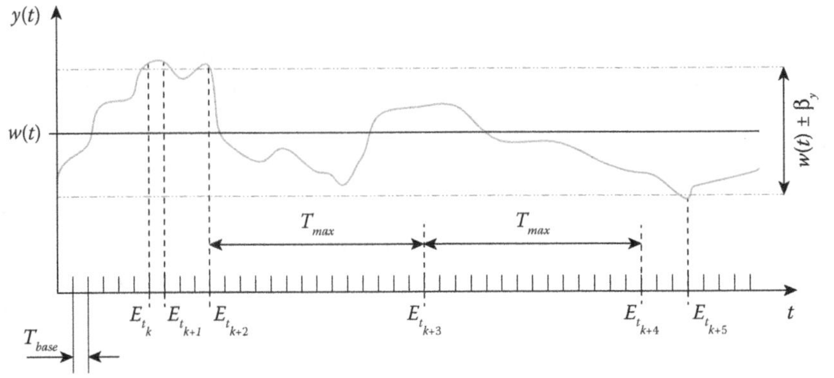


**FIGURE 8.3**
Example of event-based sampling.

blocks is neglected, $T_T \ll T_{base}$. Furthermore, for distributed event-based control systems, no information dropout is considered.

### 8.3.2 Sensor Deadband Approach for Event-Based GPC

The core idea of this section consists of applying the sensor deadband approach presented previously to MPC algorithms [37,39]. The main steps of the algorithm can be summarized as follows: a sampling rate range is defined which delimits when an event can be detected. Then, different GPC controllers are designed, one for each sampling period in that set. Finally, the proposed algorithm is focused on selecting the adequate GPC controller when an event is generated, trying to maintain a trade-off between control performance and control signal changes [38].

In this approach, event detection is based on the level-crossing sampling technique with additional time limits. With this idea, the control actions are only calculated when the process output is outside a certain band around the setpoint such as described above. Thus, while the process output stays within this band, there are no new events, and therefore new control actions are not calculated. Nevertheless, an additional event logic is included for stability issues in such a way that if the temporal difference between two consecutive events is larger than a maximum time period, a new event is generated. First, the typical benefits of any event-based control scheme are achieved, such as an important reduction in actuations. This can be useful for the life of mechanical and electromechanical actuators, reduction of transmissions in NCS, or economical savings (energy consumption due to changes in the actuators) [37]. On the contrary, the proposed control scheme allows one to use standard GPC controllers without modifying the original control algorithm; only external changes are required in the control scheme, which is valuable from a practical point of view.

Therefore, this section is devoted to presenting the proposed event-based GPC algorithm, which uses a variable sampling period to take into account the occurrence of events. The main ideas of the algorithm and the controller structure are first described and then the event detection and resampling procedures are detailed in sequence.

### 8.3.3 GPC Algorithm

As pointed out above, the proposed control structure is based on the use of the GPC algorithm as the feedback controller. Specifically, a set of GPC controllers is used to implement the proposed strategy, one for each sampling time $T_f$, $f \in [1, n_{max}]$, following the ideas from the previous section. Each individual controller in that set is implemented as a classical GPC algorithm. The GPC controller consists of applying a control sequence that minimizes a multistage cost function of the form [10]:

$$J^f = \sum_{j=N_1^f}^{N_2^f} \delta^f [\hat{y}^f(k+j|k) - w(k+j)]^2 + \sum_{j=1}^{N_u^f} \lambda^f [\Delta u^f(k+j-1)]^2, \tag{8.3}$$

where $\hat{y}^f(k+j|k)$ is an optimum $j$ step-ahead prediction of the system output on data up to discrete time $k$, $\Delta u^f(k+j-1)$ are the future control increments, and $w(k+j)$ is the future reference trajectory, considering all signals with a sampling time $T_f$ $(k = kT_f, k \in Z^+)$. Moreover, the tuning parameters are the minimum and maximum prediction horizons, $N_1^f$ and $N_2^f$; the control horizon, $N_u^f$; and the future error and control weighting factors, $\delta^f$ and $\lambda^f$, respectively [10]. The objective of GPC is to compute the future control sequence $u^f(k), u^f(k+1), \ldots, u^f(k+N_u^f-1)$ in such a way that the future plant output $y^f(k+j)$ is driven close to $w(k+j)$. This is accomplished by minimizing $J^f$.

In order to optimize the cost function, the optimal prediction of $y^f(k+j)$ for $j = N_1^f \dots N_2^f$ is obtained using a single-input single-output (SISO) linear model of the plant described by

$$A^f(z^{-1})y^f(k) = z^{-d^f}B^f(z^{-1})u^f(k) + \frac{\varepsilon^f(k)}{\Delta}, \quad (8.4)$$

where $\varepsilon^f(k)$ is the zero mean white noise, $\Delta = 1 - z^{-1}$; $d^f$ is the discrete delay of the model; and $A^f$ and $B^f$ are the following polynomials in the backward shift operator $z^{-1}$:

$$A^f(z^{-1}) = 1 + a_1^f z^{-1} + a_2^f z^{-2} + \dots + a_{na}^f z^{-na},$$

$$B^f(z^{-1}) = b_0^f + b_1^f z^{-1} + b_2^f z^{-2} + \dots + b_{nb}^f z^{-nb}.$$

Using this model, it is possible to obtain a relationship between the vector of predictions $\mathbf{y^f} = [\hat{y}^f(k + N_1^f|k), \hat{y}^f(k + N_1^f + 1|k), \dots, \hat{y}^f(k + N_2^f|k)]^T$, the vector of future control actions $\mathbf{u^f} = [\Delta u^f(k), \Delta u^f(k+1), \dots, \Delta u^f(k + N_u^f - 1)]^T$, and the free response of the process $\mathbf{f^f}$ [10]:

$$\mathbf{y^f} = \mathbf{G}\mathbf{u^f} + \mathbf{f^f},$$

where $\mathbf{G}$ is the step-response matrix of the system.

In this case, following the idea presented in [35], the prediction horizons are defined as $N_1^f = d^f + 1$ and $N_2^f = d^f + N^f$, and the weighting factor $\delta^f = 1$. Thus, the control tuning parameters are $N^f, N_u^f$, and $\lambda^f$. The minimum of $J^f$, assuming that there are no constraints on the control and output signals, can be found by making the gradient of $J^f$ equal to zero, which leads to

$$\mathbf{u^f} = (\mathbf{G^T G} + \lambda^f \mathbf{I})^{-1}\mathbf{G^T}(\mathbf{w} - \mathbf{f^f}), \quad (8.5)$$

where $\mathbf{I}$ is the identity matrix, and $\mathbf{w}$ is a vector of future references.

Notice that according to the receding horizon approach, the control signal that is actually sent to the process is the first element of vector $\mathbf{u^f}$, given by

$$\Delta u^f(k) = \mathbf{K^f}(\mathbf{f^f} - \mathbf{w}), \quad (8.6)$$

where $\mathbf{K^f}$ is the first row of matrix $(\mathbf{G^T G} + \lambda^f \mathbf{I})^{-1}\mathbf{G^T}$. This has a clear meaning if there are no future predicted errors; that is, if $\mathbf{w} - \mathbf{f^f} = \mathbf{0}$, then there is no control move since the objective will be fulfilled with the free evolution of the process. However, in the other case, there will be an increment in the control action proportional (with the factor $\mathbf{K^f}$) to that future error. Detailed formulation of classical GPC is given in [10]. When constraints are taken into account, an optimization problem must be solved using a quadratic cost function with linear inequality and equality constraints [10].

### 8.3.4 Signal Sampling and Resampling Techniques

Such as is described above, the computation of a new control action in the proposed event-based algorithm is made with a variable sampling period $T_f$. Thus, the past values of the process variables and of the control signals have to be available for sampling with that of sampling period, $T_f$. Therefore, a resampling of the corresponding signals is required.

The problem of updating the control law when there are changes in the sampling frequency can be solved by swapping between controllers working in parallel, each with a different sampling period. This is a simple task that implements swapping with signal reconstruction techniques [16]. However, using multiple controllers in parallel results in being computationally inefficient. In the approach proposed in this chapter, the algorithm shifts from one controller to another only when the sampling period changes and does not permit the controllers to operate in parallel. This requires a change in the controller parameters as well as the past values of the output and control variables required to compute the next control signal, according to the new sampling period. The control law is precomputed offline, as was explained in Section 8.3.3, generating the GPC gain $\mathbf{K^f}$ for each sampling time, $T_f$. Then the past control and output signals for that sampling period $T_f$ are calculated by using the procedure described in the following subsections.

#### *8.3.4.1 Resampling of Process Output*

As discussed previously, the controller block only receives the new state of the process output when a new event is generated. This information is stored in the controller block and is resampled to generate a vector $y^b$ including the past values of the process output with $T_{base}$ samples. The resampling of the process output is performed by using a linear approximation between two consecutive events, and afterward this linear approximation is sampled with the $T_{base}$ sampling period, resulting in $y^b(k)$ with $k = 0, T_{base}, 2T_{base}, 3T_{base}, \dots$ . Then, this resampled process output signal, $y^b$, is used to calculate the necessary past information to compute the new control action, such as described in the following steps:

- Suppose that the last value of the process output was received at time instant $t_s = k$ and is stored in $y^b(k)$ with a sampling time of $T_{base}$. Then, a new event is detected and the sampling period changes to $T_f = fT_{base}$. Thus, the new time instant is given by $t_k = k + f$, and a new value of the process output is received as $y^b(k+f) = y^f_{T_f}$.

- Using this information, a linear approximation is performed between these two samples, $y^b(k)$ and

$y^b(k+f)$, and the resulting linear equation is sampled with $T_{base}$ in order to update the process output variable $y^b$ as follows:

$$y^b(k+i) = y^b(k) - \frac{y^b(k) - y^b(k+f)}{f} i,$$

where $i = 1, \ldots, f$.

- Once the process output signal is resampled, the required past information must be obtained according to the new sampling time $T_f$, resulting in a new signal, $y_p^f$, with the past information of the process output in every $T_f$ sample. Thus, this signal is calculated taking values from the resampled process output signal, $y^b$, by moving backward in that vector with decrements of $T_f = fT_{base}$:

$$y_p^f(i) = y^b(k - jf), \quad (8.7)$$

with $i = P_y, \ldots, 1$ and $j = P_y - i$, where $P_y$ is the required number of past values.

- When there is not enough past information, the last available value is taken to fill the signal $y_p^f$.

Hence, the vector $y^f$ is obtained as a result, which contains the past process information with the new sampling period $T_f$ to be used in the calculation of the current control action [39].

#### 8.3.4.2 Reconstruction of Past Control Signals

The procedure is similar to that described for the resampling of the process output. There is a control signal, $u^b$, which is always used to store the control signal values every $T_{base}$ samples. This is done by reconstructing the control signal calculated between two consecutive events. However, in this case, there is not any linear approximation between two consecutive samples, since a ZOH is being used. Thus, the signal $u^b$ will be obtained by keeping constant the control values between two consecutive events. Nevertheless, the procedure for the control signal is done in the opposite way than for the process output. First, the required past information is calculated, and afterward the signal $u^b$ is updated:

- Let's consider that a new event is generated, that results in a new sampling period $T_f = fT_{base}$. Now, the past information for the new sampling period, $T_f$, is first calculated from the past values in $u^b$ and stored in a variable called $u_p^f$. Afterward, this information, together with the past process output data given by (8.7), will be used to calculate the new control action, $u_{T_f}^f$.
- The required past information each $T_f$ samples is calculated from values of the $u^b$ signal. As the ZOH is being considered for the control signal calculation, the past values of the new control signal, $u_p^f$, must be reconstructed according to this fact. That is, the reconstructed past values must be kept constant between samples. However, since the values in $u^b$ are sampled with $T_{base}$, the past values of $u_p^f$ are reconstructed using the average of the past $f$ values in $u^b$, as follows:

$$u_p^f(i) = \sum_{l=0}^{f-1} u^b(j-l), \quad (8.8)$$

for $i = P_u, \ldots, 1$, with $j = k - 1 - (P_u - i)f$, and $u_p^f$ and $P_u$ being the past values of $u^f$ and the number of required past values, respectively. Note that with this procedure, the new reconstructed control signal and the original base signal have the same values of the integral of the control signal in the interval.

- Then, the actual control action for the new sampling time, $u_{T_f}^f$, is calculated from the GPC control law (8.6) by using the reconstructed past control signals and the resampled past process outputs given by (8.8) and (8.7), respectively.
- Once the new control action has been calculated, $u_{T_f}^f$, the $u^b$ signal is updated by keeping constant the values between the two consecutive events. Therefore,

$$u^b(j) = u_{T_f}^f,$$

for $j = k, \ldots, k+f$.

### 8.3.5 Tuning Method for Event-Based GPC Controllers

The set of GPC controllers available at the controller block have to be properly tuned so that the control system satisfies all the design and performance requirements independently of the event-based framework. First, the maximum sampling time, $T_{max}$, has to be selected. This parameter is tuned according to the desired closed-loop time constant, $\tau_{cl}$, and ensuring the control system stability in the discretization process, where [6]

$$T_{max} \leq \frac{\tau_{cl}}{4}. \quad (8.9)$$

Then, the $n_{max}$ design parameter is established, which determines the number of GPC controllers to be considered and the faster sampling period as $T_{base} = T_{max}/n_{max}$, with $n_{max} \in Z^+$. In this way, the current sampling period $T_f$ will be given as $T_f = fT_{base}$, $f \in [1, n_{max}]$, such as is discussed above.

The next step consists of tuning the GPC parameters, $\lambda^f$, $N_2^f$, and $N_u^f$. The $\lambda^f$ parameter is designed to achieve the desired closed-loop performance for the fastest GPC controller with a $T_{base}$ sampling period, and this same value is used for the rest of the GPC controllers (notice that the algorithms are focused to use the fastest GPC controller when events are detected, and thus the tuning rules have to be mainly focused on that controller). The prediction horizon is calculated in continuous time, $N_2^c$, according to the typical GPC tuning rules (capturing the settling time of the process)[10]. Then, the prediction horizon for each GPC controller, $N_2^f$, is calculated in samples dividing the continuous-time value by the corresponding sampling period $T_f$:

$$N_2^f = \frac{N_2^c}{T_f}, \quad f \in [1, n_{max}]. \tag{8.10}$$

The control horizon is set in samples for the slowest sampling period $T_{max}$ according to typical rules [10]. Then, this same value is used for the rest of the GPC controllers:

$$N_u^f = N_u^{n_{max}}, \quad f \in [1, n_{max}]. \tag{8.11}$$

The reason to keep the same control horizon for all the GPC controllers is that having large control horizons does not improve the control performance. Besides, large control horizon values increase the computational burden considerably, what must be considered during MPC controllers design stage [50].

## 8.4 Actuator Deadband Approach

The main idea of this approach is to develop a control structure where the control signal is updated in an asynchronous manner [36]. The main goal is to reduce the number of control signal updates, saving system resources, while retaining good control performance.

In the sensor deadband approach presented in the previous section, the controller event always produces a new control action, which is transmitted to the actuator node. Most event-based control systems are focused on this idea, where the sampling events are generated by the event generator as a function of the process output. In this configuration, the control action is transmitted to the actuator even if its change is small and transmission is not justified. Therefore, this section will focus on the actuator deadband approach, which tries to face these drawbacks regarding control signal changes. The actuator deadband can be understood as a constraint on control signal increments $\Delta u(k)$:

$$|\Delta u(k)| = |u(k-1) - u(k)| \geqslant \beta_u, \tag{8.12}$$

where $\beta_u$ is the proposed deadband.

The idea of this approach consists in including this virtual deadband on the control signal (the input signal for the actuator) into a control system. The introduced deadband, $\beta_u$, will be used as an additional tuning parameter for control system design, to adjust the number of actuator events (transmissions from controller to actuator). If the deadband is set to a small value, the controller becomes more sensitive and produces more actuator events. Alternatively, when $\beta_u$ is set higher, the number of events is reduced, which is expected to result in worse control performance. In this way, it is possible to establish a desired trade-off between control performance and the number of actuator events/transmissions.

### 8.4.1 Actuator Deadband Approach for Event-Based GPC

So far, and as described in the introduction of this chapter, research efforts in event-based MPC have been focused on compensation for network issues and not on the control algorithm itself, which can greatly influence the amount of information exchanged throughout the network. Motivated by this fact, a methodology that allows a reduction of the number of actuator events is developed. This is possible to be included in the GPC by adding a virtual actuator deadband to constrain the number of changes in the control signal over the control horizon. This solution has been considered at the control synthesis stage and influences many previously discussed aspects of network limitations, such as transmission delay, quantization effect, and multi-packet transmission. At the same time, using the introduced actuator deadband, event-based controllers can reduce the number of actuator events.

The idea presented in this section consists of applying virtual constraints on the control signal. The formulated constraints introduce a deadband that disallows small control signal changes. When the required control signal increment is insignificant (smaller than an established band), it is suppressed because its effect on the controlled variable is not important. However, when the increment in the control signal is significant (bigger than the deadband), its optimal value is transmitted to the actuator and applied to the controlled process. In this way, the introduced virtual deadband is included in the optimization technique, which provides an optimal solution for the control law. The presence of the deadband in the optimization procedure can significantly reduce the number of events affecting the actuator, at the same time guaranteeing the optimality of the solution. The virtual deadband is modeled and included in the developed algorithm exploring a hybrid system design framework. The resulting optimization function contains continuous and discrete variables, and thus mixed integer quadratic programming (MIQP) has to be used to solve the formulated optimization problem. The presented idea is tested by simulation for a set of common industrial systems, and it is shown that a large reduction in the number of control events can be obtained. This can significantly improve the actuator's life span as well as reduce the number of communications between the controller and the actuator, which is important from an NCS point of view.

### 8.4.2 Classical GPC with Constraints

The minimum of the cost function $J$ (of the form referred in Equation 8.3), assuming there are no constraints on the control signals, can be found by making the gradient of $J$ equal to zero. Nevertheless, most physical processes are subjected to constraints, and the optimal solution can be obtained by minimizing the quadratic function:

$$J(\mathbf{u}) = (\mathbf{G}\mathbf{u} + \mathbf{f} - \mathbf{w})^T(\mathbf{G}\mathbf{u} + \mathbf{f} - \mathbf{w}) + \lambda \mathbf{u}^T\mathbf{u}, \quad (8.13)$$

where $\mathbf{G}$ is a matrix containing the coefficient of the input-output step response, $\mathbf{f}$ is a free response of the system, $\mathbf{w}$ is a future reference trajectory, and $\mathbf{u} = [\Delta u(t), \ldots, \Delta u(t+N-1)]$.

Equation 8.13 can be written in quadratic function form:

$$J(\mathbf{u}) = \frac{1}{2}\mathbf{u}^T\mathbf{H}\mathbf{u} + \mathbf{b}^T\mathbf{u} + \mathbf{f}_0, \quad (8.14)$$

where $\mathbf{H} = 2(\mathbf{G}^T\mathbf{G} + \lambda\mathbf{I})$, $\mathbf{b}^T = 2(\mathbf{f} - \mathbf{w})^T\mathbf{G})$, $\mathbf{f}_0 = (\mathbf{f} - \mathbf{w})^T(\mathbf{f} - \mathbf{w})$. The obtained quadratic function is minimized subject to system constraints, and a classical quadratic programming (QP) problem must be solved. The constraints acting on a process can originate from amplitude limits in the control signal, slew rate limits of the actuator, and limits on the output signals, and can be expressed in the shortened form as $\mathbf{R}\mathbf{u} \leqslant \mathbf{r}$ [10]. Many types of constraints can be imposed on the controlled process variable and can be expressed in a similar manner. Those constraints are used to force the response of the process with the goal to obtain certain characteristics [20]. In the following sections, the deadband formulation and the modification of the QP problem by a MIQP optimization problem is considered.

#### 8.4.2.1 Actuator Deadband: Problem Statement

Taking advantage of the constraints handling mechanisms in MPC controllers described above, the actuator deadband

$$|\Delta u(k)| \geqslant \beta_u,$$

can be included in the optimization procedure. Compensation techniques for actuator issues were explored in previous works about predictive control. Some examples can found in [55], where actuator nonlinearity compensation was analyzed based on the real experiments on gantry crane control problems. In that study, the external pre-compensator for the process input nonlinearity was studied, using the inverse function combined with the multivariable MPC. The proposed control architecture allows one to reduce oscillations in the controlled variable caused by the actuator nonlinearity, but it introduces a permanent offset in the reference tracking. On the other hand, the MPC algorithm was also used to compensate for quantization effects in control signal values. This topic was investigated in [45], where the control problem for a particular discrete-time linear system subject to quantization of the control set was analyzed. The main idea consists of applying a finite quantized set for the control signal, where the optimal solution is obtained by solving an online optimization problem. Authors consider the design of stabilizing control laws, which optimize a given cost index based on the state and input evolution on a finite receding horizon in a hybrid control framework.

The presence of the actuator deadband defines three regions for possible solutions of the control law, and the following nonlinear function of the actuator deadband can be used:

$$x(k) = \begin{cases} \Delta u(k) & : \quad \Delta u(k) \geqslant \beta_u \\ 0 & : \quad -\beta_u \leqslant \Delta u(k) \leqslant \beta_u \, . \\ \Delta u(k) & : \quad \Delta u(k) \leqslant -\beta_u \end{cases} \quad (8.15)$$

Most actuators in industry exhibit a backlash and other types of nonlinearities. However, controllers are normally designed without taking into account the actuator nonlinearities. Because of the predictive nature of GPC, which will be used for the developed system, the actuator deadband can be dealt with. The deadband can be treated by imposing constraints on the controller in order to generate control signals outside the deadband. Taking into account that GPC can handle different types of linear constraints, the proposed system input nonlinearity cannot be introduced directly into the optimization procedure. Therefore, the feasible region generated by this type of constraint is nonconvex and requires special techniques to solve it. The detailed description of input constraints modeling and its integration into the optimization procedure are detailed in the next sections.

In this study, it is assumed that the process output is sampled in synchronous mode with constant sampling time, and the controlled system is updated in an asynchronous way, making use of the actuator deadband. This assumption is made to highlight the influence of the control signal deadband on the controlled variable. However, the presented idea can be combined with many MPC algorithms, where nonuniform sampling techniques are used to monitor the controller variable [7,8,25,26,30,47,54].

### 8.4.3 Formulation of the Actuator Deadband Using a Hybrid Dynamical Systems Framework

The methodology presented in this section consists of including the actuator virtual deadband into the GPC design framework. The deadband nonlinearity can be handled together with other constraints on the controlled process. This approach is reasonable because the deadband nonlinearity has a nonsmooth nature, and it is characterized by discrete dynamics. Therefore, it can be expressed mathematically by a series of *if-then-else* rules [62]. The hybrid design framework developed by [9] allows one to translate those *if-then-else* rules into a set of linear logical constraints. The resulting formulation consists of a system containing continuous and discrete components, which is known as a mixed logical dynamic (MLD) system.

In accordance with (8.15), the virtual deadband forces the control signal increment to be outside the band, and in the opposite case ($-\beta_u \leq \Delta u(k) \leq \beta_u$), the control signal is kept unchanged, $\Delta u(k) = 0$.

This formulation has the typical form used to describe constraints in the GPC algorithm, which are imposed on the decision variable. To incorporate a deadzone within the GPC design framework, the input constraints have to be reconfigured. Let us introduce two logical variables, $\varphi_1$ and $\varphi_2$, to determine a condition on the control signal increments, $\Delta u(k)$. So, these logical variables are used to describe the different stages of the control signal with respect to the deadband:

$$[\Delta u(k) \geqslant \beta_u] \rightarrow [\varphi_2 = 1], \tag{8.16}$$

$$[\Delta u(k) \leqslant \beta_u] \rightarrow [\varphi_2 = 0], \tag{8.17}$$

$$[\Delta u(k) \geqslant -\beta_u] \rightarrow [\varphi_1 = 0], \tag{8.18}$$

$$[\Delta u(k) \leqslant -\beta_u] \rightarrow [\varphi_1 = 1]. \tag{8.19}$$

To make this solution more general, minimal $m$ and maximal $M$ values for the control signal increments are included in the control system design procedure, resulting in

$$M = max\{\Delta u(k)\}, \quad and \quad m = min\{\Delta u(k)\}.$$

In this way, it is possible to determine the solution region based on binary variables. Figure 8.4 shows the virtual deadband in a graphical form, where each region can be distinguished. The introduced auxiliary logical variables, $\varphi_1$ and $\varphi_2$, define three sets of possible solutions depending on their values:

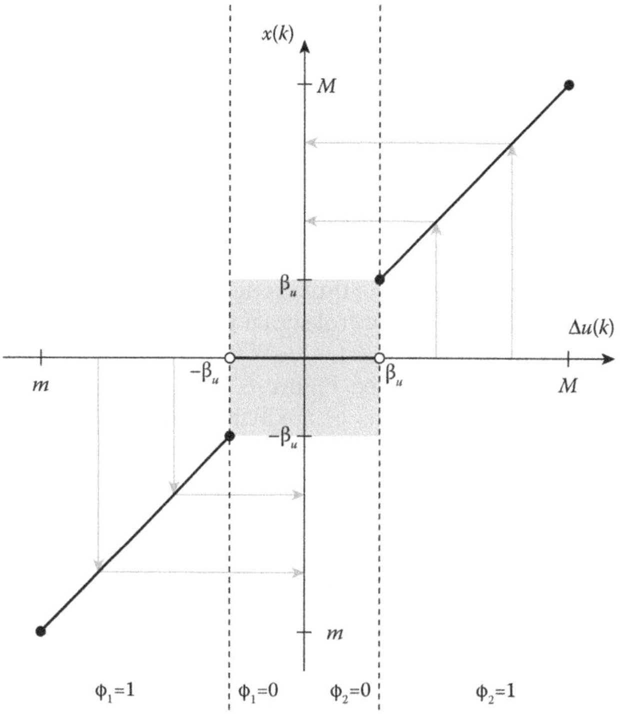


**FIGURE 8.4**
Control signal increments with deadband.

$$x(k) = \begin{cases} \Delta u(k) & : \ \Delta u(k) \geqslant \beta_u & \varphi_2 = 1 \\ 0 & : \ \Delta u(k) \leqslant \beta_u & \varphi_2 = 0 \\ 0 & : \ \Delta u(k) \geqslant -\beta_u & \varphi_1 = 0 \\ \Delta u(k) & : \ \Delta u(k) \leqslant -\beta_u & \varphi_1 = 1 \end{cases}. \tag{8.20}$$

For the overall performance of the optimization process, it is important to model the deadband with few auxiliary variables to reduce the number of logical conditions that must be verified. Thus, the proposed logic determined by (8.20) can be translated into a set of *mixed-integer linear inequalities* involving both continuous variables, $\Delta u \in \mathbb{R}$, and logical variables $\varphi_i \in \{0, 1\}$. Finally, a set of mixed-integer linear inequalities constraints for the actuator deadband are established as

$$\Delta u - (m - \beta_u)(1 - \varphi_2) \geqslant \beta_u, \tag{8.21}$$

$$\Delta u - (M - \beta_u)\varphi_2 \leqslant \beta_u, \tag{8.22}$$

$$\Delta u - (m + \beta_u)\varphi_1 \geqslant -\beta_u, \tag{8.23}$$

$$\Delta u - (M + \beta_u)(1 - \varphi_1) \leqslant -\beta_u, \tag{8.24}$$

$$\Delta u \geqslant m\varphi_1 \tag{8.25}$$

$$\Delta u \leqslant M\varphi_2. \tag{8.26}$$

To guarantee a unique solution for the optimization procedure, an additional condition must be considered. The objective of this action is a separation of the solution regions, which delimits the optimal solution to one region for each time instant:

$$\varphi_1 + \varphi_2 \leqslant 1. \tag{8.27}$$

With this restriction, the three zones that cover all possible solutions for the control signal increments can easily be determined: $\{(\varphi_1 = 1 \wedge \varphi_2 = 0), (\varphi_1 = 0 \wedge \varphi_2 = 1), (\varphi_1 = 0 \wedge \varphi_2 = 0)\}$ (see Figure 8.4 for details). These different configurations of auxiliary variables limit the solution area, assuring a solution singularity. The particular solution that can be expressed as constraints on the control signal increments:

$$\begin{array}{ccc} \varphi_1 = 1 \wedge \varphi_2 = 0 & \varphi_1 = 0 \wedge \varphi_2 = 1 & \varphi_1 = 0 \wedge \varphi_2 = 0 \\ \Delta u \geqslant m & \Delta u \geqslant \beta_u & \Delta u \geqslant m \\ \Delta u \leqslant \beta_u & \Delta u \leqslant M & \Delta u \leqslant \beta_u \\ \Delta u \geqslant m & \Delta u \geqslant -\beta_u & \Delta u \geqslant -\beta_u \\ \Delta u \leqslant -\beta_u & \Delta u \leqslant M & \Delta u \leqslant M \\ \Delta u \geqslant m & \Delta u \geqslant 0 & \Delta u \geqslant 0 \\ \Delta u \leqslant 0 & \Delta u \leqslant M & \Delta u \leqslant 0 \end{array}.$$

Hence, to obtain an optimal solution, the quadratic objective function, subject to the designed input constraints (Equations 8.21 through 8.27), has to be solved. In this way, for each time step, new control actions are obtained, avoiding control signal increments smaller than $|\beta_u|$.

### 8.4.4 MIQP-Based Design for Control Signal Deadband

In this section, the reformulated hybrid input constraints presented in the previous section are integrated into the GPC optimization problem, where the resulting formulation belongs to an MIQP optimization problem. First, for simplicity's sake, the case for the control horizon $N_u = 1$ will be introduced, and subsequently, an extension for any value of control horizon will be shown. The previous set of constraints (Equations 8.21 through 8.27) can be rewritten in the following forms:

$$\begin{aligned} \Delta u - (M - \beta_u)\varphi_2 &\leqslant \beta_u, \\ \Delta u + (M + \beta_u)\varphi_1 &\leqslant M, \\ \Delta u - M\varphi_2 &\leqslant 0, \\ -\Delta u + (m + \beta_u)\varphi_1 &\leqslant \beta_u, \\ -\Delta u - (m - \beta_u)\varphi_2 &\leqslant -m, \\ -\Delta u + m\varphi_1 &\leqslant 0, \\ \varphi_1 + \varphi_2 &\leqslant 1, \end{aligned}$$

where these constraints can be written in a matrix form for the control horizon $N_u = 1$ as follows:

$$\begin{bmatrix} 1 & 0 & -(M-\beta_u) \\ 1 & (M+\beta_u) & 0 \\ 1 & 0 & -M \\ -1 & (m+\beta_u) & 0 \\ -1 & 0 & -(m-\beta_u) \\ -1 & m & 0 \\ 0 & 1 & 1 \end{bmatrix} \begin{bmatrix} \Delta u \\ \varphi_1 \\ \varphi_2 \end{bmatrix} \leqslant \begin{bmatrix} \beta_u \\ M \\ 0 \\ \beta_u \\ -m \\ 0 \\ 1 \end{bmatrix}. \tag{8.28}$$

In the case where the control horizon is $N_u > 1$, the corresponding matrix becomes

$$\underbrace{\begin{bmatrix} 1\mathbf{D}_I & 0\mathbf{D}_I & -(M-\beta_u)\mathbf{D}_I \\ 1\mathbf{D}_I & (M+\beta_u)\mathbf{D}_I & 0\mathbf{D}_I \\ 1\mathbf{D}_I & 0\mathbf{D}_I & -M\mathbf{D}_I \\ -1\mathbf{D}_I & (m+\beta_u)\mathbf{D}_I & 0\mathbf{D}_I \\ -1\mathbf{D}_I & 0\mathbf{D}_I & -(m-\beta_u)\mathbf{D}_I \\ -1\mathbf{D}_I & m\mathbf{D}_I & 0\mathbf{D}_I \\ 0\mathbf{D}_I & 1\mathbf{D}_I & 1\mathbf{D}_I \end{bmatrix}}_{C} \underbrace{\begin{bmatrix} \Delta u\mathbf{d}_I \\ \varphi_1\mathbf{d}_I \\ \varphi_2\mathbf{d}_I \end{bmatrix}}_{x}$$

$$\leqslant \underbrace{\begin{bmatrix} \beta_u\mathbf{d}_I \\ M\mathbf{d}_I \\ 0\mathbf{d}_I \\ \beta_u\mathbf{d}_I \\ -m\mathbf{d}_I \\ 0\mathbf{d}_I \\ 1\mathbf{d}_I \end{bmatrix}}_{\rho},$$

where

$$\mathbf{D}_I = \begin{bmatrix} 1_1 & 0 & \dots & 0 \\ 0 & 1_2 & \ddots & 0 \\ \vdots & \ddots & \ddots & 0 \\ 0 & \dots & 0 & 1_{N_u} \end{bmatrix},$$

is a diagonal matrix $(N_u \times N_u)$ of ones, and

$$\mathbf{d}_I = \begin{bmatrix} 1_1 \\ \vdots \\ 1_{N_u} \end{bmatrix},$$

is a vector of ones with size $(N_u \times 1)$. The previous matrices that contain linear inequality constraints can be expressed in a general form as

$$\mathbf{C}x \leqslant \rho, \tag{8.29}$$

with $x = [x_c, x_d]^T$, where $x_c$ represents the continuous variables $\Delta u$, and $x_d$ are those of the logical variables $\varphi_i$. Introducing the matrix $\mathbf{Q}_{(3N_u \times 3N_u)}$ and $\mathbf{l}_{(3N_u \times 1)}$ defined as

$$\mathbf{Q} = \begin{bmatrix} \mathbf{H} & \underline{\mathbf{O}} & \underline{\mathbf{O}} \\ \underline{\mathbf{O}} & \underline{\mathbf{O}} & \underline{\mathbf{O}} \\ \underline{\mathbf{O}} & \underline{\mathbf{O}} & \underline{\mathbf{O}} \end{bmatrix}; \mathbf{l} = \begin{bmatrix} \mathbf{b} \\ \bar{\bar{O}} \\ \bar{\bar{O}} \end{bmatrix}, \quad (8.30)$$

where $\underline{\mathbf{O}} = N_u \times N_u$ and $\bar{\bar{O}} = N_u \times 1$ both of zeros. Now, the GPC optimization problem is expressed as

$$\min_{x} x^T \mathbf{Q} x + \mathbf{l}^T x, \quad (8.31)$$

subject to (8.29), where

$$x = \begin{bmatrix} x_c \\ x_d \end{bmatrix}$$

$$x_c \in \mathbb{R}^{nc}, \qquad x_d \in \{0, 1\}^{n_d},$$

which is a MIQP optimization problem [9]. The optimization problem involves a quadratic objective function and a set of mixed linear inequalities. The logical variables appear linearly with the optimization variables, but there is no product between the logical variables $\varphi_i$ and the optimization variable $\Delta u$.

The classical set of constraints represented by $\mathbf{Ru} \leqslant \mathbf{r}$ (see Section 8.4.2) can also be included in the optimization procedure. To fulfill this goal, it is necessary to introduce an auxiliary matrix $\check{\boldsymbol{R}}$ of the form

$$\check{\boldsymbol{R}} = \begin{bmatrix} R & \overline{O} & \overline{O} \end{bmatrix},$$

where $\overline{O}$ is a matrix of zeros with the same dimensions as $\check{\boldsymbol{R}}$. Finally, all constraints that must be considered in the optimization procedure are grouped in

$$\begin{bmatrix} C \\ \check{R} \end{bmatrix} x \leqslant \begin{bmatrix} \rho \\ r \end{bmatrix}.$$

## 8.5 Complete Event-Based GPC

The complete event-based GPC control scheme considers both sensor and actuator deadbands at the same time. To realize such an event-based control structure, it is necessary to build the control structure introduced for the sensor deadband approach. Additionally, developed controllers consider the actuator deadband in the optimization procedure. In this way, an event-based GPC controller manages two sources of events related to process output and input, respectively. Thus, the complete event-based GPC control structure has two additional tuning parameters $\beta_y$ and $\beta_u$, which determine the deadband for the sensor and the actuator, respectively. Each of these tuning parameters can be used to obtain the desired trade-off between control performance and the number of events for the sensor and the actuator, independently. In this configuration, the process output is sampled using the intelligent sensor, where the deadband sampling logic is implemented. When one of the conditions becomes true, the event generator transmits current process output to the controller node. The usage of the sensor deadband allows one to reduce the process output events $E_y$. The received information in the event-based controller node triggers the event detector to calculate the time elapsed since the last event. The obtained time value is used as the current sampling time, and the corresponding controller is selected to calculate a new control signal. Because the virtual actuator deadband is also used in such a configuration, the corresponding constraints on the control signal are active for all controllers from the set. In this way, the obtained control signal takes into account the deadband and makes the reduction of process input events $E_u$ possible.

The complete event-based GPC merges the advantages of both previously introduced methods. In the resulting configuration, the process input and output events can be tuned independently using the actuator or the sensor deadband, respectively. The developed event-based control structure can satisfy the particular requirements, which are considered under the algorithm development procedure.

## 8.6 Applications

This section presents a simulation study for an analyzed event-based GPC algorithm, applied to the greenhouse temperature control problem. The application scenario under consideration allows us to highlight the most important features and properties of the event-based control strategy.

### 8.6.1 Greenhouse Process Description and System Models

Control problems in greenhouses are mainly focused on fertigation and climate systems. The fertigation control problem is usually solved providing the amount of water and fertilizers required by the crop. The climate control problem consists in keeping the greenhouse temperature and humidity in specific ranges despite disturbances. Adaptive and feedforward controllers are commonly used for the climate control problem. Therefore, fertigation and climate systems can be represented as event-based control problems where control actions

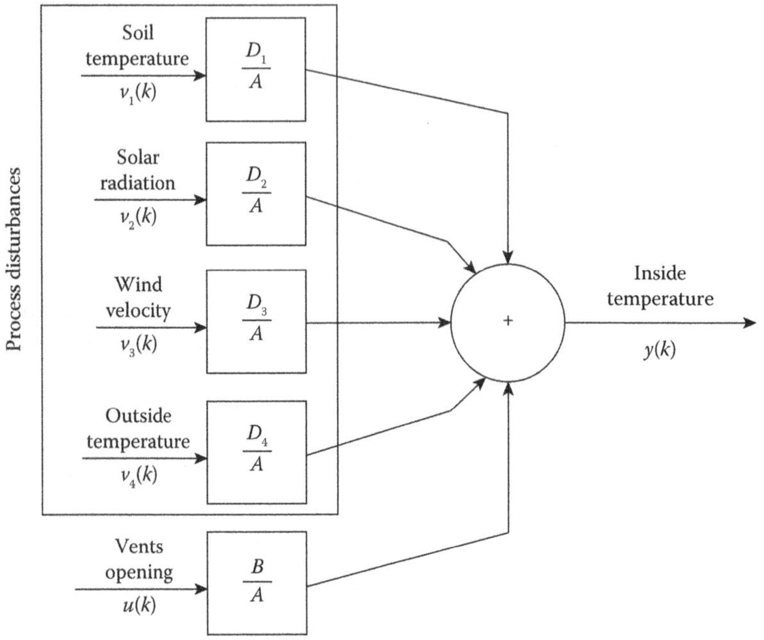


**FIGURE 8.5**
Greenhouse process model with disturbances for diurnal temperature control.

will be calculated and performed when required by the system, for instance, when water is required by the crop or when ventilation must be closed due to changes in outside weather conditions. Furthermore, such as discussed above, with event-based control systems, a new control signal is generated only when a change is detected in the system. That is, the control signal commutations are produced only when events occur. This fact is very important for the actuator life and from an economical point of view (reducing the use of electricity or fuel), especially in greenhouses where actuators are commonly composed of mechanical devices controlled by relays [17,21,27,53,56].

The experimental data used in this work were obtained from a greenhouse located at the Experimental Station of the CAJAMAR Foundation "Las Palmerillas" in Almería, Spain.* It measures an average height of 3.6 m and has covered surface area of 877 $m^2$. It is provided with natural ventilation. Additionally, it is equipped with a measurement system and software adapted to carry out the experiments of process identification and implementing different climatic control strategies. All greenhouse climatic variables were measured with a sampling period of 1 minute.

*Hierarchic Predictive Control of Processes in Semicontinuous Operation (Project DPI2004-07444-C04-04), Universidad de Almeria, Almeria, Spain. http://aer.ual.es/CJPROS/engindex.php.

The greenhouse interior temperature problem was considered as a MISO (multi-input single-output) system, where soil temperature, $v_1(k)$, solar radiation, $v_2(k)$, wind velocity, $v_3(k)$, outside temperature, $v_4(k)$, and vents-opening percentage, $u(k)$, were the input variables and the inside temperature, $y(k)$, was the output variable (see Figure 8.5). It has to be pointed out that only the vents-opening variable can be controlled, whereas the rest of the variables are considered measurable disturbances. Additionally, wind velocity is characterized by rapid changes in its dynamics, whereas solar radiation is a combination of smooth (solar cycle) and fast dynamics (caused by passing clouds). Soil and outside temperature were characterized by slow changes.

Considering all of the abovementioned process properties, the CARIMA model can be expressed as follows:

$$A(z^{-1})y(k) = z^{-d}B(z^{-1})u(k-1) + \sum_{i=1}^{4} z^{-d_{D_i}} D_i(z^{-1})v_i(k) + \frac{\varepsilon(k)}{\Delta}. \quad (8.32)$$

Many experiments were carried out over several days where a combination of PRBS and step-based input signals were applied at different operating points. It was observed that the autoregressive model with external input (ARX), using Akaike's Information Criterion (AIC), presented better dynamic behavior adjustment to the real system. This fact was confirmed by cross-correlation and residual analysis, obtaining the model's best fit of 92.53%. The following discrete-time polynomials were obtained as the estimation results around 25°C (see Figure 8.5) [40,41]:

$$
\begin{aligned}
A(z^{-1}) &= 1 - 0.3682z^{-1} + 0.0001z^{-2},\\
B(z^{-1})z^{-d} &= (-0.0402 - 0.0027z^{-1})z^{-1},\\
D_1(z^{-1})z^{-d_{D1}} &= (0.1989 + 0.0924z^{-1} + 0.1614z^{-2})z^{-2},\\
D_2(z^{-1})z^{-d_{D2}} &= (0.0001 + 0.0067z^{-1} + 0.0002z^{-2})z^{-1},\\
D_3(z^{-1})z^{-d_{D3}} &= (-0.0002 - 0.3618z^{-1} + 0.0175z^{-2})z^{-1},\\
D_4(z^{-1})z^{-d_{D4}} &= (0.0525 + 0.3306z^{-1} + 0.0058z^{-2})z^{-1}.
\end{aligned}
$$

For the controller configurations, which do not consider measurable disturbances, a typical input–output configuration based on the previous model is used with the following structure:

$$A(z^{-1})y(k) = z^{-d}B(z^{-1})u(k-1) + \frac{\varepsilon(k)}{\Delta}. \tag{8.33}$$

### 8.6.2 Performance Indexes

To compare the classical GPC and event-based GPC approaches, specific performance indexes for this type of control strategies were considered [46]. As a first measure, the integrated absolute error (IAE) is used to evaluate the control accuracy, as follows:

$$IAE = \int_0^\infty |e(t)|dt,$$

where $e(t)$ is the difference between the reference $r(t)$ and the process output $y(t)$. The IAEP compares the event-based control with the time-based control used as a reference:

$$IAEP = \int_0^\infty |y_{EB}(t) - y_{TB}(t)|dt,$$

where $y_{TB}(t)$ is the response of the time-based classical GPC. An efficiency measure index for event-based control systems can be defined as

$$NE = \frac{IAEP}{IAE}.$$

### 8.6.3 Sensor Deadband Approach

The event-based GPC control structure with sensor deadband was implemented with $T_{base} = 1$ minute, $T_{max} = 4$, $n_{max} = 4$, and thus $T_f \in [1,2,3,4]$. The control horizon was selected to $N_u^{n_{max}} = 5$ samples for all GPC controllers such as described by (8.11). The prediction horizon was set to $N_2^c = 20$ minutes in continuous time, and the rule (8.10) was used to calculate the equivalent prediction horizon for each controller with a different sampling period. Finally, the control weighting factor was adjusted to $\lambda^f = 1$ to achieve the desired closed-loop dynamics for the faster GPC controller with $T_1 = T_{base} = 1$. Due to the physical limitation of the actuator, the event-based GPC controller considers constraints on the control signal $0 < u(k) < 100\%$. Taking into account all design parameters, the control structure is composed of four controllers for each possible sampling frequency.

Furthermore, the $\beta_y$ effect parameter was again evaluated for different values, $\beta_y = [0.1, 0.2, 0.5, 0.75, 1]$. A periodic GPC with a sampling time of 1 minute was also simulated for comparison. The simulation was made for 19 days, and the numerical results are summarized in Table 8.1. It can be seen how the proposed algorithm allows one to have similar performance to the periodic GPC, but with an important reduction in events (changes in the control signal, which, in this case, are vent movements), especially where $\beta_y = 0.1$ and $\beta_y = 0.2$. Since a graphical result for 19 days will not allow one to see the results properly, representative days have been selected, which are shown in Figures 8.6 and 8.7, respectively.

These figures show the obtained results for the event-based GPC control structure with different values of $\beta_y$ and for the periodic GPC controller with a sampling time of 1 minute. As observed in Table 8.1, the classical implementation obtains the best control performance. However, the event-based GPC control structure with $\beta = 0.1$ and $\beta = 0.2$ presents almost the same performance results as the classical configuration but with an important reduction in actuations.

**TABLE 8.1**
Summary of the Performance Indexes for Event-Based GPC with Sensor Deadband Approach

| | IAE | IAEP | $E_y$ | NE | Δ IAE [%] | $\Delta E_y$ [%] |
|---|---|---|---|---|---|---|
| TB | 2275 | — | 5173 | — | — | — |
| $\beta_y = 0.1$ | 2292 | 715 | 4383 | 0.31 | 0.7 | −15.3 |
| $\beta_y = 0.2$ | 2343 | 727 | 3829 | 0.31 | 3.0 | −26.0 |
| $\beta_y = 0.5$ | 2799 | 1013 | 2656 | 0.36 | 23.0 | −48.7 |
| $\beta_y = 0.75$ | 3283 | 1457 | 2161 | 0.44 | 44.3 | −58.0 |
| $\beta_y = 1$ | 3391 | 1648 | 1828 | 0.49 | 86.3 | −64.67 |

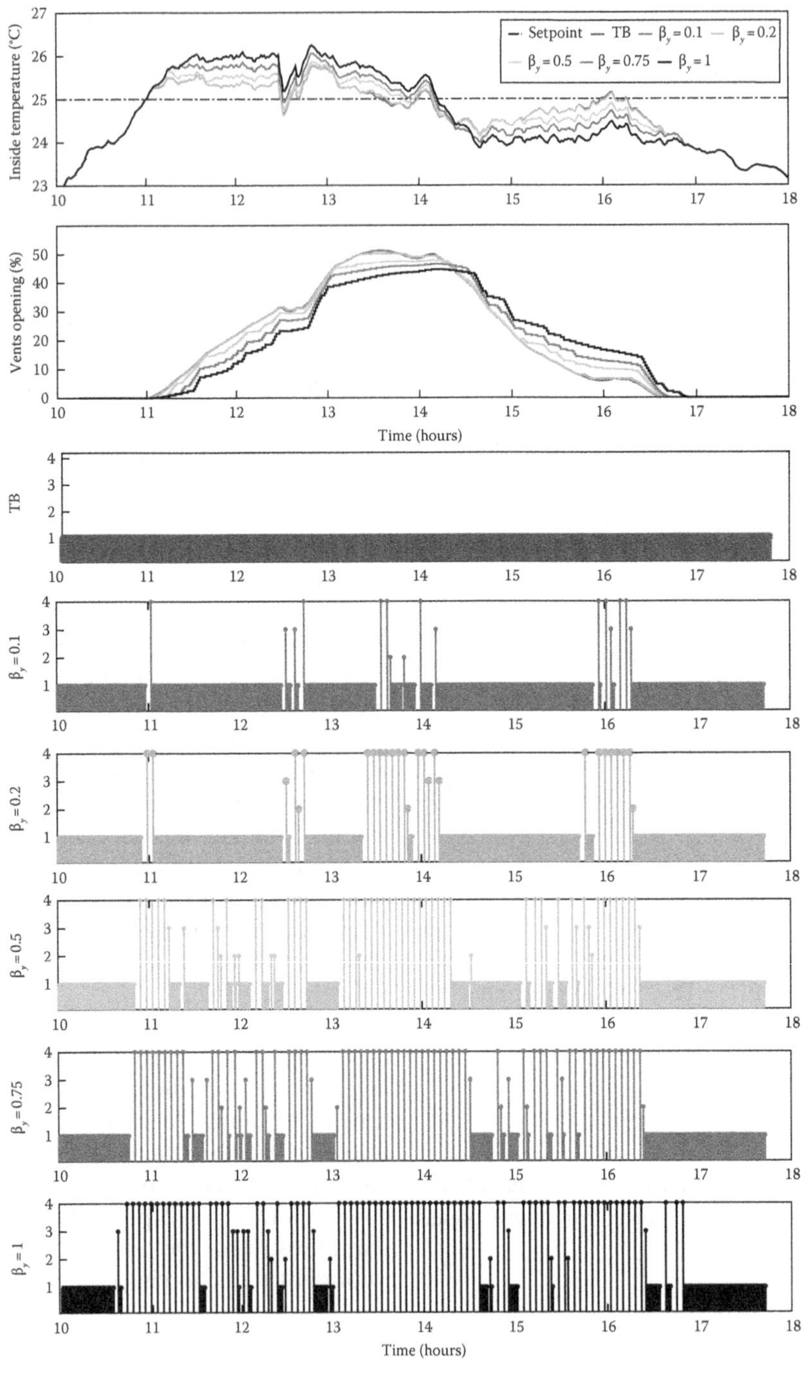


**FIGURE 8.6**
Control results for the second day for event-based GPC with sensor deadband.

The event-based control structure with $\beta = 0.5$ presents a promising trade-off between control performance and the number of actuations for this process. The evolution of events is shown on the bottom graph, where the signal amplitude corresponds to the controller, which calculates the control signal for this event. From the obtained results for the event-based GPC control structure with $\beta = 0.75$ and $\beta = 1$, it can be deduced that for the process, where some tolerance in setpoint tracking is permitted, an important reduction of actuations can be obtained. In the case of the greenhouse interior temperature control process, this reduction can be translated to economic savings, reducing the electromechanical actuator wastage, and saving electrical energy.

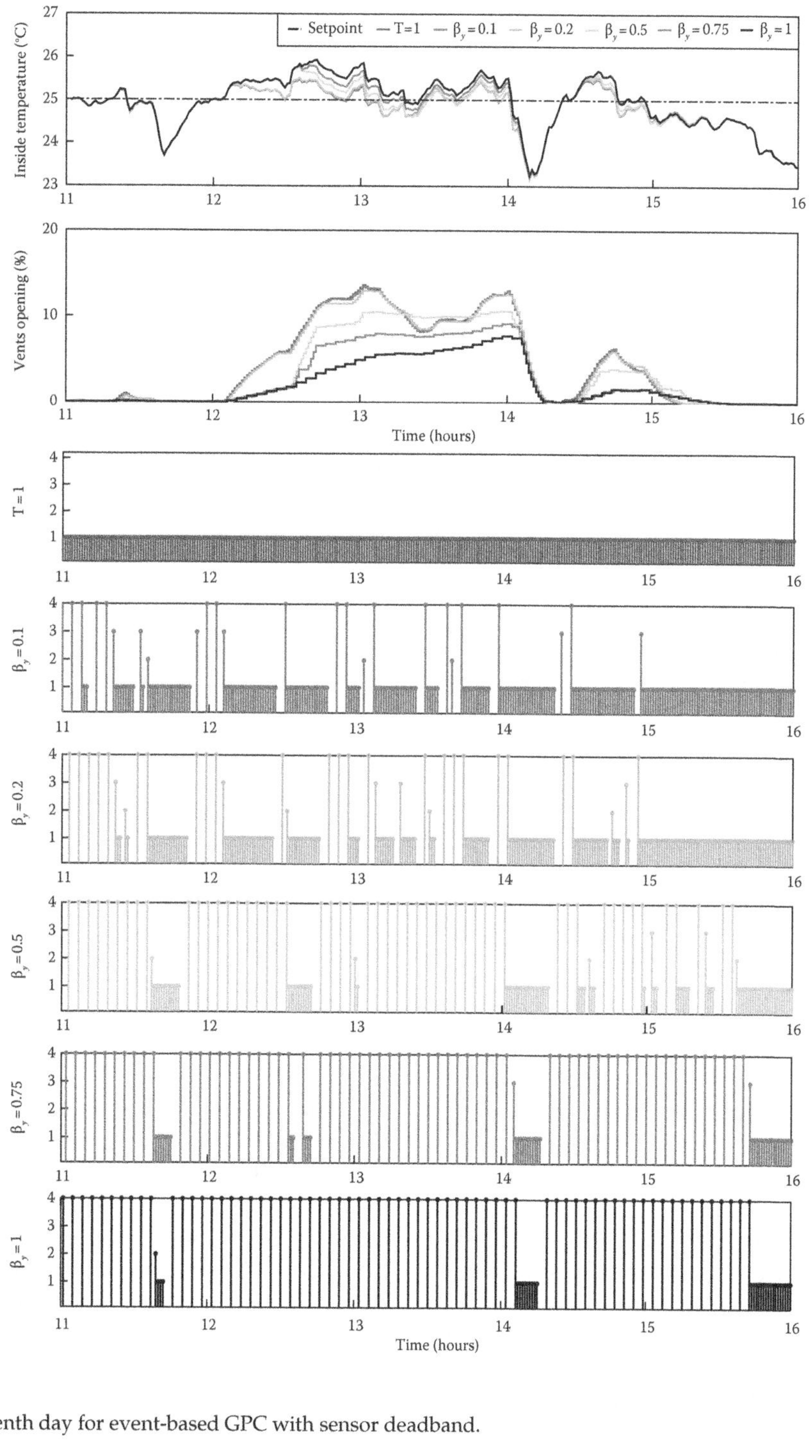


**FIGURE 8.7**
Control results for the seventh day for event-based GPC with sensor deadband.

### 8.6.4 Actuator Deadband Approach

For this event-based GPC configuration, simulations were performed for the following system parameters: the prediction horizon was set to $N_2 = 10$, the control horizon was set to $N_u = 5$, and the control weighting factor was adjusted to $\lambda = 1$ to achieve the desired closed-loop dynamics. The minimum and maximum control signal increments of the vents opening percentage were set to $m = -20\%$ and $M = 20\%$. Additionally, due to the physical limitation of the actuator, the event-based GPC controller considered constraints on the control signal $0 < u(k) < 100\%$. In this configuration, the greenhouse process controlled variable is sampled with a fixed nominal sampling time of 1 minute, and the process is updated in the event-based mode

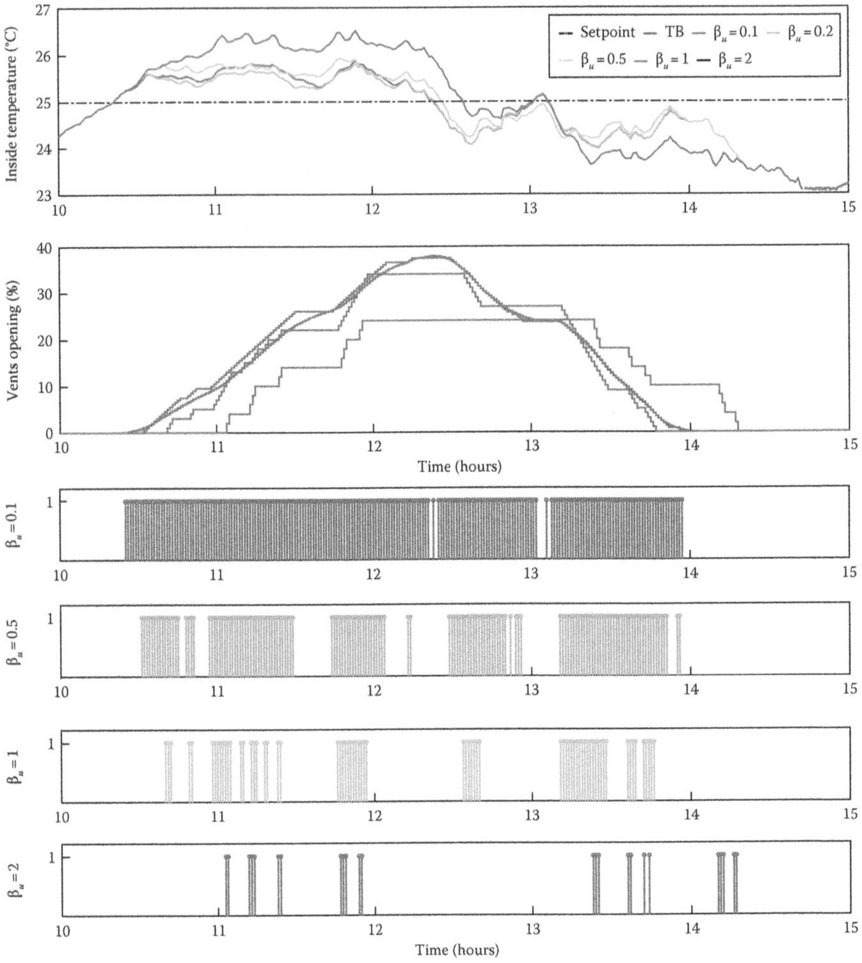


**FIGURE 8.8**
Simulation results for the eighth day for event-based GPC with actuator deadband.

taking into account the actuator deadband in MIQP optimization. The actuator virtual deadband was set to $\beta_u = [0.1, 0.5, 1, 2]$ to check its influence on the control performance.

Figures 8.8 and 8.9 show control results for a clear day and for a day with passing clouds, respectively. It can be observed that for both days, the event-based controller with small values of actuator deadband $\beta_u = 0.1$ and $\beta_u = 0.5$ obtained almost the same performance as the time-based GPC. Control results for the eighth day, where the deadband was set to $\beta_u = 2$, are characterized by poor control performance due to the response offset introduced by the actuator deadband and the low variability of the controlled variable. Control results for the 15th day for the same deadband obtained a performance comparable with other deadband configurations, despite disturbance dynamics. The bottom graphs of Figures 8.8 and 8.9 show event occurrence for different values of $\beta_u$. It can be observed that the number of events depends on the actuator deadband value and the disturbance dynamics that affect the greenhouse interior temperature. It can be observed that for the eighth day and the central part of the day between 12 and 14 hours, the number of events is reduced for all deadband values. Due to process inertia in this time range, a process controlled by an event-based GPC with an actuator deadband needs only a few control signal updates to achieve the desired setpoint. However, a time-based GPC process is updated every sampling instant, even if the error is very small. The same characteristic can be observed in Figure 8.9, where between 14 and 16 hours the event-based controllers produce fewer process updates, and the resulting control performance is kept at a good level.

Table 8.2 collects performance indexes for the event-based GPC with an actuator deadband. As can be seen, even a small actuator deadband $\beta_u = 0.1$ results in an important reduction of process input events $E_u$, where savings of 9.8% are achieved compared with time-based GPC. In this case, the $IAE$ increases by only 0.1%,

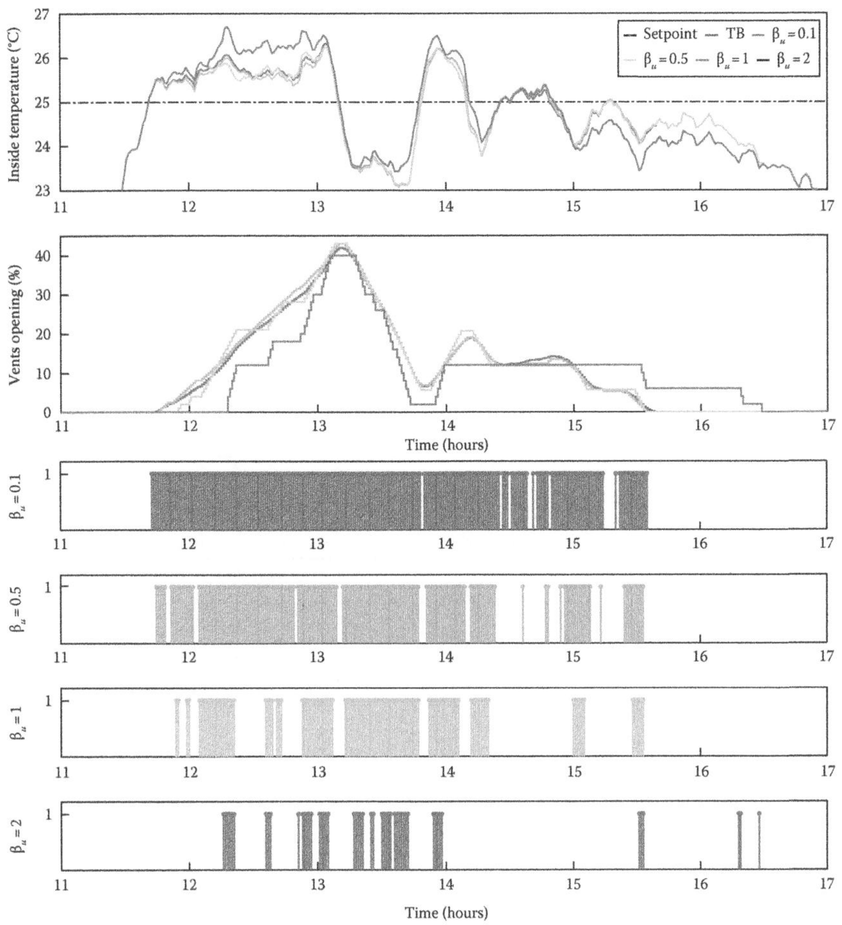


**FIGURE 8.9**
Simulation results for the 15th day for event-based GPC with actuator deadband.

**TABLE 8.2**
Summary of the Performance Indexes for Event-Based GPC with Actuator Deadband Approach

| | IAE | IAEP | $E_u$ | NE | Δ IAE [%] | $\Delta E_u$ [%] |
|---|---|---|---|---|---|---|
| TB | 2275 | — | 5173 | — | — | — |
| $\beta_u = 0.1$ | 2277 | 12 | 4665 | 0.01 | 0.1 | −9.8 |
| $\beta_u = 0.5$ | 2352 | 300 | 2694 | 0.13 | 3.4 | −47.9 |
| $\beta_u = 1$ | 2637 | 845 | 1517 | 0.32 | 15.9 | −74.5 |
| $\beta_u = 2$ | 4457 | 2894 | 453 | 0.65 | 95.9 | −91.0 |

and $IAEP$ values for individual days are close to zero, which means that the obtained control performance is the same as a time-based technique. Interesting results are obtained for $\beta_u = 1$, where $IAE$ increases by about 15.9%, but an events reduction of the order of 74% is obtained. It is observed that a direct relationship can be established between the actuator deadband value and the control performance. When the deadband increases, the control performance decreases. A similar relationship can be found between the number of process input events and the deadband value. However, for $\beta_u = 2$, the obtained control results are characterized by important precision loss, and for this reason, the optimal deadband value depends on the individual process characteristics. This facts allows one to use a $\beta_u$ as an additional tuning parameter to establish a relationship between control performance and the number of system events. Overall, good control performance, even for a large value of $\beta_u$, is achieved because the actuator deadband is considered in the optimization procedure. This fact is verified by the $NE$ measurement. On the contrary, disturbance dynamics have a strong influence on the control performance and the number of events that are obtained for individual simulation days.

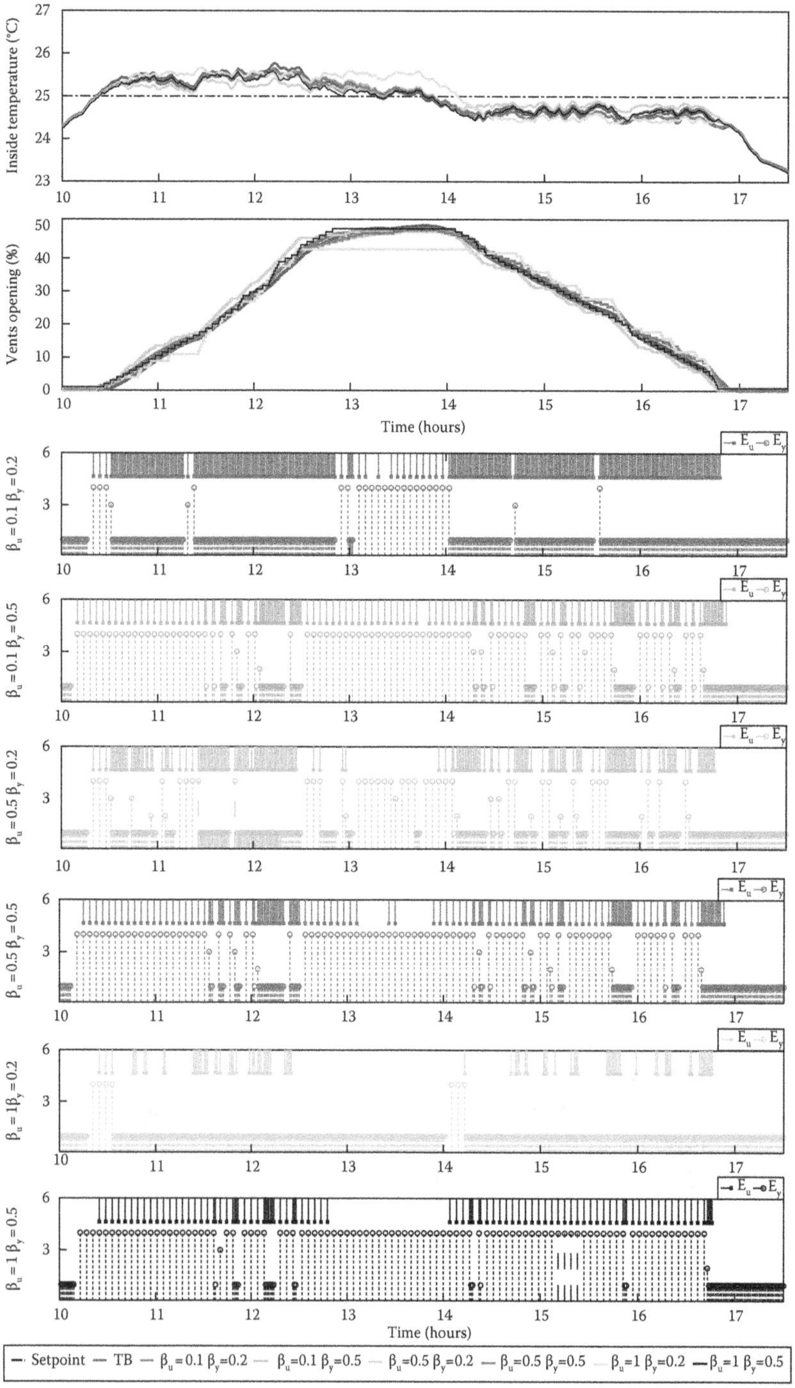


**FIGURE 8.10**
Control results for the 14th day for the complete event-based GPC scheme.

### 8.6.5 Complete Event-Based GPC

The last analyzed configuration considers actuator and sensor deadbands at the same time and forms the complete event-based GPC control structure. In this case, the control system configuration is as follows: $T_{base} =$ 1 minute, $T_{max} = 4$, $n_{max} = 4$, and thus $T_f \in [1, 2, 3, 4]$, the control horizon was selected to $N_u^{n_{max}} = 5$ samples, the prediction horizon was set to $N_2^c = 20$, and the control signal weighting factor was adjusted to $\lambda^f = 1$. Taking into account all the design requirements, the control structure is composed of four controllers for each possible sampling frequency. Additionally, each controller considers actuator deadband constraints in

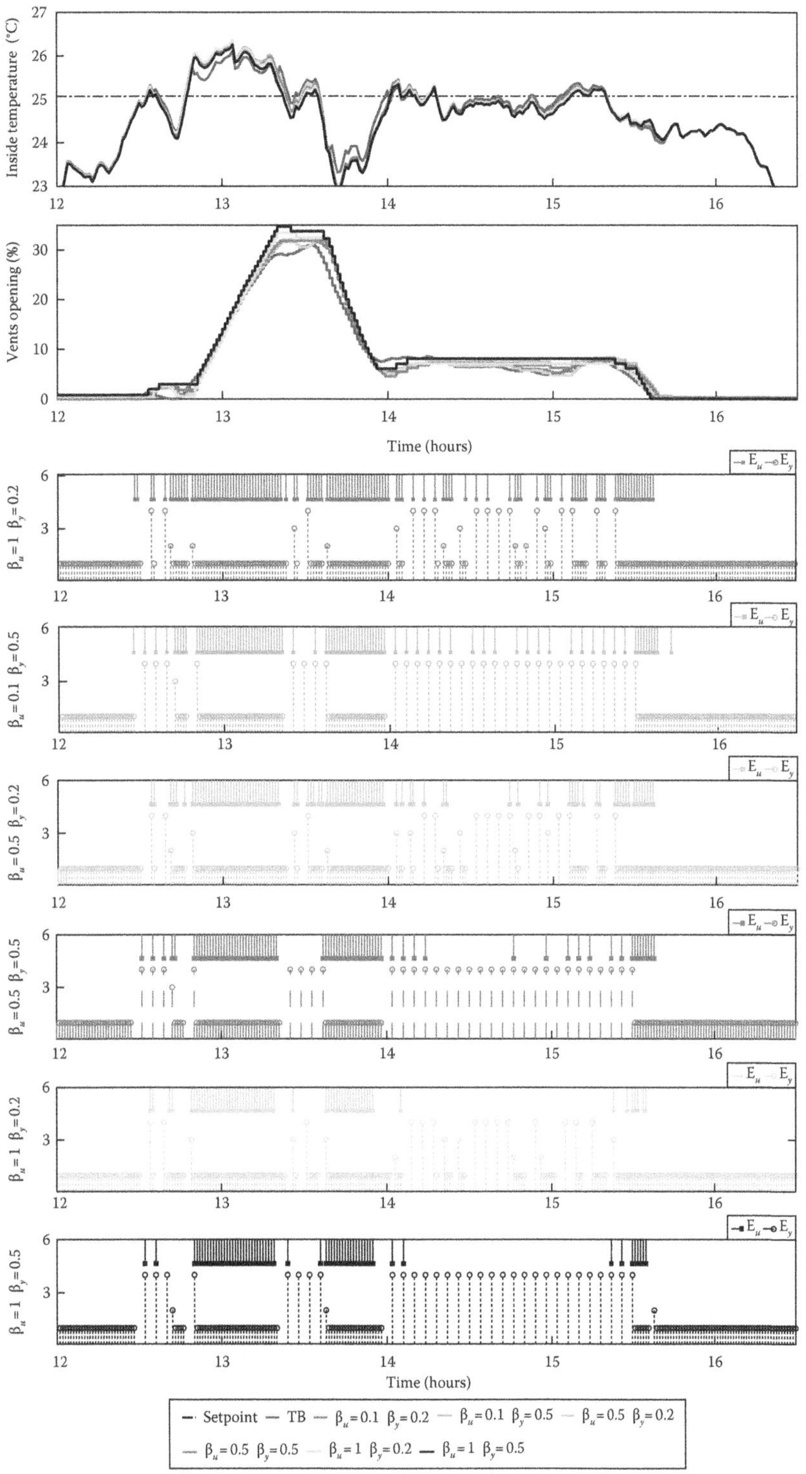


**FIGURE 8.11**
Control results for the 16th day for the complete event-based GPC scheme.

MIQP optimization. In this configuration, the actuator virtual deadband was set to $\beta_u = [0.1, 0.5, 1]$, and the sensor deadband $\beta_y = [0.2, 0.5]$. The selected deadband values were chosen considering previous simulation results in such a way that only those with good control performance were selected to check its influence on the control performance for complete event-based GPC.

Six different simulations have been performed for all combinations of sensor and actuator deadbands.

**TABLE 8.3**
Performance Indexes for the Complete Event-Based GPC Scheme

| | TB | | $\beta_u = 0.1$ | | | | | | $\beta_u = 0.5$ | | | | | | $\beta_u = 1$ | | | | | |
|---|---|---|---|---|---|---|---|---|---|---|---|---|---|---|---|---|---|---|---|---|
| | | | $\beta_y = 0.2$ | | | $\beta_y = 0.5$ | | | $\beta_y = 0.2$ | | | $\beta_y = 0.5$ | | | $\beta_y = 0.2$ | | | $\beta_y = 0.5$ | | |
| **Day** | **IAE** | $E_y = E_u$ | **IAE** | $E_y$ | $E_u$ | **IAE** | $E_y$ | $E_u$ | **IAE** | $E_y$ | $E_u$ | **IAE** | $E_y$ | $E_u$ | **IAE** | $E_y$ | $E_u$ | **IAE** | $E_y$ | $E_u$ |
| 1 | 140 | 137 | 135 | 166 | 166 | 136 | 158 | 158 | 153 | 166 | 165 | 137 | 158 | 157 | 122 | 166 | 136 | 164 | 137 | 133 |
| 2 | 78 | 240 | 94 | 186 | 169 | 100 | 121 | 112 | 108 | 175 | 136 | 101 | 121 | 103 | 104 | 197 | 76 | 142 | 121 | 86 |
| 3 | 96 | 272 | 103 | 224 | 214 | 115 | 134 | 134 | 107 | 195 | 154 | 112 | 136 | 125 | 144 | 285 | 89 | 138 | 128 | 100 |
| 4 | 106 | 272 | 110 | 229 | 228 | 120 | 163 | 161 | 114 | 214 | 176 | 120 | 160 | 145 | 118 | 246 | 100 | 151 | 136 | 107 |
| 5 | 16 | 81 | 20 | 80 | 69 | 22 | 62 | 57 | 21 | 77 | 55 | 21 | 64 | 57 | 21 | 77 | 44 | 21 | 62 | 48 |
| 6 | 207 | 424 | 197 | 370 | 357 | 211 | 265 | 262 | 204 | 370 | 300 | 212 | 261 | 245 | 185 | 381 | 158 | 224 | 219 | 173 |
| 7 | 146 | 326 | 141 | 298 | 292 | 163 | 195 | 192 | 137 | 294 | 230 | 160 | 194 | 183 | 163 | 331 | 123 | 175 | 152 | 129 |
| 8 | 84 | 176 | 91 | 186 | 186 | 97 | 119 | 118 | 101 | 172 | 145 | 97 | 121 | 116 | 100 | 196 | 92 | 131 | 107 | 97 |
| 9 | 277 | 385 | 250 | 352 | 338 | 260 | 266 | 255 | 261 | 343 | 286 | 259 | 264 | 237 | 234 | 353 | 196 | 273 | 247 | 201 |
| 10 | 103 | 321 | 109 | 244 | 236 | 123 | 157 | 154 | 114 | 231 | 173 | 124 | 154 | 132 | 127 | 267 | 92 | 150 | 138 | 99 |
| 11 | 98 | 311 | 105 | 257 | 249 | 117 | 144 | 141 | 101 | 221 | 172 | 115 | 141 | 129 | 157 | 331 | 88 | 141 | 130 | 102 |
| 12 | 53 | 181 | 68 | 159 | 156 | 79 | 88 | 87 | 76 | 146 | 115 | 80 | 89 | 80 | 80 | 164 | 62 | 125 | 98 | 69 |
| 13 | 107 | 311 | 112 | 265 | 263 | 127 | 154 | 154 | 110 | 231 | 182 | 125 | 153 | 142 | 151 | 312 | 101 | 152 | 132 | 104 |
| 14 | 131 | 361 | 129 | 332 | 330 | 163 | 175 | 174 | 114 | 271 | 209 | 161 | 171 | 162 | 198 | 382 | 104 | 176 | 136 | 118 |
| 15 | 152 | 211 | 150 | 213 | 212 | 153 | 165 | 164 | 159 | 210 | 191 | 154 | 165 | 156 | 148 | 207 | 148 | 192 | 159 | 143 |
| 16 | 91 | 161 | 102 | 143 | 137 | 104 | 113 | 109 | 119 | 142 | 117 | 104 | 113 | 98 | 107 | 145 | 86 | 145 | 112 | 88 |
| 17 | 117 | 271 | 119 | 242 | 240 | 124 | 170 | 169 | 128 | 239 | 196 | 122 | 168 | 163 | 119 | 248 | 108 | 149 | 140 | 115 |
| 18 | 83 | 281 | 95 | 216 | 206 | 103 | 133 | 129 | 107 | 214 | 166 | 103 | 134 | 116 | 103 | 230 | 71 | 146 | 129 | 89 |
| 19 | 189 | 451 | 178 | 368 | 363 | 210 | 267 | 261 | 182 | 360 | 267 | 208 | 268 | 246 | 215 | 439 | 154 | 203 | 210 | 164 |
| $\Sigma$ | 2275 | 5173 | 2307 | 4530 | 4411 | 2525 | 3049 | 2991 | 2416 | 4271 | 3435 | 2516 | 3035 | 2792 | 2597 | 4957 | 2028 | 2998 | 2693 | 2165 |
| | | $\Delta$ [%] | 1.4 | −12.4 | −14.7 | 11.0 | −41.1 | −42.2 | 6.2 | −17.4 | −33.6 | 10.6 | −41.3 | −46.0 | 14.2 | −4.2 | −60.8 | 31.8 | −47.9 | −58.1 |

The obtained results for selected days and analyzed event-based GPC configuration are shown in Figures 8.10 and 8.11. As can be observed, of all the combinations, the complete event-based GPC obtains good control performance comparable with the time-based (TB) GPC controller. For the 14th day, characterized by disturbances with low changing dynamics, it can be observed that the application of sensor and actuator deadbands allows one to achieve the desired trade-off between control accuracy and the number of system events. The bottom graphs in Figure 8.10 show the event occurrence, where actuator $E_u$ and sensor $E_y$ events are shown. It can be observed that deadband values have an important influence on the number of the event, as in the previous configuration. However, in this case, the number of input and output events are reduced simultaneously. The same characteristic is obtained for the second analyzed day. On this day, special attention should be given to the time period between 13 and 14 hours, when transients in solar radiation occur. Analyzing the event occurrence, it can be observed that the event-based GPC controller works with the fastest sampling, and for each output, the event controller produces a new control signal value.

This behavior is the same for all deadband values used in the event-based GPC structure, due to an important error between the inside temperature and the reference. The opposite effect can be found between 14 and 15 hours, when the process reaches a steady state due to low disturbance variation. In this time period, the process output is monitored with slow sampling, and the majority of the event-based controllers uses a $T_{max}$ sampling frequency. The number of system update events $E_u$ depends on the actuator deadband value, and for this time period the value is considerably reduced. The event-based GPC controller configuration with $\beta_u = 1$ and $\beta_y = 0.5$ produces only four system updates and keeps the control performance close to the TB configuration.

These results are confirmed by the performance indexes used for evaluation of the complete event-based GPC controller presented in Table 8.3. The analyzed configuration is characterized by relatively good control performance and obtains minimum and maximum $IAE$ of between 1.4% and 33.6% higher than for the TB configuration. The best trade-off between control accuracy and the number of events is obtained for $\beta_y = 0.5$ with $\beta_u = 0.5$. For this case, $IAE$ increases by about 10.6%, where output $E_y$ and input $E_u$ events were reduced by about 41.3% and 46%, respectively. As in previous event-based GPC configurations, deadband values determine overall control performance, and their selection should be carried out individually for each controlled process.

## 8.7 Summary

This chapter presented a straightforward algorithm to implement a GPC controller with event-based capabilities. The core idea of the proposed control scheme was focused on the sensor and actuator deadbands. Three configurations for an event-based GPC control structure were developed considering a sensor deadband approach, an actuator deadband approach, as well as the complete event-based scheme. In the developed event-based GPC configurations, the deadband values were used as additional tuning parameters to achieve the desired performance.

The first strategy considers the sensor deadband approach, where the resulting control algorithm allows one to use standard GPC controllers in such a way that only resampling techniques have to be used to calculate the past process output and control signal values when events are detected. Then, this resampled information was used to calculate the control law for the current sampling period. Simple tuning rules were also proposed for GPC controllers to be used within this event-based framework, where such tuning rules were tested for an important number of industrial process models. Furthermore, it was shown how it is possible to achieve an important reduction in the control signal changes without losing too much performance quality. The second approach takes into account the actuator deadband in such a way that only control signal changes higher than an established deadband limit are considered. The resulting control algorithm uses a MIQP optimization to compute control actions for the controlled process, and small control changes are suppressed to zero in accordance with the established deadband. The developed event-based control system realizes process updates only when an important change is detected in the controlled process. Otherwise, the control signal values remain unchanged. The advantages presented by this approach were obtained at the cost of a small control performance deterioration and of solving a computationally costly online optimization algorithm. Furthermore, this proposal can be applied to control any system that can be controlled with a classical GPC algorithm. The last configuration for the event-based GPC controller considered the sensor deadband and the actuator deadband simultaneously. The resulting complete event-based GPC control scheme was characterized by a good trade-off between the control performance and the number of events. For this configuration, both deadband values were adjusted independently in order to obtain the desired performance and the number of events for input and output of the process.

Finally, the chapter concluded with a simulation study, where the event-based control schemes were evaluated for the greenhouse temperature control problem. In this case, the simulations were carried out considering real data, where it was possible to see the advantages of the proposed event-based GPC controllers. For each evaluated event-based GPC configuration, it was possible to reduce the control signal changes, which in this case were associated with actuator wastage and economic costs, maintaining acceptable performance results.

## Acknowledgments

This work has been partially funded by the following projects and institutions: DPI2011-27818-C02-01/02 (financed by the Spanish Ministry of Economy and Competitiveness and ERDF funds); the UNED through postdoctoral scholarship; and supported by Cajamar Foundation.

## Bibliography

[1] P. Albertos, M. Vallés, and A. Valera. Controller transfer under sampling rate dynamic changes. In *Proceedings of the European Control Conference*, Cambridge, UK, 2003.

[2] A. Anta and P. Tabuada. To sample or not to sample: Self-triggered control for nonlinear systems. *IEEE Transactions on Automatic Control*, 55(9):2030–2042, 2010.

[3] K. E. Årzén. A simple event-based PID controller. In *Proceedings of 14th IFAC World Congress*, Beijing, China, 1999.

[4] K. J. Åström. Event based control. In *Analysis and Design of Nonlinear Control Systems* (A. Astolfi and L. Marconi, Eds.), pp. 127–148. Springer-Verlag, Berlin, Germany, 2007.

[5] K. J. Åström and B. M. Bernhardsson. Comparison of Riemann and Lebesgue sampling for first order stochastic systems. In *Proceedings of the 41st IEEE Conference on Decision and Control*, Las Vegas, NV, 2002.

[6] K. J. Åström and B. Wittenmark. *Computer Controlled Systems: Theory and Design*. Prentice Hall, Englewood Cliffs, NJ, 1997.

[7] A. Bemporad, S. Di Cairano, E. Henriksson, and K. H. Johansson. Hybrid model predictive control based on wireless sensor feedback: An experimental study. *International Journal of Robust and Nonlinear Control*, 20(2):209–225, 2010.

[8] A. Bemporad, S. Di Cairano, and J. Júlvez. Event-based model predictive control and verification of integral continuous hybrid automata. In *Hybrid Systems: Computation and Control* (J. Hespanha and A. Tiwari, Eds.), Lecture Notes in Computer Science. Springer-Verlag, Berlin, Germany, 2006.

[9] A. Bemporad and M. Morari. Control of systems integrating logic, dynamics, and constraints. *Automatica*, 35(3):407–427, 1999.

[10] E. F. Camacho and C. Bordóns. *Model Predictive Control*. Springer-Verlag, London, 2007.

[11] Y. Can, Z. Shan-an, K. Wan-zeng, and L. Li-ming. Application of generalized predictive control in networked control system. *Journal of Zhejiang University*, 7(2):225–233, 2006.

[12] A. Cervin and K. J. Åström. On limit cycles in event-based control systems. In *Proceedings of the 46th IEEE Conference on Decision and Control*. New Orleans, LA, 2007.

[13] A. Cervin, D. Henriksson, B. Lincoln, J. Eker, and K. E. Årzén. How does control timing affect performance? *IEEE Control Systems Magazine*, 23(3):16–30, 2003.

[14] D. Chen, N. Xi, Y. Wang, H. Li, and X. Tang. Event based predictive control strategy for teleoperation via Internet. In *Proceedings of the Conference on Advanced Inteligent Mechatronics*. Xi'an, China, 2008.

[15] S. Durand and N. Marchand. Further results on event-based PID controller. In *Proceedings of the European Control Conference*. Budapest, Hungary, 2009.

[16] M. S. Fadali and A. Visioli. *Digital Control Engineering—Analysis and Design*. Academic Press, Burlington, VT, 2009.

[17] I. Farkas. Modelling and control in agricultural processes. *Computers and Electronics in Agriculture*, 3(49):315–316, 2005.

[18] P. J. Gawthrop and L. Wang. Intermittent model predictive control. *Journal of Systems and Control Engineering*, 221(7):1007–1018, 2007.

[19] G. Guo. Linear systems with medium-access constraint and Markov actuator assignment. *IEEE Transaction on Circuits and Systems*, 57(11):2999–3010, 2010.

[20] J. L. Guzmán, M. Berenguel, and S. Dormido. Interactive teaching of constrained generalized predictive control. *IEEE Control Systems Magazine*, 25(2):52–66, 2005.

[21] J. L. Guzmán, F. Rodríguez, M. Berenguel, and S. Dormido. Virtual lab for teaching greenhouse climatic control. In *Proceedings of the 6th IFAC World Congress*, Prague, Czech Republic, 2005.

[22] W. P. M. H. Heemels, J. H. Sandee, and P. P. J. Van Den Bosch. Analysis of event-driven controllers for linear systems. *International Journal of Control*, 4(81):571–590, 2008.

[23] T. Henningsson and A. Cervin. Comparison of LTI and event-based control for a moving cart with quantized position measurements. In *Proceedings of the European Control Conference*, Budapest, Hungary, 2009.

[24] T. Henningsson, E. Johannesson, and A. Cervin. Sporadic event-based control of first-order linear stochastic systems. *Automatica*, 44(11):2890–2895, 2008.

[25] W. Hu, G. Liu, and D. Rees. Event-driven networked predictive control. *IEEE Transactions on Industrial Electronics*, 54(3):1603–1613, 2007.

[26] J. Jugo and M. Eguiraun. Experimental implementation of a networked input-output sporadic control system. In *Proceedings of the IEEE International Conference on Control Applications*, Yokohama, Japan, 2010.

[27] R. King and N. Sigrimis. Computational intelligence in crop production. *Computers and Electronics in Agriculture—Special Issue on Intelligent Systems in Crop Production*, 31(1):1–3, 2000.

[28] E. Kofman and J. Braslavsky. Level crossing sampling in feedback stabilization under data rate constraints. In *Proceedings of the 45th IEEE International Conference on Decision and Control*, San Diego, CA, 2006.

[29] D. Lehmann and J. Lunze. Event-based control: A state-feedback approach. In *Proceedings of the European Control Conference*, Budapest, Hungary, 2009.

[30] G. Liu, Y. Xia, J. Chen, D. Rees, and W. Hu. Networked predictive control of systems with random network delays in both forward and feedback channels. *IEEE Transaction on Industrial Electronics*, 54(3):1603–1613, 2007.

[31] J. Liu, D. Muñoz de la Peña, and P. D. Christofides. Distributed model predictive control of nonlinear systems subject to asynchronous and delayed measurements. *Automatica*, 46(1):52–61, 2010.

[32] N. Marchand. Stabilization of Lebesgue sampled systems with bounded controls: The chain of integrators case. In *Proceedings of the 17th World Congress of IFAC*, Seoul, Korea, 2008.

[33] M. Miskowicz. Send-on-delta concept: An event-based data reporting strategy. *Sensors*, 6(1):49–63, 2006.

[34] M. Nagahara and D. E. Quevedo. Spare representation for packetized predictive networked control. In *Proceedings of the 18th IFAC World Congress*, Milano, Italy, 2011.

[35] J. E. Normey-Rico and E. F. Camacho. *Control of Dead-Time Processes*. Springer-Verlag. London, 2007.

[36] A. Pawlowski, A. Cervin, J. L. Guzmán, and M. Berenguel. Generalized predictive control with actuator deadband for event-based approaches. *IEEE Transactions on Industrial Informatics*, 10(1):523–537, 2014.

[37] A. Pawlowski, I. Fernández, J. L. Guzmán, M. Berenguel, F. G. Acién, and J. E. Normey-Rico. Event-based predictive control of pH in tubular photobioreactors. *Computers and Chemical Engineering*, 65:28–39, 2014.

[38] A. Pawlowski, J. L. Guzmán, J. E. Normey-Rico, and M. Berenguel. Improving feedforward disturbance compensation capabilities in generalized predictive control. *Journal of Process Control*, 22(3):527–539, 2012.

[39] A. Pawlowski, J. L. Guzmán, J. E. Normey-Rico, and M. Berenguel. A practical approach for generalized predictive control within an event-based framework. *Computers and Chemical Engineering*, 41(6):52–66, 2012.

[40] A. Pawlowski, J. L. Guzmán, F. Rodríguez, M. Berenguel, and J. E. Normey-Rico. Predictive control with disturbance forecasting for greenhouse diurnal temperature control. In *Proceedings of the 18th World Congress of IFAC*, Milan, Italy, 2011.

[41] A. Pawlowski, J. L. Guzmán, F. Rodríguez, M. Berenguel, and J. Sánchez. Application of time-series methods to disturbance estimation in predictive control problems. In *Proceedings of the IEEE Symposium on Industrial Electronics*, Bari, Italy, 2010.

[42] A. Pawlowski, J. L. Guzmán, F. Rodríguez, M. Berenguel, J. Sánchez, and S. Dormido. Event-based control and wireless sensor network for greenhouse diurnal temperature control: A simulated case study. In *Proceedings of the 13th IEEE Conference on Emerging Technologies and Factory Automation*, Hamburg, Germany, 2008.

[43] A. Pawlowski, J. L. Guzmán, F. Rodríguez, M. Berenguel, J. Sánchez, and S. Dormido. The influence of event-based sampling techniques on data transmission and control performance. In *Proceedings of the 14th IEEE International Conference on Emerging Technologies and Factory Automation*, Mallorca, Spain, 2009.

[44] A. Pawlowski, J. L. Guzmán, F. Rodríguez, M. Berenguel, J. Sánchez, and S. Dormido. Study of event-based sampling techniques and their influence on greenhouse climate control with wireless sensor networks. In *Javier Silvestre-Blanes Edt. Factory Automation*, pages 289–312. InTech, Vukovar, Croatia, 2010.

[45] B. Picasso, S. Pancanti, A. Bemporad, and A. Bicchi. Receding-horizon control of LTI systems with quantized inputs. In *Analysis and Design of Hybrid Systems* (G. Engell and Zaytoon, Eds.), pages 259–264. Elsevier, Oxford, UK, 2003.

[46] J. Ploennigs, V. Vasyutynskyy, and K. Kabitzsch. Comparative study of energy-efficient sampling approaches for wireless control networks. *IEEE Transactions on Industrial Informatics*, 6(3):416–424, 2010.

[47] D. E. Quevedo, J. Østergaard, and D. Nešić. Packetized predictive control of stochastic systems over bit-rate limited channels with packet loss. *IEEE Transactions on Automatic Control*, 56(12):2855–2868, 2011.

[48] M. Rabi and J .S. Baras. Level-triggered control of a scalar linear system. In *Proceedings of the 15th IEEE Mediterranean Conference on Control and Automation*, Athens, Greece, 2007.

[49] M. Rabi and K. H. Johansson. Event-triggered strategies for industrial control over wireless networks. In *Proceedings of the 4th Annual International Conference on Wireless Internet*, Maui, HI, 2008.

[50] J. A. Rossiter. *Model Based Predictive Control: A Practical Approach*. CRC Press, Boca Raton, FL, 2003.

[51] J. Sánchez, A. Visoli, and S. Dormido. A two-degree-of-freedom PI controller based on events. *Journal of Process Control*, 21(4):639–651, 2011.

[52] J. H. Sandee, W. P. M. H. Heemels, and P. P. J. van den Bosch. Event-driven control as an opportunity in the multidisciplinary development of embedded controllers. In *Proceedings of the American Control Conference*, Portland, OR, 2005.

[53] N. Sigrimis, P. Antsaklis, and P. Groumpos. Control advances in agriculture and the environment. *IEEE Control System Magazine*, 21(5):8–12, 2001.

[54] M. Srinivasarao, S. C. Patwardhan, and R. D. Gudi. Nonlinear predictive control of irregularly sampled multirate systems using blackbox observers. *Journal of Process Control*, 17(1):17–35, 2007.

[55] S. W. Su, H. Nguyen, and R. Jarman. Model predictive control of gantry crane with input nonlinearity compensation. *International Journal of Aerospace and Mechanical Engineering*, 4(1):34–38, 2010.

[56] G. van Straten. What can systems and control theory do for agriculture? In *Proceedings of the 2nd IFAC International Conference AGRICONTROL*, Osijek, Croatia, 2007.

[57] P. Varutti, T. Faulwasser, B. Kern, M. Kogel, and R. Findeisen. Event based reduced attention predictive control for nonlinear uncertain systems. In *IEEE International Symposium on Computer-Aided Control System Design*. Yokohama, Japan, 2010.

[58] P. Varutti, B. Kern, T. Faulwasser, and R. Findeisen. Event-based model predictive control for networked control systems. In *Proceedings of the 48th IEEE Conference on Decision and Control*, Shanghai, China, 2009.

[59] V. Vasyuntynskyy and K. Kabitzsch. Simple PID control algorithm adapted to deadband sampling. In *Proceedings of the 12th IEEE Conference on Emerging Technologies and Factory Automation*, Patras, Greece, 2007.

[60] V. Vasyutynskyy, A. Luntovskyy, and K. Kabitzsch. Limit cycles in PI control loops with absolute deadband sampling. In *Proceedings of the 18th Crimean Conference on Microwave and Telecommunication Technology*, Sevastopol, Crimea, 2008.

[61] Z. Yao, Y. Sun, and N. H. El-Farra. Resource-aware scheduled control of distributed process systems over wireless sensor network. In *American Control Conference*. Baltimore, MD, 2010.

[62] H. Zabiri and Y. Samyudia. A hybrid formulation and design of model predictive control for system under actuator saturation and backlash. *Journal of Process Control*, 16(7):693–709, 2006.

# 9

# *Model-Based Event-Triggered Control of Networked Systems*

**Eloy Garcia**
*Infoscitex Corp.*
*Dayton, OH, USA*

**Michael J. McCourt**
*University of Florida*
*Shalimar, FL, USA*

**Panos J. Antsaklis**
*University of Notre Dame*
*Notre Dame, IN, USA*

**CONTENTS**



**ABSTRACT** The model-based event-triggered (MB-ET) control framework is presented in this chapter. This framework makes explicit use of existing knowledge of the plant dynamics to enhance the performance of the system; it also makes use of event-based communication strategies to use network communication resources more wisely. This chapter presents extensive results that consider uncertain continuous-time linear and nonlinear systems. Particular emphasis is placed on uncertain nonlinear discrete-time systems subject to external disturbances where only output measurements are available. Additional topics concerning the MB-ET control framework are also described. Multiple examples are shown through this chapter in order to illustrate the advantages and functionality of this approach.

## 9.1 Introduction

In control systems, the success of control methodologies in stabilizing and providing desired performance, in the presence of parameter uncertainties and external disturbances, has been mainly due to the use of continuous feedback information. Closed-loop feedback controllers have the property to change the response of dynamical systems and, when properly designed, are able to provide desired system behavior under a class of uncertainties and disturbances.

Robustness to parameter uncertainties and external disturbances is deteriorated when continuous feedback is not available. The main reason to implement a feedback loop and provide sensor measurements to the controller is to obtain some information about a dynamical system, which can never be perfectly described and its dynamics can never be fully known. In the absence of uncertainties and disturbances, there will be no need for feedback information (assuming the initial conditions are known or after the first transmitted feedback measurement) since an open-loop, or feed-forward, control input that provides a desired system response can be obtained based on the, known with certainty, plant parameters.

Design, analysis, and implementation of closed-loop feedback control systems have translated into successful control applications in many different areas. However, recent control applications make use of limited bandwidth communication networks to transmit information, including feedback measurements from sensor nodes to controller nodes. The use of either wired or wireless communication networks has been rapidly increasing in scenarios involving different subsystems distributed over large areas, such as in many industrial processes. Networked systems are also common in automobiles and aircraft where each vehicle may contain hundreds of individual systems, sensors, and controllers exchanging information through a digital communication channel. Thus, it becomes important at a fundamental level to study the effects of uncertainties and disturbances on systems that operate without continuous feedback measurements.

In this chapter, we present the MB-ET control framework for networked control systems (NCSs). The MB-ET control framework makes explicit use of existing knowledge of the plant dynamics, encapsulated in the mathematical model of the plant, to enhance the performance of the system; it also makes use of event-based communication strategies to use network communication resources more wisely.

The performance of an NCS depends on the performance of the communication network in addition to traditional control systems performance measures. The bandwidth of the communication network used by the control system is of major concern, because other control and data acquisition systems will typically be sharing the same digital communication network.

It turns out that stability margins, controller robustness, and other stability and performance measures may be significantly improved when knowledge of the plant dynamics is explicitly used to predict plant behavior. Note that the plant model is always used to design controllers in standard control design. The difference here is that the plant model is used explicitly in the controller implementation to great advantage. This is possible today because existing inexpensive computation power allows the simulation of the model of the plant in real time.

The MB-ET control framework represents an extension of a class of networked systems called model-based networked control systems (MB-NCSs). In MB-NCSs, a nominal model of the plant dynamics is used to estimate the state of the plant during the intervals of time that feedback measurements are unavailable to the controller node. It has been common practice in MB-NCSs to reduce network usage by implementing periodic updates [1]. However, time-varying stochastic update intervals have been analyzed as well [2]. In the MB-ET control framework, the update instants are now decided based on the current conditions and response of the underlying dynamical system.

The contents of the present chapter are as follows. In Section 9.2 we describe the MB-NCS architecture and the conditions to stabilize a linear system under such a configuration. The MB-ET control framework is also introduced in this section. Stabilization of uncertain linear systems using MB-ET control is discussed in Section 9.3. Continuous-time nonlinear systems are considered in Section 9.4. In Section 9.5, we study discrete-time nonlinear systems that are subject to parameter uncertainties and external disturbances using dissipativity tools and MB-ET control. Additional extensions to the MB-ET control framework are described in Section 9.6. Finally, concluding remarks are made in Section 9.7.

## 9.2 The Model-Based Approach for Networked Systems

### 9.2.1 Model-Based Networked Control Systems

One of the main problems to be addressed when considering an NCS is the limited bandwidth of the network. In point-to-point control systems, it is possible to send continuous measurements and control inputs. Bandwidth and dynamic responses of a plant are closely related. The faster the dynamics of the plant, the larger is its bandwidth. This usually translates to large frequency content on the controlling signal and a continuous exchange of information between the plant and the controller. In the case of discrete-time plants, the controller acts at spaced instants of time, and transmission of continuous signals is not required. However, some discrete-time systems may have a fast internal sampling, which results in large bandwidth requirements in terms of the network characteristics. In this section, we describe the MB-NCS architecture, and we derive necessary and sufficient stabilizing conditions for linear time-invariant systems when periodic updates are implemented. In the following sections, we use similar architectures for NCS, but we implement event-triggered control techniques to determine the transmission time instants.

We consider the control of a linear time-invariant dynamical system where the state sensors are connected to controllers/actuators via a network. The main goal is to reduce the number of transmitted messages over the network using knowledge of the plant dynamics. Specifically, the controller uses an explicit model of the plant that approximates the plant dynamics and makes possible the stabilization of the plant even under slow network conditions. Although in principle, we can use the same framework to study the problem of packet dropouts in NCSs, the aim here is to purposely avoid frequent broadcasting of unnecessary sensor measurements so as to reduce traffic in the network as much as possible, which in turn reduces the presence of the problems associated with high network load such as packet collisions and network-induced delays. The main idea is to update the state of the model using the actual state of the plant provided by the sensor. The rest of the time the control action is based on a plant model that is incorporated in the controller/actuator and is running open loop for a period of $h$ seconds. The control architecture is shown in Figure 9.1.

In our control architecture, having knowledge of the plant at the actuator side enables us to run the plant in open loop, while the update of the model state provides the closed-loop information needed to overcome model

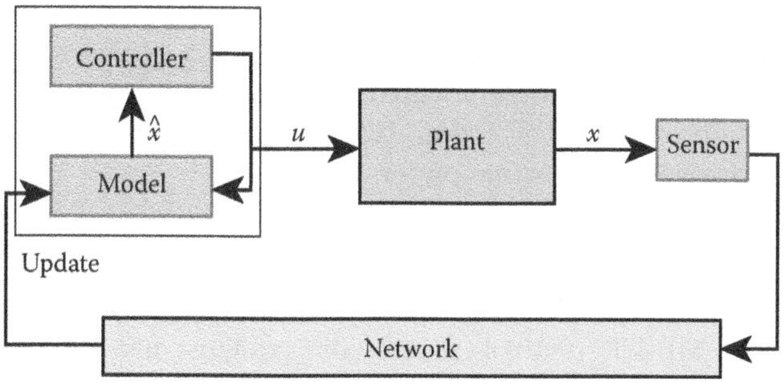


**FIGURE 9.1**
Model-based networked control systems architecture.

uncertainties and plant disturbances. In this section, we provide necessary and sufficient conditions for stability that result in a maximum update time, which depends mainly not only on the model inaccuracies but also on the designed control gain.

If all the states are available for measurement, then the sensor can send this information through the network to update the state of the model. The original plant may be open-loop unstable. We assume that the frequency at which the network updates the state in the controller is constant and that the compensated model is stable, which is typical in control systems. For simplicity, we also assume that the transportation delay is negligible, which is completely justifiable in most of the popular network standards like CAN bus or Ethernet. The case of network delays and periodic updates is addressed in [1]. In Section 9.3.2, we consider network delays when using event-based communication. The goal is to find the largest constant update period at which the network must update the model state in the controller for stability—that is, we are seeking an upper bound for $h$ the update time. Consider the control system of Figure 9.1 where the plant and the model are described, respectively, by

$$\dot{x} = Ax + Bu, \tag{9.1}$$

$$\dot{\hat{x}} = \hat{A}\hat{x} + \hat{B}u, \tag{9.2}$$

where $x, \hat{x} \in \mathbb{R}^n$, $u \in \mathbb{R}^m$. The control input is given by $u = K\hat{x}$. The state error is defined as

$$e = x - \hat{x}, \tag{9.3}$$

and represents the difference between the plant state and the model state. The modeling error matrices $\tilde{A} = A - \hat{A}$ and $\tilde{B} = B - \hat{B}$ represent the difference between the plant and the model. The update time instants are denoted by $t_i$, where

$$t_i - t_{i-1} = h, \tag{9.4}$$

for $i = 1, 2, \ldots$ (in this section, $h$ is a constant). The choice of $h$, being a constant, is simple to implement and also results in a simple analysis procedure as shown below. However, event-triggered updates can potentially bring better benefit, and they will be addressed in the following sections.

Since the model state is updated at every time instant $t_i$, then $e(t_i) = 0$, for $i = 1, 2, \ldots$. This resetting of the state error at every update time instant is a key characteristic of our control system. Define the augmented state vector $z(t)$ and the augmented state matrix $\Lambda$:

$$z(t) = \begin{bmatrix} x(t) \\ e(t) \end{bmatrix}, \qquad \Lambda = \begin{bmatrix} A + BK & -BK \\ \tilde{A} + \tilde{B}K & \hat{A} - \tilde{B}K \end{bmatrix}. \tag{9.5}$$

Thus, for $t \in [t_i, t_{i+1})$, we have that $u = K\hat{x}$ and using the error definition (9.3), the augmented system dynamics can be described by

$$\dot{z}(t) = \begin{bmatrix} \dot{x}(t) \\ \dot{e}(t) \end{bmatrix} = \begin{bmatrix} A+BK & -BK \\ \tilde{A}+\tilde{B}K & \hat{A}-\tilde{B}K \end{bmatrix} \begin{bmatrix} x(t) \\ e(t) \end{bmatrix} = \Lambda z(t), \quad (9.6)$$

for $t \in [t_i, t_{i+1})$, with updates given by

$$z(t_i) = \begin{bmatrix} x(t_i) \\ 0 \end{bmatrix} = \begin{bmatrix} x(t_i^-) \\ 0 \end{bmatrix}, \quad (9.7)$$

where $t_i - t_{i-1} = h$.

During the time interval $t \in [t_i, t_{i+1})$, the system response is given by

$$z(t) = \begin{bmatrix} x(t) \\ e(t) \end{bmatrix} = e^{\Lambda(t-t_i)} \begin{bmatrix} x(t_i) \\ 0 \end{bmatrix} = e^{\Lambda(t-t_i)} z(t_i)\ , \quad (9.8)$$

and at time instants $t_i$ we have that $z(t_i) = \begin{bmatrix} x(t_i) \\ 0 \end{bmatrix}$, that is, the error is reset to zero. We can represent this by

$$z(t_i) = \begin{bmatrix} I & 0 \\ 0 & 0 \end{bmatrix} z(t_i^-)\ . \quad (9.9)$$

Due to the periodicity of the state updates, the extended state $z(t)$ can be expressed in terms of the initial condition $x(t_0)$ as follows:

$$z(t) = e^{\Lambda(t-t_i)} \left( \begin{bmatrix} I & 0 \\ 0 & 0 \end{bmatrix} e^{\Lambda h} \begin{bmatrix} I & 0 \\ 0 & 0 \end{bmatrix} \right) z_0\ , \quad (9.10)$$

for $t \in [t_i, t_{i+1})$, where $h = t_i - t_{i-1}$ and $z_0 = \left[ \begin{smallmatrix} x(t_0) \\ e(t_0) \end{smallmatrix} \right]$.

**Theorem 9.1**

The system described by (9.6) with updates (9.7) is globally exponentially stable around the solution $z = \left[ \begin{smallmatrix} x \\ e \end{smallmatrix} \right] = \left[ \begin{smallmatrix} 0 \\ 0 \end{smallmatrix} \right]$ if and only if the eigenvalues of $\left( \left[ \begin{smallmatrix} I & 0 \\ 0 & 0 \end{smallmatrix} \right] e^{\Lambda h} \left[ \begin{smallmatrix} I & 0 \\ 0 & 0 \end{smallmatrix} \right] \right)$ are strictly inside the unit circle.

**PROOF** See [1] for the proof. ■

The MB-NCS architecture has been used to deal with different problems commonly encountered in NCSs.

In [1] the output feedback scenario was addressed by implementing a state observer at the sensor node. Network-induced delays were also considered within the MB-NCS architecture in [1]. The case where network delays are greater than the update period $h$ was studied in [3].

An important extension of this work considered time-varying updates [2]. In that case, two stochastic scenarios were studied; in the first, the assumption is that transmission times are identically independently distributed; in the second, transmission times are driven by a finite Markov chain. In both cases, conditions were derived for almost sure (probability 1) and mean square stability. Nonlinear systems have also been considered using the MB-NCS configuration. Different authors have provided stability conditions and stabilizing update rates for nonlinear MB-NCSs with and without uncertainties and for nonlinear MB-NCSs with time delays [4,5]. Different authors have dealt with similar problems in different types of applications. Motivated by activities that involve human operators, the authors of [6] point out that typically, a human operator scans information intermittently and operates the controlled system continuously; the intermittent characteristic in this case refers to the same situation presented in [1]—that is, a single measurement is used to update the internal model and generate the control input. For a skillful operator, the information is scanned less frequently. Between update intervals, the control input is generated the same way as in MB-NCSs—that is, an imperfect model of the system is used to generate an estimate of the state, and periodic measurements are used to update the state of this model. In the output feedback case, a stochastic estimator is implemented with the assumption that the statistical properties of the measurement noise are known. In both cases, the authors provide conditions for stability based on the length of the sampling interval. The authors of [7] also use a model that produces the input for the plant (possibly nonlinear) and considers a network in both sides of the plant. The actuator is assumed to have an embedded computer that decodes and resynchronizes the large packets sent from the controller that contain the predicted control input obtained by the model. In [8], Hespanha et al. use differential pulse-code modulation techniques together with a model of the plant dynamics to reduce the amount of bandwidth needed to stabilize the plant. Both the sensor and the controller/actuator have a model of the plant, which is assumed to be exact, and they are run simultaneously. Stabilizing conditions for MB-NCSs under different quantization schemes were presented in [9]. Typical static quantizers such as uniform quantizers and logarithmic quantizers were considered in [9]. Also, the design of MB-NCSs using dynamic quantizers was also addressed in the same paper.

### 9.2.2 Model-Based Event-Triggered Control

The MB-ET control framework makes use of the MB-NCS architecture described in Section 9.2.1 and the event-triggered control paradigm largely discussed in this book. The main goal in the MB-ET control framework is to adjust the update intervals based on the current state error and to send a measurement to update the state of the model only when it is necessary. This means that the update time intervals are no longer constant. The update time intervals are nonperiodic and depend on the current plant response. One key difference of the MB-ET control approach with respect to traditional event-triggered control techniques is that the transmitted measurement does not remain constant between update intervals. The transmitted measurement is used to update the state of the model $\hat{x}$, and, similar to Section 9.2.1, the model state is used to compute the control input defined as $u = K\hat{x}$.

The application of event-triggered control to the MB-NCS produces many advantages compared to periodic implementation. For instance, nonlinearities and inaccuracies that affect the system and are difficult or impossible to model and may change over time or under different physical characteristics (temperature, different load in a motor, etc.) may be handled more efficiently by tracking the state error than by updating the state at a constant rate. Also, the method presented here is robust under random packet loss. If the sensor sends data but the model state is not updated because of a packet dropout, the state error will grow rapidly above the threshold. In this case, the following event is triggered sooner, in general, than in the case when the previous measurement was successfully received. However, under a fixed transmission rate, if a packet is lost, the model will need to wait until the next update time to receive the feedback data, thus compromising the stability of the system.

The implementation of an event-based rule in a MB-NCS represents a very intuitive way to save network resources. It also considers the performance of the closed-loop real system. The implementation of the model to generate an estimate of the plant state and using that estimate to control the plant result in significant savings in bandwidth. It is also clear that the accuracy of the estimate depends on factors such as the size of the model uncertainties. One of the results in the present chapter is that these error-based updates provide more independence from model uncertainties when performing this estimation. When the error is small, which means that the state of the model is an accurate estimation of the real state, then we save energy, effort, and bandwidth by electing not to send measurements for all the time intervals in which the error remains small. The MB-ET control architecture is similar to the one shown in Figure 9.1. One difference is that copies of the nominal model parameters $\hat{A}$, $\hat{B}$ and of the control gain $K$ are now implemented in the sensor node in order to obtain the state model $\hat{x}$ at the sensor node and be able to compute the state error. The state error is compared to a threshold to determine if an event is triggered. When an event is triggered, the current measurement of the state of the plant is transmitted, and the sensor model is updated using the same state measurement.

The MB-ET control framework addressed in this chapter was introduced in [10]. This approach was used for the control of networked systems subject to quantization and network-induced delays [11]. In [12], updates of model states and model parameters are considered. The MB-ET approach has also been used for the stabilization of coupled systems [13], coordinated multi-agent tracking [14], and synchronization of multi-agent systems with general linear dynamics [15].

A similar model-based approach has been developed by Lunze and Lehman [16]. In their approach, the model is assumed to match the dynamics of the system exactly; however, the system is subject to unknown input disturbances. The main idea of the approach in [16] is the same as in this chapter, that is, to use the nominal model to generate estimates of the current state of the system. Since the system is subject to an unknown disturbance and the model is executed with zero input disturbance, then a difference between plant and model states is expected, and the sensor updates transmitted over a digital communication network are used to reset this difference between the states of the plant and of the model. The same authors have extended this approach to consider the output feedback, quantization, and network delay cases. The authors of [17] also discussed a model-based approach for stabilization of linear discrete-time systems with input disturbances and using periodic event-triggered control.

The work in [18] also offers an approach to the problem of reducing the bandwidth utilization by making use of a plant model; here, the update of the model is event driven. The model is updated when any of the states differ from the computed value for more than a certain threshold. Some stability and performance conditions are derived as functions of the plant, threshold, and magnitude of the plant-model mismatch.

## 9.3 Linear Systems

The main advantage that the event-triggered feedback strategy offers compared with the common periodic-update implementation is that the time interval between updates can be considerably increased, especially when the state of the system is close to its equilibrium point, thus releasing the communication network for other tasks. We will assume in this section that the communication delay is negligible. The approach in this section is to compute the norm of the state error and compare it to a positive threshold in order to decide if an update of the state of the model is needed. When the model is updated, the error is equal to zero; when it becomes greater than the threshold, the next update is sent.

We consider a state-dependent threshold similar to [19] where the norm of the state error is compared to a function of the norm of the state of the plant; in this way, the threshold value is not fixed anymore, and, in particular, it can be reduced as we approach the equilibrium point of the system, assuming that the zero state is the equilibrium of the system. Traditional event-triggered control techniques [19–21] consider systems controlled by static gains that generate piecewise constant inputs due to the fact that the update is held constant in the controller. The main difference in this section is that we use a model-based controller (i.e., a model of the system and a static gain); the model provides an estimate of the state between updates, and the model-based controller provides a control input for the plant that does not remain constant between measurement updates.

### 9.3.1 MB-ET Control of Uncertain Linear Systems

Let us consider the plant and model described by (9.1) and (9.2). Let use define the state error $e = \hat{x} - x$. Using the control input $u = K\hat{x}$, we obtain

$$\dot{x} = (A + BK)x + BKe, \tag{9.11}$$

and

$$\dot{\hat{x}} = (\hat{A} + \hat{B}K)\hat{x}, \tag{9.12}$$

for $t \in [t_i, t_{i+1})$. We choose a control gain $K$ that renders the closed-loop model (9.12) globally asymptotically stable. We proceed to choose a quadratic Lyapunov function $V = x^T Px$, where $P$ is symmetric positive definite and is the solution of the closed-loop model Lyapunov equation

$$(\hat{A} + \hat{B}K)^T P + P(\hat{A} + \hat{B}K) = -Q, \tag{9.13}$$

where $Q$ is a symmetric positive definite matrix. Let us first analyze the case when $\hat{B} = B$ for simplicity. Also assume that the next bound on the uncertainty holds: $\|\tilde{A}^T P + P\tilde{A}\| \leq \Delta < \underline{q}$, where $\tilde{A} = A - \hat{A}$ and $\underline{q} = \underline{\sigma}(Q)$ is the smallest singular value of $Q$ in the model Lyapunov equation (9.13). This bound can be seen as a measure of how close A and $\hat{A}$ should be.

The next theorem provides conditions on the error and its threshold value so the networked system is asymptotically stable. The error threshold is defined as a function of the norm of the state and $\Delta$ which is a bound on the uncertainty in the state matrix $A$. Similarly, the occurrence of an error event leads the sensor to send the current measurement of the state of the plant that is used in the controller to update the state of the model.

**Theorem 9.2**

Consider the system (9.1) with input $u = K\hat{x}$. Let the feedback be based on error events using the following relation:

$$\|e\| > \frac{\sigma(\underline{q} - \Delta)}{b} \|x\|, \tag{9.14}$$

where $b = \|K^T \hat{B}^T P + P\hat{B}K\|$, $0 < \sigma < 1$, and $\|\tilde{A}^T P + P\tilde{A}\| \leq \Delta < \underline{q}$. Let the model be updated when (9.14) is first satisfied. Then the system is globally asymptotically stable.

**PROOF** Proof is given in [10]. ∎

### 9.3.2 MB-ET Control of Uncertain Linear Systems over Uncertain Networks

In this section, we design stabilizing thresholds taking into account the availability not of the real variables but only of quantized measurements. Additionally, we design stabilizing thresholds using the model-based event-triggered framework for networked systems affected by both quantization and time delays.

The measured variables have to be quantized in order to be represented by a finite number of bits, so as to be used in processor operations and to be transmitted through a digital communication network. It becomes necessary to study the effects of quantization error on networked systems and on any computer-implemented control application because of the reasons just mentioned. In addition, we want to emphasize two important implications of quantization in event-triggered control. First, an important step in event-triggered control strategies is that the model-plant state error is set to zero at the update instants. When using quantization, this is no longer the case, because we use the quantized measurement of the plant state to update the state of the model, and this measurement is not, in general, the same as the real state of the plant. Second, in traditional event-triggered control techniques, the updates are triggered by comparing the norm of the state, which is not exactly available due to quantization errors, to the norm of the state error, which is not exactly available since it is a function of the real state of the plant. The problem in those approaches is that stability of the system is directly related to nonquantized measurements that are assumed to be known with certainty.

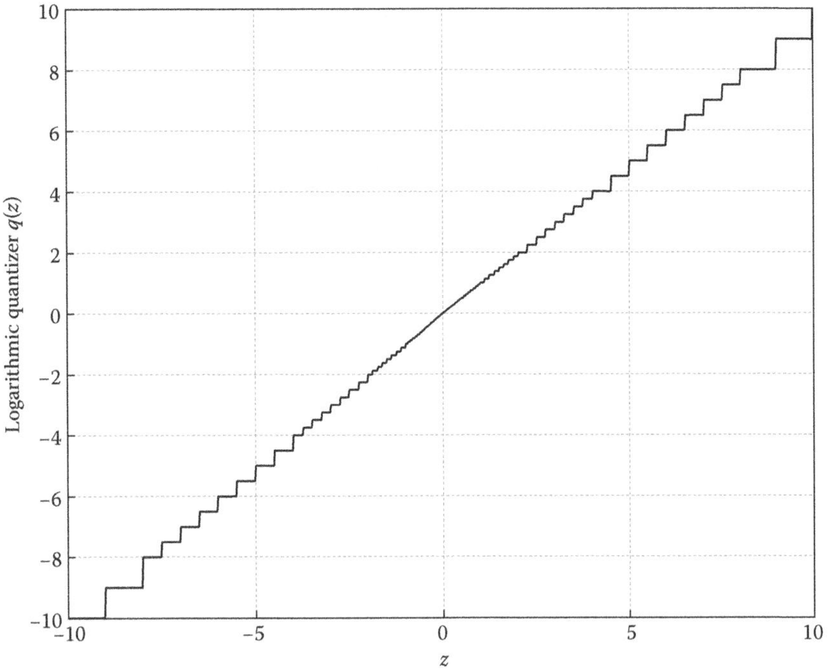


**FIGURE 9.2**
Logarithmic quantizer function.

The aim in this section is to find triggering conditions based on the available quantized variables that also ensure asymptotic stability in the presence of quantization errors. The type of quantizer that we are going to use in this section is the logarithmic quantizer. A logarithmic quantizer function is shown in Figure 9.2.

We define the logarithmic quantizer as a function $q : \mathbb{R}^n \to \mathbb{R}^n$ with the following property:

$$\|z - q(z)\| \leq \delta \|z\|, \ z \in \mathbb{R}^n, \ \delta > 0. \tag{9.15}$$

Using the logarithmic quantizer defined in (9.15), we have that at the update instants $t_i, i = 1, 2, \ldots,$ the state of the model is updated using the quantized measurement, that is,

$$q(x(t_i)) \to \hat{x}(t_i). \tag{9.16}$$

Define the quantized model-plant state error:

$$e_q(t) = \hat{x}(t) - q(x(t)), \tag{9.17}$$

where $q(x(t))$ is the quantized value of $x(t)$ at any time $t \geq 0$ using the logarithmic quantizer (9.15). Note that $q(x)$ and $e_q$ are the available variables that can be used to compute the triggering condition. Also note that $e_q(t_i) = 0$, that is, the quantized model-plant state error, is set to zero at the update instants according to the update (9.16).

Consider a stable closed-loop nominal model, and define the Lyapunov function $V = x^T P x$, where $P$ is a symmetric and positive definite matrix and is the solution of the closed-loop model Lyapunov equation (9.13). Let $\tilde{B} = B - \hat{B}$.

**Theorem 9.3**

Consider system (9.1) with control input $u = K\hat{x}$. Assume that there exists a symmetric positive definite solution $P$ for the model Lyapunov equation (9.13) and that $\|\tilde{B}\| \leq \beta$ and $\|(\tilde{A} + \tilde{B}K)^T P + P(\tilde{A} + \tilde{B}K)\| \leq \Delta < \underline{q}$. Consider the relation

$$\|e_q\| > \frac{\sigma\eta}{\delta + 1} \|q(x)\|, \tag{9.18}$$

where $\eta = \frac{\underline{q} - \Delta}{b}$, $0 < \sigma < \sigma' < 1$, $b = 2\|P\hat{B}K\| + 2\beta\|PK\|$, and let the model be updated when (9.18) holds. Then,

$$\|e\| \leq \sigma'\eta \|x\|, \tag{9.19}$$

is always satisfied, and the system is asymptotically stable when

$$\delta \leq (\sigma' - \sigma)\eta. \tag{9.20}$$

**PROOF** See [11] for the proof. ■

We now discuss stability thresholds that consider quantization and time delays. The quantized model-plant state error was defined in (9.17). Consider also the nonquantized model-plant state error $e(t) = \hat{x}(t) - x(t)$. At the update instants $t_i$, we update the model in the sensor node using the quantized measurement of the state. At this instant, we have $e_q(t_i) = 0$ at the sensor node. When considering network delays, we can reset the quantized model-state error only at the sensor node. The model-plant state error at the update instants is given by

$$e(t_i) = \hat{x}(t_i) - x(t_i) = q(x(t_i)) - x(t_i). \tag{9.21}$$

It is clear that this error cannot be set to zero at the update instants as the quantized model-plant state error, due to the existence of quantization errors when measuring the state of the plant. Using the logarithmic quantizer (9.15), we have that

$$\|e(t_i)\| = \|q(x(t_i)) - x(t_i)\| \leq \delta \|x(t_i)\|. \tag{9.22}$$

Theorem 9.4 provides conditions for asymptotic stability of the control system using quantization in the presence of network-induced delays. In this case, the admissible delays are also a function of the quantization parameter $\delta$. That is, if we are able to quantize more finely, the system is still stable in the presence of longer delays.

**Theorem 9.4**

Consider system (9.1) with control input $u = K\hat{x}$. The event-triggering condition is computed using quantized data and using error events according to (9.18). The model is updated using quantized measurements of the state of the plant. Assume that there exists a symmetric positive definite solution $P$ for the model Lyapunov equation (9.13) and a small enough $\delta$, $0 < \delta < 1$, such that $\frac{2\delta}{1-\delta} < \frac{\sigma\eta}{\delta+1}$. Assume also that $B = \hat{B}$ and the following bounds are satisfied: $\|\tilde{A}\| \leq \Delta_A$ and $\|\tilde{A}^T P + P\tilde{A}\| \leq \Delta < q$. Then there exists an $\epsilon(\delta) > 0$ such that for all network delays $\tau_N \in [0, \epsilon]$, the system is asymptotically stable. Furthermore, there exists a time $\tau > 0$ such that for any initial condition the inter-execution times $\{t_{i+1} - t_i\}$ implicitly defined by (9.18) with $\sigma < 1$ are lower bounded by $\tau$, that is, $t_{i+1} - t_i \geq \tau$, for all $i = 1, 2, \ldots$.

**PROOF** See [11] for proof. ■

### 9.3.3 Examples

**EXAMPLE 9.1: MB-ET control of a linear system** In this example, we use the inverted pendulum on a moving cart dynamics (linearized dynamics) described in example 2E in [22].

The linearized dynamics can be expressed using the state vector $x = [y\ \theta\ \dot{y}\ \dot{\theta}]^T$, where $y$ represents the displacement of the cart with respect to some reference point, and $\theta$ represents the angle that the pendulum rod makes with respect to the vertical.

The matrices corresponding to the state-space representation (9.1) are given by

$$A = \begin{bmatrix} 0 & 0 & 1 & 0 \\ 0 & 0 & 0 & 1 \\ 0 & -mg/M & 0 & 0 \\ 0 & (M+m)/Ml & 0 & 0 \end{bmatrix}, \quad B = \begin{bmatrix} 0 \\ 0 \\ 1/M \\ -1/Ml \end{bmatrix}, \tag{9.23}$$

where the nominal parameters of the model are given by $\hat{m} = 0.1$, $\hat{M} = 1$, and $\hat{l} = 1$. The real parameters represent values close to the nominal parameters but not exactly the same due to uncertainties in the measurements and specification of these parameters. The physical parameter values are given by $m = 0.09978$, $M = 1.0016$, $l = 0.9994$. Also, $\hat{g} = g = 9.8$. The input $u$ represents the external force applied to the cart. The open-loop plant and model dynamics are unstable. In this example, we use the following gain $K = [0.5379\ 25.0942\ 1.4200\ 7.4812]$.

Figure 9.3 shows the position $y(t)$ and the angle $\theta(t)$; the second subplot shows the corresponding velocities. The norm of the state converging to the origin and the threshold used to trigger events are shown in Figure 9.4, while the second subplot shows the time instants where events are triggered and communication is established.

**EXAMPLE 9.2: MB-ET control of a linear system with delays and quantization** In this example, we consider the instrument servo (DC motor driving an inertial load) dynamics from example 6A in [22]:

$$\begin{bmatrix} \dot{e} \\ \dot{\omega} \end{bmatrix} = \begin{bmatrix} 0 & 1 \\ 0 & -\alpha \end{bmatrix} \begin{bmatrix} e \\ \omega \end{bmatrix} + \begin{bmatrix} 0 \\ \beta \end{bmatrix} u, \tag{9.24}$$

where $e = \theta - \theta_r$ represents the error between the current position $\theta$ and the desired position $\theta_r$, where the desired position is assumed to be constant, $\omega$ is the angular velocity, and $u$ is the applied voltage. The parameters $\alpha$ and $\beta$ represent constants that depend on the physical parameters of the motor and load.

The nominal parameters are $\hat{\alpha} = 1$, $\hat{\beta} = 3$. The real parameters are given by $\alpha = 1.14$, $\beta = 2.8$. The model

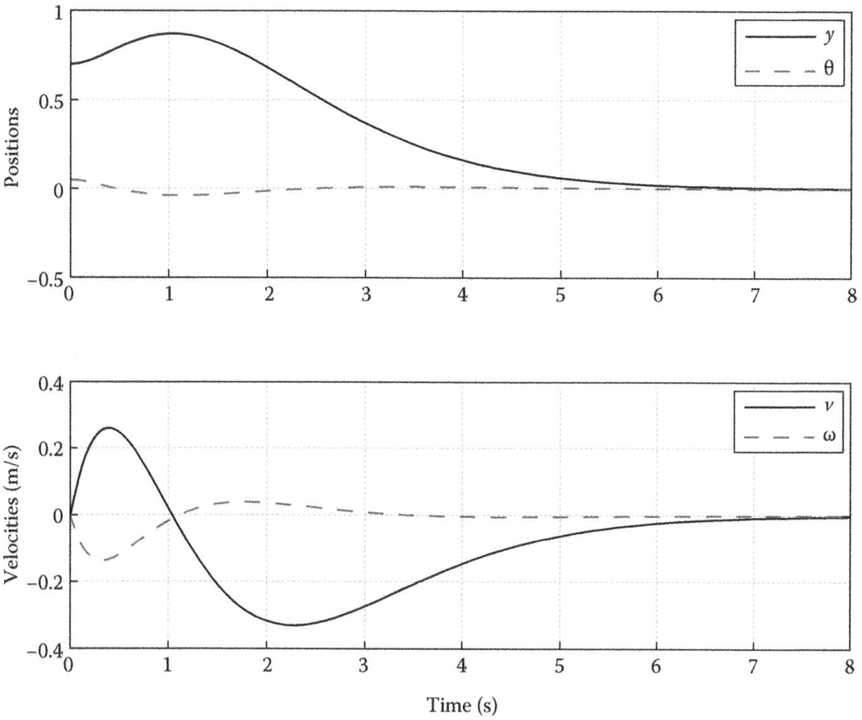


**FIGURE 9.3**
Positions and velocities of the inverted pendulum on the moving cart example using the MB-ET control framework.

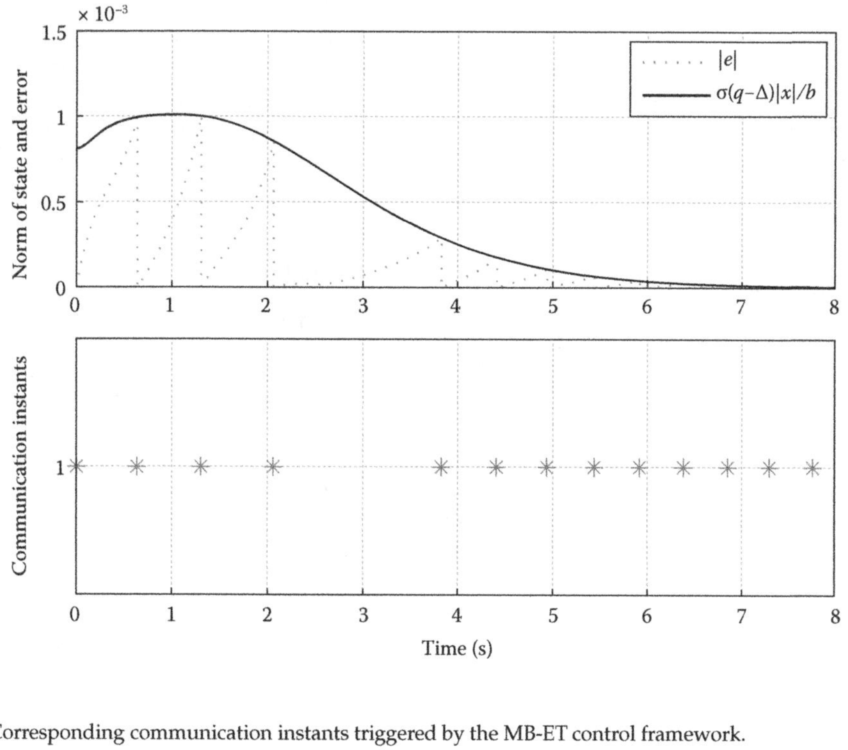


**FIGURE 9.4**
Error and state norms. Corresponding communication instants triggered by the MB-ET control framework.

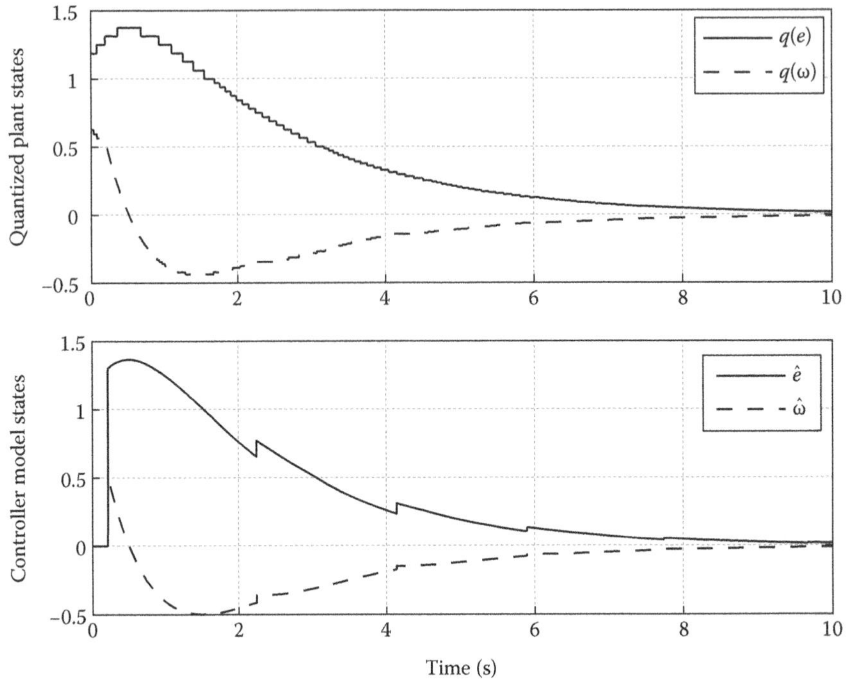


**FIGURE 9.5**
Quantized versions of the plant states and model states available in the controller node.

parameters are used for controller design; in this case the control gain is given by $K = [-0.33\ -0.33]$.

The parameters used in the MB-ET with quantization and delay framework are $\sigma = 0.6$, $\sigma' = 0.8$, $\sigma^\tau = 0.99$. The quantization parameter is $\delta = 0.08$. With these parameters we have that the admissible delays are bounded by $\epsilon = 0.468$. The system's initial conditions are $x(0) = [1.2\, 0.65]^T$; the model state is initialized to the zero vector.

Figure 9.5 shows the quantized plant states and the model states at the controller node. The norm of the state converging to the origin and the threshold used to trigger events are shown in Figure 9.6, where $y_q = \frac{\sigma\eta}{\delta+1}$ and $y' = \sigma'\eta$. The third subplot shows the time instants where events are triggered and communication is established. Because of time delays, the model in the controller node is updated after some time-varying delay $0 \leq \tau_N \leq 0.468$. This feature can be seen by looking at the second subplot of Figure 9.5 and at the third subplot of Figure 9.6 which show that the updates of the model at the controller node occur after the triggering time instants but not exactly at those instants.

## 9.4 Nonlinear Systems

In this section, we present results concerning the MB-ET control framework for nonlinear systems. The model of the system is also nonlinear. This section considers continuous-time systems with state feedback. Stability of the networked system using MB-ET control is based on Lyapunov analysis, and a lower bound on the interevent time intervals is obtained using the Gronwall–Bellman inequality. The next section considers discrete-time nonlinear systems with external disturbance and output feedback. Dissipativity methods are used for stability analysis in that case.

### 9.4.1 Analysis

Let us consider continuous-time nonlinear systems represented by

$$\dot{x} = f(x) + f_u(u). \tag{9.25}$$

The nonlinear model of the plant dynamics is given by

$$\dot{\hat{x}} = \hat{f}(\hat{x}) + \hat{f}_u(u), \tag{9.26}$$

where $x, \hat{x} \in \mathbb{R}^n$, $u \in \mathbb{R}^m$. The controller and the state error are given by $u = h(\hat{x})$ and $e = x - \hat{x}$, respectively. Thus, we have that

$$\begin{aligned} \dot{x} &= f(x) + f_u(h(\hat{x})) = f(x) + g(\hat{x}) \\ \dot{\hat{x}} &= \hat{f}(\hat{x}) + \hat{f}_u(h(\hat{x})) = \hat{f}(\hat{x}) + \hat{g}(\hat{x}). \end{aligned} \tag{9.27}$$

Let us consider plant-model uncertainties that can be characterized as follows:

$$\begin{aligned} \hat{f}(\xi) &= f(\xi) + \delta_f(\xi) \\ \hat{g}(\xi) &= g(\xi) + \delta_g(\xi). \end{aligned} \tag{9.28}$$

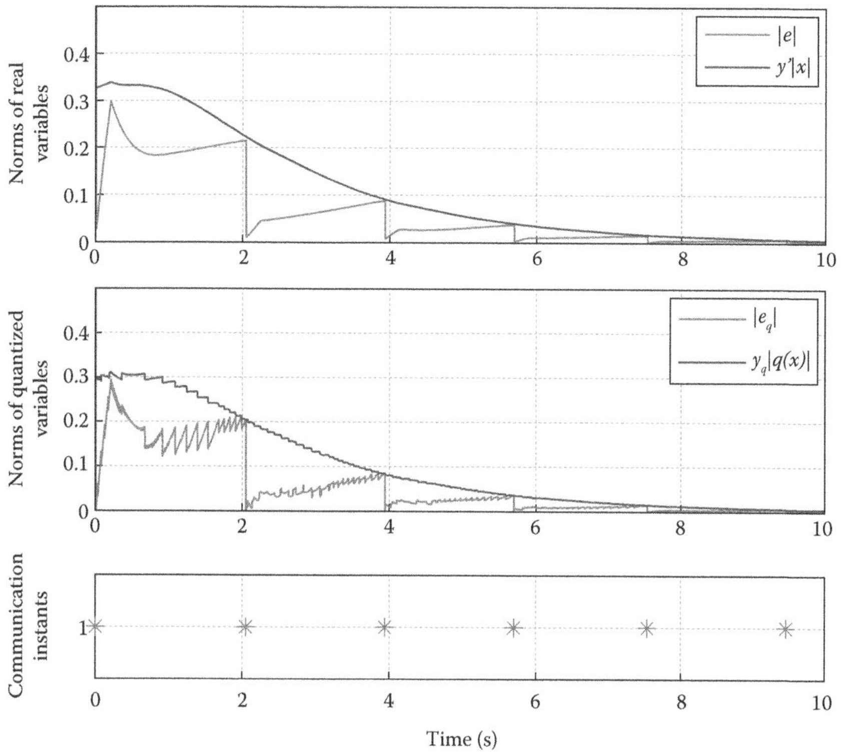


**FIGURE 9.6**
Norms of nonquantized and quantized error and plant state. Communication time instants.

Then, the model dynamics can be expressed as follows:

$$\dot{\hat{x}} = f(\hat{x}) + \delta_f(\hat{x}) + g(\hat{x}) + \delta_g(\hat{x}) = f(\hat{x}) + g(\hat{x}) + \delta(\hat{x}). \tag{9.29}$$

Assume that zero is the equilibrium point of the system $\dot{x} = f(x)$ and that $g(0) = 0$ and $\delta(0) = 0$. Also assume that both $f$ and $\delta$ are locally Lipschitz with constants $K_f$ and $K_\delta$, respectively; that is,

$$\begin{aligned} \|f(x) - f(y)\| &\leq K_f \|x - y\| \\ \|\delta(x) - \delta(y)\| &\leq K_\delta \|x - y\|, \end{aligned} \tag{9.30}$$

for $x, y \in \Omega \subseteq \mathbb{R}^n$.

Also assume that the non-networked closed-loop model is locally exponentially stable:

$$\|\hat{x}(t)\| \leq \hat{\beta} \|\hat{x}(t_0)\| e^{-\hat{\alpha}(t-t_0)}, \tag{9.31}$$

for $\hat{\alpha}, \hat{\beta} > 0$.

Let us consider the following assumptions on the non-networked closed-loop system dynamics.

**Assumption** *There exist positive constants $\alpha, \beta$ and class $\mathcal{K}$ functions $\alpha_1, \alpha_2$ such that*

$$\begin{aligned} \alpha_1(\|x\|) \leq V(x) \leq \alpha_2(\|x\|) \\ \frac{\partial V(x)}{\partial x} f(x, g(x+e)) \leq -\alpha \|x\| + \beta \|e\|. \end{aligned} \tag{9.32}$$

**Theorem 9.5**

The nonlinear model-based event-triggered control system (9.25) with $u = h(\hat{x})$ is asymptotically stable if the events are triggered when

$$\|e\| > \frac{\sigma\alpha}{\beta} \|x\|, \tag{9.33}$$

where $0 < \sigma < 1$. Furthermore, there exists a constant $\tau > 0$ such that the interevent time intervals are lower bounded by $\tau$; that is, $t_{i+1} - t_i \geq \tau$.

**PROOF** Note that at the event time instants, the model state is updated using the plant state, so we have that $e(t_i) = 0$. Also, because of the threshold (9.33), the error satisfies $\|e(t)\| \leq \frac{\sigma\alpha}{\beta} \|x\|$, for $t \geq 0$. Then, we can write

$$\dot{V} \leq -\alpha \|x\| + \beta \tfrac{\sigma\alpha}{\beta} \|x\| \leq (\sigma - 1)\alpha \|x\|, \tag{9.34}$$

and the nonlinear system (9.25) is asymptotically stable.

In order to establish a bound on the interevent time intervals, let us analyze the state error dynamics:

$$\dot{e} = \dot{x} - \dot{\hat{x}} = f(x) - f(\hat{x}) - \delta(\hat{x}). \tag{9.35}$$

Thus, the response of the state error during the interval $t \in [t_i, t_{i+1})$ is given by

$$e(t) = e(t_i) + \int_{t_i}^{t} (f(x(s)) - f(\hat{x}(s)) - \delta(\hat{x}(s)))ds, \quad (9.36)$$

where $e(t_i) = 0$, since an update has taken place at time $t_i$, and the error is reset to zero at that time instant. Then, we can write

$$\begin{aligned}\|e(t)\| &\leq \int_{t_i}^{t} \|f(x(s)) - f(\hat{x}(s))\| ds + \int_{t_i}^{t} \|\delta(\hat{x}(s))\| ds \\ &\leq K_f \int_{t_i}^{t} \|x(s) - \hat{x}(s)\| ds + K_\delta \int_{t_i}^{t} \|\hat{x}(s)\| ds \\ &\leq K_f \int_{t_i}^{t} \|e(s)\| ds + K_\delta \hat{\beta} \int_{t_i}^{t} \|\hat{x}(t_i)\| e^{-\hat{\alpha}(s-t_i)} ds \\ &\leq K_f \int_{t_i}^{t} \|e(s)\| ds + \frac{K_\delta \hat{\beta} \|\hat{x}(t_i)\|}{\hat{\alpha}} (1 - e^{-\hat{\alpha}(t-t_i)}),\end{aligned} \quad (9.37)$$

for $t \in [t_i, t_{i+1})$. We now make use of the Gronwall–Bellman inequality [23] to solve the remaining integral in (9.37).

We obtain

$$\begin{aligned}\|e(t)\| \leq & \frac{K_\delta K_f \hat{\beta} \|\hat{x}(t_i)\|}{\hat{\alpha}} \int_{t_i}^{t} (1 - e^{-\hat{\alpha}(s-t_i)}) e^{K_f(t-s)} ds \\ & + \frac{K_\delta \hat{\beta} \|\hat{x}(t_i)\|}{\hat{\alpha}} (1 - e^{-\hat{\alpha}(t-t_i)}) \\ \leq & \frac{K_\delta \hat{\beta} \|\hat{x}(t_i)\|}{\hat{\alpha}} \Big(K_f \int_{t_i}^{t} (e^{K_f(t-s)} - e^{K_f(t-s) - \hat{\alpha}(s-t_i)}) ds \\ & + 1 - e^{-\hat{\alpha}(t-t_i)}\Big) \\ \leq & \frac{K_\delta \hat{\beta} \|\hat{x}(t_i)\|}{\hat{\alpha}} \Big(e^{K_f(t-t_i)} - 1 \\ & + \frac{K_f}{K_f + \hat{\alpha}} (e^{-\hat{\alpha}(t-t_i)} - e^{K_f(t-t_i)}) + 1 - e^{-\hat{\alpha}(t-t_i)}\Big) \\ \leq & \frac{K_\delta \hat{\beta} \|\hat{x}(t_i)\|}{\hat{\alpha}} \Big(1 - \frac{K_f}{K_f + \hat{\alpha}}\Big) (e^{K_f(t-t_i)} - e^{-\hat{\alpha}(t-t_i)}) \\ \leq & \frac{K_\delta \hat{\beta} \|\hat{x}(t_i)\|}{K_f + \hat{\alpha}} (e^{K_f(t-t_i)} - e^{-\hat{\alpha}(t-t_i)}),\end{aligned} \quad (9.38)$$

for $t \in [t_i, t_{i+1})$. Let $\tau = t - t_i$, and we can see that the time $\tau > 0$ that it takes for the last expression to be equal to $\frac{\sigma\alpha}{\beta} \|x\|$ is less than or equal to the time it takes the norm of the error to grow from zero at time $t_i$ and reach the value $\frac{\sigma\alpha}{\beta} \|x\|$. That is, by establishing the relationship

$$\frac{K_\delta \hat{\beta} \|\hat{x}(t_i)\|}{K_f + \hat{\alpha}} (e^{K_f \tau} - e^{-\hat{\alpha}\tau}) = \frac{\sigma\alpha}{\beta} \|x(t)\|, \quad (9.39)$$

we guarantee that

$$\|e(t)\| \leq \frac{\sigma\alpha}{\beta} \|x\|, \quad (9.40)$$

which ensures that no event is generated before $t_i + \tau$, where $t_i$ represents the time instant corresponding to the latest event. Note that the solution $\tau$ of (9.39) is positive for any $x(t) \neq 0$. Also, if we have that $x(t_i)$=0 for some $t_i > 0$, then an event is triggered at $t = t_i$, and we have that $\hat{x}(t_i) = x(t_i) = 0$ and $e(t) = 0$ for $t \geq t_i$, because $\delta(0) = 0$ and $g(0) = 0$. Thus, the comparison (9.33) becomes $0 > \frac{\sigma\alpha}{\beta} \cdot 0$, which does not hold, and it is not necessary to generate any further event. ■

### 9.4.2 Examples

**EXAMPLE 9.3** Consider the continuous-time nonlinear system described by

$$\dot{x} = ax^2 - x^3 + bu, \quad (9.41)$$

where $a$ ($|a| < 1$) and $b$ are the unknown system parameters. The corresponding nonlinear model is given by

$$\dot{\hat{x}} = \hat{a}\hat{x}^2 - \hat{x}^3 + \hat{b}u. \quad (9.42)$$

Let us consider the following parameter values: $a = 0.9$, $b = 1$, $\hat{a} = 0.2$, and $\hat{b} = 0.4$. Selecting the control input $u = 2\hat{x}$ and the Lyapunov function $\frac{1}{2}x^2$, the following parameters can be obtained [24]: $\alpha_1(s) = \alpha_2(s) = \frac{1}{2}s^2$, $\alpha(s) = 0.84s$, and $\beta(s) = 2.66s^2$. Results of simulations are shown in Figure 9.7 for $\sigma = 0.8$. The top part of the figure shows the norms of the state and the error; the bottom part shows the communication time instants.

**EXAMPLE 9.4** Consider the following system [25]:

$$\begin{aligned}\dot{x}_1 &= b_1 u_1 \\ \dot{x}_2 &= b_2 u_2 \\ \dot{x}_3 &= a x_1 x_2,\end{aligned} \quad (9.43)$$

and the corresponding model is given by

$$\begin{aligned}\dot{\hat{x}}_1 &= \hat{b}_1 u_1 \\ \dot{\hat{x}}_2 &= \hat{b}_2 u_2 \\ \dot{\hat{x}}_3 &= \hat{a}\hat{x}_1\hat{x}_2.\end{aligned} \quad (9.44)$$

The system and model parameters are as follows: $a = 1$, $b_1 = 1$, $b_2 = 1$, $\hat{a} = 0.8$, $\hat{b}_1 = 1.5$, and $\hat{b}_2 = 1.4$. Let $\sigma = 0.6$. The control inputs are given by

$$\begin{aligned}u_1 &= -\hat{x}_1\hat{x}_2 - 2\hat{x}_2\hat{x}_3 - \hat{x}_1 - \hat{x}_3 \\ u_2 &= 2\hat{x}_1\hat{x}_2\hat{x}_3 + 3\hat{x}_3^2 - \hat{x}_2.\end{aligned} \quad (9.45)$$

Using the Lyapunov function $V(x) = \frac{1}{2}(x_1 + x_3)^2 + \frac{1}{2}(x_2 - x_3^2)^2 + x_3^2$, the following parameters can be obtained: $\alpha(s) = 91446s^2$, $\beta(s) = 147190s^2$. Figure 9.8 shows the states of the system, the norms of the error and state, and the communication time instants.

## 9.5 MB-ET Control of Discrete-Time Nonlinear Systems Using Dissipativity

Common approaches for control and analysis of NCSs often focus on state-based methods for both linear and nonlinear systems. These approaches ignore the strong

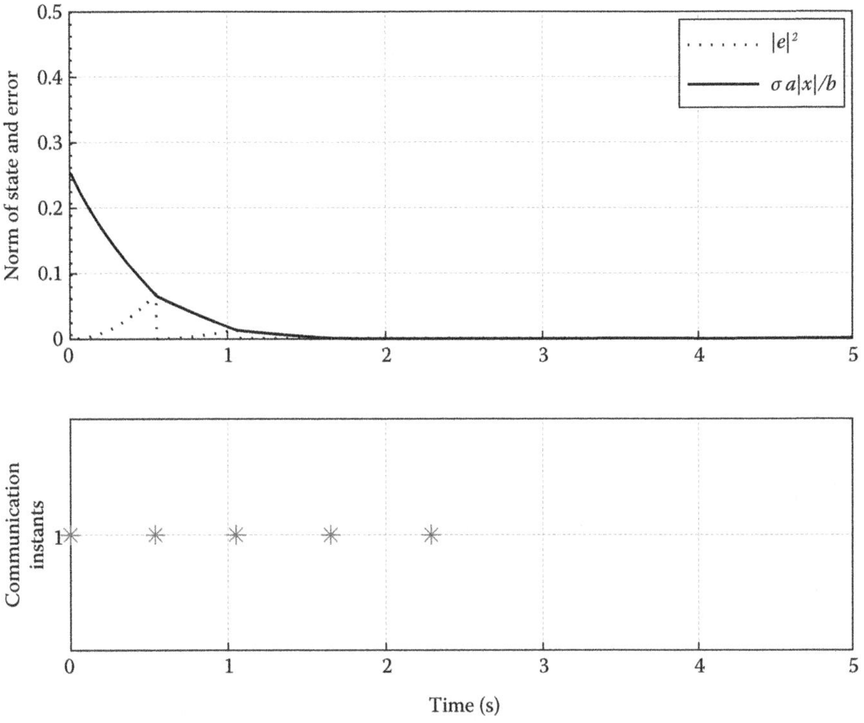


**FIGURE 9.7**
Error and state norms in Example 9.3. Corresponding communication time instants.

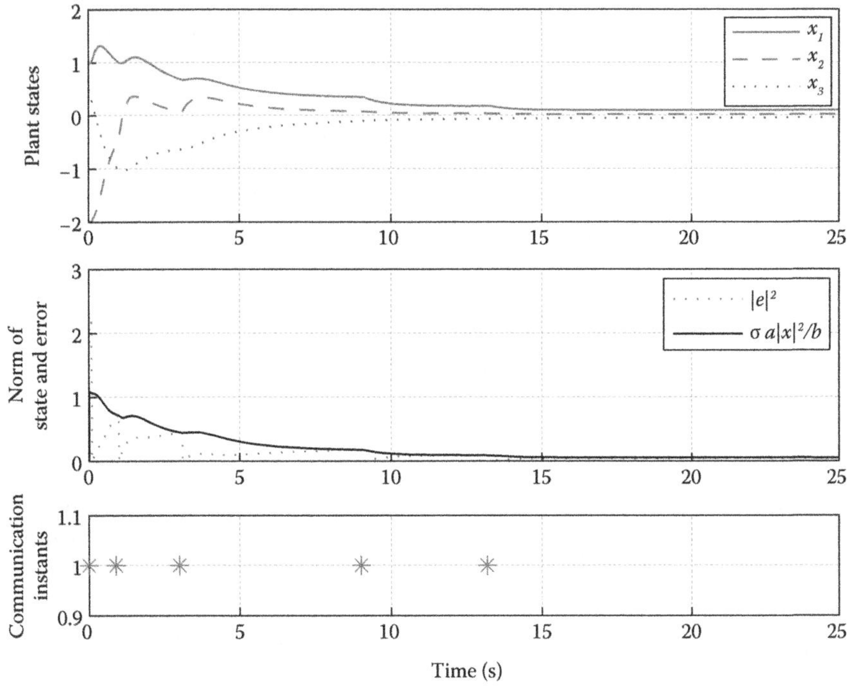


**FIGURE 9.8**
Top: states of the system in Example 9.4. Middle: error and state norms. Bottom: communication time instants.

tradition in nonlinear control to use output feedback methods to analyze feedback systems. These methods include the passivity theorem and the small-gain theorem, among others. While these analysis methods are not directly applicable to systems in the model-based framework, they are applicable to a modified version. Specifically, a signal-equivalent feedback system is derived for analyzing systems using output feedback. While passivity and finite-gain analysis are attractive options, this section focuses on the more general framework of dissipativity theory.

Dissipativity, in general, can be applied to continuous-time and to discrete-time nonlinear systems. However, the case of continuous-time systems with output feedback in the model-based framework would require a state observer in order to update the state of the model. On the other hand, discrete-time systems with output feedback can be implemented using the model-based approach without the need for a state observer by using an input–output representation of the system dynamics, as shown in this section.

This section considers the control of a discrete-time nonlinear dissipative system over a limited-bandwidth network. It is assumed that an appropriate controller has been designed for the system ignoring the effects of the network. At this point, an event-triggered aperiodic network communication law is used that takes advantage of the model-based framework to reduce communication. With the aperiodic communication, it is important to consider robustness to both model uncertainties and external disturbances due to the absence of feedback measurements at every time $k$. The goal of this section is to present a boundedness result that provides a constructive bound on the system output of a nonlinear system with output feedback. The system is subject to both external disturbances and model uncertainties; in addition, there is a lack of feedback measurements for extended intervals of time in order to reduce network usage. More details on this approach can be found in [35].

### 9.5.1 Problem Formulation

The systems of interest are single-input single-output (SISO) nonlinear discrete-time systems. The dynamics of these plants can be captured by the output model:

$$y(k) = f_{io}(y(k-1), \dots, y(k-n), u(k), \dots, u(k-m)), \quad (9.46)$$

where $y(k)$ is the current output and $u(k)$ is the current input. The current system output is a function of the $n$ previous outputs and the $m$ previous inputs. For a given plant, a model can be developed for use in the model-based framework according to the dynamics

$$\hat{y}(k) = \hat{f}_{io}(\hat{y}(k-1), \dots, \hat{y}(k-n), u(k), \dots, u(k-m)), \quad (9.47)$$

where the nonlinear function $\hat{f}_{io}(\cdot)$ represents the available model of the system function $f_{io}(\cdot)$. While the systems are controlled using only output feedback, internal models may be used for analysis purposes. These models are particularly useful for demonstrating dissipativity for a given system. As with the output models, these state-space models need not perfectly represent the system dynamics. The state-space dynamics of the plant may be given by

$$\begin{aligned} x(k+1) &= f(x(k), u(k)) \\ y(k) &= h(x(k), u(k)), \end{aligned} \quad (9.48)$$

where $x(k) \in \mathbb{R}^n$, $u(k) \in \mathbb{R}^m$, and $y(k) \in \mathbb{R}^p$. The plant may be modeled by the state-space model:

$$\begin{aligned} \hat{x}(k+1) &= \hat{f}(\hat{x}(k), u(k)) \\ \hat{y}(k) &= \hat{h}(\hat{x}(k), u(k)). \end{aligned} \quad (9.49)$$

The approach in this section makes use of the MB-ET control framework, but the approach acts on the system output instead of on the state. More details on event-triggered control for output feedback can be found in [26–30]. An intelligent sensor is implemented to compare the current system output measurement $y(k)$ to the estimated output value $\hat{y}(k)$ and only transmit the new output value when a triggering condition is satisfied. In this case, the triggering condition is when the error $e(k)$ between the current output and the estimated output,

$$e(k) = \hat{y}(k) - y(k), \quad (9.50)$$

grows above some positive threshold, $e(k) > \alpha > 0$. At this point, the current value and previous $n$ values of the output, based on the dimension of the system (9.46), are sent across the network. The model output $\hat{y}(k)$ is updated to equal $y(k)$, and the output error (9.50) is zero. This approach requires a copy of the model to be present at the sensor node to have continuous access to $\hat{y}(k)$. This additional computation is incurred in order to reduce the total network communication. Assuming no delay in updating the output, the error is always bounded:

$$|e(k)| \leq \alpha. \quad (9.51)$$

It should be noted that the reduction in network traffic is significant compared to the case in which a measurement of $y(k)$ is sent at every sampling instant. This is true even when the order of the system $n$ is large compared to the interupdate intervals. In this case, nearly every sample of the output is transmitted eventually. The average data rates are still significantly reduced when considering that bandwidth can be lost due to packet overhead and the minimum size of payload for each packet. As the minimum payload in a packet is typically much larger than a single measurement, by saving measurements and sending them all at the same time, a larger portion of the payload can be utilized, similar to the approaches in [31,32].

Feedback systems with periodic feedback typically have a low sensitivity to disturbances and unmodeled dynamics. This property is not guaranteed when considering aperiodic communication. This approach explicitly considers a plant-input disturbance $w_1(k)$ and a controller-input disturbance $w_2(k)$. Both signals are assumed to have bounded magnitude for all time $k$, but this magnitude may not go to zero. These signals can capture unmodeled dynamics as well as error introduced by discretization. These disturbances are unknown but have magnitude bounded by

$$|w_1(k)| \leq W_1(k) + c_1, \tag{9.52}$$

where the signal $W_1(k) > 0$ is an $\ell_2$ signal, and $c_1 \geq 0$ is a constant. Likewise, for $\ell_2$ signal $W_2(k) > 0$ and positive constant $c_2$,

$$|w_2(k)| \leq W_2(k) + c_2. \tag{9.53}$$

An example of such a disturbance and the appropriate bounds is given in Figure 9.9.

With all components of the problem formulation provided, the complete feedback system can be given in Figure 9.10. The switch indicates the aperiodic communication over the network. The input to the controller $u_c$ is the estimated output $\hat{y}$ with an added disturbance in $w_2$. The output of the controller $y_c$ is the calculated control effort. The actual control applied to the plant $u_p$ has some additional noise.

One of the novel components of this approach is in reformulating the MB framework into a traditional feedback problem. This is done by representing the output

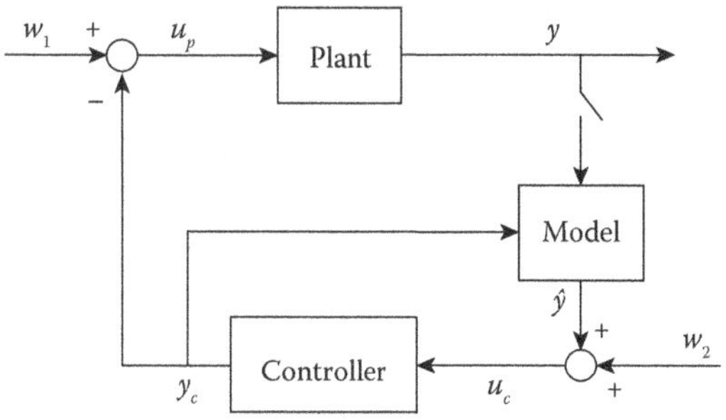


**FIGURE 9.10**
The feedback of the plant and combined controller/model. The two disturbance signals are $w_1$ and $w_2$, and the switch indicates the aperiodic communication over the network.

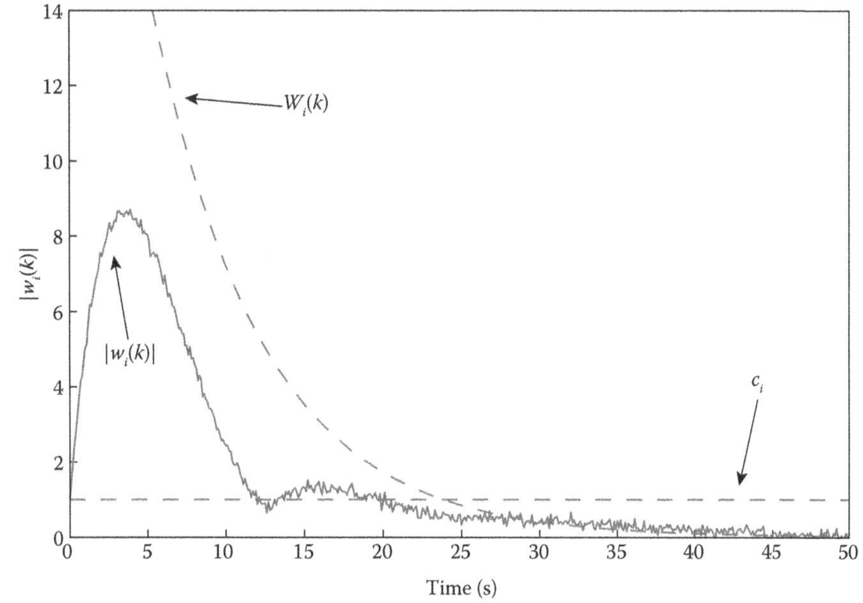


**FIGURE 9.9**
An example of an allowable disturbance $w(k)$ that is bounded by the sum of an $\ell_2$ signal $W(k) > 0$ and a constant $c > 0$.

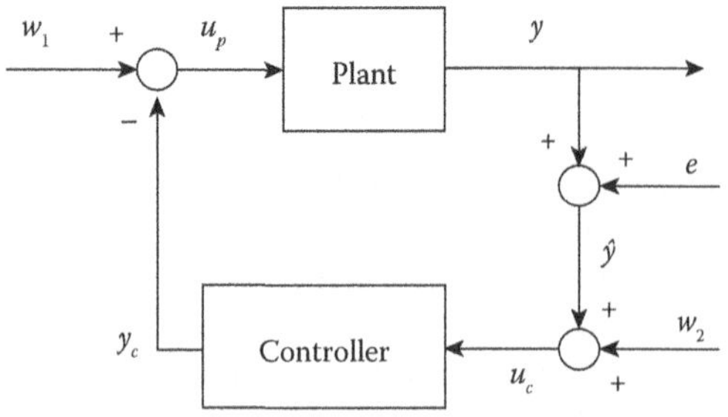


**FIGURE 9.11**
The feedback of the plant and the controller. While the model is not present, it is implicitly considered through the input of the model error $e$.

of the model $\hat{y}(k)$ as the sum of the plant output $y(k)$ and the error $e(k)$, which is now treated as an external input. In this signal-equivalent representation, the plant mapping from input $u_p(k)$ to output $y(k)$ and controller mapping from input $u_c(k)$ to output $y_c(k)$ are directly interconnected as in Figure 9.11.

In the absence of disturbances, the input to the plant is the output of the controller, and the input to the controller is the predicted plant output $\hat{y}$. This is consistent with the MB-NCS framework presented earlier in this chapter and consistent with the definition of the output error, $\hat{y}(k) = e(k) + y(k)$. The error $e(k)$ is still present in the absence of external disturbances due to model uncertainties. While this forms a standard negative feedback interconnection, it is not useful to show $\ell_2$ stability, as the error signal $e(k)$ is not an $\ell_2$ signal.

### 9.5.2 Dissipativity Theory

The current approach uses the quadratic form of dissipativity that is typically referred to as QSR dissipativity. This form is named after the parameters (matrices Q, S, and R) that define the relative importance of weighting the system input and output. In order to use dissipativity theory, some preliminaries will need to be defined.

The space of all finite energy discrete-time signals is referred to as $\ell_2$. The corresponding $\ell_2$ norm is given by

$$||w(k)||_{\ell_2} = \sqrt{\sum_{k=0}^{\infty} w^T(k)w(k)}. \tag{9.54}$$

A function $w(k)$ is in $\ell_2$ if it has finite $\ell_2$ norm. The related signal space is the superset that includes all signals with finite energy on any finite time interval. This is referred to as the extended $\ell_2$ space, or $\ell_{2e}$. The truncated $\ell_2$ norm is defined by the following:

$$||w_K(k)||_{\ell_2} = \sqrt{\sum_{k=0}^{K-1} w^T(k)w(k)}. \tag{9.55}$$

Signals in $\ell_{2e}$ have finite $\ell_2$ norm for all $K < \infty$. The systems of interest in this section map input signals $u(k) \in \ell_{2e}$ to signals $y(k) \in \ell_{2e}$. This assumption disallows inputs with finite escape time as well as systems that produce outputs with finite escape time. A system is $\ell_2$ stable if $u(k) \in \ell_2$ implies that $y(k) \in \ell_2$ for all $u(k) \in U$. An important special case of this stability is finite-gain $\ell_2$ stability, where the size of the system output can be bounded by an expression involving the size of the input. Specifically, a system is finite-gain $\ell_2$ stable if there exists a $\gamma$ and $\beta$ such that

$$||y_K(k)||_{\ell_2} \leq \gamma ||u_K(k)||_{\ell_2} + \beta, \tag{9.56}$$

$\forall K > 0$ and $\forall u(k) \in U$. The $\ell_2$ gain of the system is the smallest $\gamma$ such that there exists a $\beta$ to satisfy the inequality.

Dissipativity is an energy-based property of dynamical systems. This property relates energy stored in a system to the energy supplied to the system. The energy stored in the system is defined by an energy storage function $V(x)$. As a notion of energy, this function must be positive definite—that is, it must satisfy $V(x) > 0$ for $x \neq 0$ and $V(0) = 0$. The supplied energy is captured by an energy supply rate $\omega(u, y)$. A system is dissipative if it only stores and dissipates energy, with respect to the specific energy supply rate, and does not generate energy on its own.

**DEFINITION 9.1** A nonlinear discrete-time system (9.48) is dissipative with energy supply rate $\omega(u, y)$ if there exists a positive definite energy storage function $V(x)$ such that the following inequality holds:

$$V(x(k_2+1)) - V(x(k_1)) \leq \sum_{k=k_1}^{k_2} \omega(u(k), y(k)), \tag{9.57}$$

for all times $k_1$ and $k_2$, such that $k_1 \leq k_2$.

A particularly useful form of dissipativity with additional structure is the quadratic form, QSR dissipativity.

**DEFINITION 9.2** A discrete-time system (9.48) is QSR dissipative if it is dissipative with respect to the supply rate

$$\omega(u, y) = \begin{bmatrix} y \\ u \end{bmatrix}^T \begin{bmatrix} Q & S \\ S^T & R \end{bmatrix} \begin{bmatrix} y \\ u \end{bmatrix}, \tag{9.58}$$

where $Q = Q^T$ and $R = R^T$.

The QSR dissipative framework generalizes many areas of nonlinear system analysis. The property of passivity can be captured when $Q = R = 0$ and $S = \frac{1}{2}I$, where $I$ is the identity matrix. Systems that are finite-gain $\ell_2$ stable can be represented by $S = 0$, $Q = -\frac{1}{\gamma}I$, and $R = \gamma I$, where $\gamma$ is the gain of the system. The following theorems give stability results for QSR dissipative systems and dissipative systems in feedback.

**Theorem 9.6**

A discrete-time system is finite-gain $\ell_2$ stable if it is QSR dissipative with $Q < 0$.

**PROOF** The system being QSR dissipative implies that there exists a positive definite storage function $V(x)$ such that,

$$V(x(k_2+1)) - V(x(k_1)) \leq \sum_{k=k_1}^{k_2} \left[y^T Q y + 2y^T S u + u^T R u\right],$$

for all $k_2 \geq k_1$. The substitutions $k_1 = 0$ and $k_2 = K$ will be made. As $Q < 0$, there exists a real number $q > 0$ such that $Q \leq -qI$. Similarly, the matrices $S$ and $R$ can be bounded by their largest singular values ($s$ and $r$), $S \leq sI$ and $R \leq rI$, which gives

$$V(x(K+1)) - V(x(0)) \leq \sum_{k=0}^{K} \left[-q\,||y||^2 + 2s\,||y||\,||u|| + r\,||u||\right].$$

Rearranging terms and completing the square to remove the cross term yields

$$V(x(K+1)) - V(x(0)) \leq \sum_{k=0}^{K} \left[-\frac{1}{2q}(2s\,||u|| - q\,||y||)^2 - \frac{q}{2}\,||y||^2 + \frac{4s^2+2qr}{2q}\,||u||^2\right].$$

The squared term and $V(x(K+1))$ can be removed without violating the inequality. The remaining terms can be rearranged to give

$$\sum_{k=0}^{K} ||y||^2 \leq \sum_{k=0}^{K} \frac{4s^2+2qr}{q^2}\,||u||^2 + \frac{2}{q}V(x(0)).$$

Letting $K \to \infty$ and evaluating the summations on the norms yields the $\ell_2$ norms:

$$||y||_{\ell_2}^2 \leq \frac{4s^2+2qr}{q^2}\,||u||_{\ell_2}^2 + \frac{2}{q}V(x(0)).$$

The square root of this expression can be taken, and the fact that $\sqrt{a^2+b^2} \leq |a| + |b|$ can be used to show

$$||y||_{\ell_2} \leq \frac{\sqrt{4s^2+2qr}}{q}\,||u||_{\ell_2} + \sqrt{\frac{2}{q}V(x(0))}.$$

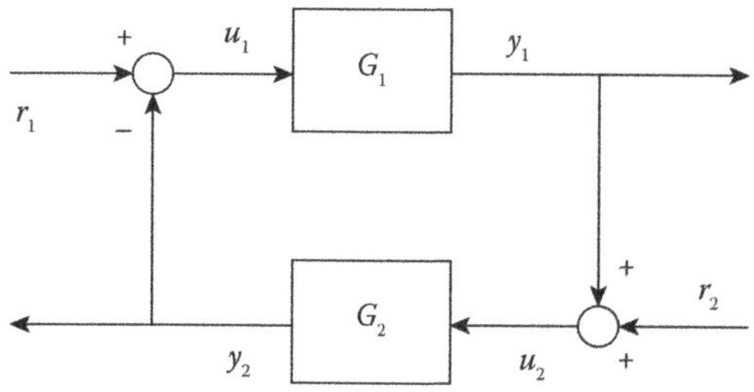


**FIGURE 9.12**
The feedback interconnection of systems $G_1$ and $G_2$.

This shows $\ell_2$ stability with $\gamma = \frac{\sqrt{4s^2+2qr}}{q}$ and $\beta = \sqrt{\frac{2}{q}V(x(0))}$. ■

**Theorem 9.7**

Consider the feedback interconnection of two QSR dissipative systems (Figure 9.12). System $G_1$ is dissipative with respect to $Q_1$, $S_1$, $R_1$ and system $G_2$ with respect to $Q_2$, $S_2$, $R_2$. The feedback interconnection is $\ell_2$ stable if there exists a positive constant $a$ such that the following matrix is negative definite:

$$\tilde{Q} = \begin{bmatrix} Q_1 + aR_2 & -S_1 + aS_2^T \\ -S_1^T + aS_2 & R_1 + aQ_2 \end{bmatrix} < 0. \tag{9.59}$$

**PROOF** Each system being QSR dissipative implies the existence of positive definite storage functions $V_1$ and $V_2$ that satisfy

$$V_1(x_1(k_2+1)) - V_1(x_1(k_1)) \leq \sum_{k=k_1}^{k_2} \begin{bmatrix} y_1 \\ u_1 \end{bmatrix}^T \begin{bmatrix} Q_1 & S_1 \\ S_1^T & R_1 \end{bmatrix} \begin{bmatrix} y_1 \\ u_1 \end{bmatrix},$$

$$V_2(x_2(k_2+1)) - V_2(x_2(k_1)) \leq \sum_{k=k_1}^{k_2} \begin{bmatrix} y_2 \\ u_2 \end{bmatrix}^T \begin{bmatrix} Q_2 & S_2 \\ S_2^T & R_2 \end{bmatrix} \begin{bmatrix} y_2 \\ u_2 \end{bmatrix},$$

where $x_i$, $u_i$, and $y_i$ are the state, input, and output of the $i$th system, respectively. The signal relationships in the feedback loop can be given by

$$u_1 = r_1 - y_2,$$
$$u_2 = r_2 + y_1.$$

A total energy storage function for the loop can be defined as

$$V(x) = V_1(x_1) + aV_2(x_2),$$

for a positive constant $a > 0$, where

$$x = \begin{bmatrix} x_1 \\ x_2 \end{bmatrix}, \quad u = \begin{bmatrix} u_1 \\ u_2 \end{bmatrix}, \quad \text{and} \quad y = \begin{bmatrix} y_1 \\ y_2 \end{bmatrix}.$$

Looking at the change in $V(x)$ over a time interval yields

$$V(x(k_2+1)) - V(x(k_1)) \leq \sum_{k=k_1}^{k_2} \begin{bmatrix} y_1 \\ u_1 \end{bmatrix}^T \begin{bmatrix} Q_1 & S_1 \\ S_1^T & R_1 \end{bmatrix} \begin{bmatrix} y_1 \\ u_1 \end{bmatrix} + a \sum_{k=k_1}^{k_2} \begin{bmatrix} y_2 \\ u_2 \end{bmatrix}^T \begin{bmatrix} Q_2 & S_2 \\ S_2^T & R_2 \end{bmatrix} \begin{bmatrix} y_2 \\ u_2 \end{bmatrix}.$$

The signal relationships can be substituted and the expression simplified to yield

$$V(x(k_2+1)) - V(x(k_1)) \leq \begin{bmatrix} y \\ u \end{bmatrix}^T \begin{bmatrix} \tilde{Q} & \tilde{S} \\ \tilde{S}^T & \tilde{R} \end{bmatrix} \begin{bmatrix} y \\ u \end{bmatrix},$$

where

$$\tilde{Q} = \begin{bmatrix} Q_1 + aR_2 & aS_2^T - S_1 \\ aS_2 - S_1^T & R_1 + aQ_2 \end{bmatrix},$$

$$\tilde{S} = \begin{bmatrix} S_1 & aR_2 \\ -R_1 & aS_2 \end{bmatrix}, \quad \text{and} \quad \tilde{R} = \begin{bmatrix} R_1 & 0 \\ 0 & aR_2 \end{bmatrix}.$$

This shows that the feedback of two QSR dissipative systems is again QSR dissipative. If $\tilde{Q} < 0$, Theorem 9.6 can be applied to demonstrate $\ell_2$ stability. ∎

QSR dissipativity can be used to assess the stability of a single system as well as systems in feedback. From a control design perspective, the QSR parameters of a given plant can be determined and used to find bounds on stabilizing QSR parameters of a potential controller. More details about general dissipativity can be found in [33], while the case of QSR dissipativity can be found in [34].

### 9.5.3 Output Boundedness Result

The main result in this approach is a tool that allows the size of the system output to be bounded when operating nonlinear discrete-time systems in the network configuration described in the previous section. As discussed previously, there are two issues with traditional stability for this network setup. The first is due to the aperiodic control updates. Between update events, the feedback system is temporarily operating as an open loop. With even small model mismatch between the actual plant and the model, the outputs between the two can drift significantly over time. Typically, the system output does not go to zero and, thus, cannot be bounded as in finite-gain $\ell_2$ stability. The second issue with traditional notions of stability is that this work allows nonvanishing input disturbances. Traditional dissipativity theory shows stability for disturbances that are in $\ell_2$ (i.e., the disturbance must converge to zero asymptotically). This approach generalizes existing results to disturbances that may not go to zero but do have an ultimate bound.

While notions of asymptotic stability or finite-gain $\ell_2$ stability are appealing, they are simply not achievable in this framework. Instead this is relaxed to a boundedness result. As this work considers systems described by an input–output relationship, the notion of $\ell_2$ stability is relaxed to a bound on the output as time goes to infinity. With output error and disturbances that are nonvanishing, the output may fluctuate over a large range. Due to fluctuations in the system input, it may be difficult to find an ultimate bound on the size of the output for all time. Instead, this framework considers an average bound on the squared system output.

**DEFINITION 9.3** A nonlinear system is average output squared bounded if after time $\bar{k}$, there exists a constant $b$ such that the following bound on the output holds for all times $k_1$ and $k_2$ larger than ($\bar{k} \leq k_1 < k_2$):

$$\frac{1}{(k_2 - k_1)} \sum_{k=k_1}^{k_2-1} y^T(k) y(k) \leq b. \tag{9.60}$$

This form of boundedness is a practical form of stability on the system output. While the output does not necessarily converge to zero, it is bounded on average with a known bound as time goes to infinity. It is important to note that this concept is not useful for an arbitrarily large bound $b$. However, the concept is informative for a small, known bound. The notion should be restricted to being used in the case when the bound is constructive and preferably when the bound can be made arbitrarily small by adjusting system parameters.

The following boundedness theorem can be applied to the analysis of a plant and a controller in the model-based framework. The plant and the model of the plant must be QSR dissipative with respect to parameters $Q_P$, $S_P$, and $R_P$. Although the plant dynamics are not known exactly, sufficient testing can be done to verify that the dissipative rate bounds the actual dissipative behavior of the system. The model-stabilizing QSR dissipative controller has been designed with parameters $Q_C$, $S_C$, and $R_C$.

**Theorem 9.8**

Consider a plant and controller in the MB-NCS framework where model mismatch may exist between the plant and the model. The network structure contains event-triggered, aperiodic updates and nonvanishing disturbances. This feedback system is average output squared bounded if there exists a positive constant $a$ such that the following matrix is negative definite

$$\tilde{Q} = \begin{bmatrix} Q_P + aR_C & aS_C^T - S_P \\ aS_C - S_P^T & R_P + aQ_C \end{bmatrix} < 0. \tag{9.61}$$

**PROOF** The plant and controller being QSR dissipative implies the existence of positive storage functions $V_P$ and $V_C$, such that

$$\Delta V_P(x_P) \leq \begin{bmatrix} y \\ u_P \end{bmatrix}^T \begin{bmatrix} Q_P & S_P \\ S_P^T & R_P \end{bmatrix} \begin{bmatrix} y \\ u_P \end{bmatrix},$$

and a similar bound on $\Delta V_C$. A total energy storage function can be defined, $V(x) = V_P(x_P) + aV_C(x_C)$, where $x = [x_P^T\ x_C^T]^T$. The total energy storage function has the dissipative property

$$\Delta V(x) \leq \begin{bmatrix} y \\ y_C \\ w_1 \\ (w_2+e) \end{bmatrix}^T \begin{bmatrix} \tilde{Q} & \tilde{S} \\ \tilde{S}^T & \tilde{R} \end{bmatrix} \begin{bmatrix} y \\ y_C \\ w_1 \\ (w_2+e) \end{bmatrix},$$

where

$$\tilde{Q} = \begin{bmatrix} Q_P + aR_C & aS_C^T - S_P \\ aS_C - S_P^T & R_P + aQ_C \end{bmatrix},$$

$$\tilde{S} = \begin{bmatrix} S_P & aR_C \\ -R_P & aS_C \end{bmatrix}, \text{ and } \tilde{R} = \begin{bmatrix} R_P & 0 \\ 0 & aR_C \end{bmatrix}.$$

Due to $\tilde{Q} < 0$ (9.61), there exists a constant $q > 0$ such that $\tilde{Q} \leq -qI$. As $\tilde{S}$ and $\tilde{R}$ are constant matrices, the largest singular values, $s$ and $r$, respectively, may be used to show the following bound on $\Delta V$:

$$\begin{aligned}\Delta V(x) \leq & -q[y^Ty + y_c^Ty_c] + 2s[y^Tw_1 + y_c^T(w_2+e)] \\ & + r[w_1^Tw_1 + (w_2+e)^T(w_2+e)].\end{aligned}$$

Completing the square can be used to remove the cross terms:

$$\begin{aligned}\Delta V(x) \leq & -\frac{q}{2}[y^Ty + y_c^Ty_c] \\ & + \frac{(4s^2+2qr)}{2q}[w_1^Tw_1 + w_2^Tw_2 + e^Te].\end{aligned}$$

Summing this inequality from $k_1$ to $k_2$ yields the following:

$$\begin{aligned}V(x(k_2)) \leq & V(x(k_1)) - \frac{q}{2}\sum_{k=k_1}^{k_2-1}[y^Ty + y_c^Ty_c] \\ & + \frac{(4s^2+2qr)}{2q}\sum_{k=k_1}^{k_2-1}[w_1^Tw_1 + w_2^Tw_2 + e^Te].\end{aligned}$$

The effect of the nonvanishing disturbances $w_1$ and $w_2$ can be bounded by constants $\epsilon_1$ and $\epsilon_2$ after some time $\bar{k}$, $|w_i(k)| \leq \epsilon_i$, for $k \geq \bar{k}$. Additionally, $|e(k)| < \alpha$ for all $k$. A single bound can be defined $\epsilon^2 = \epsilon_1^2 + \epsilon_2^2 + \alpha^2$. At this point, either the average output squared is bounded by the following:

$$\frac{1}{(k_2-k_1)}\sum_{k=k_1}^{k_2-1} y^Ty \leq \frac{(4s^2+2qr)\epsilon^2}{q^2(1-\delta)}, \tag{9.62}$$

where $0 < \delta < 1$, or not bounded by it,

$$\frac{1}{(k_2-k_1)}\sum_{k=k_1}^{k_2-1} y^Ty > \frac{(4s^2+2qr)\epsilon^2}{q^2(1-\delta)}. \tag{9.63}$$

When it is larger than this quantity, it is possible to show

$$\begin{aligned}V(x(k_2)) \leq & V(x(k_1)) - \frac{q\delta}{2}\sum_{k=k_1}^{k_2-1} y^Ty \\ & - \frac{(4s^2+2qr)}{2q}\sum_{k=k_1}^{k_2-1}[\epsilon^2 - w_1^Tw_1 - w_2^Tw_2 - e^Te].\end{aligned}$$

This can be used to show a bound on $y$:

$$\sum_{k=k_1}^{k_2-1} y^Ty \leq \frac{2}{q\delta}V(x(k_1)). \tag{9.64}$$

As the sum of $y^Ty$ is bounded, the average is also bounded. Either (9.62) or (9.64) holds, which shows that the average squared system output is bounded, satisfying Definition 9.3. Furthermore, the parameter $\delta$ can be adjusted to vary (and potentially lower) the relative size of the two bounds. As (9.64) is a fixed bound, the average output will be continually shrinking. Asymptotically, bound (9.62) will hold as $\delta \to 0$. ■

One important takeaway is that the bound on the system output is constructive. The bounds can be made smaller by adjusting the values of controller, which changes $q$, $s$, and $r$. The bounds also depend on the value of the output error threshold $\alpha$, which can be made small. The effect of the nonvanishing disturbances may be significant depending on $\epsilon_1$ and $\epsilon_2$. When these disturbances are vanishing, the bound on the output depends mainly on $\alpha$, which may be made arbitrarily small. The value of $\alpha$ may be chosen to trade off decreased communication with reduced output error.

### 9.5.4 Example

The following example was chosen to be LTI for ease of following, but the results apply to nonlinear systems as well. The QSR parameters for each system were found using state-space models, but the NCS is simulated using the equivalent input–output representation for the plant, controller, and model. The system to be controlled is unstable and uncertain with the model given by

$$\hat{A} = \begin{bmatrix} 1.05+\delta_1 & -1 \\ 0 & 0.85+\delta_2 \end{bmatrix}, \hat{B} = \begin{bmatrix} 1 \\ 0 \end{bmatrix},$$
$$\hat{C} = \begin{bmatrix} 0.5+\delta_3 & 1 \end{bmatrix}, \hat{D} = 1. \quad (9.65)$$

The model can be shown QSR dissipative ($Q_P = 0.1$, $S_P = 0.15$, and $R_P = 0$) by using the storage function:

$$\hat{V}(\hat{x}) = \hat{x}^T \begin{bmatrix} 0.12 & 0.36 \\ 0.36 & 30 \end{bmatrix} \hat{x}. \quad (9.66)$$

An example of a stabilizing controller is given by $A_C = 0.3$, $B_C = 0.8$, $C_C = 0.7$, and $D_C = 1$. This controller is QSR dissipative ($Q_C = -0.2$, $S_C = 0.15$, and $R_C = -0.3$), which can be shown using storage function, $V_c(x_c) = 0.23{x_u}^2$. The controller can be shown to stabilize the model by evaluating (9.61) with $a = 1$:

$$\tilde{Q} = \begin{bmatrix} -0.2 & 0 \\ 0 & -0.2 \end{bmatrix} < 0. \quad (9.67)$$

As discussed earlier, the actual plant is dissipative with respect to the same QSR parameters ($Q_P$, $S_P$, and $R_P$). The plant is assumed to be unknown but similar to model (9.65) where $\delta_1 = 0.09$, $\delta_2 = -0.07$, and $\delta_3 = 0.18$, which gives dynamics

$$A = \begin{bmatrix} 1.14 & -1 \\ 0 & 0.78 \end{bmatrix}, B = \begin{bmatrix} 1 \\ 0 \end{bmatrix}, C = \begin{bmatrix} 0.68 & 1 \end{bmatrix}, D = 1. \quad (9.68)$$

The QSR parameters can be verified for the plant using the same storage function.

By assumption, the controller also stabilizes the plant and satisfies the inequality for boundedness. This MB-NCS was simulated with input–output models for the plant, model, and controller. The external disturbances $w_1(k)$ and $w_2(k)$ for this example are shown in Figure 9.13.

The disturbances are bounded by 0.1 after time $k = 50$. With this magnitude of disturbance, it is not possible to guarantee that the output error stays less than 0.1. The threshold for the output error was chosen to be 0.2. These systems were simulated, and the system outputs are shown in Figure 9.14. The evolution of the output error over time is shown in the first subplot of Figure 9.15. This plot shows the error after each update takes place, that is, when the error is reset to zero. As a result, the error is always bounded as stated in (9.51). The second subplot shows the time instants at which output measurements are sent from the sensor node to the controller node. The rest of the time the networked system operates in an open loop.

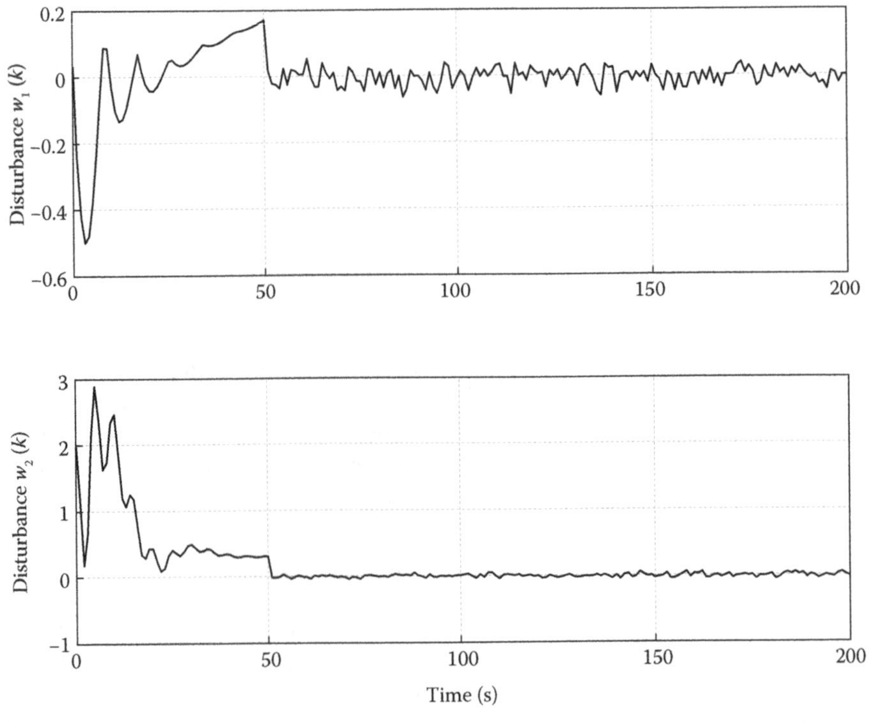


**FIGURE 9.13**
The nonvanishing disturbances $w_1(k)$ and $w_2(k)$.

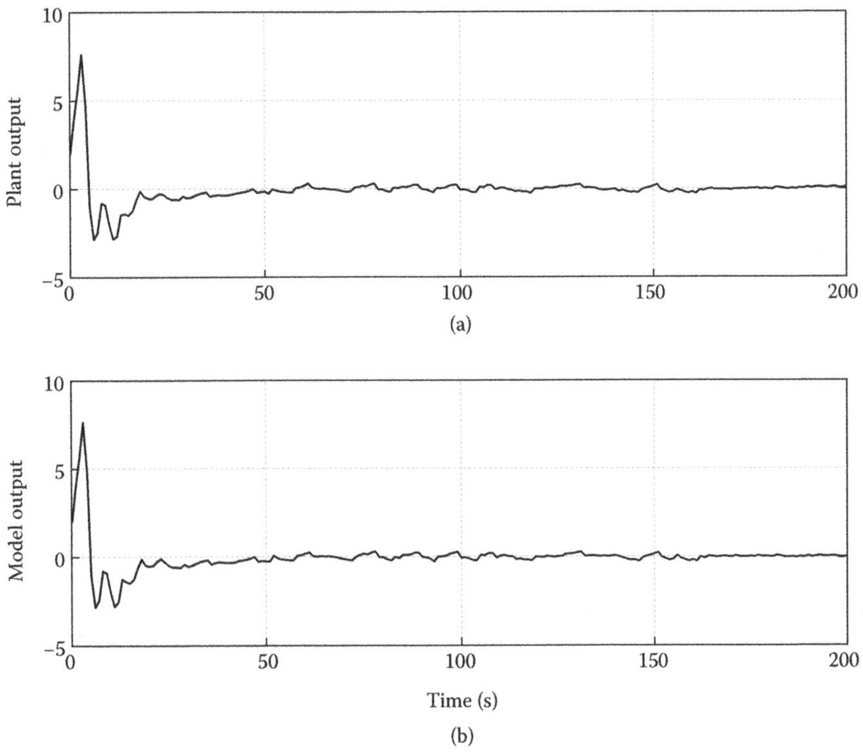


**FIGURE 9.14**
The output of the plant (a) and the model (b). The model tracks the plant closely within the update threshold.

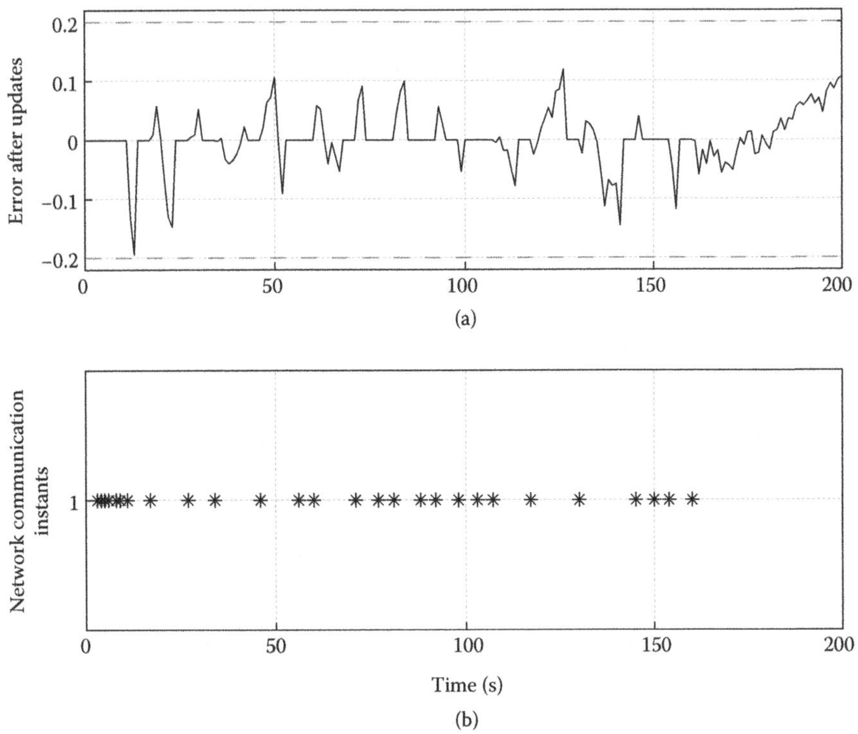


**FIGURE 9.15**
The output error (a) and communication instants (b) indicated by asterisks. The communication instants align with the pretransmission error growing to above 0.2.

Figure 9.14 shows that the outputs of the MB-NCS are bounded, as expected. Clearly, the outputs do not converge to zero, but the model output tracks the plant output closely. In this example, communication is relatively constant initially as the system responds to the large disturbances. After this point, the communication rate drops significantly. While the error stays bounded by 0.2, the communication rate is reduced by 86.5%. For comparison, a simulation was run with the output being transmitted at every time instant. For this case, after $k = 50$, the output error was as large as 0.165. The periodic output feedback provides a small reduction in output error, with an average data transmission rate that is approximately 7.4 times higher.

## 9.6 Additional Topics on Model-Based Event-Triggered Control

This section briefly describes additional problems and applications where the MB-ET control framework has been implemented.

### 9.6.1 Parameter Estimation

The first topic discussed here is concerned with the implementation of parameter estimation algorithms. The main idea is to estimate the current parameters of the real system, because in many problems, the dynamics of the system may change over time due to age, use, or nature of the physical plant. In many other situations, the initial given model is simply outdated or inaccurate. Better knowledge of the plant dynamics will provide an improvement in the control action over the network (i.e., we can achieve longer periods of time without need for feedback). Improved estimates of the plant parameters also lead to the update of the controller gain, so it can better respond to the dynamics of the real plant being controlled.

Let us consider the model-based networked architecture in Figure 9.1, and let the plant and model dynamics be represented in state-space form:

$$x(k+1) = Ax(k) + Bu(k), \tag{9.69}$$

$$\hat{x}(k+1) = \hat{A}\hat{x}(k) + \hat{B}u(k), \tag{9.70}$$

where $x, \hat{x} \in \mathbb{R}^n$, $u \in \mathbb{R}^m$. The control input is defined as $u(k) = K\hat{x}(k)$. The state of the model is updated at times $k_i$, when an event is triggered and a measurement transmitted from the sensor node to the controller node. The parameters of the model may or may not be updated at the same time instants depending on whether new estimated parameters are available.

For discrete-time linear systems of the form (9.69), it is possible to estimate the elements of the matrices $A$ and $B$ using a linear Kalman filter. In order to show this simple idea, let us focus on second-order autonomous systems with unknown time-invariant parameters $a_{ij}$. (The idea can be easily extended to higher-order systems with deterministic inputs.) The system can be written as follows:

$$\begin{bmatrix} x_1(k+1) \\ x_2(k+1) \end{bmatrix} = \begin{bmatrix} a_{11} & a_{12} \\ a_{21} & a_{22} \end{bmatrix} \begin{bmatrix} x_1(k) \\ x_2(k) \end{bmatrix}. \tag{9.71}$$

We do not know the values of the parameters, and we only receive measurements of the states $x(0), \ldots, x(k)$. At any given step, due to the iterative nature of the Kalman filter, we only need $x(k)$ and $x(k-1)$. Now we rewrite (9.71) as

$$\begin{bmatrix} \bar{x}_1(k) \\ \bar{x}_2(k) \end{bmatrix} = \begin{bmatrix} x_1(k-1) & x_2(k-1) & 0 & 0 \\ 0 & 0 & x_1(k-1) & x_2(k-1) \end{bmatrix} \times \begin{bmatrix} \hat{a}_{11}(k) \\ \hat{a}_{12}(k) \\ \hat{a}_{21}(k) \\ \hat{a}_{22}(k) \end{bmatrix} = \hat{C}(k)\hat{a}(k), \tag{9.72}$$

where $\hat{a}_{ij}$ represents the estimated values of the real parameters $a_{ij}$, and $\bar{x}_i(k)$ are the estimates of the state based on the estimated parameters and on measurements of the real plant state $x_i(k-1)$. Equation 9.72 represents the output equation of our filter. The state equation is described by

$$\begin{bmatrix} \hat{a}_{11}(k+1) \\ \hat{a}_{12}(k+1) \\ \hat{a}_{21}(k+1) \\ \hat{a}_{22}(k+1) \end{bmatrix} = \begin{bmatrix} 1 & 0 & 0 & 0 \\ 0 & 1 & 0 & 0 \\ 0 & 0 & 1 & 0 \\ 0 & 0 & 0 & 1 \end{bmatrix} \begin{bmatrix} \hat{a}_{11}(k) \\ \hat{a}_{12}(k) \\ \hat{a}_{21}(k) \\ \hat{a}_{22}(k) \end{bmatrix} = I\hat{a}(k). \tag{9.73}$$

It can be seen that the systems (9.73) and (9.72) are linear time-varying ones. Thus, we can use a linear filter to obtain estimates of the parameters $a_{ij}$. Details related to convergence of the estimated variables and to the stability of the adaptive networked system using the MB-ET control framework can be found in [12].

The case of stochastic systems described by

$$\begin{aligned} x(k+1) &= Ax(k) + Bu(k) + w(k) \\ y(k) &= x(k) + v(k), \end{aligned} \tag{9.74}$$

was also considered in [12], where the input and measurement noises, $w(k)$ and $v(k)$, are white, Gaussian, uncorrelated, and zero-mean, and have known covariance matrices $Q$ and $R$, respectively. The model of the system is still given by (9.70) and since we only measure $y(k)$, the error is now given by $e(k) = \hat{x}(k) - y(k)$. In this case, a nonlinear Kalman filter, such as the extended

Kalman filter (EKF), can be used for the estimation of states and parameters. The challenges and difficulties of using the EKF for parameter estimation and its implementation in the MB-ET control architecture have been documented in [12].

### 9.6.2 Optimal MB-ET Control

Event-triggered control techniques have been used to reduce communication between sensor and controller nodes. In general, the objective of the event-triggered control system is to achieve stability in the absence of continuous feedback measurements. Optimal control problems in NCS can be formulated in order to optimize network usage, in addition to system response and control effort.

The MB-ET control framework can be used to address the optimal control-scheduling problem. Here, we consider the design of optimal control laws and optimal thresholds for communication in the presence of *plant-model mismatch* by appropriately weighting the system performance, the control effort, and the communication cost. An approximate solution to this challenging problem is to optimize the performance of the nominal control system, which can be unstable in general, and to ensure robust stability for a given class of model uncertainties and for lack of feedback for extended intervals of time. The optimal scheduling problem is approximated by minimizing the state error and also considering the cost that needs to be paid every time we decide to send a measurement to update the model and reset the state error.

Let us consider system (9.69) and model (9.70). Assume that $\hat{B} = B$, that is, plant-model mismatch is only present in the state matrix $A$. The state error is defined as $e(k) = \hat{x}(k) - x(k)$. The following cost function weights the performance of the system, the control effort, and the price to transmit measurements from the sensor node to the controller node:

$$\min_{u,\beta} J = x^T(N)Q_N x(N) + \sum_{k=0}^{N-1} \left[x^T(k)Qx(k) + u^T(k)Ru(k) + S\beta(k)\right], \quad (9.75)$$

where $Q$ and $Q_N$ are real, symmetric, and positive semidefinite matrices, and $R$ is a real, symmetric, and positive definite matrix. The weight $S$ penalizes network communication, and $\beta(k)$ is a binary decision variable defined as follows:

$$\beta(k) = \begin{cases} 0 & \text{sensor does not send a measurement at time } k \\ 1 & \text{sensor sends a measurement at time } k \end{cases}. \quad (9.76)$$

The approximate solution to this problem offered in [36] consists of designing the optimal controller by solving the nominal linear quadratic regulator problem—that is, the control input is given by

$$u^*(N-i) = -[B^T P(i-1)B + R]^{-1} B^T P(i-1) \times \hat{A}x(N-i), \quad (9.77)$$

where $P(i)$ is recursively computed using the following:

$$P(i) = [\hat{A} + BK(N-i)]^T P(i-1)[\hat{A} + BK(N-i)] + Q + K^T(N-i)RK(N-i). \quad (9.78)$$

Now, the optimal scheduling problem can be seen as the minimization of the deviation of the system performance from the nominal performance by also considering the cost that needs to be paid by updating the model and resetting the state error. The error dynamics are given by

$$e(k+1) = \hat{x}(k+1) - x(k+1) = \hat{A}e(k) - \tilde{A}x(k), \quad (9.79)$$

where $\tilde{A} = A - \hat{A}$. Since the error matrix $\tilde{A}$ is not known, we use the nominal error dynamics:

$$e(k+1) = \hat{A}e(k).$$

Furthermore, when the sensor decides to send a measurement update, which makes $\beta(k) = 1$, we reset the error to zero. Then the complete nominal error dynamics can be represented by the following equation:

$$e(k+1) = (1 - \beta(k))\hat{A}e(k). \quad (9.80)$$

It is clear that in the nominal case, once we update the model, the state error is equal to zero for the remaining time instants. However, in a real problem, the state-error dynamics are disturbed by the state of the real system, which is propagated by means of the model uncertainties as expressed in (9.79). Then, using the available model dynamics, we implement the optimal control input and the optimal scheduler that result from the following optimization problem:

$$\begin{gathered} \min_\beta J = e^T(N)Q_N e(N) + \textstyle\sum_{k=0}^{N-1} e^T(k)Qe(k) + S\beta(k) \\ \text{subject to:} \\ e(k+1) = (1-\beta(k))\hat{A}e(k) \\ \beta(k) \in \{0,1\}. \end{gathered} \quad (9.81)$$

In order to solve the problem (9.81) we can use dynamic programming in the form of lookup tables. The main reason for using dynamic programming is that although the error will be finely quantized, the decision variable $\beta(k)$ takes only two possible values, which reduces the amount of computations performed by the dynamic programming algorithm. The sensor operations at time $k$ are reduced to measure the real state, compute and quantize the state error, and determine if the current measurement needs to be transmitted by looking at the corresponding table entries that are computed offline. The table size depends only on the horizon $N$ and the error quantization levels.

### 9.6.3 Distributed Systems

Many control applications consider the interactions of different subsystems or agents. The increased use of communication networks has made possible the transmission of information among different subsystems in order to apply improved control strategies using information from distant subsystem nodes. The MB-ET control architecture has been extended to consider this type of control system, see [13]. In the control of distributed coupled systems, the improvement in the use of network resources is obtained by implementing models of other subsystems within each local controller.

The dynamics of a set of $N$ coupled subsystems, $i = 1, \ldots, N$, can be represented by

$$\dot{x}_i = A_i x_i + B_i u_i + \sum_{j=1, j \neq i}^{N} A_{ij} x_j, \tag{9.82}$$

and the nominal models are described by

$$\dot{\hat{x}}_i = \hat{A}_i \hat{x}_i + \hat{B}_i \hat{u}_i + \sum_{j=1, j \neq i}^{N} \hat{A}_{ij} \hat{x}_j, \tag{9.83}$$

where $x_i, \hat{x}_i \in \mathbb{R}^{ni}$, and $u_i, \hat{u}_i \in \mathbb{R}^{mi}$. In this framework, each node or subsystem implements copies of the models of all subsystems, including the model corresponding to its own local dynamics in order to generate estimates of the states of all subsystems in the network. Note that the subsystems can have different dynamics and different dimensions as well. The dimensions $mi$ and $ni$ can be all different in general. Note also that each node has access to its local state $x_i$ at all times, which is used to compute the local subsystem control input defined by

$$u_i = K_i x_i + \sum_{j=1, j \neq i}^{N} K_{ij} \hat{x}_j, \tag{9.84}$$

while the model control inputs, $\hat{u}_i$, are given by

$$\hat{u}_i = K_i \hat{x}_i + \sum_{j=1, j \neq i}^{N} K_{ij} \hat{x}_j, \tag{9.85}$$

where $K_i$ and $K_{ij}$ are the stabilizing control gains to be designed. The local state is also used to compute the local state error, which is given by $e_i = \hat{x}_i - x_i$.

By measuring its local error, each subsystem is able to decide the appropriate times $t_i$ at which it should broadcast the current measured state $x_i$ to all other subsystems so that they can update the state of their local models, $\hat{x}_i$, corresponding to the state $x_i$. At the same time, the subsystem that transmitted its state needs to update its own local model corresponding to $x_i$ which makes the local state error equal to zero, that is, $e(t_i) = 0$.

A decentralized MB-ET control technique to achieve asymptotic stability of the overall uncertain system was presented in [13]. This approach is decentralized since each subsystem only needs its own local state and its own local state error to decide when to broadcast information to the rest of the agents.

## 9.7 Conclusions

This chapter offered a particular version of the event-triggered control method. Here, a nominal model of the system is implemented in the controller node to estimate the plant state or the plant output between event time instants. Results concerning the MB-ET control of linear systems were summarized in this chapter. Then, a more detailed analysis of two MB-ET control techniques for nonlinear systems were presented. The first case considered the event-triggered control of continuous-time nonlinear systems. The second case addressed discrete-time nonlinear systems with external disturbance using output feedback control and dissipativity control theory.

## Bibliography

[1] L. A. Montestruque and P. J. Antsaklis. On the model-based control of networked systems. *Automatica*, 39(10):1837–1843, 2003.

[2] L. A. Montestruque and P. J. Antsaklis. Stability of model-based control of networked systems with time varying transmission times. *IEEE Transactions on Automatic Control*, 49(9):1562–1572, 2004.

[3] E. Garcia and P. J. Antsaklis. Model-based control of continuous-time and discrete-time systems with large network induced delays. In *Proceedings of the 20th Mediterranean Conference on Control and Automation, IEEE Publ.*, Piscataway, NJ, pp. 1129–1134, 2012.

[4] I. G. Polushin, P. X. Liu, and C. H. Lung. On the model-based approach to nonlinear networked control systems. *Automatica*, 44(9):2409–2414, 2008.

[5] X. Liu. A model based approach to nonlinear networked control systems. PhD thesis, University of Alberta, 2009.

[6] K. Furuta, M. Iwase, and S. Hatakeyama. Internal model and saturating actuation in human operation from human adaptive mechatronics. *IEEE Transactions on Industrial Electronics*, 52(5):1236–1245, 2005.

[7] A. Chaillet and A. Bicchi. Delay compensation in packet-switching networked controlled systems. In *Proceedings of the 47th IEEE Conference on Decision and Control, IEEE Publ.*, Piscataway, NJ, pp. 3620–3625, 2008.

[8] J. P. Hespanha, A. Ortega, and L. Vasudevan. Towards the control of linear systems: With minimum bit-rate. In *Proceedings of the International Symposium on the Mathematical Theory of Networks and Systems*, pp. 1–15, 2002.

[9] L. A. Montestruque and P. J. Antsaklis. Static and dynamic quantization in model-based networked control systems. *International Journal of Control*, 80(1):87–101, 2007.

[10] E. Garcia and P. J. Antsaklis. Model-based event-triggered control with time-varying network delays. In *Proceedings of the 50th IEEE Conference on Decision and Control – European Control Conference, IEEE Publ.*, Piscataway, NJ, pp. 1650–1655, 2011.

[11] E. Garcia and P. J. Antsaklis. Model-based event-triggered control for systems with quantization and time-varying network delays. *IEEE Transactions on Automatic Control*, 58(2):422–434, 2013.

[12] E. Garcia and P. J. Antsaklis. Parameter estimation and adaptive stabilization in time-triggered and event-triggered model-based control of uncertain systems. *International Journal of Control*, 85(9):1327–1342, 2012.

[13] E. Garcia and P. J. Antsaklis. Decentralized model-based event-triggered control of networked systems. In *Proceedings of the American Control Conference, IEEE Publ.*, Piscataway, NJ, pp. 6485–6490, 2012.

[14] E. Garcia, Y. Cao, and D. W. Casbeer. Model-based event-triggered multi-vehicle coordinated tracking using reduced-order models. *Journal of the Franklin Institute*, 351(8):4271–4286, 2014.

[15] E. Garcia, Y. Cao, and D. W. Casbeer. Cooperative control with general linear dynamics and limited communication: Centralized and decentralized event-triggered control strategies. In *Proceedings of the American Control Conference, IEEE Publ.*, Piscataway, NJ, pp. 159–164, 2014.

[16] J. Lunze and D. Lehmann. A state-feedback approach to event-based control. *Automatica*, 46(1): 211–215, 2010.

[17] W. P. M. H. Heemels and M. C. F. Donkers. Model-based periodic event-triggered control for linear systems. *Automatica*, 49(3):698–711, 2013.

[18] J. K. Yook, D. M. Tilbury, and N. R. Soparkar. Trading computation for bandwidth: Reducing communication in distributed control systems using state estimators. *IEEE Transactions on Control Systems Technology*, 10(4):503–518, 2002.

[19] P. Tabuada. Event-triggered real-time scheduling of stabilizing control tasks. *IEEE Transactions on Automatic Control*, 52(9):1680–1685, 2007.

[20] P. Tabuada and X. Wang. Preliminary results on state-triggered scheduling of stabilizing control tasks. In *Proceedings of the 45th IEEE Conference on Decision and Control, IEEE Publ.*, Piscataway, NJ, pp. 282–287, 2006.

[21] X. Wang and M. D. Lemmon. Event-triggering in distributed networked control systems. *IEEE Transactions on Automatic Control*, 56:586–601, 2011.

[22] B. Friedland. *Control System Design: An Introduction to State-Space Methods*, Dover Publications, New York, 2005.

[23] H. K. Khalil, *Nonlinear Systems*, 3rd edition. Prentice Hall, Upper Saddle River, NJ, 2002.

[24] R. Postoyan, A. Anta, A. Nesic, and P. Tabuada. A unifying Lyapunov-based framework for the event-triggered control of nonlinear systems. In *Proceedings of the 50th IEEE Conference on Decision and Control, IEEE Publ.*, Piscataway, NJ, pp. 2559–2564, 2011.

[25] A. Anta and P. Tabuada. To sample or not to sample: Self-triggered control for nonlinear systems. *IEEE Transactions on Automatic Control*, 55(9):2030–2042, 2010.

[26] D. Lehmann and J. Lunze. Event-based output-feedback control. In *Proceedings of the 19th Mediterranean Conference on Control Automation, IEEE Publ.*, Piscataway, NJ, pp. 982–987, 2011.

[27] H. Yu and P. J. Antsaklis. Event-triggered output feedback control for networked control systems using passivity: Time-varying network induced

delays. In *Proceedings of the 50th IEEE Conference on Decision and Control, IEEE Publ.*, Piscataway, NJ, pp. 205–210, 2011.

[28] L. Li and M. D. Lemmon. Event-triggered output feedback control of finite horizon discrete-time multidimensional linear processes. In *Proceedings of the 49th IEEE Conference on Decision and Control, IEEE Publ.*, Piscataway, NJ, pp. 3221–3226, 2010.

[29] E. Garcia and P. J. Antsaklis. Output feedback model-based control of uncertain discrete-time systems with network induced delays. In *Proceedings of the 51st IEEE Conference on Decision and Control, IEEE Publ.*, Piscataway, NJ, pp. 6647–6652, 2012.

[30] M. C. F. Donkers and W. P. M. H. Heemels. Output-based event-triggered control with guaranteed Linf-gain and improved and decentralized event-triggering. *IEEE Transactions on Automatic Control*, 57(6):1362–1376, 2012.

[31] D. Georgiev and D. M. Tilbury. Packet-based control. In *Proceedings of the American Control Conference, IEEE Publ.*, Piscataway, NJ, pp. 329–336, 2004.

[32] D. E. Quevedo, E. I. Silva, and G. C. Goodwin. Packetized predictive control over erasure channels. In *Proceedings of the American Control Conference, IEEE Publ.*, Piscataway, NJ, pp. 1003–1008, 2007.

[33] J. C. Willems. Dissipative dynamical systems, Part I: General theory. *Archive for Rational Mechanics and Analysis*, 45(5):321–351, 1972.

[34] D. J. Hill and P. J. Moylan. Stability results of nonlinear feedback systems. *Automatica*, 13:377–382, 1977.

[35] M. McCourt, E. Garcia, and P. J. Antsaklis. Model-based event-triggered control of nonlinear dissipative systems. In *Proceedings of the American Control Conference, IEEE Publ.*, Piscataway, NJ, pp. 5355–5360, 2014.

[36] E. Garcia and P. J. Antsaklis. Optimal model-based control with limited communication. In *The 19th World Congress of the International Federation of Automatic Control*, pp. 10908–10913, 2014.

# 10

# *Self-Triggered and Team-Triggered Control of Networked Cyber-Physical Systems*

**Cameron Nowzari**
*University of Pennsylvania*
*Philadelphia, PA, USA*

**Jorge Cortés**
*University of California*
*San Diego, CA, USA*

**CONTENTS**



**ABSTRACT** This chapter presents extensions to event-triggered control called self-triggered and team-triggered control. Unlike event-triggered controllers that prescribe an action when an appropriate "event" has occurred, self-triggered controllers prescribe a time in the future at which an action should take place. The benefit of this is that continuous or periodic monitoring of the event is no longer required. This is especially useful in networked scenarios where the events of subsystems may depend on the state of other subsystems in the network because persistent communication is not required. However, self-triggered algorithms can generally yield conservative decisions on when actions should take place to properly guarantee the desired stability results. Consequently, we also introduce the team-triggered control approach that combines ideas from both event- and self-triggered control to help overcome these limitations.

## 10.1 Introduction

This chapter describes triggered control approaches for the coordination of networked cyber-physical systems. Given the coverage of the other chapters of this book, our focus is on self-triggered control and a novel approach we term *team-triggered control.*

The basic idea behind triggered approaches for controlled dynamical systems is to opportunistically select when to execute certain actions (e.g., update the actuation signal, sense some data, communicate some information) in order to efficiently perform various control tasks. Such approaches trade computation for sensing, actuation, or communication effort and give rise to real-time controllers that do not need to perform these actions continuously, or even periodically, in order for the system to function according to a desired level of performance. Triggered approaches for control become even more relevant in the context of networked cyber-physical systems, where computation, actuation, sensing, and communication capabilities might not be collocated. The successful operation of networked systems critically relies on the acquisition and transmission of information across different subsystems, which adds an additional layer of complexity for controller design and analysis. In fact, a given triggered controller might be implementable over some networks and not over others, depending on the requirements imposed by the triggers on the information flow across the individual subsystems.

In event-triggered control, the focus is on detecting events during the system execution that are relevant from the point of view of task completion in order to trigger appropriate actions. Event-triggered strategies result in good performance but require continuous access to some state or output to be able to monitor the possibility of an event being triggered at any time. This can be especially costly to implement when executed over networked systems (for instance, if the availability of information relies on continuous agent-to-agent communication). Other chapters in this book alleviate this issue by considering periodic or data-sampled event-triggered control, where the triggers are only checked when information is available, but this may still be inefficient if the events are not triggered often.

In self-triggered control, the emphasis is instead on developing tests that rely only on current information available to the decision maker to schedule future actions (e.g., when to measure the state again, when to recompute the control input and update the actuator signals, or when to communicate information). To do so, this approach relies critically on abstractions about how the system might behave in the future given the information available now. Self-triggered strategies can be made robust to uncertainties (since they can naturally be accounted for in the abstractions) and more easily amenable to distributed implementation. However, they often result in conservative executions because of the inherent overapproximation about the state of the system associated with the given abstractions. Intuitively, it makes sense to believe that the more accurate a given abstraction describes the system behavior, the less conservative the resulting self-triggered controller implementation. This is especially so in the context of networked systems, where individual agents need to maintain various abstractions about the state of their neighbors, the environment, or even the network.

The team-triggered approach builds on the strengths of event- and self-triggered control to synthesize a unified approach for controlling networked systems in real time, which combines the best of both worlds. The approach is based on agents making promises to one another about their future states and being responsible for warning each other if they later decide to break them. This is reminiscent of event-triggered implementations. Promises can be broad, from tight state trajectories to loose descriptions of reachability sets. With the information provided by promises, individual agents can autonomously determine what time in the future fresh information is needed to maintain a desired level of performance. This is reminiscent of self-triggered implementations. A benefit of the strategy is that because of the availability of the promises, agents do not require continuous state information about their neighbors. Also, because of the extra information provided by promises about what other agents plan to do, agents can operate more efficiently and less conservatively compared to self-triggered strategies where worst-case conditions must always be considered.

### 10.1.1 Related Work

The other chapters of this book provide a broad panoramic view of the relevant literature on general event- and self-triggered control of dynamical systems [1–3], so we focus here on networked systems understood in a broad sense. The predominant paradigm is that of a single plant that is stabilized through a decentralized triggered controller over a sensor–actuator network (see e.g., [4–6]). Fewer works have considered scenarios where multiple plants or agents together are the subject of the overall control design, where the synthesis of appropriate triggers raises novel challenges due to the lack of centralized decision making and the local agent-to-agent interactions. Exceptions include consensus via event-triggered [7–9] or self-triggered control [7], rendezvous [10], collision avoidance while performing point-to-point reconfiguration [11], distributed optimization [12–14], model predictive control [15], and model-based event-triggered control [16,17]. The works in [7,18] implement self-triggered communication schemes to perform distributed control where agents assume worst-case conditions for other agents when deciding when new information should be obtained. Distributed strategies based on event-triggered communication and control are explored in [19], where each agent has an a priori determined local error tolerance, and once it violates it, the agent broadcasts its updated state to its neighbors. The same event-triggered approach is taken in [20] to implement gradient control laws that achieve distributed optimization. In the interconnected system considered in [16], each subsystem helps neighboring subsystems by monitoring their estimates and ensuring that they stay within some performance bounds. The approach requires different subsystems to have synchronized estimates of one another even though they do not communicate at all times. In [21,22], agents do not have continuous availability of information from neighbors and instead decide when to broadcast new information to them, which is more in line with the work we present here.

### 10.1.2 Notation

Here we collect some basic notation used throughout the paper. We let $\mathbb{R}$, $\mathbb{R}_{\geq 0}$, and $\mathbb{Z}_{\geq 0}$ be the sets of real, nonnegative real, and nonnegative integer numbers, respectively. We use $\|\cdot\|$ to denote the Euclidean distance. Given $p$ and $q \in \mathbb{R}^d$, $[p,q] \subset \mathbb{R}^d$ is the closed line segment with extreme points $p$ and $q$, and $\overline{B}(p,r) = \{q \in \mathbb{R}^d \mid \|q-p\| \leq r\}$ is the closed ball centered at $p$ with radius $r \in \mathbb{R}_{\geq 0}$. Given $v \in \mathbb{R}^d \setminus \{0\}$, $\text{unit}(v)$ is the unit vector in the direction of $v$. A function $\alpha : \mathbb{R}_{\geq 0} \to \mathbb{R}_{\geq 0}$ belongs to class $\mathcal{K}$ if it is continuous, strictly increasing, and $\alpha(0) = 0$. A function $\beta : \mathbb{R}_{\geq 0} \times \mathbb{R}_{\geq 0} \to \mathbb{R}_{\geq 0}$ belongs to class $\mathcal{KL}$ if for each fixed $r$, $\beta(r,s)$ is nonincreasing with respect to $s$ and $\lim_{s\to\infty} \beta(r,s) = 0$, and for each fixed $s$, $\beta(\cdot,s) \in \mathcal{K}$. For a set $S$, we let $\partial S$ denote its boundary, and $\mathbb{P}^{cc}(S)$ denote the collection of compact and connected subsets of $S$.

## 10.2 Self-Triggered Control of a Single Plant

In this section, we review the main ideas behind the self-triggered approach for the control of a single plant. Our starting point is the availability of a controller that, when implemented in continuous time, asymptotically stabilizes the system and is certified by a Lyapunov function. For a constant input, the decision maker can compute the future evolution of the system given the current state (using abstractions of the system behavior), and determine what time in the future the controller should be updated. For simplicity, we present the discussion for a linear control system, although the results are valid for more general scenarios, such as nonlinear systems for which a controller is available that makes the closed-loop system input-to-state stable. The exposition closely follows [23].

### 10.2.1 Problem Statement

Consider a linear control system

$$\dot{x} = Ax + Bu, \tag{10.1}$$

with $x \in \mathbb{R}^n$ and $u \in \mathbb{R}^m$, for which there exists a linear feedback controller $u = Kx$ such that the closed-loop system

$$\dot{x} = (A + BK)x,$$

is asymptotically stable. Given a positive definite matrix $Q \in \mathbb{R}^{n\times n}$, let $P \in \mathbb{R}^{n\times n}$ be the unique solution to the Lyapunov equation $(A+BK)^T P + P(A+BK) = -Q$. Then, the evolution of the Lyapunov function $V_c(x) = x^T Px$ along the trajectories of the closed-loop system is

$$\dot{V}_c = x^T((A+BK)^T P + P(A+BK))x = -x^T Qx.$$

Consider now a sample-and-hold implementation of the controller, where the input is not updated continuously, but instead at a sequence of to-be-determined times $\{t_\ell\}_{\ell\in\mathbb{Z}_{\geq 0}} \subset \mathbb{R}_{\geq 0}$:

$$u(t) = Kx(t_\ell), \quad t \in [t_\ell, t_{\ell+1}). \tag{10.2}$$

Such an implementation makes sense in practical scenarios given the inherent nature of digital systems. Under this controller implementation, the closed-loop system can be written as

$$\dot{x} = (A + BK)x + BK\,e,$$

where $e(t) = x(t_\ell) - x(t)$, $t \in [t_\ell, t_{\ell+1})$, is the state error. Then, the objective is to determine the sequence of times $\{t_\ell\}_{\ell\in\mathbb{Z}_{\geq 0}}$ to guarantee some desired level of performance for the resulting system. To make this concrete, define the function

$$V(t, x_0) = x(t)^T Px(t),$$

for a given initial condition $x(0) = x_0$ (here, $t \mapsto x(t)$ denotes the evolution of the closed-loop system using (10.2)). We define the performance of the system by means of a function $S : \mathbb{R}_{\geq 0} \times \mathbb{R}^n \to \mathbb{R}_{\geq 0}$ that upper bounds the evolution of $V$. Then, the sequence of times $\{t_\ell\}$ can be implicitly defined as the times at which

$$V(t, x_0) \leq S(t, x_0), \tag{10.3}$$

is not satisfied. As stated, this is an event-triggered condition that updates the actuator signal whenever $V(t_\ell, x_0) = S(t_\ell, x_0)$. Assuming solutions are well defined, it is not difficult to see that if the performance function satisfies $S(t,x_0) \leq \beta(t,|x_0|)$, for some $\beta \in \mathcal{KL}$, then the closed-loop system is globally uniformly asymptotically stable. Moreover, if $\beta$ is an exponential function, the system is globally uniformly exponentially stable.

Therefore, one only needs to guarantee the lack of Zeno behavior. We do this by choosing the performance function $S$ so that the interevent times $t_{\ell+1} - t_\ell$ are lower bounded by some constant positive quantity. This can be done in a number of ways. For the linear system (10.1), it turns out that it is sufficient to select $S$ satisfying $\dot{V}(t_\ell) < \dot{S}(t_\ell)$ at the event times $t_\ell$. (This fact is formally stated in Theorem 10.1.) To do so, choose $R \in \mathbb{R}^{n\times n}$ positive definite such that $Q - R$ is also positive definite. Then, there exists a Hurwitz matrix $A_s \in \mathbb{R}^{n\times n}$ such that the Lyapunov equation

$$A_s^T P + PA_s = -R,$$

holds. Consider the hybrid system

$$\dot{x}_s = A_s x_s, \quad t \in [t_\ell, t_{\ell+1}),$$
$$x_s(t_\ell) = x(t_\ell),$$

whose trajectories we denote by $t \mapsto x_s(t)$, and define the performance function $S$ by

$$S(t) = x_s^T(t) P x_s(t).$$

### 10.2.2 Self-Triggered Control Policy

Under the event-triggered controller implementation described above, the decision maker needs access to the exact state at all times in order to monitor the condition (10.3). Here we illustrate how the self-triggered approach, instead, proceeds to construct an abstraction that captures the system behavior given the available information and uses it to determine what time in the future the condition might be violated (eliminating the need to continuously monitor the state).

For the particular case considered here, where the dynamics (10.1) are linear and deterministic, and no disturbances are present, it is a possible to exactly compute, at any given time $t_\ell$, the future evolution of the state, and hence, how long $t_{\ell+1} - t_\ell$ will elapse until the condition (10.3) is enabled again. In general, this is not possible, and one instead builds an abstraction that over-approximates this evolution, resulting in a trigger that generates interevent times that lower bound the interevent times generated by (10.3). This explains why the self-triggered approach generally results in more conservative implementations than the event-triggered approach, but comes with the benefit of not requiring continuous state information.

Letting $y = [x^T, e^T]^T \in \mathbb{R}^n \times \mathbb{R}^n$, we write the continuous-time dynamics as

$$\dot{y} = Fy, \quad t \in [t_\ell, t_{\ell+1}),$$

where

$$F = \begin{bmatrix} A + BK & BK \\ -A - BK & -BK \end{bmatrix}.$$

With a slight abuse of notation, we let $y_\ell = [x^T(t_\ell), 0^T]^T$ be the state $y$ at time $t_\ell$. Note that $e(t_\ell) = 0$, for all $\ell \in \mathbb{Z}_{\geq 0}$, by definition of the update times. With this notation, we can rewrite

$$S(t) = (Ce^{F_s(t-t_\ell)}y_\ell)^T P (Ce^{F_s(t-t_\ell)}y_\ell),$$
$$V(t) = (Ce^{F(t-t_\ell)}y_\ell)^T P (Ce^{F(t-t_\ell)}y_\ell),$$

where

$$F_s = \begin{bmatrix} A_s & 0 \\ 0 & 0 \end{bmatrix}, \quad C = \begin{bmatrix} I & 0 \end{bmatrix}.$$

The condition (10.3) can then be rewritten as

$$f(t, y_\ell) = y_\ell^T (e^{F^T(t-t_\ell)} C^T P C e^{F(t-t_\ell)} - e^{F_s^T(t-t_\ell)} C^T P C e^{F_s(t-t_\ell)}) y_\ell \leq 0.$$

What is interesting about this expression is that it clearly reveals the important fact that, with the information available at time $t_\ell$, the decision maker can determine the next time $t_{\ell+1}$ at which (10.3) is violated by computing $t_{\ell+1} = h(x(t_\ell))$ as the time for which

$$f(h(x(t_\ell)), y_\ell) = 0. \tag{10.4}$$

The following result from [23] provides a uniform lower bound $t_{\min}$ on the interevent times $\{t_{\ell+1} - t_\ell\}_{\ell \in \mathbb{Z}_{\geq 0}}$.

**Theorem 10.1: Lower bound on interevent times for self-triggered approach**

Given the system (10.1) with controller (10.2) and controller updates given by the self-triggered policy (10.4), the interevent times are lower bounded by

$$t_{\min} = \min\{t \in \mathbb{R}_{>0} \mid \det(M(t)) = 0\} > 0,$$

where

$$M(t) = \begin{bmatrix} I & 0 \end{bmatrix} \left(e^{F^t} C^T P C e^{Ft} - e^{F_s^t} C^T P C e^{F_s t}\right) \begin{bmatrix} I \\ 0 \end{bmatrix}.$$

Note that the above result can also be interpreted in the context of a periodic controller implementation: any period less than or equal to $t_{\min}$ results in a closed-loop system with asymptotic stability guarantees.

**REMARK 10.1: Triggered sampling versus triggered control** We must make an important distinction here between triggered *control* and triggered *sampling* (or more generally, any type of data acquisition). Due to the deterministic nature of the general linear system discussed in this section, the answers to the questions

- *When should a sample of the state be taken?*
- *When should the control signal to the system be updated?*

are identical. In general, if there are disturbances in the system dynamics, the answers to these questions will not coincide. This might be further complicated in cases when the sensor and the actuator are not collocated, and the sharing of information between them is done via communication. The simplest implementation in this case would be to update the control signal each time a sample is taken; however, in cases where computing or applying a control signal is expensive, it may be more beneficial to use a hybrid strategy combining self-triggered sampling with event-triggered control, as in [24].

## 10.3 Distributed Self-Triggered Control

In this section, we expand our previous discussion with a focus on networked cyber-physical systems. Our main objective is to explain and deal with the issues that come forward when pushing the triggered approach from controlling a single plant, as considered in Section 10.2, to coordinating multiple cooperating agents. We consider networked systems whose interaction topology is modeled by a graph that captures how information is transmitted (via, for instance, sensing or communication). In such scenarios, each individual agent does not have access to all the network information at any given time, but instead interacts with a set of neighboring agents. In our forthcoming discussion, we pay attention to the following challenges specific to these networked problems. First, the trigger designs now need to be made for individuals, as opposed to a single centralized decision maker. Triggers such as those of (10.3) or (10.4) rely on global information about the network state, which the agents do not possess in general. Therefore, the challenge is finding ways in which such conditions can be allocated among the individuals. This brings to the forefront the issue of access to information and the need for useful abstractions about the behavior of other agents. Second, the design of independent triggers that each agent can evaluate with the information available to it naturally gives rise to asynchronous network executions, in which events are triggered by different agents at different times. This "active" form of asynchronism poses challenges for the correctness and convergence analysis of the coordination algorithms, and further complicates ruling out the presence of Zeno behavior. Before getting into the description of triggered control approaches, we provide a formal description of the model for networked cyber-physical systems and the overall team objective.

### 10.3.1 Network Model and Problem Statement

We begin by describing the model for the network of agents. In order to simplify the general discussion, we make a number of simplifying assumptions, such as linear agent dynamics or time-invariant interaction topology. However, when considering more specific classes of problems, it is often possible to drop some of these assumptions.

Consider $N$ agents whose interaction topology is described by an undirected graph $\mathcal{G}$. The fact that $(i, j)$ belongs to the edge set $E$ models the ability of agents $i$ and $j$ to communicate with one another. The set of all agents that $i$ can communicate with is given by its set of neighbors $\mathcal{N}(i)$ in the graph $\mathcal{G}$. The state of agent $i \in \{1, \ldots, N\}$, denoted $x_i$, belongs to a closed set $\mathcal{X}_i \subset \mathbb{R}^{n_i}$. We assume that each agent has access to its own state at all times. According to our model, agent $i$ can access $x^i_{\mathcal{N}} = (x_i, \{x_j\}_{j\in\mathcal{N}(i)})$ when it communicates with its neighbors. The network state $x = (x_1, \ldots, x_N)$ belongs to $\mathcal{X} = \prod_{i=1}^N \mathcal{X}_i$. We consider linear dynamics for each $i \in \{1, \ldots, N\}$:

$$\dot{x}_i = A_i x_i + B_i u_i, \tag{10.5}$$

with block diagonal $A_i \in \mathbb{R}^{n_i \times n_i}$, $B_i \in \mathbb{R}^{n_i \times m_i}$, and $u_i \in \mathcal{U}_i$. Here, $\mathcal{U}_i \subset \mathbb{R}^{m_i}$ is a closed set of allowable controls for agent $i$. We assume that the pair $(A_i, B_i)$ is controllable with controls taking values in $\mathcal{U}_i$. Letting $u = (u_1, \ldots, u_N) \in \prod_{i=1}^N \mathcal{U}_i$, the dynamics of the entire network is described by

$$\dot{x} = Ax + Bu, \tag{10.6}$$

with $A = \text{diag}\,(A_1, \ldots, A_N) \in \mathbb{R}^{n\times n}$ and $B = \text{diag}\,(B_1, \ldots, B_N) \in \mathbb{R}^{n\times m}$, where $n = \sum_{i=1}^N n_i$, and $m = \sum_{i=1}^N m_i$.

The goal of the network is to drive the agents' states to some desired closed set of configurations $D \subset \mathcal{X}$ and ensure that it stays there. Depending on how the set $D$ is defined, this goal can capture different coordination tasks, including deployment, rendezvous, and formation control (see, e.g., [25]). Our objective here is to design provably correct triggered coordination strategies that agents can implement to achieve this goal. With this in mind, given the agent dynamics, the communication graph $\mathcal{G}$, and the set $D$, our starting point is the availability of a distributed, continuous control law that drives the system asymptotically to $D$. Formally, we assume that a continuous map $u^* : \mathcal{X} \to \mathbb{R}^m$ and a continuously differentiable function $V : \mathcal{X} \to \mathbb{R}$, bounded from below exist such that $D$ is the set of minimizers of $V$, and for all $x \notin D$,

$$\nabla_i V(x)\,(A_i x_i + B_i u_i^*(x)) \leq 0, \quad i \in \{1,\ldots,N\}, \tag{10.7}$$

$$\sum_{i=1}^{N} \nabla_i V(x)\,(A_i x_i + B_i u_i^*(x)) < 0. \tag{10.8}$$

We assume that both the control law $u^*$ and the gradient $\nabla V$ are distributed over $\mathcal{G}$. By this, we mean that for each $i \in \{1,\ldots,N\}$, the $i$th component of each of these objects only depends on $x^i_{\mathcal{N}}$, rather than on the full network state $x$. For simplicity, and with a slight abuse of notation, we write $u_i^*(x^i_{\mathcal{N}})$ and $\nabla_i V(x^i_{\mathcal{N}})$ to emphasize this fact when convenient. This property has the important consequence that agent $i$ can compute $u_i^*$ and $\nabla_i V$ with the exact information it can obtain through communication on $\mathcal{G}$. We note that here we only consider the goal of achieving asymptotic stability of the system, with no specification on the performance $S$ as we did in Section 10.2. For specific problems, one could reason with similar performance function requirements.

From an implementation viewpoint, the controller $u^*$ requires continuous agent-to-agent communication and continuous updates of the actuator signals. The triggered coordination strategies that we review next seek to select the time instants at which information should be acquired in an opportunistic way, such that the resulting implementation still enjoys certificates on correctness and performance similar to that of the original controller. The basic general idea is to design implementations that guarantee that the time derivative of the Lyapunov function $V$ introduced in Section 10.3.1 along the solutions of the network dynamics (10.6) is less than or equal to 0 at all times, even though the information available to the agents is partial and possibly inexact.

**REMARK 10.2: Triggered sampling, communication, and control** In light of Remark 10.1, the consideration of a networked cyber-physical system adds a third independent question that each agent of the network must answer:

- *When should I sample the (local) state of the system?*
- *When should I update the (local) control input to the system?*
- *When should I communicate with my neighbors?*

To answer these questions, the decision maker has now changed from a single agent with complete information on a single plant to multiple agents with partial information about the network. For simplicity, and to emphasize the network aspect, we assume that agents have continuous access to their own state and that control signals are updated each time communications with neighbors occur. We refer the interested reader to [16,20] where these questions are considered independently in specific classes of systems.

### 10.3.2 Event-Triggered Communication and Control

In this section, we see how an event-triggered implementation of the continuous control law $u^*$ might be realized using (10.7).

Let $t_\ell$ for some $\ell \in \mathbb{Z}_{\geq 0}$ be the last time at which all agents have received information from their neighbors. We are then interested at what time $t_{\ell+1}$ the agents should communicate again. In between updates, the simplest estimate that an agent $i$ maintains about a neighbor $j \in \mathcal{N}(i)$ is the zero-order hold given by

$$\widehat{x}^i_j(t) = x_j(t_\ell), \qquad t \in [t_\ell, t_{\ell+1}). \tag{10.9}$$

Recall that agent $i$ always has access to its own state. Therefore, $\widehat{x}^i_{\mathcal{N}}(t) = (x_i(t), \{\widehat{x}^i_j(t)\}_{j\in\mathcal{N}(i)})$ is the information available to agent $i$ at time $t$. The implemented controller is then given by

$$\widehat{x}^i_j(t) = x_j(t_\ell), \qquad u_i^{\text{event}}(t) = u_i^*(\widehat{x}^i_{\mathcal{N}}(t_\ell)),$$

for $t \in [t_\ell, t_{\ell+1})$. We only consider the zero-order hold here for simplicity, although one could also implement more elaborate schemes for estimating the state of neighboring agents given the sampled information available, and use this for the controller implementation.

The time $t_{\ell+1}$ at which an update becomes necessary is determined by the first time after $t_\ell$ that the time derivative of $V$ along the trajectory of (10.6) with $u = u^{\text{event}}$ is no longer negative. Formally, the event for when agents should request updated information is

$$\frac{d}{dt}V(x(t_{\ell+1})) = \sum_{i=1}^{N} \nabla_i V(x(t_{\ell+1})) \times \left(A_i x_i(t_{\ell+1}) + B_i u_i^{\text{event}}(t_\ell)\right) = 0. \tag{10.10}$$

Two reasons may cause (10.10) to be satisfied. One reason is that the zero-order hold control $u^{\text{event}}(t)$ for $t \in [t_\ell, t_{\ell+1})$ has become too outdated, causing an update at time $t_{\ell+1}$. Until (10.10) is satisfied, it is not necessary to update state information because inequality (10.8) implies that $\frac{d}{dt}V(x(t)) < 0$

for $t \in [t_\ell, t_{\ell+1})$ if $x(t_\ell) \notin D$ given that all agents have exact information at $t_\ell$ and $\frac{d}{dt}V(x(t_\ell)) < 0$ by continuity of $\frac{d}{dt}V(x)$. The other reason is simply that $x(t_{\ell+1}) \in D$, and thus the system has reached its desired configuration.

Unfortunately, (10.10) cannot be checked in a distributed way because it requires global information. Instead, one can define a local event that implicitly defines when a single agent $i \in \{1, \ldots, N\}$ should update its information. Letting $t^i_\ell$ be some time at which agent $i$ receives updated information, $t^i_{\ell+1} \geq t^i_\ell$ is the first time such that

$$\nabla_i V(x(t^i_{\ell+1})) \left( A_i x_i(t^i_{\ell+1}) + B_i u_i^{\text{event}}(t^i_\ell) \right) = 0. \quad (10.11)$$

This means that as long as each agent $i$ can ensure the local event (10.11) has not yet occurred, it is guaranteed that (10.10) has also not yet occurred. Note that this is a sufficient condition for (10.10) not being satisfied, but it is not necessary. There may be other ways of distributing the global trigger (10.10) for more specific classes of problems. Although this can now be monitored in a distributed way, the problem with this approach is that each agent $i \in \{1, \ldots, N\}$ needs to have continuous access to information about the state of its neighbors $\mathcal{N}(i)$ in order to evaluate $\nabla_i V(x) = \nabla_i V(x^i_{\mathcal{N}})$ and check condition (10.11). This requirement may make the event-triggered approach impractical when this information is only available through communication. Note that this does not mean it is impossible to perform event-triggered communication strategies in a distributed way, but it requires a way for agents to use only their own information to monitor some meaningful event. An example problem where this can be done is provided in [26,27], where agents use event-triggered broadcasting to solve the consensus problem.

### 10.3.3 Self-Triggered Communication and Control

In this section, we design a self-triggered communication and control strategy of the controller $u^*$ by building on the discussion in Section 10.3.2. To achieve this, the basic idea is to remove the requirement on continuous availability of information to check the test (10.11) by providing agents with possibly inexact information about the state of their neighbors.

To do so, we begin by introducing the notion of reachability sets. Given $y \in \mathcal{X}_i$, let $\mathcal{R}_i(s, y)$ be the set of reachable points under (10.5) starting from $y$ in $s$ seconds:

$$\mathcal{R}_i(s, y) = \Big\{ z \in \mathcal{X}_i \mid \exists\, u_i : [0, s] \to \mathcal{U}_i \text{ such that } z = e^{A_i s} y + \int_0^s e^{A_i (s-\tau)} B_i u_i(\tau) d\tau \Big\}.$$

Assuming here that agents have exact knowledge about the dynamics and control sets of their neighboring agents, each time an agent receives state information, it can construct sets that are guaranteed to contain their neighbors' states in the future. Formally, if $t_\ell$ is the time at which agent $i$ receives state information $x_j(t_\ell)$ from its neighbor $j \in \mathcal{N}(i)$, then

$$\mathbf{X}^i_j(t, x_j(t_\ell)) = \mathcal{R}_j(t - t_\ell, x_j(t_\ell)) \subset \mathcal{X}_j, \quad (10.12)$$

is guaranteed to contain $x_j(t)$ for all $t \geq t_\ell$. We refer to these sets as *guaranteed sets*. These capture the notion of abstractions of the system behavior that we discussed in Section 10.2 when synthesizing a self-triggered control policy for a single plant. For simplicity, we let $\mathbf{X}^i_j(t) = \mathbf{X}^i_j(t, x_j(t_\ell))$ when the starting state $x_j(t_\ell)$ and time $t_\ell$ do not need to be emphasized. We denote by $\mathbf{X}^i_{\mathcal{N}}(t) = (x_i(t), \{\mathbf{X}^i_j(t)\}_{j \in \mathcal{N}(i)})$ the information available to an agent $i$ at time $t$.

With the guaranteed sets in place, we can now provide a test using inexact information to determine when new, up-to-date information is required. Let $t^i_\ell$ be the last time at which agent $i$ received updated information from its neighbors. Until the next time $t^i_{\ell+1}$ information is obtained, agent $i$ uses the zero-order hold estimate and control:

$$\hat{x}^i_j(t) = x_j(t^i_\ell), \qquad u_i^{\text{self}}(t) = u_i^*(\hat{x}^i_{\mathcal{N}}(t^i_\ell)),$$

for $t \in [t^i_\ell, t^i_{\ell+1})$ and $j \in \mathcal{N}(i)$. As noted before, the consideration of zero-order hold is only made to simplify the exposition. At time $t^i_\ell$, agent $i$ computes the next time $t^i_{\ell+1} \geq t^i_\ell$ at which information should be acquired via

$$\sup_{y_{\mathcal{N}} \in \mathbf{X}^i_{\mathcal{N}}(t^i_{\ell+1})} \nabla_i V(y_{\mathcal{N}}) \left( A_i x_i(t^i_{\ell+1}) + B_i u_i^{\text{self}}(t^i_\ell) \right) = 0. \quad (10.13)$$

By (10.7) and the fact that $\mathbf{X}^i_j(t^i_\ell) = \{x_j(t^i_\ell)\}$, at time $t^i_\ell$ we have

$$\sup_{y_{\mathcal{N}} \in \mathbf{X}^i_{\mathcal{N}}(t^i_\ell)} \nabla_i V(y_{\mathcal{N}}) \left( A_i x_i(t^i_\ell) + B_i u_i^{\text{self}}(t^i_\ell) \right) = \nabla_i V(x^i_{\mathcal{N}}(t^i_\ell)) \left( A_i x_i(t^i_\ell) + B_i u_i^{\text{self}}(t^i_\ell) \right) \leq 0.$$

If all agents use this triggering criterium for updating information, it is guaranteed that $\frac{d}{dt}V(x(t)) \leq 0$ at all times, because for each $i \in \{1, \ldots, N\}$, the true state $x_j(t)$ is guaranteed to be in $\mathbf{X}^i_j(t)$ for all $j \in \mathcal{N}(i)$ and $t \geq t^i_\ell$.

The condition of (10.13) is appealing because it can be solved by agent $i$ with the information it possesses at time $t^i_\ell$. Once determined, agent $i$ schedules that, at time $t^i_{\ell+1}$, it will request updated information from its neighbors. The term *self-triggered* captures the fact that each agent is now responsible for deciding when it requires new information. We refer to $t^i_{\ell+1} - t^i_\ell$ as the *self-triggered request time* of agent $i \in \{1, \ldots, N\}$. Due to the conservative way in which $t^i_{\ell+1}$ is determined, it is possible that $t^i_{\ell+1} = t^i_\ell$ for some $i$, which would mean that continuous information updates are necessary. (It should be noted that this cannot happen for all $i \in \{1, \ldots, N\}$ unless the network state is already in $D$.) For a given specific task (target set $D$, Lyapunov function $V$, and controller $u^*$), one might be able to discard this issue. In other cases, this can be dealt with by introducing a dwell time such that a minimum amount of time must pass before an agent can request new information. We do not enter into the details of this here, but instead show how it can be done for a specific example in Section 10.5.

The main convergence result for the distributed self-triggered approach is provided below. We defer the sketch of the proof to Section 10.4 as it becomes a direct consequence of Proposition 10.1.

**Proposition 10.1**

For the network model and problem setup described in Section 10.3.1, if each agent executes the self-triggered algorithm presented above, the state of the entire network asymptotically converges to the desired set $D$ for all Zeno-free executions.

Note that the result is formulated for non-Zeno executions. As discussed above, dwell times can be used to rule out Zeno behavior, although we do not go into details here. The problem with the distributed self-triggered approach is that the resulting times are often conservative because the guaranteed sets can grow large quickly as they capture all possible trajectories of neighboring agents. It is conceivable that improvements can be made from tuning the guaranteed sets based on what neighboring agents *plan* to do rather than what they *can* do. This observation is at the core of the team-triggered approach proposed next.

## 10.4 Team-Triggered Control

Here we describe an approach that combines ideas from both event- and self-triggered control for the coordination of networked cyber-physical systems. The team-triggered strategy incorporates the reactive nature of event-triggered approaches and, at the same time, endows individual agents with the autonomy characteristics of self-triggered approaches to determine when and what information is needed. Agents make promises to their neighbors about their future states and inform them if these promises are violated later (hence, the connection with event-triggered control). These promises function as more accurate abstractions of the behavior of the other entities than guaranteed sets introduced in Section 10.3.3. This extra information allows each agent to compute the next time that an update is required (which in general is longer than in the purely self-triggered implementation described in Section 10.3) and request information from their neighbors to guarantee the monotonicity of the Lyapunov function (hence, the connection with self-triggered control). Our exposition here follows [28].

### 10.4.1 Promise Sets

Promises allow agents to predict the evolution of their neighbors more accurately than guaranteed sets, which in turn affects the overall network behavior. For our purposes here, a (control) *promise* that agent $j$ makes to agent $i$ at some time $t'$ is a subset $\mathcal{U}^i_j \subset \mathcal{U}_j$ of its allowable control set. This promise conveys the meaning that agent $j$ will only use controls $u(t) \in \mathcal{U}^i_j$ for all $t \geq t'$. Given the dynamics (10.5) of agent $j$ and state $x_j(t')$ at time $t'$, agent $i$ can compute the *state promise set*:

$$X^i_j(t) = \Big\{ z \in \mathcal{X}_j \mid \exists u_j : [t', t] \to \mathcal{U}^i_j \text{ s.t.} \\ z = e^{A_j(t-t')} x_j(t) + \int_{t'}^{t} e^{A_j(t-\tau)} B_j u_j(\tau) d\tau \Big\}, \quad (10.14)$$

for $t \geq t'$. This means that as long as agent $j$ keeps its promise at all times (i.e., $u_j(t) \in \mathcal{U}^i_j$ for all $t \geq t'$), then $x_j(t) \in X^i_j(t)$ for all $t \geq t'$. We denote by $X^i_{\mathcal{N}}(t) = (x_i(t), \{X^i_j(t)\}_{j \in \mathcal{N}(i)})$ the information available to an agent $i$ at time $t$. Since the promise $\mathcal{U}^i_j$ is a subset of agent $j$'s allowable control set $\mathcal{U}_j$, it follows that $X^i_j(t) \subset \mathbf{X}^i_j(t)$ for all $t \geq t'$ as well. This allows agent $i$ to operate with better information about agent $j$ than it would by computing the guaranteed sets (10.12) defined in Section 10.3.3. Promises can be generated in different ways depending on the desired task. For simplicity, we only consider static promises $\mathcal{U}^i_j$ here, although one could also consider more complex time-varying promises [28].

In general, tight promises correspond to agents having good information about their neighbors, which at the same time may result in an increased communication effort (since the promises cannot be kept for long periods of time). Loose promises correspond to agents having to use more conservative controls due to the lack of information, while at the same time potentially being able to operate for longer periods of time without communicating (because promises are not violated). These advantages rely on the assumption that promises hold throughout the evolution. As the state of the network changes and the level of task completion evolves, agents might decide to break former promises and make new ones. We discuss this process in Section 10.4.3.

**REMARK 10.3: Controllers that operate on set-valued information** With the extra information provided by promise sets, it is conceivable to design controllers that operate on set-valued information rather than point-valued information and go beyond zero-order state holds. For simplicity, we consider only controllers that operate on points in our general exposition, but illustrate how set-valued controllers can be defined in Section 10.5. We refer the interested reader to [28] for further details.

### 10.4.2 Self-Triggered Communication and Control

With the promise sets in place, we can now provide a test using the available information to determine when new, up-to-date information is required.

Let $t^i_\ell$ be the last time at which agent $i$ received updated information from its neighbors. Until the next time $t^i_{\ell+1}$ information is obtained, agent $i$ uses the zero-order hold estimate and control:

$$\widehat{x}^i_j(t) = x_j(t^i_\ell), \qquad u^{\text{team}}_i(t) = u^*_i(\widehat{x}^i_{\mathcal{N}}(t^i_\ell)),$$

for $t \in [t^i_\ell, t^i_{\ell+1})$ and $j \in \mathcal{N}(i)$. At time $t^i_\ell$, agent $i$ computes the next time $t^i_{\ell+1} \geq t^i_\ell$ at which information should be acquired via

$$\sup_{y_{\mathcal{N}} \in X^i_{\mathcal{N}}(t^i_{\ell+1})} \nabla_i V(y_{\mathcal{N}}) \left( A_i x_i(t^i_{\ell+1}) + B_i u^{\text{team}}_i(t^i_\ell) \right) = 0. \tag{10.15}$$

By (10.7) and the fact that $X^i_j(t^i_\ell) = \{x_j(t^i_\ell)\}$, at time $t^i_\ell$ we have

$$\begin{aligned} &\sup_{y_{\mathcal{N}} \in \mathbf{X}^i_{\mathcal{N}}(t^i_\ell)} \nabla_i V(y_{\mathcal{N}}) \left( A_i x_i(t^i_\ell) + B_i u^{\text{team}}_i(t^i_\ell) \right) \\ &\quad = \nabla_i V(x^i_{\mathcal{N}}(t^i_\ell)) \left( A_i x_i(t^i_\ell) + B_i u^{\text{team}}_i(t^i_\ell) \right) \leq 0. \end{aligned}$$

If all agents use this triggering criterium for updating information, and all promises among agents are kept at all times, it is guaranteed that $\frac{d}{dt} V(x(t)) \leq 0$ at all times, because for each $i \in \{1, \ldots, N\}$, the true state $x_j(t)$ is guaranteed to be in $X^i_j(t)$ for all $j \in \mathcal{N}(i)$ and $t \geq t^i_\ell$.

As long as (10.15) has not yet occurred for all agents $i \in \{1, \ldots, N\}$ for some time $t$, and the promises have not been broken, assumptions (10.7) and the continuity of the LHS of (10.15) in time guarantee

$$\frac{d}{dt} V(x(t)) \leq \sum_{i=1}^{N} \mathcal{L}_i V^{\text{sup}}(X^i_{\mathcal{N}}(t)) < 0. \tag{10.16}$$

The condition (10.15) is again appealing because it can be solved by agent $i$ with the information it possesses at time $t^i_\ell$. In general, it gives rise to a longer interevent time than the condition (10.13), because the agents employ more accurate information about the future behavior of their neighbors provided by the promises. However, as in the self-triggered method described in Section 10.3.3 and depending on the specific task, it is possible that $t^i_{\ell+1} = t_\ell$ for some agent $i$, which means this agent would need continuous information from its neighbors. If this is the case, the issue can be dealt with by introducing a dwell time such that a minimum amount of time must pass before an agent can request new information. We do not enter into the details of this here, but instead refer to [28] for a detailed discussion. Finally, we note that the conclusions laid out above only hold if promises among agents are kept at all times, which we address next.

### 10.4.3 Event-Triggered Information Updates

Agent promises may need to be broken for a variety of reasons. For instance, an agent might receive new information from its neighbors, causing it to change its former plans. Another example is given by an agent that made a promise that it is not able to keep for as long as it anticipated. Consider an agent $i \in \{1, \ldots, N\}$ that has sent a promise $\mathcal{U}^j_i$ to a neighboring agent $j$ at some time $t_{\text{last}}$, resulting in the state promise set $x_i(t) \in X^j_i(t)$. If agent $i$ ends up breaking its promise at time $t^* > t_{\text{last}}$, then it is responsible for sending a new promise $\mathcal{U}^j_i$ to agent $j$ at time $t_{\text{next}} = t^*$. This implies that agent $i$ must keep track of promises made to its neighbors and monitor them in case they are broken. This mechanism is implementable because each agent only needs information about its own state and the promises it has made to determine whether the trigger is satisfied. We also note that different notions of "broken promise" are acceptable: this might mean that $u_i(t^*) \notin \mathcal{U}^j_i$ or the more flexible $x_i(t^*) \notin X^j_i(t^*)$. The latter has an advantage in that it is not absolutely required that agent $i$ use controls only in $\mathcal{U}_i^j$, as long as the promise set that agent $j$ is operating with is still valid.

Reminiscent of the discussion on dwell times in Section 10.4.2, it may be possible for promises to broken arbitrarily fast, resulting in Zeno executions. As before, it is possible to implement a dwell time such that a minimum amount of time must pass before an agent can generate a new promise. However, in this case, some additional consideration must be given to the fact that agents might not always be operating with correct information in this case. For instance, if an agent $j$ has broken a promise to agent $i$ but is not allowed to send a new one yet due to the dwell time, agent $i$ might be operating with false information. Such a problem can be addressed using small warning messages that alert agents of this situation. Another issue that is important when considering the transmission of messages is that communication might be unreliable, with delays, packet drops, or noise. We refer to [28] for details on how these dwell times can be implemented while ensuring stability of the entire system in the presence of unreliable communication.

### 10.4.4 Convergence Analysis of the Team-Triggered Coordination Policy

The combination of the self-triggered communication and control policy with the event-triggered information updates gives rise to the team-triggered coordination policy, which is formally presented in Algorithm 10.1. The self-triggered information request in this strategy is executed by an agent anytime new information is received, whether it was actively requested by the agent, or was received from some neighbor due to the breaking of a promise.

**Algorithm 10.1**: Team-triggered coordination policy

*(Self-triggered communication and control)*

At any time $t$ agent $i \in \{1, \ldots, N\}$ receives new promise(s) $\mathcal{U}_j^i$ and position information $x_j(t)$ from neighbor(s) $j \in \mathcal{N}(i)$, agent $i$ performs:

update $\hat{x}_j^i(t') = x_j(t)$ for $t' \geq t$
compute and maintain promise set $X_j^i(t')$ for $t' \geq t$
compute own control $u_i^{\text{team}}(t') = u_i^*(\hat{x}_{\mathcal{N}}^i(t))$ for $t' \geq t$
compute first time $t^* \geq t$ such that (10.15) is satisfied
schedule information request to neighbors in $t^* - t$ seconds

*(Respond to information request)*

At any time $t$ a neighbor $j \in \mathcal{N}(i)$ requests information, agent $i$ performs:

send current state information $x_i(t)$ to agent $j$
send new promise $\mathcal{U}_i^j$ to agent $j$

*(Event-triggered information update)*

At all times $t$, agent $i$ performs:

**if** there exists $j \in \mathcal{N}(i)$ such that $x_i(t) \notin X_i^j(t)$ **then**
    send current state information $x_i(t)$ to agent $j$
    send new promise $\mathcal{U}_i^j$ to agent $j$
**end if**

It is worth mentioning that the self-triggered approach described in Section 10.3.3 is a particular case of the team-triggered approach, where the promises are simply the guaranteed (or reachable) sets described in (10.12). In this scenario, promises can never be broken so the event-triggered information updates never occur. Therefore, the class of team-triggered strategies contains the class of self-triggered strategies.

The following statement provides the main convergence result regarding distributed controller implementations synthesized via the team-triggered approach.

**Proposition 10.2**

For the network model and problem setup described in Section 10.3.1, if each agent executes the team-triggered coordination policy described in Algorithm 10.1, the state of the entire network asymptotically converges to the desired set $D$ for all Zeno-free executions.

Note that the result is only formulated for non-Zeno executions for simplicity of presentation. As discussed, the absence of Zeno behavior can be guaranteed via dwell times for the self- and event-triggered updates. In the following, we provide a proof sketch of this result that highlights its key components. We refer the interested reader to our work [28] that provides a detailed technical discussion that also accounts for the presence of dwell times in the event- and self-triggered updates.

Our first observation is that the evolution of the Lyapunov function $V$ along the trajectories of the team-triggered coordination policy is nonincreasing by design. This is a consequence of the analysis of Section 10.4.2, which holds under the assumption that promises are never broken, together with the event-triggered updates described in Section 10.4.3, which guarantee that the assumption holds. Then, the main challenge to establish the asymptotic convergence properties of the team-triggered coordination policy is in dealing with its discontinuous nature. More precisely, since the information possessed by any given agent are sets for each of its neighbors, promises shrink to singletons when updated information is received. To make this point precise, let us formally denote by

$$S_i = \mathbb{P}^{cc}(\mathcal{X}_1) \times \cdots \times \mathbb{P}^{cc}(\mathcal{X}_{i-1}) \times \mathcal{X}_i \times \mathbb{P}^{cc}(\mathcal{X}_{i+1}) \times \cdots \times \mathbb{P}^{cc}(\mathcal{X}_N),$$

the state space where the variables that agent $i \in \{1, \ldots, N\}$ possesses live. Note that this set allows us to capture the fact that each agent $i$ has perfect information about itself. Although agents only have information about their neighbors, the above space considers agents having promise information about all other agents to facilitate the analysis. This is only done to allow for a simpler technical presentation and does not impact the validity of the arguments. The space that the state of the entire network lives in is then $S = \prod_{i=1}^{N} S_i$. The information possessed by all agents of the network at some time $t$ is collected in

$$\left(X^1(t), \ldots, X^N(t)\right) \in S,$$

where $X^i(t) = (X_1^i(t), \ldots, X_N^i(t)) \in S_i$.

The team-triggered coordination policy corresponds to a discontinuous map of the form $S \times \mathbb{Z}_{\geq 0} \to S \times \mathbb{Z}_{\geq 0}$. The discontinuity makes it difficult to use standard stability methods to analyze the convergence properties of the network. Our approach to this problem consists of defining a set-valued map $M : S \times \mathbb{Z}_{\geq 0} \rightrightarrows S \times \mathbb{Z}_{\geq 0}$ with good continuity properties so that the trajectories of

$$(Z(t_{\ell+1}), \ell + 1) \in M(Z(t_\ell), \ell),$$

contain the trajectories of the team-triggered coordination policy. Although this "overapproximation procedure" enlarges the set of trajectories to consider in the analysis, the gained benefit is that of having a set-valued map with suitable continuity properties that are amenable to set-valued stability analysis.

We conclude this section by giving an idea of how to define the set-valued map $M$, which is essentially guided by the way information updates can happen under the team-triggered coordination policy. Given $(Z, \ell) \in S \times \mathbb{Z}_{\geq 0}$, we define the $(N+1)$th component of all the elements in $M(Z, \ell)$ to be $\ell + 1$. (This corresponds to evolution in time.) The $i$th component of the elements in $M(Z, \ell)$ is the set given by one of following possibilities: (1) the case when agent $i$ does not receive any new information in this step, in which case the promise sets available to agent $i$ are simply propagated forward with respect to the promises made to it; (2) the case when agent $i$ has received information from at least one neighbor $j$, in which case the promise set $X_j^i$ becomes an exact point $\{x_j\}$ and all other promise sets are propagated forward as in (1). Lumping together in a set the two possibilities for each agent gives rise to a set-valued map with good continuity properties: formally, $M$ is closed. (A set-valued map $T : X \rightrightarrows Y$ is closed if $x_k \to x$, $y_k \to y$, and $y_k \in T(x_k)$ imply that $y \in T(x)$.) The proof then builds on this construction, combined with the monotonic evolution of the Lyapunov function along the trajectories of the team-triggered coordination policy, and employs a set-valued version of LaSalle Invariance Principle, see [25], to establish the asymptotic convergence result stated in Proposition 10.2.

## 10.5 Illustrative Scenario: Optimal Robot Deployment

In this section, we illustrate the above discussion on self- and team-triggered design of distributed coordination algorithms in a scenario involving the optimal deployment of a robotic sensor network. We start by presenting the original problem formulation from [29], where the authors provide continuous-time and periodic distributed algorithms to solve the problem. The interested reader is referred to [25] for additional references and a general discussion on coverage problems.

Consider a group of $N$ agents moving in a convex polygon $S \subset \mathbb{R}^2$ with positions $P = (p_1, \ldots, p_N)$. For simplicity, consider single-integrator dynamics

$$\dot{p}_i = u_i, \tag{10.17}$$

where $\|u_i\| \leq v_{\max}$ for all $i \in \{1, \ldots, N\}$ for some $v_{\max} > 0$. The network objective is to achieve optimal deployment as measured by the following locational optimization function $\mathcal{H}$. The agent performance at a point $q$ of agent $i$ degrades with $\|q - p_i\|^2$. Assume a density $\phi : S \to \mathbb{R}$ is available, with $\phi(q)$ reflecting the likelihood of an event happening at $q$. Consider then the minimization of the function

$$\mathcal{H}(P) = E_\phi\left[\min_{i \in \{1,\ldots,N\}} \|q - p_i\|^2\right]. \tag{10.18}$$

This formulation is applicable in scenarios where the agent closest to an event of interest is the one responsible for it. Examples include servicing tasks, spatial sampling of random fields, resource allocation, and event detection.

With regard to the problem statement in Section 10.3.1, $\mathcal{H}$ plays the role of the function $V$, and the set $D$ corresponds to the set of critical points of $\mathcal{H}$. Let us next describe the distributed continuous-time controller $u^*$. In order to do so, we need to recall some notions from computational geometry. A partition of $S$ is a collection of $N$ polygons $\mathcal{W} = \{W_1, \ldots, W_N\}$ with disjoint interiors whose union is $S$. The Voronoi partition $\mathcal{V}(P) = \{V_1, \ldots, V_N\}$ of $S$ generated by the points $P$ is

$$V_i = \{q \in S \mid \|q - p_i\| \leq \|q - p_j\|, \ \forall j \neq i\}.$$

When the Voronoi regions $V_i$ and $V_j$ share an edge, $p_i$ and $p_j$ are (Voronoi) neighbors. We denote the neighbors of agent $i$ by $\mathcal{N}(i)$. For $Q \subset S$, the *mass* and *center of mass* of $Q$ with respect to $\phi$ are

$$M_Q = \int_S \phi(q)dq, \quad C_Q = \frac{1}{M_Q}\int_Q q\phi(q)dq.$$

A tuple $P = (p_1, \ldots, p_N)$ is a centroidal Voronoi configuration if $p_i = C_{V_i}$, for all $i \in \{1, \ldots, N\}$. The controller $u^*$

then simply corresponds to the gradient descent of the function $\mathcal{H}$, whose $i$th component is given by

$$-\frac{\partial \mathcal{H}}{\partial p_i} = 2M_{V_i}(C_{V_i} - p_i).$$

Remarkably, this continuous-time coordination strategy admits a periodic implementation without the need of determining any step-size in closed form. This design is based on the observation that the function $\mathcal{H}$ can be rewritten in terms of the Voronoi partition as

$$\mathcal{H}(P) = \sum_{i=1}^{N} \int_{V_i} \|q - p_i\|^2 \phi(q) dq,$$

which suggests the generalization of $\mathcal{H}$ given by

$$\mathcal{H}(P, \mathcal{W}) = \sum_{i=1}^{N} \int_{W_i} \|q - p_i\|^2 \phi(q) dq, \qquad (10.19)$$

where $\mathcal{W}$ is a partition of $S$, and the $i$th agent is responsible of the "dominance region" $W_i$. The function $\mathcal{H}$ in (10.19) is then to be minimized with respect to the locations $P$ and the dominance regions $\mathcal{W}$. The key observations now are that (1) for a fixed network configuration, the optimal partition is the Voronoi partition, that is,

$$\mathcal{H}(P, \mathcal{V}(P)) \leq \mathcal{H}(P, \mathcal{W}),$$

for any partition $\mathcal{W}$ of $S$, and (2) for a fixed partition, the optimal agent positions are the centroids, in fact,

$$\mathcal{H}(P', \mathcal{W}) \leq \mathcal{H}(P, \mathcal{W}),$$

for any $P', P$ such that $\|p'_i - C_{W_i}\| \leq \|p_i - C_{W_i}\|$, $i \in \{1, \ldots, N\}$. These two facts together naturally lead to a periodic implementation of the coordination policy $u^*$, where agents repeatedly and synchronously acquire position information from their neighbors, update their own Voronoi cell, and move toward their centroid. Both the continuous-time and periodic implementations are guaranteed to make the network state asymptotically converge to the set of centroidal Voronoi configurations.

### 10.5.1 Self-Triggered Deployment Algorithm

Here we describe a self-triggered implementation of the distributed coordination algorithm for deployment described above. There are two key elements in our exposition. The first is the notion of abstraction employed by individual agents about the behavior of their neighbors. In this particular scenario, this gives rise to agents operating with spatial partitions under uncertain information, which leads us to the second important element: the design of a controller that operates on set-valued information rather than point-valued information. Our exposition here follows [18].

*Abstraction via maximum velocity bound:* The notion of abstraction employed is based on the maximum velocity bound for (10.17). In fact, an agent knows that any other neighbor cannot travel farther than $v_{\max}\Delta t$ from its current position in $\Delta t$ seconds. Consequently, the reachable set of an agent $i \in \{1, \ldots, N\}$ is

$$\mathcal{R}_i(s, p_i) = \overline{B}(p_i, s v_{\max}) \cap S.$$

If $t_\ell$ denotes the time at which agent $i$ has just received position information $p_j(t_\ell)$ from a neighboring agent $j \in \mathcal{N}(i)$, then the guaranteed set that agent $i$ maintains about agent $j$ is

$$\mathbf{X}^i_j(t) = \overline{B}(p_j(t_\ell), (t - t_\ell)v_{\max}) \cap S,$$

which has the property that $p_j(t) \in \mathbf{X}^i_j(t)$ for $t \geq t_\ell$. Figure 10.1a shows an illustration of this concept. Agent $i$ stores the data in $\mathcal{D}^i(t) = (\mathbf{X}^i_1(t), \ldots, \mathbf{X}^i_N(t)) \subset S^N$ (if agent $i$ does not have information about some agent $j$, we set $\mathbf{X}^i_j(t) = \emptyset$) and the entire memory of the network is $\mathcal{D} = (\mathcal{D}^1, \ldots, \mathcal{D}^N) \subset S^{N^2}$.

*Design of controller operating on set-valued information:* Agents cannot compute the Voronoi partition exactly with the data structure described above. However, they can still determine lower and upper bounds on their exact Voronoi cell, as we describe next. Given a set of regions $D_1, \ldots, D_N \subset S$, each containing a site $p_i \in D_i$, the guaranteed Voronoi diagram of $S$ generated by

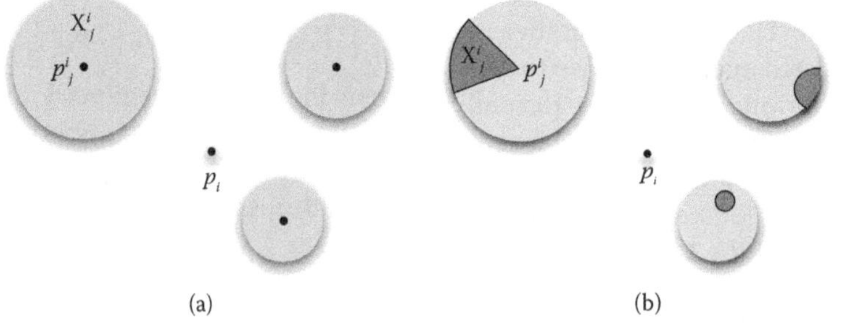


**FIGURE 10.1**
(a) Guaranteed sets and (b) promise sets kept by agent $i$. In (b), dark regions correspond to the promise sets given the promises $\mathcal{U}^i_j$.

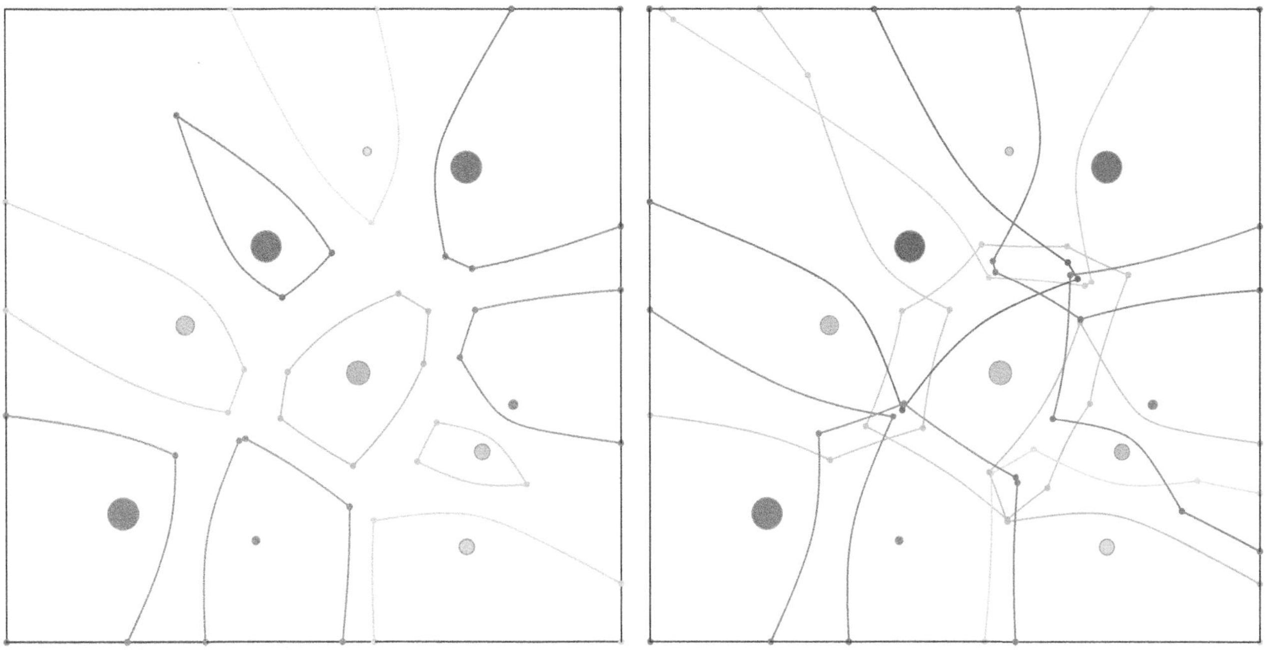

(a)

(b)

**FIGURE 10.2**
(a) Guaranteed and (b) dual guaranteed Voronoi diagrams.

$D = (D_1, \dots, D_N)$ is the collection $g\mathcal{V}(D_1, \dots, D_N) = \{gV_1, \dots, gV_N\}$,

$$gV_i = \{q \in S \mid \max_{x \in D_i} \|q - x\| \leq \min_{y \in D_j} \|q - y\| \text{ for all } j \neq i\}.$$

The guaranteed Voronoi diagram is not a partition of $S$, see Figure 10.2a. The important fact for our purposes is that, for any collection of points $p_i \in D_i$, $i \in \{1, \dots, N\}$, the guaranteed Voronoi diagram is contained in the Voronoi partition, that is, $gV_i \subset V_i$, $i \in \{1, \dots, N\}$. Each point in the boundary of $gV_i$ belongs to the set

$$\Delta_{ij}^{g} = \{q \in S \mid \max_{x \in D_i} \|q - x\| = \min_{y \in D_j} \|q - y\|\}, \quad (10.20)$$

for some $j \neq i$. Note that $\Delta_{ij}^{g} \neq \Delta_{ji}^{g}$. Agent $p_j$ is a guaranteed Voronoi neighbor of $p_i$ if $\Delta_{ij}^{g} \cap \partial gV_i$ is not empty nor a singleton. The set of guaranteed Voronoi neighbors of agent $i$ is $g\mathcal{N}_i(D)$. In our scenario, the "uncertain" regions for each agent correspond to the guaranteed sets.

One can also introduce the concept of a dual guaranteed Voronoi diagram of $S$ generated by $D_1, \dots, D_N$, which is the collection of sets $dg\mathcal{V}(D_1, \dots, D_N) = \{dgV_1, \dots, dgV_N\}$ defined by

$$dgV_i = \Big\{q \in S \mid \min_{x \in D_i} \|q - x\| \leq \max_{y \in D_j} \|q - y\| \text{ for all } j \neq i\Big\}.$$

A similar discussion can be carried out for this notion, cf. [18]. The most relevant fact for our design is that for any collection of points $p_i \in D_i$, $i \in \{1, \dots, N\}$, the dual guaranteed Voronoi diagram contains the Voronoi partition, that is, $V_i \subset dgV_i$, $i \in \{1, \dots, N\}$.

The notions of guaranteed and dual guaranteed Voronoi cells allow us to synthesize a controller building on the observation that, if

$$\|p_i - C_{gV_i}\| > \text{bnd}_i \equiv \text{bnd}(gV_i, dgV_i) = 2\,\text{cr}(dgV_i)\Big(1 - \frac{M_{gV_i}}{M_{dgV_i}}\Big), \quad (10.21)$$

then moving toward $C_{gV_i}$ from $p_i$ strictly decreases the distance $\|p_i - C_{V_i}\|$ (here, $\text{cr}(Q)$ denotes the circumradius of $Q$). With this in place, an agent can actually move closer to the centroid of its Voronoi cell (which guarantees the decrease of $\mathcal{H}$), even when it does not know the exact locations of its neighbors. Informally, each agent moves as follows:

> *[Informal description]:* The agent moves toward the centroid of its guaranteed Voronoi cell until it is within distance $\text{bnd}_i$ of it.

Note that, as time elapses without new information, the bound $\text{bnd}_i$ grows until it becomes impossible to satisfy (10.21). This gives a natural triggering condition for when agent $i$ needs updated information from its neighbors, which we describe informally as follows:

> *[Informal description]:* The agent decides that up-to-date location information is required if its computed bound in (10.21) is larger than a design parameter $\varepsilon$ and the distance to the centroid of its guaranteed cell.

The role of $\varepsilon$ is guaranteeing the lack of Zeno behavior: when an agent $i$ gets close to $C_{gV_i}$, this requires $\text{bnd}_i$ to be small. The introduction of the design parameter is critical to ensure Zeno behavior does not occur. Note that each agent can compute the future evolution of $\text{bnd}_i$ without updated information about its neighbors. This means that each agent can schedule exactly when the condition above will be satisfied, corresponding to a self-triggered implementation.

The overall algorithm is the result of combining the motion control law and the update decision policy with a procedure to acquire up-to-date information about other agents. We do not enter into the details of the latter here and instead refer the reader to [18]. The self-triggered deployment algorithm is formally presented in Algorithm 10.2.

**Algorithm 10.2**: Self-triggered deployment algorithm

Agent $i \in \{1, \ldots, N\}$ performs:

```
set D = D^i
compute L = gV_i(D) and U = dgV_i(D)
compute q = C_L and r = bnd(L, U)
(Update decision)
if r ≥ max{||q − p_i||, ε} then
    reset D^i by acquiring updated location information from Voronoi neighbors
    set D = D^i
    set L = gV(D) and U = dgV(D)
    set q = C_L and r = bnd(L, U)
end if
(Motion control)
if ||p_i − q|| > r then
    set u_i = v_max unit(q − p_i)
else
    set u_i = 0
end if
```

One can establish the same convergence guarantees for this algorithm as in the periodic or continuous case. In fact, for $\varepsilon \in [0, \text{diam}(S))$, the agents' positions starting from any initial network configuration in $S^n$ converge to the set of centroidal Voronoi configurations.

### 10.5.2 Team-Triggered Deployment Algorithm

Here we describe a team-triggered implementation of the distributed coordination algorithm for optimal deployment. Given our previous discussion, the only new element that we need to specify with respect to the exposition of the self-triggered implementation in Section 10.5.1 is the notion of promises employed by the agents. For each $j \in \{1, \ldots, N\}$, given the deployment controller $u_j^* : \prod_{i \in \mathcal{N}(j) \cup \{j\}} \mathbb{P}^{cc}(\mathcal{X}_i) \to \mathcal{U}_j$ and $\delta_j > 0$, the ball-radius control promise generated by agent $j$ for agent $i$ at time $t$ is

$$\mathcal{U}_i^j = \overline{B}(u_j(X_{\mathcal{N}}^j(t)), \delta_j) \cap \mathcal{U}_j \qquad t' \geq t, \tag{10.22}$$

for some $\delta_j > 0$. This promise is a ball of radius $\delta_j$ in the control space $\mathcal{U}_j$ centered at the control signal used at time $t$. Here we consider constant $\delta_j$ for simplicity, but one can consider time-varying functions instead, cf. [28]. Agent $i$ stores the information provided by the promise sets in $\mathcal{D}^i(t) = (X_1^i(t), \ldots, X_N^i(t)) \subset S^N$. The event-triggered information updates, by which agents inform other agents if they decide to break their promises, take care of making sure that this information is always accurate; that is, $p_i(t) \in X_i^j(t)$ for all $i \in \{1, \ldots, N\}$ and $j \in \mathcal{N}(i)$. The team-triggered deployment algorithm is formally presented in Algorithm 10.3 and has the same motion control law, update decision policy, and procedure to acquire up-to-date information about other agents as the self-triggered coordination one.

**Algorithm 10.3**: Team-triggered deployment algorithm

Agent $i \in \{1, \ldots, N\}$ performs:

```
set D = D^i
compute L = gV_i(D) and U = dgV_i(D)
compute q = C_L and r = bnd(L, U)
(Self-triggered update decision)
if r ≥ max{||q − p_i||, ε} then
    reset D^i by acquiring updated location information from Voronoi neighbors
    set D = D^i
    set L = gV(D) and U = dgV(D)
    set q = C_L and r = bnd(L, U)
end if
(Event-triggered update decision)
if p_i ∉ X_i^j for any j ∈ N(i) then
    send updated position information p_i to agent j
    send new promise U_i^j to agent j
end if
(Motion control)
if ||p_i − q|| > r then
    set u_i = v_max unit(q − p_i)
else
    set u_i = 0
end if
```

### 10.5.3 Simulations

Here we provide several simulations to illustrate the performance of the self-triggered and team-triggered deployment algorithms. We consider a network of $N = 8$ agents, moving in a 4 m $\times$ 4 m square, with a maximum velocity $v_{\max} = 1$ m/s. The density $\phi$ is a sum of two Gaussian functions:

$$\phi(x) = 1.2e^{-\|x-q_1\|^2} + e^{-\|x-q_2\|^2},$$

with $q_1 = (1, 1.5)$ and $q_2 = (2.5, 3)$. We evaluate the total power consumed by communication among the agents during the execution of each algorithm. To do this, we adopt the following model [30] for quantifying the total power $\mathcal{P}_i$ used by agent $i \in \{1, \ldots, 8\}$ to communicate, in $dBmW$ power units:

$$\mathcal{P}_i = 10 \log_{10} \left[ \sum_{j \in \{1,\ldots,N\}, i \neq j}^{n} \beta 10^{0.1 P_{i \to j} + \alpha \|p_i - p_j\|} \right], \quad (10.23)$$

where $\alpha > 0$ and $\beta > 0$ depend on the characteristics of the wireless medium, and $P_{i \to j}$ is the power received by $j$ of the signal transmitted by $i$ in units of $dBmW$. In our simulations, these values are set to 1.

For the promises made among agents in the team-triggered strategy, we use $\delta_i = 2\lambda v_{\max}$ in (10.22), where $\lambda \in [0, 1]$ is a parameter that captures the "tightness" of promises. Note that $\lambda = 0$ corresponds to exact promises, meaning trajectories can be computed exactly, and $\lambda = 1$ corresponds to no promises at all (i.e., recovers the self-triggered algorithm because promises become the entire allowable control set).

Figures 10.3 and 10.4 compare the execution of the periodic, self-triggered, and team-triggered deployment strategies (the latter for two different tightnesses of

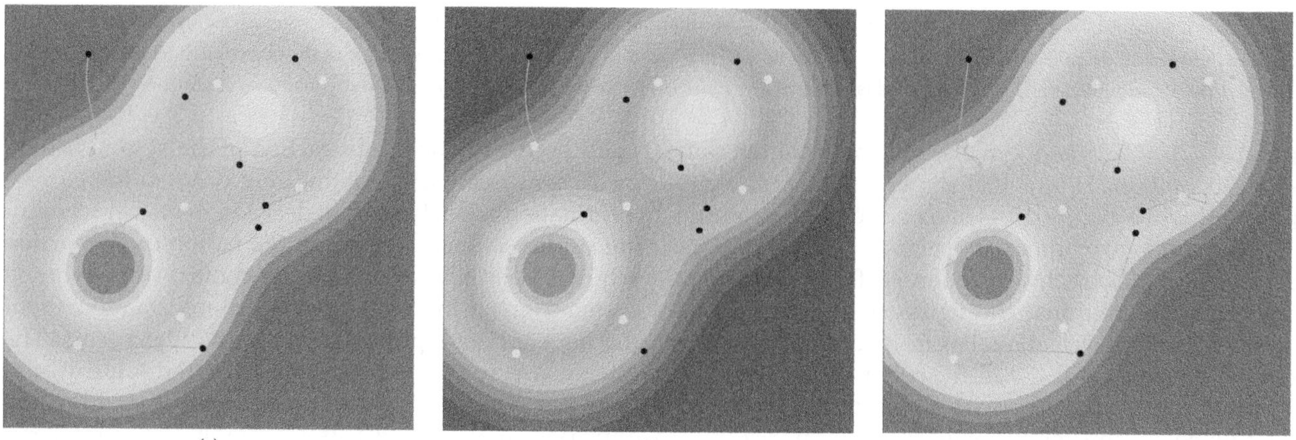

(a) (b) (c)

**FIGURE 10.3**
Plot (a) is an execution of the communicate-at-all-times strategy [29], plot (b) is an execution of the self-triggered strategy, and plot (c) is an execution of the team-triggered strategy with $\lambda = 0.5$. Black dots correspond to initial positions, and light gray dots correspond to final positions.

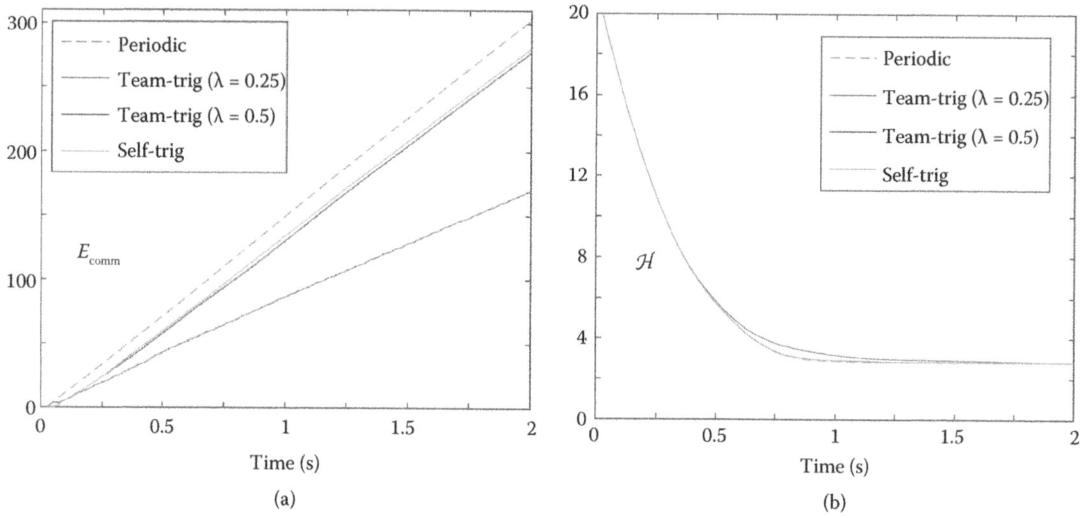


**FIGURE 10.4**
Plot (a) shows the total communication energy $E_{\text{comm}}$ in Joules of executions of the communicate-at-all-times strategy, the self-triggered strategy, and the team-triggered strategy (for two different tightnesses of promises). Plot (b) shows the evolution of the objective function $\mathcal{H}$. All simulations have the same initial condition.

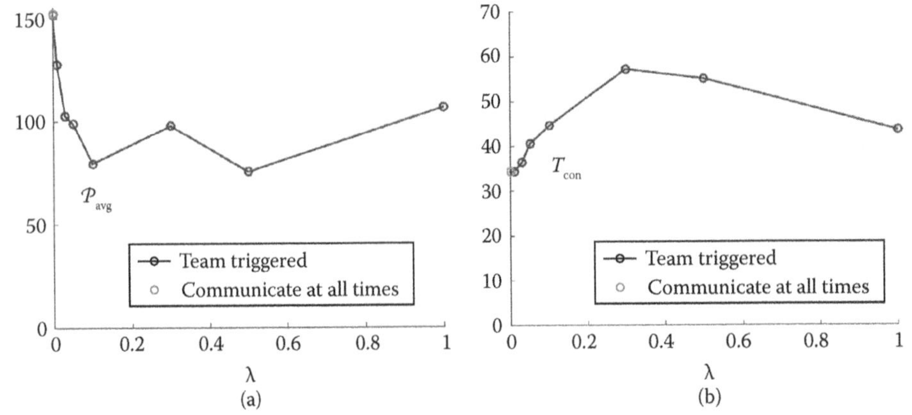


**FIGURE 10.5**
Average (a) communication power network consumption $\mathcal{P}_{avg}$ and (b) time to convergence $T_{con}$ for the team-triggered deployment strategy with varying tightness of promise. Time to convergence is the time to reach 99% of the final convergence value of the objective function. For small $\lambda$, the amount of communication energy required is substantially reduced, while the time to convergence only increases slightly.

promises, $\lambda = 0.25$ and $\lambda = 0.5$) starting from the same initial condition. Figure 10.3 shows the algorithm trajectories, while Figure 10.4a shows the total communication energy over time, and Figure 10.4b shows the evolution of the objective function $\mathcal{H}$. These plots reveal the important role that the tightness of promises has on the algorithm execution. For instance, in Figure 10.4, one can see that for $\lambda = 0.25$, the communication energy is reduced by half compared to the periodic strategy without compromising the network performance.

In Figure 10.5, we explore what happens for various different tightnesses $\lambda$ for the team-triggered strategy and compare them against the periodic and the self-triggered strategies. Figure 10.5a shows the average power consumption in terms of communication energy by the entire network for varying levels of $\lambda$. Figure 10.5b shows the time it takes each execution to reach 99% of the final convergence value of the objective function $\mathcal{H}$. As we can see from these plots, for small $\lambda > 0$, we are able to drastically reduce the amount of required communication while only slightly increasing the time to convergence.

## 10.6 Conclusions

This chapter has covered the design and analysis of triggered strategies for networked cyber-physical systems, with an emphasis on distributed interactions. At a fundamental level, our exposition aims to trade computation for less sensing, actuation, or communication effort by answering basic questions such as when to take state samples, when to update control signals, or when to transmit information. We have focused on self-triggered approaches to controller design which build on the notion of abstraction of the system behavior to synthesize triggers that rely only on the current information available to the decision maker in scheduling future actions. The role of the notion of abstraction acquires special relevance in the context of networked systems, given the limited access to information about the network state possessed by individual agents. The often conservative nature of self-triggered implementations has motivated us to introduce the team-triggered approach, which allows us to combine the best components of event- and self-triggered control. The basic idea is to have agents cooperate with each other in providing better, more accurate abstractions to each other via promise sets. Such sets can then be used by individual agents to improve the overall network performance. The team-triggered approach opens numerous venues for future research. Among them, we highlight the robustness under disturbances in the dynamics and sensor noise, more general models for individual agents, methods that tailor the generation of agent promises to the specific task at hand, methods for the systematic design of controllers that operate on set-valued information models, analytic guarantees on performance improvements with respect to self-triggered strategies (as observed in Section 10.5.3), the impact of evolving topologies on the generation of agent promises, and the application to the synthesis of distributed strategies to other coordination problems.

## Acknowledgments

This research was partially supported by NSF Award CNS-1329619.

## Bibliography


[1] M. Velasco, P. Marti, and J. M. Fuertes. The self triggered task model for real-time control systems. In *Proceedings of the 24th IEEE Real-Time Systems Symposium*, pp. 67–70, Cancun, Mexico, December 3–5, 2003.

[2] R. Subramanian and F. Fekri. Sleep scheduling and lifetime maximization in sensor networks. In *Symposium on Information Processing of Sensor Networks*, pp. 218–225, Nashville, TN, April 19–21, 2006.

[3] W. P. M. H. Heemels, K. H. Johansson, and P. Tabuada. An introduction to event-triggered and self-triggered control. In *IEEE Conference on Decision and Control*, pp. 3270–3285, Maui, HI, 2012.

[4] M. Mazo Jr. and P. Tabuada. Decentralized event-triggered control over wireless sensor/actuator networks. *IEEE Transactions on Automatic Control*, 56(10):2456–2461, 2011.

[5] X. Wang and N. Hovakimyan. $L_1$ adaptive control of event-triggered networked systems. In *American Control Conference*, pp. 2458–2463, Baltimore, MD, 2010.

[6] M. C. F. Donkers and W. P. M. H. Heemels. Output-based event-triggered control with guaranteed $L_\infty$-gain and improved and decentralised event-triggering. *IEEE Transactions on Automatic Control*, 57(6):1362–1376, 2012.

[7] D. V. Dimarogonas, E. Frazzoli, and K. H. Johansson. Distributed event-triggered control for multi-agent systems. *IEEE Transactions on Automatic Control*, 57(5):1291–1297, 2012.

[8] G. Shi and K. H. Johansson. Multi-agent robust consensus-part II: Application to event-triggered coordination. In *IEEE Conference on Decision and Control*, pp. 5738–5743, Orlando, FL, December 2011.

[9] X. Meng and T. Chen. Event based agreement protocols for multi-agent networks. *Automatica*, 49(7):2125–2132, 2013.

[10] Y. Fan, G. Feng, Y. Wang, and C. Song. Distributed event-triggered control of multi-agent systems with combinational measurements. *Automatica*, 49(2):671–675, 2013.

[11] K. Kang, J. Yan, and R. R. Bitmead. Cross-estimator design for coordinated systems: Constraints, covariance, and communications resource assignment. *Automatica*, 44(5):1394–1401, 2008.

[12] P. Wan and M. D. Lemmon. Event-triggered distributed optimization in sensor networks. In *Symposium on Information Processing of Sensor Networks*, pp. 49–60, San Francisco, CA, April 13–16, 2009.

[13] S. S. Kia, J. Cortés, and S. Martínez. Distributed convex optimization via continuous-time coordination algorithms with discrete-time communication. *Automatica*, 55:254–264, 2015.

[14] D. Richert and J. Cortés. Distributed linear programming with event-triggered communication. *SIAM Journal on Control and Optimization*, 2014. Available at http://arxiv.org/abs/1405.0535.

[15] A. Eqtami, D. V. Dimarogonas, and K. J. Kyriakopoulos. Event-triggered strategies for decentralized model predictive controllers. In *IFAC World Congress*, pp. 10068–10073, Milano, Italy, August 2011.

[16] E. Garcia and P. J. Antsaklis. Model-based event-triggered control for systems with quantization and time-varying network delays. *IEEE Transactions on Automatic Control*, 58(2):422–434, 2013.

[17] W. P. M. H. Heemels and M. C. F. Donkers. Model-based periodic event-triggered control for linear systems. *Automatica*, 49(3):698–711, 2013.

[18] C. Nowzari and J. Cortés. Self-triggered coordination of robotic networks for optimal deployment. *Automatica*, 48(6):1077–1087, 2012.

[19] X. Wang and M. D. Lemmon. Event-triggered broadcasting across distributed networked control systems. In *American Control Conference*, pp. 3139–3144, Seattle, WA, June 2008.

[20] M. Zhong and C. G. Cassandras. Asynchronous distributed optimization with event-driven communication. *IEEE Transactions on Automatic Control*, 55(12):2735–2750, 2010.

[21] X. Wang and M. D. Lemmon. Event-triggering in distributed networked control systems. *IEEE Transactions on Automatic Control*, 56(3):586–601, 2011.

[22] G. S. Seybotha, D. V. Dimarogonas, and K. H. Johansson. Event-based broadcasting for multi-agent average consensus. *Automatica*, 49(1):245–252, 2013.

[23] M. Mazo Jr., A. Anta, and P. Tabuada. On self-triggered control for linear systems: Guarantees and complexity. In *European Control Conference*, pp. 3767–3772, Budapest, Hungary, August 2009.

[24] C. Nowzari and J. Cortés. Self-triggered optimal servicing in dynamic environments with acyclic structure. *IEEE Transactions on Automatic Control*, 58(5):1236–1249, 2013.

[25] F. Bullo, J. Cortés, and S. Martínez, *Distributed Control of Robotic Networks*. Applied Mathematics Series, Princeton University Press, 2009. http://coordinationbook.info.

[26] E. Garcia, Y. Cao, H. Yu, P. Antsaklis, and D. Casbeer. Decentralised event-triggered cooperative control with limited communication. *International Journal of Control*, 86(9):1479–1488, 2013.

[27] C. Nowzari and J. Cortés. Zeno-free, distributed event-triggered communication and control for multi-agent average consensus. In *American Control Conference*, pp. 2148–2153, Portland, OR, 2014.

[28] C. Nowzari and J. Cortés. Team-triggered coordination for real-time control of networked cyberphysical systems. *IEEE Transactions on Automatic Control*, 2013. Available at http://arxiv.org/abs/1410.2298.

[29] J. Cortés, S. Martínez, T. Karatas, and F. Bullo. Coverage control for mobile sensing networks. *IEEE Transactions on Robotics and Automation*, 20(2): 243–255, 2004.

[30] S. Firouzabadi. Jointly optimal placement and power allocation in wireless networks. Master's thesis, University of Maryland at College Park, 2007.

# 11

# *Efficiently Attentive Event-Triggered Systems*

**Lichun Li**
*Georgia Institute of Technology*
*Atlanta, GA, USA*

**Michael D. Lemmon**
*University of Notre Dame*
*Notre Dame, IN, USA*

**CONTENTS**



**ABSTRACT** Event-triggered systems sample the system's outputs and transmit those outputs to a controller or an actuator when the "novelty" in the sample exceeds a specified threshold. Well-designed event-triggered systems are *efficiently attentive*, which means that the intersampling intervals get longer as the system state approaches the control system's equilibrium point. When the system regulates its state to remain in the neighborhood of an equilibrium point, efficiently attentive systems end up using fewer computational and communication resources than comparable time-triggered systems. This chapter provides sufficient conditions under which a bandwidth-limited event-triggered system is efficiently attentive. Bandwidth-limited systems transmit sampled data with a finite number of bits, which are then received at the destination with a finite delay. This chapter designs a dynamic quantizer with state-dependent upper bounds on the maximum acceptable delay under which a bandwidth-limited event-triggered system is assured to be efficiently attentive.

## 11.1 Introduction

Event triggering is a recent approach to sampled-data control in which sampling occurs only if a measure of data novelty exceeds a specified threshold. Early interest in event triggering was driven by simulation results, suggesting that these constant event-triggering thresholds could greatly increase the average sampling period over periodically sampled systems with comparable performance levels [2,3,14]. Similar experimental results were reported in [11] for state-dependent event triggers enforcing input-to-state stability and in [12] for $\mathcal{L}_2$ stability. It is, however, relatively easy to construct

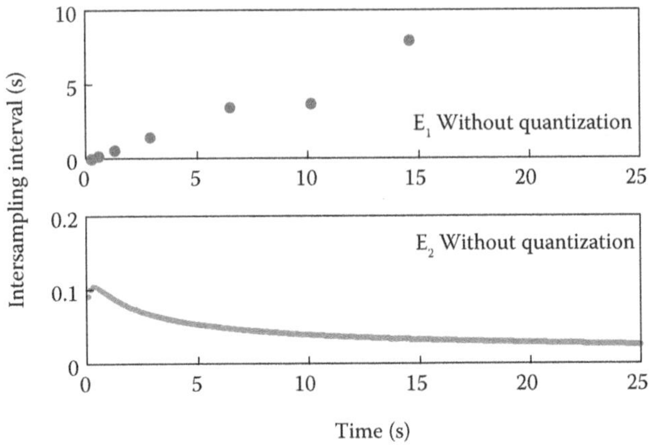


FIGURE 11.1
Intersampling intervals of event-triggered systems (11.1).

examples where an event trigger actually decreases the average sampling period of the system. Consider, for example, a cubic system

$$\dot{x} = x^3 + u; \quad x(0) = x_0, \tag{11.1}$$

where $u = -3\hat{x}_k^3$ for $t \in [s_k, s_{k+1})$, where $s_k$ is the $k$th consecutive sampling instant, and $\hat{x}_k = x(s_k)$. Let us now consider two different event triggers. The trigger $\mathbf{E}_1$ generates a sampling instant $s_{k+1}$ when $|x(t) - \hat{x}_k| = 0.5|x(t)|$, and the second trigger $\mathbf{E}_2$ generates a sampling instant when $|x(t) - \hat{x}_k| = 0.5|x^4(t)|$. For both event triggers, the system is locally asymptotically stable for all $|x_0| \leq 0.6$. But if one examines the intersampling intervals in the two plots of Figure 11.1, it should be apparent that the intersampling intervals generated by trigger $\mathbf{E}_1$ get longer as the system approaches its equilibrium. By contrast, trigger $\mathbf{E}_2$ results in intersampling intervals that get shorter as the system approaches the origin. It is obvious that trigger $\mathbf{E}_1$ conserves communication resources, while trigger $\mathbf{E}_2$ does not. The preceding examples raise an important question: namely, what are the essential characteristics of an event-triggered system that conserve its communication resources.

The event-triggered system with trigger $\mathbf{E}_1$ conserves communication resources, in the sense that as the system state approaches the equilibrium, the intersampling intervals become longer. This is a desirable property of an event-triggered system because one would conjecture that the "interesting" feedback information necessary for system stabilization should be greater when the system's state is perturbed away from the equilibrium point. We call this property *efficient attentiveness* because the control system becomes more efficient in its use of the channel as it asymptotically approaches its equilibrium point.

The intersampling interval, however, is not necessarily the best way of characterizing the information needed to stabilize an event-triggered system. This interval characterizes the rate at which sampled information is transmitted across a channel, but that information is the sampled state without any quantization error. In reality, the system's communication channel transmits information in discrete packets of a fixed size. This means that the sampled state must also be quantized with a finite number of bits which are then transmitted as a single packet across the channel. Furthermore, that packet is not instantaneously received at the other end of the channel. Transmitted packets are successfully received and decoded after a finite delay. One may, therefore, characterize the actual bandwidth needed to transmit the sampled information as a *stabilizing instantaneous bit-rate* which equals the number of bits used to encode each sampled measurement and the maximum delay with which the sampled-data system is known to be asymptotically stable. Therefore, to properly study efficiently attentive systems, one must determine conditions under which the intersampling interval increases, and the stabilizing instantaneous bit-rate decreases as the system's state asymptotically approaches the equilibrium point.

This chapter provides a sufficient condition for event-triggered systems to be efficiently attentive in the sense that the intersampling interval gets longer, and the system's instantaneous bit-rate gets smaller as the system state approaches the origin. In particular, we develop a dynamic quantization policy and identify a maximum acceptable transmission delay for which the system is assured to be locally ISS and efficiently attentive. The results consolidate earlier work in [5,6,13]. This chapter is organized as follows. Sections 11.2 and 11.3 present mathematical preliminaries and the problem statement, respectively. Sections 11.4 and 11.5 state how to properly design the threshold function and the quantization map, and the main results are presented in Sections 11.6 and 11.7 for the case without and with delay, respectively. An example to demonstrate the results is provided in Section 11.8, and a summary is given in Section 11.9 to conclude the chapter.

## 11.2 Preliminaries

Let $\mathbb{R}^n$ and $\mathbb{R}^+$ denote the $n$-dimensional Euclidean space and a set of nonnegative reals, respectively. The infinity norm in $\mathbb{R}^n$ is denoted as $\|\cdot\|$. The $\mathcal{L}$-infinity norm of a function $x(\cdot) : \mathbb{R}^+ \to \mathbb{R}^n$ is defined as $\|x\|_{\mathcal{L}_\infty} = \text{ess sup}_{t\geq 0} \|x(t)\|$. This function is said to be essentially bounded if $\|x\|_{\mathcal{L}_\infty} < \infty$.

Let $\Omega$ be a domain (open and connected subset) of $\mathbb{R}^n$, and $B_r \subset \mathbb{R}^n$ be an open ball centered at the origin with radius $r$. We say $f(\cdot) : \Omega \to \mathbb{R}^n$ is *locally Lipschitz* on $\Omega$ if each point of $\Omega$ has an open neighborhood $\Omega_0$ such that for any $x, y \in \Omega_0$, $\|f(x) - f(y)\| \leq L_0\|x - y\|$ with some constant $L_0 \geq 0$.

A function $\alpha(\cdot) : \mathbb{R}^+ \to \mathbb{R}^+$ is of class $\mathcal{K}$ if it is continuous, strictly increasing, and $\alpha(0) = 0$. The function $\alpha$ is of class $\mathcal{N}$ if it is continuous, increasing, and $\alpha(s) = \infty \Rightarrow s = \infty$. It is of class $\mathcal{M}$ if it is continuous, decreasing, and $\alpha(s) = 0 \Rightarrow s = \infty$. A function $\beta : \mathbb{R}^+ \times \mathbb{R}^+ \to \mathbb{R}^+$ is of class $\mathcal{KL}$ if $\beta(\cdot, t)$ is of class $\mathcal{K}$ for each fixed $t \geq 0$, and $\beta(r, t)$ decreases to 0 as $t \to \infty$ for each fixed $r \geq 0$.

**Lemma 11.1**

Let $g : (0, \infty) \to \mathbb{R}^+$ be a continuous function satisfying $\lim_{s \to 0} g(s) < \infty$. There must exist a class $\mathcal{N}$ function $h$ such that $g(s) \leq h(s)$, for all $s \geq 0$.

**PROOF** See the proof of Lemma 4.3 in [4]. ■

Consider a dynamical system whose state $x : \mathbb{R}^+ \to \mathbb{R}^n$ satisfies the differential equation

$$\dot{x}(t) = f(x(t), w(t)), \quad x(0) = x_0, \tag{11.2}$$

in which $x_0 \in \mathbb{R}^n$, $w : \mathbb{R}^+ \to \mathbb{R}^m$ is essentially bounded, and $f : \mathbb{R}^n \times \mathbb{R}^m \to \mathbb{R}^n$ is locally Lipschitz in $x$. A point $\bar{x} \in \mathbb{R}^n$ is an equilibrium point for system (11.2) if and only if $f(\bar{x}, 0) = 0$. The system (11.2) is *locally input-to-state stable* (ISS) if there exist $\rho_1, \rho_2 > 0$, $\gamma \in \mathcal{K}$, and $\beta \in \mathcal{KL}$, such that for all $\|x_0\| \leq \rho_1$, and $\|w\|_{\mathcal{L}_\infty} \leq \rho_2$, the system's state trajectory $x$ satisfies $\|x(t)\| \leq \beta(\|x_0\|, t) + \gamma(\|w\|_{\mathcal{L}_\infty})$ for all $t \geq 0$.

This chapter's analysis will make use of the following theorem that provides a Lyapunov-like condition for system (11.2) to be locally ISS.

**Theorem 11.1**

Let $\Omega \subset \mathbb{R}^n$ be a domain containing the origin, and $V : \Omega \to \mathbb{R}^+$ be continuous differentiable such that

$$\alpha_1(\|x\|) \leq V(x) \leq \alpha_2(\|x\|), \forall x \in \Omega,$$
$$\frac{\partial V}{\partial x} f(x, w) \leq -\alpha_3(\|x\|) + \gamma_1(\|w\|_{\mathcal{L}_\infty}), \forall x \in \Omega,$$

where $\alpha_1, \alpha_2, \alpha_3, \gamma_1 \in \mathcal{K}$. Take $r > 0$ such that $B_r \subset \Omega$, and suppose

$$\|w(t)\|_{\mathcal{L}_\infty} < \gamma_1^{-1}(\epsilon\alpha_3(\alpha_2^{-1}(\alpha_1(r)))), \text{ for some } \epsilon \in (0, 1).$$

Then, for every initial state $x(t_0)$, satisfying $\|x(t_0)\| < \alpha_2^{-1}(\alpha_1(r))$, there exist functions $\beta \in \mathcal{KL}$ and $\gamma \in \mathcal{K}$ such that the solution of (11.2) satisfies $\|x(t)\| \leq \beta(\|x(t_0)\|, t - t_0) + \gamma(\|w(t)\|_{\mathcal{L}_\infty})$ (i.e., the system (11.2) is locally ISS).

**PROOF** It can be shown that $\|x\| \geq \alpha_3^{-1}\left(\frac{\gamma_1(\|w\|_{\mathcal{L}_\infty})}{\epsilon}\right) \Rightarrow \frac{\partial V}{\partial x} f(x, w) \leq -(1 - \epsilon)\alpha_3(\|x\|)$. According to Theorem 4.18 of [4], we have Theorem 11.1. ■

## 11.3 Problem Statement

Consider a quantized event-triggered system, as shown in Figure 11.2, in which an *event detector* and a *quantizer* are co-located within the system's *sensor* with direct access to the *plant*'s state, $x$. The system *controller* is a separate subsystem whose control output, $u$, is directly applied to the plant. The sensor samples the system state at discrete time instants, $\{s_k\}_{k=1}^{\infty}$, and quantizes the sampled state so it can be transmitted across a bandwidth-limited communication channel. The *plant* is an input-to-state system satisfying the following set of differential equations:

$$\dot{x}(t) = f(x(t), u(t), w(t)), \quad x(0) = x_0, \tag{11.3}$$
$$u(t) = K(\hat{x}_k) \doteq u_k, \forall t \in [a_k, a_{k+1}), k = 0, 1, \ldots, \tag{11.4}$$

where $x(\cdot) : \mathbb{R}^+ \to \mathbb{R}^n$ is the system state, $w(\cdot) : \mathbb{R}^+ \to \mathbb{R}^q$ is an $\mathcal{L}_\infty$ disturbance with $\|w\|_{\mathcal{L}_\infty} = \bar{w} \leq c_1\|w\|_{\mathcal{L}_\infty}$ for some constant $c_1 \geq 1$, and $f : \mathbb{R}^n \times \mathbb{R}^m \times \mathbb{R}^q \to \mathbb{R}^n$ is locally Lipschitz in all three variables with $f(0, 0, 0) = 0$. The control input, $u(\cdot) : \mathbb{R}^+ \to \mathbb{R}^m$, is computed by the controller as a function of the *quantized sampled states*, $\{\hat{x}_k\}_{k=1}^{\infty}$. This quantized state, $\hat{x}_k$, is a function of the true system state, $x(s_k)$, at a sampling instant, $s_k$ for $k = 1, 2, \ldots, \infty$. The sampled state, $x(s_k)$, is quantized so it can be transmitted across the channel with a finite number of bits. The precise nature of the quantization map is discussed below. The sensor transmits the quantized sampled state at the sampling instant $s_k$,

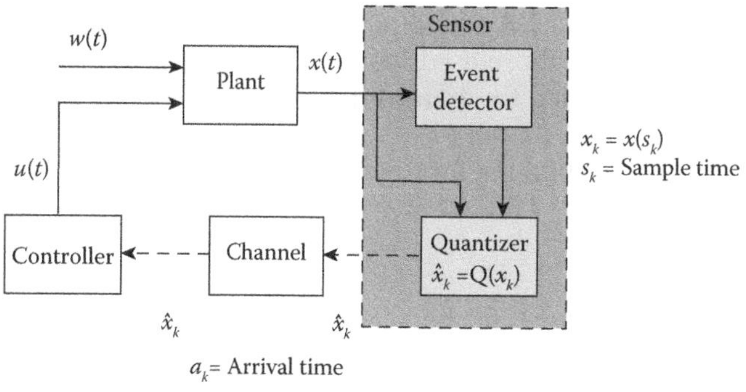


**FIGURE 11.2**
Event-triggered control system with quantization.

and that transmitted information is received by the controller at time instant $a_k$. The *delay* associated with this transmission is denoted $d_k = a_k - s_k$, and we define the $k$th intersampling interval $\tau_k = s_{k+1} - s_k$. We say the sampling, $\{s_k\}_{k=1}^{\infty}$, and arrival sequences, $\{a_k\}_{k=1}^{\infty}$, are *admissible* if $s_k \leq a_k \leq s_{k+1}$ for $k = 0, 1, \ldots, \infty$.

The sampling instants, $\{s_k\}_{k=1}^{\infty}$, can be generated in a number of ways. If the intersampling time is constant, then we obtain a traditional quantized sampled data system with a so-called *time-triggered* sampling strategy. An alternative sampling strategy generates a sampling instant $s_k$ when the difference between the current system state and the past *sampled* state, $e_{k-1}(t) = x(t) - \hat{x}_{k-1}$, exceeds a specified threshold. This chapter considers a sampling rule in which the $k$th consecutive sampling instant, $s_k$, is triggered when

$$\|e_{k-1}(t)\| \geq \theta(\|\hat{x}_{k-1}\|, \bar{w}) \doteq \theta_{k-1}, \qquad (11.5)$$

where $e_{k-1}(t) = x(t) - \hat{x}_{k-1}$. The sampling strategy described in Equation 11.5 is called a *state-dependent event trigger*. The trigger is *state dependent* because it is a function of the last sampled state.

The sampled state information, $x(s_k)$, must be transmitted across a bandwidth-limited channel. This means that the sampled state must be encoded with a finite number of bits before it is transmitted across the channel. Let $N_k$ denote the number of bits used to encode the $k$th consecutive sampled state, $x(s_k)$, and assume that these bits are transmitted in a single packet at time $s_k$ from the sensor and received at time $a_k$ by the controller. Let $\hat{x}_k$ denote the real vector *decoded* by the controller from the received packet. We assume that the *quantization error* $\delta_k = \|\hat{x}_k - x(s_k)\|$ is bounded as

$$\|\hat{x}_k - x(s_k)\| \leq \Delta(\|\hat{x}_{k-1}\|, \bar{w}) \doteq \Delta_k. \qquad (11.6)$$

The bound in (11.6) is quite general in the sense that $\Delta_k$ is a function of the system state. This allows one to consider both static and dynamic quantization strategies.

Let us now define the concept of *stabilizing instantaneous bit-rate* for the proposed event-triggered system. In particular, let the sampled state at time $s_k$ ($k = 0, 1, 2, \ldots, \infty$) be quantized with $N_k$ bits, and assume that the closed-loop system is asymptotically stable provided the transmitted information is received with a delay $d_k = a_k - s_k \leq D_k$ for all $k = 0, 1, 2, \ldots, \infty$. Then the stabilizing instantaneous bit-rate is

$$r_k = \frac{N_k}{D_k}.$$

We are interested in characterizing the asymptotic properties of the instantaneous bit-rate, $r_k$, for the event-triggered system described above.

For event-triggered systems, the quantization error and the stabilizing delay, $D_k$, will generally both be a function of the system state. This means that the stabilizing instantaneous bit-rate is a function of the system state. Since an important motivation for studying event-triggered systems is the claim that they use fewer computational and communication resources than comparable time-triggered systems, it would be valuable to identify conditions under which that claim is actually valid. As shown in the preceding example, it is quite easy to generate event-triggered systems that are not *efficient* in their use of communication resources, in that the stabilizing instantaneous bit-rate gets larger as one approaches the system's equilibrium point. This motivates the following definition.

**DEFINITION 11.1** An event-triggered system is said to be efficiently attentive if there exist functions $h_1 \in \mathcal{M}$ and $h_2 \in \mathcal{N}$, such that

1. The intersampling interval $\tau_k$ satisfies $\tau_k \geq h_1(\|\hat{x}_{k-1}\|)$ or $h_1(\|\hat{x}_k\|)$.
2. The stabilizing instantaneous bit-rate $r_k$ satisfies $r_k \leq h_2(\|\hat{x}_{k-1}\|)$ or $h_2(\|\hat{x}_k\|)$.

To be efficiently attentive, therefore, requires the intersampling interval to be increasing and the instantaneous bit-rate to be decreasing as the system state approaches the equilibrium point. This is because with a bandwidth-limited channel, one wants the system to access the channel less often as it gets closer to its equilibrium point, and one also wants it to consume fewer channel resources for each access. If one can guarantee that the system is efficiently attentive, then it will certainly be more efficient in its use of channel resources than comparable time-triggered systems, and one would have established formal conditions under which the claim that event-triggering is "more efficient" in its use of system resources can be formally verified.

## 11.4 Event-Trigger Design

The event-triggering function, $\theta(\|\hat{x}_{k-1}\|, \bar{w})$, in Equation 11.5 is chosen to render the closed-loop system in Equations 11.3 and 11.4 locally ISS. Such a triggering function exists when the following assumption is satisfied. This assumption asserts that there exists a state-feedback controller $K$ for which the system is locally ISS with respect to the state-feedback error $e(t) = \hat{x}_k - x(t)$ for all $t$ and the external disturbance $w$.

**ASSUMPTION 11.1** Let $\Omega, \Omega_e \subset \mathbb{R}^n, \Omega_w \subset \mathbb{R}^q$ be domains containing the origin. And $e(t), w(t)$ are essentially bounded signals. For system

$$\dot{x}(t) = f(x(t), K(x(t) + e(t)), w(t)), \qquad (11.7)$$

there exists a continuous differentiable function $V : \Omega \to \mathbb{R}$ satisfying

$$\alpha_1(\|x\|) \le V(x) \le \alpha_2(\|x\|), \forall x \in \Omega, \quad (11.8)$$

$$\frac{\partial V}{\partial x} f(x,w) \le -\alpha_3(\|x\|) + \gamma_1(\|e\|) + \gamma_2(\|w\|), \quad (11.9)$$

$$\forall x \in \Omega,\ e \in \Omega_e,\ w \in \Omega_w, \quad (11.10)$$

where $\alpha_{1,2,3}, \gamma_{1,2} \in \mathcal{K}$.

Let $e(t) = \hat{x}_k - x(t)$ denote the *gap* between the current state, $x(t)$, and the last sampled state, $\hat{x}_k$. Now assume that the gap satisfies

$$\|e(t)\| \le \xi(\|x(t)\|) \doteq \gamma_1^{-1}\left[c\alpha_3(\|x(t)\|) + \gamma_3(\bar{w})\right], \quad (11.11)$$

for some $c \in (0,1)$ and any $\gamma_3 \in \mathcal{K}$. Inserting this relation into Equation 11.9 yields

$$\frac{\partial V}{\partial x} f(x,w) \le -(1-c)\alpha_3(\|x\|) + \gamma(\bar{w}),$$

where $\gamma(\bar{w}) = \gamma_2(\bar{w}) + \gamma_3(\bar{w})$ is class $\mathcal{K}$. This relation shows that a sampled data is locally ISS with respect to the external disturbance $w$, provided the gap $e(t)$ satisfies the inequality in (11.11). One obvious way of enforcing this assumption is to require that the system samples the state when (11.11) is about to be violated. Inequality (11.11), therefore, represents a state-dependent event-triggering condition whose satisfaction assures the closed-loop system is locally ISS with respect to the input disturbances, $w$.

The event trigger in Equation 11.11 is a function of the system state $x(t)$ at time instant $t$. In actual implementations, it is more convenient to make this threshold dependent only on the past sampled state, $\hat{x}_k$. This is particularly important when we consider the impact that transmission delays and quantization have on the performance of the event-triggered system. In order to obtain practical event triggers that are only a function of the past sampled state, $\hat{x}_k$, one must assume that the event trigger $\xi(s,t)$ is locally Lipschitz with respect to the first argument. This observation leads to Assumption 11.2.

**ASSUMPTION 11.2** Let $B_{\bar{r}}$ be the smallest ball including $\Omega$. The function $\xi(s,t)$ defined in (11.11) is locally Lipschitz with respect to its first argument $s \in (0,\bar{r})$. Let $L_k^{\xi}$ be the Lipschitz constant of $\xi(s,t)$ with respect to s during time interval $[s_k, a_{k+1}]$.

This assumption requires that the function $\xi(\|x(t)\|,\bar{w})$ does not change too quickly with respect to $\|x(t)\|$ in a neighborhood of the origin. Examples of functions that satisfy Assumption 11.1 are polynomial functions with degree greater than one.

Let us then define a function $\underline{\xi}_k$ as

$$\underline{\xi}_k(s,t) = \frac{1}{L_k^{\xi}+1}\xi(s,t), \quad (11.12)$$

where $\xi$ is defined in Equation 11.11. Lemma 11.2 establishes the relationship between thresholds $\underline{\xi}_k(s,t)$ and $\xi(s,t)$.

**Lemma 11.2**

With Assumption 11.2,

$$\|e_k(t)\| \le \underline{\xi}_k(\|\hat{x}_k\|,\bar{w}) \Rightarrow \|e_k(t)\| \le \xi(\|x(t)\|,\bar{w}),$$
$$\forall t \in [s_k, a_{k+1}).$$

**PROOF** It is easy to see that $\xi(s,t)$ is strictly increasing with respect to $s$. Under Assumption 11.2, we have

$$\begin{aligned}\|e_k(t)\| &\le \frac{1}{L_k^{\xi}+1}\xi(\|\hat{x}_k\|,\bar{w})\\ &\le \frac{1}{L_k^{\xi}+1}\xi(\|x(t)\| + \|e_k(t)\|,\bar{w})\\ &\le \frac{1}{L_k^{\xi}+1}\xi(\|x(t)\|,\bar{w}) + \frac{L_k^{\xi}}{L_k^{\xi}+1}\|e_k(t)\|\\ \Rightarrow \|e_k(t)\| &\le \xi(\|x(t)\|,\bar{w}).\end{aligned}$$

The second inequality holds because $\|\hat{x}_k\| = \|x(t) - e_k(t)\| \le \|x(t)\| + \|e_k(t)\|$. The third inequality holds because $\xi(s,t)$ is locally Lipschitz over $s$ with Lipschitz constant $L_k^{\xi}$. ■

With Assumption 11.2, one can then define an event trigger $\theta(\cdot,\cdot)$ that is functionally dependent on $\hat{x}_k$ that still assures the closed-loop system is locally ISS. This event trigger is

$$\theta(\|\hat{x}_k\|,\bar{w}) = \rho_\theta\underline{\xi}_k(\|\hat{x}_k\|,\bar{w}) = \frac{\rho_\theta}{1+L_k^{\xi}}\xi(\|\hat{x}_k\|,\bar{w}) \doteq \theta_k, \quad (11.13)$$

where $\rho_\theta$ is a constant in $(0,1)$, and functions $\xi$ and $\underline{\xi}_k$ are defined in (11.11) and (11.12), respectively. Throughout the remainder of this chapter, the event trigger in Equation 11.13 will be used.

## 11.5 Quantizer Design

The event trigger derived in the preceding section guarantees that the closed-loop system is locally ISS *provided the infinite precision sampled state is used at the*

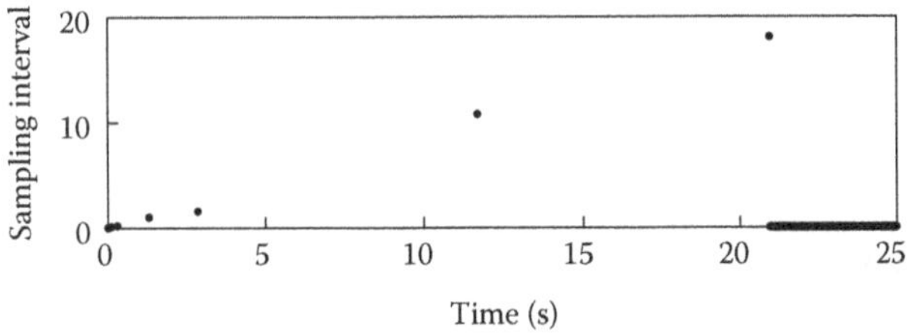


**FIGURE 11.3**
Event-triggered control system from (11.1) using a static uniform quantization level of 0.15. Note the occurrence of Zeno behavior 21 s into the simulation.

*controller*. When this information is transmitted across a bandwidth-limited channel, that information must be quantized with a finite number of bits, so there exists additional error between the current state and the last-sampled state.

One concern regarding quantization is that it may cause Zeno behavior in which the system broadcasts an infinite number of times in a finite time interval. This can occur in systems where the quantization level is larger than the state-dependent threshold function. An example of this behavior is seen in Figure 11.3 where a static uniform quantizer with a quantization level of 0.15 is used on the event-triggered example in Equation 11.1. The intersampling intervals for this simulation are shown in Figure 11.3 using event trigger $\mathbf{E}_1$ in which Zeno behavior occurs at 21 s into the simulation.

To avoid the Zeno behavior shown in Figure 11.3, one needs to guarantee that the quantization error is always less than the event trigger in Equation 11.11. The *quantization level* $\Delta_k$ for the $k$th sample is therefore defined to be

$$\delta_k \leq \Delta_k \doteq \rho_\Delta \theta(|\|\hat{x}_{k-1}\| - \theta_{k-1}|, \varpi), \qquad (11.14)$$

where $\rho_\Delta \in (0,1)$ and $\theta(s,t)$ are the threshold given in Equation 11.13. Notice that $\Delta_k$ depends on $\hat{x}_{k-1}$ instead of $\hat{x}_k$. This is because one needs $\Delta_k$ to perform the $k$th quantization, and $\hat{x}_k$, of course, is not yet available.

With quantization level $\Delta_k$, let us uniformly quantize *the uncertainty set* $\{x : \|x - \hat{x}_{k-1}\| = \theta_{k-1}\}$ of $x(s_k)$. The uncertainty set of $x(s_k)$ is the surface of, instead of all the area contained in, the hypercube centered at $\hat{x}_{k-1}$ with edge length $2\theta_{k-1}$. When the event $\|e_{k-1}(t)\| \geq \theta_{k-1}$ is triggered at $s_k$, the state $e_{k-1}(s_k)$ must satisfy $\|e_{k-1}(s_k)\| = \|x(s_k) - \hat{x}_{k-1}\| = \theta_{k-1}$.

Given the uncertainty set of $x(s_k)$ to be $\{x : \|x - \hat{x}_{k-1}\| = \theta_{k-1}\}$, and the quantization level $\Delta_k$ as in (11.14), one denotes the *quantizer* as a map $Q : \mathbb{R}^n \to \mathbb{R}^n$. Take a three-dimensional system as an example (see Figure 11.4). We first determine which facet $x(s_k)$ lies in, then uniformly divide each dimension of that facet into $J_k = \left\lceil \frac{\theta_{k-1}}{\Delta_k} \right\rceil$ mutually disjoint cells, and finally quantize

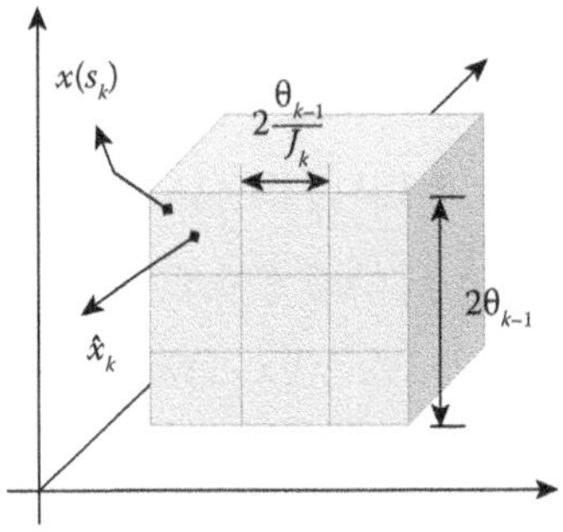


**FIGURE 11.4**
Quantization map.

$x(s_k)$ as the center of that cell which contains $x(s_k)$. Based on this idea, the quantization map $Q$ takes the form of

$$\hat{x}_k^i = Q^i(x) = \begin{cases} x^i(s_k), & \text{if } i = I_k; \\ \hat{x}_{k-1}^i - \theta_{k-1} + \left\lfloor \frac{x^i(s_k) - (\hat{x}_{k-1}^i - \theta_{k-1})}{2\theta_{k-1}/J_k} \right\rfloor \frac{2\theta_{k-1}}{J_k} + \frac{\theta_{k-1}}{J_k}, & \text{otherwise,} \end{cases} \qquad (11.15)$$

where $\hat{x}_k^i$, $Q^i$, and $x^i$ indicate the $i$th dimension of the corresponding vectors, and $I_k$ is the smallest index $i$ such that $|x^i(s_k) - \hat{x}_{k-1}^i| = \theta_{k-1}$.

The quantization map $Q$ designed in (11.15) guarantees that the infinity norm of the quantization error $\delta_k = \|x(s_k) - \hat{x}_k\| = \|e_k(s_k)\|$ is always less than the threshold function $\theta_k$ for any possible state trajectories. This assertion is formally stated and proven in the following lemma.

**Lemma 11.3**

$\delta_k < \theta_k$, for all possible state trajectories.

**PROOF** We first show that $\delta_k \leq \Delta_k$, and then show that $\Delta_k < \theta_k$.

From Equation 11.15, we see that $\delta_k \leq \frac{\theta_{k-1}}{J_k} = \frac{\theta_{k-1}}{\left\lceil \frac{\theta_{k-1}}{\Delta_k} \right\rceil} \leq \Delta_k$.

Since we quantize the uncertainty set $\{x : \|x - \hat{x}_{k-1}\| = \theta_{k-1}\}$, $\hat{x}_k$ satisfies $\|\hat{x}_k - \hat{x}_{k-1}\| = \theta_{k-1}$; hence, it is true that $\|\hat{x}_k\| \geq |\|\hat{x}_{k-1}\| - \theta_{k-1}|$. According to Equation 11.14, we have

$$\begin{aligned} \Delta_k &= \rho_\Delta \theta(|\|\hat{x}_{k-1}\| - \theta_{k-1}|, \varpi) \\ &= \rho_\Delta \frac{\rho_\theta}{1 + L_k^\xi} \xi(|\|\hat{x}_{k-1}\| - \theta_{k-1}|, \varpi) \\ &\leq \rho_\Delta \frac{\rho_\theta}{1 + L_k^\xi} \xi(\|\hat{x}_k\|, \varpi) = \rho_\Delta \theta_k < \theta_k. \end{aligned}$$

The second and the third equalities are derived from Equation 11.13, and the first inequality holds because

$\xi(\cdot,\cdot)$ is increasing in its first argument according to Equation 11.11, and the last inequality holds because $\rho_\Delta < 1$. ■

The quantization map $Q$ in (11.15) generates $N_k$ bits at the $k$th sampling. Since an $n$-dimensional box has $2n$ facets, and each facet has $n-1$ dimensions, the whole surface is divided into $2n\left\lceil\frac{\theta_{k-1}}{\Delta_k}\right\rceil^{n-1}$ parts. Therefore, the number of bits $N_k$ transmitted at the $k$th sampling is

$$N_k = \left\lceil \log_2 2n \left\lceil \frac{\theta_{k-1}}{\Delta_k} \right\rceil^{n-1} \right\rceil. \tag{11.16}$$

## 11.6 Efficient Attentiveness without Delay

While overly simplistic, it is easier to establish sufficient conditions for efficient attentiveness when transmissions are received without delay. The methods used to derive these conditions are then easily extended to handle the case when there are communication delays. This section, therefore, derives sufficient conditions for an event-triggered system to be efficiently attentive when there are zero transmission delays.

Using the approach in this section, the next section will study the case when there is delay, provide an acceptable delay, and analyze the required instantaneous bit-rate.

This section is organized as follows: Section 11.6.1 establishes that the proposed event-triggered system is locally ISS. Efficient attentiveness is then studied in Section 11.6.2. Zeno behavior is discussed in Section 11.6.3.

Assumption 11.3 is essential in guaranteeing the efficient attentiveness of an event triggered system.

**ASSUMPTION 11.3** There exists a continuous function $\bar{f}_c(\cdot)$ such that

$$\|f(x,K(x),0)\| \le \bar{f}_c(\|x\|), \forall x \in \Omega, \text{ and} \tag{11.17}$$

$$\lim_{s\to 0}\frac{\bar{f}_c(s)}{\theta(s,0)} < \infty. \tag{11.18}$$

Assumption 11.3 requires that the threshold function $\theta(\cdot,0)$ decreases more slowly than $\bar{f}_c(\cdot)$, an upper bound on the dynamics of the closed-loop system without disturbances. Take the system (11.1) as an example. The closed-loop dynamic is cubic. Event trigger 1 has linear threshold function which decreases more slowly than the closed-loop dynamic as $x$ gets close to 0; hence, the system state takes more and more time to hit the threshold function when the state gets closer and closer to the origin. Event trigger 2 has a quartic threshold function that decreases faster than the closed-loop dynamic as $x$ gets close to 0 and, hence, is triggered more and more often as the state gets closer and closer to the origin.

Suppose $\bar{f}_c(\cdot)$ is a polynomial function whose smallest degree of any term is $p$. To be efficiently attentive, the threshold function should have its smallest degree of any term no larger than $p$. For example, if the closed-loop dynamic system is linear, then the threshold function needs to have at least one term to be linear or constant.

### 11.6.1 Local Input-to-State Stability

Input-to-state stability and local input-to-state stability of event-triggered systems without delay have already been studied [1,7,9–11] without Assumption 11.3. So, Assumption 11.3 is not essential to establish stability. Although many works have discussed the stability issue in event-triggered systems, to be self-contained, we still give Lemma 11.4 followed by a brief proof.

**Lemma 11.4**

Suppose there is no delay. On Assumptions 11.1 and 11.2, the event-triggered system (11.3–11.5) with the threshold function (11.13) and the dynamic quantization level (11.14) is locally ISS.

**PROOF** Without delay, we have $s_k = a_k$ for all $k = 0, 1, \ldots$.

Apply Equation 11.4 into Equation 11.3, and compare it with Equation 11.7. We have, for the event-triggered system (11.3–11.5),

$$e(t) = e_k(t), \forall t \in [a_k, a_{k+1}), \tag{11.19}$$

where $e_k(t)$, according to the event trigger (11.5) and the threshold function (11.13), satisfies

$$\|e_k(t)\| \le \theta_k = \rho_\theta \underline{\xi}_k(\|\hat{x}_k\|, \bar{w}), \forall t \in [a_k, a_{k+1}).$$

According to Lemma 11.2,

$$\|e_k(t)\| \le \xi(\|x(t)\|, \bar{w}), \forall t \in [a_k, a_{k+1}).$$

Therefore, according to Equation 11.19, for all $t \ge 0$, along the trajectory of the system state of Equations 11.3 through 11.5,

$$e(t) \le \xi(\|x(t)\|, \bar{w}),$$

where $\xi(\cdot,\cdot)$ is defined as in (11.11).

Applying the above equation to Equation 11.10, we have

$$\begin{aligned}\dot{V} &\le -\alpha_3(\|x\|) + \gamma_1(\xi(\|x\|,\varpi)) \\ &\quad + \gamma_2(\|w\|), \forall x \in \Omega, w \in \Omega_w \\ &= -(1-c)\alpha_3(\|x\|) + \gamma_2(\|w\|) \\ &\quad + \gamma_3(\varpi), \forall x \in \Omega, w \in \Omega_w \\ &\le -(1-c)\alpha_3(\|x\|) + \gamma_2(\|w\|_{\mathcal{L}_\infty}) \\ &\quad + \gamma_3(c_1\|w\|_{\mathcal{L}_\infty}), \forall x \in \Omega,\end{aligned}$$

where $c$ is the same $c$ as defined in Equation 11.11, and $c_1$ is a constant that satisfies $\varpi \le c_1\|w\|_{\mathcal{L}_\infty}$. Together with Equation 11.8, according to Theorem 11.1, the event-triggered system (11.3–11.5) with a threshold function (11.13) and a quantization level (11.14) is locally ISS. ■

### 11.6.2 Efficiently Attentive Intersampling Interval

The preceding section focused on the performance level (local ISS) of the event-triggered system (11.3–11.5). This section concentrates on the communication resources needed to assure that performance level. Since delay is supposed to be zero in this section, we only study the intersampling interval.

Let us first study the dynamic behavior of the gap $e_k(t) = x(t) - \hat{x}_k$. Let $\tilde{u} \in \mathbb{R}^m$ be a constant control input applied to the system (11.3–11.5) during interval $[t_1, t_l)$, for some $t_1$ and $t_l$ satisfying $s_k \le t_1 < t_l$. The gap $e_k(t)$ satisfies

$$\dot{e}_k(t) = f(\hat{x}_k + e_k(t), \tilde{u}, w(t)), \forall t \in [t_1, t_l). \quad (11.20)$$

Next, we would like to analyze the dynamic behavior of $\|e_k(t)\|$. Although $\|e_k(t)\|$ may not be differentiable at every time instant during $[t_1, t_l]$, we find that $\|e_k(t)\|$ is continuous and piecewise differentiable. Let $e_k^i$ be the $i$th element of $e_k$. There must exist a sequence of time instants $t_1 < t_2 < \cdots < t_{l-1} < t_l$, such that for each small interval $[t_\ell, t_{\ell+1})$, $\|e_k(t)\| = |e_k^i(t)|$, and $e_k^i(t)$ keeps the same sign for some $i = 1, \ldots, l$. The infinity norm $\|e_k\|$ of $e_k$ is differentiable during this interval $[t_\ell, t_{\ell+1}]$ and satisfies

$$\begin{aligned}\frac{d\|e_k(t)\|}{dt} &= \frac{d|e_k^i(t)|}{dt} \le |\dot{e}_k^i(t)| = |f_i(\hat{x}_k + e_k(t), \tilde{u}, w(t))| \\ &\le \|f(\hat{x}_k + e_k(t), \tilde{u}, w(t))\| \le \bar{f}(\hat{x}_k, \tilde{u}, \varpi) + L_k^x\|e_k(t)\|,\end{aligned} \quad (11.21)$$

where

$$\bar{f}(\hat{x}_k, \tilde{u}, \varpi) = \|f(\hat{x}_k, \tilde{u}, 0)\| + L_k^w \varpi. \quad (11.22)$$

Let $L_k^x$ and $L_k^w$ be the Lipschitz constant of $f$ with respect to $x$ and $w$, respectively, during interval $[s_k, a_{k+1}]$. The first inequality holds because $\frac{d|e_k^i(t)|}{dt} = \frac{d\sqrt{e_k^i(t)^2}}{dt} = \frac{e_k^i(t)}{|e_k^i(t)|}\dot{e}_k^i(t) \le |\dot{e}_k^i(t)|$, and the last inequality holds because $f(\cdot,\cdot,\cdot)$ is locally Lipschitz in its first argument with Lipschitz constant $L_k^x$ during time interval $[s_k, a_{k+1})$.

According to the comparison principle, for all $t \in [t_\ell, t_{\ell+1}]$,

$$\|e_k(t)\| \le \frac{\bar{f}(\hat{x}_k, \tilde{u}, \varpi)}{L_k^x}(e^{L_k^x(t-t_\ell)} - 1) + \|e_k(t_\ell)\|e^{L_k^x(t-t_\ell)}. \quad (11.23)$$

Since the gap $e_k(t)$ is continuous (because all the inputs of (11.20) are bounded), so is its infinity norm $\|e_k(t)\|$. Therefore, the upper bound on $\|e_k(t_{\ell+1})\|$ computed from (11.23) can be used as the upper bound on the initial state of $\|e_k(t)\|$ for the next interval $[t_{\ell+1}, t_{\ell+2}]$, and for all $t \in [t_{\ell+1}, t_{\ell+2}]$, $\|e_k(t)\|$ satisfies

$$\|e_k(t)\| \le \frac{\bar{f}(\hat{x}_k, \tilde{u}, \varpi)}{L_k^x}(e^{L_k^x(t-t_\ell)} - 1) + \|e_k(t_\ell)\|e^{L_k^x(t-t_\ell)}.$$

By mathematical induction, it is easy to show that for all $t \in [t_1, t_l]$,

$$\|e_k(t)\| \le \frac{\bar{f}(\hat{x}_k, \tilde{u}, \varpi)}{L_k^x}(e^{L_k^x(t-t_1)} - 1) + \|e_k(t_1)\|e^{L_k^x(t-t_1)}. \quad (11.24)$$

With Equation 11.24, we are ready to find a lower bound on the intersampling interval of the event-triggered system (11.3–11.5).

**Lemma 11.5**

Suppose there is no delay. Under Assumptions 11.1 and 11.2, the event-triggered system (11.3–11.5) with the threshold function (11.13) and quantization level (11.14) has its intersampling interval $\tau_k$ satisfying

$$\tau_k \ge \frac{1}{L_k^x}\ln\left(1 + \frac{L_k^x(\theta_k - \Delta_k)}{\bar{f}(\hat{x}_k, u_k, \varpi) + L_k^x\Delta_k}\right), \forall k = 1, 2, \ldots. \quad (11.25)$$

**PROOF** If there is no delay, the control input during interval $[s_k, s_{k+1})$ is a constant and equals $u_k$. According to Equation 11.24, during interval $[s_k, s_{k+1})$, $\|e_k(s_{k+1})\|$ satisfies

$$\begin{aligned}\|e_k(s_{k+1})\| &\le \frac{\bar{f}(\hat{x}_k, u_k, \varpi)}{L_k^x}(e^{L_k^x\tau_k} - 1) + \|e_k(s_k)\|e^{L_k^x\tau_k} \\ &\le \frac{\bar{f}(\hat{x}_k, u_k, \varpi)}{L_k^x}(e^{L_k^x\tau_k} - 1) + \Delta_k e^{L_k^x\tau_k},\end{aligned}$$

where the second inequality holds because $e_k(s_k)$ is the quantization error of the $k$th quantized state, and hence satisfies $\|e_k(s_k)\| \le \Delta_k$.

According to the event trigger (11.5), the $k+1$th sampling occurs when $\|e_k(t)\| \geq \theta_k$. Therefore, we have

$$\frac{\bar{f}(\hat{x}_k, u_k, \bar{w})}{L_k^x}(e^{L_k^x \tau_k} - 1) + \Delta_k e^{L_k^x \tau_k} \geq \theta_k$$

$$\Rightarrow \tau_k \geq \frac{1}{L_k^x} \ln\left(1 + \frac{L_k^x(\theta_k - \Delta_k)}{\bar{f}(\hat{x}_k, u_k, \bar{w}) + L_k^x \Delta_k}\right). \quad \blacksquare$$

**REMARK 11.1** The intersampling intervals are strictly positive. From Lemma 11.3, we have $\Delta_k < \theta_k$. According to Equation 11.25, it is easy to conclude that the intersampling interval is strictly positive.

The lower bound (11.25) on $\tau_k$ indicates that the behavior of $\tau_k$ is strongly related with the ratio of the closed-loop dynamic $\bar{f}(\hat{x}_k, u_k, \bar{w})$ to the threshold function $\theta_k$. This relation becomes obvious if we ignore the quantization level $\Delta_k$. Assuming $\Delta_k = 0$, Equation 11.25 is simplified as

$$\tau_k \geq \frac{1}{L_k^x} \ln\left(1 + L_k^x \frac{1}{\bar{f}(\hat{x}_k, u_k, \bar{w})/\theta_k}\right), \forall k = 1, 2, \ldots.$$

If we want the lower bound to be decreasing with respect to the state, then we should require the ratio $\bar{f}/\theta_k$ to be increasing with respect to the state. But this requirement may be hard to meet during the design process. Hence, Assumption 11.3 only places requirement on the asymptotic behavior of the ratio $\bar{f}_c/\theta_k$, which assures the existence of an increasing function bounding the ratio from above, according to Lemma 11.1.

**Lemma 11.6**

Suppose there is no delay. On Assumptions 11.1 through 11.3, the event-triggered system (11.3–11.5) with threshold function (11.13) and quantization level (11.14) has efficiently attentive intersampling intervals—that is, there exists a class $\mathcal{M}$ function $h(\cdot)$ such that

$$\tau_k \geq h(\|\hat{x}_k\|).$$

**PROOF** First, we derive a more conservative lower bound on $\tau_k$ which is in a simpler form:

$$\begin{aligned}
\tau_k &\geq \frac{1}{L_k^x} \ln\left(1 + \frac{L_k^x(\theta_k - \Delta_k)}{\bar{f}(\hat{x}_k, u_k, \bar{w}) + L_k^x \Delta_k}\right) \\
&\geq \frac{\theta_k - \Delta_k}{\bar{f}(\hat{x}_k, u_k, \bar{w}) + L_k^x \theta_k} \\
&= \frac{1 - \Delta_k/\theta_k}{\bar{f}(\hat{x}_k, u_k, \bar{w})/\theta_k + L_k^x} \geq \frac{1 - \rho_\Delta}{\bar{f}(\hat{x}_k, u_k, \bar{w})/\theta_k + L_k^x} \\
&\geq \frac{1 - \rho_\Delta}{\frac{f_c(\|\hat{x}_k\|) + L_k^w \bar{w}}{\theta_k} + L_k^x}.
\end{aligned}$$

The second inequality holds because $\ln(1+x) \geq x/(1+x)$; the third inequality is derived from Lemma 11.3; and the last inequality is from Equation 11.17.

Next, we find that there are constants $\bar{L}^w$ and $\bar{L}^x$ such that $L_k^w \leq \bar{L}^w$ and $L_k^x \leq \bar{L}^x$. Since the system is locally ISS, there must exist a constant $\mu$ such that $\|x(t)\| \leq \mu$. Let $\bar{L}^w$ and $\bar{L}^x$ be the Lipschitz constant of $f$ over $w$ and $x$ for all $\|w\| \leq \bar{w}$ and $\|x\| \leq \mu$, respectively. We have $L_k^w \leq \bar{L}^w$ and $L_k^x \leq \bar{L}^x$, and the intersampling interval satisfies

$$\tau_k \geq \frac{1 - \rho_\Delta}{\frac{f_c(\|\hat{x}_k\|) + \bar{L}^w \bar{w}}{\theta_k} + \bar{L}^x}.$$

Finally, Assumption 11.3 implies that

$$\lim_{\|\hat{x}_k\| \to 0} \frac{f_c(\|\hat{x}_k\|) + \bar{L}^w \bar{w}}{\theta_k} < \infty.$$

According to Lemma 11.1, there exists a class $\mathcal{N}$ function $h_1(\cdot)$ such that

$$\frac{f_c(\|\hat{x}_k\|) + \bar{L}^w \bar{w}}{\theta_k} \leq h_1(\|\hat{x}_k\|).$$

Therefore, the intersampling interval satisfies

$$\tau_k \geq \frac{1 - \rho_\Delta}{h_1(\|\hat{x}_k\|) + \bar{L}^x} \doteq h(\|\hat{x}_k\|).$$

Since $h_1(\|\hat{x}_k\|)$ is of class $\mathcal{N}$, $h(\|\hat{x}_k\|)$ is a class $\mathcal{M}$ function. $\blacksquare$

### 11.6.3 Zeno Behavior Free

To avoid Zeno behavior, on one hand, we need to guarantee that immediately after each sampling, the event (11.5) is not triggered, that is, $\|e_k(s_k)\| < \theta_k$ for all $k$. On the other hand, we need to make sure that as the system state approaches the origin, the intersampling interval does not approach 0. The first requirement actually requires that the quantization level be smaller than the threshold for the next sampling, which is proved in Lemma 11.3 and 11.6. The second requirement is fulfilled automatically if efficient attentiveness is guaranteed. Therefore, under the same conditions of Lemma 11.6, we show the event-triggered system is Zeno behavior free.

**Corollary 11.1**

Suppose there is no delay. On Assumptions 11.1 through 11.3, the event-triggered system (11.3–11.5) with threshold function (11.13) and quantization level (11.14) is

Zeno behavior free, that is, there exists a constant $\tau_0 > 0$ such that $\tau_k \geq \tau_0$ for all $k = 1, 2, \ldots$.

**PROOF** To show there is no Zeno behavior, we need to show that there exists a strictly positive constant that is a lower bound on $\tau_k$ for all $k = 1, 2, \ldots$.

According to Lemma 11.4, the system state $x(t)$ must be bounded, and so is $\hat{x}_k$. Thus, there must exist a finite constant $\mu > 0$ such that $\|\hat{x}\| \leq \mu$.

According to Lemma 11.6, $\tau_k \geq h(\|\hat{x}_k\|)$, where $h(\cdot)$ is a class $\mathcal{M}$ function (continuous, decreasing, $h(s) = 0 \Rightarrow s = \infty$).

Therefore, $\tau_k \geq h(\|\hat{x}_k\|) \geq h(\mu) > 0$. The second inequality holds because $h(\cdot)$ is a decreasing function, and the third inequality holds because $h(s) > 0$ for all finite $s$. ■

## 11.7 Efficient Attentiveness with Delay

Let us now assume there is a positive transmission delay. With delay, the arguments used to establish efficient attentiveness are essentially unchanged. The impact of the delay is mainly on the performance level. In order to preserve local ISS, we need to bound the delay. The maximum acceptable delay is provided in Theorem 11.2. The proof of the main theorem, though more complicated, is based on the same ideas that were used when there is no delay. Interested readers can check the Appendix for the proof.

**Theorem 11.2**

Suppose the delay $d_k$ for the $k$th sampling satisfies

$$d_k \leq \min\{\bar{d}_k, T_k\} \doteq D_k, \qquad (11.26)$$

where

$$\bar{d}_k = \frac{1}{L_{k-1}^x} \ln\left(1 + \frac{L_{k-1}^x(\xi_{k-1}(\|\hat{x}_{k-1}\|, \bar{w}) - \theta_{k-1})}{\bar{f}(\hat{x}_{k-1}, u_{k-1}, \bar{w}) + L_{k-1}^x \theta_{k-1}}\right),$$

$$T_k = \frac{1}{L_k^x} \ln\left(1 + \frac{L_k^x(\theta_k - \Delta_k)}{\bar{f}(\hat{x}_k, u_{k-1}, \bar{w}) + L_k^x \Delta_k}\right).$$

The sampling and arrival sequences are admissible, that is, $d_k \leq \tau_k$ for all $k = 0, 1, \cdots$. Moreover, under Assumptions 11.1 through 11.3, the event-triggered system (11.3–11.5) with the threshold function (11.13) and the dynamic quantization level (11.14) is locally ISS, Zeno behavior free, and efficiently attentive.

**REMARK 11.2** The acceptable delay $D_k$ is strictly positive for all $k = 0, 1, \cdots$. It is easy to see from (11.13) that $\theta_{k-1} < \xi_{k-1}(\|\hat{x}_{k-1}\|, \bar{w})$; hence, $\bar{d}_k > 0$. Since $\Delta_k$ is designed to guarantee $\Delta_k < \theta_k$ as discussed in Lemma 11.3, we have $T_k > 0$. Therefore, the acceptable delay $D_k = \min\{\bar{d}_k, T_k\}$ is strictly positive.

The acceptable delay defined in (11.26) is state dependent, and is long if the system state is close to the origin. Most prior work studied constant or constant bounded delay in event-triggered systems. Theorem 11.2 suggests that the acceptable delay is state dependent and hence time varying. Moreover, from the proof of efficient attentiveness, we see that once the system is efficiently attentive, the acceptable delay is also "efficiently attentive"; that is, there exists a class $\mathcal{M}$ function bounding the acceptable delay from below. Therefore, the acceptable delay can be long if the system state is close to the origin.

Efficiently attentive event-triggered systems can share their bandwidth with other non-real-time transmission tasks. To guarantee local ISS, the communication channel should have larger bandwidth than the required instantaneous bit-rate $r_k$ for any $\hat{x}_k, \hat{x}_{k-1} \in \Omega$, where $\Omega$ is a domain defined in Assumption 11.1. Although the required channel capacity may be large if the normal operation region $\Omega$ is large, because of efficient attentiveness, the event-triggered system (11.3-11.5) may use only a portion of the channel bandwidth in an infrequent way when the system state is close to its equilibrium, and hence allows other non-real-time transmission tasks to use the channel.

## 11.8 An Example: A Rotating Rigid Spacecraft

We apply the previous technique to a rotating rigid spacecraft whose model is borrowed from [8]. Euler equations for a rotating rigid spacecraft are given by

$$\begin{aligned}\dot{x}_1 &= -x_2x_3 + u_1 + w_1,\\ \dot{x}_2 &= x_1x_3 + u_2 + w_2,\\ \dot{x}_3 &= x_1x_2 + u_3 + w_3,\end{aligned}$$

where $\|w(t)\|_{\mathcal{L}_\infty} \leq 0.005$ and $x(0) = [0.6\; 0.5\; 0.4]^T$.

A nonlinear feedback law $u_i = -x_i^3$ can render the system to be ISS as proved by the ISS-Lyapunov function $V(x) = x_1^2 + \frac{1}{2}x_2^2 + \frac{1}{2}x_3^2$. If we apply the event-triggered sampling strategy and the quantizer into the system, the feedback law is $u_i(t) = -\hat{x}_{i,k}^3$, for $t \in [a_k, a_{k+1})$. The derivative of $V$ along the trajectories of the quantized event-triggered closed-loop system satisfies

$$\dot{V} \leq 2(-x_1^4 + 3x_1^3e_1 + x_1e_1^3 + x_1w_1) + \sum_{i=2}^{3}(-x_i^4 + 3x_i^3e_i + x_ie_i^3 + x_iw_i).$$

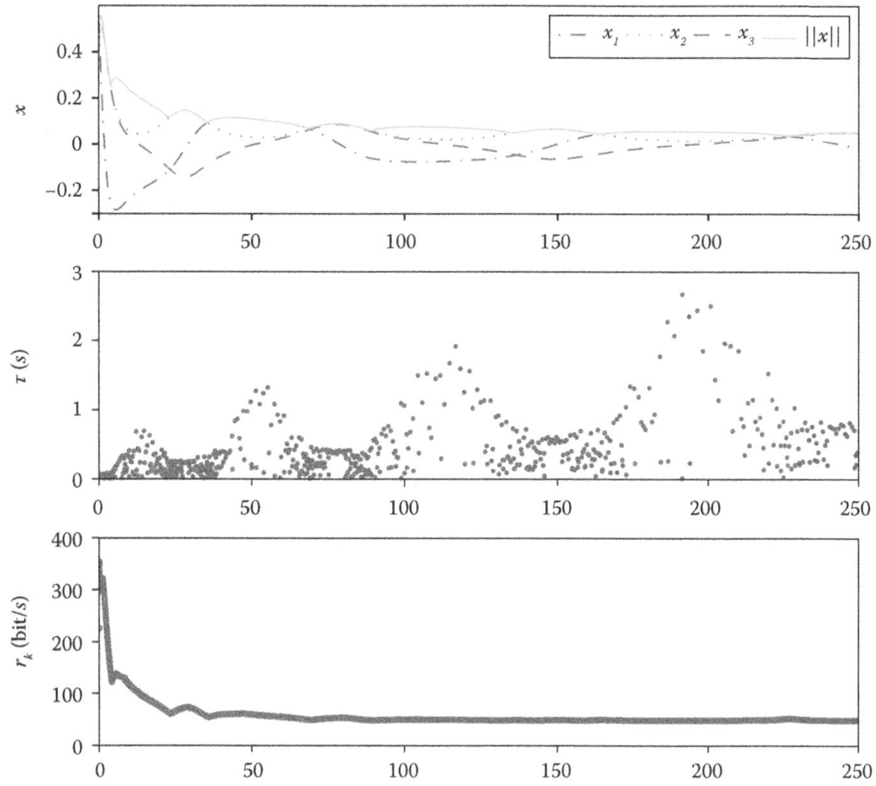


**FIGURE 11.5**
Simulation results.

If $\|e\| \leq \frac{1}{16}\|x\| + 0.1\varpi$, we have

$$\dot{V} \leq -0.249\|x\|^4 + 5.2047\varpi + 0.0075\varpi^2 + 0.004\varpi^3, \forall \|x\| \leq 1.$$

Therefore, the system is locally ISS with $\xi(s) = \frac{1}{16}s + 0.1\varpi$, whose Lipschitz constant $L_k^{\xi} = \frac{1}{16}$ for all $k$.

According to Equations 11.12 through 11.14, we design our threshold and quantization level as

$$\theta_k \doteq \theta(\|\hat{x}_k\|, \varpi) = 0.9412\rho_\theta \left(\frac{1}{16}\|\hat{x}_k\| + 0.1\varpi\right),$$

$$\Delta_k \doteq \Delta(\|\hat{x}_{k-1}\|, \varpi) = 0.9412\rho_\Delta\rho_\theta \left(\frac{1}{16}(\|\hat{x}_{k-1}\| - \theta_{k-1}) + 0.1\varpi\right),$$

where $\rho_\theta = 0.53$, and $\rho_\Delta = 0.1$.

Now, let us check whether Assumptions 11.1 through 11.3 all hold. During our design of event trigger, we have shown that Assumptions 11.1 and 11.2 hold. For Assumption 11.3, we find that $\bar{f}_c(s) = 2s^2$ with $\Omega = \{x : \|x\| < 1\}$. It is easy to see that $\lim_{s\to 0} \frac{\bar{f}_c(s)}{\theta(s,0)} = 0$, and hence Assumption 11.3 is true. According to Theorem 11.2, as long as the delay $d_k \leq D_k$ where the acceptable delay $D_k$ is defined as in (11.26), our quantized event-triggered control system is locally ISS, Zeno behavior free, and efficiently attentive.

We applied the event trigger and the quantizer into the rotating rigid spacecraft system with the delay $d_k$ for the $k$th packet to be the acceptable delay $D_k$, and ran the system for 250 s. The simulation results are given in Figure 11.5.

The top plot of Figure 11.5 shows the system state with respect to time. From this plot, we see that the system is locally ISS. The middle plot in Figure 11.5 presents the intersampling intervals with respect to time. The smallest intersampling interval is 0.1 s which is two times the simulation step size, so we can say that there is no Zeno behavior. Besides, we also find that the intersampling intervals oscillate over time. As the system state approaches the origin, the peak value of the intersampling intervals grows larger, and the sampling frequency during the trough becomes lower. Generally speaking, the intersampling interval is growing long, and the intersampling intervals are efficiently attentive. The bottom plot shows the instantaneous bit-rate with respect to time. We can see that as the system state goes to the origin, the required instantaneous bit-rate drops from about 360 bit/s to about 50 bit/s, and the curve of the instantaneous bit-rate has a similar shape as the curve of $\|x(t)\|$ (the solid line in the top plot). Thus, it is obvious that the required instantaneous bit-rate is also efficiently attentive.

## 11.9 Conclusion

As prior work showed, event-triggered systems have the potential to save communication resources while maintaining system performance, and roughly speaking, the intersampling interval is longer as the system state gets closer to the origin. This property is called *efficient*. However, not all event-triggered systems are efficiently attentive. Moreover, with capacity-limited communication channels where the effects of quantization and delay are taken into account, event-triggered systems may have Zeno behavior or become unstable. To overcome the Zeno behavior, and guarantee locally input-to-state stability and efficient attentiveness, this chapter studies how to properly design the event trigger and the quantizer, and provides a state-dependent acceptable delay. The simulation results demonstrate the main results.

## 11.10 Appendix

### 11.10.1 Proof of Admissibility of the Sampling and Arrival Sequences

First, we realize that $d_0 = 0 \leq \tau_0$.

Next, let us assume that $d_{k-1} \leq \tau_{k-1}$ holds; that is, $a_{k-1} \leq s_k$. If $d_k > \tau_k > 0$, then we have $a_{k-1} \leq s_k \leq s_{k+1} < a_k$. For interval $[s_k, s_{k+1}]$, the control input $\tilde{u} = u_{k-1}$, and the initial state $\|e_k(s_k)\|$ satisfies $\|e_k(s_k)\| \leq \Delta_k$. From Equation 11.24, we have

$$\|e_k(s_{k+1})\| \leq \frac{\bar{f}(\hat{x}_k, u_{k-1}, \bar{w})}{L_k^x}(e^{L_k^x \tau_k} - 1) + \Delta_k e^{L_k^x \tau_k}.$$

Since $\|e_k(s_{k+1})\| = \theta_k$, together with the equation above, we have $\tau_k \geq T_k$. From Equation 11.26, we further have $\tau_k \geq d_k$. This contradicts the assumption $d_k > \tau_k$.

Therefore, $d_k \leq \tau_k$ for all $k = 0, 1, \ldots, \infty$.

### 11.10.2 Proof of Local ISS

We first show that $\|e(t)\| \leq \xi(\|x(t)\|, \bar{w})$, where $e(t) = e_k(t)$, $\forall t \in [a_k, a_{k+1})$; that is, along the trajectories of the system (11.3–11.5), $\|e\| \leq \xi(\|x\|, \bar{w})$. Local ISS is, then, established using Assumption 11.1.

For all $t \in [s_k, s_{k+1}]$, we have $\|e_k(t)\| \leq \theta_k < \underline{\xi}_k(\|\hat{x}_k\|, \bar{w})$.

During interval $[s_{k+1}, a_{k+1})$, we know $\|e_k(s_{k+1})\| = \theta_k$. Because of the admissibility, the control input during this interval is $u_k$. From Equation 11.24, we have

$$\|e_k(t)\| \leq \frac{\bar{f}(\hat{x}_k, u_k, \bar{w})}{L_k^x}(e^{L_k^x(t-s_{k+1})} - 1) + \theta_k e^{L_k^x(t-s_{k+1})}.$$

Since $t - s_{k+1} \leq d_{k+1} \leq \bar{d}_{k+1}$ (11.26), we have $\|e_k(t)\| \leq \underline{\xi}_k(\|\hat{x}_k\|, \bar{w})$, for all $t \in [s_{k+1}, a_{k+1})$.

Therefore, for all $t \in [a_k, a_{k+1})$, and all $k = 0, 1, \ldots$, $\|e_k(t)\| \leq \underline{\xi}_k(\|\hat{x}_k\|, \bar{w})$. According to Lemma 11.2, $\|e_k(t)\| \leq \xi(\|x(t)\|, \bar{w}), \forall k = 0, 1, \ldots$. Let $e(t) = e_k(t)$, $\forall t \in [a_k, a_{k+1})$. We have $\|e(t)\| \leq \xi(\|x(t)\|, \bar{w})$. In other words, along the trajectories of system (11.3–11.5), $\|e\| \leq \xi(\|x\|, \bar{w})$.

Next, we show local ISS. Since $\|e\| \leq \xi(\|x\|, \bar{w})$ along the trajectories of system (11.3–11.5), according to Assumption 11.1 of system (11.3–11.5), the derivative of the ISS–Lyapunov function $V$ along the trajectories satisfies

$$\begin{aligned}
\dot{V} &\leq -\alpha_3(\|x\|) + \gamma_1(\xi(\|x\|, \bar{w})) \\
&\quad + \gamma_2(\|w\|), \forall x \in \Omega, w \in \Omega_w \\
&= -(1-c)\alpha_3(\|x\|) + \gamma_2(\|w\|) \\
&\quad + \gamma_3(\bar{w}), \forall x \in \Omega, w \in \Omega_w \\
&\leq -(1-c)\alpha_3(\|x\|) + \gamma_2(\|w\|_{\mathcal{L}_\infty}) \\
&\quad + \gamma_3(c_1\|w\|_{\mathcal{L}_\infty}), \forall x \in \Omega,
\end{aligned}$$

where $c$ is the same $c$ as defined in Equation 11.11, and $c_1$ is a constant that satisfies $\bar{w} \leq c_1\|w\|_{\mathcal{L}_\infty}$. Therefore, the system (11.3–11.5) is local ISS according to Theorem 11.1.

### 11.10.3 Proof of No Zeno Behavior

We first show that $D_k \leq \tau_k$ for all $k = 0, 1, \ldots$. From Section 11.10.1, we know that for any $d_k \in [0, D_k]$, we have $d_k \leq \tau_k$. Choose $d_k = D_k$, and we have $D_k \leq \tau_k$.

Next, we show that there exists a constant $c_1 > 0$ such that

$$\tau_k \geq D_k \geq c_1, \forall k = 0, 1, \ldots. \tag{11.27}$$

To show $\tau_k \geq c_1$, we need the following lemma.

**Lemma 11.7**

$\lim_{s \to 0} \frac{\theta_{k-1}}{\Delta_k} < \infty$.

**PROOF** If $\bar{w} \neq 0$, it is easy to show $\lim_{s \to 0} \frac{\theta_{k-1}}{\Delta_k} < \infty$.

If $\bar{w} = 0$, we first notice that $\theta(s, 0) < s$. $\theta(s, \bar{w}) = \rho_\theta \frac{1}{L^\xi + 1}\xi(s, \bar{w})$. Let $L^\theta = \rho_\theta \frac{L^\xi}{L^\xi + 1}$. Notice that $L^\theta < 1$. With Assumption 11.2 and Equation 11.13, we have $|\theta(s, \bar{w}) - \theta(t, \bar{w})| \leq L^\theta |s - t|$; hence, $\theta(s, 0) \leq L^\theta s < s$.

Let $s$ indicate $\|\hat{x}_{k-1}\|$. Since $\theta(s,0) < s$, we have

$$\frac{\theta_{k-1}}{\Delta_k} = \frac{\theta(s,0)}{\rho_\Delta \theta(s-\theta(s,0),0)} \le \frac{\theta(s,0)/\rho_\Delta}{\theta(s,0) - L^\theta \theta(s,0)} = \frac{1/\rho_\Delta}{1-L^\theta} < \infty.$$

The first inequality is derived from $\theta(s,0) - \theta(s - \theta(s,0),0) \le L^\theta \theta(s,0)$; the second inequality is derived from $L^\theta < 1$. ■

First, we show that there exists a positive constant $c_d$ such that $\bar{d}_k \ge c_d$ for all $k = 0, 1, \ldots$. From the fact that $\ln(1+x) \ge \frac{x}{1+x}$, we have

$$\bar{d}_k \ge \frac{\xi_{k-1}(\|\hat{x}_{k-1}\|, \bar{w}) - \theta_{k-1}}{\bar{f}(\hat{x}_{k-1}, u_{k-1}, \bar{w}) + L^x_{k-1}\xi_{k-1}(\|\hat{x}_{k-1}\|, \bar{w})}.$$

Since the system is local ISS, there must exist a constant $\mu$ such that $\|x(t)\| \le \mu$, and the Lipschitz constant of $f$ with respect to $x$ must satisfy $L^x_k \le \bar{L}^x$, for all $k$. Together with Equations 11.22 and 11.17,

$$\bar{d}_k \ge \frac{1/\rho_\theta - 1}{(\bar{f}_c(\|\hat{x}_{k-1}\|) + L^w_{B_\mu}\bar{w})/\theta_{k-1} + \bar{L}^x}.$$

With Assumption 11.3, we have $\lim_{\hat{x}_{k-1}\to 0} \frac{\bar{f}_c(\|\hat{x}_{k-1}\|) + \bar{L}^w\bar{w}}{\theta_{k-1}} < \infty$. For any finite $\|\hat{x}_{k-1}\|$, $\frac{\bar{f}_c(\|\hat{x}_{k-1}\|) + \bar{L}^w\bar{w}}{\theta_{k-1}} < \infty$. According to Lemma 11.1, there must exist a class $\mathcal{N}$ function $h_1$ such that $h_1(s) \ge (\bar{f}_c(s) + \bar{L}^w\bar{w})/\theta_{k-1}$; hence,

$$\bar{d}_k \ge \frac{1/\rho_\theta - 1}{h_1(\|\hat{x}_{k-1}\|) + \bar{L}^x} \ge \frac{1/\rho_\theta - 1}{h_1(\mu) + \bar{L}^x} \doteq c_d > 0. \quad (11.28)$$

Second, we show there exists a positive constant $c_t$ such that $T_k \ge c_t$ for all $k$. Since $\ln(1+x) \ge \frac{x}{1+x}$ and $\Delta_k/\theta_k \le \rho_\Delta$ (Remark 11.2),

$$T_k \ge \frac{\theta_k - \Delta_k}{\bar{f}(\hat{x}_k, u_{k-1}, \bar{w}) + L^x_k\theta_k} \ge \frac{1-\rho_\Delta}{\bar{f}(\hat{x}_k, u_{k-1}, \bar{w})/\theta_k + \bar{L}^x}.$$

Let us first look at the term $\bar{f}(\hat{x}_k, u_{k-1}, \bar{w})/\theta_k$:

$$\begin{aligned}\frac{\bar{f}(\hat{x}_k, u_{k-1}, \bar{w})}{\theta_k} &\le \frac{\bar{f}(\hat{x}_{k-1}, u_{k-1}, \bar{w}) + L^x_k\|\hat{x}_k - \hat{x}_{k-1}\|}{\theta_k} \\ &\le \frac{\bar{f}_c(\|\hat{x}_{k-1}\|) + \bar{L}^w\bar{w} + \bar{L}^x\theta_{k-1}}{\Delta_k} \\ &= \frac{\bar{f}_c(\|\hat{x}_{k-1}\|) + \bar{L}^w\bar{w}}{\theta_{k-1}}\frac{\theta_{k-1}}{\Delta_k} + \bar{L}^x\frac{\theta_{k-1}}{\Delta_k}.\end{aligned}$$

The first inequality is derived from the locally Lipschitz property of $f$. The second inequality is derived from (11.17) and the fact that $\bar{L}^x \ge L^x_k$ and $\bar{L}^w \ge L^w_k$. From the previous discussion about function $\frac{\bar{f}_c(\|\hat{x}_{k-1}\|) + \bar{L}^w\bar{w}}{\theta_{k-1}}$ and Lemma 11.7, we know there must exist a class $\mathcal{N}$ function $h_2$ such that $h_2(\|\hat{x}_{k-1}\|) \ge \frac{\bar{f}(\hat{x}_k, u_{k-1}, \bar{w})}{\theta_k}$. Therefore,

$$T_k \ge \frac{1-\rho_\Delta}{h_1(\|\hat{x}_{k-1}\|) + \bar{L}^x} \ge \frac{1-\rho_\Delta}{h_2(\mu) + \bar{L}^x} \doteq c_t > 0. \quad (11.29)$$

Therefore, $\tau_k \ge D_k \ge \min\{c_d, c_t\} > 0$ for all $k$.

### 11.10.4 Proof of Efficient Attentiveness

From Equations 11.28 and 11.29, we have $\tau_k \ge D_k \ge \bar{h}(\|\hat{x}_{k-1}\|)$, where

$$\bar{h}(\|\hat{x}_{k-1}\|) = \min\left\{\frac{1/\rho_\theta - 1}{h_1(\|\hat{x}_{k-1}\|) + \bar{L}^x}, \frac{1-\rho_\Delta}{h_2(\|\hat{x}_{k-1}\|) + \bar{L}^x}\right\}. \quad (11.30)$$

Since $h_1$ and $h_2$ are class $\mathcal{N}$ functions, $\bar{h}$ is a class $\mathcal{M}$ function. Therefore, part 1 of efficient attentiveness (Definition 11.1) is proved.

From Equation 11.16, we have

$$N_k \le 1 + \log_2 2n + (n-1)\log_2\left(1 + \frac{\theta_{k-1}}{\Delta_k}\right).$$

Lemma 11.7 shows that $\lim_{\hat{x}_{k-1}\to 0}\frac{\theta_{k-1}}{\Delta_k} < \infty$. Together with the fact that finite $\|\hat{x}_{k-1}\|$ indicates finite $\frac{\theta_{k-1}}{\Delta_k}$, according to Lemma 11.1, there exists a class $\mathcal{N}$ function $h_3$ such that $\frac{\theta_{k-1}}{\Delta_k} \le h_3(\|\hat{x}_{k-1}\|)$; hence, we have

$$N_k \le 1 + \log_2 2n + (n-1)\log_2(1 + h_3(\|\hat{x}_{k-1}\|)).$$

The instantaneous bit-rate $r_k = \frac{N_k}{D_k}$ thus satisfies

$$\begin{aligned} r_k &\le \frac{1 + \log_2 2n + (n-1)\log_2(1 + h_3(\|\hat{x}_{k-1}\|))}{\bar{h}(\|\hat{x}_{k-1}\|)} \\ &\doteq h(\|\hat{x}_{k-1}\|),\end{aligned}$$

where $\bar{h}(\cdot) \in \mathcal{M}$ is defined as in (11.30). Since $h_3(\cdot)$ is a class $\mathcal{N}$ function and $\bar{h}(\cdot)$ is a class $\mathcal{M}$ function, it is easy to see that $h(\cdot)$ is a class $\mathcal{N}$ function. This completes the proof of part 2 of efficient attentiveness.

## Bibliography


[1] A. Anta and P. Tabuada. To sample or not to sample: Self-triggered control for nonlinear systems. *IEEE Transactions on Automatic Control*, 55:2030–2042, 2010.

[2] K.-E. Årzén. A simple event-based PID controller. In *Proceedings of the 14th IFAC World Congress*, volume 18, pages 423–428, 1999.

[3] K. J. Astrom and B. M. Bernhardsson. Comparison of Riemann and Lebesgue sampling for first order stochastic systems. In *Decision and Control, 2002, Proceedings of the 41st IEEE Conference on*, volume 2, pages 2011–2016, IEEE, 2002.

[4] H. K. Khalil and J. W. Grizzle. *Nonlinear Systems*. Volume 3. Prentice Hall, New Jersey, 1992.

[5] L. Li, X. Wang, and M. Lemmon. Stabilizing bit-rates in disturbed event triggered control systems. In *The 4th IFAC Conference on Analysis and Design of Hybrid Systems*, pages 70–75, 2012.

[6] L. Li, X. Wang, and M. Lemmon. Stabilizing bit-rates in quantized event triggered control systems. In *Hybrid Systems: Computation and Control*, pages 245–254, ACM, 2012.

[7] D. Liberzon, D. Nešic, and A. R. Teel. Lyapunov-based small-gain theorems for hybrid systems. *IEEE Transactions on Automatic Control*, 59:1395–1410, 2014.

[8] W. R. Perkins and J. B. Cruz. *Engineering of Dynamic Systems*. Wiley, New York, 1969.

[9] R. Postoyan, A. Anta, D. Nesic, and P. Tabuada. A unifying lyapunov-based framework for the event-triggered control of nonlinear systems. In *Decision and Control and European Control Conference (CDC-ECC), 2011 50th IEEE Conference on*, pages 2559–2564, IEEE, 2011.

[10] A. Seuret and C. Prieur. Event-triggered sampling algorithms based on a lyapunov function. In *Decision and Control and European Control Conference (CDC-ECC), 2011 50th IEEE Conference on*, pages 6128–6133, IEEE, 2011.

[11] P. Tabuada. Event-triggered real-time scheduling of stabilizing control tasks. *IEEE Transactions on Automatic Control*, 52:1680–1685, 2007.

[12] X. Wang and M. D. Lemmon. Self-triggered feedback control systems with finite-gain $L_2$ stability. *IEEE Transactions on Automatic Control*, 54:452–467, 2009.

[13] X. Wang and M. Lemmon. Minimum attention controllers for event-triggered feedback systems. In *The 50th IEEE Conference on Decision and Control-European Control Conference (CDC-ECC)*, pages 4698–4703, IEEE, 2011.

[14] J. K. Yook, D. M. Tilbury, and N. R. Soparkar. Trading computation for bandwidth: Reducing communication in distributed control systems using state estimators. *IEEE Transactions on Control Systems Technology*, 10(4):503–518, 2002.

# 12

# *Event-Based PID Control*

**Burkhard Hensel**
*Dresden University of Technology*
*Dresden, Germany*

**Joern Ploennigs**
*IBM Research–Ireland*
*Dublin, Ireland*

**Volodymyr Vasyutynskyy**
*SAP SE, Research & Innovation*
*Dresden, Germany*

**Klaus Kabitzsch**
*Dresden University of Technology*
*Dresden, Germany*

**CONTENTS**



**ABSTRACT** The most well-known controller, especially in industry, is the PID controller. A lot of PID controller design methods are available, but most of them assume either a continuous-time control loop or equidistant sampling. When applying a PID controller in an event-based control loop without changing the control algorithm, several phenomena arise that degrade both control performance and message rate. This chapter explains these phenomena and presents the state of the art solutions to reduce or avoid these problems, including sampling scheme selection, sampling parameter adjustment, and PID algorithm optimization. Finally, tuning rules for PID controllers that are especially designed for event-based sampling are presented, allowing an adjustable trade-off between high control performance and low message rate. The focus of this chapter is on send-on-delta sampling as the most known event-based sampling scheme, but many results are valid for any event-based sampling scheme.

## 12.1 Introduction

PID controllers are very popular in automation due to their simplicity, easy tuning, and robustness [5,45]. More than 90 % of all control loops in industry use PID controllers [5,34]. PID controls are quite robust to changes of plant parameters, allowing long operation times without maintenance. They are suitable for many process types, including stable, integrating, and unstable plants, as well as plants with saturation or time delay. Practitioners can choose from a lot of tuning rules [34,71,72].

The theory behind PID controllers is well researched [43,71]. Due to the success of digital controllers, a fundamental assumption is that all devices in the control loop operate at the same equidistant and synchronized time instances. In practice, an increasing amount of easily installable wired and wireless network protocols favor event-based sampling approaches because they require less network bandwidth and energy. However, event-based sampling approaches also induce new phenomena that cannot be explained with the classical theory due to the violation of the periodic sampling assumption. As a result, practitioners need new approaches for evaluating stability and deriving parameters, advanced PID algorithms, and tuning rules.

Research in *event-based PID control* is a relatively young discipline. In 1999, Årzén presented an event-based version of the PID controller [4]. Other authors continued his work, such as Otanez et al. in 2002 [35], Sandee et al. in 2005 [49], and Vasyutynskyy et al. in 2006 [61]. Since then, an increasing amount of publications mirror the growing interest in event-based PID control.

The goal of this chapter is to synthesize this research into some practical guidelines to design and tune event-based PID controls. Section 12.2 introduces the relevant foundations of event-based sampling and control loops, as well as of the classic PID controller. For evaluating control loops, performance measures are discussed in Section 12.3. Section 12.4 throws light on typical problems in event-based PID controls. Possible solutions are discussed in Section 12.5. Send-on-delta sampling is a very common approach. We therefore present some results with focus on send-on-delta sampling in Section 12.6. Final conclusions are drawn in Section 12.7.

## 12.2 Basics

### 12.2.1 Sampled Signals

Before we dive into event-based control approaches, we shortly define our notation for periodic or event-based signals in the control loop.

All signals are defined as time series $S$ with a series of observations $s_z = s(t_z)$ and the corresponding sample times $t_z \geq 0$ for $z \in \mathbb{N}$. If the intersample interval $T_z = t_z - t_{z-1}$ is a positive constant, we speak of *uniform sampling* or a *periodic signal*.

The intersample interval $T_z$ is a random variable in embedded systems. The kind and parameters of the distribution depend on the implementation of the sampling approach [42]. Some devices sample as fast as possible. In this case, the intersample interval equals the operation cycle of the embedded device. This is common, for example, for Programmable Logic Controllers (PLC) and line-powered embedded devices such as devices using the building automation network LON [42].

Other devices use timers to sample at predefined intervals. This is very common in wireless sensor networks because it allows the nodes to go into a low-power sleep mode in-between sampling events. However, clocks in embedded systems have a variance and a drift [52], which results again in a randomly distributed intersample interval $T_z$.

For control design it is usually acceptable to simplify both cases. If the intersample intervals are very small ($T_z \to 0$), then the time series $S$ becomes similar to a continuous-time signal $s(t)$. If the clock variance and drift are significantly smaller than the sampling period, then the intersample interval can be assumed to be constant.

Most event-based sampling approaches are based on such a periodic sampling and decide for each observation if it should be transmitted. The difference of the approaches lies primarily in the filter conditions applied to the original time series. They are often based on the current sample $s_z$ and the last transmitted sample $s_l$ (transmitted at time $t_l$). Table 12.1 compares some event-based sampling approaches and their filter conditions [3,36,37,40,48]. Most common is send-on-delta (SOD) sampling as it is used in EnOcean and LON networks [33]. Other approaches are integral sampling [30] and its variation, called gradient-based sampling [41]. Both use a filter criterion based on the sampling error, while the latter also adapts the wake-up times. Model-based approaches use predictors in sender and receiver to improve the signal reconstruction in the receiver [24,50,57].

These sampling criteria may be modified with additional temporal constraints on the intertransmission interval to avoid oversampling or undersampling. A min-send-time $T_{\min}$ is used to limit the traffic load caused by oversampling if a signal changes rapidly or if the sensor is faulty. This is particularly useful when nodes sample as fast as possible such as in LON networks. As the min-send-time $T_{\min}$ defines the lower limit of the interval between messages, it has to fulfill the Nyquist–Shannon sampling theorem. The min-send-time loses its meaning in wireless sensor networks as

**TABLE 12.1**
Well-Known Sampling Approaches and Their Filter Conditions

| Approach | Condition $cond(s_z)$ | Description |
|---|---|---|
| Periodic | $(t_z - t_l) \geq T_A$ | Transmits signal with period $T_A$ |
| Send-on-delta | $\|s_z - s_l\| \geq \Delta_S$ | Transmits signal on significant change since last transmission $s_l$ [33] |
| Integral | $\sum_{j=l+1}^{z} \left\| \frac{s_{j-1}+s_j}{2} - s_l \right\| T_j \geq \Delta_I$ | Accumulates sampling error since last transmission $s_l$ and avoids steady-state errors [30,41] |
| Model-based | $\|\hat{s}_z - s_z\| \geq \Delta_P$ | Uses a signal model $\hat{s}_z$ in sender and receiver to compress and reconstruct the signal [24,50,57] |

the underlying sampling period $T_z$ should already be as large as possible for the purpose of energy efficiency.

A max-send-time $T_{max}$ defines an upper limit for the intertransmission interval and ensures that a sample is transmitted if more time passed since the last message was sent. It is primarily used to implement a periodic "watchdog signal" in the sensor that allows monitoring its correct operation. It can also be used to reduce the negative effects of event-based controls as shown in Section 12.5.2. Finally, in some cases, the max-send-timer may improve the energy efficiency of the overall communication system [21].

The resulting filtered time series $\tilde{S}$ is a subset of the original time series $S$, depending on the sampling condition $cond(s_z)$ as well as the min-send-time and the max-send-time. The intertransmission interval between two subsequent elements of the new filtered time series $\tilde{S}$ is denoted as $T_k$.

### 12.2.2 Event-Based Sampling in Control Loops

Figure 12.1 shows a basic control loop with sensor, controller, actuator, and the plant under control. A user interface may be an optional element in the control loop. The sensor observes the plant and samples the time series of plant output $y_k$ (controlled variable) where $k$ is the index of the time series. The user interface or supervisory control system provides a set-point value $r_k$

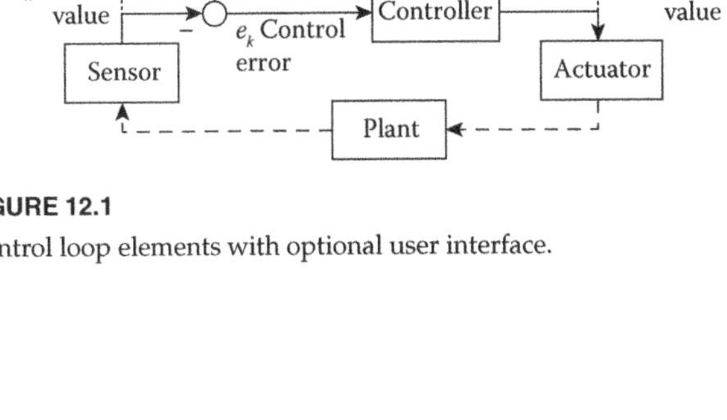

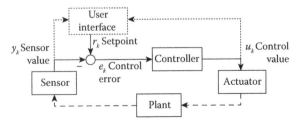

**FIGURE 12.1**
Control loop elements with optional user interface.

(reference variable). The control error $e_k$ is the difference between plant output and setpoint ($e_k = r_k - y_k$). Based on the plant output and the setpoint value, the controller computes a control value $u_k$ (manipulated variable) that is converted by the actuator to a physical action. The user interface potentially also visualizes the plant output $y_k$ or the control value $u_k$. In practice, several modules in the control loop are often combined into one device to reduce the number of devices. For example, controllers are often integrated with the actuators and the sensors are often combined with the user interface into one device [68].

The arrows in Figure 12.1 represent the data flow of the signals $y_k$, $r_k$, $e_k$, and $u_k$. The optional signal transmissions to the user interface are marked as dotted lines. It is possible that all these signals are transmitted in an event-based manner. In many applications, the setpoint changes infrequently compared to the other signals in the control loop as it is defined by the user. Therefore, event-based transmission is advantageous [16]. It is also recommendable to transmit sensor values in an event-based manner to reduce network traffic [33,40]. The actuator value may be sampled event-based to reduce along with the number of messages also the number of actuation events, as this decreases the deterioration of the actuator and can help saving control energy [36,55]. Also an event-based transmission of the control error can be found in the literature since it can be beneficial to change the triggering scheme depending on the control error [3], and it allows some theoretic investigations [6,44] as the time instance of zero control error is detected.

Independent of the signal transmission, controllers may be operated periodically or event-based [68]. In the first case, the code is run periodically, whether or not the input has changed. In the latter case, the code is only executed if the control error changes. This saves the processor load of the embedded device [4]. An event-based operation of the actuator (independent of the signal transmission scheme of the control value) may save energy.

### 12.2.3 PID Controller

The behavior of an event-based PID controller can be approximated from its continuous-time counterpart. The classic continuous-time PID controller is defined in the time domain as

$$u(t) = K_P \cdot \left( e(t) + \frac{1}{T_I} \cdot \int_0^t e(\xi)\mathrm{d}\xi + T_D \cdot \frac{\mathrm{d}e}{\mathrm{d}t} \right), \quad (12.1)$$

where $K_P$, $T_I$, and $T_D$ are proportional gain, reset time, and rate time, respectively (names according to [1]), and in the complex frequency domain (after Laplace transform) as

$$\frac{U(s)}{E(s)} = K_P \left( 1 + \frac{1}{T_I \cdot s} + T_D \cdot s \right). \quad (12.2)$$

This definition does not allow immediate changes of the control error because then the derivative action would be infinite. Therefore, the derivative part is usually filtered by a first order low-pass filter, resulting in the transfer function

$$\frac{U(s)}{E(s)} = K_P \left( 1 + \frac{1}{T_I \cdot s} + T_D \cdot \frac{s}{1 + \frac{T_D}{N_D} \cdot s} \right), \quad (12.3)$$

where $N_D$ denotes the derivative filter coefficient.

There are several approaches to approximate the ideal (continuous-time) PID controller by a discrete-time controller [48]. The following equations give one example for the proportional (P), integrative (I), and derivative (D) action, taken from [66]:

$$u_k = P_k + I_k + D_k, \quad (12.4)$$

$$P_k = K_P \cdot e_k, \quad (12.5)$$

$$I_k = I_{k-1} + \frac{K_P}{T_I} \cdot T_k \cdot e_{k-1}, \quad (12.6)$$

$$D_k = \frac{T_D}{T_D + N_D \cdot T_k} \cdot D_{k-1} + \frac{K_P \cdot T_D \cdot N_D}{T_D + N_D \cdot T_k} \cdot (e_k - e_{k-1}). \quad (12.7)$$

The start conditions are typically $D_0 = 0$ and $I_0 = 0$, if no techniques for bumpless transfer are used [58]. Section 12.5.3 will discuss several possible improvements of this algorithm.

## 12.3 Performance Measures

The main challenge of event-based controller design is the optimization of the sampling and controller parameters. For that purpose, it must be defined how an event-based control loop should be evaluated. The usual goal of continuous-time or periodic control loops is good control performance. For event-based sampling, energy criteria are also important as they consider the biggest advantage of these sampling schemes.

The criteria presented below can be used not only for evaluating PID controllers but also for other types of controllers. We show the continuous-time formulation of the control performance measures because these present the ideas of the measures more intuitively.

### 12.3.1 Basic Performance Measures

Different performance criteria have been used in the existing literature for evaluating PID controllers with event-based sampling. They can mainly be divided into control performance measures and energy consumption measures. In addition, other aspects should also be paid attention to when designing a controller, such as energy consumption of the actuator and plant, processor load, and economic considerations [19]. However, these aspects are not discussed in this chapter.

#### *Control performance measures*

Control performance usually evaluates the control error $e(t)$. As this signal changes over time, there are several possibilities for aggregating the signal into a single scalar value. The most commonly used performance measurements are:

- The $\mathcal{L}_1$ norm (integral of absolute error, IAE)

$$J_{\mathrm{IAE}} = \int_0^{T_{end}} |e(t)|\mathrm{d}t, \quad (12.8)$$

where $(0,T_{end})$ is the time range in which the error signal is evaluated is the most often used index for evaluating control performance with event-based PID controllers [3,20,35,37,39–41,44,46,47,62–69]. The normalized version of this performance measure is the mean absolute error

$$\bar{e} = \frac{J_{\mathrm{IAE}}}{T_{end}}, \quad (12.9)$$

and can be very well understood even by practitioners with less theoretical background.

- The $\mathcal{L}_2$ norm (integral of squared error, ISE)

$$J_{\mathrm{ISE}} = \int_0^{T_{end}} e^2(t)\mathrm{d}t, \quad (12.10)$$

has been used in the context of event-based PID control only in [20,22] although it is a common goal for controller optimization [34]. From the authors' point of view, ISE is better suited than IAE, because for typical applications where event-based sampling is reasonable, small errors are intentionally accepted for the advantage of reduced message rate—and ISE punishes small errors (compared with large errors) less than IAE.

- The "biased" IAE is particularly adapted to the phenomena of event-based control

$$J_{\mathrm{IAE}\Delta} = \int_0^{T_{end}} f(e)\,\mathrm{d}t, \tag{12.11}$$

$$f(e) = \begin{cases} |e| & \text{if } |e| > e_{\max} \\ 0 & \text{if } |e| \le e_{\max} \end{cases}. \tag{12.12}$$

This measure tolerates any value of the control error within a defined tolerance belt $e_{\max}$ for respecting that many event-based sampling schemes result in limit cycles, that is, small oscillations around the setpoint as shown in Section 12.4. This avoids the accumulation of long-time steady-state errors during performance evaluation. It has been used several times [6,20,60,67].

- The difference between the controlled variable $y(t)$ of the event-based control loop and the controlled variable $y_{\mathrm{ref}}$ of a periodically sampled control loop with same process and controller settings has also often been used, especially integrated as IAEP [8,37,39,46,47,60,62,64,67,68]:

$$J_{\mathrm{IAEP}} = \int_0^{T_{end}} |y_{\mathrm{ref}}(t) - y(t)|\mathrm{d}t. \tag{12.13}$$

This is a helpful measure if it is desired that the event-based control loop hardly differs from a continuous-time or discrete-time control loop. However, this is not necessarily suited for finding a reasonable trade-off between message rate and control performance as shown in Section 12.6.2. Also, the sampling efficiency, that is, the quotient of IAEP and IAE, has been used several times [37,46,47,62,64,68].

- There are some well-known performance criteria which are only applicable for evaluating step responses. Examples are settling time, overshoot, control rise time, ITAE, ITSE, or steady-state error [1]. As these measures assume setpoint steps without disturbances, their use is limited in comparison to the performance measures mentioned before. Further, limit cycles (oscillations around the setpoint, see Section 12.4) make the application difficult. For example, the settling time can only be determined with send-on-delta sampling, if the final band around the setpoint is larger than the threshold $\Delta_S$. Nevertheless, they can be helpful for some tasks, especially the overshoot. Regarding event-based PID controllers, the settling time has been evaluated in [3,6,8,11,44,67], the rise time in [3,67], and the overshoot in [3,6,8,11,20,22,23,39, 41,67].

- Other performance measures are defined in [8,11, 20,28,44,47] and robustness measures have been considered in [8,11].

#### Message rate and energy consumption measures

For event-based sampling, the message rate is also important because reducing the message rate and thus the energy consumption is the main target of event-based sampling and the reason for accepting the increased design effort and reduced control performance.

The message rate has rarely been used directly for evaluating control loops with event-based PID controllers [20,60]. More frequently, the overall number of messages $N$ which have been transmitted during one simulation has been used [3,6,8,37,41,44,65,66]. Also, the mean intertransmission interval

$$\bar{T}_{it} = \frac{T_{end}}{N}, \tag{12.14}$$

between two messages (the inverse of the message rate) has been used [20,39,41] because this interval allows comparison with the period of a periodically sampled control loop.

Several authors recommend estimation of the overall energy consumption using a weighted sum of messages and controller invocations for exact energy considerations [37,39,46], or other energy-oriented measures such as power consumption or electric current [39,40]. Also the ratio between the message number of the event-based control loop and the message number of the periodic control loop has been used [62–64,67,68].

### 12.3.2 Combination of Performance Measures

Several performance measures should be taken into account for optimizing an event-based control loop. In particular, as both the control performance and the message rate form a trade-off in many practical scenarios, at least one criterion of each category should be considered when designing event-based controllers.

The most generic type of multicriteria optimization is Pareto optimization. A solution is called Pareto optimal, if no other solution can be found which improves at least one of the performance measures without degrading any other measure. Therefore, Pareto optimization delivers a set of optimal solutions and it is up to the user to decide, which of the solutions should be realized. However, the set of Pareto optimal solutions is much smaller than the original design space, making the controller design easier.

If the control designer is interested in getting only one single solution then different approaches are suitable. One way is to optimize only one basic measure while considering constraints for all other relevant performance measures. Another possibility is the algebraic combination of the performance measures. Most performance measures mentioned before are positive real and should be minimized. For these measures, the algebraic combination can mainly be realized in two different ways:

1. Weighted sum of basic performance criteria [35,37,46,64]

$$J_{\text{sum}} = \sum_i w_i \cdot J_i. \quad (12.15)$$

The reasonable selection of the weighting factors $w_i$ may be difficult, but allows much freedom for application-specific priorities.

2. Product of basic performance criteria

$$J_{\text{Prod}} = \prod_i J_i, \quad (12.16)$$

in particular [20,22,23]

$$J_{\text{Prod}} = N \cdot J_{\text{ISE}}. \quad (12.17)$$

The advantage compared with the weighted sum is that this performance measure is optimal if none of the basic performance measures can be improved by a specific ratio without degrading the other (or product of the others for more than two used indices) by more than that ratio [22]. The disadvantage is the more difficult weighting of the performance measures (if this is necessary) as this cannot be done by simple factors. Additionally, there is a problem if one of the basic performance measures is zero or near to zero, because then the other ones have no significant influence on the comparison to other solutions.

The optimal solutions according to both strategies are always Pareto optimal. That can simply be proven by contradiction.

## 12.4 Phenomena of Event-Based PID Controllers

The simple PID control algorithm (12.4) along with event-based sampling leads to the issues presented in Figure 12.2 and described in the subsequent paragraphs [40]. The following overview explains these issues. Solutions are given in Section 12.5.

### *Approximation errors*

The control error is known by the control algorithm only in the time instances of sampling. While digital PID algorithms such as (12.4) assume usually a constant signal between two samples (zero order hold), the real controlled variable (and hence the control error, too) changes in between, thus being a source of uncertainty in the controller. In event-based sampling, the reduced amount of samples increases the uncertainty. For example, if the control error is sampled with send-on-delta

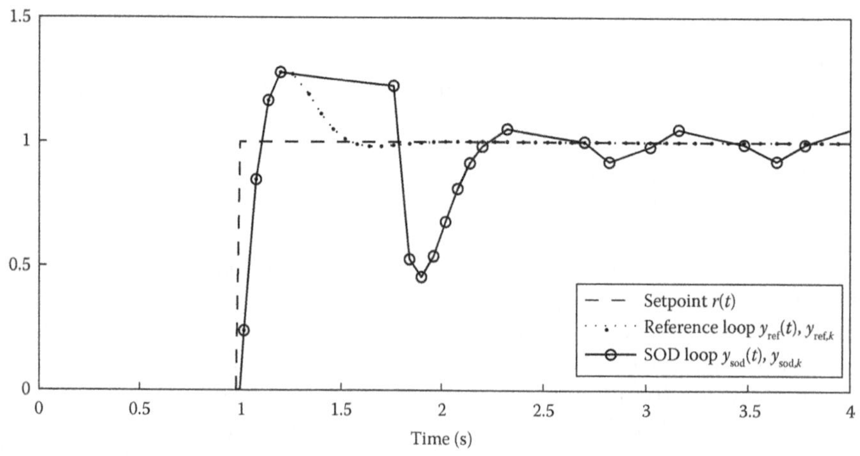


**FIGURE 12.2**
Step response of the control loop with PID algorithm (12.4) and send-on-delta sampling with $T_z = T_{\min} = 0.06$ and $T_{\max} \to \infty$. The emerging issues (e.g., sticking and limit cycles) are visible.

sampling, the uncertainty of the PID controller's integral action is limited by $\pm\Delta_S \cdot T_k$ (if $T_z \ll T_k$ and $T_{\min} \ll T_k$), that is, proportional to the intertransmission interval. Integral sampling reduces the uncertainty as also small errors accumulate and can thus cause a new event. For reaching the same mean intertransmission interval, all sampling schemes have time slots with more messages as well as with less messages than the other schemes. Therefore, all sampling schemes have advantages and disadvantages regarding accuracy of control. In general, very large intertransmission intervals ($T_k \gg T_z$, characteristic for control loops with send-on-delta sampling in the "steady state") cause large approximation errors. Although small intertransmission intervals improve the approximation of the error $e(t)$, very small intertransmission intervals can cause large differences of the derivative action compared with a periodic control loop, if the period $T_A$ of the periodic control loop is greater than the sensor sampling time $T_z$ [60]. It depends on the controller implementation and the controller and sampling parameters, how severely the approximation errors degrade the control performance as discussed in Section 12.5.

#### *Triggering of events and sticking*

Another example of event-triggering issues is that samples are not triggered, if the signal (plant output or control error) is staying inside the deadband defined by the sampling scheme such as $\Delta_S$ in send-on-delta sampling. Therefore, if not specially adapted for this case, an event-based controller is not able to control the signal within this deadband since the controller is not invoked as long as no message arrives. This can have the effect visible in Figure 12.2 at $1.2 < t < 1.8$, at the peak of the overshoot. There the event-based sensor does not produce a new message because the plant output changes just a little. The controller is then not invoked since no event arrives. The control loop achieves a temporary equilibrium, also called sticking [61], when both sensor and controller do not send any messages. The system resides in this state until some event such as a significant plant output change, load disturbance, noise, or expiry of the $T_{\max}$ timer forces the sensor to send a message. Such sticking degrades the control loop performance and is critical if the control error is large, for example, in case of large overshoots, since in that case the control error remains large for a long time.

Sticking leads further to integral action "windup." If after the sticking phase a new message arrives, the change in the integral part is

$$\Delta I_k = \frac{K_P}{T_I} \cdot T_k \cdot e_{k-1}, \tag{12.18}$$

and thus directly proportional to the duration $T_k$ of the sticking phase. A sudden, extensive change of the integral action happens at this time instance. This abrupt behavior of the integral action causes a control loop behavior that is strongly different from a continuous-time or periodic control loop. This is clearly visible in Figure 12.2 at $t \approx 1.8$.

#### *Quantization effects and limit cycles*

Event-based sampling limits the accuracy of the sampled signal and that results in non-linear quantization effects [70].

One possible effect of the quantization is a steady-state error which can be interpreted as an infinitely long sticking effect. Another possible quantization effect is limit cycles, that is, permanent oscillations around the setpoint, which appear under proper conditions in "steady-state situations" when the reference variable and the disturbances are constant. Limit cycles have already been detected by Årzén in the first publication on event-based PID control [4]. Their magnitude is mainly defined by the deadband threshold of the sampling scheme. Besides event-based sampling, another precondition for the existence of limit cycles is that the controller or the plant contains integral action [44].

In many applications, limit cycles are undesired since they increase the event rates without improving the control performance. Limit cycles may be more acceptable than a constant steady-state error because the expected value of the mean error is usually smaller.

In some applications, there are external disturbances that are practically never constant, unknown, and larger than $\Delta_S$. For example, in room temperature, control loops, outdoor air temperature variations, solar radiation, and occupant influences are severe and temporary that level crossings are generated regularly and cannot be avoided [20]. In such cases, the messages due to limit cycles can often be neglected, depending on the controller settings.

## 12.5 Solutions

The different issues related to event-based PID control need to be considered during the control design. This starts with the selection of an adequate sampling approach according to the specific design goals of the control loop as discussed in Section 12.5.1. Section 12.5.2 will discuss the selection of the sampling parameters. Special adaptations of the PID algorithms allow to avoid the critical issues. That will be explained in

Section 12.5.3. Section 12.6.2 will finally present tuning rules for the PID parameters.

### 12.5.1 Comparison of Sampling Schemes for PID Control

The first step for improving the control loop behavior is the selection of a suitable strategy for triggering events. Multiple simulation and experimental studies investigated the influence of different sampling approaches on event-based control, with the result that all sampling schemes have advantages and disadvantages.

Pawlowski et al. analyzed a greenhouse temperature control loop in a simulative study [36,37]. They explored the message rate and control loop performance of different sampling approaches (cf. Table 12.1 in Section 12.2.1). They concluded that event-based sampling allows to save about 80 % of the messages in comparison to periodic sampling. However, the parametrization of event-based sampling approaches has a large influence on the resulting control performance.

Sánchez et al. [46] analyzed the level control of a tank as an example for an industrial control process. They investigated event-based and periodically sampled PI controls in an experimental setup. They concluded that event-based approaches are convenient control strategies when the number of messages should be reduced, but that the parameter selection is an open issue. Araújo [3] realized in his thesis a simulation study of several standard process types in combination with several sampling schemes and PI control variants. The best results emerged from combining a modified PIDPLUS controller with send-on-delta sampling on the controlled variable.

The study in [40] analyzes a room temperature and an illumination control loop as two typical building automation control scenarios with different signal dynamics. The temperature control is a slow process with large dead times. On the contrary, the illumination control is a very fast process with no significant delay in comparison to the delay inflicted by the network communication and the sampling time. The study compared the event-based sampling approaches given in Table 12.1 using different performance criteria. In extension to other works, it also explores the influence of measurement noise, message losses, and transmission delays on the event-based controls. Depending on the relevance of the investigated criteria, an adequate sampling approach can be selected for a specific scenario. The criteria are defined as follows, and the results are visually presented in Figure 12.3.

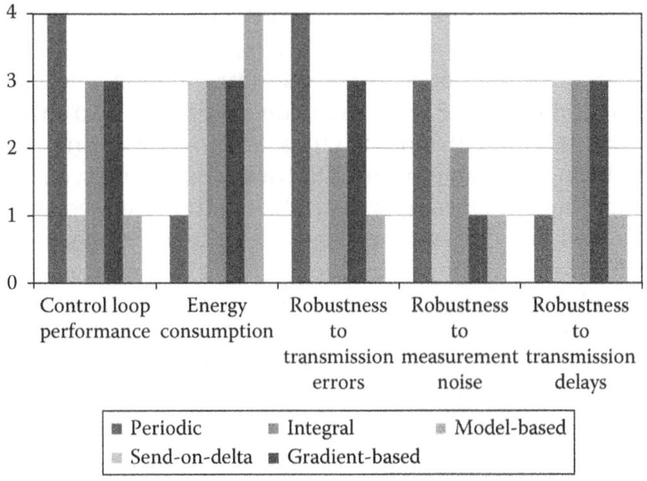


**FIGURE 12.3**
Evaluation of the adaptive sampling approaches from Table 12.1 (Rating: 4 – best suited; 3 – well suited; 2 – mean; 1 – disadvantageous).

#### *Control loop performance*

Control loop performance is related to the sampling error of event-based sampling as explained in Section 12.3. The control loop performance usually forms a trade-off with the energy consumption. A smaller sampling period $T_z$ and threshold ($\Delta_S$, $\Delta_I$, or $\Delta_P$) leads to a lower sampling error and improved control loop performance but requires more message transmissions and thus energy. In [40], $J_{\text{IAE}}$ has been used for evaluating control performance. According to these simulations, periodic sampling is the best sampling approach regarding control loop performance, followed by integral sampling and gradient-based sampling at the cost of a higher message rate.

#### *Message rate and energy consumption*

Wireless sensor nodes are often powered by batteries or energy harvesting. The energy consumption for transmitting is in many cases much higher than that for sampling [17,21,26], such that the amount of transmitted messages should be reduced to a minimum. All event-based sampling approaches can save about 80 % of the messages compared with periodic sampling without significant losses in control performance [36,37,40]. The lowest energy consumption can be reached with model-based sampling, but only if the signal is slowly changing, free of noise, and the model and algorithm are perfectly parametrized. In all other cases, integral and gradient-based sampling provide good all-around solutions, where particularly the latter is quite robust with regard to parametrizations.

#### *Robustness to transmission errors*

Wireless communication is error-prone and easily disturbed by other wireless networks, electromagnetic sources, moving people, or objects [73]. Therefore, the PID controller works on outdated information and cannot react to changes in the plant. This effect is increased for model-based sampling as the estimator model wrongly estimates the plant behavior. Periodic and gradient-based sampling are the most robust approaches because they usually transmit more samples than the other approaches.

#### Robustness to measurement noise

Measurements carry small noise added by the process disturbances or sensor hardware. This may require a filtering of the sampled signal. In the case of periodic sampling, the noise is often tolerated because it does not influence the mean sampling rate. However, in case of event-based sampling, noise may trigger the transmission condition unnecessarily and result in higher power consumption. Especially, integral and gradient-based sampling as well as model-based approaches are sensible to noise. In contrast, for send-on-delta sampling, the threshold $\Delta_S$ allows filtering noise within its tolerance belt.

#### Robustness to transmission delays

Transmission delays are problematic for the control task if they are not negligibly small compared with the dominating time constants of the control loop. Transmission delays are usually below 100 milliseconds in wired networks [38] but can be larger in wireless networks due to message caching in multihop transmissions if hops have different sleep cycles. Periodic sampling shows the worst response to large transmission delays because the numerous messages created by periodic sampling are disarranged when the delay's jitter is larger than the sampling period. As a result, the messages are not received in the correct order [54]. This adds noise at the controller's input that is boosted by the proportional and derivative part of the PID controller.

#### Alternatives

Besides considering these different triggering criteria, there are also sampling-related solutions that may help in specific cases. For example, the selection of the signal to be transmitted should be regarded. Triggering based on the control error allows for switching off the signal generation if the control error is smaller than a deadband. That can reduce the message rate in steady state [3]. On the contrary, evaluating the control error requires knowledge of the setpoint whose transmission needs additional energy in some scenarios [3,23].

If only the message rate is important and not the computational load, it is possible to invoke the controller periodically without synchronization to the transmission events [20]. This avoids sticking because the integral action is updated regularly even if no message arrives.

The performance of the event-based sampling approaches depends strongly on the parametrization and the controller. Therefore, Section 12.5.2 will investigate the sampling parametrization and Section 12.5.3 the control algorithm details.

### 12.5.2 Adjustment of the Sampling Parameters

The problems presented in Section 12.4 can be reduced by appropriate choice of the sampling parameters, in particular $T_{\max}$ and the thresholds $\Delta_S$, $\Delta_I$, or $\Delta_P$. This approach has the advantage that the PID control algorithm need not be changed.

#### Min-send-time

The setting of $T_{\min}$ should always be chosen according to the sampling theorem, that is, about 6 to 20 times smaller than the inverse of the Nyquist frequency [61]. Lower settings increase the message rate without improving the control performance significantly; higher settings endanger stability. $T_{\min}$ is only needed if the intersample interval $T_z$ is shorter than $T_{\min}$, see Section 12.2.1.

#### Max-send-time

Approximation errors and sticking are a consequence of long intertransmission intervals. Hence, reducing $T_{\max}$ reduces these problems, at the cost of an increased message rate. The differences between the event-based and continuous-time (or periodic for $T_z > 0$ or $T_{\min} > 0$) control loops diminish with decreasing max-send-time.

In [66] a rule for setting the maximum timer $T_{\max}$ has been proposed in the case of send-on-delta sampling. The max-send-time should lie between several bounds which are defined by the process time constants, the limit cycles, user-defined specifications about maximum approximation errors, and the life sign:

$$\max\left\{T_{\mathrm{eq}}, T_{\mathrm{LC}}\right\} < T_{\max} < \min\left\{\bar{T}_{\max}, T_{\mathrm{watchdog}}\right\}, \tag{12.19}$$

where $T_{\mathrm{eq}}$ is defined by

$$T_{\mathrm{eq}} = \frac{\Delta_S}{\overline{\frac{\mathrm{d}y_{\mathrm{ref}}}{\mathrm{d}t}(t)}}, \tag{12.20}$$

with the mean value $\overline{\mathrm{d}y_{\mathrm{ref}}/\mathrm{d}t}$ of the periodic reference plant output slope during the rise time of the step response. $T_{\mathrm{LC}}$ is the period of the limit cycles in steady state, $\bar{T}_{\max}$ is designed to limit the approximation errors and sticking (see [66] for details), and $T_{\mathrm{watchdog}}$ is the maximum setting to give a watchdog signal to the actuator. The watchdog time could be set to about 1 to 2 times the settling time of the closed loop [63].

If the sensor is located in another device than the controller, it is advantageous to use different max-send-timers in both devices. The max-send-time in the sensor should be as large as possible for reducing network load—as long as safety is not compromised (watchdog function) or the energy efficiency of the receiver is degraded due to timing problems [21]. The additional max-send-timer is used to invoke the controller locally if no message arrived before this timer expires. It should be as small as possible for reducing sticking and approximation errors—limited by acceptable processor load. As the controller is invoked more frequently than the sensor, it has to reuse the last received sensor value several times [22,67]. That leads to another kind of approximation error, but the sampling error is limited according to the sampling threshold.

A comparable strategy is to force a new sensor message if sticking is detected [61,63]. This increases the message rate in such situations but avoids computations with outdated data.

#### *Threshold*

For send-on-delta sampling, the threshold should be small enough that typical step responses (with step size $\Delta_r$) get a suitable resolution, that is, $\Delta_S = 0.05 \ldots 0.15 \cdot \Delta_r$ [63]. Furthermore, it should be smaller than the tolerated steady state error $\Delta_{\text{e,max}}$, or better still, $\Delta_S \leq \Delta_{\text{e,max}}/2$ [63]. However, it should not be set too small, because otherwise high-frequency noise could trigger many messages. Vasyutynskyy and Kabitzsch [62] noticed from simulations that the event rate jumps up if the standard deviation of the noise is larger than $\Delta_S/2$. Thus, $\Delta_S$ should lie between two times the noise band and the half of $\Delta_r$ [44]. Several other guidelines are discussed in [61].

Beschi et al. [8] found for send-on-delta sampling that both the mean intertransmission interval and the control performance (measured as the maximum difference between the outputs of the event-based control loop and a periodic control loop) are roughly proportional to $\Delta_S$.

For integral sampling similar rules can be used. When assuming a linear evolution of the transmitted variable between the samples, a nearly equal intertransmission interval $\bar{T}_{is}$ can be reached [39] by setting the threshold to

$$\Delta_I = \frac{1}{2} \cdot \Delta_S \cdot \bar{T}_{is}. \tag{12.21}$$

Recently, Ruiz et al. [44] presented an alternative approach for optimizing the threshold. They introduced a PI-P controller which has two parameters, the send-on-delta threshold $\Delta_S$ and an aggressiveness-related parameter $\alpha$, where $\Delta_S$ is designed for getting a desired message rate and $\alpha$ for shaping the trade-off between control performance and actuator action or robustness to modeling uncertainties.

### 12.5.3 PID Algorithms

Not only the sampling schemes and their parameters but also the PID control algorithm and its parameters can be improved to approach the problems of the simple algorithm presented by Equation 12.4. Improved algorithms are presented in this section, and improved controller parameters are discussed in Section 12.6.2. A comparison of different advanced PID control algorithms regarding several performance criteria is made in [67]. These simulations confirm that all explored algorithms have their advantages and disadvantages, and the application decides as to which strategy fits best.

#### *Integral action*

Especially the computation of integral action has been considered often in literature [48] because the integral action is the main reason of deviations between periodic and event-based control. In the simple PID algorithm defined in equation (12.4), the integral part has been computed from the last control error using a zero-order hold with

$$I_k = I_{k-1} + \frac{K_P}{T_I} \cdot T_k \cdot e_{k-1}. \tag{12.22}$$

This assumes inherently that the control error remained constant at this value $e_{k-1}$. One could intuitively assume that the implementation

$$I_k = I_{k-1} + \frac{K_P}{T_I} \cdot T_k \cdot e_k, \tag{12.23}$$

is at least equally suited, where it is assumed that the control error changed instantly after the last transmission to its current value, that is, $e_k$. The advantage is that the controller reacts immediately to the current control error. Unfortunately, this version is sensible to large control errors in combination with long intertransmission intervals. They occur in event-based sampling, for example, on setpoint changes after long steady-state periods. Then the controller assumes that the control error has been that large for the full intertransmission interval, which leads to large approximation errors. A similar case is the original implementation of Årzén [4]

$$I_k = I_{k-1} + \frac{K_P}{T_I} \cdot T_{k-1} \cdot e_{k-1}, \tag{12.24}$$

which allows a faster computation but has the same problem regarding the impact of approximation errors.

Alternatively, the control error can be estimated with a first order hold. The corresponding equation is

$$I_k = I_{k-1} + \frac{K_P}{T_I} \cdot T_k \cdot \frac{e_k + e_{k-1}}{2}. \quad (12.25)$$

This piecewise linear approximation improves the interpolation for short intertransmission intervals. However, it is also vulnerable to large control errors and long intertransmission intervals as it also contains the current control error $e_k$. Therefore, the version (12.22) is the best suited approximation. A comparison of more interpolation methods can be found in [63].

Durand and Marchand presented three other integral algorithms for send-on-delta sampling [15], repeated in the more comprehensive overview [48]. The first algorithm (already published in [14]) uses the knowledge that the signal must have been in the deadband $\pm\Delta_S$ around the setpoint up to $T_z$ before the message, because otherwise a message would have been transmitted earlier. This results in

$$I_k = I_{k-1} + \frac{K_P}{T_I} \cdot \tilde{E}_k, \quad (12.26)$$

with

$$\tilde{E}_k = (T_k - T_z) \cdot \Delta_S + T_z \cdot e_k. \quad (12.27)$$

This approximation is of course not true if $|e_{k-1}| > \Delta_S$ like in sticking situations. But in that case it is advantageous, too, as it reduces the "windup" after the next message due to the smaller change of the integral action.

The second approach uses an exponential forgetting function to limit the integral action:

$$I_k = I_{k-1} + \frac{K_P}{T_I} \cdot T_k \cdot \exp\{T_z - T_k\} \cdot e_k. \quad (12.28)$$

The best simulations regarding control performance and message rate resulted from the combination of both approaches by using

$$\tilde{E}_k = (T_k \cdot \exp\{T_z - T_k\} - T_z) \cdot \Delta_S + T_z \cdot e_k, \quad (12.29)$$

in Equation 12.26. It is also computationally the most expensive. Durand and Marchand suggest using a look-up table if the exponential function is too computationally expensive for an industrial controller.

Other examples of limited integral action can be found in [3,61,63].

### *Setpoint weighting and antireset windup*

Setpoint weighting (two degrees of freedom) can often be found in PID literature. It has the goal to avoid sudden actuator reactions on stepwise setpoint changes, as these reduce the actuator lifetime. Hence, setpoint weighting is not specific for event-based PID control [5, 58,72]. For example, Årzén [4] and Sánchez et al. [48] derived their event-based PID algorithm starting from the continuous-time PID controller

$$U(s) = K_P \cdot \Bigg(\beta \cdot R(s) - Y(s) + \frac{1}{T_I \cdot s}(R(s) - Y(s)) + \frac{T_D \cdot s}{1 + \frac{T_D}{N_D}s}(\gamma \cdot R(s) - Y(s))\Bigg). \quad (12.30)$$

and Vasyutynskyy et al. [63,69] used the derivative action

$$D_k = \frac{T_D}{T_D + N_D \cdot T_k} \cdot D_{k-1} + \frac{K_P \cdot T_D \cdot N_D}{T_D + N_D \cdot T_k} \cdot (y_k - y_{k-1}). \quad (12.31)$$

As this is not specific for event-based control, common text books about PID control can be referenced for details.

Also, antireset windup [4,23], useful for improving the integral action in the presence of actuator saturation, is independent from the sampling scheme and can be found in standard PID control literature [72].

### *Observer-based algorithm*

Another way to decrease the approximation errors as well as to avoid sticking and limit cycles is the use of a process model with an observer in the controller, which estimates periodically the current process output [65] (i.e., also without getting a message from the sensor). Based on this estimation, the controller computes the manipulated variable. If a message from the sensor containing the current process output arrives, the state of the model is updated via the observer. This allows a smaller approximation error than simply using the last transmitted value, provided that the model is accurate and disturbances change relatively seldom. A first-order observer brings acceptable control results while keeping the processing efforts in the controller small [65,67].

### *Other algorithmic extensions against limit cycles*

For avoiding limit cycles, besides the observer approach, several other algorithmic solutions have been proposed. Some authors proposed to use intentionally "biased" data (e.g., $e_k = 0$, although the true control error is not exactly zero) [3,48,59,60,63]. Another possibility is to filter the control signal [3,60]. Basis of all algorithms is a limit cycle detection strategy.

## 12.6 Special Results for Send-on-Delta Sampling

Although all sampling schemes have their own advantages and disadvantages, send-on-delta sampling [31], also called level-crossing sampling [22,27] or absolute deadband sampling [62,70], is the most commonly used approach in practice as it is standard in LON and EnOcean networks, see also [44]. Send-on-delta sampling is very easy to implement and the only parameter $\Delta_S$ is easy to understand. Therefore, it also has the strongest focus in research. For that reason, this section will present some results which have been developed only for send-on-delta sampling.

It is necessary to distinguish between *ideal* and *real* send-on-delta sampling. *Real* send-on-delta sampling fits the description above, especially containing a minimum duration between two subsequent messages due to the internal sampling period $T_z$ of the sensor or the min-send-time $T_{\min}$. In contrast, *ideal* send-on-delta sampling means that the messages are sent immediately when a level has been crossed, that is, $T_z \to 0$. Although this cannot be realized technically and would be additionally more energy-demanding as the signal to be sampled must be evaluated continuously, this assumption has been used for a lot of theoretical analyses; some of them will be discussed in this chapter.

Furthermore, many of the publications referenced in the following assume *symmetric send-on-delta sampling* (SSOD). SSOD is a subtype of *ideal* send-on-delta sampling (and hence requires that the sensor is always active). Its basic idea is that the allowed values of the event-based signals are multiples of a given threshold, including zero. Hence, it is symmetric around zero. SSOD is defined as follows [10]: The *normalized symmetric send-on-delta* map nssod : $\mathbb{R} \mapsto \mathbb{Z}$ is a nonlinear dynamical system where the input $v(t)$ and the output $v^*(t)$ are related by the expression

$$v^*(t) = \text{nssod}(v(t)) = \begin{cases} (i+1) & \text{if} \quad v(t) > (i+1) \quad \text{and} \quad v^*(t^-) = i, \\ i & \text{if} \quad v(t) \in [(i-1),(i+1)] \\ & \quad \text{and} \quad v^*(t^-) = i, \\ (i-1) & \text{if} \quad v(t) < (i-1) \quad \text{and} \quad v^*(t^-) = i, \end{cases} \tag{12.32}$$

and $v^*(t_0) = -\,\text{sgn}(v(t_0))\lfloor |v(t_0)| \rfloor$, where $t_0$ is the starting time instant, $i$ is an integer that denotes the current sampling system output, and $\lfloor s \rfloor$ denotes the integer part of $s$.

The *symmetric send-on-delta* map ssod : $\mathbb{R} \mapsto \mathbb{R}$ is a nonlinear dynamical system with an input $v(t)$ and the output $v^*(t)$ that are related by the following expression:

$$v^*(t) = \text{ssod}\,(v(t); \Delta_S, \beta) = \Delta_S \beta \,\text{nssod}\left(\frac{v(t)}{\Delta_S}\right). \tag{12.33}$$

The parameter $\beta$ allows the generalization of the results, but is usually set to 1 and can be ignored.

### 12.6.1 Stability Analyses for PID Controllers with Send-on-Delta Sampling

The basic requirement for tuning a control loop is stability, because unstable control loops are useless and often lead to damage of equipment or even compromise the safety of people. Stability analyses are hence at the center of control theory and are discussed in this section. Event-based sampling makes such considerations much more difficult than continuous-time or periodic control, but there are already some important results.

The stability of control loops with send-on-delta on the plant state and full state feedback was considered by Otanez et al. in 2002 [35]. Their result is that it can be proven using the Lyapunov criterion that the event-based system is stable if the continuous-time system with the same controller settings is stable.

A partial stability analysis for a PID and a modified PIDPLUS controller (a PID controller with limited integral action), together with a first order (or second order by an additional filter) process without time delay has been given by Araújo in 2011 [3]. He analyzed the maximum (constant) sampling period, which is much higher for the PIDPLUS controller than for the simple PID controller.

Tiberi et al. presented a stability analysis for a control loop consisting of an event-based PIDPLUS controller, a first order process without time delay, and a more complex send-on-delta triggering rule ("PI-based triggering rule") [59]. Chacón et al. performed stability analysis for a PI controller with a time-delay-free process and send-on-delta sampling based on investigations regarding limit cycles [13].

Ruiz et al. gave a stability analysis for a so-called PI-P controller (based on the Smith predictor) with a first order plus time-delay process with SSOD sampling on the control error [44]. They also formulated a general concept for analyzing stability of control loops with event-based sampling: "System stability is characterized by the absence of limit cycles and the presence of an equilibrium point around the setpoint." In conclusion, a stability proof should contain two elements: "the demonstration of the existence of the equilibrium point and determination of its reach." However, other authors showed that there also other possible approaches for investigating system stability, depending on the underlying stability definition.

Beschi et al. published several works regarding stability of control loops with SSOD.

They developed their stability analyses in a series of publications [6–8,10], of which [10] is the most generic, even if not limited to PID controllers. A variation of this approach is explained in the following paragraphs.

The process is assumed to be linear but may have any order and a time delay $\tau$. Several architectures with SSOD at different positions can be unified by combining controller and process into a single system, such that the control loop consists of a linear system and the SSOD unit. This can be formulated as follows:

$$\begin{aligned} \dot{x} &= Ax(t) + Bv^*(t) + B_o o(t), \\ w(t) &= Cx(t) + D_o o(t), \\ v(t) &= -w(t-\tau), \\ v^*(t) &= \text{ssod}\,(v(t); \Delta_S, \beta) = \Delta_S \beta \,\text{nssod}\left(\frac{v(t)}{\Delta_S}\right), \end{aligned} \tag{12.34}$$

with the initial conditions

$$\begin{aligned} x(t_0) &= x_0, \\ v^*(\xi) &= \upsilon^*(\xi), \quad \text{with} \quad \xi \in [t_0 - \tau, t_0), \end{aligned} \tag{12.35}$$

where $x(t) \in \mathbb{R}^n$ is the process state; $A \in \mathbb{R}^{n\times n}$ is the system matrix; $B \in \mathbb{R}^n$, $B_o \in \mathbb{R}^{n\times m}$, $C \in \mathbb{R}^{1\times n}$, and $D_o \in \mathbb{R}^{1\times m}$ are the other matrices and vectors describing the system properties; $w(t) \in \mathbb{R}$, $v(t) \in \mathbb{R}$, $v^*(t) \in \mathbb{R}$, and $\upsilon^*(t) \in \mathbb{R}$ are signals whose physical meaning depend on which signal in the control loop is sampled; and $x_0 \in \mathbb{R}^n$ the initial state. The external signal $o(t) \in \mathbb{R}^m$ can originate from the reference variable, disturbances, or feedforward action.

Applying the transformations

$$\begin{aligned} &\tilde{v}(t) := \frac{v(t)}{\Delta_S} && \tilde{v}^*(t) := \frac{v^*(t)}{\Delta_S \beta} && \tilde{x}(t) := \frac{x(t)}{\Delta_S}, \\ &\tilde{w}(t) := \frac{w(t)}{\Delta_S} && \tilde{o}(t) := \frac{o(t)}{\Delta_S} && \tilde{\upsilon}^*(t) := \frac{\upsilon^*(t)}{\Delta_S \beta} \\ &\tilde{x}_0(t) := \frac{x_0(t)}{\Delta_S}, \end{aligned} \tag{12.36}$$

on the event-based closed-loop system, one can obtain

$$\begin{aligned} \dot{\tilde{x}} &= A\tilde{x}(t) + B\beta\tilde{v}^*(t) + B_o\tilde{o}(t), \\ \tilde{w}(t) &= C\tilde{x}(t) + D_o\tilde{o}(t), \\ \tilde{v}(t) &= -\tilde{w}(t-\tau), \\ \tilde{v}^*(t) &= \text{nssod}(\tilde{v}(t)), \end{aligned} \tag{12.37}$$

with the initial conditions

$$\begin{aligned} \tilde{x}(t_0) &= \tilde{x}_0, \\ \tilde{v}^*(\xi) &= \tilde{\upsilon}^*(\xi), \quad \text{with} \quad \xi \in [t_0 - \tau, t_0). \end{aligned} \tag{12.38}$$

This system does not depend on $\Delta_S$ but has the same stability properties (as long as $\Delta_S$ is finite). Hence, it is possible to choose $\Delta_S$ according to the desired trade-off between message rate and precision without influencing the stability of the overall system.

For SSOD (as a subtype of *ideal* send-on-delta sampling), the difference $q(t)$ between $v(t)$ and $v^*(t)$ is limited to $\pm\Delta_S$, see also [6]. It is possible to replace (12.34) by

$$\begin{aligned} v^*(t) &= \begin{cases} v(t) + q(t) & \text{if} \quad t \geq t_0, \\ \upsilon^*(t) & \text{otherwise.} \end{cases} \\ &= v(t) + \hat{q}(t) \\ \text{with} \quad \hat{q}(t) &= \begin{cases} q(t) & \text{if} \quad t \geq t_0, \\ \upsilon^*(t) - v(t) & \text{otherwise.} \end{cases} \end{aligned} \tag{12.39}$$

So one can obtain

$$\begin{aligned} \dot{x} &= Ax(t) + Bv^*(t) + B_o o(t), \\ w(t) &= Cx(t) + D_o o(t), \\ v(t) &= -w(t-\tau), \\ v^*(t) &= v(t) + \hat{q}(t), \end{aligned} \tag{12.40}$$

with the initial conditions (12.35). This system can be transformed to

$$\begin{aligned} \dot{x} &= Ax(t) + Bv(t) + B\hat{q}(t) + B_o o(t), \\ w(t) &= Cx(t) + D_o o(t), \\ v(t) &= -w(t-\tau), \end{aligned} \tag{12.41}$$

and further by $\hat{o}(t) = (o(t)^T\, q(t))^T$ and $\hat{B}_o = (B_o\; B)$ to

$$\begin{aligned} \dot{x} &= Ax(t) + Bv(t) + \hat{B}_o\hat{o}(t), \\ w(t) &= Cx(t) + D_o o(t), \\ v(t) &= -w(t-\tau). \end{aligned} \tag{12.42}$$

Classic continuous-time stability analysis methods can be applied to this equivalent system. If this continuous-time system is stable for any bounded external signal $\hat{o}(t)$, then the event-based system is stable, too. If the event-based system is unstable then also the continuous-time system is unstable. Thus, symmetric send-on-delta sampling does not degrade the system stability compared with continuous-time control.

This result may sound surprising as the Nyquist–Shannon sampling theorem [25,28] does not hold for event-based sampling. It is also surprising because of the limit cycles which look like impending instability. In fact, limit cycles may reduce the control performance, but they have always a finite amplitude and never bring the system to instability, if the equivalent continuous-time control loop is stable. Anyhow, as limit cycles are undesirable in most cases, there are also a lot of theoretical studies on conditions for the existence of limit cycles [6–8,10,12,13,70].

The main conclusions of this section are:

1. It is possible to use continuous-time controller design methods to find controller parameters that guarantee stability of the event-based closed-loop system, at least when assuming SSOD.
2. The threshold $\Delta_S$ does not influence the stability of the system and can thus be chosen flexibly to define the trade-off between control performance and message rate.

Nevertheless, it must be noted here that the presented analysis bases on *ideal* send-on-delta sampling while for *real* send-on-delta sampling the used assumptions do not hold. Further, the controller is modeled as a continuous-time system, which does not fit to digital implementations and attenuates practical issues such as sticking and approximation errors. To the authors' knowledge, up to now there are no publications about stability analyses with *real* send-on-delta sampling.

### 12.6.2 Tuning Rules for PI Controllers with Send-on-Delta Sampling

In the first publications on event-based control, beginning with Årzen in 1999 [4], it was common to presuppose a well-tuned discrete-time (periodic) or continuous-time control loop and replace only the sampling scheme without changing the controller parameters. The main challenge of these publications was to identify and solve the new problems which arise from the event-based sampling, such as limit cycles and sticking. This often resulted in evaluating the quality of an event-based control loop by comparing it with the discrete-time or continuous-time counterpart and trying to minimize the differences (in terms of control performance and trajectories).

Appropriate tuning of the PID parameters, $K_P$, $T_I$, and $T_D$, can help improve the trade-off between the mean message rate and control performance as well as avoid limit cycles. While stability proofs and analyses on limit cycles are very important for controller design, they state nothing about how to tune the parameters inside the allowed range.

Vasyutynskyy suggested to change the parameter settings in "steady state" to avoid limit cycles [60,63] as the probability of limit cycles and sticking (and thus reduced control performance and higher message rate) rises with growing aggressiveness of the controller tuning. Araújo gave an example where limit cycles disappeared after reducing the proportional coefficient $K_P$, leading to improved control performance and message rate [3]. The problem of systematic PID parameter tuning for event-based sampling has been explored in parallel by Beschi et al. [8,11] and the authors [22,23]. Later Ruiz et al. [44] proposed a tuning rule for a PI-P controller based on the Smith predictor.

This section assumes that the send-on-delta sampling is applied to the process output. The tuning rule does not consider the optimization of the message rate for the other signals occurring in the control loop.

#### *Generic concept for evaluating the sensor energy efficiency of event-based control loops*

In [22], a generic approach for estimating the message number $N$ of a closed-loop step response after a setpoint change has been presented. This approach is the basis for the subsequent controller tuning rules. It is based on *ideal* send-on-delta sampling ($T_z \to 0$) and the assumption that $\Delta_S$ is sufficiently small in comparison to the setpoint change $\Delta_r$ (the difference between the setpoint before and after the step) that causes the step response ($\Delta_S \ll \Delta_r$).

The message rate in "steady state" is mainly determined by limit cycles and the type of disturbances. It is therefore more complex to analyze. For simplicity of derived equations, it is assumed here that the number of messages created due to limit cycles is negligible compared with the messages created during the step response. Tuning proposals with respect to disturbance rejection and limit cycles can be found in [11,23].

If the step response is free of overshoot and does not oscillate (typical for sluggish step responses), the message number of the step response can simply be approximated by

$$N \approx \frac{\Delta_r}{\Delta_S}. \tag{12.43}$$

The exact value of $N$ depends on the position of the levels relative to the setpoint values before and after the step and is bounded by

$$\left\lfloor \frac{\Delta_r}{\Delta_S} \right\rfloor \leq N \leq \left\lfloor \frac{\Delta_r}{\Delta_S} \right\rfloor + 1. \tag{12.44}$$

Thus, the more exact the approximation (12.43), the smaller the threshold $\Delta_S$, compared with the setpoint change $\Delta_r$.

As shown in Figure 12.4, it is possible to divide a typical step response with oscillations (with overshoot) in different phases—first the rising edge $P_0$ until the setpoint is reached, and then the "half cycles" $P_i$ with $i \in \mathbb{N}_+$ $(i > 0)$. For $P_0$ the number of messages can simply be estimated by

$$N(P_0) \approx \frac{\Delta_r}{\Delta_S}. \tag{12.45}$$

The other parts can be approximated by

$$N(P_i) \approx 2 \cdot \frac{|\Delta h_{a,i}|}{\Delta_S}, \tag{12.46}$$

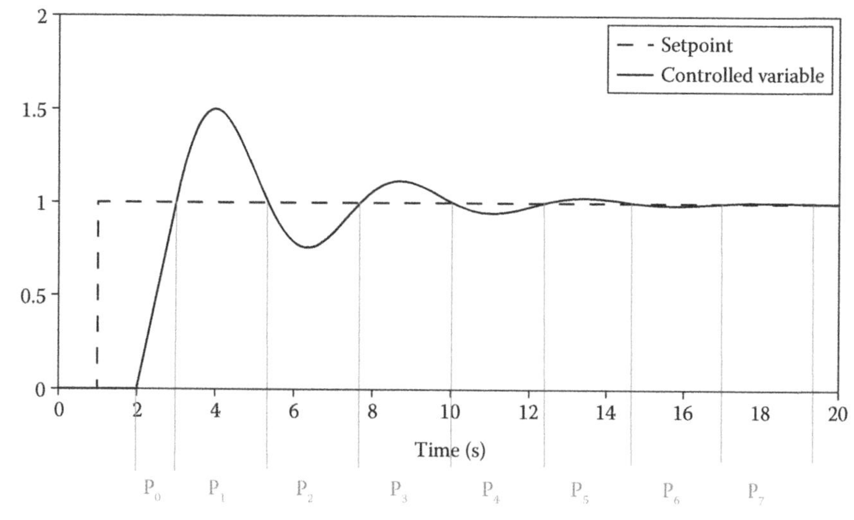


**FIGURE 12.4**
Closed-loop step response, divided into parts between the crossings with the setpoint.

where $\Delta h_{a,i} = y_{\text{extr,i}} - r_{\text{new}}$ is the absolute overshoot (the distance of each extremum $y_{\text{extr,i}}$ from the new setpoint and steady state value $r_{\text{new}}$).

For simplicity, it is assumed here that the decay ratio

$$\delta_{dr} = -\frac{\Delta h_{a,(i+1)}}{\Delta h_{a,i}}, \tag{12.47}$$

is independent of $i$. Thus,

$$\Delta h_{a,i} = \Delta h_{a,1} \cdot (-\delta_{dr})^{i-1}. \tag{12.48}$$

For PI controllers with long reset time (i.e., much proportional action and little integral action), the oscillations are usually not symmetrical around the new setpoint $r_{new}$, but the approach can be applied there, too, by modeling the step response as the superposition of a "virtual" nonoscillating step response with (12.43) and the oscillations relative to that step response with (12.46).

The relative overshoot can be defined as

$$h_r = \frac{h_{a,1}}{\Delta_r}. \tag{12.49}$$

Since for stable control loops

$$0 < \delta_{dr} < 1, \tag{12.50}$$

holds, it is possible to approximate the overall number of messages for the step response for $\Delta_S \ll \Delta_r$ as

$$N = \sum_{i=0}^{\infty} N(\mathrm{P}_i) \tag{12.51}$$

$$\approx \frac{\Delta_r}{\Delta_S} + \sum_{i=1}^{\infty} \left(2 \cdot \frac{\Delta h_{a,1}}{\Delta_S} \cdot \delta_{dr}^{i-1}\right)$$

$$= \frac{\Delta_r}{\Delta_S} \cdot \nu(h_r, \delta_{dr}), \tag{12.52}$$

with the introduction of the function

$$\nu(h_r, \delta_{dr}) := 1 + \frac{2 \cdot h_r}{1 - \delta_{dr}}, \tag{12.53}$$

which does not depend on $\Delta_S$ and $\Delta_r$.

It is important to note that the number of generated messages rises with growing overshoot and decreasing decay ratio, leading to the controller design goal that overshoot (and oscillations in general) should be as small as possible for reducing the message rate. This result is not limited to PID controllers but can be used for any type of controller using send-on-delta sampling.

For *real* send-on-delta sampling ($T_z > 0$), the message number is usually lower due to the minimum duration between two subsequent messages, especially in the rising edge $\mathrm{P}_0$. Thus, ideally, a step response should have a very steep flank at the beginning, but no oscillations. This would optimize both message rate and control performance.

### First tuning rule

In the following section, a tuning rule will be developed. It will use only proportional and integral action for several reasons. Although derivative action can improve the overall control performance of many control loops as shown by Ziegler and Nichols [74], derivative action is not often used in industry [5]. Two reasons for this fact are that derivative action amplifies high-frequency noise and that it leads to abrupt changes of the control error (e.g., after setpoint changes) to large peaks of the control signal which in many cases speeds up the wear of actuators. Both problems can be reduced (at the cost of lower effectiveness of the derivative action) by introducing a filter on the control error such as in

Equation 12.3. The large peaks after setpoint changes can also be avoided by two-degree-of-freedom approaches (i.e., setpoint weighting, see Section 12.5.3). Furthermore, derivative action increases the design effort as there is one parameter more to adjust. That makes the space of possible controller settings three dimensional.

Besides the reasons mentioned before, an additional drawback of derivative action in case of event-based sampling is that the frequency of limit cycles rises and hence the message rate.

**Assumptions of the tuning rule** For the development of the tuning rule, a process model of the form

$$G(s) = \frac{K_m}{1 + T_m s} e^{-s\tau}, \quad (12.54)$$

is assumed, with proportional action coefficient $K_m$, first order time constant $T_m$, and dead time (or time delay) $\tau$. This is a quite typical assumption [6,33,34,44]. The process type (12.54) is called FOPDT (first order plus dead time) [6], FOPTD, or FOTD (first order plus time delay) [3,13], FOLPD (first order lag plus time delay) [34], or KLT model [3]. The quotient

$$\eta = \frac{\tau}{T_m}, \quad (12.55)$$

is called "degree of difficulty" [23,51], "normalized time delay" [43], or "normalized dead time" [8]. For this process structure, there exist simple parameter identification methods [72] and a lot of PI and PID controller tuning rules [34], optimized for continuous-time control or periodic sampling.

A continuous-time PI controller $R(s)$ is used for the first analyses because this allows to abstract from the complex event-based effects. The continuous-time results will be transferred afterward to event-based controllers. For this combination of process model and controller type, there are a lot of PI controller tuning rules [34] of the form

$$K_P = \lambda \cdot \frac{T_m}{K_m \cdot \tau}, \quad (12.56)$$

$$T_I = \epsilon \cdot T_m, \quad (12.57)$$

with the process-independent constants $\lambda$ and $\epsilon$ which are given by the particular tuning rule. Since $K_P \propto \lambda$ and $T_I \propto \epsilon$, Equations 12.56 and 12.57 can be interpreted as a normalization of $K_P$ and $T_I$, respectively. According to this interpretation, each tuning rule specifies only the selection of $\lambda$ and $\epsilon$.

The open-loop transfer function becomes

$$\begin{aligned} G_o(s) &= R(s) \cdot G(s) \\ &= K_P \left(1 + \frac{1}{T_I s}\right) \cdot \frac{K_m}{1 + T_m s} e^{-s\tau} \\ &= \lambda \cdot \frac{T_m}{K_m \tau} \cdot \left(1 + \frac{1}{\epsilon \cdot T_m s}\right) \cdot \frac{K_m}{1 + T_m s} e^{-s\tau} \\ &= \frac{\lambda \cdot \left(1 + \epsilon \cdot \frac{\tau}{\eta} s\right)}{\tau \cdot s \cdot \epsilon \cdot \left(1 + \frac{\tau}{\eta} s\right)} e^{-s\tau}. \end{aligned}$$

This result contains (for constant $\lambda$, $\epsilon$, and $\eta$) only one time constant $\tau$, which can be replaced by $T_m$ or $T_I$. Changing that time constant leads only to a proportional change of the time constants of the closed-loop system, but not to a qualitative change of the system evolution in terms of overshoot, oscillations, gain or phase margins, etc. Thus, for qualitative analyses, it is enough to examine the control loop properties as a function of $\eta$ and the controller factors $\epsilon$ and $\lambda$ while it is not necessary to consider the other process parameters ($K_m$ and $T_m$).

This common approach in continuous-time systems is also applied to develop a tuning rule for event-based systems. Without loss of generality, the tuning rule which is developed below optimizes $\lambda$ and $\epsilon$, as this allows more generic results because their optimal setting depends only on $\eta$.

**Characteristics of the PI controller parameters** Figure 12.5 shows the principal influence of the PI parameters on the closed-loop step response of a continuous-time control loop with a FOLPD process. Summarizing, with increasing $\lambda$ (i.e., $K_P$) and decreasing $\epsilon$ (i.e., $T_I$) the step responses become steeper and get more significant oscillations, up to instability. The main difference between both is that large integral action (small $\epsilon$) leads to smoother signal evolutions than large proportional action. For controller tuning, it is helpful to divide the controller settings for sluggish response without overshoot, response with overshoot, and unstable response. Furthermore, there is also a relatively small area of settings for which the control loop is oscillating without overshoot, typical for nearly pure proportional control (see $T_I = 5$ in Figure 12.5b).

These typical characteristics are qualitatively shown in Figure 12.6 as a contour plot. The borders of each area depend on $\eta$. Spinner et al. [56] used comparable diagrams to explain iterative returning approaches for PI controllers.

As concluded from the estimation of the number of messages in Equation 12.53, the settings with no or only little oscillations can be expected to result in lower message rate. If it is desired to maximize the control performance without getting oscillations into the step

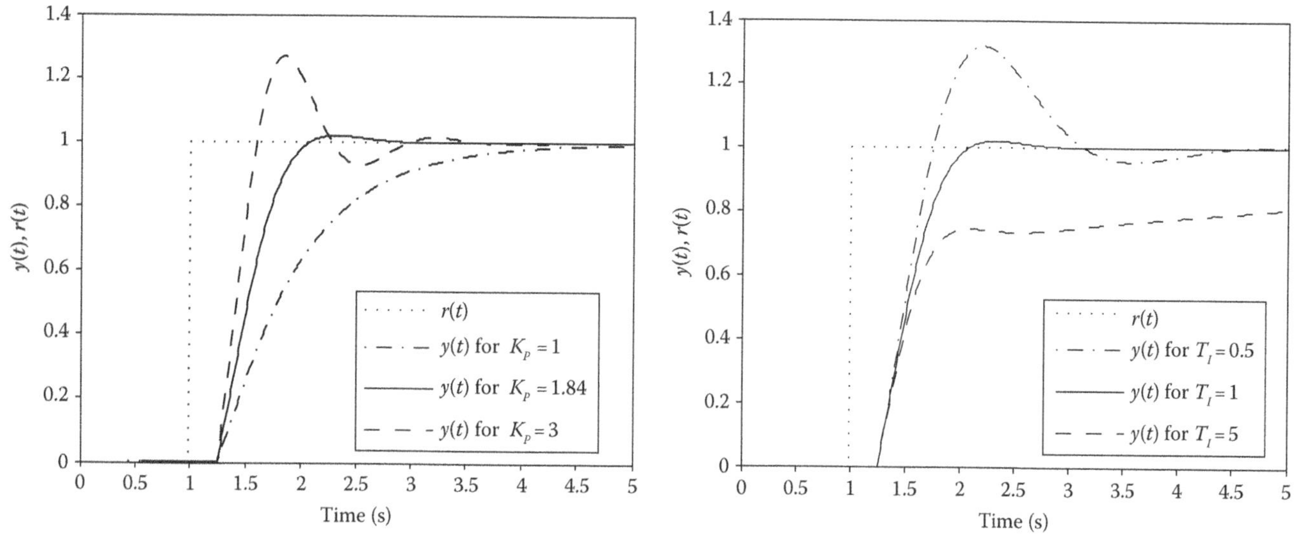


**FIGURE 12.5**
Comparison of closed-loop step responses for different settings of $K_P$ and $T_I$, respectively. The controlled variable $y(t)$ and the reference variable $r(t)$ are shown. The process parameters are $K_m = 1$, $T_m = 1$, $\tau = 0.25$. (a) Different $K_P$ for $T_I = 1$. (b) Different $T_I$ and setpoint $r(t)$ for $K_P = 1.84$.

response, the optimum must lie somewhere near the border between sluggish and oscillating step responses. If the control performance is evaluated using $J_{ISE}$, the optimum will be near to the setting with least $J_{ISE}$ under the constraint of no oscillations. This optimum is shown in each subfigure of Figure 12.6 in the form of a triangle.

If using the product $J_{Prod}$ of message number $N$ and $J_{ISE}$ as the optimization goal as defined in (12.17), the message number must be replaced with an estimation since the simulation uses a continuous-time controller. From (12.52) one can take over the approximation

$$N \approx \frac{\Delta_r}{\Delta_S} \cdot \nu. \tag{12.58}$$

This does not change the location of the optimum with respect to $\lambda$ and $\eta$ since

$$J_\nu = \nu \cdot J_{ISE} \approx \frac{\Delta_S}{\Delta_r} \cdot J_{Prod} \propto J_{Prod}. \tag{12.59}$$

For the computation of the normalized message number $\nu$, there are two main possibilities: First, it can be approximated by (12.53), or it can be approximated for $\Delta_S \ll \Delta_r$ by

$$\nu \approx \frac{1}{\Delta_r} \int_0^{T_{end}} \left| \frac{dy(\xi)}{dt} \right| d\xi, \tag{12.60}$$

what is more exact as it depends on less assumptions. When using (12.60), the optimum regarding $J_\nu$ is shown in Figure 12.6 as a square for each $\eta$.

As stated in Section 12.3.2, the Pareto optimal controller settings are very interesting, because for all *other* settings there is at least one Pareto optimal setting which improves one of both performance indices ($\nu$ or $J_{ISE}$) without degrading the other one. If the Pareto optimal points are known for a given process, all other settings are not interesting. Figure 12.6 shows the Pareto optimal region for several $\eta$. The region starts at settings with $\lambda = 0$ as there is no controller action and thus also no change of the controlled variable, resulting in $\nu = 0$ (i.e., $N = 0$). These settings are practically not reasonable, just like the sluggish settings in the upper left corner with little $\lambda$ and large $\epsilon$ which did not reach the setpoint in the simulated time and produced thus only a few messages but gave better control performance than the settings with $\lambda = 0$. These solutions would not be Pareto optimal, if the simulated time was infinitely long, because then the step responses would reach the setpoint and so the settings would have no advantage regarding $J_{ISE}$ and $\nu$ compared with the settings with faster step responses without overshoot. The practically interesting region begins at the border to settings with overshoot. Beginning there, the relevant Pareto optimal points lie on a nearly linear area beginning at oscillation-free settings (with low message rate) and ending at the point of minimum $J_{ISE}$ (marked with a plus symbol) for larger $\lambda$ and $\epsilon$ at the cost of more oscillations and thus messages.

It is possible to store the Pareto optimal settings for each $\eta$ in a look-up table and use that for controller design in a computer-aided controller design system. Since rules of thumb are very popular in practice, we give a simple linear approximation for the region $0.1 \leq \eta \leq 0.3$ which is interesting for many practical processes (e.g., for room temperature control [51]). For that purpose, a weighting factor $0 \leq \psi \leq 1$ is introduced which

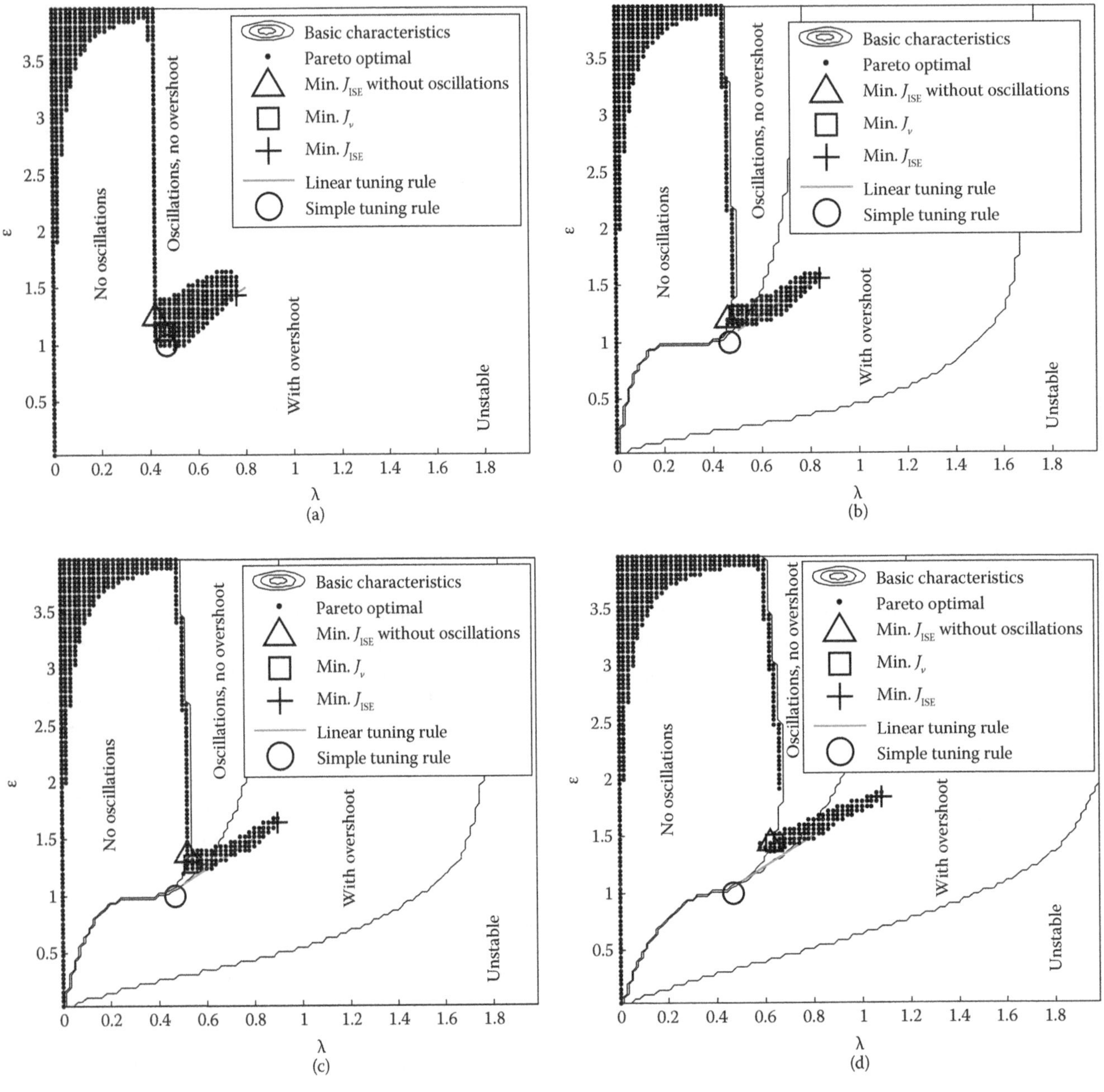


**FIGURE 12.6**
Characteristic areas as a function of $\lambda$ and $\epsilon$ for an FOLPD process. Marked are the Pareto optimal settings (dots), the point of lowest ISE without oscillations (triangle), the optimal point according to $J_\nu$ (square), the optimal point regarding $J_{ISE}$ (also with oscillations), the linear tuning rule (12.61), and the optimum according to the simple tuning rule (12.63) (circle). Analyzed by simulations with step width 0.02 and 0.04 for $\lambda$ and $\epsilon$, respectively. (a) $\eta = 0.1$. (b) $\eta = 0.3$. (c) $\eta = 0.5$, and (d) $\eta = 1$.

allows a stepless tuning between low message rate ($\psi = 0$) and high control performance $J_{ISE}$ ($\psi = 1$). The set of reasonable Pareto optimum points between them can then roughly be approximated by the following rule of thumb:

$$\begin{aligned} \lambda &= 0.5 + 0.3 \cdot \psi, \\ \epsilon &= 1.1 + 0.4 \cdot \psi. \end{aligned} \qquad (12.61)$$

These principal characteristics hold for discrete-time PI controllers, too. For event-based PI controllers, these characteristics are in principal similar if $\Delta_S \ll \Delta_r$.

#### *Simplified tuning rule*

A simpler tuning rule has been published in [22]. It is based on classic tuning rules which are targeted on little overshoot and set $\epsilon$ usually to 1 [22,34], resulting in pole-zero cancellation. The open-loop transfer function is then

$$\begin{aligned} G_o(s) &= R(s) \cdot G(s) \\ &= K_P \left(1 + \frac{1}{sT_I}\right) \cdot \frac{K_m}{1+sT_m} e^{-s\tau} \\ &= \lambda \cdot \frac{1}{\tau s} e^{-s\tau}. \end{aligned}$$

Again, $\tau$ influences only the time constants of the signal evolution but not qualitative properties such as stability, overshoot, or gain and phase margin. Here, the control loop is even independent of $\eta$. Different performance measures can thus be analyzed only as a function of $\lambda$. This is shown in Figure 12.7 for continuous-time simulations of closed-loop unit step responses without disturbances.

All settings between $0.38 \leq \lambda \leq 0.74$ are Pareto optimal regarding $\nu$ and $J_{\text{ISE}}$; other settings make no practical sense as settings with $\lambda < 0.38$ are very sluggish and settings with $\lambda > 0.74$ are very aggressive (oscillatory). Comparable to (12.61), by introducing a weighting factor $0 \leq \psi \leq 1$, the Pareto optimal points are in the area of

$$\begin{aligned} \lambda &= 0.38 + 0.36 \cdot \psi, \\ \epsilon &= 1. \end{aligned} \tag{12.62}$$

The minimum regarding $J_\nu$ is reached with $\lambda_{\text{opt}} \approx 0.468$, showing an overshoot of $h_r \approx 2.17\%$. In comparison, the minimum of the ISE is reached with $\lambda \approx 0.74$ with 24.37% overshoot. That shows that the optimum for control performance (the primary goal in discrete-time or continuous-time control) results not in a low message rate. In particular, the message rate can be reduced by 36 % if one accepts a control performance loss of only 6.8 %. This case results in the simplified tuning rule for optimizing $J_\nu$

$$\begin{aligned} \lambda &= 0.468, \\ \epsilon &= 1. \end{aligned} \tag{12.63}$$

This setting is also indicated in each subfigure of Figure 12.6 as a circle. These figures show that this is not a Pareto optimal solution although it is Pareto optimal among all solutions with $\eta = 1$.

Additional simulation results of unit step responses for event-based control loops with different settings of $\Delta_S$ using that rule are shown in [22]. The differences between the continuous-time approximation and the event-based simulation depend on $\Delta_S$ and the simulation conditions. They especially depend on the duration of the simulated "steady state" phase after the step as this phase contains limit cycles which are not modeled by the continuous-time approximation.

Also, simulations for *real* send-on-delta sampling ($T_z > 0$) can be found in [22]. If either actuator saturation is likely to appear or the frequency of limit cycles should be diminished, $K_P$ should be reduced [22,23].

Beschi et al. developed a comparable tuning rule for SSOD in parallel [8,11]. The goal of their tuning rule is the minimization of the settling time of the system step response. Also, a PI controller is used and a FOPDT process is assumed.

After transforming their results to the notation of this chapter, one can obtain

$$\begin{aligned} \lambda(\eta) &= K_1(\eta) \cdot \eta, \\ \epsilon(\eta) &= \frac{K_1(\eta)}{K_2(\eta)}, \end{aligned} \tag{12.64}$$

where $K_1(\eta)$ and $K_2(\eta)$ are polynomials whose coefficients have been found by interpolation.

The results for $\eta > 0.1$ are quite similar to the simple tuning rule (12.63), especially for $0.1 \leq \eta \leq 0.3$, see Figure 12.8. In fact, minimizing the settling time results

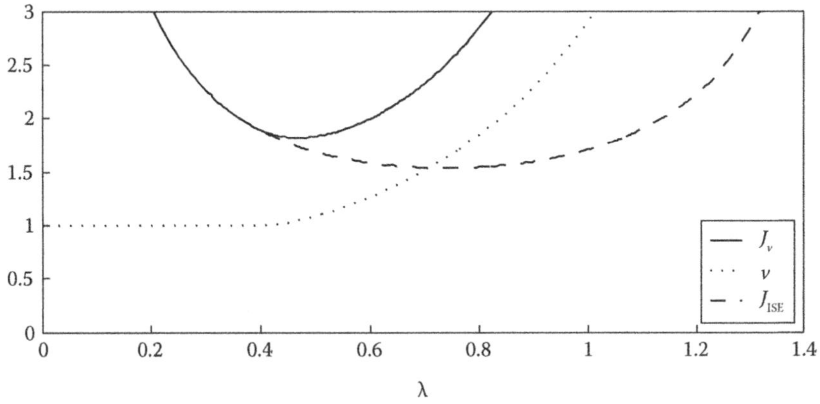


**FIGURE 12.7**
Properties of the closed-loop unit step response as a function of $\lambda$ for $K_m = T_m = \tau = 1$ and continuous-time control.

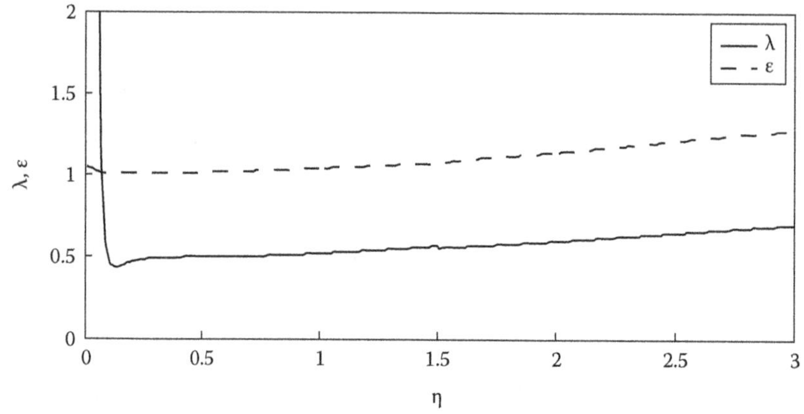


**FIGURE 12.8**
Coefficients $\lambda(\eta)$ and $\epsilon(\eta)$ computed from the tuning rule in [8].

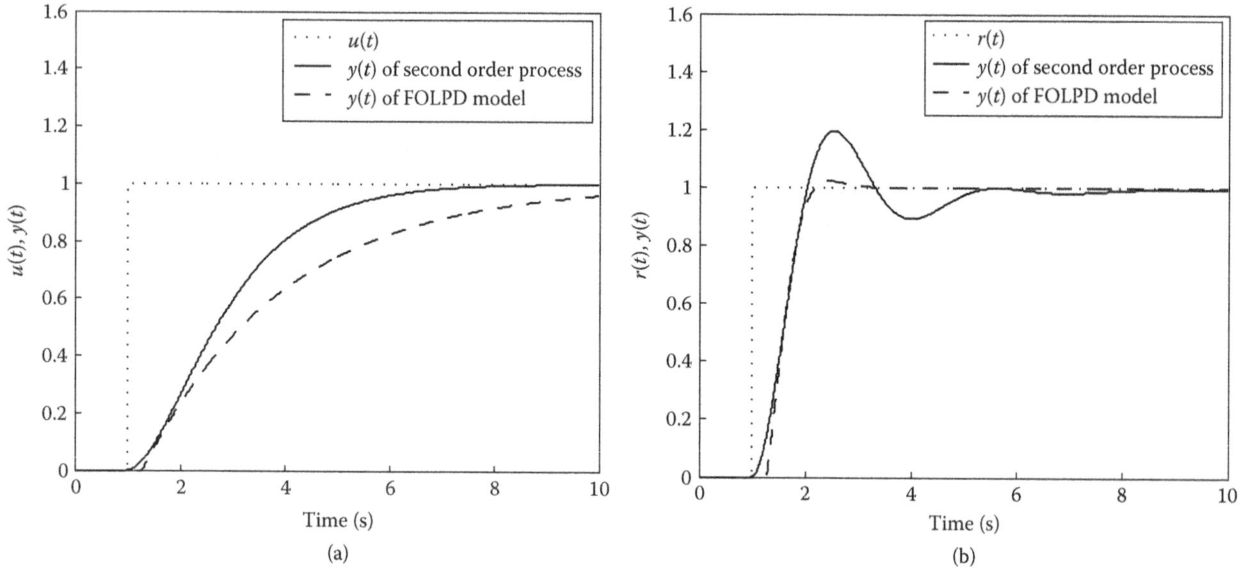


**FIGURE 12.9**
Open-loop and closed-loop step responses of a second-order process ($c = 1$) and its FOLPD model. Closed-loop step responses with $\lambda = 0.468$ and $\epsilon = 1$. (a) Open loop step responses. (b) Closed loop step responses.

in fast step responses without large overshoot, as it is also when $J_{\text{Prod}}$ is minimized.

Beschi et al. detected furthermore that the message number is more or less independent from $\eta$ but depends on the overshoot. That fits to (12.53).

### *Higher order processes*

A second order, nonoscillating process with time delay (SOPDT) can be described with the transfer function

$$G_2(s) = \frac{K_{m2}}{(1+sT_2)(1+scT_2)}e^{-s\tau_2}, \tag{12.65}$$

with $0 \leq c \leq 1$ ($T_2$ is assumed to be the larger of both time constants). There are several possibilities to approximate a SOPDT process by a FOPDT process [29,53]

$$\hat{G}_2(s) = \frac{K_m}{1+sT_m}e^{-s\tau}. \tag{12.66}$$

However, it is not the same controlling a FOPDT process or a SOPDT process, even if the latter has a FOPDT approximation with equal parameters. This has its origin in the different signal evolution of the step responses due to the different frequency response. A SOPDT process will in general result in more oscillations. If both time constants of the SOPDT process are equal ($c = 1$), the difference is most obvious (see Figure 12.9) while if one time constant of both is much smaller than the other ($c \ll 1$), the difference disappears. Figure 12.10 shows

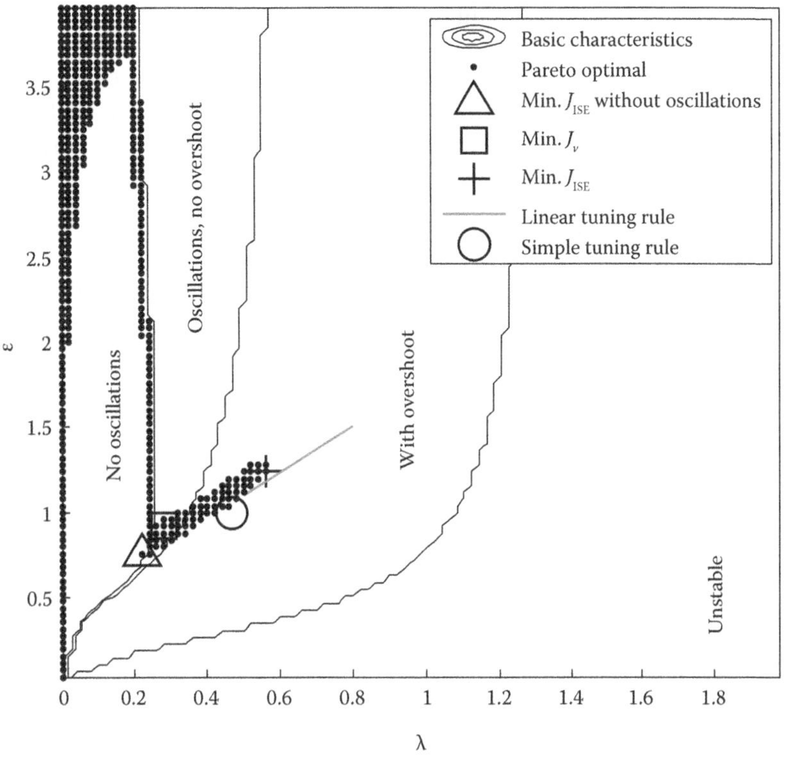


**FIGURE 12.10**
Characteristic areas as a function of $\lambda$ and $\epsilon$ for an SOLPD process (with two equal time constants, i.e., $c = 1$), which can be approximated by an FOLPD process with $\eta = 0.3$. For details, see caption of Figure 12.6.

the Pareto optimal solutions for a SOPDT process if the controller is tuned using (12.56) with the parameters from the FOPDT approximation according to [29]. It is evident that the Pareto front is significantly distanced from that of the "real" FOPDT with the same parameters (shown in Figure 12.6b). Both $\lambda$ and $\epsilon$ should be lower than for a "real" FOPDT process.

If the real process order is known, this problem can be solved by using different tuning rules for different process orders. Unfortunately, under noisy conditions, the estimation of the process order is error-prone. If the process order or $c$ is not known, the tuning rules (12.61) and (12.62) lose their meaning because the tuning parameter $\psi$ will not balance between minimum message rate and optimum control performance. In that case, the simplified tuning rule (12.63) can be used as it delivers acceptable results both for FOPDT and SOPDT processes.

## 12.7 Conclusions and Open Questions

This chapter gave an overview about the current knowledge about event-based PID control. PID control is very well accepted in industry and is thus worth exploring in event-based control. We presented typical problems of such control loops (approximation errors, sticking, and limit cycles) with their currently most accepted solutions, a comparison of several event types, parametrization, and finally some stability and tuning considerations with regard to send-on-delta sampling.

As the research in event-based PID control is a relatively young discipline, there are many unsolved questions. From a theoretical point of view, stability analyses should become more generalized. From a practical point of view, the most important problem is that message losses cannot be detected reliably before the max-send-timer expires, especially in the case of model uncertainties and unknown disturbances. This lack of knowledge increases the risk of an unstable control loop up to now [20], what is not acceptable in hard real-time applications. Sánchez et al. mention the lack of specific design or formal analysis methods for a standardized event-based PID control law as a reason as to why event-based PID control is not yet a useful alternative in some specific applications [48]. Also, the lack of results about SIMO and MIMO processes are mentioned there. As PID controllers are often used for purposes with reduced performance requirements but hard cost limits, automatic tuning procedures are of interest. First approaches in this direction can be found in [9,20,23].

For applications where disturbances and setpoints are constant over a long period, avoidance of limit cycles is of great interest, and solutions for that should become both more practical and reliable. Finally, system identification with nonuniform samples is needed for controller tuning if the sampling scheme cannot be changed for process identification. In contrast to available methods for nonuniform sampling that use only the available samples [2,18,32], the implicit knowledge about the possible range of the signal between the samples due to the sampling threshold ($\Delta_S$, $\Delta_I$ or $\Delta_P$) should be used to reduce the uncertainty about the parameters.

## Bibliography


[1] IEC 60050-351 International electrotechnical vocabulary—Part 351: Control Technology, 2006.

[2] S. Ahmed, B. Huang, and S. L. Shah. Process Parameter and Delay Estimation from Nonuniformly Sampled Data. In *Identification of Continuous-Time Models from Sampled Data* (H. Garnier and L. Wang, Eds.), pages 313–337. Springer, London, 2008.

[3] J. Araújo. Design and Implementation of Resource-Aware Wireless Networked Control Systems. *KTH Electrical Engineering*, Stockholm, Sweden, 2011. Licentiate Thesis.

[4] K. E. Årzén. A simple event-based PID controller. In *Proceedings of the 14th IFAC World Congress*, volume 18, pages 423–428, Beijing, China, July 1999.

[5] K. J. Åström and T. Hägglund. The future of PID control. *Control Engineering Practice*, 9:1163–1175, 2001.

[6] M. Beschi, S. Dormido, J. Sánchez, and A. Visioli. Characterization of symmetric send-on-delta PI controllers. *Journal of Process Control*, 22:1930–1945, 2012.

[7] M. Beschi, S. Dormido, J. Sánchez, and A. Visioli. On the stability of an event-based PI controller for FOPDT processes. In *Proceedings of the IFAC Conference on Advances in PID Control*, pages 436–441, Brescia, Italy, March 2012.

[8] M. Beschi, S. Dormido, J. Sánchez, and A. Visioli. Tuning rules for event-based SSOD-PI controllers. In *Proceedings of the 20th Mediterranean Conference on Control & Automation (MED 2012)*, pages 1073–1078, Barcelona, Spain, July 2012.

[9] M. Beschi, S. Dormido, J. Sánchez, and A. Visioli. An automatic tuning procedure for an event-based PI controller. In *Proceedings of the 52nd IEEE Conference on Decision and Control*, pages 7437–7442, Florence, Italy, December 2013.

[10] M. Beschi, S. Dormido, J. Sánchez, and A. Visioli. Stability analysis of symmetric send-on-delta event-based control systems. In *Proceedings of the 2013 American Control Conference (ACC 2013)*, pages 1771–1776, Washington, DC, June 2013.

[11] M. Beschi, S. Dormido, J. Sánchez, and A. Visioli. Tuning of symmetric send-on-delta proportional-integral controllers. *IET Control Theory and Applications*, 8(4):248–259, 2014.

[12] M. Beschi, A. Visioli, S. Dormido, and J. Sánchez. On the presence of equilibrium points in PI control systems with send-on-delta sampling. In *Proceedings of the 50th IEEE Conference on Decision and Control and European Control Conference (CDC-ECC 2011)*, pages 7843–7848, Orlando, FL, December 2011.

[13] J. Chacón, J. Sánchez, A. Visioli, L. Yebra, and S. Dormido. Characterization of limit cycles for self-regulating and integral processes with PI control and send-on-delta sampling. *Journal of Process Control*, 23:826–838, 2013.

[14] S. Durand and N. Marchand. An event-based PID controller with low computational cost. In *Proceedings of the 8th International Conference on Sampling Theory and Applications* (L. Fesquet and B. Torrésani Eds.), Marseille, France, 2009.

[15] S. Durand and N. Marchand. Further results on event-based PID controller. In *Proceedings of the 10th European Control Conference*, pages 1979–1984, Budapest, Hungary, August 2009.

[16] EnOcean. *EnOcean Equipment Profiles EEP 2.6*, June 2014.

[17] EnOcean GmbH, Oberhaching, Germany. *Dolphin Core Description V1.0*, December 2012.

[18] J. Gillberg and L. Ljung. *Frequency Domain Identification of Continuous-Time ARMA Models: Interpolation and Non-Uniform Sampling*. Technical report, Linköpings Universitet, Linköping, Sweden, 2004.

[19] B. Hensel, A. Dementjev, H.-D. Ribbecke, and K. Kabitzsch. Economic and technical influences on feedback controller design. In *Proceedings of the 39th Annual Conference of the IEEE Industrial Electronics Society (IECON)*, pages 3498–3504, Vienna, Austria, November 2013.

[20] B. Hensel and K. Kabitzsch. Adaptive controllers for level-crossing sampling: Conditions and comparison. In *Proceedings of the 39th Annual Conference of the IEEE Industrial Electronics Society (IECON)*, pages 3870–3876, Vienna, Austria, November 2013.

[21] B. Hensel and K. Kabitzsch. The energy benefit of level-crossing sampling including the actuator's energy consumption. In *Proceedings of the 17th IEEE Design, Automation & Test in Europe (DATE 2014) Conference*, pages 1–4, Dresden, Germany, March 2014.

[22] B. Hensel, J. Ploennigs, V. Vasyutynskyy, and K. Kabitzsch. A simple PI controller tuning rule for sensor energy efficiency with level-crossing sampling. In *Proceedings of the 9th International Multi-Conference on Systems, Signals & Devices*, pages 1–6, Chemnitz, Germany, March 2012.

[23] B. Hensel, V. Vasyutynskyy, J. Ploennigs, and K. Kabitzsch. An adaptive PI controller for room temperature control with level-crossing sampling. In *Proceedings of the UKACC International Conference on Control*, pages 197–204, Cardiff, UK, September 2012.

[24] A. Jain and E. Y. Chang. Adaptive sampling for sensor networks. In *Proceedings of the 1st International Workshop on Data Management for Sensor Networks (DMSN 2010)*, pages 10–16, Toronto, Canada, August 2004. ACM.

[25] A. J. Jerri. The Shannon sampling theorem—Its various extensions and applications: A tutorial review. *Proceedings of the IEEE*, 65(11):1565–1596, 1977.

[26] K. Klues, V. Handziski, C. Lu, A. Wolisz, D. Culler, D. Gay, and P. Levis. Integrating concurrency control and energy management in device drivers. In *Proceedings of the 21st ACM SIGOPS Symposium on Operating System Principles (SOSP 2007)*, pages 251–264, ACM, Stevenson, WA, 2007.

[27] E. Kofman and J. H. Braslavsky. Level crossing sampling in feedback stabilization under data-rate constraints. In *Proceedings of the 45th IEEE Conference on Decision and Control*, pages 4423–4428, San Diego, CA, December 2006.

[28] W. S. Levine, Ed. *Control System Fundamentals*. CRC Press, Boca Raton, 2000.

[29] H. Lutz and W. Wendt. *Taschenbuch der Regelungstechnik*, 8th edition. Harri Deutsch, Frankfurt am Main, 2010.

[30] M. Miskowicz. Sampling of signals in energy domain. In *Proceedings of the 10th IEEE International Conference on Emerging Technologies and Factory Automation (ETFA 2005)*, volume 1, pages 263–266, Catana, Italy, September 2005.

[31] M. Miskowicz. Send-on-delta concept: An event-based data reporting strategy. *Sensors*, 6:49–63, 2006.

[32] E. Müller, H. Nobach, and C. Tropea. Model parameter estimation from non-equidistant sampled data sets at low data rates. *The Measurement of Scientific and Technological Activities*, 9:435–441, 1998.

[33] M. Neugebauer and K. Kabitzsch. Sensor lifetime using SendOnDelta. In *Proceedings of INFORMATIK 2004—Informatik verbindet, Beiträge der 34. Jahrestagung der Gesellschaft für Informatik e.V. (GI)*, volume 2, pages 360–364, Ulm, Germany, September 2004.

[34] A. O'Dwyer. *Handbook of PI and PID Controller Tuning Rules*, 2nd edition. Imperial College Press, London, 2006.

[35] P. G. Otanez, J. R. Moyne, and D. M. Tilbury. Using deadbands to reduce communication in networked control systems. In *Proceedings of the American Control Conference 2002*, volume 4, pages 3015–3020, Anchorage, Alaska, 2002.

[36] A. Pawlowski, J. L. Guzman, F. Rodríguez, M. Berenguel, J. Sánchez, and S. Dormido. Simulation of greenhouse climate monitoring and control with wireless sensor network and event-based control. *Sensors*, 9(1):232–252, 2009.

[37] A. Pawlowski, J. L. Guzmán, F. Rodríguez, M. Berenguel, J. Sánchez, and S. Dormido. Study of event-based sampling techniques and their influence on greenhouse climate control with Wireless Sensor Networks. In *Factory Automation* (Javier Silvestre-Blanes, Ed.), pages 289–312, Rijeka, Croatia: InTech, 2010.

[38] J. Ploennigs, M. Neugebauer, and K. Kabitzsch. Diagnosis and consulting for control network performance engineering of CSMA-based networks. *IEEE Transactions on Industrial Informatics*, 4(2): 71–79, 2008.

[39] J. Ploennigs, V. Vasyutynskyy, and K. Kabitzsch. Comparison of energy-efficient sampling methods for WSNs in building automation scenarios. In *Proceedings of the 14th IEEE International Conference on Emerging Technologies and Factory Automation (ETFA 2009)*, pages 1–8, Mallorca, Spain, 2009.

[40] J. Ploennigs, V. Vasyutynskyy, and K. Kabitzsch. Comparative study of energy-efficient sampling approaches for wireless control networks. *IEEE Transactions on Industrial Informatics*, 6(3):416–424, 2010.

[41] J. Ploennigs, V. Vasyutynskyy, M. Neugebauer, and K. Kabitzsch. Gradient-based integral sampling for WSNs in building automation. In *Proceedings of the 6th European Conference on Wireless Sensor Networks (EWSN 2009)*, Cork, Ireland, February 2009.

[42] J. Plönnigs, M. Neugebauer, and K. Kabitzsch. A traffic model for networked devices in the building automation. In *Proceedings of the 5th IEEE International Workshop on Factory Communication Systems*, pages 137–145, Vienna, Austria, September 2004.

[43] K. J. Åström and T. Hägglund. *Advanced PID Control*. ISA – The Instrumentation, Systems, and Automation Society, Research Triangle Park, NC, 2006.

[44] A. Ruiz, J. E. Jiménez, J. Sánchez, and S. Dormido. A practical tuning methodology for event-based PI control. *Journal of Process Control*, 24:278–295, 2014.

[45] T. I. Salsbury. A survey of control technologies in the building automation industry. In *Proceedings of the 16th Triennial World Congress*, pages 90–100, Prague, Czech Republic, July 2005.

[46] J. Sánchez, M. Á. Guarnes, and S. Dormido. On the application of different event-based sampling strategies to the control of a simple industrial process. *Sensors*, 9:6795–6818, 2009.

[47] J. Sánchez, A. Visioli, and S. Dormido. A two-degree-of-freedom PI controller based on events. *Journal of Process Control*, 21:639–651, 2011.

[48] J. Sánchez, A. Visioli, and S. Dormido. Event-based PID Control. In *PID Control in the Third Millennium—Lessons Learned and New Approaches* (R. Vilanova and A. Visioli, Eds.), pages 495–526. Advances in Industrial Control. Springer, London, 2012.

[49] J. H. Sandee, W. P. M. H. Heemels, and P. P. J. van den Bosch. Event-driven control as an opportunity in the multidisciplinary development of embedded controllers. In *Proceedings of the American Control Conference 2005*, volume 3, pages 1776–1781, Portland, OR, June 2005.

[50] S. Santini and K. Römer. An adaptive strategy for quality-based data reduction in wireless sensor networks. In *Proceedings of the 3rd International Conference on Networked Sensing Systems (INSS 2006)*, pages 29–36, Chicago, IL, May–June 2006.

[51] Siemens Switzerland Ltd, Building Technology Group. Control technology. Technical brochure, Reference number 0-91913-en.

[52] F. Sivrikaya and B. Yener. Time synchronization in sensor networks: A survey. *IEEE Network*, 18(4): 45–50, 2004.

[53] S. Skogestad. Simple analytic rules for model reduction and PID controller tuning. *Journal of Process Control*, 13:291–309, 2003.

[54] S. Soucek and T. Sauter. Quality of service concerns in IP-based control systems. *IEEE Transactions on Industrial Electronics*, 51(6):1249–1258, 2004.

[55] Spartan Peripheral Devices. *Energy Harvesting Self Powered ME8430 Globe Terminal Unit Control Valve*. Application Sheet. Spartan Peripheral Devices, Vaudreuil, Quebec, 2014.

[56] T. Spinner, B. Srinivasan, and R. Rengaswamy. Data-based automated diagnosis and iterative retuning of proportional-integral (PI) controllers. *Control Engineering Practice*, 29:23–41, 2014.

[57] Y. S. Suh. Send-on-delta sensor data transmission with a linear predictor. *Sensors*, 7:537–547, 2007.

[58] K. K. Tan, Q.-G. Wang, C. C. Hang, and T. J. Hägglund. *Advances in PID Control*. Advances in Industrial Control. Springer, London, 1999.

[59] U. Tiberi, J. Araújo, and K. H. Johansson. On event-based PI control of first-order processes. In *Proceedings of the IFAC Conference on Advances in PID Control*, pages 448–453, Brescia, Italy, March 2012.

[60] V. Vasyutynskyy. Send-on-Delta-Abtastung in PID-Regelungen [Send-on-Delta Sampling in PID Controls]. PhD thesis, Vogt Verlag, Dresden, 246 pp., 2009.

[61] V. Vasyutynskyy and K. Kabitzsch. Implementation of PID controller with send-on-delta sampling. In *Proceedings of the International Control Conference 2006*, Glasgow, August/September 2006.

[62] V. Vasyutynskyy and K. Kabitzsch. Deadband sampling in PID control. In *Proceedings of the 5th IEEE International Conference on Industrial Informatics 2007 (INDIN 2007)*, pages 45–50, Vienna, Austria, July 2007.

[63] V. Vasyutynskyy and K. Kabitzsch. Simple PID control algorithm adapted to deadband sampling. In *Proceedings of the 12th IEEE Conference on Emerging Technologies and Factory Automation (ETFA 2007)*, pages 932–940, Patras, Greece, September 2007.

[64] V. Vasyutynskyy and K. Kabitzsch. Towards comparison of deadband sampling types. In *Proceedings of the IEEE International Symposium on Industrial Electronics*, pages 2899–2904, Vigo, Spain, June 2007.

[65] V. Vasyutynskyy and K. Kabitzsch. First order observers in event-based PID controllers. In *Proceedings of the 14th IEEE International Conference on Emerging Techonologies and Factory Automation (ETFA 2009)*, pages 1–8, Mallorca, Spain, September 2009.

[66] V. Vasyutynskyy and K. Kabitzsch. Time constraints in PID controls with send-on-delta. In *Proceedings of the 8th IFAC International Conference on Fieldbus Systems and their Applications (FeT 2009)*, pages 48–55, Ansan, Korea, May 2009.

[67] V. Vasyutynskyy and K. Kabitzsch. A comparative study of PID control algorithms adapted to send-on-delta sampling. In *Proceedings of the IEEE International Symposium on Industrial Electronics (ISIE 2010)*, pages 3373–3379, Bari, Italy, July 2010.

[68] V. Vasyutynskyy and K. Kabitzsch. Event-based control: Overview and generic model. In *Proceedings of the 8th IEEE International Workshop on Factory Communication Systems (WFCS 2010)*, pages 271–279, Nancy, France, May 2010.

[69] V. Vasyutynskyy, A. Luntovskyy, and K. Kabitzsch. Two types of adaptive sampling in networked PID control: Time-variant periodic and deadband sampling. In *Proceedings of the 17th International Crimean Conference "Microwave & Telecommunication Technology" (CriMiCo 2007)*, pages 330–331, Sevastopol, Crimea, Ukraine, September 2007.

[70] V. Vasyutynskyy, A. Luntovskyy, and K. Kabitzsch. Limit cycles in PI control loops with absolute deadband sampling. In *Proceedings of the 18th IEEE International Conference on Microwave & Telecommunication Technology (CRIMICO 2008)*, pages 362–363, Sevastopol, Ukraine, September 2008.

[71] R. Vilanova and A. Visioli, Eds. *PID Control in the Third Millennium. Lessons Learned and New Approaches*. Advances in Industrial Control. Springer, London, 2012.

[72] A. Visioli. *Practical PID Control*. Springer, London, 2006.

[73] A. Willig, K. Matheus, and A. Wolisz. Wireless technology in industrial networks. *Proceedings of the IEEE*, 93(6):1130–1151, 2005.

[74] J. G. Ziegler and N. B. Nichols. Optimum settings for automatic controllers. *Transactions of the American Society of Mechanical Engineers*, 64:759–768, 1942.

# 13

# *Time-Periodic State Estimation with Event-Based Measurement Updates*

**Joris Sijs**
*TNO Technical Sciences*
*The Hague, Netherlands*

**Benjamin Noack**
*Karlsruhe Institute of Technology*
*Karlsruhe, Germany*

**Mircea Lazar**
*Eindhoven University of Technology*
*Eindhoven, Netherlands*

**Uwe D. Hanebeck**
*Karlsruhe Institute of Technology*
*Karlsruhe, Germany*

**CONTENTS**



**ABSTRACT** To reduce the amount of data transfers in networked systems, measurements can be taken at an event on the sensor value rather than periodically in time. Yet, this could lead to a divergence of estimation results when only the received measurement values are exploited in a state estimation procedure. A solution to this issue has been found by developing estimators that perform a state update at both the event instants as well as periodically in time: when an event occurs the estimated state is updated using the measurement received, while at periodic instants the update is based on knowledge that the sensor value lies within a bounded subset of the measurement space. Several solutions for event-based state estimation will be presented in this chapter, either based on stochastic representations of random vectors, on deterministic representations of random vectors or on a mixture of the two. All solutions aim to limit the required computational resources by deriving explicit solutions for computing estimation results. Yet, the main achievement for each estimation solution is that stability of the estimation results are (not directly) dependent on the employed event sampling strategy. As such, changing the event sampling strategy does not imply to change the event-based estimator as well. This aspect is also illustrated in a case study of tracking the distribution of a chemical compound effected by wind via a wireless sensor network.

## 13.1 Introduction

Many people in our society manage their daily activities based on knowledge and information about weather conditions, traffic jams, pollution levels, energy consumptions, stock exchange, and so on. Sensor measurements are the main source of information when monitoring these surrounding processes, and a trend is to increase their amount, as they have become smaller, cheaper, and easier to use. See, for example, detailed explanations on the design of "wireless sensor networks" and its applications in [7]. As many more sensor measurements are becoming available, resource limitations of the overall system are gradually becoming the bottleneck for processing measurements automatically. The main reason for this issue is that classical signal processing solutions have been developed in a time where sensor measurements were scarce and where systems were deployed with sufficient resources for running sophisticated algorithms developed. Nowadays, these aspects are shifted, that is, there are far too many sensor measurements available for the limited resources deployed in the overall system and, moreover, it is typically not even necessary to process all sensor information available to achieve a desired performance. An example of resource-aware signal processing is presented in this chapter, where several solutions for state estimation will be presented that all aim to reduce the amount of measurement samples.

More precisely, this chapter presents a recent overview and outlook on state estimation approaches that do not require periodic measurement samples but instead were designed for exploiting a-periodic measurement samples. Some well-known a-periodic sampling strategies, known as *event-based sampling* or *event triggered sampling*, are "send-on-delta" (or Lebesgue sampling) and "integral sampling" proposed in [3,10,22,23]. When such strategies are used in estimation (and control), as will be the case in this chapter, one typically starts with the setup depicted in Figure 13.1. Note that the sensor may require access to estimation results of the estimator, for example, in case the event sampling strategy "matched sampling" of [20] is employed.

Typically, the estimator is part of a larger networked system where the output of the estimator is used by a control algorithm for computing stabilizing control actions. Advances in control theory provided several solutions for coping with the event sampled measurements directly by the controller. See, for example, the solutions on event-based control proposed in [5,8–10,12,14,16,33,37]. Yet, most of the deployed controllers used in current practices run periodically in time. Revisiting those periodic controller into an event-based one is not favorable. Mainly because the infrastructure of the control system is inflexible or because of fear for any down-time of the production facility when new controllers are tested.

The event-based estimators addressed in this chapter can serve as a solution to this problem:

- Event-based sensor measurements are sent to the estimator.
- Estimation results are computed at events *and* periodically in time.
- The time-periodic estimation results are sent to the controller.
- Stabilizing control actions are computed periodically in time.

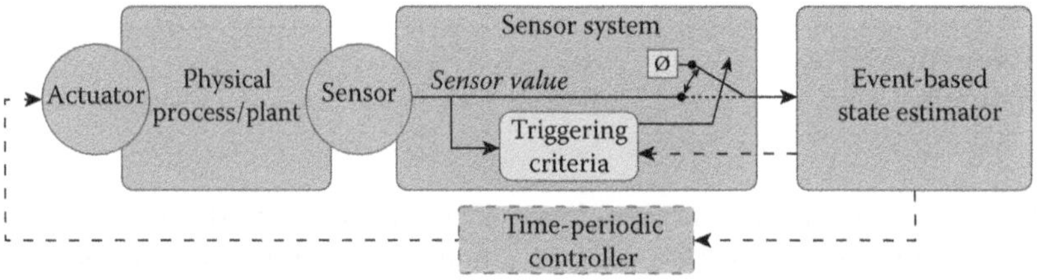


**FIGURE 13.1**
Schematic setup of event-based state estimation (and control). Therein, a sensor system studies the sensor signal on events in line with a predefined "event triggering criteria." In case of the event, the sensor signal is sampled and the corresponding measurement is sent to the estimator. In case of no event, no measurement is sent (represented by an empty set ∅ at the input of the estimator). The estimator exploits the event-triggered sensor measurements to compute time-periodic estimation results, which are possibly used by a time-periodic controller for computing stabilizing control actions.

The advantage of such an approach is twofold. First, the design of a (practical) stabilizing controller becomes independent from the employed event sampling strategy and existing control implementations might even be kept in the loop when changing from time-periodic to event-based measurements. Secondly, one can select a suitable control approach from the extensive set of time-periodic control algorithms. The purpose of this article is to assess existing event-based state estimators for time-periodic observations and apply them in an environmental monitoring application.

## 13.2 Preliminaries

$\mathbb{R}$, $\mathbb{R}_+$, $\mathbb{Z}$, and $\mathbb{Z}_+$ define the sets of real numbers, nonnegative real numbers, integer number, and nonnegative integer numbers, respectively, while, $\mathbb{Z}_{\mathbb{C}} := \mathbb{Z} \cap \mathbb{C}$, for some $\mathbb{C} \subset \mathbb{R}$. An ellipsoidal set $\mathbb{L}_{(\mu,\Sigma)} \subset \mathbb{R}^n$ centered at $\mu \in \mathbb{R}^n$ is characterized as $\mathbb{L}_{\mu,\Sigma} := \{x \in \mathbb{R}^l | (x-\mu)^\top \Sigma^{-1}(x-\mu) \leq 1\}$, for some positive definite $\Sigma \in \mathbb{R}^{n\times n}$. The Minkowski sum of two sets $\mathbb{C}_1, \mathbb{C}_2 \in \mathbb{R}^n$ is denoted as $\mathbb{C}_1 \oplus \mathbb{C}_2 := \{x + y | x \in \mathbb{C}_1, y \in \mathbb{C}_2\}$. The null-matrix and identity-matrix of suitable dimensions (clear from the context) are denoted as 0 and $I$, respectively. For a continuous-time signal $x(t)$, where $t_{\mathsf{k}} \in \mathbb{R}_+$ denotes the time of the k-th sample, let us define $x[\mathsf{k}] := x(t_{\mathsf{k}})$ and $x(t_{0:\mathsf{k}}) := (x(t_0), x(t_1), \cdots, x(t_{\mathsf{k}}))$. The $q$-th element of a vector $x \in \mathbb{R}^n$ is denoted as $\{x\}_q$, while $\{A\}_{qr}$ denotes the element of a matrix $A \in \mathbb{R}^{m\times n}$ in the $q$-th row and $r$-th column. The transpose, determinant, inverse, and trace of a matrix $A \in \mathbb{R}^{n\times n}$ are denoted as $A^\top$, $|A|$, $A^{-1}$, and $\mathrm{tr}(A)$, respectively. The minimum and maximum eigenvalue of a square matrix $A$ are denoted as $\lambda_{\min}(A)$ and $\lambda_{\max}(A)$, respectively. The $p$-norm of a vector $x \in \mathbb{R}^n$ is denoted as $\|x\|_p$.

The Delta-function $\delta : \mathbb{R}^n \to \mathbb{R}_+$ of a vector $x \in \mathbb{R}^n$ vanishes at all values of $x \neq 0$, while it is infinity when $x = 0$ and, moreover, $\int_{-\infty}^{\infty} \delta(x)dx = 1$.

The probability density function (PDF) of a random vector $x \in \mathbb{R}^n$ is denoted as $p(x)$, while $x \sim \mathcal{G}(\mu, \Sigma)$ is a short notation for stating that the PDF of $x$ is a Gaussian function with mean $\mu \in \mathbb{R}^n$ and covariance $\Sigma \in \mathbb{R}^{n\times n}$, that is, $p(x) = \mathcal{G}(x, \mu, \Sigma)$ and $\mathcal{G}(x, \mu, \Sigma) := \frac{1}{\sqrt{(2\pi)^n|\Sigma|}} e^{-0.5(x-\mu)^\top\Sigma^{-1}(x-\mu)}$. Further, for a bounded Borel set $\mathbb{C} \subset \mathbb{R}^n$ [1], the corresponding uniform PDF $\Pi_{\mathbb{C}}(x) : \mathbb{R}^n \to \mathbb{R}_+$ is characterized by $\Pi_{\mathbb{C}}(x) = 0$ if $x \notin \mathbb{C}$ and $\Pi_{\mathbb{C}}(x) = \nu^{-1}$ if $x \in \mathbb{C}$, with $\nu \in \mathbb{R}$ defined as the Lebesgue measure of $\mathbb{C}$ [15].

## 13.3 A Hybrid System Description: Event Measurements into Time-Periodic Estimates

Event-based state estimation with time-periodic estimation results deals with the setup depicted in Figure 13.1, that is, a sensor system forwarding sampled measurements to a state estimator that is possibly connected to a control implementation. Important prerequisites for the event-based state estimator in the schematic set-up of Figure 13.1 are the sampling strategies employed and the physical process considered.

**Event triggered sampling** The sensor system employs an event triggering criteria to generate a next measurement $y \in \mathbb{R}^m$ at the event instants $t_{\mathsf{e}} \in \mathbb{R}_+$, where $\mathsf{e} \in \mathbb{Z}_+$ denotes the e-th event sample. To that extent, let $\mathbb{T}_e \subset \mathbb{R}_+$ be the collection of time instants at all events, that is ,

$$\mathbb{T}_e := \{t_{\mathsf{e}} \in \mathbb{R}^n \mid \mathsf{e} \in \mathbb{Z}_+\}.$$

Further, let us introduce $\tau_{\mathsf{e}} := t_{\mathsf{e}} - t_{\mathsf{e}-1}$ as the sampling interval between two consecutive events e and $\mathsf{e}-1$. A definition of the event instants $t_{\mathsf{e}}$ is given in the next section.

**Time-periodic sampling** The control implementation employs a time-periodic sampling strategy characterized by a constant sampling interval $\tau_s \in \mathbb{R}_{>0}$. To that extent, let $\mathbb{T}_p \subset \mathbb{R}_+$ be the collection of periodic time instants, that is,

$$\mathbb{T}_p := \{\mathsf{n}\tau_s \mid \mathsf{n} \in \mathbb{Z}_+\}.$$

Notice that event instants could coincide with time-periodic instants, that is, $\mathbb{T}_e \cap \mathbb{T}_s$ might be non-empty.

**Physical process** Let us consider an autonomous, linear process discretized in time for various sampling intervals $\tau \in \mathbb{R}_{>0}$:

$$x(t) = A_\tau x(t-\tau) + B_\tau w(t), \tag{13.1}$$

$$y(t) = Cx(t) + v(t). \tag{13.2}$$

Basically, the above description could be perceived as a discretized version of a continuous time state-space model $\dot{x}(t) = Fx(t) + Ew(t)$, where $A_\tau := e^{F\tau}$ and $B_\tau := \int_0^\tau e^{F\eta} d\eta E$. The state vector is denoted as $x \in \mathbb{R}^n$ and both the process noise $w \in \mathbb{R}^l$ and measurement noise $v \in \mathbb{R}^m$ are characterized with a deterministic part and a stochastic part, that is,

$$w(t) := w_d(t) + w_s(t) \quad \text{and} \quad v(t) := v_d(t) + v_s(t).$$

Herein, $w_d(t) \in \mathbb{W}$ and $v_d(t) \in \mathbb{V}$ denote "deterministic," yet unknown, vectors taking values in the sets $\mathbb{W} \in \mathbb{R}^l$ and $\mathbb{V} \in \mathbb{R}^m$. Further, $w_s(t) \in \mathbb{R}^l$ and $v_s(t) \in \mathbb{R}^m$ denote "stochastic" random vectors taking values according to the PDFs $p(w_s(t))$ and $p(v_s(t))$. Exact characterizations of $\mathbb{W}$, $\mathbb{V}$, $p(w_s)$, and $p(v_s)$ vary per estimation approach and will be discussed in the preceding sections. Still, it is important to note that an *unbiased* noise $w(t)$ and $v(t)$ implies that both $\mathbb{W}$ and $\mathbb{V}$ have their center-of-mass collocated with the origin and that both $w_s(t)$ and $v_s(t)$ have an expected value of zero.

The purpose of the state estimator is to exploit the event measurements $y$ to compute an estimate of the state $x$. In line with the above noise characterizations, estimation results of the event-based state estimator are also represented with a deterministic and a stochastic part, that is,

$$x(t) := \hat{x}(t) + \tilde{x}_d(t) + \tilde{x}_s(t).$$

Herein, $\hat{x} \in \mathbb{R}$ is the estimated mean of $x$ and both $\tilde{x}_d, \tilde{x}_s \in \mathbb{R}^n$ are estimation errors, that is, $\tilde{x}_d + \tilde{x}_s = x - \hat{x}$. More precisely, $\tilde{x}_d(t) \in \mathbb{X}$ is a deterministic, yet unknown, vector taking values in the error-set $\mathbb{X}(t) \subset \mathbb{R}^n$ and $\tilde{x}_s(t)$ is a stochastic random vector following the error distribution $p(\tilde{x}_s(t))$.

The challenges addressed in event-based state estimation are as follows:

- Providing *stable* estimation results periodically in time when new measurements are triggered by an event sampling strategy.
- So that stability criteria of the feedback control loop do not depend on the employed sampling strategy, directly.

Such an estimator allows the sensor system to adopt different event sampling strategies depending on, for example, the expected lifetime of its battery. Moreover, a computationally efficient algorithm of the event-based state estimator is desired, to attain applicable solutions.

**REMARK 13.1** Stability of estimation results, wherein $\tilde{x}_d$ and $\tilde{x}_s$ denote the estimation errors, implies a nondivergent error-set $\mathbb{X}(t_k)$ and a nondivergent error-covariance $P(t_k)$. More precisely, there exists a constant value for both limits $\lim_{k\to\infty} \|\mathbb{X}(t_k)\| < \infty$ and $\lim_{k\to\infty} \lambda_{\max}(P(t_k)) < \infty$.

After introducing some examples and properties of event-based sampling, the chapter continues with an overview of existing estimators picking either stochastic or deterministic representations for estimating $x$. In addition, an outlook is presented toward a hybrid state estimator computing the deterministic and the stochastic part of estimation results simultaneously.

## 13.4 Event-Based Sampling

### 13.4.1 Illustrative Examples

Event-based sampling is an a-periodic sampling strategy where events are not triggered periodically in time but at instants of predefined events. Three examples are "Send-on-Delta," "Predictive Sampling," and "Matched sampling", as proposed in [3,10,20,22,29]. These strategies define that triggering the next event sample $e \in \mathbb{Z}_+$ depends on the current measurement $y(t)$ and the previously sampled $y(t_{e-n})$, for one or more $n \in \mathbb{Z}_{>1}$. More precisely, for some design parameter $\Delta(t)$ and for a predicted measurement value $\hat{y}(t)$, they define the instant of a next event $t_e$, as follows:

- Send-on-delta:
  $t_e = \inf\{t > t_{e-1} \mid \|y(t) - y(t_{e-1})\| > \Delta(t)\}.$
- Predictive sampling:
  $t_e = \inf\{t > t_{e-1} \mid \|y(t) - \hat{y}(t)\| > \Delta(t)\}.$
- Matched sampling:
  $t_e = \inf\{t > t_{e-1} \mid D_{\mathrm{KL}}(p_1(x(t))\|p_2(x(t))) > \Delta(t)\}.$

An illustrative impression of the above event sampling strategies is depicted in Figure 13.2. Matched sampling uses the Kullback–Leibler divergence for triggering new events. This divergence, denoted as $D_{\mathrm{KL}}(p_1(x)\|p_2(x)) \in \mathbb{R}_+$, is a *nonsymmetric* measure for the difference of $p_2(x)$ relative to $p_1(x)$. Therein, $p_1(x)$ is considered to be the *updated* (true) PDF of $x$ and $p_2(x)$ is a *prediction* (model) of $p_1(x)$. In line with this reasoning, let $p_2(x(t))$ denote the prediction of $x(t)$ based on the results at $t_{e-1}$, while $p_1(x(t))$ is the update of $p_2(x(t))$ with the sensor value $y(t)$. Then, their Kullback–Leibler divergence can be regarded as a measure for the relevance of $y(t)$ to the estimation results.

**REMARK 13.2** Let the estimation result at $t_{e-1}$ be the Gaussian $p(x(t_{e-1})) = \mathcal{G}(x(t_{e-1}), \hat{x}(t_{e-1}), P(t_{e-1}))$. Further, let the Kullback–Leibler divergence $D_{KL}(p_1(x(t))\|p_2(x(t)))$ correspond to $p_1(x(t)) = \mathcal{G}(x(t), \hat{x}_1(t), P_1(t))$ and $p_2(x(t)) = \mathcal{G}(x(t), \hat{x}_2(t), P_2(t))$. Then, the prediction $p_2(x(t))$ and the update $p_1(x(t))$ can be determined with the process model of (13.1) and an a-periodic Kalman filter, that is,

$$P_2(t) = A_\tau P(t_{e-1})A_\tau^\top + B_\tau \mathrm{cov}(w_s(t))B_\tau^\top;$$
$$\hat{x}_2(t) = A_\tau \hat{x}(t_{e-1});$$
$$P_1(t) = \left(P_2^{-1}(t) + C^\top(\mathrm{cov}(v_s(t)))^{-1}C\right)^{-1};$$
$$\hat{x}_1(t) = P_1^{-1}(t)\left(P_2^{-1}(t)\hat{x}_2(t) + C^\top(\mathrm{cov}(v_s(t)))^{-1}y(t)\right).$$

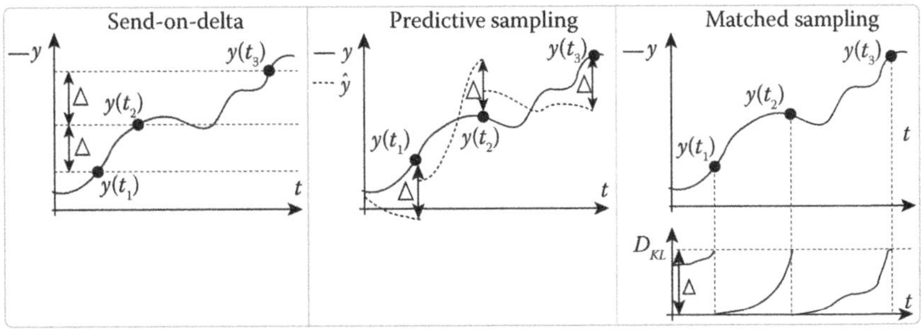


**FIGURE 13.2**
Illustrative impression for triggering event samples via send-on-delta, predictive sampling, and matched sampling. The latter one employs a Kullback–Leibler divergence denoted as $D_{KL}$.

Furthermore, the Kullback–Leibler divergence then yields:

$$D_{\mathrm{KL}}(p_1(x(t))||p_2(x(t))) := \frac{1}{2}\Big(\log|P_2(t)|\,|P_1(t)|^{-1} + \mathrm{tr}\left(P_2^{-1}(t)P_1(t)\right) - n\Big) + \frac{1}{2}\left(\hat{x}_1(t) - \hat{x}_2(t)\right)^{\top} P_2^{-1}(t)(\hat{x}_1(t) - \hat{x}_2(t)\Big).$$

The above examples will be used to derive a property of event sampling that shall be exploited in several estimation approaches presented later.

### 13.4.2 A Sensor Value Property

As new measurement samples are triggered at the instants of well-designed events, not receiving a new measurement sample at the estimator implies that no event has been triggered.* Note that this situation still contains valuable information on the sensor value, as is derived, next.

It was already shown in [30] that the triggering criteria for many of the existing event sampling approaches can be generalized into a set-criteria. To that extent, let us introduce $\mathbb{H}(\mathrm{e}, t) \subset \mathbb{R}^m$ as a set in the measurement space collecting all the sensor values that $y(t)$ is allowed to take so that no event will be triggered. Then, a generalization of event sampling has the following definition for triggering a next event $t_{\mathrm{e}}$, that is,

$$t_{\mathrm{e}} = \min\left\{t > t_{\mathrm{e}-1} | y(t) \notin \mathbb{H}(\mathrm{e}, t)\right\}. \tag{13.3}$$

Then, the sensor value property derived from the above generalization yields

**Proposition 13.1**

Let $y(t)$ be sampled with an event strategy similar to (13.3). Then, $y(t) \in \mathbb{H}(\mathrm{e}, t)$ holds for any $t \in [t_{\mathrm{e}-1}, t_{\mathrm{e}})$.

*Under the assumption that no package loss can occur.

With send-on-delta, the triggering set $\mathbb{H}(\mathrm{e}, t)$ is an $m$-dimensional ball of radius $\Delta$ centered at $y(t_{\mathrm{e}-1})$, for all $t_{\mathrm{e}-1} < t < t_{\mathrm{e}}$, while for predictive sampling the triggering set $\mathbb{H}(\mathrm{e}, t)$ is the same ball but then centered at $\hat{y}(t)$. Similarly, the triggering set $\mathbb{H}(\mathrm{e}, t)$ for matched sampling is the ellipsoidal shaped set $\mathbb{H}(\mathrm{e}, t) = \mathbb{L}_{(\mu(t), \Sigma(t))}$. Values for the center $\mu(t) \in \mathbb{R}^m$ and covariance $\Sigma(t) \in \mathbb{R}^{m\times m}$ defining the boundary of the ellipsoid are found in [20].

Proposition 13.1 formalizes the inherent measurement knowledge of event sampling, that is, *not* receiving a new measurement for any $t \in [t_{\mathrm{e}-1}, t_{\mathrm{e}})$ implies that $y(t)$ is included in $\mathbb{H}(\mathrm{e}, t)$. A characterization of $\mathbb{H}(\mathrm{e}, t)$ can be derived prior to the event instant $t_{\mathrm{e}}$ as the triggering condition is available. It is exactly this set-membership property of event sampling that gives the additional measurement information for updating estimation results.

## 13.5 Existing State Estimators

The challenge in event-based state estimation is an unknown time horizon until the next event occurs, if it even occurs at all. Solutions with a-periodic estimators, for example, [4,19,35], perform a prediction of the state $x$ periodically in time when no measurement is received. It was shown in [31] that this leads to a diverging behavior of the error-covariance $\mathrm{cov}(\tilde{x}_s)$ (unless the triggering condition depends on the error-covariance, as it is shown in [34–36]). The diverging property was proven by assuming Gaussian noise representation and no deterministic noises or state-error representations, that is, $\mathbb{W} = \emptyset$, $\mathbb{V} = \emptyset$, and $\mathbb{X} = \emptyset$. To curtail the runaway error-covariance, some alternative a-periodic estimators were proposed in [13,17] focusing on *when* to send new measurements so to minimize estimation error. Criteria developed for sending a new measurement are set to guarantee stability, implying that

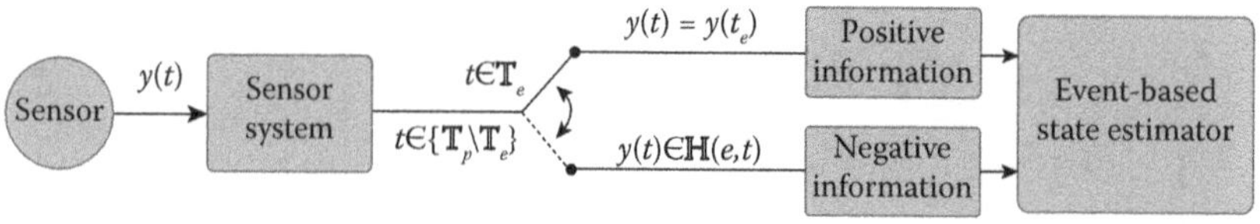


**FIGURE 13.3**
Illustrative setup of a typical event-based state estimator: a new measurement sample $y(t) = y(t_e)$ arrives at the instants of an event $t \in \mathbb{T}_e$, while the measurement information $y(t) \in \mathbb{H}(e, t)$ is implied periodically in time when no event occurs at $t \in \{\mathbb{T}_p \setminus \mathbb{T}_e\}$.

the stability of these estimators directly depends on the employed event triggering condition.

In the considered problem, a sensor is not limited to one event sampling strategy but employed strategies can be replaced depending on the situation at hand. Therefore, guaranteeing stable estimation results under these circumstances calls for a redesign of existing estimators. Key in this redesign is the additional knowledge on the sensor value that becomes available when no new measurement was sampled, that is, Proposition 13.1. An early solution able to exploit this idea was proposed in [25], where the results of Proposition 13.1 are used to perform a state update periodically in time when no event was triggered. However, the setup therein is restricted to the sampling strategy "send-on-delta" and no proof of stability was derived. The remaining three estimators of this overview do give a proof of stability and are not restricted to one specific event sampling strategy.

An overall summary of the three estimation approaches is presented, first, before continuing with more details per approach. Assumptions on the characterization of process noise, measurement noise and estimation results, that is, purely stochastic, purely deterministic or a combination of the two, form the basis for distinguishing the three event-based estimation approaches. Apart from that, all three approaches have an estimation setup as it is depicted in Figure 13.3. The estimator is able to exploit measurements from any event sampling strategy, so that updated estimation results are computed at least periodically in time. An aspect that differs is that the deterministic estimation approach checks the event criteria periodically in time, implying that $\mathbb{T}_e \subset \mathbb{T}_p$ and allowing the deterministic estimator to run periodically as well. With the stochastic and the combined estimation approaches, the event instants can occur at any time, implying that $\mathbb{T}_e$ and $\mathbb{T}_p$ will have (almost) no overlapping time instants enforcing these two estimators to run at both types of instants, that is, $\mathbb{T}_e \cap \mathbb{T}_p$. The main challenge in all three estimators is to cope with the hybrid nature of measurement information available:

- Positive information: At the instants $t \in \mathbb{T}_e$ of an event, a new measurement value $y(t) = y(t_e)$ is received.
- Negative information: At time-periodic instants $t \in \{\mathbb{T}_p \setminus \mathbb{T}_e\}$ that are not an event, the measurement information is a property that the sensor value lies within a bounded subset, that is, $y(t) \in \mathbb{H}(e, t)$.

### 13.5.1 Stochastic Representations

This section summarizes the *stochastic* event-based state estimator (sEBSE) originally developed in [30] and later applied for a target tracking application in videos in [24]. The estimator was further extended in [28] toward multiple sensor systems each having their own triggering criteria. As mentioned, the sEBSE computes updated estimation results at both event and periodic instants, that is, for all $t_k \in \mathbb{T} = \mathbb{T}_e \cup \mathbb{T}_p$. It is important to note that the sEBSE proposed in [30] does not consider deterministic noises or state errors, that is, $\mathbb{W}$, $\mathbb{V}$, and $\mathbb{X}$ are empty sets. Further, the stochastic process and measurement noise are Gaussian distributed*:

$$w_s(t_k) \sim \mathcal{G}(0, W) \quad \text{and} \quad v_s(t_k) \sim \mathcal{G}(0, V), \quad \forall t_k \in \mathbb{T}. \tag{13.4}$$

Similarly, the estimation result $x(t_k) = \hat{x}(t_k) + \tilde{x}_s(t_k)$ of this estimator is Gaussian as well, where $\hat{x}(t_k)$ is the estimated mean and the error-covariance $P(t_k) \in \mathbb{R}^{n \times n}$ characterizes $\tilde{x}_s(t_k) \sim \mathcal{G}(0, P(t_k))$. Explicit formulas for finding an approximation of $\hat{x}(t_k)$ and $P(t_k)$ are summarized next.

Let us assume that $e - 1$ events were triggered until $t_k$. Then, the new measurement information at $t_k$ is either the received measurement value $y(t_k) = y(t_e)$, when $t_k \in \mathbb{T}_e$ is an *event* instant, or it is the inherent knowledge that $y(t_k) \in \mathbb{H}(e, t_k)$, when $t_k \in \{\mathbb{T}_p \setminus \mathbb{T}_e\}$ is not an event but a time-periodic instant. This measurement information can be rewritten by introducing the bounded Borel set $\mathbb{Y}(t_k) \in \mathbb{R}^m$ as follows:

$$\mathbb{Y}(t_k) := \begin{cases} \mathbb{H}(e, t_k) & \text{if} \quad t_k \in \{\mathbb{T}_p \setminus \mathbb{T}_e\}, \\ \{y(t_e)\} & \text{if} \quad t_k \in \mathbb{T}_e. \end{cases}$$
$$\Rightarrow \quad y(t_k) \in \mathbb{Y}(t_k), \ \forall t_k \in \mathbb{T}. \tag{13.5}$$

With the above result one can derive that the PDF of $x(t_k)$, as it is to be determined by this sEBSE,

*Recall that $x \sim \mathcal{G}(\mu, \Sigma)$ is a short notation for $p(x) = \mathcal{G}(x, \mu, \Sigma)$.

yields $p(x(t_k)|y(t_0) \in \mathbb{Y}(t_0), \ldots, y(t_k) \in \mathbb{Y}(t_k))$, which is denoted as $p(x(t_k)|\mathbb{Y}(t_{0:k}))$ for brevity. An exact solution for this PDF is found by applying Bayes' rule, see [21] for more details, that is,

$$p(x(t_k)|\mathbb{Y}(t_{0:k})) = \frac{p\left(x(t_k)|\mathbb{Y}(t_{0:k-1})\right) p(y(t_k) \in \mathbb{Y}(t_k)|x(t_k))}{\int_{\mathbb{R}^n} p\left(x(t_k)|\mathbb{Y}(t_{0:k-1})\right) p(y(t_k) \in \mathbb{Y}(t_k)|x(t_k))\mathrm{d}x(t_k)}. \tag{13.6}$$

The sEBSE developed in [30] finds an single Gaussian approximation of the above equation in three steps.

**Step 1** Compute the *prediction* $p(x(t_k)|\mathbb{Y}(t_{0:k-1}))$ of (13.6) from the process model in (13.1) and the estimation result at $t_{k-1}$, that is, $p(x(t_{k-1})|\mathbb{Y}(t_{0:k-1}))$. Note that the process model is linear and that $p(x(t_{k-1})|\mathbb{Y}(t_{0:k-1})) \approx \mathcal{G}(x(t_{k-1}), \hat{x}(t_{k-1}), P(t_{k-1}))$ is Gaussian. Hence, standard (a-periodic) Kalman filtering equations can be used to find the *predicted* PDF $p(x(t_k)|\mathbb{Y}(t_{0:k-1}))$, for some sampling time $\tau_k := t_k - t_{k-1}$, that is*,

$$p(x(t_k)|\mathbb{Y}(t_{0:k-1})) := \mathcal{G}(x(t_k), \hat{x}(t_k^-), P(t_k^-)), \tag{13.7}$$

where,

$$\hat{x}(t_k^-) := A_{\tau_k}\hat{x}(t_{k-1}),$$
$$P(t_k^-) := A_{\tau_k}P(t_{k-1})A_{\tau_k}^\top + B_{\tau_k}WB_{\tau_k}^\top;$$

**Step 2** Formulate the *likelihood* $p(y(t_k) \in \mathbb{Y}(t_k)|x(t_k))$ as a summation of $N$ Gaussians and employ a sum-of-Gaussian approach proposed in [32] to solve $p(x(t_k)|\mathbb{Y}(t_{0:k}))$ in (13.6). More details on this likelihood are presented later but, for now, let us point out that this solution will result in a summation of $N$ Gaussians characterized by $N$ normalized weights $\alpha_q \in \mathbb{R}_+$, $N$ means $\hat{\theta}_q \in \mathbb{R}^n$, and $N$ covariances $\Theta_q \in \mathbb{R}^{n\times n}$:

$$p(x(t_k)|\mathbb{Y}(t_{0:k})) \approx \sum_{q\in\mathbb{Z}_{[1,N]}} \alpha_q(t_k)\mathcal{G}(x(t_k), \hat{\theta}_q(t_k), \Theta(t_k)). \tag{13.8}$$

The variables of this formula have the following expression:

$$\Theta(t_k) = \left(P^{-1}(t_k^-) + C^\top R^{-1}(t_k)C\right)^{-1},$$
$$\hat{\theta}_q(t_k) = \Theta(t_k)\left(P^{-1}(t_k^-)\hat{x}(t_k^-) + C^\top R^{-1}(t_k)\hat{y}_q(t_k)\right),$$

$$\omega_q(t_k) = e^{\left(\hat{y}_q(t_k)-C\hat{x}(t_k^-)\right)^\top \left(CP(t_k^-)C^\top+R(t_k)\right)^{-1}\left(\hat{y}_q(t_k)-C\hat{x}(t_k^-)\right)},$$

$$\alpha_q(t_k) = \omega_q(t_k)\left(\sum_{q\in\mathbb{Z}_{[1,N]}} \omega_q(t_k)\right)^{-1}.$$

Herein, $\omega_q(t_k) \in \mathbb{R}_+$, $\hat{y}_q(t_k) \in \mathbb{R}^m$, and $R(t_k) \in \mathbb{R}^{m\times m}$ are obtained from the likelihood function, which is modeled with the following weighted summation of $N$ Gaussian:

$$p\left(y(t_k) \in \mathbb{Y}(t_k)|x(t_k)\right) \approx \sum_{q\in\mathbb{Z}_{[1,N]}} \omega_q(t_k)\mathcal{G}(\hat{y}_q(t_k), Cx(t_k), R(t_k)). \tag{13.9}$$

**Step 3** Approximate the result of Step 2 from a sum of $N$ Gaussians into the desired single Gaussian $p(x(t_k)) = \mathcal{G}(x(t_k), \hat{x}(t_k)P(t_k))$. The mean $\hat{x}(t_k)$ and error-covariance $P(t_k)$ should correspond to the mean and covariance of $p(x(t_k)|\mathbb{Y}(t_{0:k}))$ in (13.8), yielding

$$\hat{x}(t_k) = \sum_{q\in\mathbb{Z}_{[1,N]}} \alpha_q(t_k)\hat{\theta}_q(t_k),$$

and

$$P(t_k) = \sum_{q\in\mathbb{Z}_{[1,N]}} \alpha_q(t_k)\Big(\Theta(t_k) + \left(\hat{x}(t_k) - \hat{\theta}_q(t_k)\right) \times \left(\hat{x}(t_k) - \hat{\theta}_q(t_k)\right)^\top\Big).$$

Stability of the above stochastic estimator sEBSE has been derived in [30].

**Theorem 13.1**

Let $\mathbb{H}(\mathsf{e}|t_k)$ be a given bounded Borel set for all $k \in \mathbb{Z}_+$ and let $(A_{\tau_s}, C)$ be an observable pair. Then, the s*EBSE* results in a stable estimate, that is, $\lim_{k\to\infty} \lambda_{\max}(P(t_k))$ exists and is bounded.

A key aspect of the sEBSE is to turn the set inclusion $y(t_k) \in \mathbb{H}(\mathsf{e}, t_k)$ into a stochastic likelihood characterized by $N$ Gaussians (see Equation 13.9). Open questions on how to come up with such a characterization are discussed in the next example and finalizes the sEBSE.

**EXAMPLE 13.1: From set inclusion to likelihood** This example finds a solution for the likelihood $p\left(y(t_k) \in \mathbb{Y}(t_k)|x(t_k)\right)$ in (13.9) by starting from the inclusion $y(t_k) \in \mathbb{Y}(t_k)$. Normally, that is, when $y(t_k)$ is an actual measurement, a likelihood is of the form $p(y(t_k)|x(t_k))$. The process model in (13.2) together with

*The notation $t_k^-$ is used to emphasize the predictive character of a variable at $t_k$.

the noise assumptions on $v_d$ and $v_s$ in (13.4) then give that such a standard likelihood is Gaussian, that is,

$$p(y(t_k)|x(t_k)) = G(y(t_k), Cx(t_k), V). \tag{13.10}$$

In the above estimator, $y(t_k)$ has been generalized into a set inclusion $y(t_k) \in \mathbb{Y}(t_k)$, which can be regarded as a quantized measurement. Results in [18] point out that the corresponding likelihood of this inclusion is found by a convolution of $p(y(t_k)|x(t_k))$ in (13.10) for all possible measurement values $y(t_k) \in \mathbb{Y}(t_k)$. In a fully stochastic description, this is similar to a convolution of the PDF in (13.10) with a uniform PDF $p_{\mathbb{Y}(t_k)}(y(t_k))$ being constant ($> 0$) for all $y(t_k) \in \mathbb{Y}(t_k)$ and 0 otherwise. The likelihood of such a quantized measurement $y(t_k) \in \mathbb{Y}(t_k)$ is then found via

$$p(y(t_k) \in \mathbb{Y}(t_k)|x(t_k)) = \int_{\mathbb{R}^m} p(y(t_k)|x(t_k)) p_{\mathbb{Y}(t_k)}(y(t_k)) dy(t_k). \tag{13.11}$$

This uniform PDF has a hybrid expression in line with $\mathbb{Y}(t_k)$ in (13.5). At instants of an event, the set is the actual measurement $\mathbb{Y}(t_k) := \{y(t_e)\}$ and the PDF $p_{\mathbb{Y}(t_k)}(y(t_k))$ is described with a delta function at $y(t_e)$. Periodically in time, the set is defined by the triggering criteria $\mathbb{Y}(t_k) = \mathbb{H}(e, t_k)$ at which $p_{\mathbb{Y}(t_k)}(y(t_k))$ will be approximated by a summation of $N$ Gaussians. As such, the following characterization is introduced:

$$p_{\mathbb{Y}(t_k)}(y(t_k)) \approx \begin{cases} \sum_{q=1}^{N} \frac{1}{N} G(y(t_k), \hat{y}_q(t_k), U(t_k)) & \text{if } t_k \in \{\mathbb{T}_p \setminus \mathbb{T}_e\}, \\ \delta(y(t_k) - y(t_e)) & \text{if } t_k \in \mathbb{T}_e. \end{cases}$$

Values of $\hat{y}_q(t_k) \in \mathbb{R}^m$, for all $q \in \mathbb{Z}_{[1,N]}$, are retrieved by taking $N$ equidistant samples in the event triggering set $\mathbb{Y}(t_k) = \mathbb{H}(e|t_k)$. Each sample represents the mean of a Gaussian $\mathcal{G}(y(t_k), \hat{y}_q(t_k), U(t_k))$, where $U \in \mathbb{R}^{m \times m}$ is a constant for each Gaussian (see Figure 13.4).

Then, substituting the above approximation and $p(y(t_k)|x(t_k))$ of (13.10) into the expression of the "set-inclusion"-likelihood in (13.11) gives a result for $p(y(t_k) \in \mathbb{Y}(t_k)|x(t_k))$ already pointed out in (13.9), where

- $N = 1$, $\hat{y}_1(t_k) = y(t_e)$ and $R(t_k) = V$ when $t_k \in \mathbb{T}_e$ is an event.
- $N \geq 1$, $\hat{y}_q(t_k)$ are the equidistantly sampled values of $\mathbb{H}(e, t_k)$ and $R(t_k) = U(t_k) + V$ when $t_k \in \{\mathbb{T}_p \setminus \mathbb{T}_e\}$ is a periodic instant.

### 13.5.2 Deterministic Representations

This section reviews the *deterministic* event-based state estimator (*d*EBSE) developed in [27]. The main difference with respect to the previous approach is that the *d*EBSE operates on time-periodic instants only, that is, $t_k \in \mathbb{T} := \mathbb{T}_p$ and $\mathbb{T}_e \subseteq \mathbb{T}_p$. As such, the sensor value is periodically checked on violation of the event triggering criteria, to generate new measurements. This simplifies the a-periodic process model in (13.1) as the sampling time is the constant $\tau_s$, that is, $\bar{A} := A_{\tau_s}$, $\bar{B} := B_{\tau_s}$, resulting in:

$$x(t_k) = \bar{A}x(t_{k-1}) + \bar{B}w(t_{k-1}); \tag{13.12}$$

$$y(t_k) = Cx(t_k) + v(t_k). \tag{13.13}$$

It is important to note that the *d*EBSE does not assume stochastic noise terms. Further, the deterministic process and measurement noise are in line with the problem formulation presented in Section 13.3:

$$w_d(t_k) \in \mathbb{W} \quad \text{and} \quad v_d(t_k) \in \mathbb{V}, \quad \forall t_k \in \mathbb{T}_p, \tag{13.14}$$

where $\mathbb{W} \subset \mathbb{R}^l$ and $\mathbb{V} \subset \mathbb{R}^m$ are unbiased. In line with these noise representations, the *d*EBSE assumes a deterministic representation of its estimation results, that is, $\tilde{x}_s(t_k) = 0$ for all $t_k \in \mathbb{T}$. The deterministic estimation

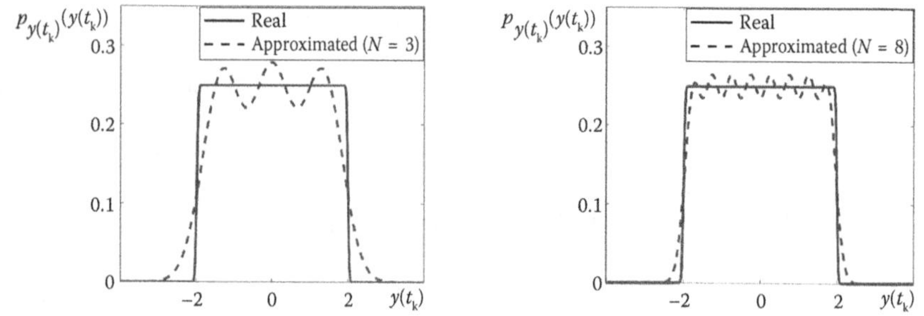


**FIGURE 13.4**
Two approximations of the uniform PDF $p_{[-2,2]}(y(t_k))$ by a summation of either 3 or 8 Gaussians having equidistantly sampled means regarding their measurement value $\hat{y}_q(t_k)$. The covariance matrix $U(t_k)$ of each individual Gaussian function is computed with the heuristic expression $U(t_k) = c\left(0.25 - 0.05e^{-\frac{4(N-1)}{15}} - 0.08e^{-\frac{4(N-1)}{180}}\right)$, where the positive scalar $c$ is equal to the Eucledian distance between two neighboring means.

error is approximated as a circular set based on some $p$-norm, that is,

$$\|\hat{x}(t_k) - x(t_k)\|_p \leq \gamma(t_k), \ \forall t_k \in \mathbb{T}_p \text{ and some } \gamma(t_k) > 0.$$

The event triggering criteria presented next follows a similar approach, for which it is assumed that $e - 1$ events were triggered until $t_k$. Then, for some known measurement value $\hat{y}^{cond}$ conditioned on the prior estimation result, the event triggering criteria of the *d*EBSE adopt a $p$-norm criteria, that is,

$$\begin{aligned} &\textit{an} \text{ event is triggered at } t_k \text{ if } \quad \|y(t_k) - \hat{y}^{cond}(t_k)\|_p > \Delta, \\ &\textit{no} \text{ event is triggered at } t_k \text{ if } \quad \|y(t_k) - \hat{y}^{cond}(t_k)\|_p \leq \Delta. \end{aligned}$$

The new measurement information at $t_k$, which is denoted with the implied measurement $z(t_k)$, is either the received measurement $y(t_k)$, when $t_k \in \mathbb{T}_e$ is an *event* instant, or it is the knowledge that $\|y(t_k) - \hat{y}^{cond}(t_k)\|_p \leq \Delta$, when $t_k \in \{\mathbb{T}_p \setminus \mathbb{T}_e\}$ is not an event instant. The latter measurement information is treated as a measurement realization via $z(t_k) = \hat{y}^{cond}(t_k) + \delta(t_k)$, where $\delta(t_k)$ is bounded by a circular set of known radius.

The implied measurement values $z(t_k)$ available at any sample instant $t_k \in \mathbb{T}$ are used to compute updated values of the estimated mean $\hat{x}(t_k)$. A value for the error bound $\gamma(t_k)$ is not actively tracked. Instead, it is proven that there exists a bound on $\gamma(t_k)$ for all $t_k \in \mathbb{T}$. Yet, before presenting this bound, let us start with the estimation procedure adopted by the *d*EBSE, which is based on the moving horizon approach presented in [2].

This moving horizon approach focuses on estimating the state at $t_{k-N}$ given all measurements from the current time instant $t_k$ until $t_{k-N}$.* As such, the method focuses on estimating $\hat{x}(t_{k-N})$ given $z(t_{k-N:k})$. The estimation result at the current instant $t_k$ is then a forward prediction of $\hat{x}(t_{k-N})$. Let us denote this forward prediction as $\hat{x}^-(t_k)$, to point out that it is still a prediction which will receive its final estimation result at $t_{k+N}$. Then, the forward prediction yields

$$\hat{x}^-(t_k) = A^N \hat{x}(t_{k-N}).$$

A solution for computing $\hat{x}(t_{k-N})$ requires the selection of several design parameters, such as $\alpha \in \mathbb{R}_+$ and $W_N \in \mathbb{R}^{n \times Nm}$. Then, the estimated mean $\hat{x}(t_{k-N})$ based on its prior result $\hat{x}(t_{k-N-1})$ yields

$$\hat{x}(t_{k-N}) = \left(\alpha I + K_N^\top K_N\right) \times \left(\alpha \bar{A} x(t_{k-N-1}) + O_N^\top W_N \begin{pmatrix} z(t_{k-N}) \\ \vdots \\ z(t_k) \end{pmatrix}\right), \tag{13.15}$$

where

$$K_N = W_N \begin{pmatrix} C & C\bar{A} & \cdots & C\bar{A}^{N+1} \end{pmatrix}^\top. \tag{13.16}$$

The employed values for $\alpha$ and $W_N$ are an indication of the confidence in measurements. A suitable value for $W_N$ depending on a positive scalar $\beta$ is found via a singular value decomposition of the observability matrix $VSU^\top := \begin{pmatrix} C & CA & \cdots & CA^{N+1} \end{pmatrix}^\top$. Then, since some singular values are likely zero, one can construct a weighted "pseudo-inverse" of this observability matrix, that is,

$$W_N = \sqrt{\beta} V S^+ U^\top. \tag{13.17}$$

The advantage of the above weight matrix selection is that $K_N = \begin{pmatrix} 0 I_{n_1} & 0 \\ 0 & \sqrt{\beta}\, I_{n_2} \end{pmatrix}$, where $n_1$ is the number of unobservable state elements and $n_2 := n - n_1$ is the number of observable state elements. With this result one can make the following generalization, which is instrumental for the stability result presented afterwards:

$$\left(\alpha I + K_N^\top K_N\right) \alpha \bar{A} = \begin{pmatrix} \bar{A}_{11} & \bar{A}_{12} \\ 0 & \frac{\alpha}{\alpha+\beta} \bar{A}_{22} \end{pmatrix}.$$

**Theorem 13.2**

Let the pair $(\bar{A}, C)$ be detectable, let $W_N$ follow (13.17), and let $\alpha > 0$ and $\beta$ be chosen such that $|\lambda_{\max}\left(\alpha(\alpha+\beta)^{-1}\bar{A}_{22}\right)| < 1$. Then, the *d*EBSE is a stable observer for the process in (13.12), that is, $\|\hat{x}(t_k) - x(t_k)\| < \epsilon$ for all $t_k \in \mathbb{T}_p$ if $\|\hat{x}(t_0) - x(t_0)\| < \eta$, for some bounded $\epsilon, \eta > 0$.

A proof of the above theorem is found in [27].

Note that the estimators in this section either address stochastic or deterministic weights. Yet, more realistic scenarios favor estimators that can address both types of noises. In most practical cases, the process and measurement noise are represented by Gaussian distributions with no deterministic part. Yet, the implied measurement information when exploiting event sampling strategies is typically modeled as additive deterministic noise on the sensor measurement $y$. An estimator able to cope with both types of noise representations is proposed in the next section.

*For clarity of expression, details on the estimation approach directly after initialization and until the $N$th sample instant are not presented. The interested reader is referred to [27] for more details.

## 13.6 A Hybrid State Estimator

This section presents a *hybrid* event-based state estimator ($h$EBSE) allowing both *stochastic* and *deterministic* representations of noise and estimation results. Yet, for clarity of exposition, the presented estimator does not consider deterministic process noises, that is, $\mathbb{W} = \emptyset$. The $h$EBSE is based on existing estimators combining stochastic and deterministic measurement noises, as they were proposed in [26] and to some extent in [11]. In line with the previous $s$EBSE, updated estimation results are computed at both event instants as well as at periodic instants, that is, for all $t_k \in \mathbb{T} = \mathbb{T}_e \cup \mathbb{T}_s$. It is important to note that the proposed estimator assumes a Gaussian distribution of the stochastic noises and an ellipsoidal set inclusion for the deterministic noise parts.* More precisely, let us introduce the following noise characteristics in line with the problem formulation of Section 13.3:

$$w_s(t_k) \sim \mathcal{G}(0, W), \; v_d(t_k) \in \mathbb{L}_{0,D} \text{ and}$$
$$v_s(t_k) \sim \mathcal{G}(0, V), \quad \forall t_k \in \mathbb{T}. \tag{13.18}$$

Herein, $W \in \mathbb{R}^{l\times l}$ is a positive definite matrix characterizing process noise, while $D, V \in \mathbb{R}^{m\times m}$ are positive definite matrices defining the ellipsoidal shaped set $\mathbb{L}_{0,D}$ and the Gaussian function $\mathcal{G}(0, V)$ for characterizing measurement noise. The $h$EBSE further defines $tx(t_k) = \hat{x}(t_k) + \tilde{x}_d(t_k) + \tilde{x}_s(t_k)$ for representing its estimation results, consisting of a stochastic and a deterministic part as introduced in Section 13.3. The estimation result has a mean $\hat{x}(t_k)$ and its estimation errors are characterized as follows:

$$\tilde{x}_d(t_k) \in \mathbb{L}_{0,X(t_k)}, \quad \tilde{x}_s(t_k) \sim \mathcal{G}(0, P(t_k)), \quad \forall t_k \in \mathbb{T}. \tag{13.19}$$

Herein, $X, P \in \mathbb{R}^{n\times n}$ are positive definite matrices defining the ellipsoidal shaped set $\mathbb{L}_{0,X(t_k)}$ and the Gaussian $\mathcal{G}(0, P(t_k))$, respectively.

**REMARK 13.3** The matrices $D$ and $X(t_k)$ are referred to as shape matrices, whereas $W$, $V$, and $P(t_k)$ are covariance matrices. Moreover, since $\tilde{x}_d$ and $\tilde{x}_s$ denote an estimation error, let us refer to $X(t_k)$ and $P(t_k)$ as the error-shape matrix and the error-covariance matrix, respectively.

Explicit formulas for finding values of $\hat{x}(t_k)$, $P(t_k)$, and $X(t_k)$ are presented, next. To that extent, it is assumed that the process model in (13.1) is available, along with the values for $W$, $V$, and $D$. The proposed $h$EBSE is presented in two stages. Firstly, the implied measurement information resulting from event triggering criteria is integrated with the deterministic part of the measurement noise $v_d$. Secondly, the state estimation formulas of the $h$EBSE are presented, which are based on the combined stochastic and set-membership estimator proposed in [26].

*Recall that $x \sim \mathcal{G}(\mu, \Sigma)$ is a short notation for $p(x) = \mathcal{G}(x, \mu, \Sigma)$ and that $\mathbb{L}_{\mu,\Sigma}\mathbb{R}^q$ is an ellipsoidal shaped set defined by $\mathbb{L}_{\mu,\Sigma} := \{x \in \mathbb{R}^q | (x-\mu)^\top \Sigma^{-1}(x-\mu) \leq 1\}$.

### 13.6.1 Implied Measurements

The measurement information available at any sample instant, that is, event or time-periodic, has already been derived in Sections 13.4.2 and 13.5.1. To summarize those results, let us assume that $\mathsf{e} - 1$ event were triggered until $t_k$. Then, the measurement information at $t_k$ is either a received measurement value $y(t_k) = y(t_\mathsf{e})$, when $t_k \in \mathbb{T}_e$ is an *event* instant, or it is a set-inclusion $y(t_k) \in \mathbb{H}(\mathsf{e}, t_k)$, when $t_k \in \mathbb{T}_p \setminus \mathbb{T}_e$ is not an event but a time-periodic instant. This measurement information can be rewritten as follows:

$$y(t_k) \in \begin{cases} \mathbb{H}(\mathsf{e}, t_k) & \text{if} \quad t_k \in \mathbb{T}_p \setminus \mathbb{T}_e, \\ \{y(t_\mathsf{e})\} & \text{if} \quad t_k \in \mathbb{T}_e. \end{cases} \tag{13.20}$$

Now, let us assume that the event triggering set $\mathbb{H}(\mathsf{e}, t_k)$ is (or can be approximated by) the ellipsoidal set $\mathbb{L}_{(\hat{y}_\mathsf{e}(t_k), H_\mathsf{e}(t_k))}$, for example,

$$\begin{aligned} [\hat{y}_\mathsf{e}(t_k), H_\mathsf{e}(t_k)] &:= \arg \min_{\hat{y}\in\mathbb{R}^m, H\succeq 0} \quad \mathrm{tr}(H), \\ \text{subject to:} & \qquad \mathbb{L}_{(\hat{y},H)} \supseteq \mathbb{H}(\mathsf{e}, t_k). \end{aligned} \tag{13.21}$$

Similarly, the remaining set $\{y(t_\mathsf{e})\}$ of the measurement information in (13.20) can be approximated by the ellipsoidal set $\lim_{\epsilon\downarrow 0} \mathbb{L}_{(y(t_\mathsf{e}),\epsilon I)}$. Substituting these values into (13.20) and considering $\epsilon \downarrow 0$ gives

$$y(t_k) \in \begin{cases} \mathbb{L}_{(\hat{y}_\mathsf{e}(t_k), H_\mathsf{e}(t_k))} & \text{if} \quad t_k \in \mathbb{T}_p \setminus \mathbb{T}_e, \\ \mathbb{L}_{(y(t_\mathsf{e}),\epsilon I)} & \text{if} \quad t_k \in \mathbb{T}_e. \end{cases}$$

One can turn the above set-membership into an equality by introducing the unbiased noise term $e(t_k)$, such that $e(t_k) \in \mathbb{L}_{(0,H_\mathsf{e}(t_k))}$ if $t_k \in \mathbb{T}_p \setminus \mathbb{T}_e$ and $e(t_k) \in \mathbb{L}_{(0,\epsilon I)}$ if $t_k \in \mathbb{T}_e$, resulting in the realization

$$y(t_k) + e(t_k) = \begin{cases} \hat{y}_\mathsf{e}(t_k) & \text{if} \quad t_k \in \mathbb{T}_p \setminus \mathbb{T}_e, \\ y(t_\mathsf{e}) & \text{if} \quad t_k \in \mathbb{T}_e. \end{cases} \tag{13.22}$$

Since $y(t_k) = Cx(t_k) + v_s(t_k) + v_d(t_k)$ already contains a deterministic noise term $v_d \in \mathbb{L}_{(0,D)}$, one can introduce $\tilde{v}(d)(t_k) := v_d(t_k) + e(t_k)$ satisfying the following set-inclusion, for some $\varsigma \in (0, 1)$, that is,

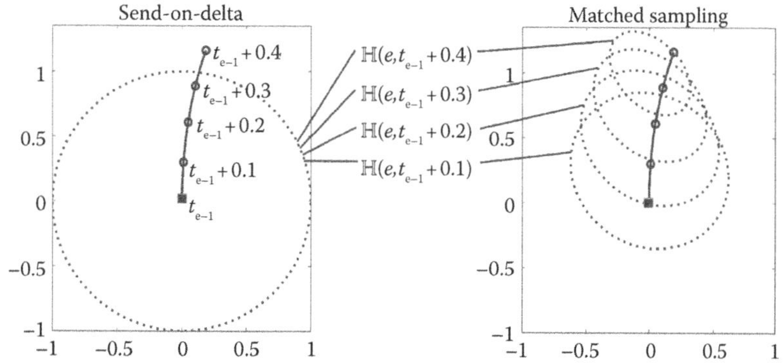


**FIGURE 13.5**
Illustrative examples of the event triggering sets $\mathbb{H}(e,t) \in \mathbb{R}^2$ that result from send-on-delta and matched sampling for the periodic sample instants $t_{e-1}+0.1$ until $t_{e-1}+0.4$ seconds. The initial measurement value is $y(t_{e-1}) = (0\ 0)^\top$ for both strategies.

$$\tilde{v}_d(t_k) \in \begin{cases} \mathbb{L}_{(0,D)} \oplus \mathbb{L}_{(0,H_e(t_k))} \subseteq \mathbb{L}_{(0,(1+\varsigma^{-1})D+(1+\varsigma)H_e(t_k))} & \text{if} \quad t_k \in \mathbb{T}_p \setminus \mathbb{T}_e, \\ \mathbb{L}_{(0,D)} \oplus \mathbb{L}_{(0,\epsilon I)} = \mathbb{L}_{(0,D)} & \text{if} \quad t_k \in \mathbb{T}_e,\ \epsilon \downarrow 0. \end{cases}$$

A suitable value for $\varsigma$ minimizes the trace of $(1+\varsigma^{-1})D + (1+\varsigma)H_e(t_k)$.

Now, combining the measurement information in (13.22) with the above noise term $\tilde{v}_d$ and the process model in (13.1) results in a new (implied) measurement $z(t_k)$ having the following measurement model and measurement realizations, for a received $y(t_e)$ and a computed $\hat{y}_e(t_k)$:

$$\text{model} \quad z(t_k) = Cx(t_k) + v_s(t_k) + \tilde{v}_d(t_k), \quad (13.23)$$

$$\text{realization} \quad z(t_k) = \begin{cases} \hat{y}_e(t_k) & \text{if} \quad t_k \in \mathbb{T}_p \setminus \mathbb{T}_e, \\ y(t_e) & \text{if} \quad t_k \in \mathbb{T}_e. \end{cases} \quad (13.24)$$

The stochastic measurement noise still follows $v_s(t_k) \sim \mathcal{G}(0,V)$, while the deterministic measurement noise $\tilde{v}_d = v_d + e$ follows $\tilde{v}_d(t_k) \in \mathbb{L}_{(0,E(t_k))}$ and

$$E(t_k) := \begin{cases} (1+\varsigma^{-1})D + (1+\varsigma)H_e(t_k) & \text{if} \quad t_k \in \mathbb{T}_p \setminus \mathbb{T}_e, \\ D & \text{if} \quad t_k \in \mathbb{T}_e. \end{cases}$$

**EXAMPLE 13.2: Ellipsoidal sets for event sampling strategies** One can derive that the event sampling strategies send-on-delta and matched sampling, as presented in Section 13.4.1, will result in an ellipsoidal shaped triggering set $\mathbb{H}(e,t_k)$. The center and shape matrix of this set directly define values for $\hat{y}_e(t_k)$ and $H_e(t_k)$ in (13.21), yielding

- Send-on-delta: $\hat{y}_e(t_k) = y(t_{e-1})$ and $H_e(t_k) = \Delta I$.
- Matched sampling: In case the sensor employs the asynchronous Kalman filter of Remark 13.2 for computing the divergence, with $\text{cov}(w_s) = W$ and $\text{cov}(v_s) = V$, then $\hat{y}_e(t_k) = Cx_2(t_k^-)$ and $H_e(t_k) = 2(\Delta - \alpha)^{-1}(V^{-1}CP_1(t_k)P_2^{-1}(t_k)P_1(t_k)CV^{-1})^{-1}$, where $\alpha = 0.5\left(\log|P_2(t_k)||P_2(t_k)|^{-1} + \text{tr}(P_2^{-1}(t_k)P_1(t_k)) - n\right)$.
- Some illustrative examples of the ellipsoidal sets for send-on-delta and matched sampling are found in Figure 13.5.

### 13.6.2 Estimation Formulas

The $h$EBSE exploits the (implied) measurement values of $z(t_k)$ as proposed in (13.23), for all $t_k \in \mathbb{T}$. The $h$EBSE aims to solve the estimation problem for the state representation $x(t_k) = \hat{x}(t_k) + \tilde{x}_d(t_k) + \tilde{x}_s(t_k)$. The underlying idea is to compute a state estimate $\hat{x}$ that minimizes the (maximum possible) mean squared error (MSE) of $\hat{x} - x$ in the presence of both stochastic and deterministic noises (or uncertainties). For clarity, let us point out that the error associated with the state estimate $\hat{x}(t)$ is composed of a stochastic and a deterministic part, that is,

$$\hat{x}(t_k) - x(t_k) = \tilde{x}_s(t_k) + \tilde{x}_d(t_k),$$

where

$$\tilde{x}_s(t_k) \sim \mathcal{G}(0, P(t_k)) \quad \text{and} \quad \tilde{x}_d(t_k) \in \mathbb{L}_{0,X(t_k)}.$$

The advantage of assuming an ellipsoidal set-inclusion for $\tilde{x}_d$ and a Gaussian distribution for $\tilde{x}_s$ is that estimation errors are characterized by $P \succ 0$ and $X \succ 0$. Since the deterministic error is nonstochastic and independent from stochastic errors, note that the considered MSE then yields

$$\begin{aligned} &\mathbb{E}\left[(\hat{x}(t_k) - x(t_k))^\top (\hat{x}(t_k) - x(t_k))\right] \\ &= \underbrace{\mathbb{E}\left[\tilde{x}_s^\top(t_k)\tilde{x}_s(t_k)\right]}_{=\text{tr}(P(t_k))} + \underbrace{\tilde{x}_d^\top(t_k)\tilde{x}_d(t_k)}_{\leq \text{tr}(X(t_k))} \\ &\leq \text{tr}\left(P(t_k) + X(t_k)\right). \end{aligned} \quad (13.25)$$

Thus, the MSE is bounded by the trace of $P(t_k) + X(t_k)$. The estimator proposed in [26] forms the basis of the *h*EBSE, as it minimizes exactly this bound. Additional to standard Kalman filtering, the estimate $\hat{x}$ is associated not only with an error-covariance $P$ but also with a error-shape matrix $X$. The values of these matrices and of the state estimate $\hat{x}$ are computed in line with standard estimation approaches, that is, with a prediction step and a measurement update.

**Step 1** Compute the predicted values* $\hat{x}(t_k^-)$, $P(t_k^-)$ and $X(t_k^-)$ given their prior results at $k-1$. Note that the process model in (13.1) is linear and that $w_s(t_k)$ is unbiased and characterized by $W \succ 0$. Moreover, the estimation error of $\hat{x}(t_k)$ is characterized by positive definite matrices $P(t_k^-)$ and $X(t_k^-)$ as well. These prerequisites are in line with standard Kalman filtering (extended with shape matrices) and it was shown in [26] that a similar prediction step can be employed here as well, that is,

$$\begin{aligned} \hat{x}(t_k^-) &= A_{\tau_k}\hat{x}(t_{k-1}),\\ P(t_k^-) &= A_{\tau_k}P(t_{k-1})A_{\tau_k}^\top + B_{\tau_k}WB_{\tau_k}^\top,\\ X(t_k^-) &= A_{\tau_k}X(t_{k-1})A_{\tau_k}^\top. \end{aligned} \tag{13.26}$$

Recall that $\tau_k := t_k - t_{k-1}$ is the sampling time. Evidently, this prediction can be computed in closed form and only differs from a Kalman filter by the additional third expression determining $X(t_k^-)$.

**Step 2** Compute the updated values $\hat{x}(t_k)$, $P(t_k)$, and $X(t_k)$, given their prediction results from Step 1 and the measurement $z(t_k)$ of (13.23). Again, the process model in (13.1) is linear and the unbiased measurement noises $\tilde{v}_d(t_k)$ and $v_s(t_k)$ are characterized by $E(t_k) \succeq 0$ and $V \succ 0$, respectively, where $\tilde{v}_d$ is a substitute of $v_d$ to include the event triggering set. As such, one can employ an unbiased update expression in line with any linear estimator, for some gain $K(t_k) \in \mathbb{R}^{n\times m}$, that is,

$$\hat{x}(t_k) = (I - K(t_k)C)\hat{x}(t_k^-) + K(t_k)z(t_k). \tag{13.27}$$

In line with the above expression, the updated error-covariance and error-shape matrices also follow standard Kalman filtering expressions, yielding

$$\begin{aligned} P(t_k) &= (I - K(t_k)C)P(t_k^-)(I - K(t_k)C)^\top\\ &\quad + K(t_k)VK(t_k)^\top, \end{aligned} \tag{13.28}$$

$$\begin{aligned} X(t_k) &= \tfrac{1}{1-\omega(t_k)}(I - K(t_k)C)X(t_k^-)(I - K(t_k)C)^\top\\ &\quad + \tfrac{1}{\omega(t_k)}K(t_k)E(t_k)K(t_k)^\top. \end{aligned} \tag{13.29}$$

*The notation $t_k^-$ is used to emphasize the predictive character of a variable at $t_k$.

The parameter $\omega(t_k) \in (0,1)$ in (13.29) guarantees that the shape matrix $X(t_k)$ corresponds to an outer ellipsoidal approximation of two ellipsoidal sets: the sets being a weighted prediction error and a measurement error, that is, $\mathbb{L}_{(0,(I-K(t_k)C)\,X(t_k^-)\,(I-K(t_k)C))}$ and $\mathbb{L}_{(0,K(t_k)\,E(t_k)\,K(t_k)^\top)}$.

Results in [26] point out that the gain $K(t_k)$ is given by

$$\begin{aligned} K(t_k) &= \left(P(t_k^-)\,C^\top + \tfrac{1}{1-\omega(t_k)}X(t_k^-)\,C^\top\right)\\ &\quad \cdot \left(C\,P(t_k^-)\,C^\top + \tfrac{1}{1-\omega(t_k)}C\,X(t_k^-)\,C^\top\right.\\ &\quad \left. + V + \tfrac{1}{\omega}E(t_k)\right)^{-1}. \end{aligned} \tag{13.30}$$

A one-dimensional convex optimization problem for $\omega^{\text{opt}} \in (0,1)$ that minimizes the posterior MSE bound in (13.25) remains to be solved, for example, with the aid of Brent's method.

**REMARK 13.4** The derived gain (13.30) embodies a systematic and consistent generalization of the standard Kalman filter for additional unknown but bounded uncertainties. Accordingly, $K$ in (13.30) reduces to the standard Kalman gain in the absence of set-membership errors, that is, $Q = 0$, $E(t_k) = 0$ implying that $X(t_k) = 0$ for all $t_k \in \mathbb{T}$. In the opposite case of vanishing error covariance matrices, the *h*EBSE yields a deterministic estimator of intersecting ellipsoidal sets.

**REMARK 13.5** The sum of the error matrices is expressed via

$$\begin{aligned} P(t_k) + X(t_k) &= \Big(\omega(t_k)\big(\omega(t_k)P(t_k^-) + X(t_k^-)\big)^{-1}\\ &\quad + (1-\omega(t_k))C^\top\\ &\quad \times \big((1-\omega(t_k))V + E(t_k)\big)^{-1}C\Big)^{-1}, \end{aligned} \tag{13.31}$$

which can be utilized to determine $\omega^{\text{opt}}$ that minimizes the right-hand side bound in (13.25). The special cases $\omega^{\text{opt}} = 0$ or $\omega^{\text{opt}} = 1$ will not be considered.

This completes the hybrid EBSE resulting in a correct description of the estimation results for including the event triggering set $\mathbb{H}(\mathrm{e}, t_k)$ into the estimation results. Before continuing with an observation case study, let us first point out the stability of estimation errors.

### 13.6.3 Asymptotic Analysis

The hybrid EBSE is said to be stable iff both $P(t_k)$ and $X(t_k)$ have finite eigenvalues for $t_k \to \infty$. Proving stability is done in two steps:

1. Introduce a $\Gamma(t_k)$, such that $P(t_k) + X(t_k) \preceq \Gamma(t_k)$ holds for all $t_k \in \mathbb{T}$.
2. Show that $\Gamma(t_k) \preceq \Sigma(t_k)$, for some $\Sigma(t_k)$ being the result of a standard, periodic Kalman filter.

Then, the proposed estimator enjoys the same stability conditions as a (standard) periodic Kalman filter would have, that is, depending on detectability and observability properties. For clarity of the presented results, it is assumed that the scalar weight $\omega$ will be constant at all sampling instants, that is, the hybrid EBSE employs $\omega(t_k) = \omega$ in (13.28), (13.29), (13.30), and (13.31) for all $t_k \in \mathbb{T}$, though the results can be generalized to weights varying in time.

Let us start with the first step where $P(t_k) + X(t_k) \preceq \Gamma(t_k)$. After this step, one can guarantee that iff $\Gamma(t_k) \succ 0$ is asymptotically stable, then both $P(t_k)$ and $X(t_k)$ shall be asymptotically stable as well. In line with the results of (13.26) and of (13.31), let us introduce the following update equation for $\Gamma(t_k)$, that is,

$$\begin{aligned}
\Gamma(t_k^-) &= A_{\tau_k}\Gamma(t_{k-1})A_{\tau_k}^\top + B_{\tau_k}WB_{\tau_k}^\top,\\
\Gamma(t_k) &\\
&= \left(\omega\left(\Gamma(t_k^-)\right)^{-1} + (1-\omega)C^\top\left((1-\omega)V + E(t_k)\right)^{-1}C\right)^{-1}.
\end{aligned} \tag{13.32}$$

**Theorem 13.3**

Consider $P(t_k)$ and $X(t_k)$ of the hybrid EBSE, for some constant $\omega \in (0,1)$, and consider $\Gamma(t_k)$ in (13.32). Further, let $\Gamma(0) := P(0) + X(0)$. Then, $P(t_k) + X(t_k) \prec \Gamma(t_k)$ holds for all $t_k \in \mathbb{T}$.

The proof of this theorem is found in the appendix.

Let us continue with the second step of this asymptotic analysis, for which we will introduce $\Sigma(t_n)$ computed via an update equation similar to a time-periodic Kalman filter, that is, for all time-periodic instants $t_n \in \mathbb{T}_p$,

$$\begin{aligned}
\Sigma(t_n^-) &= \bar{A}\Sigma(t_{n-1})\bar{A}^\top + \bar{B}W\bar{B}^\top,\\
\Sigma(t) &= \left(\omega^{-\kappa+1}\left(\Sigma(t^-)\right)^{-1}\right.\\
&\quad \left. + (1-\omega)C^\top\left((1-\omega)V + E(t_n)\right)^{-1}C\right)^{-1}.
\end{aligned} \tag{13.33}$$

The scalar $\kappa \in (0,1)$ is an upper bound on the amount of events that can occur between two consecutive periodic sample instants $(t_n)$ and $(t_{n-1})$. The constant system matrices above are defined as $\bar{A} := A_{\tau_s}$ and $\bar{B} := B_{\tau_s}$.

**Theorem 13.4**

Consider $\Gamma(t_k)$ in (13.32) and $\Sigma(t_n)$ in (13.33) and let $\Sigma(0) := \Gamma(0)$. Then, $\Gamma(t_k) \prec \Sigma(t_n)$ holds for all $t_k = t_n$ and $t_n \in \mathbb{T}_p$.

The proof of this theorem is found in the appendix. Moreover, the results in Theorem 13.4 guarantee that the hybrid EBSE proposed has asymptotically stable estimation results in case $(\bar{A}, C)$ is detectable.

## 13.7 Illustrative Case Study

Results of the stochastic and hybrid event-based state estimator are studied here in terms of estimation errors for tracking a 1D object. The process model in line with (13.1) is a double integrator, that is,

$$\begin{aligned}
x(t) &= \begin{bmatrix} 1 & \tau \\ 0 & 1 \end{bmatrix} x(t-\tau) + \begin{bmatrix} \frac{1}{2}\tau^2 \\ \tau \end{bmatrix} a(t-\tau),\\
y(t) &= \begin{bmatrix} 1 & 0 \end{bmatrix} x(t) + v_s(t).
\end{aligned}$$

The state vector $x(t)$ combines the object's position and speed. Further, $a(t) = \frac{1}{30}t \cdot \cos(\frac{1}{10}t)$ denotes the object's acceleration, while only the position is measured in $y(t)$. Since acceleration is assumed unknown, the process model in (13.1) is characterized with a process noise $w(t) := a(t)$. During the simulation, the acceleration is bounded by $|a(t)| \leq 0.9$, due to which a suitable covariance in line with [6] is $\text{cov}(a(t)) = 1.1$, resulting in an unbiased distribution $p(w(t))$ with covariance $W = 1.1$. Further, the sampling time is $\tau_s = 0.1$ seconds and the sensor noise covariance is set to $V = 2 \cdot 10^{-3}$. The object's true position, speed, and acceleration are depicted in Figure 13.6.

The *h*EBSE of Section 13.6 combining stochastic and set-membership measurement information is compared to the *s*EBSE presented in 13.5.1 limited to stochastic representations. Both estimators start with the initial estimation results $\hat{x}(0) = (0.1 \quad 0.1)^\top$ and $P(0) = 0.01 \cdot I$,

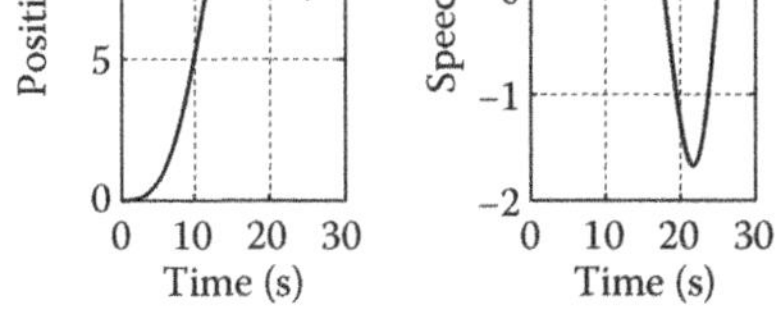

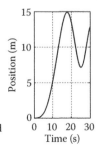

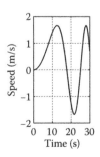

**FIGURE 13.6**
The position, speed, and acceleration of the tracked object.

while $X(0) = 0$ is chosen as the initial ellipsoidal shape matrix for the *h*EBSE. Next, the measurement information of both EBSEs is characterized.

### *h*EBSE

Measurement information of the *h*EBSE is represented by the implied measurement $z(t) = Cx(t) + v_s(t) + \tilde{v}_d(t)$. Note that the original measurement is only affected by stochastic noise and that $v_d(t) \in \emptyset$, due to which $\tilde{v}_d(t) = e(t)$ is characterized by an ellipsoidal approximation $e(t) \in \mathbb{L}_{0,E}$ of the event triggering set. In case of an event instant $t \in \mathbb{T}_e$, the measurement $y(t_e)$ is received and one obtains that $z(t) = y(t_e)$, that is, $e(t_e) \in \emptyset$ and $E(t) = 0$. At periodic time instants $t \in \mathbb{T}_p$, one has the information that $y(t) \in \mathbb{H}(e, t)$. This ellipsoidal set $\mathbb{H}(e, t)$ can be characterized with a "mass"-center, yielding an estimate of $y(t)$ and an ellipsoidal error-set resulting in an characterization of $e(t) \in \mathbb{L}_{0,E(t)}$ via $E(t) = H_e(t)$. Suitable realizations of $z(t)$ and $E(t)$ for the two employed event strategies were already given in Example 13.2, where $\Phi(t) := 2(\Delta - \alpha)^{-1}(V^{-1}CP_1(t)P_2^{-1}(t)P_1(t)CV^{-1})^{-1}$, that is,

Send-on-delta:

$$z(t) = y(t_e), \qquad E(t) = 0, \qquad \forall t \in \mathbb{T}_e,$$
$$z(t) = y(t_{e-1}), \qquad E(t) = \Delta^2, \qquad \forall t \in \mathbb{T}_p,$$

Matched sampling:

$$z(t) = y(t_e), \qquad E(t) = 0, \qquad \forall t \in \mathbb{T}_e,$$
$$z(t) = CA_{t-t_{e-1}}\hat{x}(t_{e-1}), \quad E(t) = \Phi(t), \quad \forall t \in \mathbb{T}_p.$$

### *s*EBSE

Measurement information of the *s*EBSE is represented by a single Gaussian PDF, that is, $p(y(t)) = \mathcal{G}(y(t), \hat{y}(t), R(t))$ for some (estimated) measurement value $\hat{y}(t)$ and covariance matrix $R(t) = V(t) + U(t)$. Herein, $V(t)$ is the covariance of the stochastic measurement noise $v_s(t)$, while $U(t)$ is a covariance due to any implied measurement information as it is treated as an additional stochastic noise. In case of an event instant $t \in \mathbb{T}_e$ the measurement $y(t_e)$ is received and one obtains that $\hat{y}(t) = y(t_e)$ and $U(t) = 0$. At periodic time instants $t \in \mathbb{T}_p$, one has the information that $y(t) \in \mathbb{H}(e, t)$, which is then turned into a particular value for $\hat{y}(t)$ and $U(t)$. A suitable characterization of $\hat{y}(t)$ and $U(t)$ for the two employed event sampling strategies were already given in Example 13.2, where $\Phi(t) := 2(\Delta - \alpha)^{-1}(V^{-1}CP_1(t)P_2^{-1}(t)P_1(t)CV^{-1})^{-1}$, that is,

Send-on-delta:

$$\hat{y}(t) = y(t_e), \qquad U(t) = 0, \qquad \forall t \in \mathbb{T}_e,$$
$$\hat{y}(t) = y(t_{e-1}), \qquad U(t) = \frac{3}{4}\Delta^2, \qquad \forall t \in \mathbb{T}_p,$$

Matched sampling:

$$\hat{y}(t) = y(t_e), \qquad U(t) = 0, \qquad \forall t \in \mathbb{T}_e,$$
$$\hat{y}(t) = CA_{t-t_{e-1}}\hat{x}(t_{e-1}), \quad U(t) = \frac{1}{4}\Phi(t), \quad \forall t \in \mathbb{T}_p.$$

Figure 13.7 until Figure 13.10 depict the actual squared estimation error, that is, $\|\hat{x}(t) - x(t)\|_2^2$, in comparison to the modeled estimation error, that is, $\mathrm{tr}(P(t))$ for the *s*EBSE and $\mathrm{tr}(P(t)) + \mathrm{tr}(X(t))$ for the *h*EBSE. The results depicted were obtained after averaging the outcome of 1000 runs of the considered simulation case study.

Figures 13.7 and 13.8 depict the estimation results of the *h*EBSE and the *s*EBSE, respectively, when matched sampling is employed as the event sampling strategy. Although it is not pointed out in the figures, it is worth mentioning that the *h*EBSE triggered a total amount

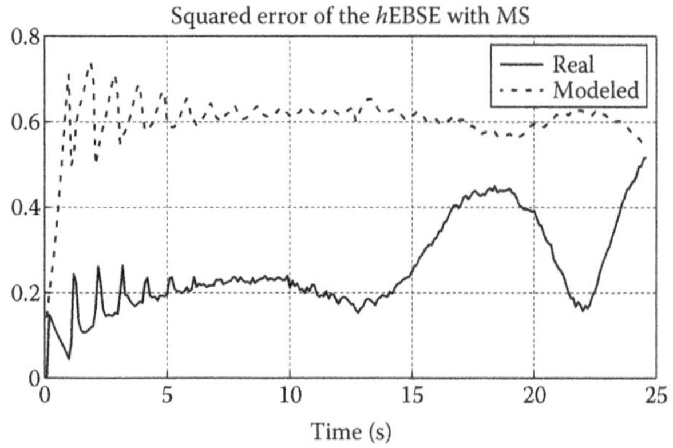


**FIGURE 13.7**
Simulation results for *matched sampling* (MS) in combination with the *h*EBSE allowing stochastic and set-membership representations. The real squared estimation error $\|\hat{x}(t) - x(t)\|_2^2$ is depicted versus the modeled (bound) of the estimation error $\mathrm{tr}(P(t)) + \mathrm{tr}(X(t))$.

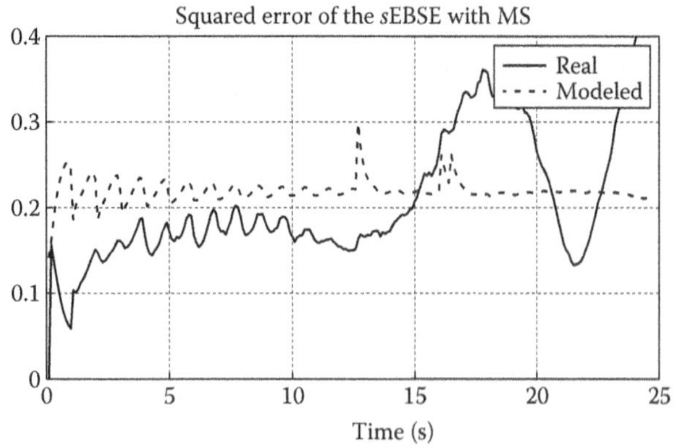


**FIGURE 13.8**
Simulation results for *matched sampling* (MS) in combination with the *s*EBSE limited to stochastic representations. The real squared estimation error $\|\hat{x}(t) - x(t)\|_2^2$ is depicted versus the modeled (bound) of the estimation error $\mathrm{tr}(P(t))$.

of 31 events (on average), while the *s*EBSE triggered 40 events (on average). Further, the real squared estimation error of both EBSEs considered is comparable. Hence, the *h*EBSE has similar estimation results with fewer events triggered, due to which less measurement samples are required saving communication resources. Yet, the main advantage of the *h*EBSE is the modeled bound on the estimation error. Figure 13.7 indicates that this modeled bound is conservative when the *h*EBSE is employed, which is not the case for the *s*EBSE depicted in Figure 13.8. This means that the *h*EBSE gives a better guarantee that the real estimation error stays within the bound as it is computed by the estimator. Such a property is important when estimation results are used for control purposes.

Figures 13.9 and 13.10 depict the estimation results of the *h*EBSE and the *s*EBSE, respectively, when send-on-delta is employed as the event sampling strategy. Since this event sampling strategy does not depend on previous estimation results but merely on the previous measurement sample, both estimators received the events at the same time instants giving a total of 115 events. Note that this is an increase of events by a factor of 3 to 4 when compared to the EBSEs in combination with matched sampling. Yet, this increase of events and thus of measurement samples is not reflected in a corresponding

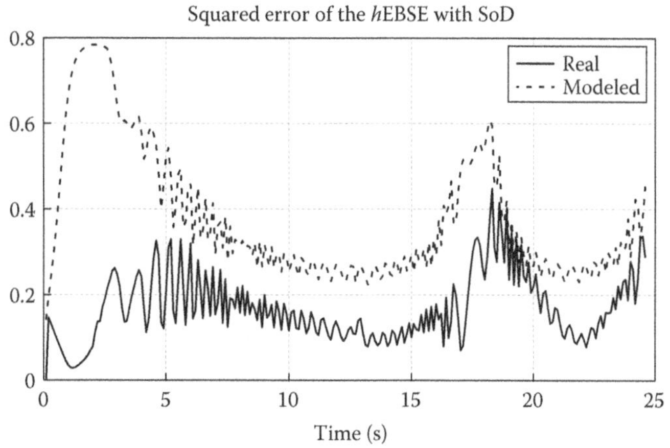


**FIGURE 13.9**
Simulation results for *send-on-delta* (SoD) in combination with the *h*EBSE allowing stochastic and set-membership representations. The real squared estimation error $\|\hat{x}(t) - x(t)\|_2^2$ is depicted versus the modeled (bound) of the estimation error $\text{tr}(P(t)) + \text{tr}(X(t))$.

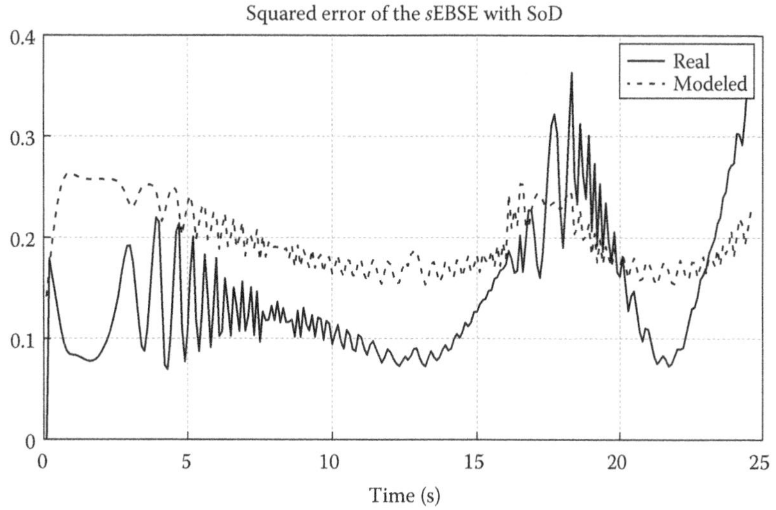


**FIGURE 13.10**
Simulation results for *send-on-delta* (SoD) in combination with the *s*EBSE limited to stochastic representations. The real squared estimation error $\|\hat{x}(t) - x(t)\|_2^2$ is depicted versus the modeled (bound) of the estimation error $\text{tr}(P(t))$.

decrease of estimation errors. Further, similar conclusions can be drawn from the estimation results with send-on-delta when comparing Figures 13.9 and 13.10. Again, the squared estimation error of the two considered EBSEs is comparable and the main advantage of the *h*EBSE is in the improved bound of the modeled estimation error.

Therefore, a fair conclusion of the *h*EBSE is that similar estimation errors are achieved when compared to *s*EBSE, although the *modeled* estimation error of the *h*EBSE is a far better bound on *real* estimation errors. Similar results are expected when comparing the *h*EBSE with the deterministic *d*EBSE, as in the latter EBSE noises should be represented as a set-membership. As such, the *h*EBSE is advantageous in networked control systems where estimation results are being used by a (stabilizing) controller.

## 13.8 Conclusions

In networked systems, high measurement frequencies may rapidly exhaust communication bandwidth and power resources when sensor data must be transmitted periodically to the state estimator. The transmission rate can significantly be reduced if an event-based strategy is employed for sampling sensor data. "send-on-delta" and "matched sampling" have been discussed as examples of such strategies. This chapter discussed several ideas to process these event sampled measurements. Typical estimation approaches perform a measurement update whenever an event is triggered, that is, at the event instants when a new measurement is received. However, observation and automation systems are mainly designed to rely on time-periodic estimation results. The time gap between events and periodic instants can simply be bridged by prediction steps, but additional knowledge then remains untapped: as long as no event is triggered, the actual sensor value does not fulfill the event-sampling criterion, thereby implying that it did not cross the edge of a particular closed set. Recent estimators do exploit this implied measurement information and three of them were discussed, here; one restricted to stochastic noise representations, one restricted to set membership noise representations, and one hybrid solution allowing both stochastic and set membership representations. The latter one is more advantageous, as measurement and process noise are typically characterized as a stochastic random vector, while the implied information results in a set membership property on the sensor value. Prospective research focuses also on unreliable networks, where delays and packet losses have to be taken into account.

## Appendix A Proof of Theorem 13.3

Let us introduce $R(t_k) := (1-\omega)C^\top\big((1-\omega)V + E(t_k)\big)^{-1}C$. Then, the update formulas of $\Gamma(t_k)$ in (13.32), yields

$$\begin{aligned}\Gamma(t_k^-) &= A_{\tau_k}\Gamma(t_{k-1})A_{\tau_k}^\top + B_{\tau_k}WB_{\tau_k}^\top,\\ \Gamma(t) &= \Big(\omega\big(\Gamma(t_k^-)\big)^{-1} + R(t_k)\Big)^{-1}, \quad \forall t_k \in \mathbb{T}.\end{aligned} \tag{13.34}$$

Similarly, let us derive the update equation for $P(t) + X(t)$ in line with the results in (13.26) and (13.31), that is,

$$\begin{aligned}&P(t_k^-) + X(t_k^-)\\ &\quad = A_{\tau_k}\left(P(t_{k-1}) + X(t_{k-1})\right)A_{\tau_k}^\top + B_{\tau_k}WB_{\tau_k}^\top,\\ &P(t_k) + X(t_k)\\ &\quad = \Big(\omega\big(\omega P(t_k^-) + X(t_k^-)\big)^{-1} + R(t_k)\Big)^{-1}, \quad \forall t_k \in \mathbb{T}.\end{aligned} \tag{13.35}$$

The inequality $P(t_k) + X(t_k) \prec \Gamma(t_k)$, for all $T_k \in \mathbb{T}$, is proven by induction: first show that $P(t_1) + X(t_1) \prec \Gamma(t_1)$ when $\Gamma(0) = P(0) + X(0)$, followed by a proof of $P(t_k) + X(t_k) \prec \Gamma(t_k)$ iff $P(t_{k-1}) + X(t_{k-1}) \prec \Gamma(t_{k-1})$.

The first step of induction starts from $\Gamma(0) = P(0) + X(0)$. The prediction $\Gamma(t_1^-)$ in (13.34) and $P(t_1^-) + X(t_1^-)$ in (13.35) give that $\Gamma(t_1^-) = P(t_1^-) + X(t_1^-)$. Substituting this result in the update equation of $\Gamma(t_1)$ in (13.34) yields

$$\Gamma(t_1) = \Big(\omega\big(P(t_1^-) + X(t_1^-)\big)^{-1} + R(t_1)\Big)^{-1}.$$

Since $\omega \in (0,1)$, this latter equality further implies that

$$\Gamma(t_1) \succ \Big(\omega\big(\omega P(t_1^-) + X(t_1^-)\big)^{-1} + R(t_1)\Big)^{-1}, \tag{13.36}$$

$$= P(t_1) + X(t_1), \tag{13.37}$$

which proves the first step of induction.

The second step starts from $P(t_{k-1}) + X(t_{k-1}) \prec \Gamma(t_{k-1})$. The prediction $\Gamma(t_k^-)$ in (13.34) and $P(t_k^-) + X(t_k^-)$ in (13.35) give that $P(t_k^-) + X(t_k^-) \prec \Gamma(t_k^-)$. Substituting this result in the update equation of $\Gamma(t_k)$ in (13.34) yields

$$\Gamma(t_k) \succ \Big(\omega\big(P(t_k^-) + X(t_k^-)\big)^{-1} + R(t_k)\Big)^{-1}.$$

Since $\omega \in (0,1)$, this latter equality further implies that

$$\Gamma(t_k) \succ \Big(\omega\big(\omega P(t_k^-) + X(t_k^-)\big)^{-1} + R(t_k)\Big)^{-1}, \tag{13.38}$$

$$= P(t_k) + X(t_k), \tag{13.39}$$

which proves the second step of induction and thereby Theorem 13.3.

## Appendix B Proof of Theorem 13.4

Let us introduce $R(t) := (1-\omega)C^\top((1-\omega)V + E(t))^{-1}C$. Then, the update formulas of $\Gamma(t_k)$ in (13.32), yields

$$\Gamma(t_k^-) = A_{\tau_k}\Gamma(t_{k-1})A_{\tau_k}^\top + B_{\tau_k}WB_{\tau_k}^\top,$$
$$\Gamma(t_k) = \left(\omega\left(\Gamma(t_k^-)\right)^{-1} + R(t_k)\right)^{-1}, \quad \forall t_k \in \mathbb{T}. \tag{13.40}$$

Similarly, the update formulas of $\Sigma(t_n)$ in (13.33) are as follows:

$$\Sigma(t_n^-) = \bar{A}\left(\Sigma(t_{n-1})\right)\bar{A}^\top + \bar{B}W\bar{B}^\top,$$
$$\Sigma(t_n) = \left(\omega^{-\kappa+1}\left(\Sigma(t_n^-)\right)^{-1} + R(t_n)\right)^{-1}, \quad \forall t_n \in \mathbb{T}_p. \tag{13.41}$$

The following result is instrumental for proving Theorem 13.4. To that extent, let $\kappa \in \mathbb{Z}_+$ be the amount of event instants in between the two consecutive time-periodic instants $\underline{t} - \tau_s$ and $\underline{t}$, or differently, $\underline{t} - \tau_s = t_{k-\kappa-1} < t_{k-\kappa} < t_{k-\kappa+1} < \ldots < t_{k-1} < t_k = \underline{t}$. Further, let us introduce the time instant $\underline{t} \in \mathbb{T}_p$, such that $t_k = \underline{t}$ and $t_n = \underline{t}$ for some $k, n \in \mathbb{Z}_+$. As $\tau_s \in \mathbb{R}_+$ is the sampling time, one has that $t_{n-1} = \underline{t} - \tau_s$ and $t_{k-\kappa-1} = \underline{t} - \tau_s$, that is, $t_{n-1} = t_{k-\kappa-1}$.

**Lemma 13.1**

Let us consider $\Gamma(\underline{t})$ characterized by (13.40) and $\Sigma(t)$ by (13.41), for some $\underline{t} \in \mathbb{T}_p$, while satisfying $\Gamma(\underline{t} - \tau_s) \preceq \Sigma(\underline{t} - \tau_s)$. Then, $\Gamma(\underline{t}^-) \preceq \omega^{-\kappa}\Sigma(\underline{t}^-)$ holds for any suitable $k, n$ such that $\underline{t} = t_k$ and $\underline{t} = t_n$.

**PROOF** The proof of this lemma start with the inequality that

$$\Gamma(t_k) \preceq \omega^{-1}\Gamma(t_k^-), \quad \forall t_k \in \mathbb{T}, \quad \text{see (13.40)}. \tag{13.42}$$

When substituting this result in the prediction step of (13.40), one can further derive that

$$\begin{aligned}\Gamma(t_k^-) &= \left(A_{\tau_{k-1}}\Gamma(t_{k-1})A_{\tau_{k-1}}^\top + B_{\tau_{k-1}}WB_{\tau_{k-1}}^\top\right),\\ &\preceq \omega^{-1}\left(A_{\tau_{k-1}}\Gamma(t_{k-1}^-)A_{\tau_{k-1}}^\top + B_{\tau_{k-1}}WB_{\tau_{k-1}}^\top\right),\\ &\preceq \omega^{-2}\Big(A_{\tau_{k-1}}A_{\tau_{k-2}}\Gamma(t_{k-2}^-)A_{\tau_{k-2}}^\top A_{\tau_{k-1}}^\top\\ &\quad + A_{\tau_{k-1}}B_{\tau_{k-2}}WB_{\tau_{k-2}}^\top A_{\tau_{k-1}}^\top + B_{\tau_{k-1}}WB_{\tau_{k-1}}^\top\Big).\end{aligned}$$

From the definition of $A_\tau$ and $B_\tau$ in Section 13.3, one obtains that $A_{\tau_i+\tau_{i-1}} = A_{\tau_i}A_{\tau_{i-1}}$ and $B_{\tau_i+\tau_{i-1}} = A_{\tau_i}B_{\tau_{i-1}} + B_{\tau_i}$ for any bounded $\tau_i > 0$ and $\tau_{i-1} > 0$. Substituting this result in the above inequality of $\Gamma(t_k^-)$ thus results in

$$\begin{aligned}\Gamma(t_k^-) &\preceq \omega^{-2}\Big(A_{\tau_{k-1}+\tau_{k-2}}\Gamma(t_{k-2}^-)A_{\tau_{k-1}+\tau_{k-2}}^\top\\ &\quad + B_{\tau_{k-1}+\tau_{k-2}}WB_{\tau_{k-1}+\tau_{k-2}}^\top\Big),\\ &\preceq \omega^{-\kappa}\Big(A_{\sum_{i=1}^{\kappa}\tau_{k-i}}\Gamma(t_{k-\kappa}^-)A_{\sum_{i=1}^{\kappa}\tau_{k-i}}^\top\\ &\quad + B_{\sum_{i=1}^{\kappa}\tau_{k-i}}WB_{\sum_{i=1}^{\kappa}\tau_{k-i}}^\top\Big).\end{aligned}$$

Note that $\Gamma(t_{k-\kappa}^-) = A_{\tau_{k-\kappa-1}}\Gamma(t_{k-\kappa-1})A_{\tau_{k-\kappa-1}}^\top + B_{\tau_{k-\kappa-1}}WB_{\tau_{k-\kappa-1}}^\top$, which after substituting in the above inequality gives that

$$\Gamma(t_k^-) \preceq \omega^{-\kappa}\left(A_{\delta_{k,\kappa-1}}\Gamma(t_{k-\kappa-1})A_{\delta_{k,\kappa-1}}^\top + B_{\delta_{k,\kappa-1}}WB_{\delta_{k,\kappa-1}}^\top\right) \tag{13.43}$$

where $\delta_{k,\kappa-1} := \sum_{i=1}^{\kappa+1}\tau_{k-i}$. The lemma considers time-periodic instants, that is, $t_k = \underline{t} \in \mathbb{T}_p$ and $t_{k-\kappa-1} = \underline{t} - \tau_s$. For those instants, one obtains $\delta_{k,\kappa-1} = \tau_s$ and thus $A_{\delta_{k,\kappa-1}} = A_{\tau_s} = \bar{A}$ and $B_{\delta_{k,\kappa-1}} = B_{\tau_s} = \bar{B}$. Substituting these results into (13.43) further implies that

$$\Gamma(\underline{t}^-) \preceq \omega^{-\kappa}\left(\bar{A}\Gamma(\underline{t} - \tau_s)\bar{A}^\top + \bar{B}W\bar{B}^\top\right).$$

From the fact that $\Sigma(\underline{t}^-) = \bar{A}\Sigma(\underline{t} - \tau_s)\bar{A}^\top + \bar{B}W\bar{B}^\top$, in combination with the assumption $\Gamma(\underline{t} - \tau_s) \preceq \Sigma(\underline{t} - \tau_s)$, one can then obtain that $\Gamma(\underline{t}^-) \preceq \omega^{-\kappa}\Sigma(\underline{t}^-)$, which completes the proof of this lemma. ■

Next, let us continue with the result of Theorem 13.4, which is proven by induction. The first step is to verify that $\Gamma(\tau_s) \preceq \Sigma(\tau_s)$ when $\Gamma(0) = \Sigma(0)$. Substituting the time-periodic instant $\underline{t} = \tau_s$ into Lemma 13.1 gives that the predicted covariance matrices $\Gamma(\tau_s^-)$ and $\Sigma(\tau_s^-)$ satisfy $\Gamma(\tau_s^-) \preceq \omega^{-\kappa}\Sigma(\tau_s^-)$. The result of this latter inequality implies that after the update equations of (13.40) and (13.41) one has that $\Gamma(\tau_s) \preceq \Sigma(\tau_s)$, which completes the first step.

The second step is to show that $\Gamma(t_k) \preceq \Sigma(t_n)$ holds for any $t_k = t_n \in \mathbb{T}_p$, when $\Gamma(t_k - \tau_s) \preceq \Sigma(t_n - \tau_s)$ holds. Since $t_k = t_n$ is a time-periodic instant and $\Gamma(t_k - \tau_s) \preceq \Sigma(t_n - \tau_s)$ holds, one can employ the results of Lemma 13.1 by considering $t_k = \underline{t}$ and $t_n = \underline{t}$. This lemma then states that $\Gamma(t_k^-)$ and $\Sigma(t_n^-)$ satisfy $\Gamma(t_k^-) \preceq \omega^{-\kappa}\Sigma(t_n^-)$. Substituting this result in the update equations of (13.40) and (13.41) further implies that $\Gamma(t_k) \preceq \Sigma(t_n)$, which completes the second step of induction and thereby, the proof of this theorem.

## Bibliography


[1] L. Aggoun and R. Elliot. *Measure Theory and Filtering*. Cambridge: Cambridge University Press, UK, 2004.

[2] A. Alessandri, M. Baglietto, and G. Battistelli. Receding-horizon estimation for discrete-time linear systems. *IEEE Transactions on Automatic Control*, 48(3):473–478, 2003.

[3] K. J. Åström and B. M. Bernhardsson. Comparison of Riemann and Lebesgue sampling for first order stochastic systems. In *Proceedings of the 41st IEEE Conf. on Decision and Control*, pages 2011–2016, Las Vegas, NV, 2002.

[4] F. L. Chernousko. *State Estimation for Dynamic Systems*. Boca Raton, FL: CRC Press, 1994.

[5] R. Cogill. Event-based control using quadratic approximate value functions. In *48th IEEE Conference on Decision and Control*, pages 5883–5888, Shanghai, China, 2009.

[6] R. E. Curry. *Estimation and Control with Quantized Measurements*. Clinton, MA: MIT Press, 1970.

[7] W. Dargie and C. Poellabauer. *Fundamentals of Wireless Sensor Networks: Theory and Practice*. Wiley, 2010.

[8] D. V. Dimarogonas and K. H. Johansson. Event-triggered control for multi-agent systems. In *Proceedings of the 48th IEEE Conference on Decision and Control*, pages 7131–7136, Shanghai, China, December 16–18, 2009.

[9] M. C. F. Donkers. *Networked and Event-Triggered Control Systems*. PhD thesis, Eindhoven University of Technology, 2012.

[10] W. P. M. H. Heemels, R. J. A. Gorter, A. van Zijl, P. P. J. van den Bosch, S. Weiland, W. H. A. Hendrix, and M. R. Vonder. Asynchronous measurement and control: A case study on motor synchronization. *Control Engineering Practice*, 7:1467–1482, 1999.

[11] T. Henningsson. Recursive state estimation for linear systems with mixed stochastic and set-bounded disturbances. In *Proceedings of the 47th IEEE Conference on Decision making and Control (CDC08)*, pages 678–683, Cancun, Mexico, December 9–11, 2008.

[12] T. Henningsson, E. Johannesson, and A. Cervin. Sporadic event-based control of first-order linear stochastic systems. *Automatica*, 44(11):2890–2895, 2008.

[13] O. C. Imer and T. Basar. Optimal estimation with limited measurements. In *Proceedings of the 44th IEEE Conference on Decision and Control*, pages 1029–1034, Seville, Spain, December 13–15, 2005.

[14] O. C. Imer and T. Basar. Optimal control with limited controls. In *American Control Conference*, pages 298–303, Minneapolis, MN, June 14–16, 2006.

[15] H. L. Lebesgue. Integrale, longueur, aire. PhD thesis, University of Nancy, 1902.

[16] D. Lehmann and J. Lunze. A state-feedback approach to event-based control. *Automatica*, 46:211–215, 2010.

[17] G. V. Moustakides, M. Rabi, and J. S. Baras. Multiple sampling for estimation on a finite horizon. In *Proceedings of the 45th IEEE Conference on Decision and Control*, pages 1351–1357, San Diego, CA, December 13–15, 2006.

[18] R. Mahler. General Bayes filtering of quantized measurements. In *Proceedings of the 14th International Conference on Information Fusion*, pages 346–352, Chicago, IL, July 5–8, 2011.

[19] M. Mallick, S. Coraluppi, and C. Carthel. Advances in asynchronous and decentralized estimation. In *Proceeding of the 2001 Aerospace Conference*, Big Sky, MT, March 10–17, 2001.

[20] J. W. Marck and J. Sijs. Relevant sampling applied to event-based state-estimation. In *Proceedings of the 4th International Conference on Sensor Technologies and Applications*, pages 618–624, Venice, Italy, July 18–25, 2010.

[21] K. V. Mardia, J. T. Kent, and J. M. Bibby. *Multivariate Analysis*. Academic Press, London, 1979.

[22] M. Miskowicz. Send-on-delta concept: An event-based data-reporting strategy. *Sensors*, 6:49–63, 2006.

[23] M. Miskowicz. Asymptotic effectiveness of the event-based sampling according to the integral criterion. *Sensors*, 7:16–37, 2007.

[24] P. Morerio, M. Pompei, L. Marcenaro, and C. S. Regazzoni. Exploiting an event based state estimator in presence of sparse measurements in video analytics. In *Proceedings of the 2014 International Conference on Acoustics, Speech and Signal Processing (ICASSP)*, pages 1871–1875, Florence, Italy, May 4–9, 2014.

[25] V. H. Nguyen and Y. S. Suh. Improving estimation performance in networked control systems applying the send-on-delta transmission method. *Sensors*, 7:2128–2138, 2007.

[26] B. Noack, F. Pfaff, and U. D. Hanebeck. Optimal Kalman gains for combined stochastic and set-membership state estimation. In *Proceedings of the 51st IEEE Conference on Decision and Control (CDC 2012)*, pages 4035–4040, Maui, HI, December 10–13, 2012.

[27] B. Saltik. Output feedback control of linear systems with event-triggered measurements. Master thesis, University of Technology, Eindhoven, 2013.

[28] D. Shi, T. Chen, and L. Shi. An event-triggered approach to state estimation with multiple point- and set-valued measurements. *Automatica*, 50(6):1641–1648, 2014.

[29] J. Sijs. State estimation in networked systems. PhD thesis, Eindhoven University of Technology, 2012.

[30] J. Sijs and M. Lazar. Event based state estimation with time synchronous updates. *IEEE Transactions on Automatic Control*, 57(10):2650–2655, 2012.

[31] B. Sinopoli, L. Schenato, M. Franceschetti, K. Poolla, M. Jordan, and S. Sastry. Kalman filter with intermittent observations. *IEEE Transactions on Automatic Control*, 49:1453–1464, 2004.

[32] H. W. Sorenson and D. L. Alspach. Recursive Bayesian estimation using Gaussian sums. *Automatica*, 7:465–479, 1971.

[33] C. Stocker. *Event-Based State-Feedback Control of Physically Interconnected Systems*. Berlin: Logos Verlag Berlin Gmbh, 2014.

[34] S. Trimpe and R. D'Andrea. An experimental demonstration of a distributed and event-based state estimation algorithm. In *Proceedings of the 18th IFAC World Congress*, pages 8811–8818, Milan, Italy, August 28–September 2, 2011.

[35] S. Trimpe and R. D'Andrea. Event-based state estimation with variance-based triggering. In *Proceedings of the 51st IEEE Conference on Decision and Control (CDC 2012)*, pages 6583–6590, Maui, HI, December 10–13, 2012.

[36] J. Wu, K. H. Johansson, and L. Shi. Event-based sensor data scheduling: Trade-off between communication rate and estimation quality. *IEEE Transactions on Automatic Control*, 58(4):1041–1046, 2013.

[37] Y. Xu and J. P. Hespanha. Optimal communication logics for networked control systems. In *Proceedings of the 43rd IEEE Conference on Decision and Control*, pages 3527–3532, Paradise Island, Bahamas, December 14–17, 2004.

# 14

## *Intermittent Control in Man and Machine*


**Peter Gawthrop**
*University of Melbourne*
*Melbourne, Victoria, Australia*

**Henrik Gollee**
*University of Glasgow*
*Glasgow, UK*

**Ian Loram**
*Manchester Metropolitan University*
*Manchester, UK*


**CONTENTS**

**ABSTRACT** It is now over 70 years since Kenneth J. Craik postulated that human control systems behave in an intermittent, rather than a continuous, fashion. This chapter provides a mathematical model of event-driven intermittent control, examines how this model explains some phenomena related to human motion control, and presents some experimental evidence for intermittency. Some new material related to constrained multivariable intermittent control is presented in the context of human standing, and some new material related to adaptive intermittent control is presented in the context of human balance and reaching.

We believe that the ideas presented here in a physiological context will also prove to be useful in an engineering context.

## 14.1 Introduction

Conventional sampled-data control uses a *zero-order* hold (ZOH), which produces a piecewise constant control signal (Franklin, Powell, and Emami-Naeini, 1994), and can be used to give a sampled-data implementation, which approximates a previously designed continuous-time controller. In contrast to conventional sampled-data control, intermittent control (Gawthrop and Wang, 2007) explicitly embeds the underlying continuous-time closed-loop system in a *generalized hold.* A number of versions of the generalized hold are available; this chapter focuses on the *system-matched* hold (SMH) (Gawthrop and Wang, 2011), which explicitly generates an open-loop intersample control trajectory based on the underlying continuous-time closed-loop control system. Other versions of the generalized hold include Laguerre function based holds (Gawthrop and Wang, 2007) and a "tapping" hold (Gawthrop and Gollee, 2012).

There are three areas where intermittent control has been used:

1. Continuous-time model-based predictive control (MPC) where the intermittency is associated with online optimization (Ronco, Arsan, and Gawthrop, 1999; Gawthrop and Wang, 2009a, 2010).
2. Event-driven control systems where the intersample interval is time varying and

determined by the event times (Gawthrop and Wang, 2009b, 2011).

3. Physiological control systems which, in some cases, have an event-driven intermittent character (Loram and Lakie, 2002; Gawthrop, Loram, Lakie, and Gollee, 2011). This intermittency may be due to the "computation" in the central nervous system (CNS). Although this chapter is orientated toward physiological control systems, we believe that it is more widely applicable.

Intermittent control has a long history in the physiological literature (e.g., Craik, 1947a,b; Vince, 1948; Navas and Stark, 1968; Neilson, Neilson, and O'Dwyer, 1988; Miall, Weir, and Stein, 1993a; Bhushan and Shadmehr, 1999; Loram and Lakie, 2002; Loram, Gollee, Lakie, and Gawthrop, 2011; Gawthrop et al., 2011). There is strong experimental evidence that some human control systems are intermittent (Craik, 1947a; Vince, 1948; Navas and Stark, 1968; Bottaro, Casadio, Morasso, and Sanguineti, 2005; Loram, van de Kamp, Gollee, and Gawthrop, 2012; van de Kamp, Gawthrop, Gollee, and Loram, 2013b), and it has been suggested that this intermittency arises in the CNS (van de Kamp, Gawthrop, Gollee, Lakie, and Loram, 2013a). For this reason, computational models of intermittent control are important and, as discussed below, a number of versions with various characteristics have appeared in the literature. Intermittent control has also appeared in various forms in the engineering literature including (Ronco et al., 1999; Zhivoglyadov and Middleton, 2003; Montestruque and Antsaklis, 2003; Insperger, 2006; Astrom, 2008; Gawthrop and Wang, 2007, 2009b; Gawthrop, Neild, and Wagg, 2012).

Intermittent control action may be initiated at regular intervals determined by a clock, or at irregular intervals determined by events; an event is typically triggered by an error signal crossing a threshold. Clock-driven control is discussed by Neilson et al. (1988) and Gawthrop and Wang (2007) and analysed in the frequency domain by Gawthrop (2009). Event-driven control is used by Bottaro et al. (2005); Bottaro, Yasutake, Nomura, Casadio, and Morasso (2008), Astrom (2008), Asai, Tasaka, Nomura, Nomura, Casadio, and Morasso (2009), Gawthrop and Wang (2009b), and Kowalczyk, Glendinning, Brown, Medrano-Cerda, Dallali, and Shapiro (2012). Gawthrop et al. (2011, Section 4) discuss event-driven control but with a lower limit $\Delta_{\min}$ on the time interval between events; this gives a range of behaviors including continuous, timed, and event-driven control. Thus, for example, threshold-based event-driven control becomes effectively clock driven with interval $\Delta_{\min}$ if the threshold is small compared to errors caused by relatively large disturbances. There is evidence that human control systems are, in fact, event driven (Navas and Stark, 1968; Loram et al., 2012; van de Kamp et al., 2013a; Loram, van de Kamp, Lakie, Gollee, and Gawthrop, 2014). For this reason, this chapter focuses on event-driven control.

As mentioned previously, intermittent control is based on an *underlying continuous-time design method;* in particular, the classical state-space approach is the basis of the intermittent control of Gawthrop et al. (2011). There are two relevant versions of this approach: state feedback and output feedback. State-feedback control requires that the current system state (e.g., angular position and velocity of an inverted pendulum) is available for feedback. In contrast, output feedback requires a measurement of the system output (e.g., angular position of an inverted pendulum). The classical approach to output feedback in a state-space context (Kwakernaak and Sivan, 1972; Goodwin, Graebe, and Salgado, 2001) is to use an observer (or the optimal version, a Kalman filter) to deduce the state from the system output.

Human control systems are associated with time delays. In engineering terms, it is well known that a predictor can be used to overcome time delay (Smith, 1959; Kleinman, 1969; Gawthrop, 1982). As discussed by many authors (Kleinman, Baron, and Levison, 1970; Baron, Kleinman, and Levison, 1970; McRuer, 1980; Miall, Weir, Wolpert, and Stein, 1993b; Wolpert, Miall, and Kawato, 1998; Bhushan and Shadmehr, 1999; Van Der Kooij, Jacobs, Koopman, and Van Der Helm, 2001; Gawthrop, Lakie, and Loram, 2008; Gawthrop, Loram, and Lakie, 2009; Gawthrop et al., 2011; Loram et al., 2012), it is plausible that physiological control systems have built in model-based prediction. Following Gawthrop et al. (2011), this chapter bases intermittent controller (IC) on an underlying predictive design.

The use of networked control systems leads to the "sampling period jitter problem" (Sala, 2007) where uncertainties in transmission time lead to unpredictable nonuniform sampling and stability issues (Cloosterman, van de Wouw, Heemels, and Nijmeijer, 2009). A number of authors have suggested that performance may be improved by replacing the standard ZOH by a generalized hold (Sala, 2005, 2007) or using a dynamical model of the system between samples (Zhivoglyadov and Middleton, 2003; Montestruque and Antsaklis, 2003). Similarly, event-driven control (Heemels, Sandee, and Bosch, 2008; Astrom, 2008), where sampling is determined by events rather than a clock, also leads to unpredictable nonuniform sampling. Hence, strategies for event-driven control would be expected to be similar to strategies for networked control. One particular form of event-driven control where events correspond to the system state moving beyond a fixed boundary has been called Lebesgue sampling in contrast to the so-called Riemann sampling of fixed-interval sampling (Astrom and Bernhardsson, 2002, 2003). In particular, Astrom (2008) uses a "control signal generator": essentially a dynamical model of the system between samples as advocated by Zhivoglyadov and Middleton (2003) for the networked control case.

As discussed previously, intermittent control has an interpretation which contains a generalized hold (Gawthrop and Wang, 2007). One particular form of hold is based on the closed-loop system dynamics of an underlying continuous control design: this will be called the SMH in this chapter. Insofar as this special case of intermittent control uses a dynamical model of the controlled system to generate the (open-loop) control between sample intervals, it is related to the strategies of both Zhivoglyadov and Middleton (2003) and Astrom (2008). However, as shown in this chapter, intermittent control provides a framework within to analyze and design a range of control systems with unpredictable nonuniform sampling possibly arising from an event-driven design. In particular, it is shown by Gawthrop and Wang (2011) that the SMH-based IC is associated with a separation principle similar to that of the underlying continuous-time controller, which states that the closed-loop poles of the intermittent control system consist of the control system poles and the observer system poles, and the interpolation using the system matched hold does not lead to the changes of closed-loop poles. As discussed by Gawthrop and Wang (2011), this separation principle is only valid when using the SMH. For example, intermittent control based on the standard ZOH does not lead to such a separation principle and therefore closed-loop stability is compromised when the sample interval is not fixed.

Human movement is characterized by low-dimensional goals achieved using high-dimensional muscle input (Shadmehr and Wise, 2005); in control system terms, the system has redundant actuators. As pointed out by Latash (2012), the abundance of actuators is an advantage rather than a problem. One approach to redundancy is by using the concept of *synergies* (Neilson and Neilson, 2005): groups of muscles which act in concert to give a desired action. It has been shown that such synergies arise naturally in the context of optimal control (Todorov, 2004; Todorov and Jordan, 2002) and experimental work has verified the existence of synergies *in vivo* (Ting, 2007; Safavynia and Ting, 2012). Synergies may be arranged in hierarchies. For example, in the context of posture, there is a natural three-level hierarchy with increasing dimension comprising task space, joint space, and muscle space. Thus, for example, a balanced posture could be a task requirement achievable by a range of possible joint torques each of which in turn corresponds to a range of possible muscle activation. This chapter focuses on the task space—joint space hierarchy previously examined in the context of robotics (Khatib, 1987).

In a similar way, humans have an abundance of measurements available; in control system terms, the system has redundant sensors. As discussed by Van Der Kooij, Jacobs, Koopman, and Grootenboer (1999) and Van Der Kooij et al. (2001), such sensors are utilized with appropriate sensor integration. In control system terms, sensor redundancy can be incorporated into state-space control using observers or Kalman–Bucy filters (Kwakernaak and Sivan, 1972; Goodwin et al., 2001); this is the dual of the optimal control problem. Again sensors can be arranged in a hierarchical fashion. Hence, optimal control and filtering provides the basis for a continuous-time control system that simultaneously applies sensor fusion to utilize sensor redundancy and optimal control to utilize actuator redundancy.

For these reasons, this chapter extends the single-input single-output IC of Gawthrop et al. (2011) to the multivariable case. As the formulation of Gawthrop et al. (2011) is set in the state space, this extension is quite straightforward. Crucially, the generalized hold, and in particular the SMH, remains as the heart of multivariable intermittent control.

The particular mathematical model of intermittent control proposed by Gawthrop et al. (2011) combines event-driven control action based on estimates of the controlled system state (position, velocity, etc.) obtained using a standard continuous-time state observer with continuous measurement of the system outputs. This model of intermittent control can be summarized as "continuous attention with intermittent action." However, the state estimate is only used at the event-driven sample time; hence, it would seem that it is not necessary for the state observer to monitor the controlled system all of the time. Moreover, the experimental results of Osborne (2013) suggest that humans can perform well even when vision is intermittently occluded. This chapter proposes an intermittent control model where a continuous-time observer monitors the controlled system intermittently: the periods of monitoring the system measurements are interleaved with periods where the measurement is *occluded*. This model of intermittent control can be summarized as "intermittent attention with intermittent action."

This chapter has two main parts:

1. Sections 14.2–14.4 give basic ideas about intermittent control.
2. Sections 14.5–14.13 explore more advanced topics and applications.

## 14.2 Continuous Control

Intermittent control is based on an *underlying design method* which, in this chapter, is taken to be conventional state space–based observer/state-feedback control (Kwakernaak and Sivan, 1972; Goodwin et al., 2001) with the addition of a state predictor (Fuller, 1968; Kleinman, 1969; Sage and Melsa, 1971; Gawthrop, 1976). Other control design approaches have been used in this context including pole-placement (Gawthrop and Ronco, 2002) and cascade control (Gawthrop, Lee, Halaki, and O'Dwyer, 2013b). It is also noted that many control designs can be embedded in LQ design (Maciejowski, 2007; Foo and Weyer, 2011) and thence used as a basis for intermittent control (Gawthrop and Wang, 2010).

Gawthrop et al. (2011) consider a single-input single-output formulation of intermittent control; this chapter considers a multi-input multi-output formulation. As in the single-input single-output case, this chapter considers linear time invariant systems with an $n \times 1$ vector state $\mathbf{x}$. As discussed by Gawthrop et al. (2011), the system, neuromuscular (NMS) and disturbances can be combined into a state-space model. For simplicity, the measurement noise signal $v_y$ will be omitted in this chapter except where needed. In contrast, however, this chapter is based on a multiple input, multiple output formulation. Thus, the corresponding state-space system has multiple outputs represented by the $n_y \times 1$ vector $\mathbf{y}$ and $n_o \times 1$ vector $\mathbf{y}_o$, multiple control inputs represented by the $n_u \times 1$ vector $\mathbf{u}$ and multiple unknown disturbance inputs represented by the $n_u \times 1$ vector $\mathbf{d}'$ where:

$$\begin{cases} \frac{d\mathbf{x}}{dt}(t) &= \mathbf{A}\mathbf{x}(t) + \mathbf{B}\mathbf{u}(t) + \mathbf{B}_d\mathbf{d}'(t) \\ \mathbf{y}(t) &= \mathbf{C}\mathbf{x}(t) \\ \mathbf{y}_o(t) &= \mathbf{C}_o\mathbf{x}(t) \end{cases}, \qquad (14.1)$$

$\mathbf{A}$ is an $n \times n$ matrix, $\mathbf{B}$ and $\mathbf{B}_d$ are a $n \times n_u$ matrices, $\mathbf{C}$ is a $n_y \times n$ matrix and $\mathbf{C}_o$ is a $n_o \times n$ matrix. The $n \times 1$ column vector $\mathbf{x}$ is the system state. In the multivariable context, there is a distinction between the $n_y \times n$ task vector $\mathbf{y}$ and the $n_o \times n$ observed vector $\mathbf{y}_o$: the former corresponds to control objectives, whereas the latter corresponds to system sensors and so provides information to the observer. Equation 14.1 is identical to Gawthrop et al. (2011, Equation 5) except that the scalar output $y$ is replaced by the vector outputs $\mathbf{y}$ and $\mathbf{y}_o$, the scalar input $u$ is replaced by the vector input $\mathbf{u}$ and the scalar input disturbance $d$ is replaced by the vector input disturbance $\mathbf{d}'$. Following standard practice (Kwakernaak and Sivan, 1972; Goodwin et al., 2001), it is assumed that $\mathbf{A}$ and $\mathbf{B}$ are such that the system (14.1) is *controllable* with respect to $\mathbf{u}$ and that $\mathbf{A}$ and $\mathbf{C}_o$ are such that the system (14.1) is *observable* with respect to $\mathbf{y}_o$.

As described previously (Gawthrop et al., 2011), Equation 14.1 subsumes a number of subsystems including the neuromuscular (actuator dynamics in the engineering context) and disturbance subsystems of Figure 14.1.

### 14.2.1 Observer Design and Sensor Fusion

The system states $\mathbf{x}$ of Equation 14.1 are rarely available directly due to sensor placement or sensor noise. As discussed in the textbooks (Kwakernaak and Sivan, 1972; Goodwin et al., 2001), an *observer* can be designed based on the system model (14.1) to approximately deduce the system states $\mathbf{x}$ from the measured signals encapsulated in the vector $\mathbf{y}_o$. In particular, the observer is given by

$$\frac{d\mathbf{x}_o}{dt}(t) = \mathbf{A}_o\mathbf{x}_o(t) + \mathbf{B}\mathbf{u}(t) + \mathbf{L}[\mathbf{y}_o(t) - \mathbf{v}_y(t)], \qquad (14.2)$$

where

$$\mathbf{A}_o = \mathbf{A} - \mathbf{L}\mathbf{C}_o, \qquad (14.3)$$

where the signal $\mathbf{v}_y(t)$ is the measurement noise. The $n \times n_o$ matrix $\mathbf{L}$ is the *observer gain matrix*. As discussed by, for example, Kwakernaak and Sivan (1972) and Goodwin et al. (2001), it is straightforward to design $\mathbf{L}$ using a number of approaches including pole-placement and the linear-quadratic optimization approach. The latter is used here and thus

$$\mathbf{L} = \mathbf{L}_o, \qquad (14.4)$$

where $\mathbf{L}_o$ is the observer gain matrix obtained using linear-quadratic optimization.

The observer deduces system states from the $n_o$ observed signals contained in $\mathbf{y}_o$; it is thus a particular form of sensor fusion with properties determined by the $n \times n_y$ matrix $\mathbf{L}$.

As discussed by Gawthrop et al. (2011), because the system (14.1) contains the disturbance dynamics of Figures 14.1 and 14.2, the corresponding observer deduces not only the state of the blocks labeled "System" and "NMS" in Figures 14.1 and 14.2, but also the state of block labeled "Dist."; thus, it acts as a *disturbance observer* (Goodwin et al., 2001, Chap. 14). A simple example appears in Section 14.4.1.

### 14.2.2 Prediction

Systems and controllers may contain pure time delays. Time delays are traditionally overcome using a *predictor*. The predictor of Smith (1959) [discussed by Astrom (1977)] was an early attempt at predictor design which,

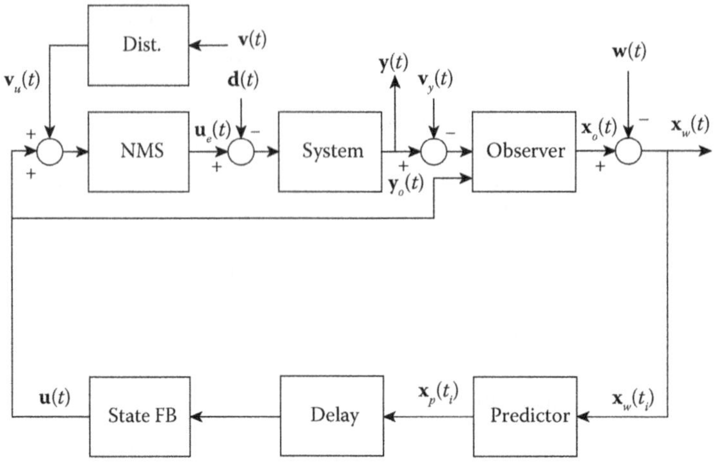


**FIGURE 14.1**
The Observer, Predictor, State-feedback (OPF) model. The block labeled "NMS" is a linear model of the neuromuscular dynamics with input $u(t)$; in the engineering context, this would represent actuator dynamics. "System" is the linear external controlled system driven by the externally observed control signal $u_e$ and disturbance $d$, and with output $y$ and associated measurement noise $v_y$. The input disturbance $v_u$ is modeled as the output of the block labeled "Dist." and driven by the external signal $v$. The block labeled "Delay" is a pure time-delay of $\Delta$ which accounts for the various delays in the human controller. The block labeled "Observer" gives an estimate $\mathbf{x}_o$ of the state $\mathbf{x}$ of the composite "NMS" and "System" (and, optionally, the "Dist.") blocks. The predictor provides an estimate of the future state error $\mathbf{x}_p(t)$ the delayed version of which is multiplied by the feedback gain vector $\mathbf{k}$ (block "State FB") to give the feedback control signal $u$. This figure is based on Gawthrop et al. (2011, Fig. 1) which is in turn based on Kleinman (1969, Fig. 2).

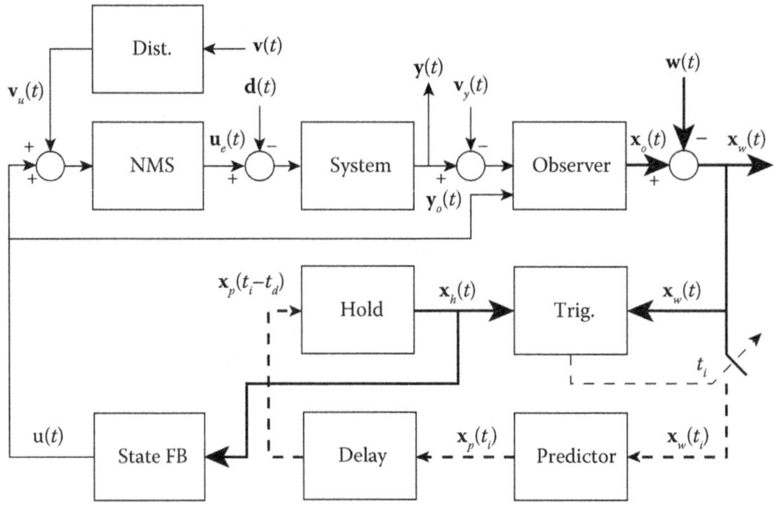


**FIGURE 14.2**
Intermittent control. This diagram has blocks in common with those of the OPF of Figure 14.1: "NMS", "Dist.", "System", "Observer", "Predictor", and "State FB", which have the same function; the continuous-time "Predictor" block of Figure 14.1 is replaced by the much simpler intermittent version here. There are three new elements: a sampling element which samples $\mathbf{x}_w$ at discrete times $t_i$; the block labeled "Hold", the system-matched hold, which provides the continuous-time input to the "State FB" block, and and the event detector block labeled "Trig.," which provides the trigger for the sampling times $t_i$. The dashed lines represent sampled signals defined only at the sample instants $t_i$. This figure is based on Gawthrop et al. (2011, Fig. 2).

however, cannot be used when the controlled system is unstable. State space–based predictors have been developed and used by a number of authors including Fuller (1968), Kleinman (1969), Sage and Melsa (1971), and Gawthrop (1976).

In particular, following Kleinman (1969), a state *predictor* is given by

$$\mathbf{x}_p(t+\Delta) = e^{\mathbf{A}\Delta}\mathbf{x}_o(t) + \int_0^{\Delta} e^{\mathbf{A}t'}\mathbf{B}\mathbf{u}(t-t')dt'. \quad (14.5)$$

Again, apart from the scalar $u$ being replaced by the vector $\mathbf{u}$ and $\mathbf{B}$ becoming an $n_u \times n$ matrix, Equation 14.5 is the same as in the single input ($n_u = 1$) case.

### 14.2.3 Controller Design and Motor Synergies

As described in the textbooks, for example, Kwakernaak and Sivan (1972) and Goodwin et al. (2001), the LQ controller problem involves minimization of

$$\int_0^{t_1} \mathbf{x}^{\mathrm{T}}(t)\mathbf{Q}_c\mathbf{x}(t) + \mathbf{u}^{\mathrm{T}}(t)\mathbf{R}_c\mathbf{u}(t)\,dt, \tag{14.6}$$

and letting $t_1 \to \infty$. $\mathbf{Q}_c$ is the $n \times n$ state-weighting matrix and $\mathbf{R}_c$ is the $n_u \times n_u$ control-weighting matrix. $\mathbf{Q}_c$ and $\mathbf{R}_c$ are used as design parameters in the rest of this chapter. As discussed previously (Gawthrop et al., 2011), the resultant state-feedback gain $\mathbf{k}$ ($n \times n_u$) may be combined with the predictor equation (14.5) to give the control signal $\mathbf{u}$

$$\mathbf{u}(t) = \mathbf{k}\mathbf{x}_w(t), \tag{14.7}$$

where

$$\mathbf{x}_w = \mathbf{x}_p(t) - \mathbf{x}_{ss}w(t). \tag{14.8}$$

As discussed by Kleinman (1969), the use of the state predictor gives a closed-loop system with no feedback delay and dynamics determined by the delay-free closed loop system matrix $\mathbf{A}_c$ given by

$$\mathbf{A}_c = \mathbf{A} - \mathbf{B}\mathbf{k}. \tag{14.9}$$

As mentioned by Todorov and Jordan (2002) and Todorov (2004), control synergies arise naturally from optimal control and are defined by the elements the $n_u \times n$ matrix $\mathbf{k}$.

A key result of state-space design in the delay free case is the *separation principle* [see Kwakernaak and Sivan (1972, section 5.3) and Goodwin et al. (2001, section 18.4)] whereby the observer and the controller can be designed separately.

### 14.2.4 Steady-State Design

As discussed in the single-input, single output case by Gawthrop et al. (2011), there are many ways to include the setpoint in the feedback controller and one way is to compute the steady-state state $\mathbf{x}_{ss}$ and control signal $\mathbf{u}_{ss}$ corresponding to the equilibrium of the ODE (14.1):

$$\frac{d\mathbf{x}}{dt} = \mathbf{0}_{n\times 1}, \tag{14.10}$$

$$\mathbf{y}_{ss} = \mathbf{C}\mathbf{x}_{ss}, \tag{14.11}$$

corresponding to a given constant value of output $\mathbf{y}_{ss}$. As discussed by Gawthrop et al. (2011), the scalars $\mathbf{x}_{ss}$ and $\mathbf{u}_{ss}$ are uniquely determined by $\mathbf{y}_{ss}$. In contrast, the multivariable case has additional flexibility; this section takes advantage of this flexibility by extending the equilibrium design in various ways.

In particular, Equation 14.11 is replaced by

$$\mathbf{y}_{ss} = \mathbf{C}_{ss}\mathbf{x}_{ss}, \tag{14.12}$$

where $\mathbf{y}_{ss}$ is a constant $n_{ss} \times m_{ss}$ matrix, $\mathbf{x}_{ss}$ is a constant $n \times m_{ss}$ matrix, and $\mathbf{C}_{ss}$ is an $n_{ss} \times m_{ss}$ matrix.

Typically, the equilibrium space defined by $\mathbf{y}_{ss}$ corresponds to the task space so that, with reference to Equation 14.1, each column of $\mathbf{y}_{ss}$ is a steady-state value of $\mathbf{y}$ (e.g., $\mathbf{y}_{ss} = \mathbf{I}_{n_y\times n_y}$) and $\mathbf{C}_{ss} = \mathbf{C}$. Further, assume that the disturbance $\mathbf{d}'(t)$ of (14.1) has $m_{ss}$ alternative constant values that form the columns of the $n_u \times m_{ss}$ matrix $\mathbf{d}_{ss}$.

Substituting the steady-state condition of Equation 14.10 into Equation 14.1 and combining with Equation 14.12 gives

$$\mathbf{S}\begin{bmatrix}\mathbf{x}_{ss}\\ \mathbf{u}_{ss}\end{bmatrix} = \begin{bmatrix}-\mathbf{B}_d\mathbf{d}_{ss}\\ y_{ss}\end{bmatrix}, \tag{14.13}$$

where

$$\mathbf{S} = \begin{bmatrix}\mathbf{A} & \mathbf{B}\\ \mathbf{C}_{ss} & \mathbf{0}_{n_{ss}\times n_u}\end{bmatrix}. \tag{14.14}$$

The matrix $\mathbf{S}$, has $n + n_{ss}$ rows $n + n_u$ columns, thus there are three possibilities:

$n_{ss} = n_u$ If $\mathbf{S}$ is full rank, Equation 14.13 has a unique solution for $\mathbf{x}_{ss}$ and $\mathbf{u}_{ss}$.

$n_{ss} < n_u$ Equation 14.13 has many solutions corresponding to a low dimensional manifold in a high dimensional space. A particular solution may be chosen to satisfy an additional criterion such as a minimum norm solution. An example is given in Section 14.7.4.

$n_{ss} > n_u$ Equation 14.13 is over-determined; a least-squares solution is possible. This case is considered in more detail in Section 14.5 and an example is given in Section 14.7.5.

Having obtained a solution for $\mathbf{x}_{ss}$, each of the $m_{ss}$ columns of the $n \times m_{ss}$ steady-state matrix $\mathbf{x}_{ss}$ can be associated with an element of a $m_{ss} \times 1$ *weighting vector* $\mathbf{w}(t)$. The error signal $\mathbf{x}_w(t)$ is then defined as the difference between the estimated state $\mathbf{x}_o(t)$ and the weighted columns of $\mathbf{x}_{ss}$ as

$$\mathbf{x}_w(t) = \mathbf{x}_o(t) - \mathbf{x}_{ss}\mathbf{w}(t). \tag{14.15}$$

Following Gawthrop et al. (2011), $\mathbf{x}_w(t)$ replaces $\mathbf{x}_o$ in the predictor equation (14.5) and the state-feedback controller remains Equation 14.7.

**Remarks**

1. In the single-input case ($n_u = 1$) setting $\mathbf{y}_{ss} = 1$ and $\mathbf{d}_{ss} = 0$ gives the same formulation as given by Gawthrop et al. (2011) and $\mathbf{w}(t)$ is the setpoint.

2. Disturbances may be unknown. Thus, using this approach requires disturbances to be estimated in some way.

3. Setpoint tracking is considered in Section 14.7.4.

4. The effect of a constant disturbance is considered in Section 14.7.5.

5. Constrained solutions are considered in Section 14.5.1.

## 14.3 Intermittent Control

Intermittent control is based on the underlying continuous-time design of Section 14.2. The purpose is to allow control computation to be performed intermittently at discrete time points—which may be determined by time (clock-driven) or the system state (event-driven)—while retaining much of the continuous-time behavior.

A disadvantage of traditional clock-driven discrete-time control (Franklin and Powell, 1980; Kuo, 1980) based on the ZOH is that the control needs to be redesigned for each sample interval. This also means that the ZOH approach is inappropriate for event-driven control. The intermittent approach avoids these issues by replacing the ZOH by the SMH. Because the SMH is based on the system state, it turns out that it does not depend on the number of system inputs $n_u$ or outputs $n_y$ and therefore the SMH described by Gawthrop et al. (2011) in the single input $n_u = 1$, single output context $n_y = 1$ context carries over to the multi-input $n_u > 1$, and multi-output $n_y > 1$ case.

This section is a tutorial introduction to the SMH-based IC in both clock-driven and event-driven cases. Section 14.3.1 looks at the various time-frames involved, Section 14.3.2 describes the SMH, and Sections 14.3.3–14.3.5 look at the observer, predictor, and feedback control, developed in the continuous-time context in Section 14.2, in the intermittent context. Section 14.3.6 looks at the event detector used for the event-driven version of intermittent control.

### 14.3.1 Time Frames

As discussed by Gawthrop et al. (2011), intermittent control makes use of three time frames:

1. **Continuous-time**, within which the controlled system (14.1) evolves is denoted by $t$.

2. **Discrete-time** points at which feedback occurs is indexed by $i$. Thus, for example, the discrete-time time instants are denoted by $t_i$ and the corresponding estimated state is $\mathbf{x}_{oi} = \mathbf{x}_o(t_i)$. The $i$th **intermittent interval** $\Delta_{ol} = \Delta_i$* is defined as

$$\Delta_{ol} = \Delta_i = t_{i+1} - t_i. \qquad (14.16)$$

This chapter distinguishes between event times $t_i$ and the corresponding sample times $t_i^s$. In particular, the model of Gawthrop et al. (2011) is extended so that sampling occurs a fixed time $\Delta_s$ after an event at time $t_i$ thus:

$$t_i^s = t_i + \Delta_s \qquad (14.17)$$

$\Delta_s$ is called the *sampling delay* in the sequel.

3. **Intermittent-time** is a continuous-time variable, denoted by $\tau$, restarting at each intermittent interval. Thus, within the $i$th intermittent interval:

$$\tau = t - t_i. \qquad (14.18)$$

Similarly, define the intermittent time $\tau^s$ after a sample by

$$\tau^s = t - t_i^s. \qquad (14.19)$$

A lower bound $\Delta_{\min}$ is imposed on each intermittent interval $\Delta_i > 0$ (14.16):

$$\Delta_i > \Delta_{\min} > 0. \qquad (14.20)$$

As discussed by Gawthrop et al. (2011) and in Section 14.4.2, $\Delta_{\min}$ is related to the Psychological Refractory Period (PRP) of Telford (1931) as discussed by Vince (1948) to explain the human response to double stimuli. As well as corresponding to the PRP explanation, the lower bound of (14.20) has two implementation advantages. Firstly, as discussed by Ronco et al. (1999), the time taken to compute the control signal (and possibly other competing tasks) can be up to $\Delta_{\min}$. It thus provides a model for a single processor bottleneck. Secondly, as discussed by Gawthrop et al. (2011), the predictor equations are particularly simple if the system time-delay $\Delta \le \Delta_{\min}$.

*Within this chapter, we will use $\Delta_{ol}$ to refer to the generic concept of intermittent interval and $\Delta_i$ to refer to the length of the $i$th interval.

### 14.3.2 System-Matched Hold

The SMH is the key component of the intermittent control. As described by Gawthrop et al. (2011, Equation 23), the SMH state $\mathbf{x}_h$ evolves in the *intermittent* time frame $\tau$ as

$$\frac{d}{d\tau}\mathbf{x}_h(\tau) = \mathbf{A}_h\mathbf{x}_h(\tau), \tag{14.21}$$

where

$$\mathbf{A}_h = \mathbf{A}_c, \tag{14.22}$$
$$\mathbf{x}_h(0) = \mathbf{x}_p(t_i^s - \Delta), \tag{14.23}$$

where $\mathbf{A}_c$ is the closed-loop system matrix (14.9) and $\mathbf{x}_p$ is given by the predictor equation (14.5). The hold state $\mathbf{x}_h$ replaces the predictor state $\mathbf{x}_p$ in the controller equation (14.7). Other holds (where $\mathbf{A}_h \neq \mathbf{A}_c$) are possible (Gawthrop and Wang, 2007; Gawthrop and Gollee, 2012).

The IC generates an open loop control signal based on the hold state $\mathbf{x}_h$ (14.21). At the intermittent sample times $t_i$, the hold state is reset to the estimated system state $\mathbf{x}_w$ generated by the observer (14.2); thus feedback occurs at the intermittent sample times $t_i$. The sample times are constrained by (14.20) to be at least $\Delta_{\min}$ apart. But, in addition to this constraint, feedback only takes place when it is needed; the event detector discussed in Section 14.3.6 provides this information.

### 14.3.3 Intermittent Observer

The IC of Gawthrop et al. (2011) uses continuous observation however, motivated by the occlusion experiments of Osborne (2013), this chapter looks a intermittent observation.

As discussed in Section 14.3.2, the predictor state $\mathbf{x}_p$ is only sampled at discrete-times $t_i$. Further, from Equation 14.5, $\mathbf{x}_p$ is a function of $\mathbf{x}_o$ at these times. Thus the only the observer performance at the discrete-times $t_i$ is important. With this in mind, this chapter proposes, in the context of intermittent control, that the continuous observer is replaced by an intermittent observer where periods of monitoring the system measurements are interleaved with periods where the measurement is *occluded*. In particular, and with reference to Figure 14.3, this chapter examines the situation where observation is occluded for a time $\Delta_{oo}$ following sampling. Such occlusion is equivalent to setting the observer gain $L = 0$ in Equation 14.2. Setting $L = 0$ has two consequences: the measured signal $y$ is ignored and the observer state evolves as the disturbance-free system.

With reference to Equation 14.23; the IC only makes use of the state estimate at the discrete time points at $t = t_i^s$ (14.17); moreover, in the event-driven case, the observer state estimate is used in Equation 14.35 to determine the event times $t_i$ and thus $t_i^s$. Hence, a good state estimate immediately after an sample at time $t_i^s$ is not required and so one would expect that occlusion ($L = 0$) would have little effect immediately after $t = t_i^s$. For this reason, define the occlusion time. $\Delta_{oo}$ as the time after $t = t_i^s$ for which the observer is open-loop $L = 0$. That is, the constant observer gain is replaced by the time varying observer gain:

$$L(t) = \begin{cases} 0 & \tau^s < \Delta_{oo} \\ L_o & \tau^s \geq \Delta_{oo} \end{cases}, \tag{14.24}$$

where $L_o$ is the observer gain designed using standard techniques (Kwakernaak and Sivan, 1972; Goodwin et al., 2001) and the intermittent time $\tau^s$ is given by (14.19).

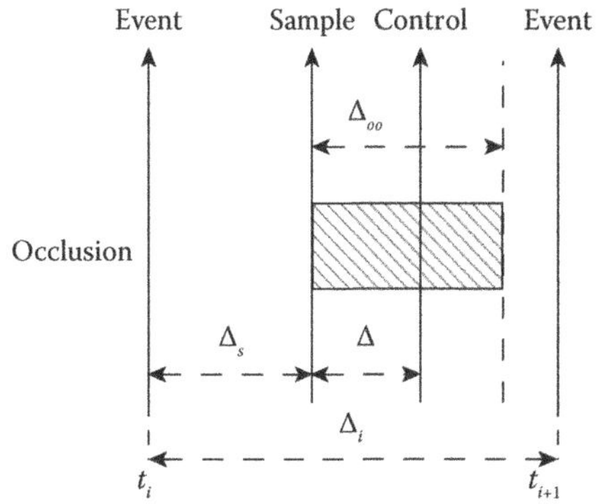


**FIGURE 14.3**
Self-occlusion. Following an event, the observer is sampled at a time $\Delta_s$ and a new control trajectory is generated. The observer is then occluded for a further time $\Delta_{oo} = \Delta_o$ where $\Delta_o$ is the internal occlusion interval. Following that time, the observer is operational and an event can be detected (14.35). The actual time between events $\Delta_i > \Delta_s + \Delta_{oo}$.

### 14.3.4 Intermittent Predictor

The continuous-time predictor of Equation 14.5 contains a convolution integral which, in general, must be approximated for real-time purposes and therefore has a speed-accuracy trade-off. This section shows that the use of intermittent control, together with the hold of Section 14.3.2, means that Equation 14.5 can be replaced by a simple exact formula.

Equation 14.5 is the solution of the differential equation (in the intermittent time $\tau$ (14.18) time frame)

$$\begin{cases} \frac{d}{d\tau}\mathbf{x}_p(\tau) & = \mathbf{A}\mathbf{x}_p(\tau) + \mathbf{B}u(\tau) \\ \mathbf{x}_p(0) & = \mathbf{x}_w(t_i^s) \end{cases}, \tag{14.25}$$

evaluated at time $\tau^s = \Delta$ where $\tau_i^s$ is given by Equation 14.17. However, the control signal $\mathbf{u}$ is not arbitrary but rather given by the hold Equation 14.21. Combining Equations 14.21 and 14.25 gives

$$\begin{cases} \frac{d}{d\tau}\mathbf{X}(\tau) & = \mathbf{A}_{ph}\mathbf{X}(\tau) \\ \mathbf{X}(0) & = \mathbf{X}_i \end{cases}, \tag{14.26}$$

where

$$\mathbf{X}(\tau) = \begin{pmatrix} \mathbf{x}_p(\tau) \\ \mathbf{x}_h(\tau) \end{pmatrix}, \tag{14.27}$$

$$\mathbf{X}_i = \begin{pmatrix} \mathbf{x}_w(t_i) \\ \mathbf{x}_p(t_i - \Delta) \end{pmatrix}, \tag{14.28}$$

and

$$\mathbf{A}_{ph} = \begin{pmatrix} \mathbf{A} & -\mathbf{Bk} \\ \mathbf{0}_{n\times n} & \mathbf{A}_h \end{pmatrix}, \tag{14.29}$$

where $\mathbf{0}$ is a zero matrix of the indicated dimensions and the hold matrix $\mathbf{A}_h$ can be $\mathbf{A}_c$ (SMH) or $\mathbf{0}$ (ZOH).

The Equation 14.26 has an explicit solution at time $\tau = \Delta$ given by

$$\mathbf{X}(\Delta) = e^{\mathbf{A}_{ph}\Delta}\mathbf{X}_i. \tag{14.30}$$

The prediction $\mathbf{x}_p$ can be extracted from (14.30) to give

$$\mathbf{x}_p(t_i) = \mathbf{E}_{pp}\mathbf{x}_w(t_i) + \mathbf{E}_{ph}\mathbf{x}_h(t_i), \tag{14.31}$$

where the $n \times n$ matrices $\mathbf{E}_{pp}$ and $\mathbf{E}_{ph}$ are partitions of the $2n \times 2n$ matrix $\mathbf{E}$:

$$\mathbf{E} = \begin{pmatrix} \mathbf{E}_{pp} & \mathbf{E}_{ph} \\ \mathbf{E}_{hp} & \mathbf{E}_{hh} \end{pmatrix}, \tag{14.32}$$

where

$$\mathbf{E} = e^{\mathbf{A}_{ph}\Delta}. \tag{14.33}$$

The intermittent predictor (14.31) replaces the continuous-time predictor (14.5); there is no convolution involved and the matrices $\mathbf{E}_{pp}$ and $\mathbf{E}_{ph}$ can be computed off-line and so do not impose a computational burden in real-time.

### 14.3.5 State Feedback

The "state-feedback" block of Figure 14.2 is implemented as

$$\mathbf{u}(t) = -\mathbf{k}\mathbf{x}_h(t). \tag{14.34}$$

This is similar to the conventional state feedback of Figure 14.1 given by Equation 14.7, but the continuous predicted state $\mathbf{x}_w(t)$ is replaced by the hold state $\mathbf{x}_h(t)$ generated by Equation 14.21.

### 14.3.6 Event Detector

The purpose of the event detector is to generate the intermittent sample times $t_i$ and thus trigger feedback. Such feedback is required when the open-loop hold state $\mathbf{x}_h$ (14.21) differs significantly from the closed-loop observer state $\mathbf{x}_w$ (14.15) indicating the presence of disturbances. There are many ways to measure such a discrepancy; following Gawthrop et al. (2011), the one chosen here is to look for a quadratic function of the error $\mathbf{e}_{hp}$ exceeding a threshold $q_t^2$:

$$E = \mathbf{e}_{hp}^{\mathrm{T}}(t)\mathbf{Q}_t\mathbf{e}_{hp}(t) - q_t^2 \geq 0, \tag{14.35}$$

where

$$\mathbf{e}_{hp}(t) = \mathbf{x}_h(t) - \mathbf{x}_w(t), \tag{14.36}$$

where $\mathbf{Q}_t$ is a positive semi-definite matrix.

### 14.3.7 The Intermittent-Equivalent Setpoint

Loram et al. (2012) introduce the concept of the *equivalent setpoint* for intermittent control. This section extends the concept and there are two differences:

1. The setpoint sampling occurs at $t_i + \Delta_s$ rather than at $t_i$ and
2. The filtered setpoint $\mathbf{w}_f$ (rather than $\mathbf{w}$) is sampled.

Define the sample time $t_i^s$ (as opposed to the event time $t_i$ and the corresponding intermittent time $\tau^s$ by

$$t_i^s = t_i + \Delta_s, \tag{14.37}$$

$$\tau^s = \tau - \Delta_s = t - t_i - \Delta_s = t - t_i^s. \tag{14.38}$$

In particular, the sampled setpoint $\mathbf{w}_s$ becomes

$$\mathbf{w}_s(t) = \mathbf{w}_f(t_i^s) \text{ for } t_i^s \leq t < t_{i+1}^s, \tag{14.39}$$

where $\mathbf{w}_f$ is the *filtered* setpoint $\mathbf{w}$. That is the sampled setpoint $\mathbf{w}_s$ is the filtered setpoint at time $t_i^s = t_i + \Delta_s$.

The equivalent setpoint $\mathbf{w}_{ic}$ is then given by

$$\mathbf{w}_{ic}(t) = \mathbf{w}_s(t - t_d) \tag{14.40}$$

$$= \mathbf{w}_f(t_i^s - t_d) \tag{14.41}$$

$$= \mathbf{w}_f(t - \tau^s - t_d) \text{ for } t_i^s \leq t < t_{i+1}^s. \tag{14.42}$$

This corresponds to the previous result (Loram et al., 2012) when $\Delta_s = 0$ and $\mathbf{w}_f(t) = \mathbf{w}(t)$.

If, however, the setpoint $\mathbf{w}(t)$ is such that $\mathbf{w}_f(t_i^s) \approx \mathbf{w}(t_s)$ (i.e., no second stimulus within the filter settling time and $\Delta_s$ is greater than the filter settling time) then Equation 14.40 may be approximated by

$$\mathbf{w}_{ic}(t) \approx \mathbf{w}(t_i^s - t_d) \text{ for } t_i^s \leq t < t_{i+1}^s \quad (14.43)$$
$$= \mathbf{w}(t - \tau^s - t_d) \text{ for } t_i^s \leq t < t_{i+1}^s. \quad (14.44)$$

As discussed in Section 14.10, the intermittent-equivalent setpoint is the basis for identification of intermittent control.

### 14.3.8 The Intermittent Separation Principle

As discussed in Section 14.3.2, the IC contains an SMH which can be views as a particular form of generalized hold (Gawthrop and Wang, 2007). Insofar as this special case of intermittent control uses a dynamical model of the controlled system to generate the (open-loop) control between sample intervals, it is related to the strategies of both Zhivoglyadov and Middleton (2003) and Astrom (2008). However, as shown in this chapter, intermittent control provides a framework within which to analyze and design a range of control systems with unpredictable nonuniform sampling possibly arising from an event-driven design.

In particular, it is shown by Gawthrop and Wang (2011), that the SMH-based IC is associated with a separation principle similar to that of the underlying continuous-time controller, which states that the closed-loop poles of the intermittent control system consist of the control system poles and the observer system poles, and the interpolation using the system matched hold does not lead to the changes of closed-loop poles. As discussed by Gawthrop and Wang (2011), this separation principle is only valid when using the SMH. For example, intermittent control based on the standard ZOH does not lead to such a separation principle and therefore closed-loop stability is compromised when the sample interval is not fixed.

As discussed by Gawthrop and Wang (2011), an important consequence of this separation principle is that the neither the design of the SMH, nor the stability of the closed-loop system in the fixed sampling case, is dependent on sample interval. It is therefore conjectured that the SMH is particularly appropriate when sample times are unpredictable or nonuniform, possibly arising from an event-driven design.

## 14.4 Examples: Basic Properties of Intermittent Control

This section uses simulation to illustrate key properties of intermittent control. Section 14.4.1 illustrates

- Timed and event-driven control (Section 14.3.6).
- The roles of the disturbance observer and series integrator (Section 14.2.1).
- The choice of event threshold (Section 14.3.6).
- The difference between control-delay and sampling delay (Section 14.3.1).
- The effect of low and high observer gain (Section 14.2.1) and
- The effect of occlusion (Section 14.3.4).

Sections 14.4.2 and 14.4.3 illustrates how the IC models two basic psychological phenomenon: the *Psychological Refractory Period* and the *Amplitude Transition Function* (ATF).

### 14.4.1 Elementary Examples

This section illustrates the basic properties of intermittent control using simple examples. In all cases, the system is given by

$$G_0(s) = \frac{1}{s^2-1} = \frac{1}{(s-1)(s+1)}$$
$$\text{Second-order unstable system,} \quad (14.45)$$
$$G_v(s) = \frac{1}{s}$$
$$\text{Simple integrator for disturbance observer.} \quad (14.46)$$

The corresponding state-space system (14.1) is

$$\mathbf{A} = \begin{pmatrix} 0 & 0 & 1 \\ 1 & 0 & 0 \\ 0 & 0 & 0 \end{pmatrix}, \quad (14.47)$$

$$\mathbf{B} = \mathbf{B}_d = \begin{pmatrix} 1 \\ 0 \\ 0 \end{pmatrix}, \quad (14.48)$$

$$\mathbf{B}_v = \begin{pmatrix} 0 \\ 0 \\ 1 \end{pmatrix}, \quad (14.49)$$

$$\mathbf{C} = \begin{pmatrix} 0 & 1 & 0 \end{pmatrix}, \quad (14.50)$$

All signals are zero except:

$$w(t) = 1 \qquad t \geq 1.1, \quad (14.51)$$
$$d(t) = 0.5 \qquad t \geq 5.1. \quad (14.52)$$

Except where stated, the intermittent control parameters are

$$\Delta_{\min} = 0.5 \quad \text{Min. intermittent interval (14.20)}$$
$$q_t = 0.1 \quad \text{Threshold (14.35)}$$
$$\Delta = 0 \quad \text{Control delay (14.5)}$$
$$\Delta_s = 0 \quad \text{Sampling delay (14.17)}$$

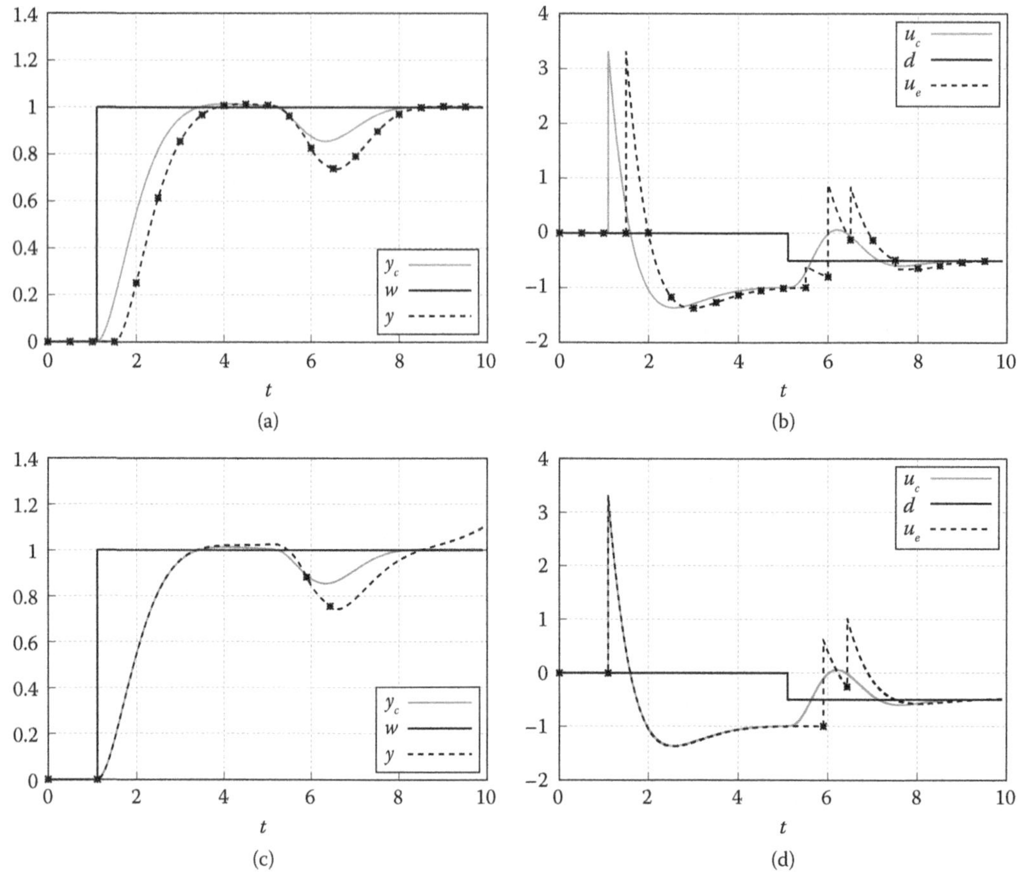


**FIGURE 14.4**
Elementary example: timed and event-driven. (a) Timed: $y$. (b) Timed: $u$. (c) Event-driven: $y$. (d) Event-driven: $u$.

Figures 14.4–14.9 are all of the same format. The left column of figures shows the system output $y$ together with the setpoint $w$ and the output $y_c$ corresponding to the underlying continuous-time design; the right column shows the corresponding control signal $u_e$ together with the negative disturbance $-d$ and the control $u_c$. In each case, the symbol (•) corresponds to an event.

Figure 14.4 contrasts timed and event driven control. In particular, Figure 14.4a and b corresponds to zero threshold ($q_t = 0$) and thus timed intermittent control with fixed interval $\Delta_{\min} = 0.5$ and Figure 14.4c and d corresponds to event-driven control. The event driven case has two advantages: the controller responds immediately to the setpoint change at time $t = 1.1$ whereas the timed case has to wait until the next sample at $t = 1.5$ and the control is only computed when required. In particular, the initial setpoint response does not need to be corrected, but the unknown disturbance means that the observer state is different from the system state for a while and so corrections need to be made until the disturbance is correctly deduced by the observer.

The simulation of Figure 14.4 includes the disturbance observer implied by the integrator of Equation 14.46; this means that the controllers are able to asymptotically eliminate the constant disturbance $d$. Figure 14.5a and b shows the effect of not using the disturbance observer. The constant disturbance $d$ is not eliminated and the IC exhibits limit cycling behavior (analyzed further by Gawthrop (2009)). As an alternative to the disturbance observer used in the simulation of Figure 14.4, a series integrator can be used by setting:

$$G_s(s) = \frac{1}{s} \quad \text{Series integrator for disturbance rejection.} \tag{14.53}$$

The corresponding simulation is shown in Figure 14.5c and d.* Although the constant disturbance $d$ is now asymptotically eliminated, the additional integrator increases both the system order and the system relative degree by one giving a more difficult system to control.

The event detector behavior depends on the threshold $q_t$ (14.35); this has already been examined in

*The system dynamics are now different; the LQ design parameter is set to $Q_c = 100$ to account for this.

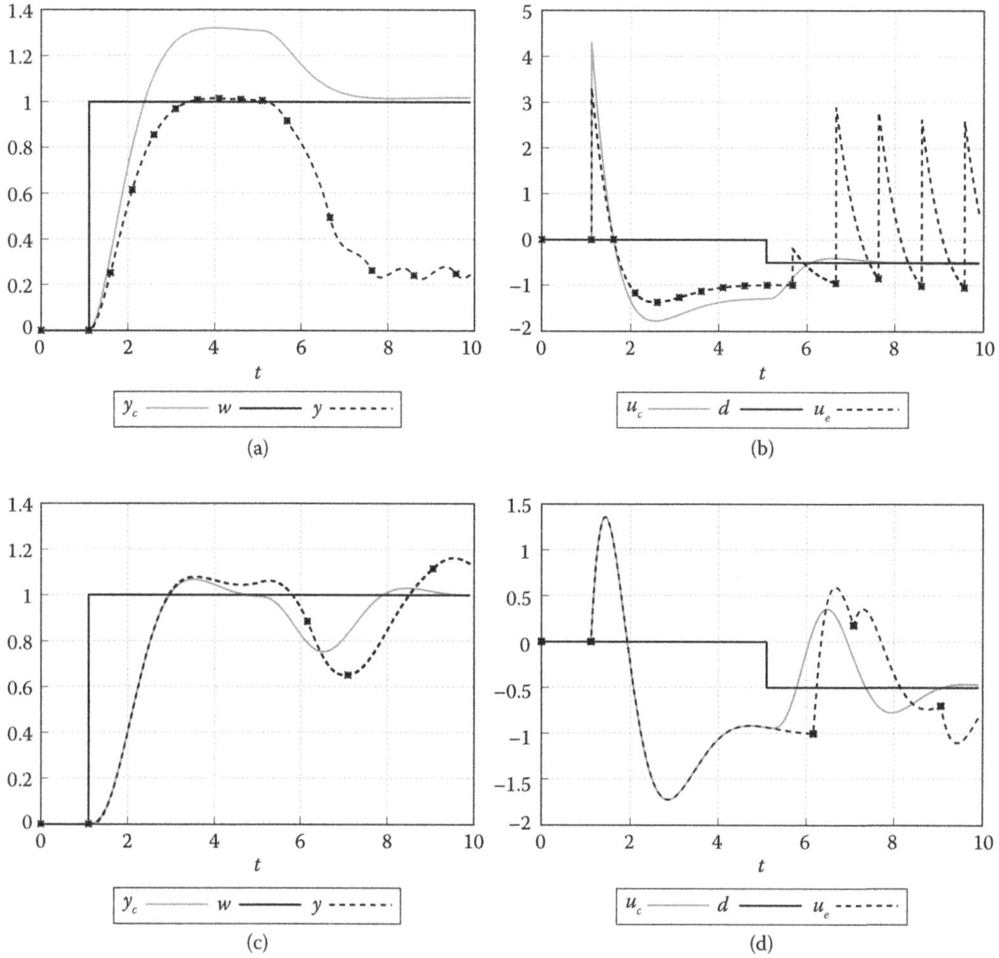


**FIGURE 14.5**

Elementary example: no disturbance observer, with and without integrator. (a) No integrator: $y$. (b) No integrator: $u$. (c) Integrator: $y$. (d) Integrator: $u$.

the simulations of Figure 14.4. Figure 14.6 shows the effect of a low ($q_t = 0.01$) and high ($q_t = 1$) threshold. As discussed in the context of Figure 14.4, the initial setpoint response does not need to be corrected, but the unknown disturbance generates events. The simulations of Figure 14.6 indicate the trade-off between performance and event rate determined by the choice of the threshold $q_t$.

The simulations of Figure 14.7 compare and contrast the two delays: control delay $\Delta$ and sample delay $\Delta_s$. In particular, Figure 14.7a and b corresponds to $\Delta = 0.4$ and $\Delta_s = 0$ but Figure 14.7c and d corresponds to $\Delta = 0$ and $\Delta_s = 0.4$. The response to the setpoint is identical as the prediction error is zero in this case; the response to the disturbance change is similar, but not identical as the prediction error is not zero in this case.

The state observer of Equation 14.2 is needed to deduce unknown states in general and the state corresponding to the unknown disturbance in particular. As discussed in the textbooks (Kwakernaak and Sivan, 1972; Goodwin et al., 2001), the choice of observer gain gives a trade-off between measurement noise and disturbance responses. The gain used in the simulations of Figure 14.4 can be regarded as medium; Figure 14.8 looks at low and high gains. As there is no measurement noise in this case, the low gain observer gives a poor disturbance response while the high gain gives an improved disturbance response.

The simulations presented in Figure 14.9 investigate the intermittent observer of Section 14.3.3. In particular, the measurement of the system output $y$ is assumed to be *occluded* for a period $\Delta_{oo}$ following a sample. Figure 14.9a and b shows simulation with $\Delta_{oo} = 0.1$ and Figure 14.9c and d shows simulation with $\Delta_{oo} = 0.5$. It can be seen that occlusion has little effect on performance for the lower value, but performance is poor for the larger value.

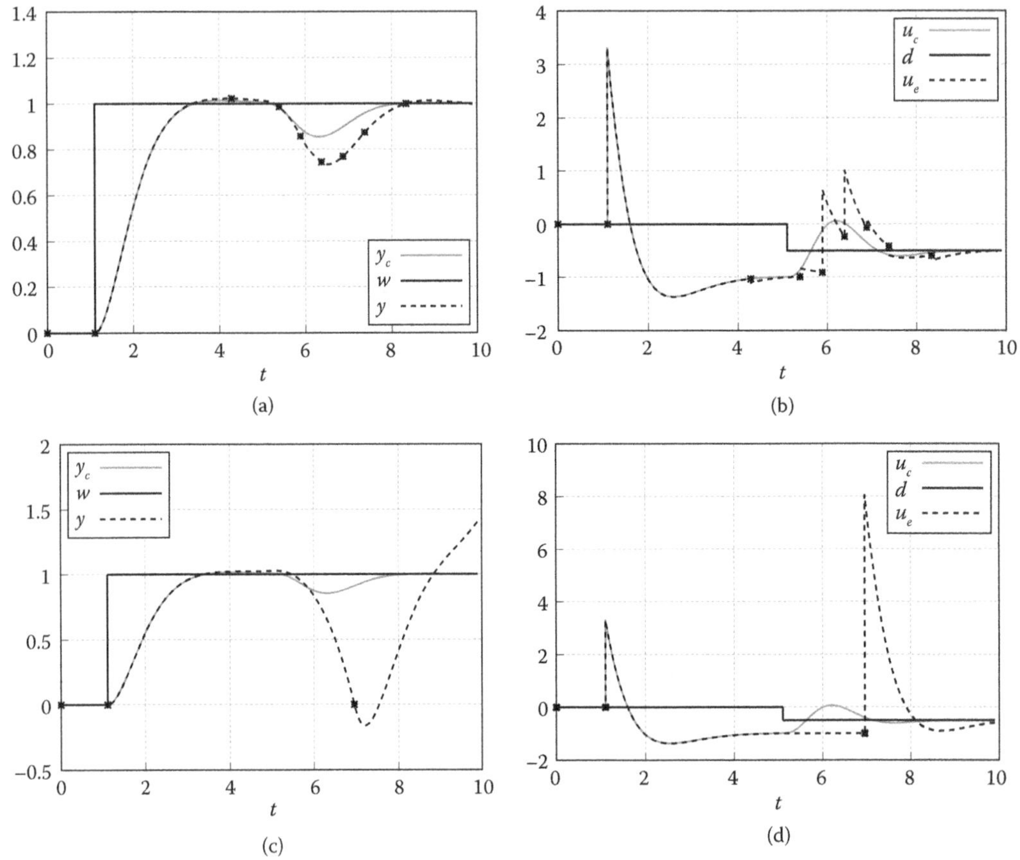


**FIGURE 14.6**
Elementary example: low and high threshold. (a) Low threshold: $y$. (b) Low threshold: $u$. (c) High threshold: $y$. (d) High threshold: $u$.

### 14.4.2 The Psychological Refractory Period and Intermittent-Equivalent Setpoint

As noted in Section 14.3.7, the intermittent sampling of the setpoint $w$ leads to the concept of the intermittent-equivalent setpoint: the setpoint that is actually used within the IC. Moreover, as noted in Section 14.3.1, there is a minimum intermittent interval $\Delta_{\text{min}}$. As discussed by Gawthrop et al. (2011), $\Delta_{\text{min}}$ is related to the *psychological refractory period* (Telford, 1931) which explains the experimental results of Vince (1948) where a second reaction time may be longer than the first. These ideas are explored by simulation in Figures 14.10–14.12. In all cases, the system is given by

$$G_0(s) = \frac{1}{s} \qquad \text{Simple integrator.} \tag{14.54}$$

The corresponding state-space system (14.1) is

$$\mathbf{A} = 0, \ \mathbf{B} = \mathbf{C} = 1. \tag{14.55}$$

All signals are zero except the signal $w_0$, which is defined as

$$w_0(t) = 1\ 0.5 \leq t \leq 1.5,\ 2.0 \leq t \leq 2.5,\ 3.0 \leq t \leq 3.2,\ 4.0 \leq t \leq 4.1, \tag{14.56}$$

and the filtered setpoint $w$ is obtained by passing $w$ through the low-pass filter $G_w(s)$ where

$$G_w(s) = \frac{1}{1 + sT_f}. \tag{14.57}$$

Except where stated, the intermittent control parameters are

| | | |
|---|---|---|
| $\Delta_{\text{min}} = 0.5$ | Min. intermittent interval (14.20) |
| $q_t = 0.1$ | Threshold (14.35) |
| $\Delta = 0$ | Control delay (14.5) |
| $\Delta_s = 0$ | Sampling delay (14.17) |

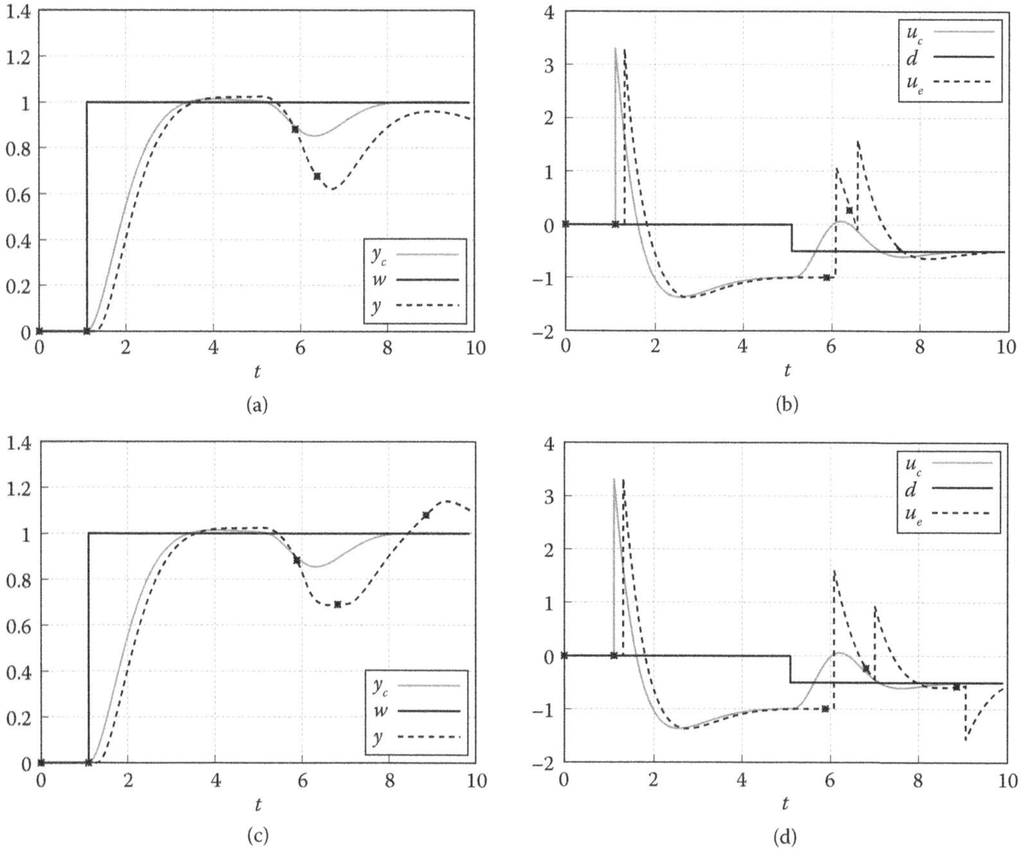


**FIGURE 14.7**
Elementary example: control-delay and sampling delay. (a) Control delay: $y$. (b) Control delay: $u$. (c) Sample delay: $y$. (d) Sample delay: $u$.

Figure 14.10a corresponds to the unfiltered setpoint with $T_f = 0$ and $w = w_0$ where $w_0$ is given by (14.56). For the first two (wider) pulses, events (•) occur at each setpoint change; but the second two (narrower) pulses, the trailing edges occur at a time less than $\Delta_{\min} = 0.5$ from the leading edges and thus the events corresponding to the trailing edges are delayed until $\Delta_{\min}$ has elapsed. Thus, the second two (narrower) pulses lead to outputs as if the pulses were $\Delta_{\min}$ wide. Figure 14.10b shows the intermittent-equivalent setpoint $w_{ic}$ superimposed on the actual setpoint $w$.

Figure 14.11a corresponds to the filtered setpoint with $T_f = 0.01$ and $w = w_0$ where $w_0$ is given by (14.56). At the event times, the setpoint has not yet reached its final value and thus the initial response is too small which is then corrected; Figure 14.11b shows the intermittent-equivalent setpoint $w_{ic}$ superimposed on the actual setpoint $w$.

The unsatisfactory behavior can be improved by delaying the sample time by $\Delta_s$ as discussed in Section 14.3.1. Figure 14.12a corresponds to Figure 14.11a except that $\Delta_s = 0.1$. Except for the short delay of $\Delta_s = 0.1$, the behavior of the first three pulses is now similar to that of Figure 14.10a. The fourth (shortest) pulse gives, however, a reduced amplitude output; this is because the sample occurs on the trailing edge of the pulse. This behavior has been observed by Vince (1948) as is related to the ATF of Barrett and Glencross (1988). Figure 14.12b shows the intermittent-equivalent setpoint $w_{ic}$ superimposed on the actual setpoint $w$. This phenomenon is further investigated in Section 14.4.3.

### 14.4.3 The Amplitude Transition Function

This section expands on the observation in Section 14.4.2, Figure 14.12, that the combination of sampling delay and a bandwidth limited setpoint can lead to narrow pulses being "missed." It turns out that the physiological equivalent of this behavior is the so-called *Amplitude Transition Function* (ATF) described by Barrett and Glencross (1988). Instead of the symmetric pulse discussed in the PRP context in Section 14.4.2, the ATF concept is based on asymmetric pulses where the step down is less than the step up leading to a nonzero final value. An example of an asymmetric pulse appears in Figure 14.13. The simulations in this section use the same

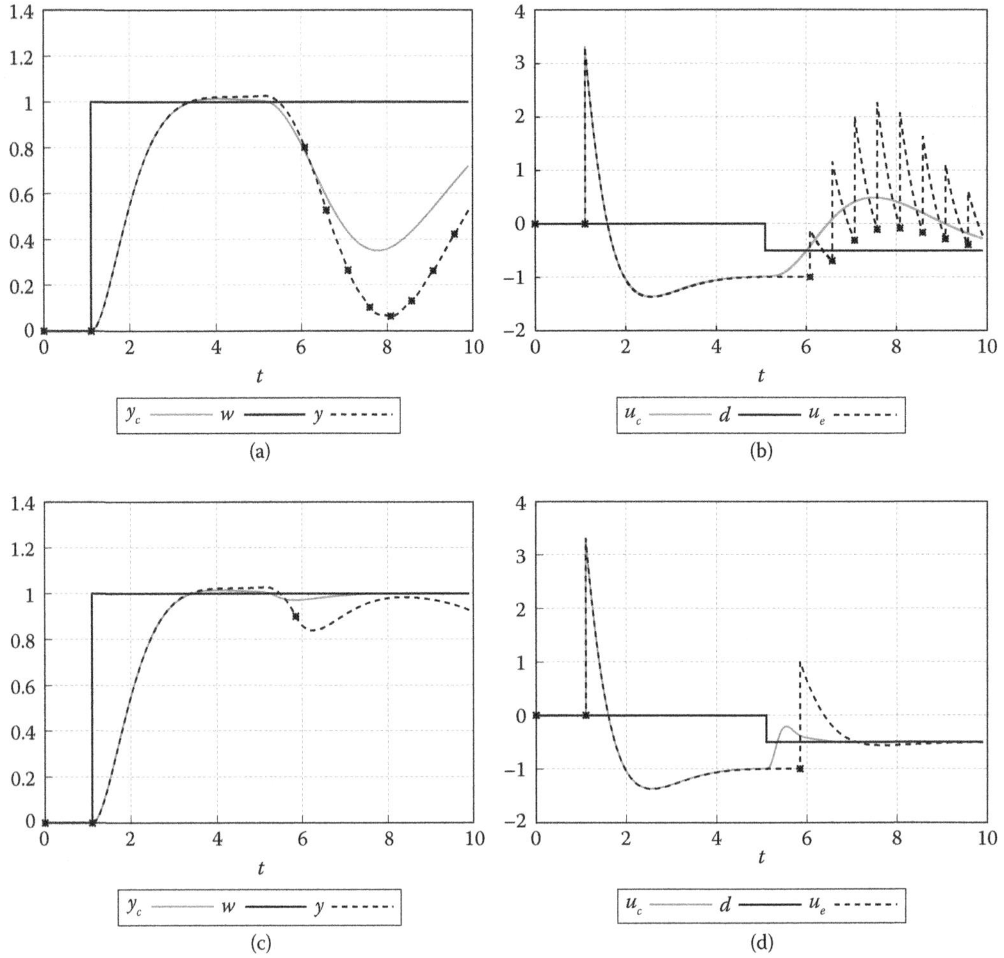


**FIGURE 14.8**
Elementary example: low and high observer gain. (a) Low observer gain: $y$. (b) Low observer gain: $u$. (c) High observer gain: $y$. (d) High observer gain: $u$.

system as in Section 14.4.2 Equations 14.54 and 14.55, but the setpoint $w_0$ of Equation 14.56 is replaced by

$$w_0(t) = \begin{cases} 0 & t < 1 \\ 1 & 1 \le t \le 1 + \Delta_p\,, \\ 0.5 & t > 1 + \Delta_p \end{cases} \tag{14.58}$$

where $\Delta_p$ is the *pulse-width*.

The system was simulated for two pulse widths: $\Delta_p =$ 200 ms (Figure 14.13a) and $\Delta_p = 100$ ms (Figure 14.13b). In each case, following Equation 14.58, the pulse was asymmetric going from 0 to 1 and back to 0.5.

At each pulse width, the system was simulated with event delay $\Delta_s = 90, 100, \ldots, 150$ ms and the control delay was set to 100 ms. Figure 14.13a shows the "usual" behavior, the 200 ms pulse is expanded to $\Delta_{ol} = 500$ ms and delayed by $\Delta + \Delta_s$. In contrast, Figure 14.13a shows the "Amplitude Transition Function" behavior: because the sampling is occurring on the downwards side of the pulse, the amplitude is reduced with increasing $\Delta_s$. Figure 14.13a is closely related to Figure 2 of Barrett and Glencross (1988).

## 14.5 Constrained Design

The design approach outlined in Sections 14.2 and 14.3 assumes that system inputs and outputs can take any value. In practice, this is not always the case and so *constraints* on both system inputs and outputs must be taken into account. There are at least three classes of constraints of interest in the context of intermittent control:

1. Constraints on the steady-state behavior of a system. These are particularly relevant in the context of multi-input ($n_u > 1$) and

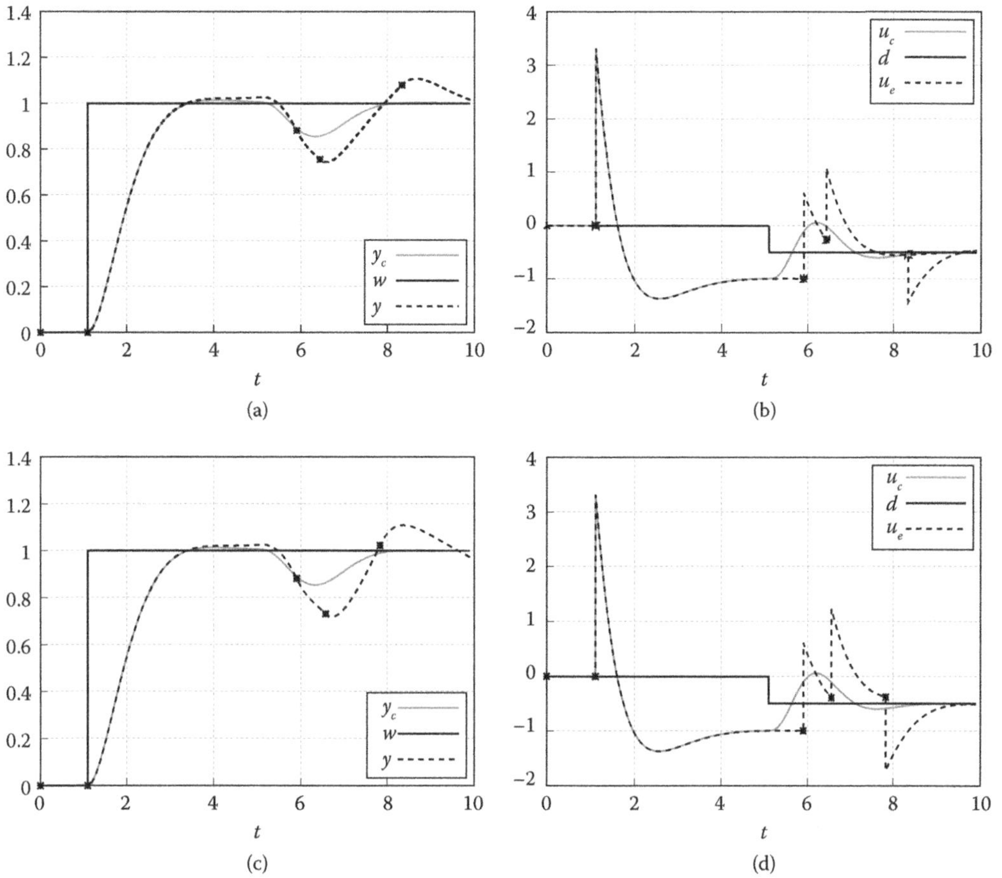


**FIGURE 14.9**
Elementary example: low and high occlusion time. (a) Low occlusion time: $y$. (b) Low occlusion time: $u$. (c) High occlusion time: $y$. (d) High occlusion time: $u$.

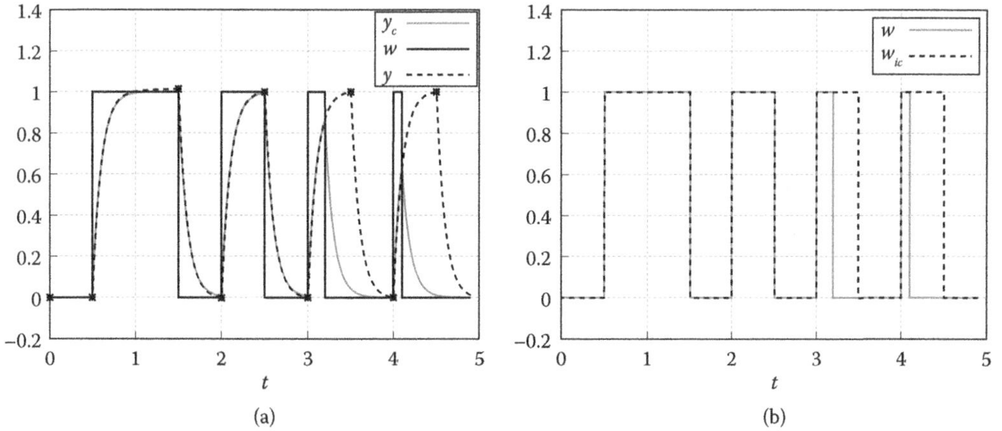


**FIGURE 14.10**
Psychological refractory period: square setpoint. (a) Intermittent control. (b) Intermittent-equivalent setpoint.

multi-output ($n_y > 1$) systems. This issue is discussed in Section 14.5.1 and is illustrated by an example in Section 14.7.

2. Amplitude constraints on the dynamical behavior of a system. This is a topic that is much discussed in the Model-based Predictive Control literature [e.g., Rawlings (2000); Maciejowski (2002); Wang (2009)]. In the context of intermittent control, constraints have been considered in the single-input

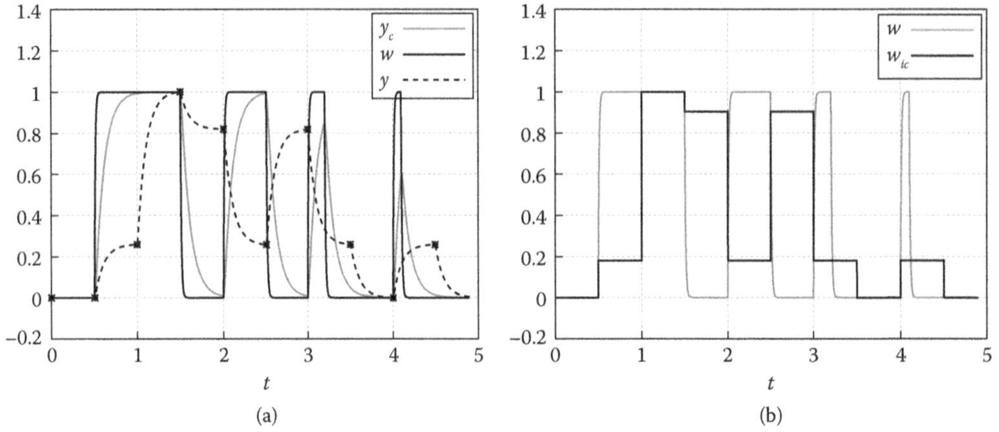


**FIGURE 14.11**
Psychological refractory period: filtered setpoint. (a) Intermittent control. (b) Intermittent-equivalent setpoint.

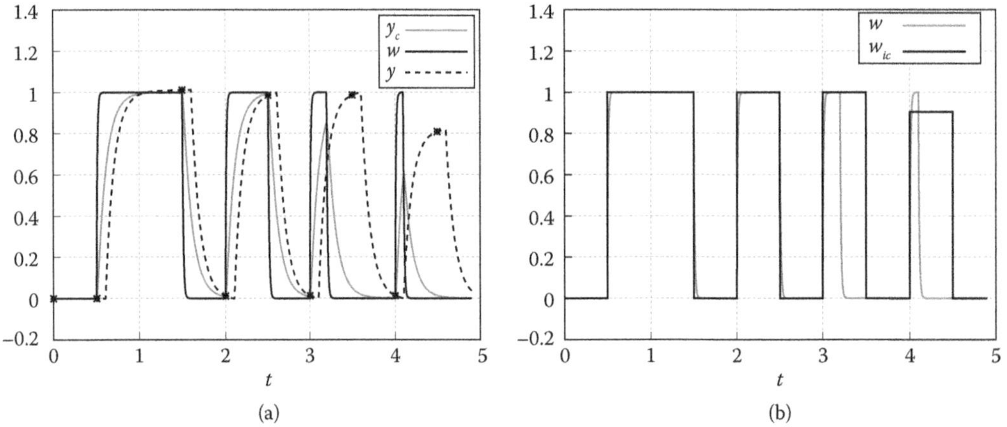


**FIGURE 14.12**
Psychological refractory period: sampling delay. (a) Intermittent control. (b) Intermittent-equivalent setpoint.

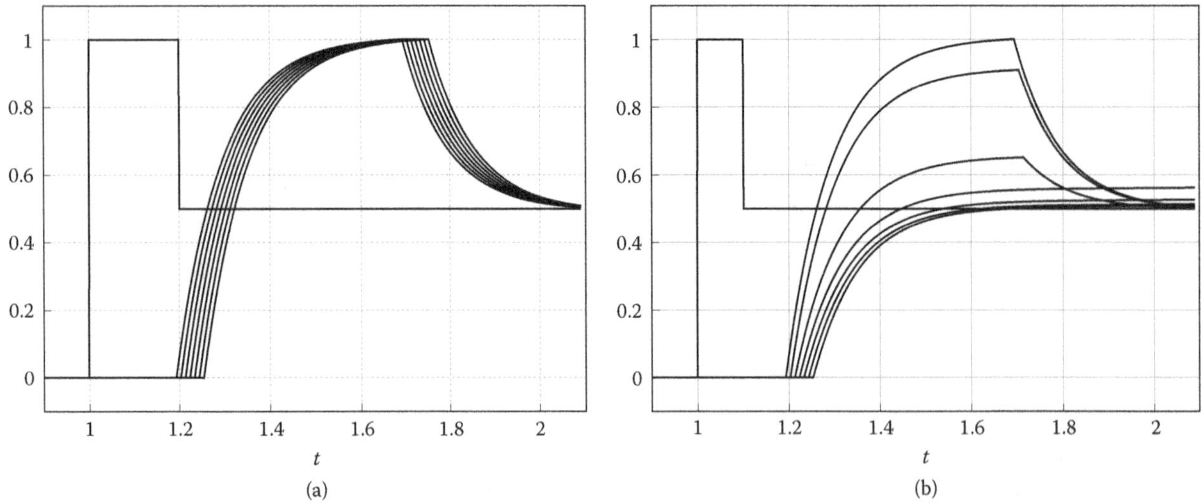


**FIGURE 14.13**
Amplitude transition function. (a) $y\&w$: pulse width 200 ms. (b) $y\&w$: pulse width 100 ms.

single-output context by Gawthrop and Wang (2009a); the corresponding multivariable case is considered in Section 14.5.2 and illustrated in Section 14.6.

3. Power constraints on the dynamical behavior of a system. This topic has been discussed by Gawthrop, Wagg, Neild, and Wang (2013c).

### 14.5.1 Constrained Steady-State Design

Section 14.2.4 considers the steady-state design of the continuous controller (CC) underlying intermittent control. In particular, Equation 14.13 gives a linear algebraic equation giving the steady-state system state $\mathbf{x}_{ss}$ and corresponding control signal $\mathbf{u}_{ss}$ yielding a particular steady-state output $\mathbf{y}_{ss}$. Although in the single-input single-output case considered by Gawthrop et al. (2011, equation 13), the solution is unique; as discussed in Section 14.2.4, the multi-input, multi-output case gives rise to more possibilities. In particular, it is not possible to exactly solve Equation 14.13 in the over-determined case where $n_{ss} > n_u$, but a least-squares solution exists. In the constrained case, this solution must satisfy two sets of constraints: an equality constraint ensuring that the equilibrium condition (14.10) holds and inequality constraints to reject physically impossible solutions.

In this context, the $n_{ss} \times n_{ss}$ *weighting matrix* $\mathbf{Q}_{ss}$ can be used to vary the relative importance of each element of $\mathbf{y}_{ss}$. In particular, define

$$\mathbf{S_Q} = \begin{bmatrix} \mathbf{A} & \mathbf{B} \\ \mathbf{Q}_{ss}\mathbf{C}_{ss} & 0_{n_{ss}\times n_u} \end{bmatrix}, \tag{14.59}$$

$$\mathbf{X}_{ss} = \begin{bmatrix} \mathbf{x}_{ss} \\ \mathbf{u}_{ss} \end{bmatrix}, \tag{14.60}$$

$$\mathbf{y}_{ss} = \begin{bmatrix} -\mathbf{B}_d\mathbf{d}_{ss} \\ \mathbf{Q}_{ss}\mathbf{y}_{ss} \end{bmatrix}, \tag{14.61}$$

and

$$\hat{\mathbf{Y}}_{ss} = \begin{bmatrix} -\mathbf{B}_d\mathbf{d}_{ss} \\ \mathbf{Q}_{ss}\hat{\mathbf{y}}_{ss} \end{bmatrix} = \mathbf{S_Q}\hat{\mathbf{X}}_{ss}. \tag{14.62}$$

This gives rise to the least-squares cost function:

$$\begin{aligned} J_{ss} &= \left(\mathbf{y}_{ss} - \hat{\mathbf{Y}}_{ss}\right)^{\mathrm{T}}\left(\mathbf{y}_{ss} - \hat{\mathbf{Y}}_{ss}\right) \\ &= \left(\mathbf{y}_{ss} - \mathbf{S_Q}\hat{\mathbf{X}}_{ss}\right)^{\mathrm{T}}\left(\mathbf{y}_{ss} - \mathbf{S_Q}\hat{\mathbf{X}}_{ss}\right). \end{aligned} \tag{14.63}$$

Differentiating with respect to $\hat{\mathbf{X}}_{ss}$ gives the weighted least-squares solution of (14.13):

$$\mathbf{S}_{\mathbf{Q}}^{\mathrm{T}}\left(\mathbf{y}_{ss} - \mathbf{S_Q}\hat{\mathbf{X}}_{ss}\right) = 0, \tag{14.64}$$

or

$$\hat{\mathbf{X}}_{ss} = \left(\mathbf{S}_{\mathbf{Q}}^{\mathrm{T}}\mathbf{S_Q}\right)^{-1}\mathbf{S}_{\mathbf{Q}}^{\mathrm{T}}\mathbf{y}_{ss}. \tag{14.65}$$

As $\mathbf{x}_{ss}$ corresponds to a steady-state solution corresponding to Equation 14.10, the solution of the least-squares problem is subject to the equality constraint:

$$\begin{bmatrix} \mathbf{A} & \mathbf{B} \end{bmatrix}\hat{\mathbf{X}}_{ss} = \mathbf{A}\mathbf{x}_{ss} + \mathbf{B}\mathbf{u}_{ss} = -\mathbf{B}_d\mathbf{d}_{ss}. \tag{14.66}$$

Furthermore, suppose that the solution must be such that the components of $\mathbf{Y}$ corresponding to $\mathbf{y}_{ss}$ are bounded above and below:

$$\hat{\mathbf{y}}_{\min} \leq \hat{\mathbf{y}} = \mathbf{C}_{ss}\hat{\mathbf{X}} \leq \hat{\mathbf{y}}_{\max}. \tag{14.67}$$

Inequality (14.67) can be rewritten as

$$\begin{bmatrix} -\mathbf{C}_{ss} \\ \mathbf{C}_{ss} \end{bmatrix}\hat{\mathbf{X}} \leq \begin{bmatrix} -\hat{\mathbf{y}}_{\min} \\ \hat{\mathbf{y}}_{\max} \end{bmatrix}. \tag{14.68}$$

The quadratic cost function (14.63) together with the linear equality constraint (14.66) and the linear inequality constraint (14.68) forms a *quadratic program* (QP), which has well-established numerical algorithms available for its solution (Fletcher, 1987).

An example of constrained steady-state optimization is given in Section 14.7.5.

### 14.5.2 Constrained Dynamical Design

MPC (Rawlings, 2000; Maciejowski, 2002; Wang, 2009) combines a quadratic cost function with linear constraints to provide optimal control subject to (hard) constraints on both state and control signal; this combination of quadratic cost and *linear* constraints can be solved by using *quadratic programming* (QP) (Fletcher, 1987; Boyd and Vandenberghe, 2004). Almost all MPC algorithms have a discrete-time framework. As a move toward a continuous-time formulation of intermittent control, the intermittent approach to MPC was introduced (Ronco et al., 1999) to reduce online computational demand while retaining continuous-time like behavior (Gawthrop and Wang, 2007, 2009a; Gawthrop et al., 2011). This section introduces and illustrates this material.*

Using the feedback control comprising the system matched hold (14.21), its initialization (14.23), and feedback (14.34) may cause state or input constraints to be violated over the intermittent interval. The key idea introduced by Chen and Gawthrop (2006) and exploited by Gawthrop and Wang (2009a) is to replace the SMH initialization (at time $t = t_i$ (14.16)) of Equation 14.23 by

$$\mathbf{x}_h(0) = \begin{cases} \mathbf{x}_p(t_i - \Delta) - \mathbf{x}_{ss}w(t_i) & \text{when constraints not violated} \\ \mathbf{U}_i & \text{otherwise} \end{cases}. \tag{14.69}$$

*Hard constraints on input *power flow* are considered by Gawthrop et al. (2013c)—these lead to *quadratically-constrained quadratic programming* (QCQP) (Boyd and Vandenberghe, 2004).

where $\mathbf{U}_i$ is the result of the online optimization to be discussed in Section 14.5.2.2.

The first step is to construct a set of equations describing the evolution of the system state $\mathbf{x}$ and the generalized hold state $\mathbf{x}_h$ as a function of the initial states and assuming that disturbances are zero.

The differential equation (14.26) has the explicit solution

$$\mathbf{X}(\tau) = \mathbf{E}(\tau)\mathbf{X}_i, \tag{14.70}$$

where

$$\mathbf{E}(\tau) = e^{\mathbf{A}_{xu}\tau}, \tag{14.71}$$

where $\tau$ is the intermittent continuous-time variable based on $t_i$.

#### 14.5.2.1 Constraints

The vector $\mathbf{X}$ (14.26) contains the system state and the state of the generalized hold; Equation 14.70 explicitly give $\mathbf{X}$ in terms of the system state $\mathbf{x}_i(t_i)$ and the hold state $\mathbf{x}_h(t_i) = \mathbf{U}_i$ at time $t_i$. Therefore, any constraint expressed at a future time $\tau$ as a linear combination of $\mathbf{X}$ can be re-expressed in terms of $\mathbf{x}_h$ and $\mathbf{U}_i$. In particular, if the constraint at time $\tau$ is expressed as

$$\Gamma_\tau \mathbf{X}(\tau) \leq \gamma_\tau, \tag{14.72}$$

where $\Gamma_\tau$ is a $2n$-dimensional row vector and $\gamma_\tau$ a scalar then the constraint can be re expressed using (14.70) in terms of the intermittent control vector $\mathbf{U}_i$ as

$$\Gamma_\tau E_u(\tau)\mathbf{U}_i \leq \gamma_\tau - \Gamma_\tau E_x(\tau)\mathbf{x}_i, \tag{14.73}$$

where $E$ has been partitioned into the two $2n \times n$ submatrices $E_x$ and $E_u$ as

$$E(\tau) = \begin{pmatrix} E_x(\tau) & E_u(\tau) \end{pmatrix}. \tag{14.74}$$

If there are $n_c$ such constraints, they can be combined as

$$\Gamma \mathbf{U}_i \leq \gamma - \Gamma_x \mathbf{x}_i, \tag{14.75}$$

where each row of $\Gamma$ is $\Gamma_\tau E_u(\tau)$, each row of $\Gamma_x$ is $\Gamma_\tau E_x(\tau)$, and each (scalar) row of $\gamma$ is $\gamma_\tau$.

Following standard MPC practice, constraints beyond the intermittent interval can be included by assuming that the control strategy will be open-loop in the future.

#### 14.5.2.2 Optimization

Following, for example, Chen and Gawthrop (2006), a modified version of the infinite-horizon LQR cost (14.6) is used:

$$J_{ic} = \int_0^{\tau_1} \mathbf{x}(\tau)^T\mathbf{Q}\mathbf{x}(\tau) + u(\tau)\mathbf{R}u(\tau)\,d\tau + \mathbf{x}(\tau_1)^T\mathbf{P}\mathbf{x}(\tau_1), \tag{14.76}$$

where the weighting matrices $\mathbf{Q}$ and $\mathbf{R}$ are as used in (14.6) and $\mathbf{P}$ is the positive-definite solution of the algebraic Riccati equation (ARE):

$$\mathbf{A}^T\mathbf{P} + \mathbf{P}\mathbf{A} - \mathbf{P}\mathbf{B}\mathbf{R}^{-1}\mathbf{B}^T\mathbf{P} + \mathbf{Q} = 0. \tag{14.77}$$

There are a number of differences between our approach to minimizing $J_{ic}$ (14.76) and the LQR approach to minimizing $J_{LQR}$ (14.6).

1. Following the standard MPC approach (Maciejowski, 2002), this is a *receding-horizon* optimization in the time frame of $\tau$ not $t$.
2. The integral is over a finite time $\tau_1$.
3. A terminal cost is added based on the steady-state ARE (14.77). In the discrete-time context, this idea is due to Rawlings and Muske (1993).
4. The minimization is with respect to the intermittent control vector $\mathbf{U}_i$ generating the control signal $u$ (14.7) through the generalized hold (14.21).

Using $\mathbf{X}$ from (14.70), (14.76) can be rewritten as

$$J_{ic} = \int_0^{\tau_1} \mathbf{X}(\tau)^T\mathbf{Q}_{xu}\mathbf{X}(\tau)\,d\tau + \mathbf{X}(\tau_1)^T\mathbf{P}_{xu}\mathbf{X}(\tau_1), \tag{14.78}$$

where

$$\mathbf{Q}_{xu} = \begin{pmatrix} \mathbf{Q} & 0_{n\times n} \\ 0_{n\times n} & \mathbf{x}_{uo}\mathbf{R}\mathbf{x}_{uo}^T \end{pmatrix}, \tag{14.79}$$

and

$$\mathbf{P}_{xu} = \begin{pmatrix} \mathbf{P} & 0_{n\times n} \\ 0_{n\times n} & 0_{n\times n} \end{pmatrix}. \tag{14.80}$$

Using (14.70), Equation 14.78 can be rewritten as

$$J_{ic} = \mathbf{X}_i^T J_{XX}\mathbf{X}_i, \tag{14.81}$$

where

$$J_{XX} = J_1 + e^{\mathbf{A}_{xu}^T\tau_1}\mathbf{P}_{xu}e^{\mathbf{A}_{xu}\tau_1}, \tag{14.82}$$

and

$$J_1 = \int_0^{\tau_1} e^{\mathbf{A}_{xu}^T\tau}\mathbf{Q}_{xu}e^{\mathbf{A}_{xu}\tau}\,d\tau. \tag{14.83}$$

The $2n \times 2n$ matrix $J_{XX}$ can be partitioned into four $n \times n$ matrices as

$$J_{XX} = \begin{pmatrix} J_{xx} & J_{xu} \\ J_{ux} & J_{uu} \end{pmatrix}. \tag{14.84}$$

**Lemma 14.1: Constrained optimization**

The minimization of the cost function $J_{ic}$ of Equation 14.76 subject to the constraints (14.75) is equivalent to the solution of the QP for the optimum value of $\mathbf{U}_i$:

$$\min_{U_i} \left\{ \mathbf{U}_i^T J_{UU} \mathbf{U}_i + \mathbf{x}_i^T J_{Ux} \mathbf{U}_i \right\}, \quad (14.85)$$

subject to $\Gamma \mathbf{U}_i \leq \gamma - \Gamma_x \mathbf{x}_i$, where $J_{UU}$ and $J_{Ux}$ are given by (14.84) and $\Gamma$, $\Gamma_x$ and $\gamma$ as described in Section 14.5.2.1.

**PROOF** See Chen and Gawthrop (2006). ■

**Remarks**

1. This optimization is dependent on the system state $\mathbf{x}$ and therefore must be accomplished at every intermittent interval $\Delta_i$.
2. The computation time is reflected in the time delay $\Delta$.
3. As discussed by Chen and Gawthrop (2006), the relation between the cost function (14.85) and the LQ cost function (14.6) means that the solution of the QP is the same as the LQ solution when constraints are not violated.

## 14.6 Example: Constrained Control of Mass–Spring System

Figure 14.14 shows a coupled mass–spring system. The five masses $m_1$–$m_5$ all have unit mass and the four springs $\kappa_1$–$\kappa_5$ all have unit stiffness. The mass positions are denoted by $y_1$–$y_5$, velocities by $v_1$–$v_5$, and the applied forces by $F_1$–$F_5$. In addition, it is assumed that the five forces $F_i$ are generated from the five control signals $u_i$ by simple integrators thus:

$$\dot{F}_i = u_i,\ i = 1, \ldots, 5. \quad (14.86)$$

This system has 15 states ($n_x = 15$), 5 inputs ($n_u = 5$), and 5 outputs ($n_y = 5$).

To examine the effect of constraints, consider the case where it is required that the velocity of the center mass ($i = 3$) is constrained above by

$$v_3 < 0.2, \quad (14.87)$$

but unconstrained below. As noted in Section 14.5.2.1, the constraints are at discrete values of intersample time $\tau$. In this case, 50 points were chosen at $\tau = 0.1, 0.2, \ldots, 5.0$. The precise choice of these points is not critical.

In addition, the system setpoint is given by

$$w_i(t) = \begin{cases} 1 & i = 3 \text{ and } 1 \leq t < 10 \\ 0 & \text{otherwise} \end{cases}. \quad (14.88)$$

Figure 14.15 shows the results of simulating the coupled mass–spring system of Figure 14.14 with constrained intermittent control with constraint given by (14.87) and setpoint by (14.88). Figure 14.15a shows the position of the first mass and Figure 14.15b the corresponding velocity; Figure 14.15c shows the position of the second mass and Figure 14.15d the corresponding velocity; Figure 14.15e shows the position of the third mass and Figure 14.15f the corresponding velocity. The fourth and fifth masses are not shown. In each case, the corresponding simulation result for the underlying continuous (unconstrained) simulation is also shown.

Note that on the forward motion of mass three, the velocity (Figure 14.15f) is constrained and this is reflected in the constant slope of the corresponding position (Figure 14.15e). However, the backward motion is unconstrained and closely approximates that corresponding to the unconstrained CC. The other masses (which have a zero setpoint) deviate more from zero whilst mass three is constrained, but are similar to the unconstrained case when mass three is not constrained.

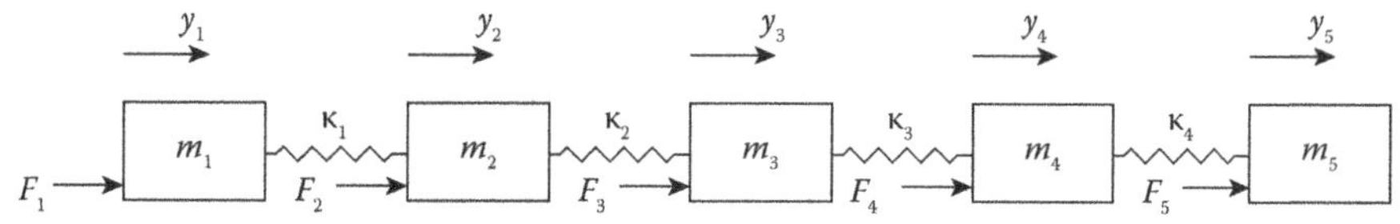


**FIGURE 14.14**
Coupled mass–spring system. The five masses $m_1$–$m_5$ all have unit mass and the four springs $\kappa_1$–$\kappa_5$ all have unit stiffness. The mass positions are denoted by $y_1$–$y_5$, velocities by $v_1$–$v_5$, and the applied forces by $F_1$–$F_5$.

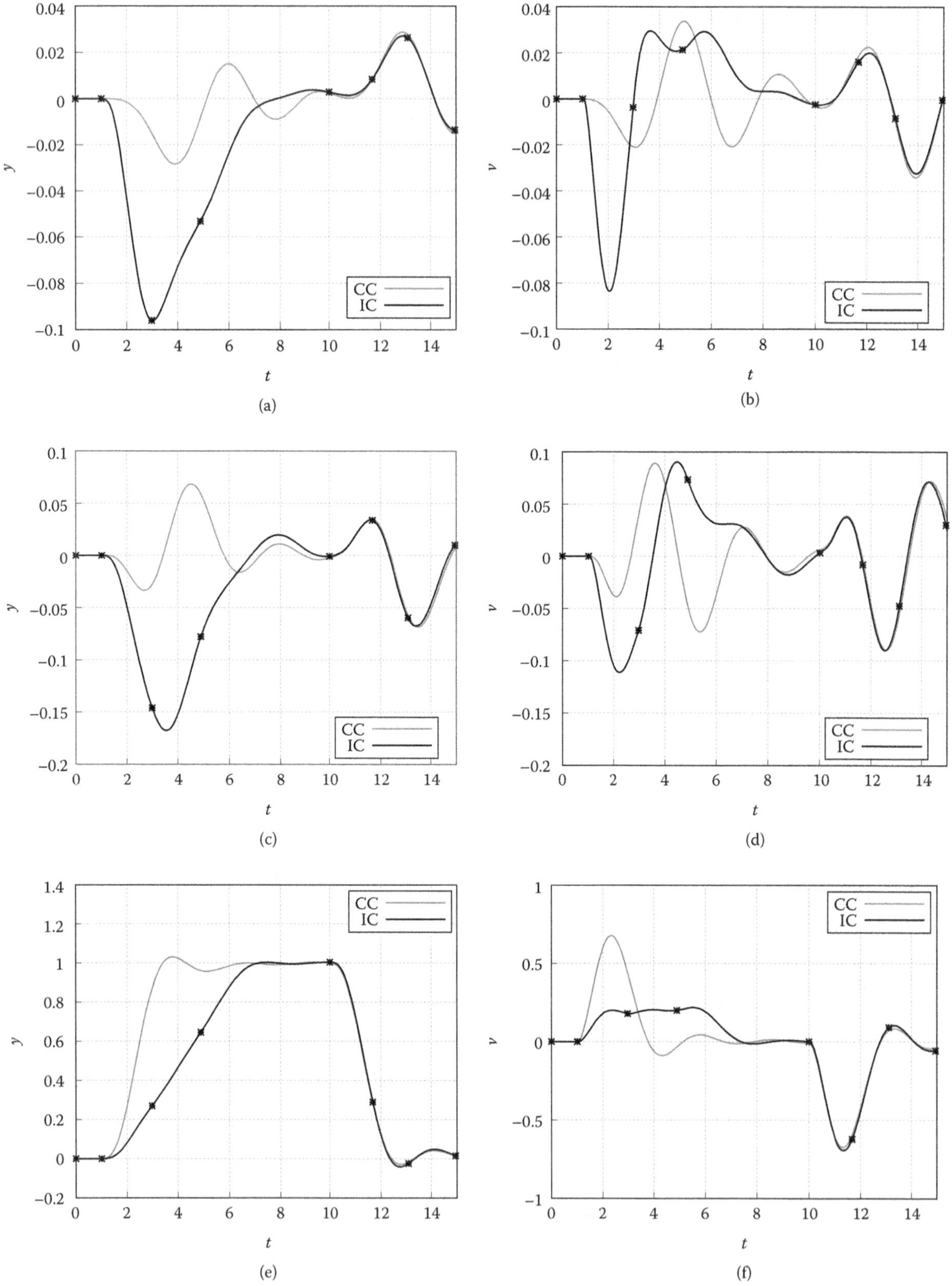


**FIGURE 14.15**
Constrained control of mass–spring system. The left-hand column shows the positions of masses 1–3 and the right-hand column the corresponding velocities. The gray line corresponds to the simulation of the underlying *unconstrained* continuous system and the black lines correspond to intermittent control; the • correspond to the intermittent sampling times $t_i$. (a) $y_1$. (b) $v_1$. (c) $y_2$. (d) $v_2$. (e) $y_3$. (f) $v_3$.

## 14.7 Examples: Human Standing

Human control strategies in the context of quiet standing have been investigated over many years by a number of authors. Early work (Peterka, 2002; Lakie, Caplan, and Loram, 2003; Bottaro et al., 2005; Loram, Maganaris, and Lakie, 2005) was based on a single inverted pendulum, single-input model of the system. More recently, it has been shown (Pinter, van Swigchem, van Soest, and Rozendaal, 2008; Günther, Grimmer, Siebert, and Blickhan, 2009; Günther, Müller, and Blickhan, 2011, 2012) that a multiple segment, multiple input model is required to model unconstrained quiet standing and this clearly has implications for the corresponding human control system. Intermittent control has been suggested as the basic algorithm by Gawthrop et al. (2011), Gawthrop et al. (2013b), and Gawthrop, Loram, Gollee, and Lakie (2014) and related algorithms have been analyzed by Insperger (2006), Stepan and Insperger (2006), Asai et al. (2009), and Kowalczyk et al. (2012).

This section uses a linear three-segment model to illustrate key features of the constrained multivariable intermittent control described in Sections 14.3 and 14.5. Section 14.7.1 describes the three-link model, Section 14.7.2 looks at a heirachical approach to muscle-level control, Section 14.7.3 looks at an intermittent explanation of quiet standing, and Sections 14.7.4 and 14.7.5 discuss tracking and disturbance rejection, respectively.

### 14.7.1 A Three-Segment Model

This section uses the linearised version of the three-link, three-joint model of posture given by Alexandrov, Frolov, Horak, Carlson-Kuhta, and Park (2005). The upper, middle, and lower links are indicated by subscripts $u$, $m$, and $l$, respectively. The linearized equations correspond to

$$\mathbf{M}\ddot{\boldsymbol{\theta}} - \mathbf{G}\boldsymbol{\theta} = \mathbf{N}\mathbf{T}, \tag{14.89}$$

where $\boldsymbol{\theta}$ is the vector of link angles given by

$$\boldsymbol{\theta} = \begin{bmatrix} \theta_l \\ \theta_m \\ \theta_u \end{bmatrix}, \tag{14.90}$$

and $\mathbf{T}$ the vector of joint torques.

The mass matrix $\mathbf{M}$ is given by

$$\mathbf{M} = b \begin{bmatrix} m_{ll} & m_{lm} & m_{lu} \\ m_{ml} & m_{mm} & m_{mu} \\ m_{ul} & m_{um} & m_{uu} \end{bmatrix}, \tag{14.91}$$

where

$$m_{ll} = m_l c_l^2 + (m_m + m_u) l_l^2 + I_l, \tag{14.92}$$
$$m_{mm} = m_m c_m^2 + m_u l_m^2 + I_m, \tag{14.93}$$
$$m_{uu} = m_u c_u^2 + I_u, \tag{14.94}$$
$$m_{ml} = m_{lm} = m_m c_m l_l + m_u l_l l_m, \tag{14.95}$$
$$m_{ul} = m_{lu} = m_u c_u l_l, \tag{14.96}$$
$$m_{um} = m_{mu} = m_u c_u l_m, \tag{14.97}$$

the gravity matrix $\mathbf{G}$ by

$$\mathbf{G} = g \begin{bmatrix} g_{ll} & 0 & 0 \\ 0 & g_{mm} & 0 \\ 0 & 0 & g_{uu} \end{bmatrix}, \tag{14.98}$$

where

$$g_{ll} = m_l c_l + (m_m + m_u) l_l, \tag{14.99}$$
$$g_{mm} = m_m c_m + m_u l_m, \tag{14.100}$$
$$g_{uu} = m_u c_u, \tag{14.101}$$

and the input matrix $\mathbf{N}$ by

$$\mathbf{N} = \begin{bmatrix} 1 & -1 & 0 \\ 0 & 1 & -1 \\ 0 & 0 & 1 \end{bmatrix}. \tag{14.102}$$

The joint angles $\phi_l, \ldots, \phi_u$ can be written in terms of the link angles as

$$\phi_l = \theta_l, \tag{14.103}$$
$$\phi_m = \theta_m - \theta_l, \tag{14.104}$$
$$\phi_u = \theta_u - \theta_m, \tag{14.105}$$

or more compactly as

$$\boldsymbol{\phi} = \mathbf{N}^{\mathrm{T}}\boldsymbol{\theta}, \tag{14.106}$$

where

$$\boldsymbol{\phi} = \begin{bmatrix} \phi_l \\ \phi_m \\ \phi_u \end{bmatrix}. \tag{14.107}$$

The values for the link lengths $l$, CoM location $c$, masses $m$, and moments of inertia (about CoM) were taken from Figure 4.1 and Table 4.1 of Winter (2009).

The model of Equation 14.89 can be rewritten as

$$\frac{d\mathbf{x}_0}{dt} = \mathbf{A}_0\mathbf{x}_0 + \mathbf{B}_0\mathbf{T}, \tag{14.108}$$
$$\mathbf{x}_0 = \begin{bmatrix} \dot{\boldsymbol{\theta}} \\ \boldsymbol{\theta} \end{bmatrix}, \tag{14.109}$$

and

$$\mathbf{A}_0 = \begin{bmatrix} \mathbf{0}_{3\times3} & -\mathbf{M}^{-1}\mathbf{G} \\ \mathbf{I}_{3\times3} & \mathbf{0}_{3\times3} \end{bmatrix}, \tag{14.110}$$
$$\mathbf{B}_0 = \begin{bmatrix} \mathbf{M}^{-1}\mathbf{N} \\ \mathbf{0}_{3\times3} \end{bmatrix}. \tag{14.111}$$

The eigenvalues of $\mathbf{A}_0$ are $\pm 2.62$, $\pm 6.54$, and $\pm 20.4$. The positive eigenvalues indicate that this system is (without control) unstable.

More sophisticated models would include nonlinear geometric and damping effects; but this model provides the basis for illustrating the properties of constrained intermittent control.

### 14.7.2 Muscle Model and Hierarchical Control

As discussed by Lakie et al. (2003) and Loram et al. (2005), the single-inverted pendulum model of balance control uses a muscle model comprising a spring and a contractile element. In this context, the effect of the spring is to counteract gravity and thus effectively slow down the toppling speed on the pendulum. This toppling speed is directly related to the maximum real part of the system eigenvalues. This is important as it reduces the control bandwidth necessary to stabilize the unstable inverted pendulum system (Stein, 2003; Loram, Gawthrop, and Lakie, 2006).

The situation is more complicated in the multiple link case as, unlike the single inverted pendulum case, the joint angles are distinct from the link angles. From Equation 14.98, the gravity matrix is diagonal in link space; on the other hand, as the muscle springs act at the joints, the corresponding stiffness matrix is diagonal in joint space and therefore cannot cancel the gravity matrix in all configurations.

The spring model used here is the multi-link extension of the model of Loram et al. (2005, Figure 1) and is given by

$$\mathbf{T}_k = \mathbf{K}_\phi \left( \boldsymbol{\phi}_0 - \boldsymbol{\phi} \right), \tag{14.112}$$

where

$$\mathbf{K}_\phi = \begin{bmatrix} k_1 & 0 & 0 \\ 0 & k_2 & 0 \\ 0 & 0 & k_3 \end{bmatrix} \text{ and } \boldsymbol{\phi}_0 = \begin{bmatrix} \phi_{l0} \\ \phi_{m0} \\ \phi_{u0} \end{bmatrix}, \tag{14.113}$$

$\mathbf{T}_k$ is the vector of spring torques at each joint, $\boldsymbol{\phi}$ contains the joint angles (14.107), and $k_1, \ldots, k_3$ are the spring stiffnesses at each joint. It is convenient to choose the control signal $\mathbf{u}$ to be

$$\mathbf{u} = \frac{d\boldsymbol{\phi}_0}{dt}, \tag{14.114}$$

and thus Equation 14.112 can be rewritten as

$$\frac{d\mathbf{T}_k}{dt} = \mathbf{K}_\phi \left( \mathbf{u} - \frac{d\boldsymbol{\phi}}{dt} \right) \tag{14.115}$$

$$= \mathbf{K}_\phi \left( \mathbf{u} - \mathbf{N}^{\mathrm{T}} \frac{d\boldsymbol{\theta}}{dt} \right). \tag{14.116}$$

Setting $\mathbf{T} = \mathbf{T}_k + \mathbf{T}_d$ where $\mathbf{T}_d$ is a disturbance torque, the composite system formed from the link dynamics (14.89) and the spring dynamics (14.115) is given by Equation 14.1 where

$$\mathbf{x} = \begin{bmatrix} \mathbf{x}_0 \\ \mathbf{T} \end{bmatrix} = \begin{bmatrix} \dot{\boldsymbol{\theta}} \\ \boldsymbol{\theta} \\ \mathbf{T} \end{bmatrix}, \tag{14.117}$$

$$\mathbf{y} = \boldsymbol{\theta}, \tag{14.118}$$

$$\mathbf{d} = \mathbf{T}_d, \tag{14.119}$$

and

$$\mathbf{A} = \begin{bmatrix} \mathbf{A}_0 & \mathbf{B}_0 \\ -\mathbf{K}_\phi \mathbf{N}^{\mathrm{T}} & \mathbf{0}_{3\times 6} \end{bmatrix} = \begin{bmatrix} \mathbf{0}_{3\times3} & -\mathbf{M}^{-1}\mathbf{G} & \mathbf{M}^{-1}\mathbf{N} \\ \mathbf{I}_{3\times3} & \mathbf{0}_{3\times3} & \mathbf{0}_{3\times3} \\ -\mathbf{K}_\phi \mathbf{N}^{\mathrm{T}} & \mathbf{0}_{3\times3} & \mathbf{0}_{3\times3} \end{bmatrix}, \tag{14.120}$$

$$\mathbf{B} = \begin{bmatrix} \mathbf{0}_{3\times3} \\ \mathbf{0}_{3\times3} \\ \mathbf{K}_\phi \end{bmatrix}, \tag{14.121}$$

$$\mathbf{B}_d = \begin{bmatrix} \mathbf{B}_0 \\ \mathbf{0}_{3\times3} \end{bmatrix} = \begin{bmatrix} \mathbf{M}^{-1}\mathbf{N} \\ \mathbf{0}_{3\times3} \\ \mathbf{0}_{3\times3} \end{bmatrix}, \tag{14.122}$$

$$\mathbf{C} = \begin{bmatrix} \mathbf{0}_{3\times3} & \mathbf{I}_{3\times3} & \mathbf{0}_{3\times3} \end{bmatrix}. \tag{14.123}$$

There are, of course, many other state-space representations with the same input–output properties, but this particular state-space representation has two useful features: first, the velocity control input of Equation 14.114 induces an integrator in each of the three inputs and secondly the state explicitly contains the joint torque due to the springs. The former feature simplifies control design in the presence of input disturbances with constant components and the latter feature allows spring preloading (in anticipation of a disturbance) to be modeled as a state initial condition. These features are used in the example of Section 14.7.5.

It has been argued (Hogan, 1984) that humans use muscle co-activation of antagonist muscles to manipulate the passive muscle stiffness and thus $\mathbf{K}_\phi$. As mentioned above, the choice of $\mathbf{K}_\phi$ in the single-link case (Loram et al., 2005, Figure 1) directly affects the toppling speed via the maximum real part of the system eigenvalues. Hence, we argue that such muscle co-activation could be used to choose the maximum real part of the system eigenvalues and thus manipulate the required closed-loop control bandwidth. However, muscle co-activation requires the flow of energy and so it makes sense to choose the minimal stiffness consistent with the required maximum real part of the system eigenvalues. Defining:

$$\mathbf{k}_{phi} = \begin{bmatrix} k_1 \\ k_2 \\ k_3 \end{bmatrix}, \tag{14.124}$$

this can be expressed mathematically as

$$\min_{\mathbf{k}_\phi} ||\mathbf{K}_\phi||, \tag{14.125}$$

subject to

$$\max\left[\Re\sigma_i\right] < \sigma_{\max}, \tag{14.126}$$

where $\sigma_i$ is the $i$th eigenvalue of $\mathbf{A}$. This is a quadratic optimization with nonlinear constraints, which can be solved by sequential quadratic programming (SQP) (Fletcher, 1987).

In the single-link case, increasing spring stiffness from zero decreases the value of the positive eigenvalue until it reaches zero, after that point the two eigenvalues form a complex-conjugate pair with zero real part. The three-link case corresponds to three eigenvalue pairs. Figure 14.16 shows how the real and imaginary parts of these six eigenvalues vary with the constraint $\sigma_{\max}$ together with the spring constants $\mathbf{k}_\phi$. Note that the spring constants and imaginary parts rise rapidly when the maximum real eigenvalue is reduced to below about 2.3.

Joint damping can be modeled by the equation:

$$\mathbf{T}_c = -\mathbf{C}_\phi \frac{d\boldsymbol{\phi}}{dt} = -\mathbf{C}_\phi \mathbf{N}^{\mathrm{T}} \frac{d\boldsymbol{\theta}}{dt}, \tag{14.127}$$

where

$$\mathbf{C}_\phi = \begin{bmatrix} c_1 & 0 & 0 \\ 0 & c_2 & 0 \\ 0 & 0 & c_3 \end{bmatrix}. \tag{14.128}$$

Setting $\mathbf{T} = \mathbf{T}_k + \mathbf{T}_c$, the matrix $\mathbf{A}$ of Equation 14.120 is replaced by

$$\mathbf{A} = \begin{bmatrix} -\mathbf{M}^{-1}\mathbf{N}\mathbf{C}_\phi\mathbf{N}^{\mathrm{T}} & -\mathbf{M}^{-1}\mathbf{G} & \mathbf{M}^{-1}\mathbf{N} \\ \mathbf{I}_{3\times 3} & \mathbf{0}_{3\times 3} & \mathbf{0}_{3\times 3} \\ -\mathbf{K}_\phi\mathbf{N}^{\mathrm{T}} & \mathbf{0}_{3\times 3} & \mathbf{0}_{3\times 3} \end{bmatrix}. \tag{14.129}$$

### 14.7.3 Quiet Standing

In the case of quiet standing, there is no setpoint tracking and no constant disturbance and thus $w = 0$ and $\mathbf{x}_{ss}$ is not computed. The spring constants were computed

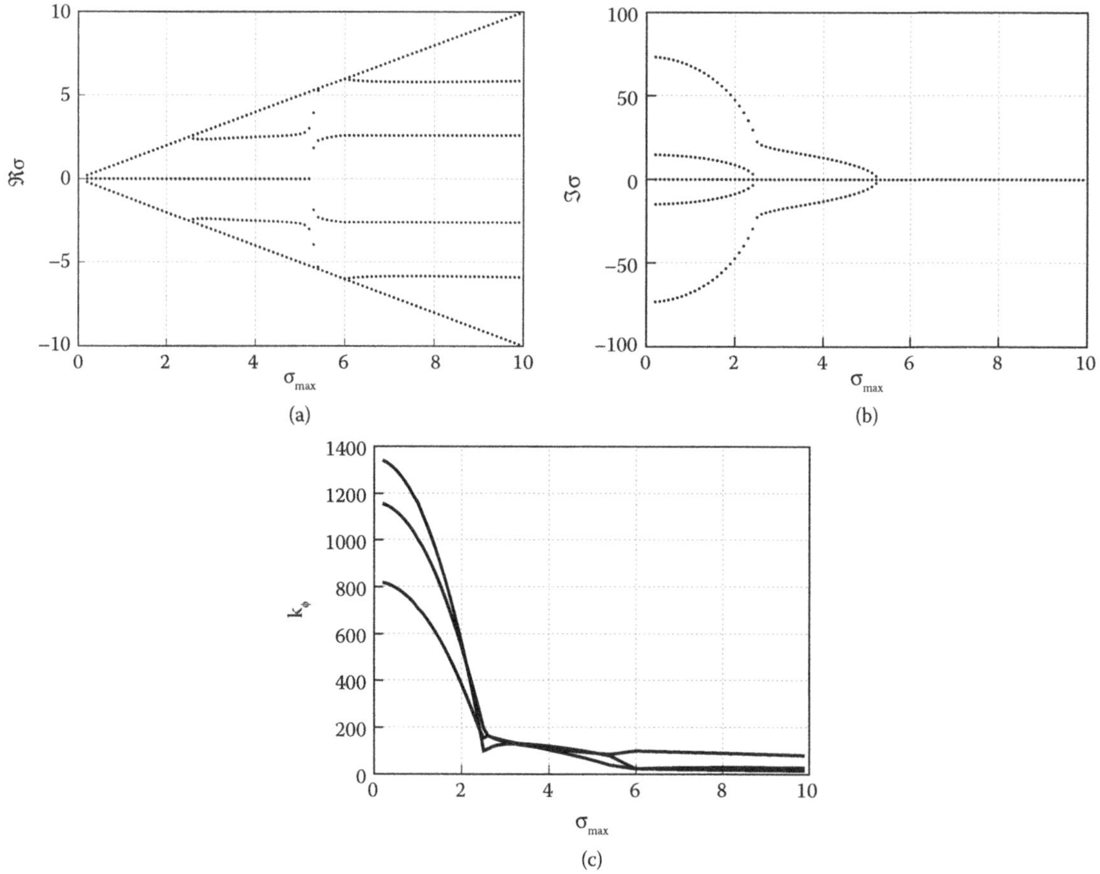


**FIGURE 14.16**
Choosing the spring constants. (a) The real parts of the nonzero eigenvalues plotted against $\sigma_{\max}$, the specified maximum real part of all eigenvalues resulting from (14.125) and (14.126). (b) The imaginary parts corresponding to (a). (c) The spring constants $\mathbf{k}_\phi$.

as in Section 14.7.2 with $\sigma_{max} = 3$. The corresponding nonzero eigenvalues of $\mathbf{A}$ are $\pm 3$, $\pm 2.38$, and $\pm j17.9$. The IC of Section 14.3, based on the continuous-time controller of Section 14.2.3, was simulated using the following control design parameters:

$$\mathbf{Q}_c = \begin{bmatrix} q_v \mathbf{I}_{3\times3} & \mathbf{0}_{3\times3} & \mathbf{0}_{3\times3} \\ \mathbf{0}_{3\times3} & q_p \mathbf{I}_{3\times3} & \mathbf{0}_{3\times3} \\ \mathbf{0}_{3\times3} & \mathbf{0}_{3\times3} & q_T \mathbf{I}_{3\times3} \end{bmatrix}, \tag{14.130}$$

where

$$q_v = 0,\ q_p = 1,\ q_T = 10, \tag{14.131}$$

$$\mathbf{R}_c = \mathbf{K}_{phi}^2. \tag{14.132}$$

The corresponding closed-loop poles are $-3.45 \pm 18.3$, $-3.95 \pm 3.82$, $-2.87 \pm 0.590$, $-5.31$, $-3.28$, and $-2.48$. The intermittent control parameters (Section 14.3) time delay $\Delta$ and minimum intermittent interval $\Delta_{min}$ were chosen as

$$\Delta = 0.1\ \text{s}, \tag{14.133}$$

$$\Delta_{min} = 0.25\ \text{s}. \tag{14.134}$$

These parameters are used in all of the following simulations.

A multisine disturbance with standard deviation 0.01 Nm was added to the control signal at the lower (ankle) joint. With reference to Equation 14.35, the threshold was set on the three segment angles so that the threshold surface (in the 9D state space) was defined as

$$\boldsymbol{\theta}^T\boldsymbol{\theta} = \mathbf{x}^T\mathbf{Q}_t\mathbf{x} = q_t^2, \tag{14.135}$$

where

$$\mathbf{Q}_t = \begin{bmatrix} \mathbf{0}_{3\times3} & \mathbf{0}_{3\times3} & \mathbf{0}_{3\times3} \\ \mathbf{0}_{3\times3} & \mathbf{I}_{3\times3} & \mathbf{0}_{3\times3} \\ \mathbf{0}_{3\times3} & \mathbf{0}_{3\times3} & \mathbf{0}_{3\times3} \end{bmatrix}, \tag{14.136}$$

Three simulations of both IC and CC were performed with event threshold $q_t = 0°$, $q_t = 0.1°$, and $q_t = 1°$ and the resultant link angles $\boldsymbol{\theta}$ are plotted against time in Figure 14.17; the black lines show the IC simulations and the gray lines the CC simulations. The three-segment model together with the spring model has nine states. Figure 14.18 shows three cross sections through this space (by plotting segment angular velocity against segment angle) for the three thresholds.

As expected, the small threshold gives smaller displacements from vertical; but the disturbance is more apparent. The large threshold gives largely self-driven behavior. This behavior is discussed in more detail by Gawthrop et al. (2014).

### 14.7.4 Tracking

As discussed in Section 14.2.4, the equilibrium state $\mathbf{x}_{ss}$ has to be designed for tracking purposes. As there are three inputs, it is possible to satisfy up to three steady-state conditions. Three possible steady-state conditions are

1. The upper link should follow a setpoint:

$$\boldsymbol{\theta}_u = w_u. \tag{14.137}$$

2. The component of ankle torque due to gravity should be zero:

$$\mathbf{T}_1 = \begin{bmatrix} 1 & 0 & 0 \end{bmatrix} \mathbf{N}^{-1}\mathbf{G}\boldsymbol{\theta} = 0. \tag{14.138}$$

3. The knee angle should follow a set point:

$$\phi_m = \theta_m - \theta_l = w_m. \tag{14.139}$$

These conditions correspond to

$$\mathbf{C}_{ss} = \begin{bmatrix} 0 & 0 & 0 & 0 & 0 & 1 & 0 & 0 & 0 \\ 0 & 0 & 0 & 27.14627 & 22.51155 & 23.74238 & 0 & 0 & 0 \\ 0 & 0 & 0 & -1 & 1 & 0 & 0 & 0 & 0 \end{bmatrix}, \tag{14.140}$$

$$\mathbf{y}_{ss} = \mathbf{I}_{3\times3}, \tag{14.141}$$

$$\mathbf{w}(t) = \begin{bmatrix} w_u(t) \\ 0 \\ w_m(t) \end{bmatrix}. \tag{14.142}$$

This choice is examined in Figures 14.19 and 14.20 by choosing the knee angle $\phi_m = \theta_m - \theta_l$. Figure 14.20 shows how the link and joint angles, and the corresponding torques, vary with $\phi_m$. Figure 14.19 shows a picture of the three links for three values of $\phi_m$. In each case, note that the upper link and the corresponding hip torque remain constant due to the first condition and that each configuration appears balanced due to condition 2.

The simulations shown in Figure 14.21 shows the tracking of a setpoint $\mathbf{w}(t)$ (14.142) using the three conditions for determining the steady state. In this example, the individual setpoint components of Equation 14.142 are

$$w_u(t) = \begin{cases} 10° & 0 < t \le 10 \\ 0° & 10 < t \le 15 \\ 0° & 10 < t \le 15 \\ 10° & 15 < t \le 15.1 \\ 0° & 15.1 < t \le 20 \\ 10° & 20 < t \le 20.25 \\ 0° & t > 20.25 \end{cases}, \tag{14.143}$$

$$w_m(t) = \begin{cases} 0° & 0 < t \le 5 \\ -20° & t > 5 \end{cases}. \tag{14.144}$$

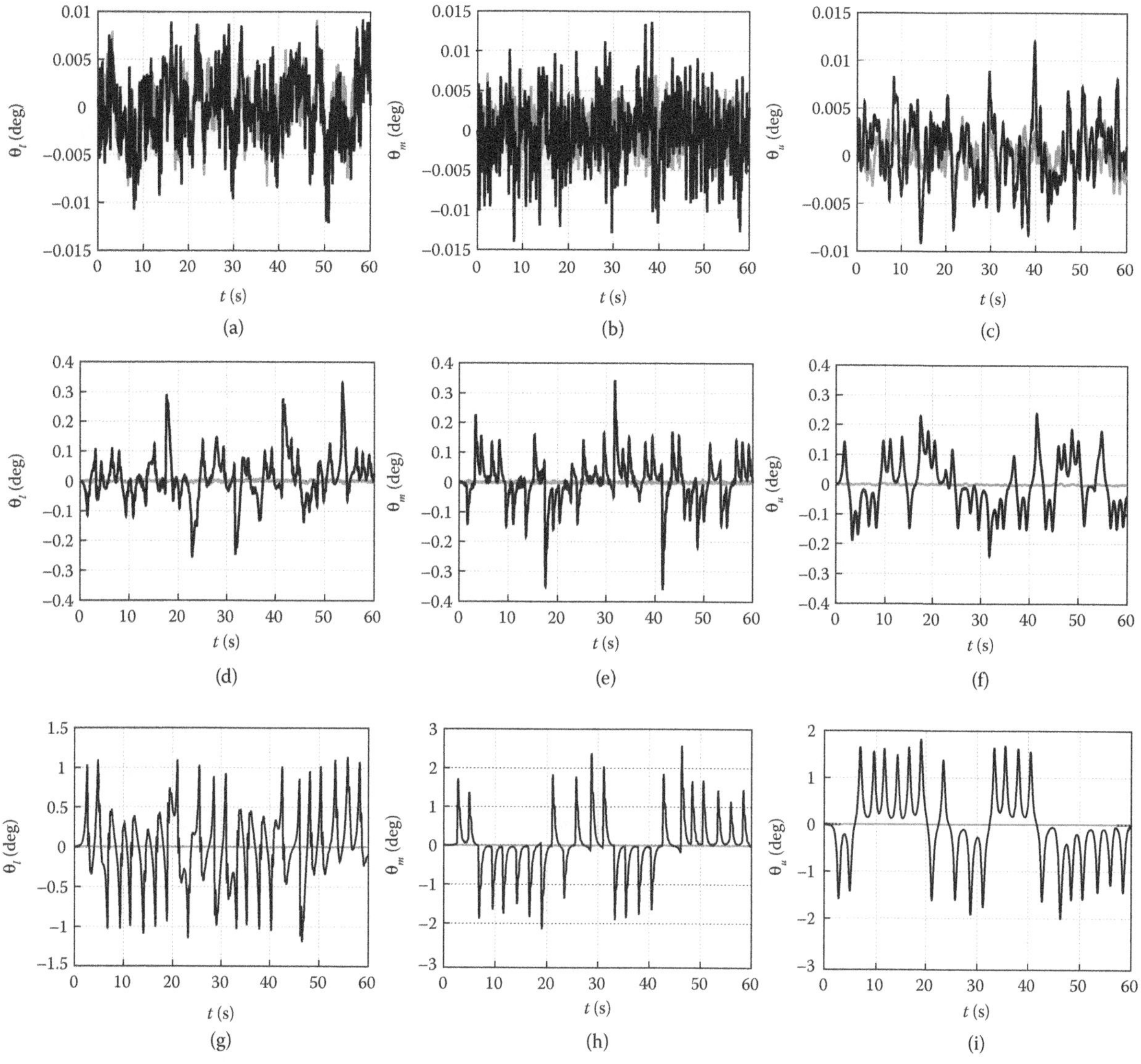


**FIGURE 14.17**
Quiet standing with torque disturbance at the lower joint. Each row shows the lower ($\boldsymbol{\theta}_l$), middle ($\boldsymbol{\theta}_m$), and upper ($\boldsymbol{\theta}_u$) link angles (deg) plotted against time $t$ (s) for a different thresholds $q_t$. Larger thresholds give larger and more regular sway angles. (a) $q_t = 0°$: $\theta_l^\circ$. (b) $q_t = 0°$: $\theta_m^\circ$. (c) $q_t = 0°$: $\theta_u^\circ$. (d) $q_t = 0.1°$: $\theta_l^\circ$. (e) $q_t = 0.1°$: $\theta_m^\circ$. (f) $q_t = 0.1°$: $\theta_u^\circ$. (g) $q_t = 1°$: $\theta_l^\circ$. (h) $q_t = 1°$: $\theta_m^\circ$. (i) $q_t = 1°$: $\theta_u^\circ$.

As a further example, only the first two conditions for determining the steady state are used; the knee is not included. These conditions correspond to:

$$\mathbf{C}_{ss} = \begin{bmatrix} 0 & 0 & 0 & 0 & 0 & 1 & 0 & 0 & 0 \\ 0 & 0 & 0 & 27.14627 & 22.51155 & 23.74238 & 0 & 0 & 0 \end{bmatrix}, \tag{14.145}$$

$$\mathbf{y}_{ss} = \mathbf{I}_{2\times 2}, \tag{14.146}$$

$$\mathbf{w}(t) = \begin{bmatrix} w_u(t) \\ 0 \end{bmatrix}. \tag{14.147}$$

The under-determined Equation 14.13 is solved using the pseudo inverse. The simulations shown in Figure 14.22 shows the tracking of a setpoint $\mathbf{w}(t)$ (14.142) using the first two conditions for determining the steady state. In this example, the individual setpoint component $w_u$ is given by (14.143). Comparing Figure 14.22b, e, and h with Figure 14.21b, e, and h, it can be seen that the knee angle is no longer explicitly controlled.

### 14.7.5 Disturbance Rejection

Detailed modeling of a human lifting and holding a heavy pole would require complicated dynamical equations. This section looks at a simple approximation to the case where a heavy pole of mass $m_p$ is held at a

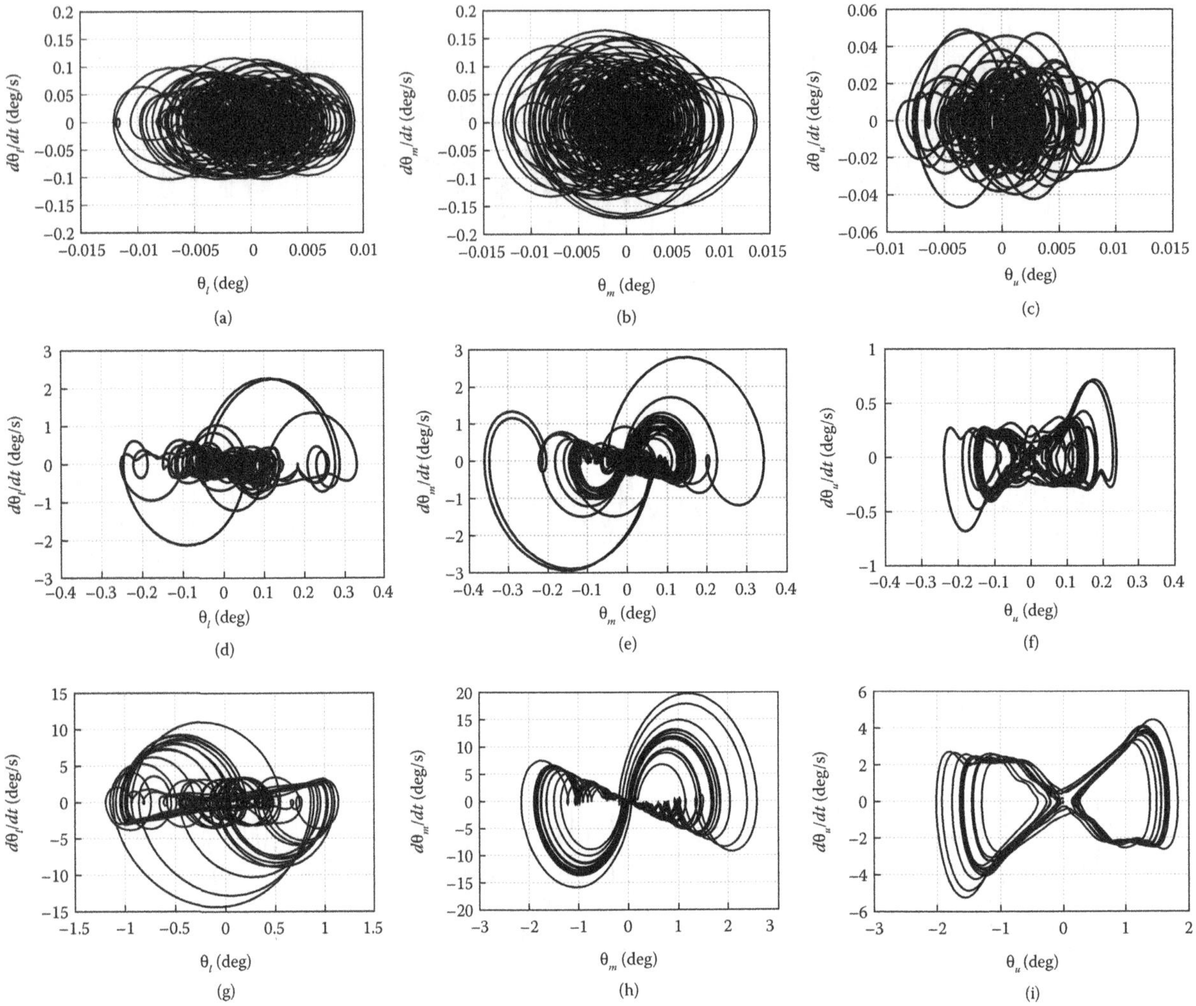


**FIGURE 14.18**

Quiet standing: phase-plane. Each plot corresponds to that of Figure 14.17 but the angular velocity is plotted against angle. Again, the increase in sway angle and angular velocity with threshold is evident. (a) $q_t = 0°$: $\theta_l^\circ$. (b) $q_t = 0°$: $\theta_m^\circ$. (c) $q_t = 0°$: $\theta_u^\circ$. (d) $q_t = 0.1°$: $\theta_l^\circ$. (e) $q_t = 0.1°$: $\theta_m^\circ$. (f) $q_t = 0.1°$: $\theta_u^\circ$. (g) $q_t = 1°$: $\theta_l^\circ$. (h) $q_t = 1°$: $\theta_m^\circ$. (i) $q_t = 1°$: $\theta_u^\circ$.

fixed distance $l_p$ to the body. In particular, the effect is modeled by

1. Adding a torque disturbance $T_d$ to the upper link where
$$T_d = gm_p l_p. \tag{14.148}$$
2. Adding a mass $m_p$ to the upper link.

In terms of the system Equation 14.1 and the three-link model of Equation 14.89, the disturbance $\mathbf{d}$ is given by

$$\mathbf{d} = N^{-1}\begin{bmatrix} 0 \\ 0 \\ T_d \end{bmatrix} \tag{14.149}$$

$$= \begin{bmatrix} 1 \\ 1 \\ 1 \end{bmatrix} T_d. \tag{14.150}$$

As discussed in Section 14.7.2, it is possible to preload the joint spring to give an initial torque. In this context, this is done by initializing the system state $\mathbf{x}$ of Equation 14.117 as

$$\mathbf{x}(0) = \begin{bmatrix} \dot{\boldsymbol{\theta}}(0) \\ \boldsymbol{\theta}(0) \\ \mathbf{T}(0) \end{bmatrix} = \begin{bmatrix} \mathbf{0}_{3\times 1} \\ \mathbf{0}_{3\times 1} \\ \kappa \mathbf{d} \end{bmatrix}, \tag{14.151}$$

$\kappa$ will be referred to as the *spring preload* and will be expressed as a percentage: thus, $\kappa = 0.8$ will be referred to as 80% preload.

There are many postures appropriate to this situation, two of which are as follows:

**Upright:** All joint *angles* are zero and the pole is balanced by appropriate joint torques (Figure 14.24 and b).

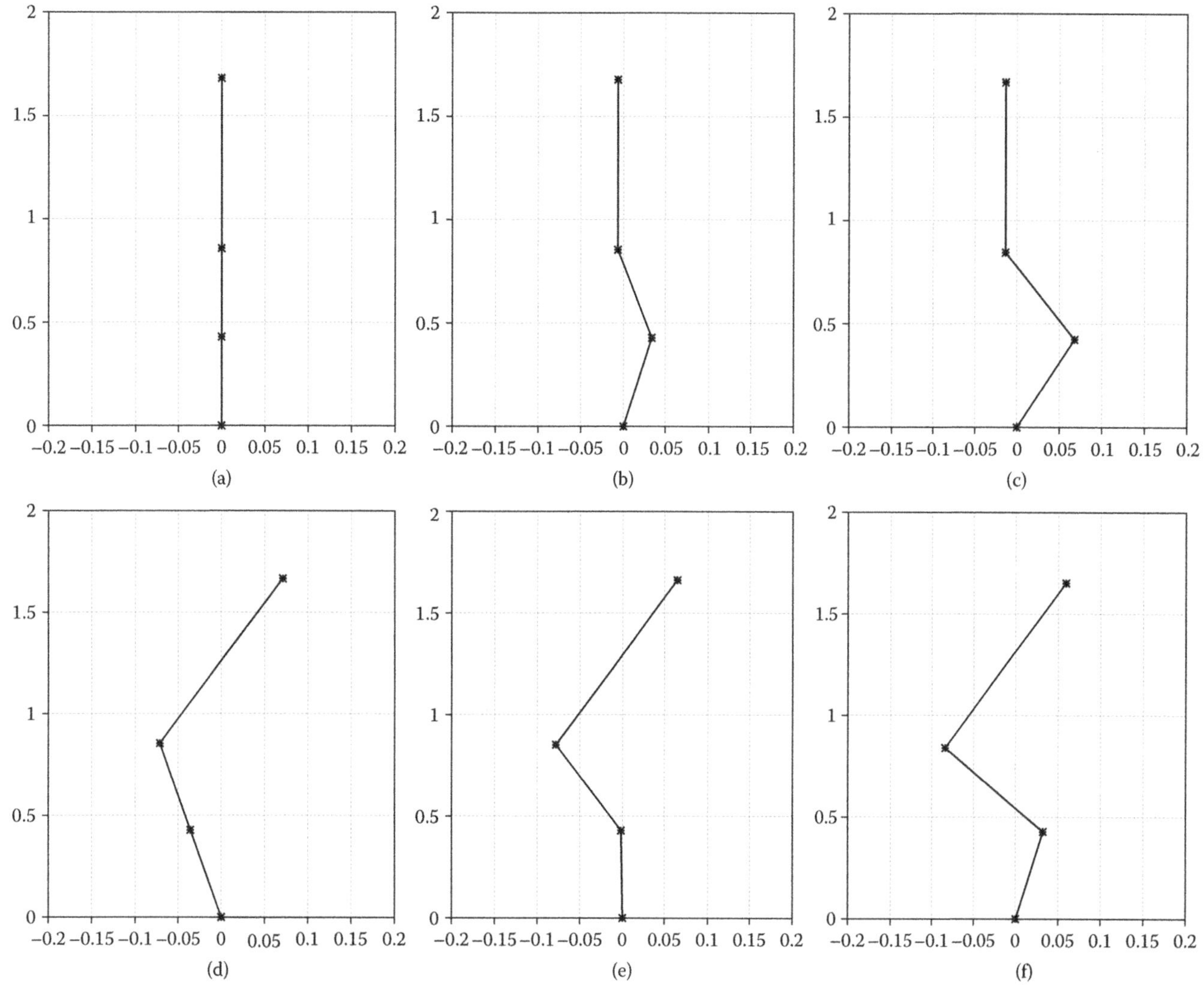


**FIGURE 14.19**
Equilibria: Link configuration. (a–c) In each configuration, the upper link is set at $\theta_u = 0°$ and the posture is balanced (no ankle torque); the knee angle is set to three possible values. (d), (e), (f) as (a), (b), (c) but $\theta_u = 10°$.

**Balanced:** All joint *torques* are zero and the pole is balanced by appropriate joint (and thus link) angles (Figure 14.24c and f).

In terms of Equation 14.12, the upright posture is specified by choosing:

$$\mathbf{C}_{ss} = \begin{bmatrix} \mathbf{0}_{3\times3} & \mathbf{I}_{3\times3} & \mathbf{0}_{3\times3} \end{bmatrix}, \tag{14.152}$$

$$\mathbf{y}_{ss} = \mathbf{0}_{3\times1}, \tag{14.153}$$

and the balanced posture is specified by choosing:

$$\mathbf{C}_{ss} = \begin{bmatrix} \mathbf{0}_{3\times3} & \mathbf{0}_{3\times3} & \mathbf{I}_{3\times3} \end{bmatrix}, \tag{14.154}$$

$$\mathbf{y}_{ss} = \mathbf{0}_{3\times1}. \tag{14.155}$$

A combination of both can be specified by choosing:

$$\mathbf{C}_{ss} = \begin{bmatrix} \mathbf{0}_{3\times3} & \mathbf{I}_{3\times3} & \mathbf{0}_{3\times3} \\ \mathbf{0}_{3\times3} & \mathbf{0}_{3\times3} & \mathbf{I}_{3\times3} \end{bmatrix}, \tag{14.156}$$

$$\mathbf{y}_{ss} = \mathbf{0}_{6\times1}, \tag{14.157}$$

$$\mathbf{Q}_{ss} = \begin{bmatrix} (1-\lambda)\mathbf{I}_{3\times3} & \mathbf{0}_{3\times3} \\ \mathbf{0}_{3\times3} & \frac{\lambda}{\mathbf{T}_d}\mathbf{I}_{3\times3} \end{bmatrix}. \tag{14.158}$$

The parameter $0 \leq \lambda \leq 1$ weights the two postures and division by $\mathbf{T}_d$ renders the equations dimensionless.

When $\mathbf{C}_{ss}$ is given by (14.156), $n_{ss} = 6$. As $n_u = 3$, $n_{ss} > n_u$ and so, as discussed in Section 14.2.4, the set of Equations 14.13 is overdetermined and the approach of Section 14.5 is used. Two situations are examined: unconstrained and constrained with hip angle and knee

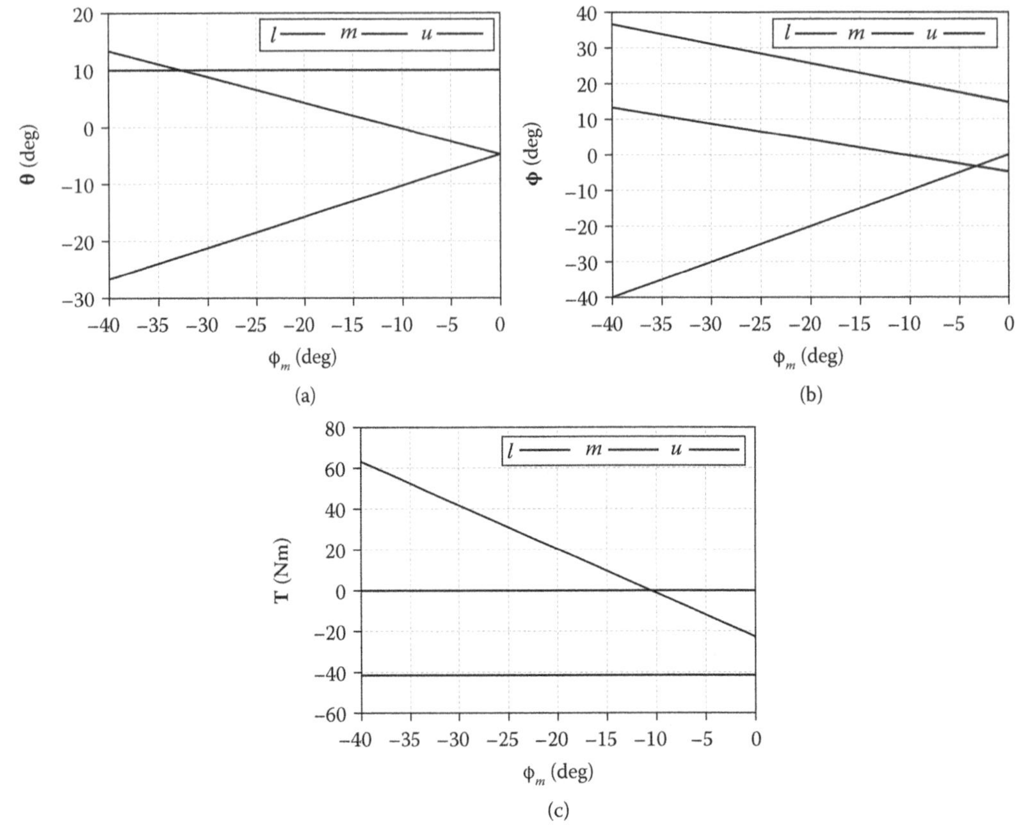


**FIGURE 14.20**

Equilibria: angles and torques. (a) The equilibrium link angles $\boldsymbol{\theta}$ are plotted against the fixed knee angle $\phi_m$ with balanced posture and the upper link at an angle of $\theta_u = 10°$. (b) and (c) as (a) but with joint angles $\boldsymbol{\phi}$ and joint torques $\mathbf{T}$, respectively. Note that the ankle torque $T_l$ is zero (balanced posture) and the waist torque $T_u$ balances the fixed $\theta_u = 10°$.

angle subject to the inequality constraints:

$$\phi_u > -0.1°, \tag{14.159}$$

$$\phi_m < 0. \tag{14.160}$$

In each case, the equality constraint (14.66) is imposed. Figure 14.23 shows how the equilibrium joint angle $\boldsymbol{\phi}$ and torque $\mathbf{T}$ vary with $\lambda$ for the two cases. As illustrated in Figure 14.24, the two extreme cases $\lambda = 0$ and $\lambda = 1$ correspond to the upright and balanced postures; other values give intermediate postures.

Figures 14.25 and 14.27 show simulation results for the two extreme cases of $\lambda$ for the unconstrained case with $m_p = 5$kg and $l_p = 0.5$m. In each case, the initial link angles are all zero ($\boldsymbol{\theta}_l = \boldsymbol{\theta}_m = \boldsymbol{\theta}_u = 0$) and the disturbance torque $\mathbf{T}_d = g$ is applied at $t = 0$. Apart from the equilibrium vector $\mathbf{x}_{ss}$, the control parameters are the same in each case.

In the case of Figure 14.25, the steady-state torques are $\mathbf{T}_l = \mathbf{T}_m = \mathbf{T}_u = -\mathbf{T}_d$ to balance $\mathbf{T}_d = gm_pl_p$; in the case of Figure 14.27, the links balance the applied torque by setting $\boldsymbol{\theta}_l = \boldsymbol{\theta}_m = 0$ and $\boldsymbol{\theta}_u = \dfrac{-\Delta}{gc_um_u}$.

## 14.8 Intermittency Induces Variability

Variability is an important characteristic of human motor control: when repeatedly exposed to identical excitation, the response of the human operator is different for each repetition. This is illustrated in Figure 14.29: Figure 14.29a shows a periodic input disturbance (periodicity 10 s), while Figure 14.29c and e shows the corresponding output signal of a human controller for different control aims. It is clear that in both cases, the control signal is different for each 10 s period of identical disturbance.

In the frequency domain, variability is represented by the observation that, when the system is excited at a range of discrete frequencies (as shown in Figure 14.29b, the disturbance signal contains frequency components at 0.1, 0.2, . . . , 10 Hz), the output response contains information at both the excited and the nonexcited frequencies (Figure 14.29d and f). The response at the nonexcited frequencies (at which the excitation signal is zero) is termed the remnant.

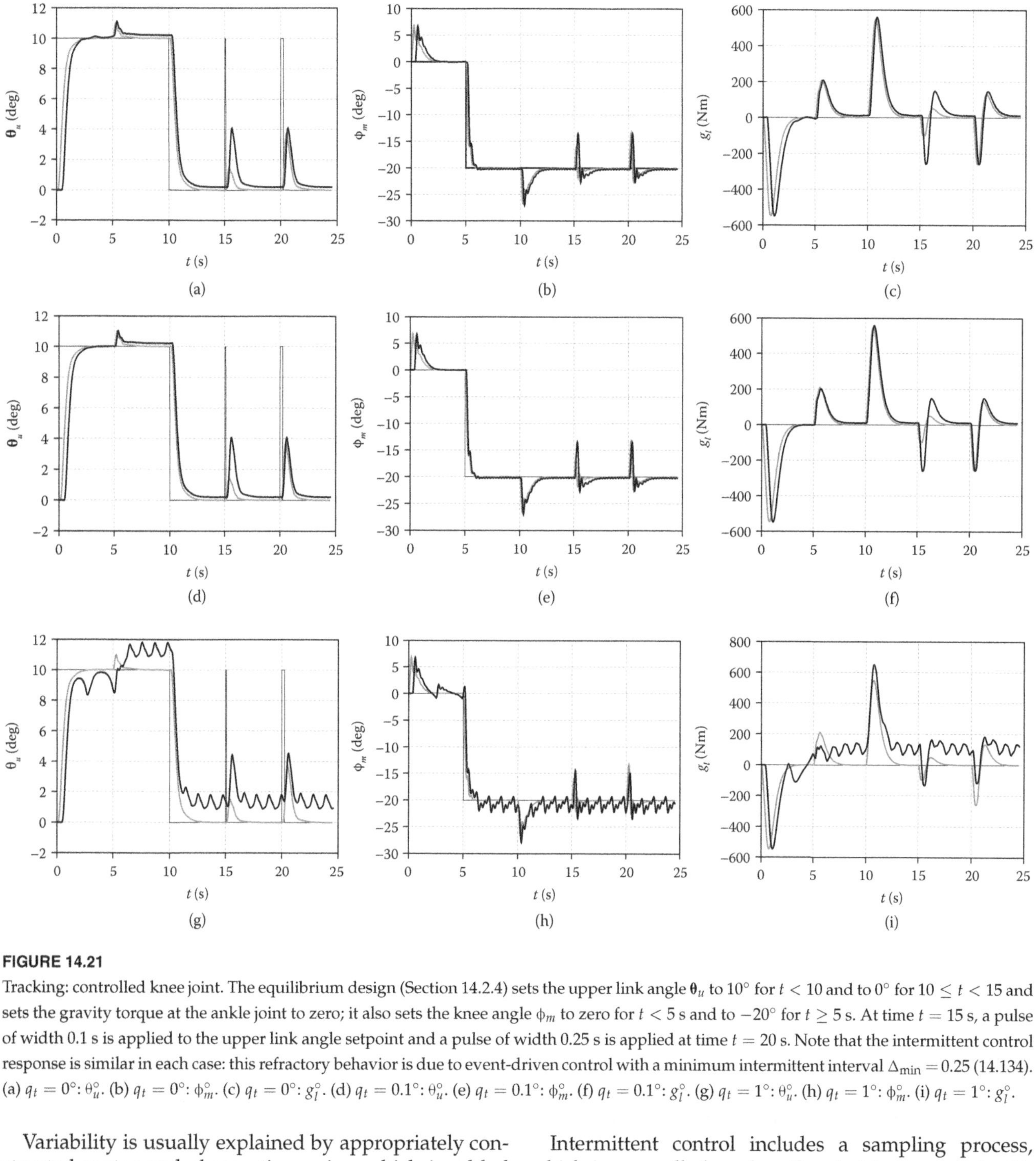


**FIGURE 14.21**

Tracking: controlled knee joint. The equilibrium design (Section 14.2.4) sets the upper link angle $\boldsymbol{\theta}_u$ to 10° for $t < 10$ and to 0° for $10 \leq t < 15$ and sets the gravity torque at the ankle joint to zero; it also sets the knee angle $\phi_m$ to zero for $t < 5$ s and to $-20°$ for $t \geq 5$ s. At time $t = 15$ s, a pulse of width 0.1 s is applied to the upper link angle setpoint and a pulse of width 0.25 s is applied at time $t = 20$ s. Note that the intermittent control response is similar in each case: this refractory behavior is due to event-driven control with a minimum intermittent interval $\Delta_{\min} = 0.25$ (14.134). (a) $q_t = 0°$: $\theta_u^\circ$. (b) $q_t = 0°$: $\phi_m^\circ$. (c) $q_t = 0°$: $g_l^\circ$. (d) $q_t = 0.1°$: $\theta_u^\circ$. (e) $q_t = 0.1°$: $\phi_m^\circ$. (f) $q_t = 0.1°$: $g_l^\circ$. (g) $q_t = 1°$: $\theta_u^\circ$. (h) $q_t = 1°$: $\phi_m^\circ$. (i) $q_t = 1°$: $g_l^\circ$.

Variability is usually explained by appropriately constructed motor and observation noise, which is added to a linear continuous-time model of the human controller (signals $v_u$ and $v_y$ in Figure 14.1; Levison, Baron, and Kleinman, 1969; Kleinman et al., 1970). While this is currently the prominent model in human control, its physiological basis is not fully established. This has led to the idea that the remnant signal might be based on structure rather than randomness (Newell, Deutch, Sosnoff, and Mayer-Kress, 2006).

Intermittent control includes a sampling process, which is generally based on thresholds associated with a trigger (see Figure 14.2). This nonuniform sampling process leads to a time-varying response of the controller. It has been suggested that the remnant can be explained by event-driven intermittent control without the need for added noise (Mamma, Gollee, Gawthrop, and Loram, 2011; Gawthrop, Gollee, Mamma, Loram, and Lakie, 2013a), and that this sampling process introduces variability (Gawthrop et al. 2013b).

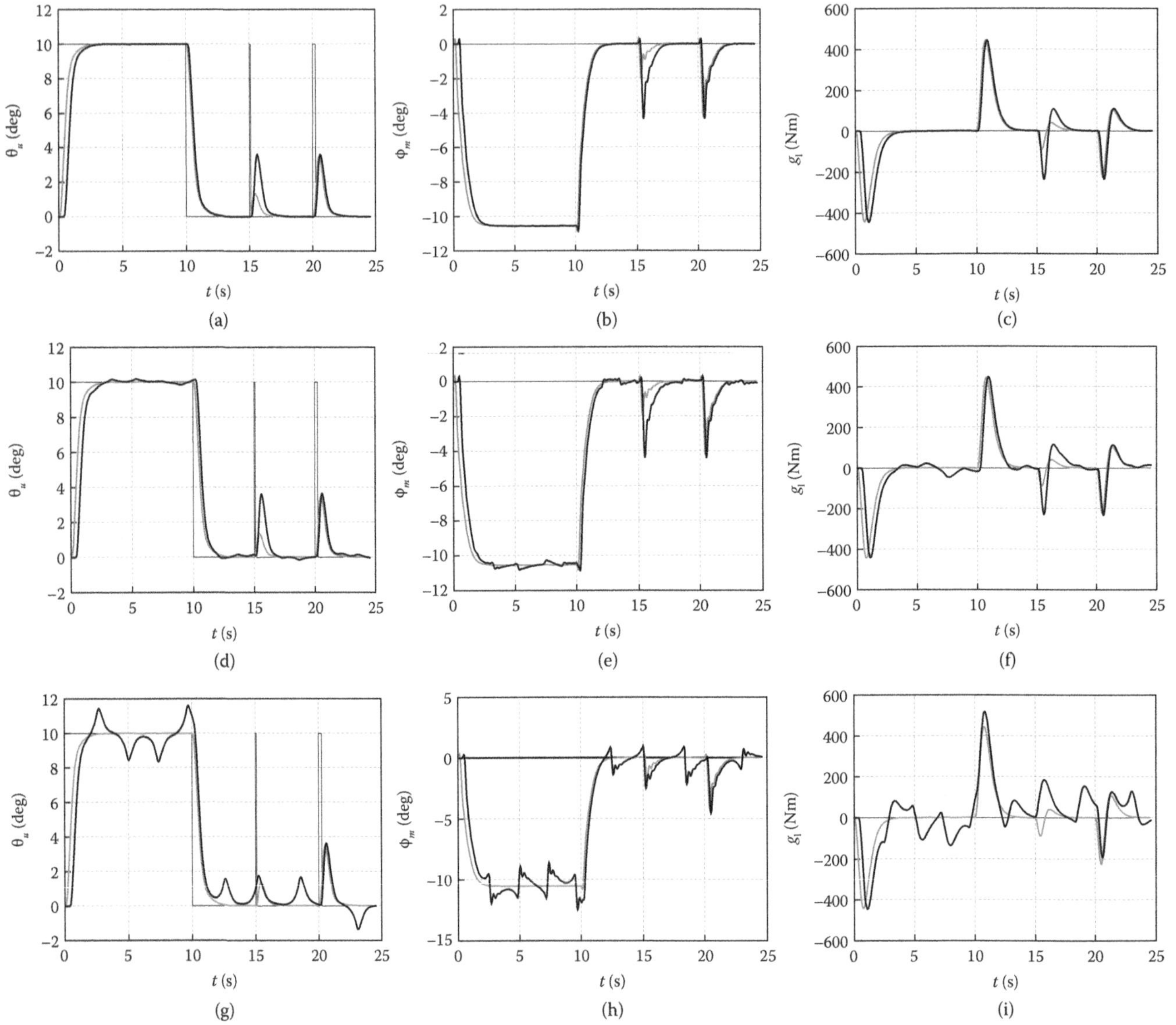


**FIGURE 14.22**

Tracking: free knee joint. This figure corresponds to Figure 14.21 except that the knee joint angle $\phi_m$ is not constrained. (a) $q_t = 0°$: $\theta_u^o$. (b) $q_t = 0°$: $\phi_m^o$. (c) $q_t = 0°$: $g_l^o$. (d) $q_t = 0.1°$: $\theta_u^o$. (e) $q_t = 0.1°$: $\phi_m^o$. (f) $q_t = 0.1°$: $g_l^o$. (g) $q_t = 1°$: $\theta_u^o$. (h) $q_t = 1°$: $\phi_m^o$. (i) $q_t = 1°$: $g_l^o$.

In this section, we will discuss how intermittency can provide an explanation for variability which is based on the controller structure and does not require a random process. Experimental data from a visual-manual control task will be used as an illustrative example.

### 14.8.1 Experimental Setup

In this section, experimental data from a visual-manual control task are used in which the participant were asked to use a sensitive, contactless, uniaxial joystick to sustain control of an unstable second-order system whose output was displayed as a dot on a oscilloscope (Loram et al., 2011). The controlled system represented an inverted pendulum with a dynamic response similar to that of a human standing [Load 2 of Table 1 in Loram, Lakie, and Gawthrop (2009)]:

$$\begin{cases} \frac{d\mathbf{x}}{dt}(t) = \begin{bmatrix} -0.0372 & 1.231 \\ 1 & 0 \end{bmatrix} \mathbf{x}(t) \\ \qquad + \begin{bmatrix} 6.977 \\ 0 \end{bmatrix} (\mathbf{u}(t) - \mathbf{d}'(t)) \\ \mathbf{y}(t) = \begin{bmatrix} 0 & 1 \end{bmatrix} \mathbf{x}(t) \\ \mathbf{y}_o(t) = \begin{bmatrix} 1 & 0 \\ 0 & 1 \end{bmatrix} \mathbf{x}(t) \end{cases} . \tag{14.161}$$

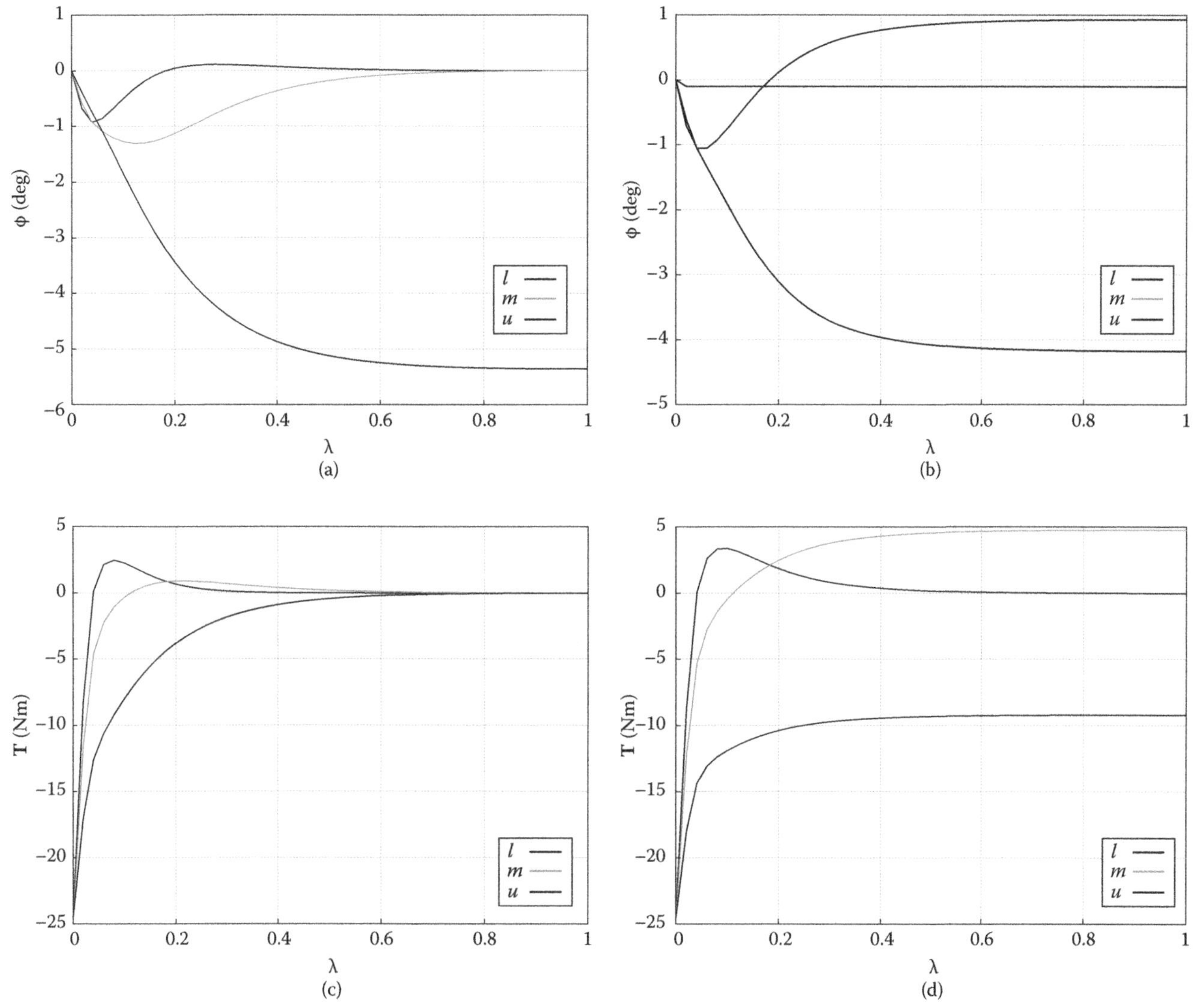


**FIGURE 14.23**

Equilibria. For a constant disturbance torque $\mathbf{T}_d$ acting on the upper link, the plots show how link angle and joint torque vary with the weighting factor $\lambda$. $\lambda = 0$ gives an upright posture (zero link and joint angles—Figure 14.24a) and $\lambda = 1$ gives a balanced posture (zero joint torques—Figure 14.24e). $\lambda = 0.5$ gives an intermediate posture (Figure 14.24c). The left column is the unconstrained case, the right column is the constrained case where hip angle $\phi_u > -0.1°$ and knee angle $\phi_m < 0$; the former constraint becomes active as $\lambda$ increases and the knee and hip joint torques are no longer zero at $\lambda = 1$. (a) Joint angle $\phi$ (deg)—unconstrained. (b) Joint angle $\phi$ (deg)—constrained. (c) Joint torque $\mathbf{T}$ (Nm)—unconstrained. (d) Joint torque $\mathbf{T}$ (Nm)—constrained.

The external disturbance signal, $d(t)$, applied to the load input, was a multi-sine consisting of $N_f = 100$ discrete frequencies $\omega_k$, with resolution $\omega_0 = 2\pi f_0$, $f_0 = 0.1$ Hz (Pintelon and Schoukens, 2001)

$$d(t) = \sum_{k=1}^{N_f} a_k \cos(\omega_k t + \phi_k) \qquad \text{with } \omega_k = 2\pi k f_0. \tag{14.162}$$

The signal $d(t)$ is periodic with $T_0 = 1/f_0 = 10$ s. To obtain an unpredictable excitation, the phases $\phi_k$ are random values taken from a uniform distribution on the open interval $(0, 2\pi)$, while $a_k = 1$ for all $k$ to ensure that all frequencies are equally excited.

We considered two control priorities using the instructions "keep the dot as close to the center as possible" ("cc," prioritizing position), and "while keeping the dot on screen, wait as long as possible before intervening" ("mi," minimizing intervention).

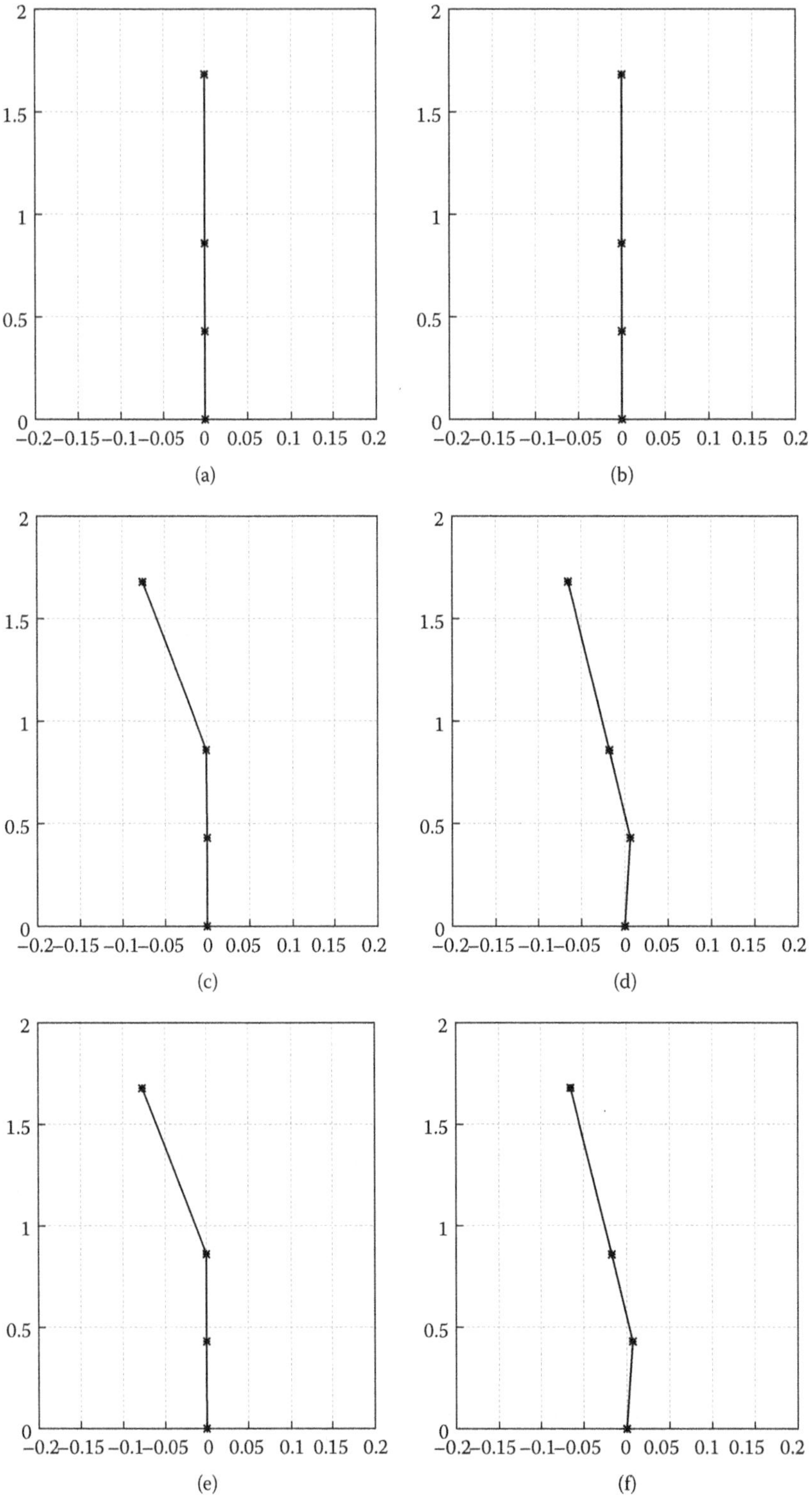


**FIGURE 14.24**
Equilibria: link configurations. (a), (c), (e) unconstrained; (b), (d), (f) constrained where hip angle $\phi_u > -0.1°$ and knee angle $\phi_m < 0$. (a) Upright ($\lambda = 0$)—unconstrained. (b) Upright ($\lambda = 0$)—constrained. (c) Intermediate ($\lambda = 0.5$)—unconstrained. (d) Intermediate ($\lambda = 0.5$)—constrained. (e) Balanced ($\lambda = 1$)—unconstrained. (f) Balanced ($\lambda = 1$)—constrained.

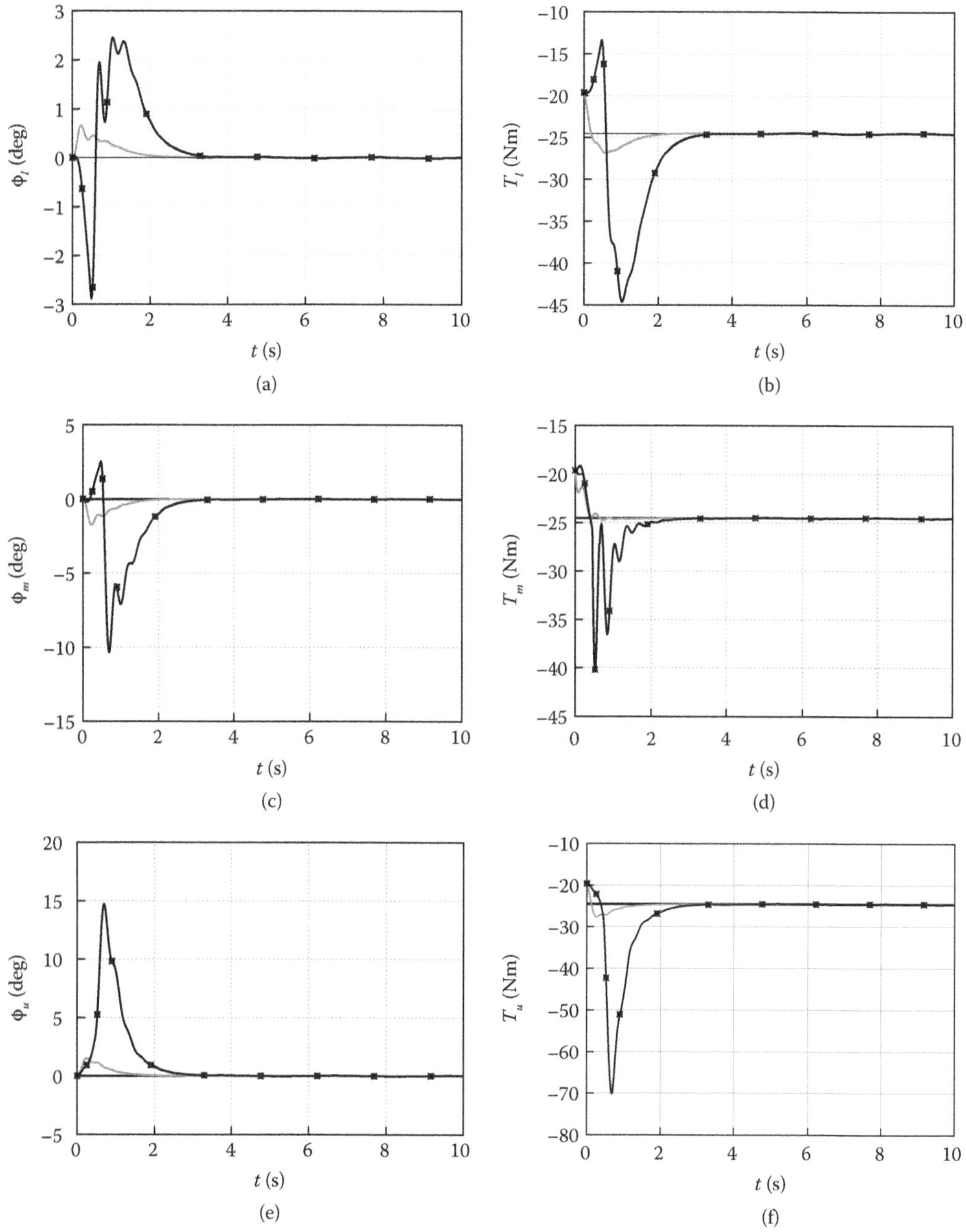


**FIGURE 14.25**

Pole-lifting simulation—constrained: upright posture ($\lambda = 0$). The spring preload $\kappa$ (14.151) is 80%. Note that the steady-state link angles are zero and the steady-state torques are all $-T_d$ to balance $T_d = gm_pl_p = 24.5$ Nm (14.148). The dots correspond to the sample times $t_i$. The intervals $\Delta_i$ (14.16) are irregular and greater than the minimum $\Delta_{\min}$. (a) Lower joint: angle $\theta_l$ (deg). (b) Lower joint: torque $T_l$ (Nm). (c) Middle joint: angle $\theta_m$ (deg). (d) Middle joint: torque $T_m$ (Nm). (e) Upper joint: angle $\theta_u$ (deg). (f) Upper joint: torque $T_u$ (Nm).

### 14.8.2 Identification of the Linear Time-Invariable (LTI) Response

Using previously established methods discussed in Section 14.9 and by Gollee, Mamma, Loram, and Gawthrop (2012), the design parameters (i.e., LQ design weightings and mean time-delay, $\Delta$) for an optimal, continuous-time linear predictive controller (PC) (Figure 14.1) are identified by fitting the complex frequency response function (FRF) relating $d$ to $u_e$ at the excited frequencies. The linear fit to the experimental data is shown in Figure 14.30a and b for the two different experimental instructions ("cc" and "mi"). Note that

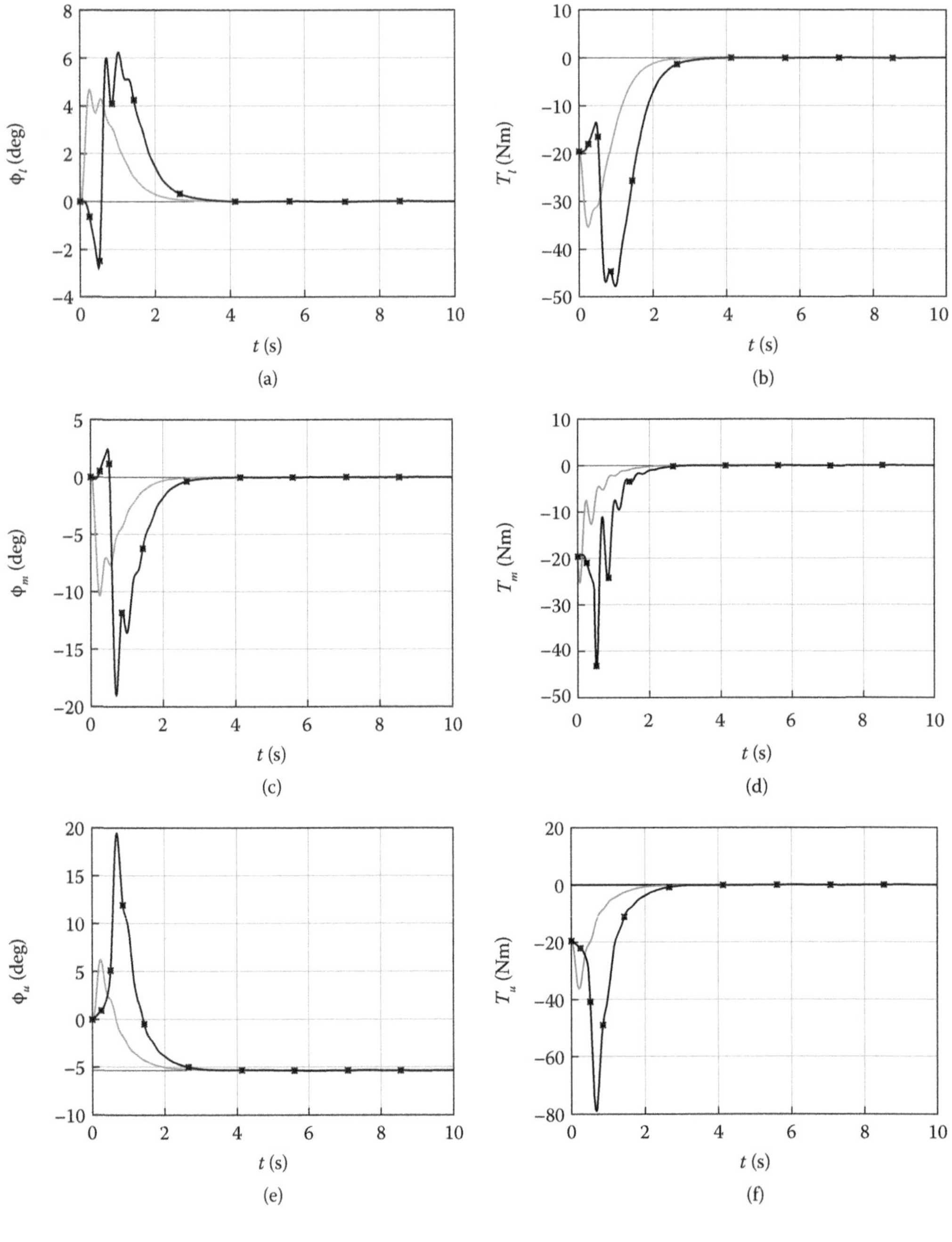


**FIGURE 14.26**
Pole-lifting simulation—unconstrained steady state: balanced posture ($\lambda = 1$). The spring preload $\kappa$ (14.151) is 80%. The steady-state joint torques are zero. (a) Lower joint: angle $\theta_l$ (deg). (b) Lower joint: torque $T_l$ (Nm). (c) Middle joint: angle $\theta_m$ (deg). (d) Middle joint: torque $T_m$ (Nm). (e) Upper joint: angle $\theta_u$ (deg). (f) Upper joint: torque $T_u$ (Nm).

the PC only fits the excited frequency components; its response at the nonexcited frequencies (bottom plots) is zero.

### 14.8.3 Identification of the Remnant Response

The controller design parameters (i.e., the LQ design weightings) obtained when fitting the LTI response, are used as the basis to model the response at the nonexcited (remnant) frequencies. First, the standard approach of adding noise to a continuous PC is demonstrated. Following this, it is shown that event-driven IC can approximate the experimental remnant response, by adjusting the threshold parameters associated with the event trigger.

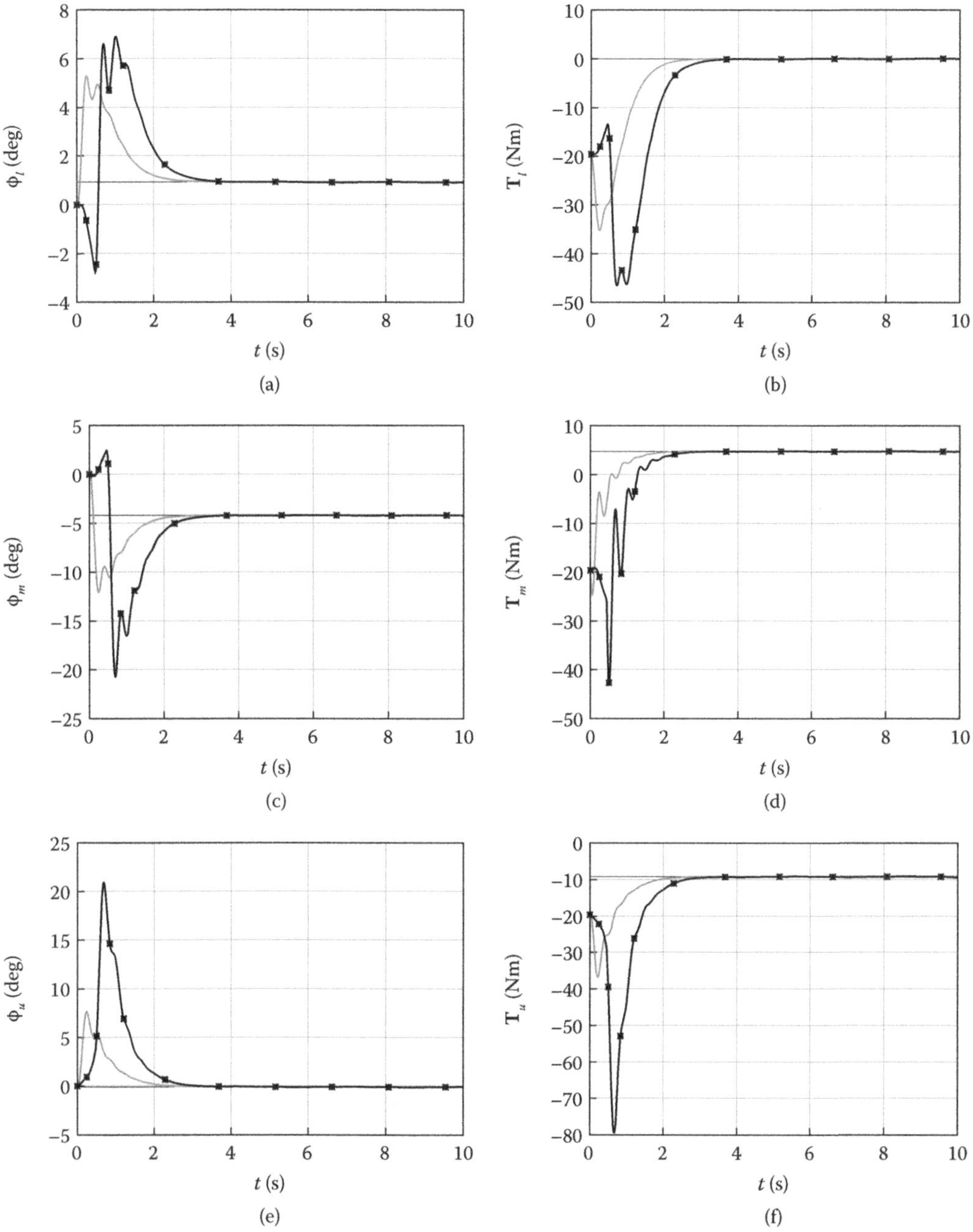


**FIGURE 14.27**
Pole-lifting simulation—constrained steady state: balanced posture ($\lambda = 1$). The spring preload $\kappa$ (14.151) is 80%. Due to the constraints, the steady-state joint torques $\mathbf{T}_u$ and $\mathbf{T}_m$ are not zero, but the upper (hip) joint is now constrained. (a) Lower joint: angle $\theta_l$ (deg). (b) Lower joint: torque $\mathbf{T}_l$ (Nm). (c) Middle joint: angle $\theta_m$ (deg). (d) Middle joint: torque $\mathbf{T}_m$ (Nm). (e) Upper joint: angle $\theta_u$ (deg). (f) Upper joint: torque $\mathbf{T}_u$ (Nm).

#### 14.8.3.1 Variability by Adding Noise

For the PC, noise can be injected either as observation noise, $v_y$, or as noise added to the input, $v_u$. The noise spectrum is obtained by considering the measured response $u_e$ at nonexcited frequencies and, using the corresponding loop transfer function (see Section 14.9.1.1), calculating the noise input ($v_u$ or $v_y$) required to generate this. The calculated noise signal is then interpolated at the excited frequencies.

Results for added input noise ($v_u$) are shown in Figure 14.31. As expected, the fit at the nonexcited frequencies is nearly perfect (Figure 14.31a and b, bottom panels). Notably, the added input noise also improves the fit at the excited frequencies (Figure 14.31a and b, top panels).

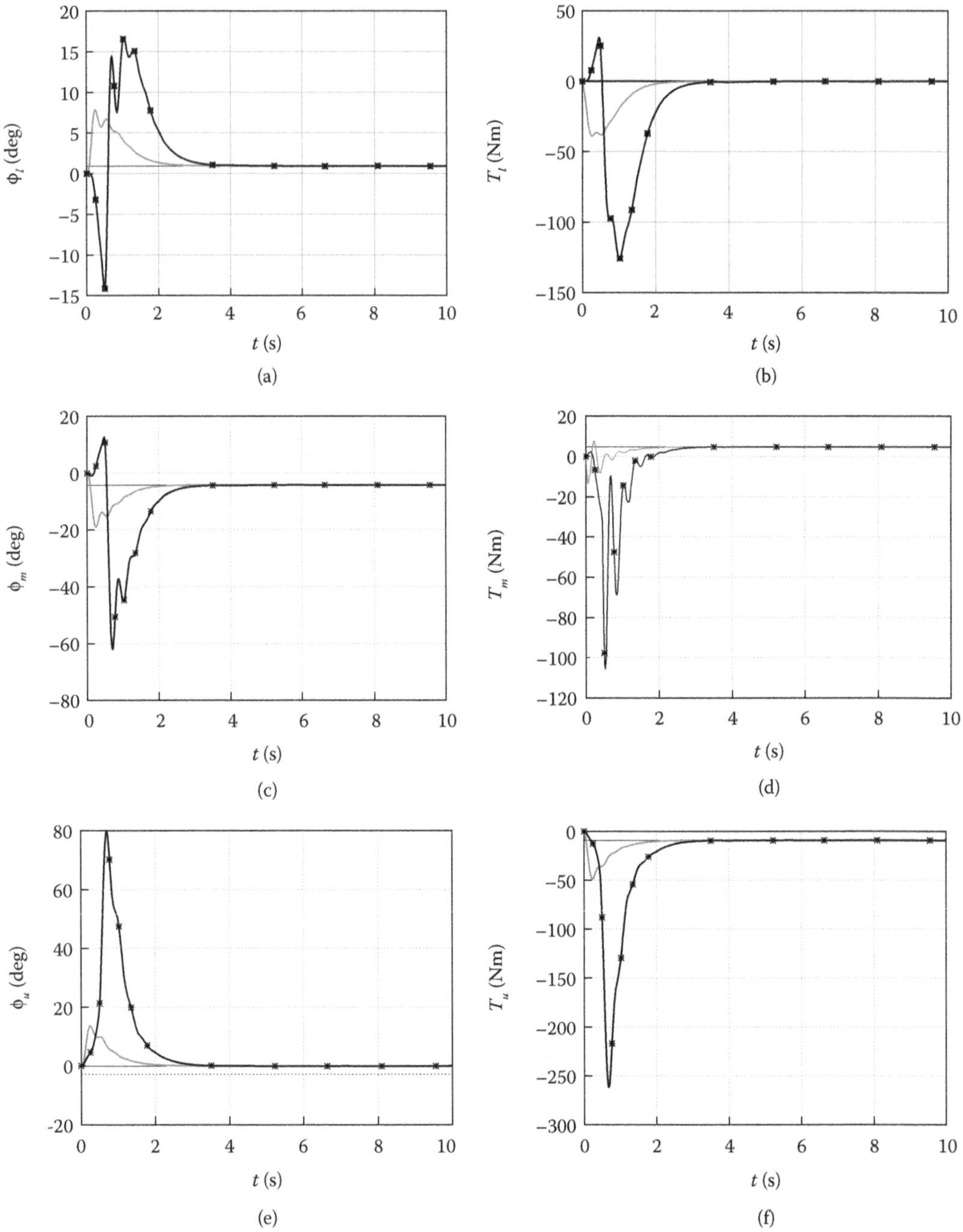


**FIGURE 14.28**
Pole-lifting simulation—constrained steady state but with 0% preload. This is the same as Figure 14.27 except that the spring preload is 0%. (a) Lower joint: angle $\theta_l$ (deg). (b) Lower joint: torque $T_l$ (Nm). (c) Middle joint: angle $\theta_m$ (deg). (d) Middle joint: torque $T_m$ (Nm). (e) Upper joint: angle $\theta_u$ (deg). (f) Upper joint: torque $T_u$ (Nm).

The spectra of the input noise $v_u$ are shown in Figure 14.31c and d. It can be observed that the noise spectra are dependent on the instructions given ("cc" or "mi"), with no obvious physiological basis to explain this difference.

#### 14.8.3.2 *Variability by Intermittency*

As an alternative explanation, a noise-free event-driven IC is considered (cf. Figure 14.2). The same design parameters as for the PC are used, with the time-delay

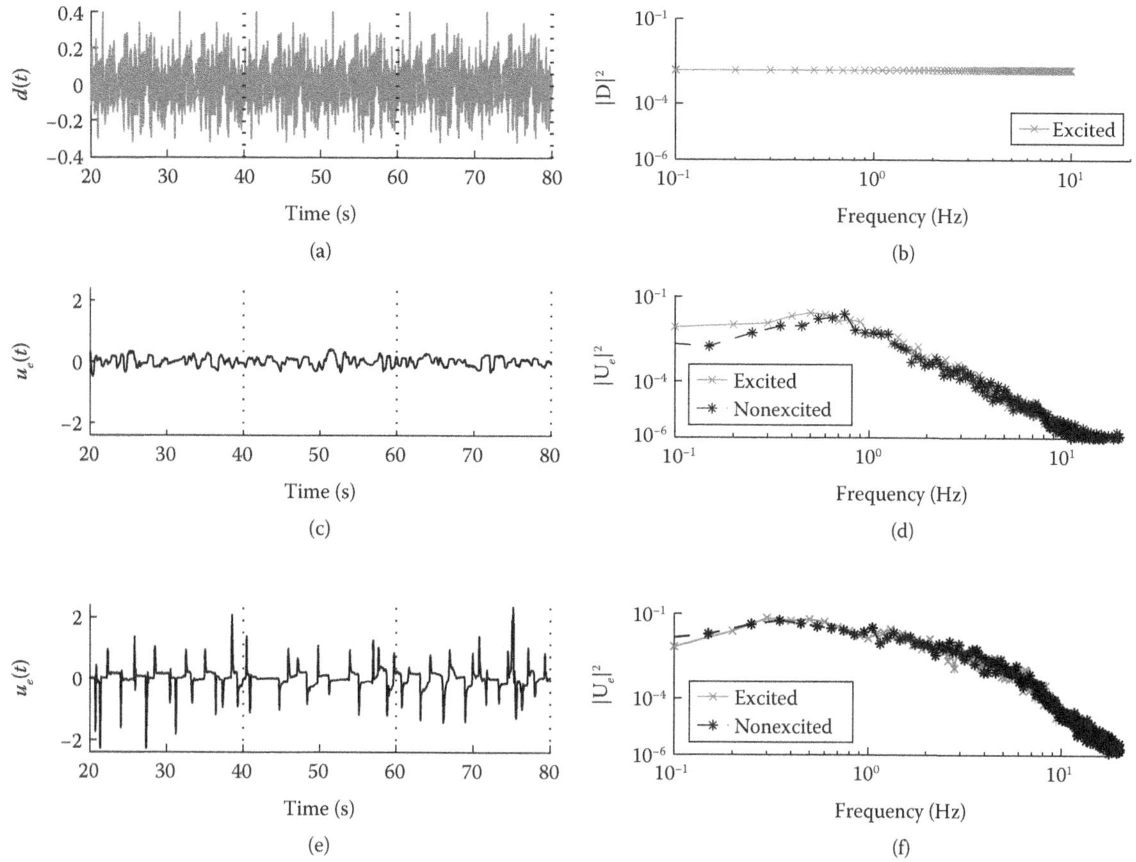


**FIGURE 14.29**
Experimental data showing variability during human motor control. Plots on the left side show time-domain signals, plots on the right side depict the corresponding frequency-domain signals. "cc"—keep close to center (position control), "mi"—minimal intervention minimize control). (a) Disturbance signal. (b) Disturbance signal. (c) Control signal (cc). (d) Control signal (cc). (e) Control signal (mi). (f) Control signal (mi).

set to a minimal value of $\Delta^{\min} = 0.1$ s and a corresponding minimal intermittent interval, $\Delta_{ol}^{\min} = 0.1$ s.

Variations in the loop delay are now the result of the event thresholds, cf. Equation 14.35. In particular, we consider the first two elements of the state prediction error $e_{hp}$, corresponding to the velocity ($e_{hp}^{v}$) and position ($e_{hp}^{p}$) states, and define an ellipsoidal event detection surface given by

$$\left(\frac{e_{hp}^{p}}{\theta^{p}}\right)^2 + \left(\frac{e_{hp}^{v}}{\theta^{v}}\right)^2 > 1, \qquad (14.163)$$

where $\theta^{p}$ and $\theta^{v}$ are the thresholds associated with the corresponding states.

To find the threshold values that resulted in simulation which best approximates the experimental remnant, both thresholds were varied between 0 (corresponding to clock-driven IC) and 3, and the threshold combination that resulted in the best least-squares fit at all frequencies (excited and nonexcited) was selected as the optimum. The resulting fit is shown in Figure 14.32a and b. For both instructions, the event-driven IC can both explain the remnant signal and improve the fit at excited frequencies.

The corresponding thresholds (for "cc": $\theta^{v} = 2.0$, $\theta^{p} = 0.2$, for "mi": $\theta^{v} = 2.1$, $\theta^{p} = 0.8$) reflect the control priorities for each instruction: for "cc" position control should be prioritized, resulting in a small value for the position threshold, while the velocity is relatively unimportant. For "mi," the control intervention should be minimal, which is associated with large thresholds on both velocity and position.

Figure 14.32c and d shows the distributions of the open-loop intervals for each condition, together with an approximation by a series of weighted Gaussian distributions (McLachlan and Peel, 2000). For

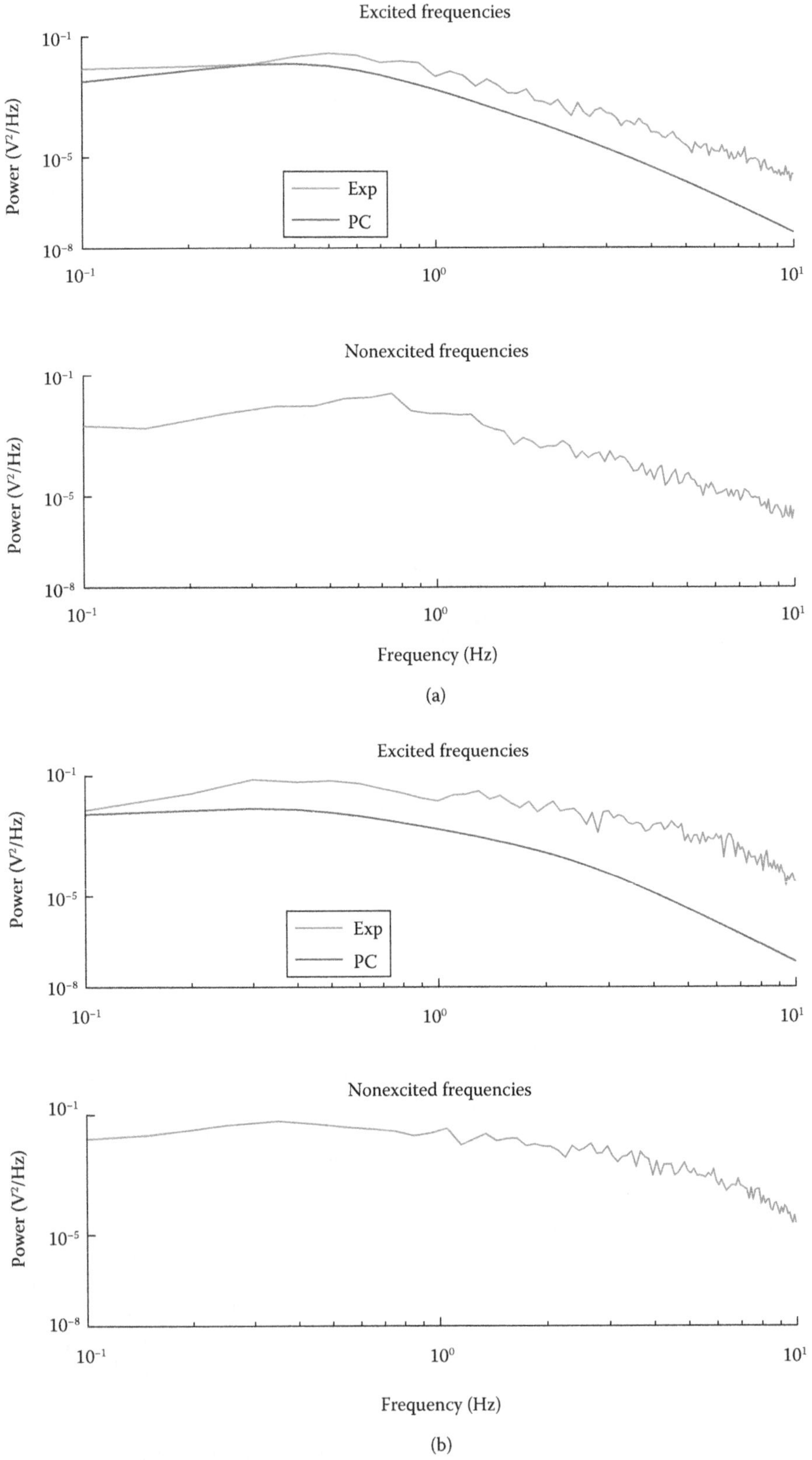


**FIGURE 14.30**
Example individual result for identification of LTI response for two experimental instructions. The continuous predictive controller can fit the excited frequencies [top graphs in (a) and (b)], but cannot explain the experimental response at nonexcited frequencies [bottom graphs in (a) and (b)]. (a) Control signal (instruction "cc"). (b) Control signal (instruction "ol").

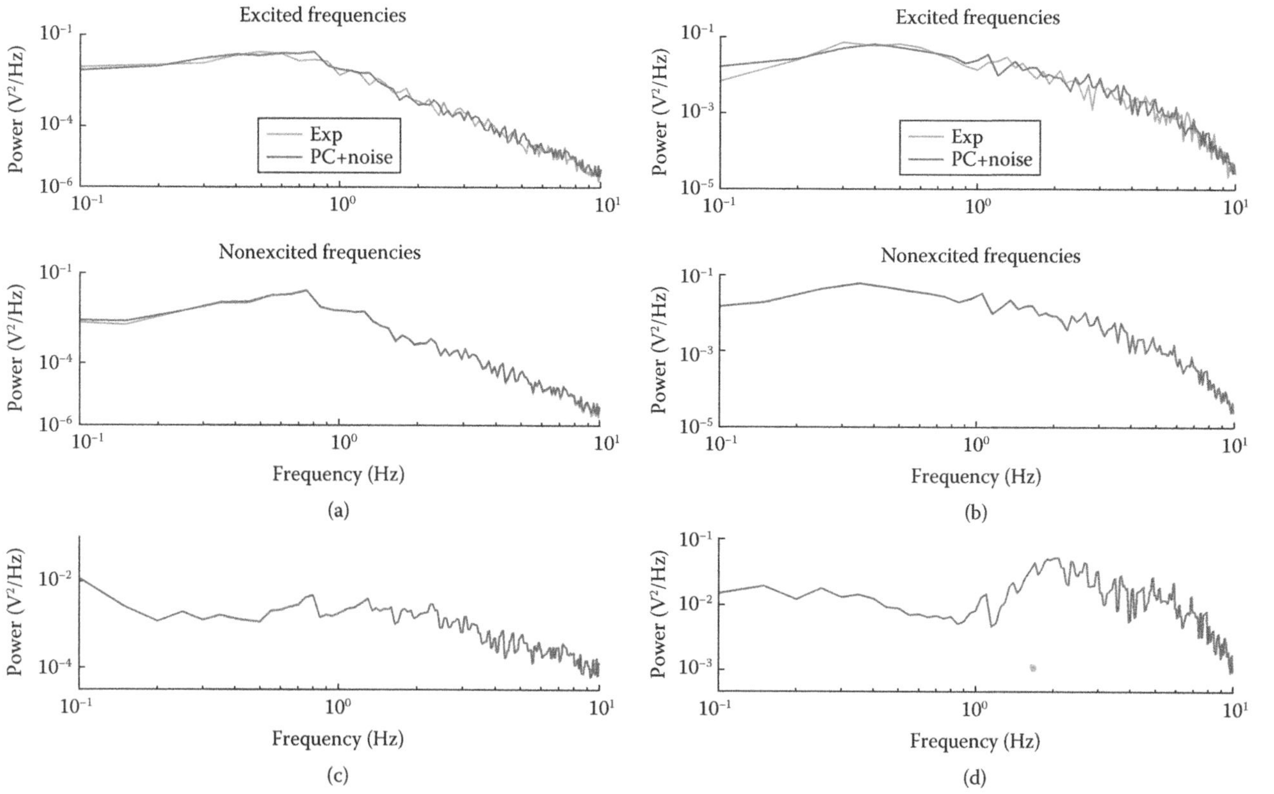


**FIGURE 14.31**

Variability as a result of colored input noise added to a predictive continuous controller. The top graphs show the resulting fit to experimental data at excited and nonexcited frequencies. The bottom graphs show the input noise added. (a) Control signal (instruction "cc"). (b) Control signal (instruction "mi"). (c) Added input noise $v_u$ (instruction "cc"). (d) Added input noise $v_u$ (instruction "mi").

position control ("cc"), open-loop intervals are clustered around a modal interval of approximately 1 s, with all $\Delta_{ol} > 0.5$ s. For the minimal intervention condition ("mi"), the open-loop intervals are clustered around a modal interval of approximately 2 s, and all $\Delta_{ol} > 1$ s. This corresponds to the expected behavior of the human operator where more frequent updates of the intermittent control trajectory are associated with the more demanding position control instruction, while the instruction to minimize intervention results in longer intermittent intervals. Thus, the identified thresholds not only result in IC models which approximate the response at excited and nonexcited frequency, but also reflect the underlying control aims.

### 14.8.4 Conclusion

The hypothesis that variability is the result of a continuous control process with added noise (PC with added noise) requires that the remnant is explained by a nonparametric input noise component. In comparison, IC introduces variability as a result of a small number of threshold parameters that are clearly related to underlying control aims.

## 14.9 Identification of Intermittent Control: The Underlying Continuous System

This section, together with Section 14.10, addresses the question of how intermittency can be identified when observing closed-loop control. In this section, it is discussed how intermittent control can masquerade as continuous control, and how the underlying continuous system can be identified. Section 14.10 addresses the question how intermittency can be detected in experimental data.

System identification provides one approach to hypothesis testing and has been used by Johansson, Magnusson, and Akesson (1988) and Peterka (2002) to test the nonpredictive hypothesis and by Gawthrop et al. (2009) to test the nonpredictive and predictive hypotheses. Given time-domain data from an sustained

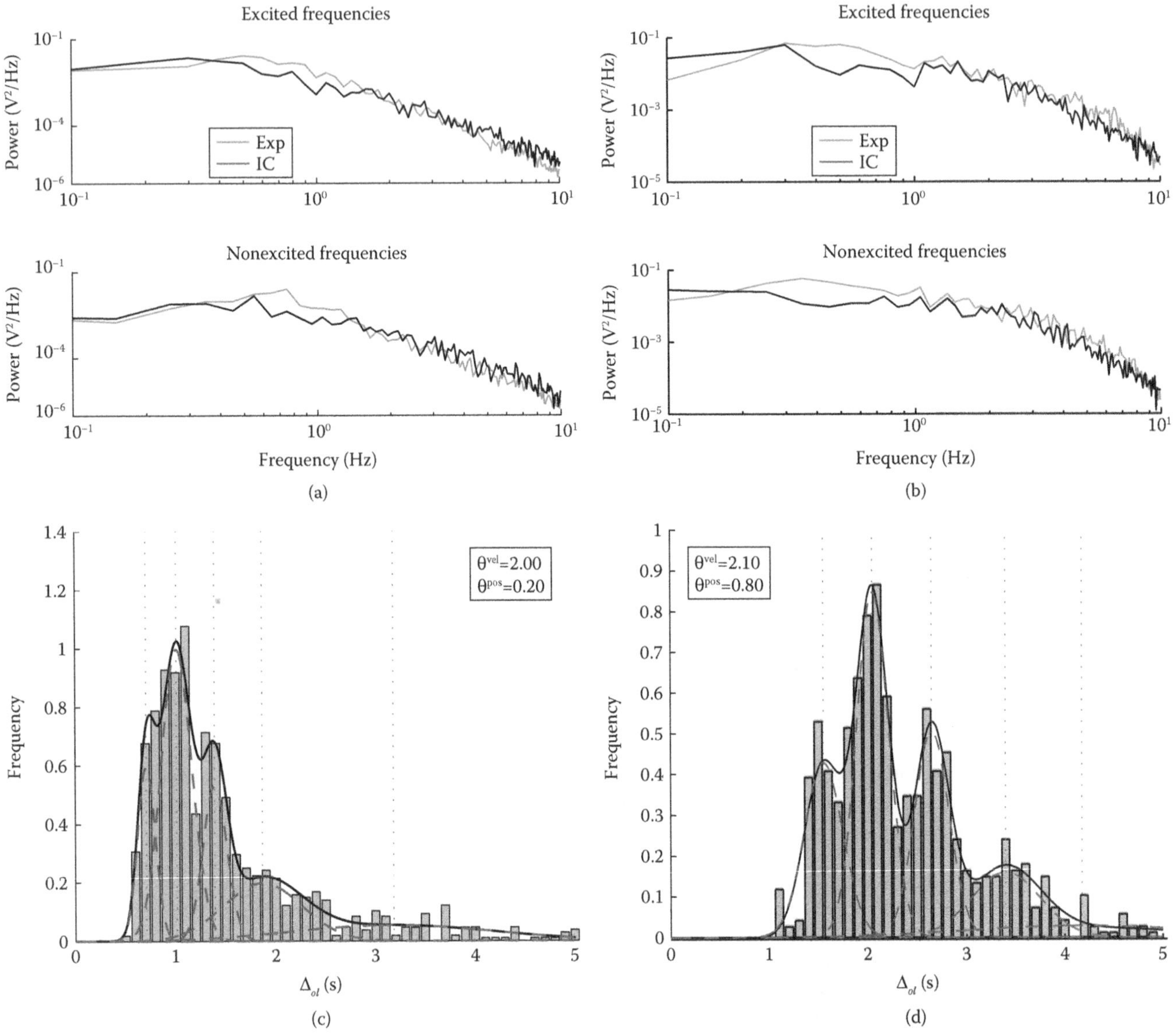


**FIGURE 14.32**
Variability resulting from event-driven IC. The top graphs show the resulting fit to experimental data at excited and nonexcited frequencies. The bottom graphs show the distribution of intermittent intervals, together with the optimal threshold values. (a) Control signal (instruction "cc"). (b) Control signal (instruction "ol"). (c) $\Delta_{ol}$ (instruction "cc"). (d) $\Delta_{ol}$ (instruction "ol").

control task which is excited by an external disturbance signal, a two-stage approach to controller estimation can be used in order to perform the parameter estimation in the frequency domain: firstly, the FRF is estimated from measured data, and secondly, a parametric model is fitted to the frequency response using nonlinear optimization (Pintelon and Schoukens, 2001; Pintelon, Schoukens, and Rolain, 2008). This approach has two advantages: firstly, computationally expensive analysis of long time-domain data sets can be reduced by estimation in the frequency domain, and secondly, advantageous properties of a periodic input signal (as advocated by Pintelon et al. (2008)) can be exploited.

In this section, first the derivation of the underlying frequency responses for a predictive continuous time controller and for the intermittent, clock-driven controller (i.e., $\Delta_{ol}$ = const) is discussed. The method is limited to clock-driven IC since frequency analysis tools are readily available only for this case (Gawthrop, 2009). The two-stage identification procedure is then outlined, followed by example results from a visual-manual control task.

The material in this section is partially based on Gollee et al. (2012).

### 14.9.1 Closed-Loop Frequency Response

As a prerequisite for system identification in the frequency domain, this section looks at the frequency response of closed-loop system corresponding to the underlying predictive continuous design method as well as that of the IC with a fixed intermittent interval (Gawthrop, 2009).

#### 14.9.1.1 *Predictive Continuous Control*

The system equations (14.1) can be rewritten in transfer function form as

$$y_o(s) = G(s)u(s) \quad \text{with } G(s) = C\left[sI - A\right]^{-1} B, \tag{14.164}$$

where $I$ is the $n \times n$ unit matrix and $s$ denotes the complex Laplace operator.

Transforming Equations 14.2 and 14.5 into the Laplace domain, assuming that disturbances $v_u$, $v_u$, and $w$ are zero:

$$\mathbf{x}_o(s) = (sI - A_o)^{-1}\left(Bu(s) + Ly_o(s)\right) \quad \text{(Observer)}, \tag{14.165}$$

$$x_p(s) = e^{A\Delta}\hat{x}(s) + (sI - A)^{-1}\left(I - e^{-(sI-A)\Delta}\right)Bu(s) \quad \text{(Predictor)}, \tag{14.166}$$

$$u(s) = -ke^{-s\Delta}x_p(s) \quad \text{(Controller)}, \tag{14.167}$$

where $I$ is the $n \times n$ unit matrix.

Equations 14.165 through 14.167 can be rewritten as

$$u(s) = -e^{-s\Delta}[H_y(s)y(s) + (H_1(s) + H_2(s))u(s)], \tag{14.168}$$

where

$$H_y(s) = ke^{A\Delta}\left(sI - A_o\right)^{-1} L, \tag{14.169}$$

$$H_1(s) = ke^{A\Delta}\left(sI - A_o\right)^{-1} B, \tag{14.170}$$

and

$$H_2(s) = k\left(sI - A\right)^{-1}(I - e^{-(sI-A)\Delta})Be^{s\Delta}. \tag{14.171}$$

It follows that the controller transfer function $H(s)$ is given by

$$H(s) = \frac{H_y(s)}{1 + H_1(s) + H_2(s)}, \tag{14.172}$$

where

$$\frac{u(s)}{y_o(s)} = -e^{-s\Delta}H(s). \tag{14.173}$$

With Equations 14.164 and 14.173, the system loop-gain $L(s)$ and closed-loop transfer function $T$ are given by

$$L(s) = e^{-s\Delta}G(s)H(s), \tag{14.174}$$

$$T(s) = \frac{u(s)}{d(s)} = \frac{L(s)}{1 + L(s)}. \tag{14.175}$$

Equation 14.175 gives a parameterized expression relating $u(s)$ and $d(s)$.

#### 14.9.1.2 *Intermittent Control*

The sampling operation in Figure 14.2 makes it harder to derive a (continuous-time) frequency response and so the details are omitted here. For the case where the intermittent interval is assumed to be constant, the basic result derived by Gawthrop (2009) apply and can be encapsulated as the following theorem*:

**Theorem**

The continuous-time system (14.1) controlled by an IC with generalized hold gives a closed-loop system where the Fourier transform $\mathbf{U}$ of the control signal $u(t)$ is given in terms of the Fourier transform $\mathbf{X^d}(j\omega)$ by

$$\mathbf{U} = F(j\omega, \theta)\left[\mathbf{X^d}(j\omega)\right]^S, \tag{14.176}$$

where

$$F(j\omega, \theta) = \mathbf{H}(j\omega)\mathbf{S_z}(e^{j\omega}), \tag{14.177}$$

$$\mathbf{H}(j\omega) = \frac{1}{\Delta_{ol}}\mathbf{k}\left[j\omega I - A_c\right]^{-1}\left[I - e^{-(j\omega I - A_c)\Delta_{ol}}\right], \tag{14.178}$$

$$\mathbf{S_z}(e^{j\omega}) = [I + \mathbf{G_z}(e^{j\omega})]^{-1}, \tag{14.179}$$

$$\mathbf{G_z}(e^{j\omega}) = \left[e^{j\omega}I - A_x\right]^{-1} B_x, \tag{14.180}$$

$$\mathbf{X^d}(j\omega) = \mathbf{G}(j\omega)\mathbf{d}(j\omega), \tag{14.181}$$

$$\mathbf{G}(j\omega) = [j\omega I - A]^{-1}B. \tag{14.182}$$

The sampling operator is defined as

$$\left[\mathbf{X^d}(j\omega)\right]^S = \sum_{k=-\infty}^{\infty} \mathbf{X^d}(j\omega - kj\omega_{ol}), \tag{14.183}$$

*This is a simplified version of (Gawthrop, 2009, Theorem 1) for the special case considered in this section.

where the *intermittent sampling-frequency* is given by $\omega_{ol} = 2\pi/\Delta_{ol}$.

As discussed in Gawthrop (2009), the presence of the sampling operator $[\mathbf{X}^{\mathrm{d}}(j\omega)]^{s}$ means that the interpretation of $F(j\omega,\theta)$ is not quite the same as that of the closed-loop transfer function $T(s)$ of (14.175), as the sample process generates an infinite number of frequencies which can lead to aliasing. As shown in Gawthrop (2009), the (bandwidth limited) observer acts as an anti-aliasing filter, which limits the effect of $[\mathbf{X}^{\mathrm{d}}(j\omega)]^{s}$ to higher frequencies and makes $F(j\omega,\theta)$ a valid approximation of $\mathbf{U}$. $F(j\omega,\theta)$ will therefore be treated as equivalent to $T(j\omega)$ in the rest of this section.

### 14.9.2 System Identification

The aim of the identification procedure is to derive an estimate for the closed-loop transfer function of the system. Our approach follows the two-stage procedure of Pintelon and Schoukens (2001) and Pintelon et al. (2008). In the first step, the frequency response transfer function is estimated based on measured input–output data, resulting in a nonparametric estimate. In the second step, a parametric model of the system is fitted to the estimated frequency response using an optimization procedure.

#### 14.9.2.1 System Setup

To illustrate the approach, we consider the visual-manual control task described in Section 14.8.1, where the subject is asked to sustain control of an unstable second-order load using a joystick, with the instruction to keep the load as close to the center as possible ("cc").

#### 14.9.2.2 Nonparametric Estimation

In the first step, a nonparametric estimate of the closed-loop FRF is derived, based on observed input–output data. The system was excited by a multi-sine disturbance signal (Equation 14.162). The output $u(t)$ of a linear system which is excited by $d(t)$ then only contains information at the same discrete frequencies $\omega_k$ as the input signal. If the system is nonlinear or noise is added, the output will contain a remnant component at nonexcited frequencies; as discussed in Section 14.8. several periods of $d(t)$ were used.

The time-domain signals $d(t)$ and $u(t)$ over one period $T_0$ of the excitation signal were transformed into the frequency domain. If the input signal has been applied over $N_p$ periods, then the frequency-domain data for the $l$th period can be denoted as $d^{[l]}(j\omega_k)$ and $u^{[l]}(j\omega_k)$, respectively, and the FRF can be estimated as

$$\hat{T}^{[l]}(j\omega_k) = \frac{u^{[l]}(j\omega_k)}{d^{[l]}(j\omega_k)}, \qquad k = 1,2,\ldots,N_f, \tag{14.184}$$

where $N_f$ denotes the number of frequency components in the excitation signal. An estimate of the FRF over all $N_p$ periods is obtained by averaging,

$$\hat{T}(j\omega_k) = \frac{1}{N_p}\sum_{l=1}^{N_p}\hat{T}^{[l]}(j\omega_k), \qquad k = 1,2,\ldots,N_f. \tag{14.185}$$

This approach ensures that only the periodic (deterministic) features related to the disturbance signal are used in the identification, and that the identification is robust with respect to remnant components.

#### 14.9.2.3 Parametric Optimization

In the second stage of the identification procedure, a parametric description, $\tilde{T}(j\omega_k,\theta)$, is fitted to the estimated FRF of Equation 14.185. The parametric FRF approximates the closed-loop transfer function (Equation 14.175), which depends in the case of predictive control, on the loop transfer function $L(j\omega_k,\theta)$, Equation 14.174, parameterized by the vector $\theta$, while for the IC this is approximated by $F(j\omega,\theta)$, Equation 14.176,

$$\tilde{T}(j\omega_k,\theta) = \begin{cases} \frac{L(j\omega_k,\theta)}{1+L(j\omega_k,\theta)} & \text{for PC} \\ F(j\omega_k,\theta) & \text{for IC} \end{cases}. \tag{14.186}$$

We use an indirect approach to parameterize the controller, where the controller and observer gains are derived from optimized design parameters using the standard LQR approach of Equation 14.6. This allows the specification of boundaries for the design parameters, which guarantee a nominally stable closed-loop system. As described in Section 14.2.3, the feedback gain vector $k$ can then be obtained by choosing the elements of the matrices $Q_c$ and $R_c$ in (14.6), and nominal stability can be guaranteed if these matrices are positive definite. As the system model is second order, we choose to parameterize the design using two positive scalars, $q_v$ and $q_p$,

$$R_c = 1 \qquad Q_c = \begin{bmatrix} q_v & 0 \\ 0 & q_p \end{bmatrix}, \text{ with } q_v, q_p > 0, \tag{14.187}$$

related to relative weightings of the velocity ($q_v$) and position ($q_p$) states.

The observer gain vector $L$ is obtained by applying the same approach to the dual system $[A^{\mathrm{T}}, C^{\mathrm{T}}, B^{\mathrm{T}}]$. It was found that the results are relatively insensitive to observer properties which was therefore parameterized by a single positive variable, $q_o$,

$$R_o = 1 \qquad Q_o = q_o\, BB^{\mathrm{T}} \text{ with } q_o > 0, \tag{14.188}$$

where $R_o$ and $Q_o$ correspond to $R_c$ and $Q_c$ in Equation 14.6 for the dual system.

The controller can then be fully specified by the positive parameter vector $\theta = [q_v, q_p, q_o, \Delta]$ (augmented by $\Delta_{ol}$ for intermittent control).

The optimization criterion $J$ is defined as the mean squared difference between the estimated FRF and its parametric fit

$$J(\theta) = \frac{1}{N_f}\sum_{k=1}^{N_f}\left[\hat{T}(j\omega_k) - \tilde{T}(j\omega_k, \theta)\right]^2. \tag{14.189}$$

This criterion favors lower frequency data since $|T(j\omega)|$ tends to be larger in this range.

The parameter vector is separated into two parts: time-delay parameters,

$$\theta_\Delta = \begin{cases} [\Delta] & \text{for PC} \\ [\Delta, \Delta_{ol}] & \text{for IC} \end{cases}, \tag{14.190}$$

and controller design parameters

$$\theta_c = [q_v, q_p, q_o], \tag{14.191}$$

such that $\theta = [\theta_\Delta, \theta_c]$. The time-delay parameters are varied over a predefined range, with the restriction that $\Delta_{ol} > \Delta$ for IC. For each given set of time-delay parameters, a corresponding set of optimal controller design parameters $\theta_c^*$ is found which solves the constrained optimization problem

$$\theta_c^* = \arg\min_{\theta_c} J([\theta_\Delta, \theta_c]), \qquad \theta_c > 0, \tag{14.192}$$

which was solved using the SQP algorithm (Nocedal and Wright, 2006) (MATLAB Optimization Toolbox, The MathWorks, Inc., Natick, MA).

The optimal cost function for each set of time-delay parameters, $J^*(\theta_\Delta)$, was calculated, and the overall optimum, $J^*$ determined. For analysis, the time-delay parameters corresponding to the optimal cost are determined, with $\Delta$ and $\Delta_{ol}$ combined for the IC to give the effective time-delay,

$$\Delta_e = \Delta + 0.5\Delta_{ol}. \tag{14.193}$$

### 14.9.3 Illustrative Example

Results from identifying the experimental data from one subject are used to illustrate the approach.

An extract of the time-domain data is shown in Figure 14.33b. The top plot shows the multi-sine

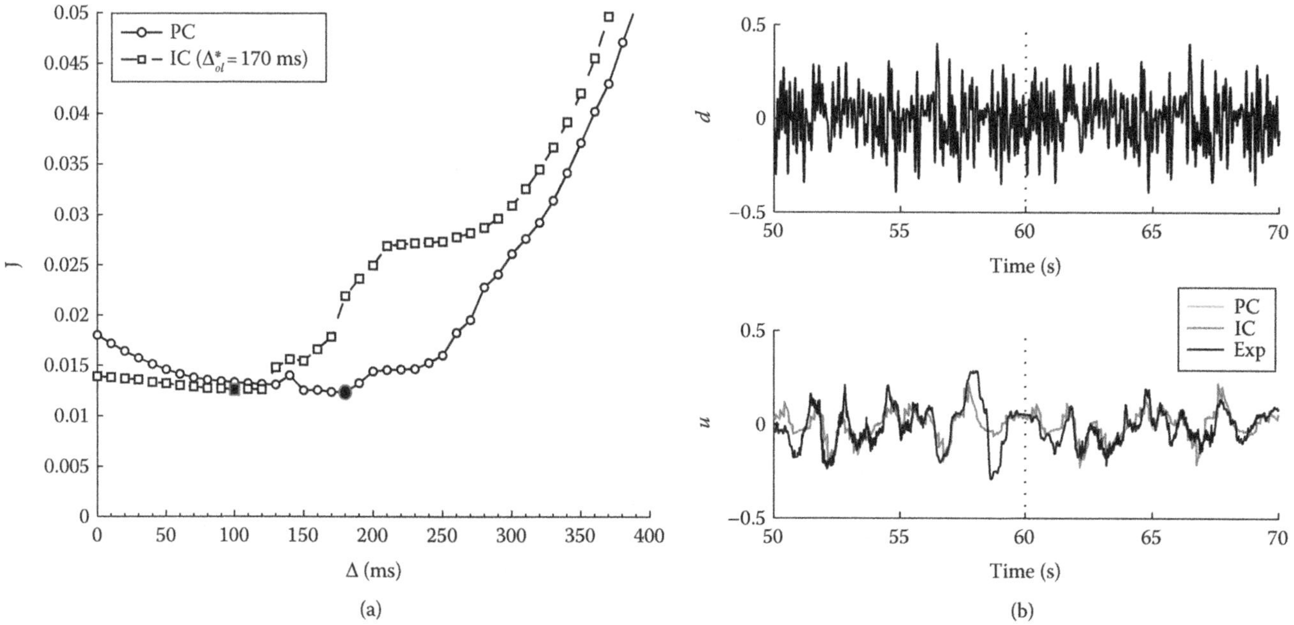


**FIGURE 14.33**
Illustrative experimental results. (a) The optimization cost (Equation 14.189) as a function of the time delay for predictive continuous control (PC) and intermittent control (IC), together with the value of the intermittent interval corresponding to the smallest cost ($\Delta^*_{ol}$). (b)–(d) Comparisons between predictive continuous control (PC), intermittent control (IC), and experimental data (Exp). Plots for PC and IC in (c) and (d) are derived analytically (Equation 14.186). (e) Comparison of the analytical FRF for the IC with the FRF derived from time-domain simulation data. (a) Cost function. (b) Time-domain signals.

*(Continued)*

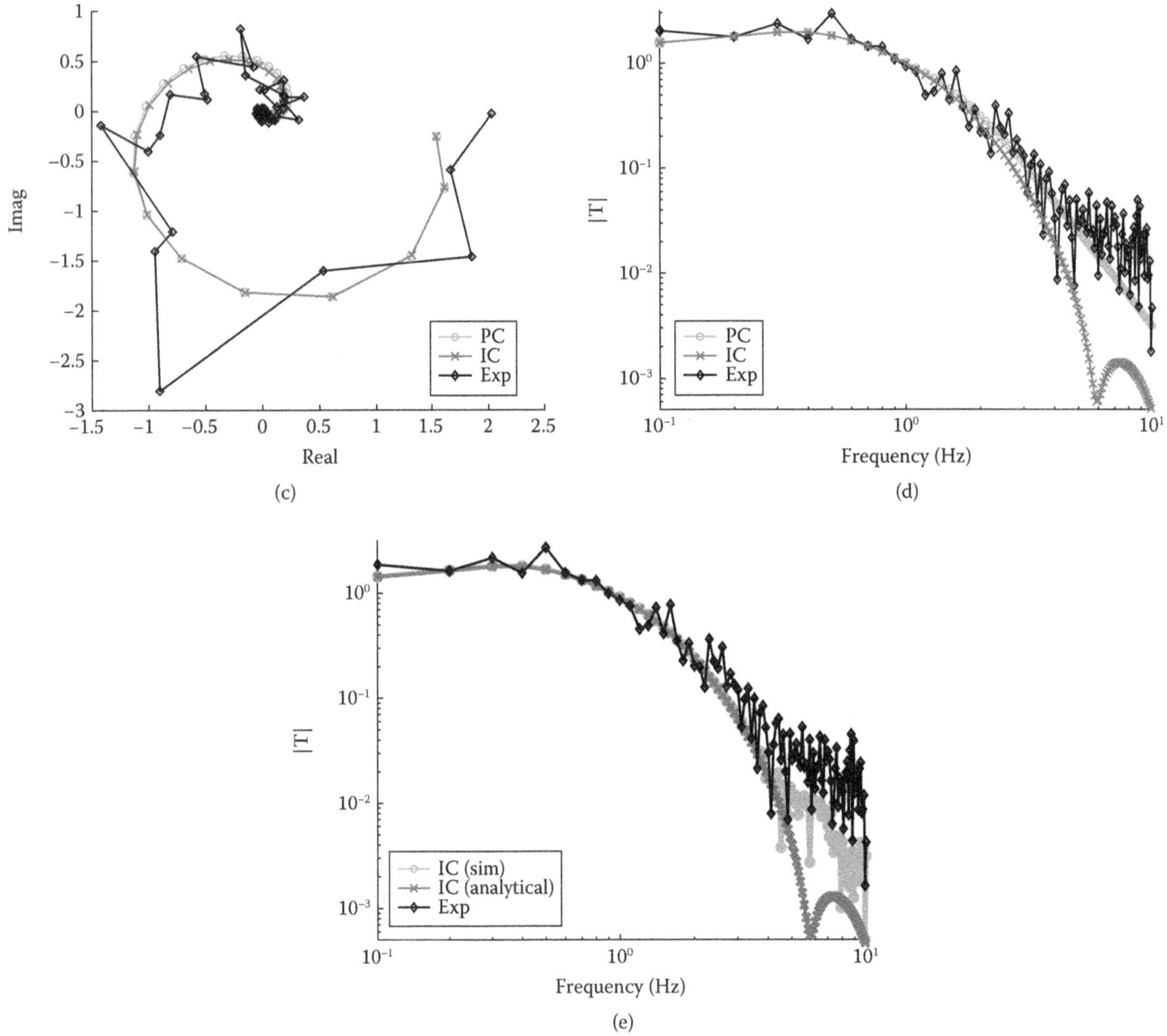


**FIGURE 14.33 (Continued)**

Illustrative experimental results. (a) The optimization cost (Equation 14.189) as a function of the time delay for predictive continuous control (PC) and intermittent control (IC), together with the value of the intermittent interval corresponding to the smallest cost ($\Delta_{ol}^{*}$). (b)–(d) Comparisons between predictive continuous control (PC), intermittent control (IC), and experimental data (Exp). Plots for PC and IC in (c) and (d) are derived analytically (Equation 14.186). (e) Comparison of the analytical FRF for the IC with the FRF derived from time-domain simulation data. (c) Nyquist plot of the FRF. (d) Bode magnitude plot of the FRF. (e) Bode magnitude plot of FRF for IC.

disturbance input over two 10 s periods, and the bottom plot depicts the corresponding measured control signal response (thin dark line). From this response, the experimental FRF was estimated in stage 1 of the identification (dark solid lines in Figure 14.33c–e).

Stage 2 of the procedure aimed to find the controller design parameters which resulted in the best fit to the experimental FRF. The corresponding cost functions for the predictive continuous and for the IC are shown in Figure 14.33a, with the minima indicated by solid markers. The estimated FRF (Equation 14.185) and their parametric fits (Equation 14.186) are shown in Figure 14.33c. It is clear that both the PC and IC are able to fit the experimental FRF equally well. The resulting controller parameters (summarized in Table 14.1) are very similar for both control architectures. This is confirmed by time-domain simulations using the estimated controllers (Figure 14.33b) where the PC and IC responses are difficult to distinguish.

Although the Nyquist plot of Figure 14.33c suggests that the PC and IC responses are virtually identical, further analysis shows that this is only the case at lower frequencies (at which most of the signal power lies). The Bode plot of the frequency response (Figure 14.33d)

**TABLE 14.1**
Estimated Controller Design Parameters

| | PC | IC |
|---|---|---|
| $q_p$ | 0.99 | 1.07 |
| $q_v$ | 0.00 | 0.00 |
| $q_o$ | 258.83 | 226.30 |
| $\Delta$ | 180 ms | 95 ms |
| $\Delta_{ol}$ | — | 170 ms |
| $\Delta_e$ | — | 180 ms |

shows that the PC and IC are indistinguishable only for frequencies up to around 2–3 Hz. This is also the frequency range up to which the PC and IC provide a good approximation to the experimental FRF.

The controller frequency responses shown in Figure 14.33c and d are based on the analytically derived expressions. For the IC, the sampling operator means that the theoretical response is only a valid approximation at lower frequencies, with aliasing evident at higher frequencies. A comparison of the analytical response with the response derived from simulated time-domain data (Figure 14.33e) shows that the simulated frequency response of the IC at higher frequency is in fact closer to the experimental data than the analytical response.

### 14.9.4 Conclusions

The results illustrate that continuous, predictive, and intermittent controllers can be equally valid descriptions of a sustained control task. Both approaches allow fitting the estimated nonparametric frequency responses with comparable quality. This implies that experimental data can be equally well explained using the PC and the IC hypotheses. This result is particularly interesting as it means that experimental results showing good fit for continuous predictive control models, dating back to at least those of Kleinman et al. (1970), do not rule out an intermittent explanation. A theoretical explanation for this result is given in Gawthrop et al. (2011, Section 4.3) where the masquerading property of intermittent control is discussed: As shown there (and illustrated in the results here), the frequency response of an intermittent controller and that of the corresponding PC are indistinguishable at lower frequency and only diverge at higher frequencies where aliasing occurs. Thus, the responses of the predictive and the intermittent controllers are difficult to distinguish, and therefore both explanations appear to be equally valid.

## 14.10 Identification of Intermittent Control: Detecting Intermittency

As discussed in Section 14.3 and by Gawthrop et al. (2011), the key feature distinguishing intermittent control from continuous control is the open-loop interval $\Delta_{ol}$ of Equations 14.16 and 14.20. As noted in Sections 14.3.1 and 14.4.2, the open-loop interval provides an explanation of the PRP of Telford (1931) as discussed by Vince (1948) to explain the human response to double stimuli. Thus, "intermittency" and "refractoriness" are intimately related. Within this interval, the control trajectory is open loop but is continuously time varying according to the basis of the generalized hold. The length of the intermittent interval gives a trade-off between continuous control (zero intermittent interval) and intermittency. Continuous control maximizes the frequency bandwidth and stability margins at the cost of reduced flexibility, whereas intermittent control provides time in the loop for optimization and selection (van de Kamp et al., 2013a; Loram et al., 2014) at the cost of reduced frequency bandwidth and reduced stability margins. The rationale for intermittent control is that it confers online flexibility and adaptability. This rationale has caused many investigators to consider whether intermittent control is an appropriate paradigm for understanding biological motor control (Craik, 1947a,b; Vince, 1948; Bekey, 1962; Navas and Stark, 1968; Neilson et al., 1988; Miall et al., 1993a; Hanneton, Berthoz, Droulez, and Slotine, 1997; Neilson and Neilson, 2005; Loram and Lakie, 2002). However, even though intermittent control was first proposed in the physiological literature in 1947, there has not been an adequate methodology to discriminate intermittent from continuous control and to identify key parameters such as the open-loop interval $\Delta_{ol}$. Within the biological literature, four historic planks of evidence (discontinuities, frequency constancy, coherence limit, and PRF) have provided evidence of intermittency in human motor control (Loram et al., 2014).

1. The existence of discontinuities within the control signal has been interpreted as sub-movements or serially planned control sequences (Navas and Stark, 1968; Poulton, 1974; Miall et al., 1993a; Miall, Weir, and Stein, 1986; Hanneton et al., 1997; Loram and Lakie, 2002).

2. Constancy in the modal rate of discontinuities, typically around 2–3 per second, has been interpreted as evidence for a central process with a well-defined timescale (Navas and

Stark, 1968; Poulton, 1974; Lakie and Loram, 2006; Loram et al., 2006).

3. The fact that coherence between unpredicted disturbance or set-point and control signal is limited to a low maximum frequency, typically of 1–2 Hz, below the mechanical bandwidth of the feedback loop has been interpreted as evidence of sampling (Navas and Stark, 1968; Loram et al., 2009, 2011).
4. The PRF has provided direct evidence of open-loop intervals but only for discrete movements and serial reaction time (e.g., push button) tasks and has not been demonstrated for sustained sensori-motor control (Vince, 1948; Pashler and Johnston, 1998; Hardwick, Rottschy, Miall, and Eickhoff, 2013).

Since these features can be reproduced by a CC with tuned parameters and filtered additive noise (Levison et al., 1969; Loram et al., 2012), this evidence is circumstantial. Furthermore, there is no theoretical requirement for regular sampling nor for discontinuities in control trajectory. Indeed, as historically observed by Craik (1947a,b), humans tend to smoothly join control trajectories following practice. Therefore, the key methodological problem is to demonstrate that on-going control is sequentially open loop even when the control trajectory is smooth and when frequency analysis shows no evidence of regular sampling.

### 14.10.1 Outline of Method

Using the intermittent-equivalent setpoint of Sections 14.3.7 and 14.4.2, we summarize a method to distinguish intermittent from continuous control (Loram et al., 2012). The identification experiment uses a specially designed paired-step set-point sequence. The corresponding data analysis uses a conventional ARMA model to relate the theoretically derived equivalent set-point (of Section 14.3.7) to the control signal. The method sequentially and iteratively adjusts the timing of the steps of this equivalent set-point to optimize the linear time invariant fit. The method has been verified using realistic simulation data and was found to robustly distinguish not only between continuous and intermittent control but also between event-driven intermittent and clock-driven intermittent control (Loram et al., 2012). This identification method is applicable for machine and biological applications. For application to humans the set-point sequence should be unpredictable in the timing and direction of steps. This method proceeds in three stages. Stages 1 and 2 are independent of model assumptions and quantify refractoriness, the key feature discriminating intermittent from continuous control.

#### *14.10.1.1 Stage 1: Reconstruction of the Set-Point*

With reference to Figure 14.34, this stage takes the known set-point and control output signals and reconstructs the set-point step times to form that sequence with a linear-time invariant response which best matches the control output. This is implemented as an optimization process in which the fit of a general linear time series model (zero-delay ARMA) is maximized by adjusting the trial set of step times. The practical algorithmic steps are stated by Loram et al. (2012). The output from stage 1 is an estimate of the time delay for each step stimulus.

#### *14.10.1.2 Stage 2: Statistical Analysis of Delays*

Delays are classified according to step (1 or 2, named reaction-time* 1 (RT1) and reaction-time 2 (RT2), respectively) and interstep interval (ISI). A significant difference in delay, RT2 vs. RT1, is *not* explained by a linear-time-invariant model. The reaction time properties, or refractoriness, are quantified by

1. The size of ISI for which RT2 > RT1. This indicates the temporal separation required to eliminate interference between successive steps.
2. The difference in delay (RT2 − RT1).

#### *14.10.1.3 Stage 3: Model-Based Interpretation*

For controllers following the generalized continuous (Figure 14.1) and intermittent (Figure 14.2) structures, the probability of a response occurring, the mean delay and the range of delays can be predicted for each inter-step-interval (Figure 14.35 and Appendix C of Loram et al. (2012)). For a CC (Figure 14.1), all delays equal the model delay ($\Delta$). Intermittent control is distinguished from continuous control by increased delays for RT2 vs. RT1 for ISIs less than the open-loop interval ($\Delta_{ol}$). Clock (zero threshold) triggered intermittent control is distinguished from threshold triggered intermittent control by the range of delays for RT1 and RT2 and by the increased mean delay for ISIs greater than the open-loop interval ($\Delta_{ol}$) (Figure 3). If the results of Stage 1–2 analysis conform to these patterns (Figure 14.35), the open-loop interval ($\Delta_{ol}$) can be estimated. Simulation also shows that the sampling delay ($\Delta_s$) can be identified from the ISI at which the delay RT2 is maximal (Figure 14.37) (van de Kamp et al., 2013a,b). Following verification by simulation (Loram et al., 2012), the method has been applied to human visually guided pursuit tracking and to whole

*In the physiological literature, "delay" is synonymous with "reaction time."

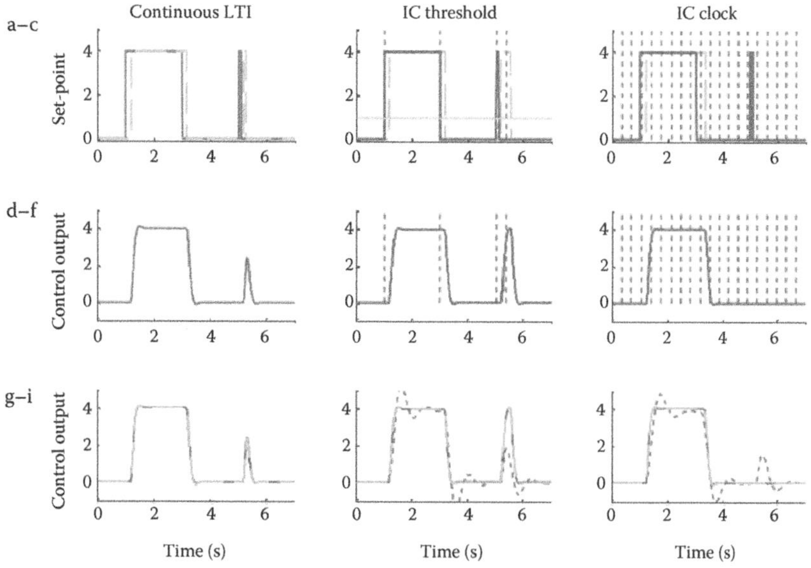


**FIGURE 14.34**
Reconstruction of the set-point. Example responses to set-point step-sequence (a–c) Solid: two paired steps (long and short interstep interval) are applied to the set-point of each of three models: continuous linear time invariant, threshold triggered intermittent control (unit threshold), and clock triggered (zero threshold) intermittent control (cols 1–3, respectively). Dashed: Set-point adjusted: time of each step follows preceding trigger by one model time delay (Δ). (d–f) Solid: Control output (ue). Gray vertical dashed: event trigger times. (g–i) Solid: control output (ue). Dash-dotted: ARMA (LTI) fit to set-point (solid in a–c). Dashed: ARMA (LTI) fit to adjusted set-point (dashed in a–c). (From I. D. Loram, et al., *Journal of the Royal Society Interface*, 9(74):2070–2084, 2012. With permission.)

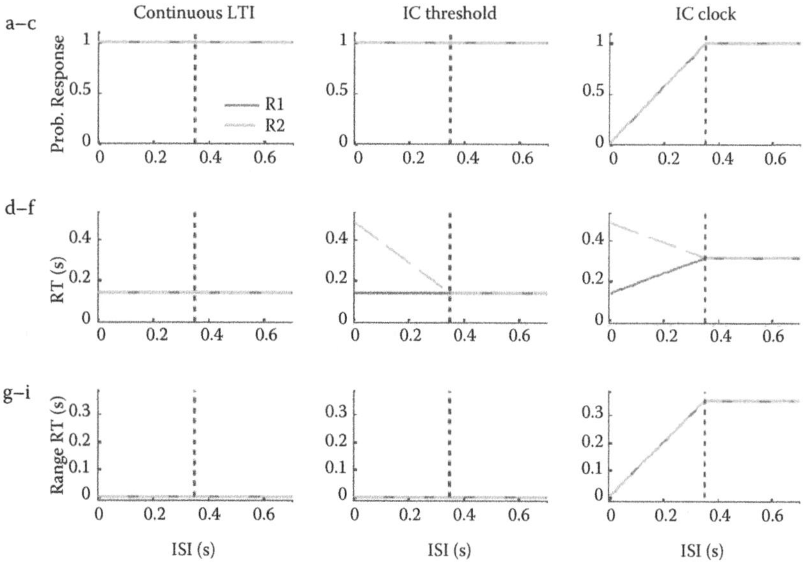


**FIGURE 14.35**
Reconstruction of the set-point. Predicted delays for varying interstep intervals. For three models, continuous linear time invariant, threshold triggered intermittent control, and clock triggered intermittent control (cols 1–3, respectively), the following is shown as a function of interstep interval (ISI): (a–c) The predicted probability of response. (d–f) The mean response delay. (g–i) The range of response delays. Responses 1 and 2 (R1, R2) are denoted by solid and dashed lines, respectively. For these calculations, the open-loop interval ($\Delta_{ol}$) is 0.35 s (vertical dashed line) and feedback time-delay (td) is 0.14 s. (From I. D. Loram, et al., *Journal of the Royal Society Interface*, 9(74):2070–2084, 2012. With permission.)

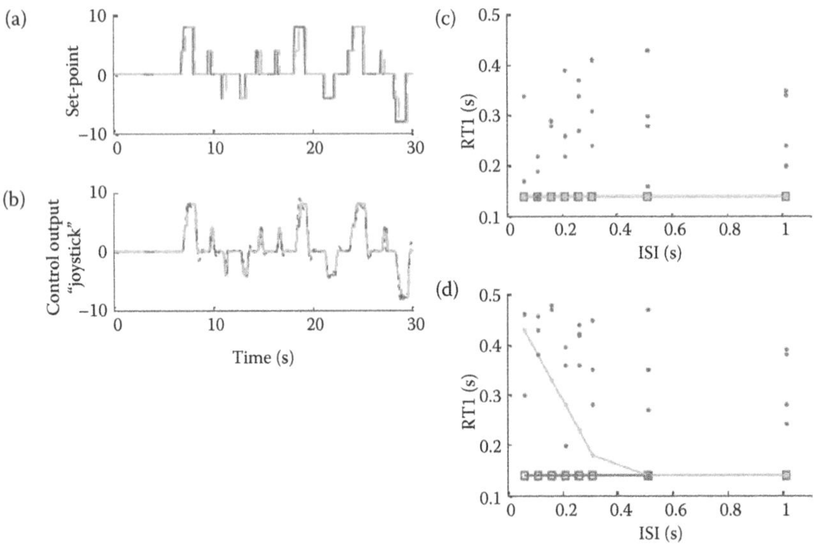


**FIGURE 14.36**
Reconstruction of the set-point. Representative stage 1 analysis. (a) Solid: set-point sequence containing eigth interstep interval (ISI) pairs with random direction (first 30 s). Dashed: adjusted set-point from step 1 analysis. After a double unidirectional step, set-point returns to zero before next pair. (b) Solid: control output (ue). Dash-dotted: ARMA (LTI) fit to set-point. Dashed: ARMA (LTI) fit to adjusted set-point. (c and d) Response times (RT1, RT2), respectively, from each of three models vs ISI. Joined square: continuous LTI. Joined dot: threshold intermittent control. Isolated dot: clock intermittent control. The system is zero order. The open-loop interval ($\Delta_{ol}$) is 0.35 s and feedback time-delay ($\Delta$) is 0.14 s. (From I. D. Loram, et al., *Journal of the Royal Society Interface*, 9(74):2070–2084, 2012. With permission.)

body pursuit tracking. In both cases, control has been shown to be intermittent rather than continuous.

### 14.10.2 Refractoriness in Sustained Manual Control

Using a uni-axial, sensitive, contactless joystick, participants were asked to control four external systems (zero, first, second-order stable, second-order unstable) using visual feedback to track as fast and accurately as possible the target which changes position discretely and unpredictably in time and direction (Figure 14.38a and b). For the zero-, first-, and second-order systems, joystick position determines system output position, velocity, and acceleration, respectively. The unstable second-order system had a time-constant equivalent to a standing human. Since the zero-order system has no dynamics requiring ongoing control, step changes in target produce discrete responses, that is, sharp responses clearly separated from periods of no response. The first- and second-order systems require sustained ongoing control of the system output position: thus, the step stimuli test responsiveness during ongoing control. The thirteen participants showed evidence of substantial open-loop interval (refractoriness) which increased with system order (0.2 to 0.5 s, 14.38 C). For first- and second-order systems, participants showed evidence of a sampling delay (0.2 to 0.25 s, 14.38 C). This evidence of refractoriness discriminates against continuous control.

### 14.10.3 Refractoriness in Whole Body Control

Control of the hand muscles may be more refined, specialized, and more intentional than control of the muscles serving the legs and trunk (van de Kamp et al., 2013a). Using online visual feedback (<100 ms delay) of a marker on the head, participants were asked to track as fast and accurately as possible a target that changes position discretely and unpredictably in time and direction (Figure 14.39a). This required head movements of 2 cm along the anterior–posterior axis and while participants were instructed not to move their feet, no other constraints or strategies were requested. The eight participants showed evidence of substantial open-loop interval (refractoriness) (0.5 s) and a sampling delay (0.3 s) (Figure 14.39b). This result extends the evidence of intermittent control from sustained manual control to integrated intentional control of the whole body.

## 14.11 Adaptive Intermittent Control

The purpose of feedback control is to reduce uncertainty (Horowitz, 1963; Jacobs, 1974). Feedback control using fixed controllers can be very successful in this role as long as sufficient bandwidth is available and the system does not contain time delays or other nonminimum phase elements (Horowitz, 1963; Goodwin et al., 2001;

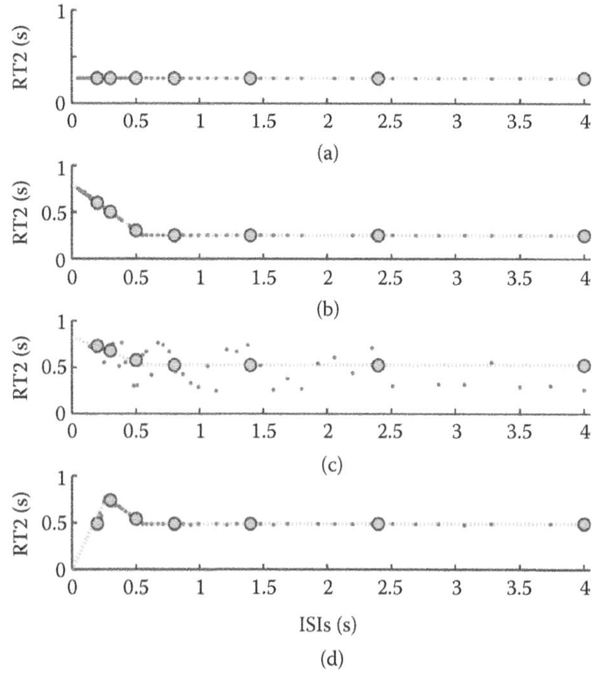


**FIGURE 14.37**
Model-based interpretation (stage 3). Parameter variants from the generalized IC model of Figure 14.2 showing several possible relationships between RT2 and inter-step interval (ISI) indicative of serial ballistic (intermittent) and continuous control behavior. The simulated system is zero order. The open-loop interval ($\Delta_{ol}$) is 0.55 s and feedback time delay ($\Delta$) is 0.25 s. For four models: (a) continuous LTI ($\Delta_{ol} = 0$), (b) externally triggered intermittent control with a prediction error threshold, (c) internally triggered intermittent control (with zero prediction error threshold, triggered to saturation), and (d) externally triggered intermittent control supplemented with a sampling delay of 0.25 s, which is associated with the ISI at the maximum delay for RT2. The joined circles represent the theoretical delays as a function of ISI, which are confirmed by the model simulations. (From C. van de Kamp, et al., *Frontiers in Computational Neuroscience*, 7(55), 2013a. With permission.)

Skogestad and Postlethwaite, 1996). However, when the system and actuators do contain such elements, an adaptive controller can be used to reduce uncertainly and thus, in time, improve controller performance. By its nature, intermittent control is a low-bandwidth controller and so adaptation is particularly appropriate in this context. Conversely, intermittent control frees computing resources that can be used for this purpose.

As discussed in the textbooks by Goodwin and Sin (1984) and by Åström and Wittenmark (1989), adaptive control of engineering systems is well-established. Perhaps, the simplest approach is to combine real-time recursive parameter estimation with a simple controller design method to give a so-called self-tuning strategy (Åström and Wittenmark, 1973; Clarke and Gawthrop, 1975, 1979, 1981; Åström and Wittenmark, 1989). However, such an approach ignores two things: the controller is based initially on incorrect parameter estimates and the quality of the parameter estimation is dependent on the controller properties. This can be formalized using the concepts of *caution* whereby the adaptive controller takes account of parameter uncertainty and *probing* whereby the controller explicitly excites the system to improve parameter estimation (Jacobs and Patchell, 1972; Bar-Shalom, 1981). Adaptive controllers that explicitly and jointly optimize controller performance and parameter estimation have been called *dual* controllers (Feldbaum, 1960; Bar-Shalom and Tse, 1974). Except in simple cases, for example, that of Astrom and Helmersson (1986), the solution to the dual control problem is impractically complex.

Since Wiener (1965) developed the idea of cybernetics, there has been a strong interest in applying both biologically-inspired and engineering-inspired ideas to adaptive control in both humans and machines; ideas arising from the biological and engineering inspired fields have been combined in various ways.

One such thread is *reinforcement learning* (Sutton and Barto, 1998), which continues to be developed both theoretically and through applications (Khan, Herrmann, Lewis, Pipe, and Melhuish, 2012). It can be argued that that "reinforcement learning is direct adaptive optimal control" (Sutton, Barto, and Williams, 1992). Artificial neural networks (ANN) have been applied to engineering control systems: see the survey of Hunt, Żbikowski, Sbarbaro, and Gawthrop (1992) and numerous textbooks (Miller, Sutton, and Werbos, 1990; Żbikowski and Hunt, 1996; Kalkkuhl, Hunt, Żbikowski, and Dzieliński, 1997). Recent work is described by Vrabie and Lewis (2009). Again, there are links between ANN methods such as back-propagation and engineering parameter estimation (Gawthrop and Sbarbaro, 1990).

Field robotics makes use of the concept of *Simultaneous Location and Mapping* (SLAM) (Durrant-Whyte and Bailey, 2006; Bailey and Durrant-Whyte, 2006). Roughly speaking, location corresponds to state estimation and mapping to parameter estimation and therefore concepts and techniques from SLAM are appropriate to adaptive (intermittent) control. In particular, the Extended Kalman Filter (EKF) and the Unscented Kalman Filter (UKF) (Julier, Uhlmann, and Durrant-Whyte, 2000; Schiff, 2012) provide the basis for SLAM and hence adaptive control.

In this section, the simplest *continuous-time* self-tuning approach (Gawthrop, 1982, 1987, 1990) is used. As indicated in the examples of Sections 14.12 and 14.13, this simple approach has interesting behaviors; nevertheless, it would be interesting to investigate more sophisticated approaches based on, for example, the EKF and UKF.

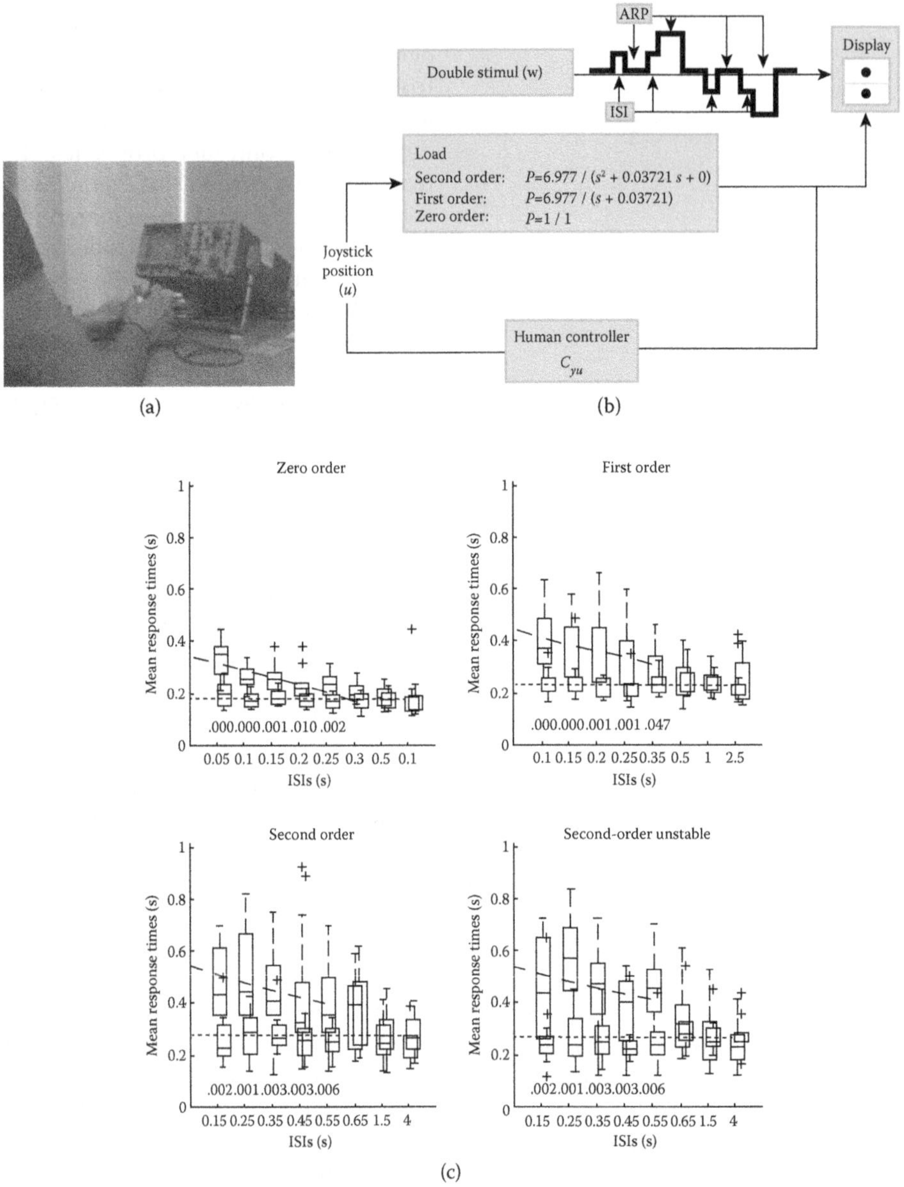


**FIGURE 14.38**

Refractoriness in sustained manual control. (a) Task setup. An oscilloscope showed real-time system output position as a small focused dot with negligible delay. Participants provided input to the system using a sensitive, uniaxial, contactless joystick. The system ran in Simulink Real-Time Windows Target within the MATLAB® environment (The MathWorks, Inc., Natick, MA). (b) Control system and experimental setup. Participants were provided with a tracking target in addition to system output. The tracking signal was constructed from four possible patterns of step sequence (uni- and reversed directional step to the left or to the right). First and second stimuli are separated by an unpredictable inter step interval (ISI), and patterns are separated by an unpredictable approximate recovery period (ARP). The participant was only aware of an unpredictable sequence of steps. (c) Group results: The four panels: zero order, first order, second order, and second-order unstable show the interparticipant mean first (RT1, black) and second (RT2, gray) response times against inter step intervals (ISIs), and $p$-values of the ANOVA's post hoc test are displayed above each ISI level dark if significant, light if not). (From C. van de Kamp, et al., *PLOS Computational Biology*, 9(1):e1002843, 2013b. With permission.)

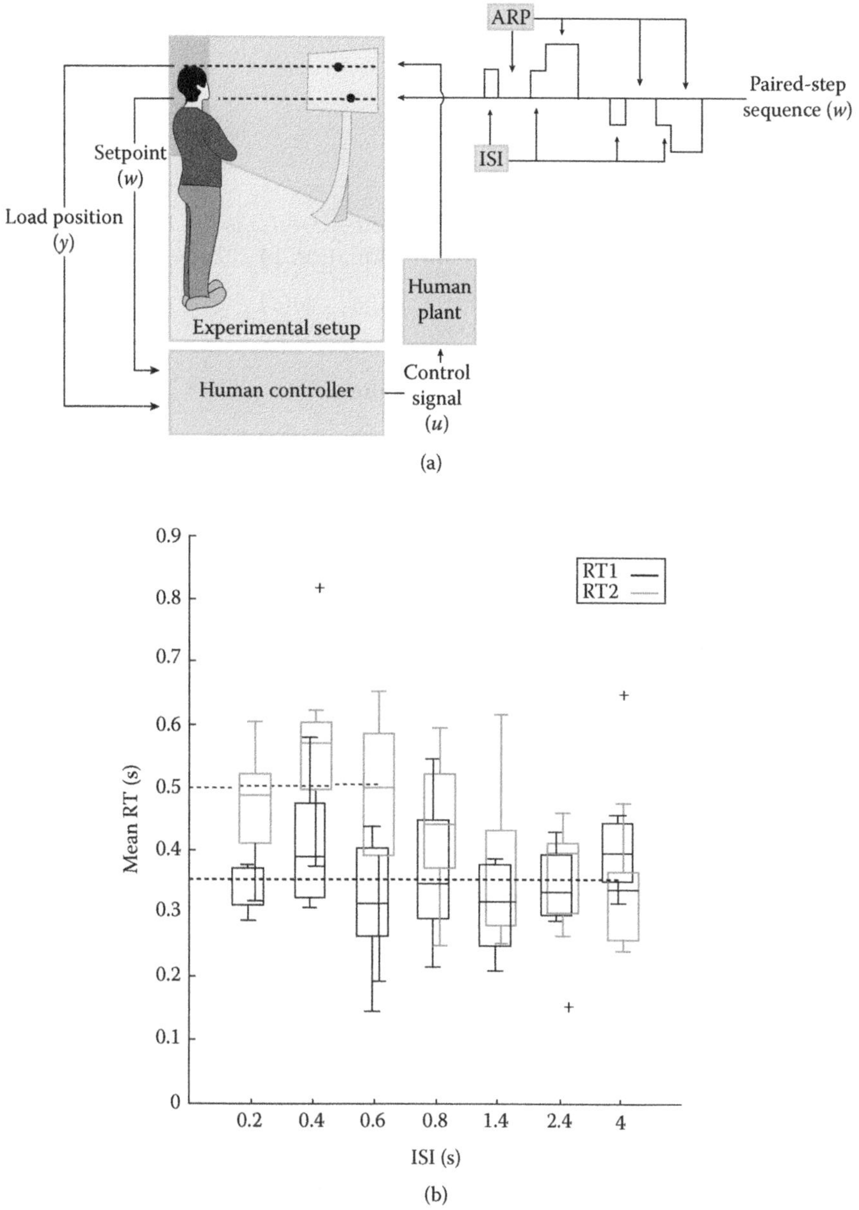


**FIGURE 14.39**

Refractoriness in whole body control. (a) The participant receives visual feed-back of the anterior–posterior head position through a dot presented on an LCD screen mounted on a trolley. Without moving their feet, participants were asked to track the position of a second dot displayed on the screen. The four possible step sequence combinations (uni- and reversed-directional step up or down) of the pursuit target are illustrated by the solid line. First and second stimuli are separated by an inter-step interval (ISI). The participant experiences an unpredictable sequence of steps. (b) Group results. Figure shows the interparticipant mean RT1 (black) and RT2 (gray) against ISI combined across the eight participants. The $p$-values of the ANOVA's post hoc test are displayed above each ISI level (black if $<0.05$, gray if not). The dotted line shows the mean RT1, the dashed line shows the regression linear fit between (interfered) RT2 and ISIs. (From C. van de Kamp, et al., *Frontiers in Computational Neuroscience*, 7(55), 2013. With permission.)

### 14.11.1 System Model

Parameter estimation is much simplified if the system can be transformed into *linear-in-the-parameters* form; the resultant model can be viewed as a *non-minimal state-space* (NMSS) representation of the system. The NMSS approach is given in discrete-time form by Young, Behzadi, Wang, and Chotai (1987) and continuous-time form by Taylor, Chotai, and Young (1998). Although a purely state-space approach to the NMSS representation is possible (Gawthrop, Wang, and Young, 2007), a polynomial approach is simpler and is presented here.

The linear-time invariant system considered in this chapter is given in Laplace transform terms by

$$\bar{y}(s) = \frac{b(s)}{a(s)}\left(\bar{u}(s) + \frac{b_\xi(s)}{a_\xi(s)}\xi_u(s)\right) + \frac{\mathbf{d}'(s)}{a(s)a_\xi(s)}, \quad (14.194)$$

where $\bar{y}(s)$, $\bar{u}(s)$, and $\xi_u(s)$ are the Laplace transformed system output, control input, and input disturbance, respectively. $\frac{b(s)}{a(s)}$ is the transfer function relating $\bar{y}(s)$ and $\bar{u}(s)$, and $\frac{b_\xi(s)}{a_\xi(s)}$ provides a transfer function model of the input disturbance. It is assumed that both transfer functions are strictly proper. The overall system initial conditions are represented by the polynomial $\mathbf{d}'(s)$.* The polynomials $a(s)$, $b(s)$, and $\mathbf{d}'(s)$ are of the form:

$$a(s) = a_0 s^n + a_1 s^{n-1} + \cdots + a_n, \quad (14.195)$$
$$b(s) = b_1 s^{n-1} + \cdots + b_n, \quad (14.196)$$
$$\mathbf{d}'(s) = \mathbf{d}'_1 s^{n-1} + \cdots + \mathbf{d}'_n, \quad (14.197)$$
$$a_\xi(s) = \alpha_0 s^{n_\xi} + \alpha_1 s^{n_\xi - 1} + \cdots + \alpha_{n_\xi}, \quad (14.198)$$
$$b_\xi(s) = \beta_0 s^{n_\xi} + \beta_1 s^{n_\xi - 1} + \cdots + \beta_{n_\xi}. \quad (14.199)$$

Finally, defining the Hurwitz polynomial $c(s)$ as

$$c(s) = c_0 s^N + c_1 s^{N-1} + \cdots + c_N, \quad (14.200)$$

where

$$N = n + n_\xi + 1. \quad (14.201)$$

Equation 14.194 may be rewritten as:

$$\frac{a(s)a_\xi(s)}{c(s)}\bar{y}(s) = \frac{b(s)a_\xi(s)}{c(s)}\bar{u}(s) + \frac{b(s)b_\xi(s)}{c(s)}\xi_u(s) + \frac{\mathbf{d}'(s)}{c(s)}. \quad (14.202)$$

For the purposes of this chapter, the polynomials $a_\xi(s)$ and $b_\xi(s)$ are defined as

$$a_\xi(s) = s, \quad (14.203)$$
$$b_\xi(s) = 1. \quad (14.204)$$

With this choice, Equation 14.202 simplifies to

$$\frac{sa(s)}{c(s)}\bar{y}(s) = \frac{sb(s)}{c(s)}\bar{u}(s) + \frac{b(s)}{c(s)}\xi_u(s) + \frac{d(s)}{c(s)}. \quad (14.205)$$

In the special case that the input disturbance is a jump to a constant value $d_\xi$ at time $t = 0^+$, then this can be modeled using Equations 14.203 and 14.204 and

$$\xi(t) = d_\xi \delta(t), \quad (14.206)$$

and

$$\xi_u(s) = d_\xi, \quad (14.207)$$

where $\delta(t - t_k)$ is the Dirac delta function.

Equation 14.202 then becomes

$$\frac{sa(s)}{c(s)}\bar{y}(s) = \frac{sb(s)}{c(s)}\bar{u}(s) + \frac{d(s)}{c(s)}, \quad (14.208)$$

where

$$d(s) = \mathbf{d}'(s) + d_\xi b(s). \quad (14.209)$$

Equation 14.205 can be rewritten in nonminimal state-space form as

$$\frac{d}{dt}\phi_y(t) = A_s\phi_y(t) - B_s y(t), \quad (14.210)$$
$$\frac{d}{dt}\phi_u(t) = A_s\phi_u(t) + B_s u(t), \quad (14.211)$$
$$\frac{d}{dt}\phi_{ic}(t) = A_s\phi_{ic}(t),\ \phi_{ic}(0) = \phi_{ic0}, \quad (14.212)$$

where

$$A_s = \begin{bmatrix} -c_1 & -c_2 & \dots & -c_{N-1} & -c_N \\ 1 & 0 & \dots & 0 & 0 \\ 0 & 1 & \dots & 0 & 0 \\ \vdots & \vdots & \dots & \vdots & \vdots \\ 0 & 0 & \dots & 0 & 0 \\ 0 & 0 & \dots & 1 & 0 \end{bmatrix}, \quad (14.213)$$

and

$$B_s = \begin{bmatrix} 1 \\ 0 \\ \vdots \\ 0 \end{bmatrix}. \quad (14.214)$$

It follows that:

$$\epsilon(t) = \theta^T\phi(t), \quad (14.215)$$

where

$$\theta = \begin{bmatrix} \mathbf{a} \\ \mathbf{b} \\ \mathbf{d} \end{bmatrix} \text{ and } \phi(t) = \begin{bmatrix} \phi_y \\ \phi_u \\ \phi_{ic} \end{bmatrix}, \quad (14.216)$$

and

$$\mathbf{a} = \begin{bmatrix} a_0 & a_1 & \dots & a_n & 0 \end{bmatrix}^T, \quad (14.217)$$
$$\mathbf{b} = \begin{bmatrix} 0 & b_1 & \dots & b_n & 0 \end{bmatrix}^T, \quad (14.218)$$
$$\mathbf{d} = \begin{bmatrix} 0 & d_1 & \dots & d_n & \dots & d_{n_c} \end{bmatrix}^T. \quad (14.219)$$

*Transfer function representations of continuous-time systems and initial conditions are discussed, for example, by Goodwin et al. (2001, Ch. 4).

### 14.11.2 Continuous-Time Parameter Estimation

As discussed by Young (1981), Gawthrop (1982, 1987), Unbehauen and Rao (1987, 1990), and Garnier and Wang (2008), least-squares parameter estimation can be performed in the continuous-time domain [as opposed to the more usual discrete-time domain as described, for example, by Ljung (1999)]. A brief outline of the method used in the following examples is given in this section:

$$e_h(t) = \hat{\theta}_u^{\mathrm{T}}\phi(t), \tag{14.220}$$

$$\begin{aligned} J(\hat{\theta}_u) &= \frac{1}{2}\int_0^t e^{\lambda(t-t')}e_h(t')^2 dt' \\ &= \hat{\theta}_u^{\mathrm{T}}\mathbf{S}(t)\hat{\theta}_u \\ &= \hat{\theta}_u^{\mathrm{T}}\mathbf{S}_{uu}(t)\hat{\theta}_u + \hat{\theta}_u^{\mathrm{T}}\mathbf{S}_{uk}(t)\theta_k + \theta_k^{\mathrm{T}}\mathbf{S}_{kk}(t)\theta_k, \end{aligned} \tag{14.221}$$

where

$$\mathbf{S}(t) = \int_0^t e^{\lambda(t-t')}\phi(t')\phi^{\mathrm{T}}(t')dt', \tag{14.222}$$

and the symmetrical matrix $\mathbf{S}(t)$ has been partitioned as

$$\mathbf{S}(t) = \begin{bmatrix} \mathbf{S}_{uu}(t) & \mathbf{S}_{uk}(t) \\ \mathbf{S}_{uk}^{\mathrm{T}}(t) & \mathbf{S}_{kk}(t) \end{bmatrix}. \tag{14.223}$$

Differentiating the cost function $J$ with respect to the vector of unknown parameters $\hat{\theta}_u$ gives

$$\frac{dJ}{d\hat{\theta}_u} = \mathbf{S}_{uu}(t)\hat{\theta}_u + \mathbf{S}_{uk}(t)\theta_k. \tag{14.224}$$

Setting the derivative to zero gives the optimal solution:

$$\hat{\theta}_u(t) = -\mathbf{S}_{uu}^{-1}(t)\mathbf{S}_{uk}(t)\theta_k. \tag{14.225}$$

Differentiating $\mathbf{S}$ (14.222) with respect to time gives

$$\frac{d\mathbf{S}}{dt} + \lambda\mathbf{S}(t) = \phi(t)\phi^{\mathrm{T}}(t). \tag{14.226}$$

### 14.11.3 Intermittent Parameter Estimation

The incremental information matrix $\tilde{\mathbf{S}}_i$ from the $i$th intermittent interval is defined as

$$\tilde{\mathbf{S}}_i = \int_{t_{i-1}}^{t_i} \phi(t')\phi^{\mathrm{T}}(t')dt'. \tag{14.227}$$

Equation 14.227 may be implemented using the differential equation (14.226) with zero initial condition at time $t_{i-1}$. The intermittent information matrix $\mathbf{S}_i$ at the $i$th intermittent interval is defined as

$$\mathbf{S}_i = \lambda_{ic}\mathbf{S}_{i-1} + \tilde{\mathbf{S}}_i. \tag{14.228}$$

Partitioning $\mathbf{S}_i$ as Equation 14.223 gives the parameter estimate of Equation 14.225.

If there is a disturbance characterized by Equations 14.203, 14.204, and 14.207, the parameters corresponding to $d(s)$ jump when the disturbance jumps. As such, a jump will give rise to an event, a new set of $d$ parameters should be estimated; this is achieved by adding a diagonal matrix to the elements of $\mathbf{S}_i$ corresponding to $d(s)$.

## 14.12 Examples: Adaptive Human Balance

As discussed by Gawthrop et al. (2014), it can be argued that the human balance control system generates ballistic control trajectories that attempt to place the unstable system at equilibrium; this leads to homoclinic orbits (Hirsch, Smale, and Devaney, 2012). However, such behavior is dependent on a good internal model. This section looks at the same ballistic balance control system as that of Gawthrop et al. (2014) but in the context of parameter adaptation.

The controlled system is given by the transfer function:

$$G(s) = \frac{b}{s^2 + a}. \tag{14.229}$$

The actual system parameters are

$$a = -1, \tag{14.230}$$

$$b = 1.1. \tag{14.231}$$

The parameters $a$ and $b$ are estimated using the intermittent parameter estimation method of Section 14.11.3 with initial values:

$$\hat{a} = -1, \tag{14.232}$$

$$\hat{b} = 1. \tag{14.233}$$

Figure 14.40a and b shows the nonadaptive controller with correct parameters of Equations 14.230 and 14.231; the behavior approximates that of the ideal ballistic controller.

Figure 14.40c and d shows the nonadaptive controller with the incorrect parameters of Equations 14.232 and 14.233; the behavior is now a limit cycle.

Figure 14.40e and f shows the adaptive controller with the initial incorrect parameters of Equations 14.232 and 14.233. Initially, the behavior corresponds to that of Figure 14.40a and b; but after about 50 s, the behavior corresponds to that of Figure 14.40c and d. The corresponding parameter estimate errors ($\hat{a} - a$ and $\hat{b} - b$) are given in Figure 14.41.

## 14.13 Examples: Adaptive Human Reaching

Repetitive reaching and pointing have been examined by a number of authors including Shadmehr and Mussa-Ivaldi (1994) (see also Shadmehr and Mussa-Ivaldi (2012)), Burdet, Tee, Mareels, Milner, Chew, Franklin, Osu, and Kawato (2006) and Tee, Franklin, Kawato, Milner, and Burdet (2010). An iterative learning control explanation of these results is given by Zhou, Oetomo, Tan, Burdet, and Mareels (2012).

As discussed by Bristow, Tharayil, and Alleyne (2006), "iterative learning control (ILC) is based on the notion that the performance of a system that executes the same task multiple times can be improved by learning from previous executions (trials, iterations, passes)." A number of survey papers are available, including those of Bristow et al. (2006), Ahn, Chen, and Moore (2007), and Wang, Gao, and III (2009), as well as a book by Xu and Tan (2003). ILC is closely related to repetitive control (Cuiyan, Dongchun, and Xianyi, 2004) and to multi-pass control (Edwards, 1974; Owens, 1977).

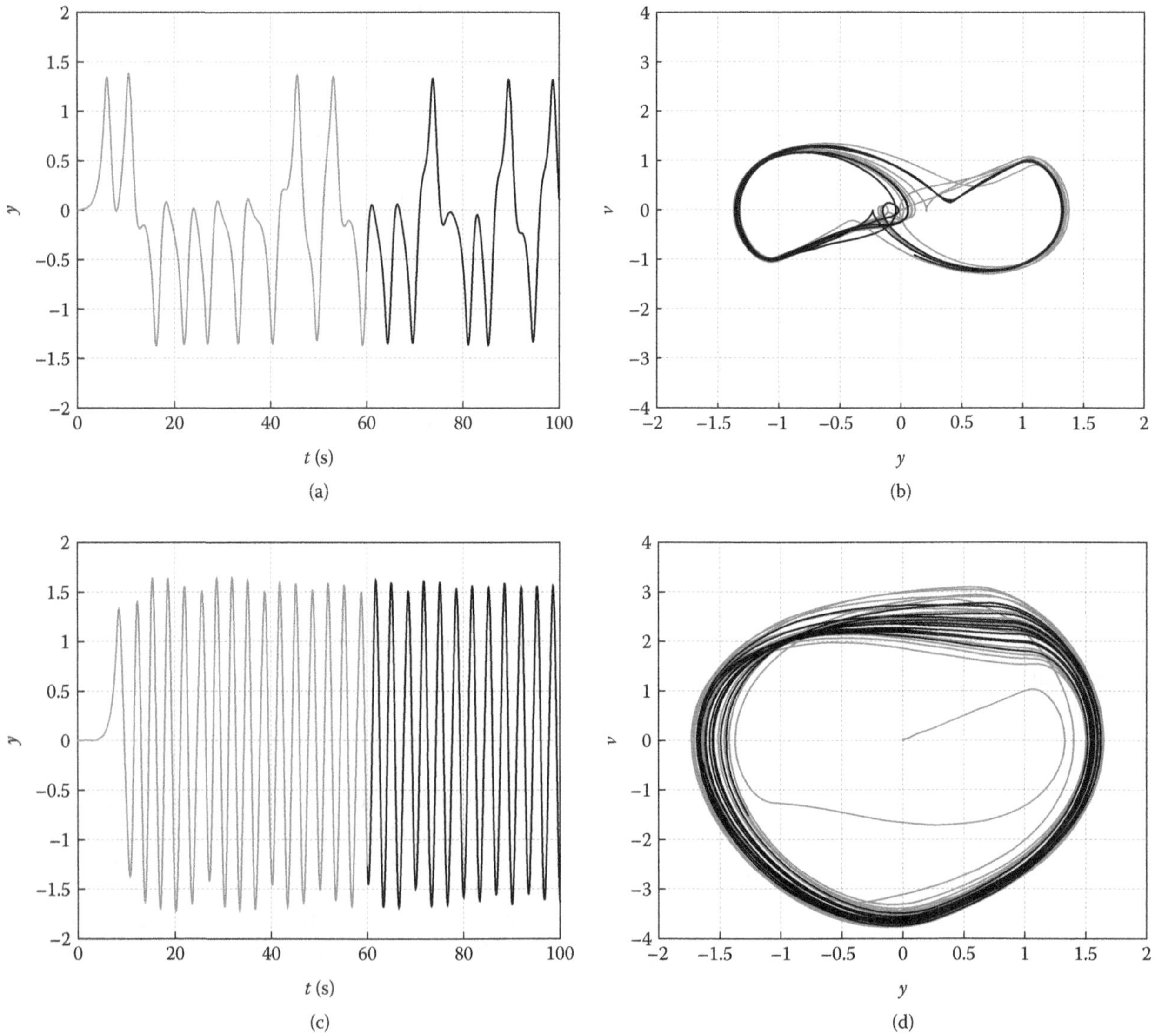


**FIGURE 14.40**

Adaptive balance control. (a) and (b) Correct parameters, no adaptation. (c) and (d) Incorrect parameters, no adaptation. (e) and (f) Incorrect parameters, with adaptation—the initial behavior corresponds to (c) and (d) and the final behavior corresponds to (a) and (b). For clarity, lines are colored grey for $t < 60$ and black for $t \geq 60$. (a) Output $y$. No adaption $\Delta b_0 = 0$. (b) Phase plane. No adaption $\Delta b_0 = 0$. (c) Output $y$. No adaption $\Delta b_0 = 0.1$. (d) Phase plane. No adaption $\Delta b_0 = 0.1$.

*(Continued)*

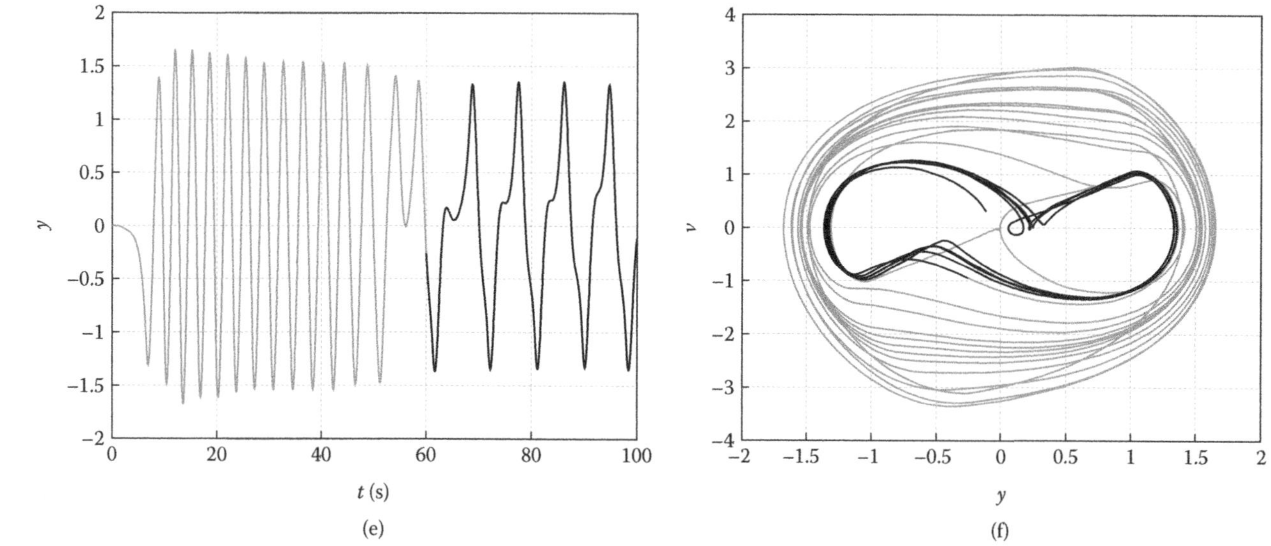


**FIGURE 14.40 (Continued)**
Adaptive balance control. (a) and (b) Correct parameters, no adaptation. (c) and (d) Incorrect parameters, no adaptation. (e) and (f) Incorrect parameters, with adaptation—the initial behavior corresponds to (c) and (d) and the final behavior corresponds to (a) and (b). For clarity, lines are colored grey for $t < 60$ and black for $t \geq 60$. (e) Output $y$. Adaption $\Delta b_0 = 0.1$. (f) Phase plane. Adaption $\Delta b_0 = 0.1$.

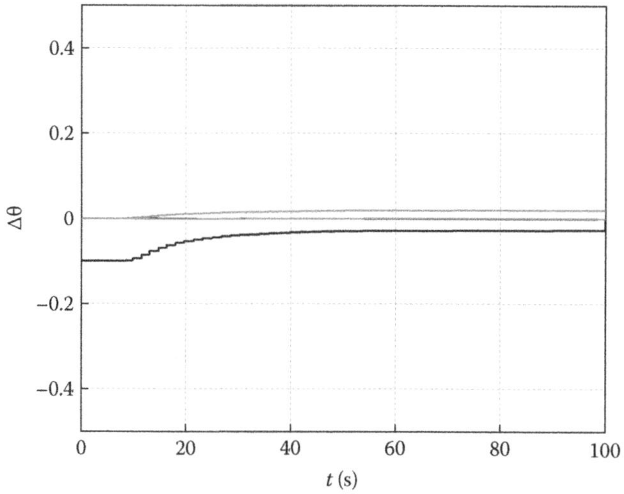


**FIGURE 14.41**
Adaptive balance control: estimated parameter. The parameter estimate errors ($\hat{a} - a$ and $\hat{b} - b$) become smaller as time increases.

This example shows how the intermittent parameter estimation method of Section 14.11.3 can be used in the context of iterative learning.

The system similar to that described in Section IV, case 3 of the paper by Zhou et al. (2012) was used. The lateral motion of the arm in the force field was described by the transfer function of Equation 14.229 with

$$a = \begin{cases} -100 & i \leq 50 \\ 0 & i > 50 \end{cases}, \tag{14.234}$$

$$b = 100. \tag{14.235}$$

The lateral target position was randomly set to $\pm 0.01$ m.

The parameters $a$ and $b$ are estimated using the intermittent parameter estimation method of Section 14.11.3 with initial values:

$$\hat{a} = 0, \tag{14.236}$$

$$\hat{b} = 200. \tag{14.237}$$

Figure 14.42 shows the system output (transverse position) $y$ and control input $u$ for five of the iterations; the sample instants are denoted by the symbol (•) and the ideal trajectory by the gray line. The initial behavior (Figure 14.42a and e) is unstable and sampling occurs at the minimum interval of 100 ms the behavior at the 50th iteration (Figure 14.42b and f) just before the parameter change is close to ideal even though the trajectory is open loop for nearly 400 ms. The behavior at the 51st iteration (Figure 14.42c and g) just after the parameter change is again poor (although stable) but has become ideal and open loop by iteration 100 (Figure 14.42d and h). The data are replotted in Figure 14.43 to show the transverse position $y$ plotted against longitudinal position.

Figure 14.44 shows the evolution of the estimated parameters with iteration number.

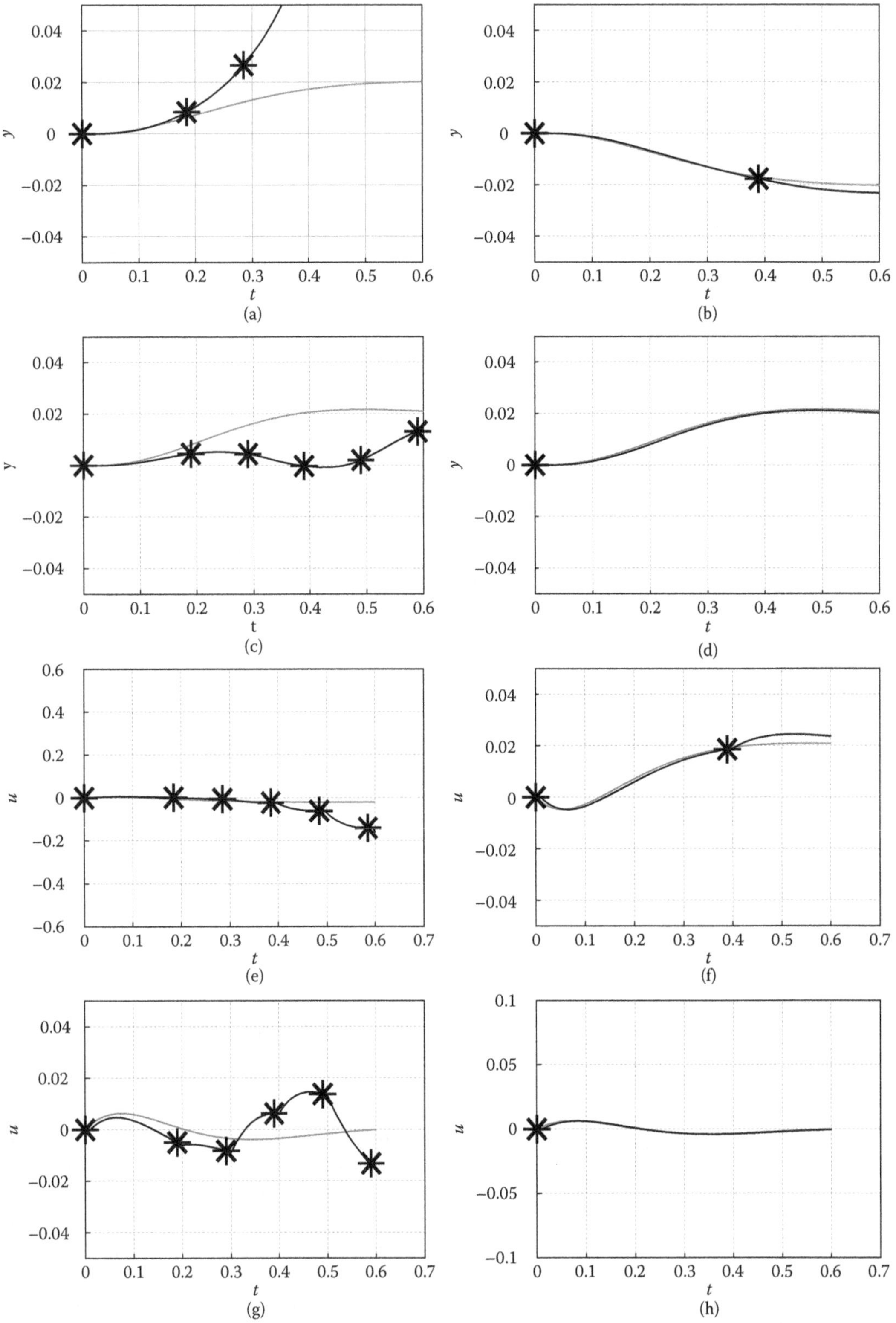


**FIGURE 14.42**

Reaching in a force-field: transverse position. The additional transverse force field is applied throughout, but the initial parameters correspond to zero force field. The sample instants are denoted by the • symbol. The behavior improves, and the intermittent interval increases, from iteration 1 to iteration 50. (a) $y$, 1st iteration. (b) $y$, 50th iteration. (c) $y$, 51st iteration. (d) $y$, 100th iteration. (e) $u$, 1st iteration. (f) $u$, 50th iteration. (g) $u$, 51st iteration. (h) $u$, 100th iteration.

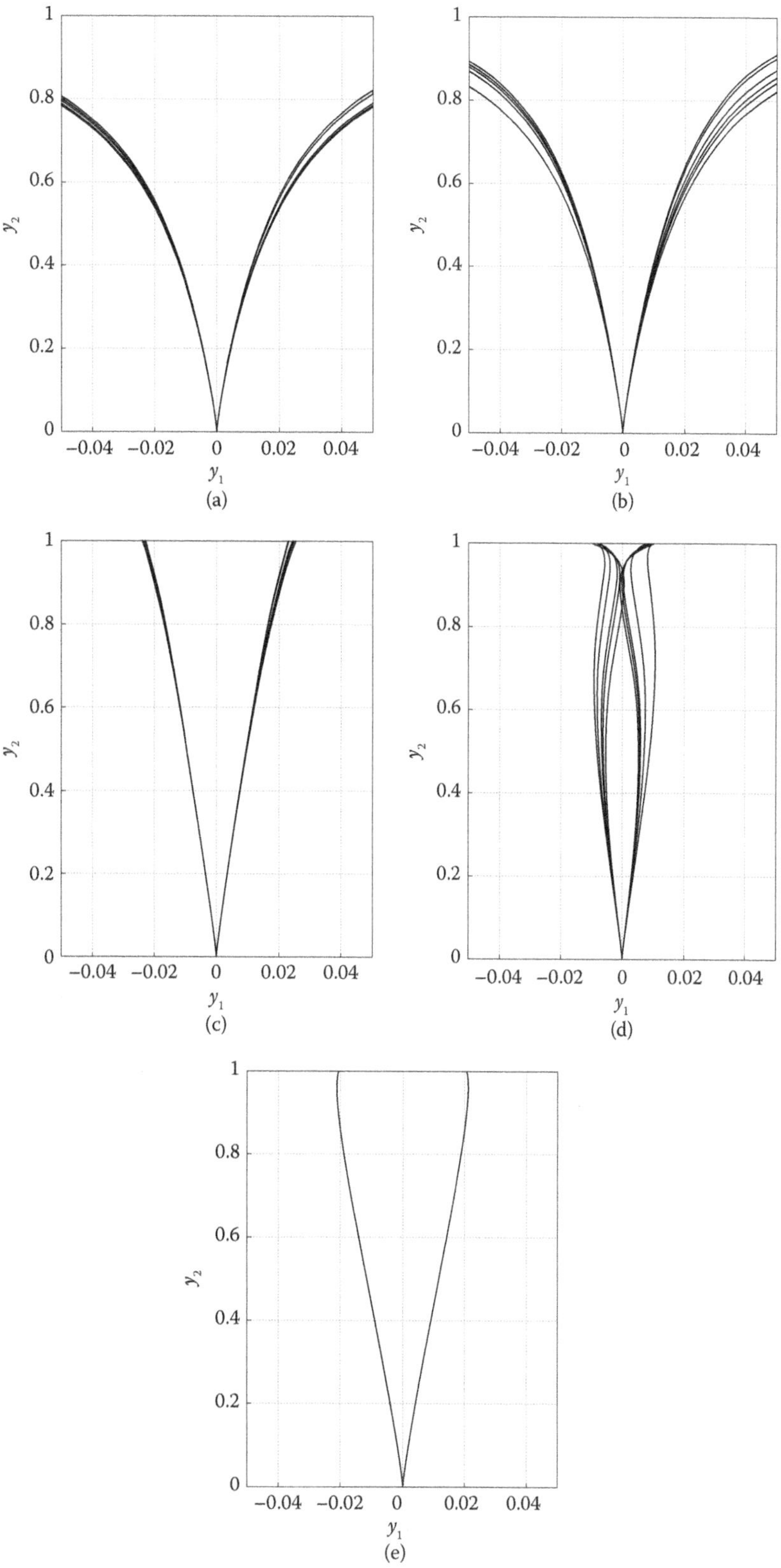


**FIGURE 14.43**

Reaching in a force-field. The data from Figure 14.42 are re-plotted against longitudinal position. (a) 1st iteration. (b) 25th iteration. (c) 50th iteration. (d) 51st iteration. (e) 100th iteration.

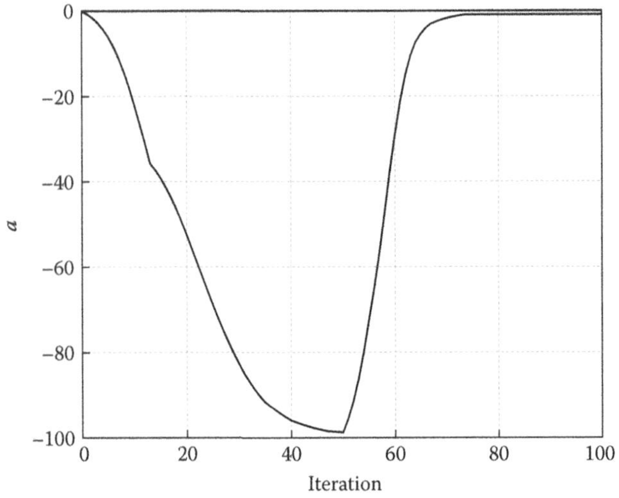


**FIGURE 14.44**
Reaching in a force-field: parameters. The transverse force field parameter is $a = -100$ for $0 \leq$ iteration $\leq 50$ and $a = 0$ for iteration $> 50$. The initial estimate is $\hat{a} = 0$.

## 14.14 Conclusion

- This chapter has given an overview of the current state-of-the-art of the event-driven Intermittent Control discussed, in the context of physiological systems, by Gawthrop et al. (2011). In particular, Intermittent Control has been shown to provide a basis for the human control systems associated with balance and motion control.

- Intermittent control arose in the context of applying control to systems and constraints, which change through time (Ronco et al., 1999). The intermittent control solution allows slow optimization to occur concurrently with a fast control action. Adaptation is intrinsic to intermittent control and yet the formal relationship between adaptive and intermittent control remains to be established. Some results of experiments with human subjects reported by Loram et al. (2011), together with the simulations of Sections 14.12 and 14.13, support the intuition that the intermittent interval somehow simplifies the complexities of dual control. A future challenge is to provide a theoretical basis formally linking intermittent and adaptive control. This basis would extend the applicability of time-varying control and would enhance investigation of biological controllers which are adaptive by nature.

- It is an interesting question as to where the event-driven intermittent control algorithm lies in the human nervous system. IC provides time within the feedback loop to use the current state to select new motor responses (control structure, law, goal, constraints). This facility provides competitive advantage in performance, adaptation, and survival and is thus likely to operate through neural structures which are evolutionarily old as well as new (Brembs, 2011). Refractoriness in humans is associated with a-modal response selection rather than sensory processing or motor execution (Pashler and Johnston, 1998). This function suggests plausible locations within premotor regions and within the slow striatal-prefrontal gating loops (Jiang and Kanwisher, 2003; Dux, Ivanoff, Asplund, and Marois, 2006; Houk, Bastianen, Fansler, Fishbach, Fraser, Reber, Roy, and Simo, 2007; Seidler, 2010; Battaglia-Mayer, Buiatti, Caminiti, Ferraina, Lacquaniti, and Shallice, 2014; Loram et al., 2014).

- It seems plausible that Intermittent Control has applications within a broader biomedical context. Some possible areas are:

  - **Rehabilitation** practice, following neuromuscular diseases such as stroke and spinal cord injury, often uses passive closed-loop learning in which movement is externally imposed by therapists or assistive technology [e.g., robotic-assisted rehabilitation (Huang and Krakauer, 2009)]. Loram et al. (2011) have shown that adaptation to parameter changes during human visual-manual control can be facilitated by using an explicitly intermittent control strategy. For successful learning, active user input should excite the system, allowing learning from the observed intermittent open-loop behavior (Loram et al., 2011).

  - **Cellular control systems** in general and gene regulatory networks in particular seem to have a intermittent nature (Albeck, Burke, Spencer, Lauffenburger, and Sorger, 2008; Balazsi, van Oudenaarden, and Collins, 2011; Liu and Jiang, 2012). It would be interesting to examine whether the intermittent control approaches of this paper are relevant in the context of cellular control systems.

- The particular intermittent control algorithm discussed within this chapter has roots and applications in control engineering (Ronco et al., 1999; Gawthrop and Wang, 2006, 2009a; Gawthrop et al., 2012) and it is hoped that this chapter will lead to further cross-fertilization of physiological and engineering research. Some possible areas are:

- **Decentralized control** (Sandell, Varaiya, Athans, and Safonov, 1978; Bakule and Papik, 2012) is a pragmatic approach to the control of large-scale systems where, for reasons of cost, convenience, and reliability, it is not possible to control the entire system by a single centralized controller. Fundamental control-theoretic principles arising from decentralized control have been available for some time (Clements, 1979; Anderson and Clements, 1981; Gong and Aldeen, 1992). More recently, following the implementation of decentralized control using networked control systems (Moyne and Tilbury, 2007), attention has focused on the interaction of communication and control theory (Baillieul and Antsaklis, 2007; Nair, Fagnani, Zampieri, and Evans, 2007) and fundamental results have appeared (Nair and Evans, 2003; Nair, Evans, Mareels, and Moran, 2004; Hespanha, Naghshtabrizi, and Xu, 2007). It would be interesting to apply the physiologically inspired approaches of this chapter to decentralized control as well as to reconsider Intermittent Control in the context of decentralized control systems.
- **Networked control systems** lead to the "sampling period jitter problem" (Sala, 2007; Moyne and Tilbury, 2007) where uncertainties in transmission time lead to unpredictable non-uniform sampling and stability issues (Cloosterman, van de Wouw, Heemels, and Nijmeijer, 2009). A number of authors have suggested that performance may be improved by replacing the standard zero-order hold by a generalized hold (Feuer and Goodwin, 1996; Sala, 2005, 2007) or using a dynamical model of the system to interpolate between samples (Zhivoglyadov and Middleton, 2003; Montestruque and Antsaklis, 2003). This can be shown to improve stability (Montestruque and Antsaklis, 2003; Hespanha et al., 2007). As shown by Gawthrop and Wang (2011), the intermittent controller has a similar feature; it therefore follows that the physiologically inspired form of intermittent controller described in this chapter has application to networked control systems.
- **Robotics** It seems likely that understanding the control mechanisms behind human balance and motion control (Loram et al., 2009; van de Kamp et al., 2013a,b) and stick balancing (Gawthrop et al., 2013b) will have applications in robotics. In particular, as discussed by van de Kamp et al. (2013a), robots, like humans, contain redundant possibilities within a multi-segmental structure. Thus, the multivariable constrained intermittent control methods illustrated in Section 14.7 and the adaptive versions illustrated in Section 14.11 may well be applicable to the control of autonomous robots.

## Acknowledgments

The work reported here is related to the linked EPSRC Grants EP/F068514/1, EP/F069022/1, and EP/F06974X/1 "Intermittent control of man and machine." PJG acknowledge the many discussions with Megan and Peter Neilson over the years, which have significantly influenced this work.

## Bibliography

H.-S. Ahn, Y. Q. Chen, and K. L. Moore. Iterative learning control: Brief survey and categorization. *IEEE Transactions on Systems, Man, and Cybernetics, Part C: Applications and Reviews*, 37(6):1099–1121, 2007. doi:10.1109/TSMCC.2007.905759.

J. G. Albeck, J. M. Burke, S. L. Spencer, D. A. Lauffenburger, and P. K. Sorger. Modeling a snap-action, variable-delay switch controlling extrinsic cell death. *PLoS Biology*, 6(12):2831–2852, 2008. doi:10.1371/journal.pbio.0060299.

A. V. Alexandrov, A. A. Frolov, F. B. Horak, P. Carlson-Kuhta, and S. Park. Feedback equilibrium control during human standing. *Biological Cybernetics*, 93:309–322, 2005. doi:10.1007/s00422-005-0004-1.

B. D. O. Anderson and D. J. Clements. Algebraic characterization of fixed modes in de-centralized control. *Automatica*, 17(5):703–712, 1981. doi:10.1016/0005-1098(81)90017-0.

Y. Asai, Y. Tasaka, K. Nomura, T. Nomura, M. Casadio, and P. Morasso. A model of postural control in quiet standing: Robust compensation of delay-induced instability using intermittent activation of feedback control. *PLoS One*, 4(7):e6169, 2009. doi:10.1371/journal.pone.0006169.

K. Astrom and B. Bernhardsson. Systems with Lebesgue sampling. In *Directions in Mathematical Systems Theory and Optimization*, A. Rantzer and C. Byrnes, Eds., pp. 1–13. Springer, 2003. doi:10.1007/3-540-36106-5.

K. J. Astrom. Frequency domain properties of Otto Smith regulators. *International Journal of Control*, 26:307–314, 1977. doi:10.1080/00207177708922311.

K. J. Astrom. Event based control. In *Analysis and Design of Nonlinear Control Systems*, A. Astolfi and L. Marconi, Eds., pp. 127–147. Springer, Heidelberg, 2008. doi:10.1007/978-3-540-74358-3.

K. J. Astrom and B. M. Bernhardsson. Comparison of Riemann and Lebesgue sampling for first order stochastic systems. In *Proceedings of the 41st IEEE Conference on Decision and Control, 2002*, volume 2, pp. 2011–2016, January 2002.

K. J. Astrom and A. Helmersson. Dual control of an integrator with unknown gain. *Computers and Mathematics with Applications*, 12(6, Part A):653–662, 1986. doi:10.1016/0898-1221(86)90052-0.

K. J. Åström and B. Wittenmark. On self-tuning regulators. *Automatica*, 9:185–199, 1973. doi:10.1016/0005-1098(73)90073-3.

K. J. Åström and B. Wittenmark. *Adaptive Control*. Addison-Wesley, Reading, MA, 1989.

T. Bailey and H. Durrant-Whyte. Simultaneous localization and mapping (SLAM): Part II. *IEEE Robotics Automation Magazine*, 13(3):108–117, 2006. doi:10.1109/MRA.2006.1678144.

J. Baillieul and P. J. Antsaklis. Control and communication challenges in networked real-time systems. *Proceedings of the IEEE*, 95(1):9–28, 2007. doi:10.1109/JPROC.2006.887290.

L. Bakule and M. Papik. Decentralized control and communication. *Annual Reviews in Control*, 36(1):1–10, 2012. doi:10.1016/j.arcontrol.2012.03.001.

G. Balazsi, A. van Oudenaarden, and J. J. Collins. Cellular decision making and biological noise: From microbes to mammals. *Cell*, 144(6):910–925, 2011. doi:10.1016/j.cell.2011.01.030.

S. Baron, D. L. Kleinman, and W. H. Levison. An optimal control model of human response part II: Prediction of human performance in a complex task. *Automatica*, 6:371–383, 1970. doi:10.1016/0005-1098(70)90052-X.

N. C. Barrett and D. J. Glencross. The double step analysis of rapid manual aiming movements. *The Quarterly Journal of Experimental Psychology Section A*, 40(2):299–322, 1988. doi:10.1080/02724988843000131.

Y. Bar-Shalom. Stochastic dynamic programming: Caution and probing. *IEEE Transactions on Automatic Control*, 26(5):1184–1195, 1981. doi:10.1109/TAC.1981.1102793.

Y. Bar-Shalom and E. Tse. Dual effect, certainty equivalence, and separation in stochastic control. *IEEE Transactions on Automatic Control*, 19(5):494–500, 1974. doi:10.1109/TAC.1974.1100635.

A. Battaglia-Mayer, T. Buiatti, R. Caminiti, S. Ferraina, F. Lacquaniti, and T. Shallice. Correction and suppression of reaching movements in the cerebral cortex: Physiological and neuropsychological aspects. *Neuroscience & Biobehavioral Reviews*, 42:232–251, 2014. doi:10.1016/j.neubiorev.2014.03.002.

G. A. Bekey. The human operator as a sampled-data system. *IRE Transactions on Human Factors in Electronics*, 3(2):43–51, 1962. doi:10.1109/THFE2.1962.4503341.

N. Bhushan and R. Shadmehr. Computational nature of human adaptive control during learning of reaching movements in force fields. *Biological Cybernetics*, 81(1):39–60, 1999. doi:10.1007/s004220050543.

A. Bottaro, M. Casadio, P. G. Morasso, and V. Sanguineti. Body sway during quiet standing: Is it the residual chattering of an intermittent stabilization process? *Human Movement Science*, 24(4):588–615, 2005. doi:10.1016/j.humov.2005.07.006.

A. Bottaro, Y. Yasutake, T. Nomura, M. Casadio, and P. Morasso. Bounded stability of the quiet standing posture: An intermittent control model. *Human Movement Science*, 27(3):473–495, 2008. doi:10.1016/j.humov.2007.11.005.

S. P. Boyd and L. Vandenberghe. *Convex Optimization*. Cambridge University Press, Cambridge, UK, 2004.

B. Brembs. Towards a scientific concept of free will as a biological trait: Spontaneous actions and decision-making in invertebrates. *Proceedings of the Royal Society B: Biological Sciences*, 278(1707):930–939, 2011. doi:10.1098/rspb.2010.2325.

D. A. Bristow, M. Tharayil, and A. G. Alleyne. A survey of iterative learning control. *IEEE Control Systems*, 26(3):96–114, 2006. doi:10.1109/MCS.2006.1636313.

E. Burdet, K. Tee, I. Mareels, T. Milner, C. Chew, D. Franklin, R. Osu, and M. Kawato. Stability and motor adaptation in human arm movements. *Biological Cybernetics*, 94:20–32, 2006. doi:10.1007/s00422-005-0025-9.

W.-H. Chen and P. J. Gawthrop. Constrained predictive pole-placement control with linear models. *Automatica*, 42(4):613–618, 2006. doi:10.1016/j.automatica.2005.09.020.

D. W. Clarke and P. J. Gawthrop. Self-tuning controller. *IEE Proceedings Part D: Control Theory and Applications*, 122(9):929–934, 1975. doi:10.1049/piee.1975.0252.

D. W. Clarke and P. J. Gawthrop. Self-tuning control. *IEE Proceedings Part D: Control Theory and Applications*, 126(6):633–640, 1979. doi:10.1049/piee.1979.0145.

D. W. Clarke and P. J. Gawthrop. Implementation and application of microprocessor-based self-tuners. *Automatica*, 17(1):233–244, 1981. doi:10.1016/0005-1098(81)90098-4.

D. J. Clements. A representation result for two-input two-output decentralized control systems. *The Journal of the Australian Mathematical Society. Series B. Applied Mathematics*, 21(1):113–127, 1979. doi:10.1017/S0334270000001971.

M. B. G. Cloosterman, N. van de Wouw, W. P. M. H. Heemels, and H. Nijmeijer. Stability of networked control systems with uncertain time-varying delays. *IEEE Transactions on Automatic Control*, 54(7):1575–1580, 2009. doi:10.1109/TAC.2009.2015543.

K. J. Craik. Theory of human operators in control systems: Part 1, the operator as an engineering system. *British Journal of Psychology*, 38:56–61, 1947a. doi:10.1111/j.2044-8295.1947.tb01141.x.

K. J. Craik. Theory of human operators in control systems: Part 2, man as an element in a control system. *British Journal of Psychology*, 38:142–148, 1947b. doi:10.1111/j.2044-8295.1948.tb01149.x.

L. Cuiyan, Z. Dongchun, and Z. Xianyi. A survey of repetitive control. In *Proceedings of the 2004 IEEE/RSJ International Conference on Intelligent Robots and Systems, 2004 (IROS 2004)*, volume 2, pp. 1160–1166, September–October 2004. doi:10.1109/IROS.2004.1389553.

H. Durrant-Whyte and T. Bailey. Simultaneous localization and mapping (SLAM): Part I. *IEEE Robotics Automation Magazine*, 13(2):99–110, 2006. doi:10.1109/MRA.2006.1638022.

P. E. Dux, J. Ivanoff, C. L. Asplund, and R. Marois. Isolation of a central bottleneck of information processing with time-resolved fMRI. *Neuron*, 52(6):1109–1120, 2006. doi:10.1016/j.neuron.2006.11.009.

J. B. Edwards. Stability problems in the control of multipass processes. *Proceedings of the Institution of Electrical Engineers*, 121(11):1425–1432, 1974. doi:10.1049/piee.1974.0299.

A. A. Feldbaum. *Optimal Control Theory*. Academic Press, New York, 1960.

A. Feuer and G. C. Goodwin. *Sampling in Digital Signal Processing and Control*. Birkhauser, Berlin, 1996.

R. Fletcher. *Practical Methods of Optimization*, 2nd edition. Wiley, Chichester, 1987.

M. Foo and E. Weyer. On reproducing existing controllers as model predictive controllers. In *Australian Control Conference (AUCC), 2011*, pp. 303–308, November 2011.

G. F. Franklin and J. D. Powell. *Digital Control of Dynamic Systems*. Addison-Wesley, Reading, MA, 1980.

G. F. Franklin, J. D. Powell, and A. Emami-Naeini. *Feedback Control of Dynamic Systems*, 3rd edition. Addison-Wesley, Reading, MA, 1994.

A. T. Fuller. Optimal nonlinear control systems with pure delay. *International Journal of Control*, 8:145–168, 1968. doi:10.1080/00207176808905662.

H. Garnier and L. Wang, Ed. *Identification of Continuous-Time Models from Sampled Data*. Advances in Industrial Control. Springer, London, 2008.

P. Gawthrop, H. Gollee, A. Mamma, I. Loram, and M. Lakie. Intermittency explains variability in human motor control. In *NeuroEng 2013: Australian Workshop on Computational Neuroscience*, Melbourne, Australia, 2013a.

P. Gawthrop, K.-Y. Lee, M. Halaki, and N. O'Dwyer. Human stick balancing: An intermittent control explanation. *Biological Cybernetics*, 107(6):637–652, 2013b. doi:10.1007/s00422-013-0564-4.

P. Gawthrop, I. Loram, H. Gollee, and M. Lakie. Intermittent control models of human standing: Similarities and differences. *Biological Cybernetics*, 108(2):159–168, 2014. doi:10.1007/s00422-014-0587-5.

P. Gawthrop, I. Loram, and M. Lakie. Predictive feedback in human simulated pendulum balancing. *Biological Cybernetics*, 101(2):131–146, 2009. doi:10.1007/s00422-009-0325-6.

P. Gawthrop, I. Loram, M. Lakie, and H. Gollee. Intermittent control: A computational theory of human control. *Biological Cybernetics*, 104(1–2):31–51, 2011. doi:10.1007/s00422-010-0416-4.

P. Gawthrop, D. Wagg, S. Neild, and L. Wang. Power-constrained intermittent control. *International Journal of Control*, 86(3):396–409, 2013c. doi:10.1080/00207179.2012.733888.

P. Gawthrop and L. Wang. The system-matched hold and the intermittent control separation principle. *International Journal of Control*, 84(12):1965–1974, 2011. doi:10.1080/00207179.2011.630759.

P. J. Gawthrop. Studies in identification and control. D.Phil. thesis, Oxford University, 1976. http://ora.ox.ac.uk/objects/uuid:90ade91d-df67-42ef-a422-0d3500331701.

P. J. Gawthrop. A continuous-time approach to discrete-time self-tuning control. *Optimal Control: Applications and Methods*, 3(4):399–414, 1982.

P. J. Gawthrop. *Continuous-time Self-tuning Control. Volume 1: Design*. Engineering Control Series, Research Studies Press, Lechworth, England, 1987.

P. J. Gawthrop. *Continuous-Time Self-tuning Control. Volume 2: Implementation*. Engineering Control Series, Research Studies Press, Taunton, England, 1990.

P. J. Gawthrop. Frequency domain analysis of intermittent control. *Proceedings of the Institution of Mechanical Engineers Pt. I: Journal of Systems and Control Engineering*, 223(5):591–603, 2009. doi:10.1243/09596518JSCE759.

P. J. Gawthrop and H. Gollee. Intermittent tapping control. *Proceedings of the Institution of Mechanical Engineers, Part I: Journal of Systems and Control Engineering*, 226(9):1262–1273, 2012. doi:10.1177/0959651812450114.

P. J. Gawthrop, M. D. Lakie, and I. D. Loram. Predictive feedback control and Fitts' law. *Biological Cybernetics*, 98(3):229–238, 2008. doi:10.1007/s00422-007-0206-9.

P. J. Gawthrop, S. A. Neild, and D. J. Wagg. Semi-active damping using a hybrid control approach. *Journal of Intelligent Material Systems and Structures*, 23(18): 965–974, 2012. doi:10.1177/1045389X12436734.

P. J. Gawthrop and E. Ronco. Predictive pole-placement control with linear models. *Automatica*, 38(3):421–432, 2002. doi:10.1016/S0005-1098(01)00231-X.

P. J. Gawthrop and D. G. Sbarbaro. Stochastic approximation and multilayer perceptrons: The gain backpropagation algorithm. *Complex System Journal*, 4: 51–74, 1990.

P. J. Gawthrop and L. Wang. Intermittent predictive control of an inverted pendulum. *Control Engineering Practice*, 14(11):1347–1356, 2006. doi:10.1016/j.conengprac.2005.09.002.

P. J. Gawthrop and L. Wang. Intermittent model predictive control. *Proceedings of the Institution of Mechanical Engineers Pt. I: Journal of Systems and Control Engineering*, 221(7):1007–1018, 2007. doi:10.1243/09596518JSCE417.

P. J. Gawthrop and L. Wang. Constrained intermittent model predictive control. *International Journal of Control*, 82:1138–1147, 2009a. doi:10.1080/00207170802474702.

P. J. Gawthrop and L. Wang. Event-driven intermittent control. *International Journal of Control*, 82(12):2235–2248, 2009b. doi:10.1080/00207170902978115.

P. J. Gawthrop and L. Wang. Intermittent redesign of continuous controllers. *International Journal of Control*, 83:1581–1594, 2010. doi:10.1080/00207179.2010.483691.

P. J. Gawthrop, L. Wang, and P. C. Young. Continuous-time non-minimal state-space design. *International Journal of Control*, 80(10):690–1697, 2007. doi:10.1080/00207170701546006.

H. Gollee, A. Mamma, I. D. Loram, and P. J. Gawthrop. Frequency-domain identification of the human controller. *Biological Cybernetics*, 106:359–372, 2012. doi:10.1007/s00422-012-0503-9.

Z. Gong and M. Aldeen. On the characterization of fixed modes in decentralized control. *IEEE Transactions on Automatic Control*, 37(7):1046–1050, 1992. doi:10.1109/9.148369.

G. C. Goodwin, S. F. Graebe, and M. E. Salgado. *Control System Design*. Prentice Hall, NJ, 2001.

G. C. Goodwin and K. S. Sin. *Adaptive Filtering Prediction and Control*. Prentice-Hall, Englewood Cliffs, NJ, 1984.

M. Günther, S. Grimmer, T. Siebert, and R. Blickhan. All leg joints contribute to quiet human stance: A mechanical analysis. *Journal of Biomechanics*, 42(16):2739–2746, 2009. doi:10.1016/j.jbiomech.2009.08.014.

M. Günther, O. Müller, and R. Blickhan. Watching quiet human stance to shake off its straitjacket. *Archive of Applied Mechanics*, 81(3):283–302, 2011. doi:10.1007/s00419-010-0414-y.

M. Günther, O. Müller, and R. Blickhan. What does head movement tell about the minimum number of mechanical degrees of freedom in quiet human stance? *Archive of Applied Mechanics*, 82(3):333–344, 2012. doi:10.1007/s00419-011-0559-3.

S. Hanneton, A. Berthoz, J. Droulez, and J. J. E. Slotine. Does the brain use sliding variables for the control of movements? *Biological Cybernetics*, 77(6):381–393, 1997. doi:10.1007/s004220050398.

R. M. Hardwick, C. Rottschy, R. C. Miall, and S. B. Eickhoff. A quantitative meta-analysis and review of motor learning in the human brain. *NeuroImage*, 67(0):283–297, 2013. doi:10.1016/j.neuroimage.2012.11.020.

W. Heemels, J. H. Sandee, and P. P. J. V. D. Bosch. Analysis of event-driven controllers for linear systems. *International Journal of Control*, 81(4):571–590, 2008. doi:10.1080/00207170701506919.

J. P. Hespanha, P. Naghshtabrizi, and Y. Xu. A survey of recent results in networked control systems. *Proceedings of the IEEE*, 95(1):138–162, 2007. doi:10.1109/JPROC.2006.887288.

M. W. Hirsch, S. Smale, and R. L. Devaney. *Differential Equations, Dynamical Systems, and an Introduction to Chaos*, 3rd edition. Academic Press, Amsterdam, 2012.

N. Hogan. Adaptive control of mechanical impedance by coactivation of antagonist muscles. *IEEE Transactions on Automatic Control*, 29(8):681–690, 1984. doi:10.1109/TAC.1984.1103644.

I. M. Horowitz. *Synthesis of Feedback Systems*. Academic Press, Amsterdam, 1963.

J. C. Houk, C. Bastianen, D. Fansler, A. Fishbach, D. Fraser, P. J. Reber, S. A. Roy, and L. S. Simo. Action selection and refinement in subcortical loops through basal ganglia and cerebellum. *Philosophical Transactions of the Royal Society of London. Series B, Biological Sciences*, 362(1485):1573–1583, 2007. doi:10.1098/rstb.2007.2063.

V. S. Huang and J. W. Krakauer. Robotic neurorehabilitation: A computational motor learning perspective. *Journal of Neuroengineering and Rehabilitation*, 6:5, 2009. doi:10.1186/1743-0003-6-5.

K. J. Hunt, R. Zbikowski, D. Sbarbaro, and P. J. Gawthrop. Neural networks for control systems—A survey. *Automatica*, 28(6):1083–1112, 1992. doi:10.1016/0005-1098(92)90053-I.

T. Insperger. Act-and-wait concept for continuous-time control systems with feedback delay. *IEEE Transactions on Control Systems Technology*, 14(5):974–977, 2006. doi:10.1109/TCST.2006.876938.

O. L. R. Jacobs. *Introduction to Control Theory*. Oxford University Press, 1974.

O. L. R. Jacobs and J. W. Patchell. Caution and probing in stochastic control. *International Journal of Control*, 16(1):189–199, 1972. doi:10.1080/00207177208932252.

Y. Jiang and N. Kanwisher. Common neural substrates for response selection across modalities and mapping paradigms. *Journal of Cognitive Neuroscience*, 15(8):1080–1094, 2003. doi:10.1162/089892903322598067.

R. Johansson, M. Magnusson, and M. Akesson. Identification of human postural dynamics. *IEEE Transactions on Biomedical Engineering*, 35(10):858–869, 1988. doi:10.1109/10.7293.

S. Julier, J. Uhlmann, and H. F. Durrant-Whyte. A new method for the nonlinear transformation of means and covariances in filters and estimators. *IEEE Transactions on Automatic Control*, 45(3):477–482, 2000. doi:10.1109/9.847726.

J. C. Kalkkuhl, K. J. Hunt, R. Zbikowski, and A. Dzielinski, Eds. *Applications of Neural Adaptive Control Technology*. Volume 17. Robotics and Intelligent Systems Series. World Scientific, Singapore, 1997.

S. G. Khan, G. Herrmann, F. L. Lewis, T. Pipe, and C. Melhuish. Reinforcement learning and optimal adaptive control: An overview and implementation examples. *Annual Reviews in Control*, 36(1):42–59, 2012. doi:10.1016/j.arcontrol.2012.03.004.

O. Khatib. A unified approach for motion and force control of robot manipulators: The operational space formulation. *IEEE Journal of Robotics and Automation*, 3(1):43–53, 1987. doi:10.1109/JRA.1987.1087068.

D. Kleinman. Optimal control of linear systems with time-delay and observation noise. *IEEE Transactions on Automatic Control*, 14(5):524–527, 1969. doi:10.1109/TAC.1969.1099242.

D. L. Kleinman, S. Baron, and W. H. Levison. An optimal control model of human response part I: Theory and validation. *Automatica*, 6:357–369, 1970. doi:10.1016/0005-1098(70)90051-8.

P. Kowalczyk, P. Glendinning, M. Brown, G. Medrano-Cerda, H. Dallali, and J. Shapiro. Modelling human balance using switched systems with linear feedback control. *Journal of the Royal Society Interface*, 9(67):234–245, 2012. doi:10.1098/rsif.2011.0212.

B. C. Kuo. *Digital Control Systems*. Holt, Reinhart and Winston, New York, 1980.

H. Kwakernaak and R. Sivan. *Linear Optimal Control Systems*. Wiley, New York, 1972.

M. Lakie, N. Caplan, and I. D. Loram. Human balancing of an inverted pendulum with a compliant linkage: Neural control by anticipatory intermittent bias. *The Journal of Physiology*, 551(1):357–370, 2003. doi:10.1113/jphysiol.2002.036939.

M. Lakie and I. D. Loram. Manually controlled human balancing using visual, vestibular and proprioceptive senses involves a common, low frequency neural process. *The Journal of Physiology*, 577(Pt 1):403–416. 2006. doi:10.1113/jphysiol.2006.116772.

M. Latash. The bliss (not the problem) of motor abundance (not redundancy). *Experimental Brain Research*, 217:1–5, 2012. doi:10.1007/s00221-012-3000-4.

W. H. Levison, S. Baron, and D. L. Kleinman. A model for human controller remnant. *IEEE Transactions on Man-Machine Systems*, 10(4):101–108, 1969. doi:10.1109/TMMS.1969.299906.

Y. Liu and H. Jiang. Exponential stability of genetic regulatory networks with mixed delays by periodically intermittent control. *Neural Computing and Applications*, 21(6):1263–1269, 2012. doi:10.1007/s00521-011-0551-4.

L. Ljung. *System Identification: Theory for the User*, 2nd edition. Information and Systems Science. Prentice-Hall, Englewood Cliffs, NJ, 1999.

I. D. Loram, P. Gawthrop, and M. Lakie. The frequency of human, manual adjustments in balancing an inverted pendulum is constrained by intrinsic physiological factors. *Journal of Physiology*, 577(1):403–416, 2006. doi:10.1113/jphysiol.2006.118786.

I. D. Loram, H. Gollee, M. Lakie, and P. Gawthrop. Human control of an inverted pendulum: Is continuous control necessary? Is intermittent control effective? Is intermittent control physiological? *The Journal of Physiology*, 589:307–324, 2011. doi:10.1113/jphysiol.2010.194712.

I. D. Loram and M. Lakie. Human balancing of an inverted pendulum: Position control by small, ballistic-like, throw and catch movements. *Journal of Physiology*, 540(3):1111–1124, 2002. doi:10.1113/jphysiol.2001.013077.

I. D. Loram, M. Lakie, and P. J. Gawthrop. Visual control of stable and unstable loads: What is the feedback delay and extent of linear time-invariant control? *Journal of Physiology*, 587(6):1343–1365, 2009. doi:10.1113/jphysiol.2008.166173.

I. D. Loram, C. N. Maganaris, and M. Lakie. Human postural sway results from frequent, ballistic bias impulses by soleus and gastrocnemius. *Journal of Physiology*, 564(Pt 1):295–311, 2005. doi:10.1113/jphysiol.2004.076307.

I. D. Loram, C. van de Kamp, H. Gollee, and P. J. Gawthrop. Identification of intermittent control in man and machine. *Journal of the Royal Society Interface*, 9(74):2070–2084, 2012. doi:10.1098/rsif.2012.0142.

I. D. Loram, C. van de Kamp, M. Lakie, H. Gollee, and P. J. Gawthrop. Does the motor system need intermittent control? *Exercise and Sport Sciences Reviews*, 42(3):117–125, 2014. doi:10.1249/JES.0000000000000018.

J. M. Maciejowski. *Predictive Control with Constraints*. Prentice Hall, Englewood Cliffs, NJ, 2002.

J. M. Maciejowski. Reverse engineering existing controllers for MPC design. In *Proceedings of the IFAC Symposium on System Structure and Control*, pp. 436–441, October 2007.

A. Mamma, H. Gollee, P. J. Gawthrop, and I. D. Loram. Intermittent control explains human motor remnant without additive noise. In *2011 19th Mediterranean Conference on Control Automation (MED)*, pp. 558–563, June 2011. doi:10.1109/MED.2011.5983113.

G. McLachlan and D. Peel. *Finite Mixture Models*. Wiley, New York, 2000.

D. McRuer. Human dynamics in man-machine systems. *Automatica*, 16:237–253, 1980. doi:10.1016/0005-1098(80)90034-5.

R. C. Miall, D. J. Weir, and J. F. Stein. Manual tracking of visual targets by trained monkeys. *Behavioural Brain Research*, 20(2):185–201, 1986. doi:10.1016/0166-4328(86)90003-3.

R. C. Miall, D. J. Weir, and J. F. Stein. Intermittency in human manual tracking tasks. *Journal of Motor Behavior*, 25:53–63, 1993a. doi:10.1080/00222895.1993.9941639.

R. C. Miall, D. J. Weir, D. M. Wolpert, and J. F. Stein. Is the cerebellum a Smith predictor? *Journal of Motor Behavior*, 25:203–216, 1993b. doi:10.1080/00222895.1993.9942050.

W. T. Miller, R. S. Sutton, and P. J. Werbos. *Neural Networks for Control*. MIT Press, Cambridge, MA, 1990.

L. A. Montestruque and P. J. Antsaklis. On the model-based control of networked systems. *Automatica*, 39(10):1837–1843, 2003. doi:10.1016/S0005-1098(03)00186-9.

J. R. Moyne and D. M. Tilbury. The emergence of industrial control networks for manufacturing control, diagnostics, and safety data. *Proceedings of the IEEE*, 95(1):29–47, 2007. doi:10.1109/JPROC.2006.887325.

G. N. Nair and R. J. Evans. Exponential stabilisability of finite-dimensional linear systems with limited data rates. *Automatica*, 39(4):585–593, 2003.

G. N. Nair, R. J. Evans, I. M. Y. Mareels, and W. Moran. Topological feedback entropy and nonlinear stabilization. *IEEE Transactions on Automatic Control*, 49(9):1585–1597, 2004.

G. N. Nair, F. Fagnani, S. Zampieri, and R. J. Evans. Feedback control under data rate constraints: An overview. *Proceedings of the IEEE*, 95(1):108–137, 2007. doi:10.1109/JPROC.2006.887294.

F. Navas and L. Stark. Sampling or Intermittency in Hand Control System Dynamics. *Biophysical Journal*, 8(2):252–302, 1968.

P. D. Neilson and M. D. Neilson. An overview of adaptive model theory: Solving the problems of redundancy, resources, and nonlinear interactions in human movement control. *Journal of Neural Engineering*, 2(3):S279–S312, 2005. doi:10.1152/jn.01144.2004.

P. D. Neilson, M. D. Neilson, and N. J. O'Dwyer. Internal models and intermittency: A theoretical account of human tracking behaviour. *Biological Cybernetics*, 58:101–112, 1988. doi:10.1007/BF00364156.

K. M. Newell, K. M. Deutsch, J. J. Sosnoff, and G. Mayer-Kress. Variability in motor output as noise: A default and erroneous proposition? In *Movement system variability* (K. Davids, S. Bennett and K.M. Newell, Eds.). Human Kinetics Publishers, Champaign, IL, 2006.

J. Nocedal and S. J. Wright. *Numerical Optimization*, 2nd edition. Springer Series in Operations Research. Springer Verlag, Berlin, 2006.

T. M. Osborne. An investigation into the neural mechanisms of human balance control. PhD thesis, School of Sport and Exercise Sciences, University of Birmingham, 2013. http://etheses.bham.ac.uk/3918/.

D. H. Owens. Stability of linear multipass processes. *Proceedings of the Institution of Electrical Engineers*, 124(11):1079–1082, 1977. doi:10.1049/piee.1977.0220.

H. Pashler and J. C. Johnston. Attentional limitations in dual-task performance. In *Attention*, H. Pashler, Ed., pp. 155–189. Psychology Press, 1998.

R. J. Peterka. Sensorimotor integration in human postural control. *The Journal of Neurophysiology*, 88(3):1097–1118, 2002.

R. Pintelon and J. Schoukens. *System Identification. A Frequency Domain Approach*. IEEE Press, New York, 2001.

R. Pintelon, J. Schoukens, and Y. Rolain. Frequency domain approach to continuous-time identification: Some practical aspects. In *Identification of Continuous-Time Models from Sampled Data*, pp. 215–248. Springer, 2008.

I. J. Pinter, R. van Swigchem, A. J. K. van Soest, and L. A. Rozendaal. The dynamics of postural sway cannot be captured using a one-segment inverted pendulum model: A PCA on segment rotations during unperturbed stance. *Journal of Neurophysiology*, 100(6):3197–3208, 2008. doi:10.1152/jn.01312.2007.

E. C. Poulton. *Tracking Skill and Manual Control*. Academic Press, New York, 1974.

J. B. Rawlings. Tutorial overview of model predictive control. *IEEE Control Systems Magazine*, 20(3):38–52, 2000.

J. B. Rawlings and K. R. Muske. The stability of constrained receding horizon control. *IEEE Transactions on Automatic Control*, 38(10):1512–1516, 1993.

E. Ronco, T. Arsan, and P. J. Gawthrop. Open-loop intermittent feedback control: Practical continuous-time GPC. *IEE Proceedings Part D: Control Theory and Applications*, 146(5):426–434, 1999. doi:10.1049/ip-cta:19990504.

S. A. Safavynia and L. H. Ting. Task-level feedback can explain temporal recruitment of spatially fixed muscle synergies throughout postural perturbations. *Journal of Neurophysiology*, 107(1):159–177, 2012. doi:10.1152/jn.00653.2011.

A. P. Sage and J. J. Melsa. *Estimation Theory with Applications to Communication and Control*. McGraw-Hill, New York, 1971.

A. Sala. Computer control under time-varying sampling period: An LMI gridding approach. *Automatica*, 41(12):2077–2082, 2005. doi:10.1016/j.automatica.2005.05.017.

A. Sala. Improving performance under sampling-rate variations via generalized hold functions. *IEEE Transactions on Control Systems Technology*, 15(4):794–797, 2007. doi:10.1109/TCST.2006.890302.

N. Sandell, P. Varaiya, M. Athans, and M. Safonov. Survey of decentralized control methods for large scale systems. *IEEE Transactions on Automatic Control*, 23(2):108–128, 1978. doi:10.1109/TAC.1978.1101704.

S. J. Schiff. *Neural Control Engineering: The Emerging Intersection between Control Theory and Neuroscience*. Computational Neuroscience. MIT Press, Cambridge, MA, 2012.

R. Seidler. Neural correlates of motor learning, transfer of learning, and learning to learn. *Exercise & Sport Sciences Reviews*, 38(1):3–9, 2010. doi:10.1097/JES.0b013e3181c5cce7.

R. Shadmehr and F. A. Mussa-Ivaldi. Adaptive representation of dynamics during learning of a motor task. *The Journal of Neuroscience*, 14(5):3208–3224, 1994.

R. Shadmehr and S. Mussa-Ivaldi. *Biological Learning and Control*. Computational Neuroscience. MIT Press, Cambridge, MA, 2012.

R. Shadmehr and S. P. Wise. *Computational Neurobiology of Reaching and Pointing: A Foundation for Motor Learning*. MIT Press, Cambridge, MA, 2005.

S. Skogestad and I. Postlethwaite. *Multivariable Feedback Control Analysis and Design*. Wiley, New York, 1996.

O. J. M. Smith. A controller to overcome dead-time. *ISA Transactions*, 6(2):28–33, 1959.

G. Stein. Respect the unstable. *IEEE Control Systems Magazine*, 23(4):12–25, 2003. doi:10.1109/MCS.2003.1213600.

G. Stepan and T. Insperger. Stability of time-periodic and delayed systems—A route to act-and-wait control. *Annual Reviews in Control*, 30(2):159–168, 2006. doi:10.1016/j.arcontrol.2006.08.002.

R. S. Sutton and A. G. Barto. *Reinforcement Learning: An Introduction*. Cambridge University Press, 1998.

R. S. Sutton, A. G. Barto, and R. J. Williams. Reinforcement learning is direct adaptive optimal control. *IEEE Control Systems*, 12(2):19–22, 1992. doi:10.1109/37.126844.

C. J. Taylor, A. Chotai, and P. C. Young. Continuous-time proportional-integral derivative-plus (PIP) control with filtering polynomials. In *Proceedings of the UKACC Conference "Control '98"*, pp. 1391–1396, Swansea, UK, September 1998.

K. Tee, D. Franklin, M. Kawato, T. Milner, and E. Burdet. Concurrent adaptation of force and impedance in the redundant muscle system. *Biological Cybernetics*, 102:31–44, 2010. doi:10.1007/s00422-009-0348-z.

C. W. Telford. The refractory phase of voluntary and associative responses. *Journal of Experimental Psychology*, 14(1):1–36, 1931. doi:10.1037/h0073262.

L. H. Ting. Dimensional reduction in sensorimotor systems: A framework for understanding muscle coordination of posture. In *Computational Neuroscience: Theoretical Insights into Brain Function, Volume 165 of Progress in Brain Research*, T. D. P. Cisek and J. F. Kalaska, Eds., pp. 299–321. Elsevier, 2007. doi:10.1016/S0079-6123(06)65019-X.

E. Todorov. Optimality principles in sensorimotor control (review). *Nature Neuroscience*, 7(9):907–915, 2004. doi:10.1038/nn1309.

E. Todorov and M. I. Jordan. Optimal feedback control as a theory of motor coordination. *Nature Neuroscience*, 5(11):1226–1235, 2002. doi:10.1038/nn963.

H. Unbehauen and G. P. Rao. *Identification of Continuous Systems*. North-Holland, Amsterdam, 1987.

H. Unbehauen and G. P. Rao. Continuous-time approaches to system identification—A survey. *Automatica*, 26(1):23–35, 1990. doi:10.1016/0005-1098(90)90155-B.

C. van de Kamp, P. Gawthrop, H. Gollee, M. Lakie, and I. D. Loram. Interfacing sensory input with motor output: Does the control architecture converge to a serial process along a single channel? *Frontiers in Computational Neuroscience*, 7(55), 2013a. doi:10.3389/fncom.2013.00055.

C. van de Kamp, P. J. Gawthrop, H. Gollee, and I. D. Loram. Refractoriness in sustained visuo-manual control: Is the refractory duration intrinsic or does it depend on external system properties? *PLOS Computational Biology*, 9(1):e1002843, 2013b. doi:10.1371/journal.pcbi.1002843.

H. Van Der Kooij, R. Jacobs, B. Koopman, and H. Grootenboer. A multisensory integration model of human stance control. *Biological Cybernetics*, 80:299–308, 1999. doi:10.1007/s004220050527.

H. Van Der Kooij, R. Jacobs, B. Koopman, and F. Van Der Helm. An adaptive model of sensory integration in a dynamic environment applied to human stance control. *Biological Cybernetics*, 84:103–115, 2001. doi:10.1007/s004220050527.

M. A. Vince. The intermittency of control movements and the psychological refractory period. *British Journal of Psychology*, 38:149–157, 1948. doi:10.1111/j.2044-8295.1948.tb01150.x.

D. Vrabie and F. Lewis. Neural network approach to continuous-time direct adaptive optimal control for partially unknown nonlinear systems. *Neural Networks*, 22(3):237–246, 2009. doi:10.1016/j.neunet.2009.03.008.

L. Wang. *Model Predictive Control System Design and Implementation Using MATLAB*, 1st edition. Springer, Berlin, 2009.

Y. Wang, F. Gao, and F. J. Doyle III. Survey on iterative learning control, repetitive control, and run-to-run control. *Journal of Process Control*, 19(10):1589–1600, 2009. doi:10.1016/j.jprocont.2009.09.006.

N. Wiener. *Cybernetics: Or the Control and Communication in the Animal and the Machine*, 2nd edition. MIT Press, Cambridge, MA, 1965.

D. A. Winter. *Biomechanics and Motor Control of Human Movement*, 4th edition. Wiley, New York, 2009.

D. M. Wolpert, R. C. Miall, and M. Kawato. Internal models in the cerebellum. *Trends in Cognitive Sciences*, 2:338–347, 1998. doi:10.1016/S1364-6613(98)01221-2.

J.-X. Xu and Y. Tan. *Linear and Nonlinear Iterative Learning Control*. Volume 291. Springer, Berlin, 2003.

P. Young. Parameter estimation for continuous-time models: A survey. *Automatica*, 17(1):23–39, 1981. doi:10.1016/0005-1098(81)90082-0.

P. C. Young, M. A. Behzadi, C. L. Wang, and A. Chotai. Direct digital and adaptive control by input-output state variable feedback pole assignment. *International Journal of Control*, 46(6):1867–1881, 1987. doi:10.1080/00207178708934021.

R. Zbikowski and K. J. Hunt, Ed. *Neural Adaptive Control Technology*. Volume 15. Robotics and Intelligent Systems Series. World Scientific, Singapore, 1996.

P. V. Zhivoglyadov and R. H. Middleton. Networked control design for linear systems. *Automatica*, 39(4):743–750, 2003. doi:10.1016/S0005-1098(02)00306-0.

S.-H. Zhou, D. Oetomo, Y. Tan, E. Burdet, and I. Mareels. Modeling individual human motor behavior through model reference iterative learning control. *IEEE Transactions on Biomedical Engineering*, 59(7):1892–1901, 2012. doi:10.1109/TBME.2012.2192437.

# Part II

# Event-Based Signal Processing

# 15

## *Event-Based Data Acquisition and Digital Signal Processing in Continuous Time*

**Yannis Tsividis**
*Columbia University*
*New York, NY, USA*

**Maria Kurchuk**
*Pragma Securities*
*New York, NY, USA*

**Pablo Martinez-Nuevo**
*Massachusetts Institute of Technology*
*Cambridge, MA, USA*

**Steven M. Nowick**
*Columbia University*
*New York, NY, USA*

**Sharvil Patil**
*Columbia University*
*New York, NY, USA*

**Bob Schell**
*Analog Devices Inc.*
*Somerset, NJ, USA*

**Christos Vezyrtzis**
*IBM*
*Yorktown Heights, NY, USA*

**CONTENTS**

**ABSTRACT** We review techniques in which uniform sampling is abandoned, with samples generated and processed instead only when there is an input event, thus causing dynamic energy use only when demanded by the information in the signal. Methods to implement event-based A/D converters and digital signal processors (DSPs) in this context are reviewed. It is shown that, compared with traditional techniques, the techniques reviewed here lead to circuits that completely avoid aliasing, respond immediately to input changes, result in lower in-band quantization error, and exhibit dynamic power dissipation that decreases when the input activity decreases. Experimental results from recent test chips, operating at kilohertz (kHz) to gigahertz (GHz) signal frequencies, are reviewed.

## 15.1 Introduction

This chapter discusses research on event-based, continuous-time (CT) signal acquisition and digital signal processing. The aim of this work is to enable use of resources only when there is a reason to do so, as dictated by the signal itself, rather than by a clock.

Among the several potential benefits of this approach is energy savings, which is important in a large number of portable applications in which long battery life is a must. An example can be found in sensor networks [1,2], in which the nodes sense one or more physical quantities (temperature, pressure, humidity, magnetic field intensity, light intensity, etc.) and perform computation and communication. Another example is biomedical devices [3–6], which can be wearable, implantable, or ingestible; in these, similar functions as above are performed. A third example is personal communication devices. Key components in such applications include analog-to-digital converters (ADCs) and digital signal processors (DSPs). In conventional systems, these utilize a fixed sampling frequency which, according to the Nyquist theorem, must be higher than twice the highest expected signal frequency. When the input properties are more relaxed (lower frequency content, bursty signals, or even periods of silence), the high sampling frequency simply wastes energy and transmission resources. To avoid this, one can use some form of nonuniform sampling [7], with the local sampling rate adapted to the properties of the signal. There have been many attempts to do this over several decades [8–14]; the techniques developed generally go by the name of adaptive sampling rate, variable sampling rate, or signal-dependent sampling. In one approach, the input is divided into segments, with the sampling rate varied from segment to segment, but kept constant within each segment [13,15]. Another approach uses a sampling rate that varies continuously in proportion to the magnitude of the slope of the signal, with the slope value obtained by a slope detector [8,9]. A similar effect is obtained by level-crossing sampling [11,12,14], illustrated in Figure 15.1a; samples are generated only when the input changes enough to cross any one of a set of amplitude levels, shown by broken lines. For signals varying less rapidly, samples are generated less frequently; the power dissipation of the associated circuitry can thus also be allowed to decrease, as illustrated in Figure 15.1b. In addition to the systems mentioned above, a variety of other systems can benefit from this approach; these include ones involving variable frequency and signals obtained from variable-speed sources, such as those encountered in Doppler-shifted signals or disk drives.

The material in this chapter expands on review papers [16–18], and uses material from [18] with copyright permission from IEEE (Copyright © 2010 IEEE).

Level-crossing sampling can be viewed as one type of event-driven or event-based sampling, which can be loosely defined as taking samples only when a significant event occurs, such as a notable change in the

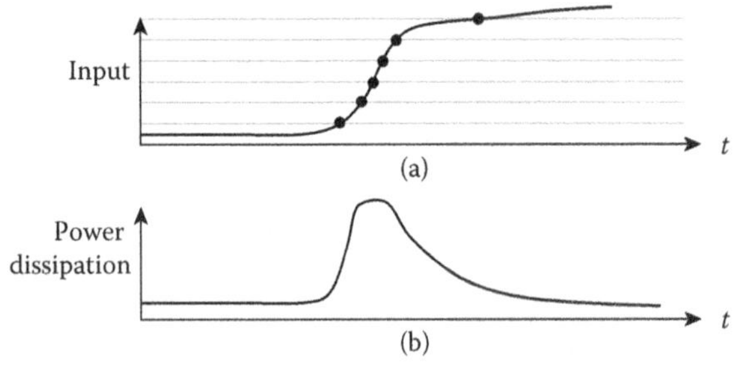


**FIGURE 15.1**
(a) Input level-crossing sampling; (b) corresponding ADC and DSP power dissipation. (From Y. Tsividis, *IEEE Transactions on Circuits and Systems II: Express Briefs*, 57(8), 577–581 © 2010 IEEE.)

quantity being monitored. Event-based and adaptive-rate control systems have been studied for a long time [8,9,19–28] and have been reviewed elsewhere [29,30], as well as in the first part of this book. Several types of event-based sampling are discussed in the above references, in the context of control. The context of this chapter is, instead, signal processing and associated data acquisition. In order to provide some focus, the discussion will concentrate mostly on level-crossing sampling in CT. However, most of the principles presented are valid for other types of event-based sampling, and for finely discretized time; comments on these will be made where appropriate. Event-based-sampled signals can be processed using event-driven digital signal processing (DSP) [31], as will be discussed later on.

Event-based systems can place demands on the accuracy with which the timing of events is recorded and processed. This shift from amplitude accuracy to timing accuracy is compatible with the advent of modern nanometer very large scale integration (VLSI) fabrication processes [32], which make higher speeds possible, but require reduced supply voltages, thus making amplitude operations increasingly difficult. Thus, the promise of the techniques discussed can be expected to improve as chip fabrication technology advances.

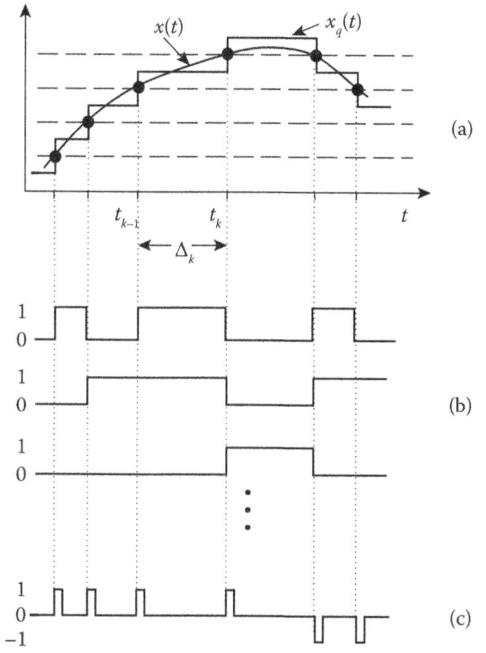


**FIGURE 15.2**
(a) Input signal ($x(t)$), threshold levels (broken lines), level-crossing samples (dots), and quantized signal ($x_q(t)$); (b) CT digital representation of quantized signal; (c) CT delta mod representation of quantized signal. (From Y. Tsividis, *IEEE Transactions on Circuits and Systems II: Express Briefs*, 57(8), 577–581 © 2010 IEEE.)

## 15.2 Level-Crossing Sampling and CT Amplitude Quantization

### 15.2.1 Level-Crossing Sampling

Level-crossing sampling [11,12,14], already introduced in Figure 15.1, is further illustrated in Figure 15.2a. A CT input signal $x(t)$ is compared to a set of discrete amplitude levels, shown by horizontal broken lines; samples, shown as dots, are taken at the instants such levels are crossed. (The waveform $x_q(t)$ in part (a), and those in parts (b) and (c), should be ignored until further notice.) The pairs $(t_k, x(t_k))$ form a representation of the signal. No sampling clock is used; the signal, in some sense, samples itself, in a way dual to conventional sampling; the sampling can be considered, in some sense, to occur along the amplitude axis, rather than the time axis. Low-frequency and low-amplitude inputs are sampled less densely in time than high-frequency, high-amplitude ones, and periods of silence are not sampled at all. However, the resulting average sampling rate can be very high if many levels are used; this can raise the power dissipation and impose excessive demands on hardware speed. Ways to reduce this problem are discussed in Section 15.2.6. Further discussion of level-crossing sampling can be found elsewhere [11,12,14], including the use of adaptive reference levels [33–36]. Related approaches and variants can be found in [37–40].

Level-crossing ADCs, due to their attention to time information, can be considered a form of time-based ADCs [41].

### 15.2.2 Relation of Level-Crossing Sampling to CT Amplitude Quantization

Consider now a CT signal passed through a CT quantizer (without sampling), as in Figure 15.3a, with decision levels as in Figure 15.2a; when a level is crossed, the output of the quantizer $x_q(t)$ goes to the closest quantized value, chosen in between the decision levels. There is no clock in the system. The quantizer's output, $x_q(t)$, is shown in Figure 15.2a. It is clear from that figure that $x_q(t)$ inherently contains all the information contained in the level-crossing samples, and that the latter can be reproduced from the former, and vice versa. It is thus seen that *level-crossing sampling, with zero-order reconstruction, can be viewed as amplitude quantization with no sampling at all.* This point of view will be seen to lead to a convenient way of deducing the spectral properties of the signals under discussion, as well to lead to appropriate hardware implementations. Unless stated otherwise, the term "level-crossing sampling" will be taken to imply also zero-order reconstruction.

### 15.2.3 Spectral Properties

Level-crossing sampling, followed by zero-order reconstruction, results in interesting spectral properties. To begin, we note that it results in no aliasing; this is shown in [42–44] and can be understood intuitively in two ways. One way is to note that high-frequency signals are automatically sampled at a higher rate than low-frequency ones; the other way is to note that the equivalent operation of CT quantization (Section 15.2.2) in Figure 15.3a involves no sampling at all, and thus no aliasing can occur. This property of level-crossing sampling in CT seems not to have been recognized in the early references on level-crossing sampling, perhaps because in those references the time was assumed discretized.

Consider now a sinusoidal input $x(t)$ as an example (Figure 15.3a); $x_q(t)$ will be periodic, and can be represented by a Fourier series, containing only harmonics, with no error spectral components in between. As many of the harmonics fall outside the baseband of the following DSP, suggested by a broken-line frequency response in Figure 15.3, the in-band error can be significantly lower than in classical systems involving uniform sampling plus quantization (Figure 15.3b), typically by 10–20 dB [31,42–44].

It is interesting to consider for a moment what it is that makes the spectral properties of the quantizer output so different from those in the conventional case, where a signal is first uniformly sampled and then quantized, as shown in Figure 15.3b. Let the frequency of the sinusoidal input be $f_{in}$; if this input is sampled at a frequency $f_s$, the output of the sampler will contain components at frequencies $mf_s \pm f_{in}$, where $m$ is an integer; these components theoretically extend to arbitrarily high frequencies. When this spectrum is now passed through the quantizer in Figure 15.3b, the nonlinearities inherent in the latter will cause harmonic distortion as well as intermodulation between the above components; in the general case, many of the intermodulation components fall in-band, as suggested by the spectrum in Figure 15.3b, and constitute what is commonly called "quantization noise." This represents extra in-band quantization error, which is absent in Figure 15.3a. (A different, but equivalent, point of view is suggested elsewhere [42,43].) Thus, by avoiding a sampling clock, one not only can achieve event-driven energy dissipation, but can also achieve lower quantization error and better spectral properties. While the above discussion uses a sinusoidal input as a simple example, more complex inputs lead to the same conclusion. The spectra of level-crossing-sampled signals are discussed in detail in another chapter in this book [45].

The use of dithering can result in further improved spectral properties for low-level signals, at the expense of an increased average sampling rate [46].

### 15.2.4 Encoding

#### 15.2.4.1 Amplitude Coding

The quantized or level-crossing-sampled signal in Figure 15.2a can be represented digitally as shown in Figure 15.2b; this can be accomplished, for example, by a flash ADC without a clock, as in Figure 15.4 [31,47]. The binary signals generated are functions of continuous

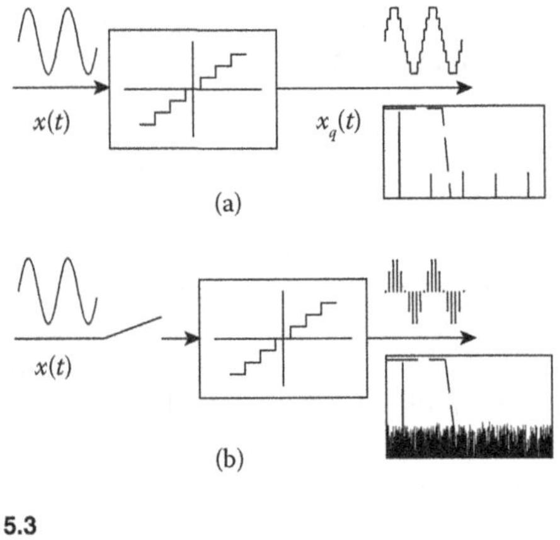


**FIGURE 15.3**
Output and its spectrum with a sinusoidal input, with (a) quantization only, and (b) uniform sampling and quantization. (From Y. Tsividis, *IEEE Transactions on Circuits and Systems II: Express Briefs*, 57(8), 577–581 © 2010 IEEE.)

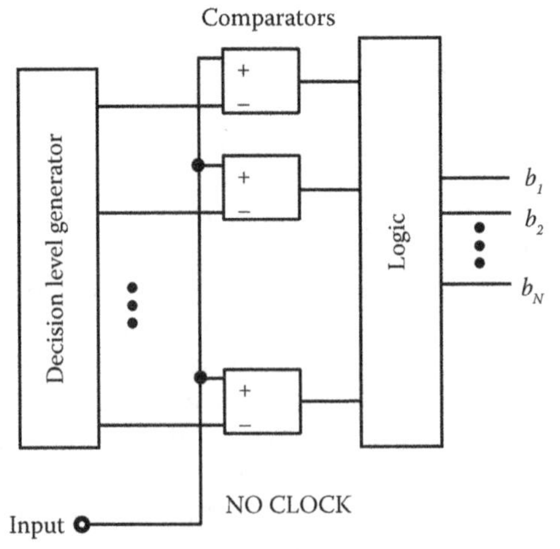


**FIGURE 15.4**
Principle of CT flash ADC.

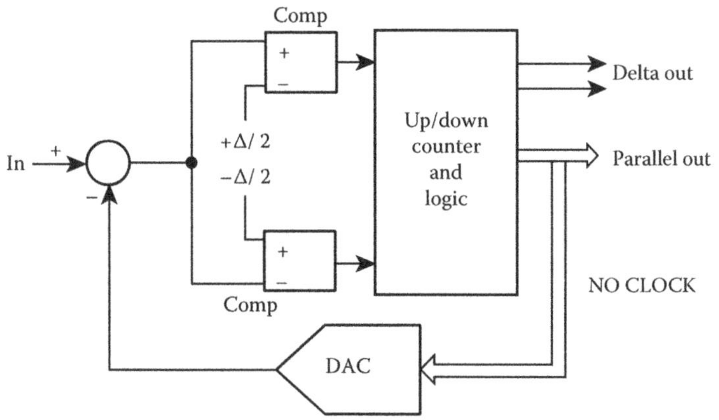


**FIGURE 15.5**
Implementation of CT asynchronous delta modulation ADC obtained from [52], with the clock removed; a parallel output is also available.

time; their transitions can occur at any time instant, as dictated by the signal itself, as shown in Figure 15.2b.

Alternatively, one can signal just the times a level is crossed, plus the direction of crossing, as shown in Figure 15.2c (the resulting waveform can then be encoded in two bits); this results in the asynchronous delta modulation [48–50]. The original signal values can be found from this signal by using an accumulator and an initial value. The accumulation operation can be performed by an analog integrator [49,50] or by a counter plus a digital-to-analog converter (DAC) [51,52]. This approach is shown in Figure 15.5; it is obtained from that in [52], with the clock removed.

For applications in which only a few quantization levels are needed, one can also use thermometer coding [53].

Several implementations of ADCs for level-crossing sampling have been presented [47,52,54–67], including some incorporated into products [68] and ones described in recent patent applications by Texas Instruments [69–71]. Some of these [52,55] use quantized time, but can be adapted to CT operation [54,72].

The maximum input rate of change that can be successfully handled depends on hardware implementation details, and on the power dissipation one is willing to accept. The input rate of change can be limited by using an analog low-pass filter in front of the ADC. Other considerations concerning the minimum and maximum sampling intervals in event-driven systems can be found in the references [19,22]. Noise immunity is improved by incorporating hysteresis [19]. If the hysteresis width is made equal to two quantization levels, one obtains the "send-on-delta" scheme [73].

In general, the design of CT ADCs is challenging; the absence of a clock makes the design of comparators especially difficult. An important consideration is the fact that the comparison time depends on the rate of change of the input; ways to minimize this problem are discussed in [74,75]. The power dissipation of CT ADCs [76] is currently significantly higher than that of clocked ADCs; however, the development efforts behind the latter is orders of magnitude larger than that devoted so far to CT ADCs, so it is too early to say what the eventual performance of CT ADCs will be.

Other ADC schemes, such as pipeline, have not so far been considered for CT implementation to the authors' knowledge.

In the above type of encoding, information on the sample time instants is not explicitly encoded. The signals in Figure 15.2b and c are functions that evolve in real time; for real-time processing, the time instants do not need to be coded, just as time does not need to be coded for real-time processing of classical analog signals.

#### 15.2.4.2 Time Coding

The event times $t_k$ can be encoded into digital words as well, and stored, if the time axis is finely quantized, resulting in pseudo-continuous operation; more conveniently, one would code the differences $\Delta_k = t_k - t_{k-1}$ [11,12,14]. This is referred to as "time coding" [12]. Time discretization needs to be sufficiently fine in order to avoid introducing errors in the signal representation, as discussed in the context of CT DSPs in Section 15.5.3.

### 15.2.5 Other Techniques

While we have focused on level-crossing sampling above, several other ways to accomplish event-based sampling are possible. For example, one can use as a criterion for sampling based not on the difference between the signal and its nearest quantization level, but rather on the integral of this difference [20,25], or on the difference between the input and a prediction of it [9,77]. A technique that relies on level-crossing sampling the derivative of the input is described in Section 15.4.

### 15.2.6 Reducing the Required Timing Resolution

CT ADCs can produce a large number of samples as compared to Nyquist-rate ADCs, particularly when the input frequency and amplitude are large. Reference [78] presents a methodology for taking advantage of the spectral properties of the CT quantization error, to adapt the resolution of a CT ADC according to the magnitude of the input slope in such a way that practically no signal degradation occurs compared to the full-resolution case. Instead of tracking fast and large deviations in the input with numerous small steps, the signal can be represented with fewer, larger steps, without causing an increase in the in-band error power. This technique is now briefly summarized.

The error waveform, $e(t) = x_q(t) - x(t)$, that results from quantization contains bell-shaped segments that vary slowly, and sawtooth-like segments that vary fast, as shown in Figure 15.6 for a sinusoidal input. The former segments occur during slowly varying portions of the signal (e.g., near the peaks of the sinusoid in the example assumed), and contain power primarily at low frequencies. The sawtooth-like segments, in contrast, occur during high-slope portions of the input and contribute primarily high-frequency error power, which is typically outside the signal bandwidth. Since fast portions of the input do not contribute significantly to the in-band error power, they can be quantized with

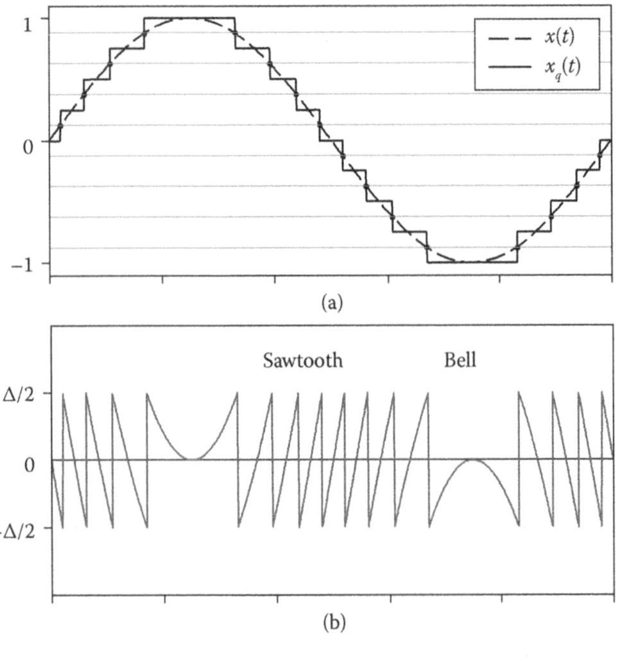


**FIGURE 15.6**
(a) Input sinusoid and its quantized version; (b) corresponding quantization error. (From M. Kurchuk and Y. Tsividis, *IEEE Transactions on Circuits and Systems I*, 57(5), 982–991 © 2010 IEEE.)

lower resolution using a variable-resolution quantizer, as illustrated in Figure 15.7. A larger quantization step for fast inputs causes an increase in the local magnitude of $e(t)$, as well as an increase in high-frequency error power, but without increasing appreciably the error power that falls in the baseband, highlighted in gray in Figure 15.7c. The sampling instants, shown in Figure 15.7b, occur less frequently and with longer intervals during the low-resolution segments, emphasized by dashed boxes; this not only leads to a decrease in the number of samples, and thus in the ensuing power dissipation, but also allows for the constraints on the processing time of the hardware to be relaxed.

Considerations involved in the judicious choice of the level-skipping algorithm are given elsewhere [78]; it is shown there that a number of difficulties can be avoided by using the approach shown in Figure 15.8, which shows high- and low-resolution transfer curves for the ADC. Note that no new decision levels and output levels need to be generated in this approach. More than two resolution settings can be used.

A generalized block diagram of a variable-resolution CT ADC is shown in Figure 15.9a, with example signals provided in Figure 15.9b. A slope magnitude detector determines whether a slope threshold has been crossed and adjusts the step of the ADC accordingly. The slope thresholds are chosen to keep the low-resolution sawtooth error power several decades outside the baseband, according to a simple criterion [78]. When a slope threshold is surpassed, the quantization step and the maximum amplitude of the error triple, as shown in Figure 15.9b. The quantized signal has larger steps for fast portions of the input and finer steps during segments of low slope.

The variable-resolution function can be realized by increasing the quantization step of an ADC or by post-processing the output of a fixed-resolution ADC; the post-processor skips an appropriate number of samples before generating a variable-resolution output. The variable-resolution output has two extra bits—one bit to indicate whether the resolution has changed, and a second bit to indicate whether that slope has increased or decreased. The slope detector function can be realized without determining the input slope by comparing the time between quantization level crossings to a threshold time. The slope detector implementation proposed in [78] consists of a few digital gates and adds insignificantly to the hardware overhead. A related approach is discussed elsewhere [55]. Time granularity reduction is considered further in [79,80].

The performance of a variable-resolution ADC has been compared with that of a fixed-resolution ADC for a sinusoidal input in the voice band in [78]. Three resolution settings were used, with a maximum resolution

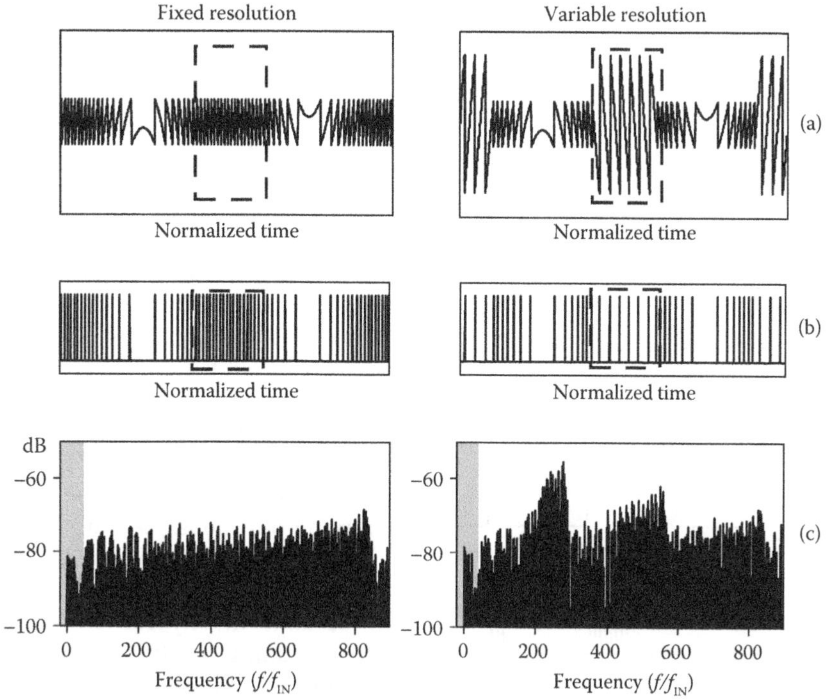


**FIGURE 15.7**
(a) Quantization error, (b) sampling instants, and (c) quantization error spectrum, with the baseband highlighted in gray, for ADCs with a fixed resolution (left) and a variable resolution (right). (From M. Kurchuk and Y. Tsividis, *IEEE Transactions on Circuits and Systems I*, 57(5), 982–991 © 2010 IEEE.)

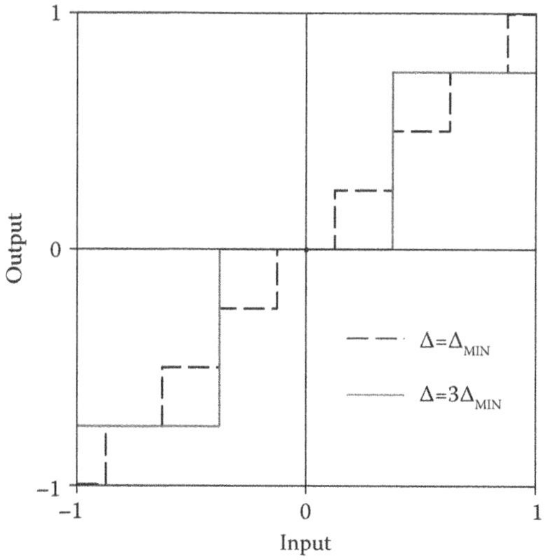


**FIGURE 15.8**
Variable-resolution ADC transfer characteristic for two resolution settings. (From M. Kurchuk and Y. Tsividis, *IEEE Transactions on Circuits and Systems I*, 57(5), 982–991 © 2010 IEEE.)

of 8 bits and a quantization step variable by factors of three. By choosing the slope thresholds appropriately, the low-resolution error was kept outside the 20 kHz bandwidth. The number of samples produced by the variable-resolution ADC, as compared with a fixed-resolution one, was found to be lower by a maximum factor of about eight. The decrease in the number of samples can be expected to result in a decrease in the power dissipation of a following CT DSP.

A method to implement variable resolution on a silicon chip, using variable comparison windows, has been presented in [75].

A very different approach [53] for reducing the timing resolution required of the hardware is now explained, with the help of Figure 15.10. Consider a signal varying very fast (e.g., in the GHz range), as shown in Figure 15.10a. The required samples, shown as dots, will be very close to each other, and the least significant bit (LSB) of the representation in Figure 15.2b will have to toggle extremely fast, as shown in Figure 15.10b. To avoid this, one can consider per-level encoding, in which a separate comparator is used for each crossing level, as shown in Figure 15.10c. This removes the previous problem, as shown in Figure 15.10d for the left-hand lobe of the signal; however, if a local extremum of the signal happens to be around a crossing level, as shown in the right-hand lobe, the resulting two samples can be very close to each other, and thus the LSB will again have to toggle extremely fast, as shown in the right part of Figure 15.10d. This problem can be avoided by using a very different type of encoding, shown in Figure 15.10e.

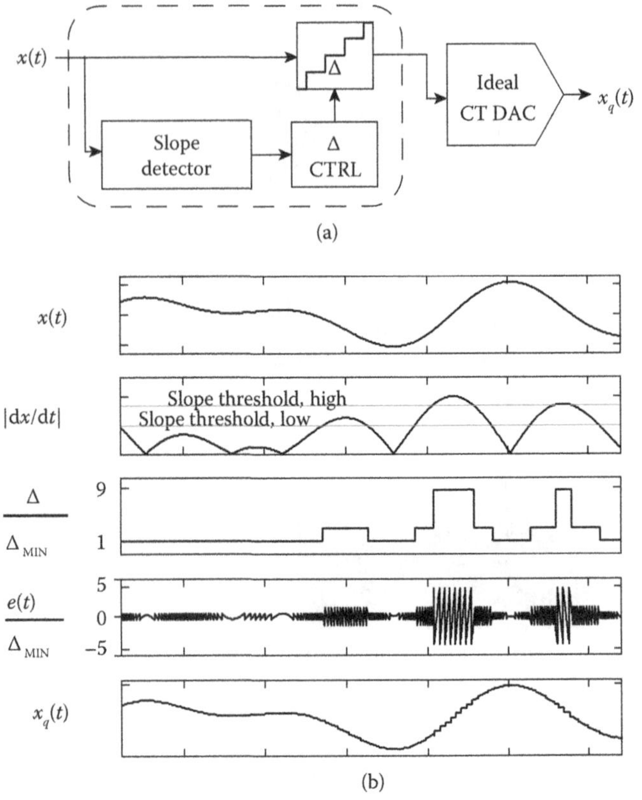


**FIGURE 15.9**
(a) Block diagram of a variable-resolution CT ADC and (b) example signals for an ADC with three resolution settings. Δ is the quantization step. (From M. Kurchuk and Y. Tsividis, *IEEE Transactions on Circuits and Systems I*, 57(5), 982–991 © 2010 IEEE.)

In it, flip-flops are used to generate two signals from the LSB: one ($R_m$) toggles on the up-going transitions of the LSB, and the other ($F_m$) on the down-going ones. These two signals now contain all information in the LSB (which can be reconstructed from them using an XOR operation) but, unlike the LSB waveform, they do not involve very narrow pulses, and thus the requirements on the following hardware are greatly reduced. The price to be paid is that each of the two signals must be processed separately, and that such processing must be carefully matched for the two paths involved.

The authors feel that the approaches described here are only the beginning, and that reduction at much higher levels should eventually be possible, depending on the method of reconstruction used.

## 15.3 Reconstruction

Reconstruction of a signal from its level-crossing samples $(t_k, x(t_k))$ need be no different from reconstruction of other types of nonuniformly sampled signals [7], and can be done in several ways. One can construct a piecewise-constant waveform from the samples, something that a quantizer inherently does (Figure 15.3a). Better performance is possible, albeit with great computational effort; if the average sampling rate exceeds twice the bandwidth of the signal, perfect reconstruction is, in theory, possible [81,82], albeit with considerable computational effort. The classical relation between quantization resolution and achievable reconstruction error no longer holds; arbitrarily small error can in principle be achieved from only a few quantization levels, provided there are enough samples to satisfy the above criterion on average sampling rate, and given sufficient computational power [83]. Intuitively, the reason for this is that, even with a few levels in Figure 15.2a, *the sample values are exact*, unlike the case in conventional, uniform sampling plus quantization.

A detailed discussion of "exact" reconstruction techniques can be found elsewhere in this book [84]. These are suitable mostly for offline applications, unless a way is found to drastically speed up the required computation time. In the present chapter, we will instead emphasize real-time applications, and thus simple piecewise-constant reconstruction will be assumed throughout.

## 15.4 Derivative Level-Crossing Sampling

As has been mentioned in Section 15.1, several applications require ultra-low-power acquisition, signal processing, and transmission. Most practical level-crossing sampling schemes are based on zero-order-hold reconstruction at the receiver [14,46,47,52,54,64,75,78,85,86]. Other schemes employ computationally intensive reconstruction techniques [7,83]; however, in applications where the receiver is on a very tight power budget, such techniques constitute a serious overhead. One can consider first-order reconstruction which might result in a smaller quantization error. Unfortunately, current first-order reconstruction techniques are noncausal: to know the signal value at a given instant between two samples, one needs to know the value of the sample following that instant. The corresponding storage need and computational effort can result in significant hardware overhead. First-order prediction techniques can be used to avoid the above noncausality, but those can result in discontinuities, and they, too, imply a significant computational overhead.

Derivative level-crossing sampling (DLCS) is an LCS technique that automatically results in piecewise-linear reconstruction in real time, with no need for storage, meant for applications in which both the transmitter and the receiver are on a tight power budget. This technique

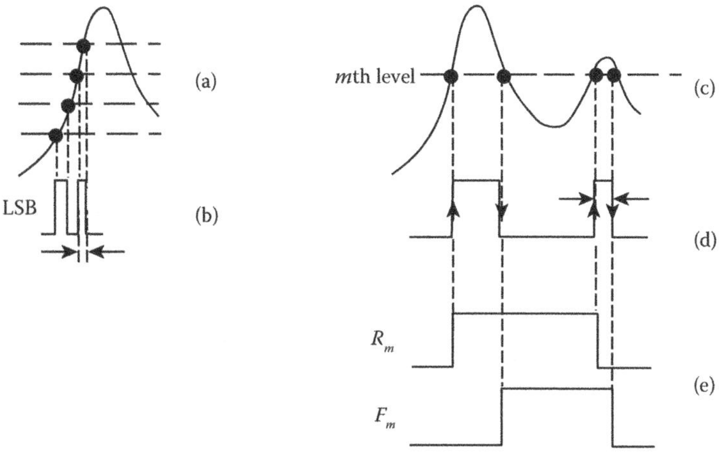


**FIGURE 15.10**
(a) Signal, threshold levels, and level-crossing samples, (b) corresponding LSB, (c) signal, threshold level, and per-level samples, (d) corresponding per-level digital signal, and (e) encoding that avoids high time-resolution requirements.

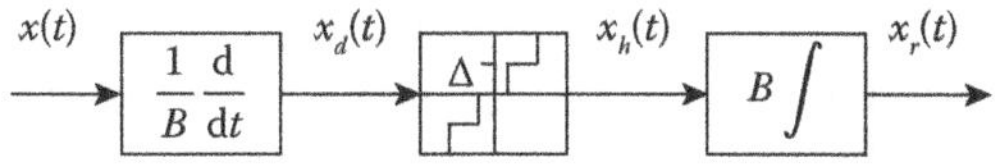


**FIGURE 15.11**
Principle of a derivative level-crossing sampling and reconstruction. (From P. Martinez-Nuevo, et al., *IEEE Transactions on Circuits and Systems II*, 62(1), 11–15 © 2015 IEEE.)

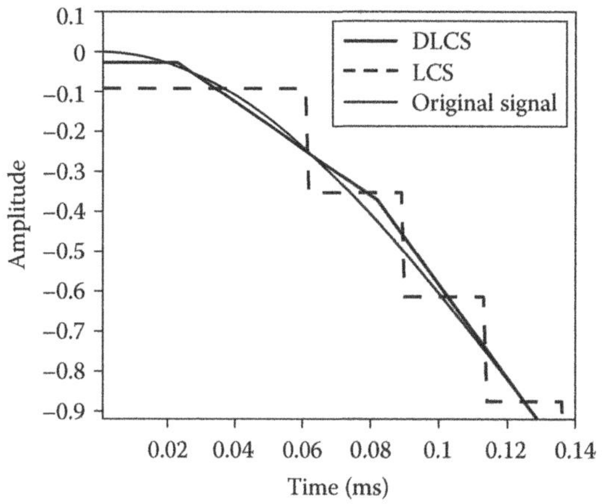


**FIGURE 15.12**
Blow-up of DLCS (first-order) and LCS (zero-order) reconstruction for a full-scale sinusoidal input signal at 2 kHz. (From P. Martinez-Nuevo, et al., *IEEE Transactions on Circuits and Systems II*, 62(1), 11–15 © 2015 IEEE.)

is discussed in detail in Reference [87], from which the material in this section is taken with permission from IEEE (copyright © 2015 by IEEE). The DLCS principle is shown in Figure 15.11. At the transmitter, the input is scaled and differentiated, and the result is level-crossing-sampled. At the receiver, the samples are zero-order-held and integrated, thus compensating for the differentiation. Thanks to integration, the scheme inherently achieves first-order reconstruction, leading to a lower reconstruction error, in real time, without the need of any linear predictor or noncausal techniques. This can be seen in Figure 15.12, which compares the output of the system to that of an LCS system with zero-order-hold reconstruction and to the original signal. As explained in Section 15.2.2, the operations of LCS and zero-order hold together are conceptually equivalent to quantization [18,43]; thus, for the purpose of analysis, the derivative can be directly quantized as shown in Figure 15.11. We assume that the input signal $x(t)$ satisfies a zero initial condition, $x(0) = 0$ (as in delta-modulated systems [50]), and is bandlimited to $B$ rad/s, bounded so that $|x(t)| \leq M$, where $M$ is a positive number; using Bernstein's inequality [88, theorem 11.1.2], we conclude that $|dx/dt|$ is bounded by $BM$. Therefore, the quantizer has an input range of $[-M, M]$.

The first-order reconstruction in DLCS shapes the quantization error with an integrator transfer function. Therefore, the high-frequency quantization artifacts are severely attenuated, whereas the low-frequency ones are not. For single-tone inputs, this results in a significantly lower mean-square error (MSE) than classical LCS of the same resolution over most of the frequency range, with the MSE increasing at low input frequencies. To improve the performance for low-frequency inputs, the adaptive-resolution (AR) DLCS, shown in Figure 15.13, has been proposed [87]. The slow-changing portions of the input derivative are quantized with a high resolution, and the fast-changing portions are quantized with a low resolution; the second derivative of the input is used to identify the rate of change of the input derivative so as to control the resolution. A high resolution around the slow-changing portions of the

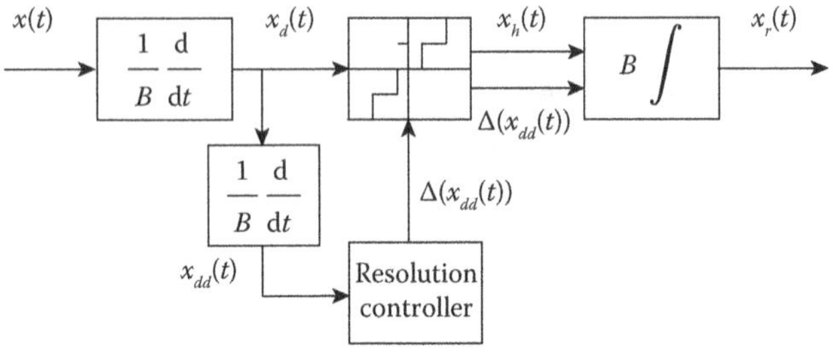


**FIGURE 15.13**
Principle of adaptive-resolution derivative level-crossing sampling and reconstruction. $\Delta(x_{dd}(t))$ denotes the variable quantization step size, which depends on the value of $x_{dd}(t)$. (From P. Martinez-Nuevo, et al., *IEEE Transactions on Circuits and Systems II*, 62(1), 11–15 © 2015 IEEE.)

derivative keeps the low-frequency quantization error components low enough such that post integration, they do not degrade the signal-to-error ratio (SER). Coarse quantization around the fast-changing regions results in a higher quantization error, but the integrator shaping significantly attenuates it during reconstruction such that it has negligible effect on the SER. A detailed discussion of spectral properties of level-crossing-sampled signals can be found in [45].

For a given SER requirement, both DLCS and AR DLCS can result in a substantial reduction in the number of samples generated, processed, and transmitted as compared with classical LCS. While improvements depend on the characteristics of the input signal, they have been consistently observed for a variety of inputs, especially for AR DLCS [87].

Both DLCS and AR DLCS exploit the varying spectral context of the input in addition to its amplitude sparsity to generate very few samples for a given MSE. These schemes have not been demonstrated in practice at the time of this writing. If they turn out to work as expected from theory, they may prove advantageous in the development of sensor networks, which consist of sensor nodes that are on a very tight power budget and spend most power in communication [2]; a reduction in the number of generated samples can substantially lower the total power dissipation by minimizing the power spent by the transmitter node in processing and transmission and that by the receiver node in reception and processing.

## 15.5 Continuous-Time Digital Signal Processing

### 15.5.1 Principle

Once a quantity has been sampled using event-driven means, it typically needs to be processed, for example, in a biomedical device in order to deduce its properties of interest, or in a local network node in preparation for transmission. If a conventional DSP were used for this, the potential advantages of the event-driven approach would be largely wasted. It has been shown that fully event-driven digital signal processing of level-crossing-sampled, or equivalently CT-quantized, signals is possible [31]. This can be done in CT [31,42–44]; or, the time can be quantized and then a clocked processor can be used [89], although then one has to deal with generating and distributing a high-frequency clock signal, and with additional frequency components in the spectrum (Section 15.5.3). In another approach, the time can be divided into frames, within which a fixed sampling rate is used [13,15,90, 91]. Here we will emphasize the CT approach, which we will refer to as CT DSP (but it is to be understood that the properties discussed also apply approximately in the case of finely discretized time). Due to its emphasis on time information, CT DSP is a form of time-domain signal processing, or time-mode signal processing [92,93].

The processing of binary signals which are functions of CT is, in a sense, "the missing category" of signal processing, as shown in Table 15.1. The first three rows under the title correspond to well-known types of processors (a representative type of the third row is switched-capacitor filters). The last category in the table is the one discussed here. If one considers the names of the first three categories, it follows by symmetry that the appropriate name for the last category is "continuous-time digital." This name is consistent with the fact that this category involves binary signals that are functions of CT (see Figure 15.2b).

CT binary signals, like the ones in Figure 15.2b, can be processed in continuous time using CT delays, CT multipliers, and CT adders. The multipliers and adders can be implemented using unclocked combinational circuits, using asynchronous digital circuitry.

Consider as a prototype the CT finite impulse response (FIR) analog structure in Figure 15.14a. The coefficient multipliers, the summer, and the delay lines are all CT. An input $x(t)$ can in principle be processed by

**TABLE 15.1**
Signal Processing Categories

| Time | Amplitude | Category |
|---|---|---|
| Continuous | Continuous | Classical analog |
| Discrete | Discrete | Classical digital |
| Discrete | Continuous | Discrete-time analog |
| Continuous | Discrete | Continuous-time digital |

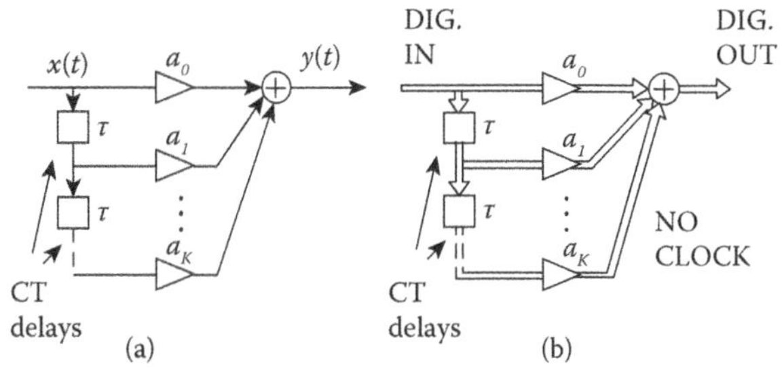


**FIGURE 15.14**
(a) CT FIR analog filter; (b) corresponding CT digital filter. (From Y. Tsividis, *IEEE Transactions on Circuits and Systems II: Express Briefs*, 57(8), 577–581 © 2010 IEEE.)

this structure, resulting in an output

$$y(t) = \sum_{k=0}^{K} a_k x(t - k\tau), \quad (15.1)$$

where τ is the delay of each delay element. Taking the Laplace transform of both sides of this equation, we find

$$Y(s) = \sum_{k=0}^{K} a_k X(s) e^{-sk\tau}. \quad (15.2)$$

Thus the transfer function, $H(s) = Y(s)/X(s)$, is

$$H(s) = \sum_{k=0}^{K} a_k e^{-sk\tau}. \quad (15.3)$$

Using $s = j\omega$, with $\omega = 2\pi f$ the radian frequency, we obtain the frequency response:

$$H(j\omega) = \sum_{k=0}^{K} a_k \left(e^{j\omega\tau}\right)^{-k}, \quad (15.4)$$

which can be seen to be periodic with period $2\pi/\tau$. It is thus seen that the frequency response is identical to that of a corresponding discrete-time filter. Thus, well-established synthesis techniques developed for discrete-time filters can directly be used for CT digital filters.

We now note that if, instead of using $x(t)$ as the input, we feed (in theory only) the structure of Figure 15.14a with a quantized version $x_q(t)$ of the input (see Figure 15.2a), the structure will produce a corresponding piecewise-constant output. Finally, if we instead have the CT digital representation of $x_q(t)$, shown in Figure 15.2b, we can replace all elements in Figure 15.14a by corresponding CT digital ones, as shown in Figure 15.14b. A rigorous proof of this is given elsewhere [42–44]. Now all signals in the processor are 0s and 1s, but they are functions of continuous, rather than discrete, time; they are composed of piecewise-constant "bit waveforms" like those in Figure 15.2b. This system then is a continuous-time digital signal processor, or CT DSP. The output of this processor can be converted to analog form using a CT D/A converter (DAC), which is basically a binary-weighted CT adder. The result is shown in Figure 15.15; this structure is meant to be shown in principle only; other, more efficient architectures will be discussed below. It can be shown that the frequency response of the entire system consisting of the CT A/D converter, the CT DSP, and the CT D/A converter is still given by (15.4) [42–44].

For the delay segments, one can use a cascade of digital delay elements operating without a clock, as shown in Figure 15.16; in the simplest case, the elemental delay elements shown can be CMOS inverters (possibly slowed down) [31], but better performance is possible using special techniques [80,94,95]. A number of such elements will need to be used, as opposed to just a single slow one, when the delay chain must be agile enough to handle many signal transitions within each delay interval. The delay can be automatically tuned using digital calibration and a time reference; this can be a clock, perhaps already used elsewhere in the system, or can be turned off after tuning [96]. In the absence of a clock, an RC time constant involving an internal

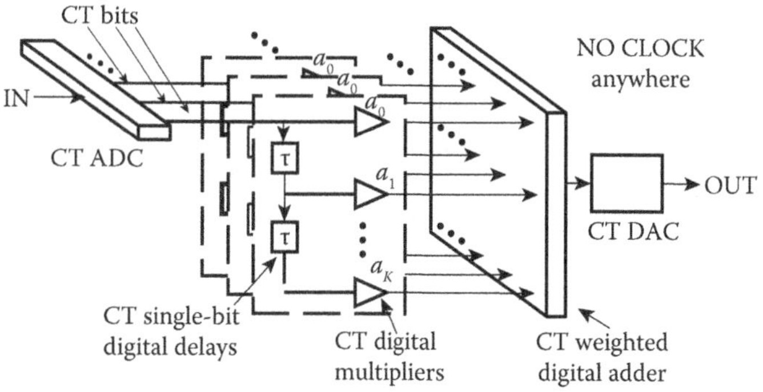


**FIGURE 15.15**
CT ADC, CT DSP, and CT ADC (principle only). (From Y. Tsividis, *Proceedings of the IEEE International Conference on Acoustics, Speech, and Signal Processing*, vol. 2, pp. 589–592 © 2004, IEEE.)

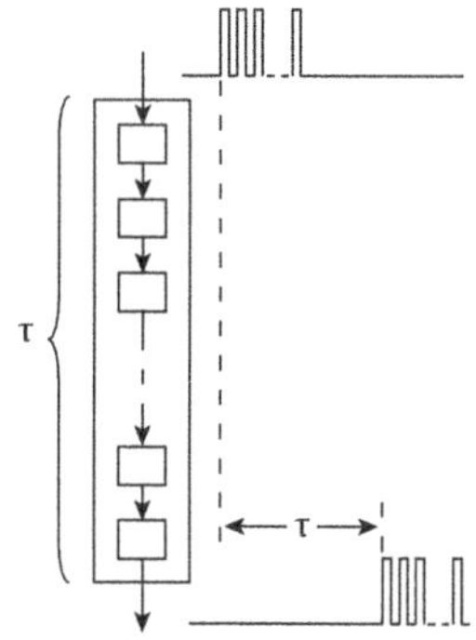


**FIGURE 15.16**
CT digital delay line composed of CT elemental delay elements.

temperature-insensitive capacitor and an external low-temperature-coefficient trimmable resistor can be used as a time reference. Such techniques have been used for decades by the industry for the automatic calibration of integrated analog filters [97].

The output of the adder in Figure 15.15 can involve arbitrarily small intervals between the signal transitions in the various paths, due to the unclocked arrival of various delayed versions of the input. If the logic cannot handle such intervals, they will be missing in the output code. Fortunately, such narrow pulses involve very little energy which falls mostly outside the baseband. Thus, as the input frequency is raised beyond the speed limits of the hardware, the degradation observed is graceful. Design considerations for the system of Figure 15.15 are discussed elsewhere [43,54]. It is very important to avoid glitches by using handshaking throughout [54]; an example will be seen in Section 15.7. Time jitter must also be carefully considered [54,94,98,99], to make sure that the noise power it contributes to the output is below that expected from the quantization distortion components. The challenging design of the adder can be largely bypassed by using a semi-digital approach employing multiple small D/A converters as shown in Figure 15.17, similar to that proposed for clocked systems [100].

It should be mentioned that the block diagrams in this section illustrate the principle of CT DSP and may not necessarily correspond to the system's microarchitecture, which has recently evolved. In [96], data are separated from the timing information inside each delay segment. When a sample enters the segment, the data is stored in a memory and remains idle while the timing goes through multiple delay cells. When the timing pulse finally exits the segment, it picks up the corresponding data and transfers it to the next segment, and this process is repeated. This timing/data decomposition is not only significantly more energy efficient, but it is also very scalable; to build a similar CT DSP for different data width, one must only change the size of the segment memories but can leave the delay part unchanged. A test chip based on this approach is described in Section 15.7.2.

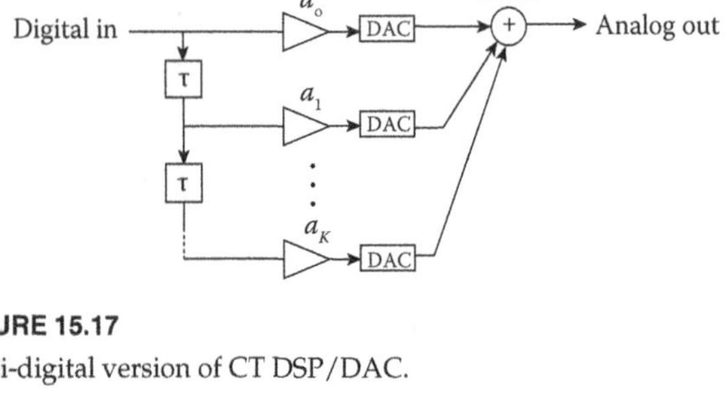


**FIGURE 15.17**
Semi-digital version of CT DSP/DAC.

As already mentioned, the frequency response in (15.4) is periodic; to attenuate the undesired frequency lobes, an output or input CT analog filter can be used, as is the case in conventional systems. Note, though, that the periodicity in the frequency response has nothing to do with aliasing; in fact, there is no aliasing, as has been discussed in Section 15.2.3. This is illustrated in Figure 15.18, where a CT DSP is compared with a classical, discrete-time one. It is seen that a signal applied at a certain input frequency results in a single frequency component at the same frequency at the output, no matter whether the input frequency is in the baseband or not. Such a property is, of course, absent in the case of classical, discrete-time systems involving uniform sampling.

Advantages occur in the time domain as well. CT DSPs react immediately to input changes, unlike classical systems which may not catch such changes until the following sampling instant; this makes them especially well suited for use in very fast digital control loops. Thus, application of CT DSP in DC-DC converters has made possible the reduction of the required very large filter capacitance, and the associated cost, by a factor of three to five [101,102].

Small mismatches in the CT delays cause small errors in the frequency response [103]; as will be seen in

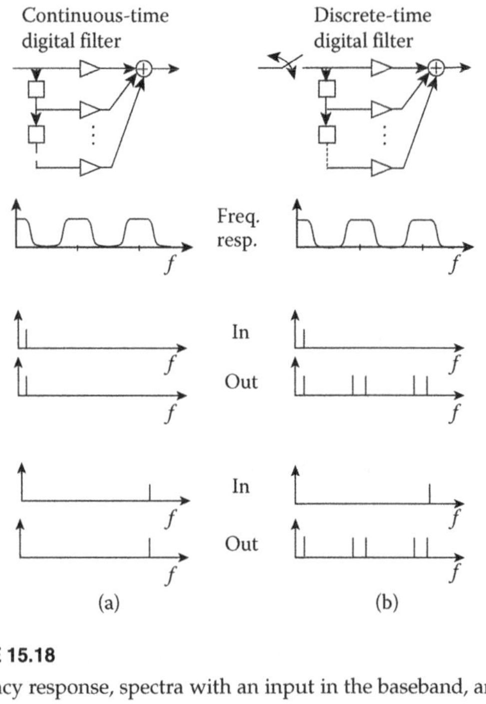


**FIGURE 15.18**
Frequency response, spectra with an input in the baseband, and spectra with an input in a higher lobe of the frequency response, for (a) a CT digital filter, and (b) a discrete-time digital filter preceded by a uniform sampler. (From Y. Tsividis, *IEEE Transactions on Circuits and Systems II: Express Briefs*, 57(8), 577–581 © 2010 IEEE.)

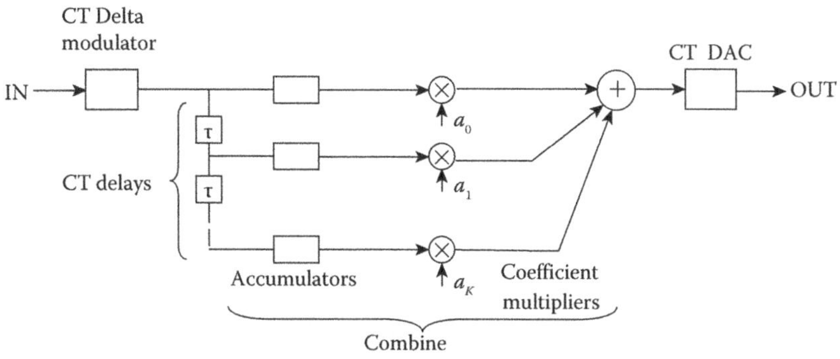


**FIGURE 15.19**
Delta mod CT ADC/DSP/DAC.

Section 15.7, such errors are not significant in practice. The CT delays can also be made different from each other on purpose, to modify the frequency response (e.g., change the size of a lobe compared to the one in the baseband), but analytical techniques for this are mostly lacking. The optimization of CT DSPs using delay adjustment has been considered in [104].

If an asynchronous delta modulator like the one in Figure 15.5 is used at the input, the processing can be done in delta-mod, as is the case for discrete-time digital filters [105]. The accumulators, required to recover the original signal from the delta mod stream, can be combined with the coefficient multipliers [54,72]. Such an implementation is shown in Figure 15.19.

The computer-aided simulation of CT DSPs can be time-consuming. Special techniques have been proposed for significantly reducing simulation time [106].

Infinite impulse-response CT DSPs have been considered elsewhere [44,107], but have not been adequately tested so far.

Several applications for CT DSPs have been proposed, including their embedding in ADCs to enhance their performance [93,108]. A single-bit version of a CT DSP has been used in communications chips [68,109,110].

### 15.5.2 "Continuous-Time" versus "Asynchronous"

A CT DSP uses asynchronous logic blocks and handshake techniques borrowed from asynchronous logic (see Section 15.7). Nevertheless, it is not appropriate to call it an asynchronous DSP. Asynchronous DSPs (see, e.g., [111]) are discrete-time systems, processing sequences of data; the time intervals between samples do not usually represent important information. In contrast, in a CT DSP the timing information is critical and must be carefully preserved. Thus, all signals in Figure 15.20 are the same to an asynchronous DSP; however, they are all distinct to a CT DSP, since the times of occurrence of each transition (or the time intervals between them) are an integral part of the CT digital signal representation.

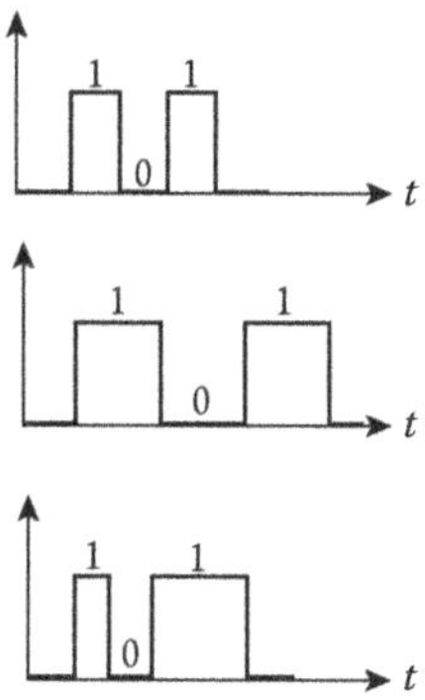


**FIGURE 15.20**
Three time waveforms. In the context of asynchronous processing, all three represent the same signal; in the context of CT DSP, they represent three distinct signals.

### 15.5.3 Time Discretization

CT DSPs are best suited to real-time applications, as their output cannot be stored. If storage is desired, the output of such processors must be sampled. This is considered elsewhere in this book [45]. In another approach, the entire operation can be replaced by pseudo-continuous operation, in which a high-frequency clock is involved throughout and the time axis becomes finely quantized [89]. As mentioned in Section 15.2.4, it is possible to quantize the time from the very beginning, in the A/D converter [11,12,14,52]. This results in spectral components in between the harmonics in Figure 15.3a. With coarse time quantization, the signal spectrum begins to resemble the one in Figure 15.3b.

Time quantization requires a high-frequency clock (with an ensuing effect on power dissipation), which may make the design of such systems more straightforward, but which one would ideally rather avoid, if true event-driven operation is sought. In fact, to reach the performance of CT DSPs in terms of spectral properties and transient response speed, the clock would have to be infinitely fast, which is infinitely difficult to implement. But this would be equivalent to not quantizing the time at all, which is relatively easy; it just implies CT operation and can be achieved without a clock, as has already been discussed.

## 15.6 Versatility of CT DSP in Processing Time-Coded Signals

In this chapter, the name "time coding" [12] has been used in relation to signals derived using level-crossing sampling. However, other types of time coding exist. An example is pulse-time modulation, defined in [112] as a class of techniques that encode the sample values of an analog signal onto the time axis of a digital signal; this class includes pulse-width modulation and pulse position modulation. Another time-coded signal can be produced by click modulation [113]. A different technique of coding a signal using only timing is described in [114]. (We also note, in passing, that implementing filter coefficient values using only timing is discussed in [32].) Yet another example is asynchronous sigma-delta modulation, which uses a self-oscillating loop involving an integrator and a Schmidt trigger, shown in Figure 15.21. This system is usually attributed to Kikkert [115], but was in fact proposed earlier by Sharma [116]; for this reason, we refer to it as the Sharma–Kikkert asynchronous sigma-delta modulator. The circuit has been further analyzed in [117], and the recovery of its output has been discussed in [118]. This system produces a self-oscillating output even in the absence of input and is thus not strictly even driven. Nevertheless, this signal can be processed with a CT DSP [119–121]. In fact, most time-coded binary signals, event-driven or not, asynchronous or synchronous, can be processed by a CT DSP. This is because a CT DSP "looks at all time."

## 15.7 Experimental Results

This section summarizes the design and measurements of several test chips designed in order to verify the principles of CT data acquisition and DSP.

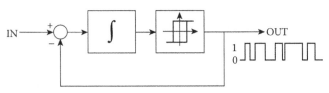

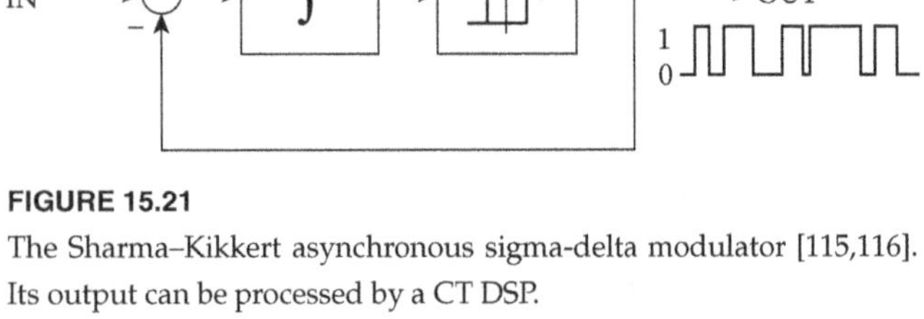


**FIGURE 15.21**
The Sharma–Kikkert asynchronous sigma-delta modulator [115,116]. Its output can be processed by a CT DSP.

### 15.7.1 A Voice-Band CT ADC/DSP/DAC

We begin with a 8b, 15th-order, finite-impulse response (FIR) continuous-time ADC, continuous-time DSP, and continuous-time DAC, collectively referred to as CT ADC/DSP/DAC, targeting speech processing [54]. A block diagram of the chip is shown in Figure 15.22. The input is applied to terminals INH and INL. The CT ADC is a variation of the asynchronous delta modulator (Section 15.2.4) [48–50,52]. The ADC's output consists of two signals (CHANGE, UPDN) that indicate the time and direction of the input change (which is equivalent to the signal in Figure 15.2c). Each of these signals takes the values of 1 or 0, and the two together are called a token. The CT ADC uses two asynchronous comparators to continuously monitor the input in order to generate signal-driven tokens.

The comparator levels are determined by the input and the signals as provided by a feedback CT DAC. Due to continuous tracking, the thresholds only move up (if dictated by the INC signal) or down (if dictated by the DEC signal) one level at a time. To coordinate the movements of tokens into the CT DSP, asynchronous handshaking (ACK signals) is used. Hysteresis is used in the comparators for noise immunity.

The CT ADC is followed by a CT DSP, composed of 15 CT delay elements, each a serial chain of individual asynchronous delay cells [94], and accumulator/multiplier blocks to do the token summation and weighting. The delay segments are implemented along the lines of Figure 15.16. The delay taps provide an 8b representation of the (now) delayed input signal, and the weighting is done by the multiplication by the programmable filter coefficients. The outputs of all the accumulator/multiplier blocks are added using an asynchronous carry-save adder to create the final filter output, which goes to the CT DAC.

A photograph of the fabricated chip implementing the system of Figure 15.22 is shown in Figure 15.23. The 8b CT ADC/DSP/DAC was fabricated in a UMC 90 nm CMOS process, operating with a 1V supply. The active area (sum of outlined areas in Figure 15.23) is 0.64 mm$^2$, with the 15 delay elements dominating the area (0.42 mm$^2$). The power consumption also reflects this dominance: when configured as a low-pass filter and processing a full-scale 1 kHz input tone, 60% of the total power consumption is dissipated by the delay elements.

In order to demonstrate some important properties of CT DSPs, we show in Figure 15.24 a frequency response and output spectra of the CT ADC/DSP/DAC when configured as a low-pass filter; see the caption for details. As expected from Section 15.5.1, the frequency response is periodic. In the spectral results, however, it can be seen that the applied 2 kHz input tone does not

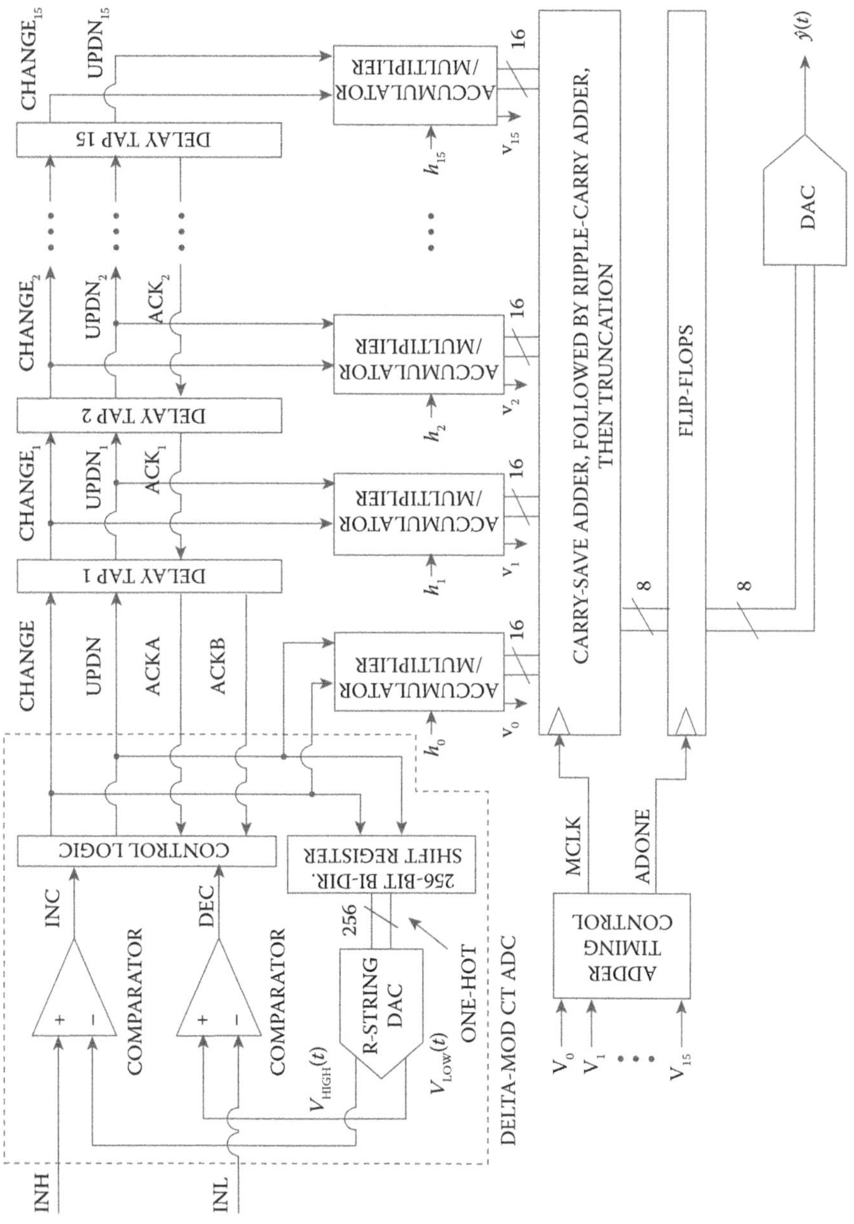


**FIGURE 15.22**
Block diagram of entire CT ADC/DSP/DAC chip. (From B. Schell and Y. Tsividis, *IEEE Journal of Solid-State Circuits*, 43(11), 2472–2481 © 2008 IEEE.)

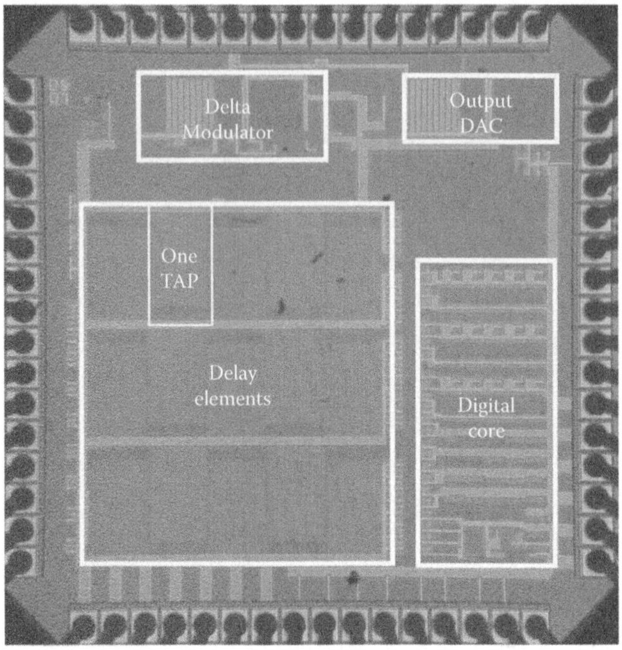


**FIGURE 15.23**
Chip photograph of 8b CT ADC/DSP/DAC. The total area of the outlined blocks is 0.64 mm$^2$. (From B. Schell and Y. Tsividis, *IEEE Journal of Solid-State Circuits*, 43(11), 2472–2481 © 2008 IEEE.)

result in an alias tone in the second lobe of the response. Similarly, a 39 kHz input tone, which places it in this second lobe, does not produce an alias tone in the baseband. This verifies the behavior illustrated in Figure 15.18.

Figure 15.25 shows an input speech signal plotted versus time, together with the measured instantaneous power consumption of the CT ADC/DSP/DAC. As can be seen, when the input is quiet (no activity), the power consumption drops to a baseline level (needed for circuit biasing). When the input becomes more active, the power consumption automatically increases, due to the processing of more tokens per second. This result stresses the utility of CT processing for signals that have varying levels of activity; the power scaling is inherent in the CT nature of the circuit. Such dynamics are possible in conventional circuits, but they require a control of some sort to detect the measure of activity of the input and perform power management; here this benefit is automatic. The actual power dissipation levels in Figure 15.25 are rather high in this early test chip; lower-power CT DSPs have now been designed (see below). A method to implement sample rate reduction based on Section 15.2.6, and thus reduce power dissipation, has been described in [75]; several other circuit techniques for power reduction are described in that reference.

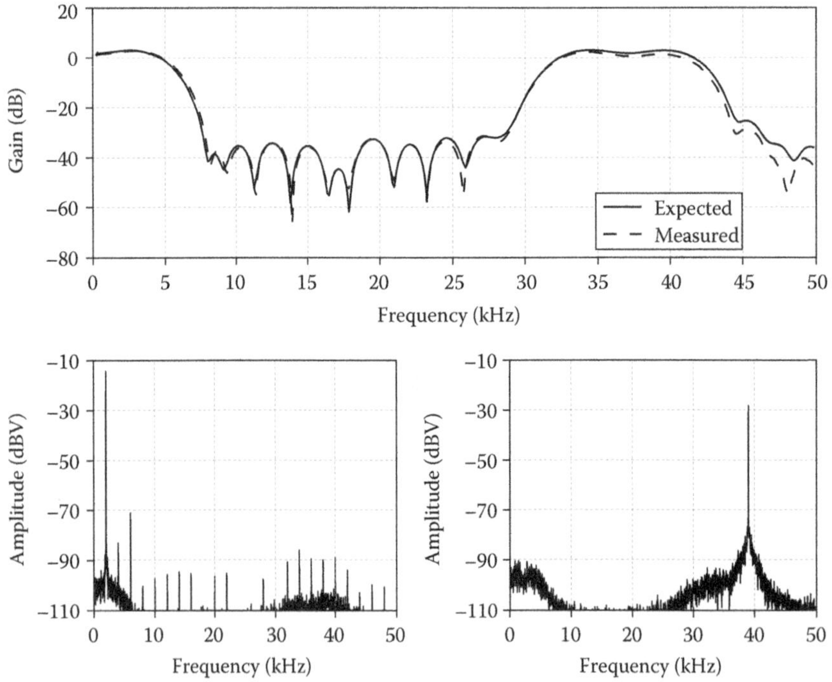


**FIGURE 15.24**
Measurement results from the 8-bit CT ADC/DSP/DAC chip, with DSP set to a low-pass filter using element delays of 27 μs [1/(37 kHz)]. Top: Frequency response. Bottom left: Output spectrum for a full-scale 2 kHz input. Bottom right: Output spectrum for 39 kHz input at −14 dB relative to full scale. No aliasing is observed. (From B. Schell and Y. Tsividis, *IEEE Journal of Solid-State Circuits*, 43(11), 2472–2481 © 2008 IEEE.)

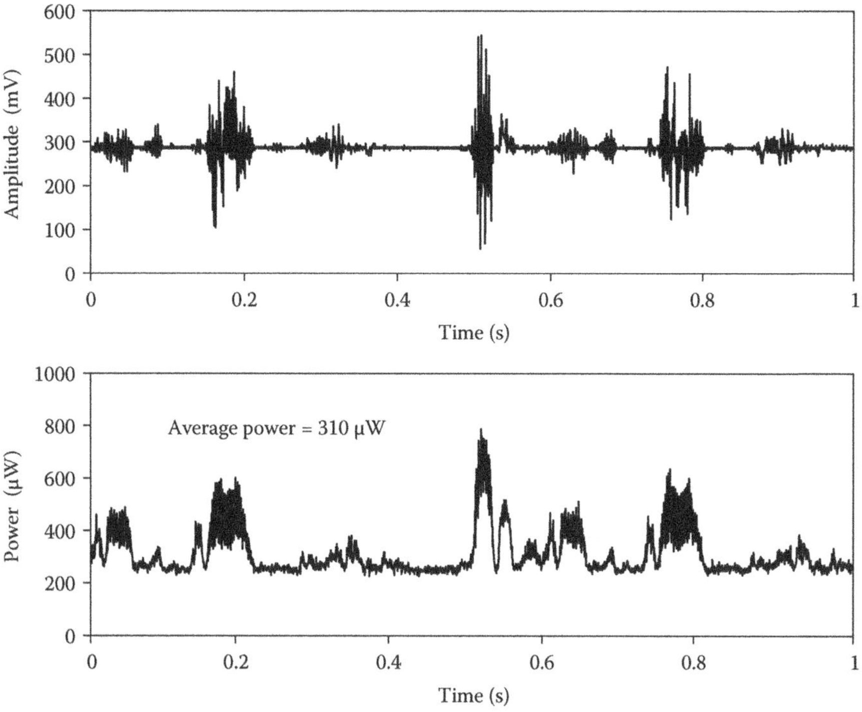


**FIGURE 15.25**
Plot of input speech signal (top) and instantaneous power consumption (bottom) of the CT ADC/DSP/DAC chip. (From B. Schell and Y. Tsividis, *IEEE Journal of Solid-State Circuits*, 43(11), 2472–2481 © 2008 IEEE.)

### 15.7.2 A General-Purpose CT DSP

A more recent general-purpose CT DSP test chip [96] is based on an entirely new microarchitecture. It provides a highly programmable CT DSP core (it does not contain an ADC or DAC), which can be connected to various ADCs and process signals of different sample rates, synchronous or asynchronous. The core consists of asynchronous (i.e., clockless) digital components, along with a calibrated delay line, allowing it to observe and respond to signals in CT. As a result, it is capable of processing signals in a wide variety of practical formats, such as PCM, PWM, ΣΔ, and others, both synchronous and asynchronous, without requiring any internal design changes. This property is not possible for synchronous DSP systems. The filter's frequency response is programmed only once, and is not affected by changes in the ADC sample rate. The chip also accommodates a larger data width than the one in [54]; it supports 8-bit-wide signals, and can be extended to arbitrary width, while the chip in [54] only supports 1-bit wide signals (resulting from asynchronous Δ modulation). The chip exhibits an SER which, for certain inputs, exceeds that of clocked systems.

Figure 15.26 shows a top-level view. The chip can receive input from a variety of ADCs at any sample rate up to 20 Msamples/s, and implements a 15th-order FIR filter. Fifteen delay segments generate delayed input versions; all copies, along with the input itself, are then weighted with proper coefficients by asynchronous multipliers and summed by a 16-way adder to form the digital output. The latter is sent to a DAC to create the analog output. Tuning is required for all CT DSPs, to ensure that the delay of the segments is equal to the desired value; this chip is the first CT DSP to incorporate on-chip automated tuning.

A "sample" sent to the chip, shown as $IN_D$ in the figure, consists of 8 (or more) data bits representing the magnitude information. Timing information is bundled with the data field; each new and valid sample is indicated by a transition on the timing signal $IN_R$. A critical requirement of the design is that timing, that is, spacing between consecutive samples, is preserved inside the CT DSP, by using a calibrated delay line, which is divided into a series of delay segments, as shown in the figure. The time interval between successive samples is maintained by the CT DSP core so as not to distort the CT signal processing function. All communication internal to the DSP employs asynchronous handshaking to indicate the timing and validity of samples; one example is a delay segment sending the latest sample to the next segment and to a multiplier.

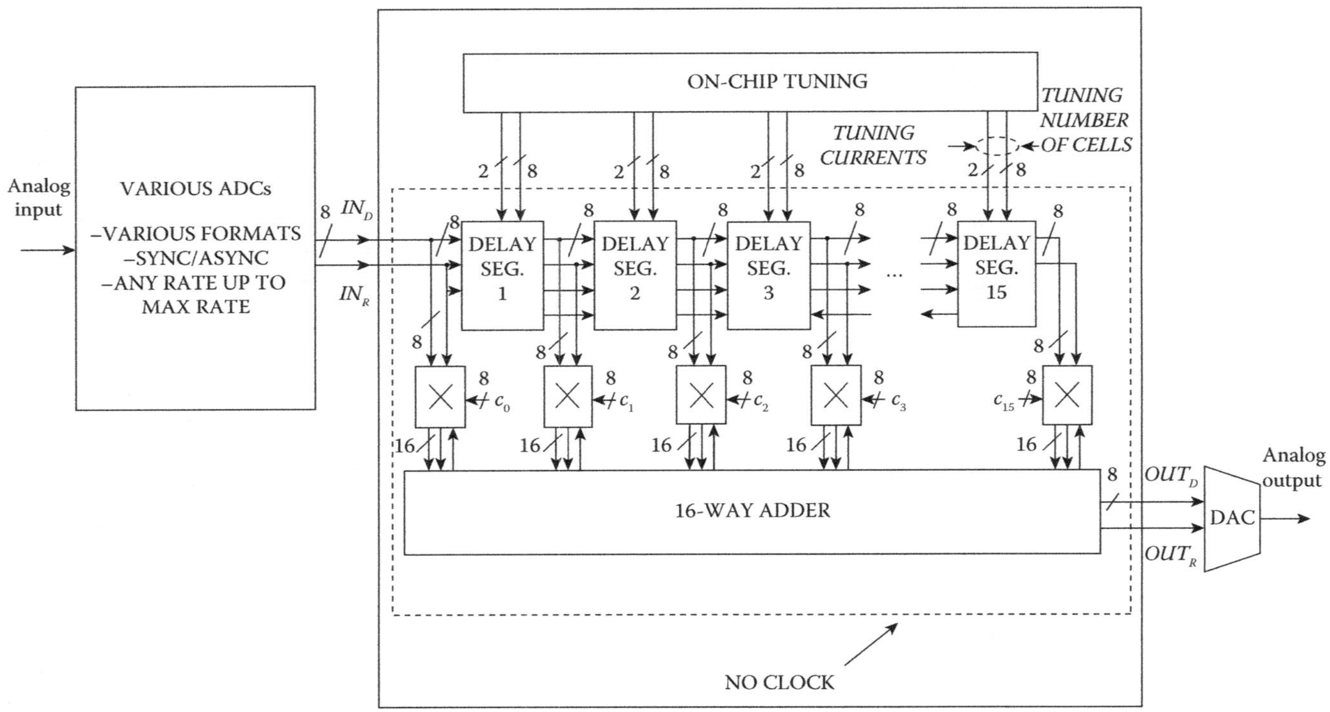


**FIGURE 15.26**
Top-level view of the flexible CT digital FIR filter as part of a ADC/DAC/DSP system. (From C. Vezyrtzis, et al., *IEEE Journal of Solid-State Circuits*, 49(10), 2292–2304 © 2014 IEEE.)

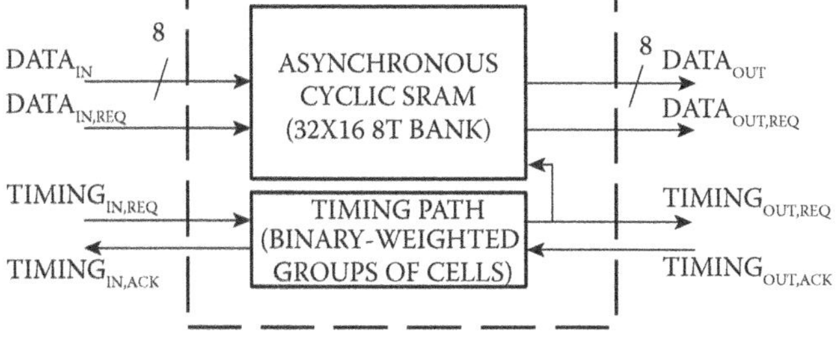


**FIGURE 15.27**
Top-level view of a delay segment, showing the segment's organization and the data and timing paths. (From C. Vezyrtzis, et al., *IEEE Journal of Solid-State Circuits*, 49(10), 2292–2304 © 2014 IEEE.)

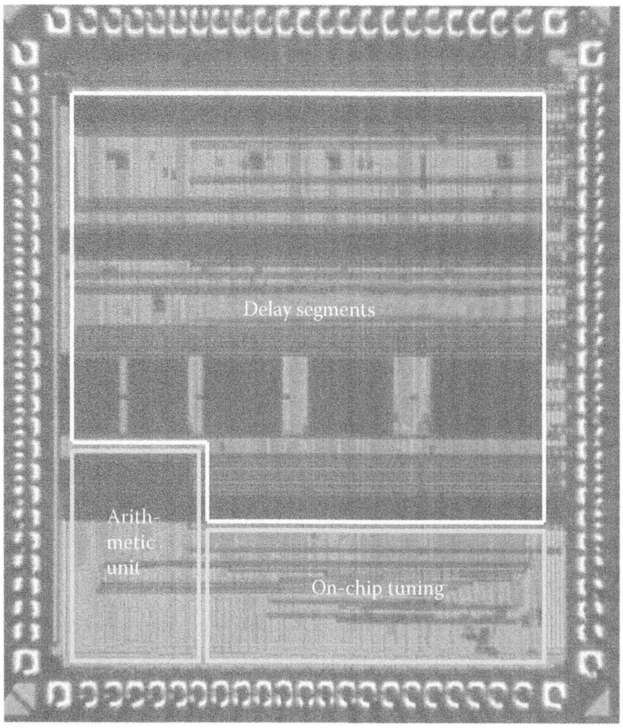


**FIGURE 15.28**
Chip photograph of the general purpose CT DSP. The core area is 5 mm$^2$. (From C. Vezyrtzis, et al., *IEEE Journal of Solid-State Circuits*, 49(10), 2292–2304 © 2014 IEEE.)

In more detail, the chip uses a calibrated pipelined delay line to provide the desired delay within the segments. Each segment is partitioned into a large number of "delay cells," as shown at the bottom of Figure 15.27. Each cell provides only a small unit of delay, and can hold at most one distinct sample at any time. This organization serves two purposes. First, the segment can hold a large number of samples inside it at a given time; this is needed when the chip is processing an input with sample spacing much smaller than the segment delay. Second, the segment can be programmed to a wide range of delay values, by selecting any combination of the different groups of delay cells (two cells, four cells, etc.).

In an early prototype CD DSP [54], data moved continually inside each segment, from cell to cell. This resulted in unnecessary data movement and wasted energy. In contrast, in the chip described here [96], this movement is entirely avoided. Instead, the timing pulse is separated from the data, and only that pulse is passed through the delay segments, as has already been mentioned in Section 15.5.1 and as shown in Figure 15.27.

Figure 15.28 shows the die photograph of the chip, as implemented in a 0.13 μm IBM CMOS technology. Due to the significant programmability and testability features incorporated in this chip, it occupies a much larger area than the one in [54]. As in that reference, the delay segments occupy the majority of the chip area, with the arithmetic (multipliers, adder) and on-chip tuning equally sharing the remaining area. A recently proposed approach [122] can be applied to the delay to significantly reduce its average power consumption, by dynamically varying its pipeline granularity to tailor it to the needs of the input sample stream.

Figure 15.29 shows the frequency response of the chip. In this experiment, the chip was programmed only once, before the measurements, and was then fed with PCM signals of three different sample rates. Despite the fact that no adjustments were made to the chip during these changes, the filter maintains its frequency response. This indicates that the response of the CT DSP is fully decoupled from the input data rate, as expected theoretically. Similar behavior was observed with other signal formats used for testing, namely, synchronous and asynchronous PWM and ΣΔ signals, and for a wide range of frequency responses. The latter were set by the chip's automated tuning blocks, controlled via a simple user interface.

The chip's power consumption grows linearly with the input rate, confirming the theoretically expected event-driven nature of CT DSPs. During idle times, the total power reduces to a baseline value resulting from leakage and circuit biasing, which is significantly reduced compared to that in [54].

### 15.7.3 A CT ADC/DSP/DAC Operating at GHz Frequencies

CT DSPs have also been tested at radio frequencies, by designing and testing a programmable 6-tap, 3-bit ADC/DSPDAC operating at frequencies up to 3.2 GHz. The chip is intended for ultra-wide-band communications [53]; it can provide flexibility in adapting

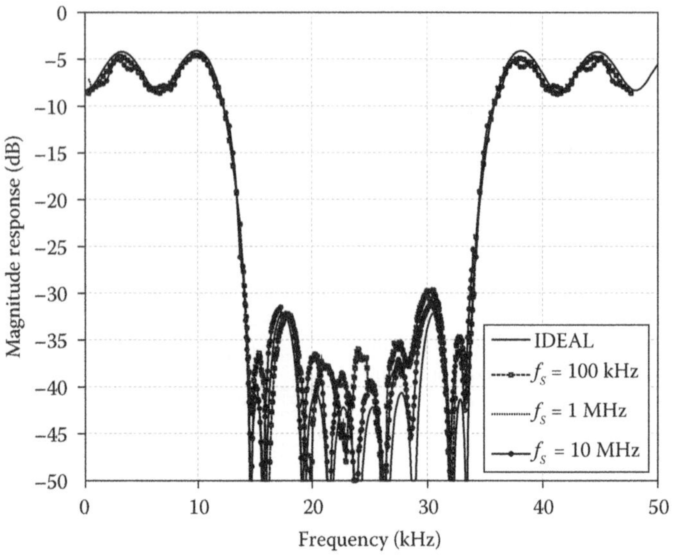


**FIGURE 15.29**
Measured frequency responses demonstrating independence from input sampling rate. PCM input, FIR filter low-pass response following automatic tuning. (From C. Vezyrtzis, et al., *IEEE Journal of Solid-State Circuits*, 49(10), 2292–2304 © 2014 IEEE.)

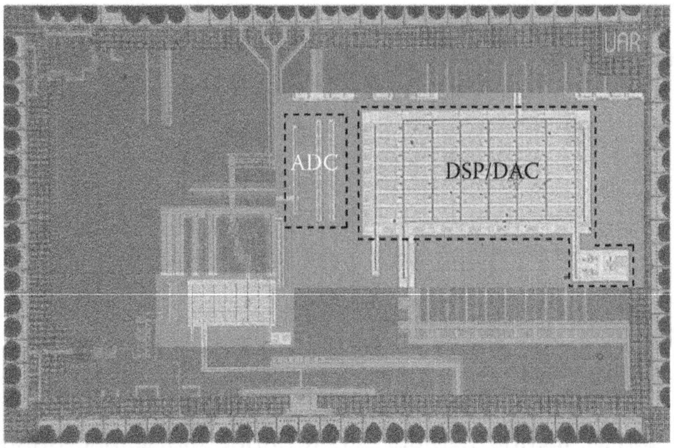


**FIGURE 15.30**
Chip photograph of the radio frequency CT ADC/DSP/DAC. The highlighted core area is 0.08 mm$^2$. (From M. Kurchuk, et al., *IEEE Journal of Solid-State Circuits*, 47(9), 2164–2173 © 2012 IEEE.)

a receiver's frequency response to the signal and interference spectrum at its input. The chip has been fabricated in 65 nm CMOS technology, and is shown in Figure 15.30. It can be seen that the core on this chip is very compact; thanks to the small number of threshold levels used and the high frequency of operation, which implies small CT delays, this area is only 0.08 mm$^2$. The chip uses the coding approach in Figure 15.10e. The power dissipation of this chip varies between 1.1 and 10 mW, depending on input activity. The chip frequency response is fully programmable. Representative frequency responses, nonlinearity measurements, and output spectra, as well as hardware design information, can be found in [53].

## 15.8 Conclusions

In contrast to conventional data acquisition and digital signal processing, event-based operations offer potential for significant advantages, if the sampling strategy is adapted to the properties of the input signal. As has been shown, true event-driven operation without using a clock is possible, both for A/D and D/A converters, and for digital signal processors. The concept of continuous-time digital signal processing has been reviewed. The known advantages of CT ADC/DSP/DAC systems in comparison to classical systems using uniform sampling and discrete time operation are the following:

- No aliasing
- No spectral components unrelated to the signal
- Lower in-band error
- Faster response to input changes
- Lower EMI emissions due to the absence of clock
- Power dissipation that inherently decreases with decreasing input activity

CT DSPs are as programmable as conventional DSPs of the same topology. However, they are best suited for real-time applications, unless their output is sampled so that it can be stored. Thus they are complementary to conventional, clocked DSPs, rather than competing with them.

The principles presented have been verified on test chips, and the results are encouraging. Further research will be necessary in order to develop more power-efficient CT ADCs and CT delay lines, to reduce the number of events needed per unit time, and to investigate other processor topologies, notably including IIR ones.

## Acknowledgment

This work was supported by National Science Foundation Grants CCF-07-01766, CCF-0964606, and CCF-1419949.

## Bibliography


[1] M. Neugebauer and K. Kabitzsch. A new protocol for a low power sensor network. In *Proceedings of the IEEE International Conference on Performance, Computing, and Communications*, Phoenix, pp. 393–399, 2004.

[2] L. M. Feeney and M. Nilsson. Investigating the energy consumption of a wireless network interface in an ad hoc networking environment. In *Proceedings of the Twentieth Annual Joint Conference of the IEEE Computer and Communications Societies*, volume 3, pp. 1548–1557, 2001.

[3] A. Yakovlev, S. Kim, and A. Poon. Implantable biomedical devices: Wireless powering and communication. *IEEE Communications Magazine*, 50(4):152–159, 2012.

[4] M. D. Linderman, G. Santhanam, C. T. Kemere, V. Gilja, S. O'Driscoll, B. M. Yu, A. Afshar, S. I. Ryu, K. V. Shenoy, and T. H. Meng. Signal processing challenges for neural prostheses. *IEEE Signal Processing Magazine*, 25(1):18–28, 2008.

[5] J. Lee, H.-G. Rhew, D. Kipke, and M. Flynn. A 64 channel programmable closed-loop deep brain stimulator with 8 channel neural amplifier and logarithmic ADC. In *Proceedings of the IEEE Symposium on VLSI Circuits*, pp. 76–77, 2008.

[6] R. M. Walker, H. Gao, P. Nuyujukian, K. Makinwa, K. Shenoy, T. Meng, and B. Murmann. A 96-channel full data rate direct neural interface in 0.13 μm CMOS. In *Proceedings of the IEEE Symposium on VLSI Circuits*, pp. 144–145, 2011.

[7] F. Marvasti. *Nonuniform Sampling: Theory and Practice*. Springer, New York, 2001.

[8] P. Ellis. Extension of phase plane analysis to quantized systems. *IRE Transactions on Automatic Control*, 4(2):43–54, 1959.

[9] R. Dorf, M. Farren, and C. Phillips. Adaptive sampling frequency for sampled-data control systems. *IRE Transactions on Automatic Control*, 7(1):38–47, 1962.

[10] J. Mitchell and W. McDaniel Jr. Adaptive sampling technique. *IEEE Transactions on Automatic Control*, 14(2):200–201, 1969.

[11] J. W. Mark and T. D. Todd. A nonuniform sampling approach to data compression. *IEEE Transactions on Communications*, 29(1):24–32, 1981.

[12] J. Foster and T.-K. Wang. Speech coding using time code modulation. In *Proceedings of the IEEE Southeastcon'91*, pp. 861–863, 1991.

[13] P. E. Luft and T. I. Laakso. Adaptive control of sampling rate using a local time-domain sampling theorem. In *Proceedings of the IEEE International Symposium on Circuits and Systems*, volume 3, pp. 145–148, 1994.

[14] N. Sayiner, H. V. Sorensen, and T. R. Viswanathan. A level-crossing sampling scheme for a/d conversion. *IEEE Transactions on Circuits and Systems II: Analog and Digital Signal Processing*, 43(4):335–339, 1996.

[15] W. R. Dieter, S. Datta, and W. K. Kai. Power reduction by varying sampling rate. In *Proceedings of the ACM International Symposium on Low power Electronics and Design*, pp. 227–232, 2005.

[16] Y. Tsividis. Event-driven, continuous-time ADCs and DSPs for adapting power dissipation to signal activity. In *Proceedings of the 2010 IEEE International Symposium on Circuits and Systems*, pp. 3581–3584, 2010.

[17] Y. P. Tsividis. Event-driven data acquisition and continuous-time digital signal processing. In *Proceedings of the IEEE Custom Integrated Circuits Conference*, pp. 1–8, 2010.

[18] Y. Tsividis. Event-driven data acquisition and digital signal processing: A tutorial. *IEEE Transactions on Circuits and Systems II: Express Briefs*, 57(8): 577–581, 2010.

[19] R. Tomovic and G. Bekey. Adaptive sampling based on amplitude sensitivity. *IEEE Transactions on Automatic Control*, 11(2):282–284, 1966.

[20] D. Ciscato and L. Martiani. On increasing sampling efficiency by adaptive sampling. *IEEE Transactions on Automatic Control*, 12(3):318–318, 1967.

[21] T. Hsia. Comparisons of adaptive sampling control laws. *IEEE Transactions on Automatic Control*, 17(6):830–831, 1972.

[22] K.-E. Årzén. A simple event-based PID controller. In *Proceedings of the 14th IFAC World Congress*, volume 18, pp. 423–428, 1999.

[23] M. Velasco, J. Fuertes, and P. Marti. The self triggered task model for real-time control systems. In *Work-in-Progress Session of the 24th IEEE Real-Time Systems Symposium (RTSS03)*, volume 384, pp. 67–70, 2003.

[24] J. Sandee, W. Heemels, and P. Van den Bosch. Event-driven control as an opportunity in the multidisciplinary development of embedded controllers. In *Proceedings of the American Control Conference*, pp. 1776–1781, 2005.

[25] M. Miskowicz. Asymptotic effectiveness of the event-based sampling according to the integral criterion. *Sensors*, 7(1):16–37, 2007.

[26] K. J. Aström. Event based control. In *Analysis and Design of Nonlinear Control Systems* (A. Astolfi and L. Marconi, Eds.), pp. 127–147. Springer, Berlin-Heidelberg, 2008.

[27] A. Anta and P. Tabuada. To sample or not to sample: Self-triggered control for nonlinear systems. *IEEE Transactions on Automatic Control*, 55(9): 2030–2042, 2010.

[28] M. Miskowicz. Efficiency of event-based sampling according to error energy criterion. *Sensors*, 10(3):2242–2261, 2010.

[29] S. Dormido, J. Sánchez, and E. Kofman. Muestreo, control y comunicación basados en eventos. *Revista Iberoamericana de Automática e Informática Industrial RIAI*, 5(1):5–26, 2008.

[30] M. Miskowicz. Event-based sampling strategies in networked control systems. In *Proceedings of IEEE International Workshop on Factory Communication Systems*, pp. 1–10, 2014.

[31] Y. Tsividis. Continuous-time digital signal processing. *Electronics Letters*, 39(21):1551–1552, 2003.

[32] Y. Tsividis. Signal processors with transfer function coefficients determined by timing. *IEEE Transactions on Circuits and Systems*, 29(12):807–817, 1982.

[33] K. M. Guan and A. C. Singer. Opportunistic sampling by level-crossing. In *Proceedings of the IEEE International Conference on Acoustics, Speech, and Signal Processing*, volume 3, pp. 1513–1516, 2007.

[34] K. M. Guan, S. S. Kozat, and A. C. Singer. Adaptive reference levels in a level-crossing analog-to-digital converter. *EURASIP Journal on Advances in Signal Processing*, 2008:183, 2008.

[35] K. M. Guan and A. C. Singer. Sequential placement of reference levels in a level-crossing analog-to-digital converter. In *Proceedings of the IEEE Conference on Information Sciences and Systems*, pp. 464–469, 2008.

[36] K. Kozmin, J. Johansson, and J. Delsing. Level-crossing ADC performance evaluation toward ultrasound application. *IEEE Transactions on Circuits and Systems I: Regular Papers*, 56(8):1708–1719, 2009.

[37] M. Greitans, R. Shavelis, L. Fesquet, T. Beyrouthy. Combined peak and level-crossing sampling scheme. In *Proceedings of the International Conference on Sampling Theory and Applications*, pages 5–8, 2011.

[38] L. C. Gouveia, T. J. Koickal, and A. Hamilton. An asynchronous spike event coding scheme for programmable analog arrays. *IEEE Transactions on Circuits and Systems I*, 58(4):791–799, 2011.

[39] T. J. Yamaguchi, M. Abbas, M. Soma, T. Aoki, Y. Furukawa, K. Degawa, S. Komatsu, and K. Asada. An equivalent-time and clocked approach for continuous-time quantization. In *Proceedings of the IEEE International Symposium on Circuits and Systems*, pp. 2529–2532, 2011.

[40] T. Marisa, T. Niederhauser, A. Haeberlin, J. Goette, M. Jacomet, and R. Vogel. Asynchronous ECG time sampling: Saving bits with Golomb-Rice encoding. *Computing in Cardiology*, 39:61–64, 2012.

[41] S. Naraghi, M. Courcy, and M. P. Flynn. A 9-bit, 14 $\mu$w and 0.06 mm$^2$ pulse position modulation ADC in 90 nm digital CMOS. *IEEE Journal of Solid-State Circuits*, 45(9):1870–1880, 2010.

[42] Y. Tsividis. Digital signal processing in continuous time: A possibility for avoiding aliasing and reducing quantization error. In *Proceedings of the IEEE International Conference on Acoustics, Speech, and Signal Processing*, volume 2, pp. 589–592, 2004.

[43] Y. Tsividis. Mixed-domain systems and signal processing based on input decomposition. *IEEE Transactions on Circuits and Systems I*, 53(10):2145–2156, 2006.

[44] B. Schell and Y. Tsividis. Analysis and simulation of continuous-time digital signal processors. *Signal Processing*, 89(10):2013–2026, 2009.

[45] Y. Chen, M. Kurchuk, N. T. Thao, and Y. Tsividis. Spectral analysis of continuous-time ADC and DSP. In *Event-Based Control and Signal Processing* (M. Miskowicz, Ed.), pp. 409–420. CRC Press, Boca Raton, FL 2015.

[46] T. Wang, D. Wang, P. J. Hurst, B. C. Levy, and S. H. Lewis. A level-crossing analog-to-digital converter with triangular dither. *IEEE Transactions on Circuits and Systems I*, 56(9):2089–2099, 2009.

[47] F. Akopyan, R. Manohar, and A. B. Apsel. A level-crossing flash asynchronous analog-to-digital converter. In *Proceedings of the IEEE International Symposium on Asynchronous Circuits and Systems*, pp. 12–22, 2006.

[48] H. Inose, T. Aoki, and K. Watanabe. Asynchronous delta-modulation system. *Electronics Letters*, 2(3):95–96, 1966.

[49] P. Sharma. Characteristics of asynchronous delta-modulation and binary-slope-quantized-pcm systems. *Electronic Engineering*, 40(479):32–37, 1968.

[50] R. Steele. *Delta Modulation Systems*. Pentech Press, London, 1975.

[51] N. Jayant. Digital coding of speech waveforms: PCM, DPCM, and DM quantizers. *Proceedings of the IEEE*, 62(5):611–632, 1974.

[52] E. Allier, G. Sicard, L. Fesquet, and M. Renaudin. A new class of asynchronous A/D converters based on time quantization. In *Proceedings of the International Symposium on Asynchronous Circuits and Systems*, pp. 196–205, 2003.

[53] M. Kurchuk, C. Weltin-Wu, D. Morche, and Y. Tsividis. Event-driven GHz-range continuous-time digital signal processor with activity-dependent power dissipation. *IEEE Journal of Solid-State Circuits*, 47(9):2164–2173, 2012.

[54] B. Schell and Y. Tsividis. A continuous-time ADC/DSP/DAC system with no clock and with activity-dependent power dissipation. *IEEE Journal of Solid-State Circuits*, 43(11):2472–2481, 2008.

[55] M. Trakimas and S. Sonkusale. A 0.8 V asynchronous ADC for energy constrained sensing applications. In *Proceedings of the IEEE Custom Integrated Circuits Conference*, pp. 173–176, 2008.

[56] V. Majidzadeh, A. Schmid, and Y. Leblebici. Low-distortion switched-capacitor event-driven analogue to-digital converter. *Electronics Letters*, 46(20):1373–1374, 2010.

[57] Y. Li, D. Zhao, M. N. van Dongen, and W. A. Serdijn. A 0.5 V signal-specific continuous-time level-crossing ADC with charge sharing. In *Proceedings of the IEEE Biomedical Circuits and Systems Conference*, pp. 381–384, 2011.

[58] T. A. Vu, S. Sudalaiyandi, M. Z. Dooghabadi, H. A. Hjortland, O. Nass, T. S. Lande, and S.-E. Hamran. Continuous-time CMOS quantizer for ultra-wideband applications. In *Proceedings of the IEEE International Symposium on Circuits and Systems*, pp. 3757–3760, 2010.

[59] S. Araujo-Rodrigues, J. Accioly, H. Aboushady, M. Louërat, D. Belfort, and R. Freire. A clockless 8-bit folding A/D converter. In *Proceedings of the IEEE Latin American Symposium on Circuits and Systems*, 2010.

[60] D. Chhetri, V. N. Manyam, and J. J. Wikner. An event-driven 8-bit ADC with a segmented resistor-string DAC. In *Proceedings of the European Conference on Circuit Theory and Design*, pp. 528–531, 2011.

[61] R. Grimaldi, S. Rodriguez, and A. Rusu. A 10-bit 5 kHz level-crossing ADC. In *Proceedings of the European Conference on Circuit Theory and Design*, pp. 564–567, 2011.

[62] R. Agarwal and S. R. Sonkusale. Input-feature correlated asynchronous analog to information converter for ECG monitoring. *IEEE Transactions on Biomedical Circuits and Systems*, 5(5):459–467, 2011.

[63] Y. Hong, I. Rajendran, and Y. Lian. A new ECG signal processing scheme for low-power wearable ECG devices. In *Proceedings of the Asia Pacific Conference on Postgraduate Research in Microelectronics and Electronics*, pp. 74–77, 2011.

[64] Y. Li, D. Zhao, and W. A. Serdijn. A sub-microwatt asynchronous level-crossing ADC for biomedical

applications. *IEEE Transactions on Biomedical Circuits and Systems*, 7(2):149–157, 2013.

[65] W. Tang, A. Osman, D. Kim, B. Goldstein, C. Huang, B. Martini, V. A. Pieribone, and E. Culurciello. Continuous time level crossing sampling ADC for bio-potential recording systems. *IEEE Transactions on Circuits and Systems I*, 60(6):1407, 2013.

[66] D. Kościelnik and M. Miskowicz. Event-driven successive charge redistribution schemes for clockless analog-to-digital conversion. In *Design, Modeling and Testing of Data Converters* (P. Carbone, S. Kiaey, and F. Xu, Eds.), Springer, pp. 161–209, 2014.

[67] X. Zhang and Y. Lian. A 300-mV 220-nW event-driven ADC with real-time QRS detection for wearable ECG sensors. *IEEE Transactions on Biomedical Circuits and Systems*, 8(6):834–843, 2014.

[68] [Online] Novelda AS. https://www.novelda.no, accessed on July 10, 2014.

[69] G. Thiagarajan, U. Dasgupta, and V. Gopinathan. Asynchronous analog-to-digital converter. US Patent application 2014/0062734 A1, March 6, 2014.

[70] G. Thiagarajan, U. Dasgupta, and V. Gopinathan. Asynchronous analog-to-digital converter having adaptive reference control. US Patent application 2014/0062735 A1, March 6, 2014.

[71] G. Thiagarajan, U. Dasgupta, and V. Gopinathan. Asynchronous analog-to-digital converter having rate control. US Patent application 2014/0062751 A1, March 6, 2014.

[72] Y. W. Li, K. L. Shepard, and Y. Tsividis. A continuous-time programmable digital FIR filter. *IEEE Journal of Solid-State Circuits*, 41(11): 2512–2520, 2006.

[73] M. Miskowicz. Send-on-delta concept: An event-based data reporting strategy. *Sensors*, 6(1):49–63, 2006.

[74] K. Kozmin, J. Johansson, and J. Kostamovaara. A low propagation delay dispersion comparator for a level-crossing A/D converter. *Analog Integrated Circuits and Signal Processing*, 62(1):51–61, 2010.

[75] C. Weltin-Wu and Y. Tsividis. An event-driven clockless level-crossing ADC with signal-dependent adaptive resolution. *IEEE Journal of Solid-State Circuits*, 48(9):2180–2190, 2013.

[76] V. Balasubramanian, A. Heragu, and C. Enz. Analysis of ultralow-power asynchronous ADCs. In *Proceedings of the IEEE International Symposium on Circuits and Systems*, pp. 3593–3596, 2010.

[77] Y. S. Suh. Send-on-delta sensor data transmission with a linear predictor. *Sensors*, 7(4):537–547, 2007.

[78] M. Kurchuk and Y. Tsividis. Signal-dependent variable-resolution clockless A/D conversion with application to continuous-time digital signal processing. *IEEE Transactions on Circuits and Systems I*, 57(5):982–991, 2010.

[79] D. Brückmann, T. Feldengut, B. Hosticka, R. Kokozinski, K. Konrad, and N. Tavangaran. Optimization and implementation of continuous time DSP-systems by using granularity reduction. In *Proceedings of the IEEE International Symposium on Circuits and Systems*, pp. 410–413, 2011.

[80] D. Brückmann, K. Konrad, and T. Werthwein. Concepts for hardware efficient implementation of continuous time digital signal processing. In *Event-Based Control and Signal Processing* (M. Miskowicz, Ed.), pp. 421–440. CRC Press, Boca Raton, FL 2015.

[81] J. Yen. On nonuniform sampling of bandwidth-limited signals. *IRE Transactions on Circuit Theory*, 3(4):251–257, 1956.

[82] F. J. Beutler. Error-free recovery of signals from irregularly spaced samples. *Siam Review*, 8(3): 328–335, 1966.

[83] C. Vezyrtzis and Y. Tsividis. Processing of signals using level-crossing sampling. In *Proceedings of the IEEE International Symposium on Circuits and Systems*, pp. 2293–2296, 2009.

[84] N. T. Thao. Event-based data acquisition and reconstruction—Mathematical background. In *Event-Based Control and Signal Processing* (M. Miskowicz, Ed.), pp. 379–408. CRC Press, Boca Raton, FL 2015.

[85] W. Tang, C. Huang, D. Kim, B. Martini, and E. Culurciello. 4-channel asynchronous bio-potential recording system. In *Proceedings of the IEEE International Symposium on Circuits and Systems*, pp. 953–956, 2010.

[86] M. Trakimas and S. R. Sonkusale. An adaptive resolution asynchronous ADC architecture for data compression in energy constrained sensing applications. *IEEE Transactions on Circuits and Systems I: Regular Papers*, 58(5):921–934, 2011.

[87] P. Martinez-Nuevo, S. Patil, and Y. Tsividis. Derivative level-crossing sampling. *IEEE Transactions on Circuits and Systems II*, 62(1):11–15, 2015.

[88] R. P. Boas. *Entire Functions.* Academic Press, London, 1954.

[89] F. Aeschlimann, E. Allier, L. Fesquet, and M. Renaudin. Asynchronous FIR filters: Towards a new digital processing chain. In *Proceedings of the IEEE International Symposium on Asynchronous Circuits and Systems*, pp. 198–206, 2004.

[90] S. M. Qaisar, L. Fesquet, M. Renaudin. Computationally efficient adaptive rate sampling and filtering. In *Proceedings of the European Signal Processing Conference*, volume 7, pp. 2139–2143, 2007.

[91] S. M. Qaisar, R. Yahiaoui, and T. Gharbi. An efficient signal acquisition with an adaptive rate A/D conversion. In *Proceedings of the IEEE International Conference on Circuits and Systems*, pp. 124–129, 2013.

[92] V. Dhanasekaran, M. Gambhir, M. M. Elsayed, E. Sánchez-Sinencio, J. Silva-Martinez, C. Mishra, L. Chen, and E. Pankratz. A 20MHz BW 68dB DR CT $\Delta\Sigma$ ADC based on a multi-bit time-domain quantizer and feedback element. In *Digest, IEEE International Solid-State Circuits Conference*, pp. 174–175, 2009.

[93] H. Pakniat and M. Yavari. A time-domain noise-coupling technique for continuous-time sigma-delta modulators. *Analog Integrated Circuits and Signal Processing*, 78(2):439–452, 2014.

[94] B. Schell and Y. Tsividis. A low power tunable delay element suitable for asynchronous delays of burst information. *IEEE Journal of Solid-State Circuits*, 43(5):1227–1234, 2008.

[95] K. Konrad, D. Brückmann, N. Tavangaran, J. Al-Eryani, R. Kokozinski, and T. Werthwein. Delay element concept for continuous time digital signal processing. In *Proceedings of the IEEE International Symposium on Circuits and Systems*, pp. 2775–2778, 2013.

[96] C. Vezyrtzis, W. Jiang, S. M. Nowick, and Y. Tsividis. A flexible, event-driven digital filter with frequency response independent of input sample rate. *IEEE Journal of Solid-State Circuits*, 49(10):2292–2304, 2014.

[97] Y. P. Tsividis. Integrated continuous-time filter design – an overview. *IEEE Journal of Solid-State Circuits*, 29(3):166–176, 1994.

[98] N. Tavangaran, D. Brückmann, T. Feldengut, B. Hosticka, R. Kokozinski, K. Konrad, and R. Lerch. Effects of jitter on continuous time digital systems with granularity reduction. In *Proceedings of the European Signal Processing Conference*, pp. 525–529, 2011.

[99] T. Redant and W. Dehaene. Joint estimation of propagation delay dispersion and time of arrival in a 40-nm CMOS comparator bank for time-based receivers. *IEEE Transactions on Circuits and Systems II: Express Briefs*, 60(2):76–80, 2013.

[100] D. K. Su and B. A. Wooley. A CMOS oversampling D/A converter with a current-mode semidigital reconstruction filter. *IEEE Journal of Solid-State Circuits*, 28(12):1224–1233, 1993.

[101] Z. Zhao and A. Prodic. Continuous-time digital controller for high-frequency DC-DC converters. *IEEE Transactions on Power Electronics*, 23(2): 564–573, 2008.

[102] Z. Zhao, V. Smolyakov, and A. Prodic. Continuous-time digital signal processing based controller for high-frequency DC-DC converters. In *Proceedings of the IEEE Applied Power Electronics Conference*, pp. 882–886, 2007.

[103] M. T. Ozgun and M. Torlak. Effects of random delay errors in continuous-time semi-digital transversal filters. *IEEE Transactions on Circuits and Systems II*, 61(1):183–190, 2014.

[104] D. Brückmann and K. Konrad. Optimization of continuous time filters by delay line adjustment. In *Proceedings of the IEEE International Midwest Symposium on Circuits and Systems*, pp. 1238–1241, 2010.

[105] G. B. Lockhart. Digital encoding and filtering using delta modulation. *Radio and Electronic Engineer*, 42(12):547–551, 1972.

[106] A. Ratiu, D. Morche, A. Arias, B. Allard, X. Lin-Shi, and J. Verdier. Efficient simulation of continuous time digital signal processing RF systems. In *Proceedings of the International Conference on Sampling Theory and Applications*, pp. 416–419, 2013.

[107] D. Brückmann. Design and realization of continuous-time wave digital filters. In *Proceedings of the IEEE International Symposium on Circuits and Systems*, pp. 2901–2904, 2008.

[108] D. Hand and M.-W. Chen. A non-uniform sampling ADC architecture with embedded alias-free asynchronous filter. In *Proceedings of the IEEE*

*Global Communications Conference*, pp. 3707–3712, 2012.

[109] S. Sudalaiyandi, M. Z. Dooghabadi, T.-A. Vu, H. A. Hjortland, Ø. Nass, T. S. Lande, and S. E. Hamran. Power-efficient CTBV symbol detector for UWB applications. In *Proceedings of the IEEE International Conference on Ultra-Wideband*, volume 1, pp. 1–4, 2010.

[110] S. Sudalaiyandi, T.-A. Vu, H. A. Hjortland, O. Nass, and T. S. Lande. Continuous-time single-symbol IR-UWB symbol detection. In *Proceedings of the IEEE International SOC Conference*, pp. 198–201, 2012.

[111] G. M. Jacobs and R. W. Brodersen. A fully asynchronous digital signal processor using self-timed circuits. *IEEE Journal of Solid-State Circuits*, 25(6):1526–1537, 1990.

[112] L. W. Couch, M. Kulkarni, and U. S. Acharya. *Digital and Analog Communication Systems*, 6th edition. Prentice Hall, Upper Saddle River, 1997.

[113] B. Logan. Click modulation. *AT&T Bell Laboratories Technical Journal*, 63(3):401–423, 1984.

[114] R. Kumaresan and Y. Wang. On representing signals using only timing information. *The Journal of the Acoustical Society of America*, 110(5):2421–2439, 2001.

[115] C. Kikkert and D. Miller. Asynchronous delta sigma modulation. In *Proceedings of the IREE*, 36:83–88, 1975.

[116] P. Sharma. Signal characteristics of rectangular-wave modulation. *Electronic Engineering*, 40(480): 103–107, 1968.

[117] E. Roza. Analog-to-digital conversion via duty-cycle modulation. *IEEE Transactions on Circuits and Systems II*, 44(11):907–914, 1997.

[118] A. A. Lazar and L. T. Tóth. Time encoding and perfect recovery of bandlimited signals. In *Proceedings of the IEEE International Conference on Acoustics, Speech, and Signal Processing*, pp. 709–712, 2003.

[119] A. Can, E. Sejdic, and L. F. Chaparro. An asynchronous scale decomposition for biomedical signals. In *Proceedings of the IEEE Signal Processing in Medicine and Biology Symposium*, pp. 1–6, 2011.

[120] N. Tavangaran, D. Brückmann, R. Kokozinski, and K. Konrad. Continuous time digital systems with asynchronous sigma delta modulation. In *Proceedings of the European Signal Processing Conference*, pp. 225–229, 2012.

[121] A. Can-Cimino, E. Sejdic, and L. F. Chaparro. Asynchronous processing of sparse signals. *Signal Processing*, 8(3):257–266, 2014.

[122] C. Vezyrtzis, Y. Tsividis, and S. M. Nowick. Designing pipelined delay lines with dynamically-adaptive granularity for low-energy applications. In *International Conference on Computer-Aided Design*, pp. 329–336, 2012.

# 16

# *Event-Based Data Acquisition and Reconstruction—Mathematical Background*

**Nguyen T. Thao**

*City College of New York*
*New York, NY, USA*

**CONTENTS**

**ABSTRACT** Event-based signal acquisition differs from traditional data acquisition in the features that one tries to capture from a continuous-time signal. But as a common point, one obtains a discrete description of the input signal in the form of a sequence of real values. Although these values are traditionally uniform samples of the input, they can be in the event-based approach nonuniform samples, and more generally innerproducts of the input with known kernel functions. But the same theoretical question of input reconstruction from these discrete values as in Shannon sampling theorem is posed. We give a tutorial on the difficult mathematical background behind this question and indicate practical methods of signal recovery based on sliding-window digital signal processing.

Event-based continuous-time data acquisition and signal processing has been discussed in Chapter 15. They promote the idea to acquire the characteristics of a continuous-time signal by detecting the time instants of precise amplitude-related events and to process the result. Advantages of this method compared with traditional pulse-code modulation (PCM) include the absence of aliasing, improved spectral properties, faster response to input changes, and power savings. This different type of data acquisition however escapes from the setting of Shannon sampling theorem, and the theoretical question of signal reconstruction from the discrete events needs to be mathematically revisited.

For illustration, a concrete example of event-based data acquisition proposed in Chapter 15 [1] is continuous-time amplitude quantization [2]. As in Figure 16.1, this operation can be viewed as providing a piecewise constant approximation of the input based on its crossings with predetermined levels. Fundamentally, the exact information about the input signal that is acquired by the quantizer is the time instants of the crossings (together with the corresponding level indices). One wonders about the potential for these crossings to uniquely characterize the input and about the possibility of better reconstructions than what continuous-time quantization proposes, with the ultimate question of perfect reconstruction.

Answers to these questions can be found in some substantial literature on the nonuniform sampling of bandlimited signals [3–7]. A number of well-known surveys have also appeared, including [8] on the various aspects and applications of nonuniform sampling and [9,10] on the more mathematical aspects. This literature may however still remain of difficult access to researchers and engineers without a special inclination for applied mathematics. Indeed, it often requires advanced knowledge on functional and complex analysis that may not be part of the common background. In this chapter, we propose to walk through this knowledge by motivating the most theoretical notions involved, while keeping a connection with the constraints of real implementations.

From a signal processing viewpoint, the main technical difficulty in this topic is the absence of simultaneous time invariance and linearity in the operations. Event-driven operations such as in continuous-time quantization are by construction time invariant but not linear. Meanwhile, analyzing this type of processing from the perspective of nonuniform sampling implies the construction of operators that are linear but not time invariant. In either case, the usual tools of signal processing such as Fourier analysis collapse and more general theories need to be involved. The appealing point of the nonuniform sampling approach is the preservation of linearity, a property that is at least commonly understood in finite-dimensional vector spaces where linear operations are described by matrices. We take advantage of this common background to progressively lead the reader to nonuniform sampling in the infinite-dimensional space of square-integrable bandlimited signals.

## 16.1 Linear Operator Approach to Sampling and Reconstruction

### 16.1.1 Linear Approach to Sampling

Figure 16.1 gives an example where a continuous-time signal $x(t)$ satisfies the relations $x(t_n) = x_n$ for a known sequence of (nonuniform) instants $(t_n)_{n\in\mathbb{Z}}$ and a known sequence of amplitude values $(x_n)_{n\in\mathbb{Z}}$. We say that $(t_n, x_n)_{n\in\mathbb{Z}}$ are samples of $x(t)$. Assuming that $x(t)$ belongs to a known space of bandlimited continuous-time signals $\mathcal{B}$, one can formalize the knowledge that $x(t_n) = x_n$ for all $n \in \mathbb{Z}$ by defining the mapping

$$\begin{array}{cccc} \mathcal{S}: & \mathcal{B} & \longrightarrow & \mathcal{D} \\ & u(t) & \mapsto & \mathbf{u} := \big(u(t_n)\big)_{n\in\mathbb{Z}} \end{array}, \qquad (16.1)$$

where $\mathcal{D}$ is a vector space of real discrete-time sequences and by writing that the image of $x(t)$ by this mapping is $\mathbf{x} := (x_n)_{n\in\mathbb{Z}}$. We use the short notation $\mathcal{S}x(t) = \mathbf{x}$. A point that needs to be emphasized is that $\mathcal{S}$ is specifically defined for the given sampling instants $(t_n)_{n\in\mathbb{Z}}$. In this approach, we are no longer concerned about how these instants have been obtained (by the event-driven system, such as the continuous-time quantizer), but only about what would happen if another bandlimited input was sampled at these *same* instants. In that sense, $\mathcal{S}$ is a virtual mapping. But, $\mathcal{S}$ is mathematically completely defined, and if the equation $\mathbf{x} = \mathcal{S}u(t)$ has a unique solution in $u(t)$, we will know (at least theoretically) that the samples $(t_n, x_n)_{n\in\mathbb{Z}}$ uniquely characterize $x(t)$. This is a minimal requirement for the perfect reconstruction of $x(t)$ to be possible from these samples.

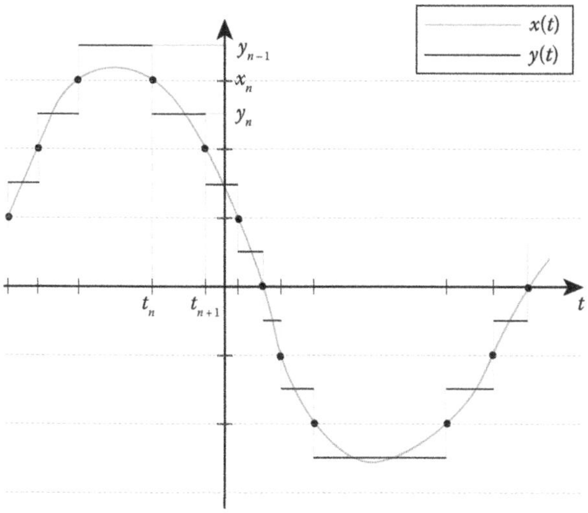


**FIGURE 16.1**
Continuous-time amplitude quantization.

The first important feature of this approach is that $\mathcal{S}$ is linear, as can be easily verified. According to standard mathematical terminology, $\mathcal{S}$ is called a *linear operator*, or simply an *operator* from $\mathcal{B}$ to $\mathcal{D}$. As a result, it is easy to see that $\mathbf{x} = \mathcal{S}u(t)$ has a unique solution in $u(t)$ *if and only if* the null space of $\mathcal{S}$ defined by

$$\text{Null}(\mathcal{S}) := \big\{u(t) \in \mathcal{B} : \mathcal{S}u(t) = 0\big\}, \qquad (16.2)$$

is reduced to the zero signal. This is because $\mathbf{x} = \mathcal{S}u(t)$ if and only if $\mathbf{x} = \mathcal{S}(u(t)+v(t))$ for all $v(t) \in \text{Null}(\mathcal{S})$, since $\mathcal{S}(u(t)+v(t)) = \mathcal{S}u(t) + \mathcal{S}v(t)$ by linearity of $\mathcal{S}$. Interestingly, the property that $\text{Null}(\mathcal{S}) = \{0\}$ (here 0 denotes the zero function) does not involve the amplitude sequence $\mathbf{x} = (x_n)_{n\in\mathbb{Z}}$, but only depends on $\mathcal{S}$ and hence on the instants $(t_n)_{n\in\mathbb{Z}}$. In fact, this property implies that the equation $\mathbf{u} = \mathcal{S}u(t)$ has a unique solution in $u(t)$ for any $\mathbf{u}$ in the range of $\mathcal{S}$, which is the linear subspace of $\mathcal{D}$ defined by

$$\text{Range}(\mathcal{S}) := \{\mathcal{S}u(t) : u(t) \in \mathcal{B}\}. \qquad (16.3)$$

It is said in this case that $\mathcal{S}$ is *one-to-one*, or *injective*. In conclusion, $x(t)$ can be uniquely reconstructed from its samples $(t_n, x_n)_{n\in\mathbb{Z}}$ if and only if $\mathcal{S}$ is injective. It will be naturally desirable to find the condition on $(t_n)_{n\in\mathbb{Z}}$ for this property to be realized.

### 16.1.2 Linear Approach to Reconstruction

Assuming that $\mathcal{S}$ has been shown to be injective, the second difficulty is to find a formula to reconstruct $x(t)$ from $\mathbf{x} = \mathcal{S}x(t)$. For any $\mathbf{u} \in \text{Range}(\mathcal{S})$, we know from the previous section that there exists a unique function $u(t) \in \mathcal{B}$ such that $\mathbf{u} = \mathcal{S}u(t)$. By calling $\mathcal{S}^{-1}\mathbf{u}$ the unique solution, we obtain a mapping $\mathcal{S}^{-1}$ from $\text{Range}(\mathcal{S})$ back to $\mathcal{B}$ that is easily shown to be linear as well. As a particular case, we will have $x(t) = \mathcal{S}^{-1}\mathbf{x}$. The difficulty here is not just the analytical derivation of $\mathcal{S}^{-1}$ but the characterization of its domain $\text{Range}(\mathcal{S})$. It is easier to search for a linear mapping from the whole space $\mathcal{D}$ to $\mathcal{B}$ that coincides with $\mathcal{S}^{-1}$ on $\text{Range}(\mathcal{S})$. In this chapter, we call in general a *reconstruction operator* any linear function that maps $\mathcal{D}$ into $\mathcal{B}$. Although $\text{Range}(\mathcal{S})$ is not known explicitly, a reconstruction operator $\mathcal{R}$ coincides with $\mathcal{S}^{-1}$ on $\text{Range}(\mathcal{S})$ if and only if $u(t) = \mathcal{R}\mathbf{u}$ whenever $\mathbf{u} = \mathcal{S}u(t)$. Denoting by $\mathcal{RS}$ the operator such that $\mathcal{RS}u(t) = \mathcal{R}(\mathcal{S}u(t))$, this is equivalent to writing that

$$\mathcal{RS} = \mathcal{I},$$

where $\mathcal{I}$ is the identity mapping of $\mathcal{B}$. It is said in this case that $\mathcal{R}$ is a *left inverse* of $\mathcal{S}$. The advantage of working on the whole space $\mathcal{D}$ is that $\mathcal{R}$ can be uniquely characterized by its action on the shifted impulse sequences

$\boldsymbol{\delta}_k := (\delta_{n-k})_{n\in\mathbb{Z}}$, where $\delta_n$ is equal to 1 at $n = 0$ and 0 when $n \neq 0$. Indeed, for any $\mathbf{u} = (u_n)_{n\in\mathbb{Z}} \in \mathcal{D}$, we have $\mathbf{u} = \sum_{k\in\mathbb{Z}} u_k \boldsymbol{\delta}_k$ and the application of $\mathcal{R}$ on both sides of this equality yields by linearity

$$\mathcal{R}\,\mathbf{u} = \sum_{n\in\mathbb{Z}} u_n\, r_n(t), \tag{16.4}$$

where

$$r_n(t) := \mathcal{R}\boldsymbol{\delta}_n. \tag{16.5}$$

We call $\{r_n(t)\}_{n\in\mathbb{Z}}$ the *impulse responses* of $\mathcal{R}$. All of them are bandlimited functions in $\mathcal{B}$. The convergence of the above summation requires certain constraints on the signal spaces and operators that will be presented in Section 16.1.3. By composing $\mathcal{S}$ with $\mathcal{R}$, we then have

$$\mathcal{R}\mathcal{S}\, u(t) = \sum_{n\in\mathbb{Z}} u(t_n)\, r_n(t). \tag{16.6}$$

We can then conclude that $\mathcal{R}$ is a left inverse of $\mathcal{S}$ if and only if its impulse responses $\{r_n(t)\}_{n\in\mathbb{Z}}$ satisfy the equation

$$u(t) = \sum_{n\in\mathbb{Z}} u(t_n)\, r_n(t), \tag{16.7}$$

for all $u(t) \in \mathcal{B}$. In this case, we call $\{r_n(t)\}_{n\in\mathbb{Z}}$ the *reconstruction functions*. This gives an equation similar to Shannon sampling theorem, except that $r_n(t)$ is not the shifted version of a single function $r(t)$. The difficulty is of course to find functions $r_n(t)$ that satisfy (16.7) for all $u(t) \in \mathcal{B}$.

### 16.1.3 Functional Setting

Until now, we have remained vague about the signal spaces. For an infinite summation such as in (16.4) to be defined, the signal spaces need to be at least equipped with some metric to define the notion of convergence. This also will be needed when dealing with the sensitivity of reconstructions to errors in samples. In this chapter, we assume that $\mathcal{B}$ is the space of square-integrable signals bandlimited by a maximum frequency $\Omega_o$. In this space, the magnitude of an error signal $e(t)$ can be measured by $\|e(t)\|$ such that*

$$\|u(t)\|^2 := \frac{1}{T_o}\int_{-\infty}^{\infty} |u(t)|^2\, dt, \tag{16.8}$$

where $T_o := \frac{\pi}{\Omega_o}$ is the Nyquist period of the signals of $\mathcal{B}$. Meanwhile, we take $\mathcal{D}$ to be the space of square-summable sequences [often denoted by $\ell^2(\mathbb{Z})$] in which the magnitude of an error sequence $\mathbf{e} = (e_n)_{n\in\mathbb{Z}}$ is measured by $\|\mathbf{e}\|_{\mathcal{D}}$ such that

$$\|\mathbf{u}\|_{\mathcal{D}}^2 := \sum_{n\in\mathbb{Z}} |u_n|^2. \tag{16.9}$$

These space restrictions imply constraints on the sampling operator $\mathcal{S}$ and hence the sampling instants $(t_n)_{n\in\mathbb{Z}}$. Indeed, without any restrictions on these instants, $\mathcal{S}$ does not necessarily map $\mathcal{B}$ into $\mathcal{D}$ [meaning that $\mathbf{u} = (u(t_n))_{n\in\mathbb{Z}}$ is not necessarily square summable when $u(t)$ is square integrable and bandlimited, which would typically happen when the values of $(t_n)_{n\in\mathbb{Z}}$ are too dense in time]. In fact, it is desirable to require the stronger condition that $\mathcal{S}$ be continuous with respect to the norms $\|\cdot\|$ and $\|\cdot\|_{\mathcal{D}}$, so that $\|\mathcal{S}u(t)\|_{\mathcal{D}}$ is not only finite for any $u(t) \in \mathcal{B}$ but can be made arbitrarily small by making the norm $\|u(t)\|_{\mathcal{D}}$ small enough. Given the linearity of $\mathcal{S}$, this is realized if and only if there exists a constant $\beta > 0$ such that

$$\|\mathcal{S}u(t)\|_{\mathcal{D}} \leq \beta\, \|u(t)\|, \tag{16.10}$$

for all $u(t) \in \mathcal{B}$. It is said in this case that $\mathcal{S}$ is *bounded*. The smallest possible value of $\beta$ is called the norm of $\mathcal{S}$ and is equal to†

$$\|\mathcal{S}\| := \sup_{u(t)\in\mathcal{B}\backslash\{0\}} \frac{\|\mathcal{S}u(t)\|_{\mathcal{D}}}{\|u(t)\|} = \sup_{\substack{u(t)\in\mathcal{B}\\ \|u(t)\|=1}} \|\mathcal{S}u(t)\|_{\mathcal{D}}, \tag{16.11}$$

where $\mathcal{B}\backslash\{0\}$ denotes the set $\mathcal{B}$ without its zero element. Then, $\mathcal{S}$ is bounded if and only if $\|\mathcal{S}\| < \infty$. A task will be to find the condition on $(t_n)_{n\in\mathbb{Z}}$ for this property to be achieved.

Naturally, these space restrictions also imply constraints on any considered reconstruction operator $\mathcal{R}$. Not only must the impulse responses $\{r_n(t)\}_{n\in\mathbb{Z}}$ be functions of $\mathcal{B}$, but it is also desirable that $\mathcal{R}$ be continuous with respect to $\|\cdot\|_{\mathcal{D}}$ and $\|\cdot\|$. This is again ensured by imposing that $\mathcal{R}$ be bounded such that $\|\mathcal{R}\mathbf{u}\| \leq \gamma\, \|\mathbf{u}\|_{\mathcal{D}}$ for some $\gamma > 0$ and any $\mathbf{u} \in \mathcal{D}$. It can be seen that this condition simultaneously guarantees the unconditional convergence of the sum in (16.4).

### 16.1.4 Generalized Sampling

The operator $\mathcal{S}$ of (16.1) can be seen as a particular case of operators of the type

$$\begin{array}{rccl} \mathcal{S}: & \mathcal{B} & \longrightarrow & \mathcal{D} \\ & u(t) & \mapsto & \mathbf{u} = (S_n u(t))_{n\in\mathbb{Z}} \end{array}, \tag{16.12}$$

where $S_n$ is for each $n$ a linear functional of $\mathcal{B}$, that is, a continuous linear function from $\mathcal{B}$ to $\mathbb{R}$. Most of the

*Contrary to the default mathematical practice, we have included the factor $\frac{1}{T_o}$ in the definition of $\|\cdot\|^2$ so that $\|u(t)\|$ is homogeneous to an amplitude.

†The supremum of a function is its smallest upper bound. When this value is attained by the function, it is the same as the maximum of the function. The infimum of a function is defined in a similar manner.

material covered until now is actually not specific to the definition $S_n u(t) = u(t_n)$ implied by (16.1) and can be restated in the general case of (16.12). One obtains a linear functional of $\mathcal{B}$ by forming, for example, the mapping

$$S_n u(t) = \frac{1}{T_o} \int_{-\infty}^{\infty} s_n(t)\, u(t)\, dt, \tag{16.13}$$

where $s_n(t)$ is some signal in $\mathcal{B}$. In fact, *any* linear functional of $\mathcal{B}$ can be put in this form due to Riesz representation theorem. Using the standard inner-product notation

$$\langle x, y \rangle := \frac{1}{T_o} \int_{-\infty}^{\infty} x(t)\, y(t)\, dt, \tag{16.14}$$

we now call a sampling operator any mapping of the form

$$\begin{array}{rccc} \mathcal{S}: & \mathcal{B} & \longrightarrow & \mathcal{D} \\ & u(t) & \mapsto & \mathbf{u} = \big(\langle s_n, u\rangle\big)_{n\in\mathbb{Z}} \end{array}. \tag{16.15}$$

In this generalized context, we call each value $\langle s_n, u\rangle$ a *sample* of $u(t)$. In the same way, a reconstruction operator $\mathcal{R}$ is uniquely defined by the family of its impulse responses $\{r_n(t)\}_{n\in\mathbb{Z}}$ and the operator $\mathcal{S}$ is uniquely defined by the family of functions $\{s_n(t)\}_{n\in\mathbb{Z}}$, which we call the kernel functions of $\mathcal{S}$.

The composed operator $\mathcal{RS}$ of (16.6) yields the more general expression

$$\mathcal{RS}\, u(t) = \sum_{n\in\mathbb{Z}} \langle s_n, u\rangle\, r_n(t). \tag{16.16}$$

By injecting the integral expression of $\langle s_n, u\rangle$ in Equation 16.16 and interchanging the summation and the integral, we obtain

$$\mathcal{RS}\, u(t) = \int_{-\infty}^{\infty} h(t,\tau)\, u(\tau)\, d\tau, \tag{16.17}$$

where

$$h(t,\tau) := \frac{1}{T_o} \sum_{n\in\mathbb{Z}} r_n(t)\, s_n(\tau). \tag{16.18}$$

This gives the classic kernel representation of $\mathcal{RS}$ as a linear (but not time-invariant) transformation.

### 16.1.5 Examples of Sampling Operator

We derive the kernel functions of important examples of sampling operator. A basic class of operators is achieved with kernel functions $s_n(t)$ of the type $s_n(t) = s(t-t_n)$, where $s(t)$ is a fixed bandlimited function of $\mathcal{B}$. It is easy to see from (16.14) that

$$\big\langle s(t-t_n), u(t)\big\rangle = \tfrac{1}{T_o}(\bar{s} * u)(t_n), \tag{16.19}$$

where $*$ is the convolution product and $\bar{s}(t) := s(-t)$. Defining

$$\mathrm{s}_o(t) := \mathrm{sinc}(\tfrac{t}{T_o}), \tag{16.20}$$

where

$$\mathrm{sinc}(x) := \frac{\sin(\pi x)}{\pi x},$$

it is easy to see that

$$\langle s_n, u\rangle = u(t_n) \qquad \text{with} \qquad s_n(t) := \mathrm{s}_o(t-t_n). \tag{16.21}$$

This is because $\frac{1}{T_o}\mathrm{s}_o(t)$ is even and is the impulse response of the ideal low-pass filter of cutoff frequency $\Omega_o$, so that $\frac{1}{T_o}(\bar{\mathrm{s}}_o * u)(t) = (\frac{1}{T_o}\mathrm{s}_o * u)(t) = u(t)$ for any $u(t) \in \mathcal{B}$. This leads to the initial sampling operator $\mathcal{S}$ of (16.1) which we call the *direct sampling operator* at the instants $(t_n)_{n\in\mathbb{Z}}$ to distinguish it from the other upcoming operators. Sampling the derivative $u'(t)$ of $u(t)$ can also be achieved in this manner. Indeed, as $u'(t)$ also belongs to $\mathcal{B}$, $u'(t_n) = \frac{1}{T_o}(\mathrm{s}_o * u')(t_n) = \frac{1}{T_o}(\mathrm{s}_o' * u)(t_n)$ by integration by part. As $\mathrm{s}_o'(t)$ is an odd function, it follows from (16.19) that $\big\langle -\mathrm{s}_o'(t-t_n), u(t)\big\rangle = u'(t_n)$. Denoting by $u^{(i)}(t)$ the $i$th derivative of $u(t)$, we have more generally

$$\langle s_n, u\rangle = u^{(i)}(t_n) \qquad \text{with} \qquad s_n(t) := (-1)^i \mathrm{s}_o^{(i)}(t-t_n).$$

There are also interesting cases where $s_n(t)$ is not the shifted version of a single function $s(t)$. For example,

$$\langle s_n, u\rangle = \int_{t_n}^{t_{n+1}} u(t)\, dt \qquad \text{with} \qquad s_n(t) := \mathrm{s}_o(t) * p_n(t), \tag{16.22}$$

where $p_n(t)$ is the rectangular function defined in Figure 16.2a. This result is easily derived from the general identity

$$\langle x, f * y\rangle = \langle \bar{f} * x, y\rangle, \tag{16.23}$$

since $\langle s_n, u\rangle = \langle \mathrm{s}_o * p_n, u\rangle = \langle p_n, \mathrm{s}_o * u\rangle = \langle p_n, T_o\, u\rangle = \int_{t_n}^{t_{n+1}} u(t)\, dt$. This corresponds to the sampling of an integrated version of $u(t)$ followed by the differentiation of consecutive samples. This type of sampling has been notably used in [11].

### 16.1.6 Sampling–Reconstruction System and Optimization

Usually, the type of sampling operator $\mathcal{S}$ is imposed by the application, and the freedom of design for

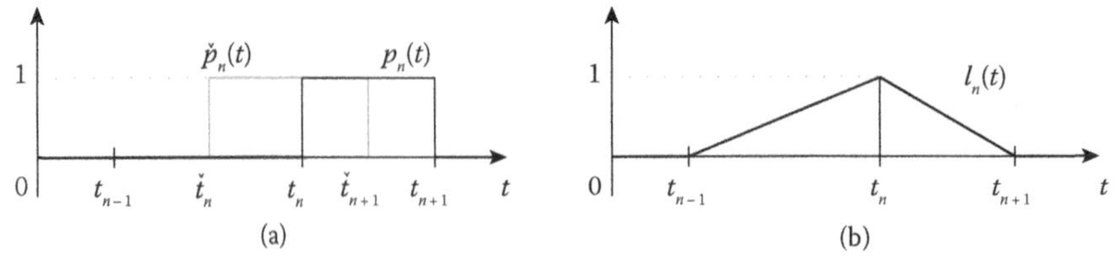


**FIGURE 16.2**
(a) Rectangular functions $p_n(t)$ and $\check{p}_n(t)$ with $\check{t}_n := \frac{1}{2}(t_{n-1}+t_n)$. (b) Triangular function $l_n(t)$.

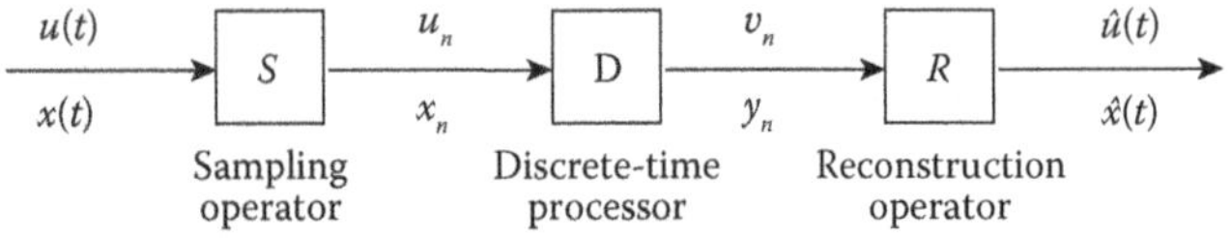


**FIGURE 16.3**
Framework of $\mathcal{R}\mathbf{D}\mathcal{S}$ system for the approximate reconstruction of $x(t)$ from $(x_n)_{n\in\mathbb{Z}} = \mathcal{S}x(t)$.

the reconstruction operator $\mathcal{R}$ is limited due to the constraints of analog circuit implementation. As left inverses of $\mathcal{S}$ are most likely not rigorously implementable in practice, the goal is at least to minimize the error $\|\hat{u}(t)-u(t)\|$ of the reconstructed signal $\hat{u}(t) := \mathcal{R}\mathcal{S}u(t)$. As the sampled signal $\mathbf{u} = \mathcal{S}u(t)$ is discrete in time, we consider the larger framework of sampling–reconstruction systems that include an intermediate discrete-time processor $\mathbf{D}$ as shown in Figure 16.3. In this case, $\hat{u}(t) := \mathcal{R}\mathbf{D}\mathcal{S}u(t)$. Since $\hat{u}(t)-u(t) = (\mathcal{R}\mathbf{D}\mathcal{S}-\mathcal{I})u(t)$ and $\mathcal{R}\mathbf{D}\mathcal{S}-\mathcal{I}$ is linear, the error $\|\hat{u}(t)-u(t)\|$ is proportional to $\|u(t)\|$. It is then desirable to optimize $\mathbf{D}$ and $\mathcal{R}$ toward the minimization of the coefficient of proportionality for the given sampling operator $\mathcal{S}$. This is formalized as the minimization of $\|\mathcal{R}\mathbf{D}\mathcal{S}-\mathcal{I}\|$, where, for any linear operator $\mathcal{A}$ of $\mathcal{B}$,

$$\|\mathcal{A}\| := \sup_{u(t)\in\mathcal{B}\setminus\{0\}} \frac{\|\mathcal{A}u(t)\|}{\|u(t)\|} = \sup_{\substack{u(t)\in\mathcal{B}\\ \|u(t)\|=1}} \|\mathcal{A}u(t)\|.$$

Once $\mathbf{D}$ and $\mathcal{R}$ have been designed, one approximately reconstructs $x(t)$ with $\hat{x}(t) = \mathcal{R}\mathbf{D}\mathbf{x}$, where $\mathbf{x} = \mathcal{S}x(t)$.

### 16.1.7 Revisiting Continuous-Time Quantization

Consider again continuous-time quantization presented in the introduction as an example of event-driven signal approximation. Assume that $x(t) \in \mathcal{B}$ is quantized in amplitude as illustrated in Figure 16.1, to give a piecewise constant approximation $y(t)$. The best reconstruction of $x(t)$ that can be performed from $y(t)$ is to remove the out-of-band content of $y(t)$. This leads to the signal

$$\hat{x}(t) = \varphi_o(t) * y(t),$$

where $\varphi_o(t)$ is the impulse response of the bandlimitation filter, explicitly equal to

$$\varphi_o(t) := \tfrac{1}{T_o}\,\mathrm{sinc}\left(\tfrac{t}{T_o}\right). \tag{16.24}$$

Interestingly, the transformation from $x(t)$ to $\hat{x}(t)$ can be modeled as an $\mathcal{R}\mathbf{D}\mathcal{S}$ system as described in Figure 16.3. The quantized signal $y(t)$ is a piecewise constant function of the form $y(t) = \sum_{n\in\mathbb{Z}} y_n\, p_n(t)$, where $y_n$ is the constant value of $y(t)$ in the interval $[t_n, t_{n+1})$ and $p_n(t)$ is the rectangular function of Figure 16.2a. We formally write $y(t) = \mathcal{P}\mathbf{y}$, where $\mathbf{y} = (y_n)_{n\in\mathbb{Z}}$ and

$$\mathcal{P}\mathbf{u} := \sum_{n\in\mathbb{Z}} u_n\, p_n(t). \tag{16.25}$$

This is basically the zero-order hold operation at the nonuniform clock instants $(t_n)_{n\in\mathbb{Z}}$. The bandlimited signal $\hat{x}(t)$ is then equal to $\mathcal{R}\mathbf{y}$ where

$$\mathcal{R}\mathbf{u} := \varphi_o(t) * \mathcal{P}\mathbf{u}. \tag{16.26}$$

$\mathcal{R}$ can be equivalently presented as the reconstruction operator of impulse responses

$$r_n(t) := \varphi_o(t) * p_n(t).$$

At each instant $t_n$, the input $x(t)$ must cross the quantization level lying between $y_{n-1}$ and $y_n$ in amplitude. When quantization is uniform as in the example of Figure 16.1, we must have $x(t_n) = x_n$, where $x_n := \frac{1}{2}(y_{n-1}+y_n)$. The sequence $\mathbf{x} = (x_n)_{n\in\mathbb{Z}}$ is then equal to $\mathcal{S}x(t)$ with the direct sampling operator of (16.1) and $\mathbf{y} = \mathbf{D}\mathbf{x}$, where $\mathbf{D}$ can be described as the IIR filter of recursive equation

$$y_n = 2x_n - y_{n-1}. \tag{16.27}$$

We have thus established the three relations $\mathbf{x} = \mathcal{S}x(t)$, $\mathbf{y} = \mathbf{D}\mathbf{x}$, and $\hat{x}(t) = \mathcal{R}\mathbf{y}$ of the $\mathcal{R}\mathbf{D}\mathcal{S}$ system of Figure 16.3.

The introduction raised the theoretical question of possible reconstruction improvements after amplitude quantization. In the equivalent $\mathcal{R}\mathbf{D}\mathcal{S}$ system model, $\mathcal{S}$ is imposed by the quantizer since it is the direct sampling operator at the instants $(t_n)_{n\in\mathbb{Z}}$, where the input $x(t)$

of interest crosses the quantization thresholds. Naturally, little improvement will be expected if the sequence $(t_n)_{n\in\mathbb{Z}}$ is such that $\mathcal{S}$ is not injective. So assume that the injectivity of $\mathcal{S}$ is effective. Possible improvements will depend on how much freedom of design one has on the operators $\mathbf{D}$ and $\mathcal{R}$. In the basic implementation of amplitude quantization, the zero-order hold function $\mathcal{P}$ is limited to operate on discrete-valued sequences $(y_n)_{n\in\mathbb{Z}}$. This greatly limits the freedom of design of the operator $\mathbf{D}$, which is to output these sequences. By allowing zero-order hold operations on real-valued sequences, we will see under reasonable conditions that there exists a discrete-time operator $\mathbf{D}$ such that $\mathcal{R}\mathbf{D}\mathcal{S} = \mathcal{I}$, thus allowing the perfect reconstruction of $x(t)$ from $\mathbf{x} = \mathcal{S}x(t)$.

## 16.2 The Case of Periodic Bandlimited Signals

The previous section left a number of difficult unanswered questions, such as the condition for $\mathcal{S}$ to be injective and a formula for finding a left inverse $\mathcal{R}$. These questions are surprisingly easy to answer in the context of periodic bandlimited signals, assuming a period that is a multiple of their Nyquist period. This is because the signals of a given bandwidth and a given period form a vector space of finite dimension. The sampling–reconstruction problem is then brought to the familiar ground of linear algebra in finite dimension. Although new issues appear when taking the period to the limit of infinity, the periodic case provides some useful basic intuition while also permitting easy and rigorous computer simulations.

### 16.2.1 Space of Periodic Bandlimited Signals

A signal $x(t)$ that is $T_p$ periodic and bandlimited yields a finite Fourier series expansion of the form

$$x(t) = \sum_{k=-L}^{L} X_k\, e^{j2\pi kt/T_p}, \tag{16.28}$$

for some positive integer $L$. We call $\mathcal{B}$ the space of all signals of this form for a given $L$. This is a vector space of dimension $K := 2L+1$. Any signal $x(t)$ of $\mathcal{B}$ is uniquely defined by the $K$-dimensional vector X of its Fourier coefficients $(X_{-L}, \ldots, X_L)$, which we formally write as

$$\mathrm{X} = \mathcal{F}x(t).$$

The mapping $\mathcal{F}$ can be thought of as a discrete-frequency Fourier transform. It satisfies the generalized Parseval's equality, which states that

$$\frac{1}{T_p}\int_0^{T_p} x(t)\, y(t)\, \mathrm{d}t = \sum_{k=-L}^{L} \mathrm{X}_k^* \mathrm{Y}_k,$$

where $\mathrm{X} := \mathcal{F}x(t)$ and $\mathrm{Y} := \mathcal{F}y(t)$ for any $x(t), y(t) \in \mathcal{B}$, and $\mathrm{X}_k^*$ denotes the complex conjugate of $\mathrm{X}_k$. In short notation, we write

$$\langle x, y\rangle = \mathrm{X}^*\mathrm{Y}, \tag{16.29}$$

where

$$\langle x, y\rangle := \frac{1}{T_p}\int_0^{T_p} x(t)\, y(t)\, \mathrm{d}t,$$

X and Y are seen as column vectors and $\mathbf{M}^*$ denotes the conjugate transpose of a matrix $\mathbf{M}$ also called the *adjoint* of $\mathbf{M}$, such that $(\mathbf{M}^*)_{i,j}$ is the complex conjugate of $\mathbf{M}_{j,i}$ for all $i, j$.

### 16.2.2 Sampling Operator

We show how generalized sampling introduced in Section 16.1.4 is formulated in finite dimension. In its generality, a sampling operator is any linear function $\mathcal{S}$ that maps a signal $x(t)$ of $\mathcal{B}$ into a vector $\mathbf{x}$ of $N$ values. Like any linear mapping in finite dimension, $\mathcal{S}$ can be represented by a matrix. We call $\mathbf{S}$ the matrix that maps the $K$ components of $x(t)$ in the Fourier basis $\{e^{j2\pi kt/T_p}\}_{-L\le k\le L}$ into $\mathbf{x} = \mathcal{S}x(t) \in \mathbb{R}^N$. This matrix is rectangular of size $N\times K$. It can be concisely defined as

$$\mathbf{S} := \mathcal{S}\mathcal{F}^{-1}. \tag{16.30}$$

Then, $\mathcal{S}x(t) = \mathbf{S}\mathrm{X}$ where $\mathrm{X} := \mathcal{F}x(t)$. Let $\mathrm{S}_n$ be the $n$th column vector of $\mathbf{S}^*$ so that

$$\mathbf{S}^* = \left[\mathrm{S}_1\, \mathrm{S}_2 \cdots \mathrm{S}_N\right]. \tag{16.31}$$

As $\mathrm{S}_n^*$ is the $n$th row vector of $\mathbf{S}$, $\mathbf{S}\mathrm{X}$ is the $N$-dimensional vector of components $(\mathrm{S}_n^*\mathrm{X})_{1\le n\le N}$. Defining

$$s_n(t) := \mathcal{F}^{-1}\mathrm{S}_n, \tag{16.32}$$

it follows from (16.29) that $\mathrm{S}_n^*\mathrm{X} = \langle s_n, x\rangle$. Thus, any sampling operator can be put in the form

$$\begin{array}{rccl} \mathcal{S}: & \mathcal{B} & \longrightarrow & \mathbb{R}^N \\ & x(t) & \mapsto & \mathbf{x} = \left(\langle s_n, x\rangle\right)_{1\le n\le N}. \end{array} \tag{16.33}$$

### 16.2.3 "Nyquist Condition" for Direct Sampling

We saw in Section 16.1.1 that a necessary and sufficient condition for a linear mapping $\mathcal{S}$ to allow the theoretical recovery of $x(t)$ from $\mathcal{S}x(t)$ is that $\mathcal{S}$ be injective. In the finite-dimensional case of (16.33), it is a basic result of linear algebra that $\mathcal{S}$ is injective if and only if the dimension of its range, also called the rank of $\mathcal{S}$, is equal to the dimension $K$ of $\mathcal{B}$. One immediately concludes that

$$N \geq K, \tag{16.34}$$

is a necessary condition for $\mathcal{S}$ to be injective. The sufficiency of this condition of course depends on $\mathcal{S}$. Assuming that $N \geq K$, $\mathcal{S}$ is injective if and only if its matrix $\mathbf{S}$ has full rank $K$. We are going to show that this is always realized with the direct sampling operator

$$\mathcal{S}x(t) = \big(x(t_1), x(t_2), \ldots, x(t_N)\big), \tag{16.35}$$

for $N$ distinct instants $t_1, t_2, \ldots, t_N$ in $[0, T_p)$. From (16.30), $\mathbf{S}\mathrm{X} = \mathcal{S}x(t)$ where $x(t) = \mathcal{F}^{-1}\mathrm{X}$. From (16.28), we have

$$x(t_n) = \sum_{k=-L}^{L} e^{j2\pi k t_n/T_p}\, X_k.$$

Since $x(t_n)$ is the $n$th component of $\mathbf{S}\mathrm{X}$, the entries of the matrix $\mathbf{S}$ are

$$\mathbf{S}_{n,k} = e^{j2\pi k t_n/T_p}, \tag{16.36}$$

for $n = 1, \ldots, N$ and $k = -L, \ldots, L$. The first $K$ row vectors of $\mathbf{S}$ form the Vandermonde matrix $\{u_n^k\}_{1\leq n\leq K, -L\leq k\leq L}$ where $u_n := e^{j2\pi t_n/T_p}$. Its determinant is in magnitude equal to $\prod_{1\leq n<m\leq K} |u_m - u_n|$. Since $t_1, \ldots, t_K$ are distinct in $[0, T_p)$, then $u_1, \ldots, u_K$ are distinct on the unit circle. This makes the determinant nonzero and hence $\mathbf{S}$ of rank $K$. Thus, with direct sampling, $x(t)$ can be recovered from $\mathcal{S}x(t)$ *if and only if* the number $N$ of samples is at least equal to the number of Nyquist periods within one period of $x(t)$. This constitutes the Nyquist condition for the nonuniform sampling of periodic signals. Note that there is no condition on the location of the sampling instants, only a condition on their number. We say that the sampling is *critical* when $N = K$. This implies the idea that no sample can be dropped for the reconstruction to be possible. The term *oversampling* is used when $N > K$.

Returning to general sampling operators of the type (16.33), $N \geq K$ is also sufficient for $\mathcal{S}$ to be injective when $s_n(t) = s(t - t_n)$, where $t_1, \ldots, t_N$ are distinct in $[0, T_p)$ and $s(t)$ is a function of $\mathcal{B}$ whose Fourier coefficients $S_{-L}, \ldots, S_L$ are nonzero. This is because similarly to (16.19),

$$\big\langle s(t - t_n), u(t)\big\rangle = \tfrac{1}{T_p}(\bar{s} \circledast u)(t_n), \tag{16.37}$$

where $(x \circledast y)(t) := \int_0^{T_p} x(\tau)y(t-\tau)\mathrm{d}\tau$, and the mapping $u(t) \mapsto (\bar{s} \circledast u)(t)$ is invertible in $\mathcal{B}$ as shown in Appendix 16.2.5. The operator $\mathcal{S}$ such that $\langle s_n, x\rangle = \int_{t_n}^{t_{n+1}} x(t)\,\mathrm{d}t$ for all $n = 1, \ldots, N$ with $t_{N+1} := t_1 + T_p$ can also be shown to be injective when $N \geq K$.

### 16.2.4 Sampling Pseudoinverse

Once a sampling operator $\mathcal{S}$ of the form (16.33) has been shown to be injective, the next question is to find a left inverse $\mathcal{R}$. In the present context of periodic signals, a reconstruction operator $\mathcal{R}$ is any linear mapping from $\mathbb{R}^N$ to $\mathcal{B}$. Consider the reconstruction operator $\mathcal{S}^*$ defined by

$$\mathcal{S}^*\mathbf{x} = \sum_{n=1}^{N} x_n\, s_n(t), \tag{16.38}$$

for all $\mathbf{x} \in \mathbb{R}^N$. There is no reason for $\mathcal{S}^*$ to be a left inverse of $\mathcal{S}$. However, it can be shown that the composed operator $\mathcal{S}^*\mathcal{S}$ is systematically invertible in $\mathcal{B}$ given the injectivity of $\mathcal{S}$. This is easily seen by matrix representation. By linearity of $\mathcal{F}$, $\mathcal{F}\mathcal{S}^*\mathbf{x} = \sum_{n=1}^{N} x_n\, \mathcal{F}s_n(t) = \sum_{n=1}^{N} x_n\, \mathrm{S}_n$ from the relation (16.32). Therefore, $\mathcal{F}\mathcal{S}^*$ is the $K\times N$ matrix $\big[\mathrm{S}_1\ \mathrm{S}_2 \cdots \mathrm{S}_N\big]$. It follows from (16.31) that

$$\mathbf{S}^* = \mathcal{F}\mathcal{S}^*. \tag{16.39}$$

With (16.30), we obtain that $\mathcal{S}^*\mathcal{S} = \mathcal{F}^{-1}(\mathbf{S}^*\mathbf{S})\mathcal{F}$. From the injective assumption of $\mathcal{S}$, $\mathbf{S}$ is an $N\times K$ matrix with $N \geq K$ and full rank $K$. It is a basic result of linear algebra that $\mathbf{S}^*\mathbf{S}$ is also of rank $K$. But since $\mathbf{S}^*\mathbf{S}$ is a $K\times K$ matrix, it is invertible. This implies the invertibility of $\mathcal{S}^*\mathcal{S}$. We can then define the reconstruction operator

$$\mathcal{S}^+ := (\mathcal{S}^*\mathcal{S})^{-1}\mathcal{S}^*. \tag{16.40}$$

Since $\mathcal{S}^+\mathcal{S} = (\mathcal{S}^*\mathcal{S})^{-1}\mathcal{S}^*\mathcal{S} = \mathcal{I}$ where $\mathcal{I}$ is the identity operator of $\mathcal{B}$, then $\mathcal{S}^+$ is a left inverse of $\mathcal{S}$. The operator $\mathcal{S}^+$ is called the pseudoinverse of $\mathcal{S}$. By applying $(\mathcal{S}^*\mathcal{S})^{-1}$ on both sides of (16.38), we have

$$\mathcal{S}^+\mathbf{x} = \sum_{n=1}^{N} x_n\, \tilde{s}_n(t) \quad \text{where} \quad \tilde{s}_n(t) := (\mathcal{S}^*\mathcal{S})^{-1} s_n(t). \tag{16.41}$$

We show in Figure 16.4 examples of functions $\{s_n(t)\}_{1\leq n\leq N}$ and $\{\tilde{s}_n(t)\}_{1\leq n\leq N}$ for the direct sampling operator of (16.35). According to (16.37), this operator is obtained by taking $s_n(t) = s_o(t - t_n)$ where $s_o(t)$ is such that $\frac{1}{T_p}(\bar{s}_o \circledast u)(t) = u(t)$. As shown in Appendix 16.2.5, this is achieved by the function

$$s_o(t) := \sum_{k=-L}^{L} e^{j2\pi k t/T_p}. \tag{16.42}$$

One can also show that $s_o(t)$ is a $T_p$-periodized sinc function. Note from Figure 16.4 that contrary to $s_n(t)$, $\tilde{s}_n(t)$ is not the shifted version of a single function. This makes the derivation of $\tilde{s}_n(t)$ difficult in practice. In the particular case where $N = K$, $\mathcal{S}$ is invertible and hence

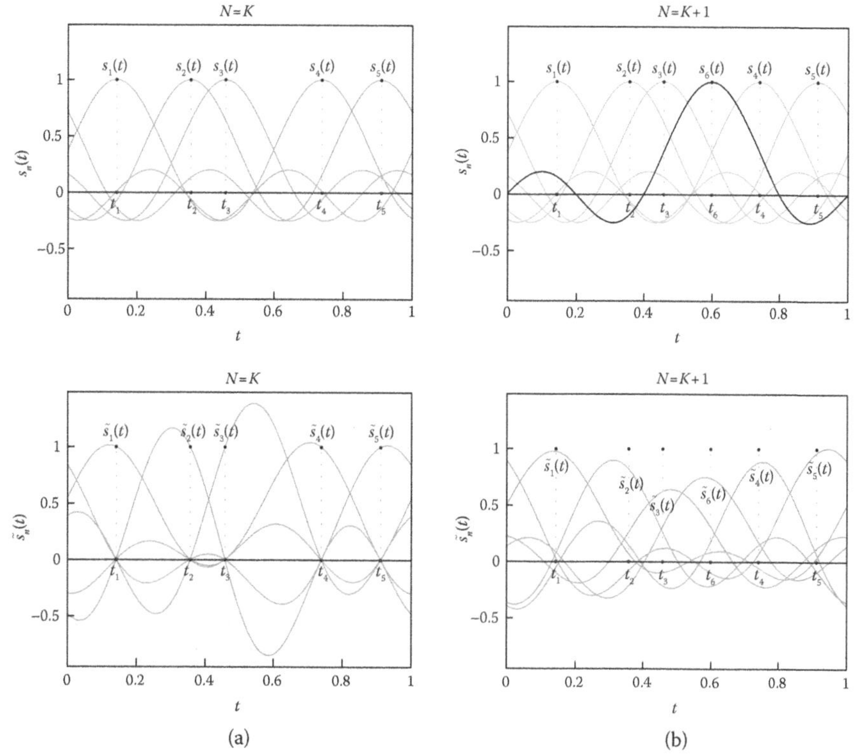


**FIGURE 16.4**
Examples of functions $s_n(t) := s_o(t-t_n)$ and $\tilde{s}_n(t)$: (a) case $N = K = 5$ (critical sampling); (b) case $N = K+1 = 6$ (oversampling). The sampling instants of the case $N = K+1$ are the sampling instants $t_1, \ldots, t_5$ of the case $N = K$, plus one additional instant $t_6$ (time index ordering is not necessary here).

$\mathcal{S}^+ = \mathcal{S}^{-1}$. Since $\tilde{s}_n(t) = \mathcal{S}^+\boldsymbol{\delta}_n$ where $\boldsymbol{\delta}_n$ is the $N$-dimensional vector whose $n$th component is 1 and the others 0, then $\mathcal{S}\tilde{s}_n(t) = \boldsymbol{\delta}_n$, which amounts to writing that

$$\tilde{s}_n(t_k) = \begin{cases} 1, & k = n \\ 0, & k = 1, \ldots, N \text{ and } k \neq n \end{cases}. \tag{16.43}$$

We say that the functions $\tilde{s}_n(t)$ are interpolation functions at the sampling instants $t_1, \ldots, t_N$. This property is confirmed by the figure in the case $N = K$. This interpolating property, however, no longer holds when $N > K$. This can also be observed in the figure in the case $N = K+1$.

### 16.2.5 Appendix

#### 16.2.5.1 Fourier Coefficients of Circular Convolution

Given the Fourier coefficient formula $X_k = \frac{1}{T_p}\int_0^{T_p} x(t)\, e^{-j2\pi kt/T_p}dt$, it is easy to verify that the $k$th Fourier coefficient of $z(t) := x(t) \circledast y(t)$ is $Z_k = T_p\, X_k Y_k$. Consider the operator $\mathcal{A}u(t) := (\bar{s} \circledast u)(t)$ for $s(t) \in \mathcal{B}$. Let $\mathrm{U} = (U_k)_{-L \leq k \leq L}$ and $\mathrm{V} = (V_k)_{-L \leq k \leq L}$ be the vectors of Fourier coefficients of $u(t)$ and $\mathcal{A}u(t)$, respectively. Then, $V_k = T_p S_k^* U_k$ for each $k = -L, \ldots, L$.

The operator $\mathcal{A}$ is invertible in $\mathcal{B}$ if and only if the mapping $\mathrm{U} \mapsto \mathrm{V}$ is invertible in $\mathbb{C}^N$. This is in turn equivalent to the fact that $S_k$ is nonzero for all $k = -L, \ldots, L$.

Next, $\mathcal{A}u(t) = T_p\, u(t)$ for all $u(t) \in \mathcal{B}$ if and only if $T_p S_k^* U_k = T_p U_k$ for each $k = -L, \ldots, L$ and all $U_k \in \mathbb{C}$. This is in turn equivalent to having $S_k = 1$ for all $k = -L, \ldots, L$, and hence $s(t) = \mathcal{F}^{-1}(1, \ldots, 1)$ which gives the function $s_o(t)$ of (16.42).

## 16.3 Sampling and Reconstruction Operators in Hilbert Space

While the periodic case brings some useful intuition to the problem of nonuniform sampling, its extension to aperiodic bandlimited signals is the major challenge in this topic. In the periodic case, the injectivity of $\mathcal{S}$ immediately led to a left inverse. This was based on the introduction of the adjoint operator (16.38) and followed by the operator construction of (16.40). With a sampling operator $\mathcal{S}x(t) = \big(\langle s_n, x\rangle\big)_{n\in\mathbb{Z}}$ in the aperiodic case, one can naturally generalize (16.38) into

$$\mathcal{S}^*\mathbf{u} = \sum_{n\in\mathbb{Z}} u_n s_n(t). \qquad (16.44)$$

Under the injectivity of $\mathcal{S}$, one is then tempted to define the pseudoinverse

$$\mathcal{S}^+ := (\mathcal{S}^*\mathcal{S})^{-1}\mathcal{S}^*. \qquad (16.45)$$

Unfortunately, the injectivity of $\mathcal{S}$ is not sufficient to make $\mathcal{S}^*\mathcal{S}$ invertible in infinite dimension. In this section, we give the reader basic tools from the theory of Hilbert spaces that permit an extension of the properties of interest in finite dimension to the infinite-dimensional case. We refer the reader to [12–14] for further details on the mathematical notions invoked in this section.

### 16.3.1 Adjoint Operator

To understand the properties of $\mathcal{S}^*\mathcal{S}$, one needs to understand the fundamental concept behind the construction of $\mathcal{S}^*$. We know the explicit definition of the adjoint $\mathbf{S}^*$ for a $N\times K$ matrix $\mathbf{S}$ in terms of the matrix coefficients. A generalization of this notion to operators would require a definition that does not involve matrix coefficients. Note that $\mathbf{x}^*(\mathbf{S}^*\mathbf{y}) = (\mathbf{S}\mathbf{x})^*\mathbf{y}$ for all $\mathbf{x}\in\mathbb{C}^K$ and $\mathbf{y}\in\mathbb{C}^N$ seen as column vectors. This relation also uniquely characterizes $\mathbf{S}^*$ as each of its coefficients can be retrieved by choosing $\mathbf{x}$ and $\mathbf{y}$ with single nonzero coefficients. Using the notation $\langle\mathbf{x},\mathbf{y}\rangle_P := \mathbf{x}^*\mathbf{y}$ for any $\mathbf{x}$ and $\mathbf{y}$ of $\mathbb{C}^P$, $\mathbf{S}^*$ can also be defined as the unique $K\times N$ matrix such that

$$\langle\mathbf{x},\mathbf{S}^*\mathbf{y}\rangle_K = \langle\mathbf{S}\mathbf{x},\mathbf{y}\rangle_N, \qquad (16.46)$$

for all $\mathbf{x}\in\mathbb{C}^K$ and $\mathbf{y}\in\mathbb{C}^N$.

The generalization of this notion of adjoint to an operator $\mathcal{S}$ from $\mathcal{B}$ to $\mathcal{D}$ is possible when $\mathcal{B}$ and $\mathcal{D}$ are equipped with the structure of Hilbert spaces. The space $\mathcal{B}$ is indeed a Hilbert space with respect to the function $\langle\cdot,\cdot\rangle$ defined in (16.14). This means that $\langle\cdot,\cdot\rangle$ is an *inner product* in $\mathcal{B}$, and $\mathcal{B}$ is complete with respect to the norm $\|\cdot\|$ of (16.8) which is induced by $\langle\cdot,\cdot\rangle$ in the sense that $\|x(t)\| = \sqrt{\langle x,x\rangle}$. The same can be said of the space $\mathcal{D}$ with the norm $\|\cdot\|_{\mathcal{D}}$ of (16.9) induced by the inner product

$$\langle\mathbf{x},\mathbf{y}\rangle_{\mathcal{D}} := \sum_{n\in\mathbb{Z}} x_n y_n.$$

Under these circumstances and given our initial assumption that $\mathcal{S}$ is bounded as stated in (16.10), it is shown that there exists a unique operator $\mathcal{S}^*$ from $\mathcal{D}$ to $\mathcal{B}$ that satisfies the relation

$$\langle u,\mathcal{S}^*\mathbf{v}\rangle = \langle\mathcal{S}u,\mathbf{v}\rangle_{\mathcal{D}}, \qquad (16.47)$$

for all $u\in\mathcal{B}$ and $\mathbf{v}\in\mathcal{D}$. This is in fact the original definition of the adjoint $\mathcal{S}^*$ of $\mathcal{S}$. It is also known that $\mathcal{S}^*$ is bounded. It is easy to verify that $\mathcal{S}^*$ yields the expression of (16.44). Indeed, defining $\mathcal{R}\mathbf{u} := \sum_{n\in\mathbb{Z}} u_n s_n(t)$, we have $\langle u(t),\mathcal{R}\mathbf{v}\rangle = \langle u(t),\sum_{n\in\mathbb{Z}} v_n\, s_n(t)\rangle = \sum_{n\in\mathbb{Z}}\langle u(t), s_n(t)\rangle\, v_n = \langle\mathcal{S}u(t),\mathbf{v}\rangle_{\mathcal{D}}$. So $\mathcal{R} = \mathcal{S}^*$. Due to the symmetry of the relation (16.47), $\mathcal{S}^*$ naturally yields $\mathcal{S}$ as its adjoint operator, that is, $(\mathcal{S}^*)^* = \mathcal{S}$.

### 16.3.2 Orthogonality in Hilbert Space

We will see that adjoint operators yield special properties connected to the notion of orthogonality induced by inner products. Two signals $x(t)$ and $y(t)$ of $\mathcal{B}$ are said to be orthogonal and symbolized by $x(t)\perp y(t)$ when $\langle x,y\rangle = 0$. Because the norm $\|\cdot\|$ is induced by the inner product $\langle\cdot,\cdot\rangle$, there is some outstanding interaction between orthogonality and metric properties. The most basic interaction lies in the Pythagorean theorem, which states that $\|z(t)\|^2 = \|x(t)\|^2 + \|y(t)\|^2$ when $z(t) = x(t) + y(t)$ and $x(t)\perp y(t)$. In this case, $\|z(t)\|$ is larger than or equal to both $\|x(t)\|$ and $\|y(t)\|$. The second outstanding interaction is in the notion of orthogonal projection. If $\mathcal{V}$ is a *closed* linear subspace of $\mathcal{B}$ (meaning that $\mathcal{V}$ includes all its limit points with respect to $\|\cdot\|$) and $x(t)\in\mathcal{B}$, then there exists a unique $y(t)\in\mathcal{V}$ that minimizes the distance $\|y(t)-x(t)\|$. This signal $y(t)$ is denoted by $\mathcal{P}_{\mathcal{V}}x(t)$, is also the unique element of $\mathcal{V}$ such that $x(t)-y(t)\perp\mathcal{V}$ (meaning that $x(t)-y(t)$ is orthogonal to all elements of $\mathcal{V}$), and is called the orthogonal projection of $x(t)$ onto $\mathcal{V}$. As $x(t) = y(t) + (x(t)-y(t))$, one obtains from the Pythagorean theorem that $\|\mathcal{P}_{\mathcal{V}}x(t)\| \le \|x(t)\|$ (Bessel's inequality). All these properties can be restated in the space $\mathcal{D}$ with the inner product $\langle\cdot,\cdot\rangle_{\mathcal{D}}$ and its induced norm $\|\cdot\|_{\mathcal{D}}$.

More properties of interest are about orthogonal space decompositions. For any subset $\mathcal{A}\subset\mathcal{B}$, let $\mathcal{A}^\perp$ denote the set of all elements of $\mathcal{B}$ that are orthogonal to $\mathcal{A}$. By continuity of each argument of $\langle\cdot,\cdot\rangle$ with respect to $\|\cdot\|$ (as a result of Cauchy–Schwarz inequality), it is shown that $\mathcal{A}^\perp$ is a closed subspace of $\mathcal{B}$ and is equal to $\overline{\mathcal{A}}^\perp$, where $\overline{\mathcal{A}}$ is the set of all the limits points of $\mathcal{A}$ and is called the closure of $\mathcal{A}$. By orthogonal projection, it is then shown that $\mathcal{B} = \mathcal{V}\oplus\mathcal{V}^\perp$ for any closed subspace $\mathcal{V}$ of $\mathcal{B}$ (where the addition symbol $\oplus$ simply recalls that $\mathcal{V}\cap\mathcal{V}^\perp = \{0\}$). One obtains from this that $(\mathcal{V}^\perp)^\perp = \mathcal{V}$ [by saying that $\mathcal{B} = \mathcal{V}^\perp\oplus(\mathcal{V}^\perp)^\perp$ since $\mathcal{V}^\perp$ is closed and noticing that $(\mathcal{V}^\perp)^\perp\supset\mathcal{V}$]. This is again valid in $\mathcal{D}$ with respect to $\langle\cdot,\cdot\rangle_{\mathcal{D}}$.

These properties imply some outstanding consequence on the relation between $\mathcal{S}$ and $\mathcal{S}^*$. Defining

$\text{Null}(\mathcal{S}^*) := \{\mathbf{u} \in \mathcal{D} : \mathcal{S}^*\mathbf{u} = 0\}$ and $\text{Range}(\mathcal{S}^*) := \{\mathcal{S}^*\mathbf{u} : \mathbf{u} \in \mathcal{D}\}$ similarly to (16.2) and (16.3), one obtains the basic identities

$$\text{Range}(\mathcal{S}^*)^\perp = \text{Null}(\mathcal{S}), \qquad (16.48)$$

$$\text{Range}(\mathcal{S})^\perp = \text{Null}(\mathcal{S}^*). \qquad (16.49)$$

This is proved as follows. If $u(t) \in \text{Null}(\mathcal{S})$, (16.47) implies that $\langle u, \mathcal{S}^*\mathbf{v}\rangle = \langle \mathcal{S}u, \mathbf{v}\rangle_{\mathcal{D}} = 0$ for all $\mathbf{v} \in \mathcal{D}$. Hence, $u(t) \in \text{Range}(\mathcal{S}^*)^\perp$. Conversely, if $u(t) \in \text{Range}(\mathcal{S}^*)^\perp$, one can see from the identities

$$\|\mathcal{S}u(t)\|_{\mathcal{D}}^2 = \langle \mathcal{S}u, \mathcal{S}u\rangle_{\mathcal{D}} = \langle u, \mathcal{S}^*\mathcal{S}u\rangle, \qquad (16.50)$$

that $\|\mathcal{S}u(t)\|_{\mathcal{D}} = 0$ since $\mathcal{S}^*\mathcal{S}u(t) \in \text{Range}(\mathcal{S}^*)$. Hence, $u(t) \in \text{Null}(\mathcal{S})$. This proves (16.48). The second relation is shown in a similar manner.

### 16.3.3 Injectivity of $\mathcal{S}$ and Surjectivity of $\mathcal{S}^*$

The pending question has been how to recognize that $\mathcal{S}$ is injective. From (16.48), this is effective if and only if $\{0\} = \text{Range}(\mathcal{S}^*)^\perp$, which is also equal to $\overline{\text{Range}(\mathcal{S}^*)}^\perp$. Since $\{0\}^\perp = \mathcal{B}$ and $\overline{\text{Range}(\mathcal{S}^*)}$ is a closed subspace, then

$$\mathcal{S} \text{ is injective} \iff \overline{\text{Range}(\mathcal{S}^*)} = \mathcal{B}. \qquad (16.51)$$

In practice, the above property means that any signal $u(t) \in \mathcal{B}$ can be approximated with arbitrary precision with respect to the norm $\|\cdot\|$ by a function of the type

$$v(t) = \mathcal{S}^*\mathbf{v} = \sum_{n\in\mathbb{Z}} v_n\, s_n(t), \qquad (16.52)$$

with $\mathbf{v} \in \mathcal{D}$. It is said in this case that $\{s_n(t)\}_{n\in\mathbb{Z}}$ is *complete* in $\mathcal{B}$.

In the case where $\mathcal{S}$ is the direct sampling operator at instants $(t_n)_{n\in\mathbb{Z}}$, we recall that the functions $s_n(t)$ of (16.52) are the shifted sinc functions $s_0(t-t_n)$. When $\mathcal{S}$ is injective, we know from the previous section that any $u(t) \in \mathcal{B}$ can be arbitrarily approached by linear combinations of these functions with coefficients in $\mathcal{D}$. One could see in this a generalized version of Shannon sampling theorem where the sinc functions are shifted in a nonuniform manner. However, a true generalization would require $u(t)$ to be exactly achieved by a fixed linear combination of the sinc functions with coefficients in $\mathcal{D}$. For this to be realized, one needs the stronger property that

$$\text{Range}(\mathcal{S}^*) = \mathcal{B}.$$

It is said in this case that $\mathcal{S}^*$ is *onto*, or *surjective*. With any operator $\mathcal{S}$, this is the required property to guarantee the existence of an exact expansion

$$x(t) = \sum_{n\in\mathbb{Z}} \tilde{x}_n\, s_n(t), \qquad (16.53)$$

with $(\tilde{x}_n)_{n\in\mathbb{Z}}$ in $\mathcal{D}$. The next theorem derived from [15] gives necessary and sufficient conditions for $\mathcal{S}^*$ to be surjective.

**Theorem 16.1**

Let $\mathcal{S}$ be a bounded operator from $\mathcal{B}$ to $\mathcal{D}$. The following propositions are equivalent:

(i) $\mathcal{S}^*$ is surjective ($\text{Range}(\mathcal{S}^*) = \mathcal{B}$).

(ii) $\mathcal{S}$ is injective and $\text{Range}(\mathcal{S})$ is closed in $\mathcal{D}$.

(iii) $\mathcal{S}^*\mathcal{S}$ is invertible.

(iv) There exists a constant $\alpha > 0$ such that for all $u(t)$ in $\mathcal{B}$,

$$\alpha\, \|u(t)\| \le \|\mathcal{S}u(t)\|_{\mathcal{D}}. \qquad (16.54)$$

The equivalence between (ii) and (iv) can be found in [15, theorem 1.2]. The equivalence between (i) and (ii) is easily established from (16.51) and the fact from [15, lemma 1.5] that $\text{Range}(\mathcal{S}^*)$ is closed in $\mathcal{B}$ if and only if $\text{Range}(\mathcal{S})$ is closed in $\mathcal{D}$. It is clear that (iii) implies (i). Assume now (i) and (ii). From [15, Lemma 1.4], $\text{Null}(\mathcal{S}^*\mathcal{S}) = \text{Null}(\mathcal{S})$, so $\mathcal{S}^*\mathcal{S}$ is injective due to (ii). Since $\text{Range}(\mathcal{S})$ is closed from (ii), it follows from (16.49) that $\mathcal{D} = \text{Range}(\mathcal{S}) \oplus \text{Null}(\mathcal{S}^*)$. With (i), $\mathcal{B} = \text{Range}(\mathcal{S}^*) = \mathcal{S}^*(\mathcal{D}) = \mathcal{S}^*(\text{Range}(\mathcal{S}) \oplus \text{Null}(\mathcal{S}^*)) = \mathcal{S}^*(\text{Range}(\mathcal{S})) = \mathcal{S}^*\mathcal{S}(\mathcal{B})$. So $\mathcal{S}^*\mathcal{S}$ is injective and surjective, which implies (iii).

The largest possible value of $\alpha$ in (16.54) is equal to

$$\gamma(\mathcal{S}) := \inf_{u(t)\in\mathcal{B}\setminus\{0\}} \frac{\|\mathcal{S}u(t)\|_{\mathcal{D}}}{\|u(t)\|} = \inf_{\substack{u(t)\in\mathcal{B}\\ \|u(t)\|=1}} \|\mathcal{S}u(t)\|_{\mathcal{D}}, \qquad (16.55)$$

which is called the *minimum modulus* of $\mathcal{S}$. Property (iv) is equivalent to stating that $\gamma(\mathcal{S}) > 0$.

### 16.3.4 Frame

Since Theorem 16.1 is based on the assumption that $\mathcal{S}$ is bounded, the two inequalities (16.10) and (16.54) are needed to ensure the surjectivity of $\mathcal{S}^*$. By squaring their members, these inequalities are equivalent to

$$A\,\|u(t)\|^2 \le \sum_{n\in\mathbb{Z}} |\langle s_n, u\rangle|^2 \le B\,\|u(t)\|^2, \qquad (16.56)$$

for all $u(t) \in \mathcal{B}$, where both $A = \alpha^2$ and $B = \beta^2$ are positive. It is said in this case that the family of functions

$\{s_n(t)\}_{n\in\mathbb{Z}}$ is a *frame* of $\mathcal{B}$. The constants $A$ and $B$ are called the frame bounds. This condition is equivalent to writing that there exist positive constants $A$ and $B$ such that

$$A \leq \gamma(\mathcal{S})^2 \qquad \text{and} \qquad \|\mathcal{S}\|^2 \leq B. \tag{16.57}$$

Under this condition, we know that $x(t)$ yields an exact expansion of the type (16.53). The current drawback of this expression is that it is not a function of the sample sequence $\mathbf{x} = \mathcal{S}x(t)$. This is fixed by the equivalence between the surjectivity of $\mathcal{S}^*$ and the invertibility of $\mathcal{S}^*\mathcal{S}$ in Theorem 16.1. The latter property allows the definition of the pseudoinverse $\mathcal{S}^+$ in (16.45), which provides us with a left inverse of $\mathcal{S}$. Hence, when $\mathbf{x} = \mathcal{S}x(t)$,

$$x(t) = \mathcal{S}^+\mathbf{x} = \sum_{n\in\mathbb{Z}} x_n\, \tilde{s}_n(t), \tag{16.58}$$

where $\{\tilde{s}_n(t)\}_{n\in\mathbb{Z}}$ are the impulse responses of $\mathcal{S}^+$. Since $\{s_n(t)\}_{n\in\mathbb{Z}}$ are the impulse responses of $\mathcal{S}^*$ and $\mathcal{S}^+ = (\mathcal{S}^*\mathcal{S})^{-1}\mathcal{S}^*$, then

$$\tilde{s}_n(t) := (\mathcal{S}^*\mathcal{S})^{-1}s_n(t), \tag{16.59}$$

for all $n \in \mathbb{Z}$. The invertibility of $\mathcal{S}^*\mathcal{S}$ also permits the derivation of a sequence $\tilde{\mathbf{x}} = (\tilde{x}_n)_{n\in\mathbb{Z}}$ satisfying (16.53) in terms of $\mathbf{x} = \mathcal{S}x(t)$. Indeed, note that $\mathcal{S}^+$ can be expressed as

$$\mathcal{S}^+ = \mathcal{S}^*\big[\mathcal{S}(\mathcal{S}^*\mathcal{S})^{-2}\mathcal{S}^*\big], \tag{16.60}$$

where $\mathcal{A}^2 := \mathcal{A}\mathcal{A}$ for any operator $\mathcal{A}$ of $\mathcal{B}$. Hence, $x(t) = \mathcal{S}^+\mathbf{x} = \mathcal{S}^*\tilde{\mathbf{x}}$ with

$$\tilde{\mathbf{x}} := \big[\mathcal{S}(\mathcal{S}^*\mathcal{S})^{-2}\mathcal{S}^*\big]\mathbf{x}. \tag{16.61}$$

Given that the initial knowledge is the family of functions $\{s_n(t)\}_{n\in\mathbb{Z}}$ and the sequence of samples $(x_n)_{n\in\mathbb{Z}}$, both reconstruction formulas (16.53) and (16.58) require some transformation to obtain either $\{\tilde{s}_n(t)\}_{n\in\mathbb{Z}}$ or $(\tilde{x}_n)_{n\in\mathbb{Z}}$. The difference between these two formulas, however, is that (16.58) involves a continuous-time transformation (see Equation 16.59), while (16.53) requires a discrete-time transformation (see Equation 16.61). Even though both $\mathcal{S}$ and $\mathcal{S}^*$ involve continuous-time signals (either as inputs or outputs), the global operator $\mathcal{S}(\mathcal{S}^*\mathcal{S})^{-2}\mathcal{S}^*$ of (16.61) has pure discrete-time inputs and outputs. This difference is of major importance for practical implementations.

### 16.3.5 Tight Frame

In general, the derivation of $\mathcal{S}^*\mathcal{S}$ and its inverse is a major difficulty. But there is a special case where this is trivial. It is when

$$\|\mathcal{S}u(t)\|_{\mathcal{D}} = \alpha\|u(t)\|, \tag{16.62}$$

for some constant $\alpha > 0$ and for all $u(t) \in \mathcal{B}$. Let us show in this case that

$$\mathcal{S}^*\mathcal{S} = \alpha^2\mathcal{I}. \tag{16.63}$$

By squaring the members of (16.62), one obtains $\alpha^2\langle u,u\rangle = \langle \mathcal{S}u, \mathcal{S}u\rangle_{\mathcal{D}} = \langle u, \mathcal{S}^*\mathcal{S}u\rangle$ so that $\langle u, \mathcal{A}u\rangle = 0$ with $\mathcal{A} := \mathcal{S}^*\mathcal{S} - \alpha^2\mathcal{I}$. This operator is easily seen to be self-adjoint, meaning that $\langle u, \mathcal{A}v\rangle = \langle \mathcal{A}u, v\rangle$ for all $u(t), v(t) \in \mathcal{B}$. It is known in this case that $\mathcal{A}$ must be the zero operator [14, corollary 2.14], which leads to (16.63). The pseudoinverse of $\mathcal{S}$ is then simply

$$\mathcal{S}^+ = \frac{1}{\alpha^2}\,\mathcal{S}^*.$$

The condition (16.62) is equivalent to the frame condition (16.75) with $A = B = \alpha^2$. The frame $\{s_n(t)\}_{n\in\mathbb{Z}}$ is said in this case to be *tight*.

### 16.3.6 Neumann Series

Another equivalent criterion can be added in Theorem 16.1, more specifically on the invertibility of $\mathcal{S}^*\mathcal{S}$. For any operator $\mathcal{A}$ of $\mathcal{B}$, the Neumann series

$$\mathcal{A}^{-1} = \sum_{m=0}^{\infty}(\mathcal{I} - \mathcal{A})^m, \tag{16.64}$$

is known to converge when

$$\|\mathcal{I} - \mathcal{A}\| < 1. \tag{16.65}$$

In (16.64), the notation $\mathcal{A}^m$ is recursively defined by $\mathcal{A}^{m+1} := \mathcal{A}\mathcal{A}^m$ with $\mathcal{A}^0 = \mathcal{I}$. The expansion (16.64) is nothing but a generalization of the geometric series $\sum_{m=0}^{\infty} u^m = (1-u)^{-1}$, or equivalently $\sum_{m=0}^{\infty}(1-u)^m = u^{-1}$. Although (16.65) is only a sufficient condition for $\mathcal{A}$ to be invertible, one obtains the following necessary and sufficient condition for $\mathcal{S}^*\mathcal{S}$ to be invertible.

**Proposition 16.1**

Let $\mathcal{S}$ be a bounded sampling operator. Then, $\mathcal{S}^*\mathcal{S}$ is invertible *if and only if* there exists $c > 0$ such that $\|\mathcal{I} - c\,\mathcal{S}^*\mathcal{S}\| < 1$.

Let us justify this result. When $\mathcal{S}^*\mathcal{S}$ is invertible, we know that the frame condition (16.56) is satisfied. As the central member in this condition is equal to $\langle u, \mathcal{S}^*\mathcal{S}u\rangle$ according to (16.50), it is easy to verify that $|\langle u, \mathcal{A}u\rangle| \leq \gamma\|u(t)\|^2$ where $\mathcal{A} := \mathcal{I} - c\,\mathcal{S}^*\mathcal{S}$,

$c := \frac{2}{A+B} > 0$, and $\gamma := \frac{B-A}{B+A} < 1$ [3]. As $\mathcal{A}^* = \mathcal{A}$, it is known that $\|\mathcal{A}\| = \sup_{\|u\|=1} |\langle u, \mathcal{A}u\rangle|$ [14, proposition 2.13]. Therefore, $\|\mathcal{A}\| \leq \gamma$. This proves that $\|\mathcal{I} - c\,\mathcal{S}^*\mathcal{S}\| < 1$. Conversely, if this inequality is satisfied for some $c > 0$, $c\,\mathcal{S}^*\mathcal{S}$ is invertible due to the Neumann series, and so is $\mathcal{S}^*\mathcal{S}$.

### 16.3.7 Critical Sampling and Oversampling

Assume that $\{s_n(t)\}_{n\in\mathbb{Z}}$ is a frame. We know that any of the propositions of Theorem 16.1 is true. Hence, $\mathcal{S}$ is injective and $\mathcal{S}^*$ is surjective. Meanwhile, no condition is implied on the injectivity of $\mathcal{S}^*$, or the surjectivity of $\mathcal{S}$. Note that these two properties are related since

$$\mathcal{D} = \text{Range}(\mathcal{S}) \oplus \text{Null}(\mathcal{S}^*), \qquad (16.66)$$

as a result of (16.49) plus the fact that Range$(\mathcal{S})$ is closed from Theorem 16.1(ii). Thus, $\mathcal{S}^*$ is injective if and only if $\mathcal{S}$ is surjective. This leads to two cases.

**Oversampling:** This is the case where $\mathcal{S}^*$ is not injective ($\mathcal{S}$ is not surjective). Since Null$(\mathcal{S}^*) \neq \{0\}$, there exists $\mathbf{u} = (u_n)_{n\in\mathbb{Z}} \in \mathcal{D}\backslash\{0\}$ such that $\mathcal{S}^*\mathbf{u} = \sum_{n\in\mathbb{Z}} u_n s_n(t) = 0$. At least one component $u_k$ of $(u_n)_{n\in\mathbb{Z}}$ is nonzero, so one can express $s_k(t)$ as a linear combination of $\{s_n(t)\}_{n\neq k}$ with square-summable coefficients. If we call $\mathcal{S}_k$ the sampling operator of kernel functions $\{s_n(t)\}_{n\neq k}$, one easily concludes that Range$(\mathcal{S}_k^*)$ = Range$(\mathcal{S}^*) = \mathcal{B}$ (since $\mathcal{S}^*$ is surjective). As $\mathcal{S}_k$ is obviously still bounded, it follows from the implication (i) $\Rightarrow$ (iv) of Theorem 16.1 that $\{s_n(t)\}_{n\neq k}$ is still a frame. This justifies the term of "oversampling" for this case. Meanwhile, as $\mathcal{S}$ is not surjective, Range$(\mathcal{S})$ does not cover the whole space $\mathcal{D}$. In other words, there are sequences $\mathbf{u} \in \mathcal{D}$ that cannot be the sampled version of any signal in $\mathcal{B}$. An intuitive way to understand this phenomenon is to think that oversampling creates samples with some interdependence which prevents them from attaining all possible values.

**Critical sampling:** This is the case where $\mathcal{S}^*$ is injective ($\mathcal{S}$ is surjective). Then, $\mathcal{S}^*$ is invertible since it is also surjective. So every signal of $\mathcal{B}$ is uniquely expanded as a linear combination of $\{s_n(t)\}_{n\in\mathbb{Z}}$ with square-summable coefficients. Hence, $\{s_n(t)\}_{n\in\mathbb{Z}}$ is a basis of $\mathcal{B}$. Under the special condition (16.56), it is said to be a *Riesz basis*. In this situation, $s_k(t)$ never belongs to Range$(\mathcal{S}_k^*)$ for any $k \in \mathbb{Z}$, and consequently, Range$(\mathcal{S}_k^*)$ never covers the whole space $\mathcal{B}$. Therefore, $\{s_n(t)\}_{n\in\mathbb{Z}}$ ceases to be a frame after the removal of any of its functions. This family is said to be an *exact frame*. The term "critical sampling" for this case is from a signal-processing perspective. Now, $\mathcal{S}$ is also invertible (since it is also surjective and injective). So $\mathcal{S}^+$ coincides with $\mathcal{S}^{-1}$ and $\tilde{s}_k(t) = \mathcal{S}^+\boldsymbol{\delta}_k = \mathcal{S}^{-1}\boldsymbol{\delta}_k$ for all $k \in \mathbb{Z}$. Then,

$$\mathcal{S}\tilde{s}_k(t) = \boldsymbol{\delta}_k, \qquad (16.67)$$

for all $k \in \mathbb{Z}$. The $n$th components of the above sequences yield the equality

$$\langle s_n, \tilde{s}_k\rangle = \begin{cases} 1, & n = k \\ 0, & n \neq k \end{cases}.$$

It is said that $\{\tilde{s}_n(t)\}_{n\in\mathbb{Z}}$ is biorthogonal to $\{s_n(t)\}_{n\in\mathbb{Z}}$.

### 16.3.8 Optimality of Pseudoinverse

We show that the pseudoinverse $\mathcal{S}^+$ is the left inverse of $\mathcal{S}$ that is the least sensitive to sample errors with respect to the norms $\|\cdot\|_{\mathcal{D}}$ and $\|\cdot\|$. Assume that the acquired sample sequence of a signal $x(t)$ of $\mathcal{B}$ is

$$\mathbf{x} = \mathcal{S}x(t) + \mathbf{e}, \qquad (16.68)$$

where $\mathbf{e}$ is some sequence of sample errors. When attempting a reconstruction of $x(t)$ by applying a left inverse $\mathcal{R}$ of $\mathcal{S}$, we obtain

$$\mathcal{R}\mathbf{x} = x(t) + \mathcal{R}\mathbf{e}. \qquad (16.69)$$

Thus, $\mathcal{R}\mathbf{e}$ is the error of reconstruction. One is concerned with its magnitude $\|\mathcal{R}\mathbf{e}\|$ with respect to the error norm $\|\mathbf{e}\|_{\mathcal{D}}$. Analytically, one wishes to minimize the norm of $\mathcal{R}$ defined by

$$\|\mathcal{R}\| := \sup_{\mathbf{e}\in\mathcal{D}\backslash\{0\}} \frac{\|\mathcal{R}\mathbf{e}\|}{\|\mathbf{e}\|_{\mathcal{D}}}, \qquad (16.70)$$

in a way similar to (16.11). Let us show that

$$\|\mathcal{R}\| \geq \|\mathcal{S}^+\| = \gamma(\mathcal{S})^{-1}. \qquad (16.71)$$

The inequality $\|\mathcal{R}\| \geq \gamma(\mathcal{S})^{-1}$ results from (16.55) and the relations

$$\begin{aligned}\sup_{\mathbf{e}\in\mathcal{D}\backslash\{0\}} \frac{\|\mathcal{R}\mathbf{e}\|}{\|\mathbf{e}\|_{\mathcal{D}}} &\geq \sup_{u(t)\in\mathcal{B}\backslash\{0\}} \frac{\|\mathcal{R}\mathcal{S}u(t)\|}{\|\mathcal{S}u(t)\|_{\mathcal{D}}} \\ &= \sup_{u(t)\in\mathcal{B}\backslash\{0\}} \frac{\|u(t)\|}{\|\mathcal{S}u(t)\|_{\mathcal{D}}} \\ &= \left(\inf_{u(t)\in\mathcal{B}\backslash\{0\}} \frac{\|\mathcal{S}u(t)\|_{\mathcal{D}}}{\|u(t)\|}\right)^{-1}.\end{aligned}$$

This implies in particular that $\|\mathcal{S}^+\| \geq \gamma(\mathcal{S})^{-1}$ since $\mathcal{S}^+$ is a left inverse of $\mathcal{S}$. So what remains to be proved is that $\|\mathcal{S}^+\| \leq \gamma(\mathcal{S})^{-1}$. Let $\mathbf{e} \in \mathcal{D}\backslash\{0\}$. According to (16.66), $\mathbf{e}$ can be decomposed as $\mathbf{e} = \mathcal{S}e(t) + \mathbf{e}_0$, where $e(t) \in \mathcal{B}$ and $\mathbf{e}_0 \in$ Null$(\mathcal{S}^*)$. Now, $\mathcal{S}e(t) \perp \mathbf{e}_0$ due to (16.49). By the Pythagorean theorem, $\|\mathbf{e}\|_{\mathcal{D}} \geq \|\mathcal{S}e(t)\|_{\mathcal{D}}$. Meanwhile, $\mathcal{S}^+\mathbf{e} = \mathcal{S}^+\mathcal{S}e(t) + \mathcal{S}^+\mathbf{e}_0 = e(t)$ since Null$(\mathcal{S}^*) \subset$ Null$(\mathcal{S}^+)$. So $\|\mathcal{S}^+\mathbf{e}\| = \|e(t)\| \leq \gamma(\mathcal{S})^{-1}\|\mathcal{S}e(t)\|_{\mathcal{D}} \leq \gamma(\mathcal{S})^{-1}\|\mathbf{e}\|_{\mathcal{D}}$. This suffices to prove that $\|\mathcal{S}^+\| \leq \gamma(\mathcal{S})^{-1}$.

### 16.3.9 Condition Number of Sampling–Reconstruction System

The sensitivity of the global sampling reconstruction system to errors in the samples is more precisely evaluated as follows. Consider again Equations 16.68 and 16.69, where an error sequence $\mathbf{e}$ is added to the sample sequence $\mathcal{S}x(t)$ and a left inverse $\mathcal{R}$ of $\mathcal{S}$ is applied on $\mathbf{x} = \mathcal{S}x(t) + \mathbf{e}$ to obtain the reconstruction $\mathcal{R}\mathbf{x} = x(t) + \mathcal{R}\mathbf{e}$. While the relative error in the input $\mathbf{x}$ is $\|\mathbf{e}\|_{\mathcal{D}} / \|\mathcal{S}x(t)\|_{\mathcal{D}}$, the relative error in the reconstruction $\mathcal{R}\mathbf{x}$ is $\|\mathcal{R}\mathbf{e}\| / \|x(t)\|$. In practice, one wishes to limit the worst-case ratio between these two quantities

$$\kappa := \sup_{\mathbf{e},x(t)} \frac{\|\mathcal{R}\mathbf{e}\| / \|x(t)\|}{\|\mathbf{e}\|_{\mathcal{D}} / \|\mathcal{S}x(t)\|_{\mathcal{D}}}, \tag{16.72}$$

where $\mathbf{e}$ and $x(t)$ are taken among all nonzero signals of $\mathcal{D}$ and $\mathcal{B}$, respectively. Note that $\kappa$ cannot be less than 1 [since the ratio in (16.72) is 1 when $\mathbf{e} = \alpha\,\mathcal{S}x(t)$ in which case $\mathcal{R}\mathbf{e} = \alpha\, x(t)$]. It is easy to see that

$$\kappa = \|\mathcal{R}\|\;\|\mathcal{S}\|. \tag{16.73}$$

According to (16.71),

$$\kappa \geq \kappa(\mathcal{S}) := \frac{\|\mathcal{S}\|}{\gamma(\mathcal{S})}, \tag{16.74}$$

and $\kappa = \kappa(\mathcal{S})$ when $\mathcal{R}$ is chosen to be the pseudoinverse $\mathcal{S}^{+}$. The value $\kappa(\mathcal{S})$ is intrinsic to the operator $\mathcal{S}$ and is called the condition number of $\mathcal{S}$. To have $\kappa(\mathcal{S})$ as small as possible, one needs to have $\gamma(\mathcal{S})$ and $\|\mathcal{S}\|$ as close as possible to each other in ratio. They are equal when and only when the condition (16.62) is met, and hence when the frame $\{s_n(t)\}_{n\in\mathbb{Z}}$ is tight. When only frame bounds $A$ and $B$ in (16.56) are known, we obtain from (16.57) that

$$\kappa(\mathcal{S}) \leq \sqrt{\frac{B}{A}}.$$

Again, $A$ and $B$ should be as close as possible to each other in ratio.

## 16.4 Direct Sampling of Aperiodic Bandlimited Signals

For the reconstruction of a signal $x(t) \in \mathcal{B}$ from $\mathbf{x} = \mathcal{S}x(t)$ to be realistically possible, we saw in the previous section that it is desirable for the kernel functions $\{s_n(t)\}_{n\in\mathbb{Z}}$ of $\mathcal{S}$ to be a frame. In this section, we concentrate on the direct sampling operator $\mathcal{S}u(t) = \big(u(t_n)\big)_{n\in\mathbb{Z}}$. In this case, the kernel functions are such that $\langle s_n, u\rangle = u(t_n)$, and hence the frame condition (16.56) takes the form

$$A\,\|u(t)\|^2 \;\leq\; \sum_{n\in\mathbb{Z}} \big|u(t_n)\big|^2 \;\leq\; B\,\|u(t)\|^2, \tag{16.75}$$

for some constants $B \geq A > 0$ and all $u(t) \in \mathcal{B}$. We recall from (16.21) that $s_n(t) = s_o(t - t_n)$ where $s_o(t) := \mathrm{sinc}(\frac{t}{T_o})$. The main question is what sampling instants $(t_n)_{n\in\mathbb{Z}}$ can guarantee this condition and what formulas of reconstruction are available.

### 16.4.1 Oversampling

We saw in the periodic case of Section 16.2.3 that $\mathcal{S}$ is injective if and only if the number of samples within one period of $x(t)$ is at least the number of Nyquist periods within this time interval. As an aperiodic signal is often thought of as a periodic signal of infinite period, one would conjecture that $\mathcal{S}$ is injective when the number of samples is at least the number of Nyquist periods within any time interval centered about 0 of large enough size. This property is achieved when there exist two constants

$$T_s < T_o \qquad \text{and} \qquad \mu \geq 0,$$

such that

$$|t_n - nT_s| \leq \mu T_s, \tag{16.76}$$

for all $n$. It was actually proved in [3] that this guarantees the frame property of (16.75) under the additional condition that

$$t_{n+1} - t_n \geq d > 0, \tag{16.77}$$

for all $n \in \mathbb{Z}$ and some constant $d$.

Note that conditions (16.76) and (16.77) remain valid after the suppression of any sample. Indeed, if we remove any given instant $t_k$ by defining $t'_n = t_n$ for all $n < k$ and $t'_n = t_{n+1}$ for all $n \geq k$, the sequence $t'_n$ satisfies (16.76) with $\mu' = \mu + 1$ and (16.77) with $d' = d$. Then, $\big(x(t'_n)\big)_{n\in\mathbb{Z}}$ satisfies (16.75) for some bounds $B' > A' > 0$. This is a situation of oversampling according to Section 16.3.7. This can be in fact repeated any finite number of times. Thus, $x(t)$ can still be recovered from $\big(x(t_n)\big)_{n\in\mathbb{Z}}$ after dropping any finite subset of samples.

### 16.4.2 Critical Sampling

It was shown in [7,16,17] that (16.75) is also achieved under (16.76) and (16.77) with

$$T_s = T_o \qquad \text{and} \qquad \mu < \frac{1}{4}.$$

It was also shown that $\{s_n(t)\}_{n\in\mathbb{Z}}$ is an exact frame. This is the case of critical sampling described in

Section 16.3.7, where $\mathcal{S}$ is invertible. From (16.67), we have

$$\tilde{s}_n(t_k) = \begin{cases} 1, & k = n \\ 0, & k \neq n \end{cases}. \quad (16.78)$$

similar to (16.43). This is the generalization in infinite dimension of the interpolation functions obtained in Section 16.2.4 in the periodic case with $N = K$. The case $\mu = 0$ corresponds to uniform sampling and brings us back to Shannon sampling theorem.

### 16.4.3 Lagrange Interpolation

In the above case of critical sampling, the reconstruction functions $\{\tilde{s}_n(t)\}_{n\in\mathbb{Z}}$ of (16.58) were also shown in [7,16,17] to coincide with the Lagrange interpolation functions

$$r_n(t) = \prod_{\substack{i=-\infty \\ i\neq n}}^{\infty} \frac{t - t_i}{t_n - t_i}. \quad (16.79)$$

It is easy to verify by construction that these functions satisfy (16.78).

In the oversampling case $T_s < T_o$ with any finite value $\mu$, the left inverses of $\mathcal{S}$ are no longer unique and the impulse responses $\{\tilde{s}_n(t)\}_{n\in\mathbb{Z}}$ of $\mathcal{S}^+$ need not satisfy the interpolation property of (16.78) as was demonstrated in the periodic case (see Figure 16.4b). It was shown in [7] that a left inverse $\mathcal{R}$ of $\mathcal{S}$ can still be found with impulse responses $\{\tilde{r}_n(t)\}_{n\in\mathbb{Z}}$ satisfying (16.78) (generalized Lagrange interpolation functions). We will, however, skip the details of these functions as they involve exponential functions with little potential for practical implementations.

When $T_s < T_o$ and $\mu < \frac{1}{4}$, however, it is interesting to see that the operator $\mathcal{R}$ of impulse responses $\{r_n(t)\}_{n\in\mathbb{Z}}$ given in (16.79) still satisfies $\mathcal{RS}u(t) = u(t)$ for any $u(t) \in \mathcal{B}$. The rigorous way to explain this is that $\mathcal{RS} = \mathcal{I}$ in the larger space $\mathcal{B}'$ of signals with no frequency content beyond $\Omega_o' := \frac{\pi}{T_s}$, which is indeed larger than $\Omega_o = \frac{\pi}{T_o}$. One simply needs to think of $T_o' := T_s$ as the Nyquist period in $\mathcal{B}'$. In this view, $\mathcal{R}$ is not just the pseudoinverse of $\mathcal{S}$ in $\mathcal{B}'$ but is also its exact inverse (since the sampling is critical with respect to signals in $\mathcal{B}'$). The considered input $x(t)$ just happens to be in the subspace $\mathcal{B}$ of $\mathcal{B}'$, which is not a problem. What prevents $\mathcal{R}$ from being rigorously a left inverse of $\mathcal{S}$ in $\mathcal{B}$ is that the functions $\{r_n(t)\}_{n\in\mathbb{Z}}$ belong to $\mathcal{B}'$ and have no reason to be in $\mathcal{B}$. A consequence of practical impact is when some random error sequence $\mathbf{e}$ is added to $\mathcal{S}x(t)$ as in (16.68). Even though $x(t)$ is in $\mathcal{B}$, the reconstruction error $\mathcal{R}\mathbf{e}$ shown in (16.69) belongs to $\mathcal{B}'$ and most likely yields frequency oscillations beyond the baseband of $x(t)$.

### 16.4.4 Weighted Sampling

We started Section 16.4 with the intuition that perfect reconstruction should be possible when the number of samples is at least the number of Nyquist periods within any time interval centered about 0 of large enough size. Condition (16.76) imposes that $(t_n)_{n\in\mathbb{Z}}$ is *uniformly dense* according to the terminology of [9, equation 34]. Intuitively, it seems, however, that perfect reconstruction should be achievable with a nonuniform density of sampling instants as long as it remains larger than the Nyquist density at any point. The only feature of the direct sampling operator that prevents this is in fact the uniform norm of its kernel functions. With the only assumption that

$$t_{n+1} - t_n \leq d' < T_o, \quad (16.80)$$

for all $n \in \mathbb{Z}$ and some constant $d'$, it was shown in [18, equation 4.6] that there exist constants $B \geq A > 0$ such that

$$A\,\|u(t)\|^2 \leq \sum_{n\in\mathbb{Z}} c_n |u(t_n)|^2 \leq B\,\|u(t)\|^2, \quad (16.81)$$

for all $u(t) \in \mathcal{B}$, where

$$c_n := \frac{t_{n+1} - t_{n-1}}{2T_o}. \quad (16.82)$$

Specific values of the bounds are $A = \left(1 - \frac{T_m}{T_o}\right)^2$ and $B = \left(1 + \frac{T_m}{T_o}\right)^2$ where

$$T_m := \sup_{n\in\mathbb{Z}}(t_{n+1} - t_n). \quad (16.83)$$

The weight $c_n$ can be interpreted as an estimate of the local normalized sampling period. Thus, the frame condition (16.56) is achieved by the kernel functions $\{\sqrt{c_n}\, s_0(t - t_n)\}$ just based on condition (16.80). Hence, the sampling operator $\mathcal{S}'$ defined by these kernel functions is such that $\mathcal{S}'^*\mathcal{S}'$ is invertible. Maintaining $\mathcal{S}$ as the direct sampling operator at instants $(t_n)_{n\in\mathbb{Z}}$, it is easy to derive that $\mathcal{S}'^*\mathcal{S}' = \mathcal{S}^*\mathbf{C}\mathcal{S}$ where $\mathbf{C}$ is the diagonal operator of $\mathcal{D}$ such that the $n$th component of $\mathbf{Cu}$ is

$$(\mathbf{Cu})_n = c_n u_n. \quad (16.84)$$

Since $(\mathcal{S}^*\mathbf{C}\mathcal{S})^{-1}\mathcal{S}^*\mathbf{C}\mathcal{S} = \mathcal{I}$, the operator

$$\mathcal{R} := (\mathcal{S}^*\mathbf{C}\mathcal{S})^{-1}\mathcal{S}^*\mathbf{C},$$

is a left inverse of $\mathcal{S}$. Interestingly, this works without the requirement that $\mathcal{S}$ be bounded, and worse, when $\mathcal{S}x(t)$ is not even square summable (and hence not in $\mathcal{D}$). This is because there is no constraint on how large the density of $(t_n)_{n\in\mathbb{Z}}$ can be.

## 16.5 Constrained Reconstruction

The reconstruction $x(t) = \mathcal{R}\mathbf{x}$ from $\mathbf{x} = \mathcal{S}x(t)$ is the most delicate part of real implementations. The first reason is the complicated definition of the impulse responses of a left inverse $\mathcal{R}$ as can be seen, for example, in (16.59) when $\mathcal{R} = \mathcal{S}^+$. As a second reason, these impulse responses must be produced in continuous-time implementations by analog circuits with limited precision and freedom from a mathematical viewpoint. Since the input to a reconstruction operator is discrete in time, an alternative approach is to look for left inverses of $\mathcal{S}$ in the form $\mathcal{R}\mathbf{D}$ where $\mathbf{D}$ is an operator from $\mathcal{D}$ to $\mathcal{D}$, which we will simply call a *discrete-time operator*. In this way, computational requirements on $\mathcal{R}$ can be relaxed and partly moved to $\mathbf{D}$ which in practice is implemented by pure digital circuits. A more rational way to proceed is to seek the condition on imposed operators $\mathcal{S}$ and $\mathcal{R}$, for $\mathcal{R}\mathbf{D}\mathcal{S} = \mathcal{I}$ to be achievable with some discrete-time operator $\mathbf{D}$. When this is realized, we will say that $(\mathcal{S},\mathcal{R})$ is an admissible sampling–reconstruction pair. When $\mathbf{D}$ can be found explicitly, then the $\mathcal{R}\mathbf{D}\mathcal{S}$ system of Figure 16.3 achieves perfect reconstruction.

### 16.5.1 Condition for $(\mathcal{S},\mathcal{R})$ to Be Admissible

Let us show the following result.

**Proposition 16.2**

Let $\mathcal{S}$ and $\mathcal{R}$ be bounded sampling and reconstruction operators, respectively. The following statements are equivalent:

(i) The equality $\mathcal{R}\mathbf{D}\mathcal{S} = \mathcal{I}$ can be achieved with some discrete-time operator $\mathbf{D}$.

(ii) $\mathcal{S}$ is injective and $\mathcal{R}$ is surjective.

(iii) The inequality $\|\mathcal{I}-\mathcal{R}\mathbf{C}\mathcal{S}\| < 1$ can be achieved with some discrete-time operator $\mathbf{C}$.

Assume (i). It is easy to see that $\mathrm{Null}(\mathcal{S}) \subset \mathrm{Null}(\mathcal{R}\mathbf{D}\mathcal{S}) = \{0\}$ and $\mathrm{Range}(\mathcal{R}) \supset \mathrm{Range}(\mathcal{R}\mathbf{D}\mathcal{S}) = \mathcal{B}$. This implies (ii). Conversely, assume (ii). In the same way the injectivity of $\mathcal{S}$ guarantees the existence of a left inverse $\check{\mathcal{S}}$ of $\mathcal{S}$, it can be shown that the surjectivity of $\mathcal{R}$ implies the existence of a *right inverse* of $\mathcal{R}$, that is, a sampling operator $\check{\mathcal{R}}$ such that $\mathcal{R}\check{\mathcal{R}} = \mathcal{I}$ [one can show that the restriction of $\mathcal{R}$ to any subspace of $\mathcal{D}$ complementary to $\mathrm{Null}(\mathcal{R})$ is invertible and take for $\check{\mathcal{R}}$ its inverse]. With the discrete-time operator

$$\mathbf{D} := \check{\mathcal{R}}\check{\mathcal{S}}, \tag{16.85}$$

we obtain $\mathcal{R}\mathbf{D}\mathcal{S} = (\mathcal{R}\check{\mathcal{R}})(\check{\mathcal{S}}\mathcal{S}) = \mathcal{I}$, which proves (i). The implication (i) $\Rightarrow$ (iii) is trivial since $\|\mathcal{I}-\mathcal{R}\mathbf{D}\mathcal{S}\| = 0$. Finally, assume (iii). We know from the Neumann series that $\mathcal{R}\mathbf{C}\mathcal{S}$ is invertible. With the discrete-time operator

$$\mathbf{D} := \mathbf{C}\mathcal{S}(\mathcal{R}\mathbf{C}\mathcal{S})^{-2}\mathcal{R}\mathbf{C}, \tag{16.86}$$

we obtain $\mathcal{R}\mathbf{D}\mathcal{S} = \mathcal{R}\mathbf{C}\mathcal{S}(\mathcal{R}\mathbf{C}\mathcal{S})^{-2}\mathcal{R}\mathbf{C}\mathcal{S} = \mathcal{I}$, which proves (i).

We have a number of remarks.

1. The condition (ii) for $(\mathcal{S},\mathcal{R})$ to be admissible is quite weak as $\mathcal{S}$ and $\mathcal{R}$ only need to satisfy separate properties and need not have any connection between each other, at least from a theoretical viewpoint. For $\mathcal{R}$ to be surjective, it is necessary and sufficient that its impulse responses $\{r_n(t)\}_{n\in\mathbb{Z}}$ form a frame. Indeed, calling $\mathcal{S}'$ the sampling operator whose kernel functions are $\{r_n(t)\}_{n\in\mathbb{Z}}$, we know from Sections 16.3.3 and 16.3.4 that these functions form a frame if and only if $\mathcal{S}'^*$ is surjective. Now $\mathcal{S}'^*$ is nothing but $\mathcal{R}$. Meanwhile, the mere injectivity of $\mathcal{S}$ does not require its kernel functions to form a frame.

2. When condition (ii) is satisfied, it is easy to show that $\mathcal{R}\mathbf{D}\mathcal{S} = \mathcal{I}$ if and only if $(\mathcal{S}\mathcal{R})\mathbf{D}(\mathcal{S}\mathcal{R}) = \mathcal{S}\mathcal{R}$. As a result, $\mathcal{R}\mathbf{D}\mathcal{S} = \mathcal{I}$ if and only if $\mathbf{D}$ is a *generalized inverse* of the discrete-time operator $\mathcal{S}\mathcal{R}$ [19, definition 1.38]. The idea of generalized inversion can also be qualitatively seen as follows. To obtain $\mathcal{R}\mathbf{D}\mathcal{S} = \mathcal{I}$, one needs to have $u(t) = \mathcal{R}\tilde{\mathbf{u}}$ where $\tilde{\mathbf{u}} := \mathbf{D}\mathbf{u}$ and $\mathbf{u} := \mathcal{S}u(t)$, for any given $u(t) \in \mathcal{B}$. The second relation $\tilde{\mathbf{u}} := \mathbf{D}\mathbf{u}$ then appears as an "inversion" of the equation $\mathbf{u} = \mathcal{S}\mathcal{R}\tilde{\mathbf{u}}$ that results from the first and third relations. One cannot talk about the inverse of $\mathcal{S}\mathcal{R}$ as in the most general case, $\mathcal{S}\mathcal{R}$ is neither injective nor surjective. Hence, the term "inversion" can only be understood here in a generalized sense.

3. The condition (iii) for $(\mathcal{S},\mathcal{R})$ to be admissible is mainly interesting for its practical contribution. Once a discrete-time operator $\mathbf{C}$ has been found to achieve $\|\mathcal{I}-\mathcal{R}\mathbf{C}\mathcal{S}\| < 1$, (16.86) gives a concrete way to obtain a generalized inverse $\mathbf{D}$ of $\mathcal{S}\mathcal{R}$.

4. Once condition (ii) is known to be satisfied, (16.85) gives another method to find a generalized inverse $\mathbf{D}$ of $\mathcal{S}\mathcal{R}$. However, it is likely to be less practical than (16.86), as separate generalized inverse $\check{\mathcal{S}}$ and $\check{\mathcal{R}}$ to $\mathcal{S}$ and $\mathcal{R}$ need to be found [we will see later that the square power in (16.86) can be eliminated when dealing with numerical approximations]. Meanwhile, (16.85) has its theoretical appeal. Specifically, (16.85) yields as a particular solution the Moore–Penrose pseudoinverse $(\mathcal{S}\mathcal{R})^+$ of $\mathcal{S}\mathcal{R}$ [19]. This is achieved by taking $\check{\mathcal{S}} = \mathcal{S}^+$ and $\check{\mathcal{R}} = \mathcal{R}^+$ where

$$\mathcal{R}^+ := \mathcal{R}^*(\mathcal{R}\mathcal{R}^*)^{-1}.$$

This operator is well defined from the mere condition that $\mathcal{R}$ is surjective from (ii). Indeed, the sampling operator $\mathcal{S}'$ introduced in the first remark above is such that $\mathcal{S}'^* = \mathcal{R}$ and hence $\mathcal{R}^* = (\mathcal{S}'^*)^* = \mathcal{S}'$. Then $\mathcal{R}\mathcal{R}^* = \mathcal{S}'^*\mathcal{S}'$, which is invertible by Theorem 16.1 since $\mathcal{S}'^* = \mathcal{R}$ is surjective.

5. In the case where $\mathcal{R} = \mathcal{S}^*$, the operator of (16.61) can be seen as the particular case of (16.85) with $\check{\mathcal{S}} = \mathcal{S}^+$ and $\check{\mathcal{R}} = (\mathcal{S}^*)^+$, and of (16.86) with $\mathbf{C} = \mathbf{I}$.

### 16.5.2 Known Cases Where $\|\mathcal{I} - \mathcal{R}\mathbf{C}\mathcal{S}\| < 1$

With the direct sampling operator $\mathcal{S}$ at instants $(t_n)_{n\in\mathbb{Z}}$ under the mere sampling condition (16.80), a number of reconstruction operators $\mathcal{R}$ have been found such that $\|\mathcal{I} - \mathcal{R}\mathbf{C}\mathcal{S}\| < 1$ with some simple, discrete-time operator $\mathbf{C}$. In this case, we recall from the previous section that $\mathcal{R}\mathbf{D}$ is a left inverse of $\mathcal{S}$ with $\mathbf{D}$ given in (16.86).

With the operator

$$\mathcal{R}\mathbf{u} := \varphi_o(t) * \check{\mathcal{P}}\mathbf{u}, \tag{16.87}$$

where

$$\check{\mathcal{P}}\mathbf{u} := \sum_{n\in\mathbb{Z}} u_n \check{p}_n(t),$$

and $\check{p}_n(t)$ is the rectangular function described in Figure 16.2a, it was shown in [18, equation 5.7] that $\|\mathcal{I} - \mathcal{R}\mathcal{S}\| \le \frac{T_m}{T_o}$, where $T_m$ is given in (16.83). This operator $\mathcal{R}$ is similar to that of (16.26) except that the $n$th impulse response $\check{p}_n(t)$ of $\check{\mathcal{P}}$ is the rectangular function on $[\check{t}_n, \check{t}_{n+1})$ instead of $[t_n, t_{n+1})$, where $\check{t}_n := \frac{1}{2}(t_{n-1}+t_n)$ (see Figure 16.2a). Under the condition (16.80), we obtain $\|\mathcal{I} - \mathcal{R}\mathcal{S}\| < 1$. This is a case where $\mathbf{C}$ is simply identity in $\mathcal{D}$. This is actually the result that enabled the proof of (16.81).

From a practical viewpoint, the operator $\mathcal{R}$ of (16.87) may be somewhat inconvenient as it introduces the extra time instants $(\check{t}_n)_{n\in\mathbb{Z}}$ in the processing. Meanwhile, if one tries to analyze $\|\mathcal{I}-\mathcal{R}\mathcal{S}\|$ with $\mathcal{R}$ from (16.26), one will face some obstacle as this operator holds its input value $y_n$ at $t_n$ during the interval $[t_n, t_{n+1}]$ before bandlimitation, causing a global signal delay in the operator $\mathcal{R}\mathcal{S}$. This can be fixed by feeding into $\mathcal{R}$ at $t_n$ the average between $y_n$ and $y_{n+1}$, instead of $y_n$. This leads us to consider the analysis of $\|\mathcal{I}-\mathcal{R}\mathbf{C}\mathcal{S}\|$, where $\mathbf{C}$ is the discrete-time operator such that the $n$th component of $\mathbf{C}\mathbf{y}$ is equal to

$$(\mathbf{C}\mathbf{y})_n := \tfrac{1}{2}(y_n+y_{n+1}). \tag{16.88}$$

By slightly modifying the mathematical arguments of [18], we prove in Appendix 16.5.4 the following result.

**Proposition 16.3**

Let $\mathcal{S}$ be the direct sampling operator at $(t_n)_{n\in\mathbb{Z}}$, $\mathcal{R}$ be the operator defined in (16.26), and $\mathbf{C}$ be the discrete-time operator defined in (16.88). Then, $\|\mathcal{I}-\mathcal{R}\mathbf{C}\mathcal{S}\| \le \frac{T_m}{T_o}$.

Again, one obtains $\|\mathcal{I}-\mathcal{R}\mathbf{C}\mathcal{S}\| < 1$ under (16.80).

As a final example, it was shown in [20, equation 48] that $\|\mathcal{I}-\mathcal{R}\mathcal{S}\| \le (\frac{T_m}{T_o})^2$ with the operator

$$\mathcal{R}\mathbf{u} := \varphi_o(t) * \mathcal{L}\mathbf{u}, \tag{16.89}$$

where

$$\mathcal{L}\mathbf{u} := \sum_{n\in\mathbb{Z}} u_n\, l_n(t),$$

and $l_n(t)$ is the triangular function shown in Figure 16.2b. In other words, $\mathcal{R}$ is the linear interpolator at instants $(t_n)_{n\in\mathbb{Z}}$ followed by bandlimitation. Once again, one obtains $\|\mathcal{I}-\mathcal{R}\mathcal{S}\| < 1$ under (16.80).

Given that $\|\mathcal{A}^*\| = \|\mathcal{A}\|$ for any bounded operator $\mathcal{A}$ of $\mathcal{B}$, and $(\mathcal{R}\mathbf{C}\mathcal{S})^* = \mathcal{S}^*\mathbf{C}^*\mathcal{R}^*$, one obtains $\|\mathcal{I}-\mathcal{S}^*\mathbf{C}^*\mathcal{R}^*\| = \|\mathcal{I}-\mathcal{R}\mathbf{C}\mathcal{S}\|$. This identity allows to consider the reconstruction operator $\mathcal{R}' = \mathcal{S}^*$ (sinc reconstruction) with the sampling operator $\mathcal{S}' = \mathcal{R}^*$, where $\mathcal{R}$ is any of the operators listed above [11]. The adjoint $\mathbf{C}^*$ can be viewed as the conjugate transpose of $\mathbf{C}$ seen as an infinite matrix. When $\mathcal{R}$ is defined by (16.26), the kernel functions of the sampling operator $\mathcal{S}' = \mathcal{R}^*$ coincide with those of (16.22) (up to a scaling factor $T_o$).

### 16.5.3 Experiments

We return to the problem of Section 16.1.7 where continuous-time amplitude quantization followed by bandlimitation was modeled as an $\mathcal{R}\mathbf{D}\mathcal{S}$ system. We recall in this case that $\mathcal{S}$ is the direct sampling operator at the instants $(t_n)_{n\in\mathbb{Z}}$ where $x(t)$ crosses the quantization levels, $\mathcal{R}$ performs a zero-order hold followed by a bandlimitation as described in (16.26), and $\mathbf{D}$ is defined by the recursive equation (16.27). By nature of bandlimited quantization, this $\mathcal{R}\mathbf{D}\mathcal{S}$ system only performs an approximate reconstruction of the input. When the zero-order hold is extended to real-valued discrete-time inputs (as opposed to just discrete-amplitude discrete-time inputs in standard quantization), we know from the previous section that $\mathcal{R}\mathbf{D}\mathcal{S} = \mathcal{I}$ can be achieved with the operator $\mathbf{D}$ of (16.86) by the operator $\mathbf{C}$ of (16.88) under condition (16.80). We plot in Figure 16.5a the result of a simulation performed with periodic signals, where the bandlimited input $x(t)$ is the black solid curve, the quantizer output is the piecewise constant signal, and its bandlimited version $\hat{x}(t)$ is the gray dashed curve. For reference, the gray horizontal lines represent the quantization levels. According to the $\mathcal{R}\mathbf{D}\mathcal{S}$ model, the quantizer output is also the zero-order hold of the sequence $(y_n)_{n\in\mathbb{Z}}$ output by the operator $\mathbf{D}$ of (16.27). We show in Figure 16.5b the new piecewise-constant signal obtained when the operator $\mathbf{D}$ of (16.86) is used instead with $\mathbf{C}$ from (16.88), and

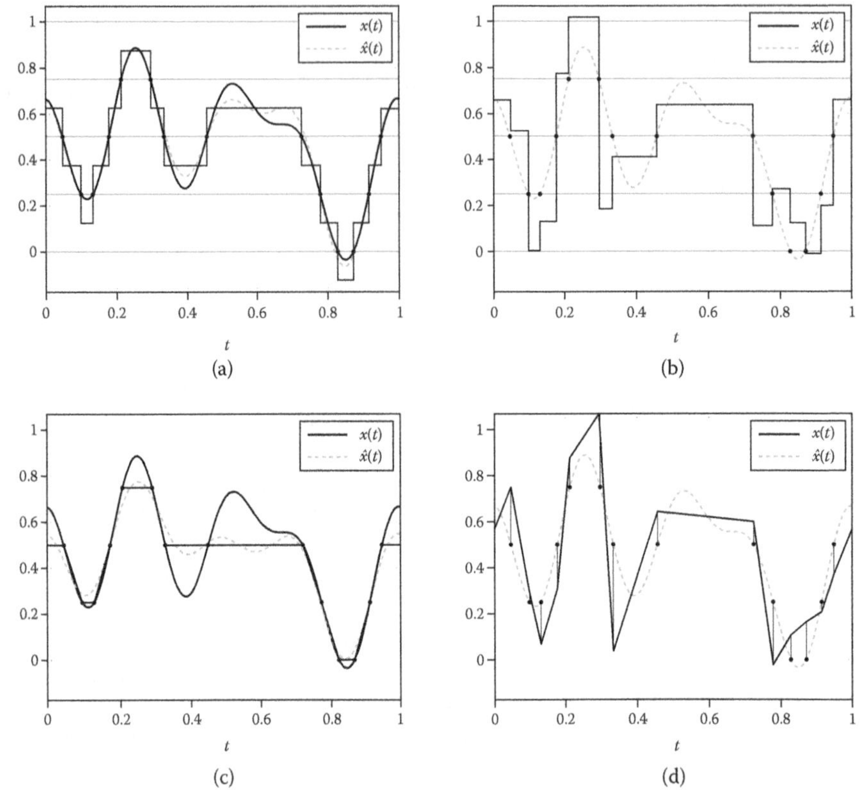


**FIGURE 16.5**

Piecewise constant and piecewise linear approximations of periodic bandlimited input $x(t)$ (black solid curves) from nonuniform samples (black dots), and bandlimited version $\hat{x}(t)$ of approximations (gray dashed curves): (a) continuous-time amplitude quantization; (b) bandlimited-optimal piecewise-constant approximation; (c) linear interpolation; and (d) bandlimited-optimal piecewise-linear approximation.

see that its bandlimited version $\hat{x}(t)$ also represented by a gray dashed line coincides exactly with $x(t)$. This leads to the nontrivial conclusion that the piecewise constant values of a continuous-time quantizer can be modified such that the input is perfectly recovered after bandlimitation.

To emphasize the generality of the method, we show in Figure 16.5c and d an experiment with the same input and sampling operator but where $\mathcal{R}$ is the linear interpolator followed by bandlimitation as described in (16.89). In Figure 16.5c, **D** is taken to be identity while in Figure 16.5d, it is replaced by the operator **D** of (16.86) with **C** = **I**. Perfect reconstruction is again confirmed in part (d) of the figure. The piecewise linear curve in this figure represents the linear interpolation of the sequence **y** output by **D**.

### 16.5.4 Appendix

The proof of Proposition 16.3 uses the following result as a core property.

**Lemma 16.1**

Let $v(t)$ be continuously differentiable on $[0,T]$ such that $v(T) = -v(0)$. Then $\int_0^T |v(t)|^2 \mathrm{d}t \leq \frac{T^2}{\pi^2} \int_0^T |v'(t)|^2 \mathrm{d}t$, where $v'(t)$ denotes the derivative of $v(t)$.

This is shown as follows. Let $w(t)$ be the $2T$-periodic function such that $w(t) = v(t)$ for $t \in [0,T)$ and $w(t) = -v(t-T)$ for $t \in [T,2T)$. The function $w(t)$ is continuous everywhere, differentiable everywhere except at multiples of $T$, and has a zero average. By Wirtinger's inequality, $\int_0^{2T} |w(t)|^2 \mathrm{d}t \leq \frac{(2T)^2}{(2\pi)^2} \int_0^{2T} |w'(t)|^2 \mathrm{d}t$. By dividing each member of this inequality by 2, one proves the lemma.

For the proof of Proposition 16.3, one then proceeds in a way similar to [18]. Using Parseval's equality for the Fourier transform, it is easy to see in the Fourier domain that for any square-integrable function $u(t)$, $\|\varphi_0(t) * u(t)\| \leq \|u(t)\|$, where $\varphi_0(t)$ is defined in (16.24). Let $u(t) \in \mathcal{B}$. Since $\varphi_0(t) * u(t) = u(t)$, it follows from (16.26) that

$u(t) - \mathcal{RCS}u(t) = \varphi_o(t) * (u(t) - \mathcal{PCS}u(t))$. Then, $\|u(t)-\mathcal{RCS}u(t)\| \leq \|u(t)-\mathcal{PCS}u(t)\|$. From (16.25) and (16.88), $\mathcal{PCS}u(t) = \sum_{n\in\mathbb{Z}} y_n p_n(t)$, where $y_n := \frac{1}{2}(u(t_n)+u(t_{n+1}))$. As $\sum_{n\in\mathbb{Z}} p_n(t) = 1$, one can write $u(t)-\mathcal{RCS}u(t) = \sum_{n\in\mathbb{Z}} v_n(t)p_n(t)$, where $v_n(t) := u(t) - y_n$. Then, $\|u(t)-\mathcal{PCS}u(t)\|^2 = \frac{1}{T_o}\sum_{n\in\mathbb{Z}} \int_{t_n}^{t_{n+1}} |v_n(t)|^2 dt$. Note that $v_n(t_{n+1}) = \frac{1}{2}(u(t_{n+1})-u(t_n)) = -v_n(t_n)$. From the above lemma, $\int_{t_n}^{t_{n+1}} |v_n(t)|^2 dt \leq \frac{(t_{n+1}-t_n)^2}{\pi^2}\int_{t_n}^{t_{n+1}} |u'(t)|^2 dt$. Hence, $\|u(t)-\mathcal{RCS}u(t)\|^2 \leq \frac{T_m^2}{\pi^2}\|u'(t)\|^2$. By Bernstein's inequality, $\|u'(t)\| \leq \Omega_0\|u(t)\|$. Since $\frac{\Omega_o}{\pi} = \frac{1}{T_o}$, then $\|u(t)-\mathcal{RCS}u(t)\| \leq \frac{T_m}{T_o}\|u(t)\|$. This proves Proposition 16.3.

## 16.6 Deviation from Uniform Sampling

It is, in general, difficult to have analytical formulas for the precise behavior of an operator of the type $\mathcal{RS}$. The mathematical derivations of the previous section are limited to specific kernel and reconstruction functions and are mostly based on inequalities that may not be tight. For example, $\|\mathcal{I}-\mathcal{RS}\|$ can only be shown to be upper bounded by 1 when $T_m = T_o$ with the operators $\mathcal{S}$ and $\mathcal{R}$ used in [18], while it can be shown by Fourier analysis to be equal to $1-\frac{2}{\pi} < 0.37$ when the sampling is uniform at the Nyquist rate, as will be shown in this section. While inequalities have been useful to prove sufficient conditions, more refined estimates may be needed for design optimizations. With the lack of analytical formulas, one mainly need to rely on numerical experiments. However, it may be useful to be guided by the intuition that views nonuniform sampling as a perturbation of uniform sampling. In the latter case, quantities can be derived by Fourier transform and are expected to continuously evolve as the operators experience variations such as deviations on the sampling instants.

### 16.6.1 Sensitivity to Sampling Errors

The approach of nonuniform sampling as a perturbation of uniform sampling is first useful to understand the sensitivity of sampling to errors. We recall from Section 16.3.9 that this sensitivity is measured by the condition number $\kappa(\mathcal{S})$. This quantity is, however, difficult to derive in the nonuniform case. So let us see what can be said in the uniform case. We include all the sampling operators presented in Section 16.1.5 with the assumption that $t_n = nT_s$ for all $n$ where $T_s \leq T_o$. Regardless of $\mathcal{S}$, the kernel functions of $\mathcal{S}$ take the form $s_n(t) = s(t-nT_s)$ where $s(t)$ is some function in $\mathcal{B}$. This is true even when $s_n(t)$ is not originally of the form $s(t-t_n)$ under general sampling instants $(t_n)_{n\in\mathbb{Z}}$. Consider the example (16.22), where $s_n(t) = s_o(t) * p_n(t)$ and $p_n(t)$ is described in Figure 16.2a. The function $s_n(t)$ is not the shifted version of a single function, except when $t_n = nT_s$, in which case $s_n(t) = s(t-nT_s)$, where $s(t) := s_o(t) * p(t)$ and $p(t)$ is the rectangular function supported by $[0, T_s)$. Regardless of the function $s(t) \in \mathcal{B}$, $\gamma(\mathcal{S})$ and $\|\mathcal{S}\|$ can be derived in terms of the Fourier transform $S(\omega)$ of $s(t)$ defined as

$$S(\omega) = \int_{-\infty}^{\infty} s(t)\, e^{-j\omega t}\, dt.$$

We show in Appendix 16.6.3 that

$$\gamma(\mathcal{S}) = \frac{1}{\sqrt{T_o T_s}} \inf_{|\omega|<\Omega_o} |S(\omega)|, \tag{16.90}$$

and

$$\|\mathcal{S}\| = \frac{1}{\sqrt{T_o T_s}} \sup_{|\omega|<\Omega_o} |S(\omega)|,$$

assuming that $S(\omega)$ is continuous in $(-\Omega_o, \Omega_o)$. The frame $\{s_n(t)\}_{n\in\mathbb{Z}}$ is tight when $S(\omega)$ is constant in $(-\Omega_o, \Omega_o)$. This is achieved when and only when $s(t)$ is proportional to the sinc function $s_o(t)$, since $s(t) \in \mathcal{B}$. In the case of direct sampling where $s(t) = s_o(t)$, $S(\omega) = T_o$ for all $\omega \in (-\Omega_o, \Omega_o)$. In this situation,

$$\gamma(\mathcal{S}) = \|\mathcal{S}\| = \sqrt{\rho}, \tag{16.91}$$

where $\rho$ is the oversampling ratio

$$\rho := \frac{T_o}{T_s}.$$

Several factors can contribute to increase the ratio $\kappa(\mathcal{S}) = \|\mathcal{S}\|/\gamma(\mathcal{S})$ from its minimal value 1. Obviously, in-band distortions in the signal $s(t)$ are a first factor as they contribute to create a gap between the infimum and supremum values of $|S(\omega)|$ over the interval $(\Omega_o, \Omega_o)$. But more generally, any distortion of the operator $\mathcal{S}$ will have a similar effect. While Fourier analysis can no longer be used when sampling uniformity is lost, one can simply return to the original definitions of $\gamma(\mathcal{S})$ and $\|\mathcal{S}\|$ in (16.55) and (16.11) as the infimum and the supremum values of $\|\mathcal{S}u(t)\|_{\mathcal{D}}$ for functions $u(t) \in \mathcal{B}$ of norm 1. Any distortion of $\mathcal{S}$ is likely to separate these two values apart.

In the case of direct sampling, having $s_n(t) = s_o(t-t_n)$ where the values $t_n$ are deviated from the uniform sampling instants $nT_s$ is a particular way to distort the operator $\mathcal{S}$ and hence increase $\kappa(\mathcal{S})$. There is a simple scenario in which $\kappa(\mathcal{S})$ can be made arbitrarily large. Start with $t_n = nT_o$ for all $n$, but deviate $t_0$ toward $t_1 = T_o$. The minimum modulus $\gamma(\mathcal{S})$ is expected to continuously vary with $t_0$. When $t_0$ reaches $t_1$, the situation is intuitively similar to uniform sampling at the Nyquist period $T_o$ in which the sample at instant 0 has been dropped. According to Section 16.4.2, $\{s_o(t-nT_o)\}_{n\in\mathbb{Z}}$ is an exact frame, so $\{s_o(t-nT_o)\}_{n\neq 0}$ ceases to be a frame. Since $\|\mathcal{S}\|$ is only expected to experience a small variation when $t_0$ moves from 0 to $t_1$, then $\gamma(\mathcal{S})$ must become 0. Thus, $\kappa(\mathcal{S})$ is expected to grow infinitely as $t_0$ tends to $t_1$. Qualitatively, this gives the general sense that $\kappa(\mathcal{S})$ grows with the degree of nonuniformity.

### 16.6.2 Analysis of $\|\mathcal{I}-\mathcal{RS}\|$

When the available reconstruction operator $\mathcal{R}$ is not a left inverse of $\mathcal{S}$, one wishes to analyze $\|\mathcal{I}-\mathcal{RS}\|$ so that $\mathcal{R}$ can still be approximately adopted as a left inverse when this number is small enough, or $\mathcal{R}\mathbf{D}$ can be made a left inverse of $\mathcal{S}$ according to the method of Section 16.5.1 (in the case $\mathbf{C} = \mathbf{I}$) when this number is less than 1. Again, initial clues are obtained in the case of uniform sampling. Assume that the impulse responses of $\mathcal{R}$ are of the type $r_n(t) = r(t-nT_s)$, where $r(t) \in \mathcal{B}$ and make the same assumptions on $\mathcal{S}$ as in the previous section under uniform sampling. The function $h(t,\tau)$ of (16.18) then becomes

$$h(t,\tau) := \frac{1}{T_o} \sum_{n\in\mathbb{Z}} r(t-nT_s)\, \bar{s}(nT_s-\tau). \qquad (16.92)$$

With the bandlimitation of $r(t)$ and $s(t)$, we show in Appendix 16.6.3 that

$$h(t,\tau) = h_s(t-\tau), \qquad (16.93)$$

where

$$h_s(t) := \frac{1}{T_o T_s}\,(r * \bar{s})(t).$$

It then follows from (16.17) that $\mathcal{RS}$ is linear and time-invariant of impulse response $h_s(t)$, that is,

$$\mathcal{RS}x(t) = h_s(t) * x(t). \qquad (16.94)$$

In a way similar to (16.90), it can be shown that

$$\|\mathcal{I}-\mathcal{RS}\| = \sup_{|\omega|<\Omega_o} \big|1 - H_s(\omega)\big|, \qquad (16.95)$$

where $H_s(\omega)$ is the Fourier transform of $h_s(t)$.

Consider, for example, the case studied in [18] where $\mathcal{S}$ is the direct sampling operator at $(t_n)_{n\in\mathbb{Z}}$ and $\mathcal{R}$ is defined by (16.87). When $t_n = nT_s$ for all $n \in \mathbb{Z}$, the above equations are applicable with $s(t) = s_o(t)$ and $r(t) = \varphi_o(t) * \check{p}(t)$, where $\check{p}(t)$ is the rectangular function supported by $[-\frac{T_s}{2}, \frac{T_s}{2})$. In this case, it is easy to derive that $H_s(\omega) = \mathrm{sinc}(\frac{1}{2\rho}\frac{\omega}{\Omega_o})$. With this, $\|\mathcal{I}-\mathcal{RS}\| = 1-\mathrm{sinc}(\frac{1}{2\rho})$. Its maximum is obtained with $\rho = 1$, that is, $T_s = T_o$, and is equal to $1-\frac{2}{\pi} \simeq 0.36$. This nonzero value is due to the inband distortion of the function $\check{p}(t)$. Note in this condition that the analysis of [18] can only give the upper bound $\|\mathcal{I}-\mathcal{RS}\| \leq 1$ (see Section 16.5.2).

In general, one should try to have $\|\mathcal{I}-\mathcal{RS}\|$ as small as possible under uniform sampling. This value is exactly zero when $H_s(\omega) = 1$ for all $\omega \in (-\Omega_o, \Omega_o)$, which is achieved when $R(\omega) = \frac{T_o T_s}{S^*(\omega)}$ in this frequency range. As the sampling instants are progressively deviated, it is expected that $\|\mathcal{I}-\mathcal{RS}\|$ continuously grows from its initial value. When $\|\mathcal{I}-\mathcal{RS}\|$ becomes too large for $\mathcal{R}$ to be considered as a satisfactory left inverse, the next concern is whether $\|\mathcal{I}-\mathcal{RS}\|$ can be maintained less than 1 for the reason explained above.

In spite of the lack of analytical formulas for the growth of $\|\mathcal{I}-\mathcal{RS}\|$ with sampling deviations, this intuitive approach may still be productive in the design process. With the direct sampling operator $\mathcal{S}$ at uniform instants $t_n = nT_s$ with $T_s \leq T_o$, we know from (16.91) that $\|\mathcal{S}u(t)\| = \sqrt{\rho}\|u(t)\|$ for all $u(t) \in \mathcal{B}$, and hence $\mathcal{S}^*\mathcal{S} = \rho\mathcal{I}$ from (16.63). Therefore, one obtains $\|\mathcal{I}-\mathcal{RS}\| = 0$ with $\mathcal{R} := \frac{1}{\rho}\mathcal{S}^*$. One can also write that $\|\mathcal{I}-\mathcal{S}^*\mathbf{C}\mathcal{S}\| = 0$ where $\mathbf{C} = \frac{1}{\rho}\mathbf{I}$. Assume now that the sampling period varies slowly in time. Even if the variation is extremely slow, $\|\mathcal{I}-\mathcal{S}^*\mathbf{C}\mathcal{S}\|$ is expected to grow as a function of the peak-to-peak amplitude of variation. Now, instead of having $\mathbf{C}$ with a diagonal constantly equal to $\frac{1}{\rho} = \frac{T_s}{T_o}$, it is intuitive that $\|\mathcal{I}-\mathcal{S}^*\mathbf{C}\mathcal{S}\|$ will be maintained small if this constant is adaptively adjusted to the "local" sampling period. A natural way to do so is to choose $\mathbf{C}$ as defined by (16.82) and (16.84). This brings us to the idea of weighted sampling discussed in Section 16.4.4.

### 16.6.3 Appendix

#### 16.6.3.1 Proof of (16.90)

Let $u(t) \in \mathcal{B}\backslash\{0\}$. The $n$th component of $\mathcal{S}u(t)$ is $\langle s(t-nT_s), u(t)\rangle = \frac{1}{T_o}\,(\bar{s} * u)(nT_s)$. Thus,

$$\|\mathcal{S}u(t)\|^2_{\mathcal{D}} = \frac{1}{T_o^2} \sum_{n\in\mathbb{Z}} \big|(\bar{s} * u)(nT_s)\big|^2. \qquad (16.96)$$

Since $(\bar{s} * u)(t)$ is bandlimited by $\Omega_o = \frac{\pi}{T_o} \leq \frac{\pi}{T_s}$, Shannon sampling theorem applied at the sampling period $T_s$ yields $(\bar{s} * u)(t) = \sum_{n\in\mathbb{Z}} (\bar{s} * u)(nT_s)\,\mathrm{sinc}(\frac{t}{T_s}-n)$. The family $\{\mathrm{sinc}(\frac{t}{T_s}-n)\}_{n\in\mathbb{Z}}$ is known to be orthogonal with respect to $\langle\cdot,\cdot\rangle$. By Parseval's equality,

$$\|(\bar{s} * u)(t)\|^2 = \|\mathrm{sinc}(\tfrac{t}{T_s})\|^2 \sum_{n\in\mathbb{Z}} \big|(\bar{s} * u)(nT_s)\big|^2. \quad (16.97)$$

Since $\int_{-\infty}^{\infty} |\text{sinc}(t)|^2 dt = 1$, then $\|\text{sinc}(\frac{t}{T_s})\|^2 = \frac{1}{T_o}\int_{\infty}^{\infty} |\text{sinc}(\frac{t}{T_s})|^2\, dt = \frac{T_s}{T_o}$. From the above two equations and Parseval's equality for the Fourier transform, we obtain

$$\frac{\|\mathcal{S}u(t)\|^2_{\mathcal{D}}}{\|u(t)\|^2} = \frac{1}{T_o T_s}\frac{\|(\bar{s}*u)(t)\|^2}{\|u(t)\|^2} = \frac{1}{T_o T_s}\frac{\int_{-\infty}^{\infty}|S(\omega)|^2\,|U(\omega)|^2\,d\omega}{\int_{-\infty}^{\infty}|U(\omega)|^2\,d\omega}.$$

Define $S_m := \inf_{|\omega|<\Omega_o} |S(\omega)|$. It is clear that $\frac{\|\mathcal{S}u(t)\|^2_{\mathcal{D}}}{\|u(t)\|^2} \geq \frac{S_m^2}{T_o T_s}$. By assumption on $S(\omega)$, it is possible to find for any $\epsilon > 0$ an interval $I \subset (-\Omega_o, \Omega_o)$ of nonzero measure such that $|S(\omega)| \leq S_m + \epsilon$ for all $\omega \in I$. By taking $U(\omega)$ equal to 1 for $\omega \in I$ and 0 otherwise, we find $\frac{\|\mathcal{S}u(t)\|^2_{\mathcal{D}}}{\|u(t)\|^2} \leq \frac{(S_m+\epsilon)^2}{T_o T_s}$. This shows that $\gamma(\mathcal{S})^2 = \frac{S_m^2}{T_o T_s}$ and hence the first relation of (16.90). The second relation is obtained in a similar manner.

#### 16.6.3.2 Justification of (16.93)

Let $\delta(t)$ denote the Dirac impulse. For any given $\theta, \tau \in \mathbb{R}$, $r(t-\theta)\,\bar{s}(\theta-\tau) = \big(r(t)*\delta(t-\theta)\big)\bar{s}(\theta-\tau) = r(t) * \big(\delta(t-\theta)\,\bar{s}(\theta-\tau)\big) = r(t) * \big(\delta(t-\theta)\,\bar{s}(t-\tau)\big)$. From (16.92), we can then write $h_s(t,\tau) = \frac{1}{T_o}\, r(t) * s_\tau(t)$ where $s_\tau(t) := \bar{s}(t-\tau)\sum_{n\in\mathbb{Z}}\delta(t-nT_s)$. The Fourier transform of $s_\tau(t)$ is that of $\bar{s}(t-\tau)$ periodized by $\Omega_s := \frac{2\pi}{T_s}$ and divided by $T_s$ [21]. For any given $\tau$, $\bar{s}(t-\tau)$ is bandlimited by $\Omega_o$ [since this is true for $s(t)$]. Since $2\Omega_o = \frac{2\pi}{T_o} \leq \Omega_s$, the portion of $s_\tau(t)$ that lies in $[-\Omega_o, \Omega_o]$ in the frequency domain is equal to $\frac{1}{T_s}\,\bar{s}(t-\tau)$. Since $r(t)$ is bandlimited by $\Omega_o$, then $r(t) * s_\tau(t) = r(t) * \big(\frac{1}{T_s}\,\bar{s}(t-\tau)\big) = \frac{1}{T_s}\,(r*\bar{s})(t-\tau)$.

## 16.7 Reconstruction by Successive Approximations

For given sampling and reconstruction operators $\mathcal{S}$ and $\mathcal{R}$, we saw in Section 16.5 under what condition on these operators the $\mathcal{R}\mathbf{D}\mathcal{S}$ system of Figure 16.3 can achieve perfect reconstruction. The major difficulty, however, is the implementation of the intermediate discrete-time processor $\mathbf{D}$. In the favorable case where a discrete-time operator $\mathbf{C}$ can be found to achieve the inequality $\|\mathcal{I}-\mathcal{R}\mathbf{C}\mathcal{S}\| < 1$, an available solution for $\mathbf{D}$ is given by (16.86). But the central difficulty is the inversion of $\mathcal{R}\mathbf{C}\mathcal{S}$ in this solution. The Neumann series (16.64) provides a method for the numerical evaluation of $(\mathcal{R}\mathbf{C}\mathcal{S})^{-1}$. Given the constraint of finite computation, one will of course only consider a finite number of terms in this series. The smaller $\|\mathcal{I}-\mathcal{R}\mathbf{C}\mathcal{S}\|$ is compared to 1, the faster the series converges and, therefore, the smaller is the number of terms one can afford to consider. We study in this section various aspects of the approximate implementation of $\mathbf{D}$.

### 16.7.1 Basic Algorithm

To simplify the analysis, we start by assuming that $\|\mathcal{I}-\mathcal{R}\mathcal{S}\| < 1$. According to Section 16.5.1, $\mathcal{R}\mathbf{D}\mathcal{S} = \mathcal{I}$ with $\mathbf{D} := \mathcal{S}(\mathcal{R}\mathcal{S})^{-2}\mathcal{R}$ (case where $\mathbf{C} = \mathbf{I}$). This operator could be approximated by injecting a truncated Neumann series for $(\mathcal{R}\mathcal{S})^{-1}$. This is, however, complicated due to the square power. Now, it is not necessary to implement an operator $\hat{\mathbf{D}}$ that approximates $\mathbf{D}$, but sufficient to find $\hat{\mathbf{D}}$ so that $\mathcal{R}\hat{\mathbf{D}}$ approximates $\mathcal{R}\mathbf{D}$. We have $\mathcal{R}\mathbf{D} = \mathcal{R}\mathcal{S}(\mathcal{R}\mathcal{S})^{-2}\mathcal{R} = (\mathcal{R}\mathcal{S})^{-1}\mathcal{R}$. We then obtain from the Neumann series of $(\mathcal{R}\mathcal{S})^{-1}$ that $\mathcal{R}\mathbf{D} \simeq \mathcal{Q}_p\mathcal{R}$ for $p$ large enough, where

$$\mathcal{Q}_p := \sum_{m=0}^{p-1}(\mathcal{I}-\mathcal{R}\mathcal{S})^m. \qquad (16.98)$$

Next, note that

$$(\mathcal{I}-\mathcal{R}\mathcal{S})\,\mathcal{R} = \mathcal{R} - \mathcal{R}\mathcal{S}\mathcal{R} = \mathcal{R}\,(\mathbf{I}-\mathcal{S}\mathcal{R}), \qquad (16.99)$$

where $\mathbf{I}$ is the identity operator of $\mathcal{D}$. By induction, one finds that $(\mathcal{I}-\mathcal{R}\mathcal{S})^m\mathcal{R} = \mathcal{R}(\mathbf{I}-\mathcal{S}\mathcal{R})^m$ for all $m \geq 0$ [22, p. 598]. By defining

$$\mathbf{D}_p := \sum_{m=0}^{p-1}(\mathbf{I}-\mathcal{S}\mathcal{R})^m, \qquad (16.100)$$

one obtains $\mathcal{Q}_p\mathcal{R} = \mathcal{R}\mathbf{D}_p$. Thus,

$$\mathcal{R}\mathbf{D} \simeq \mathcal{R}\mathbf{D}_p.$$

Since $\mathcal{R}\mathbf{D}\mathcal{S} = \mathcal{I}$, $x^{(p)}(t) := \mathcal{R}\mathbf{D}_p\mathbf{x}$ is an approximation of $x(t)$ with $\mathbf{x} := \mathcal{S}x(t)$. Computationally,

$$x^{(p)}(t) = \mathcal{R}\mathbf{y}^{(p)}, \qquad (16.101)$$

where

$$\mathbf{y}^{(p)} := \mathbf{D}_p\mathbf{x}. \qquad (16.102)$$

For the implementation of $\mathbf{D}_p$, note that

$$\mathbf{D}_{i+1} = (\mathbf{I}-\mathcal{S}\mathcal{R})\mathbf{D}_i + \mathbf{I},$$

with $\mathbf{D}_0$ equal to the zero operator. Therefore, $\mathbf{y}^{(i)}$ satisfies the recursive relation

$$\mathbf{y}^{(i+1)} = (\mathbf{I}-\mathcal{S}\mathcal{R})\,\mathbf{y}^{(i)} + \mathbf{x}, \qquad (16.103)$$

starting from $\mathbf{y}^{(0)} = 0$. The operator $\mathbf{D}_p$ is then implemented in practice by iterating the above equation $p$ times.

Note that the expression of $\mathbf{D}_p$ in (16.100) looks like $\mathcal{Q}_p$ in (16.98) and thus like the partial sum of a Neumann series. One would be initially tempted to claim that $\mathbf{D}_p$ tends to the inverse of $\mathcal{SR}$ when $p$ goes to $\infty$. The problem is that $\mathcal{SR}$ is only invertible when restricted to Range$(\mathcal{S})$. One cannot either rigorously state that $\mathbf{D}_p$ converges to a generalized inverse of $\mathcal{SR}$. Indeed, we will see in the next section that $\mathbf{y}^{(p)} := \mathbf{D}_p\mathbf{x}$ systematically diverges with $p$ whenever $\mathbf{x}$ is not in Range$(\mathcal{S})$.

### 16.7.2 Reconstruction Estimates with Sampling Errors

We have assumed until now $\mathbf{x}$ to be exactly the samples of a bandlimited signal $x(t)$. As considered in Section 16.3.8, one should, however, expect in practice to have

$$\mathbf{x} = \mathcal{S}x(t) + \mathbf{e},$$

where $\mathbf{e}$ is a vector of sampling errors. Thanks to the invertibility of $\mathcal{RS}$ in $\mathcal{B}$, it can be shown that the error sequence $\mathbf{e}$ can be uniquely decomposed as

$$\mathbf{e} = \mathcal{S}e(t) + \mathbf{e}_0, \qquad (16.104)$$

such that $e(t) \in \mathcal{B}$ and $\mathbf{e}_0 \in$ Null$(\mathcal{R})$. Indeed, by applying $\mathcal{R}$ to this equality, we find $\mathcal{R}\mathbf{e} = \mathcal{RS}e(t)$, which leads to the unique solution $e(t) = (\mathcal{RS})^{-1}\mathcal{R}\mathbf{e}$ and $\mathbf{e}_0 = \mathbf{e} - \mathcal{S}e(t)$. Conversely, these two signals do satisfy (16.104) and $\mathcal{R}\mathbf{e}_0$ is easily verified to be 0. Note by the way that this proves that $\mathcal{D} =$ Range$(\mathcal{S}) \oplus$ Null$(\mathcal{R})$, which is a generalization of (16.66). We can then write

$$\mathbf{x} = \check{\mathbf{x}} + \mathbf{e}_0,$$

where

$$\check{\mathbf{x}} := \mathcal{S}\check{x}(t) \quad \text{and} \quad \check{x}(t) := x(t) + e(t).$$

Since $\mathcal{R}\mathbf{e}_0 = 0$, then $(\mathbf{I}-\mathcal{SR})\,\mathbf{e}_0 = \mathbf{e}_0$ and by induction $(\mathbf{I}-\mathcal{SR})^m\mathbf{e}_0 = \mathbf{e}_0$ for all $m \geq 0$. Hence, $\mathbf{D}_p\mathbf{e}_0 = p\,\mathbf{e}_0$ and the vector $\mathbf{y}^{(p)}$ of (16.102) yields the expression

$$\mathbf{y}^{(p)} = \check{\mathbf{y}}^{(p)} + p\,\mathbf{e}_0, \qquad (16.105)$$

where

$$\check{\mathbf{y}}^{(p)} := \mathbf{D}_p\check{\mathbf{x}}.$$

If $\mathbf{e} \notin$ Range$(\mathcal{S})$, then $\mathbf{e}_0 \neq 0$ and the component $p\,\mathbf{e}_0$ makes $\mathbf{y}^{(p)}$ *diverge* with $p$. However, this component is immediately canceled after application of $\mathcal{R}$ so that the output of the system $x^{(p)}(t)$ defined in (16.101) yields the final expression

$$x^{(p)}(t) = \mathcal{R}\mathbf{y}^{(p)} = \mathcal{R}\check{\mathbf{y}}^{(p)} = \mathcal{R}\mathbf{D}_p\check{\mathbf{x}} = \mathcal{R}\mathbf{D}_p\mathcal{S}\,\check{x}(t).$$

We conclude that $x^{(p)}(t)$ tends to $\check{x}(t)$ when $p$ goes to infinity. The error $e(t)$ contained in the signal $\check{x}(t) = x(t) + e(t)$ results qualitatively from the component of the sample error $\mathbf{e}$ that is undistinguishable from the samples of a bandlimited signal. Now, the convergence of $x^{(p)}(t)$ toward $\check{x}(t)$ is only valid as long as the growing error $p\,\mathbf{e}_0$ contained in the intermediate signal $\mathbf{y}^{(p)}$ remains within the tolerance of the hardware. In typical implementations, the value of $p$ remains, however, relatively small due to computation complexity limitations, and $\mathbf{y}^{(p)}$ may keep magnitudes similar to $\check{\mathbf{y}}^{(p)}$ assuming a relatively small error vector $\mathbf{e}_0$.

### 16.7.3 Estimate Improvement by Contraction

The Neumann series actually includes some deeper mechanisms of sequence convergence. Inspired by (16.103), consider two estimates of $x(t)$ of the form

$$y(t) = \mathcal{R}\mathbf{y} \quad \text{and} \quad y'(t) = \mathcal{R}\mathbf{y}', \qquad (16.106)$$

where $\mathbf{y}$ and $\mathbf{y}'$ are related to each other by

$$\mathbf{y}' = (\mathbf{I}-\mathcal{SR})\,\mathbf{y} + \mathbf{x}, \qquad (16.107)$$

with $\mathbf{x} = \mathcal{S}x(t)$. Let us show that

$$\|y'(t)-x(t)\| \leq \gamma\,\|y(t)-x(t)\|, \qquad (16.108)$$

where

$$\gamma := \|\mathcal{I}-\mathcal{RS}\|.$$

By operating $\mathcal{R}$ on both sides of (16.107) and applying the identity (16.99) backward, one finds that

$$y'(t) = (\mathcal{I}-\mathcal{RS})\,y(t) + \mathcal{R}\,\mathbf{x}. \qquad (16.109)$$

Since $\mathbf{x} = \mathcal{S}x(t)$, note that $x(t) = (\mathcal{I}-\mathcal{RS})\,x(t) + \mathcal{R}\,\mathbf{x}$. By subtracting this from the above equation, one obtains $y'(t)-x(t) = (\mathcal{I}-\mathcal{RS})(y(t)-x(t))$. This implies (16.108).

The impact of (16.108) is outstanding. Regardless of how $\mathbf{y}$ has been obtained to produce $y(t) = \mathcal{R}\mathbf{y}$ as an estimate of $x(t)$, the transformed vector $\mathbf{y}'$ of (16.107) systematically leads to a better estimate $y'(t) = \mathcal{R}\mathbf{y}'$ when $\|\mathcal{I}-\mathcal{RS})\| < 1$. The estimate improvement of (16.109) has also some intuitive interpretation, better seen from the equivalent formulation

$$y'(t) = y(t) - \mathcal{R}\big(\mathcal{S}y(t) - \mathbf{x}\big).$$

Ideally, one would wish to subtract from $y(t)$ the estimate error signal $e(t) := y(t)-x(t)$, which is of course not available. The signal $y'(t)$ is obtained by subtracting instead $\mathcal{R}(\mathcal{S}y(t) - \mathbf{x}) = (\mathcal{RS})e(t)$ since $\mathbf{x} = \mathcal{S}x(t)$. When $\mathcal{RS}$ is not too far away from identity, this still reduces the error contained in $y(t)$. This is at least mathematically guaranteed when $\|\mathcal{I}-\mathcal{RS}\| < 1$.

Returning to the iteration (16.103), the successive reconstruction estimates $x^{(i)}(t)$ of (16.101) thus satisfy the inequality $\|x^{(i+1)}(t)-x(t)\| \leq \gamma\,\|x^{(i)}(t)-x(t)\|$, which recursively leads to

$$\|x^{(i)}(t)-x(t)\| \leq \gamma^i \|x^{(0)}(t)-x(t)\|.$$

This is true regardless of the initial estimate $x^{(0)}(t) = \mathcal{R}\mathbf{y}^{(0)}$ [not just with $x^{(0)}(t) = \mathcal{R}.0 = 0$ as was used for the Neumann series]. This points to a more general design of the processor $\mathbf{D}$, where a first estimate $\mathbf{y}^{(0)}$ of $\tilde{\mathbf{x}}$ is constructed using any method (including heuristic and nonlinear methods) and fewer iterations of (16.107) are needed.

### 16.7.4 General Case of Admissible Pair $(\mathcal{S}, \mathcal{R})$

In the most general case of an admissible pair $(\mathcal{S}, \mathcal{R})$, $\|\mathcal{I}-\mathcal{R}\mathcal{S}\|$ may not be less than 1 and one may need to look for some discrete-time operator $\mathbf{C}$ such that $\|\mathcal{I}-\mathcal{R}\mathbf{C}\mathcal{S}\| < 1$. Once such an operator is found, one can redo the derivations from the beginning of Section 16.7 with the complete expression (16.86) of $\mathbf{D}$ including $\mathbf{C}$. A simpler alternative is to keep these derivations as is but replace $\mathcal{R}$ by $\mathcal{R}' := \mathcal{R}\mathbf{C}$, which does satisfy $\|\mathcal{I}-\mathcal{R}'\mathcal{S}\| < 1$. The recursive relation (16.103) then simply becomes

$$\mathbf{y}^{(i+1)} = (\mathbf{I}-\mathcal{S}\mathcal{R}\mathbf{C})\,\mathbf{y}^{(i)} + \mathbf{x}, \qquad (16.110)$$

starting from some initial guess $\mathbf{y}^{(0)}$, and the reconstructed estimate of (16.101) is changed into

$$x^{(p)}(t) := \mathcal{R}\mathbf{C}\mathbf{y}^{(p)}. \qquad (16.111)$$

As shown in Figure 16.6, the processor $\mathbf{D}$ then consists of providing an initial guess $\mathbf{y}^{(0)}$, iterating (16.110) to obtain $\mathbf{y}^{(p)}$, and finally computing $\mathbf{y} = \mathbf{C}\mathbf{y}^{(p)}$ so that $\mathcal{R}\mathbf{y} = \mathcal{R}\mathbf{C}\mathbf{y}^{(p)} = x^{(p)}(t)$.

We show in Figure 16.7a how an input $x(t)$ is approached by successive estimates $x^{(i)}(t)$ from (16.111) under the experimental conditions of Figure 16.5a and b.

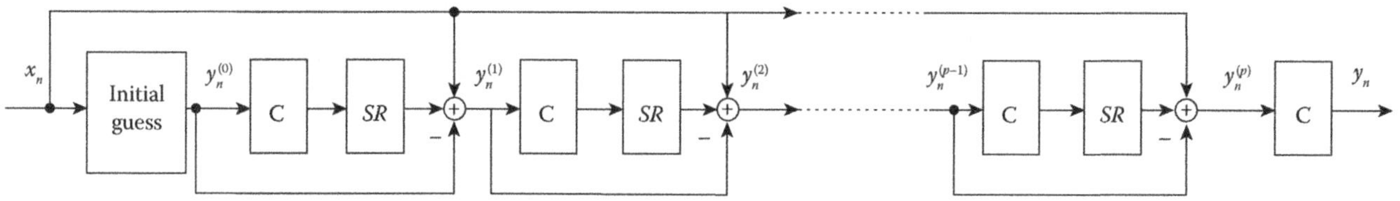


**FIGURE 16.6**
Implementation of processor $\mathbf{D}$.

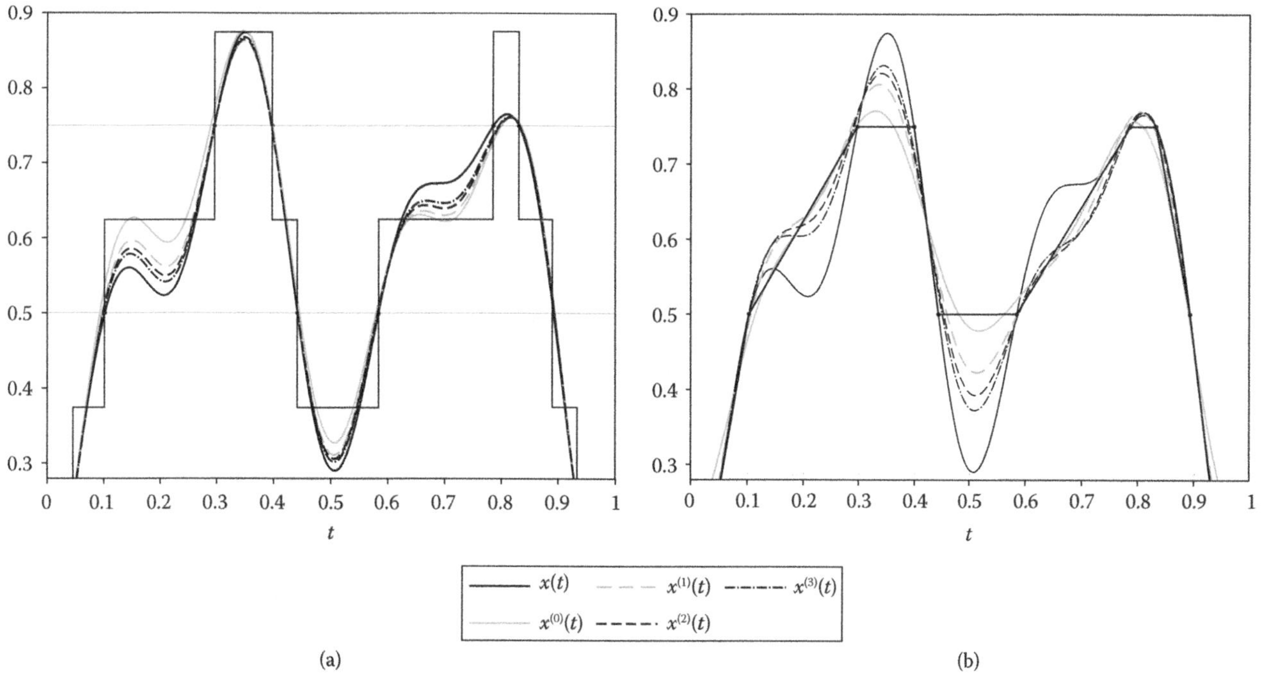


**FIGURE 16.7**
Successive approximations of input $x(t)$ by $x^{(i)}(t)$ of (16.111): (a) conditions of experiment of Figure 16.5a and (b) conditions of experiment of Figure 16.5c.

We recall that $\mathcal{S}$ is the direct sampling operator at the instants of crossings of the input with the quantization thresholds represented by dashed gray horizontal lines, $\mathcal{R}$ is the zero-order hold operator followed by bandlimitation as described in (16.26), and $\mathbf{C}$ is given in (16.88). The initial guess $\mathbf{y}^{(0)}$ is chosen so that $x^{(0)}(t) = \mathcal{R}\mathbf{C}\mathbf{y}^{(0)}$ is the bandlimited version of the quantized signal represented by the piecewise constant function in Figure 16.7a. We show in Figure 16.7b a similar experiment, where $\mathcal{R}$ is the linear interpolator at the same instants followed by bandlimitation and $\mathbf{C} = \mathbf{I}$. The initial guess $\mathbf{y}^{(0)}$ is just taken to be directly $\mathbf{x} = \mathcal{S}x(t)$.

We have seen examples where the introduction of an auxiliary operator $\mathbf{C}$ is a necessity to make $\|\mathcal{I}-\mathcal{R}\mathbf{C}\mathcal{S}\|$ less than 1. But the ultimate direction of research is find $\mathbf{C}$ so that $\|\mathcal{I}-\mathcal{R}\mathbf{C}\mathcal{S}\|$ is as small as possible. The smaller it is, the lower is the number $p$ of iterations needed to make $x^{(p)}(t)$ a good approximation of $x(t)$. Intuitively also, the closer $\|\mathcal{I}-\mathcal{R}\mathbf{C}\mathcal{S}\|$ is to 0, the closer $\mathbf{C}$ is to a generalized inverse of $\mathcal{S}\mathcal{R}$. The iterative estimation of $x(t)$ can then be seen as a systematic procedure to compensate for the errors of $\mathbf{C}$ with respect to a true generalized inverse. In the absence of analytical expressions for the generalized inversion of $\mathcal{S}\mathcal{R}$, this opens the door to heuristic designs of $\mathbf{C}$. We have seen until now examples of design where $\mathbf{C}$ is a discrete-time operator with only one or two nonzero diagonals. A direction of investigation is the minimization of $\|\mathcal{I}-\mathcal{R}\mathbf{C}\mathcal{S}\|$ using operators $\mathbf{C}$ with $N$ nonzero and nonconstant diagonals. From an implementation viewpoint, such an operator is a time-varying FIR filter with $N$ taps.

## 16.8 Discrete-Time Computation Implementations

### 16.8.1 Discrete-Time Processor D

In the proposed design of the processor $\mathbf{D}$ of Figure 16.6, the most critical part from an implementation viewpoint is the operator $\mathcal{S}\mathcal{R}$. Since

$$\mathcal{S}\mathcal{R}\mathbf{y} = \mathcal{S}\Big( \sum_{k\in\mathbb{Z}} y_k\, r_k(t)\Big) = \sum_{k\in\mathbb{Z}} y_k\, \mathcal{S}r_k(t),$$

and $\mathcal{S}r_k(t) = \big(\langle s_n, r_k\rangle\big)_{n\in\mathbb{Z}}$, then the $n$th component of the sequence $\mathcal{S}\mathcal{R}\mathbf{y}$ is

$$(\mathcal{S}\mathcal{R}\mathbf{y})_n = \sum_{k\in\mathbb{Z}} \langle s_n, r_k\rangle\, y_k.$$

The operator $\mathcal{S}\mathcal{R}$ can be seen as an infinite matrix $\mathbf{M}$ of entries $M_{n,k} := \langle s_n, r_k\rangle$ with $n, k \in \mathbb{Z}$. For practical implementation, the above summation needs to be truncated. As $\langle s_n, r_k\rangle$ tends to 0 as $t_k$ gets far away from $t_n$, $(\mathcal{S}\mathcal{R}\mathbf{y})_n$ is approximated as

$$(\mathcal{S}\mathcal{R}\mathbf{y})_n \simeq \sum_{k\in K_n} \langle s_n, r_k\rangle\, y_k, \tag{16.112}$$

where

$$K_n = \{k \in \mathbb{Z} : |t_k - t_n| \le T\},$$

for some constant $T$. This operation amounts again to an FIR filter with time-varying tap coefficients. The main issue is the computation of each coefficient $\langle s_n, r_k\rangle$. In practice, it is unrealistic to evaluate numerically the integral implied by this inner product. One would usually rely on lookup tables. However, both $s_n(t)$ and $r_n(t)$ depend on a sequence of instants $(t_k)_{k\in\mathbb{Z}}$ and, hence, are multiparameter dependent. The use of single-parameter lookup tables is fortunately possible with the sampling and reconstruction operators considered in this chapter and listed in Table 16.1. In this table, $s(t)$ and $r(t)$ are any bandlimited functions, $s_o(t) = \mathrm{sinc}(\frac{t}{T_o})$, $\varphi_o(t) = \frac{1}{T_o}\mathrm{sinc}(\frac{t}{T_o})$, as defined in (16.20) and (16.24), while $p_n(t)$ and $l_n(t)$ are the rectangular and triangular functions defined in Figure 16.2. In practice, one needs to precompute densely enough samples of the function $f(t)$ given in Table 16.1 and store them in a lookup table. The last column of Table 16.1 shows what values of $f(t)$ needs to be read to obtain $\langle s_n, r_k\rangle$. Depending on the case, it also indicates some additional algebraic operations to be performed.

We now show how Table 16.1 is derived. Case (i) is easily obtained from the identity (16.19), which implies that $\langle s_n, r_k\rangle = \frac{1}{T_o}(\bar{s} * r_k)(t_n) = \frac{1}{T_o}(\bar{s} * r)(t_n - t_k)$. With the same identity, we obtain in case (ii), $\langle s_n, r_k\rangle = \langle s_n, \varphi_o * p_k\rangle = \frac{1}{T_o}(\bar{s} * \varphi_o * p_k)(t_n) = \frac{1}{T_o}(\bar{s} * p_k)(t_n)$, since $s(t)$ is bandlimited. Given the description of the rectangular function $p_n(t)$ in Figure 16.2a, one easily finds with any function $h(t)$ that [23]

$$(h * p_k)(t) = \int_{t-t_{k+1}}^{t-t_k} h(t)\,\mathrm{d}t = f(t-t_k) - f(t-t_{k+1}), \tag{16.113}$$

where

$$f(t) := \int_0^t h(\tau)\,\mathrm{d}\tau.$$

This leads to the result of $\langle s_n, r_k\rangle$ shown in the table. Case (iii) is obtained in a similar manner. With the identity (16.23) and the above result (16.113), we obtain in case (iv), $\langle s_n, r_k\rangle = \langle s_o * p_n, \varphi_o * p_k\rangle = \langle p_n, \bar{s}_o * \varphi_o * p_k\rangle = \langle p_n, s_o * p_k\rangle = \int_{t_n}^{t_{n+1}} (\varphi_o * p_k)(t)\mathrm{d}t = \int_{t_n}^{t_{n+1}} g(t-t_k)\mathrm{d}t - \int_{t_n}^{t_{n+1}} g(t-t_{k+1})\mathrm{d}t$, where $g(\tau) := \int_0^\tau \varphi_o(\tau')\,\mathrm{d}\tau'$. One then easily finds the result of $\langle s_n, r_k\rangle$ given in the table. Case (v) is like case (ii) except that $p_n(t)$ is replaced

**TABLE 16.1**

Lookup-Table Calculation of $\langle s_n, r_k\rangle$

| | Sampling Kernel Functions $s_n(t)$ | Reconstruction Functions $r_n(t)$ | Lookup-Table Function $f(t)$ | $\langle s_n, r_k\rangle$ |
|---|---|---|---|---|
| (i) | $s(t-t_n)$ | $r(t-t_n)$ | $\frac{1}{T_o}(\bar{s} * r)(t)$ | $f(t_n-t_k)$ |
| (ii) | $s(t-t_n)$ | $(\varphi_o * p_n)(t)$ | $\frac{1}{T_o}\int_0^t \bar{s}(\tau)\,d\tau$ | $f(t_n-t_k) - f(t_n-t_{k+1})$ |
| (iii) | $(s_o * p_n)(t)$ | $r(t-t_n)$ | $\int_0^t \bar{r}(\tau)\,d\tau$ | $f(t_k-t_n) - f(t_k-t_{n+1})$ |
| (iv) | $(s_o * p_n)(t)$ | $(\varphi_o * p_n)(t)$ | $\int_0^t\int_0^\tau \varphi_o(\tau')\,d\tau' d\tau$ | $f(t_{n+1}-t_k) - f(t_n-t_k)$ $-f(t_{n+1}-t_{k+1}) + f(t_n-t_{k+1})$ |
| (v) | $s(t-t_n)$ | $(\varphi_o * l_n)(t)$ | $\frac{1}{T_o}\int_0^t (t-\tau)\,\bar{s}(\tau)\,d\tau$ | $(f(t_n-t_{k+1}) - f(t_n-t_k))/(t_{k+1}-t_k)$ $-(f(t_n-t_k) - f(t_n-t_{k-1}))/(t_k-t_{k-1})$ |

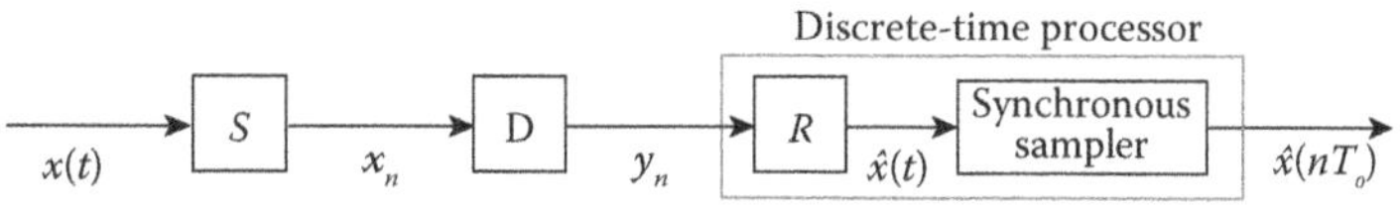


**FIGURE 16.8**

Conversion of nonuniform samples into standard PCM signal.

by $l_n(t)$. Thus, $\langle s_n, r_k\rangle = \frac{1}{T_o}(\bar{s} * l_k)(t_n)$. We show in Appendix 16.8.4 that for any function $h(t)$,

$$(h * l_k)(t) = \frac{f(t-t_{k+1}) - f(t-t_k)}{t_{k+1}-t_k} - \frac{f(t-t_k) - f(t-t_{k-1})}{t_k - t_{k-1}}, \tag{16.114}$$

where

$$f(t) := \frac{1}{T_o}\int_0^t (t-\tau)\,h(\tau)\,d\tau.$$

This leads to the result of $\langle s_n, r_k\rangle$ in the case (v) of the table.

### 16.8.2 Synchronous Sampling of Reconstructed Signal

An important application is the conversion of nonuniform samples into a PCM signal for interface compatibility with standard digital systems. This can be done by resampling the output $\hat{x}(t)$ of the $\mathcal{RDS}$ system at the Nyquist period $T_o$ to obtain the sequence $(\hat{x}(nT_o))_{n\in\mathbb{Z}}$. In practice, one does not necessarily need to produce physically the continuous-time signal before resampling it. It is more efficient to directly and digitally compute $(\hat{x}(nT_o))_{n\in\mathbb{Z}}$ from the output $(y_n)_{n\in\mathbb{Z}}$ of the processor **D** (see Figure 16.8). Let us concentrate on the computation of $\hat{x}(t_d)$ at a given discrete instant $t_d$. As $\hat{x}(t) = \mathcal{R}\mathbf{y}$, then $\hat{x}(t_d) = \sum_{n\in\mathbb{Z}} y_n r_n(t_d)$. Again, for finite computation complexity, one calculates the approximation

$$\hat{x}(t_d) \simeq \sum_{n\in K_d} y_n r_n(t_d), \tag{16.115}$$

where $K_d = \{k \in \mathbb{Z} : |t_k - t_d| \le T\}$ for some constant $T$. With the examples of reconstruction operator considered in the previous sections, $r_n(t_d)$ can be again obtained from the lookup table of a one-argument function $f(t)$. When $r_n(t) = r(t-t_n)$ for some function $r(t)$, $r_n(t_d) = r(t_d-t_n)$, which requires a table for the function $f(t) = r(t)$. When $r_n(t) = \varphi_o(t) * p_n(t)$ as in the case (ii) of Table 16.1, one obtains from (16.113) that $r_n(t_d) = f(t_d-t_k) - f(t_d-t_{k+1})$, where $f(\tau) := \frac{1}{T_o}\int_0^\tau \varphi_o(t)\,dt$. This technique was first introduced in [23]. When $r_n(t) = \varphi_o(t) * l_n(t)$ as in the case (v) of Table 16.1, one then applies (16.114) with $h(t) = \varphi_o(t)$ and $t = t_d$.

### 16.8.3 Lagrange Interpolation

In the previous section, we proposed to convert nonuniform samples into a PCM signal by resampling the output of an $\mathcal{RDS}$ system. With the direct sampling operator $\mathcal{S}x(t) = (x(t_n))_{n\in\mathbb{Z}}$ under the conditions (16.76) and (16.77) with $T_s \le T_o$ and $\mu < \frac{1}{4}$, we recall from Section 16.4.3 that $x(t) \in \mathcal{B}$ can be exactly reconstructed by the Lagrange interpolation formula

$$x(t) = \sum_{n\in\mathbb{Z}} x(t_n)\,r_n(t), \tag{16.116}$$

where $\forall\{r_n(t)\}_{n\in\mathbb{Z}}$ are given in (16.79). In this case, an alternative method of PCM conversion is to resample an estimate $\hat{x}(t)$ of $x(t)$ obtained from a finite-complexity approximation of the above formula. A standard way to approximate Lagrange interpolation is to take $\hat{x}(t)$ such that for each $k$,

$$\forall t\in[t_k,t_{k+1}],\quad \hat{x}(t)=\sum_{n=k-K+1}^{k+K} x(t_n)\,\hat{r}_{k,n}(t),\quad (16.117)$$

where

$$\hat{r}_{k,n}(t)=\prod_{\substack{i=k-K+1\\ i\neq n}}^{k+K}\frac{t-t_i}{t_n-t_i},$$

and $K$ is a truncation integer parameter. Using the rectangular function $p_n(t)$ defined in Figure 16.2a, one can write more concisely that $\hat{x}(t)=\sum_{(k,n)\in I}x(t_n)\,\hat{r}_{k,n}(t)\,p_k(t)$ for all $t\in\mathbb{R}$, where $I$ is the set of all integer pairs $(k,n)$ such that $k-K+1\leq n\leq k+K$. Since this is equivalent to $n-K\leq k\leq n+K-1$, we also have

$$\forall t\in\mathbb{R},\quad \hat{x}(t)=\sum_{n\in\mathbb{Z}}x(t_n)\,\hat{r}_n(t),\qquad (16.118)$$

where

$$\hat{r}_n(t)=\sum_{k=n-K}^{n+K-1}\hat{r}_{k,n}(t)\,p_k(t).$$

We show in Figure 16.9 examples of such functions $\hat{r}_n(t)$ for various values of $K$ and sampling configurations. With $K=1$, one obtains $\hat{r}_n(t)=\frac{t-t_{n-1}}{t_n-t_{n-1}}p_{n-1}(t)+\frac{t-t_{n+1}}{t_n-t_{n+1}}p_n(t)=l_n(t)$, where $l_n(t)$ is the triangular function of Figure 16.2b, and thus recognizes linear interpolation.

The theory that led to the perfect reconstruction (16.116) does not apply when $\mu\geq\frac{1}{4}$ [7]. Interestingly, it is still possible to show that the estimate $\hat{x}(t)$ of (16.117) uniformly converges to $x(t)$ when $K$ goes to infinity regardless of $\mu\geq 0$ with a large enough oversampling ratio $\rho=\frac{T_\rho}{T_s}$. Using basic properties of Lagrange polynomials from [24, pp. 186–192] and extending derivations from [25,26] to the present case, we show in Section 16.8.4 that at large enough $K$,

$$|\hat{x}(t)-x(t)|\leq\frac{x_m}{\sqrt{\pi}}K^{2\mu-\frac{1}{2}}\left(\frac{\pi}{2\rho}\right)^{2K},\qquad (16.119)$$

where $x_m$ is the input maximum amplitude. From the above upper bound, $\hat{x}(t)$ is guaranteed to converge uniformly to $x(t)$ when the oversampling ratio $\rho$ is more than $\frac{\pi}{2}\simeq 1.57$, *regardless* of $\mu$. Note that the assumption $\rho>\frac{\pi}{2}$ is only a sufficient condition permitting an analytical proof of convergence and has nothing of a necessary condition.

### 16.8.4 Appendix

#### 16.8.4.1 Proof of (16.114)

Let $\rho_{a,b,c}(t)$ be the triangular function shown in dashed line in Figure 16.10d. In the convolution expression

$$(h*l_k)(t)=\int_{-\infty}^{\infty}h(\tau)\,l_k(t-\tau)\,d\tau,$$

$l_k(t-\tau)=\rho_{a,b,c}(\tau)$ with $(a,b,c)=(t-t_{k+1},t-t_k,t-t_{k-1})$. The key is to expand the three-parameter function $\rho_{a,b,c}(\tau)$ in terms of a one-parameter function. Consider first the two-parameter function $\rho_{a,b}(\tau)$ shown in Figure 16.10c, that is zero on $(-\infty,a]$, linear on $[a,b]$ of slope 1, and constant on $[b,\infty)$. It is easy to see from Figure 16.10d that

$$\rho_{a,b,c}(\tau)=\frac{\rho_{a,b}(\tau)}{b-a}-\frac{\rho_{b,c}(\tau)}{c-b}.$$

Now, let $\rho_t(\tau)$ be the one-parameter function equal to $|t-\tau|$ when $0\leq\tau\leq t$ or $t\leq\tau<0$, and 0 otherwise. This function is illustrated in Figure 16.10a. By comparing the parts (b) and (c) of Figure 16.10, it is easy to see that

$$\rho_{a,b}(\tau)=\rho_a(\tau)-\rho_b(\tau)+(b-a)\,u(\tau),$$

where $u(\tau)$ is the unit step function. By injecting this into the above expression of $\rho_{a,b,c}(\tau)$, we obtain

$$\rho_{a,b,c}(\tau)=\frac{\rho_a(\tau)-\rho_b(\tau)}{b-a}-\frac{\rho_b(\tau)-\rho_c(\tau)}{c-b}.$$

Defining $f(t):=\int_{-\infty}^{\infty}h(\tau)\rho_t(\tau)\,d\tau$, it is clear that

$$\int_{-\infty}^{\infty}h(\tau)\,\rho_{a,b,c}(\tau)\,d\tau=\frac{f(a)-f(b)}{b-a}-\frac{f(b)-f(c)}{c-b}.$$

Then, $(h*l_k)(t)$ is obtained by replacing $(a,b,c)$ in the above expression by $(t-t_{k+1},t-t_k,t-t_{k-1})$. Meanwhile, $f(t)$ yields the simpler expression in (16.114) using the above definition of $\rho_t(\tau)$.

#### 16.8.4.2 Derivation of the Bound (16.119)

Without loss of generality, consider the function $\hat{x}(t)$ of (16.117) for $t\in[t_0,t_1]$ ($k$=0). The first crucial step is a result of [24] stating that

$$\hat{x}(t)-x(t)=\frac{1}{(2K)!}x^{(2K)}(u)\,w(t),$$

where

$$w(t)=\prod_{i=-K+1}^{K}(t-t_i),$$

$x^{(n)}(t)$ is the $n$th derivative of $x(t)$ and $u$ is some value in $[t_{-K+1},t_K]$. From (16.76), $t_i^-\leq t_i\leq t_i^+$, where $t_i^\pm:=$

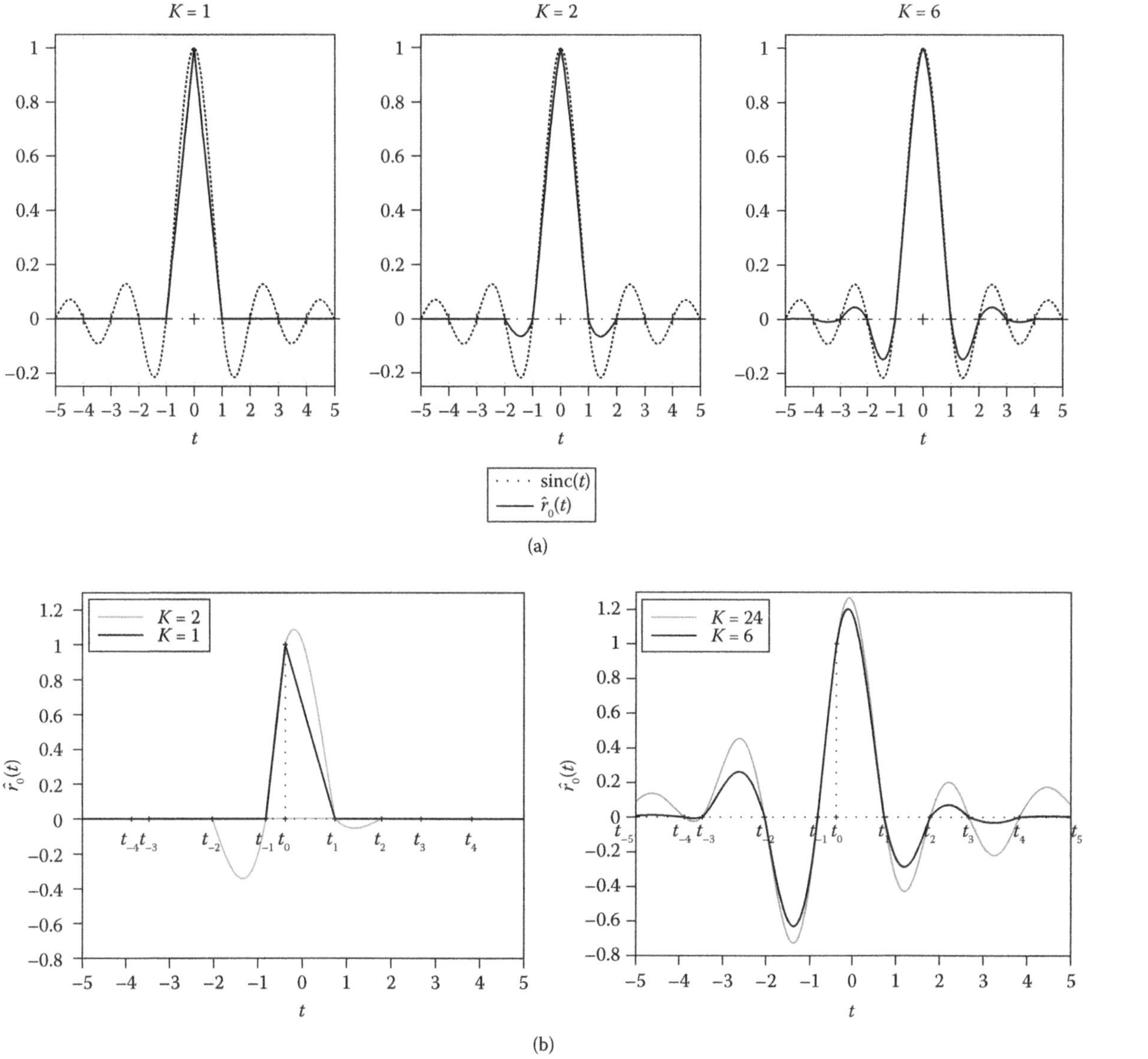


**FIGURE 16.9**
Truncated Lagrange interpolation function $\hat{r}_0(t)$ given in (16.118) for various values of $K$: (a) uniform sampling $t_n = kT_s$ (with $T_s = 1$) and (b) nonuniform sampling.

$(i \pm \mu)T_s$. Since $t \in [t_0, t_1]$, then $t_i^- \leq t_i \leq t \leq t_j \leq t_j^+$ for all integers $i \leq 0$ and $j \geq 1$. Then, $|w(t)| \leq v_-(t)\, v_+(t)$, where

$$v_-(t) := \prod_{i=-K+1}^{0} (t - t_i^-),$$

and

$$v_+(t) := \prod_{j=1}^{K} (t_j^+ - t).$$

The zeros of $v_-(t)$ and the zeros of $v_+(t)$ can be seen to be symmetric about $\frac{T_s}{2}$. The product $v_-(t)\, v_+(t)$ then reaches a maximum within $[t_0^-, t_1^+]$ at $\frac{T_s}{2}$. One finds that

$$v_-\left(\tfrac{T_s}{2}\right) = v_+\left(\tfrac{T_s}{2}\right) = \prod_{j=1}^{K} \left(t_j^+ - \tfrac{1}{2}T_s\right)$$
$$= T_s^K \prod_{j=1}^{K} \left(j+\mu-\tfrac{1}{2}\right) = T_s^K \frac{\Gamma(K+\mu+\frac{1}{2})}{\Gamma(\mu+\frac{1}{2})},$$

given the relation $\Gamma(t+1) = t\,\Gamma(t)$ satisfied by the gamma function $\Gamma$. By Bernstein inequality, $|x^{(2K)}(u)| \leq x_m \Omega_o^{2K}$. Then, for any $t \in [t_0, t_1] \subset [t_0^-, t_1^+]$,

$$|\hat{x}(t) - x(t)| \leq \frac{x_m}{\Gamma(\mu+\frac{1}{2})^2} (T_s\Omega_o)^{2K} \frac{\Gamma(K+\mu+\frac{1}{2})^2}{(2K)!}. \quad (16.120)$$

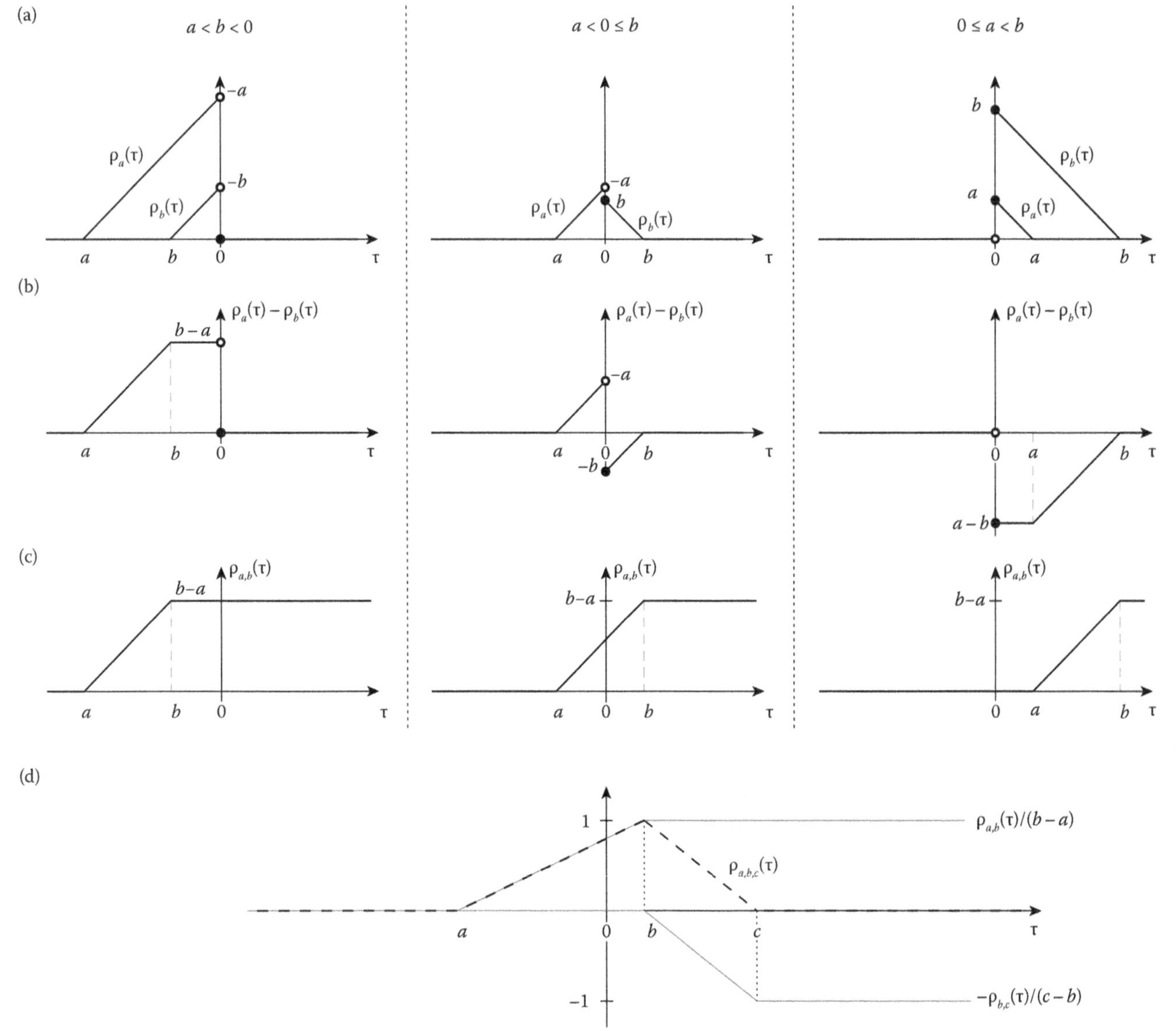


**FIGURE 16.10**
Decomposition of triangular function $\rho_{a,b,c}(\tau)$: (a) $\rho_a(\tau)$ and $\rho_b(\tau)$; (b) $\rho_a(\tau) - \rho_b(\tau)$; (c) $\rho_{a,b}(\tau)$; and (d) $\rho_{a,b,c}(\tau)$.

Using Stirling formula $\Gamma(t+1) \sim \sqrt{2\pi t}(\frac{t}{e})^t$ at large $t$, $n! = \Gamma(n+1)$ and $\beta := \mu - \frac{1}{2}$, we have at large $K$

$$\frac{\Gamma(K+\mu+\frac{1}{2})^2}{(2K)!} = \frac{\Gamma(K+\beta+1)^2}{(2K)!} \sim \frac{2\pi(K+\beta)}{\sqrt{2\pi 2K}} \frac{\left(\frac{K+\beta}{e}\right)^{2K+2\beta}}{\left(\frac{2K}{e}\right)^{2K}}$$
$$\sim \frac{2\pi(K+\beta)}{\sqrt{2\pi 2K}} \left(\frac{K+\beta}{e}\right)^{2\beta} \left(1+\frac{\beta}{K}\right)^{2K} \frac{1}{2^{2K}}$$
$$\sim \frac{\sqrt{\pi} K^{2\mu-\frac{1}{2}}}{2^{2K}},$$

using the limit $(1+\frac{\beta}{K})^K \sim e^\beta$. One obtains (16.119) after injecting the above result into (16.120), the inequality $\Gamma(\mu + \frac{1}{2}) \geq \Gamma(\frac{1}{2}) = \sqrt{\pi}$ and the relation $\Omega_o = \frac{\pi}{T_o}$.

## Acknowledgments

The author thanks Yannis Tsividis for motivating the elaboration of this chapter and for his useful comments, and Sinan Güntürk for his help on the mathematical aspects of operator theory.

## Bibliography

[1] Y. Tsividis, M. Kurchuk, P. Martinez-Nuevo, S. Nowick, B. Schell, and C. Vezyrtzis. Event-based data acquisition and digital signal processing in

continuous time. In *Event-Based Control and Signal Processing* (M. Miskowicz, Ed.), pp. 353–378. CRC Press, Boca Raton, FL 2015.

[2] J. W. Mark and T. Todd. A nonuniform sampling approach to data compression. *IEEE Transactions on Communications*, 29:24–32, 1981.

[3] R. Duffin and A. Schaeffer. A class of nonharmonic Fourier series. *Transactions of the American Mathematical Society*, 72:341–366, 1952.

[4] J. L. Yen. On nonuniform sampling of bandwidth-limited signals. *IEEE Transactions on Circuit Theory*, CT-3:251–257, 1956.

[5] F. J. Beutler. Error-free recovery of signals from irregularly spaced samples. *SIAM Review*, 8:328–335, 1966.

[6] H. Landau. Sampling, data transmission, and the Nyquist rate. *Proceedings of the IEEE*, 55:1701–1706, 1967.

[7] K. Yao and J. Thomas. On some stability and interpolatory properties of nonuniform sampling expansions. *IEEE Transactions on Circuit Theory*, 14:404–408, 1967.

[8] F. Marvasti. *Nonuniform Sampling: Theory and Practice*. New York: Kluwer, 2001.

[9] J. Benedetto. Irregular sampling and frames. In *Wavelets: A Tutorial in Theory and Applications*, C. K. Chui, Ed., pp. 445–507. Academic Press, Boston, MA, 1992.

[10] H. G. Feichtinger and K. Gröchenig. Theory and practice of irregular sampling. In *Wavelets: Mathematics and Applications*, J. Benedetto, Ed., pp. 318–324. CRC Press, Boca Raton, FL, 1994.

[11] A. Lazar and L. T. Tóth. Perfect recovery and sensitivity analysis of time encoded bandlimited signals. *IEEE Transactions on Circuits and Systems I: Regular Papers*, 51:2060–2073, 2004.

[12] M. Vetterli, V. K. Goyal, and J. Kovacevic. *Foundations of Signal Processing*. Cambridge University Press, Cambridge, UK, 2014.

[13] A. W. Naylor and G. R. Sell. *Linear Operator Theory in Engineering and Science, volume 40 of Applied Mathematical Sciences*, 2nd edition. Springer-Verlag, New York, NY, 1982.

[14] J. B. Conway. *A course in Functional Analysis, volume 96 of Graduate Texts in Mathematics*, 2nd edition. Springer-Verlag, New York, NY, 1990.

[15] C. S. Kubrusly. *Spectral Theory of Operators on Hilbert Spaces*. Birkhäuser/Springer, New York, NY, 2012.

[16] R. Paley and N. Wiener. *Fourier Transforms in the Complex Domain, volume 19 of American Mathematical Society Colloquium Publications*. American Mathematical Society, Providence, RI, 1987. Reprint of the 1934 original.

[17] N. Levinson. *Gap and Density Theorems, volume 26 of American Mathematical Society Colloquium Publications*. American Mathematical Society, New York, NY, 1940.

[18] K. Gröchenig. Sharp results on irregular sampling of bandlimited functions. In *Probabilistic and Stochastic Methods in Analysis, with Applications (Il Ciocco, 1991), volume 372 of NATO Advanced Science Institutes Series C: Mathematical and Physical Sciences* (J.S.Byrnes, J.L.Byrnes, K.A.Hargreaves, and K. Berry, Eds.), pp. 323–335. Kluwer Academic Publishers, Dordrecht, 1992.

[19] A. A. Boichuk and A. M. Samoilenko. *Generalized Inverse Operators and Fredholm Boundary-Value Problems*. VSP, Utrecht, 2004. Translated from the Russian by P. V. Malyshev and D. V. Malyshev.

[20] K. Gröchenig. Reconstruction algorithms in irregular sampling. *Mathematics of Computation*, 59(199):181–194, 1992.

[21] A. V. Oppenheim and R. W. Schafer. *Discrete-Time Signal Processing*. Prentice-Hall Signal Processing Series, Upper Saddle River, NJ, 2011.

[22] J. Benedetto and S. Scott. Frames, irregular sampling, and a wavelet auditory model. In *Nonuniform Sampling: Theory and Practice* (F. Marvasti, Ed.), pp. 585–617. Kluwer, New York, NY, 2001.

[23] D. Hand and M.-W. Chen. A non-uniform sampling ADC architecture with embedded alias-free asynchronous filter. In *Global Communications Conference (GLOBECOM), 2012 IEEE*, Anaheim, CA, pp. 3707–3712, December 2012.

[24] P. Henrici. *Elements of Numerical Analysis*. Wiley, New York, NY, 1964.

[25] J. J. Knab. System error bounds for Lagrange polynomial estimation of band-limited functions. *IEEE Transactions on Information Theory*, 21:474–476, 1975.

[26] Z. Cvetković. Single-bit oversampled A/D conversion with exponential accuracy in the bit rate. *IEEE Transactions on Information Theory*, 53:3979–3989, 2007.

# 17

# *Spectral Analysis of Continuous-Time ADC and DSP*

**Yu Chen**
*Columbia University*
*New York, NY, USA*

**Maria Kurchuk**
*Pragma Securities*
*New York, NY, USA*

**Nguyen T. Thao**
*City College of New York*
*New York, NY, USA*

**Yannis Tsividis**
*Columbia University*
*New York, NY, USA*

**CONTENTS**



**ABSTRACT** This chapter analyzes the error introduced in continuous-time amplitude quantization. Spectral features of the quantization error added on bandlimited signals are highlighted. The effects of synchronization and time coding on the error spectrum are discussed. The error at the output of a continuous-time DSP is also analyzed.

## 17.1 Introduction

The traditional view that quantization amounts to an additive and independent source of white noise has mainly resulted from the culture of discrete-time signals; to produce those from a continuous-time input, both input sampling and quantization are performed. In contrast to this, in continuous-time (CT) analog-to-digital converter (ADC) only quantization is performed [1], as shown in Figure 17.1. In this chapter, we show how amplitude quantization in continuous time can be analyzed as a memoryless (nonlinear) transformation of continuous waveforms. This yields spectral results that are harmonic based, rather than noise oriented. The resulting spectrum is discussed in Section 17.2.

While pure continuous-time processing of signals has advantages as presented in Chapter 15, one still needs to consider and evaluate the effects of an eventual sampling of the signals thus obtained for interface compatibility with standard data processing (including storage, transmission, and off-line computation). Section 17.3 addresses this issue.

As mentioned in [1], one can store the nonuniformly spaced event times $t_k$ by finely quantizing the time axis, which results in "pseudocontinuous" operation. A basic question when doing this is how to decide what the resolution of time quantization should be, such that the accuracy of the resulting signal is comparable with that of a purely continuous-time system. Section 17.4 discusses this problem.

Finally, the results obtained for CT quantizers are extended to the output of a CT DSP in Section 17.5.

## 17.2 Continuous-Time Amplitude Quantization

Continuous-time amplitude quantization (Figure 17.1) converts the input $x(t)$ to a piecewise constant signal $x_q(t)$ and allows further digitizing. Basically, $x_q(t)$ is an instantaneous transformation of $x(t)$, which can be expressed as

$$x_q(t) = Q(x(t)), \tag{17.1}$$

where $Q$ is the scalar function of quantization. The error signal $e(t) = x_q(t) - x(t)$ is then also a memoryless function of $x(t)$

$$e(t) = E(x(t)), \tag{17.2}$$

where $E$ is the error function of $Q$ defined by

$$E(x) = Q(x) - x. \tag{17.3}$$

The above error signal $e(t)$ is defined with the implicit assumptions that it does not contain a signal component or a constant, and the quantization does not add delay. In the more general case, the error has to be defined as follows. The mean square error (MSE) is obtained by the following minimization algorithm over $a$, $b$, and $\tau$, which compensates for the gain, DC offset, and time delay of the quantization, respectively:

$$MSE = \min_{a,b,\tau}\{\overline{(x_q(t) - ax(t-\tau) - b)^2}\}, \tag{17.4}$$

Once the three parameters are determined, the error signal can then be defined as

$$e(t) = x_q(t) - ax(t-\tau) - b. \tag{17.5}$$

For simplicity, this chapter assumes that quantizers have a unity gain, zero DC shift, and zero time delay, so that $e(t) = x_q(t) - x(t)$. However, the principles presented are also valid for other cases.

After reviewing basic knowledge on the error function $E(x)$ of standard scalar quantizers, we will tackle the spectral analysis of $e(t)$ successively when $x(t)$ is a single tone and when it is composed of two tones.

### 17.2.1 Quantization Characteristics

Quantization consists of approximating an input number $x$ by the nearest value from a predetermined finite set of quantum levels. In general, these levels may not be uniformly spaced, as is the case in signal companding. In this chapter, we concentrate on the simpler but standard case of uniform spacing. Figure 17.2a and b shows the two typical transfer functions $Q$ of uniform scalar quantizers [2]. They are both staircase functions. The mid-rise quantizer of Figure 17.2a has a discontinuity at $x = 0$. Meanwhile, the mid-tread quantizer of Figure 17.2b has a quantum level at 0 and is preferred in event-driven systems, as it can tolerate small input imperfections around zero (DC offset, noise) and maintain a zero output in the absence of a signal. An odd number of quantum levels is however necessary to achieve symmetry around the origin. With a quantization step size $\Delta$ and an $N$-bit resolution of quantum levels, the single-ended full-scale amplitude of the mid-tread quantizer is

$$A_{FS} = \left(2^{N-1} - \frac{1}{2}\right)\Delta,$$

(while it is $2^{N-1}\Delta$ with the mid-rise quantizer).

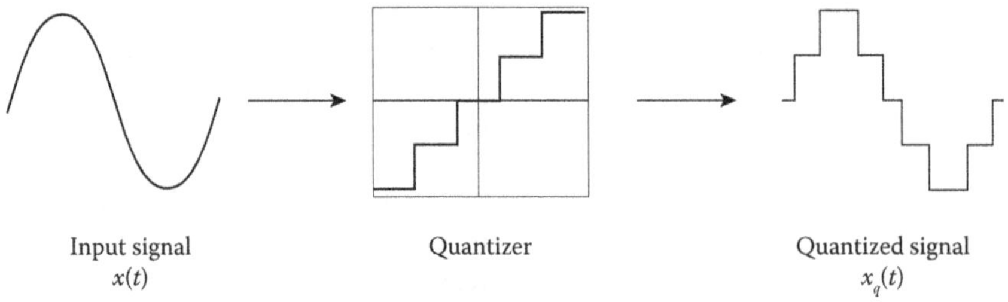


**FIGURE 17.1**
A continuous-time amplitude quantization operation.

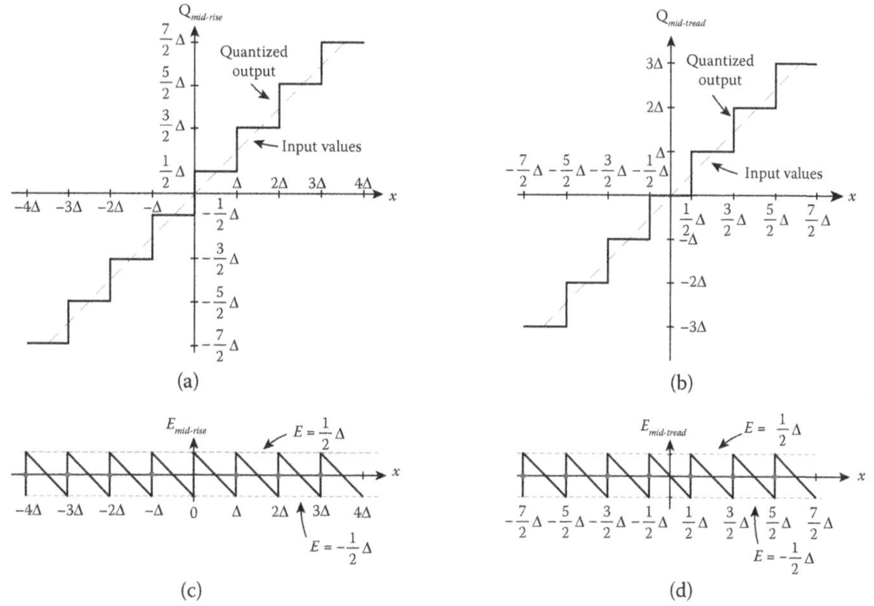


**FIGURE 17.2**
Quantization characteristics and error functions: (a) characteristic of mid-rise quantizer; (b) characteristic of mid-tread quantizer; (c) quantization error of mid-rise quantizer; and (d) quantization error of mid-tread quantizer.

Figure 17.2c and d shows the respective error functions $E(x)$. They are both odd periodic sawtooth functions of period $\Delta$. When $x$ is a random variable that is uniformly distributed over a range of a multiple of $\Delta$ in length, $E(x)$ is known to be uniformly distributed between $-\frac{\Delta}{2}$ and $\frac{\Delta}{2}$ [3]. In this case, $E(x)$ yields the classical variance value of $\frac{\Delta^2}{12}$.

When the above statistical assumption on the input is not satisfied, one needs to resort to a deterministic analysis of the error function $E(x)$. As a periodic function, $E(x)$ can be expanded in a Fourier series. With mid-tread quantizers, the expansion is [4,5]

$$E(x) = \Delta \sum_{k=1}^{\infty} (-1)^k \frac{\sin(2k\pi x/\Delta)}{k\pi}. \tag{17.6}$$

By removing $(-1)^k$ from this equation, one obtains the series corresponding to mid-rise quantizers. In the following discussion, however, we will focus on the mid-tread characteristic and choose by default the error expansion of (17.6).

### 17.2.2 Single-Tone Input

We now analyze the quantization error signal when $x(t)$ is a single-tone (sinusoidal) input. Figure 17.3 shows both the quantized output $x_q(t) = Q(x(t))$ and the quantization error signal $e(t) = E(x(t))$ for a 6-bit mid-tread quantizer with a full-scale single-tone input. As illustrated in part c of this figure, Pan and Abidi [6] divided the analysis of $e(t)$ into three different regions:

1. The bell-like pulses formed by the quantization of the sine wave around its peaks;
2. The sawtooth patterns arising from the quantization of the sine wave around its zero-crossings, where it is close to an ideal ramp;
3. The transition parts between the above two regions.

Although the amplitude values of a sinusoid are not uniformly distributed [7] (maximal density values being obtained at the peak values of the sinusoid, as expected from Figure 17.3c), the probability distribution of $e(t)$ is still modeled as uniform in $[-\frac{\Delta}{2}, \frac{\Delta}{2}]$ [3]. In the following, "power", denoted by $P$, refers to a mean-square value. With a full-scale sinusoidal input, the total error power of $\frac{\Delta^2}{12}$ resulting from this model appears to be satisfactorily accurate in practice. Given the power of a full-scale sinusoidal input of $\frac{A_{FS}^2}{2}$, the signal-to-error ratio ($SER$) of the quantized output is

$$SER = \frac{P_{signal}}{P_{error}} = \frac{A_{FS}^2/2}{\Delta^2/12}.$$

With an $N$-bit mid-tread quantizer, $A_{FS} = (2^{N-1} - \frac{1}{2})\Delta$. Thus, the $SER$ in decibels yields the expression

$$SER_{dB} = 10\log_{10} \frac{\left((2^{N-1} - 1/2)\Delta\right)^2/2}{\Delta^2/12} \approx 6.02N + 1.76 \text{ dB},$$

where the approximation assumes $2^N \gg 1$.

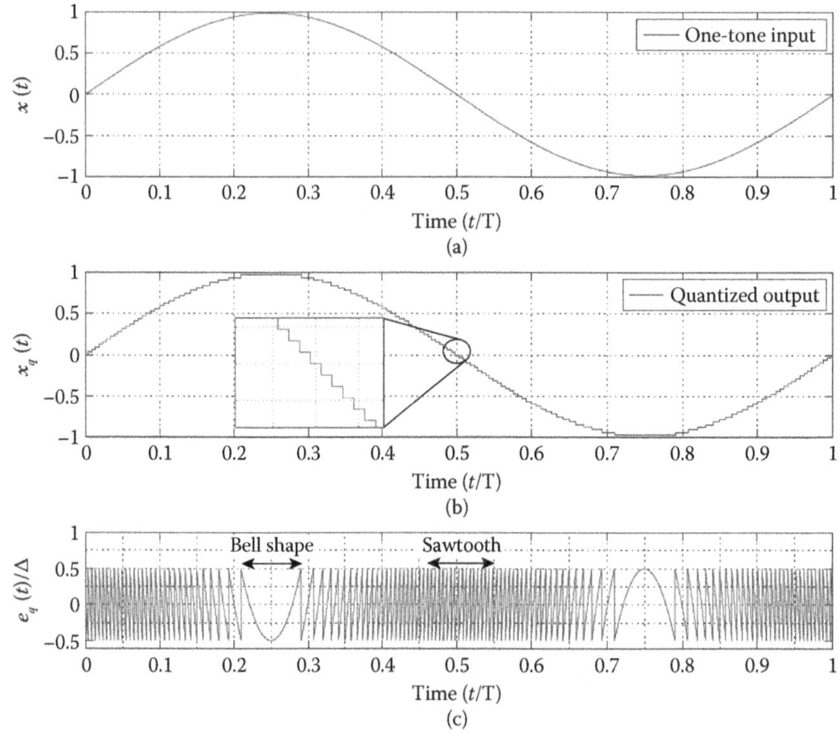


**FIGURE 17.3**
A quantized sinusoid waveform and its quantization error. The quantizer has a 6-bit resolution. (a) A full-scale sinusoidal input; (b) the quantized output waveform (the piecewise-constant waveform is more clearly shown in the zoomed-in plot.); and (c) the quantization error signal. Its amplitude is normalized to the quantization step Δ.

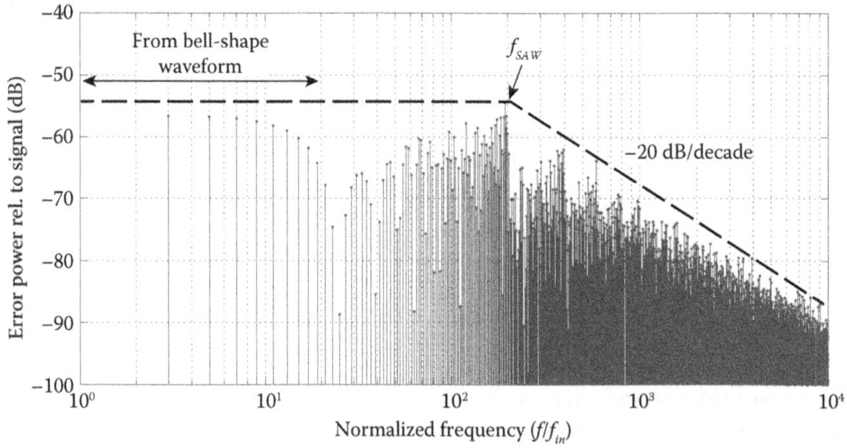


**FIGURE 17.4**
Power spectrum of the quantization error in Figure 17.3c relative to signal power. The error signal is obtained by quantizing a full-scale sinusoidal signal with a 6-bit mid-tread quantizer. The x-axis shows the frequency normalized to $f_{in}$ and is plotted on a *logarithmic* scale.

In the frequency domain, the bell-like pulses and the sawtooth waveform have very different contributions to the spectrum. Figure 17.4 shows the spectrum of the error signal $e(t)$ in Figure 17.3c. Since $e(t)$ is periodic with the same period as the input, its spectrum is composed of harmonics, with the first harmonic at the input frequency. Since the quantization error function $E(x)$ is odd-symmetric around the origin and the input signal $x(t)$ is so-called half-wave symmetry, the corresponding error signal $e(t)$ is also "half-wave symmetry," and thus the resulting discrete spectrum contains only odd-order harmonics. The bell-like pulses are periodic in the time domain at the input frequency and contribute mostly low-order harmonics. In [8], it is observed that small variations in the amplitude of the input may cause large changes in the bell-like pulses. Thus, the power of these low-order harmonics is very sensitive to the input amplitude. The fast-varying sawtooth part contributes mostly to the high-frequency part of the error power. Let the minimum sawtooth duration $T_{SAW}$ be the shortest time for the input to cross a quantization interval, and define the sawtooth frequency $f_{SAW}$ as its inverse. For a bandlimited input, $T_{SAW}$ is approximately $\frac{\Delta}{|dx/dt|_{max}}$, which implies that

$$f_{SAW} = \frac{|dx/dt|_{max}}{\Delta}. \tag{17.7}$$

For an input of the form $x(t) = A_{in} \sin(2\pi f_{in} t)$,

$$f_{SAW} = \frac{2\pi A_{in} f_{in}}{\Delta}. \tag{17.8}$$

With a full-scale input, $A_{in} = (2^{N-1} - \frac{1}{2})\Delta$ and hence

$$f_{SAW} = 2\pi\left(2^{N-1} - \frac{1}{2}\right) f_{in} \approx 2^N \pi f_{in}. \tag{17.9}$$

The last approximation assumes $2^N \gg 1$, which is satisfied in most practical cases. In the given example, $N = 6$. This leads to $f_{SAW} = 201 f_{IN}$ (indicated in Figure 17.4). The spectrum beyond $f_{SAW}$ contains the harmonics of the fundamental components below $f_{SAW}$. A qualitative observation is in order. In Figure 17.4, one can observe that the amplitude of the components up to $f_{SAW}$ can roughly be thought to have a constant envelope, while the amplitude of the components beyond $f_{SAW}$ can be seen to drop off, on average, with a $-20$ dB/decade slope (the reason for the value of this slope will be discussed below).

Assume the single-tone input has a full-scale amplitude $A_{FS}$, an arbitrary input frequency $f_{in}$ and an arbitrary phase shift $\phi$, that is:

$$x(t) = A_{FS} \sin(2\pi f_{in} t + \phi). \tag{17.10}$$

By substituting (17.10) into (17.6), the quantization error becomes:

$$e(t) = \Delta \sum_{k=1}^{\infty} (-1)^k \frac{1}{\pi k} \sin\left(\frac{2\pi k A_{FS}}{\Delta} \sin(2\pi f_{in} t + \phi)\right). \tag{17.11}$$

A complete Fourier series expansion of $e(t)$ can then be obtained by using Bessel functions and Jacobi-Anger expansion [5,9,10]. This process is involved and thus is not included in this chapter. We only highlight some important results. The term of (17.11) corresponding to $k = 1$ is

$$e_1(t) = -\frac{2\Delta}{\pi} \sum_{m=1, m\,\text{odd}}^{\infty} J_m\left(\frac{2\pi A_{FS}}{\Delta}\right) \times \sin\left[m(2\pi f_{in} t + \phi)\right],$$

where $J_m(.)$ is the Bessel function of the first kind of order $m$. This is a sum of all the odd-order harmonics of the input frequency with amplitudes $J_m(\frac{2\pi A_{FS}}{\Delta})\frac{2\Delta}{\pi}$. According to [9],

$$J_m(\beta) \approx 0, m > \beta + 1. \tag{17.12}$$

Thus, $J_m(\frac{2\pi A_{FS}}{\Delta}) = J_m(2\pi(2^{N-1} - \frac{1}{2}))$ is negligible for $m > 2^N \pi$. In other words, all the significant tones of $e_1(t)$ have frequencies below $2^N \pi f_{IN}$, which is exactly the $f_{SAW}$ defined in Equation 17.9.

The $k$th-order term of $e(t)$ in (17.11) can be expressed as

$$e_k(t) = (-1)^k \frac{2\Delta}{k\pi} \sum_{m=1, m\,\text{odd}}^{\infty} J_m\left(\frac{2k\pi A_{FS}}{\Delta}\right) \times \sin\left[m(2\pi f_{in} t + \phi)\right].$$

It has the same form as $e_1(t)$ but with a factor of $1/k$ in front of the summation and a factor of $k$ inside the argument of the Bessel function. $(-1)^k$ in front does not affect its amplitude. Using again (17.12), we find the significant tones of $e_k(t)$ are present only up to $k f_{SAW}$. The frequency range $[(k-1) f_{SAW}, k f_{SAW}]$, which we call the $k$th $f_{SAW}$ band, contains the significant tones from the $k$th and higher order terms only, that is, those resulting from $\sum_{i=k}^{\infty} e_i(t)$. However, because of the scaling factor $1/k$, the power of the tones within each $k$th $f_{SAW}$ is mostly determined by the term of the lowest order $e_k(t)$. They are approximately $k^2$ times lower than the tones within the first $f_{SAW}$ band (i.e., $[0, f_{SAW}]$). This result is better observed on the plot of Figure 17.5, plotted using a linear frequency axis. A staircase-like spectrum is observed. Each step has the same width of $f_{SAW}$. Also, the average power in each $k$th $f_{SAW}$ band is $k^2$ lower than the tones in the first $f_{SAW}$ band. This is consistent with the observed slope of $-20$ dB/decade indicated on the logarithmic scale of Figure 17.4. A zoomed-in plot around $f_{in}$ is provided to more clearly show the discrete feature of the spectrum. Only harmonics are present, while no component exists at any in-between frequency.

#### 17.2.2.1 Power within $[0, f_{SAW}]$

It was observed in [8] that the error power below $f_{SAW}$ dominates the entire quantization error power. The reason is that the tones beyond $f_{SAW}$ are coming from the higher order terms of Equation 17.11 and thus their

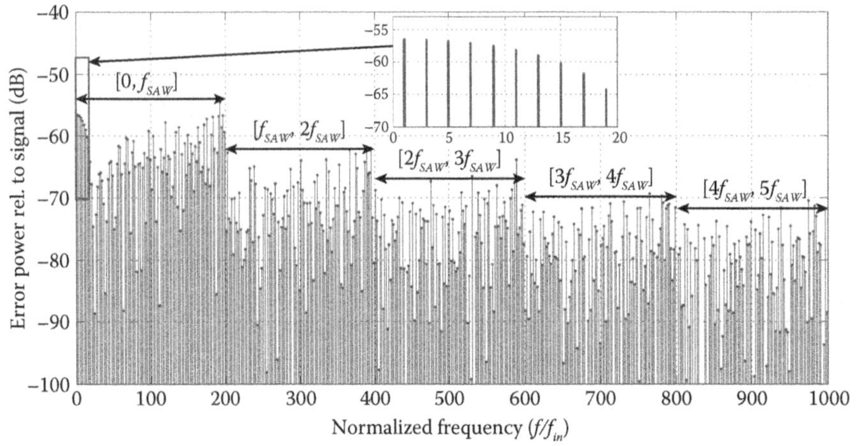


**FIGURE 17.5**
Power spectrum of the quantization error in Figure 17.3c relative to signal power. The error signal is obtained by quantizing a full-scale sinusoidal signal with a 6-bit mid-tread quantizer. The x-axis shows the frequency normalized to $f_{in}$ and is plotted on a *linear* scale. A zoomed-in plot shows a narrow spectrum around $f_{in}$ (i.e., $[0, 20f_{in}]$).

power is scaled down by $1/k^2$. From (17.11), it can be found numerically that 71% of the total error power lies in the band $[0, f_{SAW}]$. It is also interesting to estimate this number using the staircase model of Figure 17.5. According to this rough model, the power spectral density in the $k$th $f_{SAW}$ band is assumed flat and proportional to $1/k^2$. As the error tones are uniformly spaced (with a distance of $2f_{IN}$), the error power within the $k$th $f_{SAW}$ band decreases in $1/k^2$ with $k$. The ratio of power that lies in $[0, f_{SAW}]$ is then $1/\sum_{k=1}^{\infty} k^2 = 1/\frac{\pi^2}{6} \simeq 61\%$, which can be compared to the numerical result above.

### 17.2.3 Two-Tone Input

Assume now that the input is a two-tone signal $x(t) = A_{in1}\sin(2\pi f_{in1}t) + A_{in2}\sin(2\pi f_{in2}t)$ (assuming zero phase in the two). The sawtooth frequency $f_{SAW}$ can still be derived from (17.7) and yields

$$f_{SAW} = \frac{2\pi A_{in1}f_{in1} + 2\pi A_{in2}f_{in2}}{\Delta}.$$

By defining the weighted-averaged frequency

$$f_{in,avg} = \frac{A_{in1}f_{in1} + A_{in_2}f_{in2}}{A_{FS}},$$

one obtains the equivalent expression

$$f_{SAW} = \frac{2\pi A_{FS}f_{in,avg}}{\Delta},$$

which coincides with $f_{SAW}$ of a full-scale single-tone input of frequency $f_{in,avg}$ according to (17.8). Figure 17.6a plots the spectrum of the quantization error $e(t)$ resulting from this input in the case where $A_{in1} = A_{in2} = \frac{A_{FS}}{2}$ and $N = 6$. The spectrum now contains not only harmonics but also intermodulation products. The tones are uniformly spaced by a distance of $\Delta f = f_{in2} - f_{in1}$ in frequency. Given the bit resolution $N = 6$, $f_{SAW} = \pi(2^{N-1} - \frac{1}{2}) = 201f_{in,avg}$. This value is consistent with the value of $f_{SAW}$ obtained from the graphical method by finding the transition frequency between the flat part and the descending part of the spectrum. Such agreement will also be observed with more complicated inputs as will be seen in Section 17.2.4. It can be found numerically that the power below the maximum sawtooth frequency in this two-tone example is 78% of the total power of quantization error.

The total error power approximately equals to $\frac{\Delta^2}{12}$, which results in an $SER = 34.8$ dB (i.e., $10\log_{10}\frac{A_{in1}^2/2 + A_{in2}^2/2}{\Delta^2/12}$). However, if a narrow baseband is considered, only a few harmonics and intermodulation components fall in band and the resulting in-band $SER$ can be much higher. Figure 17.6b shows an example baseband, which is $[0, 2f_{in,avg}]$. A limited number of tones are evenly distributed within the baseband, with no component exists at any other frequency. The in-band $SER$ can be found by summing up the relative power of all these tones, which is 51 dB. The choice of baseband is rather arbitrary in this two-tone case, where input signal does not occupy a full band. We can choose a baseband with almost any width and find the in-band $SER$ correspondingly. Figure 17.6c shows the in-band quantization error power as a function of the ratio of baseband's upper band-edge frequency to $f_{in,avg}$. As the band-edge moves to high frequencies, more and more harmonics and intermodulation tones fall into the baseband and the resulting in-band error

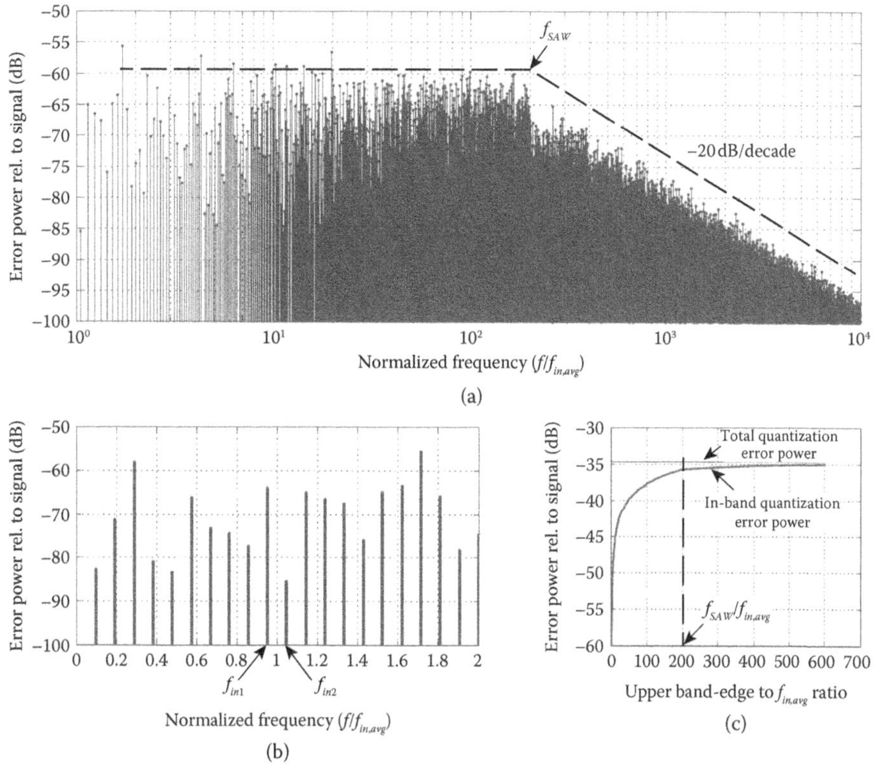


**FIGURE 17.6**

Quantization error relative to signal power in the two-tone test. The quantizer has a 6-bit resolution. Full load of the quantizer is achieved by setting the amplitudes of the two input tones as half of $A_{FS}$. (a) A wide spectrum of the quantization error plotted on a *logarithmic* scale; the frequency axis is normalized to $f_{in,avg}$. (b) A narrow baseband spectrum of the quantization error on *linear* scale; the frequency axis is normalized to $f_{in,avg}$ and shows the ranges from 0 to 2. The locations of the two input tones are shown. (c) In-band quantization error power relative to signal power, as a function of the ratio of upper band-edge frequency to $f_{in,avg}$.

power increases. Although the curve is monotonically increasing, its changing rate becomes very slow once the band-edge is larger than $f_{SAW}$. This is because the error tones beyond $f_{SAW}$ contain low power. The horizontal line on the top shows the total quantization error power relative to the signal power. It is the ultimate value of the relative error power when the baseband is infinitely wide.

### 17.2.4 Bandlimited Gaussian Inputs

We now consider a more complex class of signals, namely bandlimited signals with a Gaussian amplitude probability density function. In [11,12], it is found that the probability for such a random input to overload the quantizer is negligible (less than 1 in 10,000) as long as its root-mean-square amplitude $A_{RMS}$ is four times smaller than the quantizer's full-scale range $A_{FS}$. The spectrum of its quantization error has similar features to one-tone and two-tone inputs, that is, (1) it has a corner frequency (which we will denote by $f_{SAW,Gaussian}$); (2) beyond $f_{SAW,Gaussian}$, the spectrum follows a $-20$ dB/decade slope. $f_{SAW,Gaussian}$ can be calculated by averaging the $f_{SAW}$ contributed by each input tone [8], that is,

$$f_{SAW,Gaussian} = \overline{f_{SAW}} = \frac{2\pi A_{RMS}\overline{f_{in}}}{\Delta}.$$

Calling $f_{BW}$ the maximum frequency of input, we obtain $f_{SAW,Gaussian} = \pi A_{RMS} f_{BW}/\Delta$.

Figure 17.7 shows an example spectrum of the quantization error (solid line). Contrary to the one-tone and two-tone cases, the spectrum is continuous. $A_{RMS}$ is chosen to be $A_{FS}/4$ and thus $f_{SAW,Gaussian} = \pi 2^{N-3} f_{BW}$. For $N = 6$ in this case, $f_{SAW,Gaussian} \approx 25 f_{BW}$. This is consistent with the plot of the quantization error. Beyond $f_{SAW,Gaussian}$, a $-20$ dB/decade slope is apparent. Integrating the error power over the entire frequency axis, we find the $SER$ equals to 29 dB. In the same figure, the spectrum of the bandlimited Gaussian input is also plotted (dotted line). Its power spectrum is white up to $f_{BW}$. Beyond that, its power drops significantly (due to filtering). In this bandlimited Gaussian case, the signal band is well defined as $[0, f_{BW}]$. The bandwidth of the quantization error is much wider than that of the input signal.

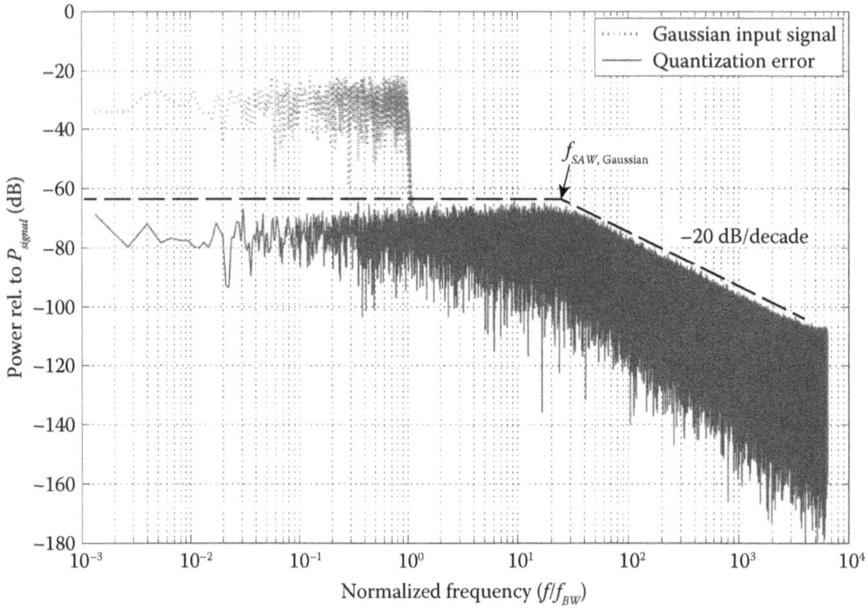


**FIGURE 17.7**
Power spectra in the bandlimited Gaussian input case. The root-mean-square amplitude $A_{RMS}$ is set as one-fourth of $A_{FS}$. The quantizer has a 6-bit resolution. The x-axis shows the frequency normalized to $f_{BW}$ and is plotted on a *logarithmic* scale. The dotted line is the power spectrum of input signal. It is generated by low-pass filtering a white Gaussian noise with a cutoff frequency of 500 Hz. The solid line shows the power spectrum of the quantization error. Both power values are normalized to $P_{signal}$, the total power of the input signal.

Integrating the error power within the signal band, we find the in-band $SER$ to be 45 dB, which is 16 dB higher than the $SER$ calculated over the entire frequency axis.

To summarize Section 17.2, $f_{SAW}$ plays an important role in the spectral characterization of any continuous-time quantized signals which are originally bandlimited. The value of this quantity can be obtained from Equation 17.7. Within $[0, f_{SAW}]$, the error power spectral density is almost flat. Beyond $f_{SAW}$, the spectral density drops with a slope of $-20$ dB/decade. The frequency range $[0, f_{SAW}]$ contains the majority of the total power of the quantization error [8,13], which is roughly in the range of 70–80%.

## 17.3 Uniform Sampling of Quantized Continuous-Time Signals

### 17.3.1 Context and Issues of Uniform Sampling

While continuous-time amplitude-quantized signals are free from aliasing, the lengths of the steps in their piecewise-constant waveform varies, and this prevents their simple integration with standard systems operating under a fixed clock rate. To resolve this issue, a uniform sampling of the continuous-time outputs is necessary. This inevitably brings back the issue of aliasing. In fact, because continuous-time quantization is an instantaneous operation, permuting the order of amplitude quantization and time sampling gives the same result [14]. Indeed, the sample of the amplitude-quantized signal $x_q(t) = Q(x(t))$ at an instant $t_k$ is $x_q(t_k) = Q(x(t_k))$, which is equal to the quantized version of the sample of $x(t)$ at $t_k$. This makes the system equivalent to a conventional ADC involving sampling and quantization, because the two operations can be interchanged. In spite of this system equivalence, there are still two strong reasons for using continuous-time systems. (1) The major part of the hardware in a continuous-time system is event-driven and thus dissipates power more efficiently compared to clocked systems. (2) Using oversampling to improve the $SER$ is straightforward with continuous-time systems. Only the sampling clock frequency of the output synchronizer needs to be increased. In conventional systems on the contrary, oversampling requires an increase of the clock rate of the entire signal processing chain.

### 17.3.2 Analysis of Aliased Quantization Error—A Staircase Model

Next, we investigate how aliasing degrades signal accuracy when a quantized continuous-time quantized signal is sampled. It is usually believed that in-band quantization error can be arbitrarily reduced by increasing sampling frequency. We will see that there is a limit to this error reduction.

Assume that a bandlimited input $x(t)$ of maximum frequency $f_{BW}$ is quantized and then uniformly sampled at a frequency $f_s \geq 2f_{BW}$. The sampled signal includes an error component due to quantization, equal to

$$e_s(t) = \sum_{k=-\infty}^{\infty} e(nT_s)\,\delta(t-kT_s),$$

where $e(t)$ is the continuous-time quantization error signal and $T_s = \frac{1}{f_s}$. We wish to evaluate the power of $e_s(t)$ that lies in the input baseband $[0, f_{BW}]$. Denoting by $E(f)$ the Fourier transform of $e(t)$, we have

$$E_s(f) = E(f) * \sum_{n=-\infty}^{\infty} \delta(f - nf_s). \tag{17.13}$$

As a result, the in-band error power is the sum of the error power in a bandwidth of $\pm f_{BW}$ around each harmonic of the sampling frequency $f_s$ [8]. This convolution operation is visually explained in Figure 17.8. As was explained in Section 17.2, $|E(f)|^2$ along the frequency axis can be reasonably modeled as a staircase function, whose steps correspond to the $f_{SAW}$ bands of successive orders. The heights of the steps represent the power spectrum density, and decrease as $1/k^2$. The dark areas represent the frequency bands of width $2f_{BW}$ centered around the multiples of $f_s$ (excluding the zero multiple). As (17.13) indicates, after $e(t)$ is sampled at frequency $f_s$, all the dark parts are shifted into the baseband $[-f_{BW}, f_{BW}]$ and added to the power originally located within it.

Assuming the aliased components are independent, we can estimate the total aliased error power $P_{al}$ from the double-sided spectra by the following equation:

$$P_{al} \approx 2\sum_{k=1}^{\infty} N_k \frac{P_{bb}}{k^2}, \tag{17.14}$$

where $P_{bb}$ is the original error power within the baseband before aliasing and $N_k$ is the number of positive multiples of $f_s$ that fall in the $k$th $f_{SAW}$ band. The factor of 2 is due to the two-sided spectrum (in signed frequency, the $k$th $f_{SAW}$ band includes the interval $[(k-1)f_{SAW}, kf_{SAW}]$ and its negative counterpart). The total quantization error power in $e_s(t)$ is

$$P_{total} = P_{al} + P_{bb}. \tag{17.15}$$

In (17.14), $P_{bb}$ is solely dependent on the input and the quantizer. Meanwhile, $N_k$ depends on the ratio of $\frac{f_s}{f_{SAW}}$.

When $\frac{f_s}{f_{SAW}} \ll 1$, the aliased error power $P_{al}$ is dominated by the part coming from the first $f_{SAW}$ band where power density is the highest. Thus, Equation 17.15 can be reduced to

$$P_{total} \approx 2N_1 P_{bb} + P_{bb}.$$

With $N_1 \gg 1$ (which is true in most oversampling scenarios), it can be further reduced to

$$P_{total} \approx 2N_1 P_{bb}.$$

The worst value of $SER_{dB}$, denoted by $SER_{dB,worst}$, is achieved at Nyquist sampling, where the entire error power is aliased into the baseband. As long as the condition of $\frac{f_s}{f_{SAW}} \ll 1$ is satisfied, doubling $f_s$ halves $N_1$ and thus decreases $P_{total}$ approximately by a factor of 2. A 3 dB improvement in $SER_{dB}$ is expected.

When $\frac{f_s}{f_{SAW}} \gg 1$, $P_{al}$ results from high-order $f_{SAW}$ bands, and is consequently negligible compared to $P_{bb}$. In this case, Equation 17.15 reduces to

$$P_{total} \approx P_{bb}.$$

The error power becomes independent of the sampling frequency and $SER_{dB}$ reaches its highest value, denoted by $SER_{dB,best}$. At infinite oversampling, the result converges to the continuous-time quantization error power.

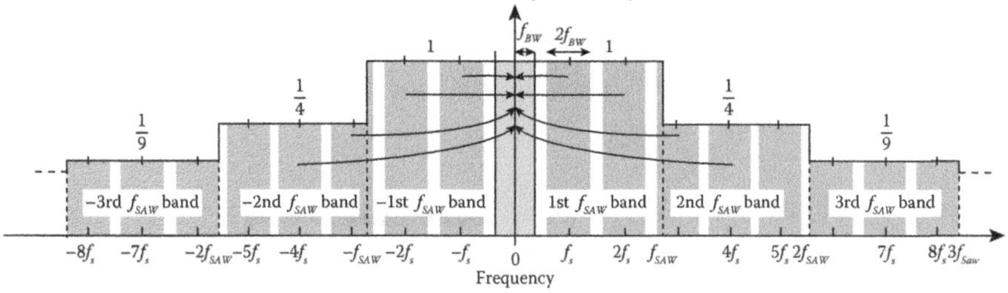


**FIGURE 17.8**

Staircase-modeled error spectrum before aliasing. The height represents the power spectrum density and is normalized to the value of the first $f_{SAW}$ band. The dark parts represent the bands which will be shifted into the baseband after sampling. In this example, $f_s$ is assumed to be a fraction of $f_{SAW}$; thus, each $f_{SAW}$ band contains multiple dark parts.

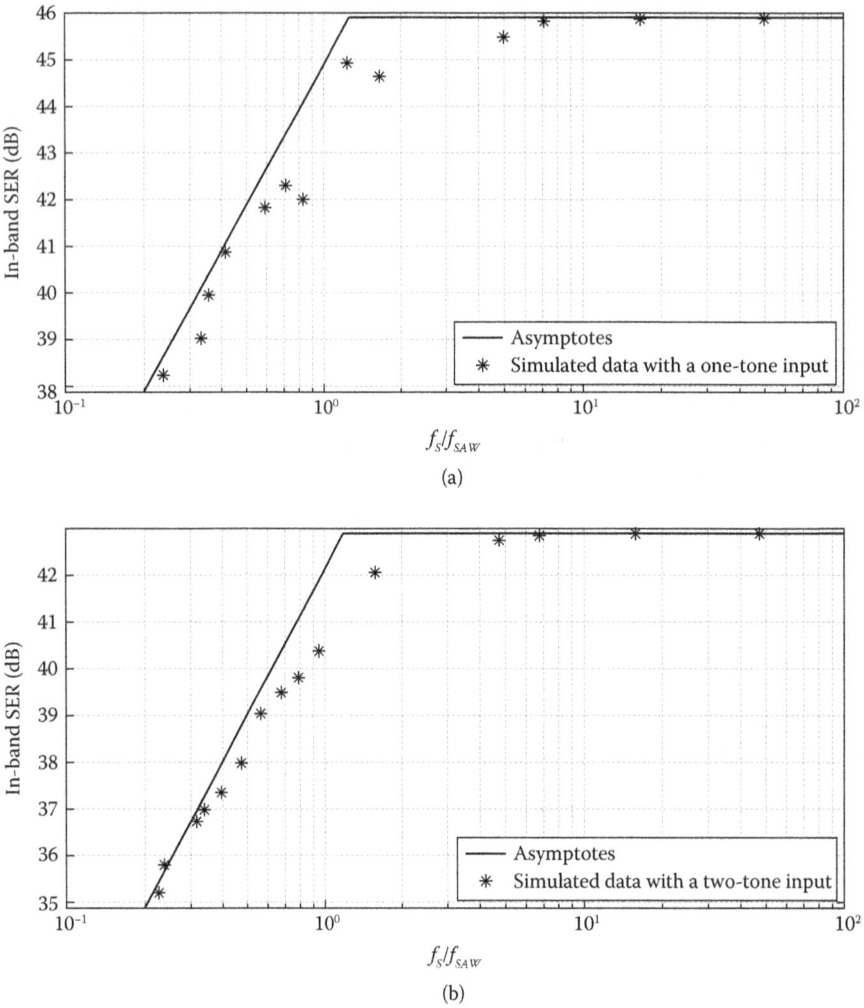


**FIGURE 17.9**
Theoretical asymptotes and simulation results of $SER_{dB}$ using the staircase model. In both cases, a 6-bit mid-tread quantizer is used. (a) One-tone test with $f_{BW} = 20f_{in}$; (b) two-tone test with $f_{BW} = 20f_{avg,in}$.

$P_{total}$ *does not* go to zero, as one would have assumed by applying the "3 dB reduction for each doubling of $f_s$" rule.

As this analysis shows, the plot of $SER_{dB}$ versus $\frac{f_s}{f_{SAW}}$ has two asymptotes. As shown in Figure 17.9, for $\frac{f_s}{f_{SAW}} \ll 1$ the plot asymptotically approaches a straight line of slope of 10 dB/decade (i.e., 3 dB/octave), starting from the lowest point of $SER_{dB,worst}$. For $\frac{f_s}{f_{SAW}} \gg 1$, $SER_{dB}$ approaches $SER_{dB,best}$.

Figure 17.9 also shows the results of MATLAB® experiments to verify the staircase model. One-tone and two-tone inputs are used with a 6-bit quantizer. The measured points satisfactorily reproduce the asymptotic trends in the regions $\frac{f_s}{f_{SAW}} \ll 1$ and $\frac{f_s}{f_{SAW}} \gg 1$, with some local deviations that can be attributed to the following factors: the model assumes a flat spectrum within each $f_{SAW}$ band, which is not exactly true; the assumption implied by the power additivity in (17.14), that is, that all the aliased components are independent is not rigorously satisfied; and Equation 17.14 assumes that each band $[nf_s - f_{BW}, nf_s + f_{BW}]$ entirely falls into one $f_{SAW}$ band of order $k$. Figure 17.8 shows that this latter assumption fails when $nf_s$ is close to the boundaries of an $f_{SAW}$ band (e.g., $n = 3$).

## 17.4 CT-ADC with Time Coding

The event times $t_k$ in CT-ADCs are nonuniformly spaced and can have arbitrary values. As mentioned in [1], time coding (a term adopted from [15]) is necessary when one wants to store the nonuniform samples.

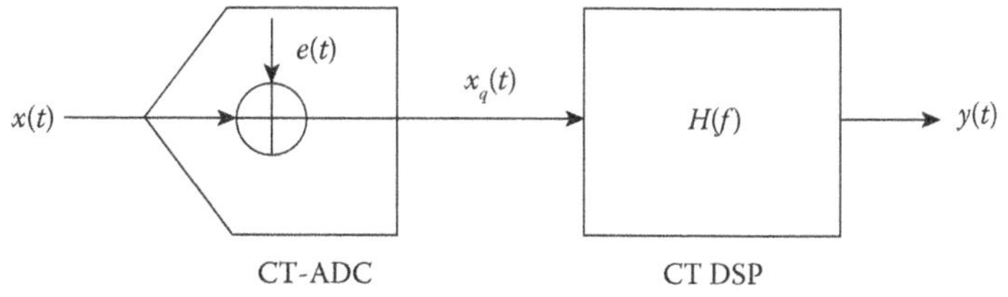


**FIGURE 17.10**
A continuous-time signal processing system composed of a CT-ADC and a CT DSP

A high-frequency clock can be used to quantize the time intervals between any two successive events $\Delta_k = t_k - t_{k-1}$ . An interesting question is how high the clock frequency should be. To answer this question, we note that the output of such a CT-ADC with time coding is equivalent to a clocked ADC with the same sampling frequency, and thus the staircase model in Figure 17.8 applies. According to the previous result that the aliased error power becomes negligible when $\frac{f_s}{f_{SAW}} \gg 1$, the time discretization needs to be sufficiently fine so that the in-band $SER$ is comparable to the continuous-time case, with no sampling. Thus, the sampling frequency has to be several times higher than $f_{SAW}$ to assure the highest in-band $SER$.

## 17.5 Error Analysis of CT DSP

The error at the output of a continuous-time DSP (CT DSP) is analyzed next. Figure 17.10 shows a signal processing system composed of a CT-ADC and a CT DSP. Mathematically, the CT-ADC (i.e., the quantizer) can be represented as a summer with an additional input $e(t)$, which is the quantization error signal. The output of the quantizer $x_q(t)$ can then be expressed as

$$x_q(t) = x(t) + e(t).$$

Since the CT-DSP is a linear-time-invariant (LTI) system, both the input and the error are processed by the same transfer function $H(f)$ of the DSP. The Fourier transform of the output of the CT DSP, $Y(f)$, can be expressed as

$$Y(f) = X(f)H(f) + E(f)H(f), \tag{17.16}$$

where $X(f)$ and $E(f)$ are the Fourier transforms of $x(t)$ and $e(t)$, respectively. The error at the output of a CT-DSP is the second term in the equation, whose power spectrum can be represented as $|E(f)|^2|H(f)|^2$.

As presented in [1], CT-DSP consists of an FIR or an IIR filter with a uniform tap delay $T_D$. This results in a periodic frequency response with a period of $1/T_D$. While the bandwidth of $X(f)$ is less than half this period, $|E(f)|^2$ remains dominant in the whole interval $[0, f_{SAW}]$, which typically covers several $1/T_D$ periods. Thus, the power spectrum $|E(f)|^2|H(f)|^2$ of the filtered quantization error signal periodically yields regions of high values at least several times outside the input baseband. When the output of the CT DSP is interfaced with standard (i.e., synchronous) data processing blocks, uniform sampling is necessary, synchronously with the clock of those blocks. To obtain an in-band $SER$ comparable with the nonsampling case, a sampling frequency at least higher than $f_{SAW}$ should be used to prevent any significant out-of-band error power being aliased into the baseband.

## Acknowledgment

This work was supported by National Science Foundation (NSF) Grants CCF-07-01766, CCF-0964606, and CCF-1419949.

## Bibliography

[1] Y. Tsividis, M. Kurchuk, P. Martinez-Nuevo, S. Nowick, S. Patil, B. Schell, and C. Vezyrtzis. Event-based data acquisition and digital signal processing in continuous time. In *Event-Based Control and Signal Processing* (M. Miskowicz, Ed.), CRC Press, Boca Raton, FL, 2015, pp. 253–278.

[2] A. Gersho and R. M. Gray. *Vector Quantization and Signal Compression*. Kluwer Academic Publishers, Boston, 1992.

[3] S. Haykin. *Communication Systems*. Wiley, New York, 2001.

[4] H. E. Rowe. *Signals and Noise in Communication Systems*. van Nostrand, New York, 1965, p. 314.

[5] N. M. Blachman. The intermodulation and distortion due to quantization of sinusoids. *IEEE Transactions on Acoustics, Speech and Signal Processing*, 33:1417–1426, 1985.

[6] H. Pan and A. Abidi. Spectral spurs due to quantization in Nyquist ADCs. *IEEE Transactions on Circuits and Systems I: Regular Papers*, 51:1422–1439, 2004.

[7] T. Claasen and A. Jongepier. Model for the power spectral density of quantization noise. *IEEE Transactions on Acoustics, Speech and Signal Processing*, 29(4):914–917, 1981.

[8] M. Kurchuk. Signal encoding and digital signal processing in continuous time. PhD thesis, Columbia University, 2011.

[9] P. Z. Peebles. *Communication System Principles*. Addison-Wesley, Advanced Book Program, MA, 1976.

[10] V. Balasubramanian, A. Heragu, and C. Enz. Analysis of ultralow-power asynchronous ADCS. In *Proceedings of 2010 IEEE International Symposium on Circuits and Systems (ISCAS)*, pp. 3593–3596, May 2010.

[11] W. R. Bennett. Spectra of quantized signals. *Bell System Technical Journal*, 27(3):446–472, 1948.

[12] N. S. Jayant and P. Noll. *Digital Coding of Waveforms: Principles and Applications to Speech and Video*. Prentice Hall, Englewood Cliffs, NJ, pp. 115–251, 1984.

[13] D. Hand and M. S. Chen. A non-uniform sampling ADC architecture with embedded alias-free asynchronous filter. *Global Communications Conference (GLOBECOM)*. IEEE, 2012.

[14] Y. Tsividis. Digital signal processing in continuous time: A possibility for avoiding aliasing and reducing quantization error. *IEEE Transactions on Acoustics, Speech, and Signal Processing*, 2:ii-589–ii-592, 2004.

[15] J. Foster and T.-K. Wang. Speech coding using time code modulation. In *IEEE Proceedings of the Southeastcon '91*, pp. 861–863, 1991.

# 18

# *Concepts for Hardware-Efficient Implementation of Continuous-Time Digital Signal Processing*

**Dieter Brückmann**
*University of Wuppertal*
*Wuppertal, Germany*

**Karsten Konrad**
*University of Wuppertal*
*Wuppertal, Germany*

**Thomas Werthwein**
*University of Wuppertal*
*Wuppertal, Germany*

**CONTENTS**



**ABSTRACT** Event-driven signal processing techniques have the potential for significant reduction of power consumption and improvement of performance. In particular, systems used for applications which deal with bursty or heavily changing traffic can profit from this special method of signal processing. Even though quite a number of implementations are already described in the literature, such techniques are however, still in a research stage. Thus, further improvements must be made until this kind of signal processing has the level of maturity required for implementation in a commercial product. The most critical point is the realization of the time-continuous delay elements, which require much more hardware than their conventional sampled data counterparts. Therefore in this contribution, several recently developed new concepts for delay element implementation will be described, which enable considerable hardware reduction. Furthermore, it will be shown that the hardware requirements can be also reduced by applying asynchronous sigma delta modulation for quantization. Finally, filter structures are proposed, which are especially well-suited for event-driven signal processing. All described, techniques enable considerable improvements with respect to a hardware-efficient implementation of CT signal processing.

## 18.1 Basics of Digital Signal Processing in Continuous Time

The basic idea behind continuous-time DSP systems is to perform signal processing only when the input signal has changed by a certain amount. This can, for example, be achieved by using a CT Analog to Digital Converter (ADC) for quantization. This converter type generates a new digital output value only when a new quantization level is reached. Since no discrete-time (DT) sampling is performed in a CT system [1–6,27], there will be no aliasing and quantization merely results in harmonic distortion. Thus, the signal-to-noise ratio (SNR) in the band of interest is better than that of the respective DT sampled data system, provided that quantizers with the same resolution are used. Note that even though the total quantization noise is the same for both systems, the noise is spread over a wider frequency range by CT signal processing, whereas in a sampled data system the total quantization noise is concentrated in the frequency range from zero to the sampling rate. Furthermore, the signal values generated by the CT-ADC contain the information about the time instants of signal changes and the actual digital data. In [4,5], it is proposed to compose the signal values to be processed of pulses at the time instants of signal changes and the respective digital data values. This pulse train can be used to trigger the subsequent CT circuitry [4,5] and to start signal processing.

The most important advantage of this kind of signal processing is the large potential for minimization of the power consumption. Since the analog input signal is only processed when it changes, no signal processing will be performed for constant input signals and thus during these time intervals the dynamic power consumption is nearly zero. In DT systems on the contrary, the dynamic power consumption is directly proportional to the sampling frequency $f_s$ of the system, which must be at least twice the maximum input frequency $f_{in,max}$, to comply with the Nyquist theorem. These distinct properties make CT signal processing interesting for a number of applications, especially for those that deal with bursty signals.

The time interval between two consecutive changes of the continuously processed signal $x_{CT}(t)$ is not predictable. Therefore, contrary to classical sampled data systems, this time distance must be also preserved. A characteristic parameter of CT systems is the minimum time interval $T_{gran}$ between two successive quantization level changes, also designated as granularity time [6]. $T_{gran}$ is determined by the resolution used for quantization and by the maximum frequency of the input signal. Assuming that a sinusoidal input signal $x(t)$ with frequency $f_{in,max}$ and amplitude $A$ is quantized by an $N$ Bit CT mid-tread ADC, $T_{gran}$ can be determined by the following relationship [6]:

$$T_{gran} = \frac{1}{2^{N-1} \cdot 2\pi A f_{in,\max}}. \tag{18.1}$$

Key components of digital signal processing systems are arithmetic units with adders and multipliers, and delay elements. In sampled data systems, the delay time $T_D$ in most cases is equal to the sampling period or it is a multiple of it. Therefore, the delay elements can be implemented by simple digital storage elements such as registers. CT systems, however, operate clockless; therefore, their arithmetic units must be implemented asynchronously. Asynchronous logic circuits are well-known [7,26] and the respective circuitry can be also used for CT systems. In classical asynchronous systems, however, only the order of the events is preserved, whereas in CT systems the time distance between successive signal values must be also retained. This results in a more complex implementation of the respective delay circuits.

A straightforward implementation of a CT delay element, as described in [5,6], consists of a cascade connection of base cells, each delaying the signal by $T_{gran}$. Depending on the total time delay $T_D$ to be realized and the granularity of the input signal, the required number $N_{D1}$ of base cells can be quite large and is given by

$$N_{D1} = \left\lceil \frac{T_D}{T_{gran}} \right\rceil. \tag{18.2}$$

Figure 18.1 shows a block diagram of such a delay element consisting of $N_{D1}$ base cells. Each cell is composed of an analog delay circuit and a digital storage element or latch consisting of an FF. Each time a new input pulse arrives at the analog circuitry of the base cell, the respective digital data word is stored in the Flip Flop. When the time period $T_{gran}$ is expired, an output pulse is generated by the delay circuit and the digital data word is transferred to the next base cell. The latch can be also composed of several FFs in parallel depending on the word length to be processed. The overall delay time $T_D$ to be implemented is determined by the system requirements. This can, for example, be the specification of a filter characteristic to be fulfilled.

As an example, let us consider an input signal quantized with a resolution of $N = 10$ bits and with a maximum frequency of $f_{in,max} = 4\,\text{kHz}$. Using (18.1) we obtain a value of $T_{gran} = 77.7\,\text{ns}$ for the granularity. Thus, for the realization of a CT delay element with $T_D = 62.5\,\mu\text{s}$ at least 806 CT base cells are required, whereas in a DT system a single latch is sufficient. For a resolution of $N = 8$ bits, the number of required base cells reduces to 202.

Even though this large number is an obvious disadvantage with respect to the implementation costs, the CT delay element offers a new degree of freedom for

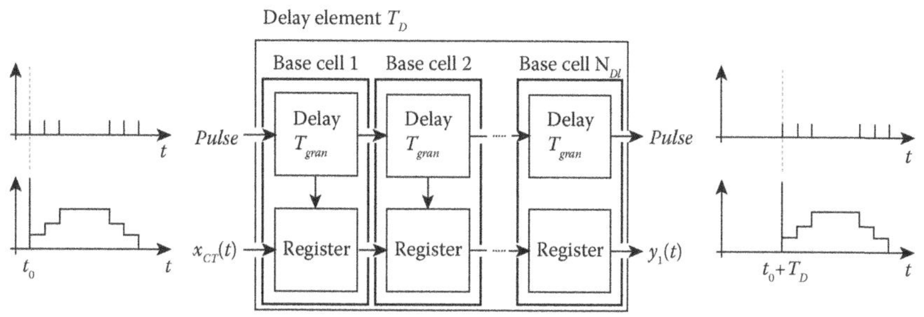


**FIGURE 18.1**
Classical CT delay element with cascaded base cells and input and output signals.

the filter design. Since the delay element usually is composed of a cascade of a large number of base cells, the delay time $T_D$ can be adjusted in a wide range with a high resolution [30]. This can be achieved either by making the time delays of the base cells smaller than $T_{gran}$ or by adding additional base cells to each delay element. It has been shown that the coefficients of a continuous-time digital filter can be optimized by applying this so-called delay line adjustment [8–11]. In the best case, the filter coefficients can be simplified by this method in such a way that no multiplications must be carried out in the filter anymore. In addition, the filter characteristic can be improved by delay line adjustment. By avoiding multipliers, the required chip area for a filter implementation can be considerably reduced. The method has been described in more detail in Section 18.4.

## 18.2 CT Delay Element Concept with Reduced Hardware Requirements

It is obvious that in CT systems the delay element is the most critical functional block. To make CT systems more competitive, it is thus essential to optimize these functional blocks. The respective objectives are power consumption and chip area reduction, without sacrificing CT signal quality. The time delay generated by a base cell should be equal or smaller than $T_{gran}$, otherwise signal information can get lost and signal quality would suffer. Of course, it is possible to make the time delay smaller than $T_{gran}$; however, this would result in an increased number $N_{D1}$ of base cells, and there with a larger chip area.

### 18.2.1 Functional Description of the New Base Cell

To meet the aforementioned requirements, a new delay element concept has been developed, which can operate with a reduced number of delay circuits per delay element. Based on this concept, hardware requirements and power consumption can be considerably reduced. In the classical delay element in Figure 18.1, one delay circuit is responsible for the transfer of one signal value through one base cell. This happens $N_{D1}$ times, until the signal value is delayed by $T_D$ and is shifted out of the delay element [28,29].

The new base cell $T_B$ shown in Figure 18.2, however, can store up to $M$ signal values at the same time. It consists of only one analog delay cell but the digital circuitry is enlarged to a shift register with $M$ FF stages. This base cell can be further extended for input signals with a word length of $N$ bits by arranging $N$ shift registers in parallel as shown in the figure. Since $M$ FFs are cascaded in each shift register, the time delay generated by the new base cell is increased to $T_{BDC}$, with $T_{BDC} = M \cdot T_{gran}$. Therefore, a signal value is now kept for the time interval $T_{BDC}$ in one base cell.

If a new signal value arrives at the base cell, it is stored in the first stages of the shift registers and the counter is reset. This counter reset together with the active input pulse triggers the delay circuit via the upper AND gate. After the time interval $T_{gran}$, the delay circuit generates an output pulse, which triggers the shift registers so that the stored signal values are shifted to the next FF stages. Furthermore, this output pulse generates a new trigger signal for the delay cell and it increments the counter, as long as the counter has not reached the value $M$-1. Thus, the delay circuit keeps on generating pulses with the time grid $T_{gran}$ until the counter has reached $M$-1 and no values are stored in the registers anymore. If all signal values are shifted out, the delay circuit stops generating pulses and goes to power down mode until the input signal changes again. Furthermore, for each signal value shifted out of a base cell, a trigger signal "pulse out" is generated by the respective AND gate. This pulse can be applied to the next cascaded base cell to trigger the delay circuit of this cell and to store the respective data value.

Since each base cell generates a delay of $M \cdot T_{gran}$, the number of base cells required for a delay cell with

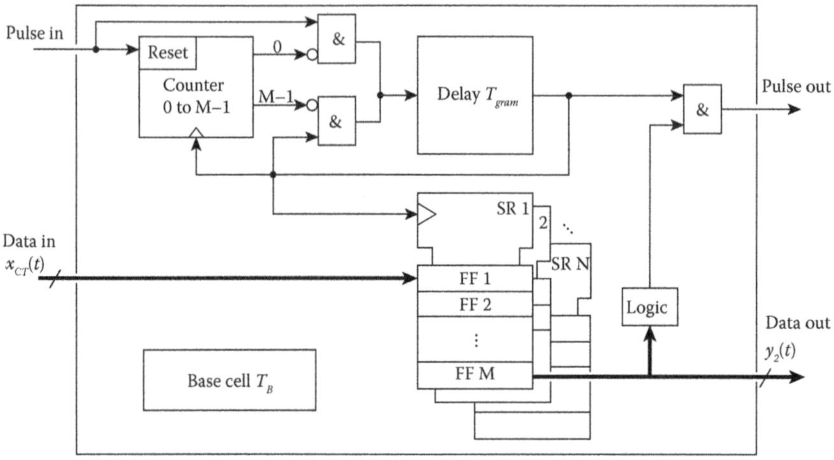


**FIGURE 18.2**
CT base cell $T_B$ with shift registers as storage elements.

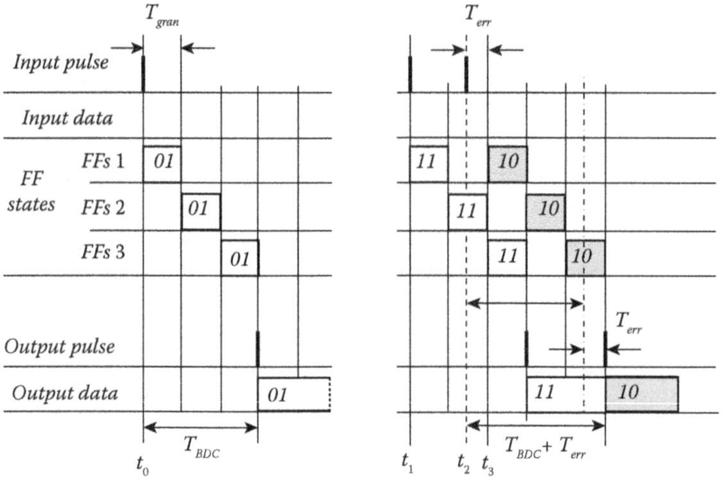


**FIGURE 18.3**
Timing diagram and FF states for a data transfer into the new base cell, with word length of 2 bits.

delay time $T_D$ can be reduced to $N_{D1}/M$, even though the total number $N_{D1}$ of digital storage elements has not changed. Note that an additional base cell is required if $N_{D1}/M$ is not an integer. This cell consists of one analog circuit and the remaining digital storage elements.

Based on this concept, the delay circuit of one base cell controls several data values stored in the respective shift registers at the same time. Thus, the number of delay circuits per delay element is reduced, which results in a smaller chip area. Furthermore, the delay circuit controls the shift of all values in one base cell simultaneously. Therefore, less delay circuits are active at the same time and the dynamic power consumption is reduced.

The timing diagram in Figure 18.3 illustrates the operation of the new base cell in more detail for a 2-bit input signal. At time instant $t_0$, when a new signal value $\{0\,1\}$ arrives at the base cell, the corresponding pulse activates the delay circuit and the signal value is stored in the first stage of the shift registers. After the time interval $T_{gran}$, the delay circuit triggers the shift registers and $\{0\ 1\}$ moves on to the second stage. This recurs until the shift registers run empty and the signal value $\{0\,1\}$ is shifted out of the base cell after time interval $T_{BDC}$.

The next signal value $\{1\ 1\}$ arrives at time instant $t_1$ and is now stored in the first register stage. Furthermore, the delay circuit is started again, and after $T_{gran}$, the value $\{1\,1\}$ is shifted to the second register stage. As long as signal values are contained in the base cell, it operates in the time grid $T_{gran}$. Thus, even though the next input value arrives at time instant $t_2 = t_1 + T_{gran} + T$, it is not stored before $t_3 = t_1 + 2T_{gran}$ in the first register stage. At the same time, the values already stored in the second stage are shifted to the third. The next pulse is generated by the delay circuit again after $T_{gran}$, causing another shift for the values in register stages 1 and 3. Thus, only one pulse is required for the shift of both signal values. With increasing number $M$ of shift register stages, the number of active delay circuits decreases, which results in lower power consumption. A possible drawback of the new base cell is that the signal itself is also affected. As can be seen in the timing diagram (Figure 18.3), the signal value arriving at time instant $t_2$ suffers an additional delay and is shifted out of the base cell not before $T_{BDC} + T_{err}$. This time shift always happens if the time distance between two consecutive signal changes is smaller than $T_{BDC}$ and if these changes are not spaced by an integer multiple of $T_{gran}$. The time shift $T_{err}$ is in the range of $0 < T_{err} < T_{gran}$ and can be considered also as a unidirectional jitter-like error. The resulting CT signal is identical to a signal sampled with frequency $1/T_{gran}$. Since sampling occurs only when more than one signal value is stored in one base cell at the same time, only the fast changing regions of the input signal will be affected.

The limit frequency $f_D$, where sampling comes into operation, is determined by the granularity used for the delay element and the number $M$ of FF stages used for the registers. Based on (18.1) the new parameter $f_D$ can be evaluated [9]. Since the time interval during which a signal value is stored in one base cell is given now by $T_{BDC} = M \cdot T_{gran}$, the time distance between two successive pulses generated by an input signal with frequency $f_D$ needs to be longer than $T_{BDC}$. Thus, the granularity of a signal with frequency $f_D$ and amplitude $A = 1$ must be equal to $M \cdot T_{gran}$, with

$$M \cdot T_{gran} = \frac{1}{2^{N-1} \cdot 2\pi f_D}. \tag{18.3}$$

Therefore, the limit frequency $f_D$ is obtained from

$$f_D = \frac{f_{in,\max}}{M}. \tag{18.4}$$

If the input signal frequency is above $f_D$, the output signal $y_2(t)$ of the delay element becomes a sampled version of the input signal with sampling frequency $1/T_{gran}$. Note, however, that even for signal frequencies larger than $f_D$, the signal is only sampled in its fast changing regions. A sinusoidal signal with maximum input frequency $f_{in,max}$ is still processed as a continuous-time signal during time intervals where the signal is only slowly changing.

### 18.2.2 Required Chip Area

In the following, the required chip area for the different delay element realizations will be determined in more detail. Each base cell of the classical CT delay element in Figure 18.1 consists of an analog delay circuit, generating the time delay $T_{gran}$, and a digital latch where the respective signal value is stored. The required number $N_{D1}$ of cascaded base cells in the delay element is obtained from (18.2). The respective chip area $A_{D1}$ for such a delay element can be thus approximated by

$$A_{D1} = N_{D1} \cdot A_{DC} + N_{D1} \cdot A_R, \tag{18.5}$$

with $A_{DC}$ the chip area of the delay circuitry and $A_R$ the chip area of the latch(es). For modern semiconductor technologies, $A_{DC}$ is considerably larger than $A_R$. Depending on the minimum feature size used for implementation, the ratio $A_{DC}/A_R$ is in the range of 13/1 to 25/1 [12]. This ratio increases with downscaling of the semiconductor technologies, since the chip area of the capacitor in the analog delay circuit cannot be decreased by the same amount as the chip area for the digital circuitry. Thus, $A_{D1}$ is mainly determined by the chip area of the analog delay circuits.

The required chip area for a delay element can be considerably reduced by using the new base cell $T_B$ in Figure 18.2. As described previously, this cell consists of one analog delay circuit and a shift register composed of $M$-cascaded storage cells, so that several values can be stored. Even though the total number of storage cells in the delay element must be still $N_{D1}$ to generate the time delay $T_D$, the number of base cells can be reduced to $N_{D2} = \lceil N_{D1}/M \rceil$. Thus, the required overall chip area $A_{D2}$ of a delay element can be now approximated by

$$A_{D2} = \left\lceil \frac{N_{D1}}{M} \right\rceil \cdot A_{DC} + N_{D1} \cdot A_R. \tag{18.6}$$

If $N_{D1}/M$ is not an integer, an additional base cell at the end of the base cell cascade within the delay element contains the remaining registers to obtain a total number of $N_{D1}$ storage elements. Using (18.5) and (18.6), the normalized chip area has been determined for different values of $M$. In Table 18.1, the respective values normalized to the chip area of the classical delay element are listed. The implementation in a standard CMOS technology had been assumed [12]. Furthermore, it has been assumed that the chip area for wiring and the control logic is nearly constant for both types of delay elements.

The improvements with respect to chip area and power consumption get larger with increasing $M$. If $M = 1$, there is no difference between the classical and the novel base cell. The improvements converge to a maximum for $M = N_{D1}$ when all storage elements are in one base cell. However, for large $M$, the limit frequency $f_D$ becomes very low, thus the signal would be sampled most of the time. The optimum for $M$ is thus between the maximum and the minimum value. The determination of the optimum number of storage elements per base cell will be discussed in Section 18.2.4.

**TABLE 18.1**
Normalized Delay Element Size for Different Values of $M$, $N_{D1} = 202$, $A_{DC}/A_R = 13$, and $N = 8$

| $M$ | 1 | 2 | 3 | 4 | 5 | 6 | 10 | 20 | 30 | 40 | 50 | 100 |
|---|---|---|---|---|---|---|---|---|---|---|---|---|
| $A_{D2}/A_{D1}$ | 1 | 0.54 | 0.38 | 0.31 | 0.26 | 0.23 | 0.17 | 0.12 | 0.1 | 0.1 | 0.09 | 0.08 |

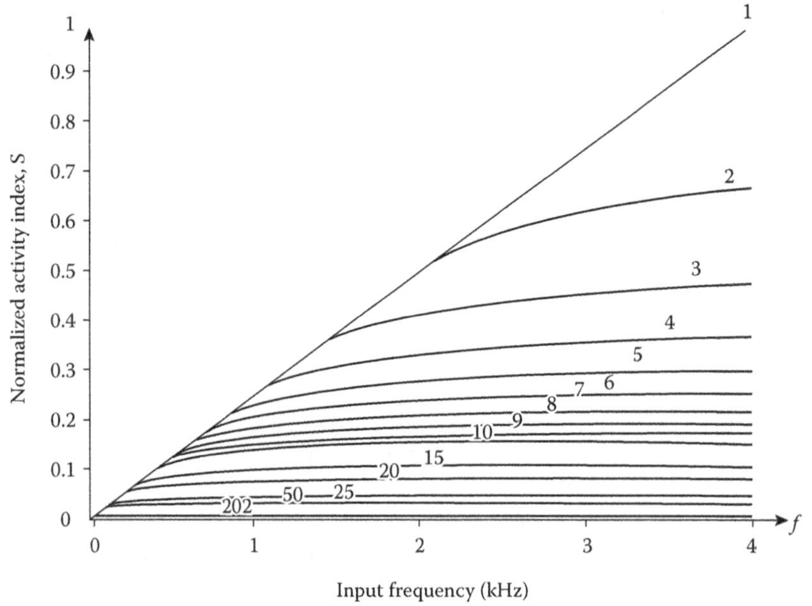


**FIGURE 18.4**
Normalized number of active analog delay cells per second versus frequency; parameter is $M$, the number of storage elements per base cell.

### 18.2.3 Reduction of Power Consumption

To compare the power consumption of the new base cell with that of a classical CT delay element, an activity index $S$ is introduced, where $S$ is equivalent to the number of active analog delay circuits per second. In the classical CT delay element, an incoming signal value has to pass all base cells of a delay element and, therefore, $N_{D1}$ delay circuits are activated for transmission. In [4], it has been shown that $(2^{N+1} - 4)f_{in}$ signal values are generated per second when time continuously quantizing a sinusoidal signal of frequency $f_{in}$ with $N$ bits. Thus, the respective activity index $S_1$ is given by

$$S_1 = (2^{N+1} - 4) \cdot f_{in} \cdot N_{D1}. \tag{18.7}$$

Since the new delay element contains several registers per base cell, the number of active analog delay circuits is reduced correspondingly. For frequencies below $f_D$, the time distance between consecutive changes is larger than $T_{BDC}$ and, therefore, the base cell will operate like the classical base cell. Thus, the respective activity index $S_2$ is equal to $S_1$. For frequencies above $f_D$, however, the shift of several signal values is initiated by only one delay circuit, and $S_2$ will be smaller than $S_1$.

Figure 18.4 shows a plot of the normalized activity index $S$ versus the input frequency for several values of $M$. The parameters are the same as in the example in Section 18.1. The curves in Figure 18.4 confirm that the savings increase fast for small values of $M$ and converge against a maximum for the case where all registers are combined in one base cell.

### 18.2.4 Optimum Number of Registers per Base Cell

To profit from the advantages of CT signal processing also for the highest signal frequency the system is designed for, the proposed delay element should operate time continuously at least during the time intervals with slower signal changes. Since the onset of sampled data operation is determined by $M \cdot T_{gran}$, the optimum value for $M$ is an important design parameter for the new base cell.

As has been shown previously, the required chip area decreases with increasing values of $M$. However, if the

**TABLE 18.2**
Maximum Number $M_{max}$ of Register Stages per Base Cell for Different Quantizer Resolutions, $N$

| $N$ | 4 | 5 | 6 | 7 | 8 | 9 | 10 | 11 | 12 | 13 | 14 | 15 |
|---|---|---|---|---|---|---|---|---|---|---|---|---|
| $M_{max}$ | 5 | 7 | 10 | 15 | 21 | 31 | 44 | 63 | 89 | 127 | 180 | 255 |

chosen $M$ is too large, the base cell operates as a sampled data system all the time and any benefit of CT signal processing is lost. To avoid permanent sampling of the signal, $M$ must be equal or smaller than a maximum value $M_{max}$. $M_{max}$ should be chosen such that the regions of a sinusoidal signal, where the amplitude approaches its minimum or maximum value, are still processed continuously in time. In [9], it had been shown that $M_{max}$ can be determined by the following relationship:

$$M_{\max} = \left\lfloor \frac{2}{T_{gran}} \cdot \left( \frac{1}{4 f_{in,\max}} - \frac{\arcsin\left(1 - \frac{Q}{2A}\right)}{2\pi f_{in,\max}} \right) - 1 \right\rfloor = \left\lfloor 2^N \cdot \left( \frac{\pi}{2} - \arcsin\left(1 - \frac{Q}{2A}\right)\right) - 1 \right\rfloor, \tag{18.8}$$

where $Q = 1/2^{N-1}$ is the quantization unit and $A = 1 - Q$ is the amplitude of the sinusoidal input signal.

With (18.8), it is now possible to evaluate the optimum number of registers per base cell for any number $N$ of bits used for quantization. In Table 18.2, $M_{max}$ is listed for different values of $N$. For the example considered previously with $N = 8$, a value of $M_{\max} = 21$ is obtained by (18.8). Therefore by using (18.6) and (18.7), the improvements with respect to chip area and power consumption can be determined. It turns out that chip area is reduced to 12%, and the number of active delay circuits is reduced to 8% of the original values.

### 18.2.5 Performance Considerations

Even though the delay element based on the new base cell $T_B$ operates like a sampled data system in the regions of fast signal changes, the advantages of CT signal processing still hold for regions with slower signal changes, if $M$ is chosen smaller than $M_{max}$. Furthermore, compared with systems with CT delay elements proposed up to now, the power consumption can be considerably reduced.

The improvement of the in-band SNR due to CT signal processing compared with a sampled data signal processing is illustrated in Figure 18.5. While the in-band SNR of the DT signal is limited by the aliasing components of the quantization noise, the SNR of the

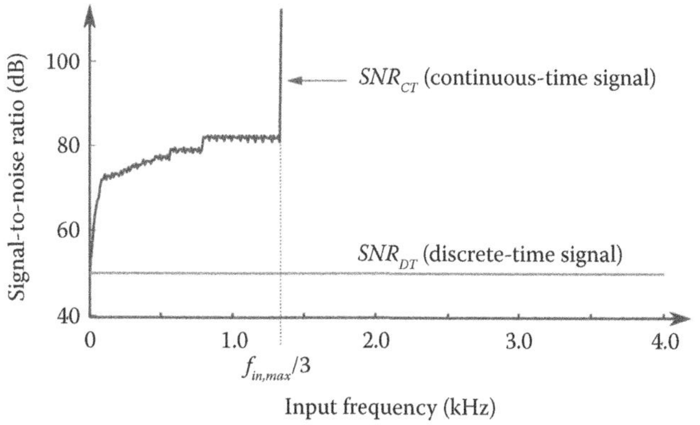


**FIGURE 18.5**
Signal-to-noise ratio (SNR) for a continuous-time system and SNR for a discrete-time system for a sampled data system, with 8-bit quantizer resolution.

CT signal is determined by the harmonics within the baseband. For frequencies above $f_{in,max}/3$, all harmonics are shifted out of the band of interest and the $SNR_{CT}$ approaches infinity.

Since the signal is partially sampled by the new delay element $T_B$, the SNR will be degraded compared with a conventional CT system. In a worst-case scenario, which is either approached for frequencies above $f_D$ and $M > M_{max}$, or for frequencies above $f_{in,max}$, the signal is completely sampled. For a sinusoidal signal with frequency $f_{in,max}$ and amplitude $A = 1$, the worst-case SNR in the baseband is given by

$$SNR_{CT,wc} = 1.76\,\text{dB} + N \cdot 6.02\,\text{dB} + 10\log_{10}(R)\,\text{dB}. \tag{18.9}$$

The oversampling factor $R$ is the ratio of the sampling frequency $1/T_{gran}$ to the baseband width, and with (18.1), the following relationship holds

$$R = \frac{1}{T_{gran} \cdot 2 f_{in,\max}} = \pi 2^{N-1}. \tag{18.10}$$

Thus, (18.9) can be written as

$$SNR_{CT,wc} = (3.72 + N \cdot 9.03)\,\text{dB}. \tag{18.11}$$

Even though the worst-case SNR approaches the SNR of a system sampled with $1/T_{gran}$, in most cases the SNR

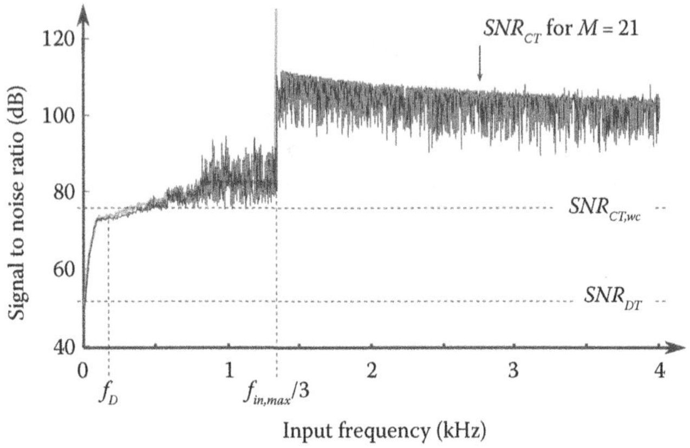


**FIGURE 18.6**
SNR for a conventional CT system and a CT system with the new base cell $T_B$, with $M = 21$ register stages and 8-bit quantizer resolution.

obtained with the new base cell is much better. This is due to the fact that major parts of the signal are still processed continuously in time.

Figure 18.6 shows a plot of the SNR versus the frequency of the input signal for $M = 21$. As can be seen, the SNR of the system with the new delay cell is much better than the worst-case $SNR_{CT,wc}$, even though it is slightly degraded compared with the SNR of the original CT signal.

## 18.3 Asynchronous Sigma Delta Modulation for CT DSP Systems

Another promising approach for optimizing CT digital signal processing arises from quantization by a time-continuous sigma delta modulator. This converter type generates a digital signal with a small word length from the analog input. Similar to delta-modulated signals [2], the respective digital signal processing must be thus designed only for the small word length. However, contrary to delta modulation, not only the deltas between successive data values are processed, even though the data stream can have a word length as small as 1 bit. Thus, no accumulators are needed to retrieve the actual signal. Furthermore, due to noise-shaping performed by the modulator, the SNR can be considerably improved in the band of interest. However, since no sampling is performed in CT systems only an Asynchronous Sigma–Delta Modulator (ASDM) [13–20] can be used for quantization.

The fundamentals of this approach are described in the following and a mathematical description is given. Furthermore, a relationship between the limit cycle frequency and the hysteresis of the quantizer is derived. Finally, simulation results are presented and a quantitative comparison with delta modulator (DM) is given.

### 18.3.1 Asynchronous Sigma–Delta Modulation

A promising approach for the reduction of implementation costs and for improving performance of time-continuous DSP systems is to use sigma–delta modulation (SDM) for quantization. SDM is a well-established method for sampled data systems and a large number of different realizations of respective converters is implemented and described in the literature [13,14]. The digital output signal has a low word length which nevertheless can have an excellent SNR in certain frequency bands. This is achieved by applying oversampling and noise-shaping techniques to improve the SNR in the band of interest. Thus, the classical sigma delta modulator is a sampled data system and requires a clock for sampling. Noise-shaping is obtained by a closed-loop configuration which results in filtering of the quantization noise. When a high-pass characteristic is implemented, the quantization noise is reduced in the base band and is shifted to the higher frequency range. While in classical SDM, sampling is done before the closed-loop system, in continuous-time configurations sampling is carried out inside the closed loop in front of the quantizer [13,14]. Both approaches, however, generate sampled data signals as output signals.

A realization which operates completely in continuous time is obtained by using ASDMs. In the respective loop configuration shown in Figure 18.7, a limit cycle is automatically generated which is modulated with the input signal [15–20]. This modulator type also converts the analog input signal into a digital signal with a small word length. The most important design parameter is the limit cycle frequency, which determines the spectral properties of the output signal and the SNR in the band of interest. Further design parameters are the order of the linear filter, and the resolution and the hysteresis of the quantizer. The configuration shown in Figure 18.7 is a closed-loop system with a linear filter with transfer function $H_L(s)$ and the nonlinear quantizer. The resolution of the quantizer is only 1 bit in the simplest case and can be modeled by a nonlinear transfer characteristic $H_{NL}(A)$, with $A$ the amplitude of the input signal.

Normally, the quantizer is implemented with a hysteresis $h$ which can be also used for optimization of the overall system performance. $x(t)$ and $y(t)$ are the input and output signals of the system, respectively. For simplicity, a simple integrator is chosen as the linear part, and its output signal is $i(t)$. For actual implementation, the ideal integrator can be replaced by a first- or second-order low-pass filter, which limits the gain below the characteristic frequency to a constant value.

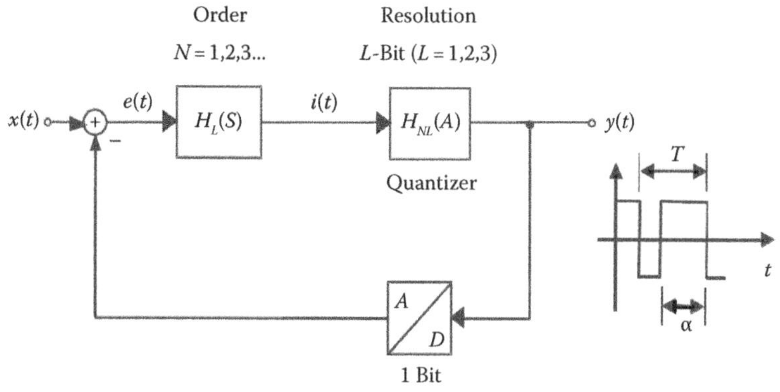


**FIGURE 18.7**
ASDM closed-loop system.

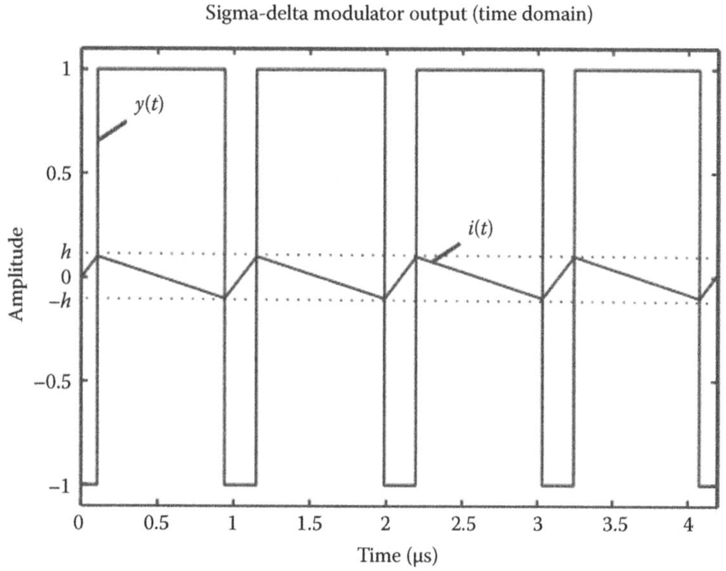


**FIGURE 18.8**
ASDM output signal $y(t)$ and integrator output $i(t)$.

The 1-bit quantizer operates as a nonlinear element with hysteresis $h$ [15,16,20].

Figure 18.8 shows the time-domain wave forms of $y(t)$ and $i(t)$ for a constant input signal with $x(t) = 0.7$ applied to the ASDM. A value of $h = 0.1$ was chosen for the hysteresis of the quantizer. As shown by the figure, the output of the ASDM is a periodic square wave.

For further processing by a continuous-time DSP system, the square wave can be transformed into a pulse train at the transitions of $y(t)$ and an additional signal corresponding to the data values. The accumulation process necessary for the DM-encoded signal is not required anymore. Thus, the ASDM output can be processed by the circuit configurations described before for CT digital signal processing.

Again the power consumption of the CT DSP system is mainly determined by the rate of change of the pulse train, which is two times the frequency of the output square wave. For minimum power consumption, this rate should be as small as possible.

Even though no clock signal is applied, an unforced periodic oscillation is generated at the output of the loop structure. The frequency of this limit cycle is maximal if the input signal is zero and changes with the amplitude of the input signal. This self-oscillation adds extra frequency components to the output spectrum. To keep the band of interest unaffected by these distorting frequency components, the limit cycle frequency $\omega_C$ must be properly chosen.

As shown in Figure 18.8, the transitions of the ASDM output signal $y(t)$ occur whenever the integrator output $i(t)$ reaches the threshold levels $h$ or $-h$ of the quantizer. ASDM systems introduce no quantization noise to the system since no sampling is performed.

In the following, the limit cycle frequency $f_c = \omega_c/2\pi$ will be determined for a system with a zero input signal and an integrator as the linear filter. As has been shown in [15], the relationship between the filter transfer function $H_L(j\omega)$ and the hysteresis $h$ for an ASDM system with input signal $x(t) = 0$ can be described by

$$\frac{4}{\pi} \cdot \sum_{n=1,3,5,\ldots}^{\infty} \frac{1}{n} \mathrm{Im}\left\{H_L\left(jn\omega_c\right)\right\} = \pm h. \tag{18.12}$$

The transfer function $H_L(j\omega)$ of an integrator with characteristic frequency $\omega_p$ is given by

$$H_L\left(j\omega\right) = \frac{1}{j\omega/\omega_p}. \tag{18.13}$$

Inserting this expression into (18.12) results in

$$-\frac{4}{\pi} \cdot \sum_{n=1,3,5,\ldots}^{\infty} \frac{1}{n} \cdot \frac{1}{n\omega_c/\omega_p} = \pm h. \tag{18.14}$$

The limit cycle frequency $\omega_c$ can be thus determined by

$$\omega_c = \frac{4\omega_p}{\pi h} \cdot \sum_{n=1,3,5,\ldots}^{\infty} \frac{1}{n^2} \tag{18.15}$$

$$= \frac{\pi}{2} \cdot \frac{\omega_p}{h}. \tag{18.16}$$

Therewith, the maximum limit cycle frequency $f_c = \omega_c/2\pi$ is given by

$$f_c = \frac{\omega_p}{4h}. \tag{18.17}$$

In [16], it was shown that for a nonzero input signal, the limit cycle frequency is lower than the value which is given by (18.17). If a sinusoidal input signal $x(t) = A \cdot \sin(2\pi f_{in} t)$ is applied, the limit cycle frequency is a function of time and is given by

$$f_{cs}(t) = \left[1 - A^2 \cdot \sin^2\left(2\pi f_{in} t\right)\right] \cdot f_c. \tag{18.18}$$

Furthermore, the average limit cycle frequency for a sinusoidal signal $x(t)$ is obtained from

$$f_0 = \left(1 - \frac{A^2}{2}\right) \cdot f_c. \tag{18.19}$$

Additional distorting frequency components are generated at the output of the ASDM by the sinusoidal input signal. These components appear around $f_0$ and can be described by Bessel functions at frequencies $f_0 \pm k \cdot f_{in}$, where $k$ is an even integer [16].

The average limit cycle frequency $f_0$ can be made smaller by decreasing $\omega_p$ and by increasing $h$ as can be concluded from (18.17) and (18.19). The lower limit of $f_0$ is determined by the distorting frequency components which are shifted into the band of interest and thus reduce the in-band SNR. Therefore, the ASDM should be designed in a way that $f_0$ is chosen large enough, so that these components are out of the band of interest. Furthermore, small values of $h$ make the hardware implementation more critical. A reasonable lower value for $h$ is 0.01.

### 18.3.2 Optimization of a First-Order Asynchronous Sigma–Delta Modulator

Since the transfer characteristic of an ideal integrator can be approximated in a real implementation only over a limited frequency range, a better choice for $H_L(j\omega)$ is a first-order low-pass filter transfer function, which is given by

$$H_L\left(j\omega\right) = \frac{1}{1 + j\omega/\omega_p}. \tag{18.20}$$

The output power spectrum of a respective ASDM with hysteresis $h = 0.01$ and characteristic frequency $\omega_p = 15\,\mathrm{kHz}$ is shown in Figure 18.9. The sinusoidal input signal with frequency $f_{in} = 3.644\,\mathrm{kHz}$ had an amplitude of $A = 0.8$. The noise of the output signal increases up to the limit cycle frequency $f_0$ and for higher frequencies it decreases again. The respective spectrum for a sampled data SDM looks different, since it increases up to half the sampling frequency.

The limit cycle frequency of this first-order modulator can be again determined by (18.17), as has been shown in [15]. Using this relationship with the parameters given above, a maximum limit cycle frequency $f_c = 375\,\mathrm{kHz}$ is obtained. Thus, with (18.19) the average limit cycle frequency for the considered sinusoid is given by $f_0 = 255\,\mathrm{kHz}$. As can be seen in Figure 18.9, there are a number of Bessel components around $f_0$. The frequency band of interest from 0 to 4 kHz is, however, only slightly affected and an SNR of more than 70 dB is obtained.

The distorting frequency components diminish completely from the baseband range and the SNR becomes even better by increasing the characteristic frequency of the low-pass filter and therewith the average value of the limit cycle frequency $f_0$. However, when increasing the limit cycle frequency, the unforced oscillations at the ASDM also increase. This results in an increased pulse rate of the change signal. Based on these considerations, it becomes clear that the choice of the limit cycle frequency is a trade-off between the SNR and the data rate.

Figure 18.10 shows a plot of the in-band signal-to-error ratio (SER) versus the limit cycle frequency $\omega_p$ and the frequency of the input signal. The amplitude of the input signal was $A = 0.8$. As can be seen from the figure,

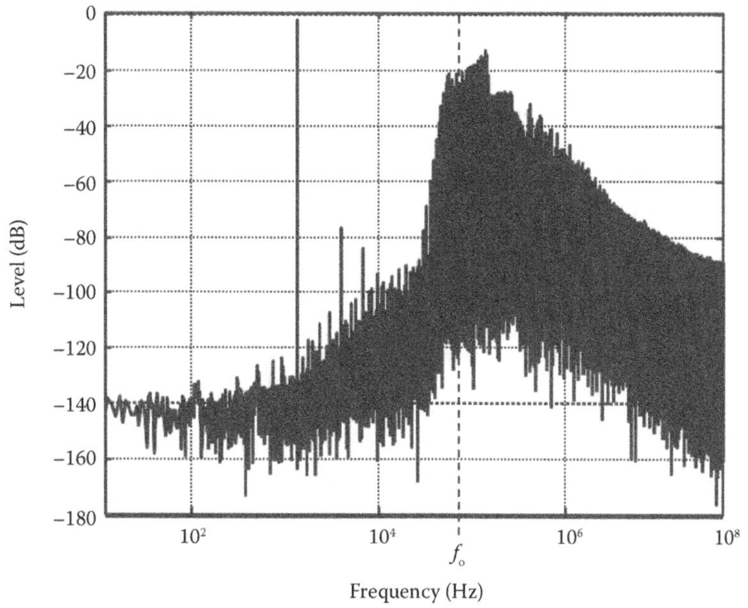


**FIGURE 18.9**
ASDM output power spectrum.

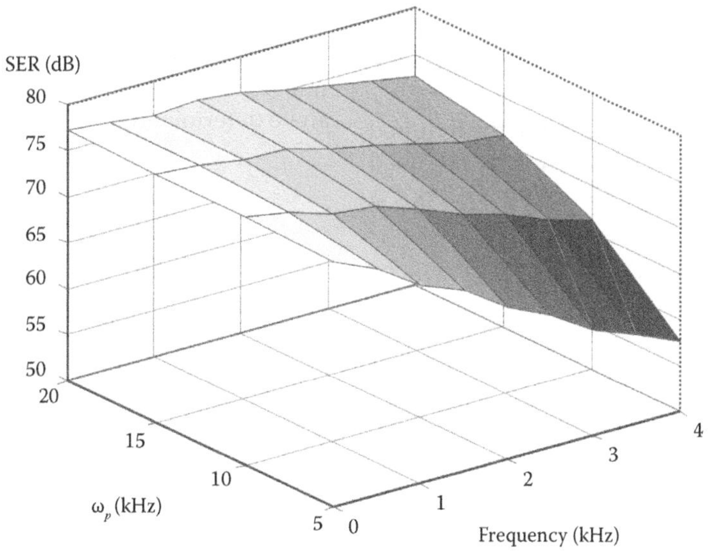


**FIGURE 18.10**
SER of the ASDM output.

the SER improves with increasing limit cycle frequency $\omega_p$ and degrades with increasing input frequency $f_{in}$. For $\omega_p \geq 15\,\text{kHz}$, the SER is better than 70 dB for all input frequencies in the range of 0–4 kHz.

The previous considerations have shown that for an ASDM system the rate of the pulse train is constant and does not depend on the input signal frequency. It is obtained from the average limit cycle frequency by

$$R_{ASDM} = 2f_0 < 2f_c. \tag{18.21}$$

The maximum rate $R_{ASDM,max} = 2 \cdot f_c$ is obtained for a zero input signal. The unwanted out-of-band frequency components and the limit cycle of the binary ADSM signal can be suppressed by decimation filters at the output of the ASDM system. For this signal processing, it is essential that the time distances between the transitions of the ASDM output are preserved. In the classical DT DSP, the required accuracy for the detection of these transitions would result in a very high oversampling rate. When using CT DSP, these time distances are, however, automatically kept by the CT delay lines with a very high accuracy. In Chapter 18, a proper architecture for an optimized CT decimation filter is proposed.

The granularity $T_{gran,ASDM}$ required for processing of the ASDM output signal can be determined from the minimum distance between adjacent signal transitions. The input signal $x(t)$ is assumed to be either a constant or a sinusoidal signal. The instantaneous duty cycle $D(t)$ of the output signal is defined as the ratio of the instantaneous pulse width $\alpha(t)$ and the limit cycle period $T(t)$ [16] and is thus given by

$$D(t) = \frac{\alpha(t)}{T(t)} = \frac{1}{2}(1 + x(t)). \quad (18.22)$$

With the instantaneous limit cycle frequency $f_{cs}(t) = 1/T(t)$ and solving for $\alpha(t)$, the following relationship is obtained:

$$\alpha(t) = \frac{1 + x(t)}{2f_{cs}(t)}. \quad (18.23)$$

With (18.18), the limit cycle frequency can be written as $f_{cs}(t) = (1 - x^2(t)) \cdot f_c$. Inserting this relationship into (18.23) gives

$$\alpha(t) = \frac{1}{(1 - x(t)) \cdot f_c}. \quad (18.24)$$

For a sinusoidal signal with amplitude $A$, the required granularity $T_{gran,ASDM}$ for CT processing is obtained from the minimum value of $\alpha(t)$ [20]. With (18.24) it is given by

$$T_{gran,ASDM} = \frac{1}{(1 + |A|) \cdot 2f_c}. \quad (18.25)$$

### 18.3.3 Performance and Hardware Comparison of ASDM with DM

To achieve about the same performance with a delta-modulated signal, a quantizer with a resolution of $N = 10$ bits is required. The respective SER versus the input frequency is shown in Figure 18.11. The in-band SER is considerably degraded for frequencies below $f_{in,max}/3$ due to the third harmonic falling into the band of interest. For $f_{in} > f_{in,max}/3$, the remaining in-band distortion component is located at the input frequency itself. Since this component is constant for input frequencies in the range of $f_{in,max} > f_{in} > f_{in,max}/3$, the respective SER is also constant.

For a delta-modulated signal, the pulse rate of the change signal increases linearly with the input frequency $f_{in}$. Since all $2^N - 1$ quantization levels used by a mid-tread quantizer are crossed two times during one period of a sinusoidal signal with maximum amplitude, the respective maximum pulse rate of the change signal is given by [6]:

$$R_{DM,\max} = \left(2^{N+1} - 4\right) \cdot f_{in}. \quad (18.26)$$

The input frequency $f_{in}^*$, for which both systems generate the same number of change pulses per unit time, can be determined by using (18.21) and (18.26):

$$f_{in}^* = \frac{f_0}{2^N - 2}. \quad (18.27)$$

Thus, for input signal frequencies smaller than $f_{in}^*$, the DM system generates fewer change pulses than

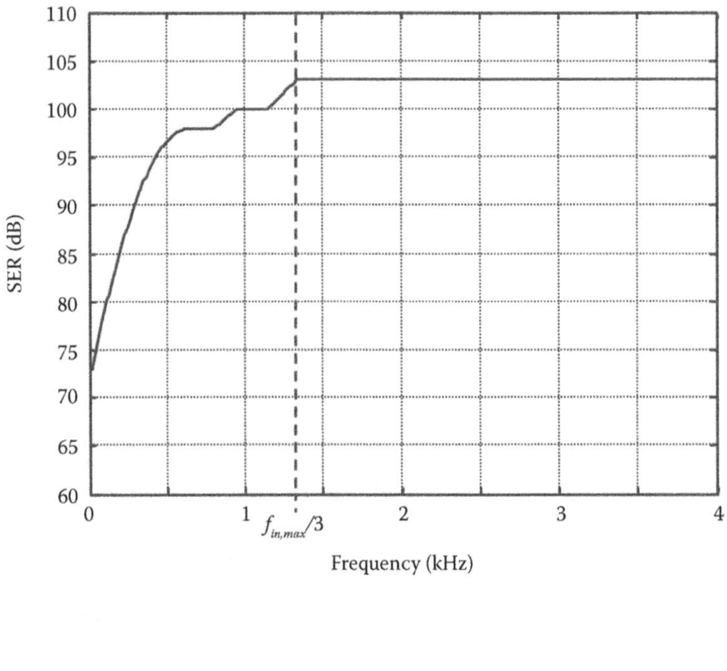


**FIGURE 18.11**
SER of the DM output.

the ASDM, whereas for frequencies larger than $f_{in}^*$, the ADSM performs better with respect to the pulse rate.

To compare the performance of the optimized ASDM and DM systems, a sinusoidal input signal with a maximum amplitude of $A = 0.8$ is applied to both systems. For DM, a resolution of $N = 10$ bits is required to fulfill the specification with an SER>70 dB. Similarly, as shown in Figure 18.9, a first-order ASDM with $\omega_p = 15$ kHz and $h = 0.01$ also fulfills the specification. With (18.16) and (18.18), a value of $f_0 = 255$ kHz is obtained for the average limit cycle frequency. Inserting the values into (18.27), we obtain $f_{in}^* = 250$ Hz. Therefore, for input signal frequencies above 250 Hz, the ASDM generates fewer change pulses and still fulfills the specification.

The relationship (18.25) also holds for a constant input signal $x(t) = A$. Thus, with $A = 0.8$ and $f_c = 375$ kHz, a very moderate value of $T_{gran,ASDM} = 740$ ns is obtained. Using (18.1), the required granularity for the DM system with $N = 10$ can be determined to be $T_{gran,DM} = 97$ ns, which is significantly smaller. Therefore, the requirements for the delay elements used for CT signal processing behind the quantizer are more relaxed for the ASDM system, the granularity can be chosen more than seven times larger. The number of required delay cells per delay element and therewith the implementation costs are reduced by the same amount.

In a classical DT SDM system, each doubling of the sampling rate results in an improvement of about 9 dB for the SNR [13]. Therefore, to achieve an in-band SNR of 70 dB for the DT SDM output, an oversampling ratio of 140 must be used. For $f_{in,max} = 4$ kHz, this results in a sampling rate of 1.2 MHz. Comparing this value with $f_c = 375$ kHz for the CT ASDM, verifies the superior performance of the latter also in comparison to a sampled data system with SDM.

Thus, it could be verified that digitization by an ASDM can result in several advantages. Compared with DM systems, the pulse rate of the change signal can be reduced for a large range of input frequencies, even though it does not depend on the input frequency itself anymore. Furthermore, it has been shown that decimation filtering behind the ASDM can be performed very efficiently by CT filters since the granularity requirements are relaxed and no accumulation is required.

## 18.4 Optimized Filter Structures for Event-Driven Signal Processing

Digital filtering is one of the most often implemented functionalities realized by digital signal processing systems. Digital filters are required either for separation of signals which have been combined or for restoration of signals that have been distorted. Plenty of structures and architectures have been proposed over the years for filter implementation. All of them have their advantages and disadvantages, and a number of parameters decide the best choice for a certain application. These parameters are, for example, cost and speed limitations, required performance, and others.

Of course, the selection of the best suited filter structure for a certain application is also an important issue for event-driven systems. The well-established methods for the design of digital sampled data filters can be also applied to CT filters. Conventional digital filters consist of registers as storage elements, adders, and possibly multipliers. In a continuous-time realization, the registers are replaced by continuous-time delay lines to eliminate the clock. When choosing the delays of the delay lines equal to the delays of the sampled data filter, which is usually one clock cycle, the transfer function remains the same, including the periodicity. Since the constraints with respect to a hardware implementation are however different, the optimization procedure can result in a different architecture best suited for a certain application. The most important difference is the considerably higher implementation cost for the time-continuous delay elements. Thus, the number of delay elements and the respective word length should be as small as possible for an optimized solution.

Furthermore, in a classical digital filter, the time delay must be either one clock period or a multiple of it. For continuous-time filters, this restriction, however, does no longer hold. Thus, an additional degree of freedom is available for the filter design by adjusting the time delays of the delay lines [10,11]. Various values can be chosen for every delay element. This additional flexibility can be used for designing filters with very simple coefficients thus avoiding multiplications. Instead, the respective operations can be exclusively realized by shift-and-add operations. Furthermore, the required number of shift-and-adds for every coefficient can be minimized by delay line adjustment. In the simplest case, the coefficients are made equal to a power of two. Furthermore, delay line adjustment can be used for improving the filter characteristic in the band of interest compared with the case with equal delays.

### 18.4.1 Continuous-Time Digital FIR Filters

The filter structure most often used for sampled data realizations is the finite impulse response filter shown in Figure 18.12, where the delays $D_i$ all have the same value [21]. The delay time in sampled data filters is usually equal to the sample time or a multiple of it.

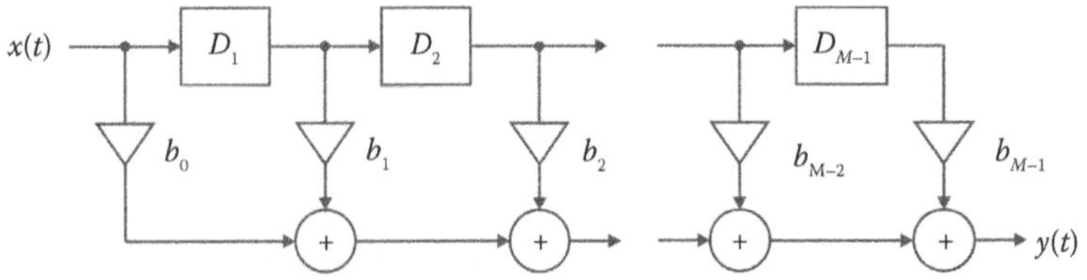


**FIGURE 18.12**
Continuous-time digital FIR-filter.

An advantage of this structure is its regularity; only Multiply and Accumulate (MAC) operations are required. Furthermore, stability is always guaranteed and the group delay of the filter can be made constant over the complete frequency range. The drawbacks of this architecture are, however, that the filter order and therewith the number of delay elements and multiplications is large compared with other structures such as recursive filters. Due to its robustness and its simplicity, the FIR structure had been also selected for the hardware implementation of continuous-time filters, a 16th-order FIR-filter is described in [5]. To minimize the hardware requirements for the delay cells, a delta modulator had been used to generate a data stream with 1-bit word length as input to the filter. Therefore, the delay elements must only be designed for this small word length. Due to the 1-bit input to the multipliers, the multipliers degenerate to accumulators, further reducing the hardware requirements. In spite of these improvements, the required chip area for the proposed FIR-filter is, however, considerably larger than that of a respective sampled data filter. Furthermore, due to delta modulation, errors are accumulated. Each error or missed value has an effect on all succeeding values. Thus, a reset for the filter must be generated at regular intervals. In [22,23], an improved realization of an FIR CT-filter of 16th order is described. The input data word is processed in parallel with a word length of 8 bits and respective $8 \times 8$ bit multipliers are implemented. The required chip area is, however, considerably larger than for the previously mentioned bit-serial implementation. Thus additional efforts are necessary to make CT-filter interesting for a commercial product.

Considerable hardware reduction and performance improvement can be obtained by applying the methods described in the preceding sections. The drawbacks of delta modulation can be avoided by using either an asynchronous $\Sigma\Delta$-modulator for quantization or by processing input signals with a larger word length. As has been shown, the size of the time-continuous delay elements can be reduced by up to 90% when using the delay element concept described in Section 18.2. Further improvements with respect to implementation costs and performance can be obtained by adjusting the values of the delay elements.

The output of the filter in Section 18.12 with coefficients $b_k$ and time delays $D_i$ is described in the time domain by the following relationship:

$$y(t) = \sum_{k=0}^{M-1} b_k \cdot x(t - D_k), \tag{18.28}$$

$$\text{with } D_k = \sum_{i=1}^{k} D_i \text{ and } D_0 = 0. \tag{18.29}$$

The respective transfer function $H(j\omega)$, with $\omega$ the frequency variable, is given by

$$H(j\omega) = \sum_{k=0}^{M-1} b_k \cdot e^{-j\omega D_k}. \tag{18.30}$$

To make the transfer function symmetrical with respect to $\omega = 0$, which is desirable for the majority of applications, the following restriction must hold for the values of the delay elements:

$$D_k = D_{M-k} \quad \text{for } k = 1 \text{ to } M-1. \tag{18.31}$$

Furthermore, to obtain a linear-phase filter, either a symmetry or an antisymmetry condition must hold for the filter coefficients:

$$b_k = \pm b_{M-1-k} \quad \text{for } k = 0 \text{ to } M-1. \tag{18.32}$$

If these relationships hold, the group delay of the filter is determined by

$$\tau_g = \frac{1}{2} \sum_{i=0}^{M-1} D_i. \tag{18.33}$$

Equation 18.33 confirms that the group delay is constant over all frequencies even for the time-continuous case. Constant group delay is a quite important property for many applications.

### 18.4.2 Optimized Continuous-Time Digital FIR-Filter

As an example, we will consider in the following a linear-phase low-pass filter which can be used, for example, for suppression of the quantization noise behind a continuous-time ΣΔ-A/D-converter [7,8]. The optimization of the filter transfer function is performed by adjusting the delay lines. When normalizing the frequency axis to the pass-band edge frequency, the pass-band extents from 0 to 1 and the stopband from 4 to 32. If the filter is implemented as a sampled data system, a 32nd-order FIR-filter is required for a minimum stop-band attenuation of 42 dB. To avoid severe performance degradation, the coefficients must be quantized at least with 12-bit resolution. The respective values are listed in Table 18.3. The pass-band droop is 0.1 dB in the range from 0 to 0.25 and increases to 1.8 dB at the pass-band edge frequency, which is acceptable for the considered application. The respective filter transfer function is shown in Figure 18.13, curve 1. To avoid multipliers and to implement each coefficient with no more than one shift-and-add operation, the values have been simplified as shown in the third column of Table 18.3. Use of these coefficient values, however, results in significant performance degradation, the stopband attenuation is reduced

**TABLE 18.3**
Coefficient Values for the 32nd-Order Linear-Phase FIR Low-Pass Filter

| Coefficient | $b_1, b_{33}$ | $b_2, b_{32}$ | $b_3, b_{31}$ | $b_4, b_{30}$ | $b_5, b_{29}$ | $b_6, b_{28}$ | $b_7, b_{27}$ | $b_8, b_{26}$ | $b_9, b_{25}$ |
|---|---|---|---|---|---|---|---|---|---|
| Value quantized to 12 bits | −0.02920 | 0.05109 | 0.07299 | 0.11679 | 0.16788 | 0.23358 | 0.31387 | 0.40146 | 0.48905 |
| Optimized for 1 shift&add | $2^{-5}$ | $2^{-4}$ | $2^{-3}-2^{-6}$ | $2^{-3}+2^{-5}$ | $2^{-2}-2^{-5}$ | $2^{-2}+2^{-5}$ | $2^{-1}-2^{-3}$ | $2^{-1}-2^{-4}$ | $2^{-1}+2^{-5}$ |

| Coefficient | $b_{10}, b_{24}$ | $b_{11}, b_{23}$ | $b_{12}, b_{22}$ | $b_{13}, b_{21}$ | $b_{14}, b_{20}$ | $b_{15}, b_{19}$ | $b_{16}, b_{18}$ | $b_{17}$ |
|---|---|---|---|---|---|---|---|---|
| Value quantized to 12 bits | 0.58394 | 0.67883 | 0.76642 | 0.84672 | 0.91241 | 0.96350 | 0.99270 | 1 |
| Optimized for 1 shift&add | $2^{-1}+2^{-3}$ | $1-2^{-2}$ | $1-2^{-2}$ | $1-2^{-3}$ | $1-2^{-4}$ | $1-2^{-5}$ | 1 | 1 |

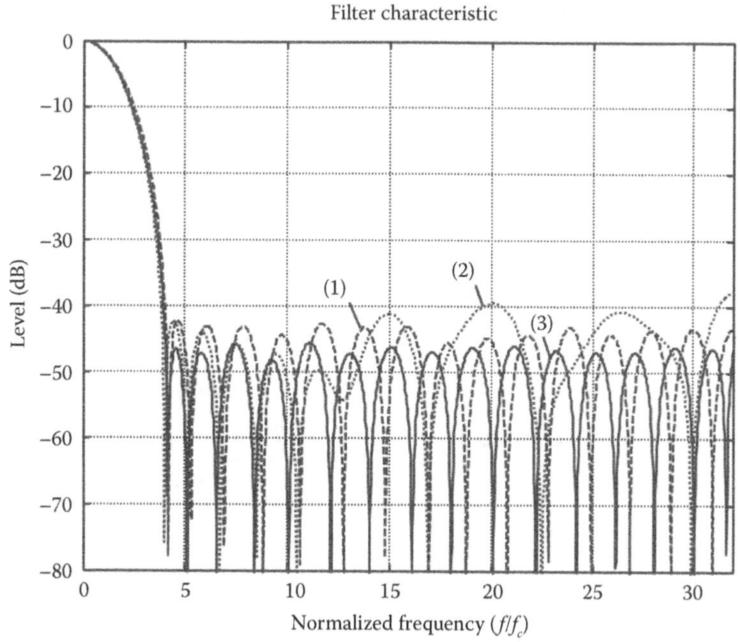


**FIGURE 18.13**
Transfer functions of the 32nd-order FIR low-pass filter. Curve (1), multiplications realized by 12-bit coefficients; curve (2), multiplications realized by 1 shift&add operation; and curve (3), optimized delays, multiplications are realized by 1 shift&add.

**TABLE 18.4**
Delay Values of the Optimized 32nd-Order Linear-Phase FIR Low-Pass Filter

| Delay | D1 | D2 | D3 | D4 | D5 | D6 | D7 | D8 |
|---|---|---|---|---|---|---|---|---|
| | D32 | D31 | D30 | D29 | D28 | D27 | D26 | D25 |
| $D_i/T_s$ | 0.24 | 1.17 | 1.03 | 1.05 | 0.98 | 1.025 | 0.985 | 0.98 |
| **Delay** | **D9** | **D10** | **D11** | **D12** | **D13** | **D14** | **D15** | **D16** |
| | **D24** | **D23** | **D22** | **D21** | **D20** | **D19** | **D18** | **D17** |
| $D_i/T_s$ | 0.975 | 1.025 | 0.98 | 0.95 | 1.005 | 0.98 | 0.985 | 0.98 |

to less than 37.9 dB. The respective transfer characteristic is shown by curve (2) in Figure 18.13.

To improve the filter characteristic, delay line adjustment has been applied. The values of the delay elements are optimized using an optimization procedure based on differential evolution. The respective transfer characteristic, which has been obtained with the simple coefficients, is shown by curve (3) in Figure 18.13. The stopband attenuation is improved to more than 45 dB by using the optimized delay values listed in Table 18.4. Note that the overall delay, the group delay, and the pass-band behavior of the filter are only slightly affected, the optimized filter is still in linear phase.

To evaluate the influence of delay tolerances, a Monte–Carlo analysis had been performed. For the simulation, a standard deviation of 0.5% for the tolerances of the delay values was assumed. The respective histogram for the minimum stopband attenuation $a_s$ is shown in Figure 18.14. The mean value of the minimum stopband attenuation is 43 dB; in more than 90% of the cases, the attenuation in the stopband is better than 42 dB confirming the good sensitivity properties of this structure with respect to tolerances of the delay elements. These mismatch requirements can be directly translated into mismatch specifications on transistor level. Thus, the strength of the proposed method had been confirmed by the example. It has been shown that the adjustment of the delay elements gives an additional degree of freedom for the optimization of continuous-time digital filters.

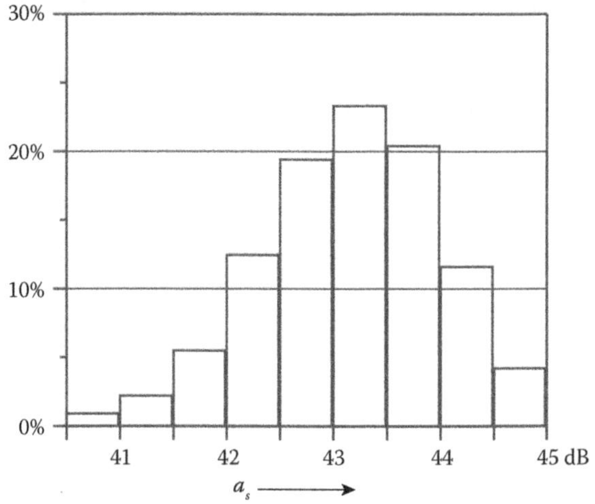


**FIGURE 18.14**
Histogram of the minimum stopband attenuation $a_s$ for a standard deviation of 0.5% for the tolerances of the delays.

### 18.4.3 Optimized Continuous-Time Recursive Low-Pass Filter

As a second example, we will consider a recursive low-pass filter. The stopband and pass-band edge frequencies are 3.4 kHz and 4.6 kHz, respectively, for a minimum stopband attenuation of 30 dB and a pass-band ripple better than 0.2 dB. The unit delay was chosen to 62.5 μs corresponding to a repetition frequency for the transfer characteristic of 16 kHz. It turned out that a bireciprocal wave digital filter [8,10,24,25] of fifth order fulfills the specification. The respective filter structure is shown in Figure 18.15. The coefficient values are truncated to 6 bits and the values of the delay elements are given in Table 18.5. The transfer characteristic is shown in Figure 18.16.

An optimization of the filter characteristic and of the coefficient values had been performed by adjusting the delay lines starting with the values in Table 18.5. The respective optimized values are also listed in the table. Even though the coefficient values are simplified, the minimum stopband attenuation is improved by about 7 dB to more than 37 dB. The respective filter characteristic is also shown in Figure 18.16.

Thus, the strength of the proposed method could be confirmed by both examples. It has been shown that the adjustment of the delay elements gives an additional degree of freedom for optimization of continuous-time digital filters.

Note that if all filter delays have the same value $T_D$, the filter transfer characteristic normally is replicated with the frequency $1/T_D$. This does not hold however anymore, if the filter characteristic is optimized by delay

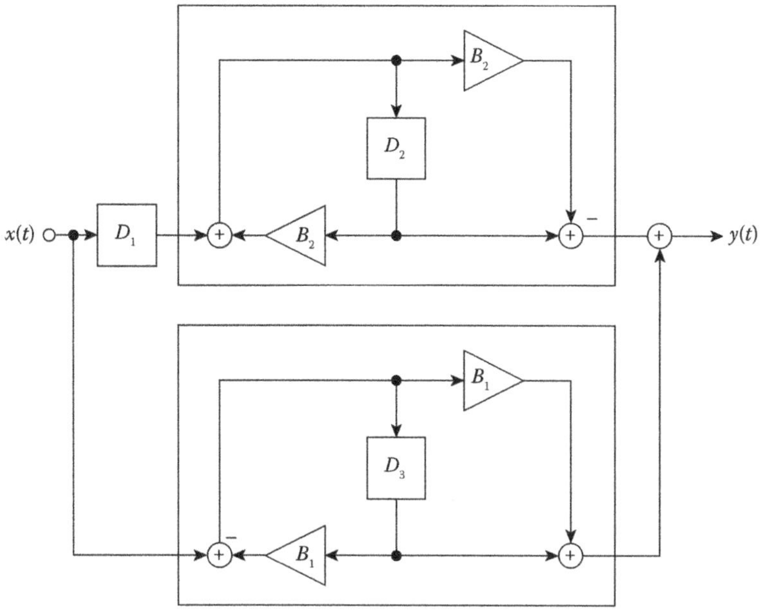


**FIGURE 18.15**
Fifth-order bireciprocal wave digital filter.

**TABLE 18.5**
Coefficient and Delay Values of the Recursive Filter

| | $B_1$ | $B_2$ | $D_1$ | $D_2$ | $D_3$ |
|---|---|---|---|---|---|
| Sampled data filter | $2^{-2}+2^{-5}$ | $-1+2^{-2}$ | T | 2T | 2T |
| CT filter | $2^{-2}$ | $-1+2^{-2}$ | $1.09\cdot T$ | $2.10\cdot T$ | $2.16\cdot T$ |

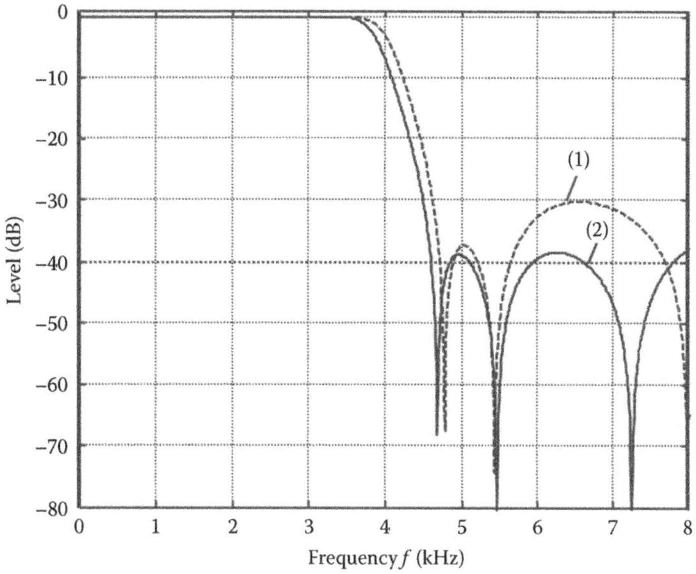


**FIGURE 18.16**
Transfer functions of the fifth-order bireciprocal wave digital filter, curve (1), sampled data filter, curve (2), optimized CT filter.

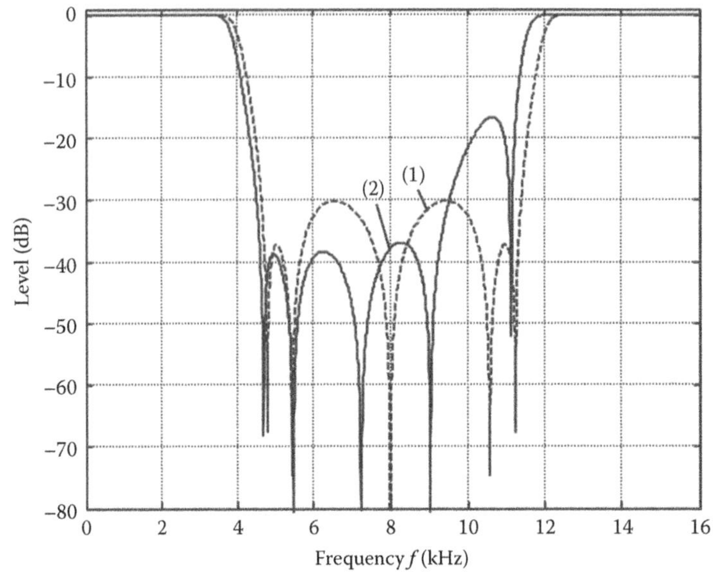


**FIGURE 18.17**
Transfer functions of the recursive wave digital filter for a wider frequency range, curve (1), sampled data filter, curve (2), optimized CT filter.

line adjustment in the band of interest. The respective filter characteristic is shown for a wider frequency band in Figure 18.17. The stopband attenuation drops in the out-of-band region from 8 kHz to 11.4 kHz to less than 20 dB.

## 18.5 Summary

Event-driven CT DSP systems can have a number of advantages compared with conventional sampled data systems. Due to the nonsampled operation, no aliasing of higher frequency components and of noise occurs and due to the asynchronous operation the power consumption can be minimized. Since the implementations proposed up to now, however, require a considerably larger chip area than the respective sampled data systems, continuous-time systems have not yet reached the maturity for commercial implementation.

Therefore, in this contribution, a number of improvements have been proposed, which result in considerable savings with respect to the required hardware and with respect to power consumption. Since the continuous-time delay elements are the most critical functional blocks with respect to chip area and power consumption, a new CT base cell has been proposed. Using this new cell, the chip area can be reduced by up to 90% at the cost of only minor performance losses. Furthermore, the drawbacks of delta modulation can be avoided by using a conventional continuous-time quantizer and performing digital signal processing with the full word length. The proposed delay element enables parallel processing with only minor additional hardware costs.

Another way to reduce implementation costs for CT DSP systems is to perform quantization by ASDM. This quantizer type can generate a 1-bit time-continuous digital signal without the drawbacks of delta modulation. This approach was analyzed in detail in Section 18.3.

In Section 18.4, filter structures that are well-suited for implementation in continuous time are analyzed. It has been shown that an additional degree of freedom for the filter optimization process is given, since the filter characteristic depends on the length of the CT delay lines. By adjusting the values of the delays, the filter characteristic can be improved and the coefficient values can be simplified. Thus, the hardware requirements for an implementation can be also reduced if the filter is properly designed. This has been verified by applying the method to a 32nd-order FIR-filter and to a fifth-order recursive filter.

The proposed methods can bring continuous-time systems an important step closer to commercial application. Event-driven systems based on continuous-time DSPs can be attractive for a number of applications including biomedical implants.

## Bibliography

[1] Y. P. Tsividis. Event-driven data acquisition a digital signal processing—A tutorial. *IEEE Transactions on Circuits and Systems—II*, 57(8):577–581, 2010.

[2] C. Weltin-Wu and Y. Tsividis. An event-driven clockless level-crossing ADC with signal-dependent adaptive resolution. *IEEE Journal of Solid-State Circuits*, 48(9):2180–2190, 2013.

[3] B. J. Rao, O. V. Krishna, and V. S. Kesav. A continuous time ADC and digital signal processing system for smart dust wireless sensor applications. *International Journal in Engineering and Technology*, 2(6):941–949, 2013.

[4] Y. P. Tsividis. Mixed-domain systems and signal processing based on input decomposition. *IEEE Transactions on Circuits and Systems—I: Regular Papers*, 53(10):2145–2156, 2006.

[5] Y. W. Lee, K. L. Shepard, and Y. P. Tsividis. A continuous-time programmable digital FIR-filter. *IEEE Journal of Solid State Circuits*, 39(21):2512–2520, 2006.

[6] B. Schell and Y. P. Tsividis. A continuous-time ADC/DSP/DAC system with no clock and with activity-dependent power dissipation. *IEEE Journal of Solid-State Circuits*, 43(11):2472–2481, 2008.

[7] F. Aeschlimann, E. Allier, L. Fesquet, and M. Renaudin. Asynchronous FIR filters, towards a new digital processing chain. In *Proceedings of the International Symposium on Asynchronous Circuits and Systems*, pp. 198–206, 2004.

[8] D. Brückmann. Design and realization of continuous-time wave digital filters. In *Proceedings ISCAS 2008*, pp. 2901–2904, Seattle, WA, 18–21 May 2008.

[9] K. Konrad, D. Brückmann, N. Tavangaran, J. Al-Eryani, R. Kokozinski, and T. Werthwein. Delay element concept for continuous time digital signal processing. In *2013 IEEE International Symposium on Circuits and Systems (ISCAS)*, pp. 2775–2778, Bejing, China, 19–23 May 2013.

[10] K. Konrad, D. Brückmann, and N. Tavangaran. Optimization of continuous time filters by delay line adjustment. In *2010, 53rd IEEE International Midwest Symposium on Circuits and Systems (MWSCAS)*, pp. 1238–1241, Seattle, WA, 1–4 August 2010.

[11] D. Brückmann, K. Konrad, and N. Tavangaran. Delay line adjustment for the optimization of digital continuous time filters. In *ICECS 2010 17th IEEE International Conference on Electronics, Circuits, and Systems*, pp. 843–846, 12–15 December 2010.

[12] J. Al-Eryani, A. Stanitzki, K. Konrad, N. Tavangaran, D. Brückmann, and R. Kokozinski. Continuous-time digital filters using delay cells with feedback. *13. ITG/GMM-Fachtagung (Analog 2013)*, Aachen, 4–6 March 2013.

[13] S. R. Norsworthy, R. Schreier, and G. C. Temes. *Delta-Sigma Data Converters, Theory, Design, and Simulation*. IEEE Press, Piscataway, NJ, 1997.

[14] O. Shoaei and W. M. Snelgrove. Optimal (bandpass) continuous-time sigma delta modulator. *Proceedings of ISCAS*, 5:489–492, 1994.

[15] S. Ouzounov, E. Roza, J. A. Hegt, G. van der Weide, and A H. M. van Roermund. Analysis and design of high-performance asynchronous sigma-delta modulators with a binary quantizer. *IEEE Journal of Solid-State Circuits*, 41(3):588–596, 2006.

[16] E. Roza. Analog-to-digital conversion via duty-cycle modulation. *IEEE Transactions on Circuits and Systems*, 44(11):907–914, 1997.

[17] P. Benabes, M. Keramat, and R. Kielbasa. A methodology for designing continuous-time sigma delta modulators. In *IEEE European Design and Test Conference (EDIC)*, pp. 46–50, March 1997.

[18] L. Breems and J. Huising. *Continuous-Time Sigma-Delta Modulation for A/D Conversion in Radio Receivers*. Springer, Heidelberg, Berlin, New York, NY, 2001.

[19] J. Daniels, W. Dehaene, M. S. J. Steyaert, and A. Wiesbauer. A/D conversion using asynchronous delta-sigma modulation and time-to-digital conversion. *IEEE Transactions on Circuits and Systems I, Regular Papers*, 57(9):2404–2412, 2010.

[20] N. Tavangaran, D. Brückmann, R. Kokozinski, and K. Konrad. Continuous time digital systems with asynchronous sigma delta modulation. In *EUSIPCO 2012*, Bucharest, Romania, pp. 225–229, 27–31 August 2012.

[21] J. G. Proakis and D. G. Manolakis. *Digital Signal Processing, Principles. Algorithms and Applications*, 4th edition. Pearson Prentice Hall, Upper Saddle River, NJ, 2007.

[22] C. Vezyrtzis, W. Jiang, S. M. Nowick, and Y. Tsividis. A flexible, clockless digital filter. In *Proceedings of the IEEE European Solid-State Circuits Conference*, pp. 65–68, September 2013.

[23] C. Vezyrtzis, W. Jiang, S. M. Nowick, and Y. Tsividis. A flexible, event-driven digital filter with frequency response independent of input sample rate. *IEEE Journal of Solid-State Circuits*, 49(10):2292–2304, 2014.

[24] A. Fettweis. Wave digital filters: Theory and practice. *Proceedings of the IEEE*, 74:270–327, 1986.

[25] D. Brückmann and L. Bouhrize. Filter stages for a high-performance reconfigurable radio receiver with minimum system delay. In *Proceedings ICASSP 2007*, Honolulu, HI, 15–20 April 2007.

[26] M. Renaudin. Asynchronous circuits and systems. A promising design alternative. *Journal of Microelectronic Engineering*, 54:133–149, 2000

[27] M. Kurchuk and Y. P. Tsividis. Signal-dependent variable-resolution clockless A/D conversion with application to continuous-time digital signal processing. *IEEE Transactions on Circuits and Systems—I: Regular Papers*, 57(5):982–991, 2010.

[28] D. Brückmann, T. Feldengut, B. Hosticka, R. Kokozinski, K. Konrad, and N. Tavangaran. Optimization and implementation of continuous time DSP-systems by using granularity reduction. In *2011 IEEE International Symposium on Circuits and Systems (ISCAS)*, pp. 410–413, 15–18 May 2011.

[29] N. Tavangaran, D. Brückmann, T. Feldengut, B. Hosticka, K. Konrad, R. Kokozinski, and R. Lerch. Effects of jitter on continuous time digital systems with granularity reduction. In *EUSIPCO 2011*, Barcelona, Spain, 29 August–2 September 2011.

[30] J. Al-Eryani, A. Stanitzki, K. Konrad, N. Tavangaran, D. Brückmann, and R. Kokozinski. Low-power area-efficient delay element with a wide delay range. In *IEEE International Conference on Electronics, Circuits and Systems (ICECS)*, Seville Spain, 9–12 December 2012.

# 19

# *Asynchronous Processing of Nonstationary Signals*


**Azime Can-Cimino**
*University of Pittsburgh*
*Pittsburgh, PA, USA*

**Luis F. Chaparro**
*University of Pittsburgh*
*Pittsburgh, PA, USA*


**CONTENTS**




**ABSTRACT** Representation and processing of nonstationary signals, from practical applications, are complicated by their time-varying statistics. Signal- and time–frequency-dependent asynchronous approaches discussed in this chapter provide efficient sampling and data storage, low-power consumption, and analog processing of such signals. The asynchronous procedures considered and contrasted are based on the level-crossing (LC) approach and the asynchronous sigma delta modulator (ASDM). LC sampling results in nonuniform sampling, and the signal reconstruction requires better localized functions than the sinc, in this case the prolate spheroidal wave functions, as well as regularization. An ASDM maps its analog input into a binary output with zero-crossing times depending on the input amplitude. Duty-cycle analysis allows development of low- and high-frequency signal decomposers, and by modifying the ASDM, a filter bank analysis is developed. These approaches are useful in denoising, compression, and decomposition of signals in biomedical and control applications.


## 19.1 Introduction

In this chapter, we discuss issues related to the asynchronous data acquisition and processing of nonstationary signals that result from practical applications. Data acquisition, storage, and processing of these signals are challenging issues; such is the case, for instance, in continuous monitoring in biomedical and sensing network applications. Moreover, low-power consumption and analog processing are important requirements for devices used in biomedical applications. The above issues are addressed by asynchronous signal processing [1,6,7,16,20,29].

Typically, signals resulting from practical applications are nonstationary, and the variation over time of the signal statistics characterizes them. As such, the representation and processing of nonstationary signals are commonly done assuming the statistics either do not change with time (stationarity) or remain constant in short time intervals (local stationarity). Better approaches result by using joint time–frequency spectral characterization that directly considers the time variability [8,32]. According to the Wold–Cramer representation [28], a nonstationary process can be thought of as the output of a time-varying system with white noise as input. It can thus be shown that the distribution of the power of a nonstationary signal is a function of time and frequency [18,19,28]. By ignoring that the frequency content is changing with time, conventional synchronous sampling collects unnecessary samples in segments where the signal is quiescent. A more appropriate approach would be to make the process signal-dependent, leading to a nonuniform sampling that can be implemented with asynchronous processes.

Power consumption and type of processing are necessary constrains for devices used to process signals from biomedical and sensor network applications. In brain–computer interfacing, for instance, the size of the devices, the difficulty in replacing batteries, possible harm to the patient from high frequencies generated by fast clocks, and the high power cost incurred in data transmission calls for asynchronous methodologies [9,13,14,25,26], and of analog rather than digital processing [38,39].

Sampling and reconstruction of nonstationary signals require a signal-dependent approach. Level-crossing (LC) sampling [13,14] is a signal-dependent sampling method such that for a given set of quantization levels, the sampler acquires a sample whenever the signal coincides with one of those levels. LC sampling is thus independent of the bandlimited signal constraint imposed by the sampling theory and has no quantization error in the amplitude. Although very efficient in the collection of significant data from the signal, LC sampling requires an *a priori* set of quantization levels and results in nonuniform sampling where for each sample we need to know not only its value but also the time at which it occurs. Quantization in the LC approach is required for the sample times rather than the amplitudes. Selecting quantization levels that depend on the signal, rather than fixed levels, as is commonly done, is an issue of research interest [10,14].

The signal dependency of LC—akin to signal sparseness in compressive sensing—allows efficient sampling for nonstationary signals but complicates their reconstruction. Reconstruction based on the Nyquist–Shannon's sampling theory requires the signal to be bandlimited. In practice, the bandlimitedness condition and Shannon's sinc interpolation are not appropriate or well-posed problem when considering nonstationary signals. The prolate spheroidal wave (PSW) or Slepian functions [35], having better time and frequency localization than the sinc function [9,40], are more appropriate for the signal reconstruction. These functions allow more compression in the sampling [9,11], and whenever the signal has band-pass characteristics, modulated PSW functions provide parsimonious representations [27,34].

Sampling and reconstruction of nonstationary signals are also possible using the asynchronous sigma delta modulator (ASDM) [2,23–25,30]. The ASDM is an analog, low-power, nonlinear feedback system that maps a bounded analog input signal into a binary output signal with zero-crossing times that depend on the amplitude of the input signal. The ASDM operation can be shown to be that of an adaptive LC sampling scheme [10]. Using duty-cycle modulation to analyze the ASDM, a multilevel representation of an analog signal in terms of localized averages—computed in windows with supports that depend on the amplitude of the signal—is obtained. Such representation allows us to obtain a signal decomposition that generalizes the Haar wavelet representation [5]. This is accomplished by latticing the time–frequency plane choosing fixed frequency ranges and allowing the time windows to be set by the signal in each of these frequency ranges. It is possible to obtain multilevel decompositions of a signal by cascading modules using ASDMs sharing the greedy characteristics of compressive sensing. We thus have two equivalent representations of a signal: one resulting from the binary output of the ASDM and the other depending on the duty-cycle modulation.

The local characterization of the input signal of an ASDM is given by an integral equation that depends on the parameters of the ASDM and on the zero-crossing times of the binary output. Rearranging the feedback system, the resulting modified ASDM can be characterized by a difference equation that can be used to obtain nonuniform samples of the input. In a way, this is equivalent to an asynchronous analog-to-digital converter that requires the quantization of the sampling times. Using the difference equation to recursively estimate input samples and the quantized sampling times, it is possible to reconstruct the analog input signal.

With the modified ASDM and latticing the joint time–frequency space into defined frequency bands and time ranges depending on the scale parameter, we obtain a bank of filter decomposition for nonstationary signals, which is similar to those obtained from wavelet representations. This provides a way to efficiently process nonstationary signals. The asynchronous approaches proposed in this chapter are especially well suited for processing signals that are sparse in time, and signals in low-power applications [6]. The different approaches are illustrated using synthetic and actual signals.

## 19.2 Asynchronous Sampling and Reconstruction

According to the Wold–Cramer decomposition [28], a nonstationary process $\xi(t)$ can be represented as the output of a linear, time-varying system with impulse response $h(t, \tau)$ and white noise $\varepsilon(t)$ as input:

$$\xi(t) = \int_{-\infty}^{\infty} h(t, \tau)\varepsilon(\tau)d\tau. \quad (19.1)$$

The white noise $\varepsilon(t)$, as a stationary process, is expressed in terms of sinusoids with random amplitudes and phases as

$$\varepsilon(t) = \int_{-\infty}^{\infty} e^{j\Omega t} dZ(\Omega), \quad (19.2)$$

where the process $Z(\Omega)$ is a random process with orthogonal increments such that

$$E[dZ(\Omega_1)dZ^*(\Omega_2)] = 0 \qquad \Omega_1 \neq \Omega_2,$$

and

$$E[|dZ(\Omega)|^2] = \frac{d\Omega}{2\pi}.$$

Replacing (19.2) into (19.1), we have

$$\xi(t) = \int_{-\infty}^{\infty} H(t, \Omega)e^{j\Omega t} dZ(\Omega), \quad (19.3)$$

where $H(t, \Omega)$ is the generalized transfer function of the linear time-varying system. Thus, the mean and variance of $\xi(t)$ are functions of time, and the Wold–Cramer evolutionary spectrum is defined as

$$S(t, \Omega) = |H(t, \Omega)|^2. \quad (19.4)$$

Accordingly, the distribution of the power of the nonstationary process $\xi(t)$ for each time $t$ is a function of the frequency $\Omega$ [18,19], and the sampling of nonstationary signals, given their time-varying spectra, requires a continuously changing sampling time. In segments where there is a great deal of variation of the signal, the sampling period should be small and it should be large in segments where the signal does not change much. Thus, sampling using a uniform sampling period—determined by the highest frequency in the signal—is wasteful in segments with low activity; that is, the sampling should be signal dependent and nonuniform. As we will see, it is important to consider not only the frequency content of the signal but also its amplitude.

### 19.2.1 LC Sampling and Reconstruction

The LC sampler [13] is signal dependent and efficient, especially for signals sparse in time. For a set of amplitude quantization levels $\{q_i\}$, an LC sampler acquires a sample whenever the amplitude of the signal coincides with one of these given quantization levels. It thus results in nonuniform sampling: more samples are taken in regions where the signal has significant activity, and fewer where the signal is quiescent. The nonuniformity of the sampling requires that the value of the sample and the time at which it occurs be kept—a more complex approach than synchronous sampling—but LC sampling has the advantage that no bandlimiting constraint is required. From the sample values obtained by the LC sampler, the signal $\xi(t)$ is approximated by a multilevel signal

$$\hat{\xi}(t) = \sum_k q_k p_k(t), \quad (19.5)$$

for a unit pulse $p_k(t) = u(t - t_k) - u(t - t_{k+1}), t_{k+1} > t_k$, where $u(t)$ is the unit-step function and $q_k$ is the quantization level that coincides with $\xi(t_k)$. Typically, the quantization levels are chosen uniformly, but better choices have been considered [10,13]. Reconstruction from the nonuniform samples $\{\xi(t_k)\}$ can be attained using the prolate spheroidal wave functions (PSWFs) $\{\phi_m(t)\}$ [9,35,40].

The PSWFs, among all orthogonal bases defined in a time-limited domain $[-T, T]$, are the ones with maximum energy concentration within a given band of frequencies $(-W, W)$. They are eigenfunctions of the integral operator:

$$\phi_n(t) = \frac{1}{\lambda_n}\int_{-T}^{T} \phi_n(x)s(t - x)dx, \quad (19.6)$$

where $s(t)$ is the sinc function and $\{\lambda_n\}$ are the corresponding eigenvalues.

Given that the PSWFs $\{\phi_n(t)\}$ are orthogonal in infinite and finite supports, the sinc function $s(t)$ can be expanded in terms of them for a sampling period $T_s$ as

$$s(t - kT_s) = \sum_{m=0}^{\infty} \phi_m(kT_s)\phi_m(t), \quad (19.7)$$

so that the Shannon's sinc interpolation [9,40] of a signal $\xi(t)$ becomes

$$\xi(t) = \sum_{m=0}^{\infty}\left[\sum_{k=-\infty}^{\infty} \xi(kT_s)\phi_m(kT_s)\right]\phi_m(t). \quad (19.8)$$

If $\xi(t)$ is finitely supported in $0 \leq t \leq (N-1)T_s$ and has a Fourier transform $\Xi(\Omega)$ essentially bandlimited in a frequency band $(-W, W)$ (i.e., most of the signal energy is in this band), then $\xi(t)$ can be expressed as

$$\xi(t) = \sum_{m=0}^{M-1} \gamma_m\phi_m(t), \quad (19.9)$$

where

$$\gamma_m = \sum_{k=0}^{N-1} \xi(kT_s)\phi_m(kT_s),$$

where the value of $M$ is chosen from the eigenvalues $\{\lambda_n\}$ indicating the energy concentration.

Using PSWFs, $N_{\ell c}$ sample values $\{\xi(t_k)\}$, acquired by an LC sampler at nonuniform times $\{t_k,\ k=0,\ldots,N_{\ell c}-1\}$, are represented by a matrix equation [11]

$$\xi(\mathbf{t_k}) = \mathbf{\Phi}(\mathbf{t_k})\gamma_M, \qquad (19.10)$$

where $\mathbf{\Phi}(\mathbf{t_k})$ is an $N_{\ell c} \times M$ matrix, in general not square as typically $M < N_{\ell c}$, with entries the PSWFs $\{\phi_i(t_k)\}$ computed at the times $\{t_k\}$ of the vector $\mathbf{t_k}$. The expansion coefficients $\{\gamma_m\}$ in the vector $\gamma_M$ are found as

$$\gamma_M = \mathbf{\Phi}^{\dagger}(\mathbf{t_k})\xi(\mathbf{t_k}), \qquad (19.11)$$

where $\dagger$ indicates the pseudoinverse. Replacing these expansion coefficients in (19.9) provides at best an approximation to the original signal. However, depending on the distribution of the time samples, the above problem could become ill-conditioned requiring regularization techniques [15,37].

The Tikhonov regularized solution of the measurement equation (19.10) can be posed as the mean-square minimization:

$$\gamma_{M_\mu} = \arg\min_{\gamma_M}\{\|\mathbf{\Phi}(\mathbf{t_k})\gamma_M - \xi(\mathbf{t_k})\|^2 + \mu\|\gamma_M\|^2\}, \qquad (19.12)$$

where $\mu$ is a trade-off between losses and smoothness of the solution, and for $\mu = 0$ the solution coincides with the pseudoinverse solution (19.11). The optimal value for the regularization parameter $\mu$ is determined *ad hoc* [11,15]. The regularized solution $\gamma_{M_\mu}$ of (19.12) is given by

$$\gamma_{M_\mu} = (\mathbf{\Phi}(\mathbf{t_k})^T\mathbf{\Phi}(\mathbf{t_k}) + \mu\mathbf{I})^{-1}\mathbf{\Phi}(\mathbf{t_k})^T\xi(\mathbf{t_k}). \qquad (19.13)$$

The above illustrates that the sampling of nonstationary signals using the LC sampler is complicated by the need to keep samples and sampling times in the nonuniform sampling, and that the signal reconstruction is in general a not well-posed problem requiring regularization with *ad hoc* parameters.

### 19.2.2 ASDM Sampling and Reconstruction

Sampling using an ASDM [10] provides a better alternative to the LC sampler. An ASDM [25,30] is a nonlinear analog feedback system that operates at low power and consists of an integrator and a Schmitt trigger (see Figure 19.1). It has been used to time-encode bounded and bandlimited analog signals into continuous-time signals with binary amplitude [25].

Given a bounded signal $x(t)$, $|x(t)| \le c$, for an appropriate value of the scale parameter $\kappa$ of the ASDM, it maps the amplitude of $x(t)$ into a binary signal $z(t)$ of amplitude $\pm b$, where $b > c$. If in $t_k \le t \le t_{k+1}$ the output of the Schmitt trigger is the binary signal $z(t) = b(-1)^{k+1}[u(t-t_k) - u(t-t_{k+1})]$, where $u(t)$ is the unit-step function, then the output $y(t)$ of the integrator is bounded, that is, $|y(t)| < \delta$. As such, the difference $y(t_{k+1}) - y(t_k)$ equals

$$\frac{1}{\kappa}\left[\int_{t_k}^{t_{k+1}} x(\tau)d\tau - (-1)^{k+1}b(t_{k+1}-t_k)\right] = 2\delta,$$

or the integral equation [25]

$$\int_{t_k}^{t_{k+1}} x(\tau)d\tau = (-1)^k[-b(t_{k+1}-t_k) + 2\kappa\delta], \qquad (19.14)$$

relating the amplitude information of $x(t)$ to the duration of the pulses in $z(t)$—or equivalently, to the zero-crossing times $\{t_k\}$. Approximating the integral in (19.14) provides an approximation of the signal using the zero-crossing times [23].

The ASDM parameters $b$ and $\kappa$ relate to the signal as follows: if $|x(t)| < c$, $b$ is chosen to be $b > c$ to guarantee together with $\kappa$ that the output of the integrator increases and decreases within bounds $\pm\,\delta$; $\kappa$ is connected with the amplitude and the maximum frequency of the signal as shown next. A sufficient condition for the reconstruction of the original signal from nonuniform samples is [22]

$$\max_k(t_{k+1} - t_k) \le T_N, \qquad (19.15)$$

where $T_N = 1/2f_{max}$ is the Nyquist sampling period for a bandlimited signal with maximum frequency $f_{max}$.

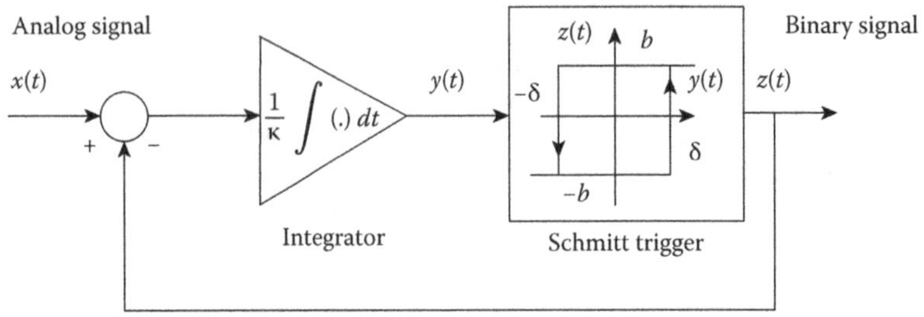


**FIGURE 19.1**
Asynchronous sigma delta modulator.

Using Equations 19.15 and 19.14, and that $|x(t)| < c$, we have that

$$\frac{\kappa}{b+c} \leq t_{k+1} - t_k \leq \frac{\kappa}{b-c} \leq \frac{1}{2f_{max}}, \quad (19.16)$$

displaying the relation between $\kappa$, the difference in amplitude $b - c$ and the maximum frequency $f_{max}$ of the signal. Because of this, the representation given by (19.14) requires $x(t)$ to be bandlimited.

The binary rectangular pulses $z(t)$, connected with the amplitude of the signal $x(t)$, are characterized by the duty cycle $0 < \alpha_k/T_k < 1$ of two consecutive pulses of duration $T_k = \alpha_k + \beta_k$, where $\alpha_k = t_{k+1} - t_k$ and $\beta_k = t_{k+2} - t_{k+1}$ [30]. Letting

$$\zeta_k = \frac{\alpha_k - \beta_k}{\alpha_k + \beta_k}, \quad (19.17)$$

we have that the duty cycle is given by $\alpha_k/T_k = (1 + \zeta_k)/2$, where, as shown next, $\zeta_k$ is the local average of $x(t)$ in $t_k \leq t \leq t_{k+2}$. Indeed, the integral equation (19.14) can be written in terms of $z(t)$ as

$$\int_{t_k}^{t_{k+1}} x(\tau)d\tau = (-1)^{k+1} \int_{t_k}^{t_{k+1}} z(\tau)d\tau + 2(-1)^k \kappa\delta,$$

which can be used to get local estimate of the signal average $\zeta_k$ in $[t_k, t_{k+2}]$. Using the above equation to find the integral in $[t_k, t_{k+2}]$ and dividing it by $T_k$, we have

$$\begin{aligned}\frac{1}{T_k}\int_{t_k}^{t_{k+2}} x(\tau)d\tau &= \frac{(-1)^{k+1}}{T_k}\left[\underbrace{\int_{t_k}^{t_{k+1}} z(t)dt}_{\alpha_k} - \underbrace{\int_{t_{k+1}}^{t_{k+2}} z(t)dt}_{\beta_k}\right] \\ &= (-1)^{k+1}\frac{\alpha_k - \beta_k}{\alpha_k + \beta_k} \\ &= (-1)^{k+1}\zeta_k, \quad (19.18)\end{aligned}$$

or the local mean of $x(t)$ in the segment $[t_k, t_{k+2}]$.

Thus, for a bounded and bandlimited analog signal, the corresponding scale parameter $\kappa$ determines the width of an appropriate window—according to the nonuniform zero-crossing times—to compute an estimate of the local average. This way, the ASDM provides either a representation of its input $x(t)$ by its binary output $z(t)$ and the integral equation, or a sequence of local averages $\{\zeta_k\}$ at nonuniform times $\{t_{2k}\}$. Different from the LC sampler, the ASDM only requires the sample times. The only drawback of this approach for general signals is the condition that the input signal be bandlimited. A possible alternative to avoid this is to lattice the time–frequency plane using arbitrary frequency windows and time windows connected to the amplitude of the input signal. This results in an asynchronous decomposition, which we discuss next.

It is of interest to remark here the following:

- Using the trapezoidal rule to approximate the integral in Equation 19.14 is possible to obtain an expression to reconstruct the original signal using zero-crossing times [10].

- A multilevel signal can be represented by zero crossings using the ASDM. Consider the signal given as the approximation from the LC sampler:

$$\hat{x}(t) = \sum_{i=-N}^{N} q_i[u(t-\tau_i) - u(t-\tau_{i+1})], \quad (19.19)$$

  where $u(t)$ is the unit-step function and $\{q_i\}$ are given quantization levels. Processing the component $\hat{x}_i(t) = q_i[u(t-\tau_i) - u(t-\tau_{i+1})]$ with the ASDM, the output would be a binary signal $z_i(t) = u(t-t_i) - 2u(t-t_{i+1}) + u(t-t_{i+2})$ for some zero-crossing times $t_{i+k}$, $k = 0, 1, 2$. Letting $t_i = \tau_i$ and $t_{i+2} = \tau_{i+1}$, $b > \max(x(t))$ and $\delta$ be a fixed number, we need to find $t_{i+1}$ and the corresponding scale parameter $\kappa_i$. If $q_i$ is considered the local average and $\alpha_i = t_{i+1} - \tau_i$ and $\beta_i = \tau_{i+1} - t_{i+1}$, we obtain the following two equations to find $t_{i+1}$ and $\kappa_i$:

$$(i) \quad q_i = (-1)^{i+1}\frac{\alpha_i - \beta_i}{\alpha_i + \beta_i}b,$$

$$(ii) \quad \kappa_i = \frac{q_i\alpha_i + (-1)^k b\alpha_i}{2\delta},$$

  or the local mean and the value of $\kappa_i$ obtained from the integral equation (19.14). Since $\alpha_i + \beta_i = \tau_{i+1} - \tau_i$ and $\alpha_i - \beta_i = 2t_{i+1} - (\tau_i + \tau_{i+1})$, we obtain from equation $(i)$

$$t_{i+1} = \frac{q_i(\tau_{i+1} - \tau_i) + (\tau_{i+1} + \tau_i)}{2(-1)^{i+1}b},$$

  that is used to obtain $\alpha_i$, which we can then use to find $\kappa_i$ in equation $(ii)$. This implies that for a multilevel signal, the ASDM is an adaptive LC sampler. For each pulse in the multilevel signal, we can find appropriate zero-crossing times to represent it exactly. In the above, the value of $\kappa_i$ adapts to the width of the pulse so that the values of $\alpha_i$ and $\beta_i$ are connected with the local average $q_i$ in the window set by $\kappa_i$.

For any signal $x(t)$, not necessarily multilevel, the second comment indicates that $x(t)$ can be approximated by a multilevel signal where the $q_i$ values are estimates of the local average for windows set by the parameter $\kappa$. The advantage of obtaining such a multilevel representation is that it can be converted into continuous-time binary signals, which can be processed in the continuous time [21,38,39]. Moreover, since the ASDM circuitry does not include a clock, it consumes low power and is suitable for low-voltage complementary metal-oxide-semiconductor (CMOS) technology [30]. These two characteristics of the ASDM make it very suitable for biomedical data acquisition systems [2,17] where low-power and analog signal processing are desirable.

One of the critical features of the ASDM that motivates further discussions is the modulation rate of its output binary signal, that is, how fast $z(t)$ crosses zero, which depends on the input signal as well as the parameters $\delta$ and $\kappa$. To see the effect of just $\kappa$ on the output, we let $x(t) = c$, $\delta = 0.5$, and $b > c$, giving

$$\alpha_k = t_{k+1} - t_k = \frac{\kappa}{c + (-1)^k b}, \quad \beta_k = \frac{\kappa}{c - (-1)^k b},$$

so that $z(t)$ is a sequence of pulses that repeat periodically. It thus can be seen that the lower the $\kappa$, the faster the $z(t)$ crosses zero and vice versa (see Figure 19.2). We thus consider $\kappa$ a scale parameter, which we will use to decompose signals. To simplify the analysis, the bias parameter of the ASDM is set to $b = 1$, which requires normalization of the input signal, so that $\max(x(t)) < 1$ and the threshold parameter can be set to $\delta = 1/2$.

## 19.3 ASDM Signal Decompositions

A nonstationary signal can be decomposed using the ASDM by latticing the time–frequency plane so that the frequency is divided into certain bands and the time into window lengths depending on the amplitude of the signal. A parameter critical in the decomposition is the scale parameter $\kappa$. If we let in (19.16) $\delta = 0.5$ and $b = c + \Delta$, for $\Delta > 0$, values of $\kappa$ are obtained as

$$\kappa \le \frac{b - c}{2f_{max}} = \frac{\Delta}{2f_{max}}, \tag{19.20}$$

so that for a chosen value of $\Delta$, we can obtain for different frequency bands with maximum frequency $f_{max}$ the corresponding scale parameters. These parameters in turn determine the width of the analysis time windows. Considering the local average $\zeta_k$, the best linear estimator of the signal in $[t_k, t_{k+2}]$ when no data are provided, the ASDM time-encoder can be thought of an optimal LC sampler.

A basic module for the proposed low-frequency decomposition is shown in Figure 19.3. The decomposer consists of a cascade of $L$ of these modules, each one consisting of a low-pass filter, an ASDM, an averager, a smoother, and an adder. The number of modules, $L$, is determined by the scale parameters used to decompose the input signal. For a scale parameter $\kappa_i$, the ASDM maps the input signal into a binary signal with sequences $\{\alpha_k\}$ and $\{\beta_k\}$, which the averager converts into local averages $\{\zeta_k\}$. The smoother is used to smooth out the multilevel signal output so that there is no discontinuity inserted by the adder when the multilevel signal is subtracted from the input signal of the corresponding module. Each of the modules operates similarly but at a different scale.

Starting with a scale factor $\kappa_1 = \Delta/2B_0$, where $B_0$ is the bandwidth of the low-pass filter of the first module, the other scales are obtained according to

$$\kappa_\ell = \frac{\kappa_1}{2^{\ell-1}} \qquad \ell = 2, \ldots, L,$$

for the $\ell$th module, by appropriately changing the bandwidth of the corresponding low-pass filter. The width of the analysis windows decreases as $\ell$ increases. For the first module, we let $f_0(t) = x(t)$, and the input to the modules beyond the first one can be written sequentially as follows:

$$\begin{aligned} f_1(t) &= x(t) - d_1(t) \\ f_2(t) &= f_1(t) - d_2(t) = x(t) - d_1(t) - d_2(t) \\ &\vdots \\ f_j(t) &= x(t) - d_1(t) - d_2(t) \cdots - d_j(t) \\ &\vdots \\ f_L(t) &= x(t) - \sum_{l=1}^{L} d_l(t), \end{aligned} \tag{19.21}$$

where the $\{d_\ell(t)\}$ are low-frequency components. We thus have the decomposition

$$x(t) = \sum_{\ell=1}^{L} d_\ell(t) + f_L(t). \tag{19.22}$$

The component $f_L(t)$ can be thought as the error of the decomposition. Considering $d_\ell(t)$ a multilevel signal, we thus have the decomposition

$$x(t) = \sum_{\ell=1}^{L} d_\ell(t) + f_L(t) \approx \sum_{\ell=1}^{L} \sum_k \zeta_{k,\ell}\, p_k(t) + f_L(t), \tag{19.23}$$

where as before $p_k(t) = u(t - t_k) - u(t - t_{k+2})$, and the averages $\{\zeta_{k,\ell}\}$ depend on the scale being used. This scale decomposition is analogous to a wavelet decomposition for analog signals.

Also similar to the wavelet analysis, a dual of the above decomposition is also possible; that is, nonstationary signals can be decomposed in terms of high-frequency components. A basic module for such a decomposer is shown in Figure 19.4. The input is

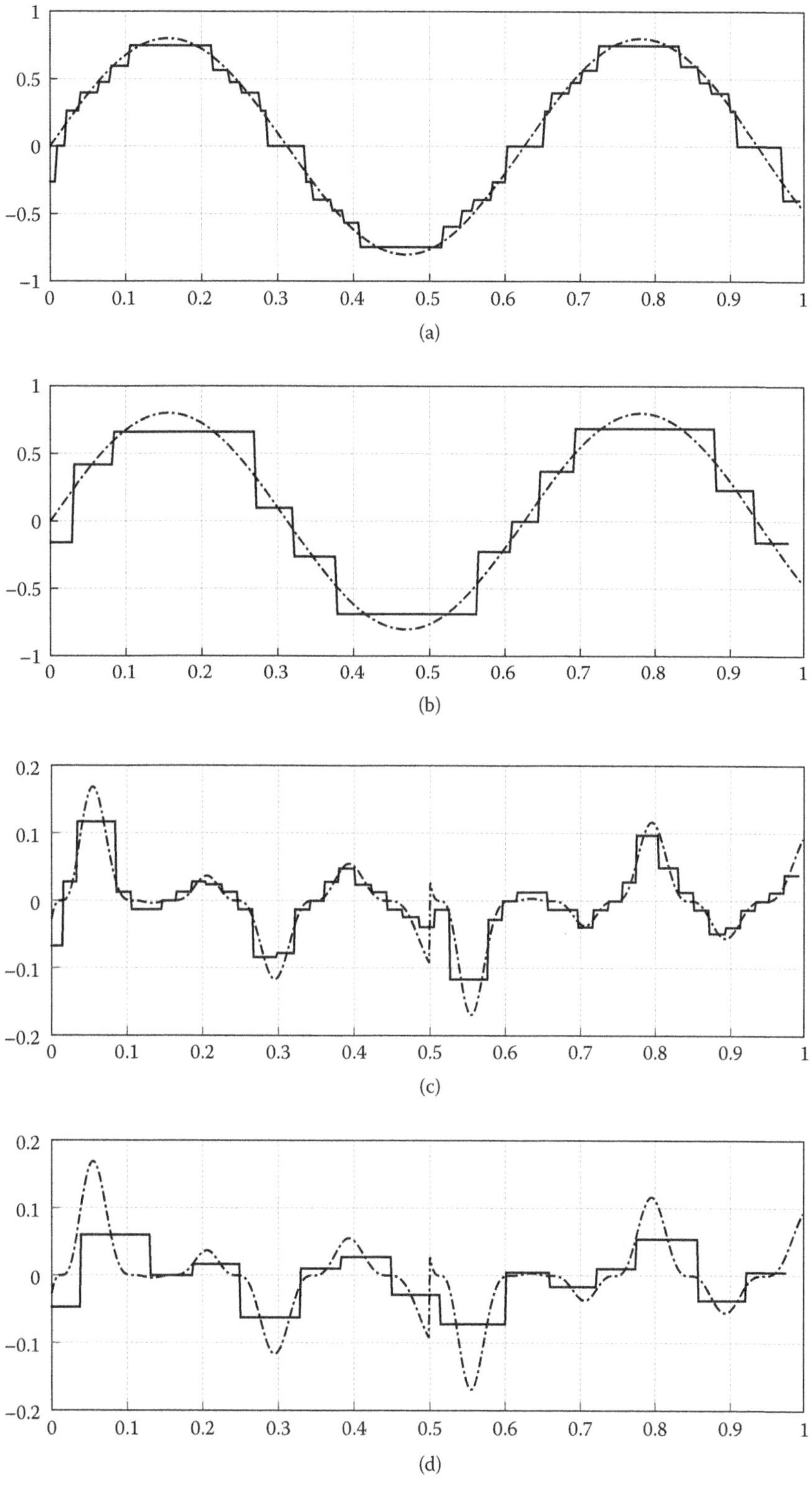


**FIGURE 19.2**
Local-average approximation of a sinusoidal signal with (a) $\kappa = 0.1$ and (b) $\kappa = 1$. Local-average approximation of a signal with (c) $\kappa = 1$ and (d) $\kappa = 2$.

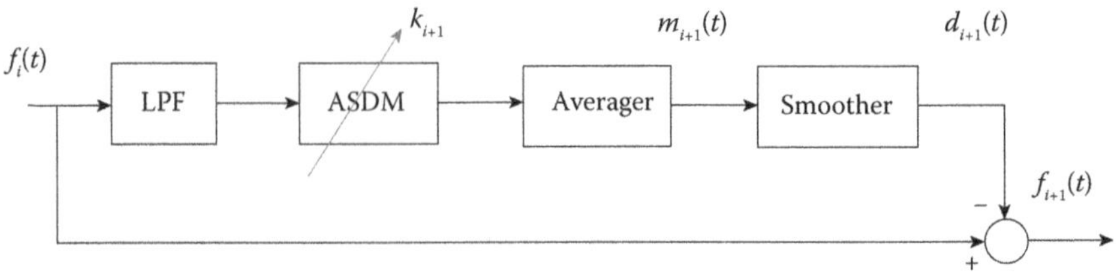


**FIGURE 19.3**
Basic module of the low-frequency decomposer.

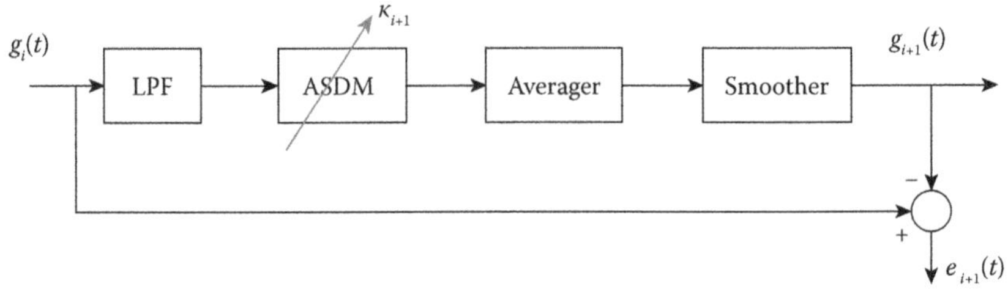


**FIGURE 19.4**
Basic module of the dual high-frequency decomposer.

now expressed in terms of high-frequency components $\{e_i(t)\}$. Letting $g_0(t) = x(t)$, we have that

$$\begin{aligned} g_1(t) &= x(t) - e_1(t) \\ g_2(t) &= x(t) - e_1(t) - e_2(t) \\ &\vdots \\ g_M(t) &= x(t) - \sum_{i=1}^{M} e_i(t). \end{aligned} \tag{19.24}$$

The decomposition in terms of high-frequency components is then given by

$$x(t) = \sum_{i=1}^{M} e_i(t) + g_M(t), \tag{19.25}$$

where $g_M(t)$ can be seen as the error of the decomposition.

In terms of signal compression, the decomposition of the input signal $x(t)$ can be interpreted in two equivalent ways:

1. Time-encoding: If the multilevel signals $\{m_\ell(t)\}$ (see Figure 19.3) were inputted into ASDMs, for each of these signals we would obtain binary signals $\{z_\ell(t)\}$ with crossing times $\{t_{\ell,k}\}$ that would permit us to reconstruct the multilevel signals. As before, these signals could then be low-pass filtered to obtain the $\{d_\ell(t)\}$ to approximate $x(t)$ as in (19.22). Thus, the array $\{t_{\ell,k}\}$, $1 \le \ell \le L$, $k \in K_\ell$, where $K_\ell$ corresponds to the number of pulses used in each decomposition module, would provide a representation of the components of the input signal. This interpretation uses time encoding [24].
2. Pulse-modulation: Each $\{z_\ell(t)\}$, the output of the ASDMs in the decomposition, provides random sequences $\{\alpha_{\ell,k}, \beta_{\ell,k}\}$ from which we can compute sequences of local averages over two-pulse widths and their length or $\{\zeta_{\ell,k}, T_{\ell,k}\}$ for $1 \le \ell \le L$, $k \in K_\ell$, where $K_\ell$ corresponds to the number of pulses used in each decomposition module. For a nondeterministic signal, the sequence $\{\zeta_{\ell,k}\}$ would be random, and as such their distributions would characterize $d_\ell(t)$ as well as the signal $x(t)$. Thus, $\{\zeta_{\ell,k}, T_{\ell,k}\}$ provides the same compression as the one provided by $\{\alpha_{\ell,k}, \beta_{\ell,k}\}$ and $\{t_{\ell,k}\}$. To obtain higher compression, we could consider the distribution of the $\{\zeta_{\ell,k}\}$ and ignore values clustered around one of the averages.

The local-average approximation obtained above for scale parameters $\{\kappa_k\}$ of the ASDM is similar to the Haar wavelet representation [36] but with the distinction that the time windows are signal dependent instead of being fixed. The pulses with duty cycle defined by the sequence $\{\alpha_k, \beta_k\}$ can be regarded as Haar wavelets with signal-dependent scaling and translation [4,7].

### 19.3.1 Simulations

To illustrate the decompositions, we consider phonocardiograph recordings of heart sounds [31,33]. The heart sounds have a duration of 1.024 s and were sampled

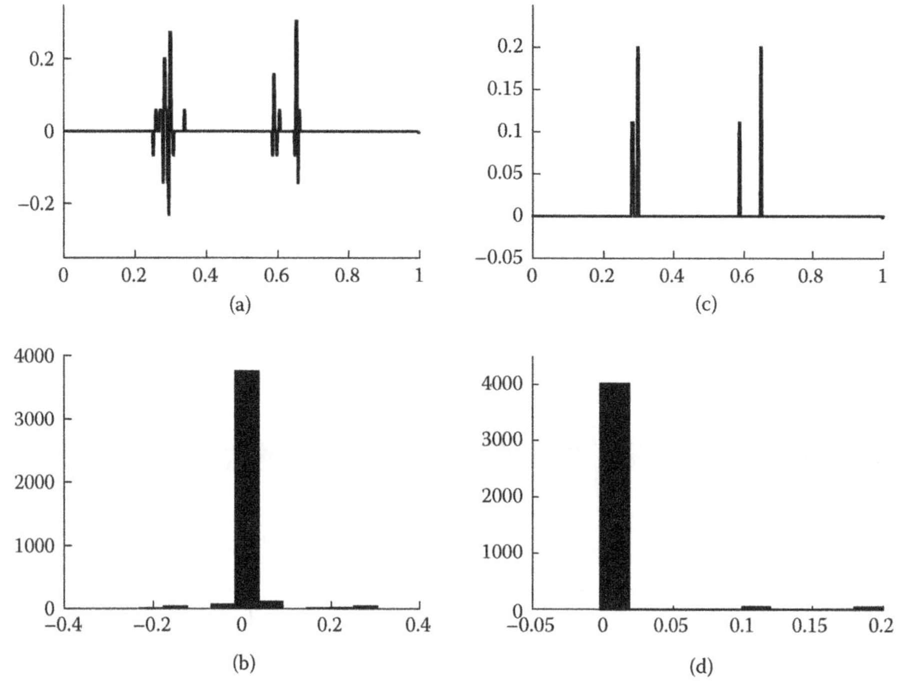


**FIGURE 19.5**
(a) Multilevel output signal from the first module and (b) its local-mean distribution; (c) multilevel output signal from the second module and (d) its local-mean distribution.

at a rate of 4000 samples per second. For the signal to approximate an analog signal, it was interpolated with a factor of 8. Two modules were needed to decompose the test signal. Figure 19.5 displays the local-mean distribution and multilevel signal obtained by the two-module decomposition. The multilevel signals obtained in the consecutive modules enable significant compression in information (mainly in the spikes and in the neural firing signals). The reconstruction of the signal from these local averages are shown in Figure 19.6b, where a root mean square error of $2.14 \times 10^{-6}$ and a compression ratio of 24.8% are obtained.

To illustrate the robustness to additive noise of the algorithms, we added a zero-mean noise to the original signal (see Figure 19.6c) giving a noisy signal with a signal-to-noise ratio (SNR) of −2.48 dB, and attempted its reconstruction. The reconstructed signal is shown in Figure 19.6d, having an SNR of 27.54 dB. We obtained the following results using performance metrics typically used in biomedical applications:

- The cross-correlation value evaluating the similarity between the reconstructed and original signal:

$$\gamma = \frac{\sum_{n=1}^{N}(x(n) - \mu_x)(\hat{x}(n) - \mu_{\hat{x}})}{\sqrt{\sum_{n=1}^{N}(x(n) - \mu_x)^2}\sqrt{\sum_{n=1}^{N}(\hat{x}(n) - \mu_{\hat{x}})^2}} \times 100$$
$$= 99\%,$$

where $N$ is the length of the signal, $x$ is the original (uncontaminated by noise) signal, and $\hat{x}$ is the reconstructed signal.

- The root mean square error:

$$\epsilon = \sqrt{\frac{1}{N}\sum_{n=1}^{N}(x(n) - \hat{x}(n))^2} = 0.0012.$$

- The maximum error:

$$\psi = \max[x(n) - \hat{x}(n)] = 0.0014.$$

## 19.4 Modified ASDM

Although the signal dependency of the ASDM provides an efficient representation of an analog signal by means of the zero-crossing times, the asynchronous structure complicates the signal reconstruction by requiring a solution to the integral equation (19.14). By extracting the integrator in the feedforward loop of the ASDM, we obtain a different configuration which we call the modified asynchronous sigma delta modulator (MASDM) (Figure 19.7). The MASDM provides a better solution for encoding and processing as we will see, while keeping the signal dependency and the low power [3]. To keep the input of the reduced model $x(t)$ now

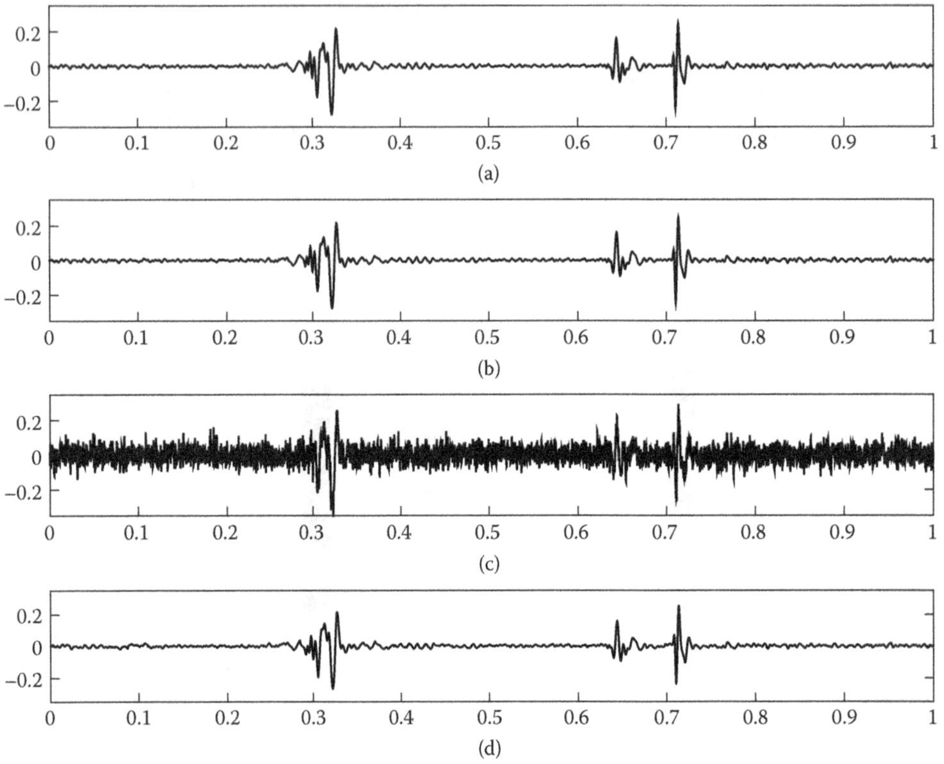

**FIGURE 19.6**
(a) Phonocardiogram recording and (b) reconstructed signal; (c) phonocardiogram recording with additive noise and (d) reconstructed signal.

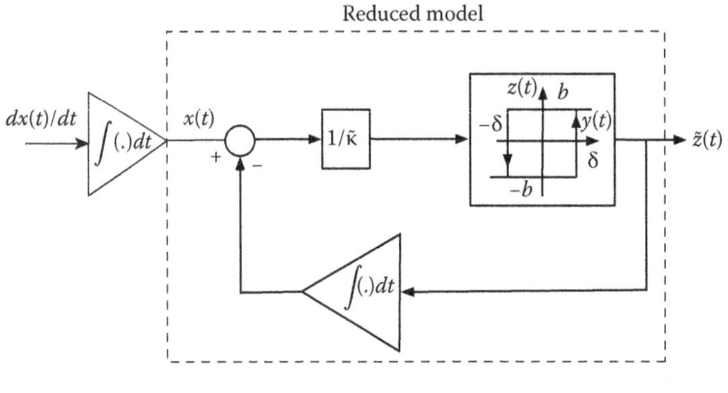


**FIGURE 19.7**
Block diagram of the modified asynchronous sigma delta modulator (MASDM).

requires that the input of the MASDM be its derivative $dx(t)/dt$.

It is very important to notice that the MASDM is equivalent to an ASDM with the input being the derivative of the signal, $dx(t)/dt$. The new configuration provides a different approach to dealing with the integral equation (19.14). Letting $b = 1$ (i.e., the input is normalized) and $\delta = 0.5$, Equation 19.14 with input $dx(t)/dt$ now gives

$$\int_{\tilde{t}_k}^{\tilde{t}_{k+1}} dx(\tau) = x(\tilde{t}_{k+1}) - x(\tilde{t}_k) = (-1)^k[-(\tilde{t}_{k+1} - \tilde{t}_k) + \tilde{\kappa}], \quad (19.26)$$

or the recursive equation

$$x(\tilde{t}_{k+1}) = x(\tilde{t}_k) + (-1)^k[-(\tilde{t}_{k+1} - \tilde{t}_k) + \tilde{\kappa}], \quad (19.27)$$

that will permit us to obtain the samples using an initial value of the signal and the zero-crossing times. The tilde notation for the zero-crossing times and the ASDM scale parameter is used to emphasize that when the input is $dx(t)/dt$ instead of $x(t)$, these values could be different.

A possible module for a low-frequency decomposer is shown in Figure 19.8. The top branch provides an estimate of the dc value of the signal by means of a

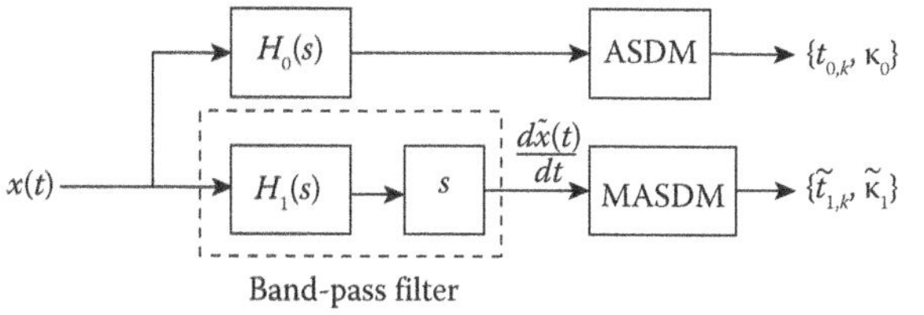


**FIGURE 19.8**
Low-frequency analysis module using the ASDM and MASDM. The filter $H_0(s)$ is a narrowband low-pass, and $H_1(s)$ is a wide-band low-pass filter.

narrowband low-pass filter $H_0(s)$ and an ASDM while the lower branch uses a wider bandwidth low-pass filter $H_1(s)$ cascaded with a derivative operator. The input to the MASDM is the derivative of the output of $H_1(s)$. This module displays two interesting features that simplify the analysis by letting the input be the derivative of the low-passed signal:

- For the branch using the MASDM, the integral equation is reduced to a first-order difference equation with input a function of zero-crossing times $\{\tilde{t}_{1,k}\}$ and a scale parameter $\tilde{\kappa}_1$. Thus, one only needs to keep the zero-crossing times and the scale parameter to recursively obtain samples of the output signal of the low-pass filter $H_1(s)$ at nonuniform times rather than solving the integral equation (19.14). The upper branch, using the ASDM, would require an averager (as in the decomposers given before) to obtain an estimate of the mean component $\bar{x}(t)$. The proposed low-frequency decomposer can thus be thought of as a nonuniform sampler that allows the recovery of the sample values recursively using Equation 19.27 and an averager for the mean component.

- The low-pass filter cascaded with the derivative operator is equivalent to a band-pass filter with a zero at zero. The proposed configuration has some overlap in the low frequencies but points to a bank-of-filters structure which we develop next.

When implementing the MASDM, the value of $\tilde{\kappa}$ depends on the maximum of the derivative:

$$c_d = \max\left|\frac{dx(t)}{dt}\right|.$$

Although this value would not be available, $c_d$ can be associated with the maximum amplitude $c$ of the input $x(t)$ under the assumption that $x(t)$ is continuous:

$$\left|\frac{dx(t)}{dt}\right| \leq \lim_{T\to 0}\frac{|x(t_i+T)|+|x(t_i)|}{T} = \lim_{T\to 0}\frac{2c}{T}.$$

Letting $T \leq 1/(2f_{max})$, where $f_{max}$ is the maximum frequency of $x(t)$, we have

$$c_d \leq \frac{2c}{T}. \quad (19.28)$$

The lack of knowledge of $f_{max}$ does not permit us to determine a value for the scale parameter $\tilde{\kappa}$ directly. However, if the bandwidth of the low-pass filter $H_1(s)$ is $B$ [rad/s], the scale parameter should satisfy

$$\tilde{\kappa} \leq \frac{\pi(1-c_d)}{B}. \quad (19.29)$$

To illustrate the nonuniform sampling using the MASDM, consider the sparse signal shown in Figure 19.9. The location of the nonuniform samples acquired from the signal is shown on the right of the figure. Notice that, as desired, more samples are taken in the sections of the signal where information is available, and fewer or none when there is no information.

### 19.4.1 ASDM/MASDM Bank of Filters

Asynchronous analysis using filter banks exhibit similar characteristics as wavelets. Indeed, compression and denoising of nonstationary signals are possible by configuring ASDMs and MASDMs into filter banks [36].

To extend the scope of asynchronous processing provided by the previous low- and high-frequency decomposers, we now consider the analysis and synthesis of nonstationary signals using a bank of filters and the ASDM and the MASDM in the analysis. In the synthesis component, we use an averager and a PSWF interpolator. The bank-of-filters structure shown in Figure 19.10 extends the low-frequency decomposer shown in Figure 19.8. We now have a set of low-pass $G_0(s)$ and band-pass filters $\{sG_i(s),\ i = 1,\ldots,M\}$ that provide bandlimited signals for processing with ASDMs to estimate the dc bias, and for MASDMs to obtain the components at other frequencies. The upper frequency of each of these filters is used to determine the corresponding scale ASDMs and MASDMs.

For each of the frequency bands in the bank of filters, the ASDMs and the MASDMs provide the zero-crossing times $\{t_{i,k}\}$ and the corresponding scale parameters $\{\kappa_i\}$ needed to recursively obtain the nonuniform samples in the synthesis component.* As indicated before, the parameters of the MASDM are set according to the derivative of the input, which can be done by using Equations 19.28 and 19.29. An averager in the synthesis part is needed to reconstruct the dc bias signal $\bar{x}_0(t)$, and the recursive equation (19.27) together with an interpolator to obtain the $\hat{x}_i(t)$. Finally, it is important to recognize

*For simplicity in Figure 19.10, we do not use the tilde notation for the zero-crossing times or the scale parameters.

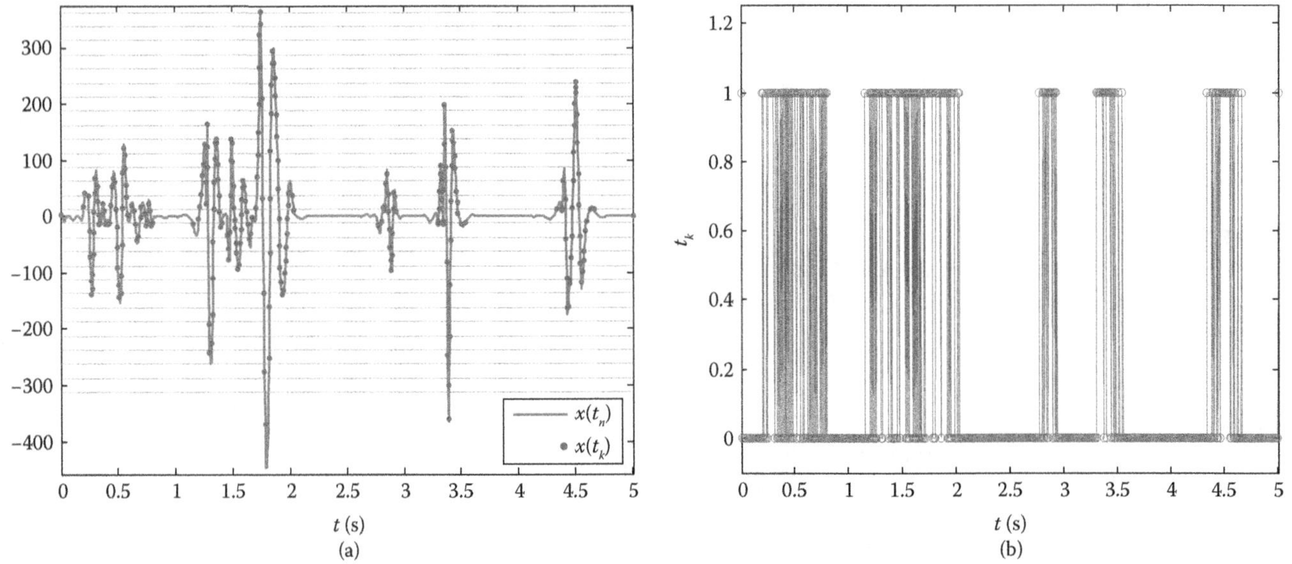


**FIGURE 19.9**
Left: nonuniform sampling using the MASDM; right: sample locations.

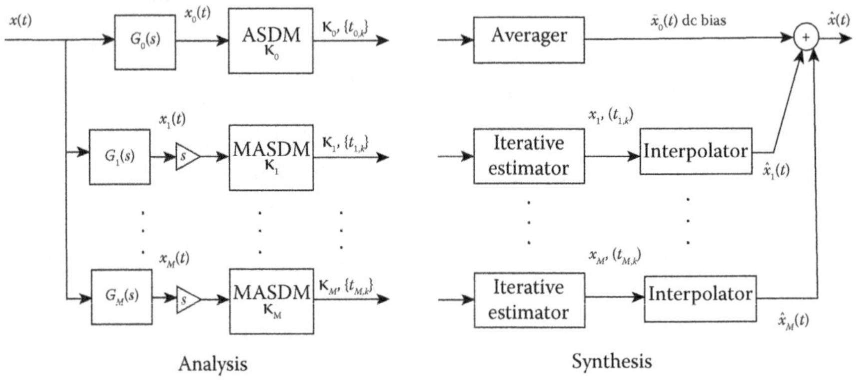


**FIGURE 19.10**
ASDM/MASDM bank of filters: analysis and synthesis components.

that one of the zeros at zero of the band-pass filters $\{sG_i(s)\}$ are used to provide the derivative input to the MASDMs.

As a side remark, the dependence of the recursion (19.27) on $\tilde{\kappa}$ discussed before can be eliminated by considering evaluating Equation 19.26 in two consecutive time intervals to get

$$\int_{\tilde{t}_k}^{\tilde{t}_{k+2}} dx(\tau) = x(\tilde{t}_{k+2}) - x(\tilde{t}_k) = t_{k+2} - 2t_{k+1} + t_k,$$

so that the new recursion is given by

$$x(\tilde{t}_{k+2}) = x(\tilde{t}_k) + \tilde{t}_{k+2} - 2\tilde{t}_{k+1} + \tilde{t}_k. \qquad (19.30)$$

This not only eliminates the value of $\tilde{\kappa}$ in the calculations but also reduces the number of sample values $\{x(\tilde{t}_k)\}$.

To recover the nonuniform samples, we need to transmit the zero crossing for each of the branches of the bank of filters (and possibly the scale parameters if they are not known *a priori* or if do not use the above remark). Lazar and Toth [25] have obtained bounds for the reconstruction error for a given number of bits used to quantize the $\{t_k\}$. They show that for a bandlimited signal, quantizing the zero-crossing times is equivalent to quantizing the signal amplitude in uniform sampling. Thus, the time elapsed between two consecutive samples is quantized according to a timer with a period of $T_c$ seconds. Theoretically, the quantization error can be minimized as much as needed by simply reducing $T_c$, or increasing the number of quantization time levels.

The filter-bank decomposer can be considered a nonuniform analog-to-digital converter, where it is the zero-crossing times that are digitized. Reconstruction of the original analog signal requires a different approach to the uniform sampling. As indicated before, the sinc basis in Shannon's reconstruction can be replaced by the PSWFs [35] having better time and frequency localization than the sinc function [40]. The procedure is similar to the one used for signal reconstruction from samples obtained by the LC sampler in Section 19.2.1.

Consider, for instance, the reconstruction of the sparse signal shown in Figure 19.9 using the samples $\{\tilde{t}_k\}$ taken by the decomposer. The sample values $\{x(\tilde{t}_k)\}$ are obtained iteratively using Equation 19.27. As in Section 19.2.1, these samples can be represented using a projection of finite dimension with PSWFs as

$$\mathbf{x}(\tilde{\mathbf{t}}_\mathbf{k}) = \mathbf{\Phi}(\tilde{\mathbf{t}}_\mathbf{k})\boldsymbol{\alpha}_\mathbf{M}, \qquad (19.31)$$

where $\boldsymbol{\alpha}_\mathbf{M}$ is the vector of the expansion coefficients associated with the PSWFs. In some cases, baseband PSWFs do not work well if the signal has band-pass characteristics. Parsimonious representations are possible using modulated PSWFs, which are obtained by multiplying the PSWFs with complex exponentials; that is, the entries of the $\mathbf{\Phi}$ matrix are then

$$\psi_{n,\tau}(t) = e^{j\omega_m t}\phi_{n,\tau}(t),$$

where $\{\phi_{n,\tau}(t)\}$ are baseband PSWFs. Modulated PSWFs keep the time–frequency concentration of the baseband PSWFs while being confined to a certain frequency interval. Appropriate values for the expansion coefficients are obtained again using Tikhonov's regularization.

Figure 19.11 compares the reconstruction errors when using PSWFs and modulated PSWFs. For comparison

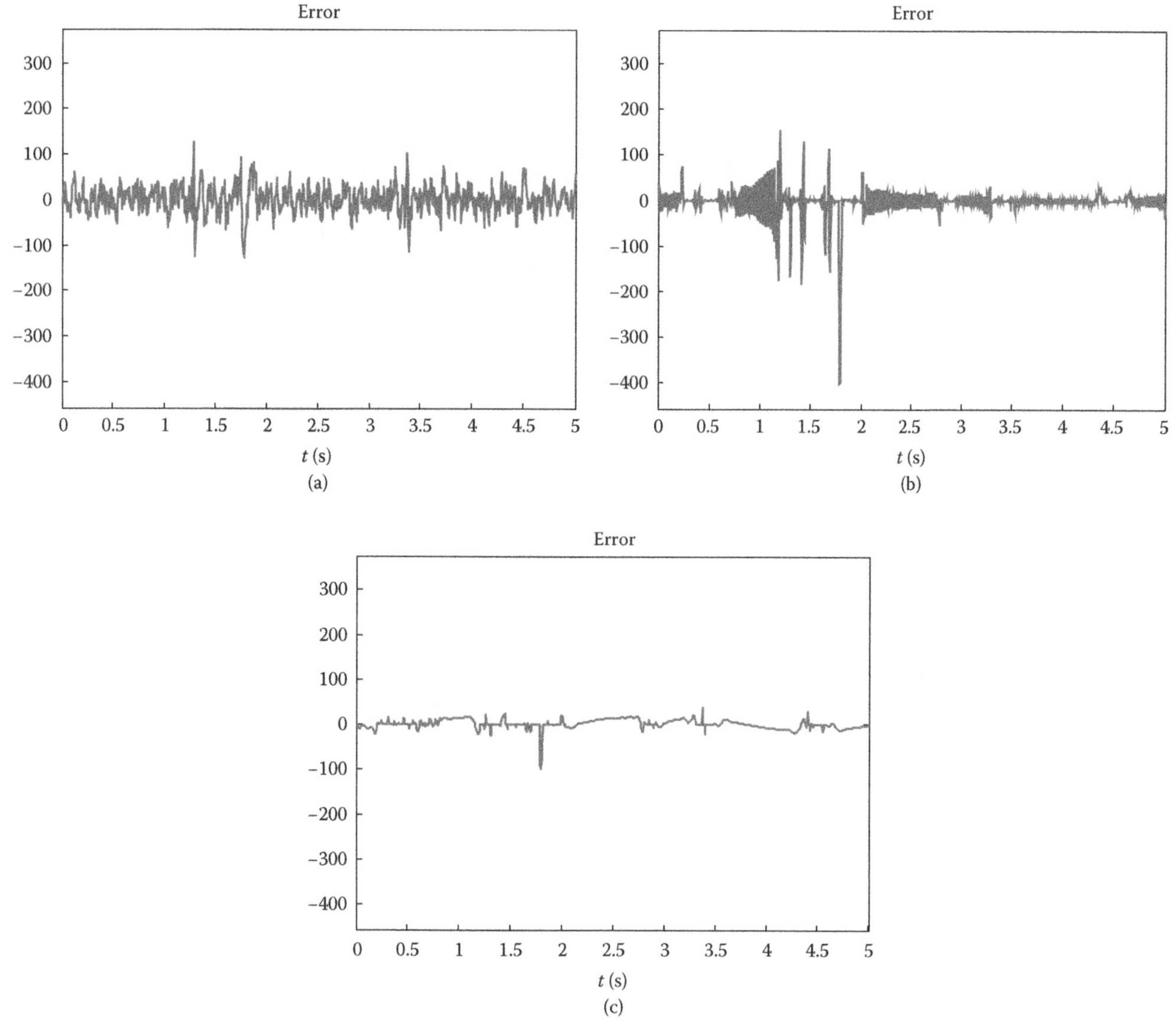


**FIGURE 19.11**
Reconstruction errors when 25% of the samples are used: (a) discrete cosine transform (DCT)-based compressive sensing (signal-to-noise ratio [SNR] = 7.45 dB); (b) baseband prolate spheroidal wave (PSW) functions (SNR = 7.11 dB); and (c) modulated PSW functions (SNR = 16.54 dB).

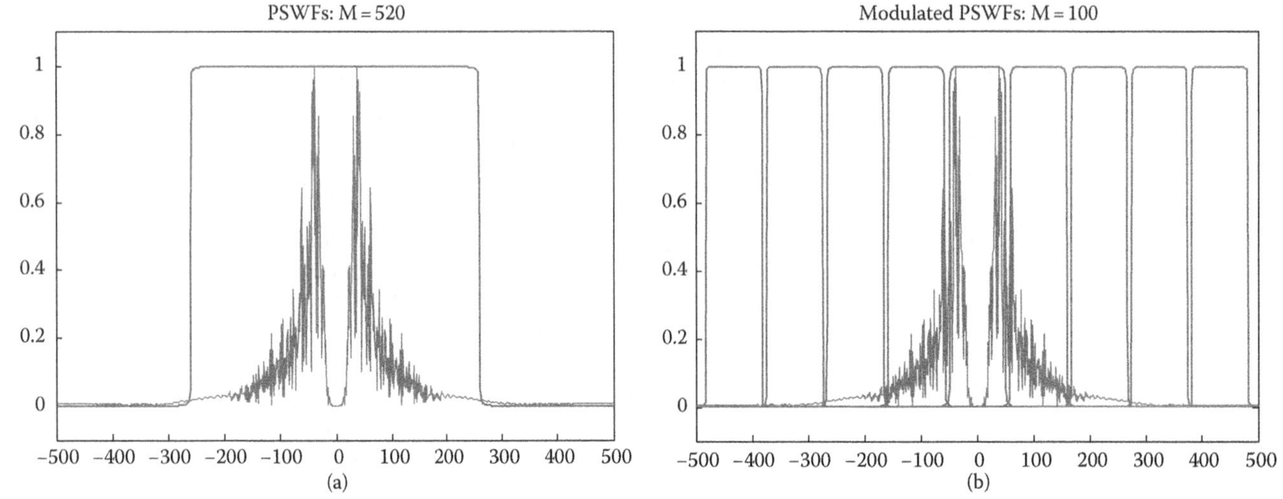


**FIGURE 19.12**
Signal spectra of the signal and of the prolate spheroidal wave (PSW) functions when using (a) 520 baseband functions and (b) 100 modulated functions.

reasons, we also include compressive sensing method using discrete cosine transform (DCT) [12]. If the PSWFs and the signal occupy the same band, the PSWFs provide an accurate and parsimonious representation. Otherwise, the representation is not very accurate as illustrated in Figure 19.11b. To improve this performance, we capture the spectrum of the signal with narrowband-modulated PSWFs (see Figure 19.12b), and the resulting recovery error is depicted in Figure 19.11c. The reconstruction using modulated PSWFs results in much improved error performance.

In addition to improved performance, the number of functions used in the reconstruction is significantly reduced when adapting their frequency support to that of the signal. As indicated in Figure 19.12b, only 100 modulated PSWFs are needed for the test signal, rather than 520 baseband PSWFs (Figure 19.12a). Hence, we conclude that the reconstruction using modulated PSWFs is better suited to reconstruct the MASDM samples.

## 19.5 Conclusions

In this chapter, we considered the asynchronous processing of nonstationary signals that typically occur in practical applications. As indicated by the Wold–Cramer spectrum of these signals, their power spectra depend jointly on time and frequency. It is thus that synchronous sampling is wasteful by using a unique sampling period, which in fact should be a function of time. As shown, the sampling of nonstationary signals needs to be signal dependent, or nonuniform. The LC sampler, commonly used, requires an *a priori* set of quantization levels and results in a set of samples and the corresponding sampling times that at best can give a multilevel approximation to the original signal. Sampling and reconstruction of nonstationary signals is then shown to be more efficiently done by using an ASDM, which provides a binary representation of a bandlimited and bounded analog signal by means of zero-crossing times that are related to the signal amplitude. Deriving a local-mean representation from the binary representation is possible to obtain an optimal LC representation using the ASDM. Moreover, studying the effect of the scale parameter used in the ASDM, and considering a latticing of the time–frequency space so that the frequency bands are determined and the time windows are set by the signal, we derive a low-frequency decomposer and its dual. The low-frequency decomposer can be seen as a modified Haar wavelet decomposition. By modifying the structure of the ASDM, we obtain a more efficient bank-of-filters structure. The interpolation of the nonuniform samples in the synthesis component is efficiently done by means of the PSW or Slepian functions, which are more compact in both time and frequency than the sinc functions used in the Shannon's interpolation. Measurement equations—similar to those obtained in compressive sensing—are typically ill-conditioned. The solution is alleviated by using Tikhonov's regularization. Once the analysis and synthesis is understood, the bank-of-filters approach is capable of providing efficient processing of nonstationary signals, comparable to the one given by wavelet packages. The proposed techniques should be useful in processing signals resulting from biomedical or control systems.

## Bibliography


[1] F. Aeschlimann, E. Allier, L. Fesquet, and M. Renaudin. Asynchronous FIR filters: Towards a new digital processing chain. In *10th International Symposium on Asynchronous Circuits and Systems, 2004*, Crete, Greece, pp. 198–206, April 19–23, 2004.

[2] E. V. Aksenov, Y. M. Ljashenko, A. V. Plotnikov, D. A. Prilutskiy, S. V. Selishchev, and E. V. Vetvetskiy. Biomedical data acquisition systems based on sigma-delta analogue-to-digital converters. In *IEEE Signal Processing in Medicine and Biology Symposium*, Istanbul, Turkey, volume 4, pp. 3336–3337, October 25–28, 2001.

[3] A. Can and L. Chaparro. Asynchronous sampling and reconstruction of analog sparse signals. In *20th European Signal Processing Conference*, Bucharest Romania, pages 854–858, August 27–31, 2012.

[4] A. Can, E. Sejdić, O. Alkishriwo, and L. Chaparro. Compressive asynchronous decomposition of heart sounds. In *IEEE Statistical Signal Processing Workshop (SSP)*, pp. 736–739, August 2012.

[5] A. Can, E. Sejdić, and L. F. Chaparro. An asynchronous scale decomposition for biomedical signals. In *IEEE Signal Processing in Medicine and Biology Symposium*, New York, NY, pp. 1–6, December 10, 2011.

[6] A. Can-Cimino, E. Sejdić, and L. F. Chaparro. Asynchronous processing of sparse signals. *IET Signal Processing*, 8(3):257–266, 2014.

[7] L. F. Chaparro, E. Sejdić, A. Can, O. A. Alkishriwo, S. Senay, and A. Akan. Asynchronous representation and processing of nonstationary signals: A time-frequency framework. *IEEE Signal Processing Magazine*, 30(6):42–52, 2013.

[8] L. Cohen. *Time–frequency Analysis: Theory and Applications*. Prentice Hall, NJ, 1995.

[9] S. Şenay, L. Chaparro, and L. Durak. Reconstruction of nonuniformly sampled time-limited signals using Prolate Spheroidal Wave functions. *Signal Processing*, 89:2585–2595, 2009.

[10] S. Şenay, L. F. Chaparro, M. Sun, and R. Sclabassi. Adaptive level–crossing sampling and reconstruction. In *European Signal Processing Conference (EUSIPCO)*, Aalborg, Denmark, pp. 1296–1300, August 23–27, 2010.

[11] S. Şenay, J. Oh, and L. F. Chaparro. Regularized signal reconstruction for level-crossing sampling using Slepian functions. *Signal Processing*, 92:1157–1165, 2012.

[12] D. L. Donoho. Compressed sensing. *IEEE Transactions on Information Therapy*, 52:1289–1306, 2006.

[13] K. Guan, S. Kozat, and A. Singer. Adaptive reference levels on a level–crossing analog-to-digital converter. *EURASIP Journal on Advances in Signal Processing*, 2008:1–11, January 2008.

[14] K. Guan and A. Singer. A level-crossing sampling scheme for non-bandlimited signals. *IEEE International Conference on Acoustics, Speech, and Signal Processing*, 3:381–383, 2006.

[15] P. Hansen and D. P. Oleary. The use of L–curve in the regularization of discrete ill–posed problems. *SIAM Journal on Scientific Computing*, 14:1487–1503, 1993.

[16] T. Hawkes and P. Simonpieri. Signal coding using asynchronous delta modulation. *IEEE Transactions on Communications*, 22(5):729–731, 1974.

[17] C. Kaldy, A. Lazar, E. Simonyi, and L. Toth. *Time Encoded Communications for Human Area Network Biomonitoring*. Technical Report 2-07, Department of Electrical Engineering, Columbia University, New York, 2007.

[18] A. S. Kayhan, A. El-Jaroudi, and L. F. Chaparro. Evolutionary periodogram for nonstationary signals. *IEEE Transactions on Signal Processing*, 42:1527–1536, 1994.

[19] A. S. Kayhan, A. El-Jaroudi, and L. F. Chaparro. Data–adaptive evolutionary spectral estimation. *IEEE Transactions on Signal Processing*, 43:204–213, 1995.

[20] D. Kinniment, A. Yakovlev, and B. Gao. Synchronous and asynchronous a-d conversion. *IEEE Transactions on VLSI Systems*, 8(2):217–220, 2000.

[21] M. Kurchuk and Y. Tsividis. Signal-dependent variable-resolution clockless A/D conversion with application to continuous-time digital signal processing. *IEEE Transactions on Circuits and Systems—I: Regular Papers*, 57(5):982–991, 2010.

[22] A. Lazar, E. Simonyi, and L. Toth. Time encoding of bandlimited signals, an overview. In *Proceedings of the Conference on Telecommunication Systems, Modeling and Analysis*, Dallas, TX, November 17–20, 2005.

[23] A. Lazar, E. Simonyi, and L. Toth. An overcomplete stitching algorithm for time decoding machines. *IEEE Transactions on Circuits and Systems*, 55:2619–2630, 2008.

[24] A. Lazar and L. Toth. Time encoding and perfect recovery of band-limited signals. In *IEEE International Conference on Acoustics, Speech, and Signal Processing*, volume 6, pp. 709–712, April 2003.

[25] A. Lazar and L. Toth. Perfect recovery and sensitivity analysis of time encoded bandlimited signals. *IEEE Transactions on Circuits and Systems*, 51:2060–2073, October 2004.

[26] F. Marvasti. *Nonuniform Sampling: Theory and Practice*. Kluwer Academic/Plenum Publishers, New York, 2001.

[27] J. Oh, S. Şenay, and L. F. Chaparro. Signal reconstruction from nonuniformly spaced samples using evolutionary Slepian transform-based POCS. *EURASIP Journal on Advances in Signal Processing*, February 2010.

[28] M. B. Priestley. *Spectral Analysis and Time Series*. Academic Press, London, 1981.

[29] M. Renaudin. Asynchronous circuits and systems: A promising design alternatives. *Microelectronic Engineering*, 54:133–149, 2000.

[30] E. Roza. Analog-to-digital conversion via duty cycle modulation. *IEEE Transactions on Circuits and Systems II: Analog to Digital Processing*, 44(11):907–914, 1997.

[31] E. Sejdić, A. Can, L. F. Chaparro, C. Steele, and T. Chau. Compressive sampling of swallowing accelerometry signals using time-frequency dictionaries based on modulated discrete prolate spheroidal sequences. *EURASIP Journal on Advances in Signal Processing*, 2012:101, 2012.

[32] E. Sejdić, I. Djurović, and J. Jiang. Time-frequency feature representation using energy concentration: An overview of recent advances. *Digital Signal Processing*, 19:153–183, 2009.

[33] E. Sejdić and J. Jiang. Selective regional correlation for pattern recognition. *IEEE Transactions on Systems, Man and Cybernetics*, 37(1):82–93, 2007.

[34] E. Sejdić, M. Luccini, S. Primak, K. Baddour, and T. Willink. Channel estimation using DPSS based frames. In *IEEE International Conference on Acoustics, Speech and Signal Processing*, pp. 2849–2952, March 2008.

[35] D. Slepian and H. Pollak. Prolate spheroidal wave functions, Fourier analysis and uncertainty. *Bell System Technical Journal*, 40:43–64, 1961.

[36] G. Strang and T. Nguyen. *Wavelets and Filter Banks*. Wellesley-Cambridge Press, Wellesley, MA, 1997.

[37] A. N. Tikhonov. Solution of incorrectly formulated problems and the regularization method. *Soviet Mathematics—Doklady*, 4:1035–1038, 1963.

[38] Y. Tsividis. Digital signal processing in continuous time: A possibility for avoiding aliasing and reducing quantization error. In *IEEE International Conference on Acoustics, Speech and Signal Processing*, volume 2, pp. 589–92, May 2004.

[39] Y. Tsividis. Mixed-domain systems and signal processing based on input decomposition. *IEEE Transactions on Circuits and Systems*, 53:2145–2156, 2006.

[40] G. Walter and X. Shen. Sampling with prolate spheroidal wave functions. *Sampling Theory in Signal and Image Processing*, 2:25–52, 2003.

# 20

# *Event-Based Statistical Signal Processing*

**Yasin Yılmaz**
*University of Michigan*
*Ann Arbor, MI, USA*

**George V. Moustakides**
*University of Patras*
*Rio, Greece*

**Xiaodong Wang**
*Columbia University*
*New York, NY, USA*

**Alfred O. Hero**
*University of Michigan*
*Ann Arbor, MI, USA*

**CONTENTS**

**ABSTRACT** In traditional time-based sampling, the sampling mechanism is triggered by predetermined sampling times, which are mostly uniformly spaced (i.e., periodic). Alternatively, in event-based sampling, some predefined events on the signal to be sampled trigger the sampling mechanism; that is, sampling times are determined by the signal and the event space. Such an alternative mechanism, setting the sampling times free, can enable simple (e.g., binary) representations in the event space. In real-time applications, the induced sampling times can be easily traced and reported with high accuracy, whereas the amplitude of a time-triggered sample needs high data rates for high accuracy.

In this chapter, for some statistical signal processing problems, namely detection (i.e., binary hypothesis testing) and parameter estimation, in resource-constrained distributed systems (e.g., wireless sensor networks), we show how to make use of the time dimension for data/information fusion, which is not possible through the traditional fixed-time sampling.

## 20.1 Introduction

Event-based paradigm is an alternative to conventional time-driven systems in control [2,13,28] and signal processing [37,43,61]. Event-based methods are *adaptive* to the observed entities, as opposed to the time-driven techniques. In signal processing, they are used for data compression [37], analog-to-digital (A/D) conversion [23,30,61], data transmission [42,43,55], imaging applications [11,26,29], detection [17,24,73], and estimation [16,75]. We also see a natural example in biological sensing systems. In many multicellular organisms, including plants, insects, reptiles, and mammals, the all-or-none principle, according to which neurons fire, that is, transmit electrical signals, is an event-based technique [19].

In signal processing applications, event-based paradigm is mainly used as a means of *nonuniform sampling*. In conventional uniform sampling, the sampling frequency is, in general, selected based on the highest expected spectral frequency. When the lower frequency content in the input signal is dominant (e.g., long periods of small change), such high-frequency sampling wastes considerable power. For many emerging applications that rely on scarce energy resources (e.g., wireless sensor networks), a promising alternative is event-based sampling, in which a sample is taken when a significant event occurs in the signal. Several closely related signal-dependent sampling techniques have been proposed, for example, level-crossing sampling [24], Lebesgue sampling [17], send-on-delta [42], time-encoding machine [29], and level-triggered sampling [73]. In these event-based sampling methods, samples are taken based on the signal amplitude instead of time, as opposed to the conventional uniform sampling. Analogous to the comparison between the Riemann and Lebesgue integrals, the amplitude-driven and conventional time-driven sampling techniques are also called Lebesgue sampling and Riemann sampling, respectively [2]. As a result, the signal is encoded in the sampling times, whereas in uniform sampling the sample amplitudes encode the signal. This yields a significant advantage in real-time applications, in which sampling times can be tracked via simple one-bit signaling. Specifically, event-based sampling, through one-bit representations of the samples, enables high-resolution recovery, which requires many bits per sample in uniform sampling. In other words, event-based sampling can save energy and bandwidth (if samples are transmitted to a receiver) in real-time applications in terms of encoding samples.

### 20.1.1 Event-Based Sampling

In level-crossing sampling, which is mostly used for A/D conversion [23,50,61], in general, uniform sampling levels in the amplitude domain are used, as shown in Figure 20.1. A/D converters based on level-crossing sampling are free of a sampling clock, which is a primary energy consumer in traditional A/D converters [61]. A version of level-crossing sampling that ignores successive crossings of the same level is used to reduce the sampling rate, especially for noisy signals [28]. This technique is called level-crossing sampling with hysteresis (LCSH) due to the hysteretic quantizer it leads to (see Figure 20.1).

Time-encoding machine is a broad event-based sampling concept, in which the signal is compared with a reference signal and sampled at the crossings [21,29]. The reference signal is possibly updated at the sampling instants (Figure 20.2). Motivated by the integrate-and-fire neuron model, a mathematical model for nerve cells, in some time-encoding machines, the signal is first integrated and then sampled. The asynchronous delta–sigma modulator, a nonlinear modulation scheme mainly used for A/D conversion, is an instance of integrate-and-fire time-encoding machine [30]. The ON–OFF time-encoding machine, which models the

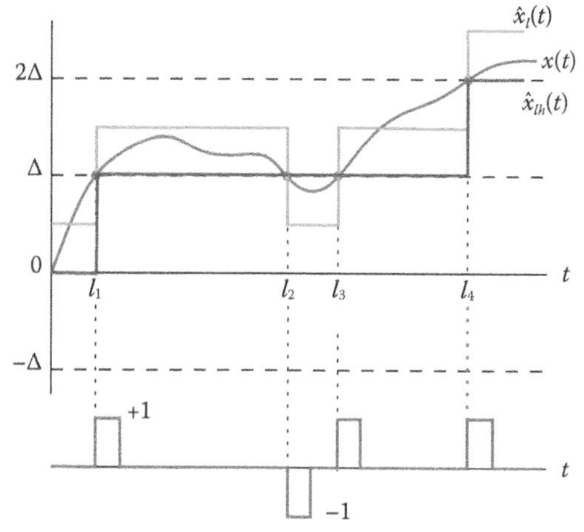


**FIGURE 20.1**
Level-crossing sampling with uniform sampling levels results in nonuniform sampling times $l_{1-4}$ and the quantized signal $\hat{x}_l(t)$. If the repeated crossings at $l_2$ and $l_3$ are discarded, $\hat{x}_{lh}(t)$ is produced by a hysteretic quantizer. One-bit encoding of the samples is shown below.

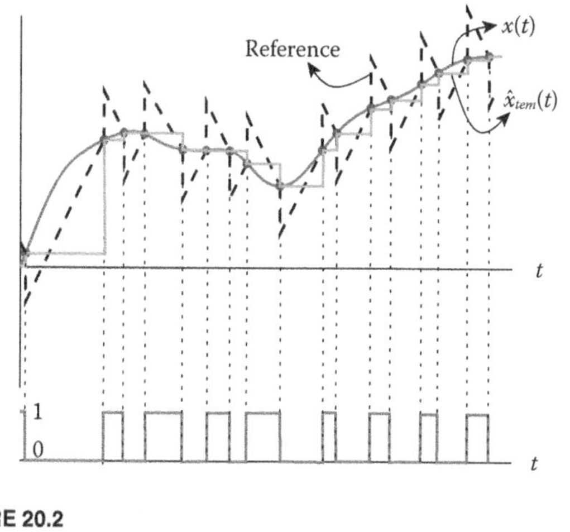


**FIGURE 20.2**
Reference signal representation of a time-encoding machine and the piecewise constant signal $\hat{x}_{tem}(t)$ resulting from the samples. At the sampling times, the reference signal switches the pair of offset and slope between $(-\delta, b)$ and $(\delta, -b)$, as in [30]. One-bit encoding of the samples is also shown below.

ON and OFF bipolar cells in the retina [39], uses two reference signals to capture the positive and negative changes in the signal. The ON–OFF time-encoding machine without integration coincides with the LCSH [29]. Hardware applications of ON–OFF time-encoding machines are seen in neuromorphic engineering [33,77] and brain–machine interfaces [3,49].

The theory of signal reconstruction from nonuniform samples applies to event-based sampling [61]. Exact reconstruction is possible if the average sampling rate is above the Nyquist rate (i.e., twice the bandwidth of the signal) [4,30]. Various reconstruction methods have been proposed in [4,30,38,70]. The similarity measures for sequences of level-crossing samples have been discussed, and an appropriate measure has been identified in [44].

### 20.1.2 Decentralized Data Collection

Decentralized data collection in resource-constrained networked systems (e.g., wireless sensor networks) is another fundamental application (in addition to A/D conversion) of sampling. In such systems, the central processor does not have access to all observations in the system due to physical constraints, such as energy and communication (i.e., bandwidth) constraints. Hence, the choice of sampling technique is of great importance to obtain a good summary of observations at the central processor. Using an adaptive sampling scheme, only the informative observations can be transmitted to the central processor. This potentially provides a better summary than the conventional (nonadaptive) sampling scheme that satisfies the same physical constraints. As a toy example to adaptive transmission, consider a bucket carrying water to a pool from a tap with varying flow. After the same number of carriages, say ten, the scheme that empties the bucket only when it is filled (i.e., adaptive to the water flow) exactly carries ten buckets of water to the pool, whereas the scheme that periodically empties the bucket (i.e., nonadaptive), in general, carries less water.

Based on such an adaptive scheme, the send-on-delta concept, for decentralized acquisition of continuous-time band-limited signals, samples and transmits only when the observed signal changes by $\pm\Delta$ since the last sampling time [42,55]. In other words, instead of transmitting at deterministic time instants, it waits for the event of $\pm\Delta$ change in the signal amplitude to sample and transmit. Although the change here is with respect to the last sample value, which is in general different from the last sampling level in level-crossing sampling, they coincide for continuous-time band-limited signals. Hence, for continuous-time band-limited signals, send-on-delta sampling is identical to LCSH (Figure 20.1). For systems in which the accumulated, instead of the current, absolute error (similar to the mean absolute error) is used as the performance criterion, an extension of the send-on-delta concept, called the integral send-on-delta, has been proposed [43]. This extension is similar to the integrate-and-fire time-encoding machine. Specifically, a $\Delta$ increase in the integral of absolute error triggers sampling (and transmission).

In essence, event-based processing aims to simplify the signal representation by mapping the real-valued amplitude, which requires infinite number of bits after the conventional time-driven sampling, to a digital value in the event space, which needs only a few bits. In most event-based techniques, including the ones discussed above, a single bit encodes the event type when an event occurs (e.g., $\pm\Delta$ change, upward/downward level crossing, reference signal crossing). In decentralized data collection, this single-bit quantization in the event space constitutes a great advantage over the infinite-bit representation of a sample taken at a deterministic time. Moreover, to save further energy and bandwidth, the number of samples (i.e., event occurrences) can be significantly reduced by increasing $\Delta$. That is, a large enough $\Delta$ value in send-on-delta sparsifies the signal with binary nonzero values, which is ideal for decentralized data collection. On the contrary, the resolution of observation summary at the central processor decreases with increasing $\Delta$, showing the expected trade-off between performance and consumption of physical resources (i.e., energy and bandwidth). In real-time reporting of the sparse signal in the event space to the central processor, only the nonzero values are sampled and transmitted when encountered*. Since event-based processing techniques, in general, first quantize the signal in terms of the events of interest, and then sample the quantized signal, they apply a *quantize-and-sample* strategy, instead of the sample-and-quantize strategy followed by the conventional time-driven processing techniques.

### 20.1.3 Decentralized Statistical Signal Processing

If, in a decentralized system, data are collected for a specific purpose (e.g., hypothesis testing, parameter estimation), then we should locally process raw observations as much as possible before transmitting to minimize processing losses at the central processor, in addition to the transmission losses due to physical constraints. For instance, in hypothesis testing, each node in the network can first compute and then report the log-likelihood ratio (LLR) of its observations, which is the sufficient statistic. Assuming independence of observations across nodes, the central processor can simply sum the reported LLRs and decide accordingly without further processing. On the contrary, if each node transmits its raw observations in a decentralized fashion, the central processor needs to process the lossy data to approximate LLR, which is in general a nonlinear function. The LLR approximation in the latter report-and-process strategy is clearly worse than the one in the former process-and-report strategy.

*Unlike compressive sensing, the binary nonzero values are simply reported in real time without any need for offline computation.

An event-based sampling technique, called level-triggered sampling, has been proposed to report the corresponding sufficient statistic in binary hypothesis testing [17] and parameter estimation [16] for continuous-time band-limited observations. The operation of level-triggered sampling is identical to that of send-on-delta sampling (i.e., $\pm\Delta$ changes in the local sufficient statistic since the last sampling time triggers a new sample), but it is motivated by the sequential probability ratio test (SPRT), the optimum sequential detector (i.e., binary hypothesis test) for independent and identically distributed (iid) observations. Without a link to event-based sampling, it was first proposed in [27] as a repeated SPRT procedure for discrete-time observations. In particular, when its local LLR exits the interval $(-\Delta, \Delta)$, each node makes a decision: null hypothesis $\mathsf{H}_0$ if it is less than or equal to $-\Delta$, and alternative hypothesis $\mathsf{H}_1$ if it is greater than or equal to $\Delta$. Then another cycle of SPRT starts with new observations. The central processor, called the fusion center, also runs SPRT by computing the joint LLR of such local decisions.

Due to the numerous advantages of digital signal processing (DSP) and digital communications over their analog counterparts, a vast majority of the existing hardware work with discrete-time signals. Although there is a significant interest in building a new DSP theory based on event-based sampling [30,44,60,64], such a theory is not mature yet, and thus it is expected that the conventional A/D converters, based on uniform sampling, will continue to dominate in the near future. Since digital communications provide reliable and efficient information transmission, with the support of inexpensive electronics, it is ubiquitous nowadays [25, page 23]. Hence, even if we perform analog signal processing and then event-based sampling on the observed continuous-time signal, we will most likely later on need to quantize time (i.e., uniformly sample the resulting continuous-time signal) for communication purposes. In that case, we should rather apply event-based sampling to uniformly sampled discrete-time observations at the nodes. This also results in a compound architecture which can perform time-driven, as well as event-driven, tasks [42,46]. As a result, level-triggered sampling with discrete-time observations (see Figure 20.3) has been considered for statistical signal processing applications [32,73–76].

In level-triggered sampling, a serious complication arises with discrete-time observations: when a sample is taken, the change since the last sampling time, in general, exceeds $\Delta$ or $-\Delta$ due to the jumps in the discrete-time signal, known as the *overshoot* problem (see Figure 20.3). Note from Figure 20.3 that the sampling thresholds are now signal dependent, as opposed to level-crossing sampling (with hysteresis), shown in

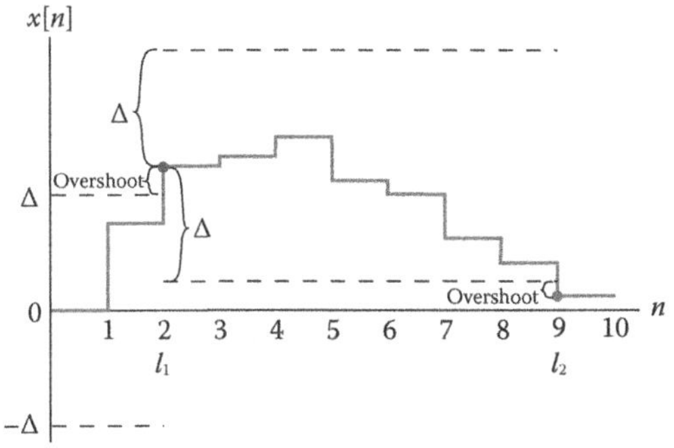


**FIGURE 20.3**
Level-triggered sampling with discrete-time observations.

Figure 20.1. Overshoots disturb the binary quantized values in terms of $\Delta$ change (i.e., the one-bit encoding in Figure 20.1) since now fractional $\Delta$ changes are possible. As a result, when sampled, such fractional values in the event space cannot be exactly encoded into a single bit. Overshoots are also observed with continuous-time band-unlimited signals (i.e., that have jumps), which are observed in practice due to noise. Hence, for practical purposes, we need to deal with the overshoot problem.

In level-triggered sampling, the overshoot problem is handled in several ways. In the first method, a single-bit encoding is still used with quantization levels that include average overshoot values in the event space. That is, for positive ($\geq\Delta$)/negative ($\leq-\Delta$) changes, the transmitted $+1/-1$ bit represents a fractional change $\bar{\theta} > \Delta/\underline{\theta} < -\Delta$, where $\bar{\theta} - \Delta/\underline{\theta} + \Delta$ compensates for the average overshoot above $\Delta$/below $-\Delta$. Examples of this overshoot compensation method are seen in the decentralized detectors of [27,74], and also in Section 20.2.2, in which the LLR of each received bit at the fusion center (FC) is computed. There are two other overshoot compensation methods in the literature, both of which quantize each overshoot value. In [73] and [75], for detection and estimation purposes, respectively, each quantized value is transmitted in a few bits via separate pulses, in addition to the single bit representing the sign of the $\Delta$ change. On the contrary, in [76], and also in Section 20.3.3.2, pulse-position modulation (PPM) is used to transmit each quantized overshoot value. Specifically, the unit time interval is divided into a number of subintervals, and a short pulse is transmitted for the sign bit at the time slot that corresponds to the overshoot value. Consequently, to transmit each quantized overshoot value, more energy is used in the former method, whereas more bandwidth is required in the latter.

In the literature, level-triggered sampling has been utilized to effectively transmit the sufficient local statistics in decentralized systems for several applications, such as spectrum sensing in cognitive radio networks [73], target detection in wireless sensor networks [76], joint spectrum sensing and channel estimation in cognitive radio networks [71], security in multiagent reputation systems [32], and power quality monitoring in power grids [31].

### 20.1.4 Outline

In this chapter, we analyze the use of event-based sampling as a means of information transmission for decentralized detection and estimation. We start with the decentralized detection problem in Section 20.2. Two challenges, namely noisy transmission channels and multimodal information sources, have been addressed via level-triggered and level-crossing sampling in Sections 20.2.2 and 20.2.3, respectively.

Then, in Section 20.3, we treat the sequential estimation of linear regression parameters under a decentralized setup. Using a variant of level-triggered sampling, we design a decentralized estimator that achieves a close-to-optimum average stopping time performance and linearly scales with the number of parameters while satisfying stringent energy and computation constraints.

Throughout the chapter, we represent scalars with lower-case letters, vectors with bold lower-case letters, and matrices with bold upper-case letters.

## 20.2 Decentralized Detection

We first consider the decentralized detection (i.e., hypothesis testing) problem, in which a number of distributed nodes (e.g., sensors), under energy and bandwidth constraints, sequentially report a summary of their discrete-time observations to an FC, which makes a decision as soon as possible satisfying some performance constraints.

### 20.2.1 Background

Existing works on decentralized detection mostly consider the fixed-sample-size approach, in which the FC makes a decision at a deterministic time using a fixed number of samples from nodes (e.g., [57,59,67]). The sequential detection approach, in which the FC at each time chooses either to continue receiving new samples or to stop and make a decision, is also of significant interest (e.g., [8,40,62]). In [17,27,73,74,76], SPRT is used both at the nodes and the FC. SPRT is the optimum sequential detector for iid observations in terms of minimizing the average sample number among all sequential tests satisfying the same error probability constraints [65]. Compared with the best fixed-sample-size detector, SPRT requires, on average, four times less samples for the same level of confidence for Gaussian signals [47, page 109].

Under stringent energy and bandwidth constraints where nodes can only infrequently transmit a single bit (which can be considered as a local decision), the optimum local decision function is the likelihood ratio test (LRT), which is nothing but a one-bit quantization of LLR, for a fixed *decision fusion* rule under the fixed-sample-size setup [57]. Similarly, the optimum fusion rule at the FC is also an LRT under the Bayesian [6] and Neyman–Pearson [58] criteria. Since SPRT, which is also a one-bit quantization of LLR with the deadband $(-\Delta, \Delta)$, is the sequential counterpart of LRT, these results readily extend to the sequential setup as a double-SPRT scheme [27].

Under relaxed resource constraints, the optimum local scheme is a multibit quantization of LLR [66], which is the necessary and sufficient statistic for the detection problem, while the optimum *data fusion* detector at the FC is still an LRT under the fixed-sample-size setup. Thanks to the event-based nature of SPRT, even its single-bit decision provides data fusion capabilities. More specifically, when it makes a decision, we know that LLR $\geq \Delta$ if $\mathsf{H}_1$ is selected, or LLR $\leq -\Delta$ if $\mathsf{H}_0$ is selected. For continuous-time bandlimited observations, we have a full precision, that is, LLR $= \Delta$ or LLR $= -\Delta$ depending on the decision, which requires infinite number of bits with LRT under the fixed-sample-size setup. The repeated SPRT structure of level-triggered sampling enables LLR tracking, that is, sequential data fusion [17,74]. For discrete-time observations, the single-bit decision at each SPRT step (i.e., one-bit representation of a level-triggered sample as in Figure 20.3) may provide high-precision LLR tracking if overshoots are small compared with $\Delta$. Otherwise, under relaxed resource constraints, each overshoot can be quantized into additional bits [73,76], resulting in a multibit quantization of the changes in LLR with the deadband $(-\Delta, \Delta)$, analogous to the multibit LLR quantization under the fixed-sample-size setup [66].

The conventional approach to decentralized detection, assuming ideal transmission channels, addresses only the noise that contaminates the observations at nodes (e.g., [17,57]). Nevertheless, in practice, the channels between nodes and the FC are noisy. Following the conventional approach, at the FC, first a communication block recovers the transmitted information bits, and then an independent signal processing block performs detection using the recovered bits. Such an independent two-step procedure inflicts performance loss due to the data-processing inequality [9]. For optimum performance, without a communication block, the received signals should be processed in a channel-aware manner [7,34].

In this section, we first design in Section 20.2.2 channel-aware decentralized detection schemes based on level-triggered sampling for different noisy channel models. We then show in Section 20.2.3 how to fuse multimodal data from disparate sources for decentralized detection.

### 20.2.2 Channel-Aware Decentralized Detection

Consider a network of $K$ distributed nodes (e.g., a wireless sensor network) and an FC, which can be one of the nodes or a dedicated processor (Figure 20.4). Each node $k$ computes the LLR $L_k[n]$ of discrete-time signal $x_k[n]$ it observes, and sends the level-triggered LLR samples to the FC, which fuses the received samples and sequentially decides between two hypotheses, $\mathsf{H}_0$ and $\mathsf{H}_1$.

Assuming iid observations $\{x_k[n]\}_n$ across time, and independence across nodes, the local LLR at node $k$ and the global LLR are given by

$$L_k[n] = \log \frac{f_{k,1}(x_k[1], \ldots, x_k[n])}{f_{k,0}(x_k[1], \ldots, x_k[n])} = \sum_{m=1}^{n} \log \frac{f_{k,1}(x_k[m])}{f_{k,0}(x_k[m])}$$

$$= \sum_{m=1}^{n} l_k[m] = L_k[n-1] + l_k[n],$$

$$L[n] = \sum_{k=1}^{K} L_k[n],$$

respectively, where $f_{k,j}, j = 0,1$ is the probability density/mass function of the observed signal at node $k$ under $\mathsf{H}_j$, and $l_k[n]$ is the LLR of $x_k[n]$.

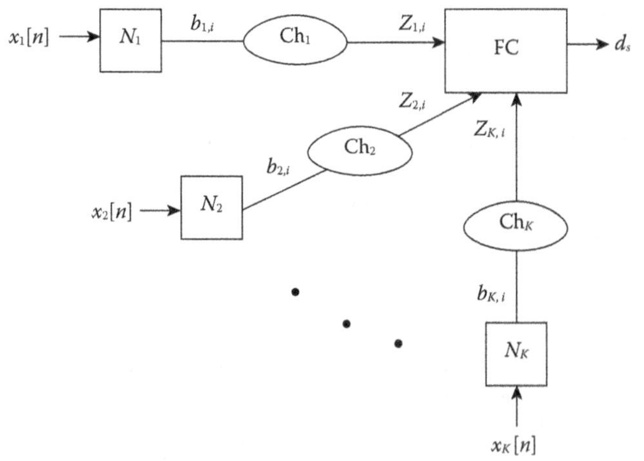


**FIGURE 20.4**
A network of $K$ nodes and a fusion center (FC). Each node $k$ processes its observations $\{x_k[n]\}_n$ and transmits information bits $\{b_{k,i}\}_i$. The FC then, upon receiving the signals $\{z_{k,i}\}$, makes a detection decision $d_S$ at the random time $S$.

#### 20.2.2.1 Procedure at Nodes

Each node $k$ samples $L_k[n]$ via level-triggered sampling at a sequence of random times $\{t_{k,i}\}_i$ that are determined by $L_k[n]$ itself. Specifically, the $i$th sample is taken when the LLR change $L_k[n] - L_k[t_{k,i-1}]$ since the last sampling time $t_{k,i-1}$ exceeds a constant $\Delta$ in absolute value, that is,

$$t_{k,i} \triangleq \min\left\{n > t_{k,i-1} : L_k[n] - L_k[t_{k,i-1}] \notin (-\Delta, \Delta)\right\},\quad t_{k,0} = 0,\ L_k[0] = 0. \tag{20.1}$$

It has been shown in [73, section IV-B] that $\Delta$ can be determined by

$$\Delta \tanh\left(\frac{\Delta}{2}\right) = \frac{1}{R}\sum_{k=1}^{K} |\mathsf{E}_j[L_k[1]]|, \tag{20.2}$$

to ensure that the FC receives messages with an average rate of $R$ messages per unit time interval.

Let $\lambda_{k,i}$ denote the LLR change during the $i$th sampling interval, $(t_{k,i-1}, t_{k,i}]$, that is,

$$\lambda_{k,i} \triangleq L_k[t_{k,i}] - L_k[t_{k,i-1}] = \sum_{n=t_{k,i-1}+1}^{t_{k,i}} l_k[n].$$

Immediately after sampling at $t_{k,i}$, as shown in Figure 20.4, an information bit $b_{k,i}$ indicating the threshold crossed by $\lambda_{k,i}$ is transmitted to the FC, that is,

$$b_{k,i} \triangleq \text{sign}(\lambda_{k,i}). \tag{20.3}$$

#### 20.2.2.2 Procedure at the FC

Let us now analyze the received signal $z_{k,i}$ at the FC corresponding to the transmitted bit $b_{k,i}$ (see Figure 20.4). The FC computes the LLR

$$\tilde{\lambda}_{k,i} \triangleq \log\frac{g_{k,1}(z_{k,i})}{g_{k,0}(z_{k,i})}, \tag{20.4}$$

of each received signal $z_{k,i}$ and approximates the global LLR, $L[n]$, as

$$\tilde{L}[n] \triangleq \sum_{k=1}^{K}\sum_{i=1}^{J_{k,n}} \tilde{\lambda}_{k,i},$$

where $J_{k,n}$ is the total number of LLR messages received from node $k$ until time $n$, and $g_{k,j}$, $j = 0, 1$, is the pdf of $z_{k,i}$ under $H_j$.

In fact, the FC recursively updates $\tilde{L}[n]$ whenever it receives an LLR message from any node. In particular, suppose that the $m$th LLR message $\tilde{\lambda}_m$ from any sensor is received at time $t_m$. Then at $t_m$, the FC performs the following update:

$$\tilde{L}[t_m] = \tilde{L}[t_{m-1}] + \tilde{\lambda}_m,$$

and uses $\tilde{L}[t_m]$ in an SPRT procedure with two thresholds $A$ and $-B$, and the following decision rule

$$d_{t_m} \triangleq \begin{cases} \mathsf{H}_1, & \text{if } \tilde{L}[t_m] \geq A, \\ \mathsf{H}_0, & \text{if } \tilde{L}[t_m] \leq -B, \\ \text{wait for } \tilde{\lambda}_{m+1}, & \text{if } \tilde{L}[t_m] \in (-B, A). \end{cases}$$

The thresholds $(A, B > 0)$ are selected to satisfy the error probability constraints

$$\mathsf{P}_0(d_S = \mathsf{H}_1) \leq \alpha \text{ and } \mathsf{P}_1(d_S = \mathsf{H}_0) \leq \beta, \tag{20.5}$$

with equalities, where $\mathsf{P}_j, j = 0, 1$, denotes the probability under $\mathsf{H}_j$, $\alpha$ and $\beta$ are the error probability bounds given to us, and

$$S \triangleq \min\{n > 0 : \tilde{L}[n] \notin (-B, A)\}, \tag{20.6}$$

is the decision time.

Comparing (20.1) with (20.6), we see that each node, in fact, applies a local SPRT with thresholds $\Delta$ and $-\Delta$ within each sampling interval. At node $k$, the $i$th local SPRT starts at time $t_{k,i-1} + 1$ and ends at time $t_{k,i}$ when the local test statistic $\lambda_{k,i}$ exceeds either $\Delta$ or $-\Delta$. This local hypothesis testing produces a local decision represented by the information bit $b_{k,i}$ in (20.3), and induces the local error probabilities

$$\alpha_k \triangleq \mathsf{P}_0(b_{k,i} = 1) \text{ and } \beta_k \triangleq \mathsf{P}_1(b_{k,i} = -1). \tag{20.7}$$

We next discuss how to compute $\tilde{\lambda}_{k,i}$, the LLR of received signal $z_{k,i}$, given by (20.4), under ideal and noisy channels.

#### 20.2.2.3 Ideal Channels

**Lemma 20.1**

Assuming ideal channels between nodes and the FC, that is, $z_{k,i} = b_{k,i}$, we have

$$\tilde{\lambda}_{k,i} = \begin{cases} \log\frac{\mathsf{P}_1(b_{k,i}=1)}{\mathsf{P}_0(b_{k,i}=1)} = \log\frac{1-\beta_k}{\alpha_k} \geq \Delta, & \text{if } b_{k,i} = 1, \\ \log\frac{\mathsf{P}_1(b_{k,i}=-1)}{\mathsf{P}_0(b_{k,i}=-1)} = \log\frac{\beta_k}{1-\alpha_k} \leq -\Delta, & \text{if } b_{k,i} = -1. \end{cases} \tag{20.8}$$

**PROOF** The equalities follow from (20.7). The inequalities can be obtained by applying a change of measure. To show the first one, we write

$$\alpha_k = \mathsf{P}_0(\lambda_{k,i} \geq \Delta) = \mathsf{E}_0[\mathbb{1}_{\{\lambda_{k,i} \geq \Delta\}}], \tag{20.9}$$

where $\mathsf{E}_j$ is the expectation under $\mathsf{H}_j, j = 0, 1$ and $\mathbb{1}_{\{\cdot\}}$ is the indicator function. Note that

$$e^{-\lambda_{k,i}} = \frac{f_{k,0}(x_k[t_{k,i-1}+1], \ldots, x_k[t_{k,i}])}{f_{k,1}(x_k[t_{k,i-1}+1], \ldots, x_k[t_{k,i}])},$$

can be used to compute the expectation integral in terms of $f_{k,1}$ instead of $f_{k,0}$, that is, to change the probability measure under which the expectation is taken from $f_{k,0}$ to $f_{k,1}$. Hence,

$$\begin{aligned}\alpha_k &= \mathsf{E}_1[e^{-\lambda_{k,i}} \mathbb{1}_{\{\lambda_{k,i} \geq \Delta\}}] \\ &\leq e^{-\Delta} \mathsf{E}_1[\mathbb{1}_{\{\lambda_{k,i} \geq \Delta\}}] = e^{-\Delta} \mathsf{P}_1(\lambda_{k,i} \geq \Delta) \\ &= e^{-\Delta}(1 - \beta_k),\end{aligned}$$

giving us the first inequality in (20.8). The second inequality follows similarly. ■

We see from Lemma 20.1 that the FC, assuming ideal channels, can compute $\tilde{\lambda}_{k,i}$, the LLR of the sign bit $b_{k,i}$ if the local error probabilities $\alpha_k$ and $\beta_k$ are available. It is also seen that $\tilde{\lambda}_{k,i}$ is, in magnitude, larger than the corresponding sampling threshold, and thus includes a constant compensation for the random overshoot of $\lambda_{k,i}$ above $\Delta$ or below $-\Delta$. The relationship of this constant compensation to the average overshoot, and the *order-1* asymptotic optimality it achieves are established in [17].

In the no-overshoot case, as with continuous-time band-limited observations, the inequalities in (20.8) become equalities since in (20.9) we can write $\alpha_k = \mathsf{P}_0(\lambda_{k,i} = \Delta)$. This shows that the LLR update in (20.8) adapts well to the no-overshoot case, in which the LLR change that triggers sampling is either $\Delta$ or $-\Delta$.

**Theorem 20.1: [17, Theorem 2]**

Consider the asymptotic regime in which the target error probabilities $\alpha, \beta \to 0$ at the same rate. If the sampling threshold $\Delta \to \infty$ is slower than $|\log \alpha|$, then, under ideal channels, the decentralized detector which uses the LLR update given by (20.8) for each level-triggered sample is order-1 asymptotically optimum, that is,

$$\frac{\mathsf{E}_j[S]}{\mathsf{E}_j[S_o]} = 1 + o(1), \; j = 0, 1, \tag{20.10}$$

where $S_o$ is the decision time of the optimum (centralized) sequential detector, SPRT, satisfying the error probability bounds $\alpha$ and $\beta$ [cf. (20.5)].

For the proof and more details on the result, see [17, Theorem 2] and the discussion therein. Using the traditional uniform sampler followed by a quantizer, a similar order-1 asymptotic optimality result cannot be obtained by controlling the sampling period with a constant number of quantization bits [73, section IV-B]. The significant performance gain of level-triggered sampling against uniform sampling is also shown numerically in [73, section V].

Order-1 is the most frequent type of asymptotic optimality encountered in the literature, but it is also the weakest. Note that in order-1 asymptotic optimality, although the average decision time ratio converges to 1, the difference $\mathsf{E}_j[S] - \mathsf{E}_j[S_o]$ may be unbounded. Therefore, stronger types of asymptotic optimality are defined. The difference remains bounded (i.e., $\mathsf{E}_j[S] - \mathsf{E}_j[S_o] = O(1)$) in order-2 and diminishes (i.e., $\mathsf{E}_j[S] - \mathsf{E}_j[S_o] = o(1)$) in order-3. The latter is extremely rare in the literature, and the schemes of that type are considered optimum per se for practical purposes.

#### 20.2.2.4 Noisy Channels

In the presence of noisy channels, one subtle issue is that since the sensors asynchronously sample and transmit the local LLR, the FC needs to first reliably detect the sampling time to update the global LLR. We first assume that the sampling time is reliably detected and focus on deriving the LLR update at the FC. We discuss the issue of sampling time detection later on.

In computing the LLR $\tilde{\lambda}_{k,i}$ of the received signal $z_{k,i}$, we make use of the local sensor error probabilities $\alpha_k, \beta_k$, and the channel parameters that characterize the statistical property of the channel.

##### *20.2.2.4.1 Binary Erasure Channels*

We first consider binary erasure channels (BECs) between sensors and the FC with erasure probabilities $\epsilon_k$, $k = 1, \ldots, K$. Under BEC, a transmitted bit $b_{k,i}$ is lost with probability $\epsilon_k$, and it is correctly received at the FC (i.e., $z_{k,i} = b_{k,i}$) with probability $1 - \epsilon_k$.

**Lemma 20.2**

Under BEC with erasure probability $\epsilon_k$, the LLR of $z_{k,i}$ is given by

$$\tilde{\lambda}_{k,i} = \begin{cases} \log \frac{\mathsf{P}_1(z_{k,i}=1)}{\mathsf{P}_0(z_{k,i}=1)} = \log \frac{1-\beta_k}{\alpha_k}, & \text{if } z_{k,i} = 1, \\ \log \frac{\mathsf{P}_1(z_{k,i}=-1)}{\mathsf{P}_0(z_{k,i}=-1)} = \log \frac{\beta_k}{1-\alpha_k}, & \text{if } z_{k,i} = -1. \end{cases} \tag{20.11}$$

**PROOF** We have $z_{k,i} = b$, $b = \pm 1$, with probability $1 - \epsilon_k$ only when $b_{k,i} = b$. Hence,

$$\mathsf{P}_j(z_{k,i} = b) = \mathsf{P}_j(b_{k,i} = b)(1 - \epsilon_k),\ j = 0, 1.$$

In the LLR expression, the $1 - \epsilon_k$ terms on the numerator and denominator cancel out, giving the result in (20.11). ■

Note that under BEC, the channel parameter $\epsilon_k$ is not needed when computing the LLR $\tilde{\lambda}_{k,i}$. Note also that in this case, a received bit bears the same amount of LLR information as in the ideal channel case [cf. (20.8)], although a transmitted bit is not always received. Hence, the channel-aware approach coincides with the conventional approach which relies solely on the received signal. Although the LLR updates in (20.8) and (20.11) are identical, the fusion rules under BEC and ideal channels are not. This is because under BEC, the decision thresholds $A$ and $B$ in (20.6), due to the information loss, are in general different from those in the ideal channel case.

#### *20.2.2.4.2 Binary Symmetric Channels*

Next, we consider binary symmetric channels (BSCs) with crossover probabilities $\epsilon_k$ between sensors and the FC. Under BSC, the transmitted bit $b_{k,i}$ is flipped (i.e., $z_{k,i} = -b_{k,i}$) with probability $\epsilon_k$, and it is correctly received (i.e., $z_{k,i} = b_{k,i}$) with probability $1 - \epsilon_k$.

**Lemma 20.3**

Under BSC with crossover probability $\epsilon_k$, the LLR of $z_{k,i}$ can be computed as

$$\tilde{\lambda}_{k,i} = \begin{cases} \log \frac{1-\hat{\beta}_k}{\hat{\alpha}_k}, & \text{if } z_{k,i} = 1, \\ \log \frac{\hat{\beta}_k}{1-\hat{\alpha}_k}, & \text{if } z_{k,i} = -1, \end{cases} \tag{20.12}$$

where $\hat{\alpha}_k = \alpha_k(1 - 2\epsilon_k) + \epsilon_k$ and $\hat{\beta}_k = \beta_k(1 - 2\epsilon_k) + \epsilon_k$.

**PROOF** Due to the nonzero probability of receiving a wrong bit, we now have

$$\begin{aligned}\mathsf{P}_j(z_{k,i} = b) &= \mathsf{P}(z_{k,i} = b | b_{k,i} = b)\mathsf{P}_j(b_{k,i} = b) \\ &\quad + \mathsf{P}(z_{k,i} = b | b_{k,i} = -b)\mathsf{P}_j(b_{k,i} = -b),\end{aligned}$$

e.g., $\mathsf{P}_0(z_{k,i} = 1) = (1 - \epsilon_k)\alpha_k + \epsilon_k(1 - \alpha_k)$,

$j = 0, 1$, $b = \pm 1$. Defining $\hat{\alpha}_k = \alpha_k(1 - 2\epsilon_k) + \epsilon_k$ and $\hat{\beta}_k = \beta_k(1 - 2\epsilon_k) + \epsilon_k$, we obtain the LLR expression given in (20.12). ■

Note that for $\alpha_k < 0.5$, $\beta_k < 0.5$, $\forall k$, which we assume true for $\Delta > 0$,

$$\hat{\alpha}_k = \alpha_k + \epsilon_k(1 - 2\alpha_k) > \alpha_k,$$

and similarly $\hat{\beta}_k > \beta_k$. Thus, $|\tilde{\lambda}_{k,i}^{BSC}| < |\tilde{\lambda}_{k,i}^{BEC}|$ from which we expect a higher performance loss under BSC than the one under BEC. Finally, note also that, unlike the BEC case, under BSC the FC needs to know the channel parameters $\{\epsilon_k\}$ to operate in a channel-aware manner.

#### *20.2.2.4.3 Additive White Gaussian Noise Channels*

Now, assume that the channel between each sensor and the FC is an additive white Gaussian noise (AWGN) channel. The received signal at the FC is given by

$$z_{k,i} = y_{k,i} + w_{k,i}, \tag{20.13}$$

where $w_{k,i} \sim \mathcal{N}_c(0, \sigma_k^2)$ is the complex white Gaussian noise, and $y_{k,i}$ is the transmitted signal at sampling time $t_{k,i}$, given by

$$y_{k,i} = \begin{cases} a, & \text{if } \lambda_{k,i} \geq \Delta, \\ b, & \text{if } \lambda_{k,i} \leq -\Delta, \end{cases} \tag{20.14}$$

where the transmission levels $a$ and $b$ are complex in general.

**Lemma 20.4**

Under the AWGN channel model in (20.13), the LLR of $z_{k,i}$ is given by

$$\tilde{\lambda}_{k,i} = \log \frac{(1 - \beta_k)e^{-c_{k,i}} + \beta_k e^{-d_{k,i}}}{\alpha_k e^{-c_{k,i}} + (1 - \alpha_k)e^{-d_{k,i}}}, \tag{20.15}$$

where $c_{k,i} = \frac{|z_{k,i}-a|^2}{\sigma_k^2}$ and $d_{k,i} = \frac{|z_{k,i}-b|^2}{\sigma_k^2}$.

**PROOF** The distribution of the received signal given $y_{k,i}$ is $z_{k,i} \sim \mathcal{N}_c(y_{k,i}, \sigma_k^2)$. The probability density function of $z_{k,i}$ under $\mathsf{H}_j$ is then given by

$$\begin{aligned} g_{k,j}(z_{k,i}) &= g_{k,j}(z_{k,i} | y_{k,i} = a)\mathsf{P}_j(y_{k,i} = a) \\ &\quad + g_{k,j}(z_{k,i} | y_{k,i} = b)\mathsf{P}_j(y_{k,i} = b), \\ \text{e.g., } g_{k,1}(z_{k,i}) &= \frac{(1 - \beta_k)e^{-\frac{|z_{k,i}-a|^2}{\sigma_k^2}} + \beta_k e^{-\frac{|z_{k,i}-b|^2}{\sigma_k^2}}}{\pi\sigma_k^2}. \end{aligned} \tag{20.16}$$

Defining $c_{k,i} \triangleq \frac{|z_{k,i}-a|^2}{\sigma_k^2}$ and $d_{k,i} \triangleq \frac{|z_{k,i}-b|^2}{\sigma_k^2}$, and substituting $g_{k,0}(z_{k,i})$ and $g_{k,1}(z_{k,i})$ into $\tilde{\lambda}_{k,i} = \log \frac{g_{k,1}(z_{k,i})}{g_{k,0}(z_{k,i})}$, we obtain (20.15). ■

If the transmission levels $a$ and $b$ are well separated, and the signal-to-noise ratio $\frac{|y_{k,i}|}{|w_{k,i}|}$ is high enough, then

$$\tilde{\lambda}_{k,i} \approx \begin{cases} \log \frac{1-\beta_k}{\alpha_k}, & \text{if } y_{k,i} = a, \\ \log \frac{\beta_k}{1-\alpha_k}, & \text{if } y_{k,i} = b, \end{cases}$$

resembling the ideal channel case, given by (20.8). Due to the energy constraints at nodes, assume a maximum transmission power $P^2$. In accordance with the above observation, it is shown in [74, Section V-C] that the antipodal signaling (e.g., $a = P$ and $b = -P$) is optimum.

*20.2.2.4.4 Rayleigh Fading Channels*

Assuming a Rayleigh fading channel model, the received signal is given by

$$z_{k,i} = h_{k,i} y_{k,i} + w_{k,i}, \quad (20.17)$$

where $h_{k,i} \sim \mathcal{N}_c(0, \sigma_{h,k}^2)$, $y_{k,i}$, and $w_{k,i}$ are as before.

**Lemma 20.5**

Under the Rayleigh fading channel model in (20.17), the LLR of $z_{k,i}$ is given by

$$\tilde{\lambda}_{k,i} = \log \frac{\frac{1-\beta_k}{\sigma_{a,k}^2} e^{-c_{k,i}} + \frac{\beta_k}{\sigma_{b,k}^2} e^{-d_{k,i}}}{\frac{\alpha_k}{\sigma_{a,k}^2} e^{-c_{k,i}} + \frac{1-\alpha_k}{\sigma_{b,k}^2} e^{-d_{k,i}}}, \quad (20.18)$$

where $c_{k,i} = \frac{|z_{k,i}|^2}{\sigma_{a,k}^2}$, $d_{k,i} = \frac{|z_{k,i}|^2}{\sigma_{b,k}^2}$, $\sigma_{a,k}^2 = |a|^2 \sigma_{h,k}^2 + \sigma_k^2$, and $\sigma_{b,k}^2 = |b|^2 \sigma_{h,k}^2 + \sigma_k^2$.

**PROOF** Given $y_{k,i}$, we have $z_{k,i} \sim \mathcal{N}_c(0, |y_{k,i}|^2 \sigma_{h,k}^2 + \sigma_k^2)$. Similar to (20.16), we can write

$$\begin{aligned} g_{k,1}(z_{k,i}) &= \frac{1-\beta_k}{\pi \sigma_{a,k}^2} e^{-c_{k,i}} + \frac{\beta_k}{\pi \sigma_{b,k}^2} e^{-d_{k,i}}, \\ g_{k,0}(z_{k,i}) &= \frac{\alpha_k}{\pi \sigma_{a,k}^2} e^{-c_{k,i}} + \frac{1-\alpha_k}{\pi \sigma_{b,k}^2} e^{-d_{k,i}}, \end{aligned} \quad (20.19)$$

where $c_{k,i} \triangleq \frac{|z_{k,i}|^2}{\sigma_{a,k}^2}$, $d_{k,i} \triangleq \frac{|z_{k,i}|^2}{\sigma_{b,k}^2}$, $\sigma_{a,k}^2 \triangleq |a|^2 \sigma_{h,k}^2 + \sigma_k^2$, and $\sigma_{b,k}^2 \triangleq |b|^2 \sigma_{h,k}^2 + \sigma_k^2$. Substituting $g_{k,0}(z_{k,i})$ and $g_{k,1}(z_{k,i})$ into $\tilde{\lambda}_{k,i} = \log \frac{g_{k,1}(z_{k,i})}{g_{k,0}(z_{k,i})}$, we obtain (20.18). ■

In this case, different messages $a$ and $b$ are expressed only in the variance of $z_{k,i}$. Hence, with antipodal signaling, they become indistinguishable (i.e., $\sigma_{a,k}^2 = \sigma_{b,k}^2$) and as a result $\tilde{\lambda}_{k,i} = 0$. This suggests that we should separate $|a|$ and $|b|$ as much as possible to decrease the uncertainty at the FC, and in turn to decrease the loss in the LLR update $\tilde{\lambda}_{k,i}$ with respect to the ideal channel case. Assuming a minimum transmission power $Q^2$ to ensure reliable detection of an incoming signal at the FC, in addition to the maximum transmission power $P^2$ due to the energy constraints, it is numerically shown in [74, Section V-D] that the optimum signaling scheme corresponds to either $|a| = P$, $|b| = Q$ or $|a| = Q$, $|b| = P$.

*20.2.2.4.5 Rician Fading Channels*

For Rician fading channels, we have $h_{k,i} \sim \mathcal{N}_c(\mu_k, \sigma_{h,k}^2)$ in (20.17).

**Lemma 20.6**

With Rician fading channels, $\tilde{\lambda}_{k,i}$ is given by (20.18), where $c_{k,i} = \frac{|z_{k,i} - a\mu_k|^2}{\sigma_{a,k}^2}$, $d_{k,i} = \frac{|z_{k,i} - b\mu_k|^2}{\sigma_{b,k}^2}$, $\sigma_{a,k}^2 = |a|^2 \sigma_{h,k}^2 + \sigma_k^2$, and $\sigma_{b,k}^2 = |b|^2 \sigma_{h,k}^2 + \sigma_k^2$.

**PROOF** Given $y_{k,i}$, the received signal is distributed as $z_{k,i} \sim \mathcal{N}_c(\mu_k y_{k,i}, |y_{k,i}|^2 \sigma_{h,k}^2 + \sigma_k^2)$. The likelihoods $g_{k,1}(z_{k,i})$ and $g_{k,0}(z_{k,i})$ are then written as in (20.19) with $\sigma_{a,k}^2 = |a|^2 \sigma_{h,k}^2 + \sigma_k^2$, $\sigma_{b,k}^2 = |b|^2 \sigma_{h,k}^2 + \sigma_k^2$, and the new definitions $c_{k,i} = \frac{|z_{k,i} - a\mu_k|^2}{\sigma_{a,k}^2}$, $d_{k,i} = \frac{|z_{k,i} - b\mu_k|^2}{\sigma_{b,k}^2}$. Finally, the LLR is given by (20.18). ■

The Rician model covers the previous two continuous channel models. Particularly, the $\sigma_{h,k}^2 = 0$ case corresponds to the AWGN model, and the $\mu_k = 0$ case corresponds to the Rayleigh model. It is numerically shown in [74, Section V-E] that depending on the values of parameters $(\mu_k, \sigma_{h,k}^2)$, either the antipodal signaling of the AWGN case or the ON–OFF type signaling of the Rayleigh case is optimum.

*20.2.2.4.6 Discussions*

Considering the unreliable detection of sampling times under continuous channels, we should ideally integrate this uncertainty into the fusion rule of the FC. In other words, at the FC, the LLR of received signal

$$z_k[n] = h_k[n] y_k[n] + w_k[n],$$

instead of $z_{k,i}$ given in (20.17), should be computed at each time instant $n$ if the sampling time of node $k$ cannot be reliably detected. In the LLR computations of Lemmas 20.4 and 20.5, the prior probabilities $\mathsf{P}_j(y_{k,i} = a)$ and $\mathsf{P}_j(y_{k,i} = b)$ are used. These probabilities are in fact conditioned on the sampling time $t_{k,i}$. Here, we need the unconditioned prior probabilities of the signal $y_k[n]$ which at each time $n$ takes a value of $a$ or $b$ or 0, that is,

$$y_k[n] = \begin{cases} a & \text{if } L_k[n] - L_k[t_{k,i-1}] \geq \Delta, \\ b & \text{if } L_k[n] - L_k[t_{k,i-1}] \leq -\Delta, \\ 0 & \text{if } L_k[n] - L_k[t_{k,i-1}] \in (-\Delta, \Delta), \end{cases}$$

instead of $y_{k,i}$ given in (20.14).

Then, the LLR of $z_k[n]$ is given by

$$\tilde{\lambda}_k[n] = \log \frac{g_{k,1}(z_k[n])}{g_{k,0}(z_k[n])},$$
$$g_{k,1}(z_k[n]) = \Big[g_{k,1}(z_k[n]|y_k[n] = a)(1-\beta_k) + g_{k,1}(z_k[n]|y_k[n] = b)\beta_k\Big]\mathsf{P}_1(y_k[n] \neq 0) + g_{k,1}(z_k[n]|y_k[n] = 0)\,\mathsf{P}_1(y_k[n] = 0),$$
$$g_{k,0}(z_k[n]) = \Big[g_{k,0}(z_k[n]|y_k[n] = a)\alpha_k + g_{k,0}(z_k[n]|y_k[n] = b)(1-\alpha_k)\Big] \times \mathsf{P}_0(y_k[n] \neq 0) + g_{k,0}(z_k[n]|y_k[n] = 0)\,\mathsf{P}_0(y_k[n] = 0),$$

where $g_{k,j}(z_k[n]|y_k[n])$ is determined by the channel model. Since the FC has no prior information on the sampling times of nodes, the probability of sampling, that is, $\mathsf{P}_j(y_k[n] \neq 0)$, can be shown to be $\frac{1}{\mathsf{E}_j[\tau_{k,i}]}$, where $\mathsf{E}_j[\tau_{k,i}]$ is the average sampling interval of node $k$ under $\mathsf{H}_j$, $j = 0, 1$.

Alternatively, a two-step procedure can be applied by first detecting a message and then using the LLR updates previously derived in Lemmas 20.4–20.6. Since it is known that most of the time $\tilde{\lambda}_k[n]$ is uninformative, corresponding to the no message case, a simple thresholding can be applied to perform LLR update only when it is informative. The thresholding step is in fact a Neyman–Pearson test (i.e., LRT) between the presence and absence of a message signal. The threshold can be adjusted to control the false alarm (i.e., type-I error) and misdetection (i.e., type-II error) probabilities. Setting the threshold sufficiently high, we can obtain a negligible false alarm probability, leaving us with the misdetection probability. Thus, if an LLR survives after thresholding, in the second step it is recomputed as in the channel-aware fusion rules obtained in Lemmas 20.4–20.6.

An information-theoretic analysis for the decentralized detectors in Sections 20.2.2.3 and 20.2.2.4 can be found in [74]. Specifically, using renewal processes, closed-form expressions for average decision time are derived under both the nonasymptotic and asymptotic regimes.

### 20.2.3 Multimodal Decentralized Detection

In monitoring of complex systems, multimodal data, such as sensor measurements, images, and texts, are collected from disparate sources. The emerging concepts of Internet of Things (IoT) and Cyber-Physical Systems (CPS) show that there is an increasing interest in connecting more and more devices with various sensing capabilities [41]. The envisioned future power grid, called Smart Grid, is a good example for such heterogeneous networks. Monitoring and managing wide-area smart grids require the integration of multimodal data from electricity consumers, such as smart home and smart city systems, as well as various electricity generators (e.g., wind, solar, coal, nuclear) and sensing devices across the grid [15].

Multisensor surveillance (e.g., for military or environmental purposes) is another application in which multimodal data from a large number of sensors (e.g., acoustic, seismic, infrared, optical, magnetic, temperature) are fused for a common statistical task [18]. An interesting multidisciplinary example is nuclear facility monitoring for treaty verification. From a data-processing perspective, using a variety of disparate information sources, such as electricity consumption, satellite images, radiation emissions, seismic vibrations, shipping manifests, and intelligence data, a nuclear facility can be monitored to detect anomalous events that violate a nuclear treaty.

Information-theoretic and machine-learning approaches to similar problems can be found in [18] and [56], respectively. We here follow a Bayesian probabilistic approach to the multimodal detection problem.

#### 20.2.3.1 Latent Variable Model

Consider a system of $K$ information sources (i.e., a network of $K$ nodes). From each source $k$, a discrete-time signal $x_k[n], n \in \mathbb{N}$, is observed, which follows the probability distribution $\mathcal{D}_k(\boldsymbol{\theta}_k)$ with the parameter vector $\boldsymbol{\theta}_k$, $k = 1, \ldots, K$. Given $\boldsymbol{\theta}_k$, the temporal observations $\{x_k[n]\}_n$ from source $k$ are assumed iid. Some information sources may be of same modality.

A latent variable vector $\boldsymbol{\phi}$ is assumed to correlate information sources by controlling their parameters (Figure 20.5). Then, the joint distribution of all observations collected until time $N$ can be written as

$$f\left(\{x_k[n]\}_{k=1,n=1}^{K,N}\right) = \int_{\chi_\phi}\int_{\chi_1}\cdots\int_{\chi_K} f\Big(\{x_k[n]\}_{k,n}\big|\{\boldsymbol{\theta}_k\},\boldsymbol{\phi}\Big) \times f\Big(\{\boldsymbol{\theta}_k\}\big|\boldsymbol{\phi}\Big)\, f\Big(\boldsymbol{\phi}\Big)\, d\boldsymbol{\theta}_1\cdots d\boldsymbol{\theta}_K\, d\boldsymbol{\phi},$$

where $\chi_\phi$ and $\chi_k$ are the supports of $\boldsymbol{\phi}$ and $\boldsymbol{\theta}_k$, $k = 1, \ldots, K$. Assuming $\{\boldsymbol{\theta}_k\}$ are independent,

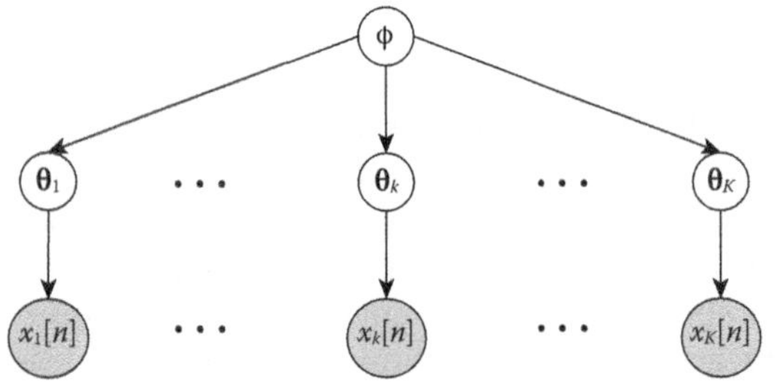


**FIGURE 20.5**
A Bayesian network of $K$ information sources linked through the latent variable vector $\boldsymbol{\phi}$. The probability distribution of the observation $x_k[n]$ is parameterized by the random vector $\boldsymbol{\theta}_k$, whose distribution is determined by $\boldsymbol{\phi}$, which is also random. The observed variables are represented by filled circles.

given $\boldsymbol{\phi}$ we have

$$\begin{aligned} &f\left(\{x_k[n]\}_{k,n}\right) \\ &= \int_{\chi_\phi}\left\{\int_{\chi_1}\prod_{n=1}^{N} f\left(x_1[n]|\boldsymbol{\theta}_1\right) f\left(\boldsymbol{\theta}_1|\boldsymbol{\phi}\right) d\boldsymbol{\theta}_1\right. \\ &\quad \times\cdots\left.\int_{\chi_K}\prod_{n=1}^{N} f\left(x_K[n]|\boldsymbol{\theta}_K\right) f\left(\boldsymbol{\theta}_K|\boldsymbol{\phi}\right) d\boldsymbol{\theta}_K\right\} f\left(\boldsymbol{\phi}\right) d\boldsymbol{\phi} \\ &= \int_{\chi_\phi} f\left(\{x_1[n]\}|\boldsymbol{\phi}\right)\cdots f\left(\{x_K[n]\}|\boldsymbol{\phi}\right) f\left(\boldsymbol{\phi}\right) d\boldsymbol{\phi}, \end{aligned} \tag{20.20}$$

where $f\left(x_k[n]|\boldsymbol{\theta}_k\right)$, $k = 1,\ldots,K$, is the probability density/mass function of the distribution $\mathcal{D}_k(\boldsymbol{\theta}_k)$. If $f\left(\boldsymbol{\theta}_k|\boldsymbol{\phi}\right)$ corresponds to the conjugate prior distribution for $\mathcal{D}_k(\boldsymbol{\theta}_k)$, then $f\left(\{x_k[n]\}_n|\boldsymbol{\phi}\right)$ can be written in closed form.

#### 20.2.3.2 Hypothesis Testing

If the latent variable vector $\boldsymbol{\phi}$ is deterministically specified under both hypotheses, that is,

$$\begin{aligned} \mathsf{H}_0 &: \boldsymbol{\phi} = \boldsymbol{\phi}_0, \\ \mathsf{H}_1 &: \boldsymbol{\phi} = \boldsymbol{\phi}_1, \end{aligned} \tag{20.21}$$

then observations $\{x_k[n]\}_k$ from different sources are independent under $\mathsf{H}_j, j = 0,1$, since $\{\boldsymbol{\theta}_k\}$ are assumed independent given $\boldsymbol{\phi}$. In that case, the global likelihood under $\mathsf{H}_j$ is given by (20.20) without the integral over $\boldsymbol{\phi}$, that is,

$$f_j(\{x_k[n]\}_{k,n}) = \prod_{k=1}^{K} f\left(\{x_k[n]\}_n|\boldsymbol{\phi} = \boldsymbol{\phi}_j\right).$$

Using $f_1(\{x_k[n]\}_{k,n})$ and $f_0(\{x_k[n]\}_{k,n})$, the global LLR at time $N$ is written as

$$L[N] = \sum_{k=1}^{K} \log\frac{f\left(\{x_k[n]\}_{n=1}^{N}|\boldsymbol{\phi} = \boldsymbol{\phi}_1\right)}{f\left(\{x_k[n]\}_{n=1}^{N}|\boldsymbol{\phi} = \boldsymbol{\phi}_0\right)} = \sum_{k=1}^{K} L_k[N]. \tag{20.22}$$

For sequential detection, SPRT can be applied by comparing $L[n]$ at each time to two thresholds $A$ and $-B$. The sequential test continues until the stopping time

$$S = \min\{n \in \mathbb{N} : L[n] \notin (-B, A)\}, \tag{20.23}$$

and makes the decision

$$d_S = \begin{cases} \mathsf{H}_1, & \text{if } L[S] \geq A, \\ \mathsf{H}_0, & \text{if } L[S] \leq -B, \end{cases} \tag{20.24}$$

at time $S$.

In a decentralized system, where all observations cannot be made available to the FC due to resource constraints, each node $k$ (corresponding to information source $k$) can compute its LLR $L_k[n]$ and transmit event-based samples of it to the FC, as will be described in Section 20.2.3.4. Then, summing the LLR messages from nodes, the FC computes the approximate global LLR $\tilde{L}[n]$ and uses it in the SPRT procedure similar to (20.23) and (20.24).

In many cases, it may not be possible to deterministically specify $\boldsymbol{\phi}$ under the hypotheses, but a statistical description may be available, that is,

$$\begin{aligned} \mathsf{H}_0 &: \boldsymbol{\phi} \sim \mathcal{D}_{\phi,0}(\boldsymbol{\theta}_{\phi,0}), \\ \mathsf{H}_1 &: \boldsymbol{\phi} \sim \mathcal{D}_{\phi,1}(\boldsymbol{\theta}_{\phi,1}). \end{aligned} \tag{20.25}$$

In such a case, to compute the likelihood under $\mathsf{H}_j$, we need to integrate over $\boldsymbol{\phi}$ as shown in (20.20). Hence, in general, the global LLR

$$\begin{aligned} &L[N] \\ &= \log\frac{\int_{\chi_\phi} f\left(\{x_1[n]\}|\boldsymbol{\phi}\right)\cdots f\left(\{x_K[n]\}|\boldsymbol{\phi}\right) f_1\left(\boldsymbol{\phi}\right) d\boldsymbol{\phi}}{\int_{\chi_\phi} f\left(\{x_1[n]\}|\boldsymbol{\phi}\right)\cdots f\left(\{x_K[n]\}|\boldsymbol{\phi}\right) f_0\left(\boldsymbol{\phi}\right) d\boldsymbol{\phi}}, \end{aligned} \tag{20.26}$$

does not have a closed-form expression. However, for a reasonable number of latent variables (i.e., entries of $\boldsymbol{\phi}$), effective numerical computation may be possible through Monte Carlo simulations. Once $L[n]$ is numerically computed, SPRT can be applied as in (20.23) and (20.24).

For decentralized detection, each node $k$ can now compute the functions of $\{x_k[n]\}_n$ included in $f\left(\{x_k[n]\}_n|\boldsymbol{\phi}\right)$ (see the example below), which has a closed-form expression thanks to the assumed conjugate prior on the parameter vector $\boldsymbol{\theta}_k$ [see (20.20)], and send event-based samples to the FC. Upon receiving such messages, the FC computes approximations to those functions; uses them in (20.26) to compute $\tilde{L}[N]$, an approximate global LLR; and applies the SPRT procedure using $\tilde{L}[N]$. Details will be provided in Section 20.2.3.4.

#### 20.2.3.3 Example

As an example to the multimodal detection scheme presented in this section, consider a system with three types of information sources: Gaussian source (e.g., real-valued physical measurements), Poisson source (e.g., event occurrences), and multinomial source (e.g., texts). We aim to find the closed-form expression of the sufficient statistic $f\left(\{x[n]\}|\boldsymbol{\phi}\right)$ for each modality. Let the discrete-time signals

$$x_g[n] \sim \mathcal{N}(\mu_\phi, \sigma^2),\ \ x_p[n] \sim \text{Pois}(\lambda_\phi),$$
$$\boldsymbol{x}_m[n] \sim \text{Mult}(1, \boldsymbol{p}_\phi),\ n \in \mathbb{N}, \tag{20.27}$$

denote the Gaussian, Poisson, and multinomial observations, respectively (see Figure 20.6). Multinomial distribution with a single trial and category probabilities $\boldsymbol{p}_\phi = [p_{\phi,1}, \ldots, p_{\phi,M}]$ is used for $\boldsymbol{x}_m[n]$, whose realization is a binary vector with an entry 1 at the index corresponding to the category observed at time $n$, and 0 at the others. The Poisson observation $x_p[n]$ denotes the number of occurrences for an event of interest in a unit time interval, where $\lambda_\phi$ is the average rate of event occurrences.

Among the parameters, only the variance $\sigma^2$ of the Gaussian model is assumed known. We assume conjugate prior distributions for the unknown parameters. Specifically, we assume a Gaussian prior on the mean

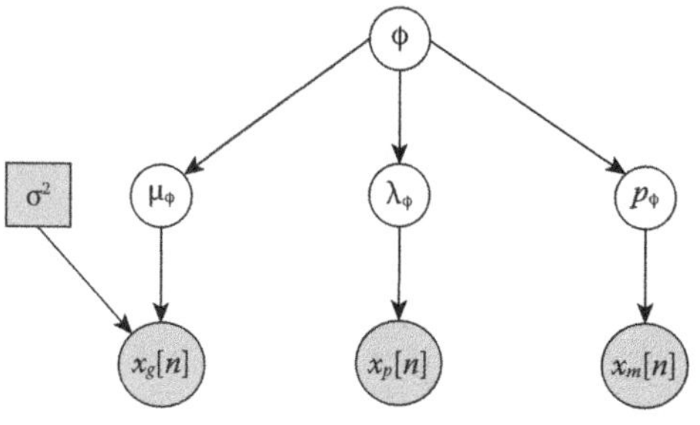


**FIGURE 20.6**
The Bayesian network considered in the example. The variance of the Gaussian source, which is a known constant, is represented by a filled square.

$\mu_\phi$ of the Gaussian model, a gamma prior on the rate parameter $\lambda_\phi$ of the Poisson model, and a Dirichlet prior on the probability vector $\boldsymbol{p}_\phi$ of the multinomial model, that is,

$$\mu_\phi \sim \mathcal{N}(\bar{\mu}_\phi, \bar{\sigma}_\phi^2),\ \ \lambda_\phi \sim \Gamma(\alpha_\phi, \beta_\phi),\ \ \boldsymbol{p}_\phi \sim \text{Dir}(\boldsymbol{\gamma}_\phi), \tag{20.28}$$

where the hyperparameters $\bar{\mu}_\phi, \bar{\sigma}_\phi^2, \alpha_\phi, \beta_\phi$, and $\boldsymbol{\gamma}_\phi$ are completely specified by the latent variable vector $\boldsymbol{\phi}$.

**Lemma 20.7**

For the example given in (20.27) and (20.28), the joint distribution of observations from each source conditioned on $\boldsymbol{\phi}$ is given by

$$f\left(\{x_g[n]\}_{n=1}^N|\boldsymbol{\phi}\right) = \frac{\exp\left(-\frac{\sum_{n=1}^N x_g[n]^2}{2\sigma^2} - \frac{\bar{\mu}_\phi^2}{2\bar{\sigma}_\phi^2} + \frac{\left(\frac{\sum_{n=1}^N x_g[n]}{\sigma^2} + \frac{\bar{\mu}_\phi}{\bar{\sigma}_\phi^2}\right)^2}{2\left(\frac{N}{\sigma^2} + \frac{1}{\bar{\sigma}_\phi^2}\right)}\right)}{(2\pi)^{N/2}\sigma^N\bar{\sigma}_\phi\sqrt{\frac{N}{\sigma^2} + \frac{1}{\bar{\sigma}_\phi^2}}}, \tag{20.29}$$

$$f\left(\{x_p[n]\}_{n=1}^N|\boldsymbol{\phi}\right) = \frac{\Gamma\left(\alpha_\phi + \sum_{n=1}^N x_p[n]\right)}{\Gamma(\alpha_\phi)\prod_{n=1}^N x_p[n]!}\frac{\beta_\phi^{\alpha_\phi}}{(\beta_\phi + N)^{\alpha_\phi + \sum_{n=1}^N x_p[n]}}, \tag{20.30}$$

$$f\left(\{\boldsymbol{x}_m[n]\}_{n=1}^N|\boldsymbol{\phi}\right) = \frac{\Gamma\left(\sum_{i=1}^M \gamma_{\phi,i}\right)}{\Gamma\left(\sum_{i=1}^M\left(\gamma_{\phi,i} + \sum_{n=1}^N x_{m,i}[n]\right)\right)} \times \prod_{i=1}^M \frac{\Gamma\left(\gamma_{\phi,i} + \sum_{n=1}^N x_{m,i}[n]\right)}{\Gamma(\gamma_{\phi,i})}, \tag{20.31}$$

where $\Gamma(\cdot)$ is the gamma function.

**PROOF** Given $\boldsymbol{\phi}$, $\{x_g[n]\}$ are iid with $\mathcal{N}(\mu_\phi, \sigma^2)$, where $\mu_\phi \sim \mathcal{N}(\bar{\mu}_\phi, \bar{\sigma}_\phi^2)$, hence

$$f\left(\{x_g[n]\}, \mu_\phi\right) = \frac{\exp\left(-\frac{\sum_{n=1}^N (x_g[n]-\mu_\phi)^2}{2\sigma^2} - \frac{(\mu_\phi - \bar{\mu}_\phi)^2}{2\bar{\sigma}_\phi^2}\right)}{(2\pi)^{\frac{N+1}{2}}\sigma^N\bar{\sigma}_\phi}.$$

After some manipulations, we can show that

$$f\left(\{x_g[n]\},\mu_\phi\right) = \frac{\exp\left(-\frac{\sum_{n=1}^N x_g[n]^2}{2\sigma^2} - \frac{\bar{\mu}_\phi^2}{2\bar{\sigma}_\phi^2} + \frac{\left(\frac{\sum_{n=1}^N x_g[n]}{\sigma^2}+\frac{\bar{\mu}_\phi}{\bar{\sigma}_\phi^2}\right)^2}{2\left(\frac{N}{\sigma^2}+\frac{1}{\bar{\sigma}_\phi^2}\right)}\right)}{(2\pi)^{N/2}\sigma^N\bar{\sigma}_\phi\sqrt{\frac{N}{\sigma^2}+\frac{1}{\bar{\sigma}_\phi^2}}}$$

$$\times \underbrace{\sqrt{\frac{\frac{1}{\bar{\sigma}_\phi^2}+\frac{N}{\sigma^2}}{2\pi}}\exp\left(-\frac{\frac{1}{\bar{\sigma}_\phi^2}+\frac{N}{\sigma^2}}{2}\left(\mu_\phi - \frac{\frac{\bar{\mu}_\phi}{\bar{\sigma}_\phi^2}+\frac{\sum_{n=1}^N x_g[n]}{\sigma^2}}{\frac{1}{\bar{\sigma}_\phi^2}+\frac{N}{\sigma^2}}\right)^2\right)}_{f(\mu_\phi|\{x_g[n]\})},$$

where from the conjugate prior property, it is known that the posterior distribution of $\mu_\phi$ is also Gaussian with mean $\frac{\frac{\bar{\mu}_\phi}{\bar{\sigma}_\phi^2}+\frac{\sum_{n=1}^N x_g[n]}{\sigma^2}}{\frac{1}{\bar{\sigma}_\phi^2}+\frac{N}{\sigma^2}}$ and variance $\frac{1}{\bar{\sigma}_\phi^2}+\frac{N}{\sigma^2}$. Hence, the result in (20.29) follows. Note that $f\left(\{x_g[n]\}_{n=1}^N|\boldsymbol{\phi}\right)$ is a multivariate Gaussian distribution, where all entries of the mean vector are $\bar{\mu}_\phi$, the diagonal entries of covariance matrix are $\bar{\sigma}_\phi^2+\sigma^2$, and the off-diagonals are $\bar{\sigma}_\phi^2$.

Similarly, for Poisson observations, we write

$$f\left(\{x_p[n]\},\lambda_\phi\right) = \prod_{n=1}^N \frac{\lambda_\phi^{x_p[n]}e^{-\lambda_\phi}}{x_p[n]!}\frac{\beta_\phi^{\alpha_\phi}}{\Gamma(\alpha_\phi)}\lambda_\phi^{\alpha_\phi-1}e^{-\beta_\phi\lambda_\phi},$$

since $\{x_p[n]\}$ are iid with $\text{Pois}(\lambda_\phi)$ given $\lambda_\phi$, and the prior is $\Gamma(\alpha_\phi,\beta_\phi)$. The posterior distribution is known to be $\Gamma(\alpha_\phi+\sum_{n=1}^N x_p[n],\beta_\phi+N)$; hence,

$$f\left(\{x_p[n]\},\lambda_\phi\right) = \underbrace{\frac{(\beta_\phi+N)^{\alpha_\phi+\sum_{n=1}^N x_p[n]}}{\Gamma\left(\alpha_\phi+\sum_{n=1}^N x_p[n]\right)}\lambda_\phi^{\alpha_\phi+\sum_{n=1}^N x_p[n]-1}e^{-(\beta_\phi+N)\lambda_\phi}}_{f(\lambda_\phi|\{x_p[n]\})}$$

$$\times \underbrace{\frac{\Gamma\left(\alpha_\phi+\sum_{n=1}^N x_p[n]\right)}{\Gamma(\alpha_\phi)\prod_{n=1}^N x_p[n]!}\frac{\beta_\phi^{\alpha_\phi}}{(\beta_\phi+N)^{\alpha_\phi+\sum_{n=1}^N x_p[n]}}}_{f(\{x_p[n]\})},$$

proving (20.30).

Finally, for the multinomial observations, $\{\boldsymbol{x}_m[n]\}$ are iid with the probability vector $\boldsymbol{p}_\phi$; the prior is $\text{Dir}(\boldsymbol{\gamma}_\phi)$; and the posterior is $\text{Dir}(\boldsymbol{\gamma}_\phi+\sum_{n=1}^N \boldsymbol{x}_m[n])$; hence,

$$f\left(\{\boldsymbol{x}_m[n]\},\boldsymbol{p}_\phi\right) = \prod_{i=1}^M p_{\phi,i}^{\sum_{n=1}^N x_{m,i}[n]}\frac{\Gamma\left(\sum_{i=1}^M\gamma_{\phi,i}\right)}{\prod_{i=1}^M\Gamma(\gamma_{\phi,i})}\prod_{i=1}^M p_{\phi,i}^{\gamma_{\phi,i}-1}$$

$$= \frac{\Gamma\left(\sum_{i=1}^M\left(\gamma_{\phi,i}+\sum_{n=1}^N x_{m,i}[n]\right)\right)}{\prod_{i=1}^M\Gamma\left(\gamma_{\phi,i}+\sum_{n=1}^N x_{m,i}[n]\right)}\prod_{i=1}^M p_{\phi,i}^{\gamma_{\phi,i}+\sum_{n=1}^N x_{m,i}[n]-1}$$

$$\times \underbrace{\frac{\Gamma\left(\sum_{i=1}^M\gamma_{\phi,i}\right)}{\Gamma\left(\sum_{i=1}^M\left(\gamma_{\phi,i}+\sum_{n=1}^N x_{m,i}[n]\right)\right)} \times \prod_{i=1}^M\frac{\Gamma\left(\gamma_{\phi,i}+\sum_{n=1}^N x_{m,i}[n]\right)}{\Gamma(\gamma_{\phi,i})}}_{f(\{\boldsymbol{x}_m[n]\}|\boldsymbol{\phi})},$$

concluding the proof. ■

In testing hypotheses that deterministically specify $\boldsymbol{\phi}$ as in (20.21), the local LLR for each modality can be computed using Lemma 20.7, for example,

$$L_g[N] = \log\frac{f\left(\{x_g[n]\}_{n=1}^N|\boldsymbol{\phi}=\boldsymbol{\phi}_1\right)}{f\left(\{x_g[n]\}_{n=1}^N|\boldsymbol{\phi}=\boldsymbol{\phi}_0\right)}.$$

Then, under a centralized setup, SPRT can applied as in (20.23) and (20.24) using the global LLR

$$L[N] = L_g[N]+L_p[N]+L_m[N], \tag{20.32}$$

or under a decentralized setup, each node reports event-based samples of its local LLR, and the FC applies SPRT using $\tilde{L}[N]=\tilde{L}_g[N]+\tilde{L}_p[N]+\tilde{L}_m[N]$.

On the contrary, while testing hypotheses that statistically specify $\boldsymbol{\phi}$ as in (20.25), the global LLR is computed using the results of Lemma 20.7 in (20.26). In this case, for decentralized detection, each node can only compute the functions of its observations that appear in the conditional joint distributions, given by (20.29)–(20.31), and do not depend on $\boldsymbol{\phi}$.

For example, the Gaussian node from (20.29) can compute

$$\frac{\sum_{n=1}^N x_g[n]^2}{2\sigma^2} \text{ and } \frac{\sum_{n=1}^N x_g[n]}{\sigma^2}, \tag{20.33}$$

and send their event-based samples to the FC, which can effectively recover such samples as will be shown next, and uses them in (20.29). Although, in this case, event-based sampling is used to transmit only some simple functions of the observations, which needs further processing at the FC, the advantages of using event-based sampling on the functions of $x_g[n]$, instead of conventional uniform sampling on $x_g[n]$ itself, is still significant. First, the error induced by using the recovered functions in the highly nonlinear expression of (20.29) is smaller than that results from using the recovered observations in (20.29) because transmission loss grows with the processing at the FC. Second, the transmission rate can be considerably lower than that of uniform sampling because only the important changes in functions are reported, censoring the uninformative observations.

The Poisson and multinomial processes are inherently event-based as each observation $x_p[n]/\boldsymbol{x}_m[n]$ marks an event occurrence after a random (e.g., exponentially distributed for Poisson process) waiting time since $x_p[n-1]/\boldsymbol{x}_m[n-1]$. Therefore, each new observation $x_p[n]/\boldsymbol{x}_m[n]$ is reported to the FC. Moreover, they take integer values ($\boldsymbol{x}_m[n]$ can be represented by the index of nonzero element); thus, no quantization error takes place.

#### 20.2.3.4 Decentralized Implementation

Due to the nonlinear processing of recovered messages at the FC [cf. (20.33)], in the decentralized testing of hypotheses with statistical descriptions of $\boldsymbol{\phi}$, we should more carefully take care of the overshoot problem.

In level-triggered sampling, the change in the signal is measured with respect to the signal value at the most recent sampling time, which possibly includes an overshoot and hence is not perfectly available to the FC even if a multibit scheme is used to quantize the overshoot. Therefore, the past quantization errors, as well as the current one, cumulatively decrease the precision of recovered signal at the FC. The accumulation of quantization errors may not be of practical interest if the individual errors are small (i.e., sufficiently large number of bits are used for quantization and/or the jumps in the signal are sufficiently small) and stay small after the processing at the FC, and the FC makes a quick decision (i.e., the constraints on detection error probabilities are not very stringent). However, causing an avalanche effect, it causes a significant problem for the asymptotic decision time performance of the decentralized detector (e.g., in a regime of large decision times due to stringent error probability constraints) even if the individual errors at the FC are small.

In [31], using of fixed reference levels is proposed to improve the asymptotic performance of level-triggered sampling, which corresponds to LCSH (see Figure 20.7). Since LCSH handles the overshoot problem better than level-triggered sampling, it suits better to the case in (20.33) where the FC performs nonlinear processing on the recovered signal. We here show that it also achieves a better asymptotic performance at the expense of much more complicated nonasymptotic performance analysis. Furthermore, we consider multiple crossings of sampling levels due to large jumps in the signal (Figure 20.7).

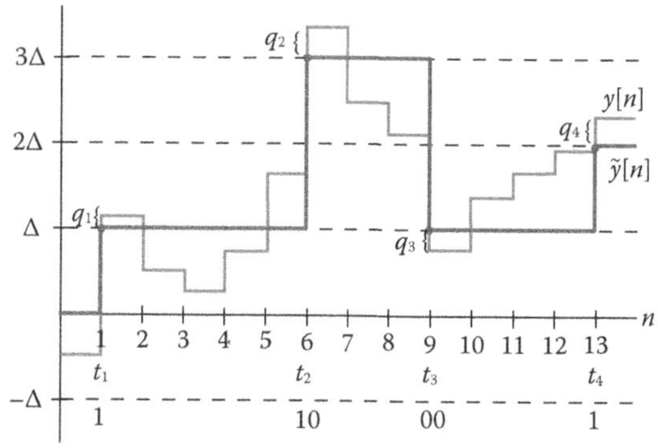


**FIGURE 20.7**
Level-crossing sampling with hysteresis applied to $y[n]$. The recovered signal $\tilde{y}[n]$ at the FC and the transmitted bits are shown. Multiple crossings are handled by transmitting additional bits. Overshoots $\{q_i\}$ take effect individually (no overshoot accumulation).

Sampling a signal $y[n]$ via LCSH with level spacing $\Delta$, a sample is taken whenever an upper or lower sampling level is crossed, as shown in Figure 20.7. Specifically, the $i$th sample is taken at time

$$t_i \triangleq \{n > t_{i-1} : |y[n] - \psi_{i-1}\Delta| \geq \Delta\}, \qquad (20.34)$$

where $\psi_{i-1}$ is the sampling level in terms of $\Delta$ that was most recently crossed. In general, $y[t_i]$ may cross multiple levels, that is, the number of level crossings

$$\eta_i \triangleq \left\lfloor \frac{|y[t_i] - \psi_{i-1}\Delta|}{\Delta} \right\rfloor \geq 1. \qquad (20.35)$$

In addition to the sign bit

$$b_{i,1} = \text{sign}(y[t_i] - \psi_{i-1}\Delta), \qquad (20.36)$$

which encodes the first crossing with its direction, we send

$$r \triangleq \left\lceil \frac{\eta_i - 1}{2} \right\rceil, \qquad (20.37)$$

more bits $b_{i,2}, \ldots, b_{i,r+1}$, where each following bit 1/0 represents a double/single crossing. For instance, the bit sequence 0110, where the first 0 denotes downward crossing (i.e., $b_{i,1} = -1$), is sent for $-7\Delta < y[t_i] - \psi_{i-1}\Delta \leq -6\Delta$.

In that way, the FC can obtain $\eta_i$ from received bits and keep track of the most recently crossed level as

$$\psi_i = \psi_{i-1} + b_{i,1}\eta_i. \qquad (20.38)$$

It approximates $y[n]$ with

$$\tilde{y}[n] = \psi_i\Delta, \; t_i \leq n < t_{i+1}. \qquad (20.39)$$

As a result, only the current overshoot causes error in $\tilde{y}[n]$; that is, overshoots do not accumulate, as opposed to level-triggered sampling. This is especially important when reporting signals that are processed further at the FC [cf. (20.33)]. It also ensures order-2 asymptotic optimality with finite number of bits per sample when used to transmit iid local LLRs from unimodal sources (i.e., the case in Theorem 20.1).

**Theorem 20.2**

Consider the decentralized detector that uses the LCSH-based transmission scheme given in (20.34)–(20.39) to report the local LLRs, $\{L_k[n]\}$, from iid nodes to the FC, and applies the SPRT procedure at the FC substituting the recovered global LLR, $\tilde{L}[n] = \sum_{k=1}^{K} \tilde{L}_k[n]$, in (20.23) and (20.24). It is order-2 asymptotically optimum, that is,

$$\mathsf{E}_j[S] - \mathsf{E}_j[S_o] = O(1),\ j = 0, 1,\ \text{ as } \alpha, \beta \to 0, \quad (20.40)$$

where $S$ and $S_o$ are the decision times of decentralized detector and the optimum (centralized) SPRT satisfying the same (type-I and type-II) error probability bounds $\alpha$ and $\beta$ [cf. (20.5)].

**PROOF** Assuming finite and nonzero Kullback–Leibler (KL) information numbers $-\mathsf{E}_0[L[1]], \mathsf{E}_1[L[1]]$, for order-2 asymptotic optimality, it suffices to show that

$$\mathsf{E}_1\left[L[S]\right] - \mathsf{E}_1\left[L[S_o]\right] = O(1),\ j = 0, 1,\ \text{ as } \alpha, \beta \to 0. \quad (20.41)$$

The proof under $\mathsf{H}_0$ follows similarly. Let us start by writing

$$\mathsf{E}_1\left[L[S]\right] = \mathsf{E}_1\left[\tilde{L}[S] + (L[S] - \tilde{L}[S])\right]. \quad (20.42)$$

Thanks to the multibit transmission scheme based on LCSH, no overshoot accumulation takes place, and thus the absolute errors satisfy

$$|\tilde{L}_k[n] - L_k[n]| < \Delta,\ \forall k, n,$$
$$|\tilde{L}[n] - L[n]| = \left|\sum_{k=1}^{K} \tilde{L}_k[n] - L_k[n]\right|$$
$$\le \sum_{k=1}^{K} |\tilde{L}_k[n] - L_k[n]| < K\Delta,\ \forall n. \quad (20.43)$$

The approximate LLR $\tilde{L}[S]$ at the stopping time exceeds $A$ or $-B$ by a finite amount, that is,

$$\tilde{L}[S] < A + C \text{ or } \tilde{L}[S] > -B - C, \quad (20.44)$$

where $C$ is a constant. Now let us analyze how the stopping threshold $A$ behaves as $\alpha, \beta \to 0$. Start with

$$\alpha = \mathsf{P}_0(\tilde{L}[S] \ge A) = \mathsf{E}_0\left[\mathbb{1}_{\{\tilde{L}[S] \ge A\}}\right],$$

where applying a change of measure using $e^{-L[S]}$ as in Lemma 20.1 we can write

$$\alpha = \mathsf{E}_1\left[e^{-L[S]}\mathbb{1}_{\{\tilde{L}[S] \ge A\}}\right]$$
$$= \mathsf{E}_1\left[e^{-\tilde{L}[S] + \tilde{L}[S] - L[S]}\mathbb{1}_{\{\tilde{L}[S] \ge A\}}\right].$$

From (20.43),

$$\alpha \le e^{-A + K\Delta}$$
$$A \le |\log \alpha| + K\Delta. \quad (20.45)$$

Combining (20.42)–(20.45), we get

$$\mathsf{E}_1\left[L[S]\right] \le |\log \alpha| + 2K\Delta + C. \quad (20.46)$$

In SPRT with discrete-time observations, due to the overshoot problem, the KL divergence at the stopping time is larger than that in the no-overshoot case [53, page 21], that is,

$$\mathsf{E}_1[L[S_o]] \ge (1 - \beta)\log\frac{1 - \beta}{\alpha} + \beta \log\frac{\beta}{1 - \alpha}$$
$$= (1 - \beta)|\log \alpha| - \beta|\log \beta| + (1 - \beta)\log(1 - \beta)$$
$$- \beta \log(1 - \alpha). \quad (20.47)$$

From (20.46) and (20.47),

$$\mathsf{E}_1\left[L[S]\right] - \mathsf{E}_1[L[S_o]]$$
$$\le 2K\Delta + C - \beta|\log \alpha| - \beta|\log \beta|$$
$$+ (1 - \beta)\log(1 - \beta) - \beta\log(1 - \alpha),$$

where the last three terms tend to zero as $\alpha, \beta \to 0$. Assuming $\alpha$ and $\beta$ tend to zero at comparable rates, the term $\beta|\log \alpha|$ also tends to zero, leaving us with the constant $2K\Delta + C$. The decentralized detector applies the SPRT procedure with a summary of observations; hence, it cannot satisfy the error probability constraints with a smaller KL divergence than that of the centralized SPRT, that is, $\mathsf{E}_1\left[L[S]\right] - \mathsf{E}_1[L[S_o]] \ge 0$. This concludes the proof. ■

In fact, the proof for (20.40) holds also for the case of multimodal sources in (20.22) and (20.32), where the local LLRs are independent but not identically distributed. Since in this non-iid case SPRT may not be optimum, we cannot claim asymptotic optimality by satisfying (20.40). However, centralized SPRT still serves as a very important benchmark; hence, (20.40) is a valuable result also for the multimodal case.

The power of Theorem 20.2 lies in the fact that the LCSH-based decentralized detector achieves order-2 asymptotic optimality by using a finite (in most cases small) number of bits per sample. Order-2 asymptotic optimality resolves the overshoot problem because it is the state-of-the-art performance in the no-overshoot case (i.e., with continuous-time band-limited observations), achieved by LCSH, which coincides with level-triggered sampling in this case. On the contrary, for order-2 asymptotic optimality with discrete-time observations, the number of bits per sample required by the level-triggered sampling-based detector tends to infinity with a reasonably low rate, $\log|\log\alpha|$ [73, Section IV-B].

In the LCSH-based detector, to avoid overshoot accumulation, the overshoot of the last sample is included toward the new sample, correlating the two samples. Consequently, samples (i.e., messages of change in the signal) that result from LCSH are neither independent nor identically distributed. As opposed to level-triggered sampling, in which samples are iid and hence form a renewal process, the statistical descriptions of samples in LCSH are quite intractable. The elegant (nonasymptotic and asymptotic) results obtained for level-triggered sampling in [74] therefore do not apply to LCSH here.

## 20.3 Decentralized Estimation

In this section, we are interested in sequentially estimating a vector of parameters (i.e., regression coefficients) $\boldsymbol{\theta} \in \mathbb{R}^p$ at a random stopping time $S$ in the following linear (regression) model:

$$x[n] = \boldsymbol{h}[n]^T\boldsymbol{\theta} + w[n],\ n \in \mathbb{N}, \tag{20.48}$$

where $x[n] \in \mathbb{R}$ is the observed sample, $\boldsymbol{h}[n] \in \mathbb{R}^p$ is the vector of regressors, and $w[n] \in \mathbb{R}$ is the additive noise. We consider the general case in which $\boldsymbol{h}[n]$ is random and observed at time $n$, which covers the deterministic $\boldsymbol{h}[n]$ case as a special case. This linear model is commonly used in many applications. For example, in system identification, $\boldsymbol{\theta}$ is the unknown system coefficients, $\boldsymbol{h}[n]$ is the (random) input applied to the system, and $x[n]$ is the output at time $n$. Another example is the estimation of wireless (multiple-access) channel coefficients, in which $\boldsymbol{\theta}$ is the unknown channel coefficients, $\boldsymbol{h}[n]$ is the transmitted (random) pilot signal, $x[n]$ is the received signal, and $w[n]$ is the additive channel noise.

In (20.48), at each time $n$, we observe the sample $x[n]$ and the vector $\boldsymbol{h}[n]$; hence, $\{(x[m], \boldsymbol{h}[m])\}_{m=1}^n$ are available. We assume $\{w[n]\}$ are i.i.d. with $\mathsf{E}[w[n]] = 0$ and $\mathsf{Var}(w[n]) = \sigma^2$. The least squares (LS) estimator minimizes the sum of squared errors, that is,

$$\hat{\boldsymbol{\theta}}_N = \arg\min_{\boldsymbol{\theta}} \sum_{n=1}^{N} (x[n] - \boldsymbol{h}[n]^T\boldsymbol{\theta})^2, \tag{20.49}$$

and is given by

$$\begin{aligned}\hat{\boldsymbol{\theta}}_N &= \left(\sum_{n=1}^{N} \boldsymbol{h}[n]\boldsymbol{h}[n]^T\right)^{-1} \sum_{n=1}^{N} \boldsymbol{h}[n]x[n] \\ &= (\boldsymbol{H}_n^T\boldsymbol{H}_n)^{-1}\boldsymbol{H}_n^T\boldsymbol{x}_n,\end{aligned} \tag{20.50}$$

where $\boldsymbol{H}_n = [\boldsymbol{h}[1],\ldots,\boldsymbol{h}[n]]^T$ and $\boldsymbol{x}_n = [x[1],\ldots,x[n]]^T$. Note that spatial diversity (i.e., a vector of observations and a regressor matrix at time $n$) can be easily incorporated in (20.48) in the same way we deal with temporal diversity. Specifically, in (20.49) and (20.50), we would also sum over the spatial dimensions.

Under the Gaussian noise, $w[n] \sim \mathcal{N}(0,\sigma^2)$, the LS estimator coincides with the minimum variance unbiased estimator (MVUE) and achieves the CRLB, that is, $\mathsf{Cov}(\hat{\boldsymbol{\theta}}_n|\boldsymbol{H}_n) = \mathrm{CRLB}_n$. To compute the CRLB, we first write, given $\boldsymbol{\theta}$ and $\boldsymbol{H}_n$, the log-likelihood of the vector $\boldsymbol{x}_n$ as

$$\begin{aligned}L_n &= \log f(\boldsymbol{x}_n|\boldsymbol{\theta},\boldsymbol{H}_n) \\ &= -\sum_{m=1}^{n} \frac{(x[m] - \boldsymbol{h}[m]^T\boldsymbol{\theta})^2}{2\sigma^2} - \frac{t}{2}\log(2\pi\sigma^2).\end{aligned} \tag{20.51}$$

Then, we have

$$\mathrm{CRLB}_n = \left(\mathsf{E}\left[-\frac{\partial^2}{\partial\boldsymbol{\theta}^2}L_n\Big|\boldsymbol{H}_n\right]\right)^{-1} = \sigma^2\boldsymbol{U}_n^{-1}, \tag{20.52}$$

where $\mathsf{E}\left[-\frac{\partial^2}{\partial\boldsymbol{\theta}^2}L_n|\boldsymbol{H}_n\right]$ is the Fisher information matrix and $\boldsymbol{U}_n \triangleq \boldsymbol{H}_n^T\boldsymbol{H}_n$ is a nonsingular matrix. Since $\mathsf{E}[\boldsymbol{x}_n|\boldsymbol{H}_n] = \boldsymbol{H}_n\boldsymbol{\theta}$ and $\mathsf{Cov}(\boldsymbol{x}_n|\boldsymbol{H}_n) = \sigma^2\boldsymbol{I}$, from (20.50) we have $\mathsf{E}[\hat{\boldsymbol{\theta}}_n|\boldsymbol{H}_n] = \boldsymbol{\theta}$ and $\mathsf{Cov}(\hat{\boldsymbol{\theta}}_n|\boldsymbol{H}_n) = \sigma^2\boldsymbol{U}_n^{-1}$; thus, from (20.52) $\mathsf{Cov}(\hat{\boldsymbol{\theta}}_n|\boldsymbol{H}_n) = \mathrm{CRLB}_n$. Note that the maximum likelihood (ML) estimator that maximizes (20.51) coincides with the LS estimator in (20.50).

In general, the LS estimator is the best linear unbiased estimator (BLUE). In other words, any linear unbiased estimator of the form $\boldsymbol{A}_n\boldsymbol{x}_n$ with $\boldsymbol{A}_n \in \mathbb{R}^{n\times t}$, where $\mathsf{E}[\boldsymbol{A}_n\boldsymbol{x}_n|\boldsymbol{H}_n] = \boldsymbol{\theta}$, has a covariance no smaller than that of the LS estimator in (20.50), that is, $\mathsf{Cov}(\boldsymbol{A}_n\boldsymbol{x}_n|\boldsymbol{H}_n) \geq \sigma^2\boldsymbol{U}_n^{-1}$ in the positive semidefinite sense. To see this result, we write $\boldsymbol{A}_n = (\boldsymbol{H}_n^T\boldsymbol{H}_n)^{-1}\boldsymbol{H}_n^T + \boldsymbol{B}_n$ for some $\boldsymbol{B}_n \in \mathbb{R}^{n\times t}$, and then $\mathsf{Cov}(\boldsymbol{A}_n\boldsymbol{x}_n|\boldsymbol{H}_n) = \sigma^2\boldsymbol{U}_n^{-1} + \sigma^2\boldsymbol{B}_n\boldsymbol{B}_n^T$, where $\boldsymbol{B}_n\boldsymbol{B}_n^T$ is a positive semidefinite matrix.

The recursive least squares (RLS) algorithm enables us to compute $\hat{\boldsymbol{\theta}}_n$ in a recursive way as follows:

$$\begin{aligned}&\hat{\boldsymbol{\theta}}_n = \hat{\boldsymbol{\theta}}_{n-1} + \boldsymbol{q}_n(x[n] - \boldsymbol{h}[n]^T\hat{\boldsymbol{\theta}}_{n-1}), \\ &\text{where } \boldsymbol{q}_n = \frac{\boldsymbol{P}_{n-1}\boldsymbol{h}[n]}{1 + \boldsymbol{h}[n]^T\boldsymbol{P}_{n-1}\boldsymbol{h}[n]} \\ &\text{and } \boldsymbol{P}_n = \boldsymbol{P}_{n-1} - \boldsymbol{q}_n\boldsymbol{h}[n]^T\boldsymbol{P}_{n-1},\end{aligned} \tag{20.53}$$

where $\boldsymbol{q}_n \in \mathbb{R}^p$ is a gain vector and $\boldsymbol{P}_n = \boldsymbol{U}_n^{-1}$. While applying RLS, we first initialize $\hat{\boldsymbol{\theta}}_0 = 0$ and $\boldsymbol{P}_0 = \delta^{-1}\boldsymbol{I}$, where 0 represents a zero vector and $\delta$ is a small number, and then at each time $n$ compute $q_n$, $\hat{\theta}_n$, and $P_n$ as in (20.53).

### 20.3.1 Background

Energy constraints are inherent to wireless sensor networks [1]. Since data transmission is the primary source of energy consumption, it is essential to keep transmission rates low in wireless sensor networks, resulting in a *decentralized* setup. Decentralized parameter estimation is a fundamental task performed in wireless sensor networks [5,10,14,35,45,48,51,52,54,68,69,78]. In *sequential* estimation, the objective is to minimize the (average) number of observations for a given target accuracy level [36]. To that end, a sequential estimator $(S, \hat{\theta}_S)$, as opposed to a traditional fixed-sample-size estimator, is equipped with a stopping rule which determines an appropriate time $S$ to stop taking new observations based on the observation history. Hence, the stopping time $S$ (i.e., the number of observations used in estimation) is a random variable. Endowed with a stopping mechanism, a sequential estimator saves not only time but also energy, both of which are critical resources. In particular, it avoids unnecessary data processing and transmission.

Decentralized parameter estimation has been mainly studied under two different network topologies. In the first one, sensors communicate to an FC that performs estimation based on the received information (e.g., [14,35,45,48,51,68]). The other commonly studied topology is called ad hoc network, in which there is no designated FC, but sensors compute their local estimators and communicate them through the network (e.g., [5,10,52,54,78]). Decentralized estimation under both network topologies is reviewed in [69]. Many existing works consider parameter estimation in linear models (e.g., [10,14,35,45,54,68]). Whereas in [5,48,51,52,69,78] a general nonlinear signal model is assumed. The majority of existing works on decentralized estimation (e.g., [10,14,35,45,48,51,52,54,68,69]) study fixed-sample-size estimation. There are a few works, such as [5,16], that consider sequential decentralized parameter estimation. Nevertheless, [5] assumes that sensors transmit real numbers, and [16] focuses on continuous-time observations, which can be seen as practical limitations.

In decentralized detection [17,73,74,76] and estimation [75], level-triggered sampling (cf. Figure 20.3), an adaptive sampling technique which infrequently transmits a few bits, for example, one bit, from sensors to the FC, has been used to achieve low-rate transmission. It has been also shown that the decentralized schemes based on level-triggered sampling significantly outperform their counterparts based on conventional uniform sampling in terms of average stopping time. We here use a form of level-triggered sampling that infrequently transmits a single pulse from sensors to the FC and, at the same time, achieves a close-to-optimum average stopping time performance [76].

The stopping capability of sequential estimators comes with the cost of sophisticated analysis. In most cases, it is not possible with discrete-time observations to find an optimum sequential estimator that attains the sequential Cramér-Rao lower bound (CRLB) if the stopping time $S$ is adapted to the complete observation history [20]. Alternatively, in [22] and more recently in [16,75], it was proposed to restrict $S$ to stopping times that are adapted to a specific subset of the complete observation history, which leads to simple optimum solutions. This idea of using a restricted stopping time first appeared in [22] with no optimality result. In [16], with continuous-time observations, a sequential estimator with a restricted stopping time was shown to achieve the sequential version of the CRLB for scalar parameter estimation. In [75], for scalar parameter estimation with discrete-time observations, a similar sequential estimator was shown to achieve the conditional sequential CRLB for the same restricted class of stopping times.

We deal with discrete-time observations in this section. In Section 20.3.2, the optimum sequential estimator that achieves the conditional sequential CRLB for a certain class of stopping times is discussed. We then develop in Section 20.3.3 a computation- and energy-efficient decentralized scheme based on level-triggered sampling for sequential estimation of vector parameters.

### 20.3.2 Optimum Sequential Estimator

In this section, we aim to find the optimal pair $(S, \hat{\theta}_S)$ of stopping time and estimator corresponding to the optimal sequential estimator. The stopping time for a sequential estimator is determined according to a target estimation accuracy. In general, the average stopping time is minimized subject to a constraint on the estimation accuracy, which is a function of the estimator covariance, that is,

$$\min_{S, \hat{\theta}_S} \mathsf{E}[S] \ \text{ s.t. } f\left(\mathsf{Cov}(\hat{\theta}_S)\right) \leq C, \tag{20.54}$$

where $f(\cdot)$ is a function from $\mathbb{R}^{p \times p}$ to $\mathbb{R}$ and $C \in \mathbb{R}$ is the target accuracy level.

The accuracy function $f$ should be a monotonic function of the covariance matrix $\mathsf{Cov}(\hat{\theta}_S)$, which is positive semidefinite, to make consistent accuracy assessments; for example, $f(\mathsf{Cov}(\hat{\theta}_S)) > f(\mathsf{Cov}(\hat{\theta}_{S'}))$ for $S < S'$ since $\mathsf{Cov}(\hat{\theta}_S) \succ \mathsf{Cov}(\hat{\theta}_{S'})$ in the positive definite sense. Two popular and easy-to-compute choices are the trace $\mathsf{Tr}(\cdot)$, which corresponds to the mean squared error (MSE), and the Frobenius norm $\|\cdot\|_F$. Before handling the problem in (20.54), let us explain why we are interested in restricted stopping times that are adapted to a subset of observation history.

#### 20.3.2.1 Restricted Stopping Time

Denote $\{\mathcal{F}_n\}$ as the filtration that corresponds to the samples $\{x[1],\ldots,x[n]\}$ where $\mathcal{F}_n = \sigma\{x[1],\ldots,x[n]\}$ is the σ-algebra generated by the samples observed up to time $n$, that is, the accumulated history related to the observed samples, and $\mathcal{F}_0$ is the trivial σ-algebra. Similarly, we define the filtration $\{\mathcal{H}_n\}$ where $\mathcal{H}_n = \sigma\{h[1],\ldots,h[n]\}$ and $\mathcal{H}_0$ is again the trivial σ-algebra. It is known that, in general, *with discrete-time observations* and an unrestricted stopping time, that is $\{\mathcal{F}_n \cup \mathcal{H}_n\}$-adapted, the sequential CRLB is not attainable under any noise distribution except for the Bernoulli noise [20].

On the contrary, in the case of *continuous-time observations with continuous paths*, the sequential CRLB is attained by the LS estimator with an $\{\mathcal{H}_n\}$-adapted stopping time that depends only on $H_S$ [16]. Moreover, in the following lemma, we show that, with discrete-time observations, the LS estimator attains the conditional sequential CRLB for the $\{\mathcal{H}_n\}$-adapted stopping times.

**Lemma 20.8**

With a monotonic accuracy function $f$ and an $\{\mathcal{H}_n\}$-adapted stopping time $S$, we can write

$$f\left(\mathsf{Cov}(\hat{\theta}_S|\mathcal{H}_S)\right) \geq f\left(\sigma^2 U_S^{-1}\right), \tag{20.55}$$

for all unbiased estimators under Gaussian noise, and for all linear unbiased estimators under non-Gaussian noise, and the LS estimator

$$\hat{\theta}_S = U_S^{-1}V_S,\ V_S \triangleq H_S^T x_S, \tag{20.56}$$

satisfies the inequality in (20.55) with equality.

**PROOF** Since the LS estimator, with $\mathsf{Cov}(\hat{\theta}_n|\mathcal{H}_n) = \sigma^2 U_n^{-1}$, is the MVUE under Gaussian noise and the BLUE under non-Gaussian noise, we write

$$\begin{aligned} f&\left(\mathsf{Cov}(\hat{\theta}_S|\mathcal{H}_S)\right) \\ &= f\left(\mathsf{E}\left[\sum_{n=1}^{\infty}(\hat{\theta}_n-\theta)(\hat{\theta}_n-\theta)^T\,\mathbb{1}_{\{n=S\}}\Big|H_n\right]\right) \\ &= f\left(\sum_{n=1}^{\infty}\mathsf{E}\left[(\hat{\theta}_n-\theta)(\hat{\theta}_n-\theta)^T\Big|H_n\right]\mathbb{1}_{\{n=S\}}\right) && (20.57) \\ &\geq f\left(\sum_{n=1}^{\infty}\sigma^2 U_n^{-1}\,\mathbb{1}_{\{n=S\}}\right) && (20.58) \\ &= f\left(\sigma^2 U_S^{-1}\right), && (20.59) \end{aligned}$$

for all unbiased estimators under Gaussian noise and for all linear unbiased estimators under non-Gaussian noise. The indicator function $\mathbb{1}_{\{A\}} = 1$ if $A$ is true, and 0 otherwise. We used the facts that the event $\{S = n\}$ is $\mathcal{H}_n$-measurable and $\mathsf{E}[(\hat{\theta}_n-\theta)(\hat{\theta}_n-\theta)^T|H_n] = \mathsf{Cov}(\hat{\theta}_n|H_n) \geq \sigma^2 U_n^{-1}$ to write (20.57) and (20.58), respectively. ■

#### 20.3.2.2 Optimum Conditional Estimator

We are interested in $\{\mathcal{H}_n\}$-adapted stopping times to use the optimality property of the LS estimator in the sequential sense, shown in Lemma 20.8.

The common practice in sequential analysis minimizes the average stopping time subject to a constraint on the estimation accuracy which is a function of the estimator covariance. The optimum solution to this classical problem proves to be intractable for even moderate number of unknown parameters [72]. Hence, it is not a convenient model for decentralized estimation. Therefore, we follow an alternative approach and formulate the problem conditioned on the observed $\{h[n]\}$ values, which yields a tractable optimum solution for any number of parameters.

In the presence of an ancillary statistic whose distribution does not depend on the parameters to be estimated, such as the regressor matrix $H_n$, the conditional covariance $\mathsf{Cov}(\hat{\theta}_n|H_n)$ can be used to assess the accuracy of the estimator more precisely than the (unconditional) covariance, which is in fact the mean of the former (i.e., $\mathsf{Cov}(\hat{\theta}_S) = \mathsf{E}[\mathsf{Cov}(\hat{\theta}_n|H_n)]$) [12,22]. Motivated by this fact, we propose to reformulate the problem in (20.54) conditioned on $H_n$, that is,

$$\min_{S,\hat{\theta}_S} \mathsf{E}[S] \text{ s.t. } f\left(\mathsf{Cov}(\hat{\theta}_S|H_S)\right) \leq C. \tag{20.60}$$

Note that the constraint in (20.60) is stricter than the one in (20.54) since it requires that $\hat{\theta}_S$ satisfies the target accuracy level for each realization of $H_S$, whereas in (20.54) it is sufficient that $\hat{\theta}_S$ satisfies the target accuracy level on average. In other words, in (20.54), even if $f(\mathsf{Cov}(\hat{\theta}_S|H_S)) > C$ for some realizations of $H_S$, we can still satisfy $f(\mathsf{Cov}(\hat{\theta}_S)) \leq C$. In fact, we can always have $f(\mathsf{Cov}(\hat{\theta}_S)) = C$ by using a probabilistic stopping rule such that we sometimes stop above $C$, that is, $f(\mathsf{Cov}(\hat{\theta}_S|H_S)) > C$, and the rest of the time at or below $C$, that is, $f(\mathsf{Cov}(\hat{\theta}_S|H_S)) \leq C$. On the contrary, in (20.60) we always have $f(\mathsf{Cov}(\hat{\theta}_S|H_S)) \leq C$; moreover, since we observe discrete-time samples, in general we have $f(\mathsf{Cov}(\hat{\theta}_S|H_S)) < C$ for each realization of $H_S$. Hence, the optimal objective value $\mathsf{E}[S]$ in (20.54) will, in general, be smaller than that in (20.60). Note that on the contrary, if we observed continuous-time processes with continuous paths, then we could always have $f(\mathsf{Cov}(\hat{\boldsymbol{\theta}}_S|\boldsymbol{H}_S)) = C$ for each realization of $\boldsymbol{H}_S$, and thus the optimal objective values of (20.60) and (20.54) would be the same.

Since minimizing $S$ also minimizes $\mathsf{E}[S]$, in (20.60) we are required to find the first time that a member of our class of estimators (i.e., unbiased estimators under Gaussian noise and linear unbiased estimators under non-Gaussian noise) satisfies the constraint $f(\mathsf{Cov}(\hat{\boldsymbol{\theta}}_S|\boldsymbol{H}_S)) \leq C$, as well as the estimator that attains this earliest stopping time. From Lemma 20.8, it is seen that the LS estimator, given by (20.56), among its competitors, achieves the best accuracy level $f(\sigma^2\boldsymbol{U}_S^{-1})$ at any stopping time $S$. Hence, for the conditional problem, the optimum sequential estimator is composed of the stopping time

$$S = \min\{n \in \mathbb{N} : f\left(\sigma^2\boldsymbol{U}_n^{-1}\right) \leq C\}, \tag{20.61}$$

and the LS estimator

$$\hat{\boldsymbol{\theta}}_S = \boldsymbol{U}_S^{-1}V_S, \tag{20.62}$$

which can be computed recursively as in (20.53). The recursive computation of $\boldsymbol{U}_n^{-1} = \boldsymbol{P}_n$ in the test statistic in (20.61) is also given in (20.53).

Note that for an accuracy function $f$ such that $f(\sigma^2\boldsymbol{U}_n^{-1}) = \sigma^2 f(\boldsymbol{U}_n^{-1})$, for example, $\mathsf{Tr}(\cdot)$ and $\|\cdot\|_F$, we can use the following stopping time:

$$S = \min\{n \in \mathbb{N} : f\left(\boldsymbol{U}_n^{-1}\right) \leq C'\}, \tag{20.63}$$

where $C' = \frac{C}{\sigma^2}$ is the relative target accuracy with respect to the noise power. Hence, given $C'$ we do not need to know the noise variance $\sigma^2$ to run the test given by (20.63). Note that $\boldsymbol{U}_n = \boldsymbol{H}_n^T\boldsymbol{H}_n$ is a nondecreasing positive semidefinite matrix, that is, $\boldsymbol{U}_n \succeq \boldsymbol{U}_{n-1}, \forall t$, in the positive semidefinite sense. Thus, from the monotonicity of $f$, the test statistic $f\left(\sigma^2\boldsymbol{U}_n^{-1}\right)$ is a nonincreasing scalar function of time. Specifically, for accuracy functions $\mathsf{Tr}(\cdot)$ and $\|\cdot\|_F$, we can show that if the minimum eigenvalue of $\boldsymbol{U}_n$ tends to infinity as $t \to \infty$, then the stopping time is finite, that is, $S < \infty$.

In the conditional problem, for any $n$, we have a simple stopping rule given in (20.63), which uses the target accuracy level $\frac{C}{\sigma^2}$ as its threshold, hence known beforehand. For the special case of scalar parameter estimation, we do not need a function $f$ to assess the accuracy of the estimator because instead of a covariance matrix we now have a variance $\frac{\sigma^2}{u_n}$, where $u_n = \sum_{m=1}^{n} h_m^2$ and $h_n$ is the scaling coefficient in (20.48). Hence, from (20.62) and (20.63), the optimum sequential estimator in the scalar case is given by

$$S = \min\left\{n \in \mathbb{N} : u_n \geq \frac{1}{C'}\right\}, \quad \hat{\theta}_S = \frac{v_S}{u_S}, \tag{20.64}$$

where $\frac{u_n}{\sigma^2}$ is the Fisher information at time $n$. That is, we stop the first time the gathered Fisher information exceeds the threshold $1/C$, which is known.

### 20.3.3 Decentralized Estimator

In this section, we propose a computation- and energy-efficient decentralized estimator based on the optimum conditional sequential estimator and level-triggered sampling. Consider a network of $K$ distributed sensors and an FC which is responsible for determining the stopping time and computing the estimator. In practice, due to the stringent energy constraints, sensors must infrequently convey low-rate information to the FC, which is the main concern in the design of a decentralized sequential estimator.

As in (20.48), each sensor $k$ observes

$$x_k[n] = \boldsymbol{h}_k[n]^T\boldsymbol{\theta} + w_k[n], \quad n \in \mathbb{N}, \; k = 1, \ldots, K, \tag{20.65}$$

as well as the regressor vector $\boldsymbol{h}_k[n] = [h_{k,1}[n], \ldots, h_{k,p}[n]]^T$ at time $n$, where $\{w_k[n]\}_{k,n}$ are independent, zero-mean, that is, $\mathsf{E}[w_k[n]] = 0, \; \forall k, n$, and $\mathsf{Var}(w_k[n]) = \sigma_k^2, \; \forall n$. Then, similar to (20.50), the weighted least squares (WLS) estimator

$$\hat{\boldsymbol{\theta}}_n = \arg\min_{\boldsymbol{\theta}} \sum_{k=1}^{K}\sum_{m=1}^{n} \frac{\left(x_k[m] - \boldsymbol{h}_k[m]^T\boldsymbol{\theta}\right)^2}{\sigma_k^2},$$

is given by

$$\begin{aligned}\hat{\boldsymbol{\theta}}_n &= \left(\sum_{k=1}^{K}\sum_{m=1}^{n} \frac{\boldsymbol{h}_k[m]\boldsymbol{h}_k[m]^T}{\sigma_k^2}\right)^{-1} \sum_{k=1}^{K}\sum_{m=1}^{n} \frac{\boldsymbol{h}_k[m]x_k[m]}{\sigma_k^2} \\ &= \bar{\boldsymbol{U}}_n^{-1}\bar{V}_n,\end{aligned} \tag{20.66}$$

where $\bar{\boldsymbol{U}}_n^k \triangleq \frac{1}{\sigma_k^2}\sum_{m=1}^{n} \boldsymbol{h}_k[m]\boldsymbol{h}_k[m]^T$, $\bar{V}_n^k \triangleq \frac{1}{\sigma_k^2}\sum_{m=1}^{n} \boldsymbol{h}_k[m] x_k[m]$, $\bar{\boldsymbol{U}}_n = \sum_{k=1}^{K} \bar{\boldsymbol{U}}_n^k$, and $\bar{V}_n = \sum_{k=1}^{K} \bar{V}_n^k$. As before, it can be shown that the WLS estimator $\hat{\boldsymbol{\theta}}_n$ in (20.66) is the BLUE under the general noise distributions. Moreover, in the Gaussian noise case, where $w_k[n] \sim \mathcal{N}(0, \sigma_k^2) \; \forall n$ for each $k$, $\hat{\boldsymbol{\theta}}_n$ is also the MVUE.

Following the steps in Section 20.3.2.2, it is straightforward to show that the optimum sequential estimator for the conditional problem in (20.60) is given by the stopping time

$$S = \min\left\{n \in \mathbb{N} : f\left(\bar{\boldsymbol{U}}_n^{-1}\right) \leq C\right\}, \tag{20.67}$$

and the WLS estimator $\hat{\boldsymbol{\theta}}_S$, given by (20.66). Note that $(S, \hat{\boldsymbol{\theta}}_S)$ is achievable only in the centralized case, where all local observations until time $n$, that is, $\{(x_k[m], \boldsymbol{h}_k[m])\}_{k=1,m=1}^{K,n}$, are available to the FC. Local processes $\{\bar{\boldsymbol{U}}_n^k\}_{k,n}$ and $\{\bar{V}_n^k\}_{k,n}$ are used to compute the stopping time and the estimator as in (20.67) and (20.66), respectively. On the contrary, in a decentralized system, the FC can compute approximations $\widetilde{\boldsymbol{U}}_n^k$ and $\widetilde{V}_n^k$ and then use these approximations to compute the stopping time and estimator as in (20.67) and (20.66), respectively.

#### *20.3.3.1 Linear Complexity*

If each sensor $k$ reports $\bar{\boldsymbol{U}}_n^k \in \mathbb{R}^{p\times p}$ and $\bar{V}_n^k \in \mathbb{R}^p$ to the FC in a straightforward way, then $O(p^2)$ terms need to be transmitted, which may not be practical, especially for large $p$, in a decentralized setup. Similarly, in the literature, the distributed implementation of the Kalman filter, which covers RLS as a special case, through its inverse covariance form, namely the information filter, requires the transmission of an $p \times p$ information matrix and an $p \times 1$ information vector (e.g., [63]).

To overcome this problem, considering $\mathsf{Tr}(\cdot)$ as the accuracy function $f$ in (20.67), we propose to transmit only the $p$ diagonal entries of $\bar{\boldsymbol{U}}_n^k$ for each $k$, yielding linear complexity $O(p)$. Using the diagonal entries of $\bar{\boldsymbol{U}}_n$, we define the diagonal matrix

$$\boldsymbol{D}_n \triangleq \operatorname{diag}\left(d_{n,1}, \ldots, d_{n,p}\right)$$
$$\text{where } d_{n,i} = \sum_{k=1}^{K}\sum_{m=1}^{n} \frac{h_{k,i}[m]^2}{\sigma_k^2}, \; i = 1, \ldots, p. \tag{20.68}$$

We further define the correlation matrix

$$\boldsymbol{R} = \begin{bmatrix} 1 & r_{12} & \cdots & r_{1p} \\ r_{12} & 1 & \cdots & r_{2p} \\ \vdots & \vdots & \ddots & \vdots \\ r_{1p} & r_{2p} & \cdots & 1 \end{bmatrix}, \tag{20.69}$$

$$\text{where } r_{ij} = \frac{\sum_{k=1}^{K} \frac{\mathsf{E}[h_{k,i}[n]h_{k,j}[n]]}{\sigma_k^2}}{\sqrt{\sum_{k=1}^{K} \frac{\mathsf{E}[h_{k,i}[n]^2]}{\sigma_k^2} \sum_{k=1}^{K} \frac{\mathsf{E}[h_{k,j}[n]^2]}{\sigma_k^2}}}, i, j = 1, \ldots, p.$$

**Proposition 20.1**

For sufficiently large $n$, we can make the following approximations:

$$\bar{\boldsymbol{U}}_n \approx \boldsymbol{D}_n^{1/2} \boldsymbol{R}\, \boldsymbol{D}_n^{1/2}$$
$$\text{and } \mathsf{Tr}\left(\bar{\boldsymbol{U}}_n^{-1}\right) \approx \mathsf{Tr}\left(\boldsymbol{D}_n^{-1}\boldsymbol{R}^{-1}\right). \tag{20.70}$$

**PROOF** The approximations are motivated from the special case where $\mathsf{E}[h_{k,i}[n]h_{k,j}[n]] = 0, \; \forall k, \; i,j = 1, \ldots, p, \; i \neq j$. In this case, by the *law of large numbers* for sufficiently large $n$, the off-diagonal elements of $\frac{\bar{\boldsymbol{U}}_n}{n}$ vanish, and thus we have $\frac{\bar{\boldsymbol{U}}_n}{n} \approx \frac{\boldsymbol{D}_n}{n}$ and $\mathsf{Tr}(\bar{\boldsymbol{U}}_n^{-1}) \approx \mathsf{Tr}(\boldsymbol{D}_n^{-1})$. For the general case where we might have $\mathsf{E}[h_{k,i}[n]h_{k,j}[n]] \neq 0$ for some $k$ and $i \neq j$, using the diagonal matrix $\boldsymbol{D}_n$ we write

$$\mathsf{Tr}\left(\bar{\boldsymbol{U}}_n^{-1}\right) = \mathsf{Tr}\left(\left(\boldsymbol{D}_n^{1/2} \underbrace{\boldsymbol{D}_n^{-1/2}\bar{\boldsymbol{U}}_n\boldsymbol{D}_n^{-1/2}}_{\boldsymbol{R}_n} \boldsymbol{D}_n^{1/2}\right)^{-1}\right), \tag{20.71}$$

$$= \mathsf{Tr}\left(\boldsymbol{D}_n^{-1/2}\boldsymbol{R}_n^{-1}\boldsymbol{D}_n^{-1/2}\right),$$
$$= \mathsf{Tr}\left(\boldsymbol{D}_n^{-1}\boldsymbol{R}_n^{-1}\right). \tag{20.72}$$

Note that each entry $r_{n,ij}$ of the newly defined matrix $\boldsymbol{R}_n$ is a normalized version of the corresponding entry $\bar{u}_{n,ij}$ of $\bar{\boldsymbol{U}}_n$. Specifically, $r_{n,ij} = \frac{\bar{u}_{n,ij}}{\sqrt{d_{n,i}d_{n,j}}} = \frac{\bar{u}_{n,ij}}{\sqrt{\bar{u}_{n,ii}\bar{u}_{n,jj}}}, \; i,j = 1, \ldots, p$, where the last equality follows from the definition of $d_{n,i}$ in (20.68). Hence, $\boldsymbol{R}_n$ has the same structure as in (20.69) with entries

$$r_{n,ij} = \frac{\sum_{k=1}^{K}\sum_{m=1}^{n} \frac{h_{k,i}[m]h_{k,j}[m]}{\sigma_k^2}}{\sqrt{\sum_{k=1}^{K}\sum_{m=1}^{n} \frac{h_{k,i}[m]^2}{\sigma_k^2} \sum_{k=1}^{K}\sum_{m=1}^{n} \frac{h_{k,j}[m]^2}{\sigma_k^2}}},$$
$$i, j = 1, \ldots, p.$$

For sufficiently large $n$, by the law of large numbers

$$r_{n,ij} \approx r_{ij} = \frac{\sum_{k=1}^{K} \frac{\mathsf{E}[h_{k,i}[n]h_{k,j}[n]]}{\sigma_k^2}}{\sqrt{\sum_{k=1}^{K} \frac{\mathsf{E}[h_{k,i}[n]^2]}{\sigma_k^2} \sum_{k=1}^{K} \frac{\mathsf{E}[h_{k,j}[n]^2]}{\sigma_k^2}}}, \tag{20.73}$$

and $\boldsymbol{R}_n \approx \boldsymbol{R}$, where $\boldsymbol{R}$ is given in (20.69). Hence, for sufficiently large $n$, we can make the approximations in (20.70) using (20.71) and (20.72). ■

Then, assuming that the FC knows the correlation matrix $\boldsymbol{R}$, that is, $\{\mathsf{E}[h_{k,i}[n]h_{k,j}[n]]\}_{i,j,k}$* and $\{\sigma_k^2\}$ [cf. (20.69)], it can compute the approximations in (20.70) if sensors report their local processes $\left\{\boldsymbol{D}_n^k\right\}_{k,n}$ to the FC, where $\boldsymbol{D}_n = \sum_{k=1}^{K} \boldsymbol{D}_n^k$. Note that each local process $\left\{\boldsymbol{D}_n^k\right\}_n$ is $p$-dimensional, and its entries at time $n$ are

*The subscripts $i$ and $j$ in the set notation denote $i = 1, \ldots, p$ and $j = i, \ldots, p$. In the special case where $\mathsf{E}[h_{k,i}[n]^2] = \mathsf{E}[h_{\ell,i}[n]^2]$, $k, \ell = 1, \ldots, K$, $i = 1, \ldots, p$, the correlation coefficients

$$\left\{\xi_{ij}^k = \frac{\mathsf{E}[h_{k,i}[n]h_{k,j}[n]]}{\sqrt{\mathsf{E}[h_{k,i}[n]^2]\mathsf{E}[h_{k,j}[n]^2]}} : i = 1, \ldots, p-1, \; j = i+1, \ldots, p\right\}_k,$$

together with $\{\sigma_k^2\}$ are sufficient statistics since $r_{ij} = \frac{\sum_{k=1}^{K} \xi_{ij}^k/\sigma_k^2}{\sum_{k=1}^{K} 1/\sigma_k^2}$ from (20.73).

given by $\left\{d^k_{n,i} = \sum_{m=1}^{n} \frac{h_{k,i}[m]^2}{\sigma_k^2}\right\}_i$ [cf. (20.68)]. Hence, we propose that each sensor $k$ sequentially reports the local processes $\{\boldsymbol{D}^k_n\}_n$ and $\{\bar{V}^k_n\}_n$ to the FC, achieving linear complexity $O(p)$. On the other side, the FC, using the information received from sensors, computes the approximations $\{\widetilde{\boldsymbol{D}}_n\}$ and $\{\widetilde{V}_n\}$, which are then used to compute the stopping time

$$\widetilde{\mathcal{S}} = \min\left\{n \in \mathbb{N} : \mathrm{Tr}\left(\widetilde{\boldsymbol{U}}_n^{-1}\right) \le \widetilde{C}\right\}, \tag{20.74}$$

and the estimator

$$\widetilde{\boldsymbol{\theta}}_{\widetilde{\mathcal{S}}} = \widetilde{\boldsymbol{U}}_{\widetilde{\mathcal{S}}}^{-1} \widetilde{V}_{\widetilde{\mathcal{S}}}, \tag{20.75}$$

similar to (20.67) and (20.66), respectively. The approximations $\mathrm{Tr}\left(\widetilde{\boldsymbol{U}}_n^{-1}\right)$ in (20.74) and $\widetilde{\boldsymbol{U}}_{\widetilde{\mathcal{S}}}$ in (20.75) are computed using $\widetilde{\boldsymbol{D}}_n$ as in (20.70). The threshold $\widetilde{C}$ is selected through simulations to satisfy the constraint in (20.60) with equality, that is, $\mathrm{Tr}\left(\mathrm{Cov}(\widetilde{\boldsymbol{\theta}}_{\widetilde{\mathcal{S}}}|\boldsymbol{H}_{\widetilde{\mathcal{S}}})\right) = C$.

#### 20.3.3.2 Event-Based Transmission

Level-triggered sampling provides a very convenient way of information transmission in decentralized systems [17,73–76]. Specifically, decentralized methods based on level-triggered sampling, transmitting low-rate information, enable highly accurate approximations and thus high-performance schemes at the FC. They significantly outperform conventional decentralized methods, which sample local processes using the traditional uniform sampling and send the quantized versions of samples to the FC [73,75].

Existing methods employ level-triggered sampling to report a scalar local process to the FC. Using a similar procedure to report each distinct entry of $\bar{\boldsymbol{U}}^k_n$ and $\bar{V}^k_n$, we need $O(p^2)$ parallel procedures, which may be prohibitive in a decentralized setup for large $p$. Hence, we use the approximations introduced in the previous subsection, achieving linear complexity $O(p)$. Data transmission and thus energy consumption also scale linearly with the number of parameters, which may easily become prohibitive for a sensor with limited battery. We address this energy efficiency issue by infrequently transmitting a single pulse with very short duration, which encodes, in time, the overshoot in level-triggered sampling [76].

We will next describe the proposed decentralized estimator based on level-triggered sampling in which each sensor nonuniformly samples the local processes $\{\boldsymbol{D}^k_n\}_n$ and $\{\bar{V}^k_n\}_n$, and transmits a single pulse for each sample to the FC, and the FC computes $\{\widetilde{\boldsymbol{D}}_n\}$ and $\{\widetilde{V}_n\}$ using received information.

##### 20.3.3.2.1 *Sampling and Recovery of $\boldsymbol{D}^k_n$*

Each sensor $k$ samples each entry $d^k_{n,i}$ of $\boldsymbol{D}^k_n$ at a sequence of random times $\{s^k_{m,i}\}_{m\in\mathbb{N}}$ given by

$$s^k_{m,i} \triangleq \min\left\{n \in \mathbb{N} : d^k_{n,i} - d^k_{s^k_{m-1,i},i} \ge \Delta^k_i\right\}, \; s^k_{0,i} = 0, \tag{20.76}$$

where $d^k_{n,i} = \sum_{p=1}^{n} \frac{h_{k,i}[p]^2}{\sigma_k^2}$, $d^k_{0,i} = 0$, and $\Delta^k_i > 0$ is a constant threshold that controls the average sampling interval. Note that the sampling times $\{s^k_{m,i}\}_m$ in (20.76) are dynamically determined by the signal to be sampled, that is, realizations of $d^k_{n,i}$. Hence, they are random, whereas sampling times in the conventional uniform sampling are deterministic with a certain period. According to the sampling rule in (20.76), a sample is taken whenever the signal level $d^k_{n,i}$ increases by at least $\Delta^k_i$ since the last sampling time. Note that $d^k_{n,i} = \sum_{p=1}^{n} \frac{h_{k,i}[p]2}{\sigma_k^2}$ is nondecreasing in $n$.

After each sampling time $s^k_{m,i}$, sensor $k$ transmits a single pulse to the FC at time

$$t^k_{m,i} \triangleq s^k_{m,i} + \delta^k_{m,i},$$

indicating that $d^k_{n,i}$ has increased by at least $\Delta^k_i$ since the last sampling time $s^k_{m-1,i}$. The delay $\delta^k_{m,i}$ between the transmission time and the sampling time is used to linearly encode the overshoot

$$q^k_{m,i} \triangleq \left(d^k_{s^k_{m,i},i} - d^k_{s^k_{m-1,i},i}\right) - \Delta^k_i, \tag{20.77}$$

and is given by

$$\delta^k_{m,i} = \frac{q^k_{m,i}}{\phi_d} \in [0,1), \tag{20.78}$$

where $\phi_d^{-1}$ is the slope of the linear encoding function, as shown in Figure 20.8, known to sensors and the FC.

Assume a global clock, that is, the time index $n \in \mathbb{N}$ is the same for all sensors and the FC, meaning that the FC knows the potential sampling times. Assume further ultra-wideband (UWB) channels between sensors and the FC, in which the FC can determine the time of flight of pulses transmitted from sensors. Then, FC can measure the transmission delay $\delta^k_{m,i}$ if it is bounded by unit time, that is, $\delta^k_{m,i} \in [0,1)$. To ensure this, from (20.78), we need to have $\phi_d > q^k_{m,i}$, $\forall k, m, i$. Assuming a bound for overshoots, that is, $q^k_{m,i} < \theta_d$, $\forall k, m, i$, we can achieve this by setting $\phi_d > \theta_d$.

Consequently, the FC can uniquely decode the overshoot by computing $q^k_{m,i} = \phi_d \delta^k_{m,i}$ (cf. Figure 20.8), using

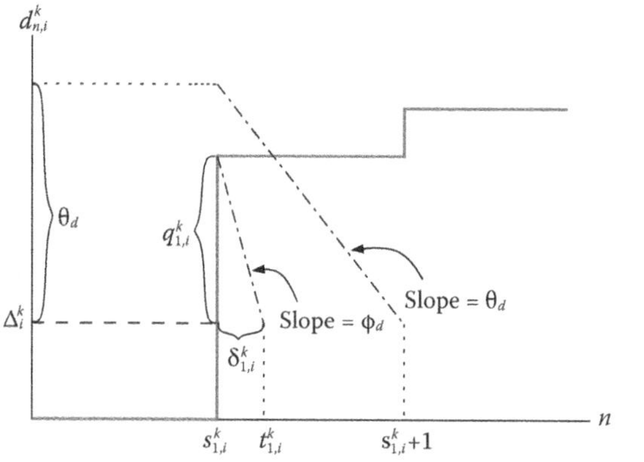


**FIGURE 20.8**

Illustration of sampling time $s^k_{m,i}$, transmission time $t^k_{m,i}$, transmission delay $\delta^k_{m,i}$, and overshoot $q^k_{m,i}$. We encode $q^k_{m,i} < \theta_d$ in $\delta^k_{m,i} = t^k_{m,i} - s^k_{m,i} < 1$ using the slope $\phi_d > \theta_d$.

which it can also find the increment occurred in $d^k_{n,i}$ during the interval $(s^k_{m-1,i}, s^k_{m,i}]$ as

$$d^k_{s^k_{m,i},i} - d^k_{s^k_{m-1,i},i} = \Delta^k_i + q^k_{m,i},$$

from (20.77). It is then possible to reach the signal level $d^k_{s^k_{m,i},i}$ by accumulating the increments occurred until the $m$th sampling time, that is,

$$d^k_{s^k_{m,i},i} = \sum_{\ell=1}^{m} \left(\Delta^k_i + q^k_{\ell,i}\right) = m\Delta^k_i + \sum_{\ell=1}^{m} q^k_{\ell,i}. \qquad (20.79)$$

Using $\{d^k_{s^k_{m,i},i}\}_m$, the FC computes the staircase approximation $\tilde{d}^k_{n,i}$ as

$$\tilde{d}^k_{n,i} = d^k_{s^k_{m,i},i}, \quad t \in [t^k_{m,i}, t^k_{m+1,i}), \qquad (20.80)$$

which is updated when a new pulse is received from sensor $k$, otherwise kept constant. Such approximate local signals of different sensors are next combined to obtain the approximate global signal $\tilde{d}_{n,i}$ as

$$\tilde{d}_{n,i} = \sum_{k=1}^{K} \tilde{d}^k_{n,i}. \qquad (20.81)$$

In practice, when the $m$th pulse in the global order regarding dimension $i$ is received from sensor $k_m$ at time $t_{m,i}$, instead of computing (20.79) through (20.81), the FC only updates $\tilde{d}_{n,i}$ as

$$\tilde{d}_{t_{m,i},i} = \tilde{d}_{t_{m-1,i},i} + \Delta^{k_m}_i + q_{m,i}, \quad \tilde{d}_{0,i} = \epsilon, \qquad (20.82)$$

and keeps it constant when no pulse arrives. We initialize $\tilde{d}_{n,i}$ to a small constant $\epsilon$ to prevent dividing by zero while computing the test statistic [cf. (20.83)].

Note that in general $\tilde{d}_{t_{m,i},i} \neq d_{s_{m,i},i}$ unlike (20.80) since all sensors do not necessarily sample and transmit at the same time. The approximations $\{\tilde{d}_{n,i}\}_i$ form $\tilde{\boldsymbol{D}}_n = \text{diag}(\tilde{d}_{n,1}, \ldots, \tilde{d}_{n,p})$, which is used in (20.74) and (20.75) to compute the stopping time and the estimator, respectively. Note that to determine the stopping time as in (20.74), we need to compute $\text{Tr}(\tilde{\boldsymbol{U}}_t^{-1})$ using (20.70) at times $\{t_m\}$ when a pulse is received from any sensor regarding any dimension. Fortunately, when the $m$th pulse in the global order is received from sensor $k_m$ at time $t_m$ regarding dimension $i_m$, we can compute $\text{Tr}(\tilde{\boldsymbol{U}}_{t_m}^{-1})$ recursively as follows:

$$\text{Tr}\left(\tilde{\boldsymbol{U}}_{t_m}^{-1}\right) = \text{Tr}\left(\tilde{\boldsymbol{U}}_{t_{m-1}}^{-1}\right) - \frac{\kappa_{i_m}(\Delta^{k_m}_{i_m} + q_m)}{\tilde{d}_{t_m,i_m}\tilde{d}_{t_{m-1},i_m}},$$

$$\text{Tr}\left(\tilde{\boldsymbol{U}}_0^{-1}\right) = \sum_{i=1}^{p} \frac{\kappa_i}{\epsilon}, \qquad (20.83)$$

where $\kappa_i$ is the $i$th diagonal element of the inverse correlation matrix $\boldsymbol{R}^{-1}$, known to the FC. In (20.83), pulse arrival times are assumed to be distinct for the sake of simplicity. In case multiple pulses arrive at the same time, the update rule will be similar to (20.83) except that it will consider all new arrivals together.

#### 20.3.3.2.2 *Sampling and Recovery of $\bar{V}^k_n$*

Similar to (20.76), each sensor $k$ samples each entry $\bar{v}^k_{n,i}$ of $\bar{V}^k_n$ at a sequence of random times $\{\rho^k_{m,i}\}_m$ written as

$$\rho^k_{m,i} \triangleq \min\left\{n \in \mathbb{N} : \left|\bar{v}^k_{n,i} - \bar{v}^k_{\rho^k_{m-1,i},i}\right| \geq \gamma^k_i\right\}, \quad \rho^k_{0,i} = 0, \qquad (20.84)$$

where $\bar{v}^k_{n,i} = \sum_{p=1}^{n} \frac{h^k_{p,i} y^k_p}{\sigma^2_k}$ and $\gamma^k_i$ is a constant threshold, available to both sensor $k$ and the FC. See (20.2) for selecting $\gamma^k_i$. Since $\bar{v}^k_{n,i}$ is neither increasing nor decreasing, we use two thresholds $\gamma^k_i$ and $-\gamma^k_i$ in the sampling rule given in (20.84).

Specifically, a sample is taken whenever $\bar{v}^k_{n,i}$ increases or decreases by at least $\gamma^k_i$ since the last sampling time. Then, after a transmission delay

$$\chi^k_{m,i} = \frac{\eta^k_{m,i}}{\phi_v},$$

where $\eta^k_{m,i} \triangleq \left|\bar{v}^k_{\rho^k_{m,i},i} - \bar{v}^k_{\rho^k_{m-1,i},i}\right| - \gamma^k_i$ is the overshoot, sensor $k$ at time

$$\tau^k_{m,i} \triangleq \rho^k_{m,i} + \chi^k_{m,i},$$

transmits a single pulse $b^k_{m,i}$ to the FC, indicating whether $\bar{v}^k_{n,i}$ has changed by at least $\gamma^k_i$ or $-\gamma^k_i$ since the last sampling time $\rho^k_{m-1,i}$. We can simply write $b^k_{m,i}$ as

$$b^k_{m,i} = \text{sign}\left(\bar{v}^k_{\rho^k_{m,i},i} - \bar{v}^k_{\rho^k_{m-1,i},i}\right). \tag{20.85}$$

Assume again that (i) there exists a global clock among sensors and the FC, (ii) the FC determines channel delay (i.e., time of flight), and (iii) overshoots are bounded by a constant, that is, $\eta^k_{m,i} < \theta_v$, $\forall k, m, i$, and we set $\phi_v > \theta_v$. With these assumptions, we ensure that the FC can measure the transmission delay $\chi^k_{m,i}$ and accordingly decode the overshoot as $\eta^k_{m,i} = \phi_v \chi^k_{m,i}$. Then, upon receiving the $m$th pulse $b_{m,i}$ regarding dimension $i$ from sensor $k_m$ at time $\tau_{m,i}$, the FC performs the following update:

$$\widetilde{v}_{\tau_{m,i},i} = \widetilde{v}_{\tau_{m-1,i},i} + b_{m,i}\left(\gamma^{k_m}_i + \eta_{m,i}\right), \tag{20.86}$$

where $\{\widetilde{v}_{n,i}\}_i$ compose the approximation $\widetilde{V}_n = [\widetilde{v}_{n,1}, \ldots, \widetilde{v}_{n,p}]^T$. Recall that the FC employs $\widetilde{V}_n$ to compute the estimator as in (20.75).

The level-triggered sampling procedure at each sensor $k$ for each dimension $i$ is summarized in Algorithm 20.1. Each sensor $k$ runs $p$ of these procedures in parallel. The sequential estimation procedure at the FC is also summarized in Algorithm 20.2. We assumed, for the sake of clarity, that each sensor transmits pulses to the FC for each dimension through a separate channel, that is, parallel architecture. On the contrary, in practice the number of parallel channels can be decreased to two by using identical sampling thresholds $\Delta$ and $\gamma$ for all sensors and for all dimensions in (20.76) and (20.84), respectively. Moreover, sensors can even employ a single channel to convey information about local processes $\{d^k_{n,i}\}$ and $\{\bar{v}^k_{n,i}\}$ by sending ternary digits to the FC. This is possible since pulses transmitted for $\{d^k_{n,i}\}$ are unsigned.

#### *20.3.3.3 Discussions*

We introduced the decentralized estimator in Section 20.3.3.2 initially for a system with infinite time precision. In practice, due to bandwidth constraints, discrete-time systems with finite precision are of interest. For example, in such systems, the overshoot $q^k_{m,i} \in \left[j\frac{\theta_d}{N}, (j+1)\frac{\theta_d}{N}\right), j = 0, 1, \ldots, N-1$, is quantized into $\hat{q}^k_{m,i} = \left(j + \frac{1}{2}\right)\frac{\theta_d}{N}$, where $N$ is the number of quantization levels. More specifically, a pulse is transmitted at time $t^k_{m,i} = s^k_{m,i} + \frac{j+1/2}{N}$, where the transmission delay $\frac{j+1/2}{N} \in (0,1)$ encodes $\hat{q}^k_{m,i}$. This transmission scheme is called pulse position modulation (PPM).

In UWB and optical communication systems, PPM is effectively employed. In such systems, $N$, which denotes the precision, can be easily made large enough so that the quantization error $|\hat{q}^k_{m,i} - q^k_{m,i}|$ becomes insignificant. Compared with conventional transmission techniques which convey information by varying the power level, frequency, and/or phase of a sinusoidal wave, PPM (with UWB) is extremely energy efficient at the expense of high bandwidth usage since only a single pulse with very short duration is transmitted per sample. Hence, PPM suits well to energy-constrained sensor network systems.

#### *20.3.3.4 Simulations*

We next provide simulation results to compare the performances of the proposed scheme with linear complexity, given in Algorithms 20.1 and 20.2, the nonsimplified version of the proposed scheme with quadratic complexity and the optimal centralized scheme. A wireless sensor network with 10 identical sensors and an FC is considered to estimate a five-dimensional deterministic vector of parameters, that is, $p = 5$. We assume i.i.d. Gaussian noise with unit variance at all sensors, that is, $w_k[n] \sim \mathcal{N}(0,1), \forall k, n$. We set the correlation coefficients $\{r_{ij}\}$ [cf. (20.73)] of the vector $\boldsymbol{h}_k[n]$ to 0 and 0.5 in Figure 20.9 to test the performance of the proposed

**Algorithm 20.1** The level-triggered sampling procedure at the $k$th sensor for the $i$th dimension

1: Initialization: $n \leftarrow 0,\ m \leftarrow 0,\ \ell \leftarrow 0,\ \lambda \leftarrow 0,\ \psi \leftarrow 0$
2: **while** $\lambda < \Delta^k_i$ **and** $\psi \in (-\gamma^k_i, \gamma^k_i)$ **do**
3: $\quad n \leftarrow n + 1$
4: $\quad \lambda \leftarrow \lambda + \frac{h_{k,i}[n]^2}{\sigma^2_k}$
5: $\quad \psi \leftarrow \psi + \frac{h_{k,i}[n] x_k[n]}{\sigma^2_k}$
6: **end while**
7: **if** $\lambda \geq \Delta^k_i$ {sample $d^k_{n,i}$} **then**
8: $\quad m \leftarrow m + 1$
9: $\quad s^k_{m,i} = n$
10: $\quad$ Send a pulse to the fusion center at time instant $t^k_{m,i} = s^k_{m,i} + \frac{\lambda - \Delta^k_i}{\phi_d}$
11: $\quad \lambda \leftarrow 0$
12: **end if**
13: **if** $\psi \notin (-\gamma^k_i, \gamma^k_i)$ {sample $\bar{v}^k_{n,i}$} **then**
14: $\quad \ell \leftarrow \ell + 1$
15: $\quad \rho^k_{\ell,i} = n$
16: $\quad$ Send $b^k_{\ell,i} = \text{sign}(\psi)$ to the fusion center at time instant $\tau^k_{\ell,i} = \rho^k_{\ell,i} + \frac{|\psi| - \gamma^k_i}{\phi_v}$
17: $\quad \psi \leftarrow 0$
18: **end if**
19: Stop if the fusion center instructs so; otherwise go to line 2.

**Algorithm 20.2** The sequential estimation procedure at the fusion center

Initialization: $\mathsf{Tr} \leftarrow \sum_{i=1}^{p} \frac{\kappa_i}{\epsilon}$, $m \leftarrow 1$, $\ell \leftarrow 1$, $\tilde{d}_i \leftarrow \epsilon \; \forall i$, $\tilde{v}_i \leftarrow 0 \; \forall i$

**while** $\mathsf{Tr} < \tilde{C}$ **do**

  Wait to receive a pulse

  **if** $m$th pulse about $d_{n,i}$ arrives from sensor $k$ at time $n$ **then**

    $q_m = \phi_d(n - \lfloor n \rfloor)$

    $\mathsf{Tr} \leftarrow \mathsf{Tr} - \frac{\kappa_i(\Delta_i^k + q_m)}{\tilde{d}_i(\tilde{d}_i + \Delta_i^k + q_m)}$

    $\tilde{d}_i = \tilde{d}_i + \Delta_i^k + q_m$

    $m \leftarrow m + 1$

  **end if**

  **if** $\ell$th pulse $b_\ell$ about $v_{n,j}$ arrives from sensor $k$ at time $n$ **then**

    $\eta_\ell = \phi_v(n - \lfloor n \rfloor)$

    $\tilde{v}_j = \tilde{v}_j + b_\ell(\gamma_j^k + \eta_\ell)$

    $\ell \leftarrow \ell + 1$

  **end if**

**end while**

Stop at time $\tilde{S} = n$

$\tilde{\boldsymbol{D}} = \mathrm{diag}(\tilde{d}_1, \ldots, \tilde{d}_p)$, $\tilde{\boldsymbol{U}}^{-1} = \tilde{\boldsymbol{D}}^{-1/2}\boldsymbol{R}^{-1}\tilde{\boldsymbol{D}}^{-1/2}$, $\tilde{V} = [\tilde{v}_1, \ldots, \tilde{v}_p]^T$

$\tilde{\boldsymbol{\theta}} = \tilde{\boldsymbol{U}}^{-1}\tilde{V}$

Instruct sensors to stop.

scheme in the uncorrelated and correlated cases. We compare the average stopping time performance of the proposed scheme with linear complexity to those of the other two schemes for different MSE values. In Figure 20.9, the horizontal axis represents the signal-to-error ratio in decibel, where $n\text{MSE} \triangleq \frac{\text{MSE}}{\|\boldsymbol{\theta}\|_2^2}$, that is, the MSE normalized by the square of the Euclidean norm of the vector to be estimated.

In the uncorrelated case, where $r_{ij} = 0$, $\forall i, j$, $i \neq j$, the proposed scheme with linear complexity nearly attains the performance of the nonsimplified scheme with quadratic complexity as seen in Figure 20.9. This result is rather expected since in this case $\bar{\boldsymbol{U}}_n \approx \boldsymbol{D}_n$ for sufficiently large $n$, where $\bar{\boldsymbol{U}}_n$ and $\boldsymbol{D}_n$ are used to compute the stopping time and the estimator in the nonsimplified and simplified schemes, respectively. Strikingly, the decentralized schemes (simplified and nonsimplified) achieve very close performances to that of the optimal centralized scheme, which is obviously unattainable in a decentralized system, thanks to the efficient information transmission through level-triggered sampling.

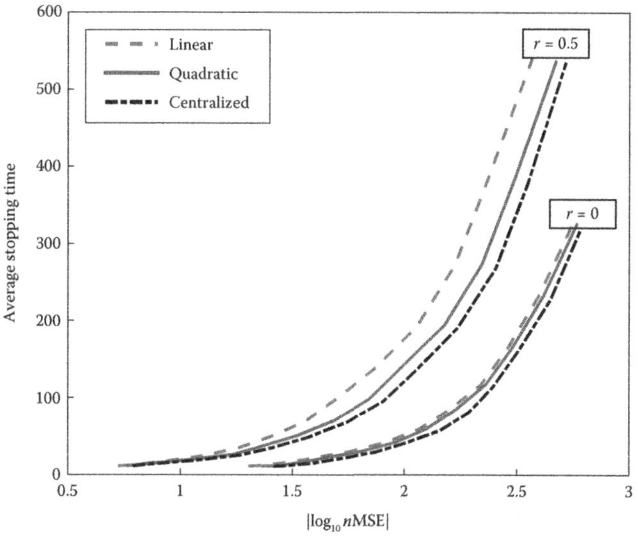


**FIGURE 20.9**
Average stopping time performances of the optimal centralized scheme and the decentralized schemes based on level-triggered sampling with quadratic and linear complexity versus normalized MSE values when scaling coefficients are uncorrelated, that is, $r_{ij} = 0, \forall i, j$, and correlated with $r_{ij} = 0.5, \forall i, j$.

It is seen in Figure 20.9 that the proposed simplified scheme exhibits an average stopping time performance close to those of the nonsimplified scheme and the optimal centralized scheme even when the scaling coefficients $\{h_{k,i}[n]\}_i$ are correlated with $r_{ij} = 0.5$, $\forall i, j$, $i \neq j$, justifying the simplification proposed in Section 20.3.3.1 to obtain linear complexity.

## 20.4 Conclusion

Event-based sampling techniques, adapting the sampling times to the signal to be sampled, provide energy- and bandwidth-efficient information transmission in resource-constrained distributed (i.e., decentralized) systems, such as wireless sensor networks. We have first designed and analyzed event-based detection schemes under challenging environments, namely noisy transmission channels between nodes and the fusion center, and multimodal observations from disparate information sources. Then, we have identified an optimum sequential estimator which lends itself to decentralized systems. For large number of unknown parameters, we have further proposed a simplified scheme with linear complexity.

## Acknowledgments

This work was funded in part by the U.S. National Science Foundation under grant CIF1064575, the U.S. Office of Naval Research under grant N000141210043, the Consortium for Verification Technology under Department of Energy National Nuclear Security Administration award number DE-NA0002534, and the Army Research Office (ARO) grant number W911NF-11-1-0391.

## Bibliography

[1] I. F. Akyildiz, W. Su, Y. Sankarasubramaniam, and E. Cayirci. A survey on sensor networks. *IEEE Communications Magazine*, 40(8):102–114, 2002.

[2] K. J. Astrom and B. M. Bernhardsson. Comparison of Riemann and Lebesgue sampling for first order stochastic systems. In *41st IEEE Conference on Decision and Control*, Las Vegas, Nevada, volume 2, pages 2011–2016, December 2002.

[3] R. Bashirullah, J. G. Harris, J. C. Sanchez, T. Nishida, and J. C. Principe. Florida wireless implantable recording electrodes (FWIRE) for brain machine interfaces. In *IEEE International Symposium on Circuits and Systems (ISCAS 2007)*, New Orleans, LA, pages 2084–2087, May 2007.

[4] F. J. Beutler. Error-free recovery of signals from irregularly spaced samples. *SIAM Review*, 8: 328–355, 1966.

[5] V. Borkar and P. P. Varaiya. Asymptotic agreement in distributed estimation. *IEEE Transactions on Automatic Control*, 27(3):650–655, 1982.

[6] Z. Chair and P. K. Varshney. Optimal data fusion in multiple sensor detection systems. *IEEE Transactions on Aerospace and Electronic Systems*, 22(1):98–101, 1986.

[7] J.-F. Chamberland and V. V. Veeravalli. Decentralized detection in sensor networks. *IEEE Transactions on Signal Processing*, 51(2):407–416, 2003.

[8] S. Chaudhari, V. Koivunen, and H. V. Poor. Autocorrelation-based decentralized sequential detection of OFDM signals in cognitive radios. *IEEE Transactions on Signal Processing*, 57(7):2690–2700, 2009.

[9] B. Chen, L. Tong, and P. K. Varshney. Channel-aware distributed detection in wireless sensor networks. *IEEE Signal Processing Magazine*, 23(4):16–26, 2006.

[10] A. K. Das and M. Mesbahi. Distributed linear parameter estimation over wireless sensor networks. *IEEE Transactions on Aerospace and Electronic Systems*, 45(4):1293–1306, 2009.

[11] D. Drazen, P. Lichtsteiner, P. Hafliger, T. Delbruck, and A. Jensen. Toward real-time particle tracking using an event-based dynamic vision sensor. *Experiments in Fluids*, 51(5):1465–1469, 2011.

[12] B. Efron and D. V. Hinkley. Assessing the accuracy of the maximum likelihood estimator: Observed versus expected fisher information. *Biometrika*, 65(3):457–487, 1978.

[13] P. Ellis. Extension of phase plane analysis to quantized systems. *IRE Transactions on Automatic Control*, 4(2):43–54, 1959.

[14] J. Fang and H. Li. Adaptive distributed estimation of signal power from one-bit quantized data. *IEEE Transactions on Aerospace and Electronic Systems*, 46(4):1893–1905, 2010.

[15] H. Farhangi. The path of the smart grid. *IEEE Power and Energy Magazine*, 8(1):18–28, 2010.

[16] G. Fellouris. Asymptotically optimal parameter estimation under communication constraints. *Annals of Statistics*, 40(4):2239–2265, 2012.

[17] G. Fellouris and G. V. Moustakides. Decentralized sequential hypothesis testing using asynchronous communication. *IEEE Transactions on Information Theory*, 57(1):534–548, 2011.

[18] J. W. Fisher, M. J. Wainwright, E. B. Sudderth, and A. S. Willsky. Statistical and information-theoretic methods for self-organization and fusion of multimodal, networked sensors. *The International Journal of High Performance Computing Applications*, 16(3):337–353, 2002.

[19] J. Fromm and S. Lautner. Electrical signals and their physiological significance in plants. *Plant, Cell & Environment*, 30(3):249–257, 2007.

[20] B. K. Ghosh. On the attainment of the Cramér-rao bound in the sequential case. *Sequential Analysis*, 6(3):267–288, 1987.

[21] D. Gontier and M. Vetterli. Sampling based on timing: Time encoding machines on shift-invariant subspaces. *Applied and Computational Harmonic Analysis*, 36(1):63–78, 2014.

[22] P. Grambsch. Sequential sampling based on the observed fisher information to guarantee the accuracy of the maximum likelihood estimator. *Annals of Statistics*, 11(1):68–77, 1983.

[23] K. M. Guan, S. S. Kozat, and A. C. Singer. Adaptive reference levels in a level-crossing analog-to-digital converter. *EURASIP Journal on Advances in Signal Processing*, 2008:183:1–183:11, 2008.

[24] K. M. Guan and A. C. Singer. Opportunistic sampling by level-crossing. In *IEEE International Conference on Acoustics, Speech and Signal Processing (ICASSP)*, Honolulu, HI, volume 3, pages III-1513–III-1516, April 2007.

[25] S. Haykin. *Communication Systems*, 4th edition. Wiley, New York, NY, 2001.

[26] M. Hofstatter, M. Litzenberger, D. Matolin, and C. Posch. Hardware-accelerated address-event processing for high-speed visual object recognition. In *18th IEEE International Conference on Electronics, Circuits and Systems (ICECS)*, Beirut, Lebanon, pages 89–92, December 2011.

[27] A. M. Hussain. Multisensor distributed sequential detection. *IEEE Transactions on Aerospace and Electronic Systems*, 30(3):698–708, 1994.

[28] E. Kofman and J. H. Braslavsky. Level crossing sampling in feedback stabilization under data-rate constraints. In *45th IEEE Conference on Decision and Control*, San Diego, CA, pages 4423–4428, December 2006.

[29] A. A. Lazar and E. A. Pnevmatikakis. Video time encoding machines. *IEEE Transactions on Neural Networks*, 22(3):461–473, 2011.

[30] A. A. Lazar and L. T. Toth. Perfect recovery and sensitivity analysis of time encoded bandlimited signals. *IEEE Transactions on Circuits and Systems I: Regular Papers*, 51(10):2060–2073, 2004.

[31] S. Li and X. Wang. Cooperative change detection for online power quality monitoring. http://arxiv.org/abs/1412.2773.

[32] S. Li and X. Wang. Quickest attack detection in multi-agent reputation systems. *IEEE Journal of Selected Topics in Signal Processing*, 8(4):653–666, 2014.

[33] P. Lichtsteiner, C. Posch, and T. Delbruck. A 128 $\times$ 128 120 db 15 $\mu$s latency asynchronous temporal contrast vision sensor. *IEEE Journal of Solid-State Circuits*, 43(2):566–576, 2008.

[34] B. Liu and B. Chen. Channel-optimized quantizers for decentralized detection in sensor networks. *IEEE Transactions on Information Theory*, 52(7):3349–3358, 2006.

[35] Z.-Q. Luo, G. B. Giannakis, and S. Zhang. Optimal linear decentralized estimation in a bandwidth constrained sensor network. In *International Symposium on Information Theory*, Adelaide, Australia, pages 1441–1445, September 2005.

[36] N. Mukhopadhyay, M. Ghosh, and P. K. Sen. *Sequential Estimation*. Wiley, New York, NY, 1997.

[37] J. W. Mark and T. D. Todd. A nonuniform sampling approach to data compression. *IEEE Transactions on Communications*, 29(1):24–32, 1981.

[38] F. Marvasti. *Nonuniform Sampling Theory and Practice*. Kluwer, New York, NY, 2001.

[39] R. H. Masland. The fundamental plan of the retina. *Nature Neuroscience*, 4(9):877–886, 2001.

[40] Y. Mei. Asymptotic optimality theory for decentralized sequential hypothesis testing in sensor networks. *IEEE Transactions on Information Theory*, 54(5):2072–2089, 2008.

[41] D. Miorandi, S. Sicari, F. De Pellegrini, and I. Chlamtac. Internet of things: Vision, applications and research challenges. *Ad Hoc Networks*, 10(7):1497–1516, 2012.

[42] M. Miskowicz. Send-on-delta concept: An event-based data reporting strategy. *Sensors*, 6:49–63, 2006.

[43] M. Miskowicz. Asymptotic effectiveness of the event-based sampling according to the integral criterion. *Sensors*, 7:16–37, 2007.

[44] B. A. Moser and T. Natschlager. On stability of distance measures for event sequences induced by level-crossing sampling. *IEEE Transactions on Signal Processing*, 62(8):1987–1999, 2014.

[45] E. J. Msechu and G. B. Giannakis. Sensor-centric data reduction for estimation with WSNs via censoring and quantization. *IEEE Transactions on Signal Processing*, 60(1):400–414, 2012.

[46] F. De Paoli and F. Tisato. On the complementary nature of event-driven and time-driven models. *Control Engineering Practice*, 4(6):847–854, 1996.

[47] H. Vincent Poor. *An Introduction to Signal Detection and Estimation*. Springer, New York, NY, 1994.

[48] A. Ribeiro and G. B. Giannakis. Bandwidth-constrained distributed estimation for wireless sensor networks-part II: Unknown probability density function. *IEEE Transactions on Signal Processing*, 54(7):2784–2796, 2006.

[49] J. C. Sanchez, J. C. Principe, T. Nishida, R. Bashirullah, J. G. Harris, and J. A. B. Fortes. Technology and signal processing for brain–machine interfaces. *IEEE Signal Processing Magazine*, 25(1):29–40, 2008.

[50] N. Sayiner, H. V. Sorensen, and T. R. Viswanathan. A level-crossing sampling scheme for a/d conversion. *IEEE Transactions on Circuits and Systems II: Analog and Digital Signal Processing*, 43(4):335–339, 1996.

[51] I. D. Schizas, G. B. Giannakis, and Z.-Q. Luo. Distributed estimation using reduced-dimensionality sensor observations. *IEEE Transactions on Signal Processing*, 55(8):4284–4299, 2007.

[52] I. D. Schizas, A. Ribeiro, and G. B. Giannakis. Consensus in ad hoc wsns with noisy links-part I: Distributed estimation of deterministic signals. *IEEE Transactions on Signal Processing*, 56(1):350–364, 2008.

[53] D. Siegmund. *Sequential Analysis, Tests and Confidence Intervals*. Springer, New York, NY, 1985.

[54] S. S. Stankovic, M. S. Stankovic, and D. M. Stipanovic. Decentralized parameter estimation by consensus based stochastic approximation. *IEEE Transactions on Automatic Control*, 56(3):531–543, 2011.

[55] Y. S. Suh. Send-on-delta sensor data transmission with a linear predictor. *Sensors*, 7(4):537–547, 2007.

[56] S. Sun. A survey of multi-view machine learning. *Neural Computing and Applications*, 23(7–8):2031–2038, 2013.

[57] R. R. Tenney and N. R. Sandell. Detection with distributed sensors. *IEEE Transactions on Aerospace and Electronic Systems*, 17(4):501–510, 1981.

[58] S. C. A. Thomopoulos, R. Viswanathan, and D. C. Bougoulias. Optimal decision fusion in multiple sensor systems. *IEEE Transactions on Aerospace and Electronic Systems*, 23(5):644–653, 1987.

[59] J. Tsitsiklis. Decentralized detection by a large number of sensors. *Mathematics of Control, Signals, and Systems*, 1(2):167–182, 1988.

[60] Y. Tsividis. Digital signal processing in continuous time: A possibility for avoiding aliasing and reducing quantization error. In *IEEE International Conference on Acoustics, Speech, and Signal Processing (ICASSP '04)*, Montreal, Quebec, Canada, volume 2, pages II-589–II-592, May 2004.

[61] Y. Tsividis. Event-driven data acquisition and digital signal processing—A tutorial. *IEEE Transactions on Circuits and Systems II: Express Briefs*, 57(8):577–581, 2010.

[62] V. V. Veeravalli, T. Basar, and H. V. Poor. Decentralized sequential detection with a fusion center performing the sequential test. *IEEE Transactions on Information Theory*, 39(2):433–442, 1993.

[63] T. Vercauteren and X. Wang. Decentralized sigma-point information filters for target tracking in collaborative sensor networks. *IEEE Transactions on Signal Processing*, 53(8):2997–3009, 2005.

[64] C. Vezyrtzis and Y. Tsividis. Processing of signals using level-crossing sampling. In *IEEE International Symposium on Circuits and Systems (ISCAS 2009)*, Taipei, Taiwan, pages 2293–2296, May 2009.

[65] A. Wald and J. Wolfowitz. Optimum character of the sequential probability ratio test. *The Annals of Mathematical Statistics*, 19(3):326–329, 1948.

[66] D. J. Warren and P. K. Willett. Optimal decentralized detection for conditionally independent sensors. In *American Control Conference*, Pittsburgh, PA, pages 1326–1329, June 1989.

[67] P. Willett, P. F. Swaszek, and R. S. Blum. The good, bad and ugly: Distributed detection of a known signal in dependent Gaussian noise. *IEEE Transactions on Signal Processing*, 48(12):3266–3279, 2000.

[68] J.-J. Xiao, S. Cui, Z.-Q. Luo, and A. J. Goldsmith. Linear coherent decentralized estimation. *IEEE Transactions on Signal Processing*, 56(2):757–770, 2008.

[69] J.-J. Xiao, A. Ribeiro, Z.-Q. Luo, and G. B. Giannakis. Distributed compression-estimation using wireless sensor networks. *IEEE Signal Processing Magazine*, 23(4):27–41, 2006.

[70] J. L. Yen. On nonuniform sampling of bandwidth-limited signals. *IRE Transactions on Circuit Theory*, 3(4):251–257, 1956.

[71] Y. Yılmaz, Z. Guo, and X. Wang. Sequential joint spectrum sensing and channel estimation for dynamic spectrum access. *IEEE Journal on Selected Areas in Communications*, 32(11):2000–2012, 2014.

[72] Y. Yılmaz, G. V. Moustakides, and X. Wang. Sequential and decentralized estimation of linear regression parameters in wireless sensor networks. *IEEE Transactions on Aerospace and Electronic Systems*, to be published. http://arxiv.org/abs/1301.5701.

[73] Y. Yılmaz, G. V. Moustakides, and X. Wang. Cooperative sequential spectrum sensing based on level-triggered sampling. *IEEE Transactions on Signal Processing*, 60(9):4509–4524, 2012.

[74] Y. Yılmaz, G. V. Moustakides, and X. Wang. Channel-aware decentralized detection via level-triggered sampling. *IEEE Transactions on Signal Processing*, 61(2):300–315, 2013.

[75] Y. Yılmaz and X. Wang. Sequential decentralized parameter estimation under randomly observed fisher information. *IEEE Transactions on Information Theory*, 60(2):1281–1300, 2014.

[76] Y. Yılmaz and X. Wang. Sequential distributed detection in energy-constrained wireless sensor networks. *IEEE Transactions on Signal Processing*, 62(12):3180–3193, 2014.

[77] K. A. Zaghloul and K. Boahen. Optic nerve signals in a neuromorphic chip I: Outer and inner retina models. *IEEE Transactions on Biomedical Engineering*, 51(4):657–666, 2004.

[78] T. Zhao and A. Nehorai. Distributed sequential Bayesian estimation of a diffusive source in wireless sensor networks. *IEEE Transactions on Signal Processing*, 55(4):1511–1524, 2007.

# 21

# *Spike Event Coding Scheme*

**Luiz Carlos Paiva Gouveia**
*University of Glasgow*
*Glasgow, UK*

**Thomas Jacob Koickal**
*Beach Theory*
*Sasthamangalam, Trivandrum, India*

**Alister Hamilton**
*University of Edinburgh*
*Edinburgh, UK*

**CONTENTS**

**ABSTRACT** In this chapter, we present and describe a spike-based signal coding scheme for general analog computation, suitable for diverse applications, including programmable analog arrays and neuromorphic systems. This coding scheme provides a reliable method for both communication and computation of signals using an asynchronous digital common access channel. Using ternary coders and decoders, this scheme codes analog signals into time domain using asynchronous digital spikes. The spike output is dependent on input signal activity and is similar for constant signals.

Systems based on this scheme present intrinsic computation: several arithmetic operations are performed without additional hardware and can be fully programmed. The design methodology and analog circuit design of the scheme are presented. Both simulations and test results from prototype chips implemented using a 3.3-V, 0.35-μm CMOS technology are presented.

## 21.1 Introduction

Electronic systems are composed of a number of interconnected components ranging from single devices to more complex circuits. Such components operate a defined transformation (transfer function) on given information (signals). There are two aspects of such systems: the signals processing themselves and the transmission of such processed signals to other components of the system. Such aspects are not always well distinguished one from another but are always present, and their designs are of fundamental importance as integrated systems grow in size and complexity.

The design of signal processing and communication techniques and methods are dependent on how the signal is represented. In other words, how the information is presented in a physical medium. A common method to classify a signal is according to their number of possible states. And in this classification system there are two extreme cases: analog and digital representations. The information encoded into analog signals has infinite states while digital representation uses a number of discrete—usually binary—possible states.

Systems designed and operating using digital representation are more robust and flexible than their analog counterparts. The discrete nature of the digital representation implies signals are less prone to electronic noise, component mismatch, and other physical restrictions, improving their information accuracy and allowing easier regeneration. This robustness allows for the easier, cheaper, and faster design of complex systems, contributing to the massive use of digital systems, that is, digital signal processors (DSPs), microcontrollers, and microprocessors.

Analog signal representation is also used despite the cited digital advantages. The main reason is that many signals are either originally presented to the system in this representation or need to be delivered as such by the system. Typical examples of this situation are most sensing systems—such as image, audio, temperature, and other sensors—and aerial signal transmission such as the wide variety of radio communications. In other words, analog circuits are used to interface with real world because most of its information—measurements and controls—are continuous variables as well. Moreover, analog designs usually present a more efficient solution than digital circuits because they are usually smaller, consume less power, and provide a higher processing speed.

Signal dynamics are also important and add another dimension to signal classification: time. As for the amplitude, timing characteristics of the signal are also presented in continuous or discrete fashion. While in the former case the signal value is valid at every instant, in the latter its value is sampled at defined instants. Therefore, it is possible to use different combinations of signal amplitude and timing representations. Systems in such a two-dimensional classification method [1] can be defined as a DTDA, a CTDA, a CTDA, or a DTCA system, where D, C, T, and A stand for discrete, continuous, time, and amplitude, respectively.

Most event-based signal processing uses CTDA representation of signals. Therefore, it combines—among other factors—the higher amplitude noise tolerance from digital signals and the more efficient bandwidth and power usage of asynchronous systems. Several event-driven communication and processing techniques and methods have been proposed and used for many years. In this chapter, we present a specific scheme—named the Asynchronous Spike Event Coding Scheme (ASEC)—and discuss some signal processing functions inherent to it. We also present an analog very-large-scale integration (aVLSI) implementation and its results. Although such a coding scheme can be useful for several applications, we refer to a set of interesting applications at the end of the chapter.

## 21.2 Asynchronous Spike Event Coding Scheme

Signals are processed in several stages in most signal processing architectures. An example is a typical signal acquisition system where an analog signal is first conditioned (filtered and amplified), converted to digital representation, and stored or transmitted. The communication flow from one stage to another is normally fixed, but flexible systems also require flexible signal flow (routing). To obtain maximum flexibility, a communication channel that is commonly available to every processing stage—or signal processing units (SPUs)—is required. A suitable channel shall be defined along with methods to access it.

For flexible-routing systems using the method proposed in this chapter, an asynchronous, digital channel is selected—due to its compatibility with event-based systems—and the Address Event Representation (AER) communication method [2] is used to control the access to it. Due to the asynchronous nature of these events, the AER protocol is an appropriate choice although other asynchronous communication protocols can be used as well.

A method to convert CTCA (analog) signals into a representation suitable for this asynchronous digital channel—and convert back—is required. In this chapter, we use a variation of delta modulation. The association of this signal conversion with the access mechanism to the channel is named the Asynchronous Spike Event Coding (ASEC) communication scheme [3]. An ASEC scheme presents a *Spike Event* coder and decoder pair as shown in Figure 21.1. This pair works on the onset of specific events: the activity of input analog signals triggers the coder operation whereas changes on the state of the channel similarly triggers the decoder operation. These spike events are then transmitted using the specified channel.

### 21.2.1 The Conversion: Ternary Spike Delta Modulation

The coding scheme used in this chapter is based on a variant of delta modulation. The one used in this work is named Ternary Spike Delta (TSD) modulation although it has been given different denominations over the years. As a delta modulation variation, general aspects of delta modulation are presented first.

The working principle of delta modulation is based on limiting the error $e(t)$ between the input (sender) signal $x_s(t)$ and an internal variable $z_s(t)$ such as

$$|e(t)| = |x_s(t) - z_s(t)| \leq e(t)_{max}, \tag{21.1}$$

using a negative feedback loop in the modulator or *coder*. In other words, these modulations work by forcing a feedback signal $z_s(t)$ to track the input signal. The error is sensed by a (normally 1-bit) quantizer system to produce the coder output $y_s(t)$ while the internal variable $z_s(t)$ is generated by the integration of this output signal, that is,

$$z_s(t) = k_i \int y_s(t)dt, \tag{21.2}$$

where $k_i$ is the integration gain.

The communication process is completed at the demodulator or *decoder* side where the signal $z_s(t)$ is replicated as $z_r(t)$

$$z_r(t) = k_i \int y_r(t)dt, \tag{21.3}$$

with $y_r(t)$ being the spikes at the decoder input.

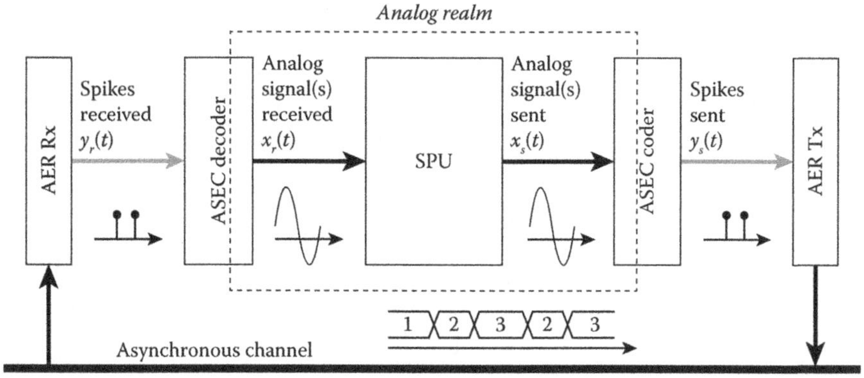


**FIGURE 21.1**
Generic ASEC–SPU interface diagram. Block diagram of the Asynchronous Spike Event Coding (ASEC) scheme interfacing with a given analog SPU and the common asynchronous digital channel. In this architecture, the realm of analog representation of the signals is limited by the coder and decoder to the SPU. This helps to reduce the signal degradation due to noise sources present in analog channels. Spikes are coded and decoded into the channel digital representation (SPU address) using the AER transmitter (AER Tx) and receiver (AER Rx) circuits. Control and registers (not shown) are used to configure the operation of the scheme.

To obtain this replication, an ideal channel $CH$ where $y_r(t) = y_s(t)$ is considered. Furthermore, both coder and decoder integrators are required to present the same initial condition and gain so that $z_r(t)$ and $z_s(t)$ signals present the same DC levels and amplitudes.

A better approximation $x_r(t)$ of the input signal is obtained by averaging the reconstructed signal $z_r(t)$. A low-pass filter (LPF) with an impulse response function $h(t)$ can be used for this purpose:

$$x_r(t) = h(z_r(t)) \approx \overline{z_r(t)} = \overline{x_s(t) - e(t)}. \tag{21.4}$$

Equation 21.4 shows that the difference between the reconstructed signal and the input signal is bounded by the maximum allowed error $e(t)_{max}$.

Modulation methods based on such principles are known as delta modulations because the feedback control updates the feedback signal by a fixed amount *delta* $(\delta)$, which is a function of the resolution of the converter. If the system is designed to provide $N_b$ bits of resolution, then

$$e(t)_{max} = \delta = \frac{\Delta x_{max}}{2^{N_b}}, \tag{21.5}$$

where $\Delta x_{max} = x_{max} - x_{min}$ is the maximum amplitude variation of the input signal. The parameter $\delta$—known as tracking or quantization step—is used in quantizer and integrator designs to limit the error to its maximum.

In the most common delta modulation version (PWM-based), the quantized output is the output of the coder itself, that is, $y_s(t) = Q(e(t))$. However, in spike-based versions, the output of the quantizer(s) trigger(s) a spike generator (SG) circuit such as $y_s(t) = SG(c(t))$ with $c(t) = Q(e(t))$.

The Ternary Spike Delta (TSD) modulation was first presented in [4]. This modulation is similar to the schemes described in [1,5,6], which are based on the principle of irregular sampling used to implement asynchronous A/D converters, for instance. A similar version was also used in [7,8] for low-bandwidth, low-power sensors.

In ternary modulations, the output presents three possible states. The transmitted signal is represented by a series of "positive" and "negative" spikes, each resulting in an increment or decrement of $\delta$, respectively. As the decrement in $z_s(t)$ is also controlled by spikes, the absence of spikes in this modulation—the third state—indicates no change in the current value of $z_s(t)$.

Due to the three-state output, it is not possible to use only one 1-bit quantizer (comparator). Another comparator is inserted in the coder loop as shown in Figure 21.2a, which presents the coder and decoder block diagrams of TSD modulation. This extra comparator also senses the error signal $e(t)$ but its comparison threshold $e_{th2}$ is different.* Both comparator outputs feed the SG block to generate proper spikes $y_s(t)$. These spikes are then transmitted simultaneously to the coder and decoder Pulse Generator (PG) blocks. Depending on these spike signals, each PG block generate control signals to define the respective integrator (INT) output.

The thresholds difference $\Delta e_{th} = |e_{th2} - e_{th1}|$ impacts the performance of the coder, and it is ideally the same as the tracking step, that is, $\Delta e_{th} = \delta$. This ideal value is used in this initial analysis but deviations from this ideal value will be considered later in this chapter.

Although the integrator can present different gains for positive and negative spikes, only those cases where the gains are symmetric are considered here. After analyzing the Figure 21.2b, one can verify the variation in the feedback signal $z_s(t)$ since the initial time $t_0$ is

$$\Delta z_s(t) = \delta(n_p - n_n), \tag{21.6}$$

where $n_p$ is the number of positive spikes and $n_n$ of negative spikes generated since $t_0$. According to Equation 21.6, the value of the signal $z_s(t)$ at any time $t$ in TSD depends only on the number of positive and negative spikes received and processed since $t_0$.

### 21.2.2 The Channel: Address Event Representation

The AER communication protocol was proposed and developed by researchers interested in replicating concepts of signal processing of biological neural systems into VLSI systems [2]. The requirement for hundreds or thousands of intercommunicating neurons with dedicated hardwired connections is impossible in such large numbers in current 2D integrated circuits (ICs). AER aims to overcome this limitation by creating virtual connections using a common physical channel.

The original AER implementation was a point-to-point asynchronous handshaking protocol for transmission of digital words using a multiple access channel common to every element in the system. The information coded in these transmitted digital words represents the identification (address) of a SDU. Depending on the implementation, either the transmitting (as used in this work) or the receiving SPU address is coded.

The asynchronous nature of the AER protocol greatly preserves the information conveyed in the time difference between events. The main sources of inaccuracy between the generated and the received time difference are timing jitter and the presence of event collisions. Because the access to the channel is asynchronous, different elements may try to access the channel simultaneously, but the AER protocol offers mechanisms to handle these spike collisions. A unique and central

*For the sake of simplicity, $e_{th1}$ and $e_{th2}$ are symmetric in this chapter.

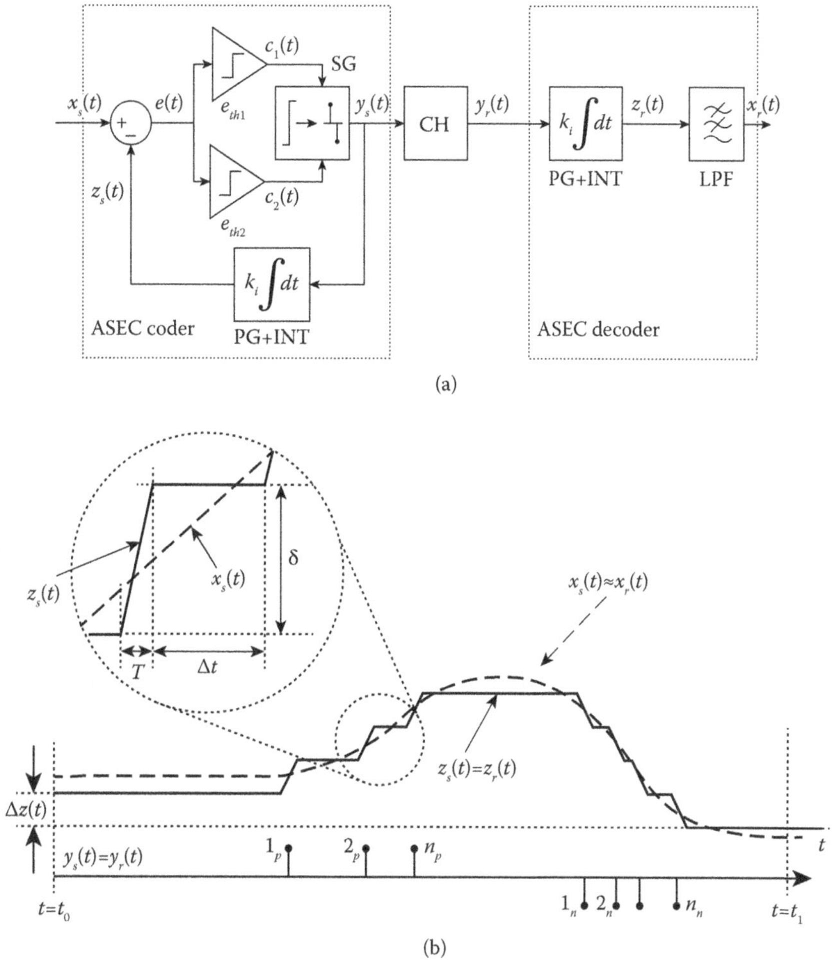


**FIGURE 21.2**
Ternary Spike Delta (TSD) modulation, modified from [3]. (a) Shows the block diagrams of coder and decoder using the channel CH. They present two comparators with different thresholds ($e_{th1}$ and $e_{th2}$), a spike generator (SG), pulse generators (PG), integrators (INT), and the decoder low-pass filter (LPF). Signal $x_s(t)$ is the coder input signal while the feedback signal $z_s(t)$ is an approximated copy of $x_s(t)$ generated from the spike output $y_s(t)$. Ideally, signals $z_r(t)$ and $y_r(t)$ are copies of signals $z_s(t)$ and $y_s(t)$, respectively, on the decoder side. Decoder output $x_r(t)$ is an approximation of $x_s(t)$ generated by the LPF. (b) Presents illustrative waveforms of related signals. The spike train $y_s(t)$ presents three states: "positive" spikes ($1_p, 2_p, \ldots, n_p$), "negative" spikes ($1_n, 2_n, \ldots, n_n$), and no activity. In the inset figure, $T$ is the fixed pulse period, $\Delta t$ is the variable inter-spike period, and $\delta$ is the fixed tracking step.

arbiter is usually used to manage collisions that are resolved by an arbiter by queuing and transmitting successively all the spike events involved in the collision. Although this process can be made relatively faster than the analog signal dynamics, it can be a source of signal distortion.

## 21.3 Computation with Spike Event Coding Scheme

The use of timing as the numerical representation for computations has been studied [9,10] and presents an important advantage due to the evolution of CMOS fabrication. While CMOS digital market (processors, memories, etc.) requirements pushes for faster circuits, many applications present well defined and fixed bandwidth. Therefore the difference between the fastest signal and the used bandwidth increases and so the dynamic range when using timing representation.*

One of the first uses of timing processing was presented in [11] where a different filter structure was presented where its filter coefficients were set by switching

*This assumption is valid when considering similar signals used in the current computations. Many of these signals are measurements of physical properties and, therefore, their dynamics are well defined. For instance, auditory signals are useful only up to 20 kHz for human hearing purposes.

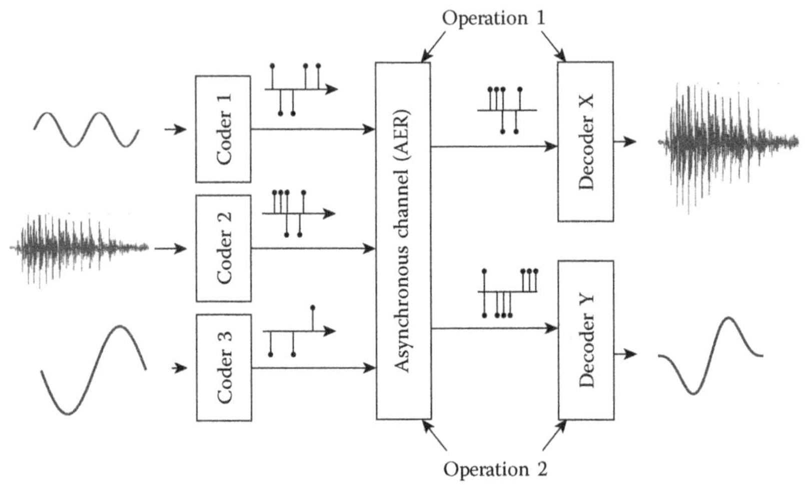


**FIGURE 21.3**
Computation within the communication scheme, modified from [3]. Coders convert analog signals into spikes which are presented to the communication channel. Arithmetic computations may be by proper configuration of the AER channel and decoders. In this example, the top decoder amplifies the audio signal from the middle coder while the bottom decoder adds the signal from top and bottom coders.

times. A multiplication of a signal by a constant (gain) using switches and a low-pass filter was shown. This kind of operation—the summation of amplified signals—is used extensively in artificial neural networks [12].

Some research groups used timing as signal representation to develop analog processing. Pulse-width modulation techniques were used in several applications. For instance it was applied to implement switched filters [9], to compute arithmetic operations [13], to convert signals, and in information storage [14]. Other modulations have been used in specific applications as voice and sound and video processing [15], but few groups are keen to use them in a more generic scope within analog processing. One example was the hybrid computation technique presented in [16]. Hybrid computation was defined as a type of processing that combines analog computation primitives with the digital signal restoration and discretization using spikes to transform a real (analog variable, the inter-spike intervals) number into an integer (discrete digital, the number of spikes) number.

The asynchronous spike event coding communication method is suitable to perform a series of generic computations. Moreover, all computations are realized with the circuits already implemented for the communication process. The idea underlying the computational properties of the communication scheme and some examples are described next.

### 21.3.1 Computational Framework

The asynchronous spike event coding scheme allows a set of arithmetic operations to be performed. For instance, in the configuration illustrated in Figure 21.3, the top decoder (decoder $X$) outputs an amplification of the audio signal while the second one (decoder $Y$) provides the summation of two inputs (sine and triangular waves) using the same communication channel. These operations are simply performed by programming parameters of the communication channel. Other operations can be implemented by combining the channel reconfigurability with the analog signal processing capability of the SPUs. The computational realm of this communication scheme enhances the computing power of the flexible system with no additional hardware.

The shared nature of the channel means it can enable multiple communications virtually at the same time: the computation operations can also be performed simultaneously. This intrinsic computation capability allows a simpler implementation than other pulse-based approaches [17]. Basic arithmetic operations will be the first examples to be presented in Section 21.3.2.

### 21.3.2 Arithmetic Operations

In this section, a set of fundamental arithmetic computations performed by the communication scheme are presented. Each operation is illustrated by simulation of the mathematical models (actual VLSI results will be presented in this chapter as well). The following results were obtained for a 4-bit resolution conversion—for easier visualization—but the working principle can be applied to any arbitrary resolution, mostly defined by the system specifications and limited by its VLSI implementation. Also, the output waveforms in the figures

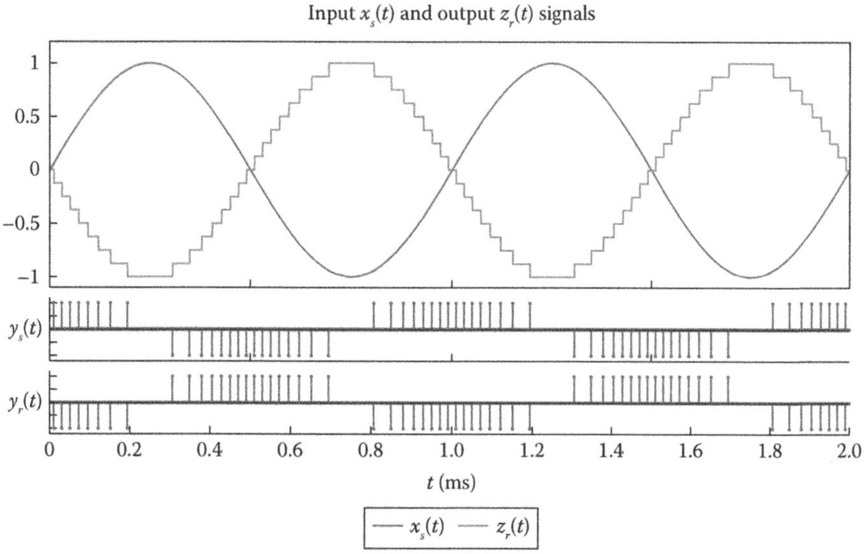


**FIGURE 21.4**
Negation operation. Snapshot of the input sine wave $x_s(t)$ and the respective negated signal $z_r(t)$. Signal negation is performed by interchanging the positive and negative incoming spikes $y_s(t)$ to generate the received spikes $y_r(t)$.

correspond to the decoder filter input $z_r(t)$ rather than the coder output $x_r(t)$ for better understanding.

#### 21.3.2.1 Negation Operation

A fundamental operation on analog signals is changing their polarity, resulting in signal negation or its opposite:

$$x_{neg}(t) = -x(t). \tag{21.7}$$

This operation is also known as the additive inverse of a number. Equation 21.6 states that if coder's and decoder's initial conditions are known, the signal value at any instant $t$ is a function of both the number of events received and their polarity. Changing the incoming spikes polarity will result in an opposite signal on the decoder output. Therefore, the negation operation is performed by setting the AER decoder to interchange the addresses of positive $y_{sp}(t)$ and negative $y_{sn}(t)$ spikes transmitted by the sender coder to form the received spikes $y_{rp}(t)$ and $y_{rn}(t)$ at the decoder.* In other words, the AER decoder performs the following address operation:

$$\begin{aligned} y_{sp}(t) &\rightarrow y_{rn}(t) \\ y_{sn}(t) &\rightarrow y_{rp}(t), \end{aligned} \tag{21.8}$$

resulting in the decoder signal $z_r(t)$ being an opposite version of the signal $z_s(t)$. If $z_r(t_0) = z_s(t_0)$, then from Equation 21.6,

$$x_{neg}(t) \approx z_r(t) = -z_s(t) = \delta(n_n - n_p) - z_s(t_0). \tag{21.9}$$

Since $z_s(t)$ and $z_r(t)$ are the quantized versions of $x_s(t)$ and $x_r(t)$, respectively, the operation of Equation 21.7 is obtained. An example using a sine wave signal is shown in Figure 21.4.

#### 21.3.2.2 Gain Operation

Another common operation in any analog processing is to change the amplitude of a signal, that is, provide a gain $G$ to the signal such as

$$x_{gain}(t) = G \times x(t). \tag{21.10}$$

This gain operation can lead to an increase (amplification) or a decrease (attenuation) of the amplitude of the input signal, with both types being possible using the ASEC scheme.

Because Equation 21.6 holds true for both coder and decoder blocks, the decoder output will be proportional to the input signal. This proportionality defines the operation gain:

$$G = \frac{\Delta x_{gain}(t)}{\Delta x(t)} \approx \frac{\Delta z_r(t)}{\Delta z_s(t)} = \frac{\delta_r}{\delta_s}. \tag{21.11}$$

where $\delta_r$ is the decoder tracking step and $\delta_s$ is its coder equivalent. Having different tracking steps, the ASEC scheme will result in the coder and decoder working with different resolutions. However, this difference would be attenuated by a proper decoder filter design. Figure 21.5a shows simulations for an amplification of a sine wave by a factor of 2 while an attenuation by the same factor is shown in Figure 21.5b.

*The coder output $y_s(t)$ as presented in Section 21.2 is actually the result of merging both positive and negative spike streams.

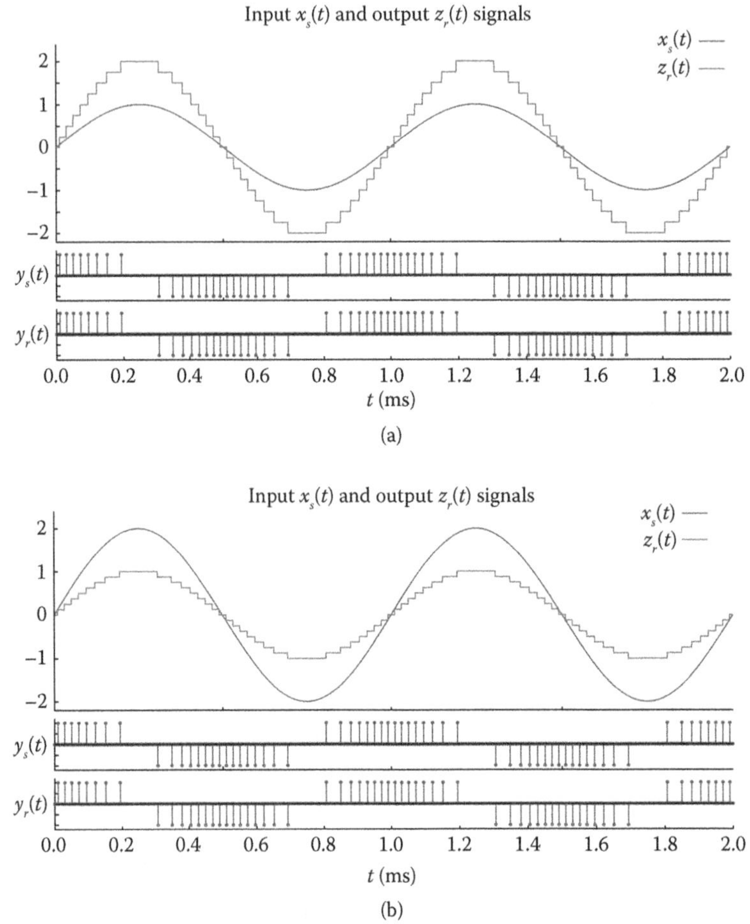


**FIGURE 21.5**
Gain operation. The input sine wave signal $x_s(t)$ is amplified by a factor of 2 in (a) and attenuated by 2 in (b), at the decoder $z_r(t)$. The gain operation is obtained by selecting the ratio of the tracking steps of the coder and the decoder. This operation will result in signal amplification by setting the tracking step in the decoder greater than in the coder. Otherwise, a signal attenuation will be produced.

#### 21.3.2.3 Modulus

Another unary function commonly presented is the modulus (or absolute value) of a signal. The modulus of a signal is the magnitude of such signal defined as

$$x_{abs}(t) = |x(t)|. \tag{21.12}$$

This function is performed by implementing the following AER control on the transmission of the spikes:

$$\begin{aligned} y_{sp}(t) \rightarrow y_{rp}(t),\ y_{sn}(t) \rightarrow y_{rn}(t)\ \ if\ \ sig(x_s(t)) = +1 \\ y_{sp}(t) \rightarrow y_{rn}(t),\ y_{sn}(t) \rightarrow y_{rp}(t)\ \ if\ \ sig(x_s(t)) = -1. \end{aligned} \tag{21.13}$$

where $sig(x_s(t))$ represents a control variable, which is a function of the signal $sig(x_s(t)) = x_s(t)/|x_s(t)|$.

This algorithm can be implemented in two different ways. The first requires an analog comparator and the second uses a digital counter. In both cases, the result in the decoder is the modulus of the signal. This operation is illustrated in Figure 21.6 for a sine wave signal.

In the first method, the original signal is compared against a reference signal representing the zero reference value. This comparator can be included in the same or in a distinct SPU. The comparator output signal $x_c(t)$ is coded and transmitted to the communication channel. The AER decoder controller reads this signal and sets $(+1)$ or resets $(-1)$ the variable $sig(x_s(t))$ on positive and negative spikes reception, respectively. If this variable is reset, the AER decoder performs the negation operation on the original signal as described before. Otherwise, no operation is performed and the output signal replicates the input signal.

The second method employs a digital counter to measure the positive and negative spikes generated by the original signal. To work properly, the initial condition of the signal should represent the zero value. If so, the counter is incremented for each positive spike and

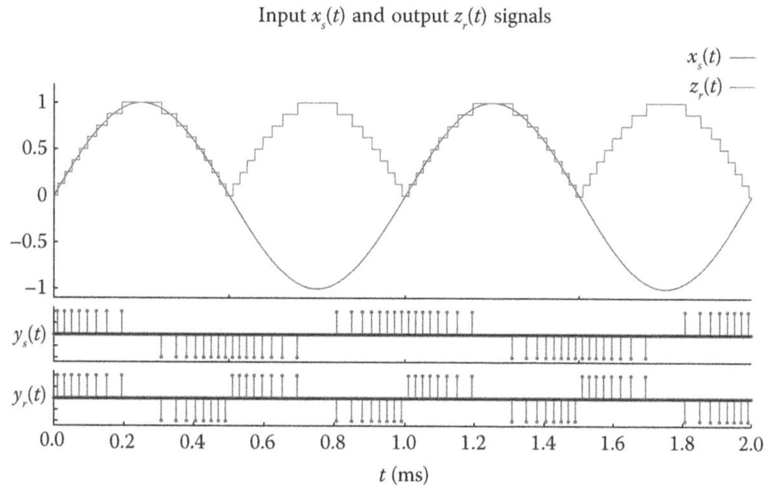


**FIGURE 21.6**
Modulus operation. In this configuration, the modulus of the input signal $x_s(t)$ is output as the result of selectively applying the negation operation. An digital counter or an analog comparator can be used to control the periods when the negation is performed.

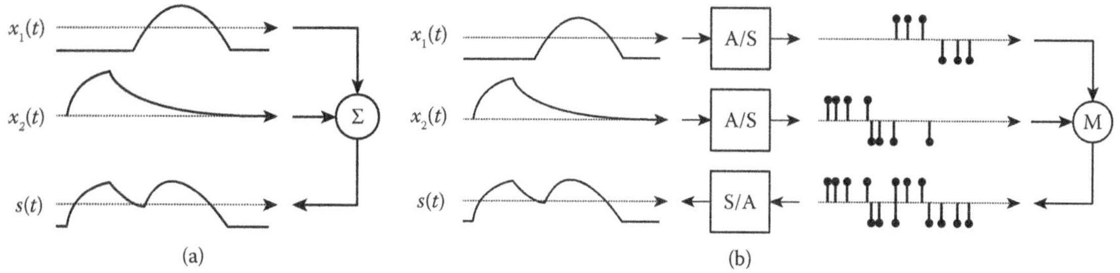


**FIGURE 21.7**
Signal summation in analog (a) and spike coding event systems (b). Analog summation can be implemented in diverse forms, the simplest being current summation according to Kirchoff's current law on a node. The summation on the proposed architecture is performed by simply *merging* the spikes generated by the input signals (operands). In (b) A/S and S/A stand for Analog-to-Spike and Spike-to-Analog conversions, respectively.

decremented for each negative spike. Every time the counter value changes from a negative value to a positive and vice-versa, the variable $sig(x_s(t))$ is updated.

#### 21.3.2.4 Summation and Subtraction

Arithmetic operations involving two or more signals are also required by any generic analog computation system. A fundamental arithmetic operation involving multiple signals is the summation of these signals. Figure 21.7 shows the conventional concept of performing summation in the analog signal domain as well as in the proposed architecture. The summation signal $x_{sum}(t)$ with $N$ signal operators is defined as

$$x_{sum}(t) = \sum_{i=1}^{N} x_i(t). \tag{21.14}$$

To implement this operation with the ASEC scheme, Equation 21.6 is considered again. From this equation, it is possible to demonstrate that the summation signal of $N$ operators is obtained at the decoder output $x_{sum}(t)$ by simply *merging* their respective output spikes

$$x_{sum}(t) \approx z_r(t) = \delta\left(\sum_{i=1}^{N} n_{rpi} - \sum_{i=1}^{N} n_{rni}\right) + z_r(t_0), \tag{21.15}$$

where $n_{rpi}$ and $n_{rni}$ are the number of positive and negative spikes, respectively, received from the $i^{th}$ operand after the initial instant $t_0$.

In this architecture, the summation is performed by routing the spikes of all operands to the same decoder, with the SPU input—after the spike-to-analog conversion—being the result. Simulation results showing the summation of two sine wave signals $x_{s1}(t)$ and $x_{s2}(t)$ are shown in Figure 21.8a, with the decoder integrator output $z_r(t)$. The signal $x_{sum}(t)$ is the theoretical result. Subtraction $x_{sub}(t)$ may be achieved by applying

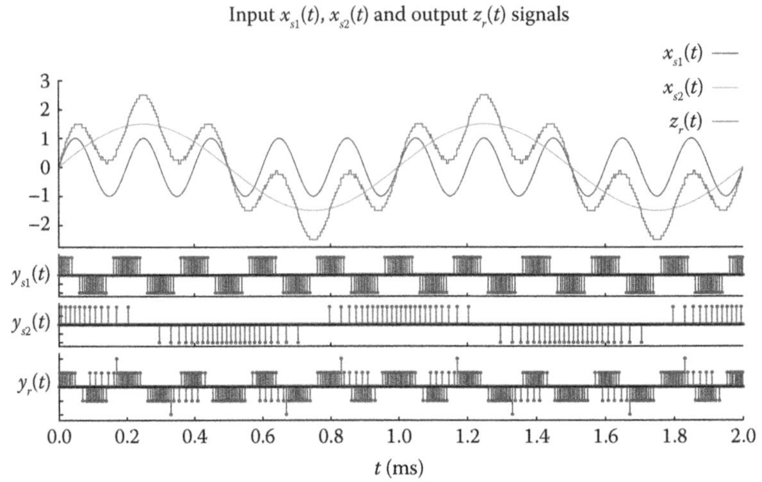


(a)

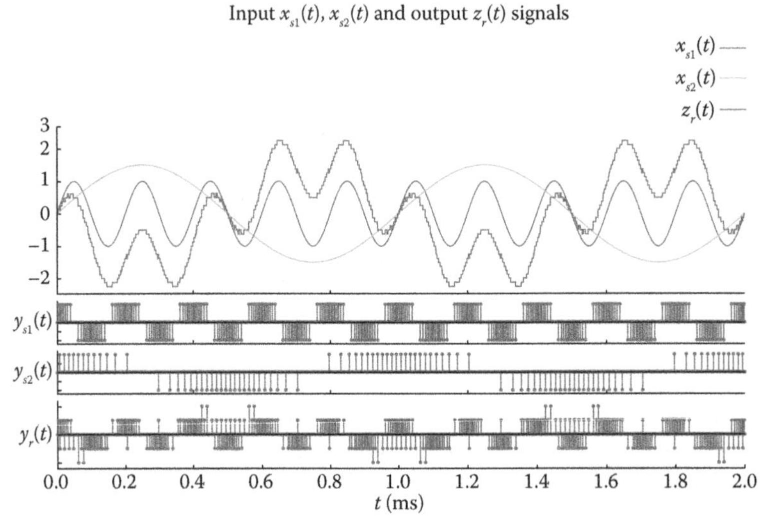


(b)

**FIGURE 21.8**

Summation and subtraction operations. (a) Shows the summation of two input sine waves, $x_1(t)$ and $x_2(t)$. Signal $z_r(t)$ is the respective summation while $x_T(t)$ is the ideal summation result. The subtraction of the same input signals $x_1(t) - x_2(t)$ is shown in (b). The summation is performed by the concatenation $y_r(t)$ of the spikes from all operands, that is, $y_1(t)$ and $y_2(t)$ in this case while subtraction also requires the negation operation. Spikes collisions are identified whenever $y_r(t)$ presents twice the amplitude in these figures.

both the summation and the negation operations outlined before, and an example is depicted in Figure 21.8b.

#### 21.3.2.5 Multiplication and Division

While some operations are performed using asynchronous spike event coding and decoding methods together with the AER decoder, the range of possible computations can be expanded when the communication scheme capabilities are combined with the functionality of the SPUs. Multiplication is an example of this cooperation. Once summation operation can be performed by the ASEC scheme, it is also possible to implemented in this architecture. This assumption is fulfilled using the following logarithmic property:

$$x_{mult}(t) = \prod_{i=1}^{N} x_i(t) = \exp\left(\sum_{i=1}^{N} \ln\, x_i\right). \qquad (21.16)$$

In other words, this property establishes that multiplication operation can be computed from summation operation if the operands $x_i(t)$ are submitted to a logarithm compression and the summation result is expanded in an exponential fashion. Although this compression and expansion are not performed by the ASEC scheme, if those operations are implemented inside

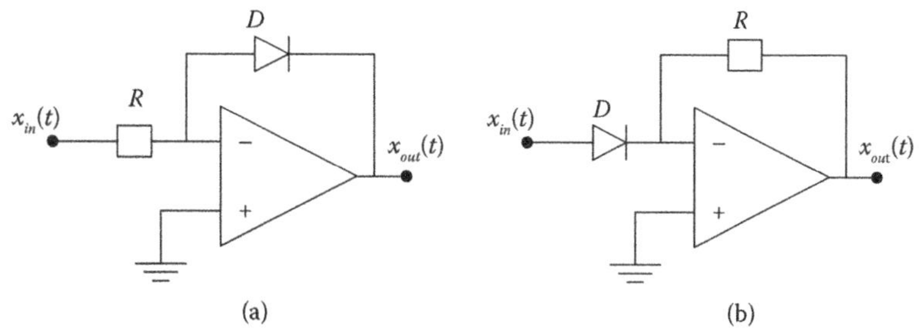


**FIGURE 21.9**
Logarithm and exponential amplifiers. These figures show traditional circuits used to realize both logarithm (a) and exponential (b) operations on analog signals. These circuits were not implemented in any chip designed for this work. These circuits can be implemented to allow multiplication and division operations.

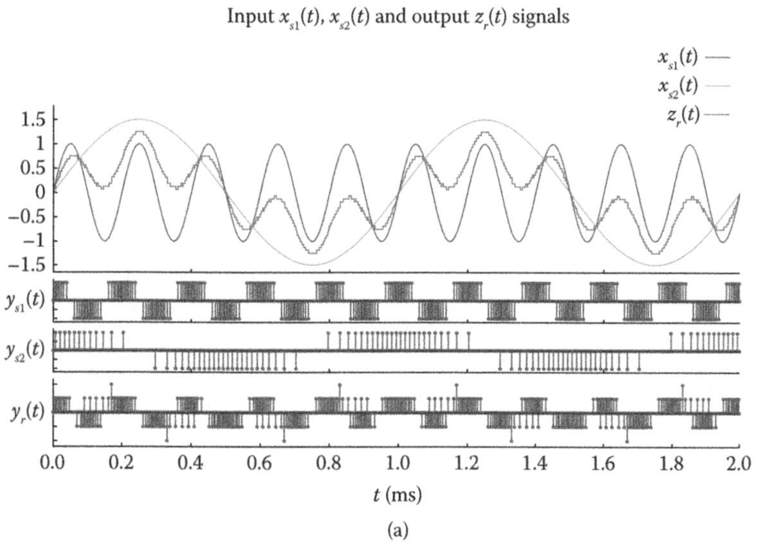


**FIGURE 21.10**
Average operation. This figure shows the input signals $x_{s1}(t)$ and $x_{s2}(t)$ and the respective average signal $z_r(t)$. This operation is performed by combining the summation and the gain operations, where the gain factor is the inverse of the number of operands.

the SPUs, both multiplication and division operations would be available in the system. Division is achieved similarly using the subtraction operation.

A conventional method to generate this logarithm compression is using the circuits presented in Figure 21.9. They are known as logarithmic and exponential amplifiers. Although they present a simple topology, the diode resistance is usually highly dependent on the temperature. These circuits were not implemented in the work described here.

#### *21.3.2.6 Average Operation*

Once the gain and the summation operations are available, another possible operation is the average of several inputs in each instant given by

$$x_{avg}(t) = \frac{1}{N}\sum_{i=1}^{N} x_i(t). \tag{21.17}$$

To implement this operation, ASEC uses the same summation and gain procedures, where the gain $G$ in 21.3.2.2 is explained here as

$$G = \frac{1}{N}. \tag{21.18}$$

Figure 21.10 presents an example of such computation using the same signals of the summation operation with the gain $G = 0.5$. This expression indicates that the minimum possible amplitude for the tracking step limits the maximum number of operators in this operation.

### 21.3.3 Signal Conversion

The main advantage of an analog system is to process analog signals using its SPUs. However, it might be of some interest to have a digital representation of the analog signals, both for storage and transmission purposes. Implementations of analog-to-digital converters (ADC) based on asynchronous modulations were presented in [16,18–20].

A simple method to provide a digital representation is to use a up/down digital counter updated on the onset of received spikes, where the counter direction is determined by the spike polarity. However, such a method is not suitable for signal storage as the timing information (time stamping) is not measured.

Realizing data conversion with the proposed scheme—and keeping the timing information for signal storage—can be achieved using time-to-digital (TDC) and digital-to-time (DTC) converters. In its simplest implementation, TDC circuits consist of a digital counter operating at a specific frequency which determines the converter resolution. Its input is an asynchronous digital signal which triggers the conversion process: read (and store) the current counter value and then reset it. The digital value represents the ratio between the interval of two consecutive inputs and the counter clock period:

$$X_i = \frac{t_i - t_{i-1}}{t_{clk}}, \tag{21.19}$$

where $X_i$ is the digital representation of the time interval $t_i - t_{i-1}$. Therefore, an AD conversion is implemented. The TDC can be used to read the output spikes $y_s(t)$ generated by a ASEC coder with input $x_s(t)$ because each spike transmitted using the ASEC scheme presents the same signal magnitude change. An extra bit is, therefore, needed to represent the polarity of the spike. For instance, the least significant bit may indicate the spike polarity.

However, with constant input signals, the ASEC coder will not generate any spikes. In this case, the digital counter can overflow and the information would be lost. One method to avoid this is to ensure the registry of these instants, storing or transmitting the maximum or minimum counter value, for instance. These values will add up until an incoming spike stops and resets the counter.

Another method, more complex, would be dynamically changing the tracking step of the ASEC coder in the same fashion as the continuously variable slope delta (CVSD) modulation. Every time the counter has overflowed, the tracking step would be reduced (halved for instance) to capture small signal variations. This approach increases the digital word size because the step size has to be stored. However, the number of generated digital words is smaller.

The complementary process can be used to perform an digital-to-analog conversion (DAC). In this case, a counter (DTC) is loaded with a digital word and then starts to count backwards until the minimum value is reached (typically zero) before reloading the next digital word. At this point, a spike with the correct polarity is transmitted to an ASEC decoder throughout the AER communication channel to (re)generate the analog signal. These applications can be used to implement complex systems that could not be fit in a simple analog system. Furthermore, this works only for nonreal-time systems.

### 21.3.4 Other Operations

#### 21.3.4.1 Shift Keying Modulations

Specific applications can also be realized with the proposed architecture. An example of those is phase-shift keying. Phase-shift keying is a digital modulation where the phase of a carrier changes according to the digital information. For instance, the binary phase-shift keying (BPSK) is defined as

$$s(t) = z_r(t) = \begin{cases} a\,sin(\omega_c t) & if \quad b_i = 0 \\ a\,sin(\omega_c t + 180^\circ) & if \quad b_i = 1, \end{cases} \tag{21.20}$$

where $\omega_c$ is the angular frequency of the carrier, $b_i$ is the bit to be transmitted, and $a$ is a function of the energy per symbol and the bit duration [21].

To implement this function using ASEC architecture, one must provide a carrier (sine wave) signal to the input of the coder and rout the resulting spikes according to the digital value to be transmitted. Whenever the input bit is zero, the carrier is replicated at the decoder output while a negation operation is performed otherwise, that is,

$$\begin{matrix} y_{sp}(t) \to y_{rp}(t),\, y_{sn}(t) \to y_{rn}(t) & if \quad b_i = 0 \\ y_{sn}(t) \to y_{rp}(t),\, y_{sp}(t) \to y_{rn}(t) & if \quad b_i = 1. \end{matrix} \tag{21.21}$$

Figure 21.11 shows the modulation result $z_r(t)$ for an input word $x_2(t) = 01011001$. Quadrature phase-shift keying (QPSK) is possible using the summation operation with another sine wave (90° phase shifted) and the fast generation of a fixed number of successive spikes to quickly change the carrier phase due to signal discontinuity. This function has a similar working principle as the function modulus described earlier. The difference is that the AER routing is controlled by an external signal $b(t)$ rather than the input signal itself.

#### 21.3.4.2 Weighted Summation

The last operation described in this section is the weighted summation. This operation, along a nonlinear transfer function, is the functional core of artificial neural networks (ANN) systems. Other applications for the weighted summation include bias compensation in statistics, center of mass calculation of a lever, among

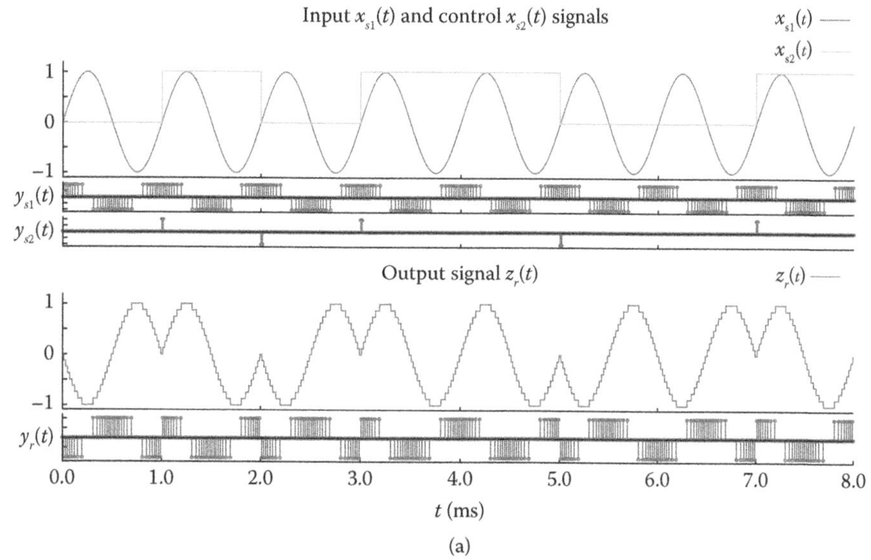


**FIGURE 21.11**
Binary phase-shift keying. This modulation is implemented by routing the spikes generated by the sine wave carrier $x_1(t)$ according to the digital input $x_2(t)$. In this example, the modulated digital word is 01011001.

others. The implementation using the proposed communication method is straightforward by combining the gain and the summation operations.* The expression for weighted summation is

$$WS(t) = \sum_{i=0}^{N} w_i(t)\ x_i(t). \tag{21.22}$$

One advantage of implementing such a computation in this architecture is the possibility of implementing massive ANNs in real time because of the system inherent scalability. On the down side, because different inputs $x_i(t)$ are multiplied by different and usually changing gains $w_i(t)$, the decoder tracking step $\delta_r$ need to be changed before receiving the correspondent spike. This results in a further delay between the instant when the AER routing computes the target address and the moment when the ASEC decoder implements the update on its output. This delay is due to setting up of new tracking step. This usually requires waiting for a DAC controlling the decoder integrator to settle to a new value.

## 21.4 ASEC aVLSI Implementation

The ASEC scheme uses the TSD modulation to perform the analog-to-time conversion and the AER protocol to route the spikes in an analog system. The ASEC block diagram presented in Figure 21.12 is the result of the rearrangement of the TSD block diagram presented in Figure 21.2a and the ASEC–SPU interface diagram presented in Figure 21.1. This rearrangement was performed to better understand the implementation of each of the blocks designed. The comparator, the spike generator, the pulse generator, and the integrator circuit design are further described in this section. Before analyzing each of these blocks individually, ASEC parameters and their designs are explained.

The first step in the design of the coder is to define the main* parameter of comparators of Figure 21.2a, that is, the thresholds $e_{th1}$ and $e_{th2}$. Absolute values of these thresholds impact on the DC level of both $z_s(t)$ and $z_r(t)$. For a null DC level error, that is, $\overline{e(t)} = 0$, these thresholds must be symmetrical ($e_{th1} = -e_{th2}$). A positive DC error is produced when $|e_{th1}| > |e_{th2}|$. Similarly, $|e_{th1}| < |e_{th2}|$ results in a negative error. More importantly, the *difference* between the comparator thresholds holds a relation with the coding resolution as

$$\Delta e_{th} = e_{th1} - e_{th2} = \delta, \tag{21.23}$$

where δ—the tracking step—is a function of the integrator gain $k_i$ as in

$$\delta = k_i T, \tag{21.24}$$

and $T$ is the duration of the pulses produced by the pulse generator block. The parameter $T$ is designed according

*This method does not work when changing the gain for constant signals.

*For the targeted applications of this scheme, the speed is not the most crucial parameter.

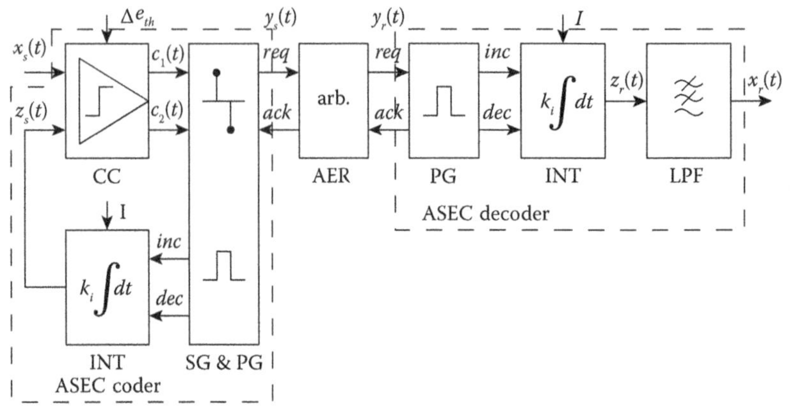


**FIGURE 21.12**
Block diagram of the ASEC implementation. CC is the *composite comparator* that performs the function of both comparators in Figure 21.2a. This block outputs the signals $c_1(t)$ and $c_2(t)$ according to the input signal $x_s(t)$ and the system parameter $\Delta e_{th} = e_{th1} - e_{th2}$. SG is the *spike generator* and PG is the *pulse generator*. PG and SG are integrated on the coder side. SG generates the AER request signal *req* and waits for the acknowledge signal *ack* from the arbiter in the AER channel. On the reception of *ack*, PG outputs pulses *inc* and *dec*. The gain $k_i$ defines the tracking step $\delta$ value of the *integrator* block INT and is set by the current $I$.

to the predicted input signals, system size, and the overall configuration because these characteristics determine the channel communication activity.

The communication channel may be overloaded when the activity of all coders exceed the channel capacity. In this implementation, the spike generator sets a minimum interval between successive output spikes as a preventive method to avoid this condition. This "refractory period" is defined as

$$\Delta t_{(min)} = \frac{1}{f_{sk(max)}} - T. \tag{21.25}$$

while the spike width is determined from

$$T = \frac{\delta}{(k+1)\,|\dot{x}_s(t)|_{(max)}}. \tag{21.26}$$

if the period is defined as a multiple of the spike "width" $\Delta t_{(min)} = kT$.

An estimation about the limitations imposed to the input signal can be derived from the definition of $\Delta t_{(min)}$. From Equation 21.26, the frequency of an input sine wave $x_s(t) = A\ \sin(\omega_{in}t)$ is

$$f_{in} = \frac{\omega_{in}}{2\pi} = \frac{1}{2\pi}\frac{\delta}{A} f_{sk(max)} = \frac{1}{2\pi}\frac{1}{2^{N_b}} f_{sk(max)}. \tag{21.27}$$

This means that once the system parameters are defined, the maximum input frequency is inversely proportional to the signal amplitude. In case the input signal frequency is greater than the value defined by Equation 21.27, the system will present slope overload [22].

The communication scheme presented in Figure 21.12 can be implemented using a small number of compact circuits for each block (comparator, spike generator, and integrators). Design methodology is presented in more detail for each of these circuits in the next subsections. The decoder low-pass filter (LPF) was not implemented on-chip, because it is an optional part of the circuit which increases the method resolution and, therefore, an offline, software-based implementation was used instead.* These ICs were designed using a 3.3 V power supply, four metal layers, a single polysilicon layer, 0.35 μm pitch CMOS process. The voltage domain was used to represent the analog signals involved in the communication process. This choice would allow an easier integration with the SPU proposed in [23].

### 21.4.1 Comparator

In the block diagram presented in Figure 21.2a, signal subtraction and comparison functions are performed by different blocks but they are implemented as a single circuit: the comparator. The design of comparators can be implemented using a preamplifier (PA) feeding a decision circuit (DC) and an output buffer (OB) [24], as shown in Figure 21.13. The preamplifier circuit provides enough gain on the input signal to reduce the impact of the mismatch on the decision circuit, which outputs a digital-like transition signaling the comparison.

*The main design parameter of the decoder filter is its pole location as it impacts the signal resolution by attenuating the undesirable out-of-band high frequency harmonics generated during the decoding process [3].

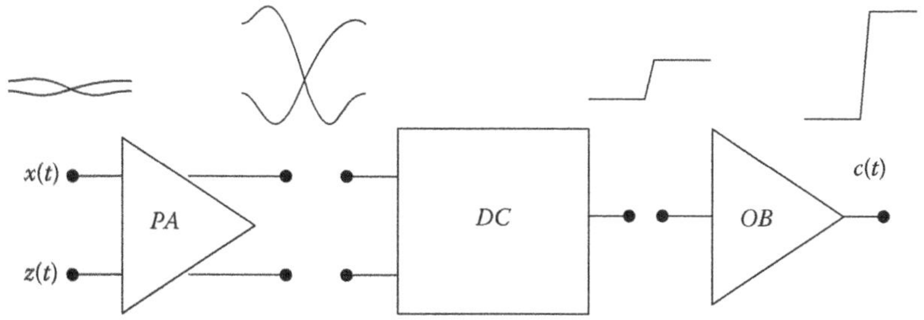


**FIGURE 21.13**
Block diagram of a typical comparator. The input signals' difference is magnified by the preamplifier (PA) and then used by the decision circuit (DC) to provide a comparison signal. An output buffer (OB) keeps the transition times independent of the load.

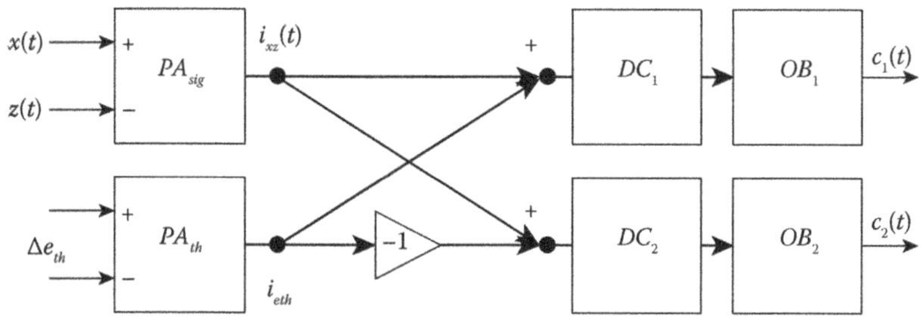


**FIGURE 21.14**
Block diagram of the compound comparator, adapted from [3]. The output current $\Delta i_{xz}$ from preamplifier $PA_A$ is a function of the inputs $x(t)$ and $z(t)$ while preamplifier $PA_B$ outputs a fixed offset current $i_{eth}$. $i_{eth}$ is both added and subtracted from $\Delta i_{xz}$, and the results are applied to the respective decision circuits ($DC$) and output buffers ($OB$).

The output buffer provides the current needed to keep the rising and falling times short for any load.

Capacitive or resistive dividers at the comparator input nodes can be used to realize the required $\Delta e_{th}$, but they impact negatively on the input impedance of the circuit [25]. $\Delta e_{th}$ can also be generated using offset comparators [25]; however, this topology suffers from low input dynamic range. An additional preamplifier, which provides a respective $I_{eth}$ on the decision circuit input, can also be used to generate programmable offset [26].

A compound comparator outputs both $c_1(t)$ and $c_2(t)$, as shown in Figure 21.14. Initially, this would require the use of four preamplifiers with two sensing the inputs $x(t)$ and $z(t)$ and two providing different offsets (two sets for each output). However, only two preamplifiers are needed. The preamplifier $PA_A$ outputs a differential current $\Delta I_{xz}(t)$ according to the difference between $x(t)$ and $z(t)$, that is, the error signal $e(t)$ in Figure 21.2a. The additional capacitive load on the input nodes $x(t)$ and $z(t)$ when using two preamplifiers is therefore avoided. The other preamplifier ($PA_B$) provides a differential current $I_{eth}$ according to voltage $\Delta e_{th}$ on its inputs. The results of adding to $(I_{xz} + I_{eth})$ and subtracting $(I_{xz} - I_{eth})$ from these currents are forwarded to the decision circuits to speed up the result.

Figure 21.15 depicts electrical schematics of the circuits used in the compound comparator. The preamplifiers are identical transconductance amplifiers. The decision circuit is a positive feedback circuit while the output buffer is a self-biased amplifier [27]. The comparator was designed to provide a hysteresis smaller than the tracking step to help avoiding excessive switching due to noise. This hysteresis is generated with the size difference between transistors $M_9 - M_{10}$ and $M_{11} - M_{12}$.

#### 21.4.1.1 Design Considerations

The system performs at its best when the tracking step $\delta$ generated at the integrator outputs is the same as the difference between the thresholds $\Delta e_{th}$ of the comparators. This is very unlikely as process mismatches [28] will certainly vary the comparator offsets $V_{os1}$ and $V_{os2}$ from the *designed* value $\Delta e_{thD}$, as shown in Figure 21.16b. Hence, the *actual* $\Delta e_{th}$ is bounded, for a $3\sigma$ variation (99.7%), by

$$\Delta e_{thD} + 3\sigma_{os} \geq \Delta e_{th} \geq \Delta e_{thD} - 3\sigma_{os}, \qquad (21.28)$$

where $\sigma_{os} = \sigma(V_{os1}) + \sigma(V_{os2})$ for uncorrelated variables and $\sigma(V_{os1})$ and $\sigma(V_{os2})$ are the standard deviations of the comparator offsets $V_{os1}$ and $V_{os2}$, respectively.

Therefore, a design margin is required to compensate for the random offsets due to process variations. If $\delta < \Delta e_{th}$, the feedback signal $z_s(t)$ is delayed and distorted, in particular in the regions close to the signal inflection

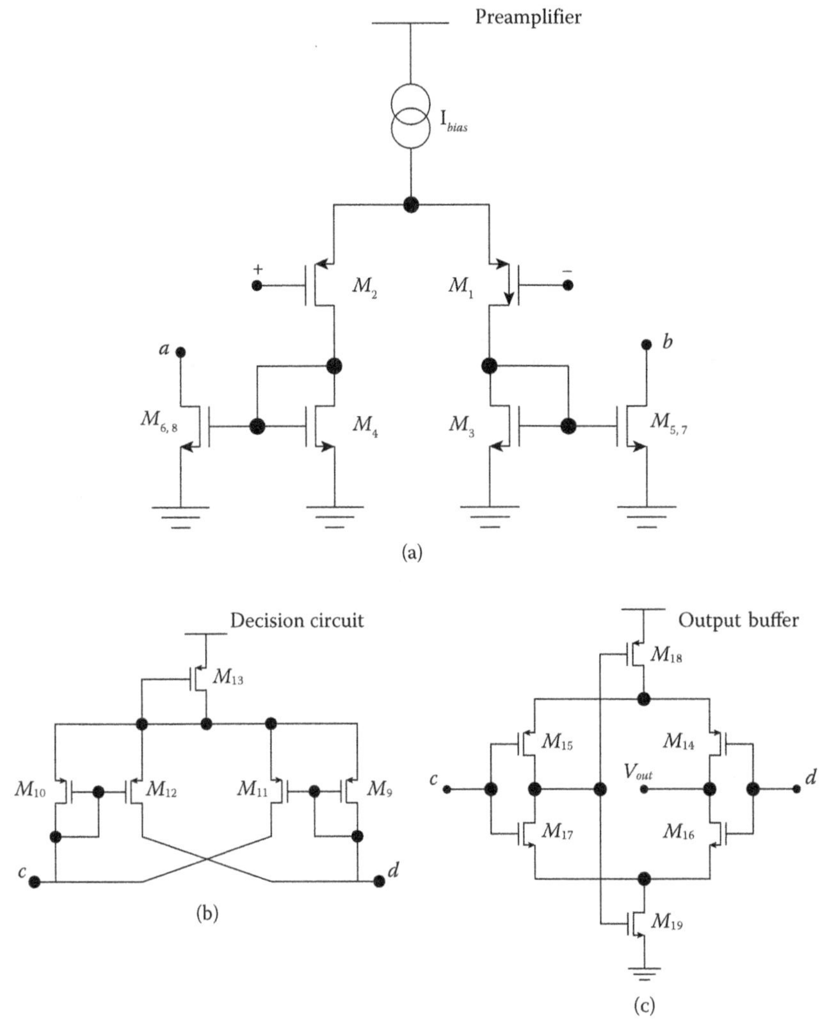


**FIGURE 21.15**
Circuit schematic for each block of the compound comparator, adapted from [3]. Transconductance preamplifier (*PA*) in (a), decision circuit (*DC*) in (b), and output stage (*OB*) in (c). In Figure 21.14, the connection of blocks $PA_{sig}$, $DC_1$, and $OB_1$ is achieved by connecting the nodes *a* and *b* to nodes *c* and *d*, respectively, where $i_a(t) - i_b(t) = i_{xy}(t)$. Similarly, for the circuit $PA_{th}$ in Figure 21.14, $i_a(t) - i_b(t) = i_{eth}$, and node *a* connects to node *d* and node *b* to node *c* to implement the negative gain.

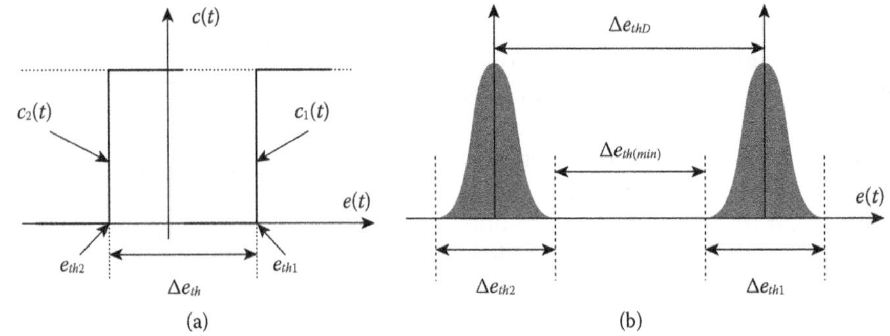


**FIGURE 21.16**
Design and mismatch effects of comparator thresholds, adapted from [3]. (a) Shows comparator transfer functions while comparator threshold variations $\Delta e_{th1}$ and $\Delta e_{th2}$, with $\Delta e_{thX} \approx 6\sigma(V_{osX})$, used to calculate the designed threshold difference $\Delta e_{thD}$, are presented in (b). For optimum design, $\Delta e_{th(min)} = \delta$.

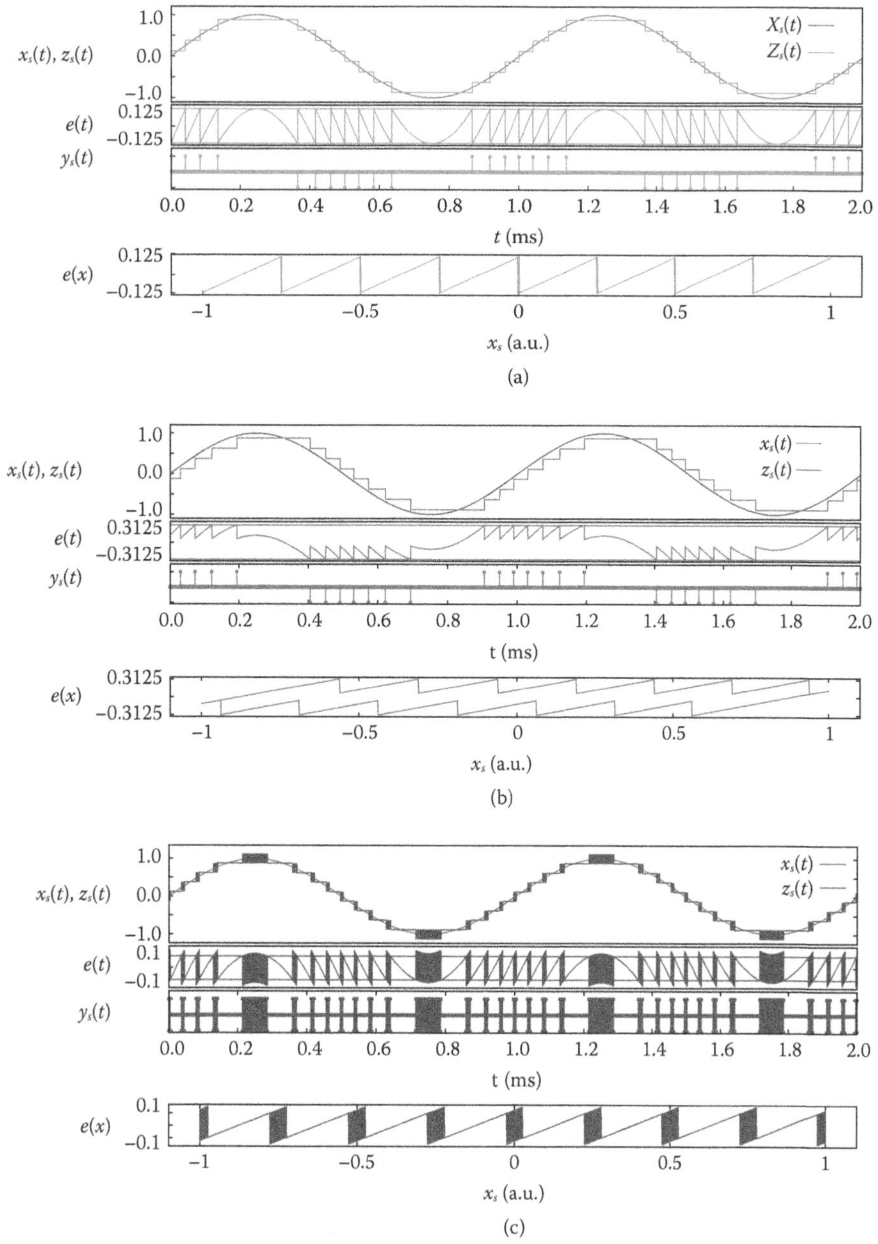


**FIGURE 21.17**

Simulation of the comparator model with mismatch. TSD model results for 3-bit resolution with different mismatch between tracking step $\delta$ and comparator threshold difference $\Delta e_{th}$. (a) Shows the case for $\Delta e_{th} = \delta$, (b) $\Delta e_{th} = 2.5 * \delta$, and (c) $\Delta e_{th} = 0.8 * \delta$, which presents excessive switching activity.

points. More critically, $z_s(t)$ will oscillate for $\delta > \Delta e_{th}$. To avoid the last case, the comparators' threshold difference is designed to meet the following safety margin

$$\Delta e_{thD} \geq \delta + 6\sigma_{os}. \tag{21.29}$$

Figure 21.17 illustrates the effects of this mismatch when $\Delta e_{th}$ is equal, greater, and smaller than $\delta$. These results were obtained from the TSD model simulation results for a 3-bit resolution. Figure 21.18 shows the influence of this mismatch type on the resolution.

### 21.4.2 Spike and Pulse Generators

The spike generator block can output positive and negative spikes according to the state of $c_1(t)$ and $c_2(t)$ signals. It is required that a positive spike is generated whenever the error $e(t) > \Delta e_{th}/2$, and similarly a negative spike is generated when $e(t) < -\Delta e_{th}/2$. The spike generator will not generate any spikes otherwise. From these spikes, corresponding pulses are generated both at the coder and the decoder. Figure 21.19a and b presents

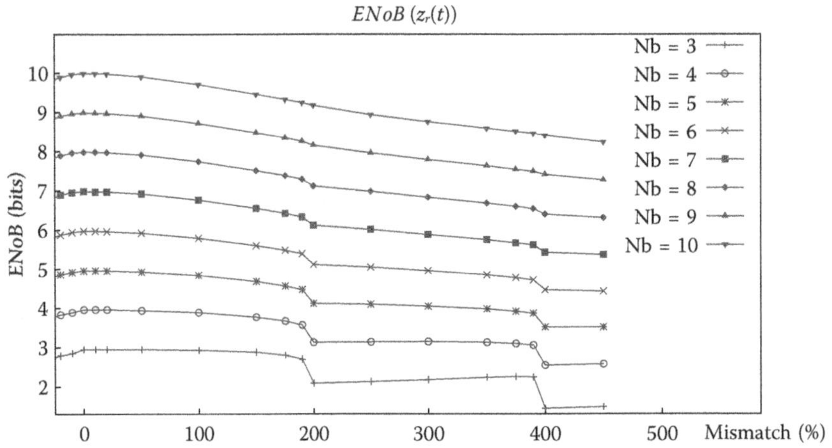


**FIGURE 21.18**
Impact of comparator mismatch on the resolution. Simulation results for TSD model with different mismatches between tracking step $\delta$ and comparator threshold difference $\Delta e_{th}$. The ENoB is measured in $z_r(t) = z_s(t)$ signal. The simulation does not include other noise sources than mismatch.

the block diagrams of the spike and pulse generators on the coder and the pulse generator on the decoder, respectively. Figure 21.19c and d shows an example of the behavior of the control signals.

The local arbiter senses which comparator outputs $c_1(t)$ and $c_2(t)$ changed first and locks its state until end of the cycle of the signal $z_s(t)$, disregarding further changes, including noise. The arbiter output then triggers the spike generator which starts the handshaking communication with the AER arbiter setting either $req_p$ or $req_n$ signal. On receipt of the *ack* signal, the pulse generator is activated and provides a pulse—*inc* or *dec*—to the integrator according to the arbiter output.

The pulse generator block includes programmable delay circuits for the generation of $T$ and $\Delta t_{(min)}$ time intervals. These blocks were implemented as a current integrator feeding a chain of inverters. This is a suitable implementation although more efficient methods have been available [29]. The control logic for this circuit was implemented using a technique developed to design asynchronous digital circuits [30]. Although the circuit in [7] presents a simpler hardware, it lacks noise tolerance provided by the arbiter.

### 21.4.3 Integrator

The integrator is the last block remaining to be analyzed. Together with the comparator threshold difference $\Delta e_{th}$, its gain $k_i$ defines the resolution of the coder. An integrator based on the switched current integration (SCI) technique [13] was designed. This implementation, used in charge pump circuits, was also used in other modulation schemes [13,31], and a unipolar version of this type of circuit driving resistors is used in steering current cells of some DAC converters [32], as shown in Figure 21.20.

The schematic of the implemented SCI circuit is presented in Figure 21.21, and it works as follows. On the detection of a positive spike, the *inc* signal is set at a high voltage level (and $\overline{inc}$ is set at low voltage). This allows for currents $0.5I$ and $1.5I$ to flow in transistors $M_{21}$ and $M_{20}$, respectively, for a fixed interval $T$. The current difference will then charge the integrating capacitor $C_{int}$. Symmetrically, the transistors $M_{22}$ and $M_{23}$, controlled by *dec* and $\overline{dec}$ signals, will provide a similar current to discharge the capacitor. The resulting current $I$ that discharges or charges the integrating capacitor $C_{int}$ is given by

$$I = C_{int}\,\frac{\delta}{T}, \tag{21.30}$$

and from Equation 21.24, the designed integrator gain may be obtained as

$$k_i = \frac{I}{C_{int}}, \tag{21.31}$$

and when there are no spikes from the spike generator, currents are driven to low-impedance nodes ($d_1 - d_4$) through $M_{24} - M_{27}$ by setting the voltages of signals $dp$ to high and $\overline{dp}$ to low levels.

This circuit operation of using simultaneously charging and discharging paths for the integration capacitor penalizes the power consumption of the system but reduces the charge injection on the integration nodes $z_s(t)$ and $z_r(t)$ [33]. Charge injection at the gate to drain capacitances of the complimentary switches will cancel each other if switches $M_{20}$ and $M_{21}$ ($M_{22}$ and $M_{23}$) have the same dimensions.

Another design approach would be switching on and off the current mirror itself instead of driving the current to low-impedance nodes. This solution would result

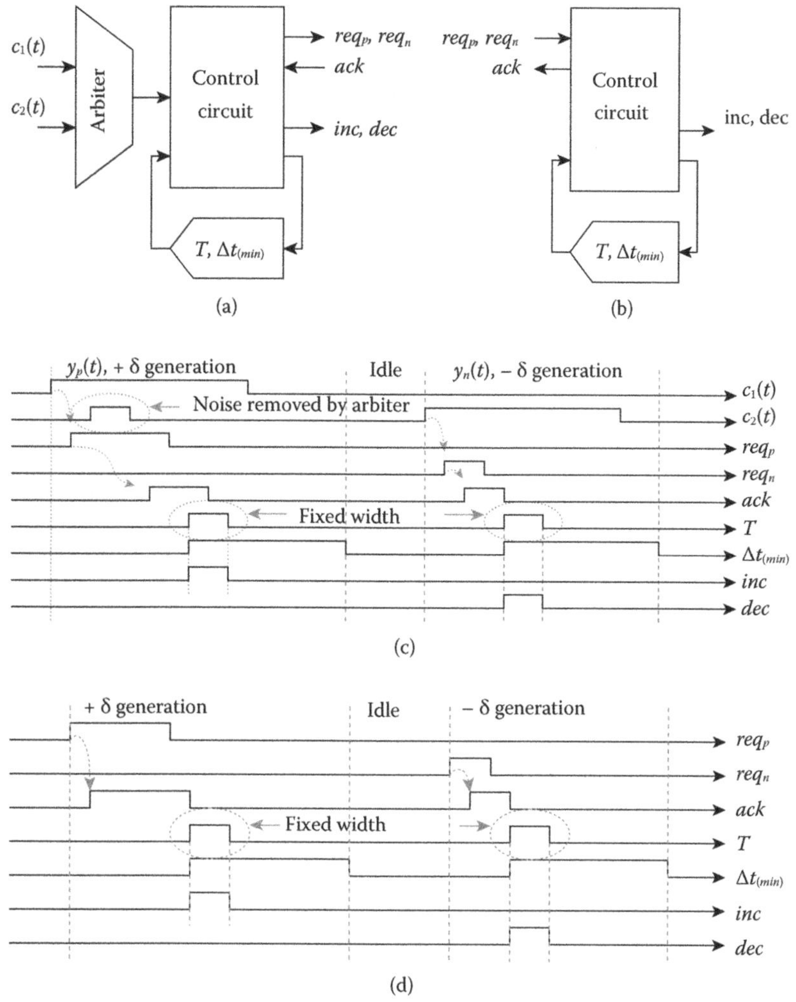


**FIGURE 21.19**

Spike (SG) and pulse (PG) generators, adapted from [3]. Coder's combined spike and pulse generators (a) and decoder's pulse (b) generator block diagrams. (c) Is an example of timing diagram of the block in the coder while (d) shows timing diagrams on the decoder. In the coder, a local arbiter selects between $c_1(t)$ and $c_2(t)$ inputs, starts the handshaking signaling (either setting the appropriate request signal $req_p$ or $req_n$), and waits for the acknowledge signal *ack* from the AER arbiter. On completion of the handshaking process, either a *inc* or *dec* pulse is generated using delay blocks for $T$ and $\Delta t_{(min)}$.

in a lower power consumption, but switching the voltage reference at the gate terminal would also result in temporary errors that would corrupt the tracking step amplitude.

The gain operation described in 21.3.2.2 relies on the fact that the tracking steps $\delta_s$ and $\delta_r$ are programmable and different. Those different tracking steps can be realized by applying different bias currents $I$ for coder and decoder integrators according to Equation 21.30.

#### 21.4.3.1 Design Considerations

According to Equation 21.30, the integrator based on the SCI technique presents three different sources of mismatches between the coder and the decoder. The first is in the pulse generation by the time delays in the pulse generator block. Because of process variations, the current source, the capacitor, and the inverters' threshold can vary and then provide different pulse width $T$. The second mismatch source is due to the size of the capacitor $C_{int}$ in each integrator. The third is the mismatch between current sources that provide the currents $1.5I$ and $0.5I$ in Figure 21.21. Moreover, the mismatch between current sources and current sink may cause a difference between the increase and the decrease tracking steps. Cascoded current mirrors were used to reduce the current mismatch.

The resulting difference may impact the functionality of the scheme. For instance, while the coder integrator will still track the input signal, the decoder integrator

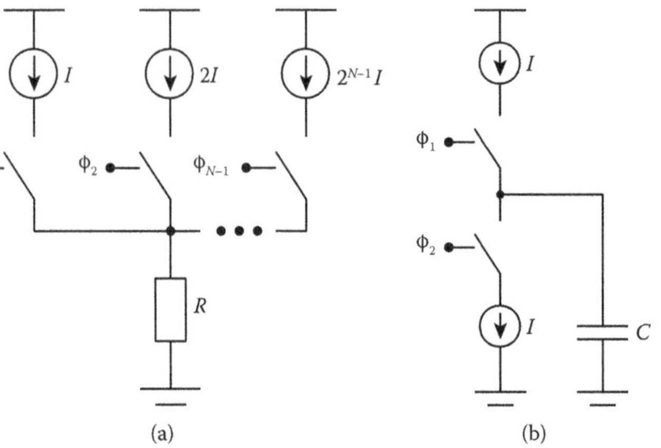


**FIGURE 21.20**
Switched current integrator (SCI) examples. (a) Unipolar version as used in current-steering DACs and (b) bipolar version used in signal modulation.

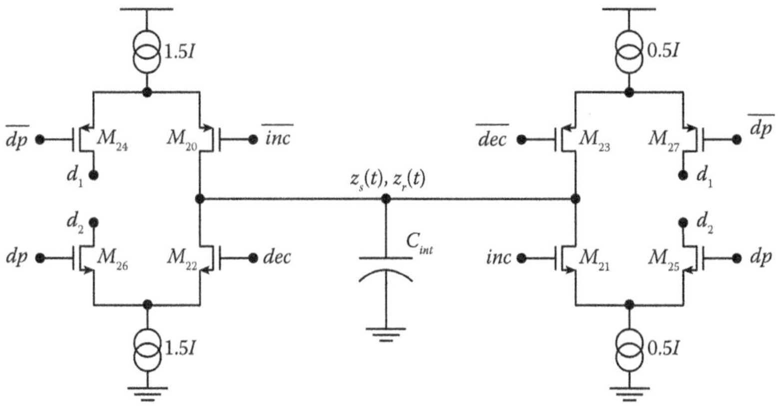


**FIGURE 21.21**
Schematic of the integrator based on an SCI technique, adapted from [3]. A positive spike triggers a voltage increment of $\delta$ according to (21.30) by turning on transistors $M_{20}$ and $M_{21}$ which then feed $1.5I$ and $0.5I$ currents to charge and discharge the capacitor $C_{int}$, respectively. For negative spikes, the complementary process decreases the capacitor voltage. In the absence of spikes, currents are drawn to low impedance nodes ($d_1 - d_2$).

might present a different behavior, normally saturating at either of the power supplies. A solution for this erratic behavior was found by changing the amplitude of the *inc* and *dec* signals. Normally, these signals present a full or half-scale amplitude.* By changing the ON state amplitude, for instance limiting the *dec* signal amplitude to a value $V_{dec}$ closer to $M_{22}$ threshold voltage, we can force the current mirror output transistor, which provides the $1.5I$ current, to leave the saturation region and enter the linear region, thus reducing the effective mirrored current. The IC provides a mechanism to change this amplitude. This calibration mechanism can thus compensate for all three mismatches sources. Figure 21.22a and b shows the decoder behavior before and after this solution. In the former, both *inc* and *dec* signals' swing amplitude is 1.65 V as designed originally while in the last the amplitude swing for the signal *inc* was reduced to 0.9 V.

### 21.4.4 Filter and AER Communication

The filter provides the averaging function in Equation 21.4 by removing high-frequency spectrum components. An ideal, unrealizable filter (sinc or brick-wall filter) would provide total rejection of out-of-band harmonics and null in-band attenuation. In practice, the cutoff frequency $\omega_p$ of a realizable low-pass filter is designed to be greater than the input signal bandwidth. The trade-off between out-of-band components suppression and signal distortion has to be considered on the filter design.

*In this implementation, the control signals were originally designed for half-scale voltage amplitude, that is, $\overline{inc}$ and $\overline{dec}$ swing from ground to $V_{dd}$ while *inc* and *dec* from ground to $V_{ss}$.

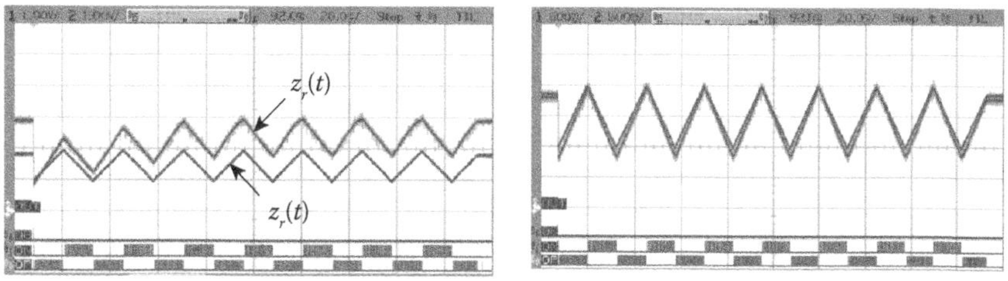

(a) (b)

**FIGURE 21.22**

SCI divergence. (a) Encoder $z_s(t)$ and decoder $z_r(t)$ integrator outputs with both *inc* and *dec* swing set to 1.65 V and (b) same signals with *inc* swing reduced to 0.9 V for compensation of integrator mismatches.

**TABLE 21.1**

Test IC Characteristics

| Parameter | | Value | Unit |
|---|---|---|---|
| Technology | | CMOS 0.35 μm 1P4M | — |
| Power Supply | | 3.3 | V |
| Area | Coder | 0.03 | $mm^2$ |
| | Decoder | 0.02 | |
| | AER | 0.025 | |
| | Total | 4.8 | |
| Power Consumption | Coder | 400 | μW |
| | Comparator | 340 | |
| | Decoder | 50 | |
| | IC | 2700 | |

The delay between the coder input $x_s(t)$ and the decoder output $x_r(t)$ is mostly impacted by the filter design. For instance, the filter will produce a 45° phase shift from the signal $z_r(t)$ when designing the pole to be $\omega_p = \omega_{in}$, for a sine wave given by $x_s(t) = A\,sin(\omega_{in} t)$. For systems designed for low frequency signals, this delay is the dominant contributor when compared against the modulator loop, $\Delta t_{(min)}$, and AER arbitration process. For instance, the conversion of a tone signal ($\omega_{in} = 2\,\pi\,4$ kHz) using an 8-bit resolution imposes $\Delta t_{(min)} \approx 300$ ns $\ll 31.2$ μs (45° phase shift).

The AER communication method was introduced to fully implement the asynchronous spike event coding scheme. Although the arbitration process was implemented inside the IC, the routing control was implemented in an external FPGA. Therefore, this IC presents two sets of control signals (request and acknowledge) and two address buses: one set from the IC to the FPGA and the other in the opposite direction.

The aVLSI implementation of the ASEC also contains four coders and four decoders. The total area of the this IC and the area of each block is presented in Table 21.1 together with the total power consumption and that for each block.

### 21.4.5 Demonstration: Speech Input

In this section, IC test results to demonstrate the communication aspects of the event coding scheme are presented. Two different input signals were used to test the communication system: a speech signal and a single-tone (sine wave) signal. Measured test results for the computation capabilities of the communication scheme presented in Section 21.2 are shown as well. A 140 ms speech signal was used to demonstrate the coding functionality. This signal was originally sampled at 44.1 kSps with an 8-bit resolution. The coding system was programmed to provide a resolution of 4 bits to demonstrate the coding properties.

Figures 21.23 and 21.24 were obtained from the implemented IC. The former shows first (a) the coder input signal $x_s(t)$ (top), the decoder integrator output $z_r(t)$ (middle), and the coder output spikes $y_s(t)$ (bottom, with the negative spikes first). The same signals are presented in detail in (b) where $x_s(t)$ and $z_r(t)$ are overlapped. This figure shows a key benefit from ASEC system as it generates no activity in the channel for signal changes smaller than $\Delta e_{th}$.

The second figure (21.24) shows signals used in the AER protocol: request and acknowledge signals and both inbound and outbound spike address buses. It is also possible to notice the difference between the 8-bit resolution input signal $x_s(t)$ and the 4-bit resolution signal $z_r(t)$ in 21.24(b).

### 21.4.6 Resolution: Single-Tone Input

A sine wave input signal with an amplitude of 2.0 $V_{pp}$ and frequency of 20 Hz ($f_{in}$) was used to measure the resolution of the system. It was generated by a DAC and it was sampled at 44.1 kSps. The coder was programmed to work with a 4-bit resolution. Figure 21.25a shows two periods of the input signal and the decoder signal $z_r(t)$, and Figure 21.25b shows the offline filtering results using a digital LPF with a cutoff frequency of 20 Hz, the same frequency as the input signal.

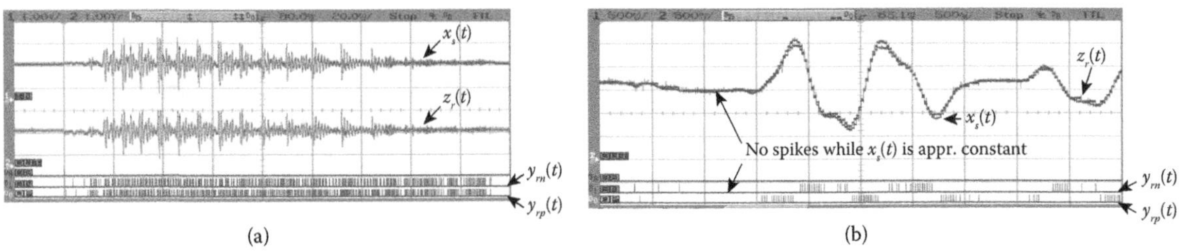


**FIGURE 21.23**

ASEC demonstration, adapted from [3]. (a) Shows the coder input $x_s(t)$ and decoder integrator output $z_r(t)$ for an audio signal with the resulting negative $y_n(t)$ and positive $y_p(t)$ spikes shown at the bottom of this figure. A detailed view of (a) is presented in (b) to show the absence of spikes during the periods when the variation on the input signal $x_s(t)$ amplitude is smaller than the coder resolution.

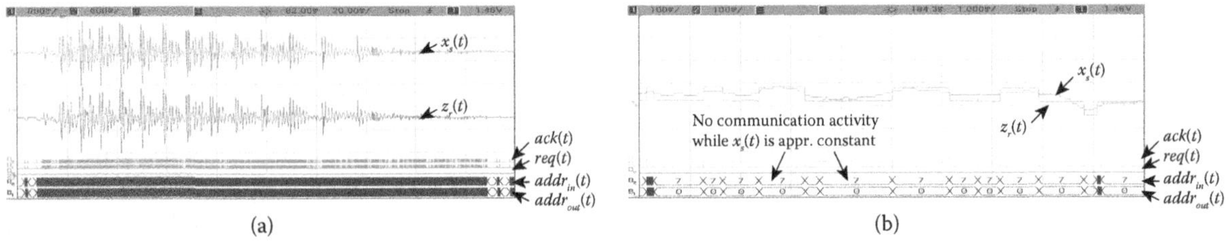


**FIGURE 21.24**

Full ASEC demonstration. (a) Shows the handshaking AER signals (request, acknowledge, and spike addresses). In (b), the difference between the 8-bit resolution input $x_s(t)$ and the 4-bit resolution output $z_r(t)$ is clear. In $addr_{out}(t)$, the absence of spikes is represented by the address 0 while the same condition is represented in the $addr_{in}(t)$ bus as number 7.

The measured total harmonic distortion (THD) for $z_r(t)$ signal is $-26$ dB leading to a resolution of approximately 4 bits. The inclusion of a filter with a cutoff frequency equal to $f_{in}$ not only realizes the signal $x_r(t)$ with a resolution of 6.35 bits but also attenuates the signal and inserts a phase shift on the signal. For comparison, when the cutoff frequency is increased to $10\times f_{in}$, the measured resolution of signal postfilter is 4.93 bits. Resolution is ultimately limited by mismatch of the components.

### 21.4.7 Computation: Gain, Negation, Summation, and BPSK

Computation properties of the ASEC were explained in Section 21.3. These explanations were illustrated with software simulations of mathematical models. Test results are also presented for some of the computations. Figure 21.26 corresponds to the gain operation while Figures 21.27 and 21.28 correspond to negation and summation (and subtraction) operations. Figure 21.29 is the test result for the BPSK. Similar inputs from the simulation were used for the IC test results.

## 21.5 Applications of ASEC-Based Systems

Some computational operations using ASEC were presented in this chapter. These included very fundamental mathematical operators such as negation, summation, and possible extension to multiplication operations. Therefore, the scope of use of this coding method—either purely ASEC-based or hybrid systems—in general signal processing is unbounded. Next, we describe two specific application fields as they are the main subjects of our research: programmable analog and mixed-mode architectures and neuromorphic systems.

If a digital system is suitable for flexible architectures due its properties, analog circuits have advantageous characteristics of being faster, smaller, and less energy demanding [35] in general. Consequently, it is desirable to develop analog architectures that also provide a high degree of flexibility experienced by FPGAs, DSPs, or digital microprocessors [36]. Similar to FPGAs, there is a wide range of potential applications for programmable analog systems, including low-power computing [37], remote sensing [38], and rapid prototyping [39].

Although digital programmable arrays mature and popular over the years, the same cannot be said about analog arrays. Among the technical challenges is the problem of sensibility of analog signals to interferences. Previous architectures relied on careful layout routing [40], using specific circuit techniques [41] or trying to limit the scope of routing [42,43]. However, many architectures have been tried in the past with no or little success.

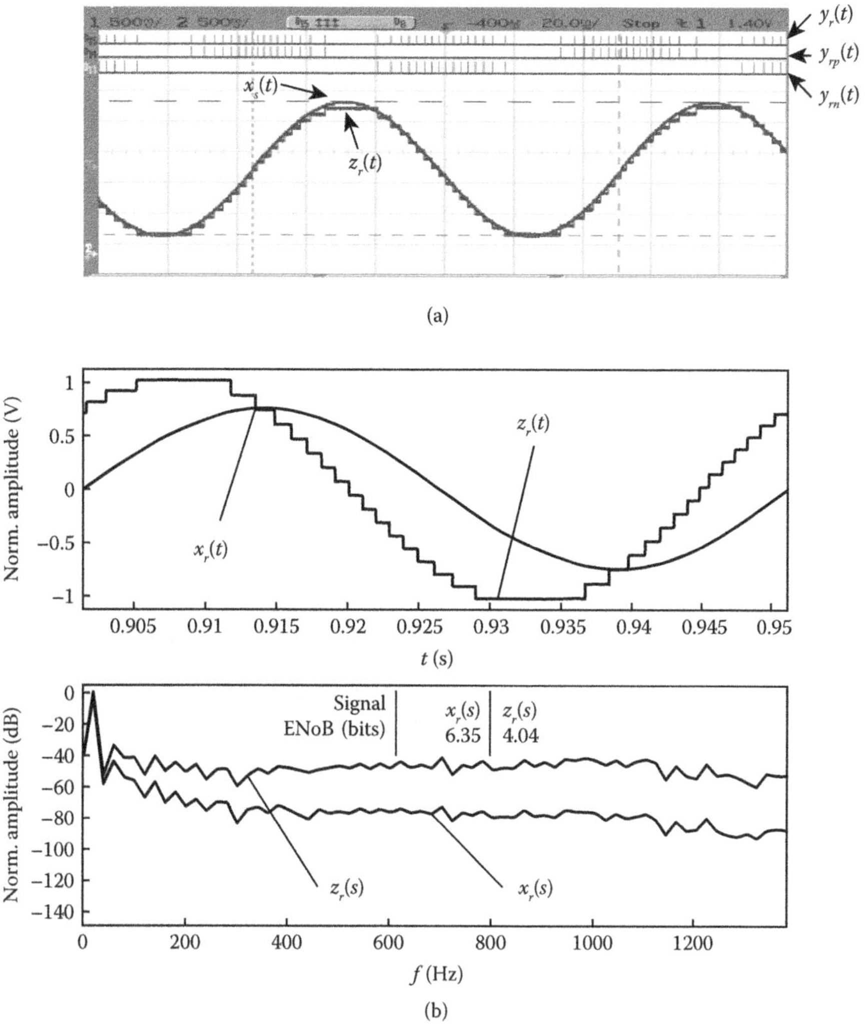


**FIGURE 21.25**
Resolution of the coding of the first implementation, adapted from [3]. (a) Oscilloscope snapshot with the input $x_s(t)$ ($2.0V_{pp}$ sine wave), decoder integrator output $z_r(t)$, negative $y_n(t)$ and positive $y_p(t)$ spikes from the coder output. (b) Reconstructed $z_r(t)$ from ADC data and decoder output $x_r(t)$ from the software filter (top). The low-pass filter cutoff frequency is the same as the input signal. The frequency spectrum $Z_r(s)$ and $X_r(s)$ of the filter input and output (bottom) is computed.

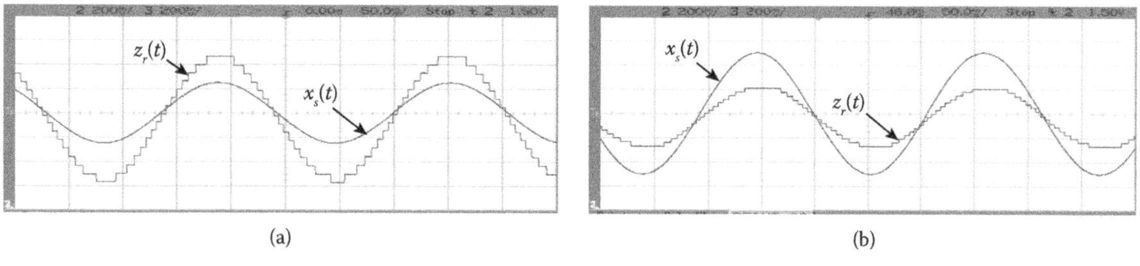


**FIGURE 21.26**
Test results for the gain operation, adapted from [34]. The input $x_s(t)$ is similar to the one in Figure 21.5. (a) Illustrates the signal amplification while (b) shows the results when the ASEC is configured to perform signal attenuation.

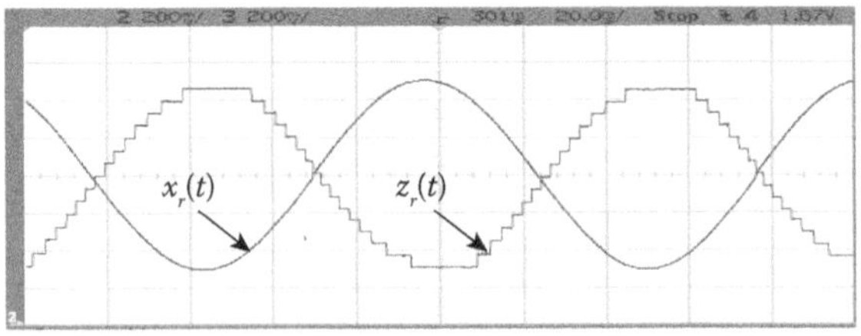


**FIGURE 21.27**
IC results for negation operation, adapted from [34]. The ASEC circuits were configured to produce a negated version of the sine wave $x_s(t)$, as shown in Figure 21.4.

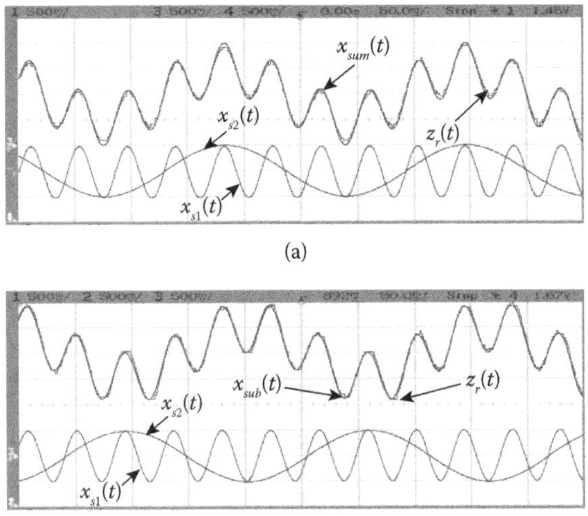


**FIGURE 21.28**
Summation and subtraction operation results from chip, adapted from [34]. This figure presents chip results similar to the simulation results in Figure 21.8. (a) Shows the summation $z_r(t)$ of two input sine waves, $x_1(t)$ and $x_2(t)$, with same amplitude but different frequencies (23 Hz and 4.7 Hz) and $x_{sum}(t)$ being the ideal summation. The subtraction of the same input signals $x_1(t) - x_2(t)$ is shown in (b), where $x_{sub}(t)$ is the ideal result. Scope DC levels were shifted to better illustration.

By using ASEC-based systems, one can improve the architecture in two areas. First, the flexibility and scalability of the system can be increased due the digital characteristics of the signal to be routed, reducing its noise susceptibility. Second, the signal dynamic range is also increased. While voltage and current dynamic ranges tend to be reduced with the evolution of the CMOS fabrication technologies, timing dynamic range tends to increase as the IC fabrication technology delivers faster and smaller transistors.

While general purpose FPAAs architectures have been continuously undermined by successive commercial failures, at least one application field has provided a

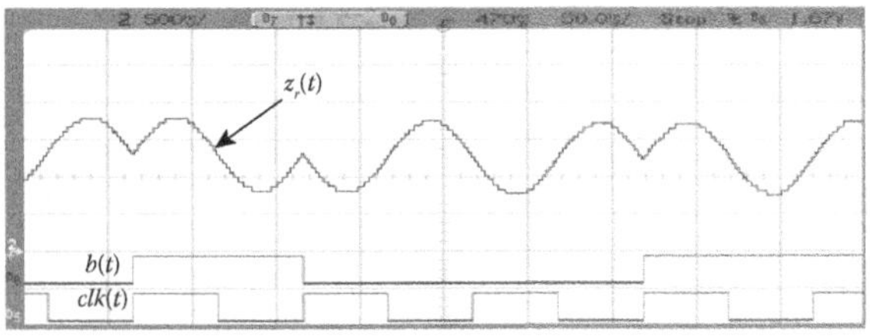


**FIGURE 21.29**
Results of the binary phase-shift keying operation from IC, adapted from [34]. In this snapshot, the input digital word $b(t)$ to be modulated by the method is 010011. $clk(t)$ frequency is the same as that of the sine carrier (not shown).

better prospectus: artificial neural networks and neuromorphic systems in particular. Artificial neural networks (ANNs) are computational architectures where each element performs a similar computational role of biological neurons and synapses. The computation is highly associated with the interconnection between the elements in the network rather than with the elements' functionality. Usually, its configuration is adaptive, and it is defined through the use of learning algorithms. Although ANN can be implemented in software, its hardware implementations (HNN) lead to greater computational performance [44]. ANNs have been designed on programmable platforms, in digital [45], analog [46,47], and mixed-mode domains [48].

Neuromorphic systems are VLSI systems designed to mimic at least some functional and computational properties of the biological nervous systems [35]. Many neuromorphic systems are spike neural networks (SNN). They are based on the concept that the information is transferred between neurons using temporal information conveyed on the onset of spikes [49]. Most neuromorphic systems are very specialized programmable arrays, where neurons, synapses, and other functional blocks act as basic blocks. However, these systems can also be implemented using generic programmable arrays, both digital—DSPs [50] and FPGAs [51,52]—and analog. For instance, a commercial analog FPAA was used to implement a I&F SNN [53] and a pulsed coupled oscillator [54]. In both ANN and neuromorphic systems, an essential signal processing operation consists of applying a weighted summation to the variables (ANN) or electric pulses (neuromorphic). This operation is readily available in ASEC-based systems as shown earlier.

In this chapter, we presented an asynchronous spike event coding scheme (ASEC) that is not only suitable for flexible analog signal routing but also presents the ability to perform a set of signal processing functions with no dedicated hardware needed. The working principle of this communication scheme was demonstrated as the association of analog-to-spike and spike-to-analog coding methods with an asynchronous digital communication channel (AER). Spike events are essentially asynchronous and robust digital signals and are easy to route on shared channels not only between CABs but also between ICs, providing improved scalability.

## Bibliography


[1] Y. Li, K. Shepard, and Y. Tsividis. Continuous-time digital signal processors. In *IEEE International Symposium on Asynchronous Circuits and Systems*, IEEE, pages 138–143, March 2005.

[2] M. Mahowald. VLSI analogs of neuronal visual processing: A synthesis of form and function. PhD thesis, California Institute of Technology, Pasadena, CA, 1992.

[3] L. C. Gouveia, T. J. Koickal, and A. Hamilton. An asynchronous spike event coding scheme for programmable analog arrays. *IEEE Transactions on Circuits and Systems I: Regular Papers*, 58(4):791–799, 2011.

[4] H. Inose, T. Aoki, and K. Watanabe. Asynchronous delta modulation system. *Electronics Letters*, 2(3):95, 1966.

[5] V. Balasubramanian, A. Heragu, and C. C. Enz. Analysis of ultra low-power asynchronous ADCs. In *Proceedings of the IEEE International Symposium on Circuits and Systems*, IEEE, pages 3593–3596, May 2010.

[6] E. Allier, G. Sicard, L. Fesquet, and M. Renaudin. A new class of asynchronous A/D converters based on time quantization. In *9th International Symposium on Asynchronous Circuits and Systems*, volume 12, pages 197–205, IEEE Computer Society, 2003.

[7] D. Chen, Y. Li, D. Xu, J. G. Harris, and J. C. Principe. Asynchronous biphasic pulse signal coding and its CMOS realization. In *2006 IEEE International Symposium on Circuits and Systems*, pages 2293–2296, IEEE, Island of Kos, 2006.

[8] M. Miskowicz. Send-on-delta concept: An event-based data reporting strategy. *Sensors*, 6(1):49–63, 2006.

[9] K. Papathanasiou, T. Brandtner, and A. Hamilton. Palmo: Pulse-based signal processing for programmable analog VLSI. *IEEE Transactions on Circuits and Systems II: Analog and Digital Signal Processing*, 49(6):379–389, 2002.

[10] V. Ravinuthula and J. G. Harris. Time-based arithmetic using step functions. In *Proceedings of the International Symposium on Circuits and Systems*, volume 1, pages I-305–I-308, May 2004.

[11] Y. Tsividis. Signal processors with transfer function coefficients determined by timing. *IEEE Transactions on Circuits and Systems*, 29(12):807–817, 1982.

[12] A. Murray. Pulse techniques in neural VLSI: A review. In *Proceedings of the IEEE International Symposium on Circuits and Systems*, volume 5, pages 2204–2207, IEEE, San Diego, CA, 2002.

[13] M. Nagata. PWM signal processing architecture for intelligent systems. *Computers & Electrical Engineering*, 23(6):393–405, 1997.

[14] A. Iwata and M. Nagata. A concept of analog-digital merged circuit architecture for future VLSI's. *Analog Integrated Circuits and Signal Processing*, 11(2):83–96, 1996.

[15] J. T. Caves, S. D. Rosenbaum, L. P. Sellars, C. H. Chan, and J. B. Terry. A PCM voice codec with on-chip filters. *IEEE Journal of Solid-State Circuits*, 14(1):65–73, 1979.

[16] R. Sarpeshkar and M. O'Halloran. Scalable hybrid computation with spikes. *Neural Computation*, 14(9):2003–2038, 2002.

[17] B. Brown and H. Card. Stochastic neural computation. I. Computational elements. *IEEE Transactions on Computers*, 50(9):891–905, 2001.

[18] E. Roza. Analog-to-digital conversion via duty-cycle modulation. *IEEE Transactions on Analog and Digital Signal Processing*, 44(11):907–914, 1997.

[19] M. Kurchuk and Y. Tsividis. Signal-dependent variable-resolution quantization for continuous-time digital signal processing. In *IEEE International Symposium on Circuits and Systems*, pages 1109–1112, IEEE, Taipei, May 2009.

[20] W. Tang and E. Culurciello. A pulse-based amplifier and data converter for bio-potentials. In *IEEE International Symposium on Circuits and Systems*, pages 337–340, IEEE, May 2009.

[21] S. G. Wilson. *Digital Modulation and Coding*. Prentice-Hall, Upper Saddle River, NJ, 1995.

[22] S. Park. Principles of sigma-delta modulation for analog-to-digital converters. *Mot. Appl. Notes APR8*, 1999.

[23] T. J. Koickal, L. C. Gouveia, and A. Hamilton. A programmable spike-timing based circuit block for reconfigurable neuromorphic computing. *Neurocomputing*, 72(16–18):3609–3616, 2009.

[24] R. Baker, H. Li, and D. Boyce. *CMOS Circuit Design, Layout, and Simulation*, 2nd edition. Wiley-IEEE Press, New York, 1997.

[25] A. A. Fayed and M. Ismail. A high speed, low voltage CMOS offset comparator. *Analog Integrated Circuits and Signal Processing*, 36(3):267–272, 2003.

[26] D. A. Yaklin. Offset comparator with common mode voltage stability, Patent No. US 5517134 A, September 1996.

[27] M. Bazes. Two novel fully complementary self-biased CMOS differential amplifiers. *IEEE Journal of Solid-State Circuits*, 26(2):165–168, 1991.

[28] J.-B. Shyu, G. C. Temes, and F. Krummenacher. Random error effects in matched MOS capacitors and current sources. *IEEE Journal of Solid-State Circuits*, 19(6):948–956, 1984.

[29] M. Kurchuk and Y. Tsividis. Energy-efficient asynchronous delay element with wide controllability. In *Proceedings of 2010 IEEE International Symposium on Circuits and Systems*, pages 3837–3840, IEEE, 2010.

[30] A. Martin. Translating concurrent programs into VLSI chips. In D. Etiemble and J.-C. Syre, Eds., *PARLE '92 Parallel Architectures and Languages Europe*, volume 605, pages 513–532–532, Springer, Berlin.

[31] D. Kośćielnik and M. Miskowicz. Asynchronous sigma-delta analog-to digital converter based on the charge pump integrator. *Analog Integrated Circuits and Signal Processing*, 55(3):223–238, 2007.

[32] J. J. Wikner and N. Tan. Modeling of CMOS digital-to-analog converters for telecommunication. *IEEE Transactions on Circuits and Systems II Analog and Digital Signal Processing*, 46(5):489–499, 1999.

[33] H. Pan. Method and system for a glitch-free differential current steering switch circuit for high speed, high resolution digital-to-analog conversion, Patent No. US 7071858 B2, June 2005.

[34] L. C. Gouveia, T. J. Koickal, and A. Hamilton. Computation in communication: Spike event coding for programmable analog arrays. In *IEEE International Symposium on Circuits and Systems*, pages 857–860, IEEE, Paris, May 2010.

[35] C. Mead. Neuromorphic electronic systems. *Proceedings of the IEEE*, 78(10):1629–1636, 1990.

[36] P. Dudek and P. J. Hicks. A CMOS general-purpose sampled-data analog processing element. *IEEE Transactions on Circuits and Systems II Analog and Digital Signal Processing*, 47(5):467–473, 2000.

[37] P. Hasler. Low-power programmable signal processing. In *Fifth International Workshop on System-on-Chip for Real-Time Applications*, pages 413–418, IEEE, Washington, DC, 2005.

[38] A. Stoica, R. Zebulum, D. Keymeulen, R. Tawel, T. Daud, and A. Thakoor. Reconfigurable VLSI architectures for evolvable hardware: From experimental field programmable transistor arrays to evolution-oriented chips. *IEEE Transactions on Very Large Scale Integration Systems*, 9(1):227–232, 2001.

[39] T. Hall and C. Twigg. Field-programmable analog arrays enable mixed-signal prototyping of embedded systems. In *48th Midwest Symposium on Circuits and Systems 2005*, Cincinnati, OH, volume 1, pages 83–86, 2005.

[40] Lattice Semiconductor Corp. ispPAC Overview, 2001.

[41] T. Hall, C. Twigg, J. Gray, P. Hasler, and D. Anderson. Large-scale field-programmable analog arrays for analog signal processing. *IEEE Transactions on Circuits and Systems—I: Regular Papers*, 52(11):2298–2307, 2005.

[42] T. Roska and L. O. Chua. The CNN universal machine: An analogic array computer. *Transactions on Circuits and Systems II Analog and Digital Signal Processing*, 40:163–173, 1993.

[43] J. Becker and Y. Manoli. A continuous-time field programmable analog array (FPAA) consisting of digitally reconfigurable Gm-cells. In *Proceedings of the IEEE International Symposium on Circuits and Systems*, Vancouver, BC, volume 1, pages I-1092–I-1095, IEEE, May 2004.

[44] J. Misra and I. Saha. Artificial neural networks in hardware: A survey of two decades of progress. *Neurocomputing*, 74(1–3):239–255, 2010.

[45] J. Liu and D. Liang. A survey of FPGA-based hardware implementation of ANNs. In *Proceedings of the International Conference on Neural Networks and Brain*, Beijing, China, volume 2, pages 915–918, IEEE, 2005.

[46] J. Maher, B. Ginley, P. Rocke, and F. Morgan. Intrinsic hardware evolution of neural networks in reconfigurable analogue and digital devices. In *IEEE Symposium on Field-Programmable Custom Computing Machines*, Napa, CA, pages 321–322, IEEE, April 2006.

[47] P. Dong, G. Bilbro, and M.-Y. Chow. Implementation of artificial neural network for real time applications using field programmable analog arrays. In *Proceedings of the IEEE International Joint Conference on Neural Networks*, Vancouver, BC, pages 1518–1524, IEEE, 2006.

[48] S. Bridges, M. Figueroa, D. Hsu, and C. Diorio. Field-programmable learning arrays. *Neural Information Processing Systems*, 15:1155–1162, 2003.

[49] W. Maass. Computing with spikes. *Special Issue on Foundations of Information Processing of TELEMATIK*, 8(1):32–36, 2002.

[50] C. Wolff, G. Hartmann, and U. Ruckert. ParSPIKE-a parallel DSP-accelerator for dynamic simulation of large spiking neural networks. In *Proceedings of the 7th International Conference on Microelectronics for Neural, Fuzzy and Bio-Inspired Systems*, Granada, Spain, pages 324–331, IEEE Computer Society, 1999.

[51] L. Maguire, T. Mcginnity, B. Glackin, A. Ghani, A. Belatreche, and J. Harkin. Challenges for large-scale implementations of spiking neural networks on FPGAs. *Neurocomputing*, 71(1–3):13–29, 2007.

[52] R. Guerrero-Rivera, A. Morrison, M. Diesmann, and T. Pearce. Programmable logic construction kits for hyper-real-time neuronal modeling. *Neural Computation*, 18(11):2651–79, 2006.

[53] P. Rocke, B. Mcginley, J. Maher, F. Morgan, and J. Harkin. Investigating the suitability of FPAAs for evolved hardware spiking neural networks. *Evolvable Systems: From Biology to Hardware*, 5216:118–129, 2008.

[54] Y. Maeda, T. Hiramatsu, S. Miyoshi, and H. Hikawa. Pulse coupled oscillator with learning capability using simultaneous perturbation and its FPAA implementation. In *ICROS-SICE International Joint Conference*, Fukuoka, Japan, pages 3142–3145, IEEE, August 2009.

# 22

# *Digital Filtering with Nonuniformly Sampled Data: From the Algorithm to the Implementation*

**Laurent Fesquet**

*University Grenoble Alpes, CNRS, TIMA*
*Grenoble, France*

**Brigitte Bidégaray-Fesquet**

*University Grenoble Alpes, CNRS, LJK*
*Grenoble, France*

**CONTENTS**

**ABSTRACT** This chapter targets the synthesis and the design of filters using nonuniformly sampled data for use in asynchronous systems. Data are sampled with a level-crossing technique. It presents several filtering strategies, namely, finite impulse-response (FIR) and infinite impulse-response (IIR) filters. The synthesis of filters in the frequency domain based on a level-crossing sampling scheme of the transfer function is also presented. Finally, an architecture for an hardware implementation of a FIR filter is given. This implementation is based on an asynchronous data-driven logic in order to benefit from the events produced by the asynchronous analog-to-digital converter.

## 22.1 Introduction

Today, our digital society exchanges data as never it has been the case in the past. The amount of data is incredibly large and the future promises that not only humans will exchange digital data but also technological equipments, robots, etc. We are close to opening the door of the Internet of things. This data orgy wastes a lot of energy and contributes to a nonecological approach of our digital life. Indeed, the Internet and the new technologies consume about 10% of the electrical power produced in the world. Design solutions already exist to enhance the energetic performances of electronic systems and circuits, of computers and their mobile applications: a lot of techniques and also a lot of publications! Nevertheless, another way to reduce energy is to rethink the sampling techniques and digital processing chains. Considering that our digital life is dictated by the Shannon theory, we produce more digital data than expected, more than necessary. Indeed, useless data induce more computation, more storage, more communications, and also more power consumption. If we disregard the Shannon theory, we can discover new sampling and processing techniques. A small set of ideas is given through examples such as filters, pattern recognition techniques, and so on, but the Pandora's Box has to be opened to drastically reduce the useless data, and mathematicians probably have a key role to play in this revolution.

Reducing the power consumption of mobile systems—such as cell phones, sensor networks, and many other electronic devices—by one to two orders of magnitude is extremely challenging, but it will be very useful to increase the system autonomy and reduce the equipment size and weight. In order to reach such a goal, the signal processing theory and the associated system architectures have to be rethought. This chapter gives the first step toward well-suited filtering techniques for nonuniform sampled signals [16].

Today signal processing systems uniformly sample analog signals (at Nyquist rate) without taking advantage of their intrinsic properties. For instance, temperature, pressure, electro-cardiograms, and speech signals significantly vary only during short moments. Thus, the digitizing system part is highly constrained due to the Shannon theory, which fixes the sampling frequency at least twice the input signal frequency bandwidth. It has been proved in [15] and [19] that analog-to-digital converters (ADCs) using a non-equi-repartition in time of samples lead to interesting power savings compared to Nyquist ADCs. A new class of ADCs called A-ADCs (for asynchronous ADCs) based on level-crossing sampling (which produces nonuniform samples in time) [2,5] and related signal processing techniques [4,18] have been developed.

This work also suggests an important change in the filter design. Like analog signals which are usually sampled uniformly in time, filter transfer functions are also usually sampled with a constant frequency step. Nonuniform sampling leads to an important reduction in the number of weight-function coefficients. Combined with a nonuniform level-crossing sampling technique performed by an A-ADC, this approach drastically reduces the computation load by minimizing the number of samples and operations, even if they are more complex.

### 22.1.1 Nonuniform Samples

We consider here nonuniform samples that are obtained through a level-crossing sampling technique. With no *a priori* knowledge of the input signal we can simply define equispaced levels in the range of the input signal. For specific applications, other distributions of levels can be defined. Contrarily to the uniform world where the sampled signal is defined by a sampling frequency and amplitudes, here each sample is the couple of an amplitude $x_j$ and a time duration $\delta t_j$, which is the time elapsed since the previous sample was taken (see Figure 22.1). This time delay can be computed with a local clock, and no global clock is needed for this purpose.

### 22.1.2 Number of Nonuniform Samples

Nonuniform sampling is especially efficient for sporadic signals. We, therefore, use here an electrocardiogram (ECG) record displayed in Figure 22.2. Its time duration is 14.2735 s. It has been recorded using 28,548 regular samples with 2000 Hz uniform sampling and displays 22 heartbeats.

Clearly all these samples are not useful for many applications. For example, to detect the main features of the signal we can use only eight levels (i.e., a 3-bit storage of amplitudes). The result is displayed in Figure 22.3, where only 1377 samples are kept to describe the signal. This number, and the induced signal compression, strongly depends on the statistical features of the signal [6], and the position and number of levels. For the same signal and 4- and 5-bit storage of amplitudes, Table 22.1 displays the number of samples and statistical data about the delays.

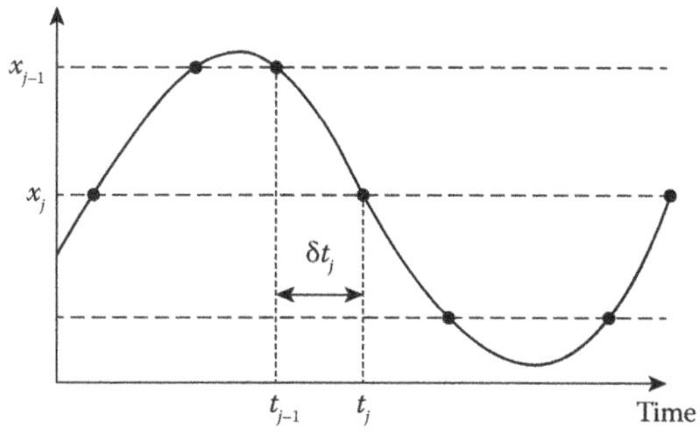


**FIGURE 22.1**
Nonuniform samples via level-crossing sampling.

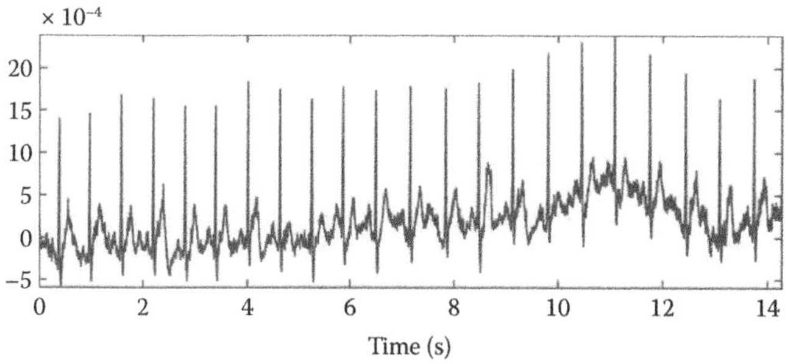


**FIGURE 22.2**
Original ECG signal composed of 28,548 regular samples.

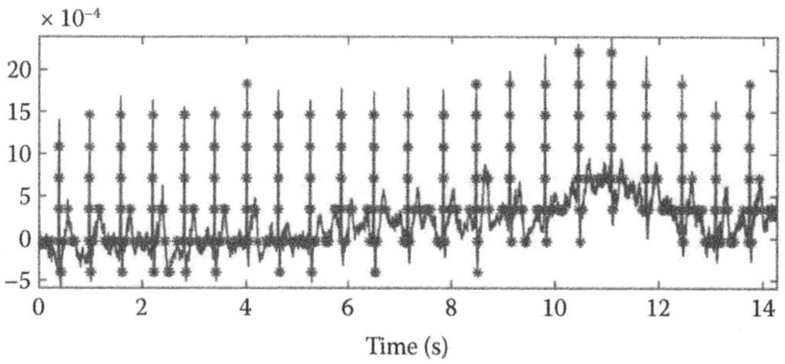


**FIGURE 22.3**
A 3-bit level-crossing sampling of the ECG signal.

## 22.2 Finite Impulse Response Filtering

Classical FIR filters are strongly based on the regular nature of both the input signal and the impulse response. However, a continuous time analog can be given that can be generalized to nonuniform input data.

### 22.2.1 Digital and Analog Convolutions

In the classical context, a $N$th order (causal) FIR filter computes the filtered signal $y_n$, from the $N+1$ previous input samples $x_i$, $i = 0, \ldots, N$:

$$y_n = \sum_{j=0}^{N} h_j x_{n-j}. \tag{22.1}$$

The filter is defined as its response to a sequence of discrete impulses. For an input signal $x(t) = \sum_{j=0}^{n} x_j \delta(t-j)$, the output at time $n$ is $y_n$. Equation 22.1 can also be cast as

$$y(n) = \int_{-\infty}^{+\infty} h(\tau)x(n-\tau)\,d\tau, \tag{22.2}$$

and $y$ is the convolution of the signal $x$ with the impulse response $h(t) = \sum_{k=0}^{N} h_k \delta(t-k)$. A real-life signal is analog and not a series of discrete impulses, but Equation 22.2 can be extended to the continuous context:

$$y(t) = \int_{-\infty}^{+\infty} h(\tau)x(t-\tau)\,d\tau = \int_{-\infty}^{+\infty} h(t-\tau)x(\tau)\,d\tau. \tag{22.3}$$

A finite number of coefficients $h_k$ is then mirrored by a finite support function $h$.

In the nonuniform context, a continuous signal is only known at times $t_j$. The continuous convolution is approximated with an algorithm that only makes use of these samples.

**TABLE 22.1**
Statistics of Uniform and Level-Crossing Samples

| | Number of Samples | Min($\delta t_j$) | Max($\delta t_j$) | Average | Median |
|---|---|---|---|---|---|
| Regular | 28,548 | 0.5 | 0.5 | 0.5 | 0.5 |
| 3-bit | 1377 | 0.0028 | 350.2 | 10.4 | 1.7 |
| 4-bit | 2414 | 0.0034 | 132.1 | 5.9 | 1.5 |
| 5-bit | 5081 | 0.0028 | 63.2 | 2.8 | 1.1 |

*Note*: Delays are given in milliseconds.

### 22.2.2 Algorithms

The signal can be approximated interpolating its nonuniform samples. Since we intend to implement the obtained algorithms on asynchronous systems, we want to keep them as simple as possible. We only describe here zeroth- and first-order interpolation, but the same derivation could be applied to higher-order interpolation of the samples.

#### 22.2.2.1 Zeroth-Order Interpolation

The signal $x$ and the impulse response $h$ are only known from their samples $(x_j, \delta t_j)$ and $(h_k, \delta\tau_k)$. We use classical filters, and $\delta\tau_k$ are all equal, but this specific feature is not taken into account to describe the algorithm, which confers symmetric roles to $x$ and $h$. Zeroth-order interpolation consists in approximating $x$ and $h$ by piecewise constant functions:

$$x(t) \simeq \sum_{j=0}^{n} x_j \chi_{[t_{j-1},t_j]} \text{ and } h(t) \simeq \sum_{k=0}^{N} h_k \chi_{[\tau_k,\tau_{k+1}]}, \quad (22.4)$$

where $\chi_I$ is the characteristic function of interval $I$. We have not chosen the same "side" for the zeroth-order interpolation of both signals. This is done on purpose, in order to have the same side, once the signal $x$ is reversed in the convolution algorithm:

$$\begin{aligned} y_n &= \int_{-\infty}^{+\infty} \sum_{j=0}^{n} x_j \chi_{[t_{j-1},t_j]}(t_n - t) \sum_{k=0}^{N} h_k \chi_{[\theta_k,\theta_{k+1}]}(t)\, dt \\ &= \sum_{j=0}^{n} x_j \sum_{k=0}^{N} h_k \int_{-\infty}^{+\infty} \chi_{[t_{j-1},t_j]}(t_n - t)\, \chi_{[\tau_k,\tau_{k+1}]}(t)\, dt \\ &= \sum_{j=0}^{n} x_j \sum_{k=0}^{N} h_k \operatorname{length}([t_n - t_j, t_n - t_{j-1}] \cap [\tau_k, \tau_{k+1}]). \end{aligned}$$

This theoretical formula would lead to the thought that $n(N+1)$ elementary contributions are needed to compute the convolution. This number can be reduced to at most $n + N + 1$ contributions, which correspond to the maximum number of nonempty intersections $[t_n - t_j, t_n - t_{j-1}] \cap [\tau_k, \tau_{k+1}]$. This value is obtained interpolating $x$ at times $\tau_k$ and $h$ at times $t_n - t_j$. Once this is done, both signals are known at the same times and both are constant on each interval. The contribution to the convolution is the product of both amplitudes by the length of the subinterval.

**REMARK 22.1** If the samples for $x$ (and $h$) are uniform, we recover the classical FIR filtering formula: $[t_n - t_j, t_n - t_{j-1}] = [n-j, n-j+1]$ and only intersects $[k, k+1]$ on a nontrivial interval if $j = n-k$. Then

$$\begin{aligned} &\sum_{j=0}^{n} a_j \sum_{k=0}^{N} h_k \operatorname{length}([n-j, n-j+1] \cap [k,k+1]) \\ &= \sum_{j=0}^{n} x_j \sum_{k=0}^{N} h_k\, \delta_{j=n-k} \\ &= \sum_{j=0}^{N} x_{n-j} h_j. \end{aligned}$$

The computation of the interval intersection lengths would use the knowledge of times $\tau_k$ and $t_n - t_j$, but only delays $\delta\tau_k$ and $\delta t_j$ are available. The algorithm introduced by Aeschlimann [3] is therefore slightly different. The subintervals are computed iteratively starting from $k = 0$ and $j = 0$, comparing the length of each sample, and subdividing the longer sample.

**Algorithm 22.1**

while $k \le N+1$ and $j \le n$

compare $\delta t_{n-j}$ and $\delta\tau_k$ and choose the smallest: $\Delta$

add $\Delta x_{n-j} h_k$ to the convolution

go to the next sample for the finished interval(s) and subdivide the other, that is,

if $\Delta = \delta t_{n-j}$, $j \leftarrow j+1$, else $\delta t_{n-j} \leftarrow \delta t_{n-j} - \Delta$

if $\Delta = \delta\tau_k$, $k \leftarrow k+1$, else $\delta\tau_k \leftarrow \delta\tau_k - \Delta$

This subdivision is done keeping the same amplitude since zeroth-order interpolation is involved. The algorithm should be performed on copies of the delay sequences, in order to be able to do the computation for the successive values of $n$.

#### 22.2.2.2 First-Order Interpolation

The principle is the same as for zeroth-order interpolation. Now $x$ and $h$ are approximated by piecewise linear functions. The support of each piece is the same as for zeroth-order interpolation, and we have to compute at most $n + N + 1$ elementary contributions. On each subinterval, the signals are linear. Their product is therefore a second-order polynomial. The Simpson quadrature formula is exact for polynomials up to degree 3, it is therefore exact in this context. On an interval of length $\Delta$ with end (left and right) values $x^\ell$ and $x^r$ for $x$ and $h^\ell$ and $h^r$ for $h$, the elementary contribution is

$$\begin{aligned} &\frac{\Delta}{6}\left(x^\ell h^\ell + x^r h^r + 4\frac{x^\ell + x^r}{2}\frac{h^\ell + h^r}{2}\right) \\ &= \frac{\Delta}{6}\left(x^\ell h^\ell + x^r h^r + (x^\ell + x^r)(h^\ell + h^r)\right). \end{aligned}$$

The differences with the zeroth-order algorithm are printed in bold font in Algorithm 22.2.

---

**Algorithm 22.2**

---

while $k \leq N+1$ and $j \leq n$

compare $\delta t_{n-j}$ and $\delta\tau_k$ and choose the smallest: $\Delta$

add **new contribution** to the convolution **with Simpson rule**

go to the next sample for the finished interval(s) and subdivide **and interpolate the amplitude** for the other, that is,

if $\Delta = \delta t_{n-j}$, $j \leftarrow j+1$, else $\delta t_{n-j} \leftarrow \delta t_{n-j} - \Delta$ **and compute interpolated value for** $x_{n-j}$

if $\Delta = \delta\tau_k$, $k \leftarrow k+1$, else $\delta\tau_k \leftarrow \delta\tau_k - \Delta$ **and compute interpolated value for** $h_k$

---

### 22.2.3 Numerical Illustration

We compare the filtering of the regular and the nonuniformly sampled ECG signal within the SPASS Toolbox [8], which is developed in MATLAB®. To be fair, we do not use the native filtering procedure in MATLAB, which is precompiled and cannot have an equivalent in the targeted architectures.

We choose a 10th-order FIR low-pass filter (MATLAB function `fir1`) with cutoff frequency 100 Hz. The regular case is the simple sum (22.1) but has to deal with the 28,548 samples and also computes 28,548 filtered samples. For the nonuniform test case, we choose the 3-bit sampled data, with 1377 samples, and also compute 1377 output samples. The input and filtered signal are displayed in Figure 22.4.

In the case of zeroth-order interpolated data (which is closer to (22.1)), the gain (compared to the regular case) is a factor of 30 in CPU time. Although the treatment of each sample is more complex, a significative gain is obtained due to the lower number of samples. It is therefore more efficient to treat directly the nonuniform samples given by the event-driven architectures. If linear interpolation is used, there is still some gain, of about seven in our example.

## 22.3 Infinite Impulse Response Filtering

There are various ways to describe IIR filters, but the state representation is the one that can be the most naturally adapted to nonuniform samples.

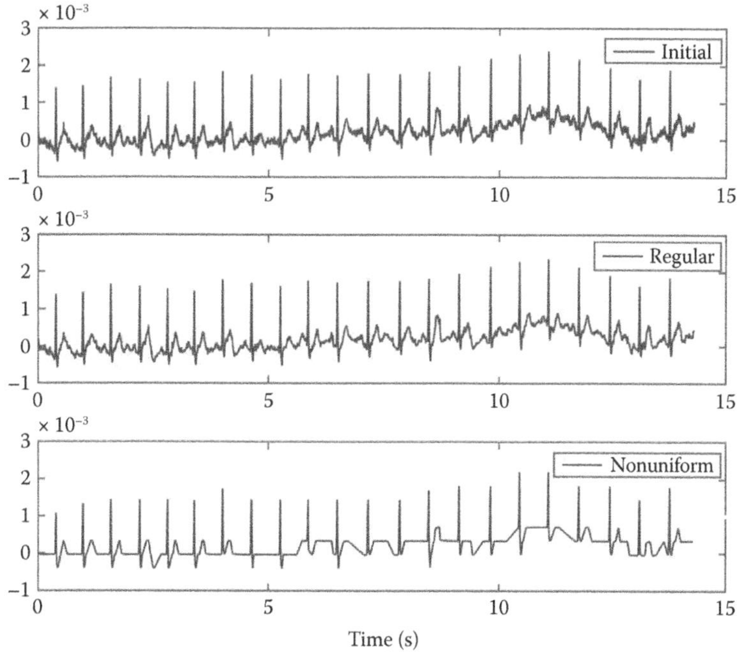


**FIGURE 22.4**
Regular and nonuniform FIR filtering.

### 22.3.1 State Representation

In the uniform world, an IIR filter is given by its transfer function in the Laplace variable:

$$\hat{h}(p) = \frac{\sum_{k=0}^{N} \alpha_k p^k}{\sum_{k=0}^{N} \beta_k p^k}. \qquad (22.5)$$

Then, the filtered signal $\hat{y}(p)$ is computed from the input signal $\hat{x}(p)$ by the relation $\hat{y}(p) = \hat{h}(p)\hat{x}(p)$. This definition is difficult to apply in the nonuniform context. Therefore, we use an equivalent formulation in the time domain which is given by the state equation. We define the state vector as $\hat{S}(p) = (\hat{s}_0(p), \ldots, \hat{s}_{N-1}(p))^t$ with coordinates

$$\hat{s}_k(p) = \frac{s^k \hat{x}(p)}{\sum_{j=0}^{N} \beta_j p^j} \quad \text{for } k = 0, \ldots, N-1,$$

(this definition is extended to $k = N$). The Laplace transform of the filtered signal $y$ can be written in terms of $\hat{S}(p)$, namely,

$$\hat{y}(p) = \sum_{k=0}^{N} \alpha_k \hat{s}_k(p).$$

The coordinates $\hat{s}_k$ may be computed step by step, thanks to the formula $\hat{s}_k(p) = p\hat{s}_{k-1}(p)$ for $k = 1, \ldots, N-1$.

Besides, the definition of $\hat{s}_0(p)$ may also be written $\sum_{k=0}^{N} \beta_k p^k \hat{s}_0(p) = \hat{x}(p)$, which if $\beta_N = 1$ (normalization condition) yields

$$\hat{s}_N(p) = -\sum_{k=0}^{N-1} \beta_k \hat{s}_k(p) + \hat{x}(p).$$

This finally leads to

$$\begin{aligned}\hat{y}(p) &= \alpha_N \hat{s}_N(p) + \sum_{k=0}^{N-1} \alpha_k \hat{s}_k(p) \\ &= \sum_{k=0}^{N-1} (\alpha_k - \alpha_N \beta_k)\hat{s}_k(p) + \alpha_N \hat{x}(p).\end{aligned}$$

These results can be cast in a system of $N$ equations, called state equations, in the time domain. Namely,

$$\begin{aligned}\frac{dS(t)}{dt} &= AS(t) + Bx(t), \\ y(t) &= CS(t) + Dx(t),\end{aligned} \qquad (22.6)$$

where

$$A = \begin{pmatrix} 0 & 1 & \cdots & 0 & 0 \\ \vdots & \vdots & \ddots & \vdots & \vdots \\ 0 & 0 & \cdots & 1 & 0 \\ 0 & 0 & \cdots & 0 & 1 \\ -\beta_0 & -\beta_1 & \cdots & -\beta_{K-2} & -\beta_{K-1} \end{pmatrix},$$

is the $N \times N$ state matrix, $B = (0 \cdots 0\ 1)^t$ is the command vector, $C = (\alpha_0 - \alpha_N\beta_0 \ \cdots \ \alpha_{N-1} - \alpha_N\beta_{N-1})$ is the observation vector, and $D = \alpha_N$ is the direct link coefficient.

The integral form of the state equation is given by

$$S(t) = e^{At}S(0) + \int_{t_0}^{t} e^{A(t-\tau)}Bx(\tau)\, d\tau. \qquad (22.7)$$

Algorithms follow from the discretization of either form, differential (22.6) or integral (22.7). The last equation will always be simply given by

$$y_n = CS_n + Dx_n.$$

### 22.3.2 Finite Difference Approximations of the Differential State Equation

#### *22.3.2.1 Euler Approximation*

The Euler method consists in writing the state equation (22.6) at time $t_{n-1}$ and using an upward approximation for the time derivative, namely,

$$\frac{S_n - S_{n-1}}{\delta t_n} = AS_{n-1} + Bx_{n-1},$$

which also reads $S_n = (I + \delta t_n A)S_{n-1} + B\delta t_n x_{n-1}$.

#### *22.3.2.2 Backward Euler Approximation*

For the backward Euler approximation, the state equation (22.6) is discretized at time $t_n$:

$$\frac{S_n - S_{n-1}}{\delta t_n} = AS_n + Bx_n,$$

that is, $S_n = (I - \delta t_n A)^{-1}(S_{n-1} + B\delta t_n x_n)$.

#### *22.3.2.3 Bilinear Approximation of an IIR Filter*

Poulton and Oksman [17] have chosen a bilinear method to approximate the time derivative in the state equation. This method consists in writing a centered approximation of Equation 22.6 at time $t_{n-1/2} = \frac{1}{2}(t_n + t_{n-1})$, that is,

$$\frac{S_n - S_{n-1}}{\delta t_n} = A\frac{S_n + S_{n-1}}{2} + B\frac{x_n + x_{n-1}}{2},$$

or

$$S_n = \left(I - \frac{\delta t_n}{2}A\right)^{-1} \times \left(\left(I + \frac{\delta t_n}{2}A\right)S_{n-1} + \frac{\delta t_n}{2}B(x_n + x_{n-1})\right).$$

#### 22.3.2.4 Runge–Kutta Schemes

We have implemented the classical fourth-order Runge–Kutta scheme (RK4), which for the state equation (22.6) and a linear approximation of $x(t_{n-1/2})$ reads

$$\begin{aligned} S_n = &\left(I + \delta t_n A + \frac{1}{2}\delta t_n^2 A^2 + \frac{1}{6}\delta t_n^3 A^3 + \frac{1}{24}\delta t_n^4 A^4\right) S_{n-1} \\ &+ \delta t_n \left(\frac{1}{2}I + \frac{1}{3}\delta t_n A + \frac{1}{8}\delta t_n^2 A^2 + \frac{1}{24}\delta t_n^3 A^3\right) Bx_{n-1} \\ &+ \delta t_n \left(\frac{1}{2}I + \frac{1}{6}\delta t_n A + \frac{1}{24}\delta t_n^2 A^2\right) Bx_n. \end{aligned}$$

We may notice that the iteration matrix that operates on $S_{n-1}$ is the fourth-order Taylor expansion of $\exp(\delta t_n A)$, which would be the exact operator.

We have also tested a third-order two-stage Runge–Kutta semi-implicit method (RK23). For the state equation (22.6), it reads

$$\begin{aligned} S_n = &(I - \delta t_n A)^{-1}\left(I - \frac{2}{3}\delta t_n A\right)^{-1} \\ &\times \left[\left(I - \frac{2}{3}\delta t_n A - \frac{1}{2}\delta t_n^2 A^2\right) S_{n-1}\right. \\ &+ \delta t_n \left(\frac{1}{2}I - \frac{1}{2}\delta t_n A\right) Bx_{n-1} \\ &\left.+ \delta t_n \left(\frac{1}{2}I - \frac{3}{2}\delta t_n A\right) Bx_n\right]. \end{aligned}$$

### 22.3.3 Quadrature of the Integral State Equation

In [13], Fontaine and Ragot choose to discretize the integral form of the state equation (22.7) directly. The integral equation is written between two asynchronous times:

$$S(t_n) = e^{A\delta t_n}S(t_{n-1}) + \int_{t_{n-1}}^{t_n} e^{A(t_n-\tau)}Bx(\tau)\,\mathrm{d}\tau.$$

The different algorithms consist in approximating the continuous signal $x(t)$ by the interpolation of order 0 or 1 of the asynchronous samples.

#### 22.3.3.1 Zeroth-Order Interpolation

In the zeroth-order interpolation $x(t) = x_{n-1}$ for $t \in ]t_{n-1}, t_n[$. Then,

$$\begin{aligned} S_n &= e^{A\delta t_n}S_{n-1} + \int_{t_{n-1}}^{t_n} e^{A(t_n-\tau)}Bx_{n-1}\,\mathrm{d}\tau \\ &= e^{A\delta t_n}S_{n-1} - A^{-1}(I - e^{A\delta t_n})Bx_{n-1}. \end{aligned}$$

If we choose $x(t) = x_n$ for $t \in ]t_{n-1}, t_n[$, we of course have

$$S_n = e^{A\delta t_n}S_{n-1} - A^{-1}(I - e^{A\delta t_n})Bx_n.$$

#### 22.3.3.2 Nearest Neighbor Interpolation

We may center time intervals in a different way and suppose that $x(t) = x_{n-1}$ for $t \in ]t_{n-1}, t_{n-1/2}[$ and $x(t) = x_n$ for $t \in ]t_{n-1/2}, t_n[$. Then,

$$\begin{aligned} S_n &= e^{A\delta t_n}S_{n-1} + \int_{t_{n-1}}^{t_{n-1/2}} e^{A(t_n-\tau)}Bx_{n-1}\,\mathrm{d}\tau \\ &\quad + \int_{t_{n-1/2}}^{t_n} e^{A(t_n-\tau)}Bx_n\,\mathrm{d}\tau \\ &= e^{A\delta t_n}S_{n-1} - A^{-1}(e^{A\delta t_n/2} - e^{A\delta t_n})Bx_{n-1} \\ &\quad - A^{-1}(I - e^{A\delta t_n/2})Bx_n. \end{aligned}$$

#### 22.3.3.3 Linear Interpolation

If $x(t)$ is approximated by a piecewise linear function $x(t) = x_{n-1} + \frac{t - t_{n-1}}{\delta t_n}(x_n - x_{n-1})$ for $t \in ]t_{n-1}, t_n[$, and the same type of computation leads to

$$\begin{aligned} S_n = &e^{A\delta t_n}S_{n-1} + A^{-1}\left[e^{A\delta t_n} + A^{-1}\frac{1}{\delta t_n}(I - e^{A\delta t_n})\right]Bx_{n-1} \\ &- A^{-1}\left[I + A^{-1}\frac{1}{\delta t_n}(I - e^{A\delta t_n})\right]Bx_n. \end{aligned}$$

The comparison of these various schemes in terms of stability and complexity has been performed in [12]. It shows that the Euler scheme has to be rejected for being unstable, and the implicit Euler scheme for being in a sense too stable (i.e., too dissipative). In the case of Runge–Kutta schemes, extra redundant samples can be taken to avoid too large inactive parts in the input signal and ensure stability.

### 22.3.4 Algorithm

In practice, the formulae cannot be easily implemented as they are given above because they need the computation of inverse matrices or exponential of matrices. In order to implement an $N$th-order filter, it is usual to decompose it in multiple first- and second-order filters that are easily implemented with classical electrical

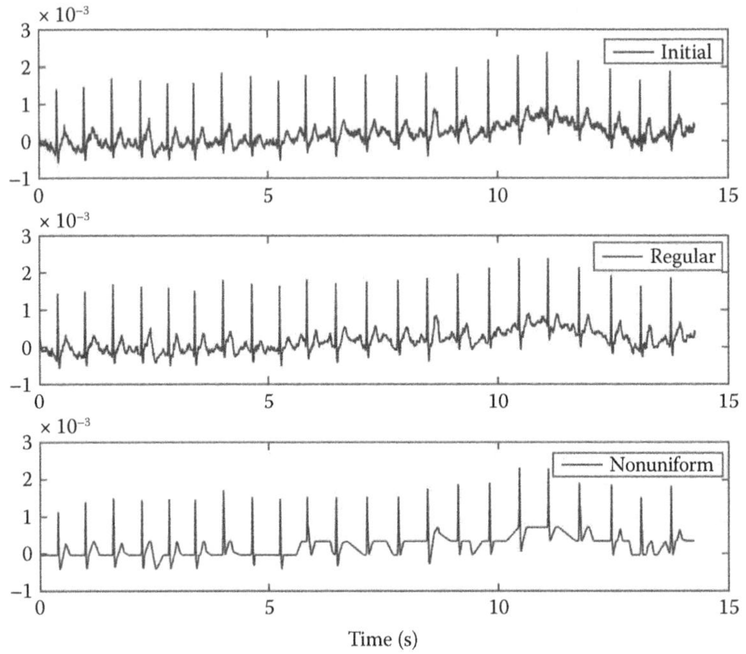


**FIGURE 22.5**
Regular and nonuniform IIR filtering.

structures. It consists in writing Equation 22.5 as a product of similar rational fractions but with $N = 1$ or 2. It is then equivalent to apply successively the induced first- and second-order filters, or the $N$th-order filter. It is equivalent from the theoretical point of view, but the succession of first- and second-order filters is easier to implement on asynchronous architecture, because we can easily give explicit values for the inverse or exponential matrices.

### 22.3.5 Numerical Illustration

The test cases in MATLAB are performed in the same framework and with the same algorithmic choices as for FIR filters. We use here a 10th-order Butterworth filter and the linearly interpolated quadrature formula to obtain the results displayed in Figure 22.5.

There is a gain of 20 in CPU time using nonuniform samples compared to regular samples. The RK4 is not stable with the 3-bit samples illustrating the stability result of [12]. The minimum number depends both on the filter (here a Butterworth one) and on the numerical scheme.

## 22.4 Nonuniform Filters in the Frequency Domain

We now want to go further from classical filtering and take advantage of the fact that filters are naturally defined in the frequency domain and should be easier to describe in this domain than in the time domain. We come back to the continuous convolution formula (22.3) and express the impulse response $h$ as the inverse Fourier transform of the analog filter transfer function $H$:

$$h(t) = \frac{1}{2\pi}\int_{-\infty}^{+\infty} H(\omega)e^{i\omega t}\,d\omega. \tag{22.8}$$

As for the previous FIR and IIR filters, we suppose that the input signal is interpolated (at most) linearly from its nonuniform samples:

$$x(t) \sim \sum_{j=1}^{n}(a_j + b_j t)\chi_{[t_{j-1},t_j]}. \tag{22.9}$$

### 22.4.1 Sampling in the Frequency Domain

The transfer function is also sampled nonuniformly and interpolated linearly with the extra feature that the transfer function is complex valued. The best way to do it is to interpolate separately the amplitude and the phase (see [9], where it is compared to interpolation in the complex plane). In practice, we define the filter for positive values of the frequency and use level-crossing sampling to define the sampling frequencies. This yields the filter samples $(\omega_k, H_k)$. Then, $H$ is interpolated. For a low-pass filter,

$$H(\omega) \sim \sum_{k=1}^{K} \Big\{ (\rho_k^0 + \rho_k^1 \omega) e^{i(\theta_k^0 + \theta_k^1 \omega)} \chi_{J_k} + (\rho_k^0 - \rho_k^1 \omega) - e^{i(\theta_k^0 - \theta_k^1 \omega)} \chi_{J_k^-} \Big\}. \tag{22.10}$$

Here, $J_k = [\omega_{k-1}, \omega_k]$ $(\omega_0 = 0)$, $J_k^- = [-\omega_k, -\omega_{k-1}]$, and

$$\rho_k^1 = \frac{|H_k| - |H_{k-1}|}{\omega_{k-1} - \omega_k}, \qquad \rho_k^0 = |H_k| - \rho_k^1 \omega_k,$$
$$\theta_k^1 = \frac{\arg(H_k) - \arg(H_{k-1})}{\omega_{k-1} - \omega_k}, \quad \theta_k^0 = \arg(H_k) - \theta_k^1 \omega_k.$$

The filtering algorithm is then obtained by "simply" computing the integral (22.3) using (22.8) and the interpolated formulae (22.9) and (22.10). The result can always be cast as

$$y(t) = \sum_{j=0}^{n} x_j \sum_{k=1}^{K} h_{jk}(t). \tag{22.11}$$

The main drawback is that the formulae for $h_{jk}(t)$ are quite complex and involve special functions such as trigonometric functions and integral trigonometric functions. The detailed formulae can be found in [9] (see also [11] for interpolation in the log-scale). Each single evaluation is therefore very costly and impossible to implement on the asynchronous systems we target as main applications. In addition, the computational cost has *a priori* an $nK$ complexity (times the computational cost of the single evaluation of a $h_{jk}(t)$).

### 22.4.2 Coefficient Number Reduction: The Ideal Filter

To reduce the cost of the previously described algorithm, we first recall that we address nonuniform samples that have been specifically chosen to reduce their number for the targeted application. Therefore, $n$ is low compared to regular sampling (by one or two decades in our applications). The next direction is to reduce $K$. It can be reduced to $K = 1$, and yield the equivalent of an infinite order scheme. Indeed, the perfect low-pass filter

$$H_0(\omega) = \begin{cases} 1 & \text{if } 0 \leq \omega \leq \omega_c, \\ 0 & \text{if } \omega > \omega_c, \end{cases} \tag{22.12}$$

would need an infinite number of filter coefficients in the time domain. With our notations, we have $K = 1$ and $H_1 = 1$. In this case, and for zeroth-order interpolation for the signal, the algorithm is quite simple [7]:

$$y(t) = \sum_{j=1}^{N} x_j (C_j(t) - C_{j-1}(t)),$$

where

$$C_j = \frac{1}{\pi} \mathrm{Si}(\omega_c (t - t_j)),$$

and Si is the integral sine equation defined by $\mathrm{Si}(x) = \int_0^x \sin(y) \frac{dy}{y}$. This has the disadvantage to be noncausal (i.e., to use all the signal samples to compute $x(t)$, even future samples). To avoid this, we can use the associated causal filter derived using the Hilbert transform of the filter transfer function:

$$y(t) = \sum_{j=1}^{N_t} x_j (C_j^c(t) - C_{j-1}^c(t)), \tag{22.13}$$

where

$$C_j^c = \frac{1}{\pi} \Big( \mathrm{Si}(\omega_c (t - t_j)) + i \,\mathrm{sgn}(t - t_j) \times \mathrm{Cin}(\omega_c (t - t_j)) \Big),$$

where Cin is the integral cosine equation defined by $\mathrm{Cin}(x) = \int_0^{|x|} (1 - \cos(y)) \frac{dy}{y}$. The details are given in [10]. In Equation 22.13, $N_t$ is the index of the last sample at time $t$.

### 22.4.3 Algorithm and Results

The special functions Si and Cin would be very difficult to implement efficiently on asynchronous hardware. Even in MATLAB their evaluation is quite slow. The best way to implement the algorithm is to create a look-up table from a previous knowledge of the range of the expressions $\omega_c(t - t_j)$. Then, values of the special function are interpolated from the look-up table. If the table takes into account the specific features (almost linear part, oscillating parts) of the special functions involved, this gives very good qualitative results with an important speed-up, at the cost of a preprocessing procedure to construct the table.

The filtering result, displayed in Figure 22.6 is again similar to the previous results (e.g., FIR filtering), which validates this approach. The cost is, of course, higher than zeroth-order FIR filtering, but it is still competitive. However, for certain applications with no *a priori* knowledge of the delays, it can be quite tricky to construct the look-up table correctly.

## 22.5 Filter Implementation Based on Event-Driven Logic

The implementation of filters with event-driven logic is a good way to fully exploit the low-power potential

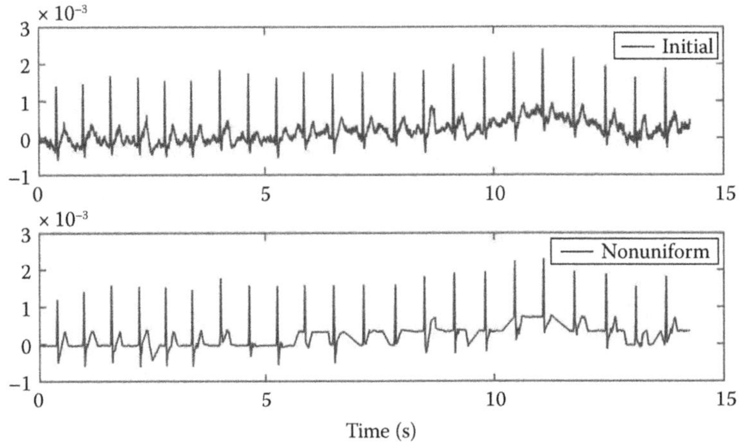


**FIGURE 22.6**
Filtering with the ideal filter.

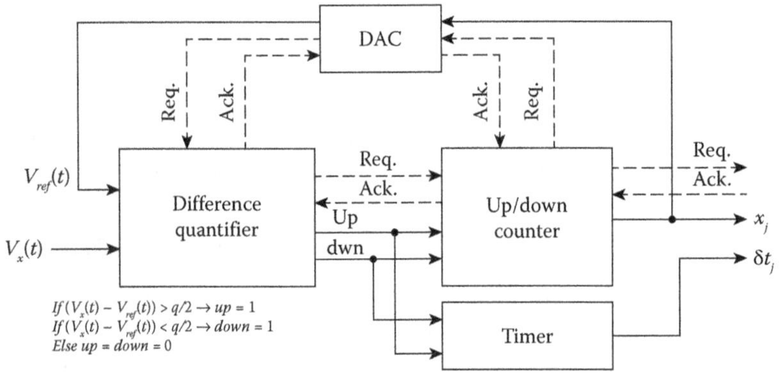


**FIGURE 22.7**
Architecture of an A-ADC.

digital filter using nonuniform sampled data. Indeed, we are not only reducing the number of samples, but we are also stopping the computation activity when no data are fed in the filter. This approach offers a good opportunity to spare energy and is particularly well suited to the nonuniform sampling schemes. Moreover, this design strategy is general enough to be applied to the different kinds of digital filters presented in the previous sections.

### 22.5.1 Analog-to-Digital Conversion for Event-Driven Technology

Many hardware strategies can be used to implement the level-crossing sampling scheme described in the previous paragraphs. One approach is a tracking loop enslaved on the analog signal to convert. For the crossing detection, the instantaneous comparison of the input analog signal $V_x$ is restricted to the two nearest levels: the one just above and the other just below. Every time a level is crossed, the two comparison levels are shifted down or up. The conversion loop, shown in Figure 22.7, is composed of four blocks: a difference quantifier, a state variable (an up/down counter and a Look-Up Table if the thresholds are not equally spaced), a digital-to-analog converter (DAC), and a timer delivering the time intervals $\delta t_j$ between samples. The $V_{ref}$ signal is changed each time a threshold is crossed and two new thresholds are defined accordingly. An example of a possible algorithm for the difference quantifier is given in Figure 22.7. This choice is very interesting for a minimization of the hardware. Moreover, only one cycle of the loop is needed to convert a sample. Notice that no external signal as a clock is used to synchronize and trigger the conversion of samples. Moreover, an asynchronous structure has been chosen for the circuit. The information transfer between each block is locally managed with a bidirectional control signaling: a request and an acknowledgment, represented by the dashed lines. This explains the name of asynchronous ADC. This block is able to simultaneously provide the magnitude of $V_x$ and the time elapsed between the last two samples. This architecture has been implemented with many FPGA and COTS (circuits on the shelf) but also in 130 nm CMOS technology from STMicroelectronics [1].

### 22.5.2 Event-Driven Logic

Unlike synchronous logic, where the synchronization is based on a global clock signal, event-driven or asynchronous logic does not need a clock to maintain the synchronization between its sub-blocks. It is considered as a data-driven logic where computation occurs only when new data arrived. Each part of an asynchronous circuit establishes a communication protocol with its neighbors in order to exchange data with them. This kind of communication protocol is known as "handshake" protocol. It is a bidirectional protocol, between two blocks called a sender and a receiver as shown in Figure 22.8. The sender starts the communication cycle by sending a request signal `'req'` to the receiver. This signal means that data are ready to be sent. The receiver starts the new computation after the detection of the `'req'` signal and sends back an acknowledgment signal `'ack'` to the sender marking the end of the communication cycle, so that a new one can start.

The main gate used in this kind of protocol is the "Muller" gate, also known as C-element. It helps—thanks to its properties—to detect a rendezvous between different signals. The C-element is in fact a state-holding gate. Table 22.2 shows its output behavior.

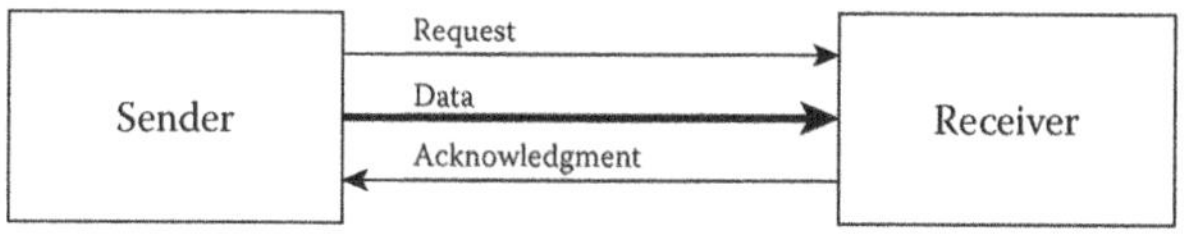


**FIGURE 22.8**
Handshake protocol used in asynchronous circuits.

**TABLE 22.2**
Output of a C-Element

| A | B | Z |
|---|---|---|
| 0 | 0 | 0 |
| 0 | 1 | Z-1 |
| 1 | 0 | Z-1 |
| 1 | 1 | 1 |

Consequently, when the output changes from `'0'` to `'1'`, we may conclude that both inputs are `'1'`. And similarly, when the output changes from `'1'` to `'0'`, we may conclude that both inputs are now set to `'0'`. This behavior could be interpreted as an acknowledgment that indicates when both inputs are `'1'` or `'0'`. This is why the C-element is extensively used in asynchronous logic and is considered as the fundamental component on which the communication protocols are based.

### 22.5.3 Micropipeline Circuits

Many asynchronous logic styles exist in the literature. It is worth mentioning that the choice of the asynchronous style affects the circuit implementation (area, speed, power, robustness, etc.). One of the most interesting styles to implement our filters with a limited area and power consumption is the micropipeline style [21]. Among all the asynchronous circuit styles, the micropipeline has the closest resemblance with the design of synchronous circuits due to the extensive use of timing assumptions [20]. Similarly as a synchronous pipeline circuit, the storage elements are controlled by control signals. Nevertheless, there is no global clock. These signals are generated by the Muller gates in the pipeline controlling the storage elements as shown in Figure 22.9.

This circuit can be seen as an asynchronous data-flow structure composed of two main blocks:

- The data path that is completely similar to the synchronous data path and locally clocked by a distributed asynchronous controller.
- The control path that is a distributed asynchronous controller able to provide the clock signal to memory elements when data are available. This feature replaces the clock tree of synchronous circuits.

This kind of circuit is easy to implement because we can adapt the synchronous flow. Indeed, the data path synthesis can be done by the existing commercial tools, and the control path insertion can be semi-automated. This is perfect for our purpose because we target event-driven

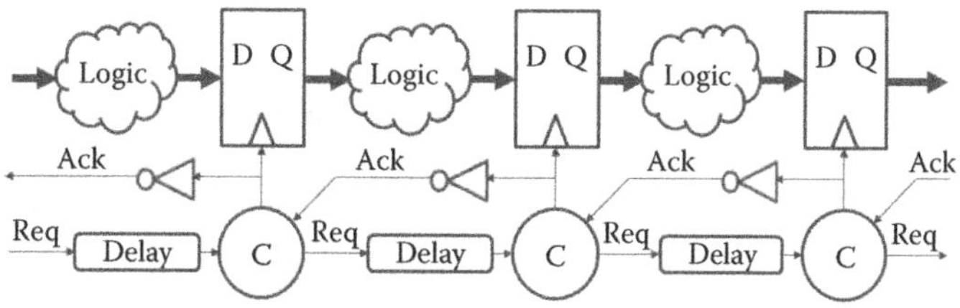


**FIGURE 22.9**
Micropipeline circuit.

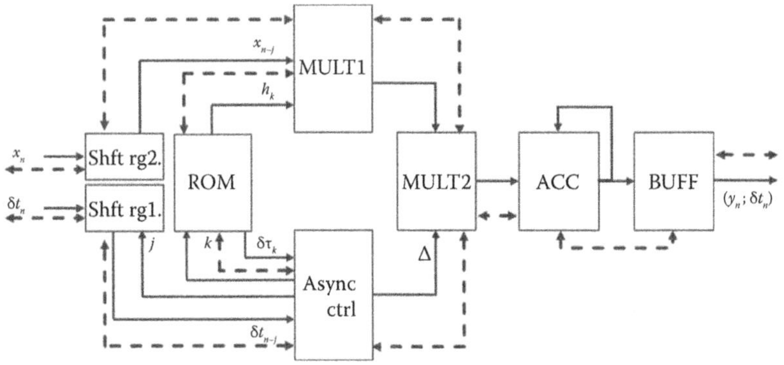


FIGURE 22.10
Architecture of FIR for nonuniform sampling.

digital signal processing. Each time new data are sampled, then the computation is activated and naturally stopped after the end of the calculation. This perfectly suits the objectives of low power while maintaining an easy circuit implementation with a similar area to their synchronous counterpart.

### 22.5.4 Implementing the Algorithm

With the nonuniform sampling scheme, the sampling instants of the FIR filter impulse response do not necessarily fit with the sampling times of the input samples. In order to avoid this, we use Algorithm 22.1 (see Section 22.2.2).

The previously proposed algorithm is implemented with the structure presented in Figure 22.10, which describes the FIR filter architecture. The dashed lines symbolize the handshake signals used for implementing the micropipeline logic. The two shift registers get the sampled signal data $(x_n, \delta t_n)$ from the A-ADC. The communication between the shift registers and the A-ADC is based on the handshake protocol as described in the previous section. The shift registers are the memory blocks of the filter. They store the input samples, magnitudes, and time intervals. The output of this register is connected to an asynchronous controller that allows selecting samples depending on the value of the selection input $j$. The ROM is used to store the impulse response coefficients. The coefficients are selected by the signal $k$. The asynchronous controller block (Async ctrl) has multiple functionalities:

- It allows determining the minimum time interval $\Delta$ among $\{\delta t_{n-j}, \delta\tau_k\}$.
- It generates the selection signals $j$ and $k$ that control the selection process in the shift registers.
- It detects the end of the convolution product round and allows starting a new one. This functionality is based on detecting the signal $k$. If $k$ reaches its maximum, this means that all the filter coefficients have been used. Thus, the convolution product is done. The asynchronous controller block generates at this point a reset signal to the other blocks, indicating that the filter is ready to start a new convolution.

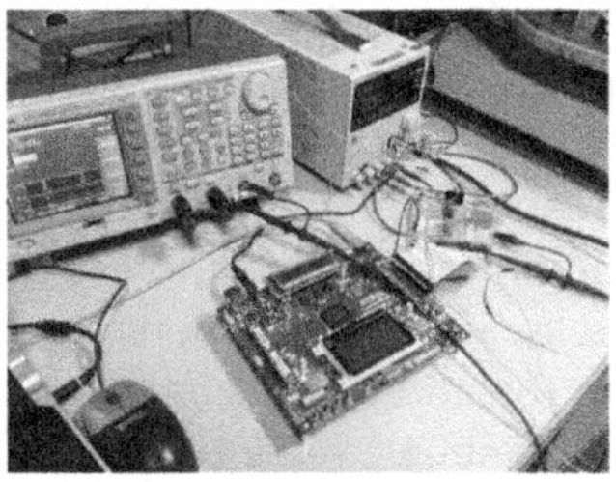

FIGURE 22.11
Nonuniform sampling and filtering experimental setup.

The two multipliers (MULT1 and MULT2) compute the intermediate values $\Delta \cdot x_{n-j} \cdot h_h$ that are accumulated in the accumulator (ACC). Finally, at the end of each convolution product cycle when the reset signal is sent, the accumulator (ACC) transfers its content to the buffer (BUFF). Then, the resulting convolution product $(y_n, \delta t_n)$ is available at the filter output.

The FIR filter has been implemented with the micropipeline logic style on an Altera FPGA board following the approach presented in [14]. The experimental setup can be seen in Figure 22.11.

## 22.6 Conclusion

In this chapter, several filtering techniques that can be applied to nonuniform sampled signals have been presented. The approaches are general and can be used with any kind of signal. We first describe the algorithm to implement a FIR filter and show that the result is convincing even if we used a small set of samples compared to the uniform approach. Then, we present the implementation of IIR filters *via* a state representation and discuss several finite difference approximations. The following section shows how nonuniform filters can be built in the frequency domain. This novel strategy for synthesizing filters is of interest for drastically reducing the coefficient number. Finally, the design architecture of such filters on a real system is presented. The technology target, asynchronous circuits, fully exploits the data time irregularity. Indeed, these circuits are event-driven and are able to react and process data only when new samples are produced. This gives a perfectly coherent system that captures new samples when the signal is evolving and processes the novel data on demand. With such an approach, it is possible to drastically reduce the system activity, the number of samples, and the data quantity sent on the output. Considering on one side today's strong requirements for many electronic equipment covering applications such as smartphones, notepads, laptops, or the emerging Internet of Things and the resulting data deluge, it becomes really urgent to stop this orgy of data that largely contributes to a colossal waste of energy.

This chapter is also a contribution to encourage people to directly process nonuniformly sampled data. Indeed, it is possible to perform spectrum analysis or pattern recognition, for instance. This leads to really interesting results demonstrating that most of the time we are using more than necessary samples. Finally, the nonuniform approach is a plus for people who target ultra-low power systems, because it helps reduce by several orders of magnitude the data volume and processing and thus the power consumption.

## Acknowledgments

This work has been partially supported by the LabEx PERSYVAL-Lab (ANR-11-LABX-0025-01).

## Bibliography


[1] E. Allier, G. Sicard, L. Fesquet, and M. Renaudin. Asynchronous level crossing analog to digital converters. *Special Issue on ADC Modelling and Testing of Measurement*, 37(4):296–309, 2005.

[2] E. Allier, G. Sicard, L. Fesquet, and M. Renaudin. A new class of asynchronous A/D converters based on time quantization. In *9th International Symposium on Asynchronous Circuits and Systems, Async'03*, pp. 196–205, IEEE, Vancouver, BC, May 2003.

[3] F. Aeschlimann. Traitement du signal echantillonne non uniformement: Algorithme et architecture. PhD thesis, INP, Grenoble, February 2006.

[4] F. Aeschlimann, E. Allier, L. Fesquet, and M. Renaudin. Asynchronous FIR filters: Towards a new digital processing chain. In *10th International Symposium on Asynchronous Circuits and Systems (Async'04)*, pp. 198–206, IEEE, Hersonisos, Crete, April 2004.

[5] F. Akopyan, R. Manohar, and A. B. Apsel. A level-crossing ash asynchronous analog-to-digital converter. In *12th IEEE International Symposium on Asynchronous Circuits and Systems (ASYNC'06)*, pp. 11–22, Grenoble, France, March 2006.

[6] B. Bidégaray-Fesquet and M. Clausel. Data driven sampling of oscillating signals. *Sampling Theory in Signal and Image Processing*, 13(2):175–187, 2014.

[7] B. Bidégaray-Fesquet and L. Fesquet. A fully non-uniform approach to FIR filtering. In *8th International Conference on Sampling Theory and Applications (SampTa'09)*, L. Fesquet and B. Torresani, Eds., pp. 1–4, Marseille, France, May 2009.

[8] B. Bidégaray-Fesquet and L. Fesquet. *SPASS 2.0: Signal Processing for ASynchronous Systems*. Software, May 2010.

[9] B. Bidégaray-Fesquet and L. Fesquet. Non-uniform filter interpolation in the frequency domain. *Sampling Theory in Signal and Image Processing*, 10:17–35, 2011.

[10] B. Bidégaray-Fesquet and L. Fesquet. *Non Uniform Filter Design*. Technical Report, Grenoble, 2015.

[11] B. Bidégaray-Fesquet and L. Fesquet. *A New Synthesis Approach for Non-Uniform Filters in the Log-Scale: Proof of Concept*. Technical Report, Grenoble, 2012.

[12] L. Fesquet and B. Bidégaray-Fesquet. IIR digital filtering of non-uniformly sampled signals via state representation. *Signal Processing*, 90:2811–2821, 2010.

[13] L. Fontaine and J. Ragot. Filtrage de signaux a echantillonnage irregulier. *Traitement du Signal*, 18:89–101, 2001.

[14] Q. T. Ho, J.-B. Rigaud, L. Fesquet, M. Renaudin, and R. Rolland. Implementing asynchronous circuits on LUT based FPGAs. In *The 12th International Conference on Field-Programmable Logic and Applications (FPL02)*, pp. 36–46, La Grande-Motte, France, 2–4 September 2005.

[15] J. W. Mark and T. D. Todd. A nonuniform sampling approach to data compression. *IEEE Transactions on Communications*, 29:24–32, 1981.

[16] F. A. Marvasti. Nonuniform sampling. Theory and practice. In *Information Technology: Transmission, Processing and Storage*, Springer, 2001.

[17] D. Poulton and J. Oksman. Filtrage des signaux a echantillonnage non uniforme. *Traitement du Signal*, 18:81–88, 2001.

[18] S. Mian Qaisar, L. Fesquet, and M. Renaudin. Adaptive rate filtering for a signal driven sampling scheme. In *International Conference on Acoustics, Speech, and Signal Processing, ICASSP 2007*, vol. 3, pp. 1465–1568, Honolulu, HI, April 2007.

[19] N. Sayiner, H. V. Sorensen, and T. R. Viswanathan. A level-crossing sampling scheme for A/D conversion. *IEEE Transactions on Circuits and Systems II*, 43:335–339, 1996.

[20] J. Sparsø and S. Furber. *Principles of Asynchronous Circuit Design: A Systems Perspective*. Springer, Boston, MA, 2001.

[21] I. E. Sutherland. Micropipelines. *Communications of the ACM*, 32:720–738, 1989.

# 23

## *Reconstruction of Varying Bandwidth Signals from Event-Triggered Samples*

**Dominik Rzepka**
*AGH University of Science and Technology*
*Kraków, Poland*

**Mirosław Pawlak**
*University of Manitoba*
*Winnipeg, MB, Canada*

**Dariusz Kościelnik**
*AGH University of Science and Technology*
*Kraków, Poland*

**Marek Miśkowicz**
*AGH University of Science and Technology*
*Kraków, Poland*

**CONTENTS**



**ABSTRACT** Level-crossing sampling is an event-based sampling scheme providing the samples whose density is dependent on the local bandwidth of the sampled signal. The use of the level crossings allows to exploit local signal properties to avoid unnecessarily fast sampling when the time-varying local bandwidth is low. In this chapter, the method of estimation of the local signal bandwidth based on counting the level crossings is proposed. Furthermore, the recovery of the original signal from the level crossings based on the methods suited for the irregular samples combined with the time warping is presented.

## 23.1 Introduction

The Shannon theorem introduced to engineering community in 1949 is a fundamental result used for digital signal processing of analog signals [1]. The main assumption under which a signal can be reconstructed from its discrete-time representation is that sampling rate is twice higher than the highest frequency component in the signal spectrum (Nyquist frequency), or twice the signal bandwidth. The class of signals with a finite bandwidth is often referred to as bandlimited. The concept of bandwidth is defined using the Fourier transform of the infinitely long signal, that is, the signal that is not vanishing in any finite time interval of $(-\infty, \infty)$. Furthermore, according to the uncertainty principle, only signals defined over infinite time interval can be exactly bandlimited. As it is known, real physical signals are always time limited, so they cannot be perfectly bandlimited and the bandlimited model is only the convenient approximation [2]. Moreover, in practice, the Nyquist frequency is determined on the basis of the finite record of signal measurements as the frequency for which higher spectral components are weak enough to be neglected. Therefore, the evaluation of the Nyquist frequency is referred not to the global but to the local signal behavior.

In the conventional signal processing system, the sampling rate corresponding to the Nyquist frequency is kept fixed when the signal is processed. Such an approach is justified in relation to signals whose spectral properties do not evolve significantly in time. However, there are some classes of signals, for example, FM signals, radar, EEG, neuronal activity, whose local spectral content is strongly varying. These signals do not change significantly their values during some long time intervals, followed by rapid aberrations over short time intervals. The maximum frequency component of the local bandwidth evaluated over finite length time intervals for such signals may change significantly and can be much lower than the Nyquist frequency defined for the global bandwidth. Intuitively, by exploiting local properties, signals can be sampled faster when the local bandwidth becomes higher and slower in regions of lower local bandwidth, which provides a potential to more efficient resource utilization.

The postulates to utilize time-varying local properties of the signal and adapt the sampling rate to changing frequency content have been proposed in several works in the past [3–9]. The objective of these approaches is to avoid unnecessarily fast sampling when the local bandwidth is low. As shown in [4], the sampling of the signal at the local Nyquist frequency preserving its perfect recovery is based on scaling the uniform sampling frequency by time-warping function that reflects varying local bandwidths. This idea, however, requires knowledge of local bandwidth to control time-varying sampling rate accordingly, and to ensure the proper signal reconstruction. The problem of triggering the sampling operations with the rate varying to local frequency content is one of the challenges of signal processing technique focused on exploiting local signal properties.

The method which can be used for this purpose is the *level-crossing sampling*, because the mean rate of level-crossing samples depends on the power spectral density of the signal [5,6,10,11]. The mean rate of level crossings is higher when the signal varies quickly and lower when it changes more slowly. The mean rate of level-crossing sampling for a bandlimited signal modeled as a stationary Gaussian process is directly proportional to the signal bandwidth, if the spectrum shape is fixed. By evaluating the number of level crossings in a time window of a finite size, the estimate of the local bandwidth can be obtained. However, as the level-crossing rate depends on the bandwidth only in the mean sense, the level-crossing instants, in general, do not match the time grid defined by the local Nyquist rate. Therefore, to recover the original signal from the level crossings, we propose to adopt methods suited for the irregular samples combined with the time warping.

The local bandwidth can be directly related to the concept of a local intensity function. This is a measure of distribution of sampling points that characterize a given signal in terms of its variability and reconstructability. In the case of samples generated from level crossings of stochastic signals, the intensity function is defined by the joint density function of the signal and its derivative. In Section 23.3.3, we provide a comprehensive introduction to the problem of estimation of the intensity function from a given set of level-crossing samples. Viewing the intensity function as the variable rate of the point process, we propose nonparametric techniques for recovering the function. The kernel nonparametric estimation method is discussed including the issue of its consistency and optimal tuning.

The signal analysis based on evolution of a local bandwidth might be used in various applications. In particular, music can be considered as possessing a local bandwidth that depends on a set of the pitches present in a given time window [9]. The concept of the local bandwidth can also be adopted to the decomposition of images into edges and textures [9]. The most promising area concerns resource-constraint applications, for example, biomedical signals processed in wireless systems with limited energy resources.

The rest of the chapter is structured as follows: Section 23.2 introduces the concept of the local bandwidth and the corresponding signal recovery problem. In Section 23.3, we develop the methodology for level-crossing sampling for signals that are modeled as Gaussian stochastic processes. Section 23.4 discusses the signal reconstruction method for non-uniformly distributed samples that occur in the level-crossing sampling strategy and any other schemes with irregular sampling.

## 23.2 Concepts of Local Bandwidth and Recovery of Signals with Varying Bandwidth

The bandwidth is a fundamental measure that defines frequency content of a signal. Intuitively, as the range of frequencies may change in time, the local bandwidth refers to the rate at which a signal varies locally. However, there is an underlying opposition between the concept of bandwidth and time locality because the former is non-local by definition [9].

### 23.2.1 Global Bandwidth

The bandwidth $\Omega$ (rad/s) (or $B = \Omega/2\pi$ [Hz]) of the finite energy signal $x(t)$ is defined using Fourier transform:

$$X(\omega) = \int_{-\infty}^{\infty} x(t)e^{-j\omega t}dt \tag{23.1}$$

as the upper frequency component in the spectrum of $x(t)$. The $\Omega$-bandlimitedness is defined as the vanishing of the Fourier transform $X(\omega)$ outside the interval $[-\Omega, \Omega]$, that is,

$$X(\omega) = 0 \quad \text{for } |\omega| > \Omega. \tag{23.2}$$

Taking into account the time-limited nature of physical signals, equation 23.2 can be replaced by a weaker requirement; the energy of the frequency components outside $[-\Omega, \Omega]$ is bounded by a certain small $\varepsilon$, that is,

$$\int_{|\omega|>\Omega} |X(\omega)|^2 d\omega < \varepsilon. \tag{23.3}$$

The perfect recovery of the ideally bandlimited signal is possible by means of the Shannon–Whittaker interpolation formula

$$x(t) = \sum_{n=-\infty}^{\infty} x(nT)\text{sinc}\left(\frac{\pi}{T}(t - nT)\right) \tag{23.4}$$

for $T \le \pi/\Omega$, where $\Omega$ is the signal bandwidth and $\text{sinc}(t) = \frac{\sin(t)}{t}$.

In practice, the number of samples is finite, but the error of reconstruction caused by such constraint decreases as the number of samples grows. In fact, for signals with sufficiently smooth Fourier spectra, it is known that the $L_2$-distance $\|x_N - x\|_2$ between $x(t)$ and the truncated version of (23.4), that is, $x_N(t) = \sum_{n=-N}^{N} x(nT)\text{sinc}\left(\frac{\pi}{T}(t - nT)\right)$ is of order $\mathcal{O}(1/N)$. See [12] for further details on the truncated error.

### 23.2.2 Local Bandwidth and Signals with Varying Bandwidth

The extension of the notion of the global bandwidth $\Omega$ to a local bandwidth $\Omega(t)$ defined at a certain time instant $t$ is not unique and can be defined in several different ways [9]. The decisive criterion of usefulness of a given definition is a possibility to formulate the sampling theorem which allows for the perfect recovery of a signal from infinite number of error-free signal samples, and perfect knowledge of local bandwidth.

Intuitively, the signal $x(t)$ bandlimited at time $t$ to the bandwidth $\Omega(t)$ can be defined as the output of an ideal low-pass filter with time-varying cutoff frequency $\Omega(t)$ for some possibly non-bandlimited signal $y(t)$ on the input. This can be represented using the Fourier transform with the variable cutting frequency, that is,

$$x(t) = \frac{1}{2\pi} \int_{-\Omega(t)}^{\Omega(t)} X(\omega)e^{j\omega t}d\omega. \tag{23.5}$$

Alternatively, we can define $\Omega(t)$ as a parameter of the following low-pass time-varying filter with the input $y(t)$ and output $x(t)$:

$$x(t) = \int_{-\infty}^{\infty} h(t - \tau, t)y(\tau)d\tau. \tag{23.6}$$

where $h(\lambda, t) = \frac{\sin(\Omega(t)\lambda)}{\pi\lambda}$ is the ideal low-pass filter response function with the time-varying bandwidth $\Omega(t)$.

Unfortunately, such representations of $x(t)$ cannot lead to the perfect reconstruction formula as was proven by Horiuchi [3]. In fact, in this case, it is possible to formulate the consistent reconstruction, for which the reconstructed signal $\tilde{x}(t) = x(t)$ only for $t = t_n$, where $\{x(t_n)\}$ are samples of $x(t)$. The reconstruction, however, is not valid for other time instants.

Since the local bandwidth definition using Fourier transform is not useful for reconstruction, the varying bandwidth reconstruction formula can be derived using a different, more general approach, based on the functional analysis [7,8]. The aim of this extension was to avoid the Gibbs phenomenon, which can occur when the signal $x(t)$ reconstructed using (23.4) is not bandlimited. However, this reconstruction scheme is characterized by relatively high complexity, so we will focus on the simpler approach yielding acceptable results.

As the notion of the bandwidth relies on the representation of a signal as a sum of sine and cosine functions with frequencies from the interval $[-\Omega, \Omega]$, the variation of the bandwidth can be, therefore, considered based on the modulation of frequencies of each signal component [4]. The frequency modulation of the sinusoidal signal is given by

$$z(t) = \sin\left(\int_{-\infty}^{t} \Omega(\tau)d\tau\right), \tag{23.7}$$

where $\Omega(t) > 0$ is an instantaneous angular frequency assumed to be a positive and continuous function of $t$. The nonlinear increase of sine phase in (23.7) can be viewed as warping the time axis, which can be formally expressed by a *warping function*

$$\gamma(t) = \int_{-\infty}^{t} \Omega(\tau)d\tau. \tag{23.8}$$

The signal $x(t)$ with varying bandwidth defined by (23.8) can now be represented as the warped version of the bandlimited signal $y(t)$

$$x(t) = y(\gamma(t)). \tag{23.9}$$

Due to the assumption that $\Omega(t)$ is positive and continuous, we have a well-defined inverse function $\alpha(t) = \gamma^{-1}(t)$ allowing to rewrite (23.9) as follows:

$$x(\alpha(t)) = y(t). \tag{23.10}$$

It is worth noting that if $\Omega(t)$ is constant, that is, $\Omega(t) = \Omega$ then $\gamma(t) = \Omega t$ and $\alpha(t) = t/\Omega$. Furthermore, we can view (23.9) as the time-varying filter with the input $y(t)$ and output $x(t)$. The corresponding transfer function of this filter can be shown to be

$$h(\lambda, t) = \frac{\delta(t - \alpha(t))}{\Omega(\alpha(t))}. \tag{23.11}$$

where $\delta(t)$ is the Dirac delta function.

In order to reconstruct the signal $x(t)$ defined in (23.9), the uniform sampling grid $\{iT, i = 0, \pm 1, \pm 2, \pm 3, \ldots\}$ should be affected by time warping, resulting in nonuniform sampling. Without loss of generality, it can be assumed that there exists a sample at time $t_0 = 0$. Then, the locations of other samples must correspond to the zeros of the time-warped function and are given by solutions of the following equation:

$$\sin(\gamma(t_n)) = 0. \tag{23.12}$$

This is equivalent to setting

$$t_n = \alpha(n\pi). \tag{23.13}$$

The formula in (23.13) defines the sampling points according to the *local Nyquist rate*, that is, to the instantaneous Nyquist rate related to the local signal bandwidth $\Omega(t)$.

The above-presented approach was introduced by Clark *et al.* [4] along with the following time-warped interpolation formula:

$$x(t) = \sum_{n=-\infty}^{\infty} x(\alpha(n\pi))\text{sinc}\left(\bar{\Omega}(t)\left(t - n\frac{\pi}{\bar{\Omega}(t)}\right)\right), \tag{23.14}$$

where $\bar{\Omega}(t) = \gamma(t)/t$ for $t \neq 0$, which is the average value of $\gamma(t)$.

Let us recall again that if $\Omega(t)$ is constant, that is, $\Omega(t) = \Omega$ then $\bar{\Omega}(t) = \Omega$ and $\alpha(t) = t/\Omega$. In this case (23.14) becomes the standard Shannon–Whittaker interpolation formula in (23.4) with $T = \pi/\Omega$.

According to the formula in (23.10), the varying bandwidth signal can be transformed into the regular bandlimited signal such that the perfectness of reconstruction formulas both in (23.4) and (23.14) is then preserved.

### 23.2.3 Reconstruction of Varying Bandwidth Signals with Nonideal Sampling Timing

To achieve the smallest possible number of samples assuring the perfect reconstruction of the signal, the samples should be taken at the rate defined by the local Nyquist rate, that is, by the instantaneous Nyquist rate related to the local signal bandwidth $\Omega(t)$ with sampling times $t_n$ given by (23.13). This would, however, require the *a priori* knowledge of the local bandwidth $\Omega(t)$ and the precise control of triggering sampling operations at instants $t_n$ according to the fluctuations of a $\Omega(t)$ in time. If this knowledge is unavailable, the local spectrum can be estimated using spectrum analyzer. However, such a procedure would require a complex hardware, and the result of estimation would be available with a considerable delay. This makes this idea impractical.

In the practically realizable level-crossing sampling, the samples are provided at times $\{t'_n\}$, such that their density in time depends on the local signal bandwidth. Due to approximated relationship with the local bandwidth, the instants $\{t'_n\}$, in general, do not match the ideal time grid $\{t_n\}$ given by (23.13), which defines the sampling according to the local Nyquist rate.

According to the local Nyquist rate (23.13), the recovery is provided by the classical Shannon theorem with the time warping, see (23.14), consequently, the recovery of the signal from the samples $\{t'_n\}$ should be obtained by means of a signal reconstruction method suited for the irregular samples combined with the time-warping function in (23.8).

The necessary condition for reconstruction of signals from irregular samples is to keep average sampling rate above the Nyquist rate [13]. Equivalently, the sampling

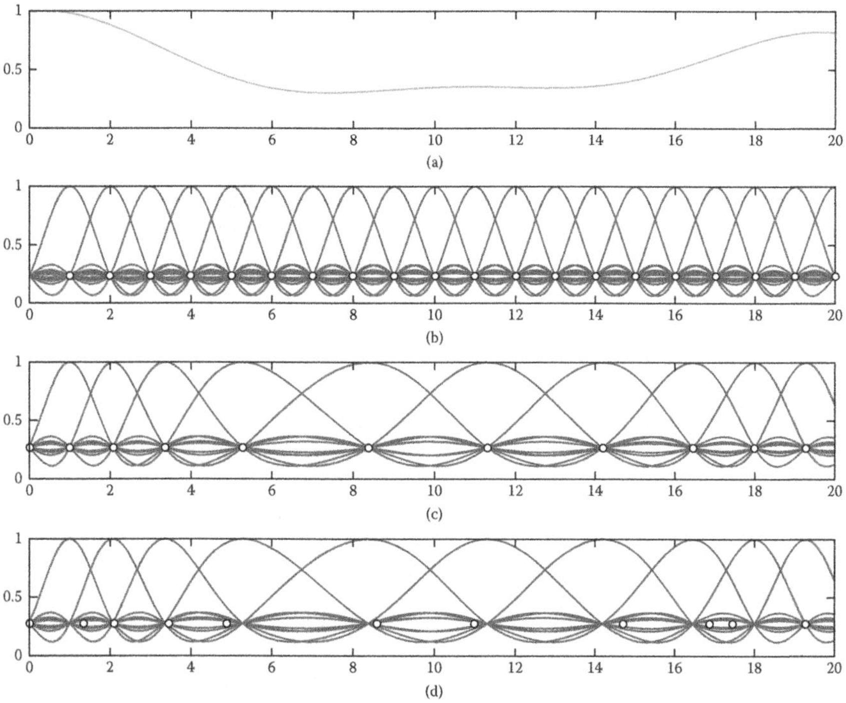


**FIGURE 23.1**
(a) Example of bandwidth function $\Omega(t)$, (b) sinc functions and time locations of uniform samples for $\Omega = \pi$, (c) sinc functions warped with $\Omega(t)$ and time locations $\{t_n\}$ of warped samples according to (23.13), and (d) sinc functions warped with by $\Omega(t)$ and the example of the sampling instants $\{t'_n\}$ irregular to $\{t_n\}$.

density, defined as

$$D(\{t_n\}) = \lim_{a\to\infty} \frac{\#\{t_n : t_n \in [-a, a]\}}{2a} \quad (23.15)$$

must be the same for irregular sampling instants $\{t'_n\}$ and time-warped instants $\{t_n\}$, that is, $D(\{t_n\}) = D(\{t'_n\})$.

Therefore, in theory, it is sufficient to sample a varying bandwidth signal at the average frequency $f_s = \bar{\Omega}/\pi$, where $\bar{\Omega} = \lim_{T\to\infty} \frac{1}{2T}\int_{-T}^{T} \Omega(t)dt$ is the average of $\Omega(t)$ over the whole time axis. This allows for the perfect recovery if $\Omega(t)$ is known exactly at each time instant. In practice, however, the recovery algorithms for irregular samples became unstable if the samples are highly irregular [14], so the closeness of the points $\{t'_n\}$ and $\{t_n\}$ is beneficial.

The reconstruction of the varying bandwidth signal using the recovery algorithm suited for the irregularly sampled constant bandwidth signal is performed in the following steps:

1. Unwarp the irregular time locations according to $u_n = \alpha(t'_n)$.
2. Reconstruct the constant-bandwidth signal $y(t)$ from the samples $y(u_n) = x(t'_n)$.
3. Warp the reconstructed signal $\tilde{y}(t)$ according to $\tilde{x}(t) = \tilde{y}(\gamma(t))$.

The methods of signal reconstruction from irregular samples are described in Section 23.4.

## 23.3 Level-Crossing Sampling of Gaussian Random Processes

### 23.3.1 Level Crossings

In the forthcoming discussion, the relationship between the local bandwidth and the level-crossing rate will be analyzed in detail based on the assumption that the input signal is a bandlimited Gaussian random process. In the classical study, Rice established the formula for the mean number of level crossings for stationary stochastic processes [15,16], see also [17]. Let

$$\theta_L = \#\{t_n \in (0,1) : x(t_n) = L\} \quad (23.16)$$

be the number of level crossings of the stochastic process $x(t)$ at the level $L$ over the interval $(0,1)$. Let $\mathbb{E}[\theta_L]$

denote the corresponding mean value of $\theta_L$. Rice's celebrated formula for $\mathbb{E}[\theta_L]$ assumes that $x(t)$ is a zero-mean, stationary, and differentiable Gaussian process with the autocorrelation function $\rho_x(\tau)$. Then, the mean number of crossings of the level $L$ by $x(t)$ is

$$\mathbb{E}[\theta_L] = \frac{1}{\pi}\sqrt{\frac{\sigma_2^2}{\sigma_0^2}}\exp\left(\frac{-L^2}{2\sigma_0^2}\right), \tag{23.17}$$

where $\sigma_0^2 = \rho_x(0)$ and $\sigma_2^2 = -\rho_x{}^{(2)}(0)$. Since $-\rho_x{}^{(2)}(\tau)$ is the autocorrelation function of the process derivative $x^{(1)}(t)$; therefore, $\sigma_0^2 = \text{Var}[x(t)]$ and $\sigma_2^2 = \text{Var}[x^{(1)}(t)]$. Hence, the factor $\sigma_2^2/\sigma_0^2$ in (23.17) is the ratio of power of $x^{(1)}(t)$ and power of $x(t)$. It is also worth noting that since $\sigma_0^2 = \int_{-\infty}^{\infty} S_x(\omega)d\omega$ and $\sigma_2^2 = \int_{-\infty}^{\infty} \omega^2 S_x(\omega)d\omega$; therefore $\sigma_0^2$ and $\sigma_2^2$ are called the spectral moments, where $S_x(\omega)$ is the power spectral density of $x(t)$. Hence, this constitutes a relation between the signal spectrum and the mean level-crossing rate. The critical question which needs to be answered is how to provide the sufficient number of level crossings to allow the reconstruction of $x(t)$.

The formula (23.17) indicates that there are three factors influencing the average rate of crossings of the level $L$ for Gaussian random processes: the value of the level $L$, the ratio of the variances $\sigma_2^2/\sigma_0^2$, respectively, of the process $x(t)$ and its derivative $x^{(1)}(t)$, and for the non-zero $L$, the variance of the process $x(t)$. The average number of level crossings $\mathbb{E}[\theta_L]$ grows with decreasing value of $|L|$, and the maximum sampling rate occurs for $L = 0$.

If the spectral density $S_x(\omega)$ of $x(t)$ is flat in the band $(\Omega_a, \Omega_b)$, that is,

$$S_x(\omega) = \begin{cases} 0 & \text{for } |\omega| \notin (\Omega_a, \Omega_b) \\ 1 & \text{for } |\omega| \in (\Omega_a, \Omega_b) \end{cases} \tag{23.18}$$

then, by virtue of (23.17), the mean number of level crossings in the unit interval is given by

$$\mathbb{E}[\theta_L] = \frac{1}{\pi\sqrt{3}}\sqrt{\frac{\Omega_b^3 - \Omega_a^3}{\Omega_b - \Omega_a}}\exp\left(-\frac{L^2}{4(\Omega_b - \Omega_a)}\right). \tag{23.19}$$

For the low-pass bandlimited Gaussian random process where $\Omega_a = 0, \Omega_b = 2\pi B$, the level-crossing rate is

$$\mathbb{E}[\theta_L] = \frac{2B}{\sqrt{3}}\exp\left(\frac{-L^2}{8\pi B}\right). \tag{23.20}$$

As follows from (23.20), the level-crossing rate for Gaussian low-pass random processes is directly proportional to the process bandwidth $B$. In fact, by expanding the exponential function we have that

$$\mathbb{E}[\theta_L] = \frac{2B}{\sqrt{3}} - \frac{L^2}{4\pi\sqrt{3}} + \mathcal{O}\left(\frac{1}{B}\right).$$

In particular, the maximum rate that is obtained for zero crossings ($L = 0$) and the flat spectrum low-pass process is

$$\mathbb{E}[\theta_0] = \frac{2B}{\sqrt{3}} \approx 1.15B < 2B. \tag{23.21}$$

If $x(t)$ is the process with non-zero mean value $m$, then the formula in (23.20) can be used to evaluate the mean rate of level crossings with respect to the shifted level $L - m$, that is, we can evaluate $\mathbb{E}[\theta_{L-m}]$. For such signals, the mean rate of crossings of the single level is too low to recover the original bandlimited Gaussian process at the Nyquist rate.

### 23.3.2 Mutilevel Crossings

The increase in the sampling rate can be achieved by triggering the sampling by multiple levels (Figure 23.2).

Under the name of the *multilevel-crossing sampling*, we mean the sampling scheme triggering samples when $x(t)$ crosses any of the levels $L_i, 1 \le i \le K$. Note that in the engineering literature, such a scheme is referred to as the level-crossing sampling [18,19]. We use the term the multilevel crossing sampling to differ it from the crossings of a single level, which has been considered in the previous section. The number of samples triggered by multilevel-crossing sampling in a unit of time is a sum of the samples triggered by each level, $L_i, 1 \le i \le K$. Sampling at level crossings gives the samples of the form $\{x(t_n) : x(t_n) = L_i\}$. Sampling the Gaussian process using multilevel crossings or crossing a single non-zero level $L_i$ makes the sampling rate dependent on the power $\sigma_0^2$ of the signal $x(t)$ appearing in (23.17).

For a given set of levels $L_1, L_2, \ldots, L_K$ and a zero mean, stationary, and differentiable Gaussian process $x(t)$, the mean rate of crossings $\mathbb{E}[\theta]$ of all levels is the sum of crossings of individual levels, that is,

$$\mathbb{E}[\theta] = \frac{1}{\pi}\sqrt{\frac{\sigma_2^2}{\sigma_0^2}}\sum_{k=1}^{K}\exp\left(\frac{-L_k^2}{2\sigma_0^2}\right). \tag{23.22}$$

In particular, if the spectral density of $x(t)$ is flat over the interval $[0, 2\pi B]$, then

$$\mathbb{E}[\theta] = \frac{2B}{\sqrt{3}}\sum_{k=1}^{K}\exp\left(\frac{-L_k^2}{8\pi B}\right). \tag{23.23}$$

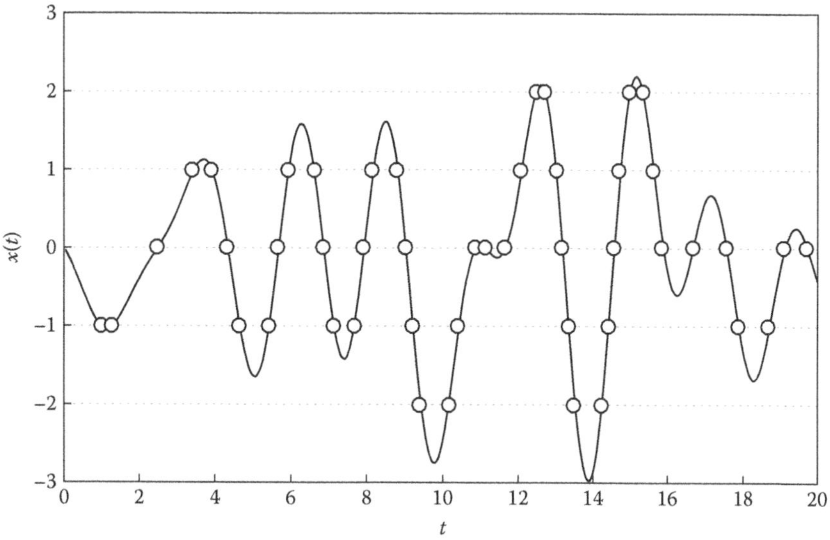


**FIGURE 23.2**
The sampling by crossings of multiple levels.

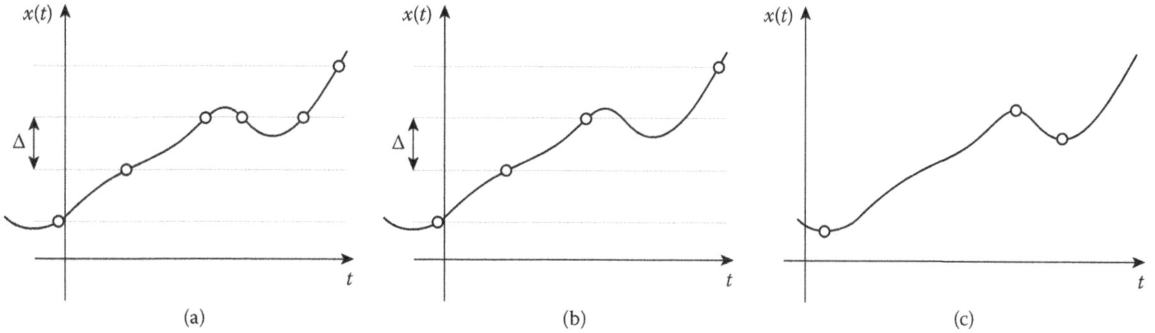


**FIGURE 23.3**
(a) Level-crossing sampling without hysteresis, (b) level-crossing sampling with hysteresis (send-on-delta sampling), and (c) extremum sampling.

The formulas in (23.22) and (23.23) refer to the level-crossing sampling without hysteresis (Figure 23.3a). By comparison, the upper bound on the mean rate $\mathbb{E}[\theta]_h$ of the level-crossing sampling with hysteresis (send-on-delta scheme, Figure 23.3b) and uniformly distributed levels for the Gaussian stationary process bandlimited to $\Omega = 2\pi B$ [11] is

$$\mathbb{E}[\theta]_h \leq \sqrt{\frac{8\pi}{3}} \frac{B\sigma_0}{\Delta} \cong 2.89 \frac{B\sigma_0}{\Delta}, \qquad (23.24)$$

where $\Delta = L_{i+1} - L_i, i = 1, \ldots, K-1$, is the distance between consecutive levels $L_i$ and $L_{i+1}$ and $\sigma_0^2 = \mathrm{Var}[\boldsymbol{x}(t)]$. The mean rate $\mathbb{E}[\theta]_h$ of the level-crossing sampling with hysteresis approaches its upper bound defined by the right side of the formula (23.24) if $\Delta \to 0$. It can be easily shown that $\mathbb{E}[\theta]_h < \mathbb{E}[\theta]$ for a given set $L_1, L_2, \ldots, L_K$ of levels.

By extending the concept of level crossings to the transforms of input signal, new sampling criteria might be formulated. For example, the zero crossings of the first derivative defines the extremum sampling (mini–max or peak sampling), which relies on capturing signal values at its local extrema (Figure 23.3c) [20–22]. Extremum sampling allows to estimate dynamically time-varying signal bandwidth similarly as level crossings. However, unlike level crossing, extremum sampling enables recovery of the original signal only at the half of Nyquist rate [22] because it is two-channel sampling, that is, the samples provide information both on signal value and on zeros of its first time derivative. Furthermore, for the low-pass bandlimited Gaussian random process (23.18) with $f_a = 0$ and $f_b = B$, the mean number of extrema in the unit interval is

$$\mathbb{E}[\theta_0]_{x'(t)} = \sqrt{2}B \approx 1.41B > B, \qquad (23.25)$$

which exceeds the minimum sufficient sampling rate for two-channel sampling and allows for reconstruction. For recovery of the signal based on irregular derivative sampling, see [23,24].

### 23.3.3 Estimation of Level-Crossing Local Intensity

The well-designed event-triggered sampling system provides the number of samples proportional to the bandwidth of a signal. However, in contrast to classic sampling paradigms, the samples obtained via level-crossing strategy are distributed irregularly. The irregularity, however, does not prohibit the recovery of the signal from samples, as it was shown in Section 23.3.1, but it has an impact on the performance of the corresponding reconstruction procedure. The reason behind this is that the sampling procedure is not adaptive as it does not take into account the local signal complexity. The latter can be captured by the concept of the local intensity of the sampled process. Hence, let

$$\theta_L(T) = \frac{\#\{t_n \in (0,T) : x(t_n) = L\}}{T} \tag{23.26}$$

be the proportion of the level-crossing points of the stochastic process $x(t)$ at the level $L$ within the time interval $(0,T)$, see (23.16) for the related notion.

The numerator in (23.26) defines a counting process that in the case when $x(t)$ is a stationary Gaussian signal behaves (as the level $L$ is increasing) as a regular Poisson point process with the constant intensity equal to $T\mathbb{E}[\theta_L]$, where $\mathbb{E}[\theta_L]$ is given by Rice's formula in (23.17) (see [17], theorem 8.4 or [25] for further details). Hence, in the case of the stationary Gaussian process the level-crossing intensity, to be denoted as $\lambda(t)$ is constant and equal to $\lambda(t) = \mathbb{E}[\theta_L]$. For a nonstationary and differentiable stochastic process $x(t)$, this is not the case and the result in [17] reveals that the relative average number of level crossings of the process $x(t)$ within $(0,T)$ is given by

$$\frac{1}{T}\int_0^T \lambda(t)dt, \tag{23.27}$$

where now $\lambda(t) = \int_{-\infty}^{\infty} |z| f_{x(t),x^{(1)}(t)}(L,z)dz$ defines the local intensity. Here, $f_{x(t),x^{(1)}(t)}(u,z)$ is the joint density of $(x(t), x^{(1)}(t))$ that, due to non-stationarity, depends on $t$. To illustrate the above discussion, let us consider a simple example concerning the local intensity $\lambda(t)$.

**EXAMPLE 23.1** Let us define a class of nonstationary stochastic processes of the form $x(t) = z(t) + \psi(t)$, where $\psi(t)$ is a differentiable deterministic function, whereas $z(t)$ is the stationary Gaussian process with zero-mean and the autocorrelation function $\rho_z(\tau)$ such that $\rho_x(0) = 1$ and $-\rho_z^{(2)}(\tau) = \sigma_2^2$. Note that the process $x(t)$ is nonstationary Gaussian with the deterministic trend and $x^{(1)}(t) = z^{(1)}(t) + \psi^{(1)}(t)$, where $\psi^{(1)}(t)$ is the derivative of $\psi(t)$. A tedious but straightforward algebra shows that the local intensity $\lambda(t)$ (as defined in (23.27)) of the process $x(t)$ at the level $L$ is given by

$$\lambda(t) = \phi(L - \psi(t))\left\{2\sigma_2\phi\left(\frac{\psi^{(1)}(t)}{\sigma_2}\right) + 2\psi^{(1)}(t)\Phi\left(\frac{\psi^{(1)}(t)}{\sigma_2}\right) - \psi^{(1)}(t)\right\}, \tag{23.28}$$

where $\phi(t)$ and $\Phi(t)$ are the pdf and CDF of the standard Gaussian random variable, respectively.

Figure 23.4 depicts the intensity $\lambda(t), 0 < t < 5$, derived in (23.28) assuming the autocorrelation function $\rho_z(\tau) = (1 - \tau^2)e^{-\tau^2}$ with the corresponding spectral moment $\sigma_2^2 = 4$. Two different trend functions $\psi_1(t) = \sin(2\pi t)$ and $\psi_2(t) = 2t$ are applied. Two level-crossing values are used, that is, $L = 0$ and $L = 2$. Note that the intensity $\lambda(t)$ corresponding to $\psi_1(t)$ is limited to the interval $(0,5)$ of the periodic function having the period 1. The $\lambda(t)$ corresponding to the linear trend $\psi_2(t)$ is the unimodal Gaussian curve confined to (0,5).

The local intensity will be used to estimate the local signal bandwidth. For example, in flat low-pass model of signal (23.18), the dependence between local bandwidth and intensity is

$$\hat{\Omega}(t) = \hat{\lambda}(t)\pi\sqrt{3}. \tag{23.29}$$

A problem of a practical importance is to estimate the unknown local intensity $\lambda(t)$ assuming only that there are $N$ level-crossing points $\{t_1, \ldots, t_N\}$ in the interval $(0,T)$. As we have already mentioned, the points $\{t_1, \ldots, t_N\}$ can be interpreted as the realization of the point process that can be described by the step function

$$N(t) = \#\{t_j : 0 < t_j \le t\} \quad \text{for } t \le T. \tag{23.30}$$

Figure 23.5 depicts the structure of $N(t)$ being a step function increasing by 1 at the time of the each level-crossing event occurring within the interval $(0,t)$.

The value of $N(t)$ can serve as a naive estimate of the expected number of points in $(0,t)$, that is, the number $\int_0^t \lambda(z)dz$. Consequently, a formal derivative $\frac{dN(t)}{dt}$ of $N(t)$ could be used as an estimate of $\frac{d}{dt}\int_0^t \lambda(z)dz = \lambda(t)$. Note the derivate $\frac{dN(t)}{dt}$ is given by

$$d(t) = \sum_j \delta(t - t_j),$$

where $\delta(t)$ is the Dirac delta function. Clearly, the estimate $d(t)$ is impractical and inconsistent since it includes

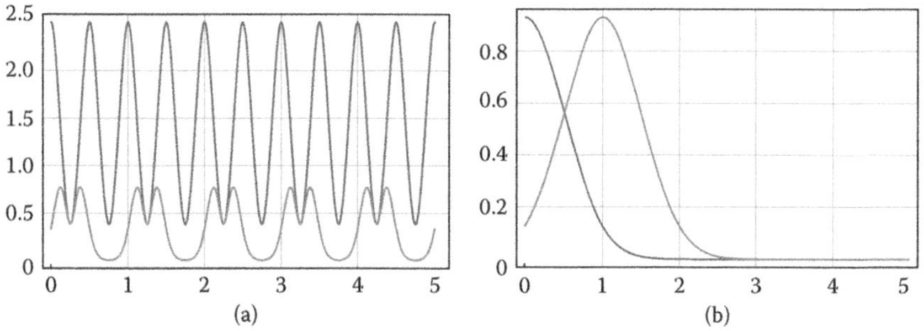


**FIGURE 23.4**
The level-crossing intensity $\lambda(t)$ of the non-stationary Gaussian stochastic processes $x(t)$ with a deterministic trend $\psi(t)$: (a) $\psi_1(t) = \sin(2\pi t)$ and (b) $\psi_2(t) = 2t$. Two values of level crossing: $L = 0$ and $L = 2$ (in solid line).

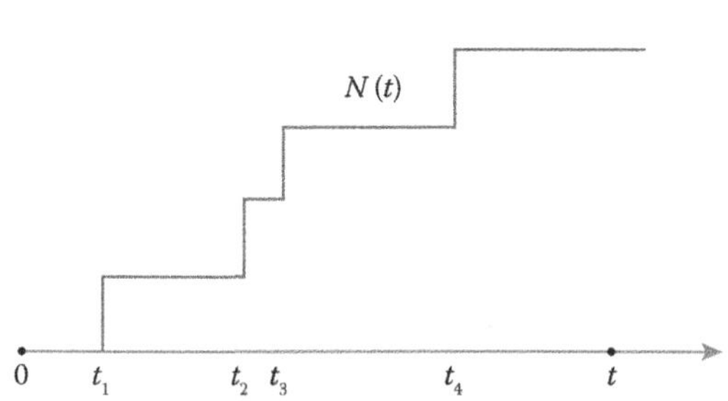


**FIGURE 23.5**
The process $N(t)$ over the interval $(0, t)$ is a step function increasing by 1 at the time $t_1$ of the each level-crossing event.

infinite spikes. A consistent estimate of $\lambda(t)$ can be obtained by a proper smoothing of the naive estimate $d(t)$. This can be accomplished by taking the convolution of $d(t)$ and the locally tuned kernel function $K_h(t)$, where $K_h(t) = h^{-1}K(t/h)$ and $h > 0$ is the smoothing parameter that controls the amount of smoothing. The kernel function $K(t)$ is assumed to be any pdf symmetric about the origin. Hence, let $\hat{\lambda}_h(t) = d(t) * K_h(t)$ be the kernel convolution estimate of $\lambda(t)$. An explicit form of this estimate is

$$\hat{\lambda}_h(t) = \sum_{n=1}^{N} K_h(t - t_n). \tag{23.31}$$

For theoretical properties of the kernel local intensity estimate in (23.31), we refer to [26,27] and the references cited therein. Some analogous analysis of the estimate $\hat{\lambda}_h(t)$ in the context of spike rate estimation for neuroscience applications was developed in [28] and the related references cited therein. It is also worth noting that if $K(t)$ is the uniform kernel, that is, $K(t) = \frac{1}{2}\mathbf{1}(|t| \le 1)$, where $\mathbf{1}(\cdot)$ is the indicator function, then $\hat{\lambda}_h(t)$ defines the binning counting method used commonly in neuroscience and biophysics.

The estimate in (23.31) is defined for a fixed level crossing $L$. In the case when we have multiple levels $\{L_i\}$, we can obtain the multiple estimates as in (23.31). The number of data points $\{t_j : 1 < j \le N\}$ depends on $L$, that is, larger $L$ makes $N$ smaller and this has a direct implication on the accuracy of $\hat{\lambda}_h(t)$. It is an interesting issue how to combine the kernel estimates corresponding to different values of level crossings to fully characterize the local bandwidth and variability of the underlying nonstationary stochastic process.

The specification of $\hat{\lambda}_h(t)$ needs the selection of the kernel function $K(t)$ and the smoothing parameter $h$. Commonly used kernels are smooth functions, and popular choices are the following: the Gaussian kernel $K(t) = \frac{1}{\sqrt{2\pi}} \exp\left(-\frac{t^2}{2}\right)$ and the compactly supported Epanechnikov kernel $K(t) = \frac{3}{2}(1 - t^2)\mathbf{1}(|t| \le 1)$. It is well known, however, that the choice of the smoothing parameter is far more important than the choice of the kernel function [29]. The data-driven choice of $h$ can be based on the approximation of the optimal $h_{ISE}$ that minimizes the integrated squared error (ISE)

$$ISE(h) = \int_0^T \left(\hat{\lambda}_h(t) - \lambda(t)\right)^2 dt. \tag{23.32}$$

The value $h_{ISE}$ cannot be determined since we do not know the form of $\lambda(t)$. Note, however, that the $ISE(h)$ can be decomposed as follows:

$$ISE(h) = \int_0^T \hat{\lambda}_h^2(t)dt - 2\int_0^T \hat{\lambda}_h(t)\lambda(t)dt + \int_0^T \lambda^2(t)dt. \tag{23.33}$$

The last term in (23.33) does not depend on $h$ and can be omitted. The first one is known up to $h$, and the only term that needs to be evaluated is the second term in (23.33). This can be estimated by the cross-validation strategy, that is, we estimate

$$\sum_{i=1}^{N} \hat{\lambda}_h^i(t_i),$$

where $\hat{\lambda}_h^i(t) = \sum_{j=1, j\neq i}^{N} K_h(t - t_j)$ is the leave-one-out estimator. Hence, $\hat{\lambda}_h^i(t)$ is the version of $\hat{\lambda}_h(t)$, where the observation $t_i$ is left out in constructing $\hat{\lambda}_h(t)$. This leads to the selection of $h_{CV}$ being the minimizer of the following criterion:

$$CV(h) = \int_0^T \left(\hat{\lambda}_h(t)\right)^2 dt - 2\sum_{i=1}^{N} \hat{\lambda}_h^i(t_i). \quad (23.34)$$

Plugging the formula in (23.31) into (23.34), we can obtain the equivalent kernel form of $CV(h)$

$$CV(h) = \sum_{i=1}^{N}\sum_{j=1}^{N} \bar{K}_h(t_i - t_j) - 4\sum_{i=1}^{N}\sum_{j=j+1}^{N} K_h(t_i - t_j),$$

where $\bar{K}_h(t) = (K_h * K_h)(t)$ is the convolution kernel, and the formula for the second term results from the symmetry of the kernel function.

It was shown in [26] that the choice $h_{CV}$ is a consistent approximation of the optimal $h_{ISE}$ if the number of points $\{t_j : 1 < j \leq N\}$ in $(0,T)$ is increasing. It is worth noting that the points $\{t_j : 1 < j \leq N\}$ constitute a correlated random process, and the aforementioned cross-validation strategy may produce a biased estimate of $h_{ISE}$. In fact, if we normalize $\lambda(t)$ such that $\lambda(t) = c\alpha(t)$, where $\int_0^T \alpha(t)dt = 1$, then for given $N$, the level-crossing points $\{t_1,\ldots,t_N\}$ have the same distribution as the order statistics corresponding to $N$ independent random variables with pdf $\alpha(t)$ on the interval $(0,T)$. This fact would suggest the modification of (23.34) with $\hat{\lambda}_h^i(t)$ replaced by its leave—$(2l+1)$—out version, that is,

$$\hat{\lambda}_h^{i,l}(t) = \sum_{j:|i-j|>l}^{N} K_h(t - t_j), \quad (23.35)$$

where $l$ is a positive integer. The choice $l$ depends on the correlation structure of $\{t_j\}$. The criterion in (23.35) takes into account the fact that the correlation between $t_i$ and $t_j$ decreases as the distance between $t_i$ and $t_j$ increases. The case $l = 0$ corresponds to the lack of correlation and then $\hat{\lambda}_h^{i,0}(t)$ is equal to $\hat{\lambda}_h^i(t)$ used in (23.34). In practical cases, the choice of $l$ in the range $1 \leq l \leq 3$ is usually sufficient, see [29]. An example of such a estimation is shown in Figure 23.6.

An interesting approach for selecting the smoothing parameter $h$, based on the Bayesian choice, was proposed in [26]. Here, one assumes that the intensity function $\lambda(t)$ is a realization of a stationary, nonnegative random process with the mean $\mu$ and the autocorrelation function $\rho_\lambda(\tau)$. Then, $\{t_1,\ldots,t_N\}$ can be viewed as a realization of double stochastic point process $N(t)$ as it is defined in (23.30). The process $N(t)$ is characterized by the property that conditioning on the realization $\lambda(t)$, the conditional process $N(t)|\lambda(t)$ is the inhomogeneous Poisson point process with the rate function $\lambda(t)$. This strategy allows the explicit evaluation of the mean square error $MSE(h) = \mathbb{E}\left[\left(\hat{\lambda}_h(t) - \lambda(t)\right)^2\right]$, where the expectation is taken now over the randomness in $\{t_1,\ldots,t_N\}$ and in $\lambda(t)$.

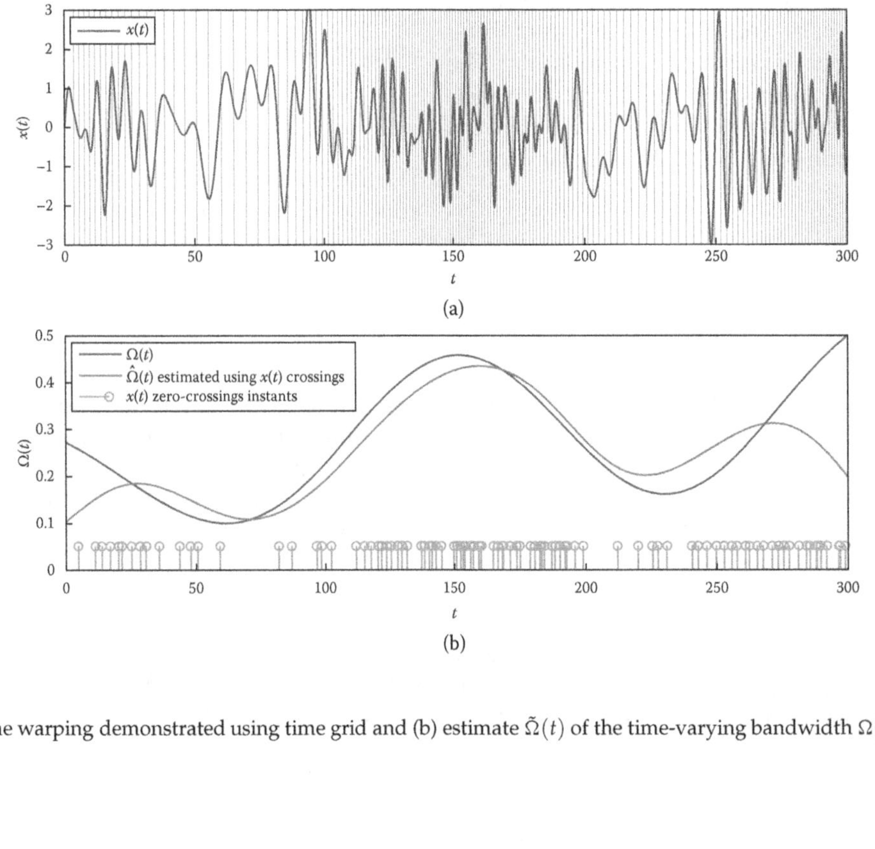


**FIGURE 23.6**
(a) Signal $x(t)$ with time warping demonstrated using time grid and (b) estimate $\tilde{\Omega}(t)$ of the time-varying bandwidth $\Omega(t)$.

The formula for the error is only a function of $\mu$ and $\rho_\lambda(\tau)$. For the uniform kernel, the formula reads as

$$MSE(h) = \rho_\lambda(0) + \frac{\mu}{2h}(1 - 2\mu\varphi(h)) + \left(\frac{\mu}{2h}\right)^2 \int_0^{2h} \varphi(z)dz, \quad (23.36)$$

where $\varphi(t) = \frac{2}{\mu^2}\int_0^t \rho_\lambda(\tau)d\tau$ [26]. Since we can easily estimate $\mu$, $\rho_\lambda(\tau)$, and consequently $\varphi(t)$, then the above expression gives an attractive strategy for selecting the smoothing parameter $h$. In fact, the computation of $h$ only needs the numerical minimization of a single variable function defined in (23.36). Let us finally mention that the localized version of the cross-validation selection method in (23.34) was examined in [28].

### 23.3.4 Estimation of Signal Parameters in Multilevel Crossing

Owing to (23.17), the level-crossing rate depends directly on the variance of the sampled Gaussian signal $x(t)$, that is, it depends on the parameter $\sigma_0^2$. The crossing rate is high for the levels close to the mean of the process $x(t)$ and gets smaller for higher values of levels. In this section, we propose a quantitative descriptor of the distribution of level crossings, that is, the frequency of visits of the Gaussian process to a level $L_i$. In fact, let $\{L_i : 1 \le i \le K\}$ be a given set of predefined levels. The probability that the level $L_i$ is selected by the process $x(t)$ can, due to (23.17), be defined as

$$g(L_i) = \frac{e^{-L_i^2/2\sigma_0^2}}{\sum_{j=1}^{K} e^{-L_j^2/2\sigma_0^2}}. \quad (23.37)$$

Since the signal samples are taken when the signal $x(t)$ crosses one of the predefined levels, this probability gives the distribution of the signal samples among the levels set $\{L_i : 1 \le i \le K\}$. The example of the distribution $g(L_i)$ in (23.37) with $K = 7$ for the zero-mean Gaussian process is shown in Figure 23.7.

In practice, the parameters $\sigma_0^2, \sigma_2^2$ defining the number of level crossings are unknown but they could be estimated from the observed number of crossings for a given set of levels $\{L_i : 1 \le i \le K\}$. Hence, let $\theta_{L_i}$ denote the number of crossings of the level $L_i$ by the observed sample path $x(t), 0 < t < 1$, of the Gaussian process with generally nonzero mean value denoted as $m$. In this case, the generalization of Rice's formula (23.17) takes the following form:

$$\mathbb{E}[\theta_L] = \frac{1}{\pi}\sqrt{\frac{\sigma_2^2}{\sigma_0^2}}\, e^{\frac{-(L-m)^2}{2\sigma_0^2}}. \quad (23.38)$$

To estimate the parameters $m$, $\sigma_2^2$, and $\sigma_0^2$, we can apply the classic least-squares method [30]. To do so, let us denote $y = \log(\mathbb{E}[\theta_L])$ and $A = \frac{1}{\pi}\sqrt{\frac{\sigma_2^2}{\sigma_0^2}}$.

Then, by a direct algebra, the formula in (23.38) can be equivalently expressed as

$$y = a + bL + cL^2, \quad (23.39)$$

where $a = \ln A - \frac{m^2}{2\sigma_0^2}$, $b = \frac{2m}{2\sigma_0^2}$, and $c = \frac{1}{2\sigma_0^2}$.

The formula in (23.39) can be viewed as a standard quadratic regression function being linear with respect to the parameters $a$, $b$, and $c$. The least-square solution

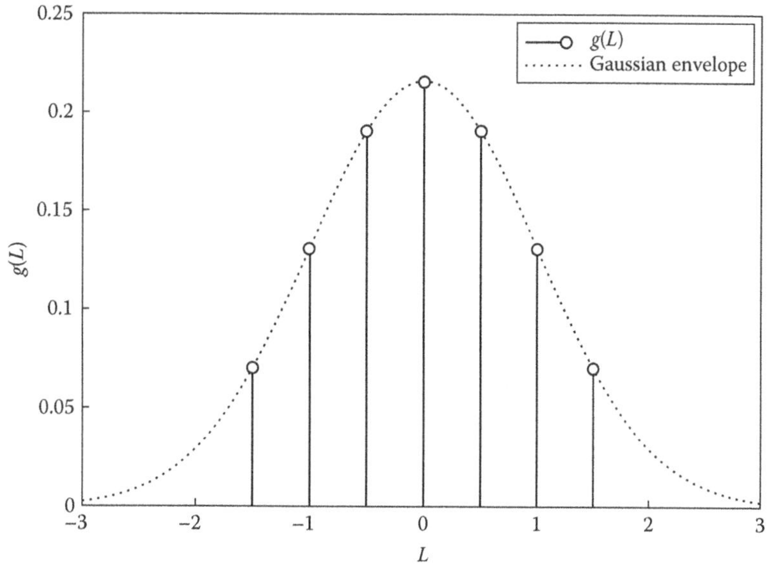


**FIGURE 23.7**
The distribution of level-crossings set for the zero-mean Gaussian process.

of this linear problem is based on the input variables $\{L_i : 1 \le i \le K\}$ and the corresponding outputs $\{y_i = \log(\theta_{L_i}) : 1 \le i \le K\}$. This produces an estimate of $a$, $b$, and $c$ as the solution of the following linear equations:

$$\begin{bmatrix} K & \Sigma_i L_i & \Sigma_i L_i^2 \\ \Sigma_i L_i & \Sigma_i L_i^2 & \Sigma_i L_i^3 \\ \Sigma_i L_i^2 & \Sigma_i L_i^3 & \Sigma_i L_i^4 \end{bmatrix} \begin{bmatrix} a \\ b \\ c \end{bmatrix} = \begin{bmatrix} \Sigma_i y_i \\ \Sigma_i L_i y_i \\ \Sigma_i L_i^2 y_i \end{bmatrix}, \quad (23.40)$$

where $\Sigma_i$ denotes summation $\Sigma_{i=1}^{K}$. It was shown in [31] that if the values of $\theta_{L_i}$ are small, the fitting error grows substantially due to the fact that $y_i = \log(\theta_{L_i})$. To prevent such a singularity of the solution, the exponential weighted version of (23.40) of the following form was proposed:

$$\begin{bmatrix} e^{2y_i} & \Sigma_i L_i e^{2y_i} & \Sigma_i L_i^2 e^{2y_i} \\ \Sigma_i L_i e^{2y_i} & \Sigma_i L_i^2 e^{2y_i} & \Sigma_i L_i^3 e^{2y_i} \\ \Sigma_i L_i^2 e^{2y_i} & \Sigma_i L_i^3 e^{2y_i} & \Sigma_i L_i^4 e^{2y_i} \end{bmatrix} \begin{bmatrix} a \\ b \\ c \end{bmatrix} = \begin{bmatrix} \Sigma_i y_i e^{2y_i} \\ \Sigma_i L_i y_i e^{2y_i} \\ \Sigma_i L_i^2 y_i e^{2y_i} \end{bmatrix}. \quad (23.41)$$

In both the aforementioned versions of the least-squares algorithm represented by (23.40) and (23.41), the searched parameter of Rice's formula in (23.38) can be easily found based on the relationships between $a, b, c$ and $m, \sigma_2^2, \sigma_0^2$.

Then, a model of bandwidth estimation presented in the previous subsections can be applied. Assuming the flat low-pass model of stationary Gaussian process, according to (23.18), we have

$$\hat{\Omega} = \sqrt{3\frac{\hat{\sigma}_2^2}{\hat{\sigma}_0^2}}, \quad (23.42)$$

where $\hat{\sigma}_2^2$ and $\hat{\sigma}_0^2$ were estimated from $\{y_i = \log(\theta_{L_i}) : 1 \le i \le K\}$. If the process is non-stationary and bandwidth is varying (but still with the assumption of spectrum flatness), then the estimation of $\Omega(t)$ must be based on the intensity functions $y_i(t) = \hat{\lambda}^{(L_i)}$, where $\{\hat{\lambda}^{(L_i)}(t), 1 \le i \le K\}$ (see Figure 23.8). Therefore, (23.41) (or (23.40), if the $y_i$ is expected to be sufficiently large) is solved for every time instants $t$, at which $\hat{\Omega}(t)$ has to be estimated.

Figure 23.9 shows the example of $\hat{\Omega}(t)$ estimation according to the scheme from Figure 23.8 for level crossings of uniformly distributed levels $L = \{-1.5, 0.75, 0, 0.75, 1.5\}$. Since the level crossings of different levels are characterized by different rates, the kernel bandwidths estimated for each level separately are different. This is not beneficial, since the kernel bandwidth should reflect the variability of the estimated signal $\Omega(t)$. To avoid this effect, the kernel bandwidth $h$ is estimated from the zero crossings, and then it used in (23.35) for each level. The bandwidth estimated from the $\{\hat{\lambda}^{(L_i)}(t), 1 \le i \le 7\}$ using Gaussian fitting (23.41) is shown in thick dashed lines. The inaccuracy of the estimation in the beginning and in the end of the shown signal stems from the edge effects. The accuracy of the estimates derived from single levels decreases for the levels with higher $|L_i|$, because the number of crossings is smaller.

## 23.4 Reconstruction of Bandlimited Signals from Nonuniform Samples

### 23.4.1 Perfect Reconstruction

As it was shown in Section 23.3, the varying bandwidth signal represented using time-warping method can be transformed into the constant bandwidth signal. The samples obtained using level crossing are irregular with respect to the time-warped sampling grid, and despite the shifting introduced by unwarping, they remain irregular with respect to the periodic grid also

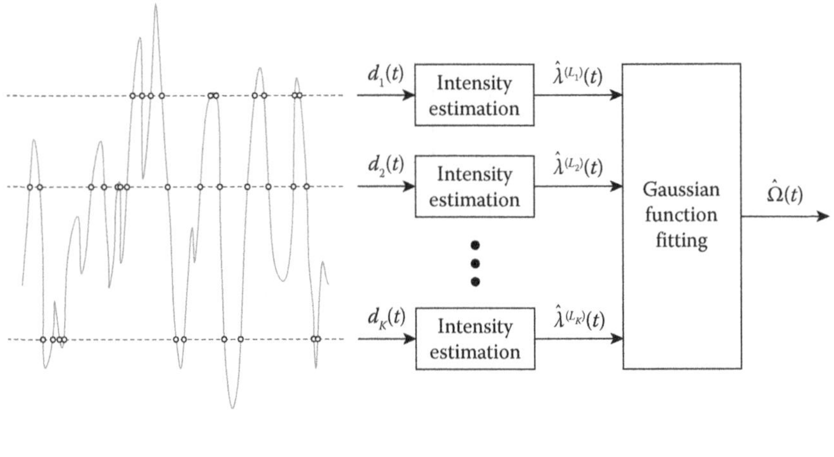


**FIGURE 23.8**
Bandwidth estimation in multilevel crossing sampling.

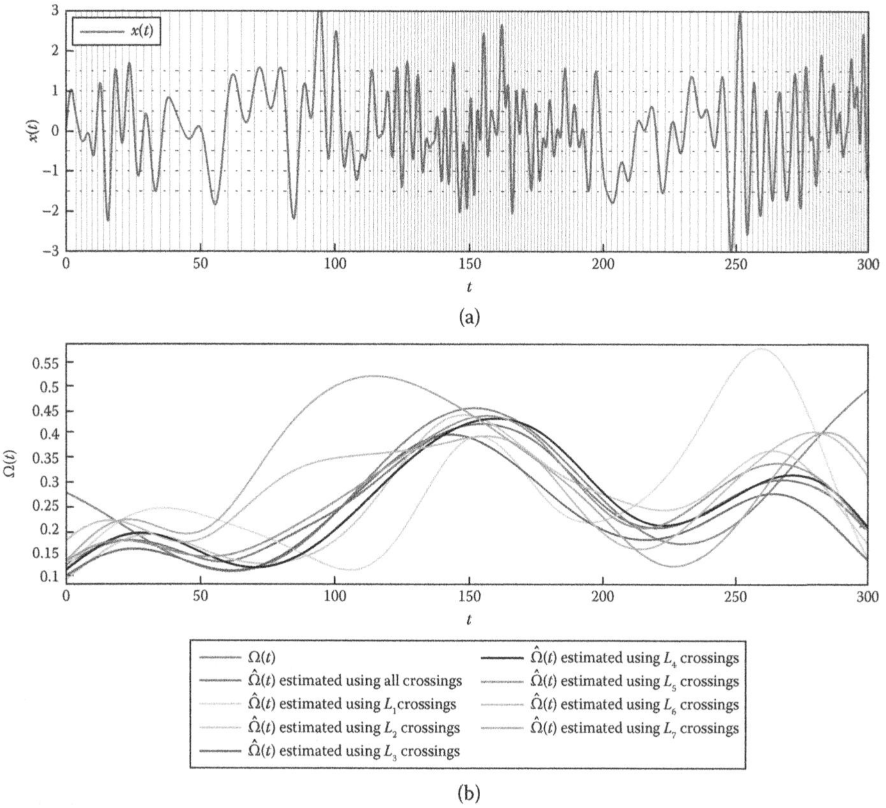


**FIGURE 23.9**
(a) Signal $x(t)$ with time warping demonstrated using time grid and (b) estimate $\tilde{\Omega}(t)$ of time-varying bandwidth $\Omega(t)$.

after the aforementioned transformation. Then, a reconstruction from the irregular samples must be applied to recover the signal from samples.

Theory of perfect reconstruction of signal from nonuniform measurements represents the sampling as an inner product of the signal $x(t)$ and the sampling function $g_n(t)$:

$$x(t_n) = \langle x(t), g_n(t)\rangle = \int_{-\infty}^{+\infty} x(t)g_n(t)dt. \qquad (23.43)$$

In the case of sampling of the bandlimited signal, the sampling function is given by $g_n(t) = \frac{\sin(\Omega(t-t_n))}{\pi(t-t_n)}$, since the samples of $x(t)$ are can be viewed as the output of ideal bandpass filter

$$\begin{aligned} x(t_n) &= \frac{1}{2\pi}\int_{-\Omega}^{\Omega} X(\omega)e^{j\omega t_n}d\omega \\ &= \int_{-\infty}^{+\infty} x(t)\frac{\Omega}{\pi}\text{sinc}(\Omega(t-t_n))dt. \end{aligned} \qquad (23.44)$$

Equation 23.44 can be interpreted in three equivalents manners: filtering $X(\omega)$ in the frequency domain, convolution of signals $x(t)$ and $\frac{\sin(\Omega t)}{\pi t}$, and inner product of these signals.

To allow the perfect reconstruction, the family of sampling functions $\{g_n(t)\}$ must constitute *a frame* or *a Riesz basis* [32]. This means that for every bandlimited signal $x(t)$ with finite energy $\langle x(t), x(t)\rangle < \infty$, there exists the coefficients $\mathbf{c} = \{\mathbf{c}[n]\}$ such that

$$x(t) = \sum_{n=-\infty}^{+\infty} \mathbf{c}[n]g_n(t). \qquad (23.45)$$

For bandlimited signals, $\{g_n(t)\}$ is a frame if the average sampling rate $\mathbb{E}\left[\frac{1}{t_{n+1}-t_n}\right]$ exceeds Nyquist rate [13] (overlapping samples $t_{n+1} = t_n$ are prohibited). The perfect reconstruction is also possible if the signal and its derivatives are sampled [23,24]. For the details of the perfect reconstruction theory, we refer to the [33] in this book, and [34,35].

### 23.4.2 Minimum Mean Squared Error Reconstruction

The perfect reconstruction requires infinite number of signal samples, which is not possible in practice. The practical recovery algorithms are aimed to minimize the reconstruction error, when signal is reconstructed from a finite number of measurements using a finite number of reconstruction functions:

$$x_N(t) = \sum_{n=1}^{N} \mathbf{c}[n] g_n(t). \tag{23.46}$$

The minimum mean squared error reconstruction can be derived [36] by solving

$$\min_{\mathbf{c}} \|x(t) - x_N(t)\|^2. \tag{23.47}$$

Denoting the error of reconstruction as $e(\mathbf{c})$, we apply the classic least-squares optimization:

$$e(\mathbf{c}) = \|x(t) - x_N(t)\|^2$$

$$= \int_{-\infty}^{+\infty} \left( x(t) - \sum_{n=1}^{N} \mathbf{c}[n] g_n(t) \right)^2 dt \tag{23.48}$$

$$= \int_{-\infty}^{+\infty} \left( x(t)^2 - 2x(t) \sum_{n=1}^{N} \mathbf{c}[n] g_n(t) dt + \sum_{n=1}^{N} \sum_{m=1}^{N} \mathbf{c}[n]\mathbf{c}[m] g_n(t) g_m(t) \right) dt \tag{23.49}$$

$$= \|x(t)\|^2 - 2 \sum_{n=1}^{N} \mathbf{c}[n] \langle x(t), g_n(t) \rangle + \sum_{n=1}^{N} \sum_{m=1}^{N} \mathbf{c}[n]\mathbf{c}[m] \langle g_n(t), g_m(t) \rangle. \tag{23.50}$$

Setting $g_n(t) = \frac{\sin(\Omega(t-t_n))}{\pi(t-t_n)}$, we obtain

$$e(\mathbf{c}) = \|x(t)\|^2 - 2 \sum_{n=1}^{N} \mathbf{c}[n] x(t_n) + \sum_{n=1}^{N} \sum_{m=1}^{N} \mathbf{c}[n]\mathbf{c}[m] \frac{\sin(\Omega(t_n - t_m))}{\pi(t_n - t_m)}. \tag{23.51}$$

To find minimum of $e(\mathbf{c})$, the gradient $\nabla e(\mathbf{c}) = 0$ is solved by

$$\frac{\partial e(\mathbf{c})}{\partial \mathbf{c}[n]} = -2x(t_n) + 2 \sum_{m=1}^{N} \mathbf{c}[m] \frac{\sin(\Omega(t_n - t_m))}{\pi(t_n - t_m)} = 0, \tag{23.52}$$

yielding

$$x(t_n) = \sum_{m=1}^{N} \mathbf{c}[m] \frac{\sin(\Omega(t_n - t_m))}{\pi(t_n - t_m)}. \tag{23.53}$$

Assembling equations for $x(t_1), x(t_2), \ldots, x(t_N)$ together gives the following liner system of equations

$$\mathbf{x} = \mathbf{G}\mathbf{c} \tag{23.54}$$

with $N \times N$ matrix $\mathbf{G}[n,m] = \frac{\sin(\Omega(t_n-t_m))}{\pi(t_n-t_m)}$ and $N \times 1$ vector $\mathbf{x}[n] = x(t_n)$. The signal $x(t)$ is approximated using coefficients $\mathbf{c}$ from the solution of the linear system (23.54), inserted into recovery equation (23.46). The analogous result can be obtained, when the signal $x(t)$ is treated as stochastic process [37].

### 23.4.3 Example of Reconstruction

The example of reconstruction is shown in Figure 23.10. The length of the examined signal was assumed to be $t_{max} = 300$. The bandwidth function $\Omega(t)$ was constructed using Shannon formula with $T_\Omega = 60$ using five uniformly distributed random samples $\Omega_n \sim \mathcal{U}(0,1)$, according to

$$\Omega(t) = \Omega_0 + \Omega_s \sum_{n=1}^{5} \Omega_n \text{sinc}\left( \frac{\pi}{T_\Omega}(t - nT_\Omega) \right). \tag{23.55}$$

Constants $\Omega_0, \Omega_s$ were set to obtain $\max \Omega(t) = 0.5$, $\min \Omega(t) = 0.1$. The signal $x(t)$ was created using 163 random samples with normal distribution, $X_n \sim \mathcal{N}(0,1)$, according to

$$x(t) = \sum_{n=1}^{163} X_n \frac{\sin(\gamma(t) - n\pi)}{\gamma(t) - n\pi} \tag{23.56}$$

where $\gamma(t)$ was calculated from (23.8). The number of samples resulted from the particular bandwidth $\Omega(t)$ obtained from (23.55) and the assumed $t_{max}$. The signal was sampled using level crossings of uniformly distributed levels $L = \{-1.5,\ 0.75,\ 0,\ 0.75,\ 1.5\}$, which yielded 511 samples.

The result of bandwidth estimation, described in Section 23.3.4, is shown in Figure 23.10b. Since the estimated bandwidth $\tilde{\Omega}(t)$ varies around the true bandwidth $\Omega(t)$, and it is better to overestimate bandwidth than to underestimate it, then a safety margin $\Omega_m = 0.04$ was added to the $\tilde{\Omega}(t)$, prior to the calculation of $\tilde{\gamma}(t)$ and $\tilde{\alpha}(t)$. Function $\tilde{\alpha}(t)$ was used for unwarping the level-crossings location, according to the procedure described in Section 23.2.3. To reconstruct the signal, the regularized variant of (23.54)

$$(\mathbf{G}^T\mathbf{G} + \varepsilon\mathbf{I})^{-1}\mathbf{G}^T\mathbf{x} = \mathbf{c} \tag{23.57}$$

was used ($\varepsilon = 0.006$). The need for regularization stems from the fact that solving (23.54) is usually ill-conditioned, which results in numerical instability [35]. Finally, the signal reconstructed using (23.46) and coefficients $\mathbf{c}$ from (23.57) was warped back, using $\tilde{\gamma}(t)$, which yielded the signal $\tilde{x}(t)$ shown in Figure 23.10a. The reconstruction error (Figure 23.10c) is low in the region where the bandwidth estimation is accurate and

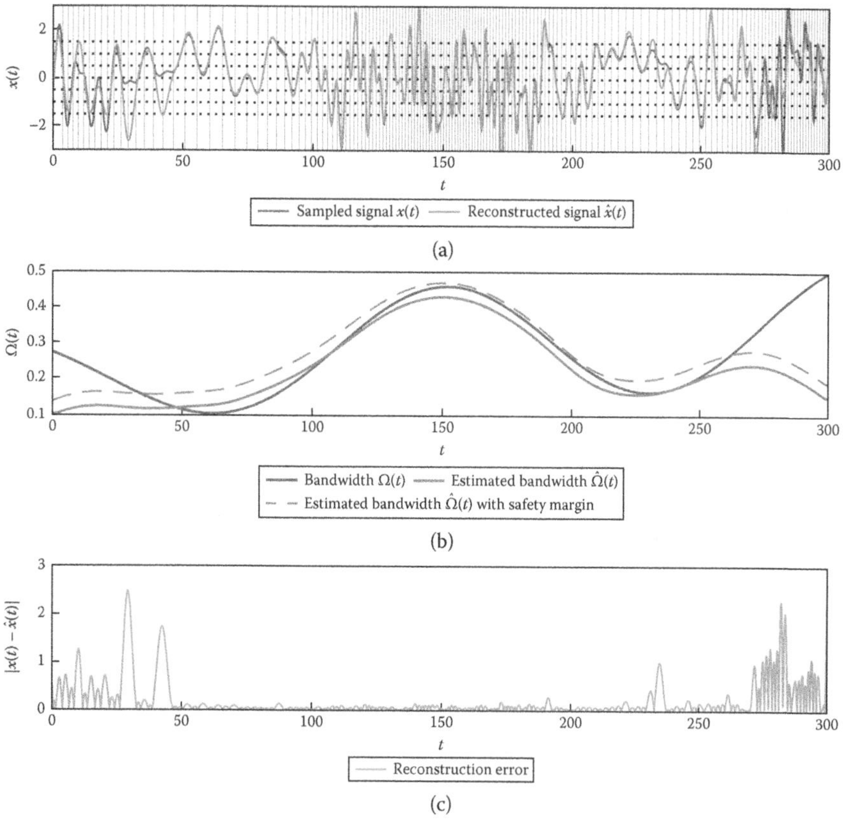


**FIGURE 23.10**
Example of signal reconstruction: (a) original and reconstructed signal, (b) estimate of bandwidth, and (c) reconstruction error.

gets worse at the edges, where the estimation suffers from edge effects.

## 23.5 Conclusions

The level-crossing sampling rate provides dynamic estimation of time-varying signal bandwidth and, therefore, allows to exploit local signal properties to avoid unnecessarily fast sampling when the local bandwidth is low. In the present study, the local bandwidth is referred to the concept of a local intensity function, which for level crossings of stochastic signals has been defined by the joint density function of the signal and its derivative. The precision of estimation was further increased for level crossings of the bandlimited Gaussian process with multiple reference levels and estimation of the local bandwidth using an algorithm for fitting the Gaussian function to the local level-crossing density. Finally, the recovery of the original signal from the level crossings by the use of the methods suited for the irregular samples combined with the time warping has been presented. The simulation results show that the proposed approach gives the successful reconstruction accuracy.

## Acknowledgments

Dominik Rzepka was supported from Dean's Grant 15.11.230.136 (AGH University of Science and Technology, Faculty of Computer Science, Electronics and Telecommunication, Kraków, Poland). Dariusz Kościelnik and Marek Miskowicz were supported by the Polish National Center of Science under grant DEC-2012/05/E/ST7/01143.

## Bibliography

[1] C. E. Shannon. Communication in the presence of noise. *Proceedings of the IRE*, 37(1):10–21, 1949.

[2] D. Slepian. On bandwidth. *Proceedings of the IEEE*, 64(3):292–300, 1976.

[3] K. Horiuchi. Sampling principle for continuous signals with time-varying bands. *Information and Control*, 13(1):53–61, 1968.

[4] J. J. Clark, M. Palmer and P. Lawrence. A transformation method for the reconstruction of functions from nonuniformly spaced samples. *IEEE Transactions on Acoustics, Speech and Signal Processing*, 33(5):1151–1165, 1985.

[5] M. Greitans and R. Shavelis. Signal-dependent sampling and reconstruction method of signals with time-varying bandwidth. In *Proceedings of SAMPTA'09, International Conference on Sampling Theory and Applications*, 2009.

[6] M. Greitans and R. Shavelis. Signal-dependent techniques for non-stationary signal sampling and reconstruction. In *Proceedings of 17th European Signal Processing Conference EUSIPCO*, 2009.

[7] Y. Hao. Generalizing sampling theory for time-varying Nyquist rates using self-adjoint extensions of symmetric operators with deficiency indices (1, 1) in Hilbert spaces. PhD Thesis, University of Waterloo, Waterloo, 2011.

[8] Y. Hao and A. Kempf. Filtering, sampling, and reconstruction with time-varying bandwidths. *IEEE Signal Processing Letters*, 17(3):241–244, 2010.

[9] D. Wei. Sampling based on local bandwidth. Master thesis, Massachusetts Institute of Technology, Boston, 2006.

[10] D. Rzepka, D. Koscielnik and M. Miskowicz. Recovery of varying-bandwidth signal for level-crossing sampling. In *Proceedings of 19th Conference of Emerging Technology and Factory Automation (ETFA)*, 2014.

[11] M. Miskowicz. Efficiency of level-crossing sampling for bandlimited Gaussian random processes. In *2006 IEEE International Workshop on Factory Communication Systems*, 2006.

[12] R. J. Marks. *Handbook of Fourier Analysis & Its Applications*. Oxford University Press, London, 2009.

[13] F. J. Beutler. Error-free recovery of signals from irregularly spaced samples. *SIAM Review*, 8(3): 328–335, 1966.

[14] D. Chen and J. Allebach. Analysis of error in reconstruction of two-dimensional signals from irregularly spaced samples. *IEEE Transactions on Acoustics, Speech and Signal Processing*, 35(2):173–180, 1987.

[15] S. O. Rice. Mathematical analysis of random noise. *Bell System Technical Journal*, 23(3):282–332, 1944.

[16] S. O. Rice. Mathematical analysis of random noise. *Bell System Technical Journal*, 24(1):46–156, 1945.

[17] G. Lindgren. *Stationary Stochastic Processes: Theory and Applications*, Chapman & Hall/CRC Texts in Statistical Science, Taylor & Francis, New York, USA, 2012.

[18] J. W. Mark and T. D. Todd. A nonuniform sampling approach to data compression. *IEEE Transactions on Communications*, 29(1):24–32, 1981.

[19] M. Miskowicz. Reducing communication by event-triggered sampling. In *Event-Based Control and Signal Processing* (M. Miskowicz, Ed.), pp. 37–58. CRC Press, Boca Raton, FL, 2015.

[20] M. Greitans, R. Shavelis, L. Fesquet, and T. Beyrouthy. Combined peak and level-crossing sampling scheme. In *9th International Conference on Sampling Theory and Applications SampTA 2011*, 2011.

[21] I. Homjakovs, M. Hashimoto, T. Hirose, and T. Onoye. Signal-dependent analog-to-digital conversion based on MINIMAX sampling. *IEICE Transactions on Fundamentals of Electronics, Communications and Computer Sciences*, 96(2):459–468, 2013.

[22] D. Rzepka and M. Miskowicz. Recovery of varying-bandwidth signal from samples of its extrema. In *Signal Processing: Algorithms, Architectures, Arrangements, and Applications (SPA), 2013*, 2013.

[23] D. Rzepka, M. Miskowicz, A. Grybos, and D. Koscielnik. Recovery of bandlimited signal based on nonuniform derivative sampling. In *Proceedings of 10th International Conference on Sampling Theory and Applications SampTA*, 2013.

[24] M. D. Rawn. A stable nonuniform sampling expansion involving derivatives. *IEEE Transactions on Information Theory*, 35(6):1223–1227, 1989.

[25] H. Cramér. On the intersections between the trajectories of a normal stationary stochastic process and a high level. *Arkiv för Matematik*, 6(4):337–349, 1966.

[26] P. Diggle. A kernel method for smoothing point process data. *Applied Statistics*, 34:138–147, 1985.

[27] P. Diggle and J. S. Marron. Equivalence of smoothing parameter selectors in density and intensity estimation. *Journal of the American Statistical Association*, 83(403):793–800, 1988.

[28] H. Shimazaki and S. Shinomoto. Kernel bandwidth optimization in spike rate estimation. *Journal of Computational Neuroscience*, 29(1–2):171–182, 2010.

[29] M. P. Wand and M. C. Jones. *Kernel Smoothing*. Chapman and Hall/CRC Press, London, UK, 1995.

[30] R. A. Caruana, R. B. Searle, T. Heller, and S. I. Shupack. Fast algorithm for the resolution of spectra. *Analytical Chemistry*, 58(6):1162–1167, 1986.

[31] H. Guo. A simple algorithm for fitting a Gaussian function. *IEEE Signal Processing Magazine*, 28:134–137, 2011.

[32] R. J. Duffin and A. C. Schaeffer. A class of non-harmonic Fourier series. *Transactions of the American Mathematical Society*, 72:341–366, 1952.

[33] N. T. Thao. Event-based data acquisition and reconstruction—mathematical background. In *Event-Based Control and Signal Processing* (M. Miskowicz, Ed.), pp. 379–408. CRC Press, Boca Raton, FL, 2015.

[34] H. G. Feichtinger and K. Gröchenig. Theory and practice of irregular sampling. In *Wavelets: Mathematics and Applications* (J. J. Benedetto and M. W. Frazier, Eds.), pp. 305–363, CRC Press, Boca Raton, FL, 1994.

[35] T. Strohmer. Numerical analysis of the non-uniform sampling problem. *Journal of Computational and Applied Mathematics*, 122(1):297–316, 2000.

[36] J. Yen. On nonuniform sampling of bandwidth-limited signals. *IRE Transactions on Circuit Theory*, 3(4):251–257, 1956.

[37] H. Choi and D. C. Munson Jr. Stochastic formulation of bandlimited signal interpolation. *IEEE Transactions on Circuits and Systems II: Analog and Digital Signal Processing*, 47(1):82–85, 2000.

# Index

## D

## E

## Q



## R

## S

## U



## V



## W



## Y



## Z

Printed and bound by CPI Group (UK) Ltd, Croydon, CR0 4YY
22/10/2024
01777611-0018